Customize this text to fit your course to a "t"!

It's your course. Teach it your way.

The probability content in Chapters 6 and 7 of this text is appropriate for those who prefer a more formal treatment of the topic. It includes an introduction to probability and probability distributions as well as a section on using simulation as a tool for estimating probabilities.

Perhaps you prefer a briefer, more informal treatment of probability, or would like to skip the text's *Graphing Calculator Explorations*. Maybe you want to pick and choose topics from this and/or other Brooks/Cole statistics texts to precisely match your syllabus. With **Cengage Learning Custom Solutions,** you can.

Cengage Learning Custom Solutions will guide you through each step of creating your perfect course. For more information about our services, visit **www.cengage.com/custom** or contact your local Brooks/ Cole representative.

Cengage Learning Custom Solutions can help you create a personalized learning solution for your statistics class. Here are some of the ways you can customize your course:

▶ Match Peck, Olsen, and Devore's text to your course syllabus by rearranging or selecting only the chapters you cover.

▶ Include your own original work such as course notes, handouts, sample tests, or additional problem sets in a professional format, working with an experienced local editor.

▶ Make classroom technology work for you and your students by utilizing our skilled Custom Media Solutions team.

▶ Draw from an extensive library of content to create a text that matches your course. With **TextChoice,** you can select content from many Brooks/Cole and other Cengage Learning best-selling titles, arrange it to fit your teaching style, and include your own work to provide your students with a complete learning solution. It's as simple as 1, 2, 3. ▼

◀ **Step 1:** Visit www.textchoice.com and select Math & Science to access our statistics content.

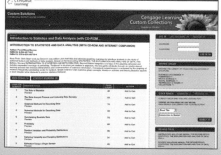

Step 2: Preview and select content. You must be a registered user. ▶

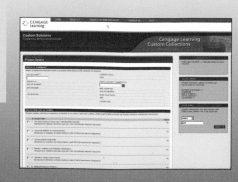

◀ **Step 3:** Finalize and submit your request. Your custom editor is available to help you build your project and make sure it meets your expectations.

ENHANCED WebAssign

Want to save time while helping students *get clear* on statistics?

Here's how: **Enhanced WebAssign.**

Enhanced WebAssign for Peck/Olsen/Devore's *Introduction to Statistics and Data Analysis*, is an easy-to-use online teaching and learning system that provides assignable homework, automatic grading, and interactive tutorial assistance. With problems and enhanced versions of examples pulled directly from the text, students get step-by-step problem-solving practice that clarifies statistics and builds conceptual understanding.

Interactive Practice for Students

As students work problems, they can link directly to

- **Read It** Relevant eBook selections from the text
- **Watch It** Videos of worked examples from the text
- **Master It** Step-by-step tutorials offering feedback at every step
- **Chat About It** Online, live help from experienced instructors

Read It

7.2 ▪ Probability Distributions for Discrete Random Variables   363

Table 7.1   Outcomes and Probabilities for Example 7.5

Outcome	Probability	x Value	Outcome	Probability	x Value
GGGG	.1296	0	GEEG	.0576	2
EGGG	.0864	1	GEGE	.0576	2
GEGG	.0864	1	GGEE	.0576	2
GGEG	.0864	1	GEEE	.0384	3
GGGE	.0864	1	EGEE	.0384	3
EEGG	.0576	2	EEGE	.0384	3
EGEG	.0576	2	EEEG	.0384	3
EGGE	.0576	2	EEEE	.0256	4

There are four different outcomes for which $x = 1$, so $p(1)$ results from summing the four corresponding probabilities:

$$p(1) = P(x = 1) = P(EGGG \text{ or } GEGG \text{ or } GGEG \text{ or } GGGE)$$
$$= P(EGGG) + P(GEGG) + P(GGEG) + P(GGGE)$$
$$= .0864 + .0864 + .0864 + .0864$$
$$= .3456$$

Watch It

Is ultrasound a reliable method for determining the gender of an unborn baby? The accompanying data on 1000 births are consistent with summary values that appeared in the online version of the *Journal of Statistics Education* ("New Approaches to Learning Probability in the First Statistics Course" [2001]).

	Ultrasound Prediction	
	Female	**Male**
Actual Gender is Female	432	48
Actual Gender is Male	130	390

"F^P" is the event that the ultrasound predicted "female"

"**P(F|F^P)**" means "probability the actual gender is female *given that* the ultrasound predicted female"

a. Use the given information to estimate the probability that a newborn baby is female, *given that* the ultrasound predicted the baby would be female.

On the go?
Study anywhere with Go Statistics™
Graphing Calculator Tutorials!

Learning statistics requires learning a new language. For instance, *z*-scores have nothing to do with snoozing, confidence intervals don't involve self-assurance, and a goodness-of-fit test isn't about trying on a new pair of jeans.

This book comes with media study tools that *do* speak your language. **Go Statistics™ Graphing Calculator Tutorials** are mini video lectures that you can download to your personal video player (PVP) or video-ready iPod® or MP3 player, or watch on your computer. The tutorials show you how to use the efficiency and convenience of a graphing calculator to solve some of the problems in this text (problem numbers are listed in the chart below). So when you're pressed for time, miss a lecture, or need extra audio and visual support to help you learn statistical techniques and prepare for tests—you'll be good to go.

Get going now!

To download **Go Statistics** tutorials to your PVP, use the access card that came with this text, or purchase an access code at **www.iChapters.com** using ISBN 10: 0-495-82811-4 / ISBN-13: 978-0-495-82811-2. The same card also lets you go online and watch (but not download) step-by-step video solutions of other text problems, identified throughout the book by a red dot.

Text Example		Related Text Problem with Downloadable Video Tutorial	Concept, Technique, or Method Explained
3.11	(pg. 97)	3.22	Frequency Histogram
3.20	(pg. 120)	3.37	Scatterplot
3.21	(pg. 122)	3.41	Time Series
4.8	(pg. 162)	4.19a	Standard Deviation from Scratch
4.11	(pg. 171)	4.30	Modified Boxplot
Exp 4.2	(pg. 196, Exploration)	4.30	Boxplot
5.3	(pg. 205)	5.5	Correlation
5.5	(pg. 213)	5.23	Calculating a Regression Equation
5.9	(pg. 226)	5.23	Calculating a Regression Equation
5.13	(pg. 239)	5.61	Nonlinear Regressions
5.14	(pg. 241)	5.61	Nonlinear Regressions
7.17	(pg. 389)	7.47	Binomial Random Variables
7.25	(pg. 407)	7.66d	*z*-scores for a Standard Normal Curve
7.29	(pg. 415)	7.83	Normal Probability Plot
9.5	(pg. 488)	9.23	Confidence Interval for Population Proportion
9.9	(pg. 502)	9.37, 9.41	Confidence Interval for Population Mean
10.11	(pg. 545)	10.30	Hypothesis Test for One Population Proportion
10.14	(pg. 555)	10.46, 10.61	Hypothesis Test for One Population Mean
11.2	(pg. 589)	11.4	Hypothesis Test for Two Population Means
11.4	(pg. 596)	11.4	Confidence Interval for Two Population Means
11.8	(pg. 613)	11.32	Hypothesis Test for Two Population Means
11.10	(pg. 623)	11.48	Hypothesis Test for Two Population Proportions
12.2	(pg. 653)	12.7	Chi-Squared Goodness-of-Fit Test
12.7	(pg. 669)	12.24	Chi-Squared Test for Independence
13.6	(pg. 715)	13.28	Scatterplot, Regression Equation, Residuals
13.13	(pg. 735)	13.55	Hypothesis Test for Regression Slope
15.4	(pg. 791)	15.20	Single Factor ANOVA
Bonus Video		4.19a (data source)	Editing Data
Bonus Video		4.26	Univariate Data

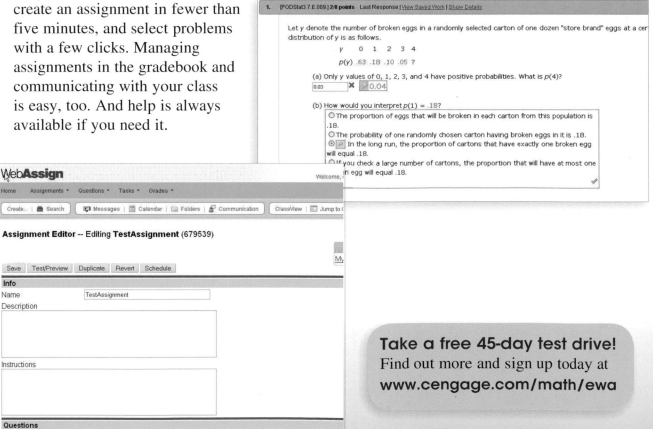

Easy Homework Management for You

Enhanced WebAssign features a simple, user-friendly interface that lets you quickly master the essential functions. You can create a course in seven easy steps, enroll students quickly, create an assignment in fewer than five minutes, and select problems with a few clicks. Managing assignments in the gradebook and communicating with your class is easy, too. And help is always available if you need it.

The Annotated Instructor's Edition contains helpful information and a variety of instructional aids.

New *Teaching Tips* precede each chapter and contain teaching suggestions for each section, recommended homework problems and activities, and information about the ancillaries that accompany the *Introduction to Statistics and Data Analysis* textbook. Some specific items in the *Tips* include

- Section-by-section summaries of important information, hints regarding topics with which students typically struggle, and suggested ways to connect statistical concepts.

- Suggested review assignments, composed of both conceptual and computational exercises. The suggested homework problems are one example of a possible subset of problems that contains both types of questions.

- Listings of assessments and teaching aids, like PowerPoint® lectures, available from the various instructor ancillaries. These ancillaries are the *Activities Manual*, the *Instructor's Resource Binder* (and CD), and the *Test Bank*. Instructions for finding the materials are given and tips for assessments, such as quizzes and tests, are also included.

- Suggested activities from the *Activities Manual*. Activities provide context and stimulate interest in the concepts. Recent educational research has shown that the best way for students to learn mathematical concepts is through carefully planned explorations or investigations, where students "develop" the concepts for themselves.

TEACHING TIPS

Intent of this chapter: After completing this chapter, students should be able to (1) explore the relationship between measures of center (mean and median), (2) describe the variability of a data set using measures of spread (range, variance, standard deviation, and interquartile range), (3) construct and describe modified boxplots, and (4) combine center, spread, and shape to find the proportion of observations within a specified range.

This chapter is a good place to emphasize that different symbols are used for parameters (values about the population) than for statistics (values calculated from a sample).

A PowerPoint® Lecture is available on the *Instructor's Resource Binder* CD.

■ Section 4.1 This section distinguishes between the population mean (μ) and sample mean ($\bar{x}$). Students should understand the relationship between the mean and median in distributions with different shapes. The activity, "Measures of Typical Value" (see below), helps students investigate this relationship.

This section also contains a brief discussion on population proportions (π) versus sample proportions (p). Stress that the *only* calculation that can be performed with categorical data is to calculate a proportion. (The use of the π and p notation for the population proportion and sample proportion, respectively, was done to maintain the convention of Greek letters for population characteristics, but other texts and the AP formula sheet use p and $\hat{p}$ for the population proportion and sample proportion, respectively. The π/p notation used in this text is acceptable on the AP exam.)

Suggested Assignment: Exercises 4.2, 4.3, 4.6, 4.10, 4.12, 4.13, 4.14
Activity: "Measures of Typical Value" (*Instructor's Resource Binder* & CD, Chapter 4 Activities Worksheets) ★

■ Section 4.2 Section 4.2 discusses measures of variability: range, interquartile range (iqr), deviation, standard deviation, and variance. It is important to do an example such as Example 4.8 so that students can "see" that the sum of deviations from the mean is zero—*always*. Discuss the difference in how population standard deviation (σ) and sample standard deviation (s) are computed.

Many students ask why we square the deviations from the mean when computing variance instead of using the apparently simple absolute values of the deviations. Historically, statistical concepts were developed using calculus. With calculus, it is easier to find derivatives and to integrate functions that contain squares rather than ab...

Students often have difficulty understan...
ing the sample standard de...
reason at thi...

TEACHING TIPS

Intent of this chapter: After completing this chapter, students should (1) understand the importance of studying statistics, (2) understand the nature and role of variability, (3) understand the data analysis process, (4) be able to identify types of data, and (5) create simple bar graphs and dot plots.

A PowerPoint® Lecture is available on the *Instructor's Resource Binder* CD.

■ Section 1.1 It is important for students to discover the impact that statistics has on their lives. A discussion of where and how statistics is used promotes the realization that statistics permeates almost all areas of our lives, and emphasizes the importance of studying statistics.

■ Section 1.2 It is well known that "Statistics is the study of variability." Many students do not realize that variability is an ever-present fact of life. Pose these questions to students: "Does every four-pound bag of sugar weigh exactly four pounds?" or "Does every 12-ounce can of soda have exactly 12 ounces in it?" The answers, of course, are "No!" and "No!" These values will vary even if ever so slightly. Students need to know that manufacturing processes (and most other processes) have "built in" variation.

The histograms in Figure 1.1 (page 5) display heights of basketball players and heights of gymnasts. Display these two histograms without identifying labels (these histograms are found on the *Instructor's Resource Binder* CD, under PowerPoint images and under Image Library) and ask students to identify which histogram is the graph for the heights of gymnasts. Use the dialogue on page 6 of the textbook to guide your class discussion.

Activity: Activity 1.1: Head Sizes—Understanding Variability (p. 22 and *Activities Manual*, p. 1), Bonus Activity 1.4: Egg Variability (*Activities Manual*, p. 8) ★, or Bonus Activity 1.5: Big Feet, Little Feet (*Activities Manual*, p. 11; this activity is similar to Activity 1.4, but requires only yardsticks.)

■ Section 1.3 Students need to understand the term "population of interest." For example, if a state wants to gather a sample of data about the math abilities of 7th grade students, then the sample should come from all schools with 7th grade students. What often happens is that a sample is taken from a small population (say a large school district in the above example), but then the results are generalized to a larger population (say the entire state). This is a common error. Samples should be drawn from the population of interest.

Suggested Assignment: Exercises 1.3, 1.4, 1.7, 1.8

■ Section 1.4 This section discusses types of variables (categorical versus numerical). Students should know that not all data that are given numerical symbols are actually numerical. Consider the variables grade level, zip code, or area code. Although these are written using numbers, it does not make sense to perform mathematical operations with them. They are categorical variables. Some students have trouble distinguishing categorical variables that are coded using numbers from numerical variables. Students should ask themselves, "Does it make sense to find a total area code?" or "Does it make sense to find an average zip code?"

For numerical variables, some students have difficulty distinguishing between discrete and continuous variables. Discrete variables are often "counts" of some item, while continuous variables are "measurements" of some variable.

★ Kathy's personal favorite

Introduction to Statistics and Data Analysis

Third Edition

Roxy Peck

California Polytechnic State University, San Luis Obispo

Chris Olsen

Thomas Jefferson High School, Cedar Rapids, IA

Jay Devore

California Polytechnic State University, San Luis Obispo

BROOKS/COLE
CENGAGE Learning™

Australia • Brazil • Japan • Korea • Mexico • Singapore • Spain • United Kingdom • United States

Introduction to Statistics and Data Analysis,
Third Edition, Enhanced Edition
Roxy Peck, Chris Olsen, Jay Devore

Senior Acquiring Sponsoring Editor: Molly Taylor

Senior Development Editor: Jay Campbell

Assistant Editor: Stefanie Beeck

Editorial Assistant: Rebecca Dashiell

Media Editor: Catie Ronquillo

Marketing Manager: Greta Kleinert

Marketing Communications Manager:
Belinda Krohmer

Project Manager, Editorial Production: Jennifer Risden

Creative Director: Rob Hugel

Art Director: Vernon Boes

Print Buyer: Becky Cross

Permissions Editor: Roberta Broyer

Production Service: Newgen

Text Designer: Stuart Paterson

Copy Editor: Nancy Dickson

Illustrator: Jade Myers; Newgen-India

Cover Designer: Stuart Paterson

Cover Image: Paul Chesley/Getty Images

Compositor: Newgen

For product information and technology assistance, contact us at
Cengage Learning Customer & Sales Support, 1-800-354-9706

For permission to use material from this text or product,
submit all requests online at **cengage.com/permissions**
Further permissions questions can be emailed to
permissionrequest@cengage.com

Library of Congress Control Number: 2006933904

ISBN-13: 978-0-495-55783-8
ISBN-10: 0-495-55783-8

Brooks/Cole
10 Davis Drive
Belmont, CA 94002-3098
USA

Cengage Learning is a leading provider of customized learning solutions with office locations around the globe, including Singapore, the United Kingdom, Australia, Mexico, Brazil, and Japan. Locate your local office at **www.cengage.com/international.**

Cengage Learning products are represented in Canada by Nelson Education, Ltd.

To learn more about Brooks/Cole, visit **www.cengage.com/brookscole**
Purchase any of our products at your local college store or at our preferred online store **www.ichapters.com.**

Printed in Canada
1 2 3 4 5 6 7 12 11 10 09 08

■ *To my nephews, Jesse and Luke Smidt, who bet I wouldn't put their names in this book.*

R. P.

■ *To my wife, Sally, and my daughter, Anna*

C. O.

■ *To Carol, Allie, and Teri.*

J. D.

■ About the Authors

ROXY PECK is Associate Dean of the College of Science and Mathematics and Professor of Statistics at California Polytechnic State University, San Luis Obispo. Roxy has been on the faculty at Cal Poly since 1979, serving for six years as Chair of the Statistics Department before becoming Associate Dean. She received an M.S. in Mathematics and a Ph.D. in Applied Statistics from the University of California, Riverside. Roxy is nationally known in the area of statistics education, and in 2003 she received the American Statistical Association's Founder's Award, recognizing her contributions to K–12 and undergraduate statistics education. She is a Fellow of the American Statistical Association and an elected member of the International Statistics Institute. Roxy has recently completed five years as the Chief Reader for the Advanced Placement Statistics Exam and currently chairs the American Statistical Association's Joint Committee with the National Council of Teachers of Mathematics on Curriculum in Statistics and Probability for Grades K–12. In addition to her texts in introductory statistics, Roxy is also co-editor of *Statistical Case Studies: A Collaboration Between Academe and Industry* and a member of the editorial board for *Statistics: A Guide to the Unknown*, 4th edition. Outside the classroom and the office, Roxy likes to travel and spends her spare time reading mystery novels. She also collects Navajo rugs and heads to New Mexico whenever she can find the time.

CHRIS OLSEN has taught statistics in Cedar Rapids, IA, for over 25 years, currently at Thomas Jefferson High School. Chris is a past member of the Advanced Placement Statistics Test Development Committee and the author of the *Teacher's Guide for Advanced Placement Statistics*. He has been a table leader at the AP Statistics reading for 6 years and since the summer of 1996 has been a consultant to the College Board. Chris leads workshops and institutes for AP Statistics teachers in the United States and internationally. Chris was the Iowa recipient of the Presidential Award for Excellence in Science and Mathematics Teaching in 1986. He was a regional winner of the IBM Computer Teacher of the Year award in 1988 and received the Siemens Award for Advanced Placement in mathematics in 1999. Chris is a frequent contributor to the AP Statistics Electronic Discussion Group and has reviewed materials for *The Mathematics Teacher*, the AP Central web site, *The American Statistician*, and the *Journal of the American Statistical Association*. He currently writes a column for *Stats* magazine. Chris graduated from Iowa State University with a major in mathematics and, while acquiring graduate degrees at the University of Iowa, concentrated on statistics, computer programming, psychometrics, and test development. Currently, he divides his duties between teaching and evaluation; in addition to teaching, he is the assessment facilitator for the Cedar Rapids, Iowa, Community Schools. In his spare time he enjoys reading and hiking. He and his wife have a daughter, Anna, who is a graduate student in Civil Engineering at Cal Tech.

JAY DEVORE earned his undergraduate degree in Engineering Science from the University of California at Berkeley, spent a year at the University of Sheffield in England, and finished his Ph.D. in statistics at Stanford University. He previously taught at the University of Florida and at Oberlin College and has had visiting appointments at Stanford, Harvard, the University of Washington, and New York University. From 1998 to 2006, Jay served as Chair of the Statistics Department at California Polytechnic State University, San Luis Obispo. The Statistics Department at Cal Poly has an international reputation for activities in statistics education. In addition to this book, Jay has written several widely used engineering statistics texts and is currently working on a book in applied mathematical statistics. He is the recipient of a distinguished teaching award from Cal Poly and is a Fellow of the American Statistical Association. In his spare time, he enjoys reading, cooking and eating good food, tennis, and travel to faraway places. He is especially proud of his wife, Carol, a retired elementary school teacher, his daughter Allison, who works for the Center for Women and Excellence in Boston, and his daughter Teri, who is finishing a graduate program in education at NYU.

Brief Contents

Sections and/or chapter numbers in color can be found at www.cengage.com/statistics/peck

Annotated Contents

Teaching Tips **T11-1**

Teaching Tips **T12-1**

Teaching Tips **T13-1**

Sections and/or chapter numbers in color can be found at www.cengage.com/statistics/peck

About This Enhanced Edition

Dear Colleagues,

Statistics is a field that interacts with and impacts many other disciplines. In fact, statistics takes its contextual scenarios from a variety of fields. An often asked question is, "Who should take statistics?" Almost everyone should! Not only will students interested in pursuing careers in science, mathematics, engineering, or medicine need statistics, but also students planning to major in business, education, economics, political science, sociology, psychology, or other social sciences. Due to the variety of reasons why students take statistics, you will often have classes composed of students with a wide range of mathematical abilities, and various levels of critical thinking skills.

As you develop your lessons, the Teaching Tips that appear in front of each chapter in this Annotated Instructor's Edition will provide you with teaching suggestions for each section, recommended homework problems and activities, and information about the ancillaries that accompany the *Introduction to Statistics and Data Analysis* textbook. These ancillaries are the *Activities Manual*, the *Instructor's Resource Binder* (and CD), and the *Test Bank*.

The *Instructor's Resource Binder* contains several components.

- **Great Resources:** This section lists books, articles, statistical software, videotapes, and other excellent resources for an introductory statistics course.
- **Chapter 1–Chapter 15:** Each section provides guides for using various software and/or videos, as well as extra examples.
- **Statistics Objectives:** This part provides learning objectives for each chapter by section.
- **Data Explorations:** This part contains extra data sets for use in each chapter.
- **Statistics Formulas Explained:** This section provides mathematical explanations for various formulas.
- **Preparing for the AP Statistics Test:** This material offers advice for preparing students for the AP Exam.
- **Activities Worksheets:** These worksheets provide good activities for each chapter, which will be acknowledged when appropriate in the Teaching Tips.

The *Instructor's Resource Binder* CD contains PowerPoint® lectures and images, the solutions manual, the electronic files for the printed *Test Bank*, the Activities Worksheets included in the *Instructor's Resource Binder*, transparency masters, and data sets.

The Teaching Tips for each chapter give you hints about which concepts students typically struggle with as well as some ideas for how to make connections among the statistical concepts. The problems listed at the end of each section are comprised of both conceptual and computational questions. The suggested homework problems are one example of a possible subset of problems that contains both types of questions.

In addition, suggested activities and other resources from the ancillaries are provided for the appropriate sections. Instructions for finding the materials are given. Tips for assessments, such as quizzes and tests, are also included.

Why should activities be part of an introductory statistics course? Regardless of whether the class is an introductory course in college or an advanced placement course in high school, activities are vital in the teaching of statistics. There are several reasons for this. First, recent educational research has shown that the best way for students to learn mathematical concepts is through carefully planned explorations or investigations, where students "develop" the concepts for themselves. The AP Statistics course, as well as a modern college-level introductory statistics course, stresses the conceptual and interpretive aspects of statistics, while capitalizing on the use of technology to perform the computations. Well-planned use of activities helps students develop an understanding of the statistical concepts necessary to be successful in this course.

In addition, although the range of mathematical achievement levels of the students in an introductory statistics course can be broad, the activities are accessible for **all** students, not just the mathematically precocious, and help them grasp the underlying and interconnected ideas in statistics. The effective use of technology for computations also makes the conceptual learning less dependent on an existing mathematical skill base. Therefore, activities make it possible for the myriad of students to succeed in this course.

An activity-based statistics course, infused with a variety of contextual settings, also promotes increased student interest in the course. For some of these students statistics will be the first occasion in their mathematical education that they will come to class eager to find out "what we will do today!" An added bonus: Students will *never* ask, "When will I ever use this?"

I would also like to stress the importance of judicious use of activities in a statistics course. Some activities are teacher-directed, while other activities require students to follow a carefully laid-out procedure to investigate a concept. No matter which type of activity is used, throughout the activity teachers should ask questions that require students to think critically and connect what they are exploring with what they currently know. Some activities, for example "Random Rectangles," are useful not only for teaching a concept (in this case, the necessity for random selection) but also for laying the groundwork for future learning (in this case, sampling distributions). This preparation allows teachers to use previous activities to connect statistical concepts for students. "Remember when we did . . ." will quickly spark the students' memories and aid in the learning process. Always summarize the activities to be sure that students understand the concepts that they investigated.

Time constraints in a college introductory course may not allow you to do all the activities suggested in this book. Therefore, college faculty will need to be selective and assign the activities that will be most advantageous for their students.

In closing, I would offer the observation that this course will undoubtedly be one of the most difficult that you will ever teach. Why? Because it requires you to develop the conceptual understanding of statistics and how these concepts are connected. For most math teachers, this takes time. However, it can be the *most* intriguing class that you will ever teach! So enjoy!

Kathy Fritz
Plano West Senior High School, Plano, Texas
Special Contributor—Teaching Tips

Kathy Fritz received her bachelor's degree in mathematics and her master's degree in psychology, concentrating on test and measurements and cognition, from the University of Houston-Victoria. She has taught mathematics for 28 years at various schools throughout Texas and has taught AP Statistics since its inaugural year, 1996. She currently teaches AP Statistics and serves as the mathematics department chair at Plano West Senior High School, where she started the AP Statistics program in 2000. Prior to that, she started the AP Statistics program at Mayde Creek High School in Katy, Texas. Kathy is a member and prior president of the Dallas Area Teachers of Statistics, a group that promotes statistics education. Kathy presents workshops for AP Statistics teachers and is lead instructor for various AP Summer Institutes. She served as an AP Exam Reader from 2002–2004 and became a Table Leader in 2005. She is also a former member of the Test Development Committee. In 2005, Kathy was voted Teacher of the Year for Plano West Senior High School, and she also received a Special Recognition Award from the College Board's Southwest Region for her work in the AP Program.

Preface

In a nutshell, statistics is about understanding the role that variability plays in drawing conclusions based on data. *Introduction to Statistics and Data Analysis,* Third Edition develops this crucial understanding of variability through its focus on the data analysis process.

An Organization That Reflects the Data Analysis Process

Students are introduced early to the idea that data analysis is a process that begins with careful planning, followed by data collection, data description using graphical and numerical summaries, data analysis, and finally interpretation of results. This process is described in detail in Chapter 1, and the ordering of topics in the first ten chapters of the book mirrors this process: data collection, then data description, then statistical inference.

The logical order in the data analysis process can be pictured as shown in the following figure.

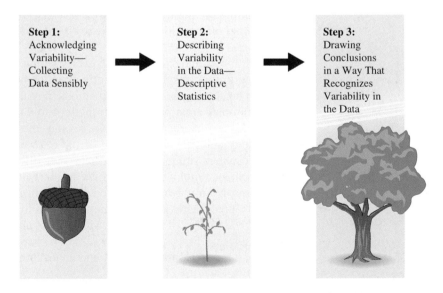

Unlike many introductory texts, *Introduction to Statistics and Data Analysis,* Third Edition is organized in a manner consistent with the natural order of the data analysis process:

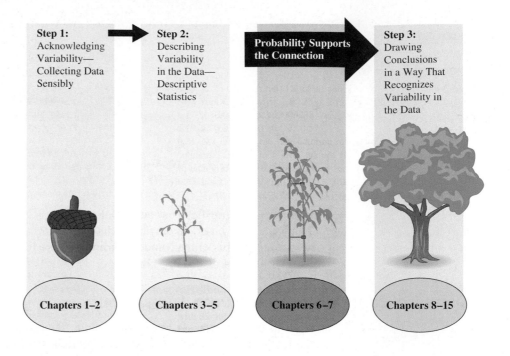

The Importance of Context and Real Data

Statistics is not about numbers; it is about data—numbers in context. It is the context that makes a problem meaningful and something worth considering. For example, exercises that ask students to compute the mean of 10 numbers or to construct a dotplot or boxplot of 20 numbers without context are arithmetic and graphing exercises. They become statistics problems only when a context gives them meaning and allows for interpretation. While this makes for a text that may appear "wordy" when compared to traditional mathematics texts, it is a critical and necessary component of a modern statistics text.

Examples and exercises with overly simple settings do not allow students to practice interpreting results in authentic situations or give students the experience necessary to be able to use statistical methods in real settings. We believe that the exercises and examples are a particular strength of this text, and we invite you to compare the examples and exercises with those in other introductory statistics texts.

Many students are skeptical of the relevance and importance of statistics. Contrived problem situations and artificial data often reinforce this skepticism. A strategy that we have employed successfully to motivate students is to present examples and exercises that involve data extracted from journal articles, newspapers, and other published sources. Most examples and exercises in the book are of this nature; they cover a very wide range of disciplines and subject areas. These include, but are not limited to, health and fitness, consumer research, psychology and aging, environmental research, law and criminal justice, and entertainment.

A Focus on Interpretation and Communication

Most chapters include a section titled "Interpreting and Communicating the Results of Statistical Analyses." These sections include advice on how to best communicate the results of a statistical analysis and also consider how to interpret statistical summaries found in journals and other published sources. A subsection titled "A Word to the Wise" reminds readers of things that must be considered in order to ensure that statistical methods are employed in reasonable and appropriate ways.

Consistent with Recommendations for the Introductory Statistics Course Endorsed by the American Statistical Association

In 2005, the American Statistical Association endorsed the report "College Guidelines in Assessment and Instruction for Statistics Education (GAISE Guidelines)," which included the following six recommendations for the introductory statistics course:

1. Emphasize statistical literacy and develop statistical thinking.
2. Use real data.
3. Stress conceptual understanding rather than mere knowledge of procedures.
4. Foster active learning in the classroom.
5. Use technology for developing conceptual understanding and analyzing data.
6. Use assessments to improve and evaluate student learning.

Introduction to Statistics and Data Analysis, Third Edition is consistent with these recommendations and supports the GAISE guidelines in the following ways:

1. **Emphasize statistical literacy and develop statistical thinking.**
   Statistical literacy is promoted throughout the text in the many examples and exercises that are drawn from the popular press. In addition, a focus on the role of variability, consistent use of context, and an emphasis on interpreting and communicating results in context work together to help students develop skills in statistical thinking.
2. **Use real data.**
   The examples and exercises from *Introduction to Statistics and Data Analysis,* Third Edition are context driven and reference sources that include the popular press as well as journal articles.
3. **Stress conceptual understanding rather than mere knowledge of procedures.**
   Nearly all exercises in *Introduction to Statistics and Data Analysis,* Third Edition are multipart and ask students to go beyond just computation. They focus on interpretation and communication, not just in the chapter sections specifically devoted to this topic, but throughout the text. The examples and explanations are designed to promote conceptual understanding. Hands-on activities in each chapter are also constructed to strengthen conceptual understanding. Which brings us to . . .
4. **Foster active learning in the classroom.**
   While this recommendation speaks more to pedagogy and classroom practice, *Introduction to Statistics and Data Analysis,* Third Edition provides 33 hands-on activities in the text and additional activities in the accompanying instructor resources that can be used in class or assigned to be completed outside of class. In addition, accompanying online materials allow students to assess their understanding and develop a personalized learning plan based on this assessment for each chapter.
5. **Use technology for developing conceptual understanding and analyzing data.**
   The computer has brought incredible statistical power to the desktop of every investigator. The wide availability of statistical computer packages such as MINITAB, S-Plus, JMP, and SPSS, and the graphical capabilities of the modern microcomputer have transformed both the teaching and learning of statistics. To highlight the role of the computer in contemporary statistics, we have included sample output

throughout the book. In addition, numerous exercises contain data that can easily be analyzed by computer, though our exposition firmly avoids a presupposition that students have access to a particular statistical package. Technology manuals for specific packages, such as MINITAB and SPSS, are available in the online materials that accompany this text.

The appearance of hand-held calculators with significant statistical and graphing capability has also changed statistics instruction in classrooms where access to computers is still limited. The computer revolution of a previous generation is now being writ small—or, possibly we should say, small*er*—for the youngest generation of investigators. There is not, as we write, anything approaching universal or even wide agreement about the proper role for the graphing calculator in college statistics classes, where access to a computer is more common. At the same time, for tens of thousands of students in Advanced Placement Statistics in our high schools, the graphing calculator is the only dependable access to statistical technology.

This text allows the instructor to balance the use of computers and calculators in a manner consistent with his or her philosophy and presents the power of the calculator in a series of Graphing Calculator Explorations. These are placed at the end of each chapter, unobtrusive to those instructors whose technology preference is the computer while still accessible to those instructors and students comfortable with graphing calculator technology. As with computer packages, our exposition avoids assuming the use of a particular calculator and presents the calculator capabilities in a generic format; specifically, we do not teach particular keystroke sequences, believing that the best source for such specific information is the calculator manual. For those using a TI graphing calculator, there is a technology manual available in the online materials that accompany this text. As much as possible, the calculator explorations are independent of each other, allowing instructors to pick and choose calculator topics that are more relevant to their particular courses.

6. **Use assessments to improve and evaluate student learning.**

Assessment materials in the form of a test bank, quizzes, and chapter exams are available in the instructor resources that accompany this text. The items in the test bank reflect the data-in-context philosophy of the text's exercises and examples.

Advanced Placement Statistics

We have designed this book with a particular eye toward the syllabus of the Advanced Placement Statistics course and the needs of high school teachers and students. Concerns expressed and questions asked in teacher workshops and on the AP Statistics Electronic Discussion Group have strongly influenced our exposition of certain topics, especially in the area of experimental design and probability. We have taken great care to provide precise definitions and clear examples of concepts that Advanced Placement Statistics instructors have acknowledged as difficult for their students. We have also expanded the variety of examples and exercises, recognizing the diverse potential futures envisioned by very capable students who have not yet focused on a college major.

Topic Coverage

Our book can be used in courses as short as one quarter or as long as one year in duration. Particularly in shorter courses, an instructor will need to be selective in deciding which topics to include and which to set aside. The book divides naturally into four major sections: collecting data and descriptive methods (Chapters 1–5), probability material (Chapters 6–8), the basic one- and two-sample inferential techniques (Chapters 9–12), and more advanced inferential methodology (Chapters 13–16). We include an early chapter (Chapter 5) on descriptive methods for bivariate numerical data. This early exposure raises questions and issues that should stimulate student interest in the subject; it is also advantageous for those teaching courses in which time constraints preclude covering advanced inferential material. However, this chapter can easily be postponed until the basics of inference have been covered, and then combined with Chapter 13 for a unified treatment of regression and correlation.

With the possible exception of Chapter 5, Chapters 1–10 should be covered in order. We anticipate that most instructors will then continue with two-sample inference (Chapter 11) and methods for categorical data analysis (Chapter 12), although regression could be covered before either of these topics. Optional portions of Chapter 14 (multiple regression) and chapter 15 (analysis of variance) and Chapter 16 (nonparametric methods) are included in the online materials that accompany this text.

A Note on Probability

The content of the probability chapters is consistent with the Advanced Placement Statistics course description. It includes both a traditional treatment of probability and probability distributions at an introductory level, as well as a section on the use of simulation as a tool for estimating probabilities. For those who prefer a briefer and more informal treatment of probability, the book *Statistics: The Exploration and Analysis of Data,* by Roxy Peck and Jay Devore, may be a more appropriate choice. Except for the treatment of probability and the omission of the Graphing Calculator Explorations, it parallels the material in this text. Please contact your sales rep for more information about this alternative and other alternative customized options available to you.

New to This Edition

There are a number of changes in the Third Edition, including the following:

- **More than 80 new examples and more than 180 new exercises that use data from current journals and newspapers are included.** In addition, more of the exercises specifically ask students to write (for example, by requiring students to explain their reasoning, interpret results, and comment on important features of an analysis).
- **Examples and exercises that make use of data sets that can be accessed online from the text website are designated by an icon in the text**, as are examples that are further illustrated in the technology manuals (MINITAB, SPSS, etc.) that are available in the online materials that accompany this text.
- **More than 90 exercises have video solutions**, presented by Brian Kotz of Montgomery College, which can be viewed online or downloaded for viewing later. These exercises are designated by an icon in the text.
- **A number of new hands-on activities have been added to the end-of-chapter activities.** These activities can be used as a chapter capstone or can be integrated at appropriate places as the chapter material is covered in class.
- **Students can now go online to test their understanding** of the material covered in each chapter **and develop a personalized learning plan** to assist them in addressing any areas of weakness.
- **A detailed description of the data analysis process now appears in Chapter 1.** Although the order of topics in the text generally mirrors the data collection process with methods of data collection covered first, two graphical displays (dotplots and bar charts) are covered in Chapter 1 so that these simple graphical analysis tools can be used in the conceptual development of experimental design and

so that students have some tools for summarizing the data they collect through sampling and experimentation in the exercises, examples, and activities of Chapter 2.

■ **A new optional section on logistic regression** is now included in Chapter 5 for those who would like more complete coverage of data analysis techniques for categorical data.

■ **Advanced topics** that are often omitted in a one-quarter or one-semester course, such as inference and variable selection methods in multiple regression (Sections 14.3 and 14.4) and analysis of variance for randomized block and two-factor designs (Sections 15.3 and 15.4), **have been moved to the online materials that accompany this text**.

■ **Coverage of distribution-free procedures for inferences** about the difference between two population or treatment means using independent samples (formerly Section 11.4) has been moved to Chapter 16. This chapter, titled "Nonparametric (Distribution-Free) Statistical Methods," also includes new material on inferences about the difference between two population or treatment means using paired samples and distribution-free analysis of variance, and is available in the online materials that accompany this text.

■ **Updated materials for instructors.** In addition to the usual instructor supplements such as a complete solutions manual and a test bank, the following are also available to instructors:

    ■ **An Instructor's Resource Binder,** which contains additional examples that can be incorporated into classroom presentations and cross-references to resources such as Fathom, Workshop Statistics, and Against All Odds. Of particular interest to those teaching Advanced Placement Statistics, the binder also includes additional data analysis questions of the type encountered on the free response portion of the Advanced Placement exam, as well as a collection of model responses.

    ■ For those who use student response systems in class, **a set of "clicker" questions** (see JoinIn™ on TurningPoint® under Instructor Resources—Media) for assessing student understanding is available.

Student Resources

■ Available Online

If your text includes a printed access card, you will have instant access to the following resources referenced throughout your text:

■ CengageNOW™ (see below for a full description of this powerful study tool).

■ Complete step-by-step instructions for MINITAB, Excel, TI-83 Graphing Calculator, JMP, and SPSS indicated by the $!$ icon throughout the text.

■ Data sets formatted for MINITAB, Excel, SPSS, SAS, JMP, TI-83, Fathom, and ASCII indicated by ● icon throughout the text.

■ Applets used in the Activities found in the text.

■ Print

Student Solutions Manual (ISBN 0-495-11876-1) by Mary Mortlock of California Polytechnic State University, San Luis Obispo.

Check your work—and your understanding—with this manual, which provides worked-out solutions to the odd-numbered problems in the text.

Activities Workbook (0-495-11883-4) by Roxy Peck.
Use this convenient workbook to take notes, record data, and cement your learning by completing textbook and bonus activities for each chapter.

■ Media

CengageNOW™
Save time, learn more, and succeed in the course with this online suite of resources (including an integrated eBook and Personalized Study plans) that give you the choices and tools you need to study smarter and get the grade. **Note:** If your text did not include a printed access card for CengageNOW, it is available for purchase online at **www.iChapters.com**.

Instructor Resources

■ Print

Annotated Instructor's Edition (0-495-55784-6)
The Annotated Instructor's Edition contains answers for all exercises, as well as an annotated table of contents with comments written by Roxy Peck.

Instructor's Solutions Manual (0-495-11879-6) by Mary Mortlock of California Polytechnic State University, San Luis Obispo.
This manual contains worked-out solutions to all of the problems in the text.

Instructor's Resource Binder (0-495-11892-3) prepared by Chris Olsen.
Includes transparencies and Microsoft® PowerPoint® slides to make lecture and class preparation quick and easy. New to this edition, we have added some Activities Worksheets authored by Carol Marchetti of Rochester Institute of Technology.

Test Bank (0-495-11880-X) by Josh Tabor of Wilson High School, Peter Flannagan-Hyde of Phoenix Country Day School, and Chris Olsen.
Includes test questions for each section of the book.

Activities Workbook (0-495-11883-4) by Roxy Peck.
Students can take notes, record data, and complete activities in this ready-to-use workbook, which includes activities from the textbook plus additional bonus activities for each chapter.

■ Media

Enhanced WebAssign (ISBN 0-495-10963-0)
Enhanced WebAssign is the most widely used homework system in higher education. Available for this title, Enhanced WebAssign allows you to assign, collect, grade, and record homework assignments via the web. This proven homework system has been enhanced to include links to the textbook sections, video examples, and problem-specific tutorials. Enhanced WebAssign is more than a homework system—it is a complete learning system for students.

CengageNOW™
CengageNOW's Personalized Study plans allow students to study smarter by diagnosing their weak areas, and helping them focus on what they need to learn. Based on responses to chapter specific pre-tests, the plans suggest a course of study for students,

including many multimedia and interactive exercises to help students better learn the material. After completing the study plan, they can take a post-test to measure their progress and understanding.

ExamView® Computerized Testing (0-495-11886-9)
Create, deliver, and customize tests and study guides (both print and online) in minutes with this easy-to-use assessment and tutorial system, which contains all questions from the Test Bank in electronic format.

JoinIn™ on TurningPoint® (0-495-11881-8)
The easiest student classroom response system to use, JoinIn features instant classroom assessment and learning.

Acknowledgments

We are grateful for the thoughtful feedback from the following reviewers that has helped to shape this text over the last two editions:

■ Reviewers of the Third Edition

Arun K. Agarwal
Grambling State University

Jacob Amidon
Finger Lakes Community College

Holly Ashton
Pikes Peak Community College

Barb Barnet
University of Wisconsin at
  Platteville

Eddie Bevilacqua
State University of New York
  College of Environmental Science
  & Forestry

Piotr Bialas
Borough of Manhattan Community
  College

Kelly Black
Union College

Gabriel Chandler
Connecticut College

Andy Chang
Youngstown State University

Jerry Chen
Suffolk Community College

Richard Chilcoat
Wartburg College

Marvin Creech
Chapman University

Ron Degges
North Dakota State University

Hemangini Deshmukh
Mercyhurst College

Ann Evans
University of Massachusetts at
  Boston
Central Carolina Community College

Guangxiong Fang
Daniel Webster College

Sharon B. Finger
Nicholls State University

Steven Garren
James Madison University

Tyler Haynes
Saginaw Valley State University

Sonja Hensler
St. Petersburg College

Trish Hutchinson
Angelo State University

Bessie Kirkwood
Sweet Briar College

Jeff Kollath
Oregon State University

Christopher Lacke
Rowan University

Michael Leitner
Louisiana State University

Zia Mahmood
College of DuPage

Art Mark
Georgoa Military College

David Mathiason
Rochester Institute of Technology

Bob Mattson
Eureka College

C. Mark Miller
York College

Megan Mocko
University of Florida

Kane Nashimoto
James Madison University

Helen Noble
San Diego State University

Broderick Oluyede
Georgia Southern University

Elaine Paris
Mercy College

Shelly Ray Parsons
Aims Community College

Judy Pennington-Price
Midway College
Hazard Community College
Jackson County High School

Michael I. Ratliff
Northern Arizona University

David R. Rauth
Duquesne University

Kevin J. Reeves
East Texas Baptist University

Robb Sinn
North Georgia College & State
University

Greg Sliwa
Broome Community College

Angela Stabley
Portland Community College

Jeffery D. Sykes
Ouachita Baptist University

Yolande Tra
Rochester Institute of Technology

Nathan Wetzel
University of Wisconsin Stevens
Point

Dr. Mark Wilson
West Virginia University Institute
of Technology

Yong Yu
Ohio State University

Toshiyuki Yuasa
University of Houston

■ Reviewers for the Second Edition

Jim Bohan
Manheim Township High School

Pat Buchanan
Pennsylvania State University

Mary Christman
American University
Iowa State University

Mark Glickman
Boston University

John Imbrie
University of Virginia

Pam Martin
Northeast Louisiana University

Paul Myers
Woodward Academy

Deanna Payton
Oklahoma State University

Michael Phelan
Chapman University

Lawrence D. Ries
University of Missouri Columbia

Alan Polansky
Northern Illinois University

Joe Ward
Health Careers High School

Additionally, we would like to express our thanks and gratitude to all who helped to make this book possible:

- Carolyn Crockett, our editor and friend, for her unflagging support and thoughtful advice for more than a decade.
- Danielle Derbenti, Beth Gershman, and Colin Blake at Brooks/Cole, for the development of all of the ancillary materials details and for keeping us on track.
- Jennifer Risden, our project manager at Brooks/Cole, and Anne Seitz at Hearthside Publishing Services, for artfully managing the myriad of details associated with the production process.
- Nancy Dickson for her careful copyediting.
- Brian Kotz for all his hard work producing the video solutions.
- Mary Mortlock for her diligence and care in producing the student and instructor solutions manuals for this book.
- Josh Tabor and Peter Flannagan-Hyde for their contributions to the test bank that accompanies the book.
- Kathy Fritz for her insightful Teaching Tips sections for the Annotated Instructor's Edition.
- Beth Chance and Francisco Garcia for producing the applet used in the confidence interval activities.
- Gary McClelland for producing the applets from *Seeing Statistics* used in the regression activities.
- Bittner Development Group for checking the accuracy of the manuscript.
- Rachel Dagdagan, a student at Cal Poly, for her help in the preparation of the manuscript.

And, as always, we thank our families, friends, and colleagues for their continued support.

Roxy Peck
Chris Olsen
Jay Devore

. . . Emphasizes Statistical Literacy and Statistical Thinking

Context Driven Applications

Real data examples and exercises throughout the text are drawn from the popular press, as well as journal articles.

■ **Exercises 2.1–2.9** ....

2.1 ▼ The article "Television's Value to Kids: It's All in How They Use It" (*Seattle Times*, July 6, 2005) described a study in which researchers analyzed standardized test results and television viewing habits of 1700 children. They found that children who averaged more than two hours of television viewing per day when they were younger than 3 tended to score lower on measures of reading ability and short term memory.
a. Is the study described an observational study or an experiment?
b. Is it reasonable to conclude that watching two or more hours of television is the cause of lower reading scores? Explain.

Page 31

Focus on Interpreting and Communicating

Chapter sections on interpreting and communicating results are designed to emphasize the importance of being able to interpret statistical output and communicate its meaning to non-statisticians. A subsection entitled "A Word to the Wise" reminds students of things that must be considered in order to ensure that statistical methods are used in reasonable and appropriate ways.

Example 3.22 Education Level and Income—Stay in School!

The time-series plot shown in Figure 3.34 appears on the U.S. Census Bureau web site. It shows the average earnings of workers by educational level as a proportion of the average earnings of a high school graduate over time. For example, we can see from this plot that in 1993 the average earnings for people with bachelor's degrees was about 1.5 times the average for high school graduates. In that same year, the average earnings for those who were not high school graduates was only about 75%

Page 123

4.5 Interpreting and Communicating the Results of Statistical Analyses

As was the case with the graphical displays of Chapter 3, the primary function of the descriptive tools introduced in this chapter is to help us better understand the variables under study. If we have collected data on the amount of money students spend on textbooks at a particular university, most likely we did so because we wanted to learn about the distribution of this variable (amount spent on textbooks) for the population of interest (in this case, students at the university). Numerical measures of center and spread and boxplots help to enlighten us, and they also allow us to communicate to

Page 186

■ A Word to the Wise: Cautions and Limitations ....

When computing or interpreting numerical descriptive measures, you need to keep in mind the following:

1. Measures of center don't tell all. Although measures of center, such as the mean and the median, do give us a sense of what might be considered a typical value for a variable, this is only one characteristic of a data set. Without additional information about variability and distribution shape, we don't really know much about the behavior of the variable.
2. Data distributions with different shapes can have the same mean and standard deviation. For example, consider the following two histograms:

Page 188

. . . Encourages Conceptual Understanding and Active Learning

Hands-on Activities in Every Chapter ▶

Thirty-three hands-on activities in the text, and additional activities in the accompanying instructor resources, can be used to encourage active learning inside or outside the classroom.

Activity 2.4   Video Games and Pain Management

Background: Video games have been used for pain management by doctors and therapists who believe that the attention required to play a video game can distract the player and thereby decrease the sensation of pain. The paper "Video Games and Health" (*British Medical Journal* [2005]:122–123) states:

"However, there has been no long term follow-up and no robust randomized controlled trials of such interventions. Whether patients eventually tire of such games is also unclear. Furthermore, it is not known whether any distracting effect depends simply on concentrating on an interactive task or whether the content of games is also an important factor as there have been no controlled trials comparing video games with other distracters. Further research should examine factors within games such as novelty, users' preferences, and relative levels of challenge and should

compare video games with other potentially distracting activities."

1. Working with a partner, select one of the areas of potential research suggested in the passage from the paper and formulate a specific question that could be addressed by performing an experiment.
2. Propose an experiment that would provide data to address the question from Step 1. Be specific about how subjects might be selected, what the experimental conditions (treatments) would be, and what response would be measured.
3. At the end of Section 2.3 there are 10 questions that can be used to evaluate an experimental design. Answer these 10 questions for the design proposed in Step 2.
4. After evaluating your proposed design, are there any changes you would like to make to your design? Explain.

Page 65

Exploration 3.3 Scaling the Histogram

When we constructed a histogram in the previous Exploration there were some numbers that we temporarily ignored in the view screen. We would like to return to those numbers now because they can seriously affect the look of a histogram. When we left the histogram the numbers in our view window were set as shown in Figure 3.43. These settings place the view window over the calculator's Cartesian system for effective viewing of the histogram from the data of Example 3.15.

We would now like to experiment a bit with the "Xscale." In all statistical graphs produced by the calculator the Xscale and Yscale choices will control the placement of the little "tick" marks on the *x* and *y* axis. In Exploration 3.2, the XScale and YScale were set at 5 and 1, respectively. The little tick marks on the *x*-axis were at multiples of 5. (Because of the data, the *x*-axis tick marks were at multiples of 5 and the *y*-axis didn't appear.) Change the Xscale value to 2 and redraw the histogram. You should see a graph similar to Figure 3.44. The *y*-axis tick marks now appear at multiples of 2, ...

Note that changing the Xscale has altered not only the tick marks but also the class intervals for the histogram. The choice of class intervals can significantly change the look and feel of the histogram. The choice of Xscale can affect judgments about the shape of the histogram. Because of this possibility it is wise to look at a histogram with varying choices of the Xscale value. If the shape appears very similar for different choices of Xscale, you can interpret and describe the shape with more confidence. However, if different Xscale choices alter the look of the histogram you should probably be more tentative.

Figure 3.43

Figure 3.44

▶ Graphing Calculator Explorations

Found at the end of most chapters, these explorations allow students to actively experience technology and promote statistical thinking.

Page 144

... Uses Technology to Develop Conceptual Understanding

Applets Allow Students to See the Concepts ▶

Within the Activities, applets are used to illustrate and promote a deeper understanding of the key statistical concepts.

Simulating Confidence Intervals

Continue generating intervals until you have seen at least 1000 intervals, and then answer the following question:
a. How does the proportion of intervals constructed that contain $\mu = 100$ compare to the stated confidence level of

Page 515

... And Analyze Data

Real Data Sets

Real data sets promote statistical analysis, as well as technology use. They are formatted for MINITAB, Excel, SPSS, SAS, JMP, TI-83, and ASCII and are indicated by the ● icon throughout the text.
▼

Example 3.14 Enrollments at Public Universities

● States differ widely in the percentage of college students who are enrolled in public institutions. The National Center for Education Statistics provided the accompanying data on this percentage for the 50 U.S. states for fall 2002.

Percentage of College Students Enrolled in Public Institutions

86	96	66	86	80	78	62	81	77	81	77	76
73	69	76	90	78	82	70	83	46	80	78	93
66	91	76	86	58	81	91	57	81	88	71	86
84	57	44	82	79	67	86	75	55	75	80	80
85	69										

The smallest observation is 44 (Rhode Island) and the largest is 96 (Alaska). It is reasonable to start the first class interval at 40 and let each interval have a width

Page 101

■ Exercises 3.22–3.34

3.22 ● Medicare's new medical plans offer a wide range of variations and choices for seniors when picking a drug plan (*San Luis Obispo Tribune*, November 25, 2005). The monthly cost for a stand-alone drug plan varies from plan to plan and from state to state. The accompanying table gives the premium for the plan with the lowest cost for each state.

State	Cost per Month (dollars)
Alabama	14.08
Alaska	20.05
Arizona	6.14
Arkansas	10.31
California	5.41
Colorado	8.62
Connecticut	7.32

State	Cost per Month (dollars)
New Hampshire	19.60
New Jersey	4.43
New Mexico	10.65
New York	4.10
North Carolina	13.27
North Dakota	1.87
Ohio	14.43
Oklahoma	10.07
Oregon	6.93
Pennsylvania	10.14
Rhode Island	7.32
South Carolina	16.57
South Dakota	1.87
Tennessee	14.08
Texas	10.31

Page 113

Step-by-Step Technology Instructions

Complete online step-by-step instructions for MINITAB, Excel, TI-83 Graphing Calculator, JMP, and SPSS are indicated by the 👣 icon throughout the text.
▼

Example 1.9 Revisiting Motorcycle Helmets

👣

Example 1.8 used data on helmet use from a sample of 1700 motorcyclists to construct a frequency distribution (Table 1.1). Figure 1.5 shows the bar chart corresponding to this frequency distribution.

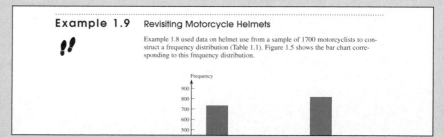

Page 17

. . . Evaluates Students' Understanding

Evaluate as You Teach Using Clickers

Using clicker content authored by Roxy Peck, evaluate your students' understanding immediately—in class—after teaching a concept. Whether it's a quick quiz, a poll to be used for in-class data, or just checking in to see if it is time to move on, our quality, tested content creates truly interactive classrooms with students' responses shaping the lecture as you teach.

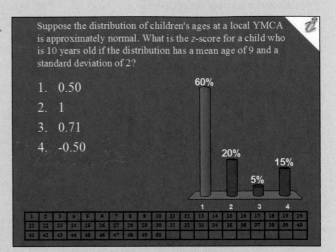

3.25 ● ▼ *USA Today* (July 2, 2001) gave the following information regarding cell phone use for men and women:

Average Number of Minutes Used per Month	Relative Frequency	
	Men	Women
0 to <200	.56	.61
200 to <400	.18	.18
400 to <600	.10	.13
600 to <800	.16	.08

a. Construct a relative frequency histogram for average number of minutes used per month for men. How would you describe the shape of this histogram?
b. Construct a relative frequency histogram for average number of minutes used per month for women. Is the distribution for average number of minutes used per month similar for men and women? Explain.
c. What proportion of men average less than 400 minutes per month?
d. Estimate the proportion of men that average less than 500 minutes per month.

Video Solutions Motivate Student Understanding

More than 90 exercises will have video solutions, presented by Brian Kotz of Montgomery College, which can be viewed online or downloaded for later viewing. These exercises will be designated by the ▼ in the text.

Page 114

Get Feedback from Roxy Peck on What You Need to Learn

CengageNOW allows students to assess their understanding and develop a personalized learning plan based on this assessment for each chapter. Pre- and post-tests include feedback authored by Roxy Peck.

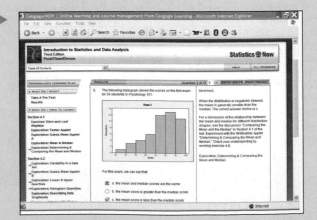

Resources for Students*

Resources for Instructors*

* See the full preface for complete descriptions.

TEACHING TIPS

Intent of this chapter:

After completing this chapter, students should (1) understand the importance of studying statistics, (2) understand the nature and role of variability, (3) understand the data analysis process, (4) be able to identify types of data, and (5) create simple bar graphs and dotplots.

A PowerPoint® Lecture is available on the *Instructor's Resource Binder* CD.

■ **Section 1.1**

It is important for students to discover the impact that statistics has on their lives. A discussion of where and how statistics is used promotes the realization that statistics permeates almost all areas of our lives, and emphasizes the importance of studying statistics.

■ **Section 1.2**

It is well known that "Statistics is the study of variability." Many students do not realize that variability is an ever-present fact of life. Pose these questions to students: "Does every four-pound bag of sugar weigh exactly four pounds?" or "Does every 12-ounce can of soda have exactly 12 ounces in it?" The answers, of course, are "No!" and "No!" These values will vary even if ever so slightly. Students need to know that manufacturing processes (and most other processes) have "built in" variation.

The histograms in Figure 1.1 (page 5) display heights of basketball players and heights of gymnasts. Display these two histograms without identifying labels (these histograms are found on the *Instructor's Resource Binder* CD, under PowerPoint images and under Image Library) and ask students to identify which histogram is the graph for the heights of gymnasts. Ask them to justify their answer. Use the dialogue on page 6 of the textbook to guide your class discussion.

Activity: Activity 1.1: Head Sizes—Understanding Variability (p. 22 and *Activities Manual*, p. 1), Bonus Activity 1.4: Egg Variability (*Activities Manual*, p. 8) ★, or Bonus Activity 1.5: Big Feet, Little Feet (*Activities Manual*, p. 11; this activity is similar to Activity 1.4, but requires only yardsticks.)

■ **Section 1.3**

Students need to understand the term "population of interest." For example, if a state wants to gather a sample of data about the math abilities of 7th grade students, then the sample should come from all schools with 7th grade students. What often happens is that a sample is taken from a small population (say a large school district in the above example), but then the results are generalized to a larger population (say the entire state). This is a common error. Samples should be drawn from the population of interest.

Suggested Assignment: Exercises 1.3, 1.4, 1.7, 1.8

■ **Section 1.4**

This section discusses types of variables (categorical versus numerical). Students should know that not all data that are given numerical symbols are actually numerical. Consider the variables grade level, zip code, or area code. Although these are written using numbers, it does not make sense to perform mathematical operations with them. They are categorical variables. Some students have trouble distinguishing categorical variables that are coded using numbers from numerical variables. Students should ask themselves, "Does it make sense to find a total area code?" or "Does it make sense to find an average zip code?"

For numerical variables, some students have difficulty distinguishing between discrete and continuous variables. Discrete variables are often "counts" of some item, while continuous variables are "measurements" of some variable.

★ Kathy's personal favorite

Bar graphs are used to display categorical data. After constructing a bar chart, students should comment on which category occurred the most (or least) often. Dotplots are used to display numerical data. They are especially suited for smaller data sets with small ranges. After constructing a dotplot, students should comment on center, shape, spread, and any unusual feature of the distribution. (For Chapter 1, students should estimate the center and discuss whether there is a little or a lot of variability. In Chapter 4, they will be able to calculate measures of center and spread.) When students are asked to compare two or more dotplots, they **must** use comparative language when describing center, spread, and shape (e.g., the median height of basketball players is *greater* than the median height of gymnasts). Just listing the medians is not comparative language (e.g., the median height of basketball players is approximately 72 inches, while the median height of gymnast is approximately 66 inches.) For any type of graph created, students **must** scale and label the axes.

Suggested Assignment: Exercises 1.11, 1.12, 1.14, 1.15, 1.21, 1.23

Activity: "Types of Data/Data Collection" (*Instructor's Resource Binder* & CD, Chapter 1 Activities Worksheets), Activity 1.2: Estimating Sizes (p. 23 and *Activities Manual* p. 3; shapes are found in the Image Library on the CD), and/or Activity 1.3: A Meaningful Paragraph (p. 24 and *Activities Manual* p. 7)

Assessment: Chapter 1 Concept Quiz (*Test Bank* and *Instructor's Resource Binder* CD)

Suggested Review Assignment: Exercises 1.27, 1.29

Assessment: Chapter 1 Test (*Test Bank* and *Instructor's Resource Binder* CD)

The Role of Statistics and the Data Analysis Process

© 2006 Jupiterimages/Brand x Pictures/trbphoto

We encounter data and conclusions based on data every day. **Statistics** is the scientific discipline that provides methods to help us make sense of data. Some people are suspicious of conclusions based on statistical analyses. Extreme skeptics, usually speaking out of ignorance, characterize the discipline as a subcategory of lying—something used for deception rather than for positive ends. However, we believe that statistical methods, used intelligently, offer a set of powerful tools for gaining insight into the world around us. Statistical methods are used in business, medicine, agriculture, social sciences, natural sciences, and applied sciences, such as engineering. The widespread use of statistical analyses in diverse fields has led to increased recognition that statistical literacy—a familiarity with the goals and methods of statistics—should be a basic component of a well-rounded educational program.

The field of statistics teaches us how to make intelligent judgments and informed decisions in the presence of uncertainty and variation. In this chapter, we consider the nature and role of variability in statistical settings, introduce some basic terminology, and look at some simple graphical displays for summarizing data.

1.1 Three Reasons to Study Statistics

Because statistical methods are used to organize, summarize, and draw conclusions from data, a familiarity with statistical techniques and statistical literacy is vital in today's society. Everyone needs to have a basic understanding of statistics, and many college majors require at least one course in statistics. There are three important reasons why statistical literacy is important: (1) to be informed, (2) to understand issues and be able to make sound decisions based on data, and (3) to be able to evaluate decisions that affect your life. Let's explore each reason in detail.

■ The First Reason: Being Informed

How do we decide whether claims based on numerical information are reasonable? We are bombarded daily with numerical information in news, in advertisements, and even in conversation. For example, here are a few of the items employing statistical methods that were part of just two weeks' news.

- The increasing popularity of online shopping has many consumers using Internet access at work to browse and shop online. In fact, the Monday after Thanksgiving has been nicknamed "Cyber Monday" because of the large increase in online purchases that occurs on that day. Data from a large-scale survey conducted in early November, 2005, by a market research firm was used to compute estimates of the percent of men and women who shop online while at work. The resulting estimates probably won't make most employers happy—42% of the men and 32% of the women in the sample were shopping online at work! (*Detroit Free Press* and *San Luis Obispo Tribune*, November 26, 2005)
- A story in the *New York Times* titled "Students Ace State Tests, but Earn D's From U.S." investigated discrepancies between state and federal standardized test results. When researchers compared state test results to the most recent results on the National Assessment of Educational Progress (NAEP), they found that large differences were common. For example, one state reported 89% of fourth graders were proficient in reading based on the state test, while only 18% of fourth graders in that state were considered proficient in reading on the federal test! An explanation of these large discrepancies and potential consequences was discussed. (*New York Times*, November 26, 2005)
- Can dogs help patients with heart failure by reducing stress and anxiety? One of the first scientific studies of the effect of therapeutic dogs found that a measure of anxiety decreased by 24% for heart patients visited by a volunteer and dog, but only by 10% for patients visited by just the volunteer. Decreases were also noted in measures of stress and heart and lung pressure, leading researchers to conclude that the use of therapeutic dogs is beneficial in the treatment of heart patients. (*San Luis Obispo Tribune*, November 16, 2005)
- Late in 2005, those eligible for Medicare had to decide which, if any, of the many complex new prescription medication plans was right for them. To assist with this decision, a program called PlanFinder that compares available options was made available online. But are seniors online? Based on a survey conducted by the *Los Angeles Times*, it was estimated that the percentage of senior citizens that go online is only between 23% and 30%, causing concern over whether providing only an online comparison is an effective way to assist seniors with this important decision. (*Los Angeles Times*, November 27, 2005)
- Are kids ruder today than in the past? An article titled "Kids Gone Wild" summarized data from a survey conducted by the Associated Press. Nearly 70% of those who participated in the survey said that people were ruder now than 20 years ago, with kids being the biggest offenders. As evidence that this is a serious problem, the author of the article also referenced a 2004 study conducted by Public Agenda, a public opinion research group. That study indicated that more than

one third of teachers had either seriously considered leaving teaching or knew a colleague who left because of intolerable student behavior. (*New York Times*, November 27, 2005)

▪ When people take a vacation, do they really leave work behind? Data from a poll conducted by Travelocity led to the following estimates: Approximately 40% of travelers check work email while on vacation, about 33% take cell phones on vacation in order to stay connected with work, and about 25% bring a laptop computer on vacation. The travel industry is paying attention—hotels, resorts, and even cruise ships are now making it easier for "vacationers" to stay connected to work. (*San Luis Obispo Tribune*, December 1, 2005)

▪ How common is domestic violence? Based on interviews with 24,000 women in 10 different countries, a study conducted by the World Health Organization found that the percentage of women who have been abused by a partner varied widely—from 15% of women in Japan to 71% of women in Ethiopia. Even though the domestic violence rate differed dramatically from country to country, in all of the countries studied women who were victims of domestic violence were about twice as likely as other women to be in poor health, even long after the violence had stopped. (*San Francisco Chronicle*, November 25, 2005)

▪ Does it matter how long children are bottle-fed? Based on a study of 2121 children between the ages of 1 and 4, researchers at the Medical College of Wisconsin concluded that there was an association between iron deficiency and the length of time that a child is bottle-fed. They found that children who were bottle-fed between the ages of 2 and 4 were three times more likely to be iron deficient than those who stopped by the time they were 1 year old. (*Milwaukee Journal Sentinel* and *San Luis Obispo Tribune*, November 26, 2005)

▪ Parental involvement in schools is often regarded as an important factor in student achievement. However, data from a study of low-income public schools in California led researchers to conclude that other factors, such as prioritizing student achievement, encouraging teacher collaboration and professional development, and using assessment data to improve instruction, had a much greater impact on the schools' Academic Performance Index. (*Washington Post* and *San Francisco Chronicle*, November 26, 2005)

To be an informed consumer of reports such as those described above, you must be able to do the following:

1. Extract information from tables, charts, and graphs.
2. Follow numerical arguments.
3. Understand the basics of how data should be gathered, summarized, and analyzed to draw statistical conclusions.

Your statistics course will help prepare you to perform these tasks.

▪ The Second Reason: Making Informed Judgments

Throughout your personal and professional life, you will need to understand statistical information and make informed decisions using this information. To make these decisions, you must be able to do the following:

1. Decide whether existing information is adequate or whether additional information is required.
2. If necessary, collect more information in a reasonable and thoughtful way.
3. Summarize the available data in a useful and informative manner.

4. Analyze the available data.
5. Draw conclusions, make decisions, and assess the risk of an incorrect decision.

People informally use these steps to make everyday decisions. Should you go out for a sport that involves the risk of injury? Will your college club do better by trying to raise funds with a benefit concert or with a direct appeal for donations? If you choose a particular major, what are your chances of finding a job when you graduate? How should you select a graduate program based on guidebook ratings that include information on percentage of applicants accepted, time to obtain a degree, and so on? The study of statistics formalizes the process of making decisions based on data and provides the tools for accomplishing the steps listed.

■ The Third Reason: Evaluating Decisions That Affect Your Life

While you will need to make informed decisions based on data, it is also the case that other people will use statistical methods to make decisions that affect you as an individual. An understanding of statistical techniques will allow you to question and evaluate decisions that affect your well-being. Some examples are:

- Many companies now require drug screening as a condition of employment. With these screening tests there is a risk of a false-positive reading (incorrectly indicating drug use) or a false-negative reading (failure to detect drug use). What are the consequences of a false result? Given the consequences, is the risk of a false result acceptable?
- Medical researchers use statistical methods to make recommendations regarding the choice between surgical and nonsurgical treatment of such diseases as coronary heart disease and cancer. How do they weigh the risks and benefits to reach such a recommendation?
- University financial aid offices survey students on the cost of going to school and collect data on family income, savings, and expenses. The resulting data are used to set criteria for deciding who receives financial aid. Are the estimates they use accurate?
- Insurance companies use statistical techniques to set auto insurance rates, although some states restrict the use of these techniques. Data suggest that young drivers have more accidents than older ones. Should laws or regulations limit how much more young drivers pay for insurance? What about the common practice of charging higher rates for people who live in urban areas?

An understanding of elementary statistical methods can help you to evaluate whether important decisions such as the ones just mentioned are being made in a reasonable way.

We hope that this textbook will help you to understand the logic behind statistical reasoning, prepare you to apply statistical methods appropriately, and enable you to recognize when statistical arguments are faulty.

1.2   The Nature and Role of Variability

Statistics is a science whose focus is on collecting, analyzing, and drawing conclusions from data. If we lived in a world where all measurements were identical for every individual, all three of these tasks would be simple. Imagine a population consisting of all students at a particular university. Suppose that *every* student took the same number of units, spent exactly the same amount of money on textbooks this semester, and favored increasing student fees to support expanding library services. For this population, there is *no* variability in the number of units, amount spent on books, or student opinion on the fee increase. A researcher studying a sample from this population to draw conclusions about these three variables would have a particularly easy task. It would not matter how many students the researcher included in the sample or how the sampled students were selected. In fact, the researcher could collect information on number of units, amount spent on books, and opinion on the fee increase by just stopping the next student who happened to walk by the library. Because there is no variability in the population, this one individual would provide complete and accurate information about the population, and the researcher could draw conclusions based on the sample with no risk of error.

The situation just described is obviously unrealistic. Populations with no variability are exceedingly rare, and they are of little statistical interest because they present no challenge! In fact, variability is almost universal. It is variability that makes life (and the life of a statistician, in particular) interesting. We need to understand variability to be able to collect, analyze, and draw conclusions from data in a sensible way. One of the primary uses of descriptive statistical methods is to increase our understanding of the nature of variability in a population.

Examples 1.1 and 1.2 illustrate how an understanding of variability is necessary to draw conclusions based on data.

Example 1.1   If the Shoe Fits

The graphs in Figure 1.1 are examples of a type of graph called a histogram. (The construction and interpretation of such graphs is discussed in Chapter 3.) Figure 1.1(a) shows the distribution of the heights of female basketball players who played at a particular university between 1990 and 1998. The height of each bar in the

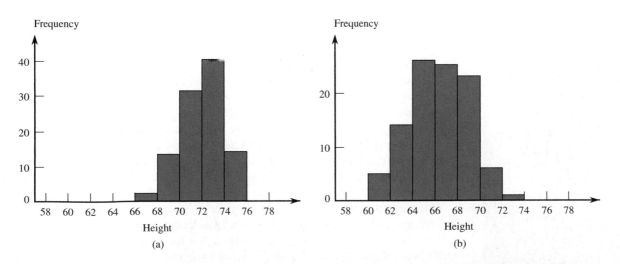

Figure 1.1 Histograms of heights (in inches) of female athletes: (a) basketball players; (b) gymnasts.

See Teaching Tips page T1-1 for class discussion of these histograms.

graph indicates how many players' heights were in the corresponding interval. For example, 40 basketball players had heights between 72 in. and 74 in., whereas only 2 players had heights between 66 in. and 68 in. Figure 1.1(b) shows the distribution of heights for members of the women's gymnastics team over the same period. Both histograms are based on the heights of 100 women.

The first histogram shows that the heights of female basketball players varied, with most heights falling between 68 in. and 76 in. In the second histogram we see that the heights of female gymnasts also varied, with most heights in the range of 60 in. to 72 in. It is also clear that there is more variation in the heights of the gymnasts than in the heights of the basketball players, because the gymnast histogram spreads out more about its center than does the basketball histogram.

Now suppose that a tall woman (5 ft 11 in.) tells you she is looking for her sister who is practicing with her team at the gym. Would you direct her to where the basketball team is practicing or to where the gymnastics team is practicing? What reasoning would you use to decide? If you found a pair of size 6 shoes left in the locker room, would you first try to return them by checking with members of the basketball team or the gymnastics team?

You probably answered that you would send the woman looking for her sister to the basketball practice and that you would try to return the shoes to a gymnastics team member. To reach these conclusions, you informally used statistical reasoning that combined your own knowledge of the relationship between heights of siblings and between shoe size and height with the information about the distributions of heights presented in Figure 1.1. You might have reasoned that heights of siblings tend to be similar and that a height as great as 5 ft 11 in., although not impossible, would be unusual for a gymnast. On the other hand, a height as tall as 5 ft 11 in. would be a common occurrence for a basketball player. Similarly, you might have reasoned that tall people tend to have bigger feet and that short people tend to have smaller feet. The shoes found were a small size, so it is more likely that they belong to a gymnast than to a basketball player, because small heights and small feet are usual for gymnasts and unusual for basketball players.

■

....

Example 1.2   Monitoring Water Quality

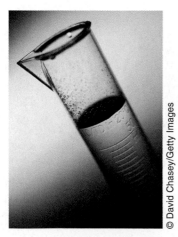

© David Chasey/Getty Images

As part of its regular water quality monitoring efforts, an environmental control board selects five water specimens from a particular well each day. The concentration of contaminants in parts per million (ppm) is measured for each of the five specimens, and then the average of the five measurements is calculated. The histogram in Figure 1.2 summarizes the average contamination values for 200 days.

Now suppose that a chemical spill has occurred at a manufacturing plant 1 mile from the well. It is not known whether a spill of this nature would contaminate groundwater in the area of the spill and, if so, whether a spill this distance from the well would affect the quality of well water.

One month after the spill, five water specimens are collected from the well, and the average contamination is 15.5 ppm. Considering the variation before the spill, would you take this as convincing evidence that the well water was affected by the spill? What if the calculated average was 17.4 ppm? 22.0 ppm? How is your reasoning related to the graph in Figure 1.2?

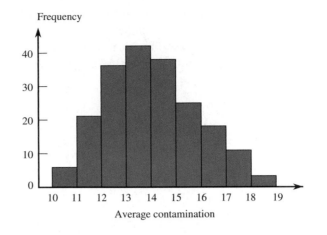

Figure 1.2 Frequency of contaminant concentration (in parts per million) in well water.

Before the spill, the average contaminant concentration varied from day to day. An average of 15.5 ppm would not have been an unusual value, so seeing an average of 15.5 ppm after the spill isn't necessarily an indication that contamination has increased. On the other hand, an average as large as 17.4 ppm is less common, and an average as large as 22.0 ppm is not at all typical of the prespill values. In this case, we would probably conclude that the well contamination level has increased.

■

In these two examples, reaching a conclusion required an understanding of variability. Understanding variability allows us to distinguish between usual and unusual values. The ability to recognize unusual values in the presence of variability is the key to most statistical procedures and is also what enables us to quantify the chance of being incorrect when a conclusion is based on sample data. These concepts will be developed further in subsequent chapters.

1.3 Statistics and the Data Analysis Process

Data and conclusions based on data appear regularly in a variety of settings: newspapers, television and radio advertisements, magazines, and professional publications. In business, industry, and government, informed decisions are often data driven. Statistical methods, used appropriately, allow us to draw reliable conclusions based on data.

Once data have been collected or once an appropriate data source has been identified, the next step in the data analysis process usually involves organizing and summarizing the information. Tables, graphs, and numerical summaries allow increased understanding and provide an effective way to present data. Methods for organizing and summarizing data make up the branch of statistics called **descriptive statistics**.

After the data have been summarized, we often wish to draw conclusions or make decisions based on the data. This usually involves generalizing from a small group of individuals or objects that we have studied to a much larger group.

For example, the admissions director at a large university might be interested in learning why some applicants who were accepted for the fall 2006 term failed to enroll at the university. The population of interest to the director consists of all accepted applicants who did not enroll in the fall 2006 term. Because this population is large and it may be difficult to contact all the individuals, the director might decide to collect data from only 300 selected students. These 300 students constitute a sample.

> **DEFINITION**
>
> The entire collection of individuals or objects about which information is desired is called the **population** of interest. A **sample** is a subset of the population, selected for study in some prescribed manner.

The second major branch of statistics, **inferential statistics**, involves generalizing from a sample to the population from which it was selected. When we generalize in this way, we run the risk of an incorrect conclusion, because a conclusion about the population is based on incomplete information. An important aspect in the development of inferential techniques involves quantifying the chance of an incorrect conclusion.

> **DEFINITION**
>
> **Descriptive statistics** is the branch of statistics that includes methods for organizing and summarizing data. **Inferential statistics** is the branch of statistics that involves generalizing from a sample to the population from which it was selected and assessing the reliability of such generalizations.

■ The Data Analysis Process

Statistics involves the collection and analysis of data. Both tasks are critical. Raw data without analysis are of little value, and even a sophisticated analysis cannot extract meaningful information from data that were not collected in a sensible way.

■ **Planning and Conducting a Statistical Study** Scientific studies are undertaken to answer questions about our world. Is a new flu vaccine effective in preventing illness? Is the use of bicycle helmets on the rise? Are injuries that result from bicycle accidents less severe for riders who wear helmets than for those who do not? How many credit cards do college students have? Do engineering students pay more for textbooks than do psychology students? Data collection and analysis allow researchers to answer such questions.

The data analysis process can be viewed as a sequence of steps that lead from planning to data collection to informed conclusions based on the resulting data. The process can be organized into the following six steps:

1. **Understanding the nature of the problem.** Effective data analysis requires an understanding of the research problem. We must know the goal of the research and what questions we hope to answer. It is important to have a clear direction before gathering data to lessen the chance of being unable to answer the questions of interest using the data collected.
2. **Deciding what to measure and how to measure it.** The next step in the process is deciding what information is needed to answer the questions of interest. In some

cases, the choice is obvious (e.g., in a study of the relationship between the weight of a Division I football player and position played, you would need to collect data on player weight and position), but in other cases the choice of information is not as straightforward (e.g., in a study of the relationship between preferred learning style and intelligence, how would you define learning style and measure it and what measure of intelligence would you use?). It is important to carefully define the variables to be studied and to develop appropriate methods for determining their values.

3. **Data collection.** The data collection step is crucial. The researcher must first decide whether an existing data source is adequate or whether new data must be collected. Even if a decision is made to use existing data, it is important to understand how the data were collected and for what purpose, so that any resulting limitations are also fully understood and judged to be acceptable. If new data are to be collected, a careful plan must be developed, because the type of analysis that is appropriate and the subsequent conclusions that can be drawn depend on how the data are collected.

4. **Data summarization and preliminary analysis.** After the data are collected, the next step usually involves a preliminary analysis that includes summarizing the data graphically and numerically. This initial analysis provides insight into important characteristics of the data and can provide guidance in selecting appropriate methods for further analysis.

5. **Formal data analysis.** The data analysis step requires the researcher to select and apply the appropriate inferential statistical methods. Much of this textbook is devoted to methods that can be used to carry out this step.

6. **Interpretation of results.** Several questions should be addressed in this final step—for example, What conclusions can be drawn from the analysis? How do the results of the analysis inform us about the stated research problem or question? and How can our results guide future research? The interpretation step often leads to the formulation of new research questions, which, in turn, leads back to the first step. In this way, good data analysis is often an iterative process.

Example 1.3 illustrates the steps in the data analysis process.

Example 1.3   A Proposed New Treatment for Alzheimer's Disease

The article "Brain Shunt Tested to Treat Alzheimer's" (*San Francisco Chronicle*, October 23, 2002) summarizes the findings of a study that appeared in the journal *Neurology*. Doctors at Stanford Medical Center were interested in determining whether a new surgical approach to treating Alzheimer's disease results in improved memory functioning. The surgical procedure involves implanting a thin tube, called a shunt, which is designed to drain toxins from the fluid-filled space that cushions the brain. Eleven patients had shunts implanted and were followed for a year, receiving quarterly tests of memory function. Another sample of Alzheimer's patients was used as a comparison group. Those in the comparison group received the standard care for Alzheimer's disease. After analyzing the data from this study, the investigators concluded that the "results suggested the treated patients essentially held their own in the cognitive tests while the patients in the control group steadily declined. However, the study was too small to produce conclusive statistical evidence." Based on these results, a much larger 18-month study was planned. That study was to include 256 patients at 25 medical centers around the country.

This study illustrates the nature of the data analysis process. A clearly defined research question and an appropriate choice of how to measure the variable of interest (the cognitive tests used to measure memory function) preceded the data collection. Assuming that a reasonable method was used to collect the data (we will see how this can be evaluated in Chapter 2) and that appropriate methods of analysis were employed, the investigators reached the conclusion that the surgical procedure showed promise. However, they recognized the limitations of the study, especially those resulting from the small number of patients in the group that received surgical treatment, which in turn led to the design of a larger, more sophisticated study. As is often the case, the data analysis cycle led to further research, and the process began anew.

■

■ **Evaluating a Research Study**   The six data analysis steps can also be used as a guide for evaluating published research studies. The following questions should be addressed as part of a study evaluation:

- What were the researchers trying to learn? What questions motivated their research?
- Was relevant information collected? Were the right things measured?
- Were the data collected in a sensible way?
- Were the data summarized in an appropriate way?
- Was an appropriate method of analysis used, given the type of data and how the data were collected?
- Are the conclusions drawn by the researchers supported by the data analysis?

Example 1.4 illustrates how these questions can guide an evaluation of a research study.

....

Example 1.4   Spray Away the Flu

The newspaper article "Spray Away Flu" (*Omaha World-Herald*, June 8, 1998) reported on a study of the effectiveness of a new flu vaccine that is administered by nasal spray rather than by injection. The article states that the "researchers gave the spray to 1070 healthy children, 15 months to 6 years old, before the flu season two winters ago. One percent developed confirmed influenza, compared with 18 percent of the 532 children who received a placebo. And only one vaccinated child developed an ear infection after coming down with influenza. . . . Typically 30 percent to 40 percent of children with influenza later develop an ear infection." The researchers concluded that the nasal flu vaccine was effective in reducing the incidence of flu and also in reducing the number of children with flu who subsequently develop ear infections.

   The researchers here were trying to find out whether the nasal flu vaccine was effective in reducing the number of flu cases and in reducing the number of ear infections in children who did get the flu. They recorded whether a child received the nasal vaccine or a placebo. (A placebo is a treatment that is identical in appearance to the treatment of interest but contains no active ingredients.) Whether or not the child developed the flu and a subsequent ear infection was also noted. These are appropriate determinations to make in order to answer the research question of interest. We typically cannot tell much about the data collection process from a newspaper article. As we will see in Section 2.3, to fully evaluate this study, we would also want to know how the participating children were selected, how it was determined that a particular child received the vaccine or the placebo, and how the subsequent diagnoses of flu and ear infection were made.

We will also have to delay discussion of the data analysis and the appropriateness of the conclusions because we do not yet have the necessary tools to evaluate these aspects of the study.

∎

Many other interesting examples of statistical studies can be found in *Statistics: A Guide to the Unknown* and in *Forty Studies That Changed Psychology: Exploration into the History of Psychological Research* (the complete references for these two books can be found in the back of the book).

∎ Exercises 1.1–1.9 .... Turn to the Expanded Answers Section for answers not shown next to exercises.

1.1 Give a brief definition of the terms *descriptive statistics* and *inferential statistics*.

1.2 Give a brief definition of the terms *population* and *sample*.

1.3 Data from a poll conducted by Travelocity led to the following estimates: Approximately 40% of travelers check work email while on vacation, about 33% take cell phones on vacation in order to stay connected with work, and about 25% bring a laptop computer on vacation (*San Luis Obispo Tribune*, December 1, 2005). Are the given percentages population values or were they computed from a sample?   computed from a sample

1.4 Based on a study of 2121 children between the ages of one and four, researchers at the Medical College of Wisconsin concluded that there was an association between iron deficiency and the length of time that a child is bottle-fed (*Milwaukee Journal Sentinel*, November 26, 2005). Describe the sample and the population of interest for this study.

1.5 The student senate at a university with 15,000 students is interested in the proportion of students who favor a change in the grading system to allow for plus and minus grades (e.g., B+, B, B−, rather than just B). Two hundred students are interviewed to determine their attitude toward this proposed change. What is the population of interest? What group of students constitutes the sample in this problem?

1.6 The supervisors of a rural county are interested in the proportion of property owners who support the construction of a sewer system. Because it is too costly to contact all 7000 property owners, a survey of 500 owners (selected at random) is undertaken. Describe the population and sample for this problem.

1.7 ▼ Representatives of the insurance industry wished to investigate the monetary loss resulting from earthquake damage to single-family dwellings in Northridge, California, in January 1994. From the set of all single-family homes in Northridge, 100 homes were selected for inspection. Describe the population and sample for this problem.

1.8 A consumer group conducts crash tests of new model cars. To determine the severity of damage to 2006 Mazda 6s resulting from a 10-mph crash into a concrete wall, the research group tests six cars of this type and assesses the amount of damage. Describe the population and sample for this problem.

1.9 A building contractor has a chance to buy an odd lot of 5000 used bricks at an auction. She is interested in determining the proportion of bricks in the lot that are cracked and therefore unusable for her current project, but she does not have enough time to inspect all 5000 bricks. Instead, she checks 100 bricks to determine whether each is cracked. Describe the population and sample for this problem.

Bold exercises answered in back   ● Data set available online but not required   ▼ Video solution available

1.4 Types of Data and Some Simple Graphical Displays

Every discipline has its own particular way of using common words, and statistics is no exception. You will recognize some of the terminology from previous math and science courses, but much of the language of statistics will be new to you.

■ Describing Data

The individuals or objects in any particular population typically possess many characteristics that might be studied. Consider a group of students currently enrolled in a statistics course. One characteristic of the students in the population is the brand of calculator owned (Casio, Hewlett-Packard, Sharp, Texas Instruments, and so on). Another characteristic is the number of textbooks purchased that semester, and yet another is the distance from the university to each student's permanent residence. A **variable** is any characteristic whose value may change from one individual or object to another. For example, *calculator brand* is a variable, and so are *number of textbooks purchased* and *distance to the university*. **Data** result from making observations either on a single variable or simultaneously on two or more variables.

A univariate data set consists of observations on a single variable made on individuals in a sample or population. There are two types of univariate data sets: categorical and numerical. In the previous example, *calculator brand* is a categorical variable, because each student's response to the query, "What brand of calculator do you own?" is a category. The collection of responses from all these students forms a categorical data set. The other two attributes, *number of textbooks purchased* and *distance to the university*, are both numerical in nature. Determining the value of such a numerical variable (by counting or measuring) for each student results in a numerical data set.

> **DEFINITION**
>
> A data set consisting of observations on a single attribute is a **univariate data set**. A univariate data set is **categorical** (or **qualitative**) if the individual observations are categorical responses. A univariate data set is **numerical** (or **quantitative**) if each observation is a number.

Example 1.5   Airline Safety Violations

© PhotoLink/Getty Images

The Federal Aviation Administration (FAA) monitors airlines and can take administrative actions for safety violations. Information about the fines assessed by the FAA appeared in the article "Just How Safe Is That Jet?" (*USA Today*, March 13, 2000). Violations that could lead to a fine were categorized as Security (S), Maintenance (M), Flight Operations (F), Hazardous Materials (H), or Other (O). Data for the variable *type of violation* for 20 administrative actions are given in the following list (these data are a subset of the data described in the article, but they are consistent with summary values given in the paper; for a description of the full data set, see Exercise 1.24):

| S | S | M | H | M | O | S | M | S | S |
| F | S | O | M | S | M | S | M | S | M |

Because *type of violation* is a categorical (nonnumerical) response, this is a categorical data set.

■

In Example 1.5, the data set consisted of observations on a single variable (*type of violation*), so this is univariate data. In some studies, attention focuses simultaneously on two different attributes. For example, both height (in inches) and weight (in pounds) might be recorded for each individual in a group. The resulting data set consists of pairs of numbers, such as (68, 146). This is called a **bivariate data set**. **Multivariate data** result from obtaining a category or value for each of two or more attributes (so bivariate data are a special case of multivariate data). For example, multivariate data would result from determining height, weight, pulse rate, and systolic blood pressure for each individual in a group. Example 1.6 illustrates a bivariate data set.

....

Example 1.6   Revisiting Airline Safety Violations

● The same article referenced in Example 1.5 ("Just How Safe Is That Jet?" *USA Today*, March 13, 2000) gave data on both the number of violations and the average fine per violation for the period 1985–1998 for 10 major airlines. The resulting bivariate data are given in the following table:

Airline	Number of Violations	Average Fine per Violation ($)
Alaska	258	5038.760
America West	257	3112.840
American	1745	2693.410
Continental	973	5755.396
Delta	1280	3828.125
Northwest	1097	2643.573
Southwest	535	3925.234
TWA	642	2803.738
United	1110	2612.613
US Airways	891	3479.237

Each of the two variables considered here is numerical (rather than categorical).

■

■ Two Types of Numerical Data ....

There are two different types of numerical data: *discrete* and *continuous*. Visualize a number line (Figure 1.3) for locating values of the numerical variable being studied. Each possible number (2, 3.125, 8.12976, etc.) corresponds to exactly one point on the number line. Now suppose that the variable of interest is the number of cylinders of an automobile engine. The possible values of 4, 6, and 8 are identified in Figure 1.4(a) by

● Data set available online

Figure 1.3 A number line.

Figure 1.4 Possible values of a variable: (a) number of cylinders; (b) quarter-mile time.

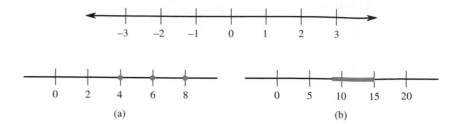

the dots at the points marked 4, 6, and 8. These possible values are isolated from one another on the line; around any possible value, we can place an interval that is small enough that no other possible value is included in the interval. On the other hand, the line segment in Figure 1.4(b) identifies a plausible set of possible values for the time it takes a car to travel one-quarter mile. Here the possible values make up an entire interval on the number line, and no possible value is isolated from the other possible values.

■

Discrete data are usually counts and continuous data are often measurements.

> ### DEFINITION
>
> A numerical variable results in **discrete** data if the possible values of the variable correspond to isolated points on the number line. A numerical variable results in **continuous** data if the set of possible values forms an entire interval on the number line.

Discrete data usually arise when each observation is determined by counting (e.g., the number of classes for which a student is registered or the number of petals on a certain type of flower).

....

Example 1.7 Calls to a Drug Abuse Hotline

● The number of telephone calls per day to a drug abuse hotline is recorded for 12 days. The resulting data set is

> 3   0   4   3   1   0   6   2   0   0   1   2

Possible values for the variable *number of calls* are 0, 1, 2, 3, . . . ; these are isolated points on the number line, so we have a sample consisting of discrete numerical data.

■

The observations on the variable *number of violations* in Example 1.6 are also an example of discrete numerical data. However, in the same example, the variable *average fine per violation* could be 3000, 3000.1, 3000.125, 3000.12476, or any other value in an entire interval, so the observations on this variable provide an example of continuous data. Other examples of continuous data result from determining task completion times, body temperatures, and package weights.

● Data set available online

In general, data are continuous when observations involve making measurements, as opposed to counting. In practice, measuring instruments do not have infinite accuracy, so possible measured values, strictly speaking, do not form a continuum on the number line. However, any number in the continuum *could* be a value of the variable. The distinction between discrete and continuous data will be important in our discussion of probability models.

▪ Frequency Distributions and Bar Charts for Categorical Data ....

An appropriate graphical or tabular display of data can be an effective way to summarize and communicate information. When the data set is categorical, a common way to present the data is in the form of a table, called a *frequency distribution*.

A **frequency distribution for categorical data** is a table that displays the possible categories along with the associated frequencies and/or relative frequencies. The **frequency** for a particular category is the number of times the category appears in the data set. The **relative frequency** for a particular category is the fraction or proportion of the observations resulting in the category. It is calculated as

$$\text{relative frequency} = \frac{\text{frequency}}{\text{number of observations in the data set}}$$

If the table includes relative frequencies, it is sometimes referred to as a **relative frequency distribution**.

....

Example 1.8   Motorcycle Helmets—Can You See Those Ears?

In 2003, the U.S. Department of Transportation established standards for motorcycle helmets. To ensure a certain degree of safety, helmets should reach the bottom of the motorcyclist's ears. The report "Motorcycle Helmet Use in 2005—Overall Results" (National Highway Traffic Safety Administration, August 2005) summarized data collected in June of 2005 by observing 1700 motorcyclists nationwide at selected roadway locations. Each time a motorcyclist passed by, the observer noted whether the rider was wearing no helmet, a noncompliant helmet, or a compliant helmet. Using the coding

$$N = \text{no helmet}$$
$$NH = \text{noncompliant helmet}$$
$$CH = \text{compliant helmet}$$

a few of the observations were

    CH   N   CH   NH   N   CH   CH   CH   N   N

There were also 1690 additional observations, which we didn't reproduce here!
In total, there were 731 riders who wore no helmet, 153 who wore a noncompliant helmet, and 816 who wore a compliant helmet.

The corresponding frequency distribution is given in Table 1.1.

Table 1.1 Frequency Distribution for Helmet Use

Helmet Use Category	Frequency	Relative Frequency
No helmet	731	.430 ←— *731/1700*
Noncompliant helmet	153	.090 ←— *153/1700*
Compliant helmet	816	.480
	1700	1.00

Total number of observations

Should total 1, but in some cases may be slightly off due to rounding

From the frequency distribution, we can see that a large number of riders (43%) were not wearing a helmet, but most of those who wore a helmet were wearing one that met the Department of Transportation safety standard.

∎

A frequency distribution gives a tabular display of a data set. It is also common to display categorical data graphically. A bar chart is one of the most widely used types of graphical displays for categorical data.

∎ Bar Charts

A **bar chart** is a graph of the frequency distribution of categorical data. Each category in the frequency distribution is represented by a bar or rectangle, and the picture is constructed in such a way that the *area* of each bar is proportional to the corresponding frequency or relative frequency.

Bar Charts

When to Use   Categorical data.

How to Construct

1. Draw a horizontal line, and write the category names or labels below the line at regularly spaced intervals.
2. Draw a vertical line, and label the scale using either frequency or relative frequency.
3. Place a rectangular bar above each category label. The height is determined by the category's frequency or relative frequency, and all bars should have the same width. With the same width, both the height and the area of the bar are proportional to frequency and relative frequency.

What to Look For

∎ Frequently and infrequently occurring categories.

Example 1.9   Revisiting Motorcycle Helmets

Example 1.8 used data on helmet use from a sample of 1700 motorcyclists to construct a frequency distribution (Table 1.1). Figure 1.5 shows the bar chart corresponding to this frequency distribution.

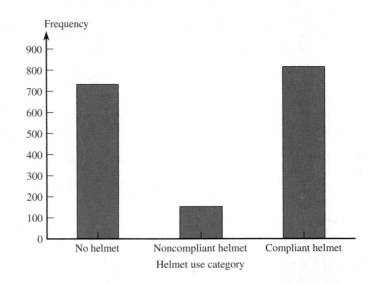

Figure 1.5 Bar chart of helmet use.

The bar chart provides a visual representation of the information in the frequency distribution. From the bar chart, it is easy to see that the compliant helmet use category occurred most often in the data set. The bar for compliant helmets is about five times as tall (and therefore has five times the area) as the bar for noncompliant helmets because approximately five times as many motorcyclists wore compliant helmets than wore noncompliant helmets.

■

■ Dotplots for Numerical Data

A dotplot is a simple way to display numerical data when the data set is reasonably small. Each observation is represented by a dot above the location corresponding to its value on a horizontal measurement scale. When a value occurs more than once, there is a dot for each occurrence and these dots are stacked vertically.

Dotplots

When to Use   Small numerical data sets.

How to Construct

1. Draw a horizontal line and mark it with an appropriate measurement scale.

(continued)

Step-by-step technology instructions available online

2. Locate each value in the data set along the measurement scale, and represent it by a dot. If there are two or more observations with the same value, stack the dots vertically.

What to Look For   Dotplots convey information about:

■ A representative or typical value in the data set.
■ The extent to which the data values spread out.
■ The nature of the distribution of values along the number line.
■ The presence of unusual values in the data set.

Example 1.10 Graduation Rates for NCAA Division I Schools in California and Texas

● *The Chronicle of Higher Education* (Almanac Issue, August 31, 2001) reported graduation rates for NCAA Division I schools. The rates reported are the percentages of full-time freshmen in fall 1993 who had earned a bachelor's degree by August 1999. Data from the two largest states (California, with 20 Division I schools; and Texas, with 19 Division I schools) are as follows:

California:	64	41	44	31	37	73	72	68	35	37
	81	90	82	74	79	67	66	66	70	63
Texas:	67	21	32	88	35	71	39	35	71	63
	12	46	35	39	28	65	25	24	22	

MINITAB, a computer software package for statistical analysis, was used to construct a dotplot of the 39 graduation rates. This dotplot is given in Figure 1.6. From the dotplot, we can see that graduation rates varied a great deal from school to school and that the graduation rates seem to form two distinguishable groups of about the same size—one group with higher graduation rates and one with lower graduation rates.

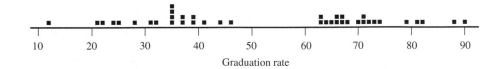

Figure 1.6 MINITAB dotplot of graduation rates.

Figure 1.7 shows separate dotplots for the California and Texas schools. The dotplots are drawn using the same scale to facilitate comparisons. From the two plots, we can see that, although both states have a high group and a low group, there are only six schools in the low group for California and only six schools in the high group for Texas.

Step-by-step technology instructions available online   ● Data set available online

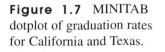
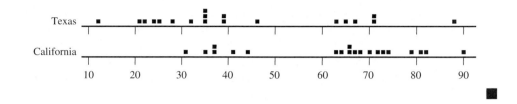

Figure 1.7 MINITAB dotplot of graduation rates for California and Texas.

■

Exercises 1.10–1.26 .... Turn to the Expanded Answers Section for answers not shown next to exercises.

1.10 Classify each of the following attributes as either categorical or numerical. For those that are numerical, determine whether they are discrete or continuous.
a. Number of students in a class of 35 who turn in a term paper before the due date   numerical (discrete)
b. Gender of the next baby born at a particular hospital
c. Amount of fluid (in ounces) dispensed by a machine used to fill bottles with soda pop   numerical (continuous)
d. Thickness of the gelatin coating of a vitamin E capsule
e. Birth order classification (only child, firstborn, middle child, lastborn) of a math major   categorical

1.11 Classify each of the following attributes as either categorical or numerical. For those that are numerical, determine whether they are discrete or continuous.
a. Brand of computer purchased by a customer
b. State of birth for someone born in the United States
c. Price of a textbook   numerical (discrete)
d. Concentration of a contaminant (micrograms per cubic centimeter) in a water sample   numerical (continuous)
e. Zip code (Think carefully about this one.)   categorical
f. Actual weight of coffee in a 1-lb can   numerical (continuous)

1.12 For the following numerical attributes, state whether each is discrete or continuous.
a. The number of insufficient-funds checks received by a grocery store during a given month   discrete
b. The amount by which a 1-lb package of ground beef decreases in weight (because of moisture loss) before purchase   continuous
c. The number of New York Yankees during a given year who will not play for the Yankees the next year   discrete
d. The number of students in a class of 35 who have purchased a used copy of the textbook   discrete

1.13 For the following numerical attributes, state whether each is discrete or continuous.
a. The length of a 1-year-old rattlesnake   continuous
b. The altitude of a location in California selected randomly by throwing a dart at a map of the state   continuous

c. The distance from the left edge at which a 12-in. plastic ruler snaps when bent sufficiently to break   continuous
d. The price per gallon paid by the next customer to buy gas at a particular station   discrete

1.14 For each of the following situations, give a set of possible data values that might arise from making the observations described.
a. The manufacturer for each of the next 10 automobiles to pass through a given intersection is noted.
b. The grade point average for each of the 15 seniors in a statistics class is determined.
c. The number of gas pumps in use at each of 20 gas stations at a particular time is determined.
d. The actual net weight of each of 12 bags of fertilizer having a labeled weight of 50 lb is determined.
e. Fifteen different radio stations are monitored during a 1-hr period, and the amount of time devoted to commercials is determined for each.

1.15 In a survey of 100 people who had recently purchased motorcycles, data on the following variables was recorded:

Gender of purchaser
Brand of motorcycle purchased
Number of previous motorcycles owned by purchaser
Telephone area code of purchaser
Weight of motorcycle as equipped at purchase

a. Which of these variables are categorical?
b. Which of these variables are discrete numerical?
c. Which type of graphical display would be an appropriate choice for summarizing the gender data, a bar chart or a dotplot?   Bar chart
d. Which type of graphical display would be an appropriate choice for summarizing the weight data, a bar chart or a dotplot?   Dotplot

1.16 ● *Spider-Man* and *Star Wars: Episode II* were the top moneymakers among the summer 2002 movie releases.

Box office totals for the top summer films in 2002 are given in the following table (*USA Today*, September 3, 2002):

Film	Box Office (millions of dollars)
Spider-Man	403.7
Star Wars: Episode II	300.1
Austin Powers in Goldmember	203.5
Signs	195.1
Men in Black II	189.7
Scooby-Doo	151.6
Lilo & Stitch	141.8
Minority Report	130.1
Mr. Deeds	124.0
XXX	123.9
The Sum of All Fears	118.5
The Bourne Identity	118.1
Road to Perdition	99.1
My Big Fat Greek Wedding	82.3
Spirit: Stallion of the Cimarron	73.2
Spy Kids 2: Island of Lost Dreams	69.1
Divine Secrets of the Ya-Ya Sisterhood	68.8
Insomnia	67.0
Stuart Little 2	61.9
Unfaithful	52.8

Use a dotplot to display these data. Write a few sentences commenting on the notable features of the dotplot.

1.17 ● The article "Fraud, Identity Theft Afflict Consumers" (*San Luis Obispo Tribune*, February 2, 2005) included the accompanying breakdown of identity theft complaints by type.

Type of Complaint	Percent of All Complaints
Credit card fraud	28%
Phone or utilities fraud	19%
Bank fraud	18%
Employment fraud	13%
Other	22%

Construct a bar chart for these data and write a sentence or two commenting on the most common types of identity theft complaints.

1.18 The U.S. Department of Education reported that 14% of adults were classified as being below a basic literacy level, 29% were classified as being at a basic literacy level, 44% were classified as being at an intermediate literacy level, and 13% were classified as being at a proficient level (*2003 National Assessment of Adult Literacy*).

a. Is the variable *literacy level* categorical or numerical?
b. Would it be appropriate to display the given information using a dotplot? Explain why or why not.
c. Construct a bar chart to display the given data on literacy level.

1.19 ● ▼ The Computer Assisted Assessment Center at the University of Luton published a report titled "Technical Review of Plagiarism Detection Software." The authors of this report asked faculty at academic institutions about the extent to which they agreed with the statement "Plagiarism is a significant problem in academic institutions." The responses are summarized in the accompanying table. Construct a bar chart for these data.

Response	Frequency
Strongly disagree	5
Disagree	48
Not sure	90
Agree	140
Strongly agree	39

1.20 ● A 2005 AP-IPSOS poll found that 21% of American adults surveyed said their child was heavier than doctors recommend. The reasons given as the most important contributing factor to the child's weight problem is summarized in the accompanying table.

Lack of exercise	38%
Easy access to junk food	23%
Genetics	12%
Eating unhealthy food	9%
Medical condition	8%
Overeating	7%

a. Construct a bar chart for the data on the most important contributing factor.
b. Do you think that it would be reasonable to combine some of these contributing factors into a single category? If so, which categories would you combine and why?

1.21 ● ▼ *USA Today* compared the average graduation rates for male football and basketball scholarship players at top-ranked universities. The graduation rates were expressed as a difference between the graduation rate for scholarship athletes and the university's overall graduation rate for men. A positive number means that the athletes had a graduation rate higher than the overall graduation rate (+2 means 2% higher) and a negative number means the athletes had a lower graduation rate than the overall graduation rate (−2 means 2% lower). Using the same scale, construct a dotplot of the graduation rate differences for football players and a dotplot for the graduation rate differences for basketball players. Comment on the similarities and differences in the two plots.

Football		Basketball	
School	Difference from overall graduation rate	School	Difference from overall graduation rate
Nebraska	+8	Butler	+29
Kansas State	+6	Kansas	+20
Arizona State	0	Dayton	+12
Penn State	−2	Stanford	+9
Georgia	−2	Xavier	−1
Purdue	−4	Marquette	−8
Iowa	−4	Michigan State	−10
Florida State	−9	Creighton	−17
Southern Cal	−10	Florida	−23
Miami	−11	Duke	−25
LSU	−12	Texas	−26
Notre Dame	−13	Arizona	−27
Ohio State	−13	Illinois	−29
Tennessee	−13	Wisconsin	−29
Virginia	−14	Notre Dame	−30
N.C. State	−14	Syracuse	−32
Oklahoma	−14	Maryland	−33
Auburn	−19	Connecticut	−38
Virginia Tech	−20	Wake Forest	−39
Colorado	−21		
Wisconsin	−21		
Florida	−24		
Pittsburgh	−24		
Texas	−26		
Michigan	−35		

1.22 ● The article "So Close, Yet So Far: Predictors of Attrition in College Seniors" (*Journal of College Student Development* [1998]: 343–348) examined the reasons that college seniors leave their college programs before graduating. Forty-two college seniors at a large public university who dropped out before graduation were interviewed and asked the main reason for discontinuing enrollment at the university. Data consistent with that given in the article are summarized in the following frequency distribution:

Reason for Leaving the University	Frequency
Academic problems	7
Poor advising or teaching	3
Needed a break	2
Economic reasons	11
Family responsibilities	4
To attend another school	9
Personal problems	3
Other	3

Summarize the given data on reason for leaving the university using a bar chart and write a few sentences commenting on the most common reasons for leaving.

1.23 ● Water quality ratings of 36 Southern California beaches were given in the article "How Safe Is the Surf?" (*Los Angeles Times*, October 26, 2002). The ratings, which ranged from A to F and which reflect the risk of getting sick from swimming at a particular beach, are given in the following list:

A+	A+	A+	F	A+	A+
A	B	A	C	C	A−
A	F	B	A+	C	A
D	F	A+	D	A+	A
D	A	A	D	A+	A+
A	A	C	F−	B	A+

a. Summarize the given ratings by constructing a relative frequency distribution and a bar chart. Comment on the interesting features of your bar chart.
b. Would it be appropriate to construct a dotplot for these data? Why or why not?   No, dotplots require numerical data.

1.24 ● The article "Just How Safe Is That Jet?" (*USA Today*, March 13, 2000) gave the following relative frequency distribution that summarized data on the type of violation for fines imposed on airlines by the Federal Aviation Administration:

Type of Violation	Relative Frequency
Security	.43
Maintenance	.39
Flight operations	.06
Hazardous materials	.03
Other	.09

Use this information to construct a bar chart for type of violation, and then write a sentence or two commenting on the relative occurrence of the various types of violation.

1.25 ▼ The article "Americans Drowsy on the Job and the Road" (Associated Press, March 28, 2001) summarized data from the 2001 Sleep in America poll. Each individual in a sample of 1004 adults was asked questions about his or her sleep habits. The article states that "40 percent of those surveyed say they get sleepy on the job and their work suffers at least a few days each month, while 22 percent said the problems occur a few days each week. And 7 percent say sleepiness on the job is a daily occurrence." Assuming that everyone else reported that sleepiness on the job was not a problem, summarize the given information by constructing a relative frequency bar chart.

1.26 "Ozzie and Harriet Don't Live Here Anymore" (*San Luis Obispo Tribune*, February 26, 2002) is the title of an article that looked at the changing makeup of America's suburbs. The article states that nonfamily households (e.g., homes headed by a single professional or an elderly widow) now outnumber married couples with children in suburbs of the nation's largest metropolitan areas. The article goes on to state:

> In the nation's 102 largest metropolitan areas, "nonfamilies" comprised 29 percent of households in 2000, up from 27 percent in 1990. While the number of married-with-children homes grew too, the share did not keep pace. It declined from 28 percent to 27 percent. Married couples without children at home live in another 29 percent of suburban households. The remaining 15 percent are single-parent homes.

Use the given information on type of household in 2000 to construct a frequency distribution and a bar chart. (Be careful to extract the 2000 percentages from the given information).

Bold exercises answered in back ● Data set available online but not required ▼ Video solution available

Activity **1.1** Head Sizes: Understanding Variability

Materials needed: Each team will need a measuring tape.

For this activity, you will work in teams of 6 to 10 people.

1. Designate a team leader for your team by choosing the person on your team who celebrated his or her last birthday most recently.

2. The team leader should measure and record the head size (measured as the circumference at the widest part of the forehead) of each of the other members of his or her team.

3. Record the head sizes for the individuals on your team as measured by the team leader.

4. Next, each individual on the team should measure the head size of the team leader. Do not share your measurement with the other team members until all team members have measured the team leader's head size.

5. After all team members have measured the team leader's head, record the different team leader head size measurements obtained by the individuals on your team.

6. Using the data from Step 3, construct a dotplot of the team leader's measurements of team head sizes. Then, using the same scale, construct a separate dotplot of the different measurements of the team leader's head size (from Step 5).

Now use the available information to answer the following questions:

7. Do you think the team leader's head size changed in between measurements? If not, explain why the measurements of the team leader's head size are not all the same.

8. Which data set was more variable—head size measurements of the different individuals on your team or the different measurements of the team leader's head size? Explain the basis for your choice.

9. Consider the following scheme (you don't actually have to carry this out): Suppose that a group of 10 people measured head sizes by first assigning each person in the group a number between 1 and 10. Then person 1 measured person 2's head size, person 2 measured person 3's head size, and so on, with person 10 finally measuring person 1's head size. Do you think that the resulting head size measurements would be more variable, less variable, or show about the same amount of variability as a set of 10 measurements resulting from a single individual measuring the head size of all 10 people in the group? Explain.

Activity 1.2   Estimating Sizes

1. Construct an activity sheet that consists of a table that has 6 columns and 10 rows. Label the columns of the table with the following six headings: (1) Shape, (2) Estimated Size, (3) Actual Size, (4) Difference (Est. − Actual), (5) Absolute Difference, and (6) Squared Difference. Enter the numbers from 1 to 10 in the "Shape" column.

2. Next you will be visually estimating the sizes of the shapes in Figure 1.8. Size will be described as the number of squares of this size

that would fit in the shape. For example, the shape

Figure 1.8 Shapes for Activity 1.2.

would be size 3, as illustrated by

You should now quickly *visually* estimate the sizes of the shapes in Figure 1.8. *Do not* draw on the figure—these are to be quick visual estimates. Record your estimates in the "Estimated Size" column of the activity sheet.

3. Your instructor will provide the actual sizes for the 10 shapes, which should be entered into the "Actual Size" column of the activity sheet. Now complete the "Difference" column by subtracting the actual value from your estimate for each of the 10 shapes.

4. What would cause a difference to be negative? positive?

5. Would the sum of the differences tell you if the estimates and actual values were in close agreement? Does a sum of 0 for the differences indicate that all the estimates were equal to the actual value? Explain.

6. Compare your estimates with those of another person in the class by comparing the sum of the absolute values of the differences between estimates and corresponding actual values. Who was better at estimating shape sizes? How can you tell?

7. Use the last column of the activity sheet to record the squared differences (e.g., if the difference for shape 1 was -3, the squared difference would be $(-3)^2 = 9$). Explain why the sum of the squared differences can also be used to assess how accurate your shape estimates were.

8. For this step, work with three or four other students from your class. For each of the 10 shapes, form a new size estimate by computing the average of the size estimates for that shape made by the individuals in your group. Is this new set of estimates more accurate than your own individual estimates were? How can you tell?

9. Does your answer from Step 8 surprise you? Explain why or why not.

Activity 1.3   A Meaningful Paragraph

Write a meaningful paragraph that includes the following six terms: **sample, population, descriptive statistics, bar chart, numerical variable,** and **dotplot**.

A "meaningful paragraph" is a coherent piece of writing in an appropriate context that uses all of the listed words. The paragraph should show that you understand the meaning of the terms and their relationship to one another. A sequence of sentences that just define the terms is *not* a meaningful paragraph. When choosing a context, think carefully about the terms you need to use. Choosing a good context will make writing a meaningful paragraph easier.

Summary of Key Concepts and Formulas

Term or Formula	Comment
Descriptive statistics	Numerical, graphical, and tabular methods for organizing and summarizing data.
Population	The entire collection of individuals or measurements about which information is desired.
Sample	A part of the population selected for study.
Categorical data	Individual observations are categorical responses (nonnumerical).
Numerical data	Individual observations are numerical (quantitative) in nature.
Discrete numerical data	Possible values are isolated points along the number line.
Continuous numerical data	Possible values form an entire interval along the number line.

Term or Formula	Comment
Bivariate and multivariate data	Each observation consists of two (bivariate) or more (multivariate) responses or values.
Frequency distribution for categorical data	A table that displays frequencies, and sometimes relative frequencies, for each of the possible values of a categorical variable.
Bar chart	A graph of a frequency distribution for a categorical data set. Each category is represented by a bar, and the area of the bar is proportional to the corresponding frequency or relative frequency.
Dotplot	A picture of numerical data in which each observation is represented by a dot on or above a horizontal measurement scale.

Chapter Review Exercises 1.27–1.31

Turn to the Expanded Answers Section for answers not shown next to exercises.

CENGAGENOW™ Know exactly what to study! Take a pre-test and receive your Personalized Learning Plan.

1.27 ● Each year, *U.S. News and World Report* publishes a ranking of U.S. business schools. The following data give the acceptance rates (percentage of applicants admitted) for the best 25 programs in the most recent survey:

16.3 12.0 25.1 20.3 31.9 20.7 30.1 19.5 36.2
46.9 25.8 36.7 33.8 24.2 21.5 35.1 37.6 23.9
17.0 38.4 31.2 43.8 28.9 31.4 48.9

Construct a dotplot, and comment on the interesting features of the plot.

1.28 ● Many adolescent boys aspire to be professional athletes. The paper "Why Adolescent Boys Dream of Becoming Professional Athletes" (*Psychological Reports* [1999]: 1075–1085) examined some of the reasons. Each boy in a sample of teenage boys was asked the following question: "Previous studies have shown that more teenage boys say that they are considering becoming professional athletes than any other occupation. In your opinion, why do these boys want to become professional athletes?" The resulting data are shown in the following table:

Response	Frequency
Fame and celebrity	94
Money	56
Attract women	29

Response	Frequency
Like sports	27
Easy life	24
Don't need an education	19
Other	19

Construct a bar chart to display these data.

1.29 ● The paper "Profile of Sport / Leisure Injuries Treated at Emergency Rooms of Urban Hospitals" (*Canadian Journal of Sports Science* [1991]: 99–102) classified noncontact sports injuries by sport, resulting in the following table:

Sport	Number of Sport Injuries
Touch football	38
Soccer	24
Basketball	19
Baseball/softball	11
Jogging/running	11
Bicycling	11
Volleyball	17
Other	47

Bold exercises answered in back   ● Data set available online but not required   ▼ Video solution available

Calculate relative frequencies and draw the corresponding bar chart.

1.30 ● Nonresponse is a common problem facing researchers who rely on mail questionnaires. In the paper "Reasons for Nonresponse on the Physicians' Practice Survey" (*1980 Proceedings of the Section on Social Statistics* [1980]: 202), 811 doctors who did not respond to the AMA Survey of Physicians were contacted about the reason for their nonparticipation. The results are summarized in the accompanying relative frequency distribution. Draw the corresponding bar chart.

Reason	Relative Frequency
1. No time to participate	.264
2. Not interested	.300
3. Don't like surveys in general	.145
4. Don't like this particular survey	.025
5. Hostility toward the government	.054
6. Desire to protect privacy	.056
7. Other reason for refusal	.053
8. No reason given	.103

1.31 ● The article "Can We Really Walk Straight?" (*American Journal of Physical Anthropology* [1992]: 19–27) reported on an experiment in which each of 20 healthy men was asked to walk as straight as possible to a target 60 m away at normal speed. Consider the following observations on cadence (number of strides per second):

0.95 0.85 0.92 1.95 0.93 1.86 1.00
0.92 0.85 0.81 0.78 0.93 0.93 1.05
0.93 1.06 1.06 0.96 0.81 0.96

Construct a dotplot for the cadence data. Do any data values stand out as being unusual?

Personal Tutor

Do you need a live tutor for homework problems?

CENGAGENOW™

Are you ready? Take your exam-prep post-test now.

Bold exercises answered in back   ● Data set available online but not required   ▼ Video solution available

TEACHING TIPS

Intent of this chapter:

After completing this chapter, students should be able to (1) distinguish between observational studies and experiments; (2) identify different sampling designs; (3) identify sources of bias in sampling; (4) describe an appropriate sampling plan given a research objective; (5) understand the principles of randomization, replication, control, and blocking in experimental design; and (6) design completely randomized and randomized block experiments.

This chapter involves a tremendous amount of new vocabulary. It is helpful for many students to see these vocabulary words presented in a variety of contextual scenarios. Be sure to review the vocabulary regularly throughout the chapter.

A PowerPoint® Lecture is available on the *Instructor's Resource Binder* CD.

■ **Section 2.1**

This section distinguishes between observational studies and experiments. It is important to stress that the *best* way to show a cause-and-effect relationship is with well-designed experiments. A common mistake is to draw a cause-and-effect conclusion from an observational study. This is inappropriate because the results of observational studies may be attributable to confounding variables. (A more in-depth discussion of confounding variables is found in Section 2.3.)

Suggested Assignment: Exercises 2.1, 2.4, 2.7, 2.9

■ **Section 2.2**

This section describes four methods for selecting samples: simple random sampling (SRS), stratified random sampling, cluster sampling, and systematic sampling. *Activity-Based Statistics* by Scheaffer, et al. (full citation listed in the *Instructors' Resource Binder* under Great Resources) contains an activity, "Random Rectangles," which demonstrates the importance of randomization in selection of a sample (SRS) from a population. You can modify this activity to demonstrate sampling using stratified, cluster, and systematic methods. This activity is brilliant because it allows you not only to demonstrate each of the sampling methods, but also to prepare students for the concept of sampling distributions. Different samples will produce different results. Thus, a discussion of the amount of variability in the values of statistics produced by the various methods can occur. Students need to realize that in practice *only* one sample is selected, and we typically do not know whether the selected sample produces a good estimate. So we need to have confidence in the method used—we want a method that tends to produce reasonable estimates. The various dotplots produced can also be used to discuss the meaning of bias. (See below for further discussion.)

A simple random sample (as defined on page 35) ". . . is a sample that is selected from a population in a way that ensures that every different possible sample of the desired size has the same chance of being selected." This definition typically doesn't make sense to students unless you provide an example. Suppose a simple random sample of 100 students were to be selected from a large high school (population of 2000). An alphabetical list of students can be numbered 1 to 2000 and, using a random number generator on a calculator or computer, 100 unique random numbers between 1 and 2000, inclusive, can be generated. (If students are using TI calculators, the random number generator needs to be seeded with a unique number for each student, say the students' ID number.) The students corresponding to those 100 numbers would comprise your sample. Using this approach, it is possible, although not likely, that all 100 students are seniors or that all 100 students are freshmen. This is the meaning of the definition—that every different possible *sample* is equally likely.

Simple random sampling ensures that every individual has an equal chance of being selected. Most of the time, sampling is performed without replacement; that is, individuals selected for a sample may not be selected again for the same sample. Further discussion of this topic is found in Section 6.5, pages 317–320.

A common mistake statistics students make is to think that small samples are never appropriate. In exam questions, students often incorrectly state that it is inappropriate to use given data because of a small sample size, missing the fact that the real problem is that the sample is not selected using randomization. The primary advantage of increasing sample size is that the variability in the values of statistics produced from different samples is reduced, but at a greater cost. Students often do not realize that sampling is expensive, and many decisions must be based on information from small samples.

When using stratified random sampling, the population is split into homogeneous groups called strata. Then random samples are selected from each stratum. In the school scenario, one possible choice of strata is grade levels. This would be an appropriate choice for strata if the grades were relatively homogeneous with respect to the characteristic of interest in a particular study. Suppose that each grade level consists of 500 students, and 25 students are randomly selected from each grade level. This is an example of a stratified random sample. (Note that every student in this example has an equal chance of being selected, but every different sample of 100 does not have an equal chance of being selected. That is, it is not possible for all 100 of the students to be seniors.) An advantage that stratified sampling has over simple random sampling is that if the strata are chosen appropriately, the variability in the values of the sample statistics produced is reduced. (If the "Random Rectangle" activity is adapted to use stratified sampling, then this reduction in variability is very evident in the dotplots.)

The cluster sampling method also requires that the population be split into groups called clusters. Unlike stratified sampling, however, clusters are not necessarily homogenous. In fact, the more heterogeneous the clusters are the better. Using the school scenario, cluster sampling can be done by randomly selecting 4 second-period classrooms (approximately 25 students each) and surveying **all** the students in each of those classrooms. Students tend to confuse stratified samples and cluster samples. To clarify, you can make the following distinction: with stratified sampling, a sample consists of *some* of the individuals from each of the strata; whereas with cluster sampling, a sample contains *all* of the individuals from every one of the selected clusters.

Systematic sampling uses a simple algorithm to select the sample. Revisiting the school scenario one more time, remember that our school consists of 2000 students. If we want a sample of 100 students, then we would divide 2000 by 100, obtaining the value 20. We would randomly select a student from the first 20 students in an alphabetical list. Beginning with (and including) that student, we would then select every 20th student on the list to be part of our sample.

This section also discusses the possible sources of bias that can occur in sampling. Students often incorrectly use the term "bias." In repeated sampling, if the value of a sample statistic tends to be too high or too low for some reason, then bias has occurred. If you use the "Random Rectangle" activity, the first step is to ask students to use their judgment to select five rectangles that they think are representative of the entire population of rectangles. A dotplot of the resulting sample means usually tends to be centered around a value that is higher than the true mean size of the population—thus there is bias introduced by the sampling method. When simple random sampling is used to select the five rectangles, the dotplot of the resulting sample means tends to be centered around the true mean size of the population—no bias is introduced by this sampling method.

Some sources of bias are selection bias (undercoverage or voluntary response), measurement or response bias, and nonresponse bias. Students tend to confuse voluntary response and nonresponse bias. If you focus on who is making the selection, it will clarify the difference between the two. In voluntary response, the individual *self-selects* to participate (for example, phone-in talk shows, magazine surveys, or surveys left on tables in a restaurant). However, in nonresponse bias, the individuals are *selected by someone else* to be part of the sample, but the individual refuses to participate or cannot be notified. Other sources of bias include convenience sampling and the wording of the questions (discussed in detail in Section 2.5).

Comprehensive coverage of this section may require three to four 50-minute classes.

Suggested Assignment: Exercises 2.11, 2.13, 2.15, 2.16, 2.21, 2.22, 2.27, 2.28, 2.30

Activity: Activity 2.1 Designing a Sampling Plan (p. 63 and *Activities Manual,* p. 13), "Sampling" (*Instructor's Resource Binder* & CD, Chapter 2 Activities Worksheets)★, and Bonus Activity 2.5: Cluster Sampling (*Activities Manual,* p. 20)★

Assessment: Chapter 2 Quiz 1 (*Test Bank* and *Instructor's Resource Binder* CD)

■ **Section 2.3**

Two basic experimental designs are typically taught in an introductory statistics class: the completely randomized design and the randomized block design. (It's important for students to understand that there are many different experimental designs *and* that some upper-level statistics courses are entirely devoted to experimental design.) In a completely randomized design, the experimental units are assigned at random to treatments (or vice versa). In a randomized block design, the experimental units are placed into blocks that are as homogeneous as possible based on an extraneous variable that could have an effect on the results of the experiment. Once placed into blocks, the experimental units in each block are then assigned at random to treatments. A matched-pairs design is a special case of blocking where individuals are paired so that they are as similar as possible with respect to some characteristic thought to affect the response. One member of each pair is assigned at random to one treatment and then the other one is assigned to the remaining treatment. (A matched-pairs design also arises when individuals receive both treatments in random order; each individual serves as a block.)

There are many variables that could potentially affect the response of an experiment. The variable whose effect we are investigating is called the explanatory variable or factor. Different values of the explanatory variable are called levels of the variable or factor. All variables other than the explanatory variable and the response variable are called extraneous variables. How can we be sure that these extraneous variables' effects on the response are not confounded with the effects of the explanatory variable? One way is to use **direct control**—that is, to keep the value of a variable constant. Suppose that we wanted to test the effects of a fertilizer on the yield of corn. Variables such as type of soil, amount of sunlight, amount of water, and amount of pests could affect the yield of corn. In a lab setting, we can use the same type of soil, provide the same amount of sunlight and water, and control the pests that occur on the plants. However, although a laboratory setting allows us to control many of these extraneous variables, it is often not feasible.

Sometimes it is not possible to directly control an extraneous variable; in that case, using this variable as a **blocking** variable will help mitigate its effects on the response. (One should only block when it is appropriate; not every experiment requires blocking.) By blocking on an extraneous variable, we are able to distinguish the effects of the explanatory variable and the blocking variable on the response. The effect of all extraneous variables that are not controlled or used as blocking variables is mitigated through **randomization**. Randomly assigning experimental units to treatments evens out the effect

★ Kathy's personal favorite

of these extraneous variables, making the treatment groups as similar as possible. (Every experiment must employ random assignment to treatments.) The goal of a well-designed experiment is to ensure that there are *no* confounding variables, which allows us to establish a cause-and-effect relationship between the explanatory variable and the response.

It is helpful to consider the design of various experiments (similar to the activities listed below) to help students understand the importance of replication, random assignment to treatments, direct control, and blocking. Students should be able to identify whether a given scenario is an experiment or an observational study. If it is an experiment, then students should be able to identify whether it is a completely randomized experiment or a randomized block experiment. Students should also be able to identify the experimental units, the explanatory variable and its levels, the blocking variable (if applicable), the treatment groups, and the response variable. In addition, students should be able to describe how experimental units are assigned to treatments, as well as discuss blinding, control groups, and other aspects of experimental design.

This section may require three to four classes.

Suggested Assignment: Exercises 2.32, 2.33, 2.34, 2.38, 2.39

Activity: Activity 2.2: An Experiment to Test the Stroop Effect (p. 64 and *Activities Manual,* p.15)★, Activity 2.3: McDonald's and the Next Billion Burgers (p. 64 and *Activities Manual,* p.17), Activity 2.4: Video Games and Pain Management (p. 64 and *Activities Manual,* p.18)★, Bonus Activity 2.6: Speed Sorting (*Activities Manual,* p. 23)★, and/or "Simple Comparative Experiments" (*Instructor's Resource Binder* & CD, Chapter 2 Activities Worksheets)

■ **Section 2.4**

This section includes additional information on experimental design. Students often mistakenly think that all experiments need a placebo. If we were testing a new asthma drug, we certainly would not want to give patients having an asthma attack a placebo! All experiments do need a comparison group, which in the given example could be a current drug on the market.

It is also helpful to use blinding in an experiment. Students should understand when an experiment is single- versus double-blind.

Suggested Assignment: Exercises 2.44, 2.46, 2.47, 2.49, 2.50

Activity: The activities listed under Section 2.3 can also be used here.

Assessment: Chapter 2 Quiz 2 (*Test Bank* and *Instructor's Resource Binder* CD) and/or Chapter 2 Concept Quiz (*Test Bank* and *Instructor's Resource Binder* CD)

■ **Section 2.5**

This section discusses the issues of survey design, including the wording and order of questions and the types of possible responses. There are interesting examples. This section complements the material in Section 2.2.

Suggested Assignment: Exercises 2.51, 2.54, 2.56

■ **Section 2.6**

Students should pay close attention to the "A Word to the Wise: Cautions and Limitations" section.

Suggested Review Assignment: Exercises 2.57, 2.59, 2.60, 2.65, 2.67, 2.70

Assessment: Chapter 2 Test (*Test Bank* and *Instructor's Resource Binder* CD)

★ Kathy's personal favorite

Chapter 2

Collecting Data Sensibly

© Wide Group/Iconica/Getty Images

Aprimary goal of statistical studies is to collect data that can then be used to make informed decisions. It should come as no surprise that the correctness of such decisions depends on the quality of the information on which they are based. The data collection step is critical to obtaining reliable information; both the type of analysis that is appropriate and the conclusions that can be drawn depend on how the data are collected. In this chapter, we first consider two types of statistical studies and then focus on two widely used methods of data collection: sampling and experimentation.

2.1 Statistical Studies: Observation and Experimentation

Two articles with the following headlines appeared in *USA Today*.

"Prayer Can Lower Blood Pressure" (August 11, 1998)
"Wired Homes Watch 15% Less Television" (August 13, 1998)

In each of these articles, a conclusion is drawn from data. In the study on prayer and blood pressure, 2391 people, 65 years or older, were followed for six years. The article states that people who attended a religious service once a week and prayed or studied the Bible at least once a day were less likely to have high blood pressure. The researcher then concluded that "attending religious services lowers blood pressure."

The second article reported on a study of 5000 families, some of whom had online Internet services. The researcher concluded that TV use was lower in homes that were wired for the Internet, compared to nonwired homes. The researcher who conducted the study cautioned that the study does not answer the question of whether higher family income could be another factor in lower TV usage.

Are these conclusions reasonable? As we will see, the answer in each case depends in a critical way on how the data were collected.

■ Observation and Experimentation

Data collection is an important step in the data analysis process. When we set out to collect information, it is important to keep in mind the questions we hope to answer on the basis of the resulting data. Sometimes we are interested in answering questions about characteristics of a single existing population or in comparing two or more well-defined populations. To accomplish this, we select a sample from each population under consideration and use the sample information to gain insight into characteristics of the population(s).

For example, an ecologist might be interested in estimating the average shell thickness of bald eagle eggs. A social scientist studying a rural community may want to determine whether gender and attitude toward abortion are related. These are examples of studies that are *observational* in nature. In these studies, we want to observe characteristics of members of an existing population or of several populations, and then use the resulting information to draw conclusions. In an observational study, it is important to obtain a sample that is representative of the corresponding population.

Sometimes the questions we are trying to answer deal with the effect of certain explanatory variables on some response and cannot be answered using data from an observational study. Such questions are often of the form, What happens when . . . ? or, What is the effect of . . . ? For example, an educator may wonder what would happen to test scores if the required lab time for a chemistry course were increased from 3 hr to 6 hr per week. To answer such questions, the researcher conducts an experiment to collect relevant data. The value of some response variable (test score in the chemistry example) is recorded under different experimental conditions (3-hr lab and 6-hr lab). In an experiment, the researcher manipulates one or more variables, called **factors**, to create the experimental conditions.

Observational Study

A study is an **observational study** if the investigator observes characteristics of a subset of the members of one or more existing populations. The goal of an observational study is usually to draw conclusions about the corresponding population or about differences between two or more populations.

Experiment

A study is an **experiment** if the investigator observes how a response variable behaves when the researcher manipulates one or more explanatory variables, sometimes also called factors. The usual goal of an experiment is to determine the effect of the manipulated factors on the response variable. In a well-designed experiment, the composition of the groups that will be exposed to different experimental conditions is determined by random assignment.

The type of conclusion that can be drawn from a research study depends on the study design. Both observational studies and experiments can be used to compare groups, but in an experiment the researcher controls who is in which group, whereas this is not the case in an observational study. This seemingly small difference is critical when it comes to the interpretation of results from the study.

Important concept

A well-designed experiment can result in data that provide evidence for a cause-and-effect relationship. This is an important difference between an observational study and an experiment. In an observational study, it is impossible to draw cause-and-effect conclusions because we cannot rule out the possibility that the observed effect is due to some variable other than the factor being studied. Such variables are called confounding variables.

> A **confounding variable** is one that is related to both group membership and the response variable of interest in a research study.

More information on confounding in Section 2.3

Consider the role of confounding variables in the following three studies:

■ The article "Panel Can't Determine the Value of Daily Vitamins" (*San Luis Obispo Tribune*, July 1, 2003) summarized the conclusions of a government advisory panel that investigated the benefits of vitamin use. The panel looked at a large number of studies on vitamin use and concluded that the results were "inadequate or conflicting." A major concern was that many of the studies were observational in nature and the panel worried that people who take vitamins might be healthier just because they tend to take better care of themselves in general. This potential confounding variable prevented the panel from concluding that taking vitamins is the cause of observed better health among those who take vitamins.

■ Studies have shown that people over age 65 who get a flu shot are less likely than those who do not get a flu shot to die from a flu-related illness during the following year. However, recent research has shown that people over age 65 who get a flu shot are also less likely than those who don't to die from *any* cause during the following year (*International Journal of Epidemiology*, December 21, 2005). This has lead to the speculation that those over age 65 who get flu shots are healthier as a group than those who do not get flu shots. If this is the case, observational studies that compare two groups—those who get flu shots and those who do not—may overestimate the effectiveness of the flu vaccine because general health differs in the two groups and so general health would be a possible confounding variable in such studies.

■ The article "Heartfelt Thanks to Fido" (*San Luis Obispo Tribune*, July 5, 2003) summarized a study that appeared in the *American Journal of Cardiology* (March 15, 2003). In this study researchers measured heart rate variability (a measure of the heart's ability to handle stress) in patients who had recovered from a heart attack. They found that heart rate variability was higher (which is good and means the heart can handle stress better) for those who owned a dog than for those who did not. Should someone who suffers a heart attack immediately go out and get a dog? Well, maybe not yet. The American Heart Association recommends additional studies to determine if the improved heart rate variability is attributable to dog ownership or due to the fact that dog owners get more exercise. If in fact dog owners do tend to get more exercise than nonowners, level of exercise is a confounding variable that would prevent us from concluding that owning a dog is the cause of improved heart rate variability.

Each of the three studies described above illustrates why potential confounding variables make it unreasonable to draw a cause-and-effect conclusion from an observational study.

Returning to the studies reported in the *USA Today* articles described at the beginning of this section, notice that both are observational studies. In the TV viewing example, the researcher merely observed the TV viewing behavior of individuals in two groups (wired and nonwired) but did not control which individuals were in which group. A possible confounding variable here is family income. Although it may be reasonable to conclude that the wired and nonwired groups differ with respect to TV viewing, it is not reasonable to conclude that the lower TV use in wired homes is caused by the fact that they have Internet service. In the prayer example, two groups were compared (those who attend a religious service at least once a week and those who do not), but the researcher did not manipulate this factor to observe the effect on blood pressure by assigning people at random to service-attending or nonservice-attending groups. As a result, cause-and-effect conclusions such as "prayer can lower blood pressure" are not reasonable based on the observed data. It is possible that some other variable, such as lifestyle, is related to both religious service attendance and blood pressure. In this case, lifestyle is an example of a potential confounding variable.

▪ Drawing Conclusions from Statistical Studies

In this section, two different types of conclusions have been described. One type involves generalizing from what we have seen in a sample to some larger population, and the other involves reaching a cause-and-effect conclusion about the effect of an explanatory variable on a response. When is it reasonable to draw such conclusions? The answer depends on the way that the data were collected. Table 2.1 summarizes the types of conclusions that can be made with different study designs.

Table 2.1 Drawing Conclusions from Statistical Studies

Study Description	Reasonable to Generalize Conclusions about Group Characteristics to the Population?	Reasonable to Draw Cause-and-Effect Conclusion?
Observational study with sample selected at random from population of interest	Yes	No
Observational study based on convenience or voluntary response sample (poorly designed sampling plan)	No	No
Experiment with groups formed by random assignment of individuals or objects to experimental conditions		
Individuals or objects used in study are volunteers or not randomly selected from some population of interest	No	Yes
Individuals or objects used in study are randomly selected from some population of interest	Yes	Yes
Experiment with groups not formed by random assignment to experimental conditions (poorly designed experiment)	No	No

As you can see from Table 2.1, it is important to think carefully about the objectives of a statistical study before planning how the data will be collected. Both observational studies and experiments must be carefully designed if the resulting data are to be useful. The common sampling procedures used in observational studies are considered in Section 2.2. In Sections 2.3 and 2.4, we consider experimentation and explore what constitutes good practice in the design of simple experiments.

■ Exercises 2.1–2.9 Turn to the Expanded Answers Section for answers not shown next to exercises.

2.1 ▼ The article "Television's Value to Kids: It's All in How They Use It" (*Seattle Times*, July 6, 2005) described a study in which researchers analyzed standardized test results and television viewing habits of 1700 children. They found that children who averaged more than two hours of television viewing per day when they were younger than 3 tended to score lower on measures of reading ability and short term memory.
a. Is the study described an observational study or an experiment?
b. Is it reasonable to conclude that watching two or more hours of television is the cause of lower reading scores? Explain.

2.2 "Fruit Juice May Be Fueling Pudgy Preschoolers, Study Says" is the title of an article that appeared in the *San Luis Obispo Tribune* (February 27, 2005). This article describes a study that found that for 3- and 4-year-olds, drinking something sweet once or twice a day doubled the risk of being seriously overweight one year later. The authors of the study state

> Total energy may be a confounder if consumption of sweet drinks is a marker for other dietary factors associated with overweight . . . (*Pediatrics*, November 2005)

Give an example of a dietary factor that might be one of the potentially confounding variables the study authors are worried about.

2.3 The article "Americans are 'Getting the Wrong Idea' on Alcohol and Health" (Associated Press, April 19, 2005) reported that observational studies in recent years that have concluded that moderate drinking is associated with a reduction in the risk of heart disease may be misleading. The article refers to a study conducted by the Centers for Disease Control and Prevention that showed that moderate drinkers, as a group, tended to be better educated, wealthier, and more active than nondrinkers. Explain why the existence of these potentially confounding factors prevent drawing the conclusion that moderate drinking is the cause of reduced risk of heart disease.

2.4 An article titled "Guard Your Kids Against Allergies: Get Them a Pet" (*San Luis Obispo Tribune*, August 28, 2002) described a study that led researchers to conclude that "babies raised with two or more animals were about half as likely to have allergies by the time they turned six."
a. Do you think this study was an observational study or an experiment? Explain.
b. Describe a potential confounding variable that illustrates why it is unreasonable to conclude that being raised with two or more animals is the cause of the observed lower allergy rate.

2.5 Researchers at the Hospital for Sick Children in Toronto compared babies born to mothers with diabetes to babies born to mothers without diabetes ("Conditioning and Hyperanalgesia in Newborns Exposed to Repeated Heel Lances," *Journal of the American Medical Association* [2002]: 857–861). Babies born to mothers with diabetes have their heels pricked numerous times during the first 36 hours of life in order to obtain blood samples to monitor blood sugar level. The researchers noted that the babies born to diabetic mothers were more likely to grimace or cry when having blood drawn than the babies born to mothers without diabetes. This led the researchers to conclude that babies who experience pain early in life become highly sensitive to pain. Comment on the appropriateness of this conclusion.

2.6 Based on a survey conducted on the DietSmart.com web site, investigators concluded that women who regularly watched *Oprah* were only one-seventh as likely to crave fattening foods as those who watched other daytime talk shows (*San Luis Obispo Tribune*, October 14, 2000).
a. Is it reasonable to conclude that watching *Oprah* causes a decrease in cravings for fattening foods? Explain.
b. Is it reasonable to generalize the results of this survey to all women in the United States? To all women who watch daytime talk shows? Explain why or why not.

2.7 ▼ A survey of affluent Americans (those with incomes of $75,000 or more) indicated that 57% would rather have more time than more money (*USA Today*, January 29, 2003).
a. What condition on how the data were collected would make the generalization from the sample to the population of affluent Americans reasonable?
b. Would it be reasonable to generalize from the sample to say that 57% of all Americans would rather have more time than more money? Explain.

2.8 Does living in the South cause high blood pressure? Data from a group of 6278 whites and blacks questioned in the Third National Health and Nutritional Examination Survey between 1988 and 1994 (see CNN.com web site article of January 6, 2000, titled "High Blood Pressure Greater Risk in U.S. South, Study Says") indicates that a greater percentage of Southerners have high blood pressure than do people in any other region of the United States. This difference in rate of high blood pressure was found in every ethnic group, gender, and age category studied. List at least two possible reasons we cannot conclude that living in the South causes high blood pressure.

2.9 Does eating broccoli reduce the risk of prostate cancer? According to an observational study from the Fred Hutchinson Cancer Research Center (see CNN.com web site article titled "Broccoli, Not Pizza Sauce, Cuts Cancer Risk, Study Finds," January 5, 2000), men who ate more cruciferous vegetables (broccoli, cauliflower, brussels sprouts, and cabbage) had a lower risk of prostate cancer. This study made separate comparisons for men who ate different levels of vegetables. According to one of the investigators, "at any given level of total vegetable consumption, as the percent of cruciferous vegetables increased, the prostate cancer risk decreased." Based on this study, is it reasonable to conclude that eating cruciferous vegetables causes a reduction in prostate cancer risk? Explain.

Bold exercises answered in back ● Data set available online but not required ▼ Video solution available

2.2 Sampling

Many studies are conducted in order to generalize from a sample to the corresponding population. As a result, it is important that the sample be representative of the population. To be reasonably sure of this, the researcher must carefully consider the way in which the sample is selected. It is sometimes tempting to take the easy way out and gather data in a haphazard way; but if a sample is chosen on the basis of convenience alone, it becomes impossible to interpret the resulting data with confidence. For example, it might be easy to use the students in your statistics class as a sample of students at your university. However, not all majors include a statistics course in their curriculum, and most students take statistics in their sophomore or junior year. The difficulty is that it is not clear whether or how these factors (and others that we might not be aware of) affect inferences based on information from such a sample.

There is no way to tell just by looking at a sample whether it is representative of the population from which it was drawn. Our only assurance comes from the method used to select the sample. ∎

There are many reasons for selecting a sample rather than obtaining information from an entire population (a **census**). Sometimes the process of measuring the characteristics of interest is destructive, as with measuring the lifetime of flashlight batteries or the sugar content of oranges, and it would be foolish to study the entire population. But the most common reason for selecting a sample is limited resources; restrictions on available time or money usually prohibit observation of an entire population.

■ Bias in Sampling

If you select a sample from a phone book, you also miss the new generation of cell phone users.

Bias in sampling is the tendency for samples to differ from the corresponding population in some systematic way. Bias can result from the way in which the sample is selected or from the way in which information is obtained once the sample has been chosen. The most common types of bias encountered in sampling situations are selection bias, measurement or response bias, and nonresponse bias.

Selection bias (sometimes also called undercoverage) is introduced when the way the sample is selected systematically excludes some part of the population of interest. For example, a researcher may wish to generalize from the results of a study to the population consisting of all residents of a particular city, but the method of selecting individuals may exclude the homeless or those without telephones. If those who are excluded from the sampling process differ in some systematic way from those who are included, the sample is virtually guaranteed to be unrepresentative of the population. If this difference between the included and the excluded occurs on a variable that is important to the study, conclusions based on the sample data may not be valid for the population of interest. Selection bias also occurs if only volunteers or self-selected individuals are used in a study, because self-selected individuals (e.g., those who choose to participate in a call-in telephone poll) may well differ from those who choose not to participate.

Measurement or response bias occurs when the method of observation tends to produce values that systematically differ from the true value in some way. This might happen if an improperly calibrated scale is used to weigh items or if questions on a survey are worded in a way that tends to influence the response. For example, a Gallup survey sponsored by the American Paper Institute (*Wall Street Journal*, May 17, 1994) included the following question: "It is estimated that disposable diapers account for less than 2 percent of the trash in today's landfills. In contrast, beverage containers, third-class mail and yard waste are estimated to account for about 21 percent of trash in landfills. Given this, in your opinion, would it be fair to tax or ban disposable diapers?" It is likely that the wording of this question prompted people to respond in a particular way.

Other things that might contribute to response bias are the appearance or behavior of the person asking the question, the group or organization conducting the study, and the tendency for people not to be completely honest when asked about illegal behavior or unpopular beliefs.

Although the terms *measurement bias* and *response bias* are often used interchangeably, the term *measurement bias* is usually used to describe systematic deviation from the true value as a result of a faulty measurement instrument (as with the improperly calibrated scale).

Nonresponse bias occurs when responses are not obtained from all individuals selected for inclusion in the sample. As with selection bias, nonresponse bias can distort results if those who respond differ in important ways from those who do not respond. Although some level of nonresponse is unavoidable in most surveys, the biasing effect on the resulting sample is lowest when the response rate is high. To minimize nonresponse bias, it is critical that a serious effort be made to follow up with individuals who do not respond to an initial request for information.

The nonresponse rate for surveys or opinion polls varies dramatically, depending on how the data are collected. Surveys are commonly conducted by mail, by phone, and by personal interview. Mail surveys are inexpensive but often have high nonresponse rates. Telephone surveys can also be inexpensive and can be implemented quickly, but they work well only for short surveys and they can also have high nonresponse rates. Personal interviews are generally expensive but tend to have better response rates. Some of the many challenges of conducting surveys are discussed in Section 2.5.

Types of Bias

Selection Bias

Tendency for samples to differ from the corresponding population as a result of systematic exclusion of some part of the population.

Measurement or Response Bias

Tendency for samples to differ from the corresponding population because the method of observation tends to produce values that differ from the true value.

Nonresponse Bias

Tendency for samples to differ from the corresponding population because data are not obtained from all individuals selected for inclusion in the sample.

Important concept

It is important to note that bias is introduced by the way in which a sample is selected or by the way in which the data are collected from the sample. Increasing the size of the sample, although possibly desirable for other reasons, does nothing to reduce bias if the method of selecting the sample is flawed or if the nonresponse rate remains high. A good discussion of the types of bias appears in the sampling book by Lohr listed in the references in the back of the book.

■ Random Sampling ....

Most of the inferential methods introduced in this text are based on the idea of random selection. The most straightforward sampling method is called simple random sampling. A **simple random sample** is a sample chosen using a method that ensures that each different possible sample of the desired size has an equal chance of being the one chosen. For example, suppose that we want a simple random sample of 10 employees chosen from all those who work at a large design firm. For the sample to be a simple random sample, each of the many different subsets of 10 employees must be equally likely to be the one selected. A sample taken from only full-time employees would not be a simple random sample of *all* employees, because someone who works part-time has no chance of being selected. Although a simple random sample may, by chance, include only full-time employees, it must be selected in such a way that each possible sample, and therefore *every* employee, has the same chance of inclusion in the sample. *It is the selection process, not the final sample, which determines whether the sample is a simple random sample.*

The letter *n* is used to denote sample size; it is the number of individuals or objects in the sample. For the design firm scenario of the previous paragraph, $n = 10$.

> **DEFINITION**
>
> A **simple random sample of size *n*** is a sample that is selected from a population in a way that ensures that every different possible sample of the desired size has the same chance of being selected.

The definition of a simple random sample implies that every individual member of the population has an equal chance of being selected. *However, the fact that every individual has an equal chance of selection, by itself, is not enough to guarantee that the sample is a simple random sample.* For example, suppose that a class is made up of 100 students, 60 of whom are female. A researcher decides to select 6 of the female students by writing all 60 names on slips of paper, mixing the slips, and then picking 6. She then selects 4 male students from the class using a similar procedure. Even though every student in the class has an equal chance of being included in the sample (6 of 60 females are selected and 4 of 40 males are chosen), the resulting sample is *not* a simple random sample because not all different possible samples of 10 students from the class have the same chance of selection. Many possible samples of 10 students—for example, a sample of 7 females and 3 males or a sample of all females—have no chance of being selected. The sample selection method described here is not necessarily a bad choice (in fact, it is an example of stratified sampling, to be discussed in more detail shortly), but it does not produce a simple random sample; this must be considered when a method is chosen for analyzing data resulting from such a sampling method.

■ **Selecting a Simple Random Sample**   A number of different methods can be used to select a simple random sample. One way is to put the name or number of each member of the population on different but identical slips of paper. The process of thoroughly mixing the slips and then selecting *n* slips one by one yields a random sample of size *n*. This method is easy to understand, but it has obvious drawbacks The mixing must be adequate, and producing the necessary slips of paper can be extremely tedious, even for populations of moderate size.

A commonly used method for selecting a random sample is to first create a list, called a **sampling frame**, of the objects or individuals in the population. Each item on the list can then be identified by a number, and a table of random digits or a random number generator can be used to select the sample. A random number generator is a procedure that produces a sequence of numbers that satisfies properties associated with the notion of randomness. Most statistics software packages include a random number generator, as do many calculators. A small table of random digits can be found in Appendix A Table 1.

When selecting a random sample, researchers can choose to do the sampling with or without replacement. **Sampling with replacement** means that after each successive item is selected for the sample, the item is "replaced" back into the population and may therefore be selected again at a later stage. In practice, sampling with replacement is rarely used. Instead, the more common method is to not allow the same item to be included in the sample more than once. After being included in the sample, an individual or object would not be considered for further selection. Sampling in this manner is called **sampling without replacement**.

Sampling without Replacement

Once an individual from the population is selected for inclusion in the sample, it may not be selected again in the sampling process. A sample selected without replacement includes *n* distinct individuals from the population.

Sampling with Replacement

After an individual from the population is selected for inclusion in the sample and the corresponding data are recorded, the individual is placed back in the population and can be selected again in the sampling process. A sample selected with replacement might include any particular individual from the population more than once.

Although these two forms of sampling are different, when the sample size *n* is small relative to the population size, as is often the case, there is little practical difference between them. In practice, the two can be viewed as equivalent if the sample size is at most 5% of the population size. (A less conservative approach that allows the sample size to be up to 10% of the population size is used in some texts.)

Example 2.1   Selecting a Random Sample of Glass Soda Bottles

Breaking strength is an important characteristic of glass soda bottles. Suppose that we want to measure the breaking strength of each bottle in a random sample of size $n = 3$ selected from four crates containing a total of 100 bottles (the population). Each crate contains five rows of five bottles each. We can identify each bottle with a number from 1 to 100 by numbering across the rows, starting with the top row of crate 1, as pictured:

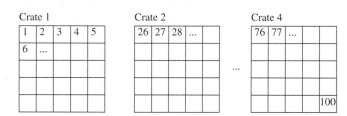

Using a random number generator from a calculator or statistical software package, we could generate three random numbers between 1 and 100 to determine which bottles would be included in our sample. This might result in bottles 15 (row 3 column 5 of crate 1), 89 (row 3 column 4 of crate 4), and 60 (row 2 column 5 of crate 3) being selected.

The goal of random sampling is to produce a sample that is likely to be representative of the population. Although random sampling does not *guarantee* that the sample will be representative, probability-based methods can be used to assess the risk of an unrepresentative sample. It is the ability to quantify this risk that allows us to generalize with confidence from a random sample to the corresponding population.

Step-by-step technology instructions available online

■ A Note Concerning Sample Size ....

It is a common misconception that if the size of a sample is relatively small compared to the population size, the sample can't possibly accurately reflect the population. Critics of polls often make statements such as, "There are 14.6 million registered voters in California. How can a sample of 1000 registered voters possibly reflect public opinion when only about 1 in every 14,000 people is included in the sample?" These critics do not understand the power of random selection!

Consider a population consisting of 5000 applicants to a state university, and suppose that we are interested in math SAT scores for this population. A dotplot of the values in this population is shown in Figure 2.1(a). Figure 2.1(b) shows dotplots of the math SAT scores for individuals in five different random samples from the population, ranging in sample size from $n = 50$ to $n = 1000$. Notice that the samples tend to reflect the distribution of scores in the population. If we were interested in using the sample to estimate the population average or to say something about the variability in SAT scores, even the smallest of the samples ($n = 50$) pictured would provide reliable information. Although it is possible to obtain a simple random sample that does not do a reasonable job of representing the population, this is likely only when the sample size is very small, and unless the population itself is small, this risk does not depend on what fraction of the population is sampled. The random selection process allows us to be confident that

Figure 2.1 (a) Dotplot of math SAT scores for the entire population. (b) Dotplots of math SAT scores for random samples of sizes 50, 100, 250, 500, and 1000.

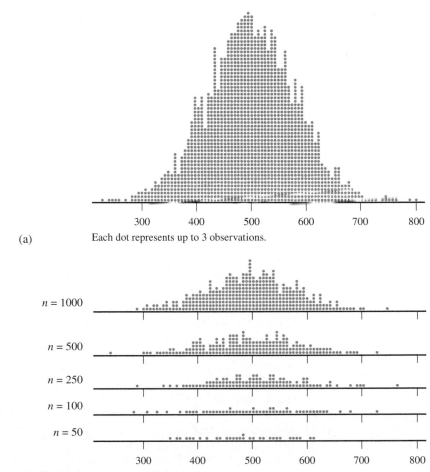

(a)   Each dot represents up to 3 observations.

(b)   Each dot represents up to 3 observations.

the resulting sample adequately reflects the population, even when the sample consists of only a small fraction of the population.

■ Other Sampling Methods ....

Simple random sampling provides researchers with a sampling method that is objective and free of selection bias. In some settings, however, alternative sampling methods may be less costly, easier to implement, or more accurate.

■ **Stratified Random Sampling**   When the entire population can be divided into a set of nonoverlapping subgroups, a method known as **stratified sampling** often proves easier to implement and more cost-effective than simple random sampling. In stratified random sampling, separate simple random samples are independently selected from each subgroup. For example, to estimate the average cost of malpractice insurance, a researcher might find it convenient to view the population of all doctors practicing in a particular metropolitan area as being made up of four subpopulations: (1) surgeons, (2) internists and family practitioners, (3) obstetricians, and (4) a group that includes all other areas of specialization. Rather than taking a random simple sample from the population of all doctors, the researcher could take four separate simple random samples— one from the group of surgeons, another from the internists and family practitioners, and so on. These four samples would provide information about the four subgroups as well as information about the overall population of doctors.

When the population is divided in this way, the subgroups are called **strata** and each individual subgroup is called a stratum (the singular of strata). Stratified sampling entails selecting a separate simple random sample from each stratum. Stratified sampling can be used instead of simple random sampling if it is important to obtain information about characteristics of the individual strata as well as of the entire population, although a stratified sample is not required to do this—subgroup estimates can also be obtained by using an appropriate subset of data from a simple random sample.

The real advantage of stratified sampling is that it often allows us to make more accurate inferences about a population than does simple random sampling. In general, it is much easier to produce relatively accurate estimates of characteristics of a homogeneous group than of a heterogeneous group. For example, even with a small sample, it is possible to obtain an accurate estimate of the average grade point average (GPA) of students graduating with high honors from a university. The individual GPAs of these students are all quite similar (a homogeneous group), and even a sample of three or four individuals from this subpopulation should be representative. On the other hand, producing a reasonably accurate estimate of the average GPA of *all* seniors at the university, a much more diverse group of GPAs, is a more difficult task. Thus, if a varied population can be divided into strata, with each stratum being much more homogeneous than the population with respect to the characteristic of interest, then a stratified random sample can produce more accurate estimates of population characteristics than a simple random sample of the same size.

■ **Cluster Sampling**   Sometimes it is easier to select groups of individuals from a population than it is to select individuals themselves. **Cluster sampling** involves dividing the population of interest into nonoverlapping subgroups, called **clusters**. Clusters are then selected at random, and all individuals in the selected clusters are included in the sample. For example, suppose that a large urban high school has 600 senior students, all of whom are enrolled in a first period homeroom. There are 24 senior homerooms, each with approximately 25 students. If school administrators wanted to select a sample of roughly 75 seniors to participate in an evaluation of the college and career placement advising available to students, they might find it much easier to select three of the senior homerooms at random and then include all the students in the selected homerooms in the sample. In this way, an evaluation survey could be administered to all students in the selected homerooms at the same time—certainly easier logistically than randomly selecting 75 students and then administering the survey to the individual seniors.

Because whole clusters are selected, the ideal situation occurs when each cluster mirrors the characteristics of the population. When this is the case, a small number of clusters results in a sample that is representative of the population. If it is not reasonable to think that the variability present in the population is reflected in each cluster, as is often the case when the cluster sizes are small, then it becomes important to ensure that a large number of clusters are included in the sample.

Be careful not to confuse clustering and stratification. Even though both of these sampling strategies involve dividing the population into subgroups, both the way in which the subgroups are sampled and the optimal strategy for creating the subgroups are different. In stratified sampling, we sample from every stratum, whereas in cluster sampling, we include only selected whole clusters in the sample. Because of this difference, to increase the chance of obtaining a sample that is representative of the population, we want to create homogeneous (similar) groups for strata and heterogeneous (reflecting the variability in the population) groups for clusters.

■ **Systematic Sampling**   **Systematic sampling** is a procedure that can be used when it is possible to view the population of interest as consisting of a list or some other sequential arrangement. A value k is specified (e.g., $k = 50$ or $k = 200$). Then one of the first k individuals is selected at random, after which every kth individual in the sequence is included in the sample. A sample selected in this way is called a **1 in k systematic sample**.

For example, a sample of faculty members at a university might be selected from the faculty phone directory. One of the first $k = 20$ faculty members listed could be selected at random, and then every 20th faculty member after that on the list would also be included in the sample. This would result in a 1 in 20 systematic sample.

The value of k for a 1 in k systematic sample is generally chosen to achieve a desired sample size. For example, in the faculty directory scenario just described, if there were 900 faculty members at the university, the 1 in 20 systematic sample described would result in a sample size of 45. If a sample size of 100 was desired, a 1 in 9 systematic sample could be used.

As long as there are no repeating patterns in the population list, systematic sampling works reasonably well. However, if there are such patterns, systematic sampling can result in an unrepresentative sample.

■ **Convenience Sampling: Don't Go There!**   It is often tempting to resort to "convenience" sampling—that is, using an easily available or convenient group to form a sample. This is a recipe for disaster! Results from such samples are rarely informative, and it is a mistake to try to generalize from a convenience sample to any larger population.

One common form of convenience sampling is sometimes called **voluntary response sampling**. Such samples rely entirely on individuals who volunteer to be a part of the sample, often by responding to an advertisement, calling a publicized telephone number to register an opinion, or logging on to an Internet site to complete a survey. It is extremely unlikely that individuals participating in such voluntary response surveys are representative of any larger population of interest.

2.10 As part of a curriculum review, the psychology department would like to select a simple random sample of 20 of last year's 140 graduates to obtain information on how graduates perceived the value of the curriculum. Describe two different methods that might be used to select the sample.

2.11 A petition with 500 signatures is submitted to a university's student council. The council president would like to determine the proportion of those who signed the petition who are actually registered students at the university. There is not enough time to check all 500 names with the registrar, so the council president decides to select a simple random sample of 30 signatures. Describe how this might be done.

2.12 During the previous calendar year, a county's small claims court processed 870 cases. Describe how a simple random sample of size $n = 50$ might be selected from the case files to obtain information regarding the average award in such cases.

2.13 The financial aid officers of a university wish to estimate the average amount of money that students spend on textbooks each term. For each of the following proposed stratification schemes, discuss whether it would be worthwhile to stratify the university students in this manner.
a. Strata corresponding to class standing (freshman, sophomore, junior, senior, graduate student)
b. Strata corresponding to field of study, using the following categories: engineering, architecture, business, other
c. Strata corresponding to the first letter of the last name: A–E, F–K, etc.

2.14 Suppose that a group of 1000 orange trees is laid out in 40 rows of 25 trees each. To determine the sugar content of fruit from a sample of 30 trees, researcher A suggests randomly selecting five rows and then randomly selecting six trees from each sampled row. Researcher B suggests numbering each tree on a map of the trees from 1 to 1000 and using random numbers to select 30 of the trees. Which selection method is preferred? Explain.

2.15 ▼ For each of the situations described, state whether the sampling procedure is simple random sampling, stratified random sampling, cluster sampling, systematic sampling, or convenience sampling.
a. All freshmen at a university are enrolled in 1 of 30 sections of a seminar course. To select a sample of freshmen at this university, a researcher selects 4 sections of the seminar course at random from the 30 sections and all students in the 4 selected sections are included in the sample. *Cluster sampling*
b. To obtain a sample of students, faculty, and staff at a university, a researcher randomly selects 50 faculty members from a list of faculty, 100 students from a list of students, and 30 staff members from a list of staff.
c. A university researcher obtains a sample of students at his university by using the 85 students enrolled in his Psychology 101 class. *Convenience sampling*
d. To obtain a sample of the seniors at a particular high school, a researcher writes the name of each senior on a slip of paper, places the slips in a box and mixes them, and then selects 10 slips. The students whose names are on the selected slips of paper are included in the sample.
e. To obtain a sample of those attending a basketball game, a researcher selects the 24th person through the door. Then, every 50th person after that is also included in the sample. *Systematic sampling*

2.16 Of the 6500 students enrolled at a community college, 3000 are part time and the other 3500 are full time. The college can provide a list of students that is sorted so that all full-time students are listed first, followed by the part-time students.
a. Select a stratified random sample that uses full-time and part-time students as the two strata and that includes 10 students from each stratum. Describe the procedure you used to select the sample, and identify the students included in your sample by placement on the sorted list.
b. Does every student at this community college have the same chance of being selected for inclusion in the sample? Explain.

2.17 Briefly explain why it is advisable to avoid the use of convenience samples.

2.18 Sometimes samples are composed entirely of volunteer responders. Give a brief description of the dangers of using voluntary response samples.

2.19 A sample of pages from this book is to be obtained, and the number of words on each selected page will be determined. For the purposes of this exercise, equations are not counted as words and a number is counted as a word only if it is spelled out—that is, *ten* is counted as a word, but *10* is not.

a. Describe a sampling procedure that would result in a simple random sample of pages from this book.
b. Describe a sampling procedure that would result in a stratified random sample. Explain why you chose the specific strata used in your sampling plan.
c. Describe a sampling procedure that would result in a systematic sample.
d. Describe a sampling procedure that would result in a cluster sample.
e. Using the process you gave in Part (a), select a simple random sample of at least 20 pages, and record the number of words on each of the selected pages. Construct a dotplot of the resulting sample values, and write a sentence or two commenting on what it reveals about the number of words on a page.
f. Using the process you gave in Part (b), select a stratified random sample that includes a total of at least 20 selected pages, and record the number of words on each of the selected pages. Construct a dotplot of the resulting sample values, and write a sentence or two commenting on what it reveals about the number of words on a page.

2.20 In 2000, the chairman of a California ballot initiative campaign to add "none of the above" to the list of ballot options in all candidate races was quite critical of a Field poll that showed his measure trailing by 10 percentage points. The poll was based on a random sample of 1000 registered voters in California. He is quoted by the Associated Press (January 30, 2000) as saying, "Field's sample in that poll equates to one out of 17,505 voters," and he added that this was so dishonest that Field should get out of the polling business! If you worked on the Field poll, how would you respond to this criticism?

2.21 A pollster for the Public Policy Institute of California explains how the Institute selects a sample of California adults ("It's About Quality, Not Quantity," *San Luis Obispo Tribune*, January 21, 2000):

> That is done by using computer-generated random residential telephone numbers with all California prefixes, and when there are no answers, calling back repeatedly to the original numbers selected to avoid a bias against hard-to-reach people. Once a call is completed, a second random selection is made by asking for the adult in the household who had the most recent birthday. It is as important to randomize who you speak to in the household as it is to randomize the household you select. If you didn't, you'd primarily get women and older people.

Comment on this approach to selecting a sample. How does the sampling procedure attempt to minimize certain types of bias? Are there sources of bias that may still be a concern?

2.22 The report "Undergraduate Students and Credit Cards in 2004: An Analysis of Usage Rates and Trends" (Nellie Mae, May 2005) estimated that 21% of undergraduates with credit cards pay them off each month and that the average outstanding balance on undergraduates' credit cards is $2169. These estimates were based on an online survey that was sent to 1260 students. Responses were received from 132 of these students. Is it reasonable to generalize the reported estimates to the population of all undergraduate students? Address at least two possible sources of bias in your answer.

2.23 A newspaper headline stated that at a recent budget workshop, nearly three dozen people supported a sales tax increase to help deal with the city's financial deficit (*San Luis Obispo Tribune*, January 22, 2005). This conclusion was based on data from a survey acknowledged to be unscientific, in which 34 out of the 43 people who chose to attend the budget workshop recommended raising the sales tax. Briefly discuss why the survey was described as "unscientific" and how this might limit the conclusions that can be drawn from the survey data.

2.24 Suppose that you were asked to help design a survey of adult city residents in order to estimate the proportion that would support a sales tax increase. The plan is to use a stratified random sample, and three stratification schemes have been proposed.

Scheme 1: Stratify adult residents into four strata based on the first letter of their last name (A–G, H–N, O–T, U–Z)
Scheme 2: Stratify adult residents into three strata: college students, nonstudents who work full time, nonstudents who do not work full time.
Scheme 3: Stratify adult residents into five strata by randomly assigning residents into one of the five strata.

Which of the three stratification schemes would be best in this situation? Explain.

2.25 The article "High Levels of Mercury Are Found in Californians" (*Los Angeles Times*, February 9, 2006) describes a study in which hair samples were tested for mercury. The hair samples were obtained from more than 6000 people who voluntarily sent hair samples to researchers at Greenpeace and The Sierra Club. The researchers found that nearly one-third of those tested had mercury levels that exceeded the concentration thought to be safe. Is it reasonable to generalize these results to the larger population of U.S. adults? Explain why or why not.

2.26 "More than half of California's doctors say they are so frustrated with managed care they will quit, retire early, or leave the state within three years." This conclusion from an article titled "Doctors Feeling Pessimistic, Study Finds" (*San Luis Obispo Tribune*, July 15, 2001) was based on a mail survey conducted by the California Medical Association. Surveys were mailed to 19,000 California doctors, and 2000 completed surveys were returned. Describe any concerns you have regarding the conclusion drawn.

2.27 ▼ Whether or not to continue a Mardi Gras Parade through downtown San Luis Obispo, CA, is a hotly debated topic. The parade is popular with students and many residents, but some celebrations have led to complaints and a call to eliminate the parade. The local newspaper conducted online and telephone surveys of its readers and was surprised by the results. The survey web site received more than 400 responses, with more than 60% favoring continuing the parade, while the telephone response line received more than 120 calls, with more than 90% favoring banning the parade (*San Luis Obispo Tribune*, March 3, 2004). What factors may have contributed to these very different results?

2.28 Based on a survey of 4113 U.S. adults, researchers at Stanford University concluded that Internet use leads to increased social isolation. The survey was conducted by an Internet-based polling company that selected its samples from a pool of 35,000 potential respondents, all of whom had been given free Internet access and WebTV hardware

in exchange for agreeing to regularly participate in surveys conducted by the polling company. Two criticisms of this study were expressed in an article that appeared in the *San Luis Obispo Tribune* (February 28, 2000). The first criticism was that increased social isolation was measured by asking respondents if they were talking less to family and friends on the phone. The second criticism was that the sample was selected only from a group that was induced to participate by the offer of free Internet service, yet the results were generalized to all U.S. adults. For each criticism, indicate what type of bias is being described and why it might make you question the conclusion drawn by the researchers.

2.29 The article "I'd Like to Buy a Vowel, Drivers Say" (*USA Today*, August 7, 2001) speculates that young people prefer automobile names that consist of just numbers and/or letters that do not form a word (such as Hyundai's XG300, Mazda's 626, and BMW's 325i). The article goes on to state that Hyundai had planned to identify the car now marketed as the XG300 with the name Concerto, until they determined that consumers hated it and that they thought XG300 sounded more "technical" and deserving of a higher price. Do the students at your school feel the same way? Describe how you would go about selecting a sample to answer this question.

2.30 The article "Gene's Role in Cancer May Be Overstated" (*San Luis Obispo Tribune*, August 21, 2002) states that "early studies that evaluated breast cancer risk among gene mutation carriers selected women in families where sisters, mothers, and grandmothers all had breast cancer. This created a statistical bias that skewed risk estimates for women in the general population." Is the bias described here selection bias, measurement bias, or nonresponse bias? Explain.

Bold exercises answered in back   ● Data set available online but not required   ▼ Video solution available

2.3 Simple Comparative Experiments

Sometimes the questions we are trying to answer deal with the effect of certain explanatory variables on some response. Such questions are often of the form, What happens when . . . ? or What is the effect of . . . ? For example, an industrial engineer may be considering two different workstation designs and might want to know whether the choice of design affects work performance. A medical researcher may want to determine how a proposed treatment for a disease compares to a standard treatment.

To address these types of questions, the researcher conducts an experiment to collect the relevant information. Experiments must be carefully planned to obtain information that will give unambiguous answers to questions of interest.

DEFINITION

An **experiment** is a planned intervention undertaken to observe the effects of one or more explanatory variables, often called **factors**, on a response variable. The fundamental purpose of the intervention is to increase understanding of the nature of the relationships between the explanatory and response variables. Any particular combination of values for the explanatory variables is called an **experimental condition** or **treatment**.

The **design** of an experiment is the overall plan for conducting an experiment.

Suppose that we are interested in determining how student performance on a first-semester calculus exam is affected by room temperature. There are four sections of calculus being offered in the fall semester. We might design an experiment in this way: Set the room temperature (in degrees Fahrenheit) to 65° in two of the rooms and to 75° in the other two rooms on test day, and then compare the exam scores for the 65° group and the 75° group. Suppose that the average exam score for the students in the 65° group was noticeably higher than the average for the 75° group. Could we conclude that the increased temperature resulted in a lower average score? Based on the information given, the answer is no because many other factors might be related to exam score. Were the sections at different times of the day? Did they have the same instructor? Different textbooks? Did the sections differ with respect to the abilities of the students? Any of these other factors could provide a plausible explanation (having nothing to do with room temperature) for why the average test score was different for the two groups. It is not possible to separate the effect of temperature from the effects of these other factors. As a consequence, simply setting the room temperatures as described makes for a poorly designed experiment. *A well-designed experiment requires more than just manipulating the explanatory variables; the design must also eliminate rival explanations or the experimental results will not be conclusive.*

Example of confounding variables

The goal is to design an experiment that will allow us to determine the effects of the relevant factors on the chosen response variable. To do this, we must take into consideration other extraneous factors that, although not of interest in the current study, might also affect the response variable.

DEFINITION

An **extraneous factor** is one that is not of interest in the current study but is thought to affect the response variable.

A well-designed experiment copes with the potential effects of extraneous factors by using random assignment to experimental conditions and sometimes also by incorporating direct control and/or blocking into the design of the experiment. Each of these strategies—random assignment, direct control, and blocking—is described in the paragraphs that follow.

A researcher can **directly control** some extraneous factors. In the calculus test example, the textbook used is an extraneous factor because part of the differences in test results might be attributed to this factor. We could control this factor directly, by requiring that all sections use the same textbook. Then any observed differences between temperature groups could not be explained by the use of different textbooks. The extraneous factor *time of day* might also be directly controlled in this way by having all sections meet at the same time.

The effects of some extraneous factors can be filtered out by a process known as **blocking**. Extraneous factors that are addressed through blocking are called *blocking factors*. Blocking creates groups (called blocks) that are similar with respect to blocking factors; then all treatments are tried in each block. In our example, we might use *instructor* as a blocking factor. If two instructors are each teaching two sections of calculus, we would make sure that for each instructor, one section was part of the 65° group and the other section was part of the 75° group. With this design, if we see a difference in exam scores for the two temperature groups, the factor *instructor* can be ruled out as a possible explanation, because both instructors' students were present in each temperature group. (Had we controlled the instructor variable by choosing to have only one instructor, that would be an example of direct control.) If one instructor taught both 65° sections and the other taught both 75° sections, we would be unable to distinguish the effect of temperature from the effect of the instructor. In this situation, the two factors (temperature and instructor) are said to be **confounded**.

Two factors are **confounded** if their effects on the response variable cannot be distinguished from one another.

If an extraneous factor is confounded with the factors defining the treatments (experimental conditions), it is not possible to draw an unambiguous conclusion about the effect of the treatment on the response. Both direct control and blocking are effective in ensuring that the controlled factors and blocking factors are not confounded with the factors that define the treatments.

We can directly control some extraneous factors by holding them constant, and we can use blocking to create groups that are similar to essentially filter out the effect of others. But what about factors, such as student ability in our calculus test example, which cannot be controlled by the experimenter and which would be difficult to use as blocking factors? These extraneous factors are handled by the use of random assignment to experimental groups—a process called **randomization**. Randomization ensures that our experiment does not systematically favor one experimental condition over any other and attempts to create experimental groups that are as much alike as possible. For example, if the students requesting calculus could be assigned to one of the four available sections using a random mechanism, we would expect the resulting groups to be similar with respect to student ability as well as with respect to other extraneous factors that we are not directly controlling or using as a basis for blocking. Note that randomization in an experiment is different from random selection of subjects. The ideal situation would be to have both random selection of subjects and random assignment of subjects to experimental conditions as this would allow conclusions from the experiment to be generalized to a larger population. For many experiments the random selection of subjects is not possible. As long as subjects are assigned at random to experimental conditions it is still possible to assess treatment effects.

To get a sense of how random assignment tends to create similar groups, suppose that 50 college freshmen are available to participate as subjects in an experiment to investigate whether completing an online review of course material before an exam improves exam performance. The 50 subjects vary quite a bit with respect to achievement, which is reflected in their math and verbal SAT scores, as shown in Figure 2.2.

Figure 2.2 Dotplots of math and verbal SAT scores for 50 freshmen.

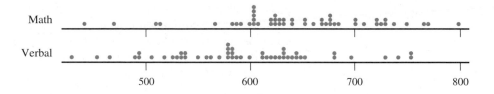

If these 50 students are to be assigned to the two experimental groups (one that will complete the online review and one that will not), we want to make sure that the assignment of students to groups does not favor one group over the other by tending to assign the higher achieving students to one group and the lower achieving students to the other.

Creating groups of students with similar achievement levels in a way that considers both verbal and math SAT scores simultaneously would be difficult, so we rely on random assignment. Figure 2.3(a) shows the math SAT scores of the students assigned to each of the two experimental groups (one shown in orange and one shown in blue) for each of three different random assignments of students to groups. Figure 2.3(b) shows the verbal SAT scores for the two experimental groups for each of the same three random assignments. Notice that each of the three random assignments produced groups that are similar with respect to *both* verbal and math SAT scores. So, if any of these three assignments were used and the two groups differed on exam performance, we could rule out differences in math or verbal SAT scores as possible competing explanations for the difference.

Figure 2.3 Dotplots for three different random assignments to two groups, one shown in orange and one shown in blue: (a) math SAT score; (b) verbal SAT score.

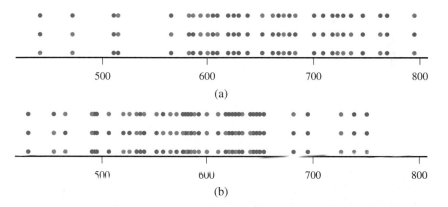

Not only will random assignment tend to create groups that are similar with respect to verbal and math SAT scores, but it will also tend to even out the groups with respect to other extraneous variables. *As long as the number of subjects is not too small, we can rely on the random assignment to produce comparable experimental groups. This is the reason that randomization is a part of all well-designed experiments.*

Not all experiments require the use of human subjects. For example, a researcher interested in comparing three different gasoline additives with respect to gasoline mileage might conduct an experiment using a single car with an empty tank. One gallon of gas with one of the additives will be put in the tank, and the car will be driven along a standard route at a constant speed until it runs out of gas. The total distance traveled on the gallon of gas could then be recorded. This could be repeated a number of times—10, for example—with each additive.

The experiment just described can be viewed as consisting of a sequence of trials. Because a number of extraneous factors (such as variations in environmental conditions like wind speed or humidity and small variations in the condition of the car) might have an effect on gas mileage, it would not be a good idea to use additive 1 for the first 10 trials, additive 2 for the next 10 trials, and so on. A better approach would be to randomly assign additive 1 to 10 of the 30 planned trials, and then randomly assign additive 2 to 10 of the remaining 20 trials. The resulting plan for carrying out the experiment might look as follows:

Trial	1	2	3	4	5	6	7	...	30
Additive	2	2	3	3	2	1	2	...	1

When an experiment can be viewed as a sequence of trials, randomization involves the random assignment of treatments to trials. *Remember that random assignment—either of subjects to treatments or of treatments to trials—is a critical component of a good experiment.* Randomization can be effective only if the number of subjects or observations in each treatment or experimental condition is large enough for each experimental group to reliably reflect variability in the population. For example, if there were only eight students requesting calculus, it is unlikely that we would get equivalent groups for comparison, even with random assignment to the four sections.

Replication is the design strategy of making multiple observations for each experimental condition. Together, replication and randomization allow the researcher to be reasonably confident of comparable experimental groups.

Key Concepts in Experimental Design

Randomization

Random assignment (of subjects to treatments or of treatments to trials) to ensure that the experiment does not systematically favor one experimental condition (treatment) over another.

Blocking

Using extraneous factors to create groups (blocks) that are similar. All experimental conditions (treatments) are then tried in each block.

Direct Control

Holding extraneous factors constant so that their effects are not confounded with those of the experimental conditions (treatments).

Replication

Ensuring that there is an adequate number of observations for each experimental condition.

To illustrate the design of a simple experiment, consider the dilemma of Anna, a waitress in a local restaurant. She would like to increase the amount of her tips, and her strategy is simple: She will write "Thank you" on the back of some of the checks before giving them to the patrons and on others she will write nothing. She plans to calculate the percentage of the tip as her measure of success (for instance, a 15% tip is common). She will compare the average percentage of the tip calculated from checks with and without the handwritten "Thank you." If writing "Thank you" does not produce higher tips, she may try a different strategy.

Anna is untrained in the art of planning experiments, but already she has taken some commonsense steps in the right direction to answer her question—Will writing "Thank you" produce the desired outcome of higher tips? Anna has defined a manageable problem, and collecting the appropriate data is feasible. It should be easy to gather data as a normal part of her work. Anna wonders whether writing "Thank you" on the customers' bills will have an effect on the amount of her tip. In the language of experimentation, we would refer to the writing of "Thank you" and the not writing of "Thank you" as **treatments** (the two experimental conditions to be compared in the experiment). The two treatments together are the possible values of the **explanatory variable**. The tipping percentage is the **response variable**. The idea behind this terminology is that the outcome of the experiment (the tipping percentage) is a *response* to the treatments *writing "Thank you"* or *not writing "Thank you."* Anna's experiment may be thought of as an attempt to explain the variability in the response variable in terms of its presumed cause, the variability in the explanatory variable. That is, as she manipulates the explanatory variable, she expects the response by her customers to vary. Anna has a good start, but now she must consider the four fundamental design principles.

Replication. Anna cannot run a successful experiment by gathering tipping information on only one person for each treatment. There is no reason to believe that any single tipping incident is representative of what would happen in other incidents, and therefore it would be impossible to evaluate the two treatments with only two subjects. To interpret the effects of a particular treatment, she must **replicate** each treatment in the experiment.

Direct Control and Randomization. There are a number of extraneous variables that might have an effect on the size of tip. Some restaurant patrons will be seated near the window with a nice view; some will have to wait for a table, whereas others may be seated immediately; and some may be on a fixed income and cannot afford a large tip. Some of these variables can be directly controlled. For example, Anna may choose to use only window tables in her experiment, thus eliminating table location as a potential confounding variable. Other variables, such as length of wait and customer income, cannot be easily controlled. As a result, it is important that Anna use randomization to decide which of the window tables will be in the "Thank you" group and which will be in the no "Thank you" group. She might do this by flipping a coin as she prepares the check for each window table. If the coin lands with the head side up, she could write "Thank you" on the bill, omitting the "Thank you" when a tail is observed.

Blocking. Suppose that Anna works on both Thursdays and Fridays. Because day of the week might affect tipping behavior, Anna should block one day of the week and make sure that observations for both treatments are made on each of the two days.

■ Evaluating an Experimental Design

The key concepts of experimental design provide a framework for evaluating an experimental design, as illustrated in the following examples.

Example 2.2 Revenge is Sweet

The article "The Neural Basis of Altruistic Punishment" (*Science*, August 27, 2004) described a study that examined motivation for revenge. Subjects in the study were all healthy, right-handed men. Subjects played a game with another player in which they could both earn money by trusting each other or one player could double-cross the other player and keep all of the money. In some cases the double cross was required by the rules of the game in certain circumstances, while in other cases the double cross was the result of a deliberate choice. The victim of a double cross was then given the opportunity to retaliate by imposing a fine, but sometimes the victim had to spend some of his own money in order to impose the fine. This study was an experiment with four experimental conditions or treatments:

1. double cross not deliberate (double cross dictated by the rules of the game) and no cost to the victim to retaliate
2. double cross deliberate and no cost to the victim to retaliate
3. double cross not deliberate and a cost to the victim to retaliate
4. double cross deliberate and a cost to the victim to retaliate

All subjects chose revenge (imposed a fine on the double-crosser) when the double cross was deliberate and retaliation was free, and 86% of the subjects chose revenge when the double cross was deliberate, even if it cost them money. Only 21% imposed a fine if the double cross was dictated by the rules of the game and was not deliberate.

Assuming that the researchers randomly assigned the subjects to the four experimental conditions, this study is an experiment that incorporated randomization, direct control (controlled sex, health, and handedness by using only healthy, right-handed males as subjects), and replication (many subjects assigned to each experimental condition).

■

Example 2.3 Subliminal Messages

The article "The Most Powerful Manipulative Messages Are Hiding in Plain Sight" (*Chronicle of Higher Education*, January 29, 1999) reported the results of an interesting experiment on priming—the effect of subliminal messages on how we behave. In the experiment, subjects completed a language test in which they were asked to construct a sentence using each word in a list of words. One group of subjects received a list of words related to politeness, and a second group was given a list of words related to rudeness. Subjects were told to complete the language test and then come into the hall and find the researcher so that he could explain the next part of the test. When each subject came into the hall, he or she found the researcher engaged in conversation. The researcher wanted to see whether the subject would interrupt the conversation. The researcher found that 63% of those primed with words related to rudeness interrupted the conversation, whereas only 17% of those primed with words related to politeness interrupted.

If we assume that the researcher randomly assigned the subjects to the two groups, then this study is an experiment that compares two treatments (primed with words related to rudeness and primed with words related to politeness). The response variable, *politeness*, has the values *interrupted conversation* and *did not interrupt conversation*. The experiment uses replication (many subjects in each treatment group) and randomization to control for extraneous variables that might affect the response.

■

Many experiments compare a group that receives a particular treatment to a **control group** that receives no treatment.

Example 2.4   Chilling Newborns? Then You Need a Control Group . . .

Researchers for the National Institute of Child Health and Human Development studied 208 infants whose brains were temporarily deprived of oxygen as a result of complications at birth (*The New England Journal of Medicine*, October 13, 2005). These babies were subjects in an experiment to determine if reducing body temperature for three days after birth improved their chances of surviving without brain damage. The experiment was summarized in a paper that stated "infants were randomly assigned to usual care (control group) or whole-body cooling." Including a control group in the experiment provided a basis for comparison of death and disability rates for the proposed cooling treatment and those for usual care. Some extraneous factors that might also affect death and disability rates, such as the duration of oxygen deprivation, could not be directly controlled, so to ensure that the experiment didn't unintentionally favor one experimental condition over the other, random assignment of the infants to the two groups was critical. Because this was a well-designed experiment, the researchers were able to use the resulting data and statistical methods that you will see in Chapter 11 to conclude that cooling did reduce the risk of death and disability for infants deprived of oxygen at birth.

■

Before proceeding with an experiment, you should be able to give a satisfactory answer to each of the following 10 questions.

1. What is the research question that data from the experiment will be used to answer?
2. What is the response variable?
3. How will the values of the response variable be determined?
4. What are the factors (explanatory variables) for the experiment?
5. For each factor, how many different values are there, and what are these values?
6. What are the treatments for the experiment?
7. What extraneous variables might influence the response?
8. How does the design incorporate random assignment of subjects to treatments (or treatments to subjects) or random assignment of treatments to trials?
9. For each extraneous variable listed in Question 7, how does the design protect against its potential influence on the response through blocking, direct control, or randomization?
10. Will you be able to answer the research question using the data collected in this experiment?

■ Exercises 2.31–2.39   Turn to the Expanded Answers Section for answers not shown next to exercises.

2.31 ▼ Based on observing more than 400 drivers in the Atlanta area, two investigators at Georgia State University concluded that people exiting parking spaces did so more slowly when a driver in another car was waiting for the space than when no one was waiting ("Territorial Defense in Parking Lots: Retaliation Against Waiting Drivers," *Journal of Applied Social Psychology* [1997]:

821–834). Describe how you might design an experiment to determine whether this phenomenon is true for your city. What is the response variable? What are some extraneous factors and how does your design control for them?

2.32 The head of the quality control department at a printing company would like to carry out an experiment to

Bold exercises answered in back   ● Data set available online but not required   ▼ Video solution available

determine which of three different glues results in the greatest binding strength. Although they are not of interest in the current investigation, other factors thought to affect binding strength are the number of pages in the book and whether the book is being bound as a paperback or a hardback.

a. What is the response variable in this experiment?
b. What factor will determine the experimental conditions?
c. What two extraneous factors are mentioned in the problem description? Are there other extraneous factors that should be considered?

2.33 A 1993 study showed that college students temporarily gained up to 9 IQ points after listening to a Mozart piano sonata. This conclusion, dubbed the Mozart effect, has since been criticized by a number of researchers who have been unable to confirm the result in similar studies. Suppose that you wanted to see whether there is a Mozart effect for students at your school.

a. Describe how you might design an experiment for this purpose.
b. Does your experimental design include direct control of any extraneous variables? Explain.
c. Does your experimental design use blocking? Explain why you did or did not include blocking in your design.
d. What role does randomization play in your design?

2.34 The following is from an article titled "After the Workout, Got Chocolate Milk?" that appeared in the *Chicago Tribune* (January 18, 2005):

> Researchers at Indiana University at Bloomington have found that chocolate milk effectively helps athletes recover from an intense workout. They had nine cyclists bike, rest four hours, then bike again, three separate times. After each workout, the cyclists downed chocolate milk or energy drinks Gatorade or Endurox (two to three glasses per hour); then, in the second workout of each set, they cycled to exhaustion. When they drank chocolate milk, the amount of time they could cycle until they were exhausted was similar to when they drank Gatorade and longer than when they drank Endurox.

The article isn't explicit about this, but in order for this to have been a well-designed experiment, it must have incorporated randomization. Briefly explain where the researcher would have needed to randomize in order for the conclusion of the experiment to be valid.

2.35 The report "Comparative Study of Two Computer Mouse Designs" (Cornell Human Factors Laboratory

Technical Report RP7992) included the following description of the subjects used in an experiment:

> Twenty-four Cornell University students and staff (12 males and 12 females) volunteered to participate in the study. Three groups of 4 men and 4 women were selected by their stature to represent the 5th percentile (female 152.1 ± 0.3 cm, male 164.1 ± 0.4 cm), 50th percentile (female 162.4 ± 0.1 cm, male 174.1 ± 0.7 cm), and 95th percentile (female 171.9 ± 0.2 cm, male 185.7 ± 0.6 cm) ranges . . . All subjects reported using their right hand to operate a computer mouse.

This experimental design incorporated direct control and blocking.

a. Are the potential effects of the extraneous variable stature (height) addressed by blocking or direct control?
b. Whether the right or left hand is used to operate the mouse was considered to be an extraneous variable. Are the potential effects of this variable addressed by blocking or direct control?   Direct control

2.36 The Institute of Psychiatry at Kings College London found that dealing with "infomania" has a temporary, but significant derogatory effect on IQ. (*Discover*, November 2005). In this experiment, researchers divided volunteers into two groups. Each subject took an IQ test. One group had to check email and respond to instant messages while taking the test while the second group took the test without any distraction. The distracted group had an average score that was 10 points lower than the average for the control group. Explain why it is important that the researchers created the two experimental groups in this study by using random assignment.

2.37 An article from the Associated Press (May 14, 2002) led with the headline "Academic Success Lowers Pregnancy Risk." The article described an evaluation of a program that involved about 350 students at 18 Seattle schools in high crime areas. Some students took part in a program beginning in elementary school in which teachers showed children how to control their impulses, recognize the feelings of others, and get what they want without aggressive behavior. Others did not participate in the program. The study concluded that the program was effective because by the time young women in the program reached age 21, the pregnancy rate among them was 38%, compared to 56% for the women in the experiment who did not take part in the program. Explain why this conclusion is valid only if the women in the experiment were randomly assigned to one of the two experimental groups.

2.38 A study in Florida is examining whether health literacy classes and using simple medical instructions that include pictures and avoid big words and technical terms can keep Medicaid patients healthier (*San Luis Obispo Tribune*, October 16, 2002). Twenty-seven community health centers are participating in the study. For 2 years, half of the centers will administer standard care. The other centers will have patients attend classes and will provide special health materials that are easy to understand. Explain why it is important for the researchers to assign the 27 centers to the two groups (standard care and classes with simple health literature) at random.

2.39 Is status related to a student's understanding of science? The article "From Here to Equity: The Influence of Status on Student Access to and Understanding of Science" (*Culture and Comparative Studies* [1999]: 577–602) described a study on the effect of group discussions on learning biology concepts. An analysis of the relationship between status and "rate of talk" (the number of on-task speech acts per minute) during group work included gender as a blocking variable. Do you think that gender is a useful blocking variable? Explain.

Bold exercises answered in back   ● Data set available online but not required   ▼ Video solution available

2.4   More on Experimental Design

The previous section covered basic principles for designing simple comparative experiments—control, blocking, randomization, and replication. The goal of an experimental design is to provide a method of data collection that (1) minimizes extraneous sources of variability in the response so that any differences in response for various experimental conditions can be more easily assessed and (2) creates experimental groups that are similar with respect to extraneous variables that cannot be controlled either directly or through blocking.

In this section, we look at some additional considerations that you may need to think about when planning an experiment.

■ Use of a Control Group

If the purpose of an experiment is to determine whether some treatment has an effect, it is important to include an experimental group that does not receive the treatment. Such a group is called a **control group**. The use of a control group allows the experimenter to assess how the response variable behaves when the treatment is not used. This provides a baseline against which the treatment groups can be compared to determine whether the treatment had an effect.

Example 2.5   Comparing Gasoline Additives

© Royalty Free/Getty Images

Suppose that an engineer wants to know whether a gasoline additive increases fuel efficiency (miles per gallon). Such an experiment might use a single car (to eliminate car-to-car variability) and a sequence of trials in which 1 gallon of gas is put in an empty tank, the car is driven around a racetrack at a constant speed, and the distance traveled on the gallon of gas is recorded.

To determine whether the additive increases gas mileage, it would be necessary to include a control group of trials where distance traveled was measured when gasoline without the additive was used. The trials would be assigned *at random* to one of the two experimental conditions (additive or no additive).

Even though this experiment consists of a sequence of trials all with the same car, random assignment of trials to experimental conditions is still important because there will always be uncontrolled variability. For example, temperature or other environmental conditions might change over the sequence of trials, the physical condition of the car might change slightly from one trial to another, and so on. Random assignment of experimental conditions to trials will tend to even out the effects of these uncontrollable factors.

■

Although we usually think of a control group as one that receives no treatment, in experiments designed to compare a new treatment to an existing standard treatment, the term *control group* is sometimes also used to describe the group that receives the current standard treatment.

Not all experiments require the use of a control group. For example, many experiments are designed to compare two or more conditions—an experiment to compare density for three different formulations of bar soap or an experiment to determine how oven temperature affects the cooking time of a particular type of cake. However, sometimes a control group is included even when the ultimate goal is to compare two or more different treatments. An experiment with two treatments and no control group might allow us to determine whether there is a difference between the two treatments and even to assess the magnitude of the difference if one exists, but it would not allow us to assess the individual effect of either treatment. For example, without a control group, we might be able to say that there is no difference in the increase in mileage for two different gasoline additives, but we wouldn't be able to tell if this was because both additives increased gas mileage by a similar amount or because neither additive had any effect on gas mileage.

■ Use of a Placebo

In experiments that use human subjects, use of a control group may not be enough to determine whether a treatment really does have an effect. People sometimes respond merely to the power of suggestion! For example, suppose a study designed to determine whether a particular herbal supplement is effective in promoting weight loss uses an experimental group that takes the herbal supplement and a control group that takes nothing. It is possible that those who take the herbal supplement and believe that they are taking something that will help them to lose weight may be more motivated and may unconsciously change their eating behavior or activity level, resulting in weight loss.

Although there is debate about the degree to which people respond, many studies have shown that people sometimes respond to treatments with no active ingredients and that they often report that such "treatments" relieve pain or reduce symptoms. So, if an experiment is to enable researchers to determine whether a treatment really has an effect, comparing a treatment group to a control group may not be enough. To address the problem, many experiments use what is called a placebo.

A **placebo** is something that is identical (in appearance, taste, feel, etc.) to the treatment received by the treatment group, except that it contains no active ingredients.

For example, in the herbal supplement experiment, rather than using a control group that received *no* treatment, the researchers might want to include a placebo group. Individuals in the placebo group would take a pill that looked just like the herbal supplement but did not contain the herb or any other active ingredient. As long as the subjects did not know whether they were taking the herb or the placebo, the placebo group would provide a better basis for comparison and would allow the researchers to determine whether the herbal supplement had any real effect over and above the "placebo effect."

■ Single-Blind and Double-Blind Experiments

Because people often have their own personal beliefs about the effectiveness of various treatments, it is desirable to conduct experiments in such a way that subjects do not know what treatment they are receiving. For example, in an experiment comparing four different doses of a medication for relief of headache pain, someone who knows that he is receiving the medication at its highest dose may be subconsciously influenced to report a greater degree of headache pain reduction. By ensuring that subjects are not aware of which treatment they receive, we can prevent the subjects' personal perception from influencing the response.

An experiment in which subjects do not know what treatment they have received is described as **single-blind**. Of course, not all experiments can be made single-blind. For example, in an experiment to compare the effect of two different types of exercise on blood pressure, it is not possible for participants to be unaware of whether they are in the swimming group or the jogging group! However, when it is possible, "blinding" the subjects in an experiment is generally a good strategy.

In some experiments, someone other than the subject is responsible for measuring the response. To ensure that the person measuring the response does not let personal beliefs influence the way in which the response is recorded, the researchers should make sure that the measurer does not know which treatment was given to any particular individual. For example, in a medical experiment to determine whether a new vaccine reduces the risk of getting the flu, doctors must decide whether a particular individual who is not feeling well actually has the flu or some other unrelated illness. If the doctor knew that a participant with flu-like symptoms had received the new flu vaccine, she might be less likely to determine that the participant had the flu and more likely to interpret the symptoms as being the result of some other illness.

There are two ways in which blinding might occur in an experiment. One involves blinding the participants, and the other involves blinding the individuals who measure the response. If participants do not know which treatment was received *and* those measuring the response do not know which treatment was given to which participant, the experiment is described as **double-blind**. If only one of the two types of blinding is present, the experiment is single-blind.

A **double-blind** experiment is one in which neither the subjects nor the individuals who measure the response know which treatment was received.

A **single-blind** experiment is one in which the subjects do not know which treatment was received but the individuals measuring the response do know which treatment was received, or one in which the subjects do know which treatment was received but the individuals measuring the response do not know which treatment was received.

■ Experimental Units and Replication

An **experimental unit** is the smallest unit to which a treatment is applied. In the language of experimental design, treatments are assigned at random to experimental units, and replication means that each treatment is applied to more than one experimental unit.

Replication is necessary for randomization to be an effective way to create similar experimental groups and to get a sense of the variability in the values of the response for individuals that receive the same treatment. As we will see in Chapters 9–15, this enables us to use statistical methods to decide whether differences in the responses in different treatment groups can be attributed to the treatment received or whether they can be explained by chance variation (the natural variability seen in the responses to a single treatment).

Be careful when designing an experiment to ensure that there is replication. For example, suppose that children in two third-grade classes are available to participate in an experiment to compare two different methods for teaching arithmetic. It might at first seem reasonable to select one class at random to use one method and then assign the other method to the remaining class. But what are the experimental units here? If treatments are randomly assigned to classes, classes are the experimental units. Because only one class is assigned to each treatment, this is an experiment with no replication, even though there are many children in each class. We would *not* be able to determine whether there was a difference between the two methods based on data from this experiment, because we would have only one observation per treatment.

One last note on replication: Don't confuse replication in an experimental design with replicating an experiment. Replicating an experiment means conducting a new experiment using the same experimental design as a previous experiment; it is a way of confirming conclusions based on a previous experiment, but it does not eliminate the need for replication in each of the individual experiments themselves.

Important concept

■ Using Volunteers as Subjects in an Experiment ....

Although the use of volunteers in a study that involves collecting data through sampling is never a good idea, it is a common practice to use volunteers as subjects in an experiment. Even though the use of volunteers limits the researcher's ability to generalize to a larger population, random assignment of the volunteers to treatments should result in comparable groups, and so treatment effects can still be assessed.

 Exercises 2.40–2.50 ·················· Turn to the Expanded Answers Section for answers not shown next to exercises.

2.40 Explain why some studies include both a control group and a placebo treatment. What additional comparisons are possible if both a control group and a placebo group are included?

2.41 Explain why blinding is a reasonable strategy in many experiments.

2.42 Give an example of an experiment for each of the following:
a. Single-blind experiment with the subjects blinded
b. Single-blind experiment with the individuals measuring the response blinded
c. Double-blind experiment
d. An experiment that is not possible to blind

2.43 ▼ Swedish researchers concluded that viewing and discussing art soothes the soul and helps relieve medical conditions such as high blood pressure and constipation (AFP International News Agency, October 14, 2005). This conclusion was based on a study in which 20 elderly women gathered once a week to discuss different works of art. The study also included a control group of 20 elderly women who met once a week to discuss their hobbies and interests. At the end of 4 months, the art discussion group was found to have a more positive attitude, to have lower blood pressure, and to use fewer laxatives than the control group.
a. Why would it be important to determine if the researchers assigned the women participating in the study at random to one of the two groups?

Bold exercises answered in back ● Data set available online but not required ▼ Video solution available

b. Explain why you think that the researchers included a control group in this study.

2.44 A novel alternative medical treatment for heart attacks seeds the damaged heart muscle with cells from the patient's thigh muscle ("Doctors Mend Damaged Hearts with Cells from Muscles," *San Luis Obispo Tribune*, November 18, 2002). Doctor Dib from the Arizona Heart Institute evaluated the approach on 16 patients with severe heart failure. The article states that "ordinarily, the heart pushes out more than half its blood with each beat. Dib's patients had such severe heart failure that their hearts pumped just 23 percent. After bypass surgery and cell injections, this improved to 36 percent, although it was impossible to say how much, if any, of the new strength resulted from the extra cells."
a. Explain why it is not reasonable to generalize to the population of all heart attack victims based on the data from these 16 patients.
b. Explain why it is not possible to say whether any of the observed improvement was due to the cell injections, based on the results of this study.
c. Describe a design for an experiment that would allow researchers to determine whether bypass surgery plus cell injections was more effective than bypass surgery alone.

2.45 ▼ The article "Doctor Dogs Diagnose Cancer by Sniffing It Out" (*Knight Ridder Newspapers*, January 9, 2006) reports the results of an experiment described in the journal *Integrative Cancer Therapies*. In this experiment, dogs were trained to distinguish between people with breast and lung cancer and people without cancer by sniffing exhaled breath. Dogs were trained to lay down if they detected cancer in a breath sample. After training, dogs' ability to detect cancer was tested using breath samples from people whose breath had not been used in training the dogs. The paper states "The researchers blinded both the dog handlers and the experimental observers to the identity of the breath samples." Explain why this blinding is an important aspect of the design of this experiment.

2.46 An experiment to evaluate whether vitamins can help prevent recurrence of blocked arteries in patients who have had surgery to clear blocked arteries was described in the article "Vitamins Found to Help Prevent Blocked Arteries" (Associated Press, September 1, 2002). The study involved 205 patients who were given either a treatment consisting of a combination of folic acid, vitamin B12, and vitamin B6 or a placebo for 6 months.

a. Explain why a placebo group was used in this experiment.
b. Explain why it would be important for the researchers to have assigned the 205 subjects to the two groups (vitamin and placebo) at random.
c. Do you think it is appropriate to generalize the results of this experiment to the population of all patients who have undergone surgery to clear blocked arteries? Explain.   No, subjects not randomly selected

2.47 Pismo Beach, California, has an annual clam festival that includes a clam chowder contest. Judges rate clam chowders from local restaurants, and the judging is done in such a way that the judges are not aware of which chowder is from which restaurant. One year, much to the dismay of the seafood restaurants on the waterfront, Denny's chowder was declared the winner! (When asked what the ingredients were, the cook at Denny's said he wasn't sure—he just had to add the right amount of nondairy creamer to the soup stock that he got from Denny's distribution center!)
a. Do you think that Denny's chowder would have won the contest if the judging had not been "blind"? Explain.
b. Although this was not an experiment, your answer to Part (a) helps to explain why those measuring the response in an experiment are often blinded. Using your answer in Part (a), explain why experiments are often blinded in this way.   So that the judging is fair

2.48 The *San Luis Obispo Tribune* (May 7, 2002) reported that "a new analysis has found that in the majority of trials conducted by drug companies in recent decades, sugar pills have done as well as—or better than—antidepressants." What effect is being described here? What does this imply about the design of experiments with a goal of evaluating the effectiveness of a new medication?

2.49 Researchers at the University of Pennsylvania suggest that a nasal spray derived from pheromones (chemicals emitted by animals when they are trying to attract a mate) may be beneficial in relieving symptoms of premenstrual syndrome (PMS) (*Los Angeles Times*, January 17, 2003).
a. Describe how you might design an experiment using 100 female volunteers who suffer from PMS to determine whether the nasal spray reduces PMS symptoms.
b. Does your design from Part (a) include a placebo treatment? Why or why not?
c. Does your design from Part (a) involve blinding? Is it single-blind or double-blind? Explain.

Bold exercises answered in back   ● Data set available online but not required   ▼ Video solution available

2.50 The article "A Debate in the Dentist's Chair" (*San Luis Obispo Tribune*, January 28, 2000) described an ongoing debate over whether newer resin fillings are a better alternative to the more traditional silver amalgam fillings. Because amalgam fillings contain mercury, there is concern that they could be mildly toxic and prove to be a health risk to those with some types of immune and kidney disorders. One experiment described in the article used sheep as subjects and reported that sheep treated with amalgam fillings had impaired kidney function.

a. In the experiment, a control group of sheep that received no fillings was used but there was no placebo group. Explain why it is not necessary to have a placebo group in this experiment.
b. The experiment compared only an amalgam filling treatment group to a control group. What would be the benefit of also including a resin filling treatment group in the experiment?
c. Why do you think the experimenters used sheep rather than human subjects?

Bold exercises answered in back ● Data set available online but not required ▼ Video solution available

---

2.5 More on Observational Studies: Designing Surveys (Optional)

Designing an observational study to compare two populations on the basis of some easily measured characteristic is relatively straightforward, with attention focusing on choosing a reasonable method of sample selection. However, many observational studies attempt to measure personal opinion or attitudes using responses to a survey. In such studies, both the sampling method and the design of the survey itself are critical to obtaining reliable information.

It would seem at first glance that a survey must be a simple method for acquiring information. However, it turns out that designing and administering a survey is not as easy as it might seem. Great care must be taken in order to obtain good information from a survey.

■ Survey Basics

A **survey** is a voluntary encounter between strangers in which an interviewer seeks information from a respondent by engaging in a special type of conversation. This conversation might take place in person, over the telephone, or even in the form of a written questionnaire, and it is quite different from usual social conversations. Both the interviewer and the respondent have certain roles and responsibilities. The interviewer gets to decide what is relevant to the conversation and may ask questions—possibly personal or even embarrassing questions. The respondent, in turn, may refuse to participate in the conversation and may refuse to answer any particular question. But having agreed to participate in the survey, the respondent is responsible for answering the questions truthfully. Let's consider the situation of the respondent.

■ The Respondent's Tasks

Our understanding of the survey process has been improved in the past two decades by contributions from the field of psychology, but there is still much uncertainty about how people respond to survey questions. Survey researchers and psychologists generally agree that the respondent is confronted with a sequence of tasks when asked a question: comprehension of the question, retrieval of information from memory, and reporting the response.

■ **Task 1: Comprehension**   Comprehension is the single most important task facing the respondent, and fortunately it is the characteristic of a survey question that is most easily controlled by the question writer. Comprehensible directions and questions are characterized by (1) a vocabulary appropriate to the population of interest, (2) simple sentence structure, and (3) little or no ambiguity. Vocabulary is often a problem. As a rule, it is best to use the simplest possible word that can be used without sacrificing clear meaning.

Simple sentence structure also makes it easier for the respondent to understand the question. A famous example of difficult syntax occurred in 1993 when the Roper organization created a survey related to the Holocaust. One question in this survey was

"Does it seem possible or does it seem impossible to you that the Nazi extermination of the Jews never happened?"

The question has a complicated structure and a double negative—"impossible . . . never happened"—that could lead respondents to give an answer opposite to what they actually believed. The question was rewritten and given a year later in an otherwise unchanged survey:

"Does it seem possible to you that the Nazi extermination of the Jews never happened, or do you feel certain that it happened?"

This question wording is much clearer, and in fact the respondents' answers were quite different, as shown in the following table (the "unsure" and "no opinion" percentages have been omitted):

Original Roper Poll		Revised Roper Poll	
Impossible	65%	Certain it happened	91%
Possible	12%	Possible it never happened	1%

It is also important to filter out ambiguity in questions. Even the most innocent and seemingly clear questions can have a number of possible interpretations. For example, suppose that you are asked, "When did you move to Cedar Rapids?" This would seem to be an unambiguous question, but some possible answers might be (1) "In 1971," (2) "When I was 23," and (3) "In the summer." The respondent must decide which of these three answers, if any, is the appropriate response. It may be possible to lessen the ambiguity with more precise questions:

1. In what year did you move to Cedar Rapids?
2. How old were you when you moved to Cedar Rapids?
3. In what season of the year did you move to Cedar Rapids?

One way to find out whether or not a question is ambiguous is to field-test the question and to ask the respondents if they were unsure how to answer a question.

Ambiguity can arise from the placement of questions as well as from their phrasing. Here is an example of ambiguity uncovered when the order of two questions differed in two versions of a survey on happiness. The questions were

1. Taken altogether, how would you say things are these days: Would you say that you are very happy, pretty happy, or not too happy?

2. Taking things altogether, how would you describe your marriage: Would you say that your marriage is very happy, pretty happy, or not too happy?

The proportions of responses to the general happiness question differed for the different question orders, as follows:

Response to General Happiness Question

	General Asked First	General Asked Second
Very happy	52.4%	38.1%
Pretty happy	44.2%	52.8%
Not too happy	3.4%	9.1%

If the goal in this survey was to estimate the proportion of the population that is generally happy, these numbers are quite troubling—they cannot both be right! What seems to have happened is that Question 1 was interpreted differently depending on whether it was asked first or second. When the general happiness question was asked after the marital happiness question, the respondents apparently interpreted it to be asking about their happiness in all aspects of their lives *except* their marriage. This was a reasonable interpretation, given that they had just been asked about their marital happiness, but it is a different interpretation than when the general happiness question was asked first. The troubling lesson here is that even carefully worded questions can have different interpretations in the context of the rest of the survey.

■ **Task 2: Retrieval from Memory**   Retrieving relevant information from memory to answer the question is not always an easy task, and it is not a problem limited to questions of fact. For example, consider this seemingly elementary "factual" question:

How many times in the past 5 years did you visit your dentist's office?

a. 0 times
b. 1–5 times
c. 6–10 times
d. 11–15 times
e. more than 15 times

It is unlikely that many people will remember with clarity every single visit to the dentist in the past 5 years. But generally, people will respond to such a question with answers consistent with the memories and facts they are able to reconstruct given the time they have to respond to the question. An individual may, for example, have a sense that he usually makes about two trips a year to the dentist's office, so he may extrapolate the typical year and get 10 times in 5 years. Then there may be three particularly memorable visits, say, for a root canal in the middle of winter. Thus, the best recollection is now 13, and the respondent will choose Answer (d), 11–15 times. Perhaps not exactly correct, but the best that can be reported under the circumstances.

What are the implications of this relatively fuzzy memory for those who construct surveys about facts? First, the investigator should understand that most factual answers are going to be approximations of the truth. Second, events closer to the time of a survey are easier to recall.

Attitude and opinion questions can also be affected in significant ways by the respondent's memory of recently asked questions. One recent study contained a survey question asking respondents their opinion about how much they followed politics. When that question was preceded by a factual question asking whether they knew the name of the congressional representative from their district, the percentage who reported they follow politics "now and then" or "hardly ever" jumped from 21% to 39%! Respondents apparently concluded that, because they didn't know the answer to the previous knowledge question, they must not follow politics as much as they might have thought otherwise. In a survey that asks for an opinion about the degree to which the respondent believes drilling for oil should be permitted in national parks, the response might be different if the question is preceded by questions about the high price of gasoline than if the question is preceded by questions about the environment and ecology.

■ **Task 3: Reporting the Response**   The task of formulating and reporting a response can be influenced by the social aspects of the survey conversation. In general, if a respondent agrees to take a survey, he or she will be motivated to answer truthfully. Therefore, if the questions aren't too difficult (taxing the respondent's knowledge or memory) and if there aren't too many questions (taxing the respondent's patience and stamina), the answers to questions will be reasonably accurate. However, it is also true that the respondents often wish to present themselves in a favorable light. This desire leads to what is known as a social desirability bias. Sometimes this bias is a response to the particular wording in a question. In 1941, the following questions were analyzed in two different forms of a survey (emphasis added):

1. Do you think the United States should *forbid* public speeches against democracy?
2. Do you think the United States should *allow* public speeches against democracy?

It would seem logical that these questions are opposites and that the proportion who would not allow public speeches against democracy should be equal to the proportion who would forbid public speeches against democracy. But only 45% of those respondents offering an opinion on Question 1 thought the United States should "forbid," whereas 75% of the respondents offering an opinion on Question 2 thought the United States should "not allow" public speeches against democracy. Most likely, respondents reacted negatively to the word *forbid*, as forbidding something sounds much harsher than not allowing it.

Some survey questions may be sensitive or threatening, such as questions about sex, drugs, or potentially illegal behavior. In this situation a respondent not only will want to present a positive image but also will certainly think twice about admitting illegal behavior! In such cases the respondent may shade the actual truth or may even lie about particular activities and behaviors. In addition, the tendency toward positive presentation is not limited to obviously sensitive questions. For example, consider the question about general happiness previously described. Several investigators have reported higher happiness scores in face-to-face interviews than in responses to a mailed questionnaire. Presumably, a happy face presents a more positive image of the respondent to the interviewer. On the other hand, if the interviewer was a clearly unhappy person, a respondent might shade answers to the less happy side of the scale, perhaps thinking that it is inappropriate to report happiness in such a situation.

It is clear that constructing surveys and writing survey questions can be a daunting task. Keep in mind the following three things:

1. Questions should be understandable by the individuals in the population being surveyed. Vocabulary should be at an appropriate level, and sentence structure should be simple.

2. Questions should, as much as possible, recognize that human memory is fickle. Questions that are specific will aid the respondent by providing better memory cues. The limitations of memory should be kept in mind when interpreting the respondent's answers.

3. As much as possible, questions should not create opportunities for the respondent to feel threatened or embarrassed. In such cases respondents may introduce a social desirability bias, the degree of which is unknown to the interviewer. This can compromise conclusions drawn from the survey data.

Constructing good surveys is a difficult task, and we have given only a brief introduction to this topic. For a more comprehensive treatment, we recommend the book by Sudman and Bradburn listed in the references in the back of the book.

Exercises 2.51–2.56 .... Turn to the Expanded Answers Section for answers not shown next to exercises.

2.51 A tropical forest survey conducted by Conservation International included the following statements in the material that accompanied the survey:

"A massive change is burning its way through the earth's environment."

"The band of tropical forests that encircle the earth is being cut and burned to the ground at an alarming rate."

"Never in history has mankind inflicted such sweeping changes on our planet as the clearing of rain forest taking place right now!"

The survey that followed included the questions given in Parts (a)–(d). For each of these questions, identify a word or phrase that might affect the response and possibly bias the results of any analysis of the responses.
a. "Did you know that the world's tropical forests are being destroyed at the rate of 80 acres per minute?"
b. "Considering what you know about vanishing tropical forests, how would you rate the problem?"
c. "Do you think we have an obligation to prevent the man-made extinction of animal and plant species?"
d. "Based on what you know now, do you think there is a link between the destruction of tropical forests and changes in the earth's atmosphere?"

2.52 Fast-paced lifestyles, where students balance the requirements of school, after-school activities, and jobs, are thought by some to lead to reduced sleep. Suppose that you are assigned the task of designing a survey that will provide answers to the accompanying questions. Write a set of survey questions that might be used. In some cases, you may need to write more than one question to adequately address a particular issue. For example, responses might be different for weekends and school nights. You may also have to

define some terms to make the questions comprehensible to the target audience, which is adolescents.

Topics to be addressed:
How much sleep do the respondents get? Is this enough sleep?
Does sleepiness interfere with schoolwork?
If they could change the starting and ending times of the school day, what would they suggest?
(Sorry, they cannot reduce the total time spent in school during the day!)   Answers will vary

2.53 Asthma is a chronic lung condition characterized by difficulty in breathing. Some studies have suggested that asthma may be related to childhood exposure to some animals, especially dogs and cats, during the first year of life ("Exposure to Dogs and Cats in the First Year of Life and Risk of Allergic Sensitization at 6 to 7 Years of Age," *Journal of the American Medical Association* [2002]: 963–972). Some environmental factors that trigger an asthmatic response are (1) cold air, (2) dust, (3) strong fumes, and (4) inhaled irritants.
a. Write a set of questions that could be used in a survey to be given to parents of young children suffering from asthma. The survey should include questions about the presence of pets in the first year of the child's life as well as questions about the presence of pets today. Also, the survey should include questions that address the four mentioned household environmental factors.
b. It is generally thought that low-income persons, who tend to be less well educated, have homes in environments where the four environmental factors are present. Mindful of the importance of comprehension, can you improve the questions in Part (a) by making your vocabulary simpler or by changing the wording of the questions?

c. One problem with the pet-related questions is the reliance on memory. That is, parents may not actually remember when they got their pets. How might you check the parents' memories about these pets?   Answers will vary.

2.54 In national surveys, parents consistently point to school safety as an important concern. One source of violence in junior high schools is fighting ("Self-Reported Characterization of Seventh-Grade Student Fights," *Journal of Adolescent Health* [1998]: 103–109). To construct a knowledge base about student fights, a school administrator wants to give two surveys to students after fights are broken up. One of the surveys is to be given to the participants, and the other is to be given to students who witnessed the fight. The type of information desired includes (1) the cause of the fight, (2) whether or not the fight was a continuation of a previous fight, (3) whether drugs or alcohol was a factor, (4) whether or not the fight was gang related, and (5) the role of bystanders.
a. Write a set of questions that could be used in the two surveys. Each question should include a set of possible responses. For each question, indicate whether it would be used on both surveys or just on one of the two.
b. How might the tendency toward positive self-presentation affect the responses of the fighter to the survey questions you wrote for Part (a)?
c. How might the tendency toward positive self-presentation affect the responses of a bystander to the survey questions you wrote for Part (a)?

2.55 Doctors are concerned about young women drinking large amounts of soda and about their decreased consumption of milk in recent years ("Teenaged Girls, Carbonated Beverage Consumption, and Bone Fractures," *Archives of Pediatric and Adolescent Medicine* [2000]: 610–613). In parts (a)–(d), construct two questions that might be included in a survey of teenage girls. Each question should include possible responses from which the respondent can select. (Note: The questions as written are vague. Your task is to clarify the questions for use in a survey, not just to change the syntax!)
a. How much "cola" beverage does the respondent consume?
b. How much milk (and milk products) is consumed by the respondent?
c. How physically active is the respondent?
d. What is the respondent's history of bone fractures?

2.56 A recent survey attempted to address psychosocial factors thought to be of importance in preventive health care for adolescents ("The Adolescent Health Review: A Brief Multidimensional Screening Instrument," *Journal of Adolescent Health* [2001]: 131–139). For each risk area in the following list, construct a question that would be comprehensible to students in grades 9–12 and that would provide information about the risk factor. Make your questions multiple-choice, and provide possible responses.
a. Lack of exercise
b. Poor nutrition
c. Emotional distress
d. Sexual activity
e. Cigarette smoking
f. Alcohol use   Answers will vary

Bold exercises answered in back   ● Data set available online but not required   ▼ Video solution available

2.6 Interpreting and Communicating the Results of Statistical Analyses

Statistical studies are conducted to allow investigators to answer questions about characteristics of some population of interest or about the effect of some treatment. Such questions are answered on the basis of data, and how the data are obtained determines the quality of information available and the type of conclusions that can be drawn. As a consequence, when describing a study you have conducted (or when evaluating a published study), you must consider how the data were collected.

The description of the data collection process should make it clear whether the study is an observational study or an experiment. For observational studies, some of the issues that should be addressed are:

1. What is the population of interest? What is the sampled population? Are these two populations the same? If the sampled population is only a subset of the population

of interest, **undercoverage** limits our ability to generalize to the population of interest. For example, if the population of interest is all students at a particular university, but the sample is selected from only those students who choose to list their phone number in the campus directory, undercoverage may be a problem. We would need to think carefully about whether it is reasonable to consider the sample as representative of the population of all students at the university. **Overcoverage** results when the sampled population is actually larger than the population of interest. This would be the case if we were interested in the population of all high schools that offer Advanced Placement (AP) Statistics but sampled from a list of all schools that offered an AP class in any subject. Both undercoverage and overcoverage can be problematic.

2. How were the individuals or objects in the sample actually selected? A description of the sampling method helps the reader to make judgments about whether the sample can reasonably be viewed as representative of the population of interest.

3. What are potential sources of bias, and is it likely that any of these will have a substantial effect on the observed results? When describing an observational study, you should acknowledge that you are aware of potential sources of bias and explain any steps that were taken to minimize their effect. For example, in a mail survey, nonresponse can be a problem, but the sampling plan may seek to minimize its effect by offering incentives for participation and by following up one or more times with those who do not respond to the first request. A common misperception is that increasing the sample size is a way to reduce bias in observational studies, but this is not the case. For example, if measurement bias is present, as in the case of a scale that is not correctly calibrated and tends to weigh too high, taking 1000 measurements rather than 100 measurements cannot correct for the fact that the measured weights will be too large. Similarly, a larger sample size cannot compensate for response bias introduced by a poorly worded question.

For experiments, some of the issues that should be addressed are:

1. What is the role of randomization? All good experiments use random assignment as a means of coping with the effects of potentially confounding variables that cannot easily be directly controlled. When describing an experimental design, you should be clear about how random assignment (subjects to treatments, treatments to subjects, or treatments to trials) was incorporated into the design.

2. Were any extraneous variables directly controlled by holding them at fixed values throughout the experiment? If so, which ones and at which values?

3. Was blocking used? If so, how were the blocks created? If an experiment uses blocking to create groups of homogeneous experimental units, you should describe the criteria used to create the blocks and their rationale. For example, you might say something like "Subjects were divided into two blocks—those who exercise regularly and those who do not exercise regularly—because it was believed that exercise status might affect the responses to the diets."

Because each treatment appears at least once in each block, the block size must be at least as large as the number of treatments. Ideally, the block sizes should be equal to the number of treatments, because this presumably would allow the experimenter to create small groups of extremely homogeneous experimental units. For example, in an experiment to compare two methods for teaching calculus to first-year college students, we may want to block on previous mathematics knowledge by using math SAT scores. If 100 students are available as subjects for this experiment, rather than creating two large groups (above-average math SAT score and below-average math SAT score), we might want to create 50 blocks of two students each, the first consisting of the two students with the highest math SAT scores, the second containing the two students with the next highest scores, and so on. We would then select one student in each block at random and assign that student to teaching method 1. The other student in the block would be assigned to teaching method 2.

■ A Word to the Wise: Cautions and Limitations ....

It is a big mistake to begin collecting data before thinking carefully about research objectives and developing a research plan, and doing so may result in data that do not enable the researcher to answer key questions of interest or to generalize conclusions based on the data to the desired populations of interest.

Clearly defining the objectives at the outset enables the investigator to determine whether an experiment or an observational study is the best way to proceed. Watch out for the following *inappropriate* actions:

1. Drawing a cause-and-effect conclusion from an observational study. Don't do this, and don't believe it when others do it!
2. Generalizing results of an experiment that uses volunteers as subjects to a larger population without a convincing argument that the group of volunteers can reasonably be considered a representative sample from the population.
3. Generalizing conclusions based on data from a sample to some population of interest. This is sometimes a sensible thing to do, but on other occasions it is not reasonable. Generalizing from a sample to a population is justified only when there is reason to believe that the sample is likely to be representative of the population. This would be the case if the sample was a random sample from the population and there were no major potential sources of bias. If the sample was not selected at random or if potential sources of bias were present, these issues would have to be addressed before a judgment could be made regarding the appropriateness of generalizing the study results.

   For example, the Associated Press (January 25, 2003) reported on the high cost of housing in California. The median home price was given for each of the 10 counties in California with the highest home prices. Although these 10 counties are a sample of the counties in California, they were not randomly selected and (because they are the 10 counties with the highest home prices) it would not be reasonable to generalize to all California counties based on data from this sample
4. Generalizing conclusions based on an observational study that used voluntary response or convenience sampling to a larger population. This is almost never reasonable.

A c t i v i t y **2.1**   Designing a Sampling Plan

Background: In this activity, you will work with a partner to develop a sampling plan.

Suppose that you would like to select a sample of 50 students at your school to learn something about how many hours per week, on average, students at your school spend engaged in a particular activity (such as studying, surfing the Internet, or watching TV).

1. Discuss with your partner whether you think it would be easy or difficult to obtain a simple random sample of students at your school and to obtain the desired information from all the students selected for the sample. Write a summary of your discussion.
2. With your partner, decide how you might go about selecting a sample of 50 students from your school

that reasonably could be considered representative of the population of interest even if it may not be a simple random sample. Write a brief description of your sampling plan, and point out the aspects of your plan that you think make it reasonable to argue that it will be representative.

3. Explain your plan to another pair of students. Ask them to critique your plan. Write a brief summary of the comments you received. Now reverse roles, and provide a critique of the plan devised by the other pair.

4. Based on the feedback you received in Step 3, would you modify your original sampling plan? If not, explain why this is not necessary. If so, describe how the plan would be modified.

A c t i v i t y **2.2**   An Experiment to Test for the Stroop Effect

Background: In 1935, John Stroop published the results of his research into how people respond when presented with conflicting signals. Stroop noted that most people are able to read words quickly and that they cannot easily ignore them and focus on other attributes of a printed word, such as text color. For example, consider the following list of words:

   green   blue   red   blue   **yellow**   red

It is easy to quickly read this list of words. It is also easy to read the words even if the words are printed in color, and even if the text color is different from the color of the word. For example, people can read the words in the list

   green   blue   red   blue   **yellow**   red

as quickly as they can read the list that isn't printed in color.

However, Stroop found that if people are asked to name the text colors of the words in the list (red, yellow, blue, green, red, green), it takes them longer. Psychologists believe that this is because the reader has to inhibit a natural response (reading the word) and produce a different response (naming the color of the text).

If Stroop is correct, people should be able to name colors more quickly if they do not have to inhibit the word response, as would be the case if they were shown the following:

1. Design an experiment to compare times to identify colors when they appear as text to times to identify colors when there is no need to inhibit a word response. Indicate how randomization is incorporated into your design. What is your response variable? How will you measure it? How many subjects will you use in your experiment, and how will they be chosen?

2. When you are satisfied with your experimental design, carry out the experiment. You will need to construct your list of colored words and a corresponding list of colored bars to use in the experiment. You will also need to think about how you will implement your randomization scheme.

3. Summarize the resulting data in a brief report that explains whether your findings are consistent with the Stroop effect.

A c t i v i t y **2.3**   McDonald's and the Next 100 Billion Burgers

Background: The article "Potential Effects of the Next 100 Billion Hamburgers Sold by McDonald's" (*American Journal of Preventative Medicine* [2005]:379–381) estimated that 992.25 million pounds of saturated fat would be consumed as McDonald's sells its next 100 billion hamburgers. This estimate was based on the assumption that the average weight of a burger sold would be 2.4 oz. This is the average of the weight of a regular hamburger

(1.6 oz.) and a Big Mac (3.2 oz.). The authors took this approach because

> "McDonald's does not publish sales and profits of individual items. Thus, it is not possible to estimate how many of McDonald's first 100 billion beef burgers sold were 1.6 oz hamburgers, 3.2 oz. Big Macs (introduced in 1968), 4.0 oz. Quarter Pounders (introduced in 1973), or other sandwiches."

This activity can be completed as an individual or as a team. Your instructor will specify which approach (individual or team) you should use.

1. The authors of the article believe that the use of 2.4 oz. as the average size of a burger sold at McDonald's is "conservative," which would result in the estimate of 992.25 million pounds of saturated fat being lower than the actual amount that would be consumed. Explain why the authors' belief might be justified.

2. Do you think it would be possible to collect data that could lead to a value for the average burger size that would be better than 2.4 oz.? If so, explain how you would recommend collecting such data. If not, explain why you think it is not possible.

Activity **2.4** Video Games and Pain Management

Background: Video games have been used for pain management by doctors and therapists who believe that the attention required to play a video game can distract the player and thereby decrease the sensation of pain. The paper "Video Games and Health" (*British Medical Journal* [2005]:122–123) states:

> "However, there has been no long term follow-up and no robust randomized controlled trials of such interventions. Whether patients eventually tire of such games is also unclear. Furthermore, it is not known whether any distracting effect depends simply on concentrating on an interactive task or whether the content of games is also an important factor as there have been no controlled trials comparing video games with other distracters. Further research should examine factors within games such as novelty, users' preferences, and relative levels of challenge and should compare video games with other potentially distracting activities."

1. Working with a partner, select one of the areas of potential research suggested in the passage from the paper and formulate a specific question that could be addressed by performing an experiment.

2. Propose an experiment that would provide data to address the question from Step 1. Be specific about how subjects might be selected, what the experimental conditions (treatments) would be, and what response would be measured.

3. At the end of Section 2.3 there are 10 questions that can be used to evaluate an experimental design. Answer these 10 questions for the design proposed in Step 2.

4. After evaluating your proposed design, are there any changes you would like to make to your design? Explain.

Summary of Key Concepts and Formulas

Term or Formula	Comment
Observational study	A study that observes characteristics of an existing population.
Simple random sample	A sample selected in a way that gives every different sample of size n an equal chance of being selected.
Stratified sampling	Dividing a population into subgroups (strata) and then taking a separate random sample from each stratum.
Cluster sampling	Dividing a population into subgroups (clusters) and forming a sample by randomly selecting clusters and including all individuals or objects in the selected clusters in the sample.
1 in k systematic sampling	A sample selected from an ordered arrangement of a population by choosing a starting point at random from the first k individuals on the list and then selecting every kth individual thereafter.

Term or Formula	Comment
Confounding variable	A variable that is related both to group membership and to the response variable.
Measurement or response bias	The tendency for samples to differ from the population because the method of observation tends to produce values that differ from the true value.
Selection bias	The tendency for samples to differ from the population because of systematic exclusion of some part of the population.
Nonresponse bias	The tendency for samples to differ from the population because measurements are not obtained from all individuals selected for inclusion in the sample.
Experiment	A procedure for investigating the effect of *experimental conditions* (which are manipulated by the experimenter) on a *response variable*.
Treatments	The experimental conditions imposed by the experimenter.
Extraneous factor	A variable that is not of interest in the current study but is thought to affect the response variable.
Direct control	Holding extraneous factors constant so that their effects are not confounded with those of the experimental conditions.
Blocking	Using extraneous factors to create experimental groups that are similar with respect to those factors, thereby filtering out their effect.
Randomization	Random assignment of experimental units to treatments or of treatments to trials.
Replication	A strategy for ensuring that there is an adequate number of observations on each experimental treatment.
Placebo treatment	A treatment that resembles the other treatments in an experiment in all apparent ways but that has no active ingredients.
Control group	A group that receives no treatment or one that receives a placebo treatment.
Single-blind experiment	An experiment in which the subjects do not know which treatment they received but the individuals measuring the response do know which treatment was received, or an experiment in which the subjects do know which treatment they received but the individuals measuring the response do not know which treatment was received.
Double-blind experiment	An experiment in which neither the subjects nor the individuals who measure the response know which treatment was received.

Chapter Review Exercises 2.57–2.70

Turn to the Expanded Answers Section for answers not shown next to exercises.

CENGAGENOW Know exactly what to study! Take a pre-test and receive your Personalized Learning Plan.

2.57 The article "Tots' TV-Watching May Spur Attention Problems" (*San Luis Obispo Tribune*, April 5, 2004) describes a study that appeared in the journal *Pediatrics*. In this study, researchers looked at records of 2500 children who were participating in a long-term health study. They found that 10% of these children had attention disorders at age 7 and that hours of television watched at age 1 and age 3 was associated with an increased risk of having an attention disorder at age 7.
a. Is the study described an observational study or an experiment?   Observational study
b. Give an example of a potentially confounding variable that would make it unwise to draw the conclusion that hours of television watched at a young age is the cause of the increased risk of attention disorder.

2.58 A study of more than 50,000 U.S. nurses found that those who drank just one soda or fruit punch a day tended to gain much more weight and had an 80% increased risk in developing diabetes compared to those who drank less than one a month. (*The Washington Post*, August 25, 2004). "The message is clear. . . . Anyone who cares about their health or the health of their family would not consume these beverages" said Walter Willett of the Harvard School of Public Health who helped conduct the study. The sugar and beverage industries said that the study was fundamentally flawed. "These allegations are inflammatory. Women who drink a lot of soda may simply have generally unhealthy lifestyles" said Richard Adamson of the American Beverage Association.
a. Do you think that the study described was an observational study or an experiment?   Observational study
b. Is it reasonable to conclude that drinking soda or fruit punch causes the observed increased risk of diabetes? Why or why not?

2.59 "Crime Finds the Never Married" is the conclusion drawn in an article from *USA Today* (June 29, 2001). This conclusion is based on data from the Justice Department's National Crime Victimization Survey, which estimated the number of violent crimes per 1000 people, 12 years of age or older, to be 51 for the never married, 42 for the divorced or separated, 13 for married individuals, and 8 for the widowed. Does being single cause an increased risk of violent crime? Describe a potential confounding variable

that illustrates why it is unreasonable to conclude that a change in marital status causes a change in crime risk.

2.60 The paper "Prospective Randomized Trial of Low Saturated Fat, Low Cholesterol Diet During the First Three Years of Life" (*Circulation* [1996]: 1386–1393) describes an experiment in which "1062 infants were randomized to either the intervention or control group at 7 months of age. The families of the 540 intervention group children were counseled to reduce the child's intake of saturated fat and cholesterol but to ensure adequate energy intake. The control children consumed an unrestricted diet."
a. The researchers concluded that the blood cholesterol level was lower for children in the intervention group. Is it reasonable to conclude that the parental counseling and subsequent reduction in dietary fat and cholesterol are the cause of the reduction in blood cholesterol level? Explain why or why not.
b. Is it reasonable to generalize the results of this experiment to all children? Explain.

2.61 The article "Workers Grow More Dissatisfied" in the *San Luis Obispo Tribune* (August 22, 2002) states that "a survey of 5000 people found that while most Americans continue to find their jobs interesting, and are even satisfied with their commutes, a bare majority like their jobs." This statement was based on the fact that only 51 percent of those responding to a mail survey indicated that they were satisfied with their jobs. Describe any potential sources of bias that might limit the researcher's ability to draw conclusions about working Americans based on the data collected in this survey.

2.62 According to the article "Effect of Preparation Methods on Total Fat Content, Moisture Content, and Sensory Characteristics of Breaded Chicken Nuggets and Beef Steak Fingers" (*Family and Consumer Sciences Research Journal* [1999]: 18–27), sensory tests were conducted using 40 college student volunteers at Texas Women's University. Give three reasons, apart from the relatively small sample size, why this sample may not be ideal as the basis for generalizing to the population of all college students.

2.63 Do ethnic group and gender influence the type of care that a heart patient receives? The following passage is from the article "Heart Care Reflects Race and Sex, Not

Symptoms" (*USA Today*, February 25, 1999, reprinted with permission):

> Previous research suggested blacks and women were less likely than whites and men to get cardiac catheterization or coronary bypass surgery for chest pain or a heart attack. Scientists blamed differences in illness severity, insurance coverage, patient preference, and health care access. The researchers eliminated those differences by videotaping actors— two black men, two black women, two white men, and two white women—describing chest pain from identical scripts. They wore identical gowns, used identical gestures, and were taped from the same position. Researchers asked 720 primary care doctors at meetings of the American College of Physicians or the American Academy of Family Physicians to watch a tape and recommend care. The doctors thought the study focused on clinical decision-making.

Evaluate this experimental design. Do you think this is a good design or a poor design, and why? If you were designing such a study, what, if anything, would you propose to do differently?

2.64 Does alcohol consumption cause increased cravings for cigarettes? Research at Purdue University suggests this is so (see CNN.com web site article "Researchers Find Link Between Cigarette Cravings and Alcohol," dated June 13, 1997). In an experiment, 60 heavy smokers and moderate drinkers were divided into two groups. One group drank vodka tonics and the other group drank virgin tonics (tonic water alone), but all subjects were told they were drinking vodka tonics. The researchers then measured the level of nicotine cravings (by monitoring heart rate, skin conductance, etc.). Those who had consumed the vodka tonics had 35% more cravings than those who did not. Assuming that the assignment of subjects to the treatment (vodka) and control groups was made at random, do you think there are any confounding factors that would make conclusions based on this experiment questionable?

2.65 An article in the *San Luis Obispo Tribune* (September 7, 1999) described an experiment designed to investigate the effect of creatine supplements on the development of muscle fibers. The article states that the researchers "looked at 19 men, all about 25 years of age and similar in weight, lean body mass, and capacity to lift weights. Ten were given creatine—25 grams a day for the first week, followed by 5 grams a day for the rest of the study. The rest were given a fake preparation. No one was told what he was getting. All the men worked out under the guidance of the same trainer. The response variable measured was gain in fat-free mass (in percent)."
a. What extraneous variables are identified in the given statement, and what strategy did the researchers use to deal with them?
b. Do you think it was important that the men participating in the experiment were not told whether they were receiving creatine or the placebo? Explain.
c. This experiment was not conducted in a double-blind manner. Do you think it would have been a good idea to make this a double-blind experiment? Explain.

2.66 The article "Heavy Drinking and Problems among Wine Drinkers" (*Journal of Studies on Alcohol* [1999]: 467–471) investigates whether wine drinkers tend to drink less excessively than those who drink beer and spirits. A sample of Canadians, stratified by province of residence and other socioeconomic factors, was selected.
a. Why might stratification by province be a good thing?
b. List two socioeconomic factors that would be appropriate to use for stratification. Explain how each factor would relate to the consumption of alcohol in general and of wine in particular.   For example, occupation or income.

2.67 Researchers at the University of Houston decided to test the hypothesis that restaurant servers who squat to the level of their customers would receive a larger tip ("Effect of Server Posture on Restaurant Tipping," *Journal of Applied Social Psychology* [1993]: 678–685). In the experiment, the waiter would flip a coin to determine whether he would stand or squat next to the table. The waiter would record the amount of the bill and of the tip and whether he stood or squatted.
a. Describe the treatments and the response variable.
b. Discuss possible extraneous factors and how they could be controlled.
c. Discuss whether blocking would be necessary.
d. Identify possible confounding variables.
e. Discuss the role of randomization in this experiment.

2.68 You have been asked to determine on what types of grasslands two species of birds, northern harriers and short-eared owls, build nests. The types of grasslands to be used include undisturbed native grasses, managed native grasses, undisturbed nonnative grasses, and managed nonnative grasses. You are allowed a plot of land 500 m square to study. Explain how you would determine where to plant the four types of grasses. What role would randomization play in this determination? Identify any confounding variables. Would this study be considered an observational study or an experiment? (Based on the article

"Response of Northern Harriers and Short-Eared Owls to Grassland Management in Illinois," *Journal of Wildlife Management* [1999]: 517–523.)

2.69 A manufacturer of clay roofing tiles would like to investigate the effect of clay type on the proportion of tiles that crack in the kiln during firing. Two different types of clay are to be considered. One hundred tiles can be placed in the kiln at any one time. Firing temperature varies slightly at different locations in the kiln, and firing temperature may also affect cracking. Discuss the design of an experiment to collect information that could be used to decide between the two clay types. How does your proposed design deal with the extraneous factor *temperature*?

2.70 A mortgage lender routinely places advertisements in a local newspaper. The advertisements are of three different types: one focusing on low interest rates, one featuring low fees for first-time buyers, and one appealing to people who may want to refinance their homes. The lender would like to determine which advertisement format is most successful in attracting customers to call for more information. Describe an experiment that would provide the information needed to make this determination. Be sure to consider extraneous factors, such as the day of the week that the advertisement appears in the paper, the section of the paper in which the advertisement appears, or daily fluctuations in the interest rate. What role does randomization play in your design?

Personal Tutor

Do you need a live tutor for homework problems?

Are you ready? Take your exam-prep post-test now.

Bold exercises answered in back ● Data set available online but not required ▼ Video solution available

Graphing Calculator Explorations

Exploration 2.1 Calculators and the Study of Statistics

You must be able to use your calculator in order to be able to analyze data. In previous math classes you may have used your calculator for graphing functions, finding solutions to equations, and arithmetic calculations. In statistics you will use your calculator differently and will also use new calculator keys and menu items. Graphing Calculator Explorations are intended to help you get maximum utility from your calculator. These explorations highlight some important features of your calculator. In order to speak to the widest possible audience, the explorations will be generic in nature, rather than showcasing a particular calculator.

Calculators vary in statistical capability and in the applications that can be downloaded from the web. The characteristics of a graphing calculator that are important for the study of statistics are:

- Capability to perform elementary statistical calculations (computing means, standard deviations, correlation coefficients, and regression equations)
- Capability to generate statistical graphs (boxplots, histograms, and scatterplots)
- Row-and-column data entry format

Exploration 2.2 Generating Random Integers

Procedures for generating random numbers have been around for a long time. The earliest techniques for generating random numbers were throwing dice, dealing cards, and selecting well-mixed numbered balls from a container. Rapidly turning discs and spinners, randomly pulsating vacuum tubes, and clicking Geiger counters have also been used to generate random numbers. Today, computers and calculators use algorithms to generate "random" numbers. These algorithms do not, strictly speaking, generate random numbers; they generate what are called "pseudo-random" numbers. For most purposes, the "random" numbers generated by today's computers and calculators are adequate, and we will refer to these numbers as random.

Generating random numbers is a built-in function of graphing calculators. To learn the appropriate keystrokes you will need to consult the manual that came with your calculator; look for "rand," or possibly "random" in the index. On most calculators a single keystroke or a short sequence of keystrokes should produce a random number, something like this: .9435974025. On some calculators the precision can be adjusted and in our discussions we will generally use four digits. When you have found the appropriate keystrokes for your calculator, generate five random numbers. Some calculators will repeat the process each time you press "enter" or "execute." Try pressing the enter/execute button four times after you get the first random number. If a random number appears each time, smile—this will save many keystrokes! Here are the numbers we obtained (yours will be different): 0.5147, 0.4058, 0.7338, 0.0440, and 0.3394. Your numbers are also in the interval $0 < 1$. This is not an accident; random number generators typically produce a random number, r, such that $0 \leq r < 1$. Unless your calculator has some additional built-in random number functions, you will have to convert each random decimal number to a more useful form—often a positive integer. To see whether your calculator has a built-in capability of generating random integers, look in your index for something like "random integer." In the discussion that follows, we will *not* assume you have a random integer capability, but if you do, please use it!

If you do not have a built-in capability to generate random integers, the method in the box below will be helpful.

Converting Random Numbers to Integers, 1, 2, 3, . . . , n

To convert a calculator-generated decimal random number r, $0 \leq r < 1$, into a random integer in the range 1, 2, 3, . . . , n, multiply r by n, add 1, and ignore the digits to the right of the decimal.

This will involve a sequence of keystrokes something like the following, where we use "*rand*" to mean the keystrokes needed to generate r:

$rand \times n + 1.$

Most scientific calculators have an "int" or "floor" function, which will truncate the decimal by "rounding down." If your calculator has this capability, you can accomplish the integer random number generation in one sequence of keystrokes:

$\text{int}(rand \times n) + 1.$

■ **Example: Generating Random Numbers** Generate five random integers between 1 and 100 for purposes of sampling from a population with 100 individuals. The keystrokes to generate a random integer between and 100 are:

$\text{int}(rand \times 100) + 1.$

(Remember, *rand* stands for the sequence of keystrokes needed to get the random number between 0 and 1 and *int* stands for the keystrokes necessary to truncate a decimal to an integer.)

The numbers we obtained (though of course your numbers will differ) are

11   55   29   38   37

We would then include the individuals identified by these numbers in the sample. (Note that in this example we are sampling from a population. If we were sampling without replacement and the random number generated by the calculator resulted in two or more random integers that were the same, we would have ignored the duplicates and generated additional random integers as needed.)

Adding 1 in the formula above isn't some sort of magic. Because the *rand* keystrokes return random numbers in the range $0 \leq r < 1$, it is possible for *rand* to deliver a 0. If you are not bothered by random integers starting at 0, you need not waste keystrokes by adding 1 each time. Also note that the arithmetic random number generators in calculators are shipped from the factory with a number called a "seed," needed to start the random number generation process. If two calculators in your class are "right out of the box" you may notice that the random numbers generated by these calculators are the same. This will not be a problem for very long; since calculator users will typically press the *rand* sequence different numbers of times, sequences will soon differ in actual classroom use.

Exploration 2.3 Randomization

■ **Random Assignment to Treatments**   The process of randomization is critical in the proper design of an experiment. Randomly assigning subjects to treatments can be accomplished using the graphing calculator's capability to generate integers from 1, 2, . . . , *n*. We will illustrate how this can be done in some common experimental situations. We will use small samples are used to illustrate the techniques, but the techniques also work for large samples.

In the first experimental situation we will assign subjects to treatments without worrying about getting equal numbers of subjects in each treatment. This experiment involves the effect of pizza on performance on a statistics exam. An instructor has decided to use three types of pizza (treatments): sausage pizza, mushroom pizza, and cheese pizza. Twelve students will take part in the experiment, each receiving a free pizza lunch for participating. The strategy for assigning treatments is very simple: Generate 12 random integers from the list {1, 2, 3}. *Before* any students are assigned, we arbitrarily assign mushroom pizza = 1, cheese pizza = 2, and sausage pizza = 3.

We now randomly generate integers between 1 and 3 using:

$$\mathrm{int}(rand \times 3) + 1.$$

Our results were (remember that yours will surely be different):

3, 1, 1, 1, 1, 3, 1, 2, 1, 3, 1, 3

These numbers are used to assign a treatment (pizza type) to each of the twelve students participating in the experiment, as shown. Entries in the table are (student, treatment number, and treatment).

1 3 (Sausage)	2 1 (Mushroom)	3 1 (Mushroom)	4 1 (Mushroom)
5 1 (Mushroom)	6 3 (Sausage)	7 1 (Mushroom)	8 2 (Cheese)
9 1 (Mushroom)	10 3 (Sausage)	11 1 (Mushroom)	12 3 (Sausage)

We can easily see some problems using this method of assignment: Mushroom pizza accounts for more than 50% of the experimental trials, and there is no replication of the cheese pizza! What if we want to have equal-sized treatment groups? We can get sample size balance among treatments by adding a rule to our procedure: "Once any treatment has 4 subjects assigned, do not assign any more to that treatment." One disadvantage of this rule is that it might be necessary to generate many random numbers before completing the assignment of subjects to treatments, but the rule is easy to understand and implement. Anticipating the need for more than 12 random numbers, we generated a sequence using int $(rand \times 3) + 1$:

1 3 3 2 1 3 3 1 1 3 3 2 1 1 3 1 2 1 1 3 2

The assignment to treatments now proceeds as shown, with the assignments appearing below the random integers:

1 3 3 2 1 3 3 1 1 3 3 2 1 1 3 1 2 1 1 3 2
M S S C M S S M M (Last three assigned Cs.)

Students 1, 5, 8, and 9 are assigned to the Mushroom group; students 4, 10, 11, and 12 are assigned to the Cheese group; students 2, 3, 6, and 7 are assigned to the Sausage group. Notice that this particular assignment was essentially done after nine random numbers were generated; the only students left to be assigned were assigned to the Cheese treatment.

■ **Random Assignment to Treatments—with Blocking** Randomization in a situation where the experimenter is using blocking to control an extraneous factor could also be implemented using random integers. Suppose the instructor in the previous example has some seats near the window and some seats that are not by the window. It is possible that students near the window might be distracted and that this might affect exam performance. Because of this, it would be reasonable to block by position in the room. Suppose that six students will participate in the experiment. We need to consider the blocking strategy as we assign treatments.

Our randomization is restricted a bit compared to what we did earlier because now we need to have each treatment represented in each block. Suppose that the six students are seated for the exam in the arrangement shown in the table below.

Near Window	Not Near Window

We now want to assign treatments at random to the students within each block. A sequence of random integers between 1 and 3 can be used for this purpose. One possible sequence is:

3 3 2 1 1 3 1 2 1 1 3 2

The treatments are assigned as shown in the table below, starting with the "Near Window" block and then, once the assignment for the "Near Window" block is completed, moving to the "Not Near Window" block. Remember that mushroom pizza = 1, cheese pizza = 2, and sausage pizza = 3.

Random Integer	Decision
3	Assign student in row one of near window block to sausage pizza treatment
3	Ignore, as there is already a student assigned to sausage pizza in the near window block
2	Assign student in row two of near window block to cheese pizza treatment
—	Assign student in row three of near window block to mushroom pizza treatment (since that is the only treatment that has not already been assigned in this block)
1	Assign student in row one of not near window block to mushroom pizza treatment
1	Ignore, as there is already a student assigned to mushroom pizza in the near window block
3	Assign student in row two of not near window block to sausage pizza treatment
—	Assign student in row three of near window block to cheese pizza treatment (since that is the only treatment that has not already been assigned in this block)

The final assignment of treatments to subjects within each block is shown in the table below. Note that each treatment appears once in each block.

Near Window	Not Near Window
Sausage	Mushroom
Cheese	Sausage
Mushroom	Cheese

Exploration 2.4 Generating Random Real Numbers

In previous Explorations, we used the *rand* function to generate random numbers between 0 and 1 and then converted them to integers. In this Exploration, we will use the *rand* function for generating random *real* numbers in some interval (a, b).

The need for random real numbers sometimes arises in sampling situations where it is not possible to make a list of the elements of the population. For example, suppose we want to study chemical residue from the use of pesticides on farms in Iowa. A map of Iowa is overlaid with a grid of 1-mile by 1-mile squares. Picking squares at random in order to measure chemical residue would be a good start, but within any randomly selected square the chemical residue measurement must be taken at some actual location in the square. To determine this location, imagine a coordinate system with the origin in the southwest corner of a square of land. Since each square is 1 mile by 1 mile, generating a random point in a square of land is easy using the *rand* function: generate two random real numbers between 0 and 1, and use them as *x*- and *y*-coordinates for the square. Suppose the *rand* function generated 0.45059, and 0.98906. We would convert this to (0.45059, 0.98906), which would then determine the location where the measurement should be made.

Suppose now that we want to select random locations from a map of the wilds of Saskatchewan, Canada in order to estimate the amount of a particular mineral in the soil. Here we can use the "natural" coordinates provided by the latitude and longitude of the map to determine location. For example, Saskatchewan's latitudes range from 49 to 60N, and longitudes from 102 to 110W. For generating random points in Saskatchewan, we can use the *rand* function and a little algebra.

We approach this problem in stages. First, consider the problem of generating a random number between 0 and *b*. One solution is to generate a random number between 0 and 1, and multiply it by the positive real number, *b*. Algebraically we see that if *rand* represents the random number generated,

$$0 \leq rand < 1$$
$$b \times (0) \leq b \times (rand) < b \times (1)$$
$$0 \leq b \times (rand) < b$$

Our ability to generate random real numbers is not limited to intervals that begin at 0. To generate a random real number between any two real numbers, *a* and *b*, we can do a bit more algebra, first multiplying a random number between 0 and 1 by $(b - a)$ and then adding *a*. The resulting number will be a random number between *a* and *b*, as shown:

$$0 \leq rand < 1$$
$$(b - a) \times (0) \leq (b - a) \times (rand) < (b - a) \times (1)$$
$$0 \leq (b - a) \times (rand) < (b - a)$$
$$0 + a \leq (b - a) \times (rand) + a < (b - a) + a$$
$$a \leq (b - a) \times (rand) + a < b$$

Saskatchewan is located between the latitudes 49 and 60N and longitudes 102 and 110W. So,

$$102 \leq (110 - 102) \times (rand) + 102 < 110$$

generates a random longitude in Saskatchewan and

$$49 \leq (60 - 49) \times (rand) + 49 < 60$$

generates a random latitude in Saskatchewan.

For example, using *rand*, we might get: 0.6217, and 0.8811. Substituting these values into the formulas above results in this random location:

$$(110 - 102) \times (0.6217) + 102 = 106.97° W$$
$$(60 - 49) \times (0.8811) + 49 = 58.69° N$$

This random location is somewhere close to Fond du Lac, on Lake Athabasca, just down the road from Uranium City!

TEACHING TIPS

Intent of this chapter:

After completing this chapter, students should be able to (1) create and interpret graphs for categorical data (bar graphs, segmented bar graphs, and pie charts), (2) create and interpret graphs for numerical data (stemplots and histograms), and (3) create and interpret graphs for bivariate, numerical data (scatterplots and time series plots).

A PowerPoint® Lecture is available on the *Instructor's Resource Binder* CD. Remember that all the graphs in the text can also be found on the *Instructor's Resource Binder* CD in the Image Library.

Note: For any type of graph created, students *must* scale and label the axes.

■ **Section 3.1**

This section discusses graphs for categorical data. Stress to students that relative frequencies should be used on the vertical axis of comparative bar graphs, especially if the sample sizes are not the same. In addition to bar graphs and pie charts, students should know how to make segmented bar graphs as shown in Example 3.6 on page 81. Segmented bar graphs are becoming more popular in the media.

Suggested Assignment: Exercises 3.2, 3.3, 3.4, 3.6, 3.13, 3.14
Activity: Activity 3.1: Locating States (p. 134 and *Activities Manual,* p. 26)

■ **Section 3.2**

Stem-and-leaf displays (also called stemplots) are used to display numerical data. Although it is not necessary to order the leaves behind each stem, doing so is helpful if you wish to find the median and quartiles (discussed in Chapter 4). Truncating is often used in creating stem-and-leaf displays; see Figure 3.11 and 3.12. Truncating is different from rounding. When truncating, digits are simply removed and their values have no effect on the remaining digits. Figures 3.13 and 3.14 show how to repeat stems in order to improve the display. When any row of leaves is inordinately long, it is helpful to repeat the stems using low (L) and high (H). This helps spread out the distribution and gives a better sense of its overall shape.

Comparative stem-and-leaf displays are especially useful when comparing two distributions. In a comparative stem-and-leaf display, the stems are listed in the middle with leaves for one distribution extending to the left and leaves for the other distribution extending to the right. Students should be able to describe the distributions displayed in stemplots. They should comment on center (an approximate typical value), spread (how much variability), shape (see Section 3.3), and unusual features. (For now, students can estimate measures of center and spread. In Chapter 4, they will be able to calculate these measures. However, it is helpful for students to begin describing graphs appropriately.) If they are asked to compare two distributions, they should use comparative language to comment on center, spread, shape, and unusual features. (Remember, listing the values is not considered a form of comparison on the AP Exam.)

Suggested Assignment: Exercises 3.16, 3.17, 3.19, 3.21
Activity: Activity 3.2: Bean Counters! (p. 134 and *Activities Manual,* p. 29)★
Assessment: Chapter 3 Quiz 1 (*Test Bank* and *Instructor's Resource Binder* CD)

■ **Section 3.3**

Histograms (graphs in which the bars touch) are used for displaying numerical data, while bar graphs (bars do not touch) are used with categorical data. With large data sets, histograms and frequency tables provide an easy way to organize and present the data.

★ Kathy's personal favorite

Histograms display discrete numerical data differently than continuous numerical data. In histograms of discrete numerical data, bars are typically centered over **each** discrete value. (*Calculator hint:* Set the minimum x-value at −0.5 to get the calculator to center bars over the values.) Notice that the overall shape of the distribution does not change when relative frequency is used on the vertical axis instead of frequency. When discrete numerical data sets have a large range, it is helpful to group discrete values together (see Example 3.13 on page 99).

In histograms of continuous numerical data, the bars extend over a class interval or class. Standard practice is that the bar includes the left-most value of the bar and all the values in between and up to but not including the right-most value of the bar. For example, if the bar is from 10 to 20, the bar includes 10 and all values up to but not including 20.

Students should know that intervals are not required to have the same width; however, if the interval widths are not equal, density is used as the height of the bar. This is calculated by the formula: $\text{density} = \dfrac{\text{relative frequency of class}}{\text{class width}}$.

Regardless of whether the data are discrete or continuous, histograms are described by discussing center, shape, spread, and unusual features (same as other graphs for numerical data). Remember, if you wish to use two separate histograms to compare two distributions, be sure to use the same scale on the horizontal axis.

Cumulative relative frequency plots (also called ogives) display percentiles (discussed in Chapter 4). Students frequently have difficulty understanding the cumulative relative frequency plot. The steeper the slope of the line (over an interval), the larger the percent of values in that interval. A flat line (over an interval) indicates that there were no values within that interval. Be sure to ask questions such as "What percent of the observations were above (or below) . . . ?" or "What value has 20% of the observations below (or above) it?"

This section also discusses the shapes of distributions (e.g., symmetric, skewed, uniform, or bimodal). The direction of skewness (positive or right versus negative or left) is a common problem for students. The *longer* tail is the direction of skewness.

The following card-sort activity can help students make connections between variables and the appropriate graphical representations: On one-third of a set of cards, write a description of a variable (e.g., "the grades on an easy statistics quiz"); on another third, write the type of variable (categorical, discrete numerical, or continuous numerical); and on the remaining third of the cards, state the type of graph with the shape this distribution would likely have (e.g., "histogram that is skewed left"). Distribute the cards to students working in groups of three or four. The group then works cooperatively to match the triplets of cards.

Suggested Assignment: Exercises 3.24, 3.25, 3.27, 3.30, 3.32, 3.33

Activity: Bonus Activity 3.4: Stretchability of Rubber Bands (*Activities Manual,* p. 35) ★

■ **Section 3.4**   This section discusses how to graph bivariate numerical data. Most students are already familiar with scatterplots and correlation. Chapter 5 provides a more in-depth discussion of scatterplots and correlation. Example 3.19 (graphs also located on *Instructor's Resource Binder* CD) provides an interesting class discussion.

This section also introduces students to time-series plots, which are used to show trends and cycles over time.

Suggested Assignment: Exercises 3.37, 3.39, 3.41, 3.43, 3.45

★ Kathy's personal favorite

Activity: "Working With Your Calculator" (*Instructor's Resource Binder* & CD, Chapter 3 Activities Worksheets) or Bonus Activity 3.3: M&M Marginal Plots (*Activities Manual*, p. 32)

Assessment: Chapter 3 Quiz 2 (*Test Bank* and *Instructor's Resource Binder* CD) and/or Chapter 3 Concept Quiz (*Test Bank* and *Instructor's Resource Binder* CD)

■ **Section 3.5**

Students should pay close attention to the "A Word to the Wise: Cautions and Limitations" section. Remember that these graphs are found on the *Instructor's Resource Binder* CD and can be used for class discussions.

Suggested Review Assignment: Exercises 3.49, 3.50, 3.52, 3.56, 3.60, 3.63

Assessment: Chapter 3 Test (*Test Bank* and *Instructor's Resource Binder* CD)

Chapter 3

Graphical Methods for Describing Data

© John Terence Turner/Taxi/Getty Images

M̲ost college students (and their parents) are concerned about the cost of a college education. *The Chronicle of Higher Education* (August, 2005) reported the average tuition and fees for 4-year public institutions in each of the 50 U.S. states for the 2003–2004 academic year. Average tuition and fees (in dollars) are given for each state:

3977	3423	3586	4010	3785	3447	5761	6176	2773	3239
3239	3323	5653	5384	4991	3686	3868	3208	5011	6230
6080	6015	5754	3754	5367	4155	4238	2724	7623	7266
3161	4885	3251	3855	6561	3200	4677	7633	5387	7482
4441	4043	3579	2902	8260	5069	4630	3172	4675	3090

Several questions could be posed about these data. What is a typical value of average tuition and fees for the 50 states? Are observations concentrated near the typical value, or does average tuition and fees differ quite a bit from state to state? Do states with low tuition and fees or states with high tuition and fees predominate, or are there roughly equal numbers of the two types? Are there any states whose average tuition and fees are somehow unusual compared to the rest? What proportion of the states have average tuition and fees exceeding $4000? exceeding $6000?

Questions such as these are most easily answered if the data can be organized in a sensible manner. In this chapter, we introduce some techniques for organizing and describing data using tables and graphs.

Improve your understanding and save time! Visit www.cengage.com/login **where you will find:**

- Step-by-step instructions for MINITAB, Excel, TI-83, SPSS, and JMP
- Video solutions to selected exercises
- Data sets available for selected examples and exercises
- Exam-prep pre-tests that build a Personalized Learning Plan based on your results so that you know exactly what to study
- Help from a live statistics tutor 24 hours a day

| 3.1 | Displaying Categorical Data: Comparative Bar Charts and Pie Charts |

■ **Comparative Bar Charts** ....

In Chapter 1 we saw that categorical data could be summarized in a frequency distribution and displayed graphically using a bar chart. Bar charts can also be used to give a visual comparison of two or more groups. This is accomplished by constructing two or more bar charts that use the same set of horizontal and vertical axes, as illustrated in Example 3.1.

Example 3.1   Perceived Risk of Smoking

The article "Most Smokers Wish They Could Quit" (*Gallup Poll Analyses,* November 21, 2002) noted that smokers and nonsmokers perceive the risks of smoking differently. The accompanying relative frequency table summarizes responses regarding the perceived harm of smoking for each of three groups: a sample of 241 smokers, a sample of 261 former smokers, and a sample of 502 nonsmokers.

Perceived Risk of Smoking	Frequency			Relative Frequency		
	Smokers	Former Smokers	Nonsmokers	Smokers	Former Smokers	Nonsmokers
Very harmful	145	204	432	.60	.78	.86
Somewhat harmful	72	42	50	.30	.16	.10
Not too harmful	17	10	15	.07	.04	.03
Not harmful at all	7	5	5	.03	.02	.01

Important concept

When constructing a comparative bar graph we use the relative frequency rather than the frequency to construct the scale on the vertical axis so that we can make meaningful comparisons even if the sample sizes are not the same. The comparative bar chart for these data is shown in Figure 3.1. It is easy to see the differences among the three groups. The proportion believing that smoking is very harmful is noticeably smaller for smokers than for either former smokers or nonsmokers, and the proportion of former smokers who believe smoking is very harmful is smaller than the proportion for nonsmokers.

To see why it is important to use relative frequencies rather than frequencies to compare groups of different sizes, consider the *incorrect* bar chart constructed using the frequencies rather than the relative frequencies (Figure 3.2). The incorrect bar chart conveys a very different and misleading impression of the differences among the three groups.

!! Step-by-step technology instructions available online.

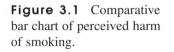

Figure 3.1 Comparative bar chart of perceived harm of smoking.

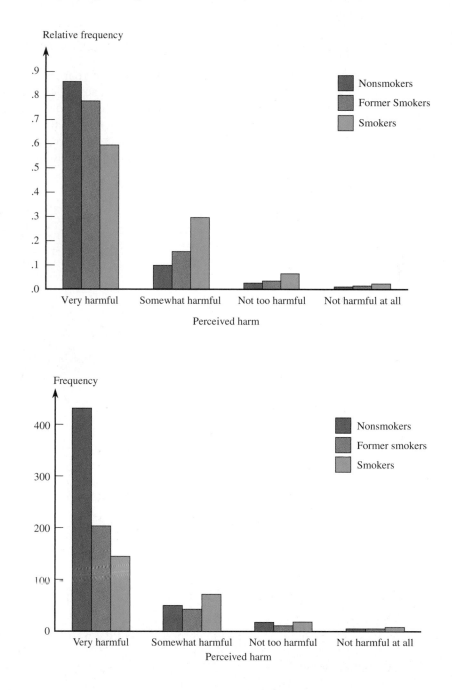

Figure 3.2 An *incorrect* comparative bar chart for the data of Example 3.1.

■ Pie Charts

A categorical data set can also be summarized using a pie chart. In a pie chart, a circle is used to represent the whole data set, with "slices" of the pie representing the possible categories. The size of the slice for a particular category is proportional to the corresponding frequency or relative frequency. Pie charts are most effective for summarizing data sets when there are not too many different categories.

Example 3.2 Life Insurance for Cartoon Characters??

The article "Fred Flintstone, Check Your Policy" (*The Washington Post,* October 2, 2005) summarized the results of a survey of 1014 adults conducted by the Life and Health Insurance Foundation for Education. Each person surveyed was asked to select which of five fictional characters, Spider-Man, Batman, Fred Flintstone, Harry Potter, and Marge Simpson, he or she thought had the greatest need for life insurance. The resulting data are summarized in the pie chart of Figure 3.3.

Figure 3.3 Pie chart of data on which fictional character most needs life insurance.

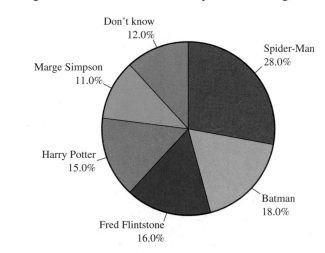

The survey results were quite different from an insurance expert's assessment. His opinion was that Fred Flintstone, a married father with a young child, was by far the one with the greatest need for life insurance. Spider-Man, unmarried with an elderly aunt, would need life insurance only if his aunt relied on him to supplement her income. Batman, a wealthy bachelor with no dependents, doesn't need life insurance in spite of his dangerous job! ■

Pie Chart for Categorical Data

When to Use Categorical data with a relatively small number of possible categories. Pie charts are most useful for illustrating proportions of the whole data set for various categories.

How to Construct

1. Draw a circle to represent the entire data set.
2. For each category, calculate the "slice" size. Because there are 360 degrees in a circle

   slice size = 360 (category relative frequency)

3. Draw a slice of appropriate size for each category. This can be tricky, so most pie charts are generated using a graphing calculator or a statistical software package.

What to Look For

■ Categories that form large and small proportions of the data set.

Example 3.3 Birds That Fish

Night herons and cattle egrets are species of birds that feed on aquatic prey in shallow water. These birds wade through shallow water, stalking submerged prey and then striking rapidly and downward through the water in an attempt to catch the prey. The article "Cattle Egrets Are Less Able to Cope with Light Refraction Than Are Other Herons" (*Animal Behaviour* [1999]: 687–694) gave data on outcome when 240 cattle egrets attempted to capture submerged prey. The data are summarized in the following frequency distribution:

Outcome	Frequency	Relative Frequency
Prey caught on first attempt	103	.43
Prey caught on second attempt	41	.17
Prey caught on third attempt	2	.01
Prey not caught	94	.39

To draw a pie chart by hand, we would first compute the slice size for each category. For the first attempt category, the slice size would be

$$\text{slice size} = (.43)(360) = 154.8 \text{ degrees}$$

155 degrees, to represent first attempt category

We would then draw a circle and use a protractor to mark off a slice corresponding to about 155°, as illustrated here in the figure shown in the margin. Continuing to add slices in this way leads to a completed pie chart.

It is much easier to use a statistical software package to create pie charts than to construct them by hand. A pie chart for the cattle egret data, created with the statistical software package MINITAB is shown in Figure 3.4(a). Figure 3.4(b) shows a pie chart constructed using similar data for 180 night herons. Although some differences

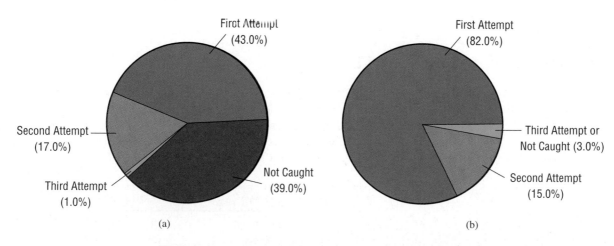

Figure 3.4 Pie charts for Example 3.3: (a) cattle egret data; (b) night heron data.

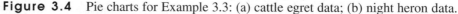
Step-by-step technology instructions available online.

Figure 3.5 Comparative bar chart for the egret and heron data.

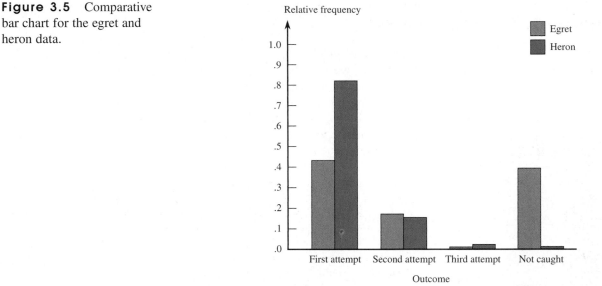

between night herons and cattle egrets can be seen by comparing the pie charts in Figures 3.4(a) and 3.4(b), it is difficult to actually compare category proportions using pie charts. A comparative bar chart (Figure 3.5) makes this type of comparison easier.

■

■ A Different Type of "Pie" Chart: Segmented Bar Charts ....

A pie chart can be difficult to construct by hand, and the circular shape sometimes makes it difficult to compare areas for different categories, particularly when the relative frequencies for categories are similar. The **segmented bar chart** (also sometimes called a stacked bar chart) avoids these difficulties by using a rectangular bar rather than a circle to represent the entire data set. The bar is divided into segments, with different segments representing different categories. As with pie charts, the area of the segment for a particular category is proportional to the relative frequency for that category. Example 3.4 illustrates the construction of a segmented bar graph.

Example 3.4 Gun Permit Denials

Between 1998 and 2004, more than 53 million background checks for gun purchases were conducted. One reason for denial of a gun permit is having a criminal history, and during this period approximately 748,000 permits were denied for this reason. The accompanying table categorizes these denials by type of criminal history (*San Francisco Chronicle,* November 2005).

Criminal History	Relative Frequency
Felony offense	.59
Domestic violence	.15
Drug-related offense	.07
Other	.19

To construct a segmented bar graph for these data, first draw a bar of any fixed width and length, and then add a scale that ranges from 0 to 1, as shown.

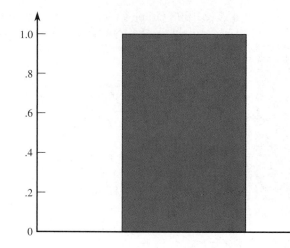

Then divide the bar into four segments, corresponding to the four possible categories in this example. The first segment, corresponding to felony offense, ranges from 0 to .59. The second segment, corresponding to domestic violence, ranges from .59 to .74 (for a length of .15, the relative frequency for this category), and so on. The segmented bar can be displayed either vertically or horizontally, as shown in Figure 3.6.

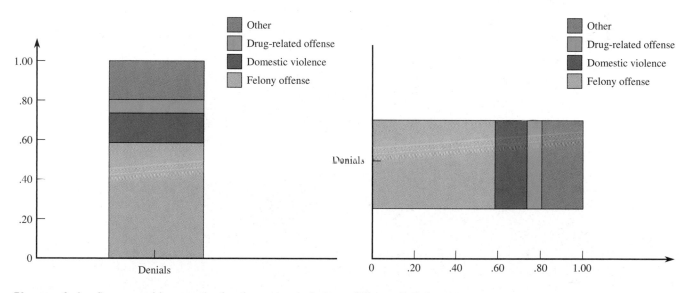

Figure 3.6   Segmented bar graphs for the gun permit data of Example 3.4.

■

■ Other Uses of Bar Charts and Pie Charts ....

As we have seen in previous examples, bar charts and pie charts can be used to summarize categorical data sets. However, they are occasionally used for other purposes, as illustrated in Examples 3.5 and 3.6.

Example 3.5   Grape Production

● The 1998 Grape Crush Report for San Luis Obispo and Santa Barbara counties in California gave the following information on grape production for each of seven different types of grapes (*San Luis Obispo Tribune,* February 12, 1999):

Type of Grape	Tons Produced
Cabernet Sauvignon	21,656
Chardonnay	39,582
Chenin Blanc	1,601
Merlot	11,210
Pinot Noir	2,856
Sauvignon Blanc	5,868
Zinfandel	7,330
Total	**90,103**

Although this table is not a frequency distribution, it is common to represent information of this type graphically using a pie chart, as shown in Figure 3.7. The pie represents the total grape production, and the slices show the proportion of the total production for each of the seven types of grapes.

Figure 3.7   Pie chart for grape production data.

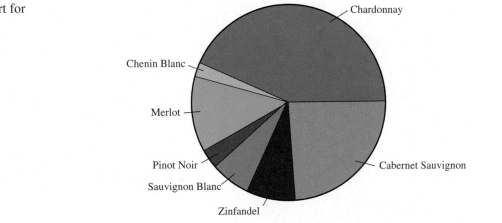

Example 3.6   Identity Theft

The article "Fraud, Identity Theft Afflict Consumers" (*San Luis Obispo Tribune,* February 2, 2005) gave the following identity theft rates for three states.

● Data set available online

State	Identity Theft (Cases per 100,000 Population)
California	122.1
Arizona	142.5
Oregon	87.8

Even though this table is not a frequency distribution, this type of information is often represented graphically in the form of a bar chart, as illustrated in Figure 3.8.

Figure 3.8 Bar chart for the identity theft rate data of Example 3.6.

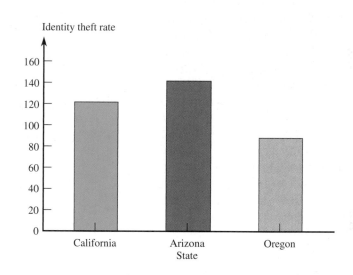

The article also gave Internet fraud rates for the same three states, again reported in cases per 100,000 population. Figure 3.9 is a comparative bar chart of the Internet fraud rates and identity theft rates for the three states.

Figure 3.9 Bar chart comparing internet fraud and identity theft rates.

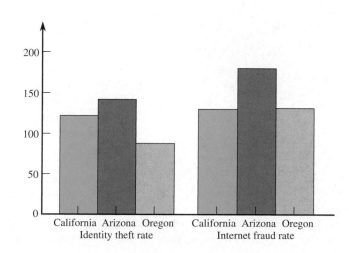

Exercises 3.1–3.14

Turn to the Expanded Answers Section for answers not shown next to exercises.

3.1 ▼ *The Chronicle of Higher Education* (August 31, 2001) published data collected in a survey of a large number of students who were college freshmen in the fall of 2001. Of those surveyed, 70.6% reported that they were attending their first-choice college, 20.8% their second-choice college, 5.5% their third-choice college, and 3.1% their fourth- or higher-choice college. Use a pie chart to display this information.

3.2 The article "Rinse Out Your Mouth" (Associated Press, March 29, 2006) summarized results from a survey of 1001 adults on the use of profanity. When asked "How many times do you use swear words in conversations?" 46% responded a few or more times per week, 32% responded a few times a month or less, and 21% responded never. Use the given information to construct a segmented bar chart.

3.3 Data from the U.S Census Bureau was used to construct the accompanying pie chart that reflects changes of residence between 1995 and 2000. Construct a bar graph for these data. Which graph is a more effective display of the data? Explain.

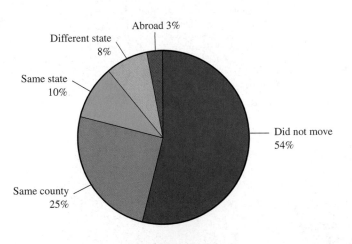

3.4 The California Healthy Kids Survey in 1999 and 2001 asked 7th, 9th, and 11th graders in San Luis Obispo County whether they smoked tobacco. In 1999, 6% of 7th graders, 17% of 9th graders, and 23% of 11th graders admitted smoking tobacco while in 2001 the corresponding figures were 3%, 13%, and 22%. Create a comparative bar chart showing bars for 7th graders together, bars for 9th graders together, and bars for 11th graders together. Comment on any interesting features of the chart.

3.5 The Institute for Highway Safety tracks which cars are most likely to be stolen by looking at theft claims per 1000 insured vehicles. For cars manufactured between 2001 and 2003, the top five were as follows:

Type of Vehicle	Claims per 1000 Vehicles
'02–'03 Cadillac Escalade EXT (luxury pickup)	20.2
'02–'03 Nissan Maxima (midsize 4-door car)	17.0
'02–'03 Cadillac Escalade (luxury SUV)	10.2
Dodge Stratus/Chrysler Sebring (midsize 4-door car)	8.3
Dodge Intrepid (large 4-door car)	7.9

Use this information to construct a pie chart.

3.6 The article "The Need to Be Plugged In" (Associated Press, December 22, 2005) described the results of a survey of 1006 adults who were asked about various technologies, including personal computers, cell phones, and DVD players. The accompanying table summarizes the responses to questions about how essential these technologies were.

	Relative Frequency		
Response	Personal Computer	Cell Phone	DVD Player
Cannot imagine living without	.46	.41	.19
Would miss but could do without	.28	.25	.35
Could definitely live without	.26	.34	.46

Construct a comparative bar chart that shows the distribution of responses for the three different technologies.

3.7 Students in California are required to pass an exit exam in order to graduate from high school. The pass rate for San Luis Obispo High School has been rising, as have the rates for San Luis Obispo County and the state of California (*San Luis Obispo Tribune,* August 17, 2004). The percentage of students that passed the test was as follows:

Year	District	Pass Rate
2002	San Luis Obispo High School	66%
2003		72%
2004		93%
2002	San Luis Obispo County	62%
2003		57%
2004		85%
2002	State of California	32%
2003		43%
2004		74%

a. Construct a comparative bar chart that allows the change in the pass rate for each group to be compared.
b. Is the change the same for each group? Comment on any difference observed.

3.8 A poll conducted by the Associated Press–Ipsos on public attitudes found that most Americans are convinced that political corruption is a major problem (*San Luis Obispo Tribune,* December 9, 2005). In the poll, 1002 adults were surveyed. Two of the questions and the summarized responses to these questions follow:

How widespread do you think corruption is in public service in America?

Hardly anyone	1%
A small number	20%
A moderate number	39%
A lot of people	28%
Almost everyone	10%
Not sure	2%

In general, which elected officials would you say are more ethical?

Democrats	36%
Republicans	33%
Both equally	10%
Neither	15%
Not sure	6%

a. For each question, construct a pie chart summarizing the data.
b. For each question construct a segmented bar chart displaying the data.
c. Which type of graph (pie chart or segmented bar graph) does a better job of presenting the data? Explain.

3.9 ▼ Poor fitness in adolescents and adults increases the risk of cardiovascular disease. In a study of 3110 adolescents and 2205 adults (*Journal of the American Medical Association,* December 21, 2005), researchers found 33.6% of adolescents and 13.9% of adults were unfit; the percentage was similar in adolescent males (32.9%) and females (34.4%), but was higher in adult females (16.2%) than in adult males (11.8%).
a. Summarize this information using a comparative bar graph that shows differences between males and females within the two different age groups.
b. Comment on the interesting features of your graphical display.

3.10 A survey of 1001 adults taken by Associated Press–Ipsos asked "How accurate are the weather forecasts in your area?" (*San Luis Obispo Tribune,* June 15, 2005). The responses are summarized in the table below.

Extremely	4%
Very	27%
Somewhat	53%
Not too	11%
Not at all	4%
Not sure	1%

a. Construct a pie chart to summarize this data.
b. Construct a bar chart to summarize this data.
c. Which of these charts—a pie chart or a bar chart best summarizes the important information? Explain.

3.11 The article "Online Shopping Grows, But Department, Discount Stores Still Most Popular with Holiday Shoppers" (*Gallup Poll Analyses,* December 2, 2002) included the data in the accompanying table. The data are based on two telephone surveys, one conducted in November 2000 and one in November 2002. Respondents were asked if they were very likely, somewhat likely, not too likely, or not at all likely to do Christmas shopping online.

| | Relative Frequency | |
Response	2000	2002
Very likely	.09	.15
Somewhat likely	.12	.14
Not too likely	.14	.12
Not at all likely	.65	.59

a. Construct a comparative bar chart. What aspects of the chart justify the claim in the title of the article that "online shopping grows?"
b. Respondents were also asked about the likelihood of doing Christmas shopping from mail order catalogues. Use the data in the accompanying table to construct a comparative bar chart. Are the changes in responses from 2000 to 2002 similar to what was observed for online shopping? Explain.

| | Relative Frequency | |
Response	2000	2002
Very likely	.12	.13
Somewhat likely	.22	.17
Not too likely	.20	.16
Not at all likely	.46	.54

3.12 The article "So Close, Yet So Far: Predictors of Attrition in College Seniors" (*Journal of College Student Development* [1998]: 343–348) examined the reasons that college seniors leave their college programs before graduating. Forty-two college seniors at a large public university who dropped out before graduation were interviewed and asked the main reason for discontinuing enrollment at the university. Data consistent with that given in the article are summarized in the following frequency distribution:

Reason for Leaving the University	Frequency
Academic problems	7
Poor advising or teaching	3
Needed a break	2
Economic reasons	11
Family responsibilities	4
To attend another school	9
Personal problems	3
Other	3

a. Would a bar chart or a pie chart be a better choice for summarizing this data set? Explain.
b. Based on your answer to Part (a), construct an appropriate graphical summary of the given data.
c. Write a few sentences describing the interesting features of your graphical display.

3.13 In a discussion of roadside hazards, the web site highwaysafety.com included a pie chart like the one shown:

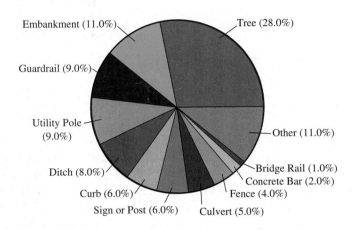

a. Do you think this is an effective use of a pie chart? Why or why not?
b. Construct a bar chart to show the distribution of deaths by object struck. Is this display more effective than the pie chart in summarizing this data set? Explain.

3.14 The article "Death in Roadwork Zones at Record High" (*San Luis Obispo Tribune*, July 25, 2001) included a bar chart similar to this one:

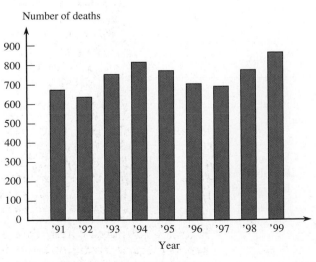

a. Comment on the trend over time in the number of people killed in highway work zones.

b. Would a pie chart have also been an effective way to summarize these data? Explain why or why not.   A pie chart would not have clearly shown the trend over time.

Bold exercises answered in back   ● Data set available online but not required   ▼ Video solution available

3.2   Displaying Numerical Data: Stem-and-Leaf Displays

A stem-and-leaf display is an effective and compact way to summarize univariate numerical data. Each number in the data set is broken into two pieces, a stem and a leaf. The **stem** is the first part of the number and consists of the beginning digit(s). The **leaf** is the last part of the number and consists of the final digit(s). For example, the number 213 might be split into a stem of 2 and a leaf of 13 or a stem of 21 and a leaf of 3. The resulting stems and leaves are then used to construct the display.

Example 3.7   Should Doctors Get Auto Insurance Discounts?

● Many auto insurance companies give job-related discounts of between 5 and 15%. The article "Auto-Rate Discounts Seem to Defy Data" (*San Luis Obispo Tribune*, June 19, 2004) included the accompanying data on the number of automobile accidents per year for every 1000 people in 40 occupations.

Occupation	Accidents per 1000	Occupation	Accidents per 1000
Student	152	Banking-finance	89
Physician	109	Customer service	88
Lawyer	106	Manager	88
Architect	105	Medical support	87
Real estate broker	102	Computer-related	87
Enlisted military	99	Dentist	86
Social worker	98	Pharmacist	85
Manual laborer	96	Proprietor	84
Analyst	95	Teacher, professor	84
Engineer	94	Accountant	84
Consultant	94	Law enforcement	79
Sales	93	Physical therapist	78
Military officer	91	Veterinarian	78
Nurse	90	Clerical, secretary	77
School administrator	90	Clergy	76
Skilled labor	90	Homemaker	76
Librarian	90	Politician	76
Creative arts	90	Pilot	75
Executive	89	Firefighter	67
Insurance agent	89	Farmer	43

♫ Step-by-step technology instructions available online   ● Data set available online

Figure 3.10 Stem-and-leaf display for accident rate per 1000 for forty occupations

```
 4 | 3
 5 |
 6 | 7
 7 | 56667889
 8 | 44567788999
 9 | 000013445689
10 | 2569
11 |
12 |
13 |
14 |                    Stem:  Tens
15 | 2                  Leaf:  Ones
```

Figure 3.10 shows a stem-and-leaf display for the accident rate data.

The numbers in the vertical column on the left of the display are the **stems**. Each number to the right of the vertical line is a **leaf** corresponding to one of the observations in the data set. The legend

Stem: Tens
Leaf: Ones

tells us that the observation that had a stem of 4 and a leaf of 3 corresponds to an occupation with an accident rate of 43 per 1000 (as opposed to 4.3 or 0.43). Similarly, the observation with the stem of 10 and leaf of 2 corresponds to 102 accidents per 1000 (the leaf of 2 is the ones digit) and the observation with the stem of 15 and leaf of 2 corresponds to 152 accidents per 1000.

The display in Figure 3.10 suggests that a typical or representative value is in the stem 8 or 9 row, perhaps around 90. The observations are mostly concentrated in the 75 to 109 range, but there are a couple of values that stand out on the low end (43 and 67) and one observation (152) that is far removed from the rest of the data on the high end.

From the point of view of an auto insurance company it might make sense to offer discounts to occupations with low accident rates—maybe farmers (43 auto accidents per 1000 farmers) or firefighters (67 accidents per 1000 firefighters) or even some of the occupations with accident rates in the 70s. The "discounts seem to defy data" in the title of the article refers to the fact that some insurers provide discounts to doctors and engineers, but not to homemakers, politicians, and other occupations with lower accident rates. Two possible explanations were offered for this apparent discrepancy. One is that it is possible that while some occupations have higher accident rates, they also have lower average cost per claim. Accident rates alone may not reflect the actual cost to the insurance company. Another possible explanation is that the insurance companies may offer the discounted auto insurance in order to attract people who would then also purchase other types of insurance such as malpractice or liability insurance.

■

The leaves on each line of the display in Figure 3.10 have been arranged in order from smallest to largest. Most statistical software packages order the leaves this way, but it is not necessary to do so to get an informative display that still shows many of the important characteristics of the data set, such as shape and spread.

Stem-and-leaf displays can be useful to get a sense of a typical value for the data set, as well as a sense of how spread out the values in the data set are. It is also easy to spot data values that are unusually far from the rest of the values in the data set. Such values are called **outliers**. The stem-and-leaf display of the accident rate data (Figure 3.10) shows an outlier on the low end (43) and an outlier on the high end (152).

> ### DEFINITION
>
> An **outlier** is an unusually small or large data value. In Chapter 4 a precise rule for deciding when an observation is an outlier is given.

Stem-and-Leaf Displays

When to Use

Numerical data sets with a small to moderate number of observations (does not work well for very large data sets)

How to Construct

1. Select one or more leading digits for the stem values. The trailing digits (or sometimes just the first one of the trailing digits) become the leaves.
2. List possible stem values in a vertical column.
3. Record the leaf for every observation beside the corresponding stem value.
4. Indicate the units for stems and leaves someplace in the display.

What to Look For   The display conveys information about

- a representative or typical value in the data set
- the extent of spread about a typical value

- the presence of any gaps in the data
- the extent of symmetry in the distribution of values
- the number and location of peaks

Example 3.8   Tuition at Public Universities

● The introduction to this chapter gave data on average tuition and fees at public institutions in the year 2004 for the 50 U.S. states. The observations ranged from a low value of 2724 to a high value of 8260. The data are reproduced here:

3977	3423	3586	4010	3785	3447	5761	6176	2773	3239
3239	3323	5653	5384	4991	3686	3868	3208	5011	6230
6080	6015	5754	3754	5367	4155	4238	2724	7623	7266
3161	4885	3251	3855	6561	3200	4677	7633	5387	7482
4441	4043	3579	2902	8260	5069	4630	3172	4675	3090

A natural choice for the stem is the leading (thousands) digit. This would result in a display with 7 stems (2, 3, 4, 5, 6, 7, and 8). Using the first two digits of a number as the stem would result in 56 stems (27, 28, . . . , 82). A stem-and-leaf display with 56 stems would not be an effective summary of the data. *In general, stem-and-leaf displays that use between 5 and 20 stems tend to work well.*

● Data set available online

If we choose the thousands digit as the stem, the remaining three digits (the hundreds, tens, and ones) would form the leaf. For example, for the first few values in the first row of data, we would have

$$3977 \rightarrow \text{stem} = 3, \text{leaf} = 977$$
$$3423 \rightarrow \text{stem} = 3, \text{leaf} = 423$$
$$3586 \rightarrow \text{stem} = 3, \text{leaf} = 586$$

The leaves have been entered in the display of Figure 3.11 in the order they are encountered in the data set. Commas are used to separate the leaves only when each leaf has two or more digits. Figure 3.11 shows that most states had average tuition and fees in the $3000 to $5000 range and that the typical average tuition and fees is around $4000. A few states have average tuition and fees at public four-year institutions that are quite a bit higher than most other states (the four states with the highest values were Vermont, Pennsylvania, New Hampshire, and New Jersey).

2	773, 724, 902
3	977, 423, 586, 785, 447, 239, 239, 323, 686, 868, 208, 754, 161, 251, 855, 200, 579, 172, 090
4	010, 991, 155, 238, 885, 677, 441, 043, 630, 675
5	761, 653, 384, 011, 754, 367, 387, 069
6	176, 230, 080, 015, 561
7	623, 266, 633, 482
8	260

Stem: Thousands
Leaf: Ones

Figure 3.11 Stem-and-leaf displays of average tuition and fees.

Truncating is different than rounding.

An alternative display (Figure 3.12) results from dropping all but the first digit of the leaf. This is what most statistical computer packages do when generating a display; little information about typical value, spread, or shape is lost in this truncation and the display is simpler and more compact.

Figure 3.12 Stem-and-leaf display of the average tuition and fees data using truncated leaves.

2	779
3	9457422368271282510
4	0912864066
5	76307330
6	12005
7	6264
8	2

Stem: Thousands
Leaf: Hundreds

▪ Repeated Stems to Stretch a Display

Sometimes a natural choice of stems gives a display in which too many observations are concentrated on just a few stems. A more informative picture can be obtained by dividing the leaves at any given stem into two groups: those that begin with 0, 1, 2, 3, or 4 (the "low" leaves) and those that begin with 5, 6, 7, 8, or 9 (the "high" leaves). Then each stem value is listed twice when constructing the display, once for the low leaves and once again for the high leaves. It is also possible to repeat a stem more than twice. For example, each stem might be repeated five times, once for each of the leaf groupings {0, 1}, {2, 3}, {4, 5}, {6, 7}, and {8, 9}.

Example 3.9   Median Ages in 2030

● The accompanying data on the Census Bureau's projected median age in 2030 for the 50 U.S. states and Washington D.C. appeared in the article "2030 Forecast: Mostly Gray" (*USA Today,* April 21, 2005). The median age for a state is the age that divides the state's residents so that half are younger than the median age and half are older than the median age.

Projected Median Age

41.0	32.9	39.3	29.3	37.4	35.6	41.1	43.6	33.7	45.4	35.6	38.7
39.2	37.8	37.7	42.0	39.1	40.0	38.8	46.9	37.5	40.2	40.2	39.0
41.1	39.6	46.0	38.4	39.4	42.1	40.8	44.8	39.9	36.8	43.2	40.2
37.9	39.1	42.1	40.7	41.3	41.5	38.3	34.6	30.4	43.9	37.8	38.5
46.7	41.6	46.4								

The ages in the data set range from 29.3 to 46.9. Using the first two digits of each data value for the stem results in a large number of stems, while using only the first digit results in a stem-and-leaf display with only three stems.

The stem-and-leaf display using single digit stems and leaves truncated to a single digit is shown in Figure 3.13. A stem-and-leaf display that uses repeated stems is shown in Figure 3.14. Here each stem is listed twice, once for the low leaves (those beginning with 0, 1, 2, 3, 4) and once for the high leaves (those beginning with 5, 6, 7, 8, 9). This display is more informative than the one in Figure 3.13, but is much more compact than a display based on two digit stems.

Figure 3.13 Stem-and-leaf display for the projected median age data.

```
2 | 9
3 | 0234556777778888899999999
4 | 000000111111222333456666    Stem: Tens
                                Leaf: Ones
```

Figure 3.14 Stem-and-leaf display for the projected median age data using repeated stems.

```
2H | 9
3L | 0234
3H | 5567777778888899999999
4L | 0000001111112223334
4H | 56666                  Stem: Tens
                           Leaf: Ones
```

■ Comparative Stem-and-Leaf Displays

Frequently an analyst wishes to see whether two groups of data differ in some fundamental way. A comparative stem-and-leaf display, in which the leaves for one group extend to the right of the stem values and the leaves for the second group extend to the left, can provide preliminary visual impressions and insights.

● Data set available online

Example 3.10 Progress for Children

● The UNICEF report "Progress for Children" (April, 2005) included the accompanying data on the percentage of primary-school-age children who were enrolled in school for 19 countries in Northern Africa and for 23 countries in Central Africa.

Northern Africa

54.6	34.3	48.9	77.8	59.6	88.5	97.4	92.5	83.9	96.9	88.9
98.8	91.6	97.8	96.1	92.2	94.9	98.6	86.6			

Central Africa

58.3	34.6	35.5	45.4	38.6	63.8	53.9	61.9	69.9	43.0	85.0
63.4	58.4	61.9	40.9	73.9	34.8	74.4	97.4	61.0	66.7	79.6
98.9										

We will construct a comparative stem-and-leaf display using the first digit of each observation as the stem and the remaining two digits as the leaf. To keep the display simple the leaves will be truncated to one digit. For example, the observation 54.6 would be processed as

$$54.6 \rightarrow \text{stem} = 5, \text{leaf} = 4 \text{ (truncated from 4.6)}$$

and the observation 34.3 would be processed as

$$34.3 \rightarrow \text{stem} = 3, \text{leaf} = 4 \text{ (truncated from 4.3)}$$

The resulting comparative stem-and-leaf display is shown in Figure 3.15.

Figure 3.15 Comparative stem-and-leaf display for percentage of children enrolled in primary school.

Central Africa			Northern Africa	
4854	3	4		
035	4	8		
838	5	49		
6113913	6			
943	7	76		
5	8	8386	Stem:	Tens
87	9	7268176248	Leaf:	Ones

From the comparative stem-and-leaf display you can see that there is quite a bit of variability in the percentage enrolled in school for both Northern and Central African countries and that the shapes of the two data distributions are quite different. The percentage enrolled in school tends to be higher in Northern African countries than in Central African countries, although the smallest value in each of the two data sets is about the same. For Northern African countries the distribution of values has a single peak in the 90s with the number of observations declining as we move toward the stems corresponding to lower percentages enrolled in school. For Central African countries the distribution is more symmetric, with a typical value in the mid 60s.

● Data set available online

Exercises 3.15–3.21 Turn to the Expanded Answers Section for answers not shown next to exercises.

3.15 ● ▼ The National Survey on Drug Use and Health, conducted in 2002 and 2003 by the Office of Applied Studies, led to the following state estimates of the total number of people (ages 12 and older) who had smoked within the last month).

State	Number of People (in thousands)
Alabama	976
Alaska	129
Arizona	1215
Arkansas	730
California	5508
Colorado	985
Connecticut	678
Delaware	174
District of Columbia	125
Florida	3355
Georgia	1779
Hawaii	217
Idaho	260
Illinois	2754
Indiana	1427
Iowa	647
Kansas	573
Kentucky	1178
Louisiana	1021
Maine	297
Maryland	1039
Massachusetts	1207
Michigan	2336
Minnesota	1122
Mississippi	680
Missouri	1472
Montana	212
Nebraska	367
Nevada	543
New Hampshire	281
New Jersey	1662
New Mexico	363
New York	4052
North Carolina	2010
North Dakota	145
Ohio	2865
Oklahoma	858

State	Number of People (in thousands)
Oregon	735
Pennsylvania	2858
Rhode Island	246
South Carolina	907
South Dakota	188
Tennessee	1343
Texas	4428
Utah	304
Vermont	134
Virginia	1487
Washington	1172
West Virginia	452
Wisconsin	1167
Wyoming	111

a. Construct a stem-and-leaf display using hundreds (of thousands) as the stems and truncating the leaves to the tens (of thousands) digit.
b. Write a few sentences describing the shape of the distribution and any unusual observations.
c. The three largest values were for California, New York, and Texas. Does this indicate that tobacco use is more of a problem in these states than elsewhere? Explain.
d. If you wanted to compare states on the basis of the extent of tobacco use, would you use the data in the given table? If yes, explain why this would be reasonable. If no, what would you use instead as the basis for the comparison?

3.16 ● The Connecticut Agricultural Experiment Station conducted a study of the calorie content of different types of beer. The calorie content (calories per 100 ml) for 26 brands of light beer are (from the web site brewery.org):

29 28 33 31 30 33 30 28 27 41 39 31 29
23 32 31 32 19 40 22 34 31 42 35 29 43

Construct a stem-and-leaf display using stems 1, 2, 3, and 4. Write a sentence or two describing the calorie content of light beers.

3.17 The stem-and-leaf display of Exercise 3.16 uses only four stems. Construct a stem-and-leaf display for these data using repeated stems 1H, 2L, 2H, . . . , 4L. For example, the first observation, 29, would have a stem of 2 and a leaf of 9. It would be entered into the display for the stem 2H, because it is a "high" 2—that is, it has a leaf that is on the high end (5, 6, 7, 8, 9).

3.18 ● Many states face a shortage of fully credentialed teachers. The percentages of teachers who are fully credentialed for each county in California were published in the *San Luis Obispo Tribune* (July 29, 2001) and are given in the following table:

County	Percentage Credentialed
Alameda	85.1
Alpine	100.0
Amador	97.3
Butte	98.2
Calaveras	97.3
Colusa	92.8
Contra Costa	87.7
Del Norte	98.8
El Dorado	96.7
Fresno	91.2
Glenn	95.0
Humbolt	98.6
Imperial	79.8
Inyo	94.4
Kern	85.1
Kings	85.0
Lake	95.0
Lassen	89.6
Los Angeles	74.7
Madera	91.0
Marin	96.8
Mariposa	95.5
Mendicino	97.2
Merced	87.8
Modoc	94.6
Mono	95.9
Monterey	84.6
Napa	90.8
Nevada	93.9
Orange	91.3
Placer	97.8
Plumas	95.0
Riverside	84.5
Sacramento	95.3

County	Percentage Credentialed
San Benito	83.5
San Bernardino	83.1
San Diego	96.6
San Francisco	94.4
San Joaquin	86.3
San Luis Obispo	98.1
San Mateo	88.8
Santa Barbara	95.5
Santa Clara	84.6
Santa Cruz	89.6
Shasta	97.5
Sierra	88.5
Siskiyou	97.7
Solano	87.3
Sonoma	96.5
Stanislaus	94.4
Sutter	89.3
Tehama	97.3
Trinity	97.5
Tulare	87.6
Tuolomne	98.5
Ventura	92.1
Yolo	94.6
Yuba	91.6

a. Construct a stem-and-leaf display for this data set using stems 7, 8, 9, and 10. Truncate the leaves to a single digit. Comment on the interesting features of the display.
b. Construct a stem-and-leaf display using repeated stems. Are there characteristics of the data set that are easier to see in the plot with repeated stems, or is the general shape of the two displays similar?

3.19 ● ▼ A report from Texas Transportation Institute (Texas A&M University System, 2005) titled "Congestion Reduction Strategies" included the accompanying data on extra travel time for peak travel time in hours per year per traveler for different sized urban areas.

Very Large Urban Areas	Extra Hours per Year per Traveler
Los Angeles, CA	93
San Francisco, CA	72
Washington DC, VA, MD	69

Very Large Urban Areas	Extra Hours per Year per Traveler
Atlanta, GA	67
Houston, TX	63
Dallas, Fort Worth, TX	60
Chicago, IL-IN	58
Detroit, MI	57
Miami, FL	51
Boston, MA, NH, RI	51
New York, NY-NJ-CT	49
Phoenix, AZ	49
Philadelphia, PA-NJ-DE-MD	38

Large Urban Areas	Extra Hours per Year per Traveler
Riverside, CA	55
Orlando, FL	55
San Jose, CA	53
San Diego, CA	52
Denver, CO	51
Baltimore, MD	50
Seattle, WA	46
Tampa, FL	46
Minneapolis, St Paul, MN	43
Sacramento, CA	40
Portland, OR, WA	39
Indianapolis, IN	38
St Louis, MO-IL	35
San Antonio, TX	33
Providence, RI, MA	33
Las Vegas, NV	30
Cincinnati, OH-KY-IN	30
Columbus, OH	29
Virginia Beach, VA	26
Milwaukee, WI	23
New Orleans, LA	18
Kansas City, MO-KS	17
Pittsburgh, PA	14
Buffalo, NY	13
Oklahoma City, OK	12
Cleveland, OH	10

a. Construct a back-to-back stem-and-leaf plot for annual delay per traveler for each of the two different sizes of urban areas.

b. Is the following statement consistent with the display constructed in part (a)? Explain.

The larger the urban areas, the greater the extra travel time during peak period travel.

3.20 ● The article "A Nation Ablaze with Change" (*USA Today,* July 3, 2001) gave the accompanying data on percentage increase in population between 1990 and 2000 for the 50 U.S. states. Also provided in the table is a column that indicates for each state whether the state is in the eastern or western part of the United States (the states are listed in order of population size):

State	Percentage Change	East/West
California	13.8	W
Texas	22.8	W
New York	5.5	E
Florida	23.5	E
Illinois	8.6	E
Pennsylvania	3.4	E
Ohio	4.7	E
Michigan	6.9	E
New Jersey	8.9	E
Georgia	26.4	E
North Carolina	21.4	E
Virginia	14.4	E
Massachusetts	5.5	E
Indiana	9.7	E
Washington	21.1	W
Tennessee	16.7	E
Missouri	9.3	E
Wisconsin	9.6	E
Maryland	10.8	E
Arizona	40.0	W
Minnesota	12.4	E
Louisiana	5.9	E
Alabama	10.1	E
Colorado	30.6	W
Kentucky	9.7	E
South Carolina	15.1	E
Oklahoma	9.7	W
Oregon	20.4	W
Connecticut	3.6	E
Iowa	5.4	E
Mississippi	10.5	E
Kansas	8.5	W

(continued)

State	Percentage Change	East/West
Arkansas	13.7	E
Utah	29.6	W
Nevada	66.3	W
New Mexico	20.1	W
West Virginia	0.8	E
Nebraska	8.4	W
Idaho	28.5	W
Maine	3.9	E
New Hampshire	11.4	E
Hawaii	9.3	W
Rhode Island	4.5	E
Montana	12.9	W
Delaware	17.6	E
South Dakota	8.5	W
North Dakota	0.5	W
Alaska	14.0	W
Vermont	8.2	E
Wyoming	8.9	W

a. Construct a stem-and-leaf display for percentage growth for the data set consisting of all 50 states. Hints: Regard the observations as having two digits to the left of the decimal place. That is, think of an observation such as 8.5 as 08.5. It will also be easier to truncate leaves to a single digit; for example, a leaf of 8.5 could be truncated to 8 for purposes of constructing the display.
b. Comment on any interesting features of the data set. Do any of the observations appear to be outliers?
c. Now construct a comparative stem-and-leaf display for the eastern and western states. Write a few sentences comparing the percentage growth distributions for eastern and western states.

3.21 ● High school dropout rates (percentages) for the period 1997–1999 for the 50 states were given in *The Chronicle of Higher Education* (August 31, 2001) and are shown in the following table:

State	Rate
Alabama	10
Alaska	7
Arizona	17
Arkansas	12
California	9
Colorado	13

State	Rate
Connecticut	9
Delaware	11
Florida	12
Georgia	13
Hawaii	5
Idaho	10
Illinois	9
Indiana	6
Iowa	7
Kansas	7
Kentucky	11
Louisiana	11
Maine	7
Maryland	7
Massachusetts	6
Michigan	9
Minnesota	6
Mississippi	10
Missouri	9
Montana	8
Nebraska	8
Nevada	17
New Hampshire	7
New Jersey	6
New Mexico	13
New York	9
North Carolina	11
North Dakota	5
Ohio	8
Oklahoma	9
Oregon	13
Pennsylvania	7
Rhode Island	11
South Carolina	9
South Dakota	8
Tennessee	12
Texas	12
Utah	9
Vermont	6
Virginia	8
Washington	8
West Virginia	8
Wisconsin	5
Wyoming	9

Note that dropout rates range from a low of 5% to a high of 17%. In constructing a stem-and-leaf display for these data, if we regard each dropout rate as a two-digit number and use the first digit for the stem, then there are only two possible stems, 0 and 1. One solution is to use repeated stems. Consider a scheme that divides the leaf range into five parts: 0 and 1, 2 and 3, 4 and 5, 6 and 7, and 8 and 9. Then, for example, stem 1 could be repeated as

   1.   with leaves 0 and 1
   1t   with leaves 2 and 3

1f   with leaves 4 and 5
1s   with leaves 6 and 7

If there had been any dropout rates as large as 18 or 19 in the data set, we would also need to include a stem 1* to accommodate the leaves of 8 and 9. Construct a stem-and-leaf display for this data set that uses stems 0f, 0s, 0*, 1., 1t, 1f, and 1s. Comment on the important features of the display.

Bold exercises answered in back   ● Data set available online but not required   ▼ Video solution available

---

3.3 Displaying Numerical Data: Frequency Distributions and Histograms

A stem-and-leaf display is not always an effective summary technique; it is unwieldy when the data set contains a large number of observations. Frequency distributions and histograms are displays that are useful for summarizing even a large data set in a compact fashion.

■ Frequency Distributions and Histograms for Discrete Numerical Data

Discrete numerical data almost always result from counting. In such cases, each observation is a whole number. As in the case of categorical data, a frequency distribution for discrete numerical data lists each possible value (either individually or grouped into intervals), the associated frequency, and sometimes the corresponding relative frequency. Recall that relative frequency is calculated by dividing the frequency by the total number of observations in the data set.

Example 3.11 Promiscuous Raccoons!

Alan and Sandy Carey/Getty Images

● The authors of the article "Behavioral Aspects of the Raccoon Mating System: Determinants of Consortship Success" (*Animal Behaviour* [1999]: 593–601) monitored raccoons in southern Texas during three mating seasons in an effort to describe mating behavior. Twenty-nine female raccoons were observed, and the number of male partners during the time the female was accepting partners (generally 1 to 4 days each year) was recorded for each female. The resulting data were as follows:

1	3	2	1	1	4	2	4	1	1	1
3	1	1	1	1	2	2	1	1	4	1
1	2	1	1	1	1	3				

The corresponding frequency distribution is given in Table 3.1. From the frequency distribution we can see that 18 of the female raccoons had a single partner. The corresponding relative frequency, 18/29 = .621, tells us that the proportion of female raccoons in the sample with a single partner was .621, or, equivalently, 62.1% of the

● Data set available online

Table 3.1 Frequency Distribution for Number of Partners

Number of Partners	Frequency	Relative Frequency
1	18	.621
2	5	.172
3	3	.103
4	3	.103
	29	.999 ← *Differs from 1 due to rounding*

females had a single partner. Adding the relative frequencies for the values of 1 and 2 gives

$$.621 + .172 = .793$$

indicating that 79.3% of the raccoons had 2 or fewer partners.

■

A histogram for discrete numerical data is a graph of the frequency distribution, and it is similar to the bar chart for categorical data. Each frequency or relative frequency is represented by a rectangle centered over the corresponding value (or range of values) and the area of the rectangle is proportional to the corresponding frequency or relative frequency.

Histogram for Discrete Numerical Data

When to Use

Discrete numerical data. Works well even for large data sets.

How to Construct

1. Draw a horizontal scale, and mark the possible values of the variable.
2. Draw a vertical scale, and mark it with either frequency or relative frequency.
3. Above each possible value, draw a rectangle centered at that value (so that the rectangle for 1 is centered at 1, the rectangle for 5 is centered at 5, and so on). The height of each rectangle is determined by the corresponding frequency or relative frequency. Often possible values are consecutive whole numbers, in which case the base width for each rectangle is 1.

What to Look For

- Central or typical value
- Extent of spread or variation
- General shape
- Location and number of peaks
- Presence of gaps and outliers

Example 3.12 Revisiting Promiscuous Raccoons

The raccoon data of Example 3.11 were summarized in a frequency distribution. The corresponding histogram is shown in Figure 3.16. Note that each rectangle in the histogram is centered over the corresponding value. When relative frequency instead of frequency is used for the vertical scale, the scale on the vertical axis is different but all essential characteristics of the graph (shape, location, spread) are unchanged.

Figure 3.16 Histogram and relative frequency histogram of raccoon data.

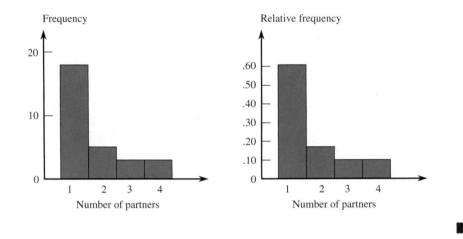

Sometimes a discrete numerical data set contains a large number of possible values and perhaps also has a few large or small values that are far away from most of the data. In this case, rather than forming a frequency distribution with a very long list of possible values, it is common to group the observed values into intervals or ranges. This is illustrated in Example 3.13

Example 3.13 Math SAT Score Distribution

Each of the 1,475,623 students who took the math portion of the SAT exam in 2005 received a score between 200 and 800. The score distribution was summarized in a frequency distribution table that appeared in the College Board report titled "2005 College Bound Seniors." A relative frequency distribution is given in Table 3.2 and the corresponding relative frequency histogram is shown in Figure 3.17. Notice that rather than list each possible individual score value between 200 and 800, the scores are grouped into intervals (200 to 249, 250 to 299, etc.). This results in a much more compact table that still communicates the important features of the data set. Also, notice that because the data set is so large, the frequencies are also large numbers. Because of these large frequencies, it is easier to focus on the relative frequencies in our interpretation. From the relative frequency distribution and histogram, we can see that while there is a lot of variability in individual math SAT scores, the majority were in the 400 to 600 range and a typical value for math SAT looks to be something in the low 500s.

Before leaving this example, take a second look at the relative frequency histogram of Figure 3.17. Notice that there is one rectangle for each score interval in the relative frequency distribution. For simplicity we have chosen to treat the very

Table 3.2 Relative Frequency Distribution of Math SAT Score

Score	Frequency	Relative Frequency
200–249	33,841	$.02 \leftarrow \dfrac{33,841}{1,475,623}$
250–299	67,117	.05
300–349	124,032	.08
350–399	166,446	.11
400–449	210,449	.14
450–499	255,904	.17
500–549	225,335	.15
550–599	188,780	.13
600–649	111,350	.08
650–699	58,762	.04
700–749	22,270	.02
750–800	11,337	.01
Total	**1,475,623**	**1.00**

Figure 3.17 Relative Frequency Histogram for the Math SAT Data

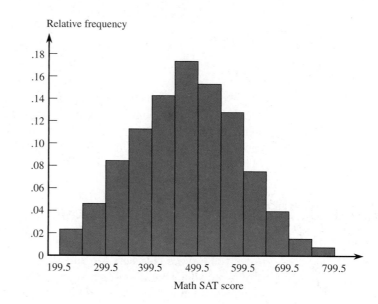

last interval, 750 to 800 as if it were 750 to 799 so that all of the score ranges in the frequency distribution are the same width. Also note that the rectangle representing the score range 400 to 449 actually extends from 399.5 to 449.5 on the score scale. This is similar to what happens in histograms for discrete numerical data where there is no grouping. For example, in Figure 3.16 the rectangle representing 2 is centered at 2 but extends from 1.5 to 2.5 on the number of partners scale.

■ Frequency Distributions and Histograms for Continuous Numerical Data

The difficulty in constructing tabular or graphical displays with continuous data, such as observations on reaction time (in seconds) or fuel efficiency (in miles per gallon), is that there are no natural categories. The way out of this dilemma is to define our own categories. For fuel efficiency data, suppose that we mark some intervals on a horizontal miles-per-gallon measurement axis, as pictured in Figure 3.18. Each data value should fall in exactly one of these intervals. If the smallest observation was 25.3 and the largest was 29.8, we might use intervals of width 0.5, with the first interval starting at 25.0 and the last interval ending at 30.0. The resulting intervals are called **class intervals**, or just **classes**. The class intervals play the same role that the categories or individual values played in frequency distributions for categorical or discrete numerical data.

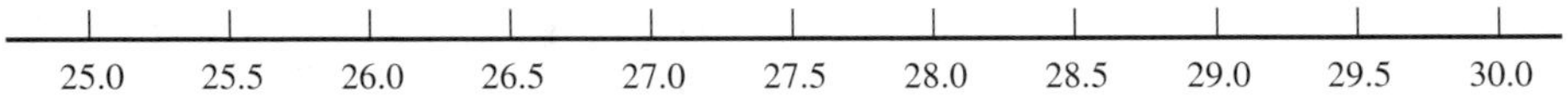

Figure 3.18 Suitable class intervals for miles-per-gallon data.

| | | | | | | | | | | |
|25.0|25.5|26.0|26.5|27.0|27.5|28.0|28.5|29.0|29.5|30.0|

There is one further difficulty we need to address. Where should we place an observation such as 27.0, which falls on a boundary between classes? Our convention is to define intervals so that such an observation is placed in the upper rather than the lower class interval. Thus, in a frequency distribution, one class might be 26.5 to <27.0, where the symbol < is a substitute for the phrase *less than*. This class will contain all observations that are greater than or equal to 26.5 and less than 27.0. The observation 27.0 would then fall in the class 27.0 to <27.5.

....

Example 3.14 Enrollments at Public Universities

● States differ widely in the percentage of college students who are enrolled in public institutions. The National Center for Education Statistics provided the accompanying data on this percentage for the 50 U.S. states for fall 2002.

Percentage of College Students Enrolled in Public Institutions

86	96	66	86	80	78	62	81	77	81	77	76
73	69	76	90	78	82	70	83	46	80	78	93
66	91	76	86	58	81	91	57	81	88	71	86
84	57	44	82	79	67	86	75	55	75	80	80
85	69										

The smallest observation is 44 (Rhode Island) and the largest is 96 (Alaska). It is reasonable to start the first class interval at 40 and let each interval have a width of 10. This gives class intervals of 40 to <50, 50 to <60, 60 to <70, 70 to <80, 80 to <90, and 90 to <100.

Table 3.3 displays the resulting frequency distribution, along with the relative frequencies.

Various relative frequencies can be combined to yield other interesting information. For example,

$$\begin{pmatrix} \text{proportion of states} \\ \text{with percent in public} \\ \text{institutions less than 60} \end{pmatrix} = \begin{pmatrix} \text{proportion in 40} \\ \text{to <50 class} \end{pmatrix} + \begin{pmatrix} \text{proportion in 50} \\ \text{to <60 class} \end{pmatrix}$$

$$= .04 + .08 = .12 \, (12\%)$$

● Data set available online

Table 3.3 Frequency Distribution for Percentage of College Students Enrolled in Public Institutions

Class Interval	Frequency	Relative Frequency
40 to <50	2	.04
50 to <60	4	.08
60 to <70	6	.12
70 to <80	14	.28
80 to <90	19	.38
90 to <100	5	.10
	50	1.00

and

$$\begin{pmatrix} \text{proportion of states} \\ \text{with percent in} \\ \text{public institutions} \\ \text{between 60 and 90} \end{pmatrix} = \begin{pmatrix} \text{proportion} \\ \text{in 60 to} \\ <70 \text{ class} \end{pmatrix} + \begin{pmatrix} \text{proportion} \\ \text{in 70 to} \\ <80 \text{ class} \end{pmatrix} + \begin{pmatrix} \text{proportion} \\ \text{in 80 to} \\ <90 \text{ class} \end{pmatrix}$$

$$= .12 + .28 + .38 = .78 \ (78\%)$$

■

There are no set rules for selecting either the number of class intervals or the length of the intervals. Using a few relatively wide intervals will bunch the data, whereas using a great many relatively narrow intervals may spread the data over too many intervals, so that no interval contains more than a few observations. Neither type of distribution will give an informative picture of how values are distributed over the range of measurement, and interesting features of the data set may be missed. In general, with a small amount of data, relatively few intervals, perhaps between 5 and 10, should be used. With a large amount of data, a distribution based on 15 to 20 (or even more) intervals is often recommended. The quantity

$$\sqrt{\text{number of observations}}$$

is often used as an estimate of an appropriate number of intervals: 5 intervals for 25 observations, 10 intervals when the number of observations is 100, and so on.

Two people making reasonable and similar choices for the number of intervals, their width, and the starting point of the first interval will usually obtain similar histograms of the data.

■ Cumulative Relative Frequencies and Cumulative Relative Frequency Plots

Rather than wanting to know what proportion of the data fall in a particular class, we often wish to determine the proportion falling below a specified value. This is easily done when the value is a class boundary. Consider the following classes and relative frequencies:

Class	0 to <25	25 to <50	50 to <75	75 to <100	100 to <125	...
Rel. freq.	.05	.10	.18	.25	.20	...

Then

proportion of observations less than 75 = proportion in one of the first three classes
$$= .05 + .10 + .18$$
$$= .33$$

Similarly,

proportion of observations less than 100 = .05 + .10 + .18 + .25 = .33 + .25 = .58

Each such sum of relative frequencies is called a **cumulative relative frequency**. Notice that the cumulative relative frequency .58 is the sum of the previous cumulative relative frequency .33 and the "current" relative frequency .25. The use of cumulative relative frequencies is illustrated in Example 3.15.

Example 3.15 Bicycle Accidents

In 1996, 50 bicycle riders died in fatal accidents in the state of New York. Table 3.4 gives a relative frequency distribution for the ages of these 50 riders (New York Department of Health). The table also includes a column of cumulative relative frequencies.

Table 3.4   Frequency Distribution with Cumulative Relative Frequencies

Age at Death	Frequency	Relative Frequency	Cumulative Relative Frequency
5 to <10	4	.08	.08
10 to <15	4	.08	.16 ◄— .08 + .08
15 to <20	7	.14	.30 ◄— .08 + .08 + .14
20 to <25	4	.08	.38
25 to <45	19	.38	.76
45 to <65	9	.18	.94
65 to <90	3	.06	1.00

The proportion of bicycle deaths that were riders less than 25 years of age is .38 (or 38%), the cumulative relative frequency for the 20 to <25 group. What about the proportion of bicycle deaths of riders younger than 35 years old? Because 35 is not the endpoint of one of the intervals in the frequency distribution, we can only estimate this proportion from the information given. The value 35 is halfway between the endpoints of the 25 to 45 interval, so it is reasonable to estimate that half of the relative frequency of .38 for this interval belongs in the 25 to 35 range. Then

estimate of proportion younger than 35 = $.08 + .08 + .14 + .08 + \frac{1}{2}(.38) = .57$

This proportion could also have been computed using the cumulative relative frequencies as

estimate of proportion younger than 35 = $.38 + \frac{1}{2}(.38) = .38 + .19 = .57$

Similarly, since 50 is one-fourth of the way from 45 to 65,

estimate of proportion younger than $50 = .76 + \frac{1}{4}(.18) = .76 + .045 = .805$

■

A **cumulative relative frequency plot** is just a graph of the cumulative relative frequencies against the upper endpoint of the corresponding interval. The pairs

(upper endpoint of interval, cumulative relative frequency)

are plotted as points on a rectangular coordinate system, and successive points in the plot are connected by a line segment. For the bicycle accident data of Example 3.15, the plotted points would be

(10, .08)   (15, .16)   (20, .30)   (25, .38)
(45, .76)   (65, .94)   (90, 1.00)

This graph displays cumulative percentages.

One additional point, the pair (lower endpoint of first interval, 0), is also included in the plot (for the bicycle accident data, this would be the point (5, 0)), and then points are connected by line segments. Figure 3.19 shows the cumulative relative frequency plot for the bicycle accident data. The cumulative relative frequency plot can be used to obtain approximate answers to questions such as

What proportion of the observations is smaller than a particular value?

and

What value separates the smallest p percent from the larger values?

For example, to determine the approximate proportion of the bicycle fatalities that were for people younger than 55, we would follow a vertical line up from 55 on the x axis and then read across to obtain the corresponding cumulative relative frequency, as illustrated in Figure 3.20(a). Approximately .84, or 84%, of the fatalities were for people younger than 55. Similarly, to find the age value that separates the youngest 30% of the fatalities from the higher ages, start at .30 on the cumulative relative frequency axis and move across and then down to find the corresponding age, as shown in Figure 3.20(b). Approximately 30% of the fatalities were for people younger than 20.

Figure 3.19 Cumulative relative frequency plot for the bicycle accident data of Example 3.15.

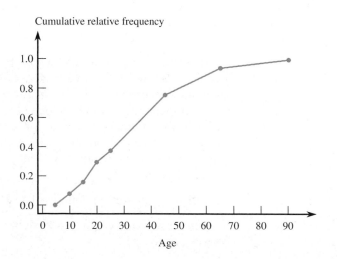

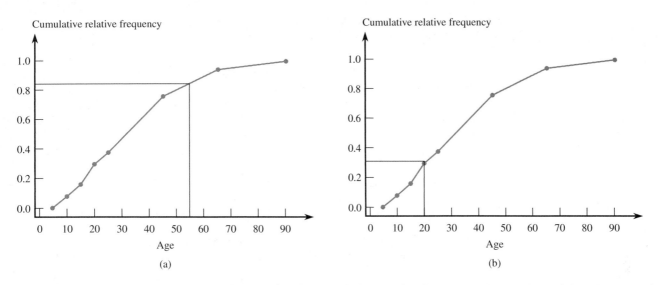

Figure 3.20 Using the cumulative relative frequency plot. (a) determining the approximate proportion of bicycle fatalities that were for people 55 or younger; (b) finding the age that separates the youngest 30% of bicycle fatalities from the oldest 70%.

■ Histograms for Continuous Numerical Data ....

When the class intervals in a frequency distribution are all of equal width, it is easy to construct a histogram using the information in a frequency distribution.

Histogram for Continuous Numerical Data When the Class Interval Widths Are Equal

When to Use

Continuous numerical data. Works well, even for large data sets.

How to Construct

1. Mark the boundaries of the class intervals on a horizontal axis.
2. Use either frequency or relative frequency on the vertical axis.
3. Draw a rectangle for each class directly above the corresponding interval (so that the edges are at the class boundaries). The height of each rectangle is the frequency or relative frequency of the corresponding class interval.

What to Look For

- Central or typical value
- Extent of spread or variation
- General shape
- Location and number of peaks
- Presence of gaps and outliers

Example 3.16 TV Viewing Habits of Children

The article "Early Television Exposure and Subsequent Attention Problems in Children" (*Pediatrics,* April 2004) investigated the television viewing habits of children in the United States. Table 3.5 gives approximate relative frequencies (read from graphs that appeared in the article) for the number of hours spent watching TV per day for a sample of children at age 1 year and a sample of children at age 3 years. The data summarized in the article were obtained as part of a large scale national survey.

Table 3.5   Relative Frequency Distribution for Number of Hours Spent Watching TV per Day

TV Hours per Day	Age 1 Year Relative Frequency	Age 3 Years Relative Frequency
0 to <2	.270	.630
2 to <4	.390	.195
4 to <6	.190	.100
6 to <8	.085	.025
8 to <10	.030	.020
10 to <12	.020	.015
12 to <14	.010	.010
14 to <16	.005	.005

Figure 3.21(a) is the relative frequency histogram for the 1-year-old children and Figure 3.21(b) is the relative frequency histogram for 3-year-old children. Notice that both histograms have a single peak with the majority of children in both age groups concentrated in the smaller TV hours intervals. Both histograms are quite stretched out at the upper end, indicating some young children watch a lot of TV.

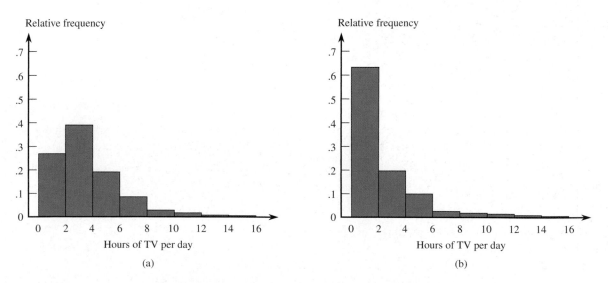

Figure 3.21   Histogram of TV hours per day: (a) 1-year-old children; (b) 3-year-old children.

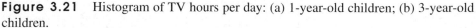

Step-by-step technology instructions available online

The big difference between the two histograms is at the low end, with a much higher proportion of 3-year-old children falling in the 0 to 2 TV hours interval than is the case for 1-year-old children. A typical number of TV hours per day for 1-year-old children would be somewhere between 2 and 4 hours, whereas a typical number of TV hours for 3-year-old children is in the 0 to 2 hour interval.

■

■ **Class Intervals of Unequal Widths**   Figure 3.22 shows a data set in which a great many observations are concentrated at the center of the set, with only a few outlying, or stray, values both below and above the main body of data. If a frequency distribution is based on short intervals of equal width, a great many intervals will be required to capture all observations, and many of them will contain no observations (0 frequency). On the other hand, only a few wide intervals will capture all values, but then most of the observations will be grouped into a few intervals. Neither choice yields an informative description of the distribution. In such situations, it is best to use a combination of wide class intervals where there are few data points and shorter intervals where there are many data points.

Figure 3.22   Three choices of class intervals for a data set with outliers: (a) many short intervals of equal width; (b) a few wide intervals of equal width; (c) intervals of unequal width.

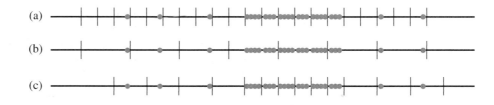

Constructing a Histogram for Continuous Data When Class Interval Widths Are Unequal

In this case, frequencies or relative frequencies should not be used on the vertical axis. Instead, the height of each rectangle, called the **density** for the class, is given by

$$\text{density} = \text{rectangle height} = \frac{\text{relative frequency of class interval}}{\text{class interval width}}$$

The vertical axis is called the **density scale**.

■

The use of the density scale to construct the histogram ensures that the area of each rectangle in the histogram will be proportional to the corresponding relative frequency. The formula for density can also be used when class widths are equal. However, when the intervals are of equal width, the extra arithmetic required to obtain the densities is unnecessary.

Example 3.17   Misreporting Grade Point Average

When people are asked for the values of characteristics such as age or weight, they sometimes shade the truth in their responses. The article "Self-Reports of Academic Performance" (*Social Methods and Research* [November 1981]: 165–185) focused on such characteristics as SAT scores and grade point average (GPA). For each student in a sample, the difference in GPA (reported–actual) was determined. Positive differences resulted from individuals reporting GPAs larger than the correct values. Most differences were close to 0, but there were some rather gross errors. Because of this, the frequency distribution based on unequal class widths shown in Table 3.6 gives an informative yet concise summary.

Table 3.6   Frequency Distribution for Errors in Reported GPA

Class Interval	Relative Frequency	Width	Density
−2.0 to <−0.4	.023	1.6	0.014
−0.4 to <−0.2	.055	.2	0.275
−0.2 to <−0.1	.097	.1	0.970
−0.1 to <0	.210	.1	2.100
0 to <0.1	.189	.1	1.890
0.1 to <0.2	.139	.1	1.390
0.2 to <0.4	.116	.2	0.580
0.4 to <2.0	.171	1.6	0.107

Figure 3.23 displays two histograms based on this frequency distribution. The histogram in Figure 3.23(a) is correctly drawn, with density used to determine the height of each bar. The histogram in Figure 3.23(b) has height equal to relative fre-

Figure 3.23   Histograms for errors in reporting GPA: (a) a correct picture (height = density); (b) an incorrect picture (height = relative frequency).

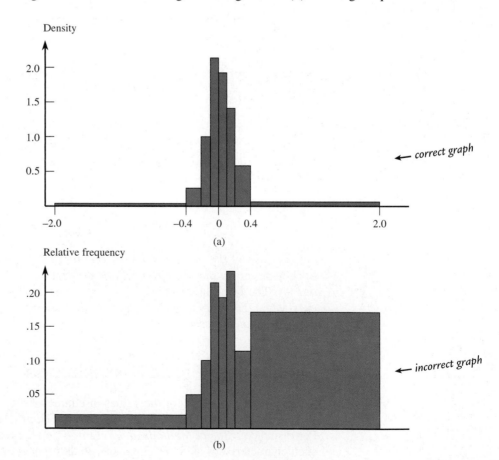

quency and is therefore not correct. In particular, this second histogram considerably exaggerates the incidence of grossly overreported and underreported values—the areas of the two most extreme rectangles are much too large. The eye is naturally drawn to large areas, so it is important that the areas correctly represent the relative frequencies.

■

■ Histogram Shapes ....

General shape is an important characteristic of a histogram. In describing various shapes it is convenient to approximate the histogram itself with a smooth curve (called a *smoothed histogram*). This is illustrated in Figure 3.24.

Figure 3.24 Approximating a histogram with a smooth curve.

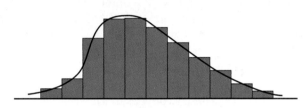

One description of general shape relates to the number of peaks, or **modes**.

A histogram is said to be **unimodal** if it has a single peak, **bimodal** if it has two peaks, and **multimodal** if it has more than two peaks.

■

These shapes are illustrated in Figure 3.25.

Figure 3.25 Smoothed histograms with various numbers of modes: (a) unimodal; (b) bimodal; (c) multimodal.

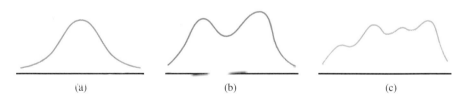

(a)   (b)   (c)

Bimodality can occur when the data set consists of observations on two quite different kinds of individuals or objects. For example, consider a large data set consisting of driving times for automobiles traveling between San Luis Obispo, California, and Monterey, California. This histogram would show two peaks, one for those cars that took the inland route (roughly 2.5 hr) and another for those cars traveling up the coast highway (3.5–4 hr). However, bimodality does not automatically follow in such situations. Bimodality will occur in the histogram of the combined groups only if the centers of the two separate histograms are far apart relative to the variability in the two data sets. Thus, a large data set consisting of heights of college students would probably not produce a bimodal histogram because the typical height for males (about 69 in.) and the typical height for females (about 66 in.) are not very far apart. Many histograms encountered in practice are unimodal, and multimodality is rather rare.

Unimodal histograms come in a variety of shapes. A unimodal histogram is **symmetric** if there is a vertical line of symmetry such that the part of the histogram to the left of the line is a mirror image of the part to the right. (Bimodal and multimodal

Figure 3.26 Several symmetric unimodal smoothed histograms.

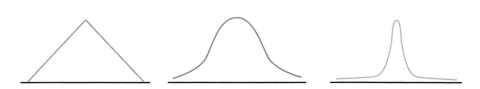

histograms can also be symmetric in this way.) Several different symmetric smoothed histograms are shown in Figure 3.26.

Proceeding to the right from the peak of a unimodal histogram, we move into what is called the **upper tail** of the histogram. Going in the opposite direction moves us into the **lower tail**.

A unimodal histogram that is not symmetric is said to be **skewed**. If the upper tail of the histogram stretches out much farther than the lower tail, then the distribution of values is **positively skewed**. If, on the other hand, the lower tail is much longer than the upper tail, the histogram is **negatively skewed**.

These two types of skewness are illustrated in Figure 3.27. Positive skewness is much more frequently encountered than is negative skewness. An example of positive skewness occurs in the distribution of single-family home prices in Los Angeles County; most homes are moderately priced (at least for California), whereas the relatively few homes in Beverly Hills and Malibu have much higher price tags.

Figure 3.27 Two examples of skewed smoothed histograms: (a) positive skew; (b) negative skew.

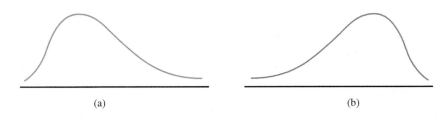

(a)   (b)

One rather specific shape, a **normal curve**, arises more frequently than any other in statistical applications. Many histograms can be well approximated by a normal curve (e.g., characteristics such as arm span and brain weight). Here we briefly mention several of the most important qualitative properties of such a curve, postponing a more detailed discussion until Chapter 7. A normal curve is both symmetric and bell shaped; it looks like the curve in Figure 3.28(a). However, not all bell-shaped curves are normal. In a normal curve, starting from the top of the bell the height of the curve decreases at a well-defined rate when moving toward either tail. (This rate of decrease is specified by a certain mathematical function.)

A curve with tails that do not decline as rapidly as the tails of a normal curve is called **heavy-tailed** (compared to the normal curve). Similarly, a curve with tails that

Figure 3.28 Three examples of bell-shaped histograms: (a) normal; (b) heavy-tailed; (c) light-tailed.

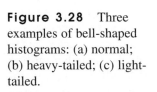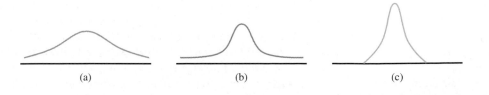

(a)   (b)   (c)

decrease more rapidly than the normal tails is called **light-tailed**. Figures 3.28(b) and 3.28(c) illustrate these possibilities. The reason that we are concerned about the tails in a distribution is that many inferential procedures that work well (i.e., they result in accurate conclusions) when the population distribution is approximately normal do poorly when the population distribution is heavy tailed.

■ Do Sample Histograms Resemble Population Histograms? ....

Sample data are usually collected to make inferences about a population. The resulting conclusions may be in error if the sample is unrepresentative of the population. So how similar might a histogram of sample data be to the histogram of all population values? Will the two histograms be centered at roughly the same place and spread out to about the same extent? Will they have the same number of peaks, and will these occur at approximately the same places?

A related issue concerns the extent to which histograms based on different samples from the same population resemble one another. If two different sample histograms can be expected to differ from one another in obvious ways, then at least one of them might differ substantially from the population histogram. If the sample differs substantially from the population, conclusions about the population based on the sample are likely to be incorrect. **Sampling variability**—the extent to which samples differ from one another and from the population—is a central idea in statistics. Example 3.18 illustrates such variability in histogram shapes.

Example 3.18 What You Should Know About Bus Drivers . . .

Table 3.7 Frequency Distribution for Number of Accidents by Bus Drivers

Number of Accidents	Frequency	Relative Frequency
0	117	.165
1	157	.222
2	158	.223
3	115	.162
4	78	.110
5	44	.062
6	21	.030
7	7	.010
8	6	.008
9	1	.001
10	3	.004
11	1	.001
	708	.998

● A sample of 708 bus drivers employed by public corporations was selected, and the number of traffic accidents in which each bus driver was involved during a 4-year period was determined ("Application of Discrete Distribution Theory to the Study of Noncommunicable Events in Medical Epidemiology," in *Random Counts in Biomedical and Social Sciences,* G. P. Patil, ed. [University Park, PA. Pennsylvania State University Press, 1970]). A listing of the 708 sample observations might look like this:

306002141...602

The frequency distribution (Table 3.7) shows that 117 of the 708 drivers had no accidents, a relative frequency of 117/708 = .165 (or 16.5%). Similarly, the proportion of sampled drivers who had 1 accident is .222 (or 22.2%). The largest sample observation was 11.

Although the 708 observations actually constituted a sample from the population of all bus drivers, we will regard the 708 observations as constituting the entire population. The first histogram in Figure 3.29, then, represents the population histogram. The other four histograms in Figure 3.29 are based on four different samples of 50 observations each

● Data set available online

Figure 3.29 Comparison of population and sample histograms for number of accidents.

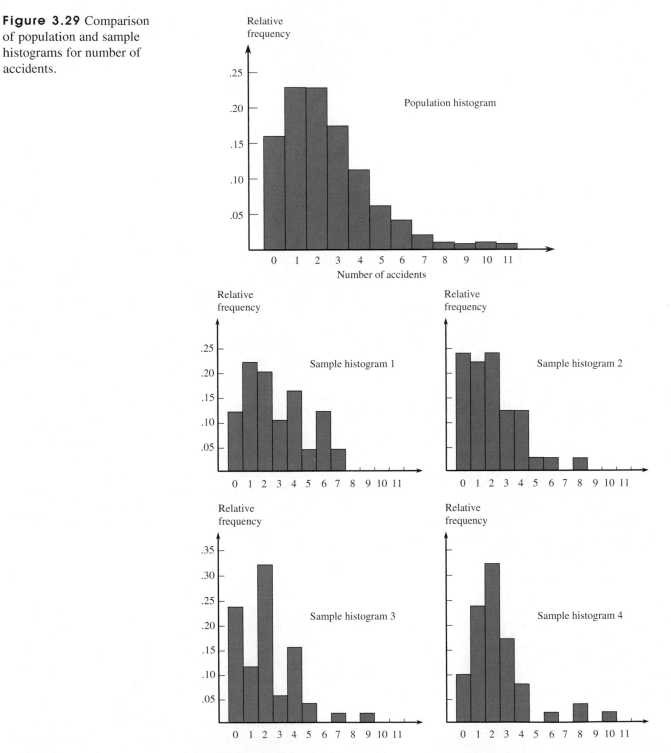

from this population. The five histograms certainly resemble one another in a general way, but some dissimilarities are also obvious. The population histogram rises to a peak and then declines smoothly, whereas the sample histograms tend to have more peaks, valleys, and gaps. Although the population data set contained an observation of 11, none of the four samples did. In fact, in the first two samples, the largest observations were 7 and 8, respectively. In Chapters 8–15 we will see how sampling variability can be described and incorporated into conclusions based on inferential methods.

■

Exercises 3.22–3.34

Turn to the Expanded Answers Section for answers not shown next to exercises.

3.22 ● Medicare's new medical plans offer a wide range of variations and choices for seniors when picking a drug plan (*San Luis Obispo Tribune,* November 25, 2005). The monthly cost for a stand-alone drug plan varies from plan to plan and from state to state. The accompanying table gives the premium for the plan with the lowest cost for each state.

State	Cost per Month (dollars)
Alabama	14.08
Alaska	20.05
Arizona	6.14
Arkansas	10.31
California	5.41
Colorado	8.62
Connecticut	7.32
Delaware	6.44
District of Columbia	6.44
Florida	10.35
Georgia	17.91
Hawaii	17.18
Idaho	6.33
Illinois	13.32
Indiana	12.30
Iowa	1.87
Kansas	9.48
Kentucky	12.30
Louisiana	17.06
Maine	19.60
Maryland	6.44
Massachusetts	7.32
Michigan	13.75
Minnesota	1.87
Mississippi	11.60
Missouri	10.29
Montana	1.87
Nebraska	1.87
Nevada	6.42

State	Cost per Month (dollars)
New Hampshire	19.60
New Jersey	4.43
New Mexico	10.65
New York	4.10
North Carolina	13.27
North Dakota	1.87
Ohio	14.43
Oklahoma	10.07
Oregon	6.93
Pennsylvania	10.14
Rhode Island	7.32
South Carolina	16.57
South Dakota	1.87
Tennessee	14.08
Texas	10.31
Utah	6.33
Vermont	7.32
Virginia	8.81
Washington	6.93
West Virginia	10.14
Wisconsin	11.42
Wyoming	1.87

a. Use class intervals of $0 to <$3, $3 to <$6, $6 to <$9 etc., to create a relative frequency distribution for these data.

b. Construct a histogram and comment on its shape.

c. Using the relative frequency distribution or the histogram, determine the proportion of the states that have a minimum monthly plan of less than $13.00 a month. About 71% of the states have a minimum monthy plan of less than $13.00 a month.

3.23 ● The following two relative frequency distributions were constructed using data that appeared in the report "Undergraduate Students and Credit Cards in 2004 (Nellie

Mae, May 2005). One relative frequency distribution is based on credit bureau data for a random sample of 1413 college students, while the other is based on the result of a survey completed by 132 of the 1260 college students who received the survey.

Credit Card Balance (dollars)—
Credit Bureau Data	**Relative Frequency**
0 to <100	.18
100 to <500	.19
500 to <1000	.14
1000 to <2000	.16
2000 to <3000	.10
3000 to <7000	.16
7000 or more	.07

Credit Card Balance (dollars)—
Survey Data	**Relative Frequency**
0 to <100	.18
100 to <500	.22
500 to <1000	.17
1000 to <2000	.22
2000 to <3000	.07
3000 to <7000	.14
7000 or more	.00

a. Construct a histogram for the credit bureau data. For purposes of constructing the histogram, assume that none of the students in the sample had a balance higher than 15,000 and that the last interval can be regarded as 7000 to < 15,000. Be sure to use the density scale when constructing the histogram.
b. Construct a histogram for the survey data. Use the same scale that you used for the histogram in part (a) so that it will be easy to compare the two histograms.
c. Comment on the similarities and differences in the histograms from Parts (a) and (b).
d. Do you think the high nonresponse rate for the survey may have contributed to the observed differences in the two histograms? Explain.

3.24 ● People suffering from Alzheimer's disease often have difficulty performing basic activities of daily living (ADLs). In one study ("Functional Status and Clinical Findings in Patients with Alzheimer's Disease," *Journal of Gerontology* [1992]: 177–182), investigators focused on

six such activities: dressing, bathing, transferring, toileting, walking, and eating. Here are data on the number of ADL impairments for each of 240 patients:

Number of impairments	0	1	2	3	4	5	6
Frequency	100	43	36	17	24	9	11

a. Determine the relative frequencies that correspond to the given frequencies.
b. What proportion of these patients had at most two impairments?   .7459
c. Use the result of Part (b) to determine what proportion of patients had more than two impairments.
d. What proportion of the patients had at least four impairments?   .1000 + .0375 + .0458 = .1833

3.25 ● ▼ *USA Today* (July 2, 2001) gave the following information regarding cell phone use for men and women:

Average Number of Minutes Used per Month	Relative Frequency	
	Men	**Women**
0 to <200	.56	.61
200 to <400	.18	.18
400 to <600	.10	.13
600 to <800	.16	.08

a. Construct a relative frequency histogram for average number of minutes used per month for men. How would you describe the shape of this histogram?
b. Construct a relative frequency histogram for average number of minutes used per month for women. Is the distribution for average number of minutes used per month similar for men and women? Explain.
c. What proportion of men average less than 400 minutes per month?   0.74
d. Estimate the proportion of men that average less than 500 minutes per month.   0.79
e. Estimate the proportion of women that average 450 minutes or more per month.   0.1775

3.26 ● U.S. Census data for San Luis Obispo County, California, were used to construct the following frequency distribution for commute time (in minutes) of working adults (the given frequencies were read from a graph that appeared in the *San Luis Obispo Tribune* [September 1, 2002] and so are only approximate):

Commute Time	Frequency
0 to <5	5,200
5 to <10	18,200
10 to <15	19,600
15 to <20	15,400
20 to <25	13,800
25 to <30	5,700
30 to <35	10,200
35 to <40	2,000
40 to <45	2,000
45 to <60	4,000
60 to <90	2,100
90 to <120	2,200

a. Notice that not all intervals in the frequency distribution are equal in width. Why do you think that unequal width intervals were used?

b. Construct a table that adds a relative frequency and a density column to the given frequency distribution.

c. Use the densities computed in Part (b) to construct a histogram for this data set. (Note: The newspaper displayed an incorrectly drawn histogram based on frequencies rather than densities!) Write a few sentences commenting on the important features of the histogram.

d. Compute the cumulative relative frequencies, and construct a cumulative relative frequency plot.

e. Use the cumulative relative frequency plot constructed in Part (d) to answer the following questions.

   i. Approximately what proportion of commute times were less than 50 min?   0.93

   ii. Approximately what proportion of commute times were greater than 22 min?   0.34

   iii. What is the approximate commute time value that separates the shortest 50% of commute times from the longest 50%?   17 mins

3.27 ● The article "Determination of Most Representative Subdivision" (*Journal of Energy Engineering* [1993]: 43–55) gave data on various characteristics of subdivisions that could be used in deciding whether to provide electrical power using overhead lines or underground lines. Data on the variable x = total length of streets within a subdivision are as follows:

1280	5320	4390	2100	1240	3060	4770	1050
360	3330	3380	340	1000	960	1320	530
3350	540	3870	1250	2400	960	1120	2120
450	2250	2320	2400	3150	5700	5220	500
1850	2460	5850	2700	2730	1670	100	5770
3150	1890	510	240	396	1419	2109	

a. Construct a stem-and-leaf display for these data using the thousands digit as the stem. Comment on the various features of the display.

b. Construct a histogram using class boundaries of 0 to 1000, 1000 to 2000, and so on. How would you describe the shape of the histogram?

c. What proportion of subdivisions has total length less than 2000? between 2000 and 4000?   .4894, .3617

3.28 Student loans can add up, especially for those attending professional schools to study in such areas as medicine, law, or dentistry. Researchers at the University of Washington studied medical students and gave the following information on the educational debt of medical students on completion of their residencies (*Annals of Internal Medicine* [March 2002]: 384–398):

Educational Debt (dollars)	Relative Frequency
0 to <5000	.427
5000 to <20,000	.046
20,000 to <50,000	.109
50,000 to <100,000	.232
100,000 or more	.186

a. What are two reasons that it would be inappropriate to construct a histogram using relative frequencies to determine the height of the bars in the histogram?

b. Suppose that no student had an educational debt of $150,000 or more upon completion of his or her residency, so that the last class in the relative frequency distribution would be 100,000 to <150,000. Summarize this distribution graphically by constructing a histogram of the educational debt data. (Don't forget to use the density scale for the heights of the bars in the histogram, because the interval widths aren't all the same.)

c. Based on the histogram of Part (b), write a few sentences describing the educational debt of medical students completing their residencies.

3.29 An exam is given to students in an introductory statistics course. What is likely to be true of the shape of the histogram of scores if:
a. the exam is quite easy?   Negatively skewed
b. the exam is quite difficult?   Positively skewed
c. half the students in the class have had calculus, the other half have had no prior college math courses, and the exam emphasizes mathematical manipulation?   Bimodal
Explain your reasoning in each case.

3.30 ● The paper "Lessons from Pacemaker Implantations" (*Journal of the American Medical Association* [1965]: 231–232) gave the results of a study that followed 89 heart patients who had received electronic pacemakers. The time (in months) to the first electrical malfunction of the pacemaker was recorded:

```
24  20  16  32  14  22   2  12  24   6  10  20
 8  16  12  24  14  20  18  14  16  18  20  22
24  26  28  18  14  10  12  24   6  12  18  16
34  18  20  22  24  26  18   2  18  12  12   8
24  10  14  16  22  24  22  20  24  28  20  22
26  20   6  14  16  18  24  18  16   6  16  10
14  18  24  22  28  24  30  34  26  24  22  28
30  22  24  22  32
```

a. Summarize these data in the form of a frequency distribution, using class intervals of 0 to <6, 6 to <12, and so on.
b. Compute the relative frequencies and cumulative relative frequencies for each class interval of the frequency distribution of Part (a).
c. Show how the relative frequency for the class interval 12 to <18 could be obtained from the cumulative relative frequencies.
d. Use the cumulative relative frequencies to give approximate answers to the following:
　　i. What proportion of those who participated in the study had pacemakers that did not malfunction within the first year? $1 - .1349 = .8651$
　　ii. If the pacemaker must be replaced as soon as the first electrical malfunction occurs, approximately what proportion required replacement between 1 and 2 years after implantation? $0.6855 - 0.1349 = 0.5506$
e. Construct a cumulative relative frequency plot, and use it to answer the following questions.
　　i. What is the approximate time at which about 50% of the pacemakers had failed?

　　ii. What is the approximate time at which only about 10% of the pacemakers initially implanted were still functioning?

3.31 The clearness index was determined for the skies over Baghdad for each of the 365 days during a particular year ("Contribution to the Study of the Solar Radiation Climate of the Baghdad Environment," *Solar Energy* [1990]: 7–12). The accompanying table summarizes the resulting data:

Clearness Index	Number of Days (frequency)
0.15 to <0.25	8
0.25 to <0.35	14
0.35 to <0.45	28
0.45 to <0.50	24
0.50 to <0.55	39
0.55 to <0.60	51
0.60 to <0.65	106
0.65 to <0.70	84
0.70 to <0.75	11

a. Determine the relative frequencies and draw the corresponding histogram. (Be careful here—the intervals do not all have the same width.)
b. Cloudy days are those with a clearness index smaller than 0.35. What proportion of the days was cloudy?   .06
c. Clear days are those for which the index is at least 0.65. What proportion of the days was clear?   .26

3.32 How does the speed of a runner vary over the course of a marathon (a distance of 42.195 km)? Consider determining both the time to run the first 5 km and the time to run between the 35 km and 40 km points, and then subtracting the 5-km time from the 35–40-km time. A positive value of this difference corresponds to a runner slowing down toward the end of the race. The histogram on page 117 is based on times of runners who participated in several different Japanese marathons ("Factors Affecting Runners' Marathon Performance," *Chance* [Fall, 1993]: 24–30). What are some interesting features of this histogram? What is a typical difference value? Roughly what proportion of the runners ran the late distance more quickly than the early distance?

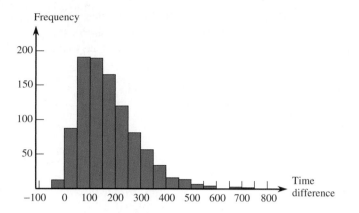

Frequency

Class Interval	I	II	III	IV	V
0 to <10	5	40	30	15	6
10 to <20	10	25	10	25	5
20 to <30	20	10	8	8	6
30 to <40	30	8	7	7	9
40 to <50	20	7	7	20	9
50 to <60	10	5	8	25	23
60 to <70	5	5	30	10	42

3.34 Using the five class intervals 100 to 120, 120 to 140, . . . , 180 to 200, devise a frequency distribution based on 70 observations whose histogram could be described as follows:

a. symmetric

b. bimodal

c. positively skewed

d. negatively skewed

3.33 Construct a histogram corresponding to each of the five frequency distributions, I–V, given in the following table, and state whether each histogram is symmetric, bimodal, positively skewed, or negatively skewed:

Bold exercises answered in back   ● Data set available online but not required   ▼ Video solution available

3.4 Displaying Bivariate Numerical Data

A bivariate data set consists of measurements or observations on two variables, x and y. For example, x might be distance from a highway and y the lead content of soil at that distance. When both x and y are numerical variables, each observation consists of a pair of numbers, such as (14, 5.2) or (27.63, 18.9). The first number in a pair is the value of x, and the second number is the value of y.

An unorganized list of bivariate data yields little information about the distribution of either the x values or the y values separately and even less information about how the two variables are related to one another. Just as graphical displays can be used to summarize univariate data, they can also help with bivariate data. The most important graph based on bivariate numerical data is a **scatterplot**.

In a scatterplot each observation (pair of numbers) is represented by a point on a rectangular coordinate system, as shown in Figure 3.30(a). The horizontal axis is identified with values of x and is scaled so that any x value can be easily located. Similarly, the vertical or y axis is marked for easy location of y values. The point corresponding

Figure 3.30 Constructing a scatterplot: (a) rectangular coordinate system; (b) point corresponding to (4.5, 15).

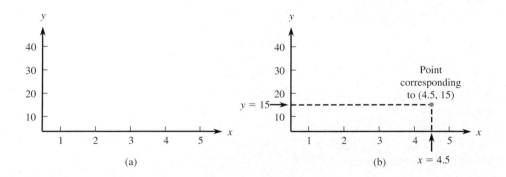

to any particular (x, y) pair is placed where a vertical line from the value on the x axis meets a horizontal line from the value on the y axis. Figure 3.30(b) shows the point representing the observation (4.5, 15); it is above 4.5 on the horizontal axis and to the right of 15 on the vertical axis.

....

Example 3.19 Olympic Figure Skating

● Do tall skaters have an advantage when it comes to earning high artistic scores in figure skating competitions? Data on x = height (in cm) and y = artistic score in the free skate for both male and female singles skaters at the 2006 Winter Olympics are shown in the accompanying table. (Data set courtesy of John Walker.)

Name	Gender	Height	Artistic
PLUSHENKO Yevgeny	M	178	41.2100
BUTTLE Jeffrey	M	173	39.2500
LYSACEK Evan	M	177	37.1700
LAMBIEL Stephane	M	176	38.1400
SAVOIE Matt	M	175	35.8600
WEIR Johnny	M	172	37.6800
JOUBERT Brian	M	179	36.7900
VAN DER PERREN Kevin	M	177	33.0100
TAKAHASHI Daisuke	M	165	36.6500
KLIMKIN Ilia	M	170	32.6100
ZHANG Min	M	176	31.8600
SAWYER Shawn	M	163	34.2500
LI Chengjiang	M	170	28.4700
SANDHU Emanuel	M	183	35.1100
VERNER Tomas	M	180	28.6100
DAVYDOV Sergei	M	159	30.4700
CHIPER Gheorghe	M	176	32.1500
DINEV Ivan	M	174	29.2500
DAMBIER Frederic	M	163	31.2500
LINDEMANN Stefan	M	163	31.0000
KOVALEVSKI Anton	M	171	28.7500
BERNTSSON Kristoffer	M	175	28.0400
PFEIFER Viktor	M	180	28.7200
TOTH Zoltan	M	185	25.1000
ARAKAWA Shizuka	F	166	39.3750
COHEN Sasha	F	157	39.0063
SLUTSKAYA Irina	F	160	38.6688
SUGURI Fumie	F	157	37.0313
ROCHETTE Joannie	F	157	35.0813
MEISSNER Kimmie	F	160	33.4625
HUGHES Emily	F	165	31.8563

● Data set available online

Name	Gender	Height	Artistic
MEIER Sarah	F	164	32.0313
KOSTNER Carolina	F	168	34.9313
SOKOLOVA Yelena	F	162	31.4250
YAN Liu	F	164	28.1625
LEUNG Mira	F	168	26.7000
GEDEVANISHVILI Elene	F	159	31.2250
KORPI Kiira	F	166	27.2000
POYKIO Susanna	F	159	31.2125
ANDO Miki	F	162	31.5688
EFREMENKO Galina	F	163	26.5125
LIASHENKO Elena	F	160	28.5750
HEGEL Idora	F	166	25.5375
SEBESTYEN Julia	F	164	28.6375
KARADEMIR Tugba	F	165	23.0000
FONTANA Silvia	F	158	26.3938
PAVUK Viktoria	F	168	23.6688
MAXWELL Fleur	F	160	24.5438

Figure 3.31(a) gives a scatterplot of the data. Looking at the data and the scatterplot, we can see that

1. Several observations have identical x values but different y values (e.g. $x = 176$ cm for both Stephane Lambiel and Min Zhang, but Lambiel's artistic score was 38.1400 and Zhang's artistic score was 31.8600). Thus, the value of y is *not* determined *solely* by the value of x but by various other factors as well.
2. At any given height there is quite a bit of variability in artistic score. For example, for those skaters with height 160 cm, artistic scores ranged from a low of about 24.5 to a high of about 39.
3. There is no noticeable tendency for artistic score to increase as height increases. There does not appear to be a strong relationship between height and artistic score.

The data set used to construct the scatter plot included data for both male and female skaters. Figure 3.31(b) shows a scatterplot of the (height, artistic score) pairs with observations for male skaters shown in blue and observations for female skaters shown in orange. Not surprisingly, the female skaters tend to be shorter than the male skaters (the observations for females tend to be concentrated toward the left side of the scatterplot). Careful examination of this plot shows that while there was no apparent pattern in the combined (male and female) data set, there may be a relationship between height and artistic score for female skaters.

Figures 3.31(c) and 3.31(d) show separate scatterplots for the male and female skaters, respectively. It is interesting to note that it appears that for female skaters, higher artistic scores seem to be associated with smaller height values, but for men there does not appear to be a relationship between height and artistic score. The relationship between height and artistic score for women is not evident in the scatterplot of the combined data.

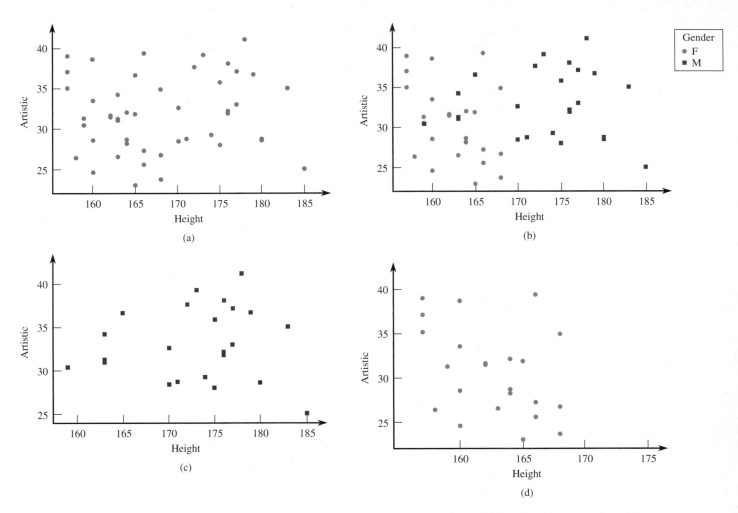

Figure 3.31 Scatterplots for the data of Example 3.19: (a) scatterplot of data; (b) scatterplot of data with observations for males and females distinguished by color; (c) scatterplot for male skaters; (d) scatterplot for female skaters.

The horizontal and vertical axes in the scatterplots of Figure 3.31 do not intersect at the point (0, 0). In many data sets, the values of *x* or of *y* or of both variables differ considerably from 0 relative to the ranges of the values in the data set. For example, a study of how air conditioner efficiency is related to maximum daily outdoor temperature might involve observations at temperatures of 80°, 82°, . . . , 98°, 100°. In such cases, the plot will be more informative if the axes intersect at some point other than (0, 0) and are marked accordingly. This is illustrated in Example 3.20.

Example 3.20 Taking Those "Hard" Classes Pays Off

● The report titled "2005 College Bound Seniors" (College Board, 2005) included the accompanying table showing the average score on the verbal section of the SAT for groups of high school seniors completing different numbers of years of study in

Step-by-step technology instructions available online   ● *Data set available online*

six core academic subjects (arts and music, English, foreign languages, mathematics, natural sciences, and social sciences and history). Figure 3.32(a) and (b) show two scatterplots of x = total number of years of study and y = average verbal SAT score. The scatterplots were produced by the statistical computer package MINITAB. In Figure 3.32(a), we let MINITAB select the scale for both axes. Figure 3.32(b) was obtained by specifying that the axes would intersect at the point (0, 0). The second plot does not make effective use of space. It is more crowded than the first plot, and such crowding can make it more difficult to see the general nature of any relationship. For example, it can be more difficult to spot curvature in a crowded plot.

Years of Study	Average Verbal SAT Score
14	446
15	453
16	459
17	467
18	484
19	501
20	545

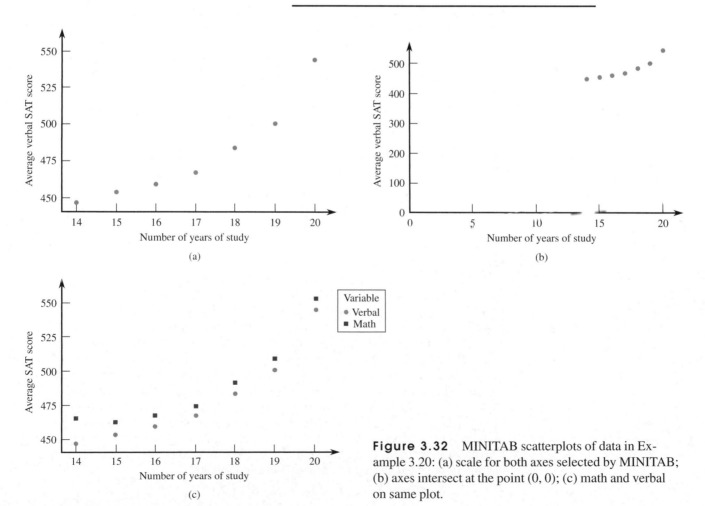

Figure 3.32 MINITAB scatterplots of data in Example 3.20: (a) scale for both axes selected by MINITAB; (b) axes intersect at the point (0, 0); (c) math and verbal on same plot.

The scatterplot for average verbal SAT score exhibits a fairly strong curved pattern, indicating that there is a strong relationship between average verbal SAT score and the total number of years of study in the six core academic subjects. Although the pattern in the plot is curved rather than linear, it is still easy to see that the average verbal SAT score increases as the number of years of study increases. Figure 3.32(c) shows a scatterplot with the average verbal SAT scores represented by orange dots and the average math SAT scores represented by blue squares. From this plot we can see that while the average math SAT scores tend to be higher than the average verbal scores at all of the values of total number of years of study, the general curved form of the relationship is similar.

■

In Chapter 5, methods for summarizing bivariate data when the scatterplot reveals a pattern are introduced. Linear patterns are relatively easy to work with. A curved pattern, such as the one in Example 3.20, is a bit more complicated to analyze, and methods for summarizing such nonlinear relationships are developed in Section 5.4.

■ Time Series Plots

Data sets often consist of measurements collected over time at regular intervals so that we can learn about change over time. For example, stock prices, sales figures, and other socio-economic indicators might be recorded on a weekly or monthly basis. A **time-series plot** (sometimes also called a time plot) is a simple graph of data collected over time that can be invaluable in identifying trends or patterns that might be of interest.

A time-series plot can be constructed by thinking of the data set as a bivariate data set, where y is the variable observed and x is the time at which the observation was made. These (x, y) pairs are plotted as in a scatterplot. Consecutive observations are then connected by a line segment; this aids in spotting trends over time.

Example 3.21 The Cost of Christmas

The article "'12 Days' of Gifts Will Run You $18,348" (*The New York Times,* December 24, 2005) described the Christmas Price Index and how it has changed over time. The Christmas Price Index is computed each year by PNC Advisors, and it reflects the cost of the giving all of the gifts described in the popular Christmas song "The Twelve Days of Christmas." For 2005, with the threat of avian flu complicating the shipment of birds such as the song's turtle doves and French hens, the Christmas Price index hit an all-time high of $18,348.87. A time-series plot of the Christmas Price Index for the past 21 years appears on the PNC web site (www.pncchristmaspriceindex.com) and the data given there were used to construct the time-series plot of Figure 3.33. The plot shows an upward trend in the index from 1984 until 1994. A dramatic drop in the cost occurred in 1995, but there has been a clear upward trend in the index since then. You can visit the web site to see individual time-series plots for each of the twelve gifts that are used to determine the Christmas Price Index (a partridge in a pear tree, two turtle doves, etc.). See if you can figure out what caused the dramatic decline in 1995.

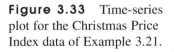

Figure 3.33 Time-series plot for the Christmas Price Index data of Example 3.21.

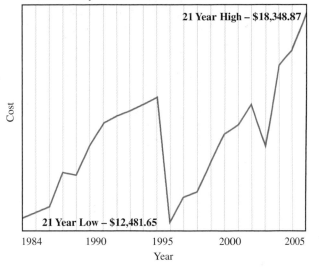

■

···

Example 3.22 Education Level and Income—Stay in School!

The time-series plot shown in Figure 3.34 appears on the U.S. Census Bureau web site. It shows the average earnings of workers by educational level as a proportion of the average earnings of a high school graduate over time. For example, we can see from this plot that in 1993 the average earnings for people with bachelor's degrees was about 1.5 times the average for high school graduates. In that same year, the average earnings for those who were not high school graduates was only about 75%

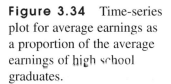

Figure 3.34 Time-series plot for average earnings as a proportion of the average earnings of high school graduates.

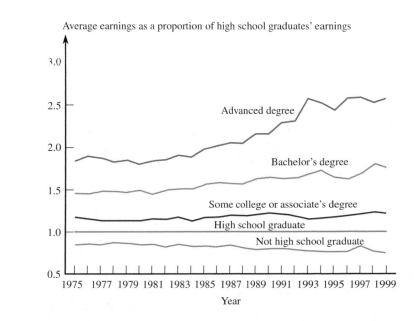

(a proportion of .75) of the average for high school graduates. The time-series plot also shows that the gap between the average earnings for high school graduates and those with a bachelor's degree or an advanced degree widened during the 1990s.

■

Exercises 3.35–3.45 Turn to the Expanded Answers Section for answers not shown next to exercises.

3.35 Does the size of a transplanted organ matter? A study that attempted to answer this question ("Minimum Graft Size for Successful Living Donor Liver Transplantation," *Transplantation* [1999]:1112–1116) presented a scatterplot much like the following ("graft weight ratio" is the weight of the transplanted liver relative to the ideal size liver for the recipient):

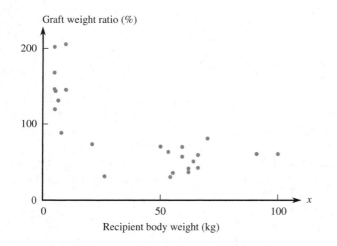

a. Discuss interesting features of this scatterplot.
b. Why do you think the overall relationship is negative?

3.36 Data on x = poverty rate (%) and y = high school dropout rate (%) for the 50 U.S. states and the District of Columbia were used to construct the following scatterplot (*Chronicle of Higher Education,* August 31, 2001):

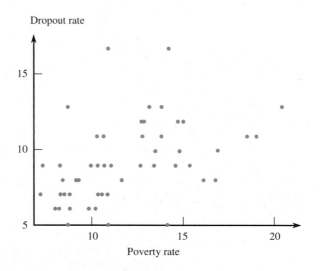

Write a few sentences commenting on this scatterplot. Would you describe the relationship between poverty rate and dropout rate as positive (y tends to increase as x increases), negative (y tends to decrease as x increases), or as having no discernible relationship between x and y?

3.37 ● ▼ A number of studies concerning muscular efficiency in humans have used competitive cyclists as subjects. It had previously been observed that variation in efficiency could not be explained by differences in cycling technique. The paper "Cycling Efficiency Is Related to the Percentage of Type I Muscle Fibers" (*Medicine and Science in Sports and Exercise* [1992]: 782–788) reported the accompanying data on x = percentage of type I (slow twitch) muscle fibers and y = cycling gross efficiency (the ratio of work accomplished per minute to caloric expenditure):

Observation	x	y
1	32	18.3
2	37	18.9
3	38	19.0
4	40	20.9
5	45	21.4
6	50	20.5
7	50	20.1
8	54	20.1
9	54	20.8
10	55	20.5
11	61	19.9
12	63	20.5
13	63	20.6
14	64	22.1
15	64	21.9
16	70	21.2
17	70	20.5
18	75	22.7
19	76	22.8

a. Construct a scatterplot of the data in which the axes intersect at the point (0, 16). How would you describe the pattern in the plot?

b. Does it appear that the value of cycling gross efficiency is determined solely by the percentage of type I muscle fibers? Explain.

3.38 ● The article "Exhaust Emissions from Four-Stroke Lawn Mower Engines" (*Journal of the Air and Water Management Association* [1997]: 945–952) reported data from a study in which both a baseline gasoline and a re-formulated gasoline were used. Consider the following observations on age (in years) and NO_x emissions (in grams per kilowatt-hour):

Engine	Age	Baseline	Reformulated
1	0	1.72	1.88
2	0	4.38	5.93
3	2	4.06	5.54
4	11	1.26	2.67
5	7	5.31	6.53
6	16	0.57	0.74
7	9	3.37	4.94
8	0	3.44	4.89
9	12	0.74	0.69
10	4	1.24	1.42

a. Construct a scatterplot of emissions versus age for the baseline gasoline.

b. Construct a scatterplot of emissions versus age for the reformulated gasoline.

c. Do age and emissions appear to be related? If so, in what way?

d. Comment on the similarities and differences in the two scatterplots of Parts (a) and (b).

3.39 ▼ One factor in the development of tennis elbow, a malady that strikes fear into the hearts of all serious players of that sport (including one of the authors), is the impact-induced vibration of the racket-and-arm system at ball contact. It is well known that the likelihood of getting tennis elbow depends on various properties of the racket used. Consider the accompanying scatterplot of x = racket resonance frequency (in hertz) and y = sum of peak-to-peak accelerations (a characteristic of arm vibration, in meters per second per second) for n = 23 different rackets ("Transfer of Tennis Racket Vibrations into the Human Forearm," *Medicine and Science in Sports and Exercise* [1992]: 1134–1140). Discuss interesting features of the data and of the scatterplot.

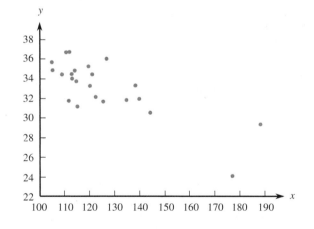

3.40 ● Stress can affect the physiology and behavior of animals, just as it can with humans (as many of us know all too well). The accompanying data on x = plasma cortisol concentration (in milligrams of cortisol per milliliter of plasma) and y = oxygen consumption rate (in milligrams per kilogram per hour) for juvenile steelhead after three 2-min disturbances were read from a graph in the paper "Metabolic Cost of Acute Physical Stress in Juvenile Steelhead" (*Transactions of the American Fisheries Society* [1987]: 257–263). The paper also included data for un-stressed fish.

x	25	36	48	59	62	72	80	100	100	137
y	155	184	180	220	280	163	230	222	241	350

a. Is the value of *y* determined solely by the value of *x*? Explain your reasoning.
b. Construct a scatterplot of the data.
c. Does an increase in plasma cortisol concentration appear to be accompanied by a change in oxygen consumption rate? Comment.

3.41 ● The National Telecommunications and Information Administration published a report titled "Falling Through the Net: Toward Digital Inclusion" (U.S. Department of Commerce, October 2000) that included the following information on access to computers in the home:

Year	Percentage of Households with a Computer
1985	8.2
1990	15.0
1994	22.8
1995	24.1
1998	36.6
1999	42.1
2000	51.0

a. Construct a time-series plot for these data. Be careful—the observations are not equally spaced in time. The points in the plot should not be equally spaced along the *x* axis.
b. Comment on any trend over time.

3.42 According to the National Association of Home Builders, the average size of a home in 1950 was 983 ft². The average size increased to 1500 ft² in 1970, 2080 ft² in 1990; and 2330 ft² in 2003 (*San Luis Obispo Tribune*, October 16, 2005).
a. Construct a time-series plot that shows how the average size of a home has changed over time.
b. If the trend of the time-series plot were to continue, what would you predict the average home size to be in 2010?   Approximately 2500 ft² by 2010.

3.43 An article that appeared in *USA Today* (September 3, 2003) included a graph similar to the one show here summarizing responses from polls conducted in 1978, 1991, and 2003 in which a sample of American adults were

asked whether or not it was a good time or a bad time to buy a house.

a. Construct a time-series plot that shows how the percentage that thought it was a good time to buy a house has changed over time.
b. Add a new line to the plot from Part (a) showing the percentage that thought it was a bad time to buy a house over time. Be sure to label the lines clearly.
c. Which graph, the given bar chart or the time-series plot, best shows the trend over time?

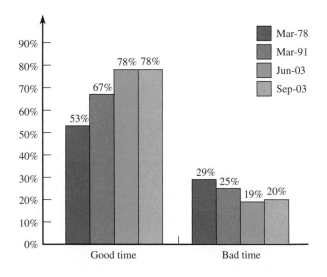

3.44 ● Some days of the week are more dangerous than others, according to Traffic Safety Facts produced by the National Highway Traffic Safety Administration. Average number of fatalities per day for each day of the week are shown in the accompanying table.

	Average Fatalities per Day (day of the week)						
	Mon	**Tue**	**Wed**	**Thurs**	**Fri**	**Sat**	**Sun**
1978–1982	103	101	107	116	156	201	159
1983–1987	98	96	99	108	140	174	140
1988–1992	97	94	97	106	139	168	135
1993–1997	97	93	96	102	129	148	127
1998–2002	99	96	98	104	129	149	130
Total	99	96	100	107	138	168	138

a. Using the midpoint of each year range (e.g., 1980 for the 1978–1982 range), construct a time-series plot that shows the average fatalities over time for each day of the week. Be sure to label each line clearly as to which day of the week it represents.

b. Write a sentence or two commenting on the difference in average number of fatalities for the days of the week. What is one possible reason for the differences?

c. Write a sentence or two commenting on the change in average number of fatalities over time. What is one possible reason for the change?

3.45 The accompanying time-series plot of movie box office totals (in millions of dollars) over 18 weeks of summer for both 2001 and 2002 appeared in *USA Today* (September 3, 2002):

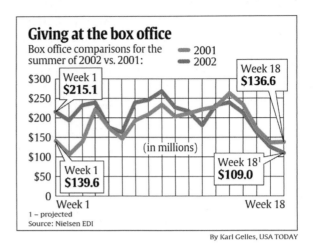

Patterns that tend to repeat on a regular basis over time are called seasonal patterns. Describe any seasonal patterns that you see in the summer box office data. Hint: Look for patterns that seem to be consistent from year to year.

Bold exercises answered in back   ● Data set available online but not required   ▼ Video solution available

---

3.5 Interpreting and Communicating the Results of Statistical Analyses

A graphical display, when used appropriately, can be a powerful tool for organizing and summarizing data. By sacrificing some of the detail of a complete listing of a data set, important features of the data distribution are more easily seen and more easily communicated to others.

■ Communicating the Results of Statistical Analyses

When reporting the results of a data analysis, a good place to start is with a graphical display of the data. A well-constructed graphical display is often the best way to highlight the essential characteristics of the data distribution, such as shape and spread for numerical data sets or the nature of the relationship between the two variables in a bivariate numerical data set.

For effective communication with graphical displays, some things to remember are:

- Be sure to select a display that is appropriate for the given type of data.
- Be sure to include scales and labels on the axes of graphical displays.
- In comparative plots, be sure to include labels or a legend so that it is clear which parts of the display correspond to which samples or groups in the data set.
- Although it is sometimes a good idea to have axes that don't cross at (0, 0) in a scatterplot, the vertical axis in a bar chart or a histogram should always start at 0 (see the cautions and limitations later in this section for more about this).

■ Keep your graphs simple. A simple graphical display is much more effective than one that has a lot of extra "junk." Most people will not spend a great deal of time studying a graphical display, so its message should be clear and straightforward.

■ Keep your graphical displays honest. People tend to look quickly at graphical displays so it is important that a graph's first impression is an accurate and honest portrayal of the data distribution. In addition to the graphical display itself, data analysis reports usually include a brief discussion of the features of the data distribution based on the graphical display.

■ For categorical data, this discussion might be a few sentences on the relative proportion for each category, possibly pointing out categories that were either common or rare compared to other categories.

■ For numerical data sets, the discussion of the graphical display usually summarizes the information that the display provides on three characteristics of the data distribution: center or location, spread, and shape.

■ For bivariate numerical data, the discussion of the scatterplot would typically focus on the nature of the relationship between the two variables used to construct the plot.

■ For data collected over time, any trends or patterns in the time-series plot would be described.

■ Interpreting the Results of Statistical Analyses

The use of graphical data displays is quite common in newspapers, magazines, and journals, so it is important to be able to extract information from such displays. For example, data on test scores for a standardized math test given to eighth-graders in 37 states, 2 territories (Guam and the Virgin Islands), and the District of Columbia were presented in the August 13, 1992, edition of *USA Today*. Figure 3.35 gives both

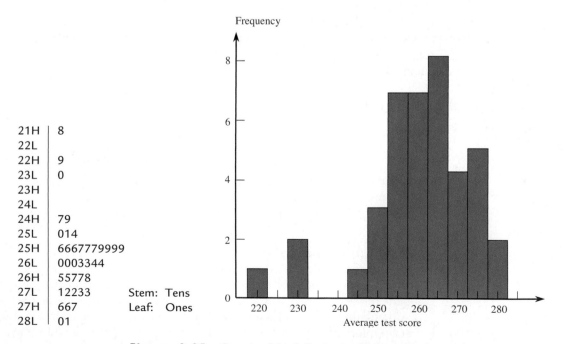

Figure 3.35   Stem-and-leaf display and histogram for math test scores.

a stem-and-leaf display and a histogram summarizing these data. Careful examination of these displays reveals the following:

1. Most of the participating states had average eighth-grade math scores between 240 and 280. We would describe the shape of this display as negatively skewed, because of the longer tail on the low end of the distribution.
2. Three of the average scores differed substantially from the others. These turn out to be 218 (Virgin Islands), 229 (District of Columbia), and 230 (Guam). These three scores could be described as outliers. It is interesting to note that the three unusual values are from the areas that are not states.
3. There do not appear to be any outliers on the high side.
4. A "typical" average math score for the 37 states would be somewhere around 260.
5. There is quite a bit of variability in average score from state to state.

How would the displays have been different if the two territories and the District of Columbia had not participated in the testing? The resulting histogram is shown in Figure 3.36. Note that the display is now more symmetric, with no noticeable outliers. The display still reveals quite a bit of state-to-state variability in average score, and 260 still looks reasonable as a "typical" average score. Now suppose that the two highest values among the 37 states (Montana and North Dakota) had been even higher. The stem-and-leaf display might then look like the one given in Figure 3.37. In this stem-and-leaf display, two values stand out from the main part of the display. This would catch our attention and might cause us to look carefully at these two states to determine what factors may be related to high math scores.

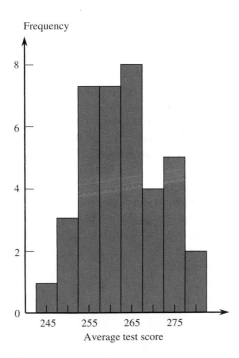

Figure 3.36 Histogram frequency for the modified math score data.

```
24H │ 79
25L │ 014
25H │ 6667779999
26L │ 0003344
26H │ 55778
27L │ 12233          Stem:  Tens
27H │ 667            Leaf:  Ones
28L │
28H │
29L │
29H │ 68
```

Figure 3.37 Stem-and-leaf display for modified math score data.

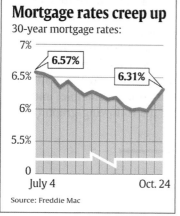

Mortgage rates creep up

30-year mortgage rates:

Source: Freddie Mac

By Quin Tian, USA TODAY

2. Be cautious of graphs with broken axes. Although it is common to see scatterplots with broken axes, be extremely cautious of time-series plots, bar charts, or histograms with broken axes. The use of broken axes in a scatterplot does not distort information about the nature of the relationship in the bivariate data set used to construct the display. On the other hand, in time-series plots, broken axes can sometimes exaggerate the magnitude of change over time. Although it is not always inadvisable to break the vertical axis in a time-series plot, it is something you should watch for, and if you see a time-series plot with a broken axis, as in the accompanying time-series plot of mortgage rates (*USA Today,* October 25, 2002), you should pay particular attention to the scale on the vertical axis and take extra care in interpreting the graph:

In bar charts and histograms, the vertical axis (which represents frequency, relative frequency, or density) should *never* be broken. If the vertical axis is broken in this type of graph, the resulting display will violate the "proportional area" principle and the display will be misleading. For example, the accompanying bar chart is similar to one appearing in an advertisement for a software product designed to help teachers raise student test scores. By starting the vertical axis at 50, the gain for students using the software is exaggerated. Areas of the bars are not proportional to the magnitude of the numbers represented—the area for the rectangle representing 68 is more than three times the area of the rectangle representing 55!

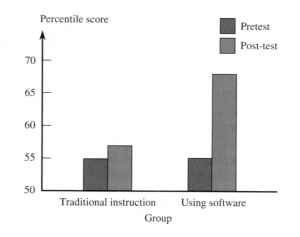

3. Watch out for unequal time spacing in time-series plots. If observations over time are not made at regular time intervals, special care must be taken in constructing the time-series plot. Consider the accompanying time-series plot, which is similar to one appearing in the *San Luis Obispo Tribune* (September 22, 2002) in an article on online banking:

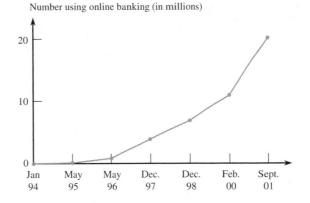

Notice that the intervals between observations are irregular, yet the points in the plot are equally spaced along the time axis. This makes it difficult to make a coherent assessment of the rate of change over time. This could have been remedied by spacing the observations differently along the time axis, as shown in the following plot:

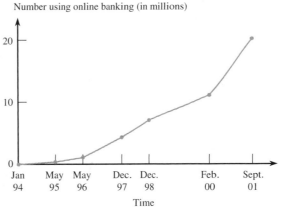

4. Be careful how you interpret patterns in scatterplots. A strong pattern in a scatterplot means that the two variables tend to vary together in a predictable way, but it does not mean that there is a cause-and-effect relationship between the two variables. We will consider this point further in Chapter 5, but in the meantime, when describing patterns in scatterplots, be careful not to use wording that implies that changes in one variable *cause* changes in the other.

5. Make sure that a graphical display creates the right first impression. For example, consider the graph at left from *USA Today* (June 25, 2001). Although this graph does not violate the proportional area principle, the way the "bar" for the "none" category is displayed makes this graph difficult to read, and a quick glance at this graph would leave the reader with an incorrect impression.

Activity **3.1** Locating States

Background: A newspaper article bemoaning the state of students' knowledge of geography claimed that more students could identify the island where the 2002 season of the TV show *Survivor* was filmed than could locate Vermont on a map of the United States. In this activity, you will collect data that will allow you to estimate the proportion of students that can correctly locate the states of Vermont and Nebraska.

1. Working as a class, decide how you will select a sample that you think will be representative of the students from your school.

2. Use the sampling method from Step 1 to obtain the subjects for this study. Subjects should be shown the accompanying map of the United States and asked to point out the state of Vermont. After the subject has given his or her answer, ask the subject to point out the state of Nebraska. For each subject, record whether or not Vermont was correctly identified and whether or not Nebraska was correctly identified.

3. When the data collection process is complete, summarize the resulting data in a table like the one shown here:

Response	Frequency
Correctly identified both states	
Correctly identified Vermont but not Nebraska	
Correctly identified Nebraska but not Vermont	
Did not correctly identify either state	

4. Construct a pie chart that summarizes the data in the table from Step 3.

5. What proportion of sampled students were able to correctly identify Vermont on the map?

6. What proportion of sampled students were able to correctly identify Nebraska on the map?

7. Construct a comparative bar chart that shows the proportion correct and the proportion incorrect for each of the two states considered.

8. Which state, Vermont or Nebraska, is closer to the state in which your school is located? Based on the pie chart, do you think that the students at your school were better able to identify the state that was closer than the one that was farther away? Justify your answer.

9. Write a paragraph commenting on the level of knowledge of U.S. geography demonstrated by the students participating in this study.

10. Would you be comfortable generalizing your conclusions in Step 8 to the population of students at your school? Explain why or why not.

Activity **3.2** Bean Counters!

Materials needed: A large bowl of dried beans (or marbles, plastic beads, or any other small, fairly regular objects) and a coin.

In this activity, you will investigate whether people can hold more in the right hand or in the left hand.

1. Flip a coin to determine which hand you will measure first. If the coin lands heads side up, start with the right hand. If the coin lands tails side up, start with the left hand. With the designated hand, reach into the bowl and grab as many beans as possible. Raise the hand over the bowl and

count to 4. If no beans drop during the count to 4, drop the beans onto a piece of paper and record the number of beans grabbed. If any beans drop during the count, restart the count. That is, you must hold the beans for a count of 4 without any beans falling before you can determine the number grabbed. Repeat the process with the other hand, and then record the following information: (1) right-hand number, (2) left-hand number, and (3) dominant hand (left or right, depending on whether you are left- or right-handed).

2. Create a class data set by recording the values of the three variables listed in Step 1 for each student in your class.

3. Using the class data set, construct a comparative stem-and-leaf display with the right-hand counts displayed on the right and the left-hand counts displayed on the left of the stem-and-leaf display. Comment on the interesting features of the display and include a comparison of the right-hand count and left-hand count distributions.

4. Now construct a comparative stem-and-leaf display that allows you to compare dominant-hand count to non-dominant-hand count. Does the display support the theory that dominant-hand count tends to be higher than non-dominant-hand count?

5. For each observation in the data set, compute the difference

$$\text{dominant-hand count} - \text{nondominant-hand count}$$

Construct a stem-and-leaf display of the differences. Comment on the interesting features of this display.

6. Explain why looking at the distribution of the differences (Step 5) provides more information than the comparative stem-and-leaf display (Step 4). What information is lost in the comparative display that is retained in the display of the differences?

Summary of Key Concepts and Formulas

Term or Formula	Comment
Frequency distribution	A table that displays frequencies, and sometimes relative and cumulative relative frequencies, for categories (categorical data), possible values (discrete numerical data), or class intervals (continuous data).
Comparative bar chart	Two or more bar charts that use the same set of horizontal and vertical axes.
Pie chart	A graph of a frequency distribution for a categorical data set. Each category is represented by a slice of the pie, and the area of the slice is proportional to the corresponding frequency or relative frequency.
Segmented bar chart	A graph of a frequency distribution for a categorical data set. Each category is represented by a segment of the bar, and the area of the segment is proportional to the corresponding frequency or relative frequency.
Stem-and-leaf display	A method of organizing numerical data in which the stem values (leading digit(s) of the observations) are listed in a column, and the leaf (trailing digit(s)) for each observation is then listed beside the corresponding stem. Sometimes stems are repeated to stretch the display.
Histogram	A picture of the information in a frequency distribution for a numerical data set. A rectangle is drawn above each possible value (discrete data) or class interval. The rectangle's area is proportional to the corresponding frequency or relative frequency.

Term or Formula	Comment
Histogram shapes	A (smoothed) histogram may be unimodal (a single peak), bimodal (two peaks), or multimodal. A unimodal histogram may be symmetric, positively skewed (a long right or upper tail), or negatively skewed. A frequently occurring shape is the normal curve.
Cumulative relative frequency plot	A graph of a cumulative relative frequency distribution.
Scatter plot	A picture of bivariate numerical data in which each observation (x, y) is represented as a point with respect to a horizontal x axis and a vertical y axis.
Time series plot	A picture of numerical data collected over time

Chapter Review Exercises 3.46–3.64

Turn to the Expanded Answers Section for answers not shown next to exercises.

CENGAGENOW Know exactly what to study! Take a pre-test and receive your Personalized Learning Plan.

3.46 Each year the College Board publishes a profile of students taking the SAT. In the report "2005 College Bound Seniors: Total Group Profile Report," the average SAT scores were reported for three groups defined by first language learned. Use the data in the accompanying table to construct a bar chart of the average verbal SAT score for the three groups.

First Language Learned	Average Verbal SAT
English	519
English and another language	486
A language other than English	462

3.47 The report referenced in Exercise 3.46 also gave average math SAT scores for the three language groups, as shown in the following table.

First Language Learned	Average Math SAT
English	521
English and another language	513
A language other than English	521

Construct a comparative bar chart for the average verbal and math scores for the three language groups. Write a few sentences describing the differences and similarities

between the three language groups as shown in the bar chart.

3.48 Each student in a sample of 227 boys and 251 girls was asked what he or she thought was most important: getting good grades, being popular, or being good at sports. The resulting data, from the paper "The Role of Sport as a Social Determinant for Children" (*Research Quarterly for Exercise and Sport* [1992]: 418–424), are summarized in the accompanying table:

Most Important	Boys Frequency	Boys Relative Frequency	Girls Frequency	Girls Relative Frequency
Grades	117	.515	130	.518
Popular	50	.220	91	.363
Sports	60	.264	30	.120

Construct a comparative bar chart for this data set. Remember to use relative frequency when constructing the bar chart because the two group sizes are not the same.

3.49 The paper "Community Colleges Start to Ask, Where Are the Men?" (*Chronicle of Higher Education*, June 28, 2002) gave data on gender for community college students. It was reported that 42% of students enrolled at community colleges nationwide were male and

58% were female. Construct a segmented bar graph for these data.

3.50 ● The article "Tobacco and Alcohol Use in G-Rated Children's Animated Films" (*Journal of the American Medical Association* [1999]: 1131–1136) reported exposure to tobacco and alcohol use in all G-rated animated films released between 1937 and 1997 by five major film studios. The researchers found that tobacco use was shown in 56% of the reviewed films. Data on the total tobacco exposure time (in seconds) for films with tobacco use produced by Walt Disney, Inc., were as follows:

223 176 548 37 158 51 299 37 11 165
74 92 6 23 206 9

Data for 11 G-rated animated films showing tobacco use that were produced by MGM/United Artists, Warner Brothers, Universal, and Twentieth Century Fox were also given. The tobacco exposure times (in seconds) for these films was as follows:

205 162 6 1 117 5 91 155 24 55 17

Construct a comparative stem-and-leaf display for these data. Comment on the interesting features of this display.

3.51 ● The accompanying data on household expenditures on transportation for the United Kingdom appeared in "Transport Statistics for Great Britain: 2002 Edition" (in *Family Spending: A Report on the Family Expenditure Survey* [The Stationary Office, 2002]). Expenditures (in pounds per week) included costs of purchasing and maintaining any vehicles owned by members of the household and any costs associated with public transportation and leisure travel.

Year	Average Transportation Expenditure	Percentage of Household Expenditures for Transportation
1990	247.20	16.2
1991	259.00	15.3
1992	271.80	15.8
1993	276.70	15.6
1994	283.60	15.1
1995	289.90	14.9
1996	309.10	15.7
1997	328.80	16.7
1998	352.20	17.0
1999	359.40	17.2
2000	385.70	16.7

a. Construct time-series plots of the transportation expense data and the percent of household expense data.
b. Do the time-series plots of Part (a) support the statement that follows? Explain why or why not. Statement: Although actual expenditures have been increasing, the percentage of the total household expenditures that was for transportation has remained relatively stable.

3.52 The web site PollingReport.com gave data from a CBS news poll conducted in December 1999. In the survey described, people were asked the following question: "All things considered, in our society today, do you think there are more advantages in being a man, more advantages in being a woman, or are there no more advantages in being one than the other?" Responses for men and for women are summarized in the following table:

Response	Relative Frequency Women	Men
Advantage in being a man	.57	.41
Advantage in being a woman	.06	.14
No advantage	.33	.40
Don't know	.04	.05

Construct a comparative bar chart for the response, and write a few sentences describing the differences in the response distribution for men and women.

3.53 The same poll described in Exercise 3.52 also asked the question, "What about salaries? These days, if a man and a woman are doing the same work, do you think the man generally earns more, the woman generally earns more, or that both earn the same amount?" The resulting data are shown in the following table:

Response	Relative Frequency Women	Men
Man earns more	.70	.59
Woman earns more	.01	.01
Both earn the same	.25	.34
Don't know	.04	.06

a. Construct a comparative bar chart that allows the responses for women and men to be compared.

Bold exercises answered in back   ● Data set available online but not required   ▼ Video solution available

b. Construct two pie charts, one summarizing the responses for women and one summarizing the responses for men.

c. Is it easier to compare the responses of women and men by looking at the comparative bar chart or the two pie charts? Explain.

d. Write a brief paragraph describing the difference between women and men with respect to the way they answered this question.

3.54 The article "The Healthy Kids Survey: A Look at the Findings" (*San Luis Obispo Tribune,* October 25, 2002) gave the accompanying information for a sample of fifth graders in San Luis Obispo County. Responses are to the question:

"After school, are you home alone without adult supervision?"

Response	Percentage
Never	8
Some of the time	15
Most of the time	16
All of the time	61

a. Summarize these data using a pie chart.

b. Construct a segmented bar chart for these data.

c. Which graphing method—the pie chart or the segmented bar chart—do you think does a better job of conveying information about response? Explain.

3.55 "If you were taking a new job and had your choice of a boss, would you prefer to work for a man or a woman?" That was the question posed to individuals in a sample of 576 employed adults (*Gallup at a Glance,* October 16, 2002). Responses are summarized in the following table:

Response	Frequency
Prefer to work for a man	190
Prefer to work for a woman	92
No difference	282
No opinion	12

a. Construct a pie chart to summarize this data set, and write a sentence or two summarizing how people responded to this question.

b. Summarize the given data using a segmented bar chart.

3.56 The accompanying stem-and-leaf display shows observations on average shower flow rate (in liters per minute) for a sample of 129 houses in Perth, Australia ("An Application of Bayes Methodology to the Analysis of Diary Records from a Water Use Study," *Journal of the American Statistical Association* [1987]: 705–711).

```
 2 | 23
 3 | 2344567789
 4 | 01356889
 5 | 00001114455666789
 6 | 000012222334445666677899999
 7 | 00012233455555668
 8 | 02233448
 9 | 012233335666788
10 | 2344455688
11 | 2335999
12 | 37
13 | 8                        Stem: Ones
14 | 36                       Leaf: Tenths
15 | 0035
16 |
17 |
18 | 9
```

a. What is the smallest flow rate in the sample?

b. If one additional house yielded a flow rate of 8.9, where would this observation be placed on the display?

c. What is a typical, or representative, flow rate?

d. Does the display appear to be highly concentrated, or quite spread out?

e. Does the distribution of values in the display appear to be reasonably symmetric? If not, how would you describe the departure from symmetry?

f. Does the data set appear to contain any outliers (observations far removed from the bulk of the data)?

3.57 ● Disparities among welfare payments by different states have been the source of much political controversy. The accompanying table reports average payment per person (in dollars) in the Aid to Families with Dependent Children Program for the 1990 fiscal year. Construct a relative frequency distribution for these data using equal interval widths. Draw the histogram corresponding to your frequency distribution.

State	Average Welfare Payment ($)
Alaska	244.90
California	218.31
Arizona	93.57
Montana	114.95

State	Average Welfare Payment ($)
Texas	56.79
Nebraska	115.15
Minnesota	171.75
Arkansas	65.96
Alabama	39.62
Illinois	112.28
Indiana	92.43
New Hampshire	164.20
Rhode Island	179.37
New Jersey	121.99
Delaware	113.66
North Carolina	91.95
Florida	95.43
Washington	160.41
Idaho	97.93
Utah	118.36
Colorado	111.20
Oklahoma	96.98
South Dakota	95.52
Iowa	129.58
Louisiana	55.81
Tennessee	65.93
Wisconsin	155.04
Ohio	115.26
Vermont	183.36
Connecticut	205.86
Pennsylvania	127.70
Maryland	132.86
South Carolina	71.91
Hawaii	187.71
Oregon	135.99
Nevada	100.25
Wyoming	113.84
New Mexico	81.87
Kansas	113.88
North Dakota	130.49
Missouri	91.93
Mississippi	40.22
Kentucky	85.21
Michigan	154.75
Maine	150.12
Massachusetts	200.99
New York	193.48
West Virginia	82.94
Virginia	97.98
Georgia	91.31

3.58 ● 2005 was a record year for hurricane devastation in the United States (*San Luis Obispo Tribune,* November 30, 2005). Of the 26 tropical storms and hurricanes in the season, 4 hurricanes hit the mainland: Dennis, Katrina, Rita, and Wilma. The U.S. insured catastrophic losses since 1989 (approximate values read from a graph that appeared in the *San Luis Obispo Tribune*, November 30, 2005) are as follows:

Year	Cost (in billions of dollars)
1989	7.5
1990	2.5
1991	4.0
1992	22.5
1993	5.0
1994	18.0
1995	9.0
1996	8.0
1997	2.6
1998	10.0
1999	9.0
2000	3.0
2001	27.0
2002	5.0
2003	12.0
2004	28.5
2005	56.8

Construct a time-series plot that shows the insured catastrophic loss over time. What do you think causes the peaks in the graph?

3.59 ● Each observation in the following data set is the number of housing units (homes or condominiums) sold during November 1992 in a region corresponding to a particular Orange County, California, ZIP code:

```
25  18  16   6  26  11  29   7   5  15  12  37
35  11  16  35  20  27  17  30  10  16  28  13
26  11  12   8   9  29   0  20  30  12  45  26
21  30  18  31   0  46  47  14  13  29  11  18
10  27   5  18  67  21  35  48  42  70  43   0
30  17  35  40  61  18  17  17
```

Construct a stem-and-leaf display, and comment on any interesting features.

3.60 Each murder committed in Utah during the period 1978–1990 was categorized by day of the week, resulting in the following frequencies: Sunday, 109; Monday, 73; Tuesday, 97; Wednesday, 95; Thursday, 83; Friday, 107; Saturday, 100.

a. Construct the corresponding frequency distribution.

b. What proportion of these murders was committed on a weekend day—that is, Friday, Saturday, or Sunday?

c. Do these data suggest that a murder is more likely to be committed on some days than on other days? Explain your reasoning.

3.61 An article in the *San Luis Obispo Tribune* (November 20, 2002) stated that 39% of those with critical housing needs (those who pay more than half their income for housing) lived in urban areas, 42% lived in suburban areas, and the rest lived in rural areas. Construct a pie chart that shows the distribution of type of residential area (urban, suburban, or rural) for those with critical housing needs.

3.62 ● Living-donor kidney transplants are becoming more common. Often a living donor has chosen to donate a kidney to a relative with kidney disease. The following data appeared in a *USA Today* article on organ transplants ("Kindness Motivates Newest Kidney Donors," June 19, 2002):

Number of Kidney Transplants

Year	Living-Donor to Relative	Living-Donor to Unrelated Person
1994	2390	202
1995	2906	400
1996	2916	526
1997	3144	607
1998	3324	814
1999	3359	930
2000	3679	1325
2001	3879	1399

a. Construct a time-series plot for the number of living-donor kidney transplants where the donor is a relative of the recipient. Describe the trend in this plot.

b. Use the data from 1994 and 2001 to construct a comparative bar chart for the type of donation (relative or unrelated). Write a few sentences commenting on your display.

3.63 ● Many nutritional experts have expressed concern about the high levels of sodium in prepared foods. The following data on sodium content (in milligrams) per frozen meal appeared in the article "Comparison of 'Light' Frozen Meals" (*Boston Globe*, April 24, 1991):

720	530	800	690	880	1050	340	810	760
300	400	680	780	390	950	520	500	630
480	940	450	990	910	420	850	390	600

Two histograms for these data are shown:

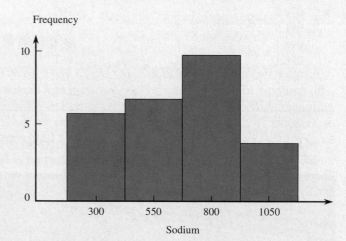

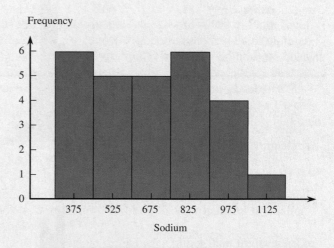

a. Do the two histograms give different impressions about the distribution of values?

b. Use each histogram to determine approximately the proportion of observations that are less than 800, and compare to the actual proportion.

3.64 ● Americium 241 (^{241}Am) is a radioactive material used in the manufacture of smoke detectors. The article "Retention and Dosimetry of Injected ^{241}Am in Beagles" (*Radiation Research* [1984]: 564–575) described a study in which 55 beagles were injected with a dose of ^{241}Am (proportional to each animal's weight). Skeletal retention of ^{241}Am (in microcuries per kilogram) was recorded for each beagle, resulting in the following data:

0.196	0.451	0.498	0.411	0.324	0.190	0.489
0.300	0.346	0.448	0.188	0.399	0.305	0.304
0.287	0.243	0.334	0.299	0.292	0.419	0.236
0.315	0.447	0.585	0.291	0.186	0.393	0.419

0.335	0.332	0.292	0.375	0.349	0.324	0.301
0.333	0.408	0.399	0.303	0.318	0.468	0.441
0.306	0.367	0.345	0.428	0.345	0.412	0.337
0.353	0.357	0.320	0.354	0.361	0.329	

a. Construct a frequency distribution for these data, and draw the corresponding histogram.

b. Write a short description of the important features of the shape of the histogram.

Personal Tutor

Do you need a live tutor for homework problems?

CENGAGENOW™

Are you ready? Take your exam-prep post-test now.

Bold exercises answered in back   ● Data set available online but not required   ▼ Video solution available

Graphing Calculator Explorations

Exploration 3.1 Using Lists on Your Calculator

Calculators and computers work their magic by storing numbers in "memory locations." To perform an addition, the computer looks in its "memory" for the two numbers, retrieves them, and adds them. In the early days of calculators (and computers) there were very few of these expensive memory cells and calculations were performed one at a time while the user entered data. The modern scientific calculator allows a very useful extension of a single memory cell: a *group* of memory cells known as a "list." Using a list, a whole set of data, complete with a "name," can be stored in the calculator and analyzed as a whole. This list capability makes the calculator a very powerful tool for analyzing and displaying data. Since all the numbers are in the calculator at the same time, graphic representations of data such as those presented in this chapter are possible.

The actual capabilities of lists and the keystrokes to use these capabilities vary from calculator to calculator, so we will not be overly specific about particular calculator keystrokes. *Your best source for learning about your calculator is the manual that came with it!* You need the manual to fully understand and realize the potential of your calculator. Your calculator may implement some of its capabilities with special keys or menus, and the menus may include functions that require additional information that must be entered in a particular order. You don't need to memorize all these details. The calculator manual is there to use as a reference.

To use your calculator for statistical analysis you must be able to manipulate lists effectively. The accompanying table describes some of the calculator features that have a "list-based" capability and that will be important to you in performing statistical analysis. In future Explorations, we will assume that you are familiar with these list features. Many of these skills will become second nature with practice.

Capability	Why It's Important
Create a list	Some lists are provided automatically for your use, already labeled "List 1 or L1," or "List 2 or L2," etc. You will want to save data in lists with more informative names than these, such as "height" or "time."
Enter data into a list	This knowledge is, of course, fundamental to all analyses you will be performing.
Delete a list from the calculator	As powerful as your calculator is, there is a limit to its memory—there will come a time when you will need to delete the old data to prepare for the new.
Insert a number in a certain location in the list Or Delete a particular number in the list	If you are like everyone else, you will eventually add an extra number you did not intend or leave out a number from where it should be. Correcting these errors is a lot faster than deleting a whole list and starting over.
Copy data from one list to another	This will give you one of those things so precious to everyone who works with a finicky calculator (or finicky fingers?): a backup copy of the data!
Perform arithmetic operations with lists	Rather than perform the same arithmetic sequence separately on a set of numbers, you can do the calculations one list at a time. For example, to change units from inches to centimeters, you can multiply *all* the numbers on the list by 2.54. Usually, this is done with a statement something like 2.54 x ListName1→ ListName2. (The equal sign, =, is sometimes used in place of the arrow symbol.)

You need to be familiar with list manipulation to do effective statistical work with your calculator. As we proceed we will be more specific—and more detailed—about how you can utilize your calculator's list capabilities. The calculator manual may have a chapter called "Using Lists." We strongly encourage you to read it!

Exploration 3.2 Setting the Statistics Window

As we move into graphical descriptions of data, we need to consider how to use your calculator to produce graphs and plots of data. If you have used your calculator to graph functions, some of what follows will be a review. If you are new to the world of graphing calculators you will need a basic understanding of how to set up your calculator's "viewing window" for displaying graphs.

The metaphor of viewing the "world" through a "window" is a good one for thinking about the calculator window. If you think of the Cartesian *x-y* axes as the calculator's "world view" and your calculator view window as a portal through which to view this Cartesian world, you will be in the right frame of mind.

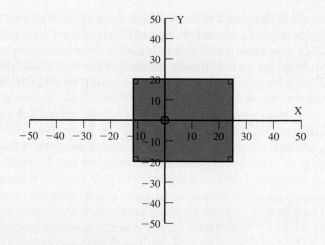

As you set up your calculator for graphing, your first problem will be, "Where in the *world* do I put my view *window*?" The quick and easy answer: You put your view window where your data is! We will illustrate how to do this by constructing a histogram, using the data from Example 3.15. To illustrate some of the problems involved in setting the view window we will begin with a slightly bad graph of a histogram and then gradually improve it. First, we want you to do something a bit strange, but trust us. Enter the function, $y = 400/x$, into your calculator. Now enter the data from Example 3.15 into your calculator in List 1. After entering the data, navigate your calculator's menu system to the histogram option. Actual keystrokes will vary among calculators, but the terms "Stat Graph" and "Stat Plot" are commonly used in calculators.

The view window is based on settings that you will manually enter into the calculator. (Your calculator may have the capability of automatically placing the view window over the appropriate position in its Cartesian system. Pretend for the moment that you are unaware of this.) To set up the view window you must navigate to your calculator's menu system, or possibly just press a "window" key. When you find the graph setup screen, it will look something like Figure 3.39. (There may be different or additional information on your screen but the numbers here will be our focus for this Exploration. Your numbers may have different values from these; for ease in following the discussion you may wish to change the values on your calculator to match those in the figure.)

The numbers on this screen determine where the viewing window is placed over the calculator's "world" coordinate system. Exit from the setup window and plot a histogram, by using a "graph" key or a menu, depending on your calculator. You should have something like what is shown in Figure 3.40. (Your graph will get better as we go.)

Notice that there actually is a histogram there, but there is a pesky function overlaying the histogram. (Suspiciously, it looks something like $y = 400/x$!) What's the problem here? First, remember that the calculator's world is the x-y coordinate system. If you have been using the calculator to plot mathematical functions and they are still "in" the calculator, both your last function and the histogram will be drawn. Oddly, this happens by design. Since a calculator user most likely will not graph a mathematical function and a histogram at the same time, the calculator graphing space is shared to save calculator memory. The solution to the histogram-and-function problem is easy: Don't graph the function. Delete the function and redraw the histogram.

We need to manually reposition the viewing window in the calculator world for a better look at the data. Return to the view window setup screen and make the follow-

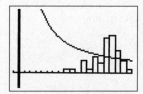

Figure 3.39

Figure 3.40

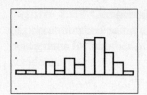

Figure 3.41

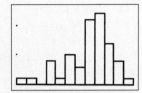

Figure 3.42

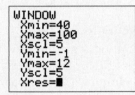

Figure 3.43

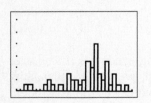

Figure 3.44

ing changes: set Xmin to 40, and Ymin to −1 and redraw the histogram. The histogram in the view window will now be positioned something like Figure 3.41.

This is certainly an improvement but the histogram is still rather small compared to the screen, sacrificing some detail. To correct this problem we will adjust the top of the view window. Return to the graph setup screen and locate the line "Ymax = 20." Change this to "Ymax = 12" and regraph the histogram. You should see a very well-spaced histogram similar to Figure 3.42.

You will find that adjusting the view window is a frequent task in creating effective statistical graphs. While each of the statistical plots has its own individuality, your construction of them will always involve positioning the view window over the Cartesian world view of your calculator. Even if the calculator has an automatic function to position statistical graphs, you will find it necessary sometimes to "improve" on the calculator's automatic choice. When you manually change your graph, keep in mind the idea of positioning a view window over a Cartesian coordinate system. This will help you organize your thoughts about how to change the view window and make the task less frustrating.

Exploration **3.3** Scaling the Histogram

When we constructed a histogram in the previous Exploration there were some numbers that we temporarily ignored in the view screen. We would like to return to those numbers now because they can seriously affect the look of a histogram. When we left the histogram the numbers in our view window were set as shown in Figure 3.43. These settings place the view window over the calculator's Cartesian system for effective viewing of the histogram from the data of Example 3.15.

We would now like to experiment a bit with the "Xscale." In all statistical graphs produced by the calculator the Xscale and Yscale choices will control the placement of the little "tick" marks on the *x* and *y* axis. In Exploration 3.2, the XScale and YScale were set at 5 and 1, respectively. The little tick marks on the *x*-axis were at multiples of 5. (Because of the data, the *x*-axis tick marks were at multiples of 5 and the *y*-axis didn't appear.) Change the Xscale value to 2 and redraw the histogram. You should see a graph similar to Figure 3.44. The *y*-axis tick marks now appear at multiples of 2, ...

Note that changing the Xscale has altered not only the tick marks but also the class intervals for the histogram. The choice of class intervals can significantly change the look and feel of the histogram. The choice of Xscale can affect judgments about the shape of the histogram. Because of this possibility it is wise to look at a histogram with varying choices of the Xscale value. If the shape appears very similar for different choices of Xscale, you can interpret and describe the shape with more confidence. However, if different Xscale choices alter the look of the histogram you should probably be more tentative.

Exploration **3.4** The Scatterplot

In this Exploration, we consider graphing a scatterplot of bivariate data. Here are the steps for creating a scatterplot:

1. Navigate your calculator's menu system to select the type of graph you want.
2. Select an appropriate viewing window.
3. Select data from two lists rather than one.

Figure 3.45a

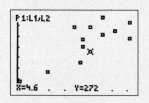

Figure 3.45b

Figure 3.46

4. Indicate which list will correspond to the horizontal axis and which will correspond to the vertical axis.

Figure 3.45a shows the selection for a scatterplot. We have selected graphing parameters and scale information as indicated Figure 3.45a. The resulting plot is shown in Figure 3.45b. The x and y scales are set so that the points will fill the view screen. Notice also that the plot is set to be made up of small squares—they are more easily seen than small dots.

One problem that frequently arises with scatterplots on calculators is the "disappearing" axis (in this case, the horizontal axis) and the lack of a discernible scale on either axis. The calculator has been hardwired to plot the x axis and y axis, but these will not be displayed if the view screen is not set up for both positive and negative values for the axes. Unfortunately the demands for axes as reference points compete with the desire for detail in a scatter plot. Although selecting a y scale from -1 to 200 would show each axis, the points would cluster close to the top of the screen, possibly hiding some detail or pattern from the data analyst.

There are no easy solutions to this problem, but use of the "trace" capability on your calculator can help a little. Pressing the trace button causes the calculator to display the coordinates of the location of a little crosshair icon. You can move this icon about the screen by pressing special "arrow" keys on your calculator. If you are displaying data rather than a function, pressing the trace button and then arrow keys may move the crosshair icon from point to point, displaying the coordinates of the points of your scatter plot. The scatter plot in Figure 3.46 shows the crosshair icon on a point (4.6, 272).

In your classroom your instructor may want you to sketch a scatter plot on a particular assignment, and most likely will want some sort of axes and scales. The scaling information is easily found in the viewing window and translated to the sketch of the plot on your paper.

TEACHING TIPS

Intent of this chapter:

After completing this chapter, students should be able to (1) explore the relationship between measures of center (mean and median), (2) describe the variability of a data set using measures of spread (range, variance, standard deviation, and interquartile range), (3) construct and describe modified boxplots, and (4) combine center, spread, and shape to find the proportion of observations within a specified range.

This chapter is a good place to emphasize that different symbols are used for parameters (values about the population) than for statistics (values calculated from a sample).

A PowerPoint® Lecture is available on the *Instructor's Resource Binder* CD.

■ Section 4.1

This section distinguishes between the population mean (μ) and sample mean ($\bar{x}$). Students should understand the relationship between the mean and median in distributions with different shapes. The activity, "Measures of Typical Value" (see below), helps students investigate this relationship.

This section also contains a brief discussion on population proportions (π) versus sample proportions (p). Stress that the *only* calculation that can be performed with categorical data is to calculate a proportion. (The use of the π and p notation for the population proportion and sample proportion, respectively, was done to maintain the convention of Greek letters for population characteristics, but other texts and the AP formula sheet use p and $\hat{p}$ for the population proportion and sample proportion, respectively. The π/p notation used in this text is acceptable on the AP exam.)

Suggested Assignment: Exercises 4.2, 4.3, 4.6, 4.10, 4.12, 4.13, 4.14

Activity: "Measures of Typical Value" (*Instructor's Resource Binder* & CD, Chapter 4 Activities Worksheets) ★

■ Section 4.2

Section 4.2 discusses measures of variability: range, interquartile range (iqr), deviation, standard deviation, and variance. It is important to do an example such as Example 4.8 so that students can "see" that the sum of deviations from the mean is zero—*always*. Discuss the difference in how population standard deviation (σ) and sample standard deviation (s) are computed.

Many students ask why we square the deviations from the mean when computing variance instead of using the apparently simple absolute values of the deviations. Historically, statistical concepts were developed using calculus. With calculus, it is easier to find derivatives and to integrate functions that contain squares rather than absolute values. Early statisticians used the easiest mathematical method that solved the problem!

Students often have difficulty understanding why we divide by $(n-1)$ when calculating the sample standard deviation. There does not seem to be a plausible, understandable reason at this level of statistics. The concept of "degrees of freedom" has sometimes been invoked. Since the deviations from the mean total zero, there are only $n-1$ deviations that are *free to vary* or give us information about variability. The last deviation from the mean must be a specific value to make the deviations equal zero, thus it is not *free to vary*.

Suggested Assignment: Exercises 4.16, 4.19, 4.20, 4.21, 4.23, 4.25, 4.26, 4.28

Activity: Several activities are provided; pick from among the following: Activity 4.1: Collecting and Summarizing Numerical Data (p. 190 and *Activities Manual*, p. 37), Activity 4.2: Airline Passenger Weights (p. 190 and *Activities Manual*, p.38), Bonus Activity

★ Kathy's personal favorite

4.4: Understanding Variability and Numerical Measures of Variability (*Activities Manual,* p. 41) ★, and Bonus Activity 4.5: Comparing Brands of Chocolate Chip Cookies (*Activities Manual,* p. 43) ★

■ **Section 4.3**

Boxplots were developed in the 1970s by John Tukey. Here's an interesting side note: When asked why he chose 1.5 iqr away from the quartiles as the value to determine outliers, Tukey answered that 1 iqr was too small and 2 iqr was too large. There is no mathematical reason for this other than Tukey's intuition, which considering how great a statistician he was, is enough for most of us! Students should *always* use modified boxplots in this course. For the AP exam, students need only to determine if an observation is an outlier, not whether it is an extreme or a mild outlier.

Boxplots are especially useful in comparing two or more distributions, but should be used with caution with small data sets ($n < 10$). When describing boxplots, students should comment on center, spread, shape, and unusual features. However, it is dangerous to infer too much about distribution shape from a boxplot, especially if the sample size is small. In addition, while the boxplot can provide some information about symmetry, approximate normality (or even modality) cannot be determined from a boxplot.

Suggested Assignment: Exercises 4.29, 4.31, 4.32, 4.33

Activity: Activity 4.3: Boxplot Shapes (p. 190 and *Activities Manual,* p.39)★ and "Quartiles, IQR, and Boxplots" (*Instructor's Resource Binder* & CD, Chapter 4 Activities Worksheets★

Assessment: Chapter 4 Quiz 1 (*Test Bank* and *Instructor's Resource Binder* CD)

■ **Section 4.4**

This section might more aptly be named "Using Center and Spread to Determine Proportions." The importance of Chebyshev's Rule is that it can be used with *any* distributions, no matter what shape. However, since students tend to have difficulty with Chebyshev's Rule and it is not part of the AP curriculum, this topic is optional. Be sure to stress that the Empirical Rule can *only* be used with distributions that are approximately normal and only gives an *approximation* of the proportion within the given interval.

Students often have trouble with percentiles. Be sure to incorporate this concept in many different situations to help them learn the meaning of percentiles. For example, you can ask students "Where is the 50th percentile?" (median) or "Where is the 75th percentile?" (third quartile).

Suggested Assignment: Exercises 4.37, 4.39, 4.41, 4.46, 4.49

Activity: "Measuring Spread (Variability)" (*Instructor's Resource Binder* & CD, Chapter 4 Activities Worksheets)★

Assessment: Chapter 4 Quiz 2 (*Test Bank* and *Instructor's Resource Binder* CD) and/or Chapter 4 Concept Quiz (*Test Bank* and *Instructor's Resource Binder* CD)

■ **Section 4.5**

Students should pay close attention to the "A Word to the Wise: Cautions and Limitations" section.

Suggested Review Assignment: Exercises 4.52, 4.53, 4.55, 4.56, 4.62, 4.66, 4.69

Assessment: Chapter 4 Test (*Test Bank* and *Instructor's Resource Binder* CD)

★ Kathy's personal favorite

4

Numerical Methods for Describing Data

© Hideji Watanabe/Sebun Photo/Getty Images

I n 2006, Medicare introduced a new prescription drug program. The article "Those Most in Need May Miss Drug Benefit Sign-Up" (*USA Today,* May 9, 2006) notes that just two weeks before the enrollment deadline only 24% of those eligible for low-income subsidies under this program had signed up. The article also gave the percentage of those eligible who had signed up in each of 49 states and the District of Columbia (information was not available for Vermont):

24	27	12	38	21	26	23	33	19	19	26	28
16	21	28	20	21	41	22	16	29	26	22	16
27	22	19	22	22	22	30	20	21	34	26	20
25	19	17	21	27	19	27	34	20	30	20	21
14	18										

What is a typical value for this data set? Is the nationwide figure of 24% representative of the individual state percentages? The enrollment percentages differ widely from state to state, ranging from a low of 12% (Arizona) to a high of 41% (Kentucky). How might we summarize this variability numerically? In this chapter we show how to calculate numerical summary measures that describe more precisely both the center and the extent of spread in a data set. In Section 4.1 we introduce the mean and the median, the two most widely used measures of the center of a distribution. The variance and the standard deviation are presented in Section 4.2 as measures of variability. In later sections we discuss several techniques for using such summary measures to describe other characteristics of the data.

4.1   Describing the Center of a Data Set

When describing numerical data, it is common to report a value that is representative of the observations. Such a number describes roughly where the data are located or "centered" along the number line, and it is called a measure of center. The two most popular measures of center are the *mean* and the *median*.

▪ The Mean

The **mean** of a set of numerical observations is just the familiar arithmetic average: the sum of the observations divided by the number of observations. It is helpful to have concise notation for the variable on which observations were made, the sample size, and the individual observations. Let

x = the variable for which we have sample data
n = the number of observations in the sample (the sample size)
x_1 = the first observation in the sample
x_2 = the second observation in the sample
$\vdots$
x_n = the nth (last) observation in the sample

For example, we might have a sample consisting of $n = 4$ observations on $x =$ battery lifetime (in hours):

$$x_1 = 5.9 \qquad x_2 = 7.3 \qquad x_3 = 6.6 \qquad x_4 = 5.7$$

Notice that the value of the subscript on x has no relationship to the magnitude of the observation. In this example, x_1 is just the first observation in the data set and not necessarily the smallest observation, and x_n is the last observation but not necessarily the largest.

The sum of $x_1, x_2, \ldots, x_n$ can be denoted by $x_1 + x_2 + \cdots + x_n$, but this is cumbersome. The Greek letter Σ is traditionally used in mathematics to denote summation. In particular, Σx denotes the sum of all the x values in the data set under consideration.*

D E F I N I T I O N

The **sample mean** of a sample of numerical observations $x_1, x_2, \ldots, x_n$, denoted by $\bar{x}$ is

$$\bar{x} = \frac{\text{sum of all observations in the sample}}{\text{number of observations in the sample}} = \frac{x_1 + x_2 + \cdots + x_n}{n} = \frac{\Sigma x}{n}$$

---

*It is also common to see Σx written as Σx_i or even as $\sum_{i=1}^{n} x_i$, but for simplicity we will usually omit the summation indices.

Example 4.1

Improving Knee Extension

● Increasing joint extension is one goal of athletic trainers. In a study to investigate the effect of a therapy that uses ultrasound and stretching (Trae Tashiro, Masters Thesis, University of Virginia, 2004) passive knee extension was measured after treatment. Passive knee extension (in degrees) is given for each of 10 participants in the study:

$$x_1 = 59 \qquad x_2 = 46 \qquad x_3 = 64 \qquad x_4 = 49 \qquad x_5 = 56$$
$$x_6 = 70 \qquad x_7 = 45 \qquad x_8 = 52 \qquad x_9 = 63 \qquad x_{10} = 52$$

The sum of these sample values is $59 + 46 + 64 + \cdots + 52 = 556$, and the sample mean passive knee extension is

$$\bar{x} = \frac{\sum x}{n} = \frac{556}{10} = 55.6$$

We would report 55.6° as a representative value of passive knee extension for this sample (even though there is no person in the sample that actually had a passive knee extension of 55.6°).

■

The data values in Example 4.1 were all integers, yet the mean was given as 55.6. It is common to use more digits of decimal accuracy for the mean. This allows the value of the mean to fall between possible observable values (e.g., the average number of children per family could be 1.8, whereas no single family will have 1.8 children).

The sample mean $\bar{x}$ is computed from sample observations, so it is a characteristic of the particular sample in hand. It is customary to use Roman letters to denote sample characteristics, as we have done with $\bar{x}$. Characteristics of the population are usually denoted by Greek letters. One of the most important of such characteristics is the population mean.

> ### DEFINITION
>
> The **population mean**, denoted by μ, is the average of all x values in the entire population.

For example, the true average fuel efficiency for all 600,000 cars of a certain type under specified conditions might be $\mu = 27.5$ mpg. A sample of $n = 5$ cars might yield efficiencies of 27.3, 26.2, 28.4, 27.9, 26.5, from which we obtain $\bar{x} = 27.26$ for this particular sample (somewhat smaller than μ). However, a second sample might give $\bar{x} = 28.52$, a third $\bar{x} = 26.85$, and so on. The value of $\bar{x}$ varies from sample to sample, whereas there is just one value for μ. We shall see subsequently how the value of $\bar{x}$ from a particular sample can be used to draw various conclusions about the value of μ.

● Data set available online

Example 4.2 illustrates how the value of $\bar{x}$ from a particular sample can differ from the value of μ and how the value of $\bar{x}$ differs from sample to sample.

Example 4.2   County Population Sizes

The 50 states plus the District of Columbia contain 3137 counties. Let x denote the number of residents of a county. Then there are 3137 values of the variable x in the population. The sum of these 3137 values is 293,655,404 (2004 Census Bureau estimate), so the population average value of x is

$$\mu = \frac{293,655,404}{3137} = 93,610.27 \text{ residents per county}$$

We used the Census Bureau web site to select three different samples at random from this population of counties, with each sample consisting of five counties. The results appear in Table 4.1, along with the sample mean for each sample. Not only are the three $\bar{x}$ values different from one another—because they are based on three different samples and the value of $\bar{x}$ depends on the x values in the sample—but also none of the three values comes close to the value of the population mean, μ. If we did not know the value of μ but had only Sample 1 available, we might use $\bar{x}$ as an *estimate* of μ, but our estimate would be far off the mark.

Table 4.1   Three Samples from the Population of All U.S. Counties (x = number of residents)

Sample 1		Sample 2		Sample 3	
County	x Value	County	x Value	County	x Value
Fayette, TX	22,513	Stoddard, MO	29,773	Chattahoochee, GA	13,506
Monroe, IN	121,013	Johnston, OK	10,440	Petroleum, MT	492
Greene, NC	20,219	Sumter, AL	14,141	Armstrong, PA	71,395
Shoshone, ID	12,827	Milwaukee, WI	928,018	Smith, MI	14,306
Jasper, IN	31,624	Albany, WY	31,473	Benton, MO	18,519
$\sum x = 208,196$		$\sum x = 1,013,845$		$\sum x = 118,218$	
$\bar{x} = 41,639.2$		$\bar{x} = 202,769.0$		$\bar{x} = 23,643.6$	

Alternatively, we could combine the three samples into a single sample with $n = 15$ observations:

$$x_1 = 22,513, \ldots, x_5 = 31,624, \ldots, x_{15} = 18,519$$
$$\sum x = 1,340,259$$
$$\bar{x} = \frac{1,340,259}{15} = 89,350.6$$

This value is closer to the value of μ but is still somewhat unsatisfactory as an estimate. The problem here is that the population of x values exhibits a lot of variability (the largest value is $x = 9,937,739$ for Los Angeles County, California, and the smallest value is $x = 52$ for Loving County, Texas, which evidently few people love). Therefore, it is difficult for a sample of 15 observations, let alone just 5, to be reasonably representative of the population. In Chapter 9 you will see how to take variability into account when deciding on a sample size.

■

Important concept

One potential drawback to the mean as a measure of center for a data set is that its value can be greatly affected by the presence of even a single *outlier* (an unusually large or small observation) in the data set.

....

Example 4.3   Number of Visits to a Class Web Site

● Forty students were enrolled in a section of a general education course in statistical reasoning during the fall quarter of 2002 at Cal Poly. The instructor made course materials, grades, and lecture notes available to students on a class web site, and course management software kept track of how often each student accessed any of the web pages on the class site. One month after the course began, the instructor requested a report that indicated how many times each student had accessed a web page on the class site. The 40 observations were:

20	37	4	20	0	84	14	36	5	331	19	0
0	22	3	13	14	36	4	0	18	8	0	26
4	0	5	23	19	7	12	8	13	16	21	7
13	12	8	42								

The sample mean for this data set is $\bar{x} = 23.10$. Figure 4.1 is a MINITAB dotplot of the data. Many would argue that 23.10 is not a very representative value for this sample, because 23.10 is larger than most of the observations in the data set—only 7 of 40 observations, or 17.5%, are larger than 23.10. The two outlying values of 84 and 331 (no, that wasn't a typo!) have a substantial impact on the value of $\bar{x}$.

Figure 4.1 A MINITAB dotplot of the data in Example 4.3

Number of Accesses

■

We now turn our attention to a measure of center that is not so sensitive to outliers —the median.

■ The Median ....

The median strip of a highway divides the highway in half, and the median of a numerical data set does the same thing for a data set. Once the data values have been listed in order from smallest to largest, the **median** is the middle value in the list, and it divides the list into two equal parts. Depending on whether the sample size n is even

♩♩ Step-by-step technology instructions available online   ● Data set available online

or odd, the process of determining the median is slightly different. When n is an odd number (say, 5), the sample median is the single middle value. But when n is even (say, 6), there are two middle values in the ordered list, and we average these two middle values to obtain the sample median.

DEFINITION

The **sample median** is obtained by first ordering the n observations from smallest to largest (with any repeated values included, so that every sample observation appears in the ordered list). Then

$$\text{sample median} = \begin{cases} \text{the single middle value if } n \text{ is odd} \\ \text{the average of the middle two values if } n \text{ is even} \end{cases}$$

Example 4.4   Web Site Data Revised

The sample size for the web site access data of Example 4.3 was $n = 40$, an even number. The median is the average of the 20th and 21st values (the middle two) in the ordered list of the data. Arranging the data in order from smallest to largest produces the following ordered list:

0	0	0	0	0	0	3	4	4	4	5	5
7	7	8	8	8	12	12	13	13	13	14	14
16	18	19	19	20	20	21	22	23	26	36	36
37	42	84	331								

The median can now be determined:

$$median = \frac{13 + 13}{2} = 13$$

Looking at the dotplot (Figure 4.1), we see that this value appears to be somewhat more typical of the data than $\bar{x} = 23.10$ is.

■

The sample mean can be sensitive to even a single value that lies far above or below the rest of the data. The value of the mean is pulled out toward such an outlying value or values. The median, on the other hand, is quite *in*sensitive to outliers. For example, the largest sample observation (331) in Example 4.3 can be increased by an arbitrarily large amount without changing the value of the median. Similarly, an increase in the second or third largest observations does not affect the median, nor would a decrease in several of the smallest observations.

This stability of the median is what sometimes justifies its use as a measure of center in some situations. For example, the article "Educating Undergraduates on Using Credit Cards" (Nellie Mae, 2005) reported that the mean credit card debt for undergraduate students in 2001 was $2327, whereas the median credit card debt was only $1770. In this case, the small percentage of students with unusually high credit card debt may be resulting in a mean that is not representative of a typical student credit card debt.

■ Comparing the Mean and the Median ....

Figure 4.2 presents several smoothed histograms that might represent either a distribution of sample values or a population distribution. Pictorially, the median is the value on the measurement axis that separates the histogram into two parts, with .5 (50%) of the area under each part of the curve. The mean is a bit harder to visualize. If the histogram were balanced on a triangle (a fulcrum), it would tilt unless the triangle was positioned at the mean. The mean is the balance point for the distribution.

Figure 4.2 The mean and the median.

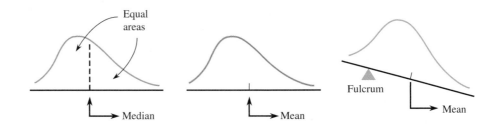

When the histogram is symmetric, the point of symmetry is both the dividing point for equal areas and the balance point, and the mean and the median are equal. However, when the histogram is unimodal (single-peaked) with a longer upper tail (positively skewed), the outlying values in the upper tail pull the mean up, so it generally lies above the median. For example, an unusually high exam score raises the mean but does not affect the median. Similarly, when a unimodal histogram is negatively skewed, the mean is generally smaller than the median (see Figure 4.3).

Figure 4.3 Relationship between the mean and the median.

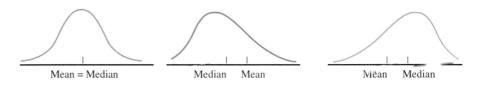

■ Trimmed Means ....

The extreme sensitivity of the mean to even a single outlier and the extreme insensitivity of the median to a substantial proportion of outliers can sometimes make both of them suspect as a measure of center. A *trimmed mean* is a compromise between these two extremes.

Which measures of center are resistant to outliers?

> ### DEFINITION
>
> A **trimmed mean** is computed by first ordering the data values from smallest to largest, deleting a selected number of values from each end of the ordered list, and finally averaging the remaining values.
>
> The **trimming percentage** is the percentage of values deleted from *each* end of the ordered list.

Sometimes the number of observations to be deleted from each end of the data set is specified. Then the corresponding trimming percentage is calculated as

$$\text{trimming percentage} = \left(\frac{\text{number deleted from each end}}{n} \right) \cdot 100$$

In other cases, the trimming percentage is specified and then used to determine how many observations to delete from each end, with

$$\text{number deleted from each end} = (\text{trimming percentage}) \cdot n$$

If the number of observations to be deleted from each end resulting from this calculation is not an integer, it can be rounded to the nearest integer (which changes the trimming percentage a bit).

....

Example 4.5 NBA Salaries

● *USA Today* (December 17, 2003) published salaries of NBA players for the 2003–2004 season. Salaries for the players of the Chicago Bulls were

Player	2003–2004 Salary
Antonio Davis	$12,000,000
Eddie Robinson	6,246,950
Jerome Williams	5,400,000
Scottie Pippen	4,917,000
Tyson Chandler	3,804,360
Marcus Fizer	3,727,000
Jay Williams	3,710,000
Eddy Curry	3,080,000
Jamal Crawford	2,578,000
Kirk Hinrich	2,098,600
Corie Blount	1,600,000
Kendall Gill	1,070,000
Chris Jefferies	840,360
Rick Brunson	813,679
Linton Johnson	366,931

A MINITAB dotplot of these data is shown in Figure 4.4. Because the data distribution is not symmetric and there are outliers, a trimmed mean is a reasonable choice for describing the center of this data set.

There are 15 observations in this data set. Deleting the two largest and the two smallest observations from the data set and then averaging the remaining values

Figure 4.4 A MINITAB dotplot of the data in Example 4.5.

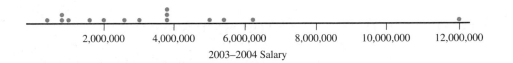

2003–2004 Salary

would result in a $\left(\dfrac{2}{15}\right)(100) = 13\%$ trimmed mean. For the Bulls salary data, the two largest salaries are \$12,000,000 and \$6,246,950 and the two smallest are \$813,679 and \$366,931. The average of the remaining 11 observations is

$$13\% \text{ trimmed mean} = \frac{5{,}400{,}000 + \cdots + 840{,}360}{11} = \frac{32{,}825{,}320}{11} = 2{,}984{,}120$$

The mean (\$3,483,525) is larger than the trimmed mean because of the unusually large values in the data set, and the trimmed mean is closer to the median salary value (\$3,080,000).

For the LA Lakers, the difference between the mean (\$4,367,343) and the 13% trimmed mean (\$2,233,061) is even more dramatic because during the 2003–2004 season one player on the Lakers earned over \$26 million and another player earned over \$13 million.

■

■ Categorical Data

The natural numerical summary quantities for a categorical data set are the relative frequencies for the various categories. Each relative frequency is the proportion (fraction) of responses that are in the corresponding category. Often there are only two possible responses (a *dichotomy*)—for example, male or female, does or does not have a driver's license, did or did not vote in the last election. It is convenient in such situations to label one of the two possible responses S (for success) and the other F (for failure). As long as further analysis is consistent with the labeling, it does not matter which category is assigned the S label. When the data set is a sample, the fraction of S's in the sample is called the *sample proportion of successes*.

> **DEFINITION**
>
> The **sample proportion of successes**, denoted by p, is
>
> $$p = \text{sample proportion of successes} = \frac{\text{number of S's in the sample}}{n}$$
>
> where S is the label used for the response designated as success.

Example 4.6 Can You Hear Me Now?

© Royalty-Free/Getty Images

It is not uncommon for a cell phone user to complain about the quality of his or her service provider. Suppose that each person in a sample of $n = 15$ cell phone users is asked if he or she is satisfied with the cell phone service. Each response is classified as S (satisfied) or F (not satisfied). The resulting data are:

$$\begin{array}{cccccccccc} S & F & S & S & S & F & F & S & S & F \\ S & S & S & F & F & & & & & \end{array}$$

This sample contains nine S's, so

$$p = \frac{9}{15} = .60$$

That is, 60% of the sample responses are S's. Of those surveyed, 60% are satisfied with their cell phone service.

■

The Greek letter π is used to denote the **population proportion of S's**.* We will see later how the value of p from a particular sample can be used to make inferences about π.

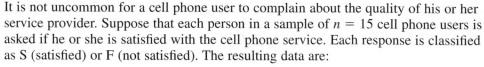

■ Exercises 4.1–4.14

Turn to the Expanded Answers Section for answers not shown next to exercises.

4.1 ● ▼ The Highway Loss Data Institute publishes data on repair costs resulting from a 5-mph crash test of a car moving forward into a flat barrier. The following table gives data for 10 midsize luxury cars tested in October 2002:

Model	Repair Cost
Audi A6	0
BMW 328i	0
Cadillac Catera	900
Jaguar X	1254
Lexus ES300	234
Lexus IS300	979
Mercedes C320	707
Saab 9-5	670
Volvo S60	769
Volvo S80	4194*

*Included cost to replace airbags, which deployed.

Compute the values of the mean and the median. Why are these values so different? Which of the mean and median is more representative of the data set, and why?

4.2 ● *USA Today* (May 9, 2006) published the accompanying average weekday circulation for the six month period ending March 31, 2006 for the top 20 newspapers in the country:

2,272,815	2,049,786	1,142,464	851,832	724,242
708,477	673,379	579,079	513,387	438,722
427,771	398,329	398,246	397,288	365,011
362,964	350,457	345,861	343,163	323,031

a. Which of the mean or the median do you think will be larger for this data set? Explain.
b. Compute the values of the mean and the median of this data set.   Mean = 683,315.2, median = 433,246.5

---

*Note that we are using the symbol π to represent a population proportion and *not* as the mathematical constant $\pi = 3.14\ldots$ Some statistics books use the symbol p for the population proportion and $\hat{p}$ for the sample proportion.

Bold exercises answered in back   ● Data set available online but not required   ▼ Video solution available

c. Of the mean and median, which does the best job of describing a typical value for this data set?
d. Explain why it would not be reasonable to generalize from this sample of 20 newspapers to the population of daily newspapers in the United States.

4.3 ● Bidri is a popular and traditional art form in India. Bidri articles (bowls, vessels, and so on) are made by casting from an alloy containing primarily zinc along with some copper. Consider the following observations on copper content (%) for a sample of Bidri artifacts in London's Victoria and Albert Museum ("Enigmas of Bidri," *Surface Engineering* [2005]: 333–339), listed in increasing order:

2.0	2.4	2.5	2.6	2.6	2.7	2.7	2.8	3.0	3.1
3.2	3.3	3.3	3.4	3.4	3.6	3.6	3.6	3.6	3.7
4.4	4.6	4.7	4.8	5.3	10.1				

a. Construct a dotplot for these data.
b. Calculate the mean and median copper content.
c. Will an 8% trimmed mean be larger or smaller than the mean for this data set? Explain your reasoning.

4.4 ● The chapter introduction gave the accompanying data on the percentage of those eligible for a low-income subsidy who had signed up for a Medicare drug plan in each of 49 states (information was not available for Vermont) and the District of Columbia (*USA Today,* May 9, 2006).

24	27	12	38	21	26	23	33
19	19	26	28	16	21	28	20
21	41	22	16	29	26	22	16
27	22	19	22	22	22	30	20
21	34	26	20	25	19	17	21
27	19	27	34	20	30	20	21
14	18						

a Compute the mean for this data set.   Mean = 23.42%
b. The article stated that nationwide, 24% of those eligible had signed up. Explain why the mean of this data set from Part (a) is not equal to 24. (No information was available for Vermont, but that is not the reason that the mean differs—the 24% was calculated excluding Vermont.)

4.5 ● The U.S. Department of Transportation reported the number of speeding-related crash fatalities for the 20 dates that had the highest number of these fatalities between 1994 and 2003 (*Traffic Safety Facts,* July 2005).

Date	Speeding-Related Fatalities	Date	Speeding-Related Fatalities
Jan 1	521	Aug 17	446
Jul 4	519	Dec 24	436
Aug 12	466	Aug 25	433
Nov 23	461	Sep 2	433
Jul 3	458	Aug 6	431
Dec 26	455	Aug 10	426
Aug 4	455	Sept 21	424
Aug 31	446	Jul 27	422
May 25	446	Sep 14	422
Dec 23	446	May 27	420

a. Compute the mean number of speeding-related fatalities for these 20 days.   Mean = 448.3 fatalities
b. Compute the median number of speeding-related fatalities for these 20 days.   Median = 446 fatalities
c. Explain why it is not reasonable to generalize from this sample of 20 days to the other 345 days of the year.

4.6 The ministry of Health and Long-Term Care in Ontario, Canada, publishes information on its web site (www.health.gov.on.ca) on the time that patients must wait for various medical procedures. For two cardiac procedures completed in fall of 2005 the following information was provided:

	Number of Completed Procedures	Median Wait Time (days)	Mean Wait Time (days)	90% Completed Within (days)
Angioplasty	847	14	18	39
Bypass surgery	539	13	19	42

a. The median wait time for angioplasty is greater than the median wait time for bypass surgery but the mean wait time is shorter for angioplasty than for bypass surgery. What does this suggest about the distribution of wait times for these two procedures?
b. Is it possible that another medical procedure might have a median wait time that is greater than the time reported for "90% completed within"? Explain.

Bold exercises answered in back   ● Data set available online but not required   ▼ Video solution available

4.7 ● ▼ Medicare's new medical plans offer a wide range of variations and choices for seniors when picking a drug plan (*San Luis Obispo Tribune,* November 25, 2005). The monthly cost for a stand-alone drug plan can vary from a low of $1.87 in Montana, Wyoming, North Dakota, South Dakota, Nebraska, Minnesota and Iowa to a high of $104.89. Here are the lowest and highest monthly premiums for stand-alone Medicare drug plans for each state:

State	$ Low	$ High
Alabama	14.08	69.98
Alaska	20.05	61.93
Arizona	6.14	64.86
Arkansas	10.31	67.98
California	5.41	66.08
Colorado	8.62	65.88
Connecticut	7.32	65.58
Delaware	6.44	68.91
District of Columbia	6.44	68.91
Florida	10.35	104.89
Georgia	17.91	73.17
Hawaii	17.18	64.43
Idaho	6.33	68.88
Illinois	13.32	65.04
Indiana	12.30	70.72
Iowa	1.87	99.90
Kansas	9.48	67.88
Kentucky	12.30	70.72
Louisiana	17.06	70.59
Maine	19.60	65.39
Maryland	6.44	68.91
Massachusetts	7.32	65.58
Michigan	13.75	65.69
Minnesota	1.87	99.90
Mississippi	11.60	70.59
Missouri	10.29	68.26
Montana	1.87	99.90
Nebraska	1.87	99.90
Nevada	6.42	64.63
New Hampshire	19.60	65.39
New Jersey	4.43	66.53
New Mexico	10.65	62.38
New York	4.10	85.02
North Carolina	13.27	65.03
North Dakota	1.87	99.90
Ohio	14.43	68.05
Oklahoma	10.07	70.79
Oregon	6.93	64.99

State	$ Low	$ High
Pennsylvania	10.14	68.61
Rhode Island	7.32	65.58
South Carolina	16.57	69.72
South Dakota	1.87	99.90
Tennessee	14.08	69.98
Texas	10.31	68.41
Utah	6.33	68.88
Vermont	7.32	65.58
Virginia	8.81	68.61
Washington	6.93	64.99
West Virginia	10.14	68.61
Wisconsin	11.42	63.23
Wyoming	1.87	99.90

Which of the following can be determined from the data? If it can be determined, calculate the requested value. If it cannot be determined, explain why not.
a. the median premium cost in Colorado
b. the number of plan choices in Virginia
c. the state(s) with the largest difference in cost between plans
d. the state(s) with the choice with the highest premium cost   Florida
e. the state for which the minimum premium cost is greatest   Alaska
f. the mean of the minimum cost of all states beginning with the letter "M" $9.0925

4.8 *Note: This exercise requires the use of a computer.* Refer to the Medicare drug plan premium data of Exercise 4.7.
a. Construct a dotplot or a stem-and-leaf display of the lowest premium cost data.
b. Based on the display in Part (a), which of the following would you expect to be the case for the lowest cost premium data?
  i. the mean will be less than the median
  ii. the mean will be approximately equal to the median   The mean would be about the same as the median.
  iii. the mean will be greater than the median
c. Compute the mean and median for the lowest cost premium data.   Mean = $9.459, median = $9.489
d. Construct an appropriate graphical display for the highest cost premium data.
e. Compute the mean and median for the highest cost premium data.   Mean = $72.85, median = $68.61

4.9 Houses in California are expensive, especially on the Central Coast where the air is clear, the ocean is blue, and the scenery is stunning. The median home price in San Luis Obispo County reached a new high in July 2004 soaring to $452,272 from $387,120 in March 2004. (*San Luis Obispo Tribune,* April 28, 2004). The article included two quotes from people attempting to explain why the median price had increased. Richard Watkins, chairman of the Central Coast Regional Multiple Listing Services was quoted as saying "There have been some fairly expensive houses selling, which pulls the median up." Robert Kleinhenz, deputy chief economist for the California Association of Realtors explained the volatility of house prices by stating: "Fewer sales means a relatively small number of very high or very low home prices can more easily skew medians." Are either of these statements correct? For each statement that is incorrect, explain why it is incorrect and propose a new wording that would correct any errors in the statement.

4.10 Consider the following statement: More than 65% of the residents of Los Angeles earn less than the average wage for that city. Could this statement be correct? If so, how? If not, why not?

4.11 ▼ A sample consisting of four pieces of luggage was selected from among those checked at an airline counter, yielding the following data on x = weight (in pounds):

$$x_1 = 33.5, x_2 = 27.3, x_3 = 36.7, x_4 = 30.5$$

Suppose that one more piece is selected; denote its weight by x_5. Find a value of x_5 such that $\bar{x}$ = sample median.   32

4.12 Suppose that 10 patients with meningitis received treatment with large doses of penicillin. Three days later, temperatures were recorded, and the treatment was considered successful if there had been a reduction in a patient's temperature. Denoting success by S and failure by F, the 10 observations are

S   S   F   S   S   S   F   F   S   S

a. What is the value of the sample proportion of successes?   .7
b. Replace each S with a 1 and each F with a 0. Then calculate $\bar{x}$ for this numerically coded sample. How does $\bar{x}$ compare to p?   .7; they are the same.
c. Suppose that it is decided to include 15 more patients in the study. How many of these would have to be S's to give $p = .80$ for the entire sample of 25 patients?

4.13 An experiment to study the lifetime (in hours) for a certain type of component involved putting 10 components into operation and observing them for 100 hr. Eight of the components failed during that period, and those lifetimes were recorded. The lifetimes of the two components still functioning after 100 hr are recorded as 100+. The resulting sample observations were

48   79   100+   35   92   86   57   100+   17   29

Which of the measures of center discussed in this section can be calculated, and what are the values of those measures?   Median = 68; 20% trimmed mean = 66.17

4.14 An instructor has graded 19 exam papers submitted by students in a class of 20 students, and the average so far is 70. (The maximum possible score is 100.) How high would the score on the last paper have to be to raise the class average by 1 point? By 2 points?

Bold exercises answered in back   ● Data set available online but not required   ▼ Video solution available

4.2 Describing Variability in a Data Set

Reporting a measure of center gives only partial information about a data set. It is also important to describe the spread of values about the center. The three different samples displayed in Figure 4.5 all have mean = median = 45. There is much variability in the first sample compared to the third sample. The second sample shows less variability than the first and more variability than the third; most of the variability in the second sample is due to the two extreme values being so far from the center.

Figure 4.5 Three samples with the same center and different amounts of variability.

Important concept

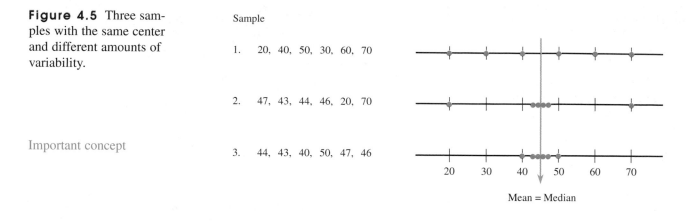

Sample

1. 20, 40, 50, 30, 60, 70

2. 47, 43, 44, 46, 20, 70

3. 44, 43, 40, 50, 47, 46

Mean = Median

The simplest numerical measure of variability is the range.

> **DEFINITION**
>
> The **range** of a data set is defined as
>
> range = largest observation − smallest observation

In general, more variability will be reflected in a larger range. However, variability is a characteristic of the entire data set, and each observation contributes to variability. The first two samples plotted in Figure 4.5 both have a range of 50, but there is less variability in the second sample.

■ Deviations from the Mean

The most common measures of variability describe the extent to which the sample observations deviate from the sample mean $\bar{x}$. Subtracting $\bar{x}$ from each observation gives a set of deviations from the mean.

> **DEFINITION**
>
> The n **deviations from the sample mean** are the differences
>
> $$(x_1 - \bar{x}), (x_2 - \bar{x}), \ldots, (x_n - \bar{x})$$

A particular deviation is positive if the corresponding x value is greater than $\bar{x}$ and negative if the x value is less than $\bar{x}$.

Example 4.7   Acrylamide Levels in French Fries

● Research by the Food and Drug Administration (FDA) shows that acrylamide (a possible cancer-causing substance) forms in high-carbohydrate foods cooked at high temperatures and that acrylamide levels can vary widely even within the same brand of food (Associated Press, December 6, 2002). FDA scientists analyzed McDonald's

● Data set available online

french fries purchased at seven different locations and found the following acryl-amide levels:

$$497 \quad 193 \quad 328 \quad 155 \quad 326 \quad 245 \quad 270$$

For this data set, $\sum x = 2014$ and $\bar{x} = 287.714$. Table 4.2 displays the data along with the corresponding deviations, formed by subtracting $\bar{x} = 287.714$ from each observation. Three of the deviations are positive because three of the observations are larger than $\bar{x}$. The negative deviations correspond to observations that are smaller than $\bar{x}$. Some of the deviations are quite large in magnitude (209.286 and -132.714, for example), indicating observations that are far from the sample mean.

Table 4.2   Deviations from the Mean for the Acrylamide Data

Observation (x)	Deviation (x − x̄)
497	209.286
193	−94.714
328	40.286
155	−132.714
326	38.286
245	−42.714
270	−17.714

■

This is why the mean is the balance point.

 In general, the greater the amount of variability in the sample, the larger the magnitudes (ignoring the signs) of the deviations. We now consider how to combine the deviations into a single numerical measure of variability. A first thought might be to calculate the average deviation, by adding the deviations together (this sum can be denoted compactly by $\sum(x - \bar{x})$) and then dividing by n. This does not work, though, because negative and positive deviations counteract one another in the summation.

Except for the effects of rounding in computing the deviations, it is always true that

$$\sum(x - \bar{x}) = 0$$

Since this sum is zero, the average deviation is always zero and so it cannot be used as a measure of variability.

■

 As a result of rounding, the value of the sum of the seven deviations in Example 4.7 is $\sum(x - \bar{x}) = .002$. If we used even more decimal accuracy in computing $\bar{x}$ the sum would be even closer to 0.

■ The Variance and Standard Deviation ....

The customary way to prevent negative and positive deviations from counteracting one another is to square them before combining. Then deviations with opposite signs but

See Teaching Tips page T4-1 for why we square deviations.

with the same magnitude, such as $+20$ and -20, make identical contributions to variability. The squared deviations are $(x_1 - \bar{x})^2, (x_2 - \bar{x})^2, \ldots, (x_n - \bar{x})^2$ and their sum is

$$(x_1 - \bar{x})^2 + (x_2 - \bar{x})^2 + \cdots + (x_n - \bar{x})^2 = \sum(x - \bar{x})^2$$

Common notation for is $\sum(x - \bar{x})^2$ is S_{xx}. Dividing this sum by the sample size n gives the average squared deviation. Although this seems to be a reasonable measure of variability, we use a divisor slightly smaller than n. (The reason for this will be explained later in this section and in Chapter 9.)

DEFINITION

The **sample variance**, denoted by s^2, is the sum of squared deviations from the mean divided by $n - 1$. That is,

$$s^2 = \frac{\sum(x - \bar{x})^2}{n - 1} = \frac{S_{xx}}{n - 1}$$

The **sample standard deviation** is the positive square root of the sample variance and is denoted by s.

A large amount of variability in the sample is indicated by a relatively large value of s^2 or s, whereas a value of s^2 or s close to 0 indicates a small amount of variability. Notice that whatever unit is used for x (such as pounds or seconds), the squared deviations and therefore s^2 are in squared units. Taking the square root gives a measure expressed in the same units as x. Thus, for a sample of heights, the standard deviation might be $s = 3.2$ in., and for a sample of textbook prices, it might be $s = \$12.43$.

....

Example 4.8   French Fries Revisited

Let's continue using the acrylamide data and the computed deviations from the mean given in Example 4.7 to obtain the sample variance and standard deviation. Table 4.3 shows the observations, deviations from the mean, and squared deviations.

Table 4.3   Deviations and Squared Deviations for the Acrylamide Data

Observation (x)	Deviation ($x - \bar{x}$)	Squared Deviation ($x - \bar{x}$)^2
497	209.286	43800.630
193	-94.714	8970.742
328	40.286	1622.962
155	-132.714	17613.006
326	38.286	1465.818
245	-42.714	1824.486
270	-17.714	313.786
		$\sum(x - \bar{x})^2 = 75{,}611.429$

Combining the squared deviations to compute the values of s^2 and s gives

$$\sum (x - \bar{x})^2 = S_{xx} = 75{,}611.429$$

and

$$s^2 = \frac{\sum (x - \bar{x})^2}{n - 1} = \frac{75{,}611.429}{7 - 1} = \frac{75{,}611.429}{6} = 12{,}601.904$$

$$s = \sqrt{12{,}601.904} = 112.258$$

■

The computation of s^2 can be a bit tedious, especially if the sample size is large. Fortunately, many calculators and computer software packages compute the variance and standard deviation upon request. One commonly used statistical computer package is MINITAB. The output resulting from using the MINITAB Describe command with the acrylamide data follows. MINITAB gives a variety of numerical descriptive measures, including the mean, the median, and the standard deviation.

Descriptive Statistics: acrylamide

Variable	N	Mean	Median	TrMean	StDev	SE Mean
acrylamide	7	287.7	270.0	287.7	112.3	42.4

Variable	Minimum	Maximum	Q1	Q3
acrylamide	155.0	497.0	193.0	328.0

The standard deviation can be informally interpreted as the size of a "typical" or "representative" deviation from the mean. Thus, in Example 4.8, a typical deviation from $\bar{x}$ is about 112.3; some observations are closer to $\bar{x}$ than 112.3 and others are farther away. We computed $s = 112.3$ in Example 4.8 without saying whether this value indicated a large or a small amount of variability. At this point, it is better to use s for comparative purposes than for an absolute assessment of variability. If we obtained a sample of acrylamide level values for a second set of McDonald's french fries and computed $s = 104.6$, then we would conclude that our original sample has more variability than our second sample.

There are measures of variability for the entire population that are analogous to s^2 and s for a sample. These measures are called the **population variance** and the **population standard deviation** and are denoted by σ^2 and σ, respectively. (We again use a lowercase Greek letter for a population characteristic.) As with s, the value of σ can be used for comparative purposes.

In many statistical procedures, we would like to use the value of σ, but unfortunately it is not usually known. Therefore, in its place we must use a value computed from the sample that we hope is close to σ (i.e., a good *estimate* of σ). We use the divisor $(n - 1)$ in s^2 rather than n because, on average, the resulting value tends to be a bit closer to σ^2. We will say more about this in Chapter 9.

An alternative rationale for using $(n - 1)$ is based on the property $\sum (x - \bar{x}) = 0$. Suppose that $n = 5$ and that four of the deviations are

$$x_1 - \bar{x} = -4 \qquad x_2 - \bar{x} = 6 \qquad x_3 - \bar{x} = 1 \qquad x_5 - \bar{x} = -8$$

Then, because the sum of these four deviations is -5, the remaining deviation must be $x_4 - \bar{x} = 5$ (so that the sum of all five is 0). Although there are five deviations, only four of them contain independent information about variability. More generally, once

See Teaching Tips
page T4-1 for explanation
of degrees of freedom.

any $(n - 1)$ of the deviations are available, the value of the remaining deviation is determined. The n deviations actually contain only $(n - 1)$ independent pieces of information about variability. Statisticians express this by saying that s^2 and s are based on $(n - 1)$ *degrees of freedom* (df).

■ The Interquartile Range

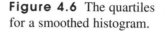

As with $\bar{x}$, the value of s can be greatly affected by the presence of even a single unusually small or large observation. The *interquartile range* is a measure of variability that is resistant to the effects of outliers. It is based on quantities called *quartiles*. The *lower quartile* separates the bottom 25% of the data set from the upper 75%, and the *upper quartile* separates the top 25% from the bottom 75%. The *middle quartile* is the median, and it separates the bottom 50% from the top 50%. Figure 4.6 illustrates the locations of these quartiles for a smoothed histogram.

Figure 4.6 The quartiles for a smoothed histogram.

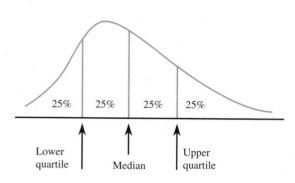

The quartiles for sample data are obtained by dividing the n ordered observations into a lower half and an upper half; if n is odd, the median is excluded from both halves. The two extreme quartiles are then the medians of the two halves. (Note: The median is only temporarily excluded for the purpose of computing quartiles. It is not excluded from the data set.)

What is the only resistant measure of spread?

> **D E F I N I T I O N** *
>
> **lower quartile** = median of the lower half of the sample
> **upper quartile** = median of the upper half of the sample
> (If n is odd, the median of the entire sample is excluded from both halves.)
>
> The **interquartile range (iqr)**, a measure of variability that is not as sensitive to the presence of outliers as the standard deviation, is given by
> **iqr = upper quartile − lower quartile**
>
> ———
> * There are several other sensible ways to define quartiles. Some calculators and software packages use an alternate definition.

The resistant nature of the interquartile range follows from the fact that up to 25% of the smallest sample observations and up to 25% of the largest sample observations can be made more extreme without affecting the value of the interquartile range.

Example 4.9   Hospital Cost-to-Charge Ratios

● The Oregon Department of Health Services publishes cost-to-charge ratios for hospitals in Oregon on its web site. The cost-to-charge ratio is computed as the ratio of the actual cost of care to what the hospital actually bills for care, and the ratio is usually expressed as a percentage. A cost-to-charge ratio of 60% means that the actual cost is 60% of what was billed. The ratios for inpatient services in 2002 at 31 hospitals in Oregon were:

68	76	60	88	69	80	75	67	71	100
63	62	71	74	64	48	100	72	65	50
72	100	63	45	54	60	75	57	74	84
83									

Figure 4.7 gives a stem-and-leaf display of the data.

Figure 4.7 MINITAB stem-and-leaf display of hospital cost-to-charge ratio data in Example 4.9.

```
Stem-and-Leaf Display: Cost-to-Charge
Stem-and-leaf of Cost-to-Charge N = 31
Leaf Unit = 1.0
   4  58
   5  04
   5  7
   6  002334
   6  5789
   7  112244
   7  556
   8  034
   8  8
   9
   9
  10  000
```

The sample size $n = 31$ is an odd number, so the median is excluded from both halves of the sample when computing the quartiles.

Ordered Data

Lower half   45 48 50 54 57 60 60 **62** 63 63 64 65 67 68 69
Median   **71**
Upper half   71 72 72 74 74 75 75 **76** 80 83 84 88 100 100 100

Each half of the sample contains 15 observations. The lower quartile is just the median of the lower half of the sample (62 for this data set), and the upper quartile is the median of the upper half (76 for this data set). This gives

lower quartile = 62
upper quartile = 76
iqr = 76 − 62 = 14

● Data set available online

The sample mean and standard deviation for this data set are 70.65 and 14.11, respectively. If we were to change the two smallest values from 45 and 48 to 25 and 28 (so that they still remain the two smallest values), the median and interquartile range would not be affected, whereas the mean and the standard deviation would change to 69.35 and 16.99, respectively.

■

The **population interquartile range** is the difference between the upper and lower population quartiles. If a histogram of the data set under consideration (whether a population or a sample) can be reasonably well approximated by a normal curve, then the relationship between the standard deviation (sd) and the interquartile range is roughly sd = iqr/1.35. A value of the standard deviation much larger than iqr/1.35 suggests a histogram with heavier (or longer) tails than a normal curve. For the hospital cost-to-charge data of Example 4.9, we had $s = 14.11$, whereas iqr/1.35 = 14/1.35 = 10.37. This suggests that the distribution of sample values is indeed heavy-tailed compared to a normal curve, as can be seen in the stem-and-leaf display.

Exercises 4.15–4.28
Turn to the Expanded Answers Section for answers not shown next to exercises.

4.15 A sample of $n = 5$ college students yielded the following observations on number of traffic citations for a moving violation during the previous year:

$$x_1 = 1 \qquad x_2 = 0 \qquad x_3 = 0 \qquad x_4 = 3 \qquad x_5 = 2$$

Calculate s^2 and s. $s^2 = 1.70, s = 1.3038$

4.16 Give two sets of five numbers that have the same mean but different standard deviations, and give two sets of five numbers that have the same standard deviation but different means.

4.17 Going back to school can be an expensive time for parents—second only to the Christmas holiday season in terms of spending (*San Luis Obispo Tribune*, August 18, 2005). Parents spend an average of $444 on their children at the beginning of the school year stocking up on clothes, notebooks, and even iPods. Of course, not every parent spends the same amount of money and there is some variation. Do you think a data set consisting of the amount spent at the beginning of the school year for each student at a particular elementary school would have a large or a small standard deviation? Explain.

4.18 The article "Rethink Diversification to Raise Returns, Cut Risk" (*San Luis Obispo Tribune*, January 21, 2006) included the following paragraph:

In their research, Mulvey and Reilly compared the results of two hypothetical portfolios and used actual data from 1994 to 2004 to see what returns they would achieve. The first portfolio invested in Treasury bonds, domestic stocks, international stocks, and cash. Its 10-year average annual return was 9.85 percent, and its volatility—measured as the standard deviation of annual returns—was 9.26 percent. When Mulvey and Reilly shifted some assets in the portfolio to include funds that invest in real estate, commodities, and options, the 10-year return rose to 10.55 percent while the standard deviation fell to 7.97 percent. In short, the more diversified portfolio had a slightly better return and much less risk.

Explain why the standard deviation is a reasonable measure of volatility and why it is reasonable to interpret a smaller standard deviation as meaning less risk.

4.19 ● The U.S. Department of Transportation reported the accompanying data on the number of speeding-related crash fatalities during holiday periods for the years from 1994 to 2003 (*Traffic Safety Facts*, July 20, 2005).

Speeding-Related Fatalities

Holiday Period	1994	1995	1996	1997	1998	1999	2000	2001	2002	2003
New Year's Day	141	142	178	72	219	138	171	134	210	70
Memorial Day	193	178	185	197	138	183	156	190	188	181
July 4th	178	219	202	179	169	176	219	64	234	184
Labor Day	183	188	166	179	162	171	180	138	202	189
Thanksgiving	212	198	218	210	205	168	187	217	210	202
Christmas	152	129	66	183	134	193	155	210	60	198

a. Compute the standard deviation for the New Year's Day data. $s = 50.058$

b. Without computing the standard deviation of the Memorial Day data, explain whether the standard deviation for the Memorial Day data would be larger or smaller than the standard deviation of the New Year's Day data.

c. Memorial Day and Labor Day are holidays that always occur on Monday and Thanksgiving always occurs on a Thursday, whereas New Year's Day, July 4th and Christmas do not always fall on the same day of the week every year. Based on the given data, is there more or less variability in the speeding-related crash fatality numbers from year to year for same day of the week holiday periods than for holidays that can occur on different days of the week? Support your answer with appropriate measures of variability.

4.20 The ministry of Health and Long-Term Care in Ontario, Canada publishes information on the time that patients must wait for various medical procedures on its web site (www.health.gov.on.ca). For two cardiac procedures completed in fall of 2005 the following information was provided:

Procedure	Number of Completed Procedures	Median Wait Time (days)	Mean Wait Time (days)	90% Completed Within (days)
Angioplasty	847	14	18	39
Bypass surgery	539	13	19	42

a. Which of the following must be true for the lower quartile of the data set consisting of the 847 wait times for angioplasty?

   i. The lower quartile is less than 14.

   ii. The lower quartile is between 14 and 18.

   iii. The lower quartile is between 14 and 39.

   iv. The lower quartile is greater than 39.

b. Which of the following must be true for the upper quartile of the data set consisting of the 539 wait times for bypass surgery?

   i. The upper quartile is less than 13.

   ii. The upper quartile is between 13 and 19.

   iii. The upper quartile is between 13 and 42.

   iv. The upper quartile is greater than 42.

c. Which of the following must be true for the 95th percentile of the data set consisting of the 539 wait times for bypass surgery?

   i. The 95th percentile is less than 13.

   ii. The 95th percentile is between 13 and 19.

   iii. The 95th percentile is between 13 and 42.

   iv. The 95th percentile is greater than 42.   The 95th percentile is greater than 42.

4.21 The accompanying table shows the low price, the high price, and the average price of homes in fifteen communities in San Luis Obispo County between January 1, 2004 and August 1, 2004 (*San Luis Obispo Tribune,* September 5, 2004).

Community	Average Price	Number Sold	Low	High
Cayucos	$937,366	31	$380,000	$2,450,000
Pismo Beach	$804,212	71	$439,000	$2,500,000
				(*continued*)

Bold exercises answered in back   ● Data set available online but not required   ▼ Video solution available

Community	Average Price	Number Sold	Low	High
Cambria	$728,312	85	$340,000	$2,000,000
Avila Beach	$654,918	16	$475,000	$1,375,000
Morro Bay	$606,456	114	$257,000	$2,650,000
Arroyo Grande	$595,577	214	$178,000	$1,526,000
Templeton	$578,249	89	$265,000	$2,350,000
San Luis Obispo	$557,628	277	$258,000	$2,400,000
Nipomo	$528,572	138	$263,000	$1,295,000
Los Osos	$511,866	123	$140,000	$3,500,000
Santa Margarita	$430,354	22	$290,000	$583,000
Atascadero	$420,603	270	$140,000	$1,600,000
Grover Beach	$416,405	97	$242,000	$720,000
Paso Robles	$412,584	439	$170,000	$1,575,000
Oceano	$390,354	59	$177,000	$1,350,000

a. Explain why the average price for the combined areas of Los Osos and Morro Bay is not just the average of $511,866 and $606,456.
b. Houses sold in Grover Beach and Paso Robles have very similar average prices. Based on the other information given, which is likely to have the higher standard deviation for price?
c. Consider houses sold in Grover Beach and Paso Robles. Based on the other information given, which is likely to have the higher median price?

4.22 ● Cost-to-charge ratios (see Example 4.9 for a definition of this ratio) were reported for the 10 hospitals in California with the lowest ratios (*San Luis Obispo Tribune,* December 15, 2002). These ratios represent the 10 hospitals with the highest markup, because for these hospitals, the actual cost was only a small percentage of the amount billed. The 10 cost-to-charge values (percentages) were

8.81	10.26	10.20	12.66	12.86	12.96
13.04	13.14	14.70	14.84		

a. Compute the variance and standard deviation for this data set.
b. If cost-to-charge data were available for all hospitals in California, would the standard deviation of this data set be larger or smaller than the standard deviation computed in

Part (a) for the 10 hospitals with the lowest cost-to-charge values? Explain.
c. Explain why it would not be reasonable to use the data from the sample of 10 hospitals in Part (a) to draw conclusions about the population of all hospitals in California.

4.23 ● In 1997 a woman sued a computer keyboard manufacturer, charging that her repetitive stress injuries were caused by the keyboard (*Genessey v. Digital Equipment Corporation*). The jury awarded about $3.5 million for pain and suffering, but the court then set aside that award as being unreasonable compensation. In making this determination, the court identified a "normative" group of 27 similar cases and specified a reasonable award as one within 2 standard deviations of the mean of the awards in the 27 cases. The 27 award amounts were (in thousands of dollars)

37	60	75	115	135	140	149	150
238	290	340	410	600	750	750	750
1050	1100	1139	1150	1200	1200	1250	1576
1700	1825	2000					

What is the maximum possible amount that could be awarded under the "2-standard deviations rule"?
$1,961,158

4.24 ● The Highway Loss Data Institute reported the following repair costs resulting from crash tests conducted in October 2002. The given data are for a 5-mph crash into a flat surface for both a sample of 10 moderately priced midsize cars and a sample of 14 inexpensive midsize cars.

Moderately Priced Midsize Cars	296	0	1085	148	1065
	0	0	341	184	370
Inexpensive Midsize Cars	513	719	364	295	305
	335	353	156	209	288
	0	0	397	243	

a. Compute the standard deviation and the interquartile range for the repair cost of the moderately priced midsize cars.   Standard deviation = $406.98, iqr = $370
b. Compute the standard deviation and the interquartile range for the repair cost of the inexpensive midsize cars.
c. Is there more variability in the repair cost for the moderately priced cars or for the inexpensive midsize cars? Justify your choice.
d. Compute the mean repair cost for each of the two types of cars.
e. Write a few sentences comparing repair cost for moderately priced and inexpensive midsize cars. Be sure to include information about both center and variability.

Bold exercises answered in back   ● Data set available online but not required   ▼ Video solution available

4.25 ● ▼ The paper "Total Diet Study Statistics on Element Results" (Food and Drug Administration, April 25, 2000) gave information on sodium content for various types of foods. Twenty-six tomato catsups were analyzed. Data consistent with summary quantities given in the paper were

Sodium content (mg/kg)

12,148	10,426	10,912	9116	13,226	11,663
11,781	10,680	8457	10,788	12,605	10,591
11,040	10,815	12,962	11,644	10,047	10,478
10,108	12,353	11,778	11,092	11,673	8758
11,145	11,495				

Compute the values of the quartiles and the interquartile range.   1st quartile = 10478, 3rd quartile = 11778, interquartile range = 1300

4.26 ● The paper referenced in Exercise 4.25 also gave summary quantities for sodium content (in milligrams per kilogram) of chocolate pudding made from instant mix:

3099 3112 2401 2824 2682 2510 2297 3959
3068 3700

a. Compute the mean, the standard deviation, and the interquartile range for sodium content of these chocolate puddings.
b. Based on the interquartile range, is there more or less variability in sodium content for the chocolate pudding data than for the tomato catsup data of Exercise 4.25?

4.27 ● Example 4.9 gave 2002 cost-to-charge ratios for inpatient services at 31 hospitals in Oregon. The same data source also provided the cost-to-charge ratios for outpatient services at these same 31 hospitals:

75 54 50 75 58 56 56 62 66 69 100 57
45 51 53 45 84 45 48 39 51 51 65 100
63 96 54 53 67 52 71

a. Use the given data to compute the quartiles and the interquartile range for the outpatient cost-to-charge ratio.
b. How does the interquartile range compare to the interquartile range for the inpatient cost-to-charge ratio from Example 4.9? Is there more variability in the inpatient or the outpatient ratio?

4.28 ● The standard deviation alone does not measure relative variation. For example, a standard deviation of $1 would be considered large if it is describing the variability from store to store in the price of an ice cube tray. On the other hand, a standard deviation of $1 would be considered small if it is describing store-to-store variability in the price of a particular brand of freezer. A quantity designed to give a relative measure of variability is the *coefficient of variation*. Denoted by CV, the coefficient of variation expresses the standard deviation as a percentage of the mean. It is defined by the formula $CV = 100\left(\dfrac{s}{\bar{x}}\right)$.
Consider two samples. Sample 1 gives the actual weight (in ounces) of the contents of cans of pet food labeled as having a net weight of 8 oz. Sample 2 gives the actual weight (in pounds) of the contents of bags of dry pet food labeled as having a net weight of 50 lb. The weights for the two samples are:

Sample 1	8.3	7.1	7.6	8.1	7.6
	8.3	8.2	7.7	7.7	7.5
Sample 2	52.3	50.6	52.1	48.4	48.8
	47.0	50.4	50.3	48.7	48.2

a. For each of the given samples, calculate the mean and the standard deviation.
b. Compute the coefficient of variation for each sample. Do the results surprise you? Why or why not?

Bold exercises answered in back   ● Data set available online but not required   ▼ Video solution available

4.3 Summarizing a Data Set: Boxplots

In Sections 4.1 and 4.2, we looked at ways of describing the center and variability of a data set using numerical measures. It would be nice to have a method of summarizing data that gives more detail than just a measure of center and spread and yet less detail than a stem-and-leaf display or histogram. A *boxplot* is one such technique. It is compact, yet it provides information about the center, spread, and symmetry or skewness of the data. We consider two types of boxplots: the skeletal boxplot and the modified boxplot.

Construction of a Skeletal Boxplot

1. Draw a horizontal (or vertical) measurement scale.
2. Construct a rectangular box with a left (or lower) edge at the lower quartile and a right (or upper) edge at the upper quartile. The box width is then equal to the iqr.
3. Draw a vertical (or horizontal) line segment inside the box at the location of the median.
4. Extend horizontal (or vertical) line segments, called whiskers, from each end of the box to the smallest and largest observations in the data set.

Example 4.10 Revisiting Hospital Cost-to-Charge Ratios

Let's reconsider the cost-to-charge data for hospitals in Oregon (Example 4.9). The ordered observations are:

Ordered Data
Lower half   45 48 50 54 57 60 60 **62** 63 63 64 65   67   68   69
Median   **71**
Upper half   71 72 72 74 74 75 75 **76** 80 83 84 88 100 100 100

To construct a boxplot of these data, we need the following information: the smallest observation, the lower quartile, the median, the upper quartile, and the largest observation. This collection of summary measures is often referred to as the **five-number summary**. For this data set (see Example 4.9), we have

> smallest observation = 45
> lower quartile = median of the lower half = 62
> median = 16th observation in the ordered list = 71
> upper quartile = median of the upper half = 76
> largest observation = 100

Figure 4.8 shows the corresponding boxplot. The median line is somewhat closer to the upper edge of the box than to the lower edge, suggesting a concentration of values in the upper part of the middle half. The upper whisker is a bit longer than the lower whisker. These conclusions are consistent with the stem-and-leaf display of Figure 4.7.

Figure 4.8 Skeletal boxplot for the cost-to-charge data of Example 4.10.

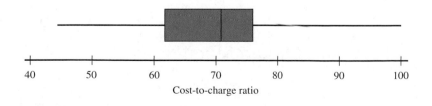

The sequence of steps used to construct a skeletal boxplot is easily modified to give information about outliers.

See Teaching Tips page T4-2
for note on John Tukey.

DEFINITION

An observation is an **outlier** if it is more than 1.5(iqr) away from the nearest quartile (the nearest end of the box).

An outlier is **extreme** if it is more than 3(iqr) from the nearest end of the box, and it is **mild** otherwise.

A **modified boxplot** represents mild outliers by shaded circles and extreme outliers by open circles, and the whiskers extend on each end to the most extreme observations that are *not* outliers.

Construction of a Modified Boxplot

1. Draw a horizontal (or vertical) measurement scale.
2. Construct a rectangular box with a left (or lower) edge is at the lower quartile and whose right (or upper) edge is at the upper quartile. The box width is then equal to the iqr.
3. Draw a vertical (or horizontal) line segment inside the box at the location of the median.
4. Determine if there are any mild or extreme outliers in the data set.
5. Draw whiskers that extend from each end of the box to the most extreme observation that is *not* an outlier.
6. Draw a solid circle to mark the location of any mild outliers in the data set.
7. Draw an open circle to mark the location of any extreme outliers in the data set.

Example 4.11 Golden Rectangles

● The accompanying data came from an anthropological study of rectangular shapes (*Lowie's Selected Papers in Anthropology*, Cora Dubios, ed [Berkeley, CA: University of California Press, 1960]: 137–142). Observations were made on the variable x = width/length for a sample of $n = 20$ beaded rectangles used in Shoshoni Indian leather handicrafts:

| .553 | .570 | .576 | .601 | .606 | .606 | .609 | .611 | .615 | .628 |
| .654 | .662 | .668 | .670 | .672 | .690 | .693 | .749 | .844 | .933 |

The quantities needed for constructing the modified boxplot follow:

median = .641   iqr = .681 − .606 = .075
lower quartile = .606   1.5(iqr) = .1125
upper quartile = .681   3(iqr) = .225

Thus,

(upper quartile) + 1.5(iqr) = .681 + .1125 = .7935
(lower quartile) − 1.5(iqr) = .606 − .1125 = .4935

♩♩ Step-by-step technology instructions available online   ● Data set available online

So 0.844 and 0.933 are both outliers on the upper end (because they are larger than 0.7935), and there are no outliers on the lower end (because no observations are smaller than 0.4935). Because

$$(\text{upper quartile}) + 3(\text{iqr}) = 0.681 + 0.225 = 0.906$$

0.933 is an extreme outlier and 0.844 is only a mild outlier. The upper whisker extends to the largest observation that is not an outlier, 0.749, and the lower whisker extends to 0.553. The boxplot is presented in Figure 4.9. The median line is not at the center of the box, so there is a slight asymmetry in the middle half of the data. However, the most striking feature is the presence of the two outliers. These two *x* values considerably exceed the "golden ratio" of 0.618, used since antiquity as an aesthetic standard for rectangles.

Figure 4.9 Boxplot for the rectangle data in Example 4.11.

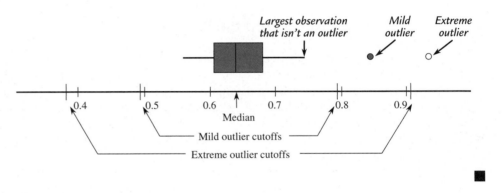

■

··

Example 4.12 State Gross Products

● *Highway Statistics 2001* (Federal Highway Administration) gave selected descriptive measures for the 50 U.S. states and the District of Columbia. The 51 observations for gross state product (in billions of dollars) and per capita state product (in dollars) are as follows:

Gross Product

110	24	134	62	1119	142	142	34	54	419	254	40
31	426	174	85	77	107	129	32	165	239	295	161
62	163	20	52	63	41	319	48	707	236	17	341
82	105	364	30	100	21	160	646	60	16	231	193
40	158	18									

Per Capita GP

25,282	39,024	28,712	24,429	34,238	35,777	43,385	45,699
103,647	28,106	33,259	33,813	28,183	35,294	29,452	29,710
29,178	27,199	29,567	25,641	32,164	38,900	30,041	34,087
22,537	29,974	22,727	31,306	36,124	34,570	39,402	27,682
38,934	31,275	26,645	30,343	34,558	31,993	30,328	30,364
26,042	28,728	29,450	32,772	28,588	27,073	34,028	33,931
22,075	30,257	37,500					

● Data set available online

Figure 4.10(a) shows a MINITAB boxplot for gross state product. Note that the upper whisker is longer than the lower whisker and that there are three outliers on the high end (from least extreme to most extreme, they are Texas, New York, and California—not surprising, because these are the three states with the largest population sizes). Figure 4.10(b) shows a MINITAB boxplot of the per capita gross state product (state gross product divided by population size). For this boxplot, the upper whisker is longer than the lower whisker and there are only two outliers on the upper end. The two outliers in this data set are Delaware and Washington, D.C., indicating that, although the three largest states have the highest gross state product, they are not the states with the highest *per capita* gross state product.

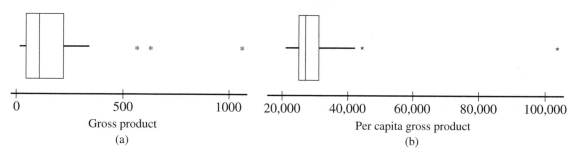

Figure 4.10 MINITAB boxplots for the data of Example 4.12: (a) gross state product; (b) per capita gross state product.

Note that MINITAB does not distinguish between mild outliers and extreme outliers in the boxplot. For the per capita gross state product data,

$$\text{lower quartile} = 28{,}183$$
$$\text{upper quartile} = 34{,}558$$
$$\text{iqr} = 34{,}558 - 28{,}183 = 9375$$

Then

$$1.5(\text{iqr}) = 9562.5$$
$$3(\text{iqr}) = 19{,}125$$

The observation for Delaware (45,699) would be a mild outlier, because it is more than 1.5(iqr) but less than 3(iqr) from the upper quartile. The observation for Washington, D.C. (103,647), is an extreme outlier, because it is more than 3(iqr) away from the upper quartile.

For the gross product data,

$$\text{lower quartile} = 41$$
$$\text{upper quartile} = 231$$
$$\text{iqr} = 231 - 41 = 190$$
$$1.5(\text{iqr}) = 285 \qquad 3(\text{iqr}) = 570$$

so the observations for Texas (646) and New York (707) are mild outliers and the observation for California (1119) is an extreme outlier.

■

With two or more data sets consisting of observations on the same variable (e.g., fuel efficiencies for four types of car or weight gains for a control group and a treatment group), **comparative boxplots** (more than one boxplot drawn using the same scale) convey initial impressions concerning similarities and differences between the data sets.

Example 4.13 NBA Salaries Revisited

The 2003–2004 salaries of NBA players published in *USA Today* (December 17, 2003) were used to construct the comparative boxplot of the salary data for five teams shown in Figure 4.11.

Figure 4.11 Comparative boxplot for salaries for five NBA teams.

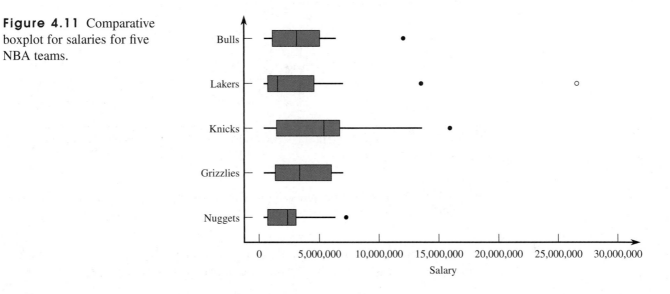

The comparative boxplot reveals some interesting similarities and differences in the salary distributions of the five teams. The minimum salary is about the same for all five teams. The median salary was lowest for the Lakers—in fact the median for the Lakers is about the same as the lower quartile for the Knicks and the Grizzlies, indicating that half of the players on the Lakers have salaries less than about $1.5 million, whereas only about 25% of the Knicks and the Grizzlies have salaries less than about $1.5 million. On the other hand, the Lakers had the player with by far the highest salary and two of the players with the highest three salaries. The Grizzlies were the only team that did not have any salary outliers.

■

Exercises 4.29–4.35
Turn to the Expanded Answers Section for answers not shown next to exercises.

4.29 Based on a large national sample of working adults, the U.S. Census Bureau reports the following information on travel time to work for those who do not work at home:

  lower quartile = 7 min
  median = 18 min
  upper quartile = 31 min

Also given was the mean travel time, which was reported as 22.4 min.

a. Is the travel time distribution more likely to be approximately symmetric, positively skewed, or negatively skewed? Explain your reasoning based on the given summary quantities.

Bold exercises answered in back   ● Data set available online but not required   ▼ Video solution available

b. Suppose that the minimum travel time was 1 min and that the maximum travel time in the sample was 205 min. Construct a skeletal boxplot for the travel time data.
c. Were there any mild or extreme outliers in the data set? How can you tell?

4.30 ● The technical report "Ozone Season Emissions by State" (U.S. Environmental Protection Agency, 2002) gave the following nitrous oxide emissions (in thousands of tons) for the 48 states in the continental U.S. states:

76	22	40	7	30	5	6	136	72	33	0
89	136	39	92	40	13	27	1	63	33	60
27	16	63	32	20	2	15	36	19	39	130
40	4	85	38	7	68	151	32	34	0	6
43	89	34	0							

Use these data to construct a boxplot that shows outliers. Write a few sentences describing the important characteristics of the boxplot.

4.31 ● ▼ A report from Texas Transportation Institute (Texas A&M University System 2005) on Congestion Reduction Strategies looked into the extra travel time (due to traffic congestion) for commute travel per traveler per year in hours for different urban areas. Below are the data for urban areas that had a population of over 3 million for the year 2002.

Urban Area	Extra Hours per Traveler per Year
Los Angeles	98
San Francisco	75
Washington DC	66
Atlanta	64
Houston	65
Dallas, Fort Worth	61
Chicago	55
Detroit	54
Miami	48
Boston	53
New York	50
Phoenix	49
Philadelphia	40

a. Compute the mean and median values for extra travel hours. Based on the values of the mean and median, is the distribution of extra travel hours likely to be approximately symmetric, positively skewed, or negatively skewed?

b. Construct a modified boxplot for these data and comment on any interesting features of the plot.

4.32 ● Shown here are the number of auto accidents per year for every 1000 people in each of 40 occupations (*Knight Ridder Tribune,* June 19, 2004):

Occupation	Accidents per 1000	Occupation	Accidents per 1000
Student	152	Banking, finance	89
Physician	109	Customer service	88
Lawyer	106	Manager	88
Architect	105	Medical support	87
Real estate broker	102	Computer-related	87
Enlisted military	99	Dentist	86
Social worker	98	Pharmacist	85
Manual laborer	96	Proprietor	84
Analyst	95	Teacher, professor	84
Engineer	94	Accountant	84
Consultant	94	Law enforcement	79
Sales	93	Physical therapist	78
Military officer	91	Veterinarian	78
Nurse	90	Clerical, secretary	77
School administrator	90	Clergy	76
Skilled laborer	90	Homemaker	76
Librarian	90	Politician	76
Creative arts	90	Pilot	75
Executive	89	Firefighter	67
Insurance agent	89	Farmer	43

a. Would you recommend using the standard deviation or the iqr as a measure of variability for this data set?
b. Are there outliers in this data set? If so, which observations are mild outliers? Which are extreme outliers?
c. Draw a modified boxplot for this data set.
d. If you were asked by an insurance company to decide which, if any, occupations should be offered a professional discount on auto insurance, which occupations would you recommend? Explain.

4.33 ● ▼ The paper "Relationship Between Blood Lead and Blood Pressure Among Whites and African Americans" (a technical report published by Tulane University School of Public Health and Tropical Medicine, 2000) gave summary quantities for blood lead level (in micrograms per deciliter) for a sample of whites and a sample of African Americans. Data consistent with the given summary quantities follow:

Whites	8.3	0.9	2.9	5.6	5.8	5.4	1.2
	1.0	1.4	2.1	1.3	5.3	8.8	6.6
	5.2	3.0	2.9	2.7	6.7	3.2	
African	4.8	1.4	0.9	10.8	2.4	0.4	5.0
Americans	5.4	6.1	2.9	5.0	2.1	7.5	3.4
	13.8	1.4	3.5	3.3	14.8	3.7	

a. Compute the values of the mean and the median for blood lead level for the sample of African Americans. Which of the mean or the median is larger? What characteristic of the data set explains the relative values of the mean and the median?

b. Construct a comparative boxplot for blood lead level for the two samples. Write a few sentences comparing the blood lead level distributions for the two samples.

4.34 ● The article "Compression of Single-Wall Corrugated Shipping Containers Using Fixed and Floating Test Platens" (*Journal of Testing and Evaluation* [1992]: 318–320) described an experiment in which several different types of boxes were compared with respect to compressive strength. Consider the following observations on four different types of boxes (these data are consistent with summary quantities given in the paper):

Type of Box	Compression Strength (lb)					
1	655.5	788.3	734.3	721.4	679.1	699.4
2	789.2	772.5	786.9	686.1	732.1	774.8
3	737.1	639.0	696.3	671.7	717.2	727.1
4	535.1	628.7	542.4	559.0	586.9	520.0

Construct a boxplot for each type of box. Use the same scale for all four boxplots. Discuss the similarities and differences in the boxplots.

4.35 ● Blood cocaine concentration (in milligrams per liter) was determined for both a sample of individuals who had died from cocaine-induced excited delirium and a sample of those who had died from a cocaine overdose without excited delirium. The accompanying data are consistent with summary values given in the paper "Fatal Excited Delirium Following Cocaine Use" (*Journal of Forensic Sciences* [1997]: 25–31):

Excited Delirium

0	0	0	0	0.1	0.1	0.1	0.1	0.2	0.2
0.3	0.3	0.3	0.4	0.5	0.7	0.7	1.0	1.5	2.7
2.8	3.5	4.0	8.9	9.2	11.7	21.0			

No Excited Delirium

0	0	0	0	0	0.1	0.1	0.1	0.1
0.2	0.2	0.2	0.3	0.3	0.3	0.4	0.5	0.5
0.6	0.8	0.9	1.0	1.2	1.4	1.5	1.7	2.0
3.2	3.5	4.1	4.3	4.8	5.0	5.6	5.9	6.0
6.4	7.9	8.3	8.7	9.1	9.6	9.9	11.0	11.5
12.2	12.7	14.0	16.6	17.8				

a. Determine the median, quartiles, and interquartile range for each of the two samples.
b. Are there any outliers in either sample? Any extreme outliers?
c. Construct a comparative boxplot, and use it as a basis for comparing and contrasting the two samples.

4.4 Interpreting Center and Variability: Chebyshev's Rule, the Empirical Rule, and z Scores

The mean and standard deviation can be combined to obtain informative statements about how the values in a data set are distributed and about the relative position of a particular value in a data set. To do this, it is useful to be able to describe how far away a particular observation is from the mean in terms of the standard deviation. For example, we might say that an observation is 2 standard deviations above the mean or that an observation is 1.3 standard deviations below the mean.

Example 4.14 Standardized Test Scores

Consider a data set of scores on a standardized test with a mean and standard deviation of 100 and 15, respectively. We can make the following statements:

1. Because $100 - 15 = 85$, we say that a score of 85 is "1 standard deviation *below* the mean." Similarly, $100 + 15 = 115$ is "1 standard deviation *above* the mean" (see Figure 4.12).
2. Because 2 standard deviations is $2(15) = 30$, and $100 + 30 = 130$ and $100 - 30 = 70$, scores between 70 and 130 are those *within* 2 standard deviations of the mean (see Figure 4.13).
3. Because $100 + (3)(15) = 145$, scores above 145 exceed the mean by more than 3 standard deviations.

Figure 4.12 Values within 1 standard deviation of the mean (Example 4.14).

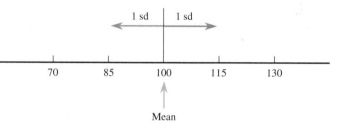

Figure 4.13 Values within 2 standard deviations of the mean (Example 4.14).

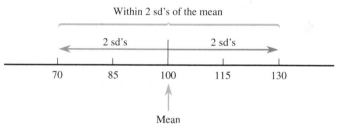

Sometimes in published articles, the mean and standard deviation are reported, but a graphical display of the data is not given. However, by using a result called Chebyshev's Rule, we can get a sense of the distribution of data values based on our knowledge of only the mean and standard deviation.

Chebyshev's Rule

Consider any number k, where $k \geq 1$. Then the percentage of observations that are within k standard deviations of the mean is at least $100\left(1 - \dfrac{1}{k^2}\right)\%$. Substituting selected values of k gives the following results.

(continued)

Number of Standard Deviations, k	$1 - \dfrac{1}{k^2}$	Percentage Within k Standard Deviations of the Mean
2	$1 - \dfrac{1}{4} = .75$	at least 75%
3	$1 - \dfrac{1}{9} = .89$	at least 89%
4	$1 - \dfrac{1}{16} = .94$	at least 94%
4.472	$1 - \dfrac{1}{20} = .95$	at least 95%
5	$1 - \dfrac{1}{25} = .96$	at least 96%
10	$1 - \dfrac{1}{100} = .99$	at least 99%

Example 4.15 Child Care for Preschool Kids

The article "Piecing Together Child Care with Multiple Arrangements: Crazy Quilt or Preferred Pattern for Employed Parents of Preschool Children?" (*Journal of Marriage and the Family* [1994]: 669–680) examined various modes of care for preschool children. For a sample of families with one such child, it was reported that the mean and standard deviation of child care time per week were approximately 36 hr and 12 hr, respectively. Figure 4.14 displays values that are 1, 2, and 3 standard deviations from the mean.

Figure 4.14 Measurement scale for child care time (Example 4.15).

0	12	24	36	48	60	72
$\bar{x} - 3s$	$\bar{x} - 2s$	$\bar{x} - s$	$\bar{x}$	$\bar{x} + s$	$\bar{x} + 2s$	$\bar{x} + 3s$

Chebyshev's Rule allows us to assert the following:

1. At least 75% of the sample observations must be between 12 and 60 hours (within 2 standard deviations of the mean).
2. Because at least 89% of the observations must be between 0 and 72, at most 11% are outside this interval. Time cannot be negative, so we conclude that at most 11% of the observations exceed 72.
3. The values 18 and 54 are 1.5 standard deviations to either side of $\bar{x}$, so using $k = 1.5$ in Chebyshev's Rule implies that at least 55.6% of the observations must be between these two values. Thus, at most 44.4% of the observations are less than 18—*not* at most 22.2%, because the distribution of values may not be symmetric.

Because Chebyshev's Rule is applicable to any data set (distribution), whether symmetric or skewed, we must be careful when making statements about the proportion above a particular value, below a particular value, or inside or outside an interval that is not centered at the mean. The rule must be used in a conservative fashion. There is another aspect of this conservatism. Whereas the rule states that at least 75% of the observations are within 2 standard deviations of the mean, in many data sets substantially more than 75% of the values satisfy this condition. The same sort of understatement is frequently encountered for other values of k (numbers of standard deviations).

Example 4.16 IQ Scores

Figure 4.15 gives a stem-and-leaf display of IQ scores of 112 children in one of the early studies that used the Stanford revision of the Binet–Simon intelligence scale (*The Intelligence of School Children*, L. M. Terman [Boston: Houghton-Mifflin, 1919]).

Summary quantities include

$$\bar{x} = 104.5 \qquad s = 16.3 \qquad 2s = 32.6 \qquad 3s = 48.9$$

Figure 4.15 Stem-and leaf display of IQ scores used in Example 4.16.

```
 6 | 1
 7 | 25679
 8 | 0000124555668
 9 | 00001123334466666778889
10 | 0001122222333566677778899999
11 | 000011223333444444477899
12 | 01111123445669
13 | 006
14 | 26                              Stem:  Tens
15 | 2                               Leaf:  Ones
```

In Figure 4.15, all observations that are within two standard deviations of the mean are shown in blue. Table 4.4 shows how Chebyshev's Rule can sometimes considerably understate actual percentages.

Table 4.4   Summarizing the Distribution of IQ Scores

k = Number of sd's	$\bar{x} \pm ks$	Chebyshev	Actual
2	71.9 to 137.1	at least 75%	*the blue* → 96% (108)
2.5	63.7 to 145.3	at least 84%	*leaves in Figure 4.15* 97% (109)
3	55.6 to 153.4	at least 89%	100% (112)

■

■ Empirical Rule

The fact that statements based on Chebyshev's Rule are frequently conservative suggests that we should look for rules that are less conservative and more precise. The most useful such rule is the **Empirical Rule**, which can be applied whenever the distribution of data values can be reasonably well described by a normal curve (distributions that are "mound" shaped).

The Empirical Rule

If the histogram of values in a data set can be reasonably well approximated by a normal curve, then

Approximately 68% of the observations are within 1 standard deviation of the mean.
Approximately 95% of the observations are within 2 standard deviations of the mean.
Approximately 99.7% of the observations are within 3 standard deviations of the mean.

The Empirical Rule makes "approximately" instead of "at least" statements, and the percentages for $k = 1$, 2, and 3 standard deviations are much higher than those of Chebyshev's Rule. Figure 4.16 illustrates the percentages given by the Empirical Rule. In contrast to Chebyshev's Rule, dividing the percentages in half is permissible, because a normal curve is symmetric.

Figure 4.16 Approximate percentages implied by the Empirical Rule.

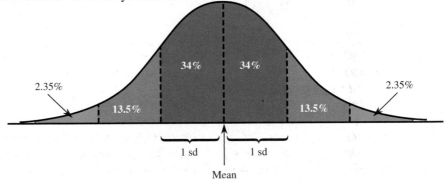

Example 4.17 Heights of Mothers and the Empirical Rule

© The Image Bank/Paul Thomas/Getty Images

One of the earliest articles to argue for the wide applicability of the normal distribution was "On the Laws of Inheritance in Man. I. Inheritance of Physical Characters" (*Biometrika* [1903]: 375–462). Among the data sets discussed in the article was one consisting of 1052 measurements of the heights of mothers. The mean and standard deviation were

$$\bar{x} = 62.484 \text{ in.} \qquad s = 2.390 \text{ in.}$$

The data distribution was described as approximately normal. Table 4.5 contrasts actual percentages with those obtained from Chebyshev's Rule and the Empirical Rule.

Table 4.5 Summarizing the Distribution of Mothers' Heights

Number of sd's	Interval	Actual	Empirical Rule	Chebyshev Rule
1	60.094 to 64.874	72.1%	68%	At least 0%
2	57.704 to 67.264	96.2%	95%	At least 75%
3	55.314 to 69.654	99.2%	99.7%	At least 89%

Clearly, the Empirical Rule is much more successful and informative in this case than Chebyshev's Rule.

■

Our detailed study of the normal distribution and areas under normal curves in Chapter 7 will enable us to make statements analogous to those of the Empirical Rule for values other than $k = 1, 2,$ or 3 standard deviations. For now, note that it is unusual to see an observation from a normally distributed population that is farther than 2 standard deviations from the mean (only 5%), and it is very surprising to see one that is more than 3 standard deviations away. If you encountered a mother whose height was 72 in., you might reasonably conclude that she was not part of the population described by the data set in Example 4.17.

■ Measures of Relative Standing

When you obtain your score after taking an achievement test, you probably want to know how it compares to the scores of others who have taken the test. Is your score above or below the mean, and by how much? Does your score place you among the top 5% of those who took the test or only among the top 25%? Questions of this sort are answered by finding ways to measure the position of a particular value in a data set relative to all values in the set. One such measure involves calculating a *z score*.

DEFINITION

The **z score** corresponding to a particular value is

$$z \text{ score} = \frac{\text{value} - \text{mean}}{\text{standard deviation}}$$

The *z* score tells us how many standard deviations the value is from the mean. It is positive or negative according to whether the value lies above or below the mean.

The process of subtracting the mean and then dividing by the standard deviation is sometimes referred to as *standardization,* and a *z* score is one example of what is called a *standardized score*.

Example 4.18 Relatively Speaking, Which Is the Better Offer?

Suppose that two graduating seniors, one a marketing major and one an accounting major, are comparing job offers. The accounting major has an offer for $45,000 per year, and the marketing student has an offer for $43,000 per year. Summary information about the distribution of offers follows:

Accounting: mean = 46,000   standard deviation = 1500
Marketing: mean = 42,500   standard deviation = 1000

Then,

$$\text{accounting } z \text{ score} = \frac{45,000 - 46,000}{1500} = -.67$$

(so \$45,000 is .67 standard deviation below the mean), whereas

$$\text{marketing } z \text{ score} = \frac{43{,}000 - 42{,}500}{1000} = .5$$

Relative to the appropriate data sets, the marketing offer is actually more attractive than the accounting offer (although this may not offer much solace to the marketing major).

◼

The z score is particularly useful when the distribution of observations is approximately normal. In this case, from the Empirical Rule, a z score outside the interval from -2 to $+2$ occurs in about 5% of all cases, whereas a z score outside the interval from -3 to $+3$ occurs only about 0.3% of the time.

▪ Percentiles

A particular observation can be located even more precisely by giving the percentage of the data that fall at or below that observation. If, for example, 95% of all test scores are at or below 650, whereas only 5% are above 650, then 650 is called the *95th percentile* of the data set (or of the distribution of scores). Similarly, if 10% of all scores are at or below 400 and 90% are above 400, then the value 400 is the 10th percentile.

> **DEFINITION**
>
> For any particular number r between 0 and 100, the **rth percentile** is a value such that r percent of the observations in the data set fall at or below that value.

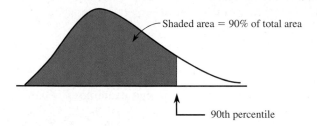

Shaded area = 90% of total area

90th percentile

Figure 4.17 Ninetieth percentile for a smoothed histogram.

Figure 4.17 illustrates the 90th percentile. We have already met several percentiles in disguise. The median is the 50th percentile, and the lower and upper quartiles are the 25th and 75th percentiles, respectively.

Example 4.19 Variable Cost as a Percentage of Operating Revenue

Variable cost as a percentage of operating revenue is an important indicator of a company's financial health. The article "A Credit Limit Decision Model for Inventory Floor Planning and Other Extended Trade Credit Arrangements" (*Decision Sciences* [1992]: 200–220) reported the following summary data for a sample of 350 corporations:

sample mean = 90.1%	sample median = 91.2%
5th percentile = 78.3%	95th percentile = 98.6%

Thus, half of the sampled firms had variable costs that were less than 91.2% of operating revenues because 91.2 is the 50th percentile. If the distribution of values was perfectly symmetric, the mean and the median would be identical. Further evidence against symmetry is provided by the two other given percentiles. The 5th percentile is 12.9 percentage points below the median, whereas the 95th percentile is only 7.4 percentage points above the median. The histogram given in the article had roughly the shape of the negatively skewed smoothed histogram in Figure 4.18 (the total area under the curve is 1).

Figure 4.18 The distribution of variable costs from Example 4.19.

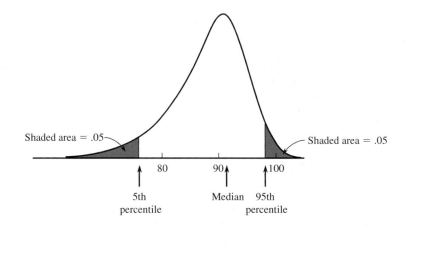

Shaded area = .05

Shaded area = .05

80   90   100

5th percentile   Median   95th percentile

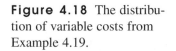

Exercises 4.36–4.50

Turn to the Expanded Answers Section for answers not shown next to exercises.

4.36 The average playing time of compact discs in a large collection is 35 min, and the standard deviation is 5 min.
a. What value is 1 standard deviation above the mean? 1 standard deviation below the mean? What values are 2 standard deviations away from the mean?
b. Without assuming anything about the distribution of times, at least what percentage of the times are between 25 and 45 min?   At least 75%
c. Without assuming anything about the distribution of times, what can be said about the percentage of times that are either less than 20 min or greater than 50 min?
d. Assuming that the distribution of times is normal, approximately what percentage of times are between 25 and 45 min? less than 20 min or greater than 50 min? less than 20 min?   Approximately 15%

4.37 ▼ In a study investigating the effect of car speed on accident severity, 5000 reports of fatal automobile accidents were examined, and the vehicle speed at impact was recorded for each one. For these 5000 accidents, the average speed was 42 mph and that the standard deviation was

15 mph. A histogram revealed that the vehicle speed at impact distribution was approximately normal.
a. Roughly what proportion of vehicle speeds were between 27 and 57 mph?   Approximately 68%
b. Roughly what proportion of vehicle speeds exceeded 57 mph?   Approximately 16%

4.38 The U.S. Census Bureau (2000 census) reported the following relative frequency distribution for travel time to work for a large sample of adults who did not work at home:

Travel Time (minutes)	Relative Frequency
0 to <5	.04
5 to <10	.13
10 to <15	.16
15 to <20	.17
20 to <25	.14
25 to <30	.05
	(continued)

Travel Time (minutes)	Relative Frequency
30 to <35	.12
35 to <40	.03
40 to <45	.03
45 to <60	.06
60 to <90	.05
90 or more	.02

a. Draw the histogram for the travel time distribution. In constructing the histogram, assume that the last interval in the relative frequency distribution (90 or more) ends at 200; so the last interval is 90 to <200. Be sure to use the density scale to determine the heights of the bars in the histogram because not all the intervals have the same width.
b. Describe the interesting features of the histogram from Part (a), including center, shape, and spread.
c. Based on the histogram from Part (a), would it be appropriate to use the Empirical Rule to make statements about the travel time distribution? Explain why or why not.
d. The approximate mean and standard deviation for the travel time distribution are 27 min and 24 min, respectively. Based on this mean and standard deviation and the fact that travel time cannot be negative, explain why the travel time distribution could not be well approximated by a normal curve.
e. Use the mean and standard deviation given in Part (d) and Chebyshev's Rule to make a statement about
   i. the percentage of travel times that were between 0 and 75 min
   ii. the percentage of travel times that were between 0 and 47 min
f. How well do the statements in Part (e) based on Chebyshev's Rule agree with the actual percentages for the travel time distribution? (Hint: You can estimate the actual percentages from the given relative frequency distribution.)

4.39 Mobile homes are tightly constructed for energy conservation. This can lead to a buildup of indoor pollutants. The paper "A Survey of Nitrogen Dioxide Levels Inside Mobile Homes" (*Journal of the Air Pollution Control Association* [1988]: 647–651) discussed various aspects of NO_2 concentration in these structures.
a. In one sample of mobile homes in the Los Angeles area, the mean NO_2 concentration in kitchens during the summer was 36.92 ppb, and the standard deviation was 11.34. Making no assumptions about the shape of the NO_2

distribution, what can be said about the percentage of observations between 14.24 and 59.60? At least 75%
b. Inside what interval is it guaranteed that at least 89% of the concentration observations will lie? (2.90, 70.94)
c. In a sample of non–Los Angeles mobile homes, the average kitchen NO_2 concentration during the winter was 24.76 ppb, and the standard deviation was 17.20. Do these values suggest that the histogram of sample observations did not closely resemble a normal curve? (Hint: What is $\bar{x} - 2s$?)

4.40 The article "Taxable Wealth and Alcoholic Beverage Consumption in the United States" (*Psychological Reports* [1994]: 813–814) reported that the mean annual adult consumption of wine was 3.15 gal and that the standard deviation was 6.09 gal. Would you use the Empirical Rule to approximate the proportion of adults who consume more than 9.24 gal (i.e., the proportion of adults whose consumption value exceeds the mean by more than 1 standard deviation)? Explain your reasoning.

4.41 A student took two national aptitude tests. The national average and standard deviation were 475 and 100, respectively, for the first test and 30 and 8, respectively, for the second test. The student scored 625 on the first test and 45 on the second test. Use z scores to determine on which exam the student performed better relative to the other test takers. The z score is larger for the second exam, so performance was better on the second exam.
4.42 Suppose that your younger sister is applying for entrance to college and has taken the SATs. She scored at the 83rd percentile on the verbal section of the test and at the 94th percentile on the math section of the test. Because you have been studying statistics, she asks you for an interpretation of these values. What would you tell her?

4.43 A sample of concrete specimens of a certain type is selected, and the compressive strength of each specimen is determined. The mean and standard deviation are calculated as $\bar{x} = 3000$ and $s = 500$, and the sample histogram is found to be well approximated by a normal curve.
a. Approximately what percentage of the sample observations are between 2500 and 3500? Approximately 68%
b. Approximately what percentage of sample observations are outside the interval from 2000 to 4000?
c. What can be said about the approximate percentage of observations between 2000 and 2500?
d. Why would you not use Chebyshev's Rule to answer the questions posed in Parts (a)–(c)?

4.44 The paper "Modeling and Measurements of Bus Service Reliability" (*Transportation Research* [1978]: 253–256) studied various aspects of bus service and presented data on travel times from several different routes. The accompanying frequency distribution is for bus travel times from origin to destination on one particular route in Chicago during peak morning traffic periods:

Class	Frequency	Relative Frequency
15 to <16	4	.02
16 to <17	0	.00
17 to <18	26	.13
18 to <19	99	.49
19 to <20	36	.18
20 to <21	8	.04
21 to <22	12	.06
22 to <23	0	.00
23 to <24	0	.00
24 to <25	0	.00
25 to <26	16	.08

a. Construct the corresponding histogram.
b. Compute (approximately) the following percentiles:
   i. 86th   iv. 95th
   ii. 15th   v. 10th
   iii. 90th

4.45 An advertisement for the "30-in. Wonder" that appeared in the September 1983 issue of the journal *Packaging* claimed that the 30-in. Wonder weighs cases and bags up to 110 lb and provides accuracy to within 0.25 oz. Suppose that a 50 oz weight was repeatedly weighed on this scale and the weight readings recorded. The mean value was 49.5 oz, and the standard deviation was 0.1. What can be said about the proportion of the time that the scale actually showed a weight that was within 0.25 oz of the true value of 50 oz? (Hint: Use Chebyshev's Rule.)

4.46 Suppose that your statistics professor returned your first midterm exam with only a z score written on it. She also told you that a histogram of the scores was approximately normal. How would you interpret each of the following z scores?
a. 2.2   d. 1.0
b. 0.4   e. 0
c. 1.8

4.47 The paper "Answer Changing on Multiple-Choice Tests" (*Journal of Experimental Education* [1980]: 18–21) reported that for a group of 162 college students, the average number of responses changed from the correct answer to an incorrect answer on a test containing 80 multiple-choice items was 1.4. The corresponding standard deviation was reported to be 1.5. Based on this mean and standard deviation, what can you tell about the shape of the distribution of the variable *number of answers changed from right to wrong*? What can you say about the number of students who changed at least six answers from correct to incorrect?

4.48 The average reading speed of students completing a speed-reading course is 450 words per minute (wpm). If the standard deviation is 70 wpm, find the z score associated with each of the following reading speeds.
a. 320 wpm   −1.86   c. 420 wpm   −0.43
b. 475 wpm   0.36   d. 610 wpm   2.29

4.49 ● ▼ The following data values are 1989 per capita expenditures on public libraries for each of the 50 states:

29.48	24.45	23.64	23.34	22.10	21.16	19.83
18.01	17.95	17.23	16.53	16.29	15.89	15.85
13.64	13.37	13.16	13.09	12.66	12.37	11.93
10.99	10.55	10.24	10.06	9.84	9.65	8.94
7.70	7.56	7.46	7.04	6.58	5.98	19.81
19.25	19.18	18.62	14.74	14.53	14.46	13.83
11.85	11.71	11.53	11.34	8.72	8.22	8.13
8.01						

a. Summarize this data set with a frequency distribution. Construct the corresponding histogram.
b. Use the histogram in Part (a) to find approximate values of the following percentiles:
i. 50th   13.158   iv. 90th   21
ii. 70th   16.25   v. 40th   11.842.
iii. 10th   6.923

4.50 The accompanying table gives the mean and standard deviation of reaction times (in seconds) for each of two different stimuli:

	Stimulus 1	Stimulus 2
Mean	6.0	3.6
Standard deviation	1.2	0.8

Bold exercises answered in back   ● Data set available online but not required   ▼ Video solution available

If your reaction time is 4.2 sec for the first stimulus and 1.8 sec for the second stimulus, to which stimulus are you reacting (compared to other individuals) relatively more quickly?

Bold exercises answered in back   ● Data set available online but not required   ▼ Video solution available

4.5 Interpreting and Communicating the Results of Statistical Analyses

As was the case with the graphical displays of Chapter 3, the primary function of the descriptive tools introduced in this chapter is to help us better understand the variables under study. If we have collected data on the amount of money students spend on textbooks at a particular university, most likely we did so because we wanted to learn about the distribution of this variable (amount spent on textbooks) for the population of interest (in this case, students at the university). Numerical measures of center and spread and boxplots help to enlighten us, and they also allow us to communicate to others what we have learned from the data.

■ Communicating the Results of Statistical Analyses

When reporting the results of a data analysis, it is common to start with descriptive information about the variables of interest. It is always a good idea to start with a graphical display of the data, and, as we saw in Chapter 3, graphical displays of numerical data are usually described in terms of center, variability, and shape. The numerical measures of this chapter can help you to be more specific in describing the center and spread of a data set.

When describing center and spread, you must first decide which measures to use. Common choices are to use either the sample mean and standard deviation or the sample median and interquartile range (and maybe even a boxplot) to describe center and spread. Because the mean and standard deviation can be sensitive to extreme values in the data set, they are best used when the distribution shape is approximately symmetric and when there are few outliers. If the data set is noticeably skewed or if there are outliers, then the observations are more spread out in one part of the distribution than in the others. In this situation, a five-number summary or a boxplot conveys more information than the mean and standard deviation do.

■ Interpreting the Results of Statistical Analyses

It is relatively rare to find raw data in published sources. Typically, only a few numerical summary quantities are reported. We must be able to interpret these values and understand what they tell us about the underlying data set.

For example, a university recently conducted an investigation of the amount of time required to enter the information contained in an application for admission into the university computer system. One of the individuals who performs this task was asked to note starting time and completion time for 50 randomly selected application forms. The resulting entry times (in minutes) were summarized using the mean, median, and standard deviation:

$$\bar{x} = 7.854$$
$$\text{median} = 7.423$$
$$s = 2.129$$

```
 4 8
 5 0234579
 6 00001234566779
 7 223556688
 8 23334
 9 002
10 011168
11 134
12 2                    Stem: Ones
13                      Leaf: Tenths
14 3
```

Figure 4.19 Stem-and-leaf display of data entry times.

What do these summary values tell us about entry times? The average time required to enter admissions data was 7.854 min, but the relatively large standard deviation suggests that many observations differ substantially from this mean. The median tells us that half of the applications required less than 7.423 min to enter. The fact that the mean exceeds the median suggests that some unusually large values in the data set affected the value of the mean. This last conjecture is confirmed by the stem-and-leaf display of the data given in Figure 4.19.

The administrators conducting the data-entry study looked at the outlier 14.3 min and at the other relatively large values in the data set; they found that the five largest values came from applications that were entered before lunch. After talking with the individual who entered the data, the administrators speculated that morning entry times might differ from afternoon entry times because there tended to be more distractions and interruptions (phone calls, etc.) during the morning hours, when the admissions office generally was busier. When morning and afternoon entry times were separated, the following summary statistics resulted:

Morning (based on $n = 20$ applications): $\bar{x} = 9.093$ median $= 8.743$ $s = 2.329$
Afternoon (based on $n = 30$ applications): $\bar{x} = 7.027$ median $= 6.737$ $s = 1.529$

Clearly, the average entry time is higher for applications entered in the morning; also, the individual entry times differ more from one another in the mornings than in the afternoons (because the standard deviation for morning entry times, 2.329, is about 1.5 times as large as 1.529, the standard deviation for afternoon entry times).

■ What to Look for in Published Data

Here are a few questions to ask yourself when you interpret numerical summary measures.

- Is the chosen summary measure appropriate for the type of data collected? In particular, watch for inappropriate use of the mean and standard deviation with categorical data that has simply been coded numerically.
- If both the mean and the median are reported, how do the two values compare? What does this suggest about the distribution of values in the data set? If only the mean or the median was used, was the appropriate measure selected?
- Is the standard deviation large or small? Is the value consistent with your expectations regarding variability? What does the value of the standard deviation tell you about the variable being summarized?
- Can anything of interest be said about the values in the data set by applying Chebyshev's Rule or the Empirical Rule?

For example, consider a study on smoking and lactation. The journal article "Smoking During Pregnancy and Lactation and Its Effects on Breast Milk Volume" (*American Journal of Clinical Nutrition* [1991]: 1011–1016) reported the following summary values for data collected on daily breast milk volume (in grams) for two groups of nursing mothers:

Group	Mean	Standard Deviation	Median
Smokers	693	110	589.5
Nonsmokers	961	120	946

Because milk volume is a numerical variable, these descriptive measures reasonably summarize the data set. For nonsmoking mothers, the mean and the median are similar, indicating that the milk volume distribution is approximately symmetric. The average milk volume for smoking mothers is 693, whereas for nonsmoking mothers the mean milk volume is 961—quite a bit higher than for the smoking mothers. The standard deviations for the two groups are similar, suggesting that, although the average milk volume may be lower for smokers, the variability in mothers' milk volume is about the same. Because the median for the smoking group is smaller than the corresponding mean, we suspect that the distribution of milk volume for smoking mothers is positively skewed.

■ A Word to the Wise: Cautions and Limitations

When computing or interpreting numerical descriptive measures, you need to keep in mind the following:

1. Measures of center don't tell all. Although measures of center, such as the mean and the median, do give us a sense of what might be considered a typical value for a variable, this is only one characteristic of a data set. Without additional information about variability and distribution shape, we don't really know much about the behavior of the variable.
2. Data distributions with different shapes can have the same mean and standard deviation. For example, consider the following two histograms:

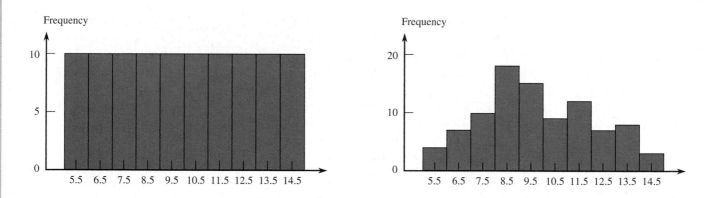

Both histograms summarize data sets that have a mean of 10 and a standard deviation of 2, yet they have different shapes.
3. Both the mean and the standard deviation are sensitive to extreme values in a data set, especially if the sample size is small. If a data distribution is skewed or if the data set has outliers, the median and the interquartile range may be a better choice for describing center and spread.
4. Measures of center and variability describe the values of the variable studied, not the frequencies in a frequency distribution or the heights of the bars in a histogram. For example, consider the following two frequency distributions and histograms:

Frequency Distribution A		Frequency Distribution B	
Value	Frequency	Value	Frequency
1	10	1	5
2	10	2	10
3	10	3	20
4	10	4	10
5	10	5	5

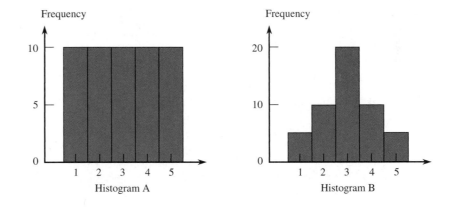

Histogram A

Histogram B

There is more variability in the data summarized by Frequency Distribution and Histogram A than in the data summarized by Frequency Distribution and Histogram B. This is because the values of the variable described by Histogram and Frequency Distribution B are more concentrated near the mean than are the values for the variable described by Histogram and Frequency Distribution A. Don't be misled by the fact that there is no variability in the frequencies in Frequency Distribution A or the heights of the bars in Histogram A.

5. Be careful with boxplots based on small sample sizes. Boxplots convey information about center, variability, and shape, but when the sample size is small, you should be hesitant to overinterpret shape information. It is really not possible to decide whether a data distribution is symmetric or skewed if only a small sample of observations from the distribution is available.

6. Not all distributions are normal (or even approximately normal). Be cautious in applying the Empirical Rule in situations where you are not convinced that the data distribution is at least approximately normal. Using the Empirical Rule in such situations can lead to incorrect statements.

7. Watch out for outliers! Unusual observations in a data set often provide important information about the variable under study, so it is important to consider outliers in addition to describing what is typical. Outliers can also be problematic—both because the values of some descriptive measures are influenced by outliers and because some of the methods for drawing conclusions from data may not be appropriate if the data set has outliers.

Activity 4.1　Collecting and Summarizing Numerical Data

In this activity, you will work in groups to collect data that will provide information about how many hours per week, on average, students at your school spend engaged in a particular activity. You will use the sampling plan designed in Activity 2.1 to collect the data.

1. With your group, pick one of the following activities to be the focus of your study:
   i. Surfing the web
   ii. Studying or doing homework
   iii. Watching TV
   iv. Exercising
   v. Sleeping

or you may choose a different activity, *subject to the approval of your instructor.*

2. Use the plan developed in Activity 2.1 to collect data on the variable you have chosen for your study.

3. Summarize the resulting data using both numerical and graphical summaries. Be sure to address both center and variability.

4. Write a short article for your school paper summarizing your findings regarding student behavior. Your article should include both numerical and graphical summaries.

Activity 4.2　Airline Passenger Weights

The article "Airlines Should Weigh Passengers, Bags, NTSB Says" (*USA Today*, February 27, 2004) states that the National Transportation Safety Board recommended that airlines weigh passengers and their bags to prevent overloaded planes from attempting to take off. This recommendation was the result of an investigation into the crash of a small commuter plane in 2003, which determined that too much weight contributed to the crash.

Rather than weighing passengers, airlines currently use estimates of average passenger and luggage weights. After the 2003 accident, this estimate was increased by 10 lbs. for passengers and 5 lbs. for luggage. Although an airplane can fly if it is somewhat overweight if all sys-

tems are working properly, if one of the plane's engines fails an overweight plane becomes difficult for the pilot to control.

Assuming that the new estimate of the average passenger weight is accurate, discuss the following questions with a partner and then write a paragraph that answers these questions.

1. What role does variability in passenger weights play in creating a potentially dangerous situation for an airline?

2. Would an airline have a lower risk of a potentially dangerous situation if the variability in passenger weight is large or if it is small?

Activity 4.3　Boxplot Shapes

In this activity, you will investigate the relationship between boxplot shapes and the corresponding five-number summary. The accompanying figure shows four boxplots, labeled A–D. Also given are four five-number summaries, labeled I–IV. Match each five-number summary to the appropriate boxplot. Note that scales are not included on the boxplots, so you will have to think about what the five-number summary implies about characteristics of the boxplot.

Five-Number Summaries

	I	II	III	IV
Minimum	40	4	0.0	10
Lower quartile	45	8	0.1	34
Median	71	16	0.9	44
Upper quartile	88	25	2.2	82
Maximum	106	30	5.1	132

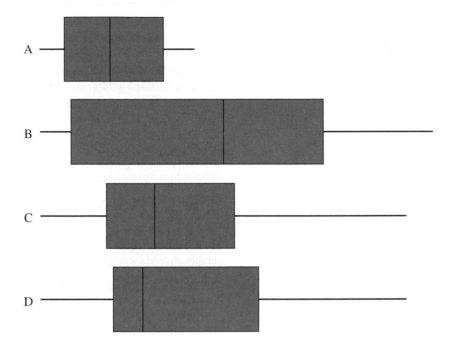

Summary of Key Concepts and Formulas

Term or Formula	Comment
$x_1, x_2, \ldots, x_n$	Notation for sample data consisting of observations on a variable x, where n is the sample size.
Sample mean, $\bar{x}$	The most frequently used measure of center of a sample. It can be very sensitive to the presence of even a single outlier (unusually large or small observation).
Population mean, μ	The average x value in the entire population.
Sample median	The middle value in the ordered list of sample observations. (For n even, the median is the average of the two middle values.) It is very insensitive to outliers.
Trimmed mean	A measure of center in which the observations are first ordered from smallest to largest, one or more observations are deleted from each end, and the remaining ones are averaged. In terms of sensitivity to outliers, it is a compromise between the mean and the median.
Deviations from the mean: $x_1 - \bar{x}, x_2 - \bar{x}, \ldots, x_n - \bar{x}$	Quantities used to assess variability in a sample. Except for rounding effects, $\sum(x - \bar{x}) = 0$.
The sample variance $s^2 = \dfrac{\sum(x - \bar{x})^2}{n - 1}$ and standard deviation $s = \sqrt{s^2}$	The most frequently used measures of variability for sample data.

gram per minute) while the participants pedaled at a specified rate on a bicycle ergometer:

12.81	14.95	15.83	15.97	17.90	18.27	18.34
19.82	19.94	20.62	20.88	20.93	20.98	20.99
21.15	22.16	22.24	23.16	23.56	35.78	36.73

a. Compute the median and the quartiles for this data set.
b. What is the value of the interquartile range? Are there outliers in this data set?
c. Draw a modified boxplot, and comment on the interesting features of the plot.

4.60 ● The risk of developing iron deficiency is especially high during pregnancy. Detecting such a deficiency is complicated by the fact that some methods for determining iron status can be affected by the state of pregnancy itself. Consider the following data on transferrin receptor concentration for a sample of women with laboratory evidence of overt iron-deficiency anemia ("Serum Transferrin Receptor for the Detection of Iron Deficiency in Pregnancy," *American Journal of Clinical Nutrition* [1991]: 1077–1081):

15.2	9.3	7.6	11.9	10.4	9.7	20.4	9.4	11.5
16.2	9.4	8.3						

Compute the values of the sample mean and median. Why are these values different here? Which one do you regard as more representative of the sample, and why?

4.61 ● The paper "The Pedaling Technique of Elite Endurance Cyclists" (*International Journal of Sport Biomechanics* [1991]: 29–53) reported the following data on single-leg power at a high workload:

244	191	160	187	180	176	174	205	211
183	211	180	194	200				

a. Calculate and interpret the sample mean and median.
b. Suppose that the first observation had been 204, not 244. How would the mean and median change?
c. Calculate a trimmed mean by eliminating the smallest and the largest sample observations. What is the corresponding trimming percentage? *191, 7.14% trimmed mean*
d. Suppose that the largest observation had been 204 rather than 244. How would the trimmed mean in Part (c) change? What if the largest value had been 284?

4.62 The paper cited in Exercise 4.61 also reported values of single-leg power for a low workload. The sample mean for $n = 13$ observations was $\bar{x} = 119.8$ (actually

119.7692), and the 14th observation, somewhat of an outlier, was 159. What is the value of $\bar{x}$ for the entire sample? *122.57*

4.63 ● The amount of aluminum contamination (in parts per million) in plastic was determined for a sample of 26 plastic specimens, resulting in the following data ("The Log Normal Distribution for Modeling Quality Data When the Mean Is Near Zero," *Journal of Quality Technology* [1990]: 105–110):

30	30	60	63	70	79	87	90	101
102	115	118	119	119	120	125	140	145
172	182	183	191	222	244	291	511	

Construct a boxplot that shows outliers, and comment on the interesting features of this plot.

4.64 ● The article "Can We Really Walk Straight?" (*American Journal of Physical Anthropology* [1992]: 19–27) reported on an experiment in which each of 20 healthy men was asked to walk as straight as possible to a target 60 m away at normal speed. Consider the following data on cadence (number of strides per second):

0.95	0.85	0.92	0.95	0.93	0.86	1.00	0.92
0.85	0.81	0.78	0.93	0.93	1.05	0.93	1.06
1.06	0.96	0.81	0.96				

Use the methods developed in this chapter to summarize the data; include an interpretation or discussion wherever appropriate. (Note: The author of the paper used a rather sophisticated statistical analysis to conclude that people cannot walk in a straight line and suggested several explanations for this.)   *Answers will vary. See solutions manual for one example.*

4.65 ● The article "Comparing the Costs of Major Hotel Franchises" (*Real Estate Review* [1992]: 46–51) gave the following data on franchise cost as a percentage of total room revenue for chains of three different types:

Budget	2.7	2.8	3.8	3.8	4.0	4.1	5.5
	5.9	6.7	7.0	7.2	7.2	7.5	7.5
	7.7	7.9	7.9	8.1	8.2	8.5	
Midrange	1.5	4.0	6.6	6.7	7.0	7.2	7.2
	7.4	7.8	8.0	8.1	8.3	8.6	9.0
First-class	1.8	5.8	6.0	6.6	6.6	6.6	7.1
	7.2	7.5	7.6	7.6	7.8	7.8	8.2
	9.6						

Construct a boxplot for each type of hotel, and comment on interesting features, similarities, and differences.

Bold exercises answered in back   ● Data set available online but not required   ▼ Video solution available

4.66 ● The accompanying data on milk volume (in grams per day) were taken from the paper "Smoking During Pregnancy and Lactation and Its Effects on Breast Milk Volume" (*American Journal of Clinical Nutrition* [1991]: 1011–1016):

Smoking	621	793	593	545	753	655
mothers	895	767	714	598	693	
Nonsmoking	947	945	1086	1202	973	981
mothers	930	745	903	899	961	

Compare and contrast the two samples.   Answers will vary. See solutions manual for one example.

4.67 The *Los Angeles Times* (July 17, 1995) reported that in a sample of 364 lawsuits in which punitive damages were awarded, the sample median damage award was $50,000, and the sample mean was $775,000. What does this suggest about the distribution of values in the sample?

4.68 ● Age at diagnosis for each of 20 patients under treatment for meningitis was given in the paper "Penicillin in the Treatment of Meningitis" (*Journal of the American Medical Association* [1984]: 1870–1874). The ages (in years) were as follows:

18 18 25 19 23 20 69 18 21 18 20 18
18 20 18 19 28 17 18 18

a. Calculate the values of the sample mean and the standard deviation.   Mean = 22.15, standard deviation = 11.366
b. Calculate the 10% trimmed mean. How does the value of the trimmed mean compare to that of the sample mean? Which would you recommend as a measure of location? Explain.
c. Compute the upper quartile, the lower quartile, and the interquartile range.
d. Are there any mild or extreme outliers present in this data set?
e. Construct the boxplot for this data set.

4.69 Suppose that the distribution of scores on an exam is closely described by a normal curve with mean 100. The 16th percentile of this distribution is 80.
a. What is the 84th percentile?   120
b. What is the approximate value of the standard deviation of exam scores?   20
c. What z score is associated with an exam score of 90?
d. What percentile corresponds to an exam score of 140?
e. Do you think there were many scores below 40? Explain.

Personal Tutor
Do you need a live tutor for homework problems?

CENGAGENOW™
Are you ready? Take your exam-prep post-test now.

Bold exercises answered in back   ● Data set available online but not required   ▼ Video solution available

Graphing Calculator Explorations

Exploration 4.1 Quartiles

In this exploration we bring up a somewhat delicate statistical point. Statisticians are not in complete agreement about the best way to calculate quartiles, and because of this you may encounter calculators or computer software packages that calculate quartiles using a different method than the one proposed in this text. These different methods lead to potentially different values for the quartiles computed by calculators and computer software, and therefore to potentially different results for the interquartile range. This *may* result in different observations being identified as outliers. This is a good news/bad news situation. The bad news: you may at first be confused by your textbook, software, and calculator giving you different "right" answers when you make boxplots. The good news outweighs the bad: unless you have a very small data set the differences among the different methods for calculating quartiles are of little practical importance.

Despite the unimportance of the differences in practice, you should be aware of what your calculator and software actually do. If your calculator and/or software uses a different definition than the text, you may be more comfortable in the knowledge that an observed discrepancy is likely to be due to the differing definitions and not an error in data entry. To check the agreement among your calculator, your computer software, and this book, use the following sets of data, designed to highlight the differences. The answers using the text definitions are in parentheses.

Test Data Set #1: 10, 20, 30, 40
(lower quartile = 15, median = 25, upper quartile = 35)

Test Data Set #2: 10, 20, 30, 40, 50
(lower quartile = 15, median = 30, upper quartile = 45)

Test Data Set #3: 10, 20, 30, 40, 50, 60
(lower quartile = 20, median = 35, upper quartile = 50)

Test Data Set #4: 10, 20, 30, 40, 50, 60, 70
(lower quartile = 20, median = 40, upper quartile = 60)

Did you obtain the same values using your calculator? If so, your calculator is probably using the same definition as the text. But, what if your calculator gives different quartiles? When analyzing real data, the method used to compute quartiles is not of great concern.

Exploration 4.2 Boxplots

Your calculator may not provide complete flexibility when graphing a boxplot. Your calculator may graph a skeletal boxplot only, a modified boxplot only, or it may offer you a choice. What if you are limited in your choices? Suppose you wish to sketch a boxplot that is different from what your calculator delivers. If your calculator presents a modified boxplot and you need to sketch a skeletal boxplot you can simply extend the whiskers of the modified boxplot to include all the outliers.

If your calculator plots only a skeletal boxplot, you have a very different problem—how can you draw the outliers on your paper without just starting from scratch? To graph the skeletal boxplot, your calculator will have already calculated the quartiles and the median of the data. To sketch a modified boxplot, you need only locate this information and then decide which, if any, of the data values are outliers.

Your calculator will have a screen that displays a statistical summary for the data. Such quantities as the minimum and maximum data element, the mean, median, and standard deviation, and the lower and upper quartile will be shown. Using this information you can do the math and see if there are any outliers. To do this you will need to calculate the following boundaries by hand, as discussed in the text:

upper quartile + 1.5 iqr = upper quartile + 1.5 (upper quartile − lower quartile)
lower quartile − 1.5 iqr = lower quartile − 1.5 (upper quartile − lower quartile)

Then look to see if any of the observations lie above or below the computed boundaries. If so, they are outliers.

If there are no outliers you are done! The skeletal boxplot and the modified boxplot will be the same. If these calculations indicate there are outliers in your data, you now must find the largest observation that is not an outlier, and the smallest observation that is not an outlier. Any observations larger and smaller than these, respectively, will be outliers. Finding the outliers "by inspection"—that is, by scanning up and down the list of data, would be tedious and error-laden. The solution? Use your calculator to sort the data, making it very easy to spot outliers by looking at either end of the data list.

Your calculator will sort the data from small to large or large to small; for the purpose of identifying outliers, it does not matter which you choose. The keystroke sequence for sorting data on your particular calculator can be found in your calculator manual; look in the chapter on "lists" or in the index under "sorting." Once the data is sorted, finding the outliers by inspection is very easy. A final suggestion before you do this: copy the numbers to a separate list and work with that list. Then, if you inadvertently mess something up, you still have the original data.

Exploration **4.3** z Scores

Despite the usefulness of standardized scores in general, and z scores in particular, there will not be a z score function on your calculator. Suppose that you wanted to compute the z scores corresponding to each observation in a data set consisting of 20 values. You might worry that this would require mind-numbing and repetitious calculations—subtracting the mean from every value and then dividing each result by the standard deviation. Fortunately, this is not the case. The list capabilities of your calculator will dramatically save time and effort. Instead of 20 sets of calculations, you can perform one calculation on the whole list of data! Moreover, the keystrokes that will work for lists mirror the keystrokes for working with individual observations.

You will first need to enter some data into a list we will refer to as List1. We will use the data in Figure 4.20(a). After entering the data, we need to find the mean and sample standard deviation. These values are $\bar{x} = 3.0$, $s = 1.58113883$. (Your calculator may not display all these digits, but it knows they are there!)

At this point you can convert all the raw scores to z scores by performing arithmetic on the whole list. You can also store the results in another list without losing your original data. Again, your required keystrokes may differ, but they will accomplish the following:

1. Subtract the mean from each value in List1 to get the deviations.
2. Divide the deviations by the standard deviation.
3. Store the results of these calculations in List2.

The sequence of keystrokes will probably be very similar to this:

$$(\text{List1} - 3.0) \div 1.58113883 \rightarrow \text{List2}$$

Some calculators may use the slash "/" to indicate division, and some may use the equal sign "=" to indicate that the results should be stored in a second list. After performing the keystrokes appropriate for your calculator, you should see the results shown in List2 of Figure 4.20(b). If the lists you have chosen are next to each other on the screen you can easily compare the raw data with the z scores.

You may be able to save a few more keystrokes by "recalling" the mean and standard deviation from the calculator memory, although this is probably a matter of style and individual taste. If you have not done any intervening statistical calculations since you calculated the mean and standard deviation, these values are still stored in the calculator memory. If you wish, you may recall those

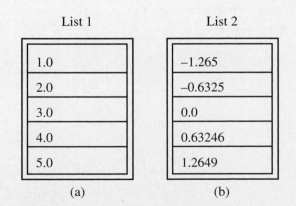

List 1

| 1.0 |
| 2.0 |
| 3.0 |
| 4.0 |
| 5.0 |

(a)

List 2

| −1.265 |
| −0.6325 |
| 0.0 |
| 0.63246 |
| 1.2649 |

(b)

Figure 4.20 Original data and z scores.

values from the calculator's memory rather than key in the numbers. These values can be inserted in your calculations in place of the actual numbers by pressing the appropriate keystroke sequence. This capability typically involves using a "VARS" key, short for "variables." (The Texas Instruments TI-80 series calculators use VARS, as does the Casio CFX-9850Ga+. This capability may exist under different names with different calculators; consult your manual!) For example, on the Texas Instruments TI-83, you would press

VARS...   Statistics... S_x...   ENTER

to recall the sample standard deviation. In the discussion to follow, the keystroke VARS will stand for the particular sequence to recall the sample mean, and VARS will stand for the particular sequence to recall the sample standard deviation.

With this symbolism in mind, the sequence of keystrokes using the recall capabilities of your calculator would transform

$$(\text{List1} - 3.0) \div 1.58113883 \rightarrow \text{List2}$$

into a keystroke sequence like this:

$$(\text{List1} - \text{VARS}\bar{x}) \div \text{VARSs} \rightarrow \text{List2}.$$

The recall capability is a very nice feature, but we have found that individuals differ in their preferences about whether to use the VARS capabilities or just key in the digits when performing arithmetic operations with lists. We encourage you to try both ways and use the method with which you are more comfortable.

TEACHING TIPS

Intent of this chapter:

After completing this chapter, students should be able to (1) calculate and interpret a correlation coefficient; (2) find the least-squares regression line (LSRL), interpret the slope of the LSRL, and use the LSRL to predict a y-value for a given x-value; (3) construct and describe scatterplots and residual plots; (4) understand when linear regression is an appropriate choice; (5) calculate and interpret the coefficient of determination and the standard deviation about the LSRL; and (6) transform nonlinear data to obtain an appropriate equation that can be used to make predictions.

A PowerPoint® Lecture is available on the *Instructor's Resource Binder* CD.

There is a lot of material in each of these sections. Therefore, you may need to spend two or three 50-minute classes on each section.

■ Section 5.1

Students typically have some experience with the concept of correlation from their algebra and precalculus classes, but this knowledge is probably superficial. The graphs in Figure 5.1 are located on the Instructor's Resource Binder CD. Ask students to identify the following: (a) which of these graphs displays the strongest positive correlation and why, and (b) which graph displays a weak correlation and why, etc. In AP Statistics, students are expected to use the calculator (or computer software) to compute the correlation coefficient. On the TI calculators, "Diagnostics" must be turned on in order for the calculator to compute and display the correlation coefficient. To turn on this capability, use the following keystrokes: [2nd] [Catalog], arrow down to "DiagnosticOn", [enter], [enter]. Once turned on, Diagnostics stays on until the calculator's memory is reset.

The "Correlation" applet (listed below) allows students to investigate the effects of adding data points to a given data set. For example, given the data set (40, 20), (50, 50), (60, 40), (70, 60), explain what happens to the correlation coefficient when each of the following points is added to the set: (50, 80), (50, 10), (55, 45), (60, 50), (80, 80), (100, 0). [Note: Remove the last data point before adding another point to the original four points.] Thus students should discover that adding a point that lies away from the trend of the original points will weaken the correlation coefficient, while adding a point that lies along the trend of the original points will strengthen the correlation coefficient (and the further away the point is from the rest of the data, the greater the effect the added point has).

The properties of the correlation coefficient (listed on page 204) are another excellent opportunity to use examples to allow students to investigate how the correlation coefficient is affected when units of measurements are altered or when the independent variable and the dependent variable are interchanged. Using a data set such as Example 5.4 (on page 238) demonstrates that just because the correlation coefficient is close to zero ($r \approx 0.038$) does not mean that no relationship exists between x and y. Students typically will want to just perform the calculations without first looking at a graphical display (in this case a scatterplot). Upon graphing the data, we see that there is a strong relationship—it's just not linear!

The second applet listed below allows students to estimate the correlation coefficients from a variety of different randomly generated scatterplots.

Suggested Assignment: Exercises 5.1, 5.2, 5.3, 5.6, 5.9, 5.10, 5.14, 5.16

Applets: Correlation Applet (*Instructor's Resource Binder* CD), Guessing Correlations (www.stat.uiuc.edu/~stat100/java/GCApplet/GCAppletFrame.html)

Activity: Activity 5.1: Exploring Correlation and Regression Technology Activity (Applets) (p. 267 and *Activities Manual,* p. 46) ★ and "Correlation" (*Instructor's Resource Binder* & CD, Chapter 5 Activities Worksheets)

★ Kathy's personal favorite

■ Section 5.2

This section discusses the least-squares regression line. Again, most students will have used a calculator to compute the LSRL, but rarely do students understand exactly why this line is the line of best fit. Using a small data set, add a trend line that is reasonable but is not the LSRL. Find the vertical deviations of each point from the LSRL, $(y - \hat{y})$. Notice that some of these deviations are negative and some are positive; therefore, we will square the deviations before adding to obtain a measure of how well the line fits the points. Next, repeat this procedure using the LSRL. Notice that the sum of the vertical deviations from the line is now zero. (Doesn't this sound familiar? See Chapter 4.) The sum of the squared deviations will be the smallest sum of squared deviations possible. The "Regdecomp" applet (listed below) is useful in showing this concept.

Students should be able to use technology to compute the LSRL. When using the TI calculators, notice that there are actually two different "LinReg" functions. Accepted statistical notation is associated with the second function, which produces the line in the form of $(a + bx)$. Students should also be able to read computer output from a regression analysis in order to find the LSRL.

In addition to computing the LSRL, students should be able to interpret the slope of the line in context, and when appropriate, interpret the intercept in context. When interpreting the slope, students often use the definition shown in the tan box on page 211, which states slope ". . . is the amount by which y increases when x increases by 1 unit." This definition is **only** correct when the points lie in a straight line and $r = 1$ (or -1). One of the most common errors made by students is not understanding that typically the LSRL is an estimate, based upon a sample, of the true LSRL for some population. Since the LSRL is an estimate, then the interpretation of the slope must indicate this estimation ". . . is the **approximate** amount by which y increases when x increases by 1 unit." The interpretation of the intercept is dependent on the context of the problem. In other words, does it make sense that the x-variable might equal zero? If so, the intercept should also be interpreted as an estimated value. Another concern is the location of the point $(0, y)$ compared to the points in the data set: If $(0, y)$ is far away from the data set points, interpreting the intercept is risky as it is equivalent to extrapolating far outside the range of the data.

Finally, students should be able to use the LSRL to make predictions for given x-values. Beware of the danger of extrapolation, as stated on page 214. Also note the important distinction made on page 217: The least-squares line of y on x should not be used to predict x-values from y-values. If one wishes to predict x-values, then a regression of x on y should be performed. The two equations are not the same.

Suggested Assignment: Exercises 5.19, 5.20, 5.21, 5.22, 5.27, 5.28, 5.30, 5.33

Applets: Regdecomp Applet (*Instructor's Resource Binder* CD)

Activity: Activity 5.1: Exploring Correlation and Regression Technology Activity (Applets) (p. 267 and *Activities Manual*, p. 46) ★

Assessment: Chapter 5 Quiz 1 (*Test Bank* and *Instructor's Resource Binder* CD)

■ Section 5.3

This section discusses ways to assess the fit of a line. One method is to create a residual plot, a scatterplot with the x-values graphed on the horizontal axis and the residuals graphed on the vertical axis. Residuals are the vertical deviations of each point from the LSRL, $(y - \hat{y})$, and should always total zero. Figure 5.16 (on page 226) illustrates that either x-values or predicted or fitted values $(\hat{y})$ can be graphed on the horizontal axis with no change in the pattern of the residual plot. On the TI calculator, when you perform the linear regression the residuals are automatically calculated and are stored in a "Resid" list.

★ Kathy's personal favorite

A common question from students is why a randomly scattered residual plot indicates that the relationship between the variables is linear. To explain this, we must discuss the theoretical regression model in Chapter 13. Amazingly, students tend to have an intuitive understanding of this model. Suppose we wanted to investigate the relationship between height and weight of a population of adult females.

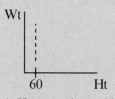

1. How much would you expect an adult woman to weigh if she were 60 inches tall? (*There is a distribution of possible weights for a woman 60 inches tall.*)

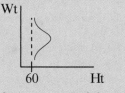

2. Are some weights more likely than others? (*Yes, the middle values.*)

So, what shape do we think this distribution is? (*Normal*)

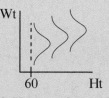

3. How much would you expect an adult woman to weigh if she were 63 inches tall? Or 66 inches tall? (*Each value of height would have a normal distribution of weights that was centered slightly higher than the previous distribution.*)

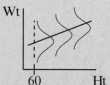

4. Where will the true line of best fit be? (*Through the means of these distributions*)

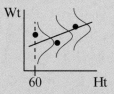

5. So if I select a random sample of women from this population, then the weights should be a random sample from these populations. Thus, if the relationship is linear, the residuals should be randomly scattered about the LSRL.

If you look long and hard enough, you can find a pattern in any residual plot. Warn students to avoid this common mistake. At first glance, if you recognize no pattern in the points, then the residuals can be considered randomly scattered.

Another consideration when assessing the fit of a line is the coefficient of determination. This measure "gives the proportion of variation in *y* that can be attributed to an approximate linear relationship between *x* and *y*" (page 228). Example 5.10 (page 229) is very helpful in aiding students' understanding and retention of this definition. In this example, we first look at the **total** amount of variation in *y* (range of motion) by calculating $\text{SSTo} = \Sigma(y - \bar{y})^2$, the sum of the squared deviations from the mean range of motion. Next, we look at the amount of variation remaining when we use x and the LSRL to predict y ($\hat{y}$) by calculating SSResid $= \Sigma(y - \hat{y})^2 = \Sigma(\text{residuals})^2$. (Why squared deviations and squared residuals? Remember, the unsquared values sum to zero.) For Example 5.10, SSTo $=$ 1564.9, meaning that a measure of the total variability in y is 1564.9; SSResid $= 1085.7$, meaning that when the LSRL is used, the remaining variability is 1085.7. By what percent did the variation in y decrease when the LSRL was used? The percent of decrease is 30.6% which is r^2. Looking back at our definition, approximately 30.6% of the variation in the range of motion (y) can be attributed to the approximate linear relationship between age (x) and range of motion (y). This is not a very high r^2; however, this is not surprising because other variables, such as weight or amount of exercise (in minutes/week), also contribute to variability in range of motion.

The standard deviation about the least-squares line (s_e) is an estimate of the standard deviation of the normal curves for the y-values at a given x-value in our theoretical model above. The value of s_e indicates the amount of spread in the y-values and gives us another tool when assessing the fit of the line. Why do we divide by ($n - 2$)? In the theoretical regression model, we have two unknown parameters, the true intercept and the true slope. Thus, the value for degrees of freedom is ($n - 2$). (Note: When calculating the standard deviation, we only have one unknown parameter, the true mean.)

The applet listed below allows students to investigate what happens to the LSRL and the correlation coefficient as data points are added to an original data set. Students may explore the effects of outliers and influential points. Students should be able to answer questions concerning what might happen to the slope (becomes steeper or less steep) or the correlation coefficient (indicating a weaker or stronger linear relationship) if a point is added or removed from the data set.

Suggested Assignment: Exercises 5.37, 5.38, 5.40, 5.42, 5.43, 5.45, 5.47, 5.48, 5.50

Applets: Regression Applet (www.stat.sc.edu/~west/javahtml/Regression.html)

Activity: Activity 5.2: Age and Flexibility (p. 268 and *Activities Manual,* p. 48) ★ and "Regression" (*Instructor's Resource Binder* & CD, Chapter 5 Activities Worksheets)

■ **Section 5.4**

Section 5.4 discusses nonlinear relationships. Although linear regression is the model that receives the most attention in an introductory statistics course, students may have experience with the other regression functions on the calculator from either algebra or precalculus classes.

On page 241, a polynomial regression curve is given by the following function: $\hat{y} = a + b_1x + b_2x^2 + b_3x^3 + \cdots + b_kx^k$. Notice that the subscript of the coefficient is the power of the x-variable and that the intercept a could also be written as b_0x^0. This is helpful in explaining why the least-squares regression equation provided on the formula pages of the AP Exam is in the form $\hat{y} = b_0 + b_1x$, where b_0 is the intercept and b_1 is the slope of the LSRL.

In the AP Statistics course, students should be able to determine if the relationship between bivariate data is linear or nonlinear. If the relationship is nonlinear, then by using one of the methods described in this section students should be able to transform the data and compute an appropriate equation for the transformed data. When expressing a transformed equation, like the one in Example 5.17 on page 248, the "$\log(y)$" is actually the predicted $\widehat{\log(y)}$. Therefore, there should be a "hat" over the entire term, $\log(y)$. In addition to finding an appropriate model using transformed data, students should also be able to use the resulting equation to make predictions.

Suggested Assignment: Exercises 5.52, 5.53, 5.56, 5.57, 5.59

Activity: Bonus Activity 5.3: Paper Towels (*Activities Manual,* p. 50) ★ and Bonus Activity 5.4: Exponential Decay (*Activities Manual,* p. 52)

Assessment: Chapter 5 Quiz 2 (*Test Bank* and *Instructor's Resource Binder* CD), Chapter 5 Quiz 3 (*Test Bank* and *Instructor's Resource Binder* CD)

■ **Section 5.5**

Logistic regression is an optional topic and is not included as part of the AP Statistics Course Description. Logistic regression is used when the dependent variable is categorical, with only two values (thought of as success or failure). While the logistic equation (given on page 256) looks complicated, it has two very simple properties: (1) for any x-value, the value of the logistic function is between 0 and 1, and (2) the graph of the logistic equation has an "S" shape. Logistic regression has many applications in medicine and the social sciences.

Suggested Assignment: Exercises 5.61, 5.63, 5.65

Assessment: Chapter 5 Quiz 3 (*Test Bank* and *Instructor's Resource Binder* CD) and/or Chapter 5 Concept Quiz (*Test Bank* and *Instructor's Resource Binder* CD)

■ **Section 5.6**

Students should pay close attention to the "What to Look for in Published Data" and the "A Word to the Wise: Cautions and Limitations" sections.

Suggested Review Assignment: Exercises 5.66, 5.67, 5.68, 5.69, 5.72, 5.74, 5.76

Assessment: Chapter 5 Test (*Test Bank* and *Instructor's Resource Binder* CD)

★ Kathy's personal favorite

© Spencer Platt/Getty Images

Chapter 5

Summarizing Bivariate Data

Unusually large brain size at age 2 to 5 years is one indicator that a child may be at risk for autism. Research described in the *Journal of the American Medical Association* (July 2003) investigated whether head circumference at age 6 to 14 months could serve as a predictor of cerebral grey matter at age 2 to 5 years. Data on head circumference (measured at age 6 to 14 months) and cerebral grey matter (measured at age 2 to 5 years) for 17 male children with autism were used to explore the relationship between these two variables.

Questions of interest are: Is there a relationship between head circumference at age 6 to 14 months and the cerebral grey matter measurement at age 2 to 5 years? If so, can a head circumference measurement taken at an early age be used to predict what the grey matter measurement will be, potentially allowing doctors to detect autism at a younger age? How accurate are such predictions of grey matter?

In this chapter, we introduce methods for describing relationships between two numerical variables and for assessing the strength of a relationship. These methods allow us to answer questions such as the ones just posed regarding the relationship between head circumference at age 6 to 14 months and the grey matter measurement at age 2 to 5 years. In Chapter 13, methods of statistical inference for drawing conclusions from data consisting of observations on two variables are developed. The techniques introduced in this chapter are also important stepping stones for analyzing data consisting of observations on three or more variables, the topic of Chapter 14.

Improve your understanding and save time! Visit www.cengage.com/login where you will find:

- Step-by-step instructions for MINITAB, Excel, TI-83, SPSS, and JMP
- Video solutions to selected exercises
- Data sets available for selected examples and exercises

- Exam-prep pre-tests that build a Personalized Learning Plan based on your results so that you know exactly what to study
- Help from a live statistics tutor 24 hours a day

5.1 Correlation

An investigator is often interested in how two or more attributes of individuals or objects in a population are related to one another. For example, an environmental researcher might wish to know how the lead content of soil varies with distance from a major highway. Researchers in early childhood education might investigate how vocabulary size is related to age. College admissions officers, who must try to predict whether an applicant will succeed in college, might use a model relating college grade point average to high school grades, and ACT or SAT scores.

Recall that a scatterplot of bivariate numerical data gives a visual impression of how strongly x values and y values are related. However, to make precise statements and draw conclusions from data, we must go beyond pictures. A **correlation coefficient** (from *co-* and *relation*) is a numerical assessment of the strength of relationship between the x and y values in a set of (x, y) pairs. In this section, we introduce the most commonly used correlation coefficient.

Figure 5.1 displays several scatterplots that show different relationships between the x and y values. The plot in Figure 5.1(a) suggests a strong *positive relationship* between x and y; for every pair of points in the plot, the one with the larger x value also has the larger y value. That is, an increase in x is inevitably paired with an increase in y. The plot in Figure 5.1(b) shows a strong tendency for y to increase as x does, but there are a few exceptions. For example, the x and y values of the two points in the extreme upper right-hand corner of the plot go in opposite directions (x increases but y decreases). Nevertheless, a plot such as this would again indicate a rather strong

Figure 5.1 Scatterplots illustrating various types of relationships: (a) positive linear relation; (b) another positive linear relation; (c) negative linear relation; (d) no relation; (e) curved relation.

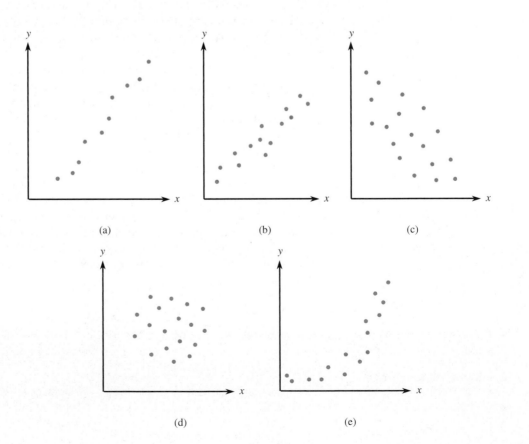

positive relationship. Figure 5.1(c) suggests that x and y are *negatively related*—as x increases, y tends to decrease. The negative relationship in this plot is not as strong as the positive relationship in Figure 5.1(b), although both plots show a well-defined linear pattern. The plot of Figure 5.1(d) indicates no strong relationship between x and y; there is no tendency for y either to increase or to decrease as x increases. Finally, as illustrated in Figure 5.1(e), a scatterplot can show evidence of a strong relationship that is curved rather than linear.

■ **Pearson's Sample Correlation Coefficient** ....

Let $(x_1, y_1), (x_2, y_2), \ldots, (x_n, y_n)$ denote a sample of (x, y) pairs. Consider replacing each x value by the corresponding z score, z_x (by subtracting $\bar{x}$ and then dividing by s_x) and similarly replacing each y value by its z score. Note that x values that are larger than $\bar{x}$ will have positive z scores and those smaller than $\bar{x}$ will have negative z scores. Also y values larger than $\bar{y}$ will have positive z scores and those smaller will have negative z scores. A common measure of the strength of linear relationship, called Pearson's sample correlation coefficient, is based on the sum of the products of z_x and z_y for each observation in the bivariate data set, $\sum z_x z_y$.

The scatterplot in Figure 5.2(a) indicates a strong positive relationship. A vertical line through $\bar{x}$ and a horizontal line through $\bar{y}$ divide the plot into four regions. In Region I, both x and y exceed their mean values, so the z score for x and the z score for y are both positive numbers. It follows that $z_x z_y$ is positive. The product of the z scores is also positive for any point in Region III, because both z scores are negative there and multiplying two negative numbers gives a positive number. In each of the other two regions, one z score is positive and the other is negative, so $z_x z_y$ is negative. But because the points generally fall in Regions I and III, the products of z scores tend to be positive. Thus, the *sum* of the products is a large positive number.

Similar reasoning for the data displayed in Figure 5.2(b), which exhibits a strong negative relationship, implies that $\sum z_x z_y$ is a large negative number. When there is no strong relationship, as in Figure 5.2(c), positive and negative products tend to counteract one another, producing a value of $\sum z_x z_y$ that is close to 0. In summary, $\sum z_x z_y$ seems to be a reasonable measure of the degree of association between x and y; it can be a large positive number, a large negative number, or a number close to 0, depending on whether there is a strong positive, a strong negative, or no strong linear relationship.

Pearson's sample correlation coefficient, denoted r, is obtained by dividing $\sum z_x z_y$ by $(n - 1)$.

D E F I N I T I O N

Pearson's sample correlation coefficient r is given by

$$r = \frac{\sum z_x z_y}{n - 1}$$

Although there are several different correlation coefficients, Pearson's correlation coefficient is by far the most commonly used, and so the name "Pearson's" is often omitted and it is referred to as simply **the correlation coefficient**.

Figure 5.2 Viewing a scatterplot according to the signs of z_x and z_y: (a) a positive relation; (b) a negative relation; (c) no strong relation.

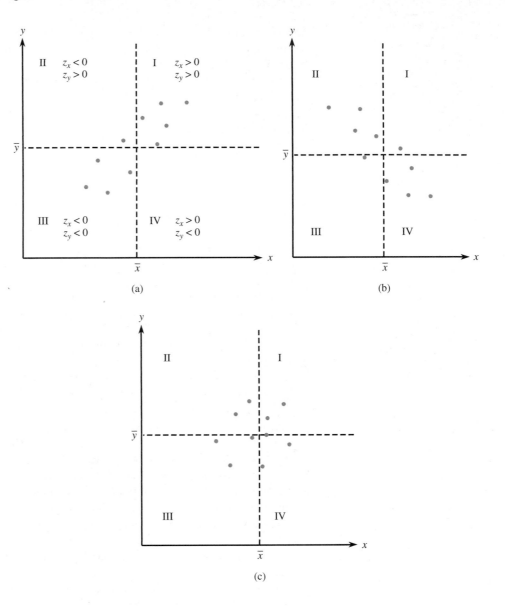

Hand calculation of the correlation coefficient is quite tedious. Fortunately, all statistical software packages and most scientific calculators can compute r once the x and y values have been input.

. . . .

Example 5.1   Graduation Rates and Student-Related Expenditures

● The web site www.collegeresults.org publishes data on U.S. colleges and universities. For the six primarily undergraduate public universities in California with enrollments between 10,000 and 20,000, six-year graduation rates and student-related expenditures per full-time student for 2003 were reported as follows:

● Data set available online

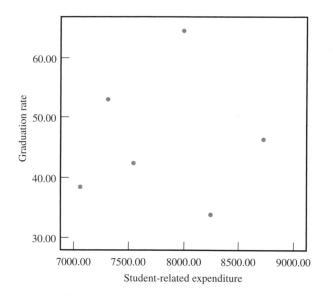

Figure 5.3 SPSS scatterplot for the data of Example 5.1.

Observation	1	2	3	4	5	6
Graduation rate	64.6	53.0	46.3	42.5	38.5	33.9
Student-related expenditure	8011	7323	8735	7548	7071	8248

Figure 5.3 is a scatterplot of these data generated using SPSS, a widely used statistics package.

Let x denote the student-related expenditure per full-time student and y denote the six-year graduation rate. It is easy to verify that

$$\bar{x} = 7822.67 \quad s_x = 622.71 \quad \bar{y} = 46.47 \quad s_y = 11.03$$

To illustrate the calculation of the correlation coefficient, we begin by computing z scores for each (x, y) pair in the data set. For example, the first observation is $(8011, 64.6)$. The corresponding z scores are

$$z_x = \frac{8011 - 7822.67}{622.71} = 0.30 \quad z_y = \frac{64.6 - 46.47}{11.03} = 1.64$$

The following table shows the z scores and the product $z_x z_y$ for each observation:

x	y	z_x	z_y	$z_x z_y$
8011	64.6	0.30	1.64	0.50
7323	53.0	−0.80	0.59	−0.48
8735	46.3	1.47	−0.02	−0.02
7548	42.5	−0.44	−0.36	0.16
7071	38.5	−1.21	−0.72	0.87
8248	33.9	0.68	−1.14	−0.78

$$\sum z_x z_y = 0.25$$

Then, with $n = 6$

$$r = \frac{\sum z_x z_y}{n - 1} = \frac{0.25}{5} = .05$$

SPSS was used to compute the correlation coefficient, producing the following computer output.

Correlations

Gradrate	Pearson Correlation	.050

Based on the scatterplot and the properties of the correlation coefficient presented in the discussion that follows this example, we conclude that there is at best a very weak positive linear relationship between student-related expenditure and graduation rate for these six universities.

■

■ Properties of *r*

See Teaching Tips on page T5-1 for tips to teach these properties.

1. *The value of r does not depend on the unit of measurement for either variable.* For example, if x is height, the corresponding z score is the same whether height is expressed in inches, meters, or miles, and thus the value of the correlation coefficient is not affected. The correlation coefficient measures the inherent strength of the linear relationship between two numerical variables.

2. *The value of r does not depend on which of the two variables is considered x.* Thus, if we had let x = graduation rate and y = student related expenditure in Example 5.1, the same value, $r = 0.05$, would have resulted.

3. *The value of r is between* -1 *and* $+1$. A value near the upper limit, $+1$, indicates a substantial positive relationship, whereas an r close to the lower limit, -1, suggests a substantial negative relationship. Figure 5.4 shows a useful way to describe the strength of relationship based on r. It may seem surprising that a value of r as extreme as $-.5$ or $.5$ should be in the weak category; an explanation for this is given later in the chapter. Even a weak correlation can indicate a meaningful relationship.

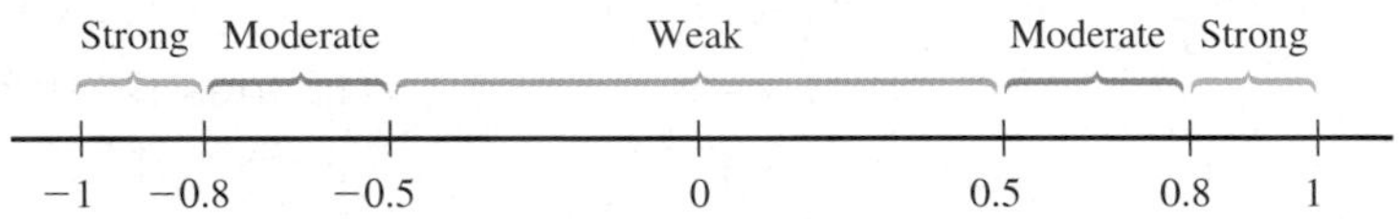

Figure 5.4 Describing the strength of a linear relationship.

4. *The correlation coefficient r* $= 1$ *only when all the points in a scatterplot of the data lie exactly on a straight line that slopes upward. Similarly, r* $= -1$ *only when all the points lie exactly on a downward-sloping line.* Only when there is a perfect linear relationship between x and y in the sample does r take on one of its two possible extreme values.

5. *The value of r is a measure of the extent to which x and y are linearly related*—that is, the extent to which the points in the scatterplot fall close to a straight line. A value of r close to 0 does not rule out *any* strong relationship between x and y; there could still be a strong relationship but one that is not linear.

Example 5.2 Nightingales: Song Repertoire Size and Body Characteristics

For male songbirds, both physical characteristics and the quality of the song play a role in a female's choice of a mate. The authors of the article "Song Repertoire Is Correlated with Body Measures and Arrival Date in Common Nightingales" (*Animal Behaviour* [2005]: 211–217) used data from $n = 20$ nightingales to reach the conclusion that there was a positive correlation between the number of different songs in a nightingale's repertoire and both body weight ($r = .53$) and wing length ($r = .47$), and a negative correlation between repertoire size and arrival date ($r = -.47$). This means that heavier birds tend to know more songs, as do birds with longer wings. The authors of the paper indicated that the observed correlation between repertoire size and body characteristics was unlikely to be due solely to the age of the bird, since all nightingales in the study were more than 3 years old and prior research indicates that repertoire size does not continue to increase with age after the third year.

The negative correlation between repertoire size and arrival date was interpreted as meaning that male nightingales who knew more songs tended to arrive at their breeding habitats earlier than those who knew fewer songs.

■

Example 5.3 Is Foal Weight Related to Mare Weight?

● Foal weight at birth is an indicator of health, so it is of interest to breeders of thoroughbred horses. Is foal weight related to the weight of the mare (mother)? The accompanying data are from the article "Suckling Behaviour Does Not Measure Milk Intake in Horses" (*Animal Behaviour* [1999]: 673–678):

Observation	1	2	3	4	5	6	7	8
Mare weight (x, in kg)	556	638	588	550	580	642	568	642
Foal weight (y, in kg)	129	119	132	123.5	112	113.5	95	104

Observation	9	10	11	12	13	14	15
Mare weight (x, in kg)	556	616	549	504	515	551	594
Foal weight (y, in kg)	104	93.5	108.5	95	117.5	128	127.5

MINITAB was used to compute the value of the correlation coefficient (Pearson's) for these data, with the following result:

Correlations (Pearson)
Correlation of Mare Weight and Foal Weight = 0.001

A correlation coefficient this close to 0 indicates no linear relationship between mare weight and foal weight. A scatterplot of the data (Figure 5.5) supports the conclusion that mare weight and foal weight are unrelated. From the correlation coefficient alone, we can conclude only that there is no *linear* relationship. We cannot rule out a more complicated curved relationship without also examining the scatterplot.

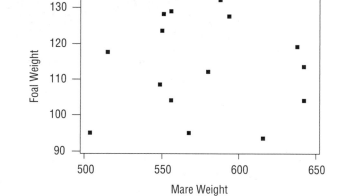

Figure 5.5 A MINITAB scatterplot of the mare and foal weight data of Example 5.3.

■

Example 5.4 Age and Marathon Times

● The article "Master's Performance in the New York City Marathon" (*British Journal of Sports Medicine* [2004]: 408–412) gave the following data on the average finishing time by age group for female participants in the 1999 marathon.

 Step-by-step technology instructions available online   ● Data set available online

Age Group	Representative Age	Average Finish Time
10–19	15	302.38
20–29	25	193.63
30–39	35	185.46
40–49	45	198.49
50–59	55	224.30
60–69	65	288.71

The scatterplot of average finish time versus representative age is shown in Figure 5.6.

Figure 5.6 Scatterplot of y = average finish time and x = age for the data of Example 5.4.

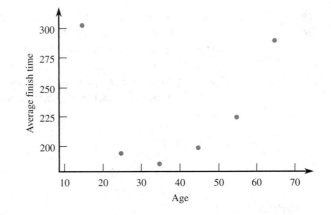

Using MINITAB to compute Pearson's correlation coefficient between age and average finish time results in the following:

Correlations: Age, Average Finish Time
Pearson correlation of Age and Average Finish Time = 0.038

This example shows the importance of interpreting r as a measure of the strength of a *linear* association. Here, r is not large, but there is a strong nonlinear relationship between age and average finish time. The point is we should not conclude that there is no relationship whatsoever simply because the value of r is small in absolute value.

■

■ The Population Correlation Coefficient ....

The sample correlation coefficient r measures how strongly the x and y values in a *sample* of pairs are linearly related to one another. There is an analogous measure of how strongly x and y are related in the entire population of pairs from which the sample was obtained. It is called the **population correlation coefficient** and is denoted ρ. (Notice again the use of a Greek letter for a population characteristic and a Roman letter for a sample characteristic.) We will never have to calculate ρ from an entire population of pairs, but it is important to know that ρ satisfies properties paralleling those of r:

ρ is the Greek letter rho.

1. ρ is a number between -1 and $+1$ that does not depend on the unit of measurement for either x or y, or on which variable is labeled x and which is labeled y.
2. $\rho = +1$ or -1 if and only if all (x, y) pairs in the population lie exactly on a straight line, so ρ measures the extent to which there is a linear relationship in the population.

In Chapter 13, we show how the sample correlation coefficient r can be used to make an inference concerning the population correlation coefficient ρ.

■ Correlation and Causation

A value of r close to 1 indicates that relatively large values of one variable tend to be associated with relatively large values of the other variable. This is far from saying that a large value of one variable *causes* the value of the other variable to be large. Correlation measures the extent of association, but *association does not imply causation*. It frequently happens that two variables are highly correlated not because one is causally related to the other but because they are both strongly related to a third variable. Among all elementary school children, the relationship between the number of cavities in a child's teeth and the size of his or her vocabulary is strong and positive. Yet no one advocates eating foods that result in more cavities to increase vocabulary size (or working to decrease vocabulary size to protect against cavities). Number of cavities and vocabulary size are both strongly related to age, so older children tend to have higher values of both variables than do younger ones. In the ABCNews.com series "Who's Counting?" (February 1, 2001), John Paulos reminded readers that correlation does not imply causation and gave the following example: Consumption of hot chocolate is negatively correlated with crime rate (high values of hot chocolate consumption tend to be paired with lower crime rates), but both are responses to cold weather.

Scientific experiments can frequently make a strong case for causality by carefully controlling the values of all variables that might be related to the ones under study. Then, if y is observed to change in a "smooth" way as the experimenter changes the value of x, a most plausible explanation would be that there is a causal relationship between x and y. In the absence of such control and ability to manipulate values of one variable, we must admit the possibility that an unidentified underlying third variable is influencing both the variables under investigation. A high correlation in many uncontrolled studies carried out in different settings can also marshal support for causality—as in the case of cigarette smoking and cancer—but proving causality is an elusive task.

■ Exercises 5.1–5.16

Turn to the Expanded Answers Section for answers not shown next to exercises.

5.1 For each of the following pairs of variables, indicate whether you would expect a positive correlation, a negative correlation, or a correlation close to 0. Explain your choice.
a. Maximum daily temperature and cooling costs
b. Interest rate and number of loan applications

c. Incomes of husbands and wives when both have full-time jobs   Positive
d. Height and IQ   No correlation
e. Height and shoe size   Positive
f. Score on the math section of the SAT exam and score on the verbal section of the same test   Positive

Bold exercises answered in back   ● Data set available online but not required   ▼ Video solution available

g. Time spent on homework and time spent watching television during the same day by elementary school children
h. Amount of fertilizer used per acre and crop yield (Hint: As the amount of fertilizer is increased, yield tends to increase for a while but then tends to start decreasing.)

5.2 Is the following statement correct? Explain why or why not.

A correlation coefficient of 0 implies that no relationship exists between the two variables under study.

5.3 Draw two scatterplots, one for which $r = 1$ and a second for which $r = -1$.

5.4 The article "That's Rich: More You Drink, More You Earn" (*Calgary Herald*, April 16, 2002) reported that there was a positive correlation between alcohol consumption and income. Is it reasonable to conclude that increasing alcohol consumption will increase income? Give at least two reasons or examples to support your answer.

5.5 ● The paper "A Cross-National Relationship Between Sugar Consumption and Major Depression?" (*Depression and Anxiety* [2002]: 118–120) concluded that there was a correlation between refined sugar consumption (calories per person per day) and annual rate of major depression (cases per 100 people) based on data from 6 countries. The following data were read from a graph that appeared in the paper:

Country	Sugar Consumption	Depression Rate
Korea	150	2.3
United States	300	3.0
France	350	4.4
Germany	375	5.0
Canada	390	5.2
New Zealand	480	5.7

a. Compute and interpret the correlation coefficient for this data set.
b. Is it reasonable to conclude that increasing sugar consumption leads to higher rates of depression? Explain.
c. Do you have any concerns about this study that would make you hesitant to generalize these conclusions to other countries?

5.6 ● Cost-to-charge ratios (the percentage of the amount billed that represents the actual cost) for 11 Oregon hospitals of similar size were reported separately for inpatient and outpatient services. The data are

Hospital	Cost-to-Charge Ratio Inpatient	Cost-to-Charge Ratio Outpatient
Blue Mountain	80	62
Curry General	76	66
Good Shepherd	75	63
Grande Ronde	62	51
Harney District	100	54
Lake District	100	75
Pioneer	88	65
St. Anthony	64	56
St. Elizabeth	50	45
Tillamook	54	48
Wallowa Memorial	83	71

a. Does there appear to be a strong linear relationship between the cost-to-charge ratio for inpatient and outpatient services? Justify your answer based on the value of the correlation coefficient and examination of a scatterplot of the data.
b. Are any unusual features of the data evident in the scatterplot?
c. Suppose that the observation for Harney District was removed from the data set. Would the correlation coefficient for the new data set be greater than or less than the one computed in Part (a)? Explain.

5.7 The following time-series plot is based on data from the article "Bubble Talk Expands: Corporate Debt Is Latest Concern Turning Heads" (*San Luis Obispo Tribune*, September 13, 2002) and shows how household debt and corporate debt have changed over the time period from 1991 (year 1 in the graph) to 2002:

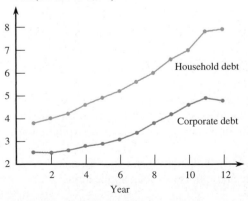

Based on the time-series plot, would the correlation coefficient between household debt and corporate debt be positive or negative? Weak or strong? What aspect of the time-series plot supports your answer?   Strong and positive.

5.8 ● Data from the U.S. Federal Reserve Board (Household Debt Service Burden, 2002) on the percentage of disposable personal income required to meet consumer loan payments and mortgage payments for selected years are shown in the following table:

Consumer Debt	Household Debt	Consumer Debt	Household Debt
7.88	6.22	6.24	5.73
7.91	6.14	6.09	5.95
7.65	5.95	6.32	6.09
7.61	5.83	6.97	6.28
7.48	5.83	7.38	6.08
7.49	5.85	7.52	5.79
7.37	5.81	7.84	5.81
6.57	5.79		

a. What is the value of the correlation coefficient for this data set? $r = 0.1178$

b. Is it reasonable to conclude in this case that there is no strong relationship between the variables (linear or otherwise)? Use a graphical display to support your answer.

5.9 ● ▼ The accompanying data were read from graphs that appeared in the article "Bush Timber Proposal Runs Counter to the Record" (*San Luis Obispo Tribune*, September 22, 2002). The variables shown are the number of acres burned in forest fires in the western United States and timber sales.

Year	Number of Acres Burned (thousands)	Timber Sales (billions of board feet)
1945	200	2.0
1950	250	3.7
1955	260	4.4
1960	380	6.8
1965	80	9.7
1970	450	11.0
1975	180	11.0
1980	240	10.2
1985	440	10.0
1990	400	11.0
1995	180	3.8

a. Is there a correlation between timber sales and acres burned in forest fires? Compute and interpret the value of the correlation coefficient.

b. The article concludes that "heavier logging led to large forest fires." Do you think this conclusion is justified based on the given data? Explain.

5.10 ● Peak heart rate (beats per minute) was determined both during a shuttle run and during a 300-yard run for a sample of $n = 10$ individuals with Down syndrome ("Heart Rate Responses to Two Field Exercise Tests by Adolescents and Young Adults with Down Syndrome," *Adapted Physical Activity Quarterly* [1995]: 43–51), resulting in the following data:

Shuttle	168	168	188	172	184	176	192
300-yd	184	192	200	192	188	180	182

Shuttle	172	188	180
300-yd	188	196	196

a. Construct a scatterplot of the data. What does the scatterplot suggest about the nature of the relationship between the two variables?

b. With x = shuttle run peak rate and y = 300-yd run peak rate, calculate r. Is the value of r consistent with your answer in Part (a)?

c. With x = 300-yd peak rate and y = shuttle run peak rate, how does the value of r compare to what you calculated in Part (b)?

5.11 The accompanying scatterplot shows observations on hemoglobin level, determined both by the standard spectrophotometric method (y) and by a new, simpler method based on a color scale (x) ("A Simple and Reliable Method for Estimating Hemoglobin," *Bulletin of the World Health Organization* [1995]: 369–373).

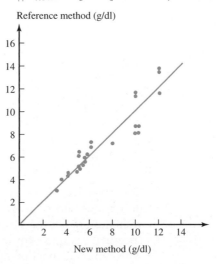

a. Does it appear that x and y are highly correlated?

b. The paper reported that $r = .9366$. How would you describe the relationship between the two variables?

c. The line pictured in the scatterplot has a slope of 1 and passes through $(0, 0)$. If x and y were always identical, all points would lie exactly on this line. The authors of the paper claimed that perfect correlation ($r = 1$) would result in this line. Do you agree? Explain your reasoning.

5.12 ● The following data on x = score on a measure of test anxiety and y = exam score for a sample of $n = 9$ students are consistent with summary quantities given in the paper "Effects of Humor on Test Anxiety and Performance" (*Psychological Reports* [1999]: 1203–1212):

x	23	14	14	0	17	20	20	15	21
y	43	59	48	77	50	52	46	51	51

Higher values for x indicate higher levels of anxiety.

a. Construct a scatterplot, and comment on the features of the plot.

b. Does there appear to be a linear relationship between the two variables? How would you characterize the relationship?

c. Compute the value of the correlation coefficient. Is the value of r consistent with your answer to Part (b)?

d. Is it reasonable to conclude that test anxiety caused poor exam performance? Explain.

5.13 According to the article "First-Year Academic Success: A Prediction Combining Cognitive and Psychosocial Variables for Caucasian and African American Students" (*Journal of College Student Development* [1999]: 599–605), there is a mild correlation between high school GPA (x) and first-year college GPA (y). The data can be summarized as follows:

$$n = 2600 \qquad \sum x = 9620 \qquad \sum y = 7436$$

$$\sum xy = 27{,}918 \qquad \sum x^2 = 36{,}168 \qquad \sum y^2 = 23{,}145$$

An alternative formula for computing the correlation coefficient that is based on raw data and is algebraically equivalent to the one given in the text is

$$r = \frac{\sum xy - \dfrac{(\sum x)(\sum y)}{n}}{\sqrt{\sum x^2 - \dfrac{(\sum x)^2}{n}}\,\sqrt{\sum y^2 - \dfrac{(\sum y)^2}{n}}}$$

Use this formula to compute the value of the correlation coefficient, and interpret this value.

5.14 An auction house released a list of 25 recently sold paintings. Eight artists were represented in these sales. The sale price of each painting appears on the list. Would the correlation coefficient be an appropriate way to summarize the relationship between artist (x) and sale price (y)? Why or why not?   No, because x, artist, is not a numerical variable.

5.15 Each individual in a sample was asked to indicate on a quantitative scale how willing he or she was to spend money on the environment and also how strongly he or she believed in God ("Religion and Attitudes Toward the Environment," *Journal for the Scientific Study of Religion* [1993]: 19–28). The resulting value of the sample correlation coefficient was $r = -.085$. Would you agree with the stated conclusion that stronger support for environmental spending is associated with a weaker degree of belief in God? Explain your reasoning.

5.16 A sample of automobiles traversing a certain stretch of highway is selected. Each one travels at roughly a constant rate of speed, although speed does vary from auto to auto. Let x = speed and y = time needed to traverse this segment of highway. Would the sample correlation coefficient be closest to .9, .3, $-.3$, or $-.9$? Explain.

Bold exercises answered in back   ● Data set available online but not required   ▼ Video solution available

5.2   Linear Regression: Fitting a Line to Bivariate Data

The objective of *regression analysis* is to use information about one variable, x, to draw some sort of conclusion concerning a second variable, y. Often an investigator wants to predict the y value that would result from making a single observation at a specified x value—for example, to predict y = product sales during a given period when amount spent on advertising is $x = \$10{,}000$. The two variables in a regression analysis play different roles: y is called the **dependent** or **response variable**, and x is referred to as the **independent**, **predictor**, or **explanatory variable**.

Scatterplots frequently exhibit a linear pattern. It is natural in such cases to summarize the relationship between the variables by finding a line that is as close as possible to the points in the plot. Before seeing how this might be done, let's review some elementary facts about lines and linear relationships.

The general form of a linear relation between x and y is $y = a + bx$. A particular relation is specified by choosing values of a and b. Thus, one such relationship is $y = 10 + 2x$; another is $y = 100 − 5x$. If we choose some x values and compute $y = a + bx$ for each value, the points in the plot of the resulting (x, y) pairs will fall exactly on a straight line.

*When interpreting the slope of the LSRL, be sure to stress that it is the **estimated or approximate** increase in y for 1-unit increase in x.*

DEFINITION

The relationship

Intercept

$$y = a + bx$$

Slope

is the equation of a straight line. The value of b, called the **slope** of the line, is the amount by which y increases when x increases by 1 unit. The value of a, called the **intercept** (or sometimes the **y-intercept** or **vertical intercept**) of the line, is the height of the line above the value $x = 0$.

The equation $y = 10 + 2x$ has slope $b = 2$, so each 1-unit increase in x is paired with an increase of 2 in y. When $x = 0$, $y = 10$, so the height at which the line crosses the vertical axis (where $x = 0$) is 10. This is illustrated in Figure 5.7(a). The slope of the line determined by $y = 100 − 5x$ is $−5$, so y increases by $−5$ (i.e., decreases by 5) when x increases by 1. The height of the line above $x = 0$ is $a = 100$. The resulting line is pictured in Figure 5.7(b).

Figure 5.7 Graphs of two lines: (a) slope $b = 2$, intercept $a = 10$; (b) slope $b = −5$, intercept $a = 100$.

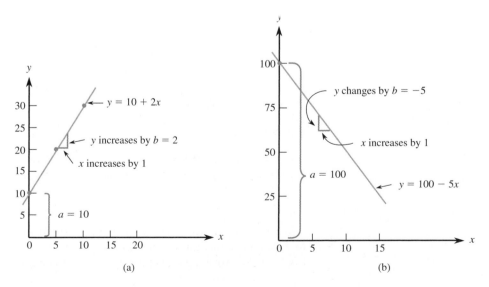

It is easy to draw the line corresponding to any particular linear equation. Choose any two *x* values and substitute them into the equation to obtain the corresponding *y* values. Then plot the resulting two (*x*, *y*) pairs as two points. The desired line is the one passing through these points. For the equation $y = 10 + 2x$, substituting $x = 5$ yields $y = 20$, whereas using $x = 10$ gives $y = 30$. The resulting two points are then (5, 20) and (10, 30). The line in Figure 5.7(a) passes through these points.

∎ Fitting a Straight Line: The Principle of Least Squares

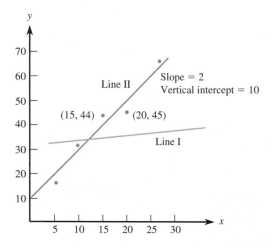

Figure 5.8 Line I gives a poor fit and Line II gives a good fit to the data.

Figure 5.8 shows a scatterplot with two lines superimposed on the plot. Line II clearly gives a better fit to the data than does Line I. In order to measure the extent to which a particular line provides a good fit, we focus on the vertical deviations from the line. For example, Line II in Figure 5.8 has equation $y = 10 + 2x$, and the third and fourth points from the left in the scatterplot are (15, 44) and (20, 45). For these two points, the vertical deviations from this line are

$$\text{3rd deviation} = y_3 - \text{height of the line above } x_3$$
$$= 44 - [10 + 2(15)]$$
$$= 4$$

and

$$\text{4th deviation} = 45 - [10 + 2(20)] = -5$$

A positive vertical deviation results from a point that lies above the chosen line, and a negative deviation results from a point that lies below this line. A particular line is said to be a good fit to the data if the deviations from the line are small in magnitude. Line I in Figure 5.8 fits poorly, because all deviations from that line are larger in magnitude (some are much larger) than the corresponding deviations from Line II.

To assess the overall fit of a line, we need a way to combine the *n* deviations into a single measure of fit. The standard approach is to square the deviations (to obtain nonnegative numbers) and then to sum these squared deviations.

> **DEFINITION**
>
> The most widely used criterion for measuring the goodness of fit of a line $y = a + bx$ to bivariate data $(x_1, y_1), \ldots, (x_n, y_n)$ is the **sum of the squared deviations** about the line
>
> $$\sum [y - (a + bx)]^2 = [y_1 - (a + bx_1)]^2 + [y_2 - (a + bx_2)]^2 + \cdots + [y_n - (a + bx_n)]^2$$
>
> The **least-squares line**, also called the **sample regression line**, is the line that minimizes this sum of squared deviations.

Fortunately, the equation of the least-squares line can be obtained without having to calculate deviations from any particular line. The accompanying box gives relatively simple formulas for the slope and intercept of the least-squares line.

The slope of the least-squares line is

$$b = \frac{\Sigma(x - \bar{x})(y - \bar{y})}{\Sigma(x - \bar{x})^2}$$

and the y intercept is

$$a = \bar{y} - b\bar{x}$$

We write the equation of the least-squares line as

$$\hat{y} = a + bx$$

Students should use $\hat{y}$ for the predicted y.

where the ^ above y indicates that $\hat{y}$ (read as y-hat) is a prediction of y resulting from the substitution of a particular x value into the equation.

Statistical software packages and many calculators can compute the slope and intercept of the least-squares line. If the slope and intercept are to be computed by hand, the following computational formula can be used to reduce the amount of time required to perform the calculations.

Calculating Formula for the Slope of the Least-Squares Line

$$b = \frac{\Sigma xy - \dfrac{(\Sigma x)(\Sigma y)}{n}}{\Sigma x^2 - \dfrac{(\Sigma x)^2}{n}}$$

....

Example 5.5   Time to Defibrillator Shock and Heart Attack Survival Rate

● Studies have shown that people who suffer sudden cardiac arrest (SCA) have a better chance of survival if a defibrillator shock is administered very soon after cardiac arrest. How is survival rate related to the time between when cardiac arrest occurs and when the defibrillator shock is delivered? This question is addressed in the paper "Improving Survival from Sudden Cardiac Arrest: The Role of Home Defibrillators" (by J. K. Stross, University of Michigan, February 2002; available at www.heartstarthome.com). The accompanying data give y = survival rate (percent) and x = mean call-to-shock time (minutes) for a cardiac rehabilitation center (where cardiac arrests occurred while victims were hospitalized and so the call-to-shock time tended to be short) and for four communities of different sizes:

Mean call-to-shock time, x	2	6	7	9	12
Survival rate, y	90	45	30	5	2

!! Step-by-step technology instructions available online   ● Data set available online

A scatterplot of these data (Figure 5.9) shows that the relationship between survival rate and mean call-to-shock time for times in the range 2–12 min could reasonably be summarized by a straight line.

The summary quantities necessary to compute the equation of the least-squares line are

$$\sum x = 36 \qquad \sum x^2 = 314 \qquad \sum xy = 729$$
$$\sum y = 172 \qquad \sum y^2 = 11{,}054$$

From these quantities, we compute

$$\bar{x} = 7.20 \qquad \bar{y} = 34.4$$

$$b = \frac{\sum xy - \dfrac{(\sum x)(\sum y)}{n}}{\sum x^2 - \dfrac{(\sum x)^2}{n}} = \frac{729 - \dfrac{(36)(172)}{5}}{314 - \dfrac{(36)^2}{5}} = \frac{-509.40}{54.80} = -9.30$$

and

$$a = \bar{y} - b\bar{x} = 34.4 - (-9.30)(7.20) = 34.4 + 66.96$$
$$= 101.36$$

The least-squares line is then

$$\hat{y} = 101.36 - 9.30x$$

This line is also shown on the scatterplot of Figure 5.9.

If we wanted to predict SCA survival rate for a community with a mean call-to-shock time of 5 minutes, we could use the point on the least-squares line above $x = 5$:

$$\hat{y} = 101.36 - 9.30(5) = 101.36 - 46.50 = 54.86$$

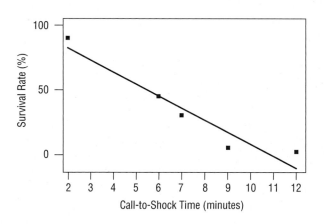

Figure 5.9   MINITAB scatterplot for the data of Example 5.5.

Predicted survival rates for communities with different mean call-to-shock times could be obtained in a similar way.

But, be careful in making predictions—the least-squares line should not be used to predict survival rate for communities much outside the range 2–12 min (the range of x values in the data set) because we do not know whether the linear pattern observed in the scatterplot continues outside this range. This is sometimes referred to as the **danger of extrapolation**.

In this example, we can see that using the least-squares line to predict survival rate for communities with mean call-to-shock times much above 12 min leads to nonsensical predictions. For example, if the mean call-to-shock time is 15 min, the predicted survival rate is negative:

$$\hat{y} = 101.36 - 9.30(15) = -38.14$$

Because it is impossible for survival rate to be negative, this is a clear indication that the pattern observed for x values in the 2–12 range does not continue outside this range. Nonetheless, the least-squares line can be a useful tool for making predictions for x values within the 2- to 12-min range.

Calculations involving the least-squares line can obviously be tedious. This is where the computer comes to our rescue. All the standard statistical packages can fit a straight line to bivariate data.

Example 5.6

Is Age Related to Recovery Time for Injured Athletes?

● How quickly can athletes return to their sport following injuries requiring surgery? The paper "Arthroscopic Distal Clavicle Resection for Isolated Atraumatic Osteolysis in Weight Lifters" (*American Journal of Sports Medicine* [1998]: 191–195) gave the following data on $x =$ age and $y =$ days after arthroscopic shoulder surgery before being able to return to their sport, for 10 weight lifters:

x	33	31	32	28	33	26	34	32	28	27
y	6	4	4	1	3	3	4	2	3	2

Figure 5.10 is a scatterplot of the data. The predominant pattern is linear, although the points in the plot spread out quite a bit about *any* line (even the least-squares line).

Figure 5.10 Scatterplot and least-squares line for the data of Example 5.6.

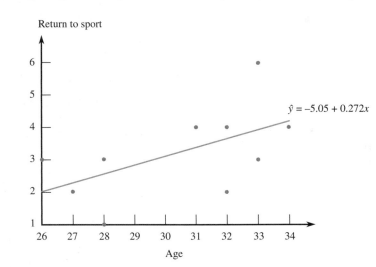

MINITAB was used to fit the least-squares line, and Figure 5.11 shows part of the resulting output. Instead of x and y, the variable labels "Return to Sport" and "Age" are used. The equation at the top is that of the least-squares line. In the rectangular table just below the equation, the first row gives information about the intercept, a, and the second row gives information concerning the slope, b. In particular, the coefficient column labeled "Coef" contains the values of a and b using more digits than in the rounded values that appear in the equation.

Figure 5.11 Partial MINITAB output for Example 5.6.

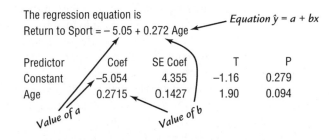

The regression equation is

Return to Sport = − 5.05 + 0.272 Age ← *Equation $\hat{y} = a + bx$*

Predictor	Coef	SE Coef	T	P
Constant	−5.054	4.355	−1.16	0.279
Age	0.2715	0.1427	1.90	0.094

Value of a   *Value of b*

● Data set available online

The least-squares line should not be used to predict the number of days before returning to practice for weight lifters with ages such as $x = 21$ or $x = 48$. These x values are well outside the range of the data, and there is no evidence to support extrapolation of the linear relationship.

■

■ Regression ....

The least-squares line is often called the **sample regression line**. This terminology comes from the relationship between the least-squares line and Pearson's correlation coefficient. To understand this relationship, we first need alternative expressions for the slope b and the equation of the line itself. With s_x and s_y denoting the sample standard deviations of the x's and y's, respectively, a bit of algebraic manipulation gives

$$b = r\left(\frac{s_y}{s_x}\right)$$

$$\hat{y} = \bar{y} + r\left(\frac{s_y}{s_x}\right)(x - \bar{x})$$

You do not need to use these formulas in any computations, but several of their implications are important for appreciating what the least-squares line does.

1. When $x = \bar{x}$ is substituted in the equation of the line, $\hat{y} = \bar{y}$ results. That is, the least-squares line passes through the *point of averages* $(\bar{x}, \bar{y})$.
2. Suppose for the moment that $r = 1$, so that all points lie exactly on the line whose equation is

$$\hat{y} = \bar{y} + \frac{s_y}{s_x}(x - \bar{x})$$

Now substitute $x = \bar{x} + s_x$, which is 1 standard deviation above $\bar{x}$:

$$\hat{y} = \bar{y} + \frac{s_y}{s_x}(\bar{x} + s_x - \bar{x}) = \bar{y} + s_y$$

That is, with $r = 1$, when x is 1 standard deviation above its mean, we predict that the associated y value will be 1 standard deviation above its mean. Similarly, if $x = \bar{x} - 2s_x$ (2 standard deviations below its mean), then

$$\hat{y} = \bar{y} + \frac{s_y}{s_x}(\bar{x} - 2s_x - \bar{x}) = \bar{y} - 2s_y$$

which is also 2 standard deviations below the mean. If $r = -1$, then $x = \bar{x} + s_x$ results in $\hat{y} = \bar{y} - s_y$, so the predicted y is also 1 standard deviation from its mean but on the opposite side of $\bar{y}$ from where x is relative to $\bar{x}$. In general, if x and y are perfectly correlated, the predicted y value associated with a given x value will be the same number of standard deviations (of y) from its mean $\bar{y}$ as x is from its mean, $\bar{x}$.
3. Now suppose that x and y are not perfectly correlated. For example, suppose $r = .5$, so the least-squares line has the equation

$$\hat{y} = \bar{y} + .5\left(\frac{s_y}{s_x}\right)(x - \bar{x})$$

Then substituting $x = \bar{x} + s_x$ gives

$$\hat{y} = \bar{y} + .5\left(\frac{s_y}{s_x}\right)(\bar{x} + s_x - \bar{x}) = \bar{y} + .5s_y$$

That is, for $r = .5$, when x lies 1 standard deviation above its mean, we predict that y will be only 0.5 standard deviation above its mean. Similarly, we can predict y when r is negative. If $r = -.5$, then the predicted y value will be only half the number of standard deviations from $\bar{y}$ that it is from $\bar{x}$ but x and the predicted y will now be on opposite sides of their respective means.

Consider using the least-squares line to predict the value of y associated with an x value some specified number of standard deviations away from $\bar{x}$. Then the predicted y value will be only r times this number of standard deviations from $\bar{y}$. In terms of standard deviations, except when $r = 1$ or -1, the predicted y will always be closer to $\bar{y}$ than x is to $\bar{x}$.

Using the least-squares line for prediction results in a predicted y that is pulled back in, or regressed, toward the mean of y compared to where x is relative to the mean of x. This regression effect was first noticed by Sir Francis Galton (1822–1911), a famous biologist, when he was studying the relationship between the heights of fathers and their sons. He found that predicted heights of sons whose fathers were above average in height were also above average (because r is positive here) but not by as much as the father's height; he found a similar relationship for fathers whose heights were below average. This regression effect has led to the term **regression analysis** for the collection of methods involving the fitting of lines, curves, and more complicated functions to bivariate and multivariate data.

The alternative form of the regression (least-squares) line emphasizes that predicting y from knowledge of x is not the same problem as predicting x from knowledge of y. The slope of the least-squares line for predicting x is $r(s_x/s_y)$ rather than $r(s_y/s_x)$ and the intercepts of the lines are almost always different. For purposes of prediction, it makes a difference whether y is regressed on x, as we have done, or x is regressed on y. *The regression line of y on x should not be used to predict x, because it is not the line that minimizes the sum of squared deviations in the x direction.*

Important concept

Turn to the Expanded Answers Section for answers not shown next to exercises.

■ Exercises 5.17–5.35

5.17 ● ▼ The article "Air Pollution and Medical Care Use by Older Americans" (*Health Affairs* [2002]: 207–214) gave data on a measure of pollution (in micrograms of particulate matter per cubic meter of air) and the cost of medical care per person over age 65 for six geographical regions of the United States:

Region	Pollution	Cost of Medical Care
North	30.0	915
Upper South	31.8	891
Deep South	32.1	968
West South	26.8	972
Big Sky	30.4	952
West	40.0	899

Bold exercises answered in back   ● Data set available online but not required   ▼ Video solution available

a. Construct a scatterplot of the data. Describe any interesting features of the scatterplot.

b. Find the equation of the least-squares line describing the relationship between y = medical cost and x = pollution. $\hat{y} = 1082.24 - 4.691x$

c. Is the slope of the least-squares line positive or negative? Is this consistent with your description of the relationship in Part (a)?   Negative, consistent with description.

d. Do the scatterplot and the equation of the least-squares line support the researchers' conclusion that elderly people who live in more polluted areas have higher medical costs? Explain.   No, relationship is negative.

5.18 ● Researchers asked each child in a sample of 411 school-age children if they were more or less likely to purchase a lottery ticket at a store if lottery tickets were visible on the counter. The percentage that said that they were *more* likely to purchase a ticket by grade level are as follows (R&J Child Development Consultants, Quebec, 2001):

Grade	Percentage That Said They Were More Likely to Purchase
6	32.7
8	46.1
10	75.0
12	83.6

a. Construct a scatterplot of y = percentage who said they were more likely to purchase and x = grade. Does there appear to be a linear relationship between x and y?

b. Find the equation of the least-squares line. $\hat{y} = -22.37 + 9.08x$

5.19 A sample of 548 ethnically diverse students from Massachusetts were followed over a 19-month period from 1995 and 1997 in a study of the relationship between TV viewing and eating habits (*Pediatrics* [2003]: 1321–1326). For each additional hour of television viewed per day, the number of fruit and vegetable servings per day was found to decrease on average by 0.14 serving.

a. For this study, what is the dependent variable? What is the predictor variable?

b. Would the least-squares line for predicting number of servings of fruits and vegetables using number of hours spent watching TV as a predictor have a positive or negative slope? Explain.

5.20 The relationship between hospital patient-to-nurse ratio and various characteristics of job satisfaction and patient care has been the focus of a number of research studies. Suppose x = patient-to-nurse ratio is the predictor variable. For each of the following potential dependent variables, indicate whether you expect the slope of the least-squares line to be positive or negative and give a brief explanation for your choice.

a. y = a measure of nurse's job satisfaction (higher values indicate higher satisfaction)

b. y = a measure of patient satisfaction with hospital care (higher values indicate higher satisfaction)

c. y = a measure of patient quality of care.

5.21 ● ▼ The accompanying data on x = head circumference z score (a comparison score with peers of the same age—a positive score suggests a larger size than for peers) at age 6 to 14 months and y = volume of cerebral grey matter (in ml) at age 2 to 5 years were read from a graph in the article described in the chapter introduction (*Journal of the American Medical Association* [2003]).

Cerebral Grey Matter (ml) 2–5 yr	Head Circumference z Scores at 6–14 Months
680	−.75
690	1.2
700	−.3
720	.25
740	.3
740	1.5
750	1.1
750	2.0
760	1.1
780	1.1
790	2.0
810	2.1
815	2.8
820	2.2
825	.9
835	2.35
840	2.3
845	2.2

a. Construct a scatterplot for these data.

b. What is the value of the correlation coefficient?

c. Find the equation of the least-squares line.

d. Predict the volume of cerebral grey matter for a child whose head circumference z score at age 12 months was 1.8.   790.68

e. Explain why it would not be a good idea to use the least-squares line to predict the volume of grey matter for a child whose head circumference z score was 3.0.

5.22 In the article "Reproductive Biology of the Aquatic Salamander *Amphiuma tridactylum* in Louisiana" (*Journal of Herpetology* [1999]: 100–105), 14 female salamanders were studied. Using regression, the researchers predicted y = clutch size (number of salamander eggs) from x = snout-vent length (in centimeters) as follows:

$$\hat{y} = -147 + 6.175x$$

For the salamanders in the study, the range of snout-vent lengths was approximately 30 to 70 cm.

a. What is the value of the y intercept of the least-squares line? What is the value of the slope of the least-squares line? Interpret the slope in the context of this problem.

b. Would you be reluctant to predict the clutch size when snout-vent length is 22 cm? Explain.   22 cm is outside the 30 to 70 cm range of x values in the data set.

5.23 ● Percentages of public school students in fourth grade in 1996 and in eighth grade in 2000 who were at or above the proficient level in mathematics were given in the article "Mixed Progress in Math" (*USA Today*, August 3, 2001) for eight western states:

State	4th grade (1996)	8th grade (2000)
Arizona	15	21
California	11	18
Hawaii	16	16
Montana	22	37
New Mexico	13	13
Oregon	21	32
Utah	23	26
Wyoming	19	25

a. Construct a scatterplot, and comment on any interesting features.

b. Find the equation of the least-squares line that summarizes the relationship between x = 1996 fourth-grade math proficiency percentage and y = 2000 eighth-grade math proficiency percentage. $\hat{y} = -3.14 + 1.52x$

c. Nevada, a western state not included in the data set, had a 1996 fourth-grade math proficiency of 14%. What would you predict for Nevada's 2000 eighth-grade math proficiency percentage? How does your prediction compare to the actual eighth-grade value of 20 for Nevada?   The predicted value of 18 is lower than the reported value of 20.

5.24 Data on high school GPA (x) and first-year college GPA (y) collected from a southeastern public research university can be summarized as follows ("First-Year Academic Success: A Prediction Combining Cognitive and Psychosocial Variables for Caucasian and African American Students," *Journal of College Student Development* [1999]: 599–605):

$$n = 2600 \qquad \sum x = 9620 \qquad \sum y = 7436$$
$$\sum xy = 27{,}918 \qquad \sum x^2 = 36{,}168 \qquad \sum y^2 = 23{,}145$$

a. Find the equation of the least-squares regression line.

b. Interpret the value of b, the slope of the least-squares line, in the context of this problem.

c. What first-year GPA would you predict for a student with a 4.0 high school GPA?   3.0716

5.25 ● Representative data on x = carbonation depth (in millimeters) and y = strength (in megapascals) for a sample of concrete core specimens taken from a particular building were read from a plot in the article "The Carbonation of Concrete Structures in the Tropical Environment of Singapore" (*Magazine of Concrete Research* [1996]: 293–300):

Depth, x	8.0	20.0	20.0	30.0	35.0
Strength, y	22.8	17.1	21.1	16.1	13.4
Depth, x	40.0	50.0	55.0	65.0	
Strength, y	12.4	11.4	9.7	6.8	

a. Construct a scatterplot. Does the relationship between carbonation depth and strength appear to be linear?

b. Find the equation of the least-squares line.

c. What would you predict for strength when carbonation depth is 25 mm?   17.5

d. Explain why it would not be reasonable to use the least-squares line to predict strength when carbonation depth is 100 mm.

5.26 The data given in Example 5.5 on x = call-to-shock time (in minutes) and y = survival rate (percent) were used to compute the equation of the least-squares line, which was

$$\hat{y} = 101.36 - 9.30x$$

The newspaper article "FDA OKs Use of Home Defibrillators" (*San Luis Obispo Tribune*, November 13, 2002)

Bold exercises answered in back   ● Data set available online but not required   ▼ Video solution available

reported that "every minute spent waiting for paramedics to arrive with a defibrillator lowers the chance of survival by 10 percent." Is this statement consistent with the given least-squares line? Explain.

5.27 An article on the cost of housing in California that appeared in the *San Luis Obispo Tribune* (March 30, 2001) included the following statement: "In Northern California, people from the San Francisco Bay area pushed into the Central Valley, benefiting from home prices that dropped on average $4000 for every mile traveled east of the Bay area." If this statement is correct, what is the slope of the least-squares regression line, $\hat{y} = a + bx$, where y = house price (in dollars) and x = distance east of the Bay (in miles)? Explain.   Slope = −4000

5.28 ● The following data on sale price, size, and land-to-building ratio for 10 large industrial properties appeared in the paper "Using Multiple Regression Analysis in Real Estate Appraisal" (*Appraisal Journal* [2002]: 424–430):

Property	Sale Price (millions of dollars)	Size (thousands of sq. ft.)	Land-to-Building Ratio
1	10.6	2166	2.0
2	2.6	751	3.5
3	30.5	2422	3.6
4	1.8	224	4.7
5	20.0	3917	1.7
6	8.0	2866	2.3
7	10.0	1698	3.1
8	6.7	1046	4.8
9	5.8	1108	7.6
10	4.5	405	17.2

a. Calculate and interpret the value of the correlation coefficient between sale price and size.
b. Calculate and interpret the value of the correlation coefficient between sale price and land-to-building ratio.
c. If you wanted to predict sale price and you could use either size or land-to-building ratio as the basis for making predictions, which would you use? Explain.
d. Based on your choice in Part (c), find the equation of the least-squares regression line you would use for predicting y = sale price. $\hat{y} = 1.333 + 0.00525x$

5.29 ● Representative data read from a plot that appeared in the paper "Effect of Cattle Treading on Erosion from

Hill Pasture: Modeling Concepts and Analysis of Rainfall Simulator Data" (*Australian Journal of Soil Research* [2002]: 963–977) on runoff sediment concentration for plots with varying amounts of grazing damage, measured by the percentage of bare ground in the plot, are given for gradually sloped plots and for steeply sloped plots.

Gradually Sloped Plots

Bare ground (%)	5	10	15	25
Concentration	50	200	250	500
Bare ground (%)	30	40		
Concentration	600	500		

Steeply Sloped Plots

Bare ground (%)	5	5	10	15
Concentration	100	250	300	600
Bare ground (%)	20	25	20	30
Concentration	500	500	900	800
Bare ground (%)	35	40	35	
Concentration	1100	1200	1000	

a. Using the data for steeply sloped plots, find the equation of the least-squares line for predicting y runoff sediment concentration using x = percentage of bare ground.
b. What would you predict runoff sediment concentration to be for a steeply sloped plot with 18% bare ground?
c. Would you recommend using the least-squares equation from Part (a) to predict runoff sediment concentration for gradually sloped plots? If so, explain why it would be appropriate to do so. If not, provide an alternative way to make such predictions.

5.30 The paper "Postmortem Changes in Strength of Gastropod Shells" (*Paleobiology* [1992]: 367–377) included scatterplots of data on x = shell height (in centimeters) and y = breaking strength (in newtons) for a sample of $n = 38$ hermit crab shells. The least-squares line was $\hat{y} = -275.1 + 244.9x$.
a. What are the slope and the intercept of this line?
b. When shell height increases by 1 cm, by how much does breaking strength tend to change?   244.9
c. What breaking strength would you predict when shell height is 2 cm?   214.7
d. Does this approximate linear relationship appear to hold for shell heights as small as 1 cm? Explain.

5.31 ● The paper "Increased Vital and Total Lung Capacities in Tibetan Compared to Han Residents of Lhasa" (*American Journal of Physical Anthropology* [1991]: 341–351) included a scatterplot of vital capacity (y) versus chest circumference (x) for a sample of 16 Tibetan natives, from which the following data were read:

x	79.4	81.8	81.8	82.3	83.7	84.3	84.3	85.2
y	4.3	4.6	4.8	4.7	5.0	4.9	4.4	5.0
x	87.0	87.3	87.7	88.1	88.1	88.6	89.0	89.5
y	6.1	4.7	5.7	5.7	5.2	5.5	5.0	5.3

a. Construct a scatterplot. What does it suggest about the nature of the relationship between x and y?
b. The summary quantities are

$$\sum x = 1368.1 \qquad \sum y = 80.9$$
$$\sum xy = 6933.48 \qquad \sum x^2 = 117,123.85 \qquad \sum y^2 = 412.81$$

Verify that the equation of the least-squares line is $\hat{y} = -4.54 + 0.1123x$, and draw this line on your scatterplot.
c. On average, roughly what change in vital capacity is associated with a 1-cm increase in chest circumference? with a 10-cm increase?   0.113, 1.13
d. What vital capacity would you predict for a Tibetan native whose chest circumference is 85 cm?   4.9994
e. Is vital capacity completely determined by chest circumference? Explain.

5.32 Explain why it can be dangerous to use the least-squares line to obtain predictions for x values that are substantially larger or smaller than those contained in the sample.

5.33 The sales manager of a large company selected a random sample of $n = 10$ salespeople and determined for each one the values of $x =$ years of sales experience and $y =$ annual sales (in thousands of dollars). A scatterplot of the resulting (x, y) pairs showed a marked linear pattern.
a. Suppose that the sample correlation coefficient is $r = .75$ and that the average annual sales is $\bar{y} = 100$. If a

particular salesperson is 2 standard deviations above the mean in terms of experience, what would you predict for that person's annual sales?   100 + 1.5(standard deviation)
b. If a particular person whose sales experience is 1.5 standard deviations below the average experience is predicted to have an annual sales value that is 1 standard deviation below the average annual sales, what is the value of r? $r = 0.67$

5.34 Explain why the slope b of the least-squares line always has the same sign (positive or negative) as does the sample correlation coefficient r.

5.35 ● The accompanying data resulted from an experiment in which weld diameter x and shear strength y (in pounds) were determined for five different spot welds on steel. A scatterplot shows a pronounced linear pattern. With $\sum(x - \bar{x}) = 1000$ and $\sum(x - \bar{x})(y - \bar{y}) = 8577$, the least-squares line is $\hat{y} = -936.22 + 8.577x$.

x	200.1	210.1	220.1	230.1	240.0
y	813.7	785.3	960.4	1118.0	1076.2

a. Because 1 lb = 0.4536 kg, strength observations can be re-expressed in kilograms through multiplication by this conversion factor: new $y = 0.4536$(old y). What is the equation of the least-squares line when y is expressed in kilograms? $\hat{y} = -424.7 + 3.891x$
b. More generally, suppose that each y value in a data set consisting of n (x, y) pairs is multiplied by a conversion factor c (which changes the units of measurement for y). What effect does this have on the slope b (i.e., how does the new value of b compare to the value before conversion), on the intercept a, and on the equation of the least-squares line? Verify your conjectures by using the given formulas for b and a. (Hint: Replace y with cy, and see what happens—and remember, this conversion will affect $\bar{y}$.)

Bold exercises answered in back   ● Data set available online but not required   ▼ Video solution available

5.3   Assessing the Fit of a Line

Once the least-squares regression line has been obtained, it is natural to examine how effectively the line summarizes the relationship between x and y. Important questions to consider are

1. Is a line an appropriate way to summarize the relationship between the two variables?
2. Are there any unusual aspects of the data set that we need to consider before proceeding to use the regression line to make predictions?
3. If we decide that it is reasonable to use the regression line as a basis for prediction, how accurate can we expect predictions based on the regression line to be?

In this section, we look at graphical and numerical methods that will allow us to answer these questions. Most of these methods are based on the vertical deviations of the data points from the regression line. These vertical deviations are called *residuals*, and each represents the difference between an actual y value and the corresponding predicted value, $\hat{y}$, that would result from using the regression line to make a prediction.

■ Predicted Values and Residuals

If the x value for the first observation is substituted into the equation for the least-squares line, the result is $a + bx_1$, the height of the line above x_1. The point (x_1, y_1) in the scatterplot also lies above x_1, so the difference

$$y_1 - (a + bx_1)$$

is the vertical deviation from this point to the line (see Figure 5.12). A point lying above the line gives a positive deviation, and a point lying below the line results in a negative deviation. The remaining vertical deviations come from repeating this process for $x = x_2$, then $x = x_3$, and so on.

Figure 5.12 Positive and negative deviations (residuals) from the least-squares line.

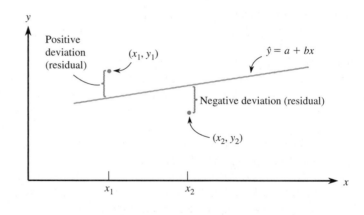

Example 5.7

How Does Range of Motion After Knee Surgery Change with Age?

● One measure of the success of knee surgery is postsurgical range of motion for the knee joint. Postsurgical range of motion was recorded for 12 patients who had surgery following a knee dislocation. The age of each patient was also recorded ("Reconstruction of the Anterior and Posterior Cruciate Ligaments After Knee Dislocation," *American Journal of Sports Medicine* [1999]: 189–194). The data are given in Table 5.1.

Table 5.1 Predicted Values and Residuals for the Data of Example 5.7

Patient	Age (x)	Range of Motion (y)	Predicted Range of Motion ($\hat{y}$)	Residual $y - \hat{y}$
1	35	154	138.07	15.93
2	24	142	128.49	13.51
3	40	137	142.42	−5.42
4	31	133	134.58	−1.58
5	28	122	131.97	−9.97
6	25	126	129.36	−3.36
7	26	135	130.23	4.77
8	16	135	121.52	13.48
9	14	108	119.78	−11.78
10	20	120	125.00	−5.00
11	21	127	125.87	1.13
12	30	122	133.71	−11.71

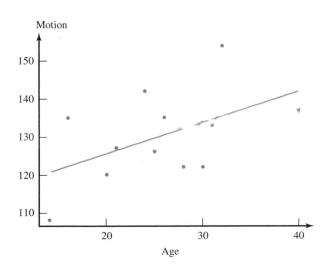

Figure 5.13 Scatterplot for the data of Example 5.7.

MINITAB was used to fit the least-squares regression line. Partial computer output follows:

Regression Analysis

The regression equation is

Range of Motion = 108 + 0.871 Age

Predictor	Coef	StDev	T	P
Constant	107.58	11.12	9.67	0.000
Age	0.8710	0.4146	2.10	0.062

s = 10.42   R-Sq = 30.6%   R-Sq(adj) = 23.7%

The resulting least-squares regression line is $\hat{y} = 107.58 + 0.871x$. A scatterplot that also includes the regression line is shown in Figure 5.13. The residuals for this data set are the signed vertical distances from the points to the regression line.

● Data set available online

For the youngest patient (patient 9 with $x_9 = 14$ and $y_9 = 108$), the corresponding predicted value and residual are

$$\text{predicted value} = \hat{y}_9 = 107.58 + .871 \text{ age} = 119.78$$
$$\text{residual} = y_9 - \hat{y}_9 = 108 - 119.78 = -11.78$$

The other predicted values and residuals are computed in a similar manner and are included in Table 5.1.

Computing the predicted values and residuals by hand can be tedious, but MINITAB and other statistical software packages, as well as many graphing calculators, include them as part of the output, as shown in Figure 5.14. The predicted values and residuals can be found in the table at the bottom of the MINITAB output in the columns labeled "Fit" and "Residual," respectively.

FIGURE 5.14 MINITAB output for the data of Example 5.7.

The regression equation is
Range of Motion = 108 + 0.871 Age

Predictor	Coef	StDev	T	P
Constant	107.58	11.12	9.67	0.000
Age	0.8710	0.4146	2.10	0.062

s = 10.42   R-Sq = 30.6%   R-Sq(adj) = 23.7%

Analysis of Variance

Source	DF	SS	MS	F	P
Regression	1	479.2	479.2	4.41	0.062
Residual Error	10	1085.7	108.6		
Total	11	1564.9			

Obs	Age	Range	Fit	SE Fit	Residual	St Resid
1	35.0	154.00	138.07	4.85	15.93	1.73
2	24.0	142.00	128.49	3.10	13.51	1.36
3	40.0	137.00	142.42	6.60	−5.42	−0.67
4	31.0	133.00	134.58	3.69	−1.58	−0.16
5	28.0	122.00	131.97	3.14	−9.97	−1.00
6	25.0	126.00	129.36	3.03	−3.36	−0.34
7	26.0	135.00	130.23	3.01	4.77	0.48
8	16.0	135.00	121.52	5.07	13.48	1.48
9	14.0	108.00	119.78	5.75	−11.78	−1.36
10	20.0	120.00	125.00	3.86	−5.00	−0.52
11	21.0	127.00	125.87	3.61	1.13	0.12
12	30.0	122.00	133.71	3.47	−11.71	−1.19

Plotting the Residuals

A careful look at residuals can reveal many potential problems. A *residual plot* is a good place to start when assessing the appropriateness of the regression line.

> **DEFINITION**
>
> A **residual plot** is a scatterplot of the (x, residual) pairs. Isolated points or a pattern of points in the residual plot indicate potential problems.

A desirable plot is one that exhibits no particular pattern, such as curvature. Curvature in the residual plot is an indication that the relationship between x and y is not linear and that a curve would be a better choice than a line for describing the relationship between x and y. This is sometimes easier to see in a residual plot than in a scatterplot of y versus x, as illustrated in Example 5.8.

Example 5.8   Heights and Weights of American Women

● Consider the accompanying data on x = height (in inches) and y = average weight (in pounds) for American females, age 30–39 (from *The World Almanac and Book of Facts*). The scatterplot displayed in Figure 5.15(a) appears rather straight. However, when the residuals from the least-squares line ($\hat{y} = 98.23 + 3.59x$) are plotted, substantial curvature is apparent (even though $r \approx .99$). It is not accurate to say that weight increases in direct proportion to height (linearly with height). Instead, average weight increases somewhat more rapidly for relatively large heights than it does for relatively small heights.

x	58	59	60	61	62	63	64	65
y	113	115	118	121	124	128	131	134

x	66	67	68	69	70	71	72
y	137	141	145	150	153	159	164

Figure 5.15 Plots for the data of Example 5.8: (a) scatterplot; (b) residual plot.

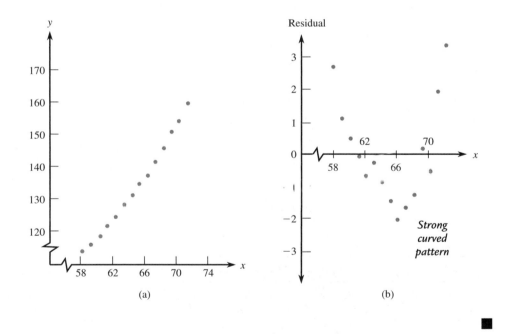

There is another common type of residual plot—one that plots the residuals versus the corresponding $\hat{y}$ values rather than versus the x values. Because $\hat{y} = a + bx$ is simply a linear function of x, the only real difference between the two types of residual plots is the scale on the horizontal axis. The pattern of points in the residual plots will be the same, and it is this pattern of points that is important, not the scale. Thus

● Data set available online

Figure 5.16 (a) Plot of residuals versus x; (b) plot of residuals versus $\hat{y}$.

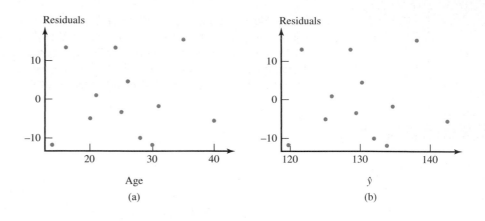

the two plots give equivalent information, as can be seen in Figure 5.16, which gives both plots for the data of Example 5.7.

It is also important to look for unusual values in the scatterplot or in the residual plot. A point falling far above or below the horizontal line at height 0 corresponds to a large residual, which may indicate some type of unusual behavior, such as a recording error, a nonstandard experimental condition, or an atypical experimental subject. A point whose x value differs greatly from others in the data set may have exerted excessive influence in determining the fitted line. One method for assessing the impact of such an isolated point on the fit is to delete it from the data set, recompute the best-fit line, and evaluate the extent to which the equation of the line has changed.

Example 5.9 Tennis Elbow

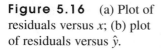

● One factor in the development of tennis elbow is the impact-induced vibration of the racket and arm at ball contact. Tennis elbow is thought to be related to various properties of the tennis racket used. The following data are a subset of those analyzed in the article "Transfer of Tennis Racket Vibrations into the Human Forearm" (*Medicine and Science in Sports and Exercise* [1992]: 1134–1140). Measurements on x = racket resonance frequency (in hertz) and y = sum of peak-to-peak accelerations (a characteristic of arm vibration in meters per second squared) are given for $n = 14$ different rackets:

Racket	Resonance (x)	Acceleration (y)	Racket	Resonance (x)	Acceleration (y)
1	105	36.0	8	114	33.8
2	106	35.0	9	114	35.0
3	110	34.5	10	119	35.0
4	111	36.8	11	120	33.6
5	112	37.0	12	121	34.2
6	113	34.0	13	126	36.2
7	113	34.2	14	189	30.0

Step-by-step technology instructions available online   ● Data set available online

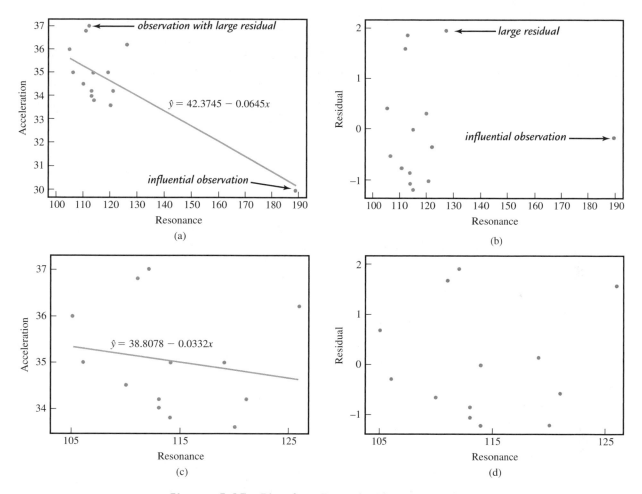

Figure 5.17 Plots from Example 5.9: (a) scatterplot for the full sample; (b) residual plot for the full sample; (c) scatterplot when the influential observation is deleted; (d) residual plot when the influential observation is deleted.

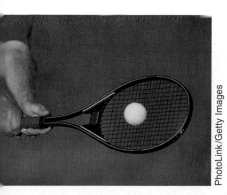

PhotoLink/Getty Images

A scatterplot and a residual plot are shown in Figures 5.17(a) and 5.17(b), respectively. One observation in the data set is far to the right of the other points in the scatterplot. Because the least-squares line minimizes the sum of squared residuals, the least-squares line is pulled down toward this discrepant point. This single observation plays a big role in determining the slope of the least-squares line, and it is therefore called an *influential observation*. Notice that an influential observation is not necessarily the one with the largest residual, because the least-squares line actually passes near this point.

Figures 5.17(c) and 5.17(d) show what happens when the influential observation is removed from the sample. Both the slope and the intercept of the least-squares line are quite different from the slope and intercept of the line with this observation included.

Deletion of the observation corresponding to the largest residual (1.955 for observation 13) also changes the values of the slope and intercept of the least-squares line, but these changes are not profound. For example, the least-squares line for the data set consisting of the 13 points that remain when observation 13 is deleted is $\hat{y} = 42.5 - .0670x$. This observation does not appear to be all that influential.

Careful examination of a scatterplot and a residual plot can help us determine the appropriateness of a line for summarizing a relationship. If we decide that a line is appropriate, the next step is to think about assessing the accuracy of predictions based on the least-squares line and whether these predictions (based on the value of x) are better in general than those made without knowledge of the value of x. Two numerical measures that are helpful in this assessment are the coefficient of determination and the standard deviation about the regression line.

■ Coefficient of Determination

Suppose that we would like to predict the price of homes in a particular city. A random sample of 20 homes that are for sale is selected, and y = price and x = size (in square feet) are recorded for each house in the sample. Undoubtedly, there will be variability in house price (the houses will differ with respect to price), and it is this variability that makes accurate prediction of price a challenge. How much of the variability in house price can be explained by the fact that price is related to house size and that houses differ in size? If differences in size account for a large proportion of the variability in price, a price prediction that takes house size into account is a big improvement over a prediction that is not based on size.

The **coefficient of determination** is a measure of the proportion of variability in the y variable that can be "explained" by a linear relationship between x and y.

> **DEFINITION**
>
> The **coefficient of determination**, denoted by r^2, gives the proportion of variation in y that can be attributed to an approximate linear relationship between x and y; $100r^2$ is the percentage of variation in y that can be attributed to an approximate linear relationship between x and y.

Variation in y can effectively be explained by an approximate straight-line relationship when the points in the scatterplot fall close to the least-squares line—that is, when the residuals are small in magnitude. A natural measure of variation about the least-squares line is the sum of the squared residuals. (Squaring before combining prevents negative and positive residuals from counteracting one another.) A second sum of squares assesses the total amount of variation in observed y values.

> **DEFINITION**
>
> The **total sum of squares**, denoted by **SSTo**, is defined as
> $$\text{SSTo} = (y_1 - \bar{y})^2 + (y_2 - \bar{y})^2 + \cdots + (y_n - \bar{y})^2 = \sum(y - \bar{y})^2$$
> The **residual sum of squares** (sometimes referred to as the error sum of squares), denoted by **SSResid**, is defined as
> $$\text{SSResid} = (y_1 - \hat{y}_1)^2 + (y_2 - \hat{y}_2)^2 + \cdots + (y_n - \hat{y}_n)^2 = \sum(y - \hat{y})^2$$

These sums of squares can be found as part of the regression output from most standard statistical packages or can be obtained using the following computational formulas:

$$SSTo = \sum y^2 - \frac{(\sum y)^2}{n}$$

$$SSResid = \sum y^2 - a \sum y - b \sum xy$$

Example 5.10 Range of Motion Revisited

Figure 5.18 displays part of the MINITAB output that results from fitting the least-squares line to the data on range of motion and age from Example 5.7. From the output,

$$SSTo = 1564.9 \text{ and } SSResid = 1085.7$$

Figure 5.18 MINITAB output for the data of Example 5.10.

Regression Analysis

The regression equation is
Range of Motion = 108 + 0.871 Age

Predictor	Coef	StDev	T	P
Constant	107.58	11.12	9.67	0.000
Age	0.8710	0.4146	2.10	0.062

S = 10.42 R-Sq = 30.6% R-Sq(adj) = 23.7%

Analysis of Variance

Source	DF	SS	MS	F	P
Regression	1	479.2	479.2	4.41	0.062
Residual Error	10	1085.7	108.6		
Total	11	1564.9			

SSResid ← 1085.7

SSTo ← 1564.9

Notice that SSResid is nearly as large as SSTo.

■

The residual sum of squares is the sum of squared vertical deviations from the least-squares line. As Figure 5.19 illustrates, SSTo is also a sum of squared vertical deviations from a line—the horizontal line at height $\bar{y}$. The least-squares line is, by definition, the one having the smallest sum of squared deviations. It follows that SSResid $\leq$ SSTo. The two sums of squares are equal only when the least-squares line *is* the horizontal line.

Figure 5.19 Interpreting sums of squares: (a) SSResid = sum of squared vertical deviations from the least-squares line; (b) SSTo = sum of squared vertical deviations from the horizontal line at height $\bar{y}$.

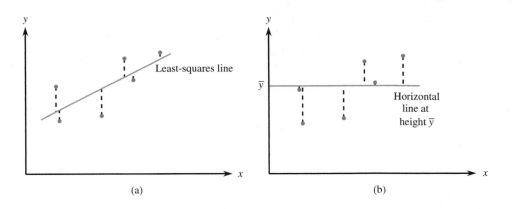

SSResid is often referred to as a measure of unexplained variation—the amount of variation in y that cannot be attributed to the linear relationship between x and y. The more the points in the scatterplot deviate from the least-squares line, the larger the value of SSResid and the greater the amount of y variation that cannot be explained by the approximate linear relationship. Similarly, SSTo is interpreted as a measure of total variation. The larger the value of SSTo, the greater the amount of variability in $y_1, y_2, \ldots, y_n$. The ratio SSResid/SSTo is the fraction or proportion of total variation that is unexplained by a straight-line relation. Subtracting this ratio from 1 gives the proportion of total variation that *is* explained:

The coefficient of determination can be computed as

$$r^2 = 1 - \frac{\text{SSResid}}{\text{SSTo}}$$

Multiplying r^2 by 100 gives the percentage of y variation attributable to the approximate linear relationship. The closer this percentage is to 100%, the more successful is the relationship in explaining variation in y.

Example 5.11 r^2 for Age and Range of Motion Data

For the data on range of motion and age from Example 5.10, we found SSTo = 1564.9 and SSResid = 1085.7. Thus

$$r^2 = 1 - \frac{\text{SSResid}}{\text{SSTo}} = 1 - \frac{1085.7}{1564.9} = .306$$

This means that only 30.6% of the observed variability in postsurgical range of motion can be explained by an approximate linear relationship between range of motion and age. Note that the r^2 value can be found in the MINITAB output of Figure 5.18, labeled "R-Sq."

The symbol r was used in Section 5.1 to denote Pearson's sample correlation coefficient. It is not coincidental that r^2 is used to represent the coefficient of determination. The notation suggests how these two quantities are related:

$$(\text{correlation coefficient})^2 = \text{coefficient of determination}$$

Thus, if $r = .8$ or $r = -.8$, then $r^2 = .64$, so 64% of the observed variation in the dependent variable can be explained by the linear relationship. Because the value of r does not depend on which variable is labeled x, the same is true of r^2. The coefficient of determination is one of the few quantities computed in a regression analysis whose value remains the same when the roles of dependent and independent variables are interchanged. When $r = .5$, we get $r^2 = .25$, so only 25% of the observed variation is explained by a linear relation. This is why a value of r between $-.5$ and $.5$ is not considered evidence of a strong linear relationship.

■ Standard Deviation About the Least-Squares Line ....

The coefficient of determination measures the extent of variation about the best-fit line *relative* to overall variation in y. A high value of r^2 does not by itself promise that the deviations from the line are small in an absolute sense. A typical observation could deviate from the line by quite a bit, yet these deviations might still be small relative to overall y variation.

Recall that in Chapter 4 the sample standard deviation

$$s = \sqrt{\frac{\sum(x - \bar{x})^2}{n - 1}}$$

was used as a measure of variability in a single sample; roughly speaking, s is the typical amount by which a sample observation deviates from the mean. There is an analogous measure of variability when a least-squares line is fit.

> **DEFINITION**
>
> The **standard deviation about the least-squares line** is given by
>
> $$s_e = \sqrt{\frac{\text{SSResid}}{n - 2}}$$

Roughly speaking, s_e is the typical amount by which an observation deviates from the least-squares line. Justification for division by $(n - 2)$ and the use of the subscript e is given in Chapter 13.

....

Example 5.12 Predicting Graduation Rates

● Consider the accompanying data on six-year graduation rate (%), student-related expenditure per full-time student, and median SAT score for the 37 primarily undergraduate public universities and colleges in the United States with enrollments between 10,000 and 20,000 (Source: College Results Online, The Education Trust).

● Data set available online

Graduation Rate	Expenditure	Median SAT
80	6244	1165
65	6757	1045
65	8011	1195
62	7224	1120
61	7229	1110
57	6355	1090
55	7650	990
53	7323	1035
52	6382	1080
49	6792	990
49	7970	1065
49	7271	1065
48	6261	1085
46	5361	1080
46	8735	948
45	8553	990
45	7874	1005
43	5908	970
43	7548	890
41	6531	1010
39	7736	980
39	7209	970
39	7355	1050
39	7071	990
38	7186	970
38	6216	1010
37	6285	1045
37	7375	930
35	5435	1010
34	8248	860
33	6401	950
30	6169	1045
29	7126	1065
28	5604	1004
28	4327	990
26	6762	865
22	4541	950

Figure 5.20 displays scatterplots of graduation rate versus student-related expenditure and graduation rate versus median SAT score. The least-squares lines and the values of r^2 and s_e are also shown.

Notice that while there is a positive linear relationship between student-related expenditure and graduation rate, the relationship is weak. The value of r^2 is only .095 (9.5%), indicating that only about 9.5% of the variability in graduation rate from university to university can be explained by student-related expenditures. The standard deviation about the regression line is $s_e = 12.0855$, which is larger than s_e for the predictor median SAT, a reflection of the fact that the points in the scatterplot of graduation rate versus student-related expenditure tend to fall farther from the regression line than is the case for the line that describes graduation rate versus

Figure 5.20 Scatterplots for the data of Example 5.12: (a) graduation rate versus student-related expenditure; (b) graduation rate versus median SAT.

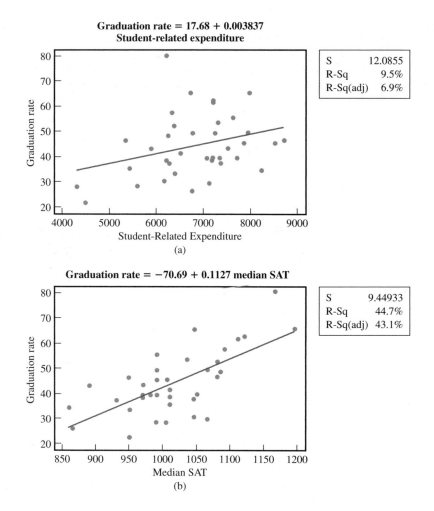

median SAT. The value of r^2 for graduation rate versus median SAT is .447 (44.7%) and $s_e = 9.44933$, indicating that the predictor median SAT does a better job of explaining variability in graduation rates and the corresponding least-squares line would be expected to produce more accurate estimates of graduation rates than would be the case for the predictor student-related expenditure.

Based on the values of r^2 and s_e, median SAT would be a better choice for predicting graduation rates than student-related expenditures. It is also possible to develop a prediction equation that would incorporate both potential predictors—techniques for doing this are introduced in Chapter 14.

5.36 ● The following table gives the number of organ transplants performed in the United States each year from 1990 to 1999 (The Organ Procurement and Transplantation Network, 2003):

Year	Number of Transplants (in thousands)
1 (1990)	15.0
2	15.7
3	16.1
4	17.6
5	18.3
6	19.4
7	20.0
8	20.3
9	21.4
10 (1999)	21.8

a. Construct a scatterplot of these data, and then find the equation of the least-squares regression line that describes the relationship between y = number of transplants performed and x = year. Describe how the number of transplants performed has changed over time from 1990 to 1999.
b. Compute the 10 residuals, and construct a residual plot. Are there any features of the residual plot that indicate that the relationship between year and number of transplants performed would be better described by a curve rather than a line? Explain.

5.37 ● The following data on x = soil depth (in centimeters) and y = percentage of montmorillonite in the soil were taken from a scatterplot in the paper "Ancient Maya Drained Field Agriculture: Its Possible Application Today in the New River Floodplain, Belize, C.A." (*Agricultural Ecosystems and Environment* [1984]: 67–84):

x	40	50	60	70	80	90	100
y	58	34	32	30	28	27	22

a. Draw a scatterplot of y versus x.
b. The equation of the least-squares line is $\hat{y} = 64.50 - 0.45x$. Draw this line on your scatterplot. Do there appear to be any large residuals?
c. Compute the residuals, and construct a residual plot. Are there any unusual features in the plot?

5.38 ● Data on pollution and cost of medical care for elderly people were given in Exercise 5.17 and are also shown here. The following data give a measure of pollution (micrograms of particulate matter per cubic meter of air) and the cost of medical care per person over age 65 for six geographic regions of the United States:

Region	Pollution	Cost of Medical Care
North	30.0	915
Upper South	31.8	891
Deep South	32.1	968
West South	26.8	972
Big Sky	30.4	952
West	40.0	899

The equation of the least-squares regression line for this data set is $\hat{y} = 1082.2 - 4.691x$, where y = medical cost and x = pollution.
a. Compute the six residuals.
b. What is the value of the correlation coefficient for this data set? Does the value of r indicate that the linear relationship between pollution and medical cost is strong, moderate, or weak? Explain.
c. Construct a residual plot. Are there any unusual features of the plot?
d. The observation for the West, (40.0, 899), has an x value that is far removed from the other x values in the sample. Is this observation influential in determining the values of the slope and/or intercept of the least-squares line? Justify your answer.

5.39 ● The following data on degree of exposure to ^{242}Cm alpha particles (x) and the percentage of exposed cells without aberrations (y) appeared in the paper "Chromosome Aberrations Induced in Human Lymphocytes by D-T Neutrons" (*Radiation Research* [1984]: 561–573):

x	0.106	0.193	0.511	0.527
y	98	95	87	85
x	1.08	1.62	1.73	2.36
y	75	72	64	55
x	2.72	3.12	3.88	4.18
y	44	41	37	40

Bold exercises answered in back ● Data set available online but not required ▼ Video solution available

Summary quantities are

$$n = 12 \qquad \sum x = 22.027 \qquad \sum y = 793$$
$$\sum x^2 = 62.600235 \qquad \sum xy = 1114.5 \qquad \sum y^2 = 57,939$$

a. Obtain the equation of the least-squares line.
b. Construct a residual plot, and comment on any interesting features.

5.40 ● Cost-to-charge ratio (the percentage of the amount billed that represents the actual cost) for inpatient and outpatient services at 11 Oregon hospitals is shown in the following table (Oregon Department of Health Services, 2002). A scatterplot of the data is also shown.

Cost-to-Charge Ratio

Hospital	Outpatient Care	Inpatient Care
1	62	80
2	66	76
3	63	75
4	51	62
5	75	100
6	65	88
7	56	64
8	45	50
9	48	54
10	71	83
11	54	100

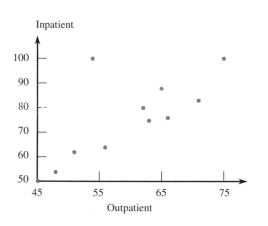

Inpatient / Outpatient

The least-squares regression line with y = inpatient cost-to-charge ratio and x = outpatient cost-to-charge ratio is $\hat{y} = -1.1 + 1.29x$.

a. Is the observation for Hospital 11 an influential observation? Justify your answer.
b. Is the observation for Hospital 11 an outlier? Explain.
c. Is the observation for Hospital 5 an influential observation? Justify your answer.
d. Is the observation for Hospital 5 an outlier? Explain.
No, residual = 4.35, which is small relative to $s_e = 12.185$.

5.41 The article "Examined Life: What Stanley H. Kaplan Taught Us About the SAT" (*The New Yorker* [December 17, 2001]: 86–92) included a summary of findings regarding the use of SAT I scores, SAT II scores, and high school grade point average (GPA) to predict first-year college GPA. The article states that "among these, SAT II scores are the best predictor, explaining 16 percent of the variance in first-year college grades. GPA was second at 15.4 percent, and SAT I was last at 13.3 percent."
a. If the data from this study were used to fit a least-squares line with y = first-year college GPA and x = high school GPA, what would the value of r^2 have been?
b. The article stated that SAT II was the best predictor of first-year college grades. Do you think that predictions based on a least-squares line with y = first-year college GPA and x = SAT II score would have been very accurate? Explain why or why not.

5.42 Exercise 5.22 gave the least-squares regression line for predicting y = clutch size from x = snout-vent length ("Reproductive Biology of the Aquatic Salamander *Amphiuma tridactylum* in Louisiana," *Journal of Herpetology* [1999]: 100–105). The paper also reported $r^2 = .7664$ and SSTo = 43,951.
a. Interpret the value of r^2.
b. Find and interpret the value of s_e (the sample size was $n = 14$).

5.43 ● ▼ The article "Characterization of Highway Runoff in Austin, Texas, Area" (*Journal of Environmental Engineering* [1998]: 131–137) gave a scatterplot, along with the least-squares line for x = rainfall volume (in cubic meters) and y = runoff volume (in cubic meters), for a particular location. The following data were read from the plot in the paper:

x	5	12	14	17	23	30	40	47
y	4	10	13	15	15	25	27	46

x	55	67	72	81	96	112	127
y	38	46	53	70	82	99	100

a. Does a scatterplot of the data suggest a linear relationship between x and y?

b. Calculate the slope and intercept of the least-squares line.   Slope = 0.827, intercept = −1.13

c. Compute an estimate of the average runoff volume when rainfall volume is 80.   65.03

d. Compute the residuals, and construct a residual plot. Are there any features of the plot that indicate that a line is not an appropriate description of the relationship between x and y? Explain.

5.44 ● The paper "Accelerated Telomere Shortening in Response to Life Stress" (*Proceedings of the National Academy of Sciences* [2004]: 17312–17315) described a study that examined whether stress accelerates aging at a cellular level. The accompanying data on a measure of perceived stress (x) and telomere length (y) were read from a scatterplot that appeared in the paper. Telomere length is a measure of cell longevity.

Perceived Stress	Telomere Length	Perceived Stress	Telomere Length
5	1.25	20	1.22
6	1.32	20	1.3
6	1.5	20	1.32
7	1.35	21	1.24
10	1.3	21	1.26
11	1	21	1.3
12	1.18	22	1.18
13	1.1	22	1.22
14	1.08	22	1.24
14	1.3	23	1.18
15	0.92	24	1.12
15	1.22	24	1.5
15	1.24	25	0.94
17	1.12	26	0.84
17	1.32	27	1.02
17	1.4	27	1.12
18	1.12	28	1.22
18	1.46	29	1.3
19	0.84	33	0.94

a. Compute the equation of the least-squares line.

b. What is the value of r^2? $r^2 = .0992$

c. Does the linear relationship between perceived stress and telomere length account for a large or small proportion of the variability in telomere length? Justify your answer.

5.45 The article "Cost-Effectiveness in Public Education" (*Chance* [1995]: 38–41) reported that for a regression of y = average SAT score on x = expenditure per pupil, based on data from $n = 44$ New Jersey school districts, $a = 766$, $b = 0.015$, $r^2 = .160$, and $s_e = 53.7$.

a. One observation in the sample was (9900, 893). What average SAT score would you predict for this district, and what is the corresponding residual?

b. Interpret the value of s_e.

c. How effectively do you think the least-squares line summarizes the relationship between x and y? Explain your reasoning.

5.46 ● The paper "Feeding of Predaceous Fishes on Out-Migrating Juvenile Salmonids in John Day Reservoir, Columbia River" (*Transactions of the American Fisheries Society* [1991]: 405–420) gave the following data on y = maximum size of salmonids consumed by a northern squawfish (the most abundant salmonid predator) and x = squawfish length, both in millimeters:

x	218	246	270	287	318	344
y	82	85	94	127	141	157

x	375	386	414	450	468
y	165	216	219	238	249

Use the accompanying output from MINITAB to answer the following questions.

The regression equation is size = −89.1 + 0.729 length

Predictor	Coef	Stdev	t ratio	p
Constant	−89.09	16.83	5.29	0.000
length	0.72907	0.04778	15.26	0.000

s = 12.56   R-sq = 96.3%   R-sq(adj) = 95.9%

Analysis of Variance

Source	DF	SS	MS	F	p
Regression	1	36736	36736	232.87	0.000
Error	9	1420	158		
Total	10	38156			

a. What maximum salmonid size would you predict for a squawfish whose length is 375 mm, and what is the residual corresponding to the observation (375, 165)?

b. What proportion of observed variation in y can be attributed to the approximate linear relationship between the two variables? $r^2 = .963$

5.47 ● The paper "Crop Improvement for Tropical and Subtropical Australia: Designing Plants for Difficult Climates" (*Field Crops Research* [1991]: 113–139) gave the following data on x = crop duration (in days) for soybeans and y = crop yield (in tons per hectare):

x	92	92	96	100	102
y	1.7	2.3	1.9	2.0	1.5
x	102	106	106	121	143
y	1.7	1.6	1.8	1.0	0.3

$$\sum x = 1060 \quad \sum y = 15.8 \quad \sum xy = 1601.1$$
$$a = 5.20683380 \quad b = -0.3421541$$

a. Construct a scatterplot of the data. Do you think the least-squares line will give accurate predictions? Explain.
b. Delete the observation with the largest x value from the sample and recalculate the equation of the least-squares line. Does this observation greatly affect the equation of the line?
c. What effect does the deletion in Part (b) have on the value of r^2? Can you explain why this is so?

5.48 A study was carried out to investigate the relationship between the hardness of molded plastic (y, in Brinell units) and the amount of time elapsed since termination of the molding process (x, in hours). Summary quantities include n = 15, SSResid = 1235.470, and SSTo = 25,321.368. Calculate and interpret the coefficient of determination.

5.49 The paper "Effects of Canine Parvovirus (CPV) on Gray Wolves in Minnesota" (*Journal of Wildlife Management* [1995]: 565–570) summarized a regression of y = percentage of pups in a capture on x = percentage of CPV prevalence among adults and pups. The equation of the least-squares line, based on n = 10 observations, was $\hat{y} = 62.9476 - 0.54975x$, with $r^2 = .57$.
a. One observation was (25, 70). What is the corresponding residual?
b. What is the value of the sample correlation coefficient?

c. Suppose that SSTo = 2520.0 (this value was not given in the paper). What is the value of s_e?

5.50 Both r^2 and s_e are used to assess the fit of a line.
a. Is it possible that both r^2 and s_e could be large for a bivariate data set? Explain. (A picture might be helpful.)
b. Is it possible that a bivariate data set could yield values of r^2 and s_e that are both small? Explain. (Again, a picture might be helpful.)
c. Explain why it is desirable to have r^2 large and s_e small if the relationship between two variables x and y is to be described using a straight line.

5.51 Some straightforward but slightly tedious algebra shows that
$$SSResid = (1 - r^2)\sum(y - \bar{y})^2$$
from which it follows that
$$s_e = \sqrt{\frac{n-1}{n-2}}\sqrt{1 - r^2}\, s_y$$
Unless n is quite small, $(n - 1)/(n - 2) \approx 1$, so
$$s_e \approx \sqrt{1 - r^2}\, s_y$$
a. For what value of r is s_e as large as s_y? What is the least-squares line in this case?
b. For what values of r will s_e be much smaller than s_y?
c. A study by the Berkeley Institute of Human Development (see the book *Statistics* by Freedman et al., listed in the back of the book) reported the following summary data for a sample of n = 66 California boys:

$r \approx .80$
At age 6, average height ≈ 46 in., standard deviation ≈ 1.7 in.
At age 18, average height ≈ 70 in., standard deviation ≈ 2.5 in.

What would s_e be for the least-squares line used to predict 18-year-old height from 6-year-old height?
d. Referring to Part (c), suppose that you wanted to predict the past value of 6-year-old height from knowledge of 18-year-old height. Find the equation for the appropriate least-squares line. What is the corresponding value of s_e?
$\hat{y} = 7.92 + .544x, s_e = 1.02$

5.4 Nonlinear Relationships and Transformations

As we have seen in previous sections, when the points in a scatterplot exhibit a linear pattern and the residual plot does not reveal any problems with the linear fit, the least-squares line is a sensible way to summarize the relationship between x and y. A linear relationship is easy to interpret, departures from the line are easily detected, and using the line to predict y from our knowledge of x is straightforward. Often, though, a scatterplot or residual plot exhibits a curved pattern, indicating a more complicated relationship between x and y. In this case, finding a curve that fits the observed data well is a more complex task. In this section, we consider two common approaches to fitting nonlinear relationships: polynomial regression and transformations.

■ Polynomial Regression

Let's reconsider the data first introduced in Example 5.4 on x = age and y = average marathon finish time:

Age Group	x = Representative Age	y = Average Finish Time
10–19	15	302.38
20–29	25	193.63
30–39	35	185.46
40–49	45	198.49
50–59	55	224.30
60–69	65	288.71

The scatterplot of these data is reproduced here as Figure 5.21. Because this plot shows a marked curved pattern, it is clear that no straight line can do a reasonable job of describing the relationship between x and y. However, the relationship can be de-

Figure 5.21 Scatterplot for the marathon data.

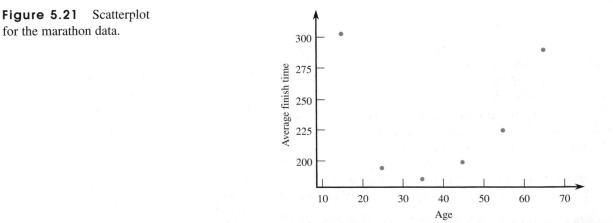

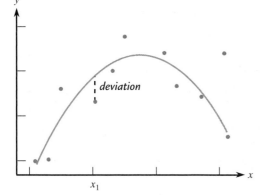

Figure 5.22 Deviation for a quadratic function.

scribed by a curve, and in this case the curved pattern in the scatterplot looks like a parabola (the graph of a quadratic function). This suggests trying to find a quadratic function of the form

$$\hat{y} = a + b_1x + b_2x^2$$

that would reasonably describe the relationship. That is, the values of the coefficients a, b_1, and b_2 in this equation must be selected to obtain a good fit to the data.

What are the best choices for the values of a, b_1, and b_2? In fitting a line to data, we used the principle of least squares to guide our choice of slope and intercept. Least squares can be used to fit a quadratic function as well. The deviations, $y - \hat{y}$, are still represented by vertical distances in the scatterplot, but now they are vertical distances from the points to a parabolic curve (the graph of a quadratic equation) rather than to a line, as shown in Figure 5.22. We then choose values for the coefficients in the quadratic equation so that the sum of squared deviations is as small as possible.

For a quadratic regression, the least squares estimates of a, b_1, and b_2 are those values that minimize the sum of squared deviations $\sum(y - \hat{y})^2$ where $\hat{y} = a + b_1x + b_2x^2$.

For quadratic regression, a measure that is useful for assessing fit is

$$R^2 = 1 - \frac{\text{SSResid}}{\text{SSTo}}$$

where $\text{SSResid} = \sum(y - \hat{y})^2$. The measure R^2 is defined in a way similar to r^2 for simple linear regression and is interpreted in a similar fashion. The notation r^2 is used only with linear regression to emphasize the relationship between r^2 and the correlation coefficient, r, in the linear case.

The general expressions for computing the least-squares estimates are somewhat complicated, so we rely on a statistical software package or graphing calculator to do the computations for us.

....

Example 5.13 Marathon Data Revisited: Fitting a Quadratic Model

For the marathon data, the scatterplot (see Figure 5.21) showed a marked curved pattern. If the least-squares line is fit to these data, it is no surprise that the line does not do a good job of describing the relationship ($r^2 = .001$ or $.1\%$ and $s_e = 56.9439$), and the residual plot shows a distinct curved pattern as well (Figure 5.23).

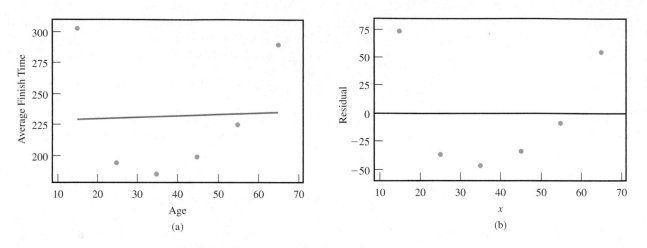

Figure 5.23 Plots for the marathon data of Example 5.13: (a) fitted line plot; (b) residual plot.

Part of the MINITAB output from fitting a quadratic function to these data is as follows:

The regression equation is
$y = 462 − 14.2\ x + 0.179\ x\text{-squared}$

Predictor	Coef	SE Coef	T	P
Constant	462.00	43.99	10.50	0.002
x	−14.205	2.460	−5.78	0.010
x-squared	0.17888	0.03025	5.91	0.010

S = 18.4813   R-Sq = 92.1%   R-Sq(adj) = 86.9%

Analysis of Variance

Source	DF	SS	MS	F	P
Regression	2	11965.0	5982.5	17.52	0.022
Residual Error	3	1024.7	341.6		
Total	5	12989.7			

The least-squares coefficients are

$$a = 462.0 \qquad b_1 = −14.205 \qquad b_2 = 0.17888$$

and the least-squares quadratic is

$$\hat{y} = 462.0 − 14.205x + 0.17888x^2$$

A plot showing the curve and the corresponding residual plot for the quadratic regression are given in Figure 5.24. Notice that there is no strong pattern in the residual plot for the quadratic case, as there was in the linear case. For the quadratic regression, $R^2 = .921$ (as opposed to .001 for the least-squares line), which means that 92.1% of the variability in average marathon finish time can be explained by an approximate quadratic relationship between average finish time and age.

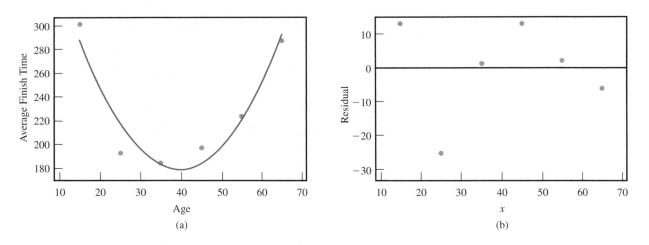

Figure 5.24   Quadratic regression of Example 5.13: (a) scatterplot; (b) residual plot.

■

Linear and quadratic regression are special cases of polynomial regression. A polynomial regression curve is described by a function of the form

$$\hat{y} = a + b_1x + b_2x^2 + b_3x^3 + \cdots + b_kx^k$$

which is called a kth-degree polynomial. The case of $k = 1$ results in linear regression ($\hat{y} = a + b_1x$) and $k = 2$ yields a quadratic regression ($\hat{y} = a + b_1x + b_2x^2$). A quadratic curve has only one bend (see Figure 5.25(a) and (b)). A less frequently encountered special case is for $k = 3$, where $\hat{y} = a + b_1x + b_2x^2 + b_3x^3$, which is called a cubic regression curve. Cubic curves have two bends, as shown in Figure 5.25(c).

Figure 5.25   Polynomial regression curves: (a) quadratic curve with $b_2 < 0$; (b) quadratic curve with $b_2 > 0$; (c) cubic curve.

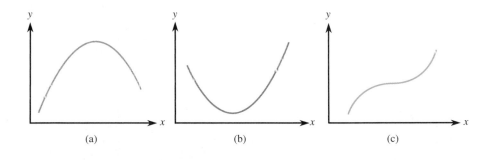

Example 5.14   Nonlinear Relationship Between Cloud Cover Index and Sunshine Index

● Researchers have examined a number of climatic variables in an attempt to understand the mechanisms that govern rainfall runoff. The article "The Applicability of Morton's and Penman's Evapotranspiration Estimates in Rainfall-Runoff Modeling"

● Data set available online

(*Water Resources Bulletin* [1991]: 611–620) reported on a study that examined the relationship between x = cloud cover index and y = sunshine index. Suppose that the cloud cover index can have values between 0 and 1. Consider the accompanying data, which are consistent with summary quantities in the article:

Cloud Cover Index (x)	Sunshine Index (y)
.2	10.98
.5	10.94
.3	10.91
.1	10.94
.2	10.97
.4	10.89
0	10.88
.4	10.92
.3	10.86

The authors of the article used a cubic regression to describe the relationship between cloud cover and sunshine. Using the given data, MINITAB was instructed to fit a cubic regression, resulting in the following output:

```
The regression equation is
y = 10.9 + 1.46 x − 7.26 x-squared + 9.23 x-cubed

Predictor       Coef      SE Coef          T        P
Constant     10.8768       0.0307     354.72    0.000
x             1.4604       0.5285       2.76    0.040
x-squared    −7.259        2.550       −2.85    0.036
x-cubed       9.234        3.346        2.76    0.040

S = 0.0315265    R-Sq = 62.0%    R-Sq(adj) = 39.3%
```

Analysis of Variance

```
Source           DF          SS           MS        F        P
Regression        3   0.0081193    0.0027064     2.72    0.154
Residual Error    5   0.0049696    0.0009939
Total             8   0.0130889
```

The least-squares cubic regression is then

$$\hat{y} = 10.8768 + 1.4604x - 7.259x^2 + 9.234x^3$$

To predict the sunshine index for a day when the cloud cover index is .45, we use the regression equation to obtain

$$\hat{y} = 10.8768 + 1.4604(.45) - 7.259(.45)^2 + 9.234(.45)^3 = 10.91$$

The least-squares cubic regression and the corresponding residual plot are shown in Figure 5.26. The residual plot does not reveal any troublesome patterns that would suggest a choice other than the cubic regression.

Figure 5.26 Plots for the data of Example 5.14: (a) least-squares cubic regression; (b) residual plot.

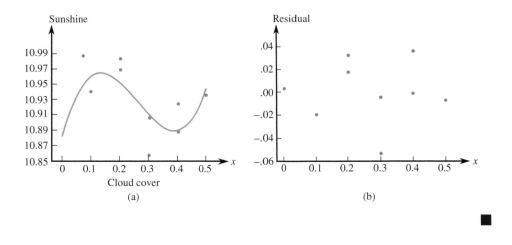

(a)

(b)

■ Transformations

An alternative to finding a curve to fit the data is to find a way to transform the x values and/or y values so that a scatterplot of the transformed data has a linear appearance. A **transformation** (sometimes called a reexpression) involves using a simple function of a variable in place of the variable itself. For example, instead of trying to describe the relationship between x and y, it might be easier to describe the relationship between $\sqrt{x}$ and y or between x and $\log(y)$. And, if we can describe the relationship between, say, $\sqrt{x}$ and y, we will still be able to predict the value of y for a given x value. Common transformations involve taking square roots, logarithms, or reciprocals.

Important concept

Example 5.15 River Water Velocity and Distance from Shore

● As fans of white-water rafting know, a river flows more slowly close to its banks (because of friction between the river bank and the water). To study the nature of the relationship between water velocity and the distance from the shore, data were gathered on velocity (in centimeters per second) of a river at different distances (in meters) from the bank. Suppose that the resulting data were as follows:

Distance	.5	1.5	2.5	3.5	4.5	5.5	6.5	7.5	8.5	9.5
Velocity	22.00	23.18	25.48	25.25	27.15	27.83	28.49	28.18	28.50	28.63

A graph of the data exhibits a curved pattern, as seen in both the scatterplot and the residual plot from a linear fit (see Figures 5.27(a) and 5.27(b)).

Let's try transforming the x values by replacing each x value by its square root. We define

$$x' = \sqrt{x}$$

The resulting transformed data are given in Table 5.2.

● Data set available online

Figure 5.27 Plots for the data of Example 5.15: (a) scatterplot of the river data; (b) residual plot from linear fit.

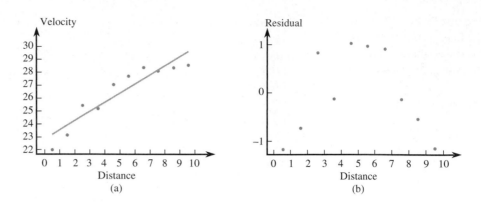

Figure 5.28(a) shows a scatterplot of y versus x' (or equivalently y versus $\sqrt{x}$). The pattern of points in this plot looks linear, and so we can fit a least-squares line using the transformed data. The MINITAB output from this regression appears below.

The residual plot in Figure 5.28(b) shows no indication of a pattern. The resulting regression equation is

$$\hat{y} = 20.1 + 3.01x'$$

Table 5.2 Original and Transformed Data of Example 5.15

Original Data		Transformed Data	
x	y	x'	y
0.5	22.00	0.7071	22.00
1.5	23.18	1.2247	23.18
2.5	25.48	1.5811	25.48
3.5	25.25	1.8708	25.25
4.5	27.15	2.1213	27.15
5.5	27.83	2.3452	27.83
6.5	28.49	2.5495	28.49
7.5	28.18	2.7386	28.18
8.5	28.50	2.9155	28.50
9.5	28.63	3.0822	28.63

Regression Analysis

The regression equation is
Velocity = 20.1 + 3.01 sqrt distance

Predictor	Coef	StDev	T	P
Constant	20.1102	0.6097	32.99	0.000
Sqrt dis	3.0085	0.2726	11.03	0.000

S = 0.6292   R-Sq = 93.8%   R-Sq(adj) = 93.1%

Analysis of Variance

Source	DF	SS	MS	F	P
Regression	1	48.209	48.209	121.76	0.000
Residual Error	8	3.168	0.396		
Total	9	51.376			

Figure 5.28 Plots for the transformed data of Example 5.15: (a) scatterplot of y versus x'; (b) residual plot resulting from a linear fit to the transformed data.

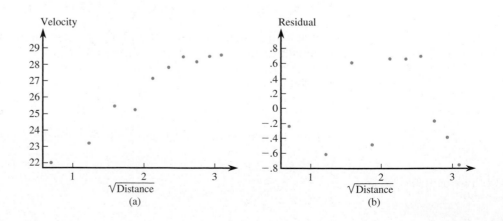

An equivalent equation is:

$$\hat{y} = 20.1 + 3.01\sqrt{x}$$

The values of r^2 and s_e (see the MINITAB output) indicate that a line is a reasonable way to describe the relationship between y and x'. To predict velocity of the river at a distance of 9 meters from shore, we first compute $x' = \sqrt{x} = \sqrt{9} = 3$ and then use the sample regression line to obtain a prediction of y:

$$\hat{y} = 20.1 + 3.01x' = 20.1 + (3.01)(3) = 29.13$$

■

In Example 5.15, transforming the x values using the square root function worked well. In general, how can we choose a transformation that will result in a linear pattern? Table 5.3 gives some guidance and summarizes some of the properties of the most commonly used transformations.

Table 5.3   Commonly Used Transformations

Transformation	Mathematical Description	Try This Transformation When
No transformation	$\hat{y} = a + bx$	The change in y is constant as x changes. A 1-unit increase in x is associated with, on average, an increase of b in the value of y.
Square root of x	$\hat{y} = a + b\sqrt{x}$	The change in y is not constant. A 1-unit increase in x is associated with smaller increases or decreases in y for larger x values.
Log of x*	$\hat{y} = a + b\log_{10}(x)$ or $\hat{y} = a + b\ln(x)$	The change in y is not constant. A 1-unit increase in x is associated with smaller increases or decreases in the value of y for larger x values.
Reciprocal of x	$\hat{y} = a + b\left(\dfrac{1}{x}\right)$	The change in y is not constant. A 1-unit increase in x is associated with smaller increases or decreases in the value of y for larger x values. In addition, y has a limiting value of a as x increases.
Log of y* (Exponential growth or decay)	$\log(\hat{y}) = a + bx$ or $\ln(\hat{y}) = a + bx$	The change in y associated with a 1-unit change in x is proportional to x.

*The values of a and b in the regression equation will depend on whether $\log_{10}$ or ln is used, but the $\hat{y}$'s and r^2 values will be identical.

....

Example 5.16 Tortilla Chips: Relationship Between Frying Time and Moisture Content

● No tortilla chip lover likes soggy chips, so it is important to find characteristics of the production process that produce chips with an appealing texture. The following

● Data set available online

data on x = frying time (in seconds) and y = moisture content (%) appeared in the article "Thermal and Physical Properties of Tortilla Chips as a Function of Frying Time" (*Journal of Food Processing and Preservation* [1995]: 175–189):

Frying time, x	5	10	15	20	25	30	45	60
Moisture content, y	16.3	9.7	8.1	4.2	3.4	2.9	1.9	1.3

Figure 5.29(a) shows a scatterplot of the data. The pattern in this plot is typical of exponential decay, with the change in y as x increases much smaller for large x values than for small x values. You can see that a 5-sec change in frying time is associated with a much larger change in moisture content in the part of the plot where the x values are small than in the part of the plot where the x values are large. Table 5.3 suggests transforming the y values (moisture content in this example) by taking their logarithms.

Figure 5.29 Plots for the data of Example 5.16: (a) scatterplot of the tortilla chip data; (b) scatterplot of transformed data with $y' = \log(y)$; (c) scatterplot of transformed data with $y' = \ln(y)$.

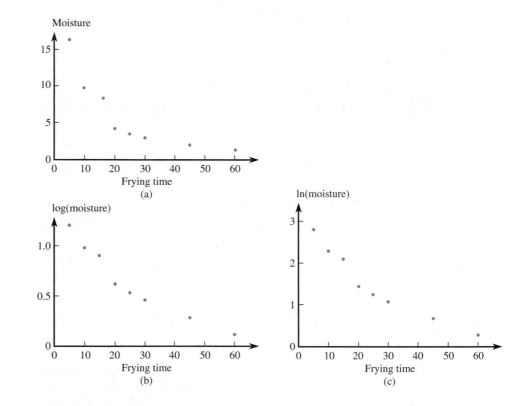

Two standard logarithmic functions are commonly used for such transformations—the common logarithm (log base 10, denoted by log or $\log_{10}$) and the natural logarithm (log base e, denoted ln). Either the common or the natural logarithm can be used; the only difference in the resulting scatterplots is the scale of the transformed y variable. This can be seen in Figures 5.29(b) and 5.29(c). These two scatterplots show the same pattern, and it looks like a line would be appropriate to describe this relationship.

Table 5.4 displays the original data along with the transformed y values using $y' = \log(y)$.

Table 5.4   Transformed Data from Example 5.16

Frying Time x	Moisture y	Log(moisture) y'
5	16.3	1.21219
10	9.7	.98677
15	8.1	.90849
20	4.2	.62325
25	3.4	.53148
30	2.9	.46240
45	1.9	.27875
60	1.3	.11394

The following MINITAB output shows the result of fitting the least-squares line to the transformed data:

Regression Analysis

The regression equation is
log(moisture) = 1.14 − 0.0192 frying time

Predictor	Coef	StDev	T	P
Constant	1.14287	0.08016	14.26	0.000
Frying t	−0.019170	0.002551	−7.52	0.000

S = 0.1246   R-Sq = 90.4%   R-Sq(adj) = 88.8%

Analysis of Variance

Source	DF	SS	MS	F	P
Regression	1	0.87736	0.87736	56.48	0.000
Residual Error	6	0.09320	0.01553		
Total	7	0.97057			

The resulting regression equation is

$$y' = 1.14 - 0.0192x$$

or, equivalently,

$$\log(y) = 1.14 - 0.0192x$$

■

■ **Fitting a Curve Using Transformations**   The objective of a regression analysis is usually to describe the approximate relationship between x and y with an equation of the form $y =$ some function of x.

If we have transformed only x, fitting a least-squares line to the transformed data results in an equation of the desired form, for example,

$$\hat{y} = 5 + 3x' = 5 + 3\sqrt{x} \qquad where\ x' = \sqrt{x}$$

or

$$\hat{y} = 4 + .2x' = 4 + .2\frac{1}{x} \qquad where\ x' = \frac{1}{x}$$

These functions specify lines when graphed using y and x', and they specify curves when graphed using y and x, as illustrated in Figure 5.30 for the square root transformation.

Figure 5.30 (a) A plot of $\hat{y} = 5 + 3x'$ where $x' = \sqrt{x}$; (b) a plot of $\hat{y} = 5 + 3\sqrt{x}$.

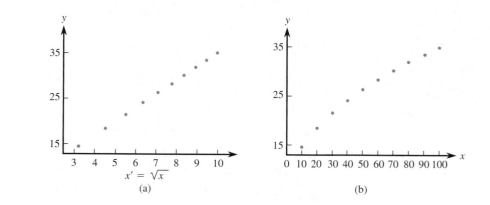

If the y values have been transformed, after obtaining the least-squares line the transformation can be undone to yield an expression of the form $y =$ some function of x (as opposed to $y' =$ some function of x). For example, to reverse a logarithmic transformation ($y' = \log(y)$), we can take the antilogarithm of each side of the equation. To reverse a square root transformation ($y' = \sqrt{y}$), we can square both sides of the equation, and to reverse a reciprocal transformation ($y' = 1/y$), we can take the reciprocal of each side of the equation. This is illustrated in Example 5.17.

Example 5.17 Revisiting the Tortilla Chip Data

For the tortilla chip data of Example 5.16, $y' = \log(y)$ and the least-squares line relating y' and x was

$$y' = 1.14 - .0192x$$

or, equivalently,

$$\log(y) = 1.14 - .0192x$$

To reverse this transformation, we take the antilog of both sides of the equation:

$$10^{\log(y)} = 10^{1.14 - .0192x}$$

Using the properties of logs and exponents

$$10^{\log(y)} = y$$
$$10^{1.14 - .0192x} = (10^{1.14})(10^{-.0192x})$$

We get

$$\hat{y} = (10^{1.14})(10^{-.0192x}) = (13.8038)(10^{-.0192x})$$

This equation can now be used to predict the y value (moisture content) for a given x (frying time). For example, the predicted moisture content when frying time is 35 sec is

$$\hat{y} = (13.8038)(10^{-.0192x}) = (13.8038)(.2128) = 2.9374$$

■

It should be noted that the process of transforming data, fitting a line to the transformed data, and then undoing the transformation to get an equation for a curved relationship between x and y usually results in a curve that provides a reasonable fit to the sample data, but it is not the least-squares curve for the data. For example, in Example 5.17, a transformation was used to fit the curve $\hat{y} = (13.8038)(10^{-.0192x})$. However, there may be another equation of the form $\hat{y} = a(10^{bx})$ that has a smaller sum of squared residuals for the *original* data than the one we obtained using transformations. Finding the least-squares estimates for a and b in an equation of this form is complicated. Fortunately, the curves found using transformations usually provide reasonable predictions of y.

■ **Power Transformations** Frequently, an appropriate transformation is suggested by the data. One type of transformation that statisticians have found useful for straightening a plot is a **power transformation**. A power (exponent) is first selected, and each original value is raised to that power to obtain the corresponding transformed value. Table 5.5 displays a "ladder" of the most frequently used power transformations. The power 1 corresponds to no transformation at all. Using the power 0 would transform every value to 1, which is certainly not informative, so statisticians use the logarithmic transformation in its place in the ladder of transformations. Other powers intermediate to or more extreme than those listed can be used, of course, but they are less frequently needed than those on the ladder. Notice that all the transformations previously presented are included in this ladder.

Table 5.5 Power Transformation Ladder

Power	Transformed Value	Name
3	(Original value)^3	Cube
2	(Original value)^2	Square
1	(Original value)	No transformation
$\frac{1}{2}$	$\sqrt{\text{Original value}}$	Square root
$\frac{1}{3}$	$\sqrt[3]{\text{Original value}}$	Cube root
0	Log(Original value)	Logarithm
-1	$\dfrac{1}{\text{Original value}}$	Reciprocal

Figure 5.31 is designed to suggest where on the ladder we should go to find an appropriate transformation. The four curved segments, labeled 1, 2, 3, and 4, represent shapes of curved scatterplots that are commonly encountered. Suppose that a scatterplot

Figure 5.31 Scatterplot shapes and where to go on the transformation ladder to straighten the plot.

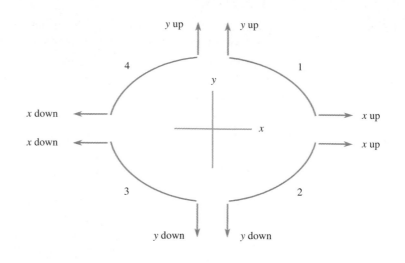

looks like the curve labeled 1. Then, to straighten the plot, we should use a power of x that is up the ladder from the no-transformation row (x^2 or x^3) and/or a power on y that is also up the ladder from the power 1. Thus, we might be led to squaring each x value, cubing each y, and plotting the transformed pairs. If the curvature looks like curved segment 2, a power up the ladder from no transformation for x and/or a power down the ladder for y (e.g., $\sqrt{y}$ or $\log(y)$) should be used.

The scatterplot for the tortilla chip data (Figure 5.29(a)) has the pattern of segment 3 in Figure 5.31. This suggests going down the ladder of transformations for x and/or for y. We found that transforming the y values in this data set by taking logarithms worked well, and this is consistent with the suggestion of going down the ladder of transformations for y.

Example 5.18 A Nonlinear Model: Iron Absorption and Polyphenol Content of Foods

● In many parts of the world, a typical diet consists mainly of cereals and grains, and many individuals suffer from a substantial iron deficiency. The article "The Effects of Organic Acids, Phytates, and Polyphenols on the Absorption of Iron from Vegetables" (*British Journal of Nutrition* [1983]: 331–342) reported the data in Table 5.6 on x = proportion of iron absorbed and y = polyphenol content (in milligrams per gram) when a particular food is consumed.

The scatterplot of the data in Figure 5.32(a) shows a clear curved pattern, which resembles the curved segment 3 in Figure 5.31. This suggests that x and/or y should be transformed by a power transformation down the ladder from 1. The authors of the article applied a square root transformation to both x and y. The resulting scatterplot in Figure 5.32(b) is reasonably straight.

● Data set available online

Table 5.6 Original and Transformed Data for Example 5.18

Vegetable	x	y	$\sqrt{x}$	$\sqrt{y}$
Wheat germ	.007	6.4	.083666	2.52982
Aubergine	.007	3.0	.083666	1.73205
Butter beans	.012	2.9	.109545	1.70294
Spinach	.014	5.8	.118322	2.40832
Brown lentils	.024	5.0	.154919	2.23607
Beetroot greens	.024	4.3	.154919	2.07364
Green lentils	.032	3.4	.178885	1.84391
Carrot	.096	.7	.309839	.83666
Potato	.115	.2	.339116	.44721
Beetroot	.185	1.5	.430116	1.22474
Pumpkin	.206	.1	.453872	.31623
Tomato	.224	.3	.473286	.54772
Broccoli	.260	.4	.509902	.63246
Cauliflower	.263	.7	.512835	.83666
Cabbage	.320	.1	.565685	.31623
Turnip	.327	.3	.571839	.54772
Sauerkraut	.327	.2	.571839	.44721

Figure 5.32 Scatterplot of data of Example 5.18: (a) original data; (b) data transformed by taking square roots.

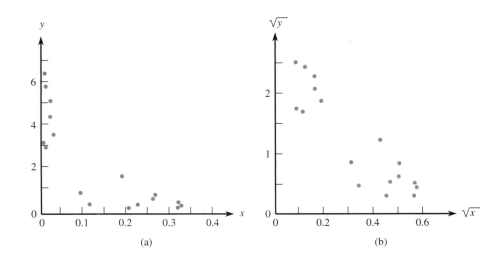

(a)　　　　　　(b)

Using MINITAB to fit a least-squares line to the transformed data ($y' = \sqrt{y}$ and $x' = \sqrt{x}$) results in the following output:

Regression Analysis

The regression equation is
sqrt y = 2.45 − 3.72 sqrt x

Predictor	Coef	StDev	T	P
Constant	2.4483	0.1833	13.35	0.000
Sqrt x	−3.7247	0.4836	−7.70	0.000

S = 0.3695   R-Sq = 79.8%   R-Sq(adj) = 78.5%

The corresponding least-squares line is

$$\hat{y}' = 2.45 - 3.72x'$$

This transformation can be reversed by squaring both sides to obtain an equation of the form y = some function of x:

$$\hat{y}'^2 = (2.45 - 3.72x')^2$$

Since $y'^2 = y$ and $x' = \sqrt{x}$ we get

$$\hat{y} = (2.45 - 3.72\sqrt{x})^2$$

■

■ Exercises 5.52–5.60

Turn to the Expanded Answers Section for answers not shown next to exercises.

5.52 ● The following data on x = frying time (in seconds) and y = moisture content (%) appeared in the paper "Thermal and Physical Properties of Tortilla Chips as a Function of Frying Time" (*Journal of Food Processing and Preservation* [1995]: 175–189):

x	5	10	15	20	25	30	45	60
y	16.3	9.7	8.1	4.2	3.4	2.9	1.9	1.3

a. Construct a scatterplot of the data. Would a straight line provide an effective summary of the relationship?
b. Here are the values of $x' = \log(x)$ and $y' = \log(y)$:

x'	0.70	1.00	1.18	1.30	1.40	1.48	1.65	1.78
y'	1.21	0.99	0.91	0.62	0.53	0.46	0.28	0.11

Construct a scatterplot of these transformed data, and comment on the pattern.
c. Based on the accompanying MINITAB output, does the least-squares line effectively summarize the relationship between y' and x'?

The regression equation is
log(moisture) = 2.02 − 1.05 log(time)

Predictor	Coef	SE Coef	T	P
Constant	2.01780	0.09584	21.05	0.000
log(time)	−1.05171	0.07091	−14.83	0.000

S = 0.0657067   R-Sq = 97.3%   R-Sq(adj) = 96.9%

Analysis of Variance

Source	DF	SS	MS	F	P
Regression	1	0.94978	0.94978	219.99	0.000
Residual Error	6	0.02590	0.00432		
Total	7	0.97569			

d. Use the MINITAB output to predict moisture content when frying time is 35 sec.   2.477
e. Do you think that predictions of moisture content using the model in Part (c) will be better than those using the model fit in Example 5.16, which used transformed y values but did not transform x? Explain.   Yes, the model in Part (c) has a lower SSResid than the model of Example 5.16.

5.53 ● ▼ The report "Older Driver Involvement in Injury Crashes in Texas" (Texas Transportation Institute, 2004) included a scatterplot of y = fatality rate (percentage of drivers killed in injury crashes) versus x = driver age. The accompanying data are approximate values read from the scatterplot.

Age	Fatality Rate	Age	Fatality Rate
40	0.75	80	2.20
45	0.75	85	3.00
50	0.95	90	3.20
55	1.05		

a. Construct a scatterplot of these data.
b. Using Table 5.5 and the ladder of transformations in Figure 5.31, suggest a transformation that might result in variables for which the scatterplot would exhibit a pattern that was more nearly linear.
c. Reexpress x and/or y using the transformation you recommended in Part (b). Construct a scatterplot of the transformed data.

Bold exercises answered in back   ● Data set available online but not required   ▼ Video solution available

d. Does the scatterplot in Part (c) suggest that the transformation was successful in straightening the plot?
e. Using the transformed variables, fit the least-squares line and use it to predict the fatality rate for 78-year-old drivers. $\hat{y} = 5.20683 - .03421x$, $\hat{y} = 1.91$

5.54 ● The paper "Aspects of Food Finding by Wintering Bald Eagles" (*The Auk* [1983]: 477–484) examined the relationship between the time that eagles spend aerially searching for food (indicated by the percentage of eagles soaring) and relative food availability. The accompanying data were taken from a scatterplot that appeared in this paper. Let *x* denote salmon availability and *y* denote the percentage of eagles in the air.

x	0	0	0.2	0.5	0.5	1.0
y	28.2	69.0	27.0	38.5	48.4	31.1
x	1.2	1.9	2.6	3.3	4.7	6.5
y	26.9	8.2	4.6	7.4	7.0	6.8

a. Draw a scatterplot for this data set. Would you describe the plot as linear or curved?
b. One possible transformation that might lead to a straighter plot involves taking the square root of both the *x* and *y* values. Use Figure 5.31 to explain why this might be a reasonable transformation.
c. Construct a scatterplot using the variables $\sqrt{x}$ and $\sqrt{y}$. Is this scatterplot straighter than the scatterplot in Part (a)?
d. Using Table 5.5, suggest another transformation that might be used to straighten the original plot.

5.55 Data on salmon availability (*x*) and the percentage of eagles in the air (*y*) were given in the previous exercise.
a. Calculate the correlation coefficient for these data.
b. Because the scatterplot of the original data appeared curved, transforming both the *x* and *y* values by taking square roots was suggested. Calculate the correlation coefficient for the variables $\sqrt{x}$ and $\sqrt{y}$. How does this value compare with that calculated in Part (a)? Does this indicate that the transformation was successful in straightening the plot?

5.56 ● The article "Organ Transplant Demand Rises Five Times as Fast as Existing Supply" (*San Luis Obispo Tribune*, February 23, 2001) included a graph that showed the number of people waiting for organ transplants each

year from 1990 to 1999. The following data are approximate values and were read from the graph in the article:

Year	Number Waiting for Transplant (in thousands)
1 (1990)	22
2	25
3	29
4	33
5	38
6	44
7	50
8	57
9	64
10 (1999)	72

a. Construct a scatterplot of the data with *y* = number waiting for transplant and *x* = year. Describe how the number of people waiting for transplants has changed over time from 1990 to 1999.
b. The scatterplot in Part (a) is shaped like segment 2 in Figure 5.31. Find a transformation of *x* and/or *y* that straightens the plot. Construct a scatterplot for your transformed variables.
c. Using the transformed variables from Part (b), fit a least-squares line and use it to predict the number waiting for an organ transplant in 2000 (Year 11).
d. The prediction made in Part (c) involves prediction for an *x* value that is outside the range of the *x* values in the sample. What assumption must you be willing to make for this to be reasonable? Do you think this assumption is reasonable in this case? Would your answer be the same if the prediction had been for the year 2010 rather than 2000? Explain.

5.57 ● A study, described in the paper "Prediction of Defibrillation Success from a Single Defibrillation Threshold Measurement" (*Circulation* [1988]: 1144–1149) investigated the relationship between defibrillation success and the energy of the defibrillation shock (expressed as a multiple of the defibrillation threshold) and presented the following data:

Bold exercises answered in back   ● Data set available online but not required   ▼ Video solution available

Energy of Shock	Success (%)
0.5	33.3
1.0	58.3
1.5	81.8
2.0	96.7
2.5	100.0

a. Construct a scatterplot of y = success and x = energy of shock. Does the relationship appear to be linear or nonlinear?
b. Fit a least-squares line to the given data, and construct a residual plot. Does the residual plot support your conclusion in Part (a)? Explain.
c. Consider transforming the data by leaving y unchanged and using either $x' = \sqrt{x}$ or $x'' = \log(x)$. Which of these transformations would you recommend? Justify your choice by appealing to appropriate graphical displays.
d. Using the transformation you recommended in Part (c), find the equation of the least-squares line that describes the relationship between y and the transformed x.
e. What would you predict success to be when the energy of shock is 1.75 times the threshold level? When it is 0.8 times the threshold level?   87.0, 52.6

5.58 ● Penicillin was administered orally to five different horses, and the concentration of penicillin in the blood was determined after five different lengths of time. The following data appeared in the paper "Absorption and Distribution Patterns of Oral Phenoxymethyl Penicillin in the Horse" (*Cornell Veterinarian* [1983]: 314–323):

x (time elapsed, hr)

| 1 | 2 | 3 | 6 | 8 |

y (penicillin concentration, mg/ml)

| 1.8 | 1.0 | 0.5 | 0.1 | 0.1 |

Construct scatterplots using the following variables. Which transformation, if any, would you recommend?
a. x and y c. x and $\sqrt{y}$
b. $\sqrt{x}$ and y d. $\sqrt{x}$ and $\sqrt{y}$
e. x and $\log(y)$ (values of $\log(y)$ are 0.26, 0, -0.30, -1, and -1)

5.59 ● The paper "Population Pressure and Agricultural Intensity" (*Annals of the Association of American Geographers* [1977]: 384–396) reported a positive association between population density and agricultural intensity. The following data consist of measures of population density (x) and agricultural intensity (y) for 18 different subtropical locations:

x	1.0	26.0	1.1	101.0	14.9	134.7
y	9	7	6	50	5	100
x	3.0	5.7	7.6	25.0	143.0	27.5
y	7	14	14	10	50	14
x	103.0	180.0	49.6	140.6	140.0	233.0
y	50	150	10	67	100	100

a. Construct a scatterplot of y versus x. Is the scatterplot compatible with the statement of positive association made in the paper?
b. The scatterplot in Part (a) is curved upward like segment 2 in Figure 5.31, suggesting a transformation that is up the ladder for x or down the ladder for y. Try a scatterplot that uses y and x^2. Does this transformation straighten the plot?
c. Draw a scatterplot that uses $\log(y)$ and x. The $\log(y)$ values, given in order corresponding to the y values, are 0.95, 0.85, 0.78, 1.70, 0.70, 2.00, 0.85, 1.15, 1.15, 1.00, 1.70, 1.15, 1.70, 2.18, 1.00, 1.83, 2.00, and 2.00. How does this scatterplot compare with that of Part (b)?
d. Now consider a scatterplot that uses transformations on both x and y: $\log(y)$ and x^2. Is this effective in straightening the plot? Explain.

5.60 ● Determining the age of an animal can sometimes be a difficult task. One method of estimating the age of harp seals is based on the width of the pulp canal in the seal's canine teeth. To investigate the relationship between age and the width of the pulp canal, researchers measured age and canal width in seals of known age. The following data on x = age (in years) and y = canal length (in millimeters) are a portion of a larger data set that appeared in the paper "Validation of Age Estimation in the Harp Seal Using Dentinal Annuli" (*Canadian Journal of Fisheries and Aquatic Science* [1983]: 1430–1441):

x	0.25	0.25	0.50	0.50	0.50	0.75	0.75	1.00
y	700	675	525	500	400	350	300	300
x	1.00	1.00	1.00	1.00	1.25	1.25	1.50	1.50
y	250	230	150	100	200	100	100	125
x	2.00	2.00	2.50	2.75	3.00	4.00	4.00	5.00
y	60	140	60	50	10	10	10	10
x	5.00	5.00	5.00	6.00	6.00
y	15	10	10	15	10

Construct a scatterplot for this data set. Would you describe the relationship between age and canal length as linear? If not, suggest a transformation that might straighten the plot.

Bold exercises answered in back   ● Data set available online but not required   ▼ Video solution available

5.5 Logistic Regression (Optional)

The correlation and regression techniques we have seen up to this point require that both variables of interest be numerical. But what if the dependent variable in a study is not numerical? This situation requires a different approach. For a dependent variable that is categorical with just two possible values (a binary variable), logistic regression can be used to describe the way in which such a dependent variable is related to a numerical predictor variable.

Example 5.19 Look Out for Those Wolf Spiders

● The paper "Sexual Cannibalism and Mate Choice Decisions in Wolf Spiders: Influence of Male Size and Secondary Sexual Characteristics" (*Animal Behaviour* [2005]: 83–94) described a study in which researchers were interested in what variables might be related to a female wolf spider's decision to kill and consume her partner during courtship or mating. The accompanying data (approximate values read from a graph in the paper) are values of x = difference in body width (female–male) and y = cannibalism, coded as 0 for no cannibalism and 1 for cannibalism for 52 pairs of courting wolf spiders.

Size Difference (mm)	Cannibalism	Size Difference (mm)	Cannibalism
−1	0	0.4	0
−1	0	0.4	0
−0.8	0	0.4	0
−0.8	0	0.4	0
−0.6	0	0.4	1
−0.6	0	0.6	0
−0.4	0	0.6	0
−0.4	0	0.6	0
−0.4	0	0.6	0
−0.4	0	0.6	0
−0.2	0	0.6	1
−0.2	0	0.6	1
−0.2	0	0.8	0
−0.2	0	0.8	0
0.0	0	0.8	1
0.0	0	0.8	1
0.0	0	0.8	1
0.0	0	1.0	0
0.0	0	1.0	0
0.0	0	1.0	1
0.2	0	1.0	1
0.2	0	1.2	0
0.2	0	1.4	0
0.2	0	1.6	1
0.2	0	1.8	1
0.2	0	2.0	1

● Data set available online

Figure 5.33 Scatterplot of the wolf spider data.

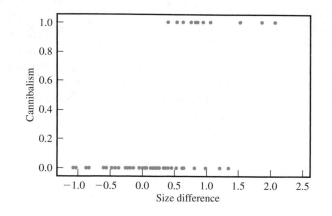

A MINITAB scatterplot of the data is shown in Figure 5.33. Note that the plot was constructed so that if two points fell in exactly the same position, one was offset a bit so that all observations would be visible. (This is called jittering.)

The scatterplot doesn't look like others we have seen before—its odd appearance is due to the fact that all y values are either 0 or 1. But, we can see from the plot that there are more occurrences of cannibalism for large x values (where the female is bigger than the male) than for smaller x values. In this situation, it makes sense to consider the probability of cannibalism (or equivalently, the proportion of the time cannibalism would occur) as being related to size difference. For example, we might focus on a single x value, say $x = 0$ where the female and male are the same size. Based on the data at hand, what can we say about the cannibalism proportion for pairs where the size difference is 0? We will return to this question after introducing the logistic regression equation.

■

A logistic regression equation is used to describe how the probability of "success" (e.g., cannibalism in the wolf spider example) changes as a numerical predictor variable, x, changes.

With p denoting the probability of success, the **logistic regression equation** is

$$p = \frac{e^{a+bx}}{1 + e^{a+bx}}$$

where a and b are constants.

The logistic regression equation looks complicated, but it has some very convenient properties. For any x value, the value of $e^{a+bx}/(1 + e^{a+bx})$ is between 0 and 1. As x changes, the graph of this equation has an "S" shape. Consider the two S-shaped

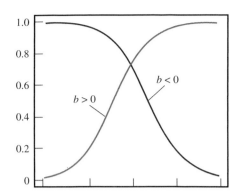

Figure 5.34   Two logistic regression curves.

curves of Figure 5.34. The blue curve starts near 0 and increases to 1 as x increases. This is the type of behavior exhibited by $p = e^{a+bx}/(1 + e^{a+bx})$ when $b > 0$. The red curve starts near 1 for small x values and then decreases as x increases. This happens when $b < 0$ in the logistic regression equation. The steepness of the curve—how quickly it rises or falls—also depends on the value of b. The farther b is from 0, the steeper the curve.

Most statistics packages, such as MINITAB and SPSS, have the capability of using sample data to compute values for a and b in the logistic regression equation to produce an equation relating the probability of success to the predictor x. An explanation of an alternate method for computing reasonable values of a and b is given later in this section.

Example 5.20 Cannibal Spiders II

MINITAB was used to fit a logistic regression equation to the wolf spider data of Example 5.19. The resulting MINITAB output is given in Figure 5.35, and Figure 5.36 shows a scatterplot of the original data with the logistic curve superimposed.

Figure 5.35   MINITAB output for the data of Example 5.20.

Response Information

Variable	Value	Count	
Cannibalism	1	11	(Event)
	0	41	
	Total	52	

Logistic Regression Table

Predictor	Coef	SE Coef	Z	P	Odds Ratio	95% CI Lower	95% CI Upper
Constant	−3.08904	0.828780	−3.73	0.000			
Size difference	3.06928	1.00407	3.06	0.002	21.53	3.01	154.05

With $a = -3.08904$ and $b = 3.06928$, the equation of the logistic regression equation is

$$p = \frac{e^{-3.08904+3.06928x}}{1 + e^{-3.08904+3.06928x}}$$

To predict or estimate the probability of cannibalism when the size difference between the female and male $= 0$, we substitute 0 into the logistic regression equation to obtain

$$p = \frac{e^{-3.08904+3.06928(0)}}{1 + e^{-3.08904+3.06928(0)}} = \frac{e^{-3.08904}}{1 + e^{-3.08904}} = .044$$

A consequence of this is that if we transform p using

$$y' = \ln\left(\frac{p}{1-p}\right)$$

we can use least squares to fit a line to the (x, y') data.

For the RatRiddance example, the transformed data are

x	p	$\dfrac{p}{1-p}$	$y' = \ln\left(\dfrac{p}{1-p}\right)$
20	0.225	0.290	−1.237
40	0.236	0.309	−1.175
60	0.398	0.661	−0.414
80	0.628	1.688	0.524
100	0.678	2.106	0.745
120	0.795	3.878	1.355
140	0.853	5.803	1.758
160	0.860	6.143	1.815
180	0.921	11.658	2.456
200	0.940	15.667	2.752

The resulting best fit line is:

$$y' = a + bx$$
$$= -1.6033 + 0.221x$$

We can check the transformed linear model fit in the customary way, checking the scatterplot and the residual plot, as shown in Figure 5.38 (a) and (b). Although there seems to be an ever-so-slight hint of curvature in the data, the linear model appears to fit quite well. In fact, the linear fit accounts for about 97% of the variation in y'.

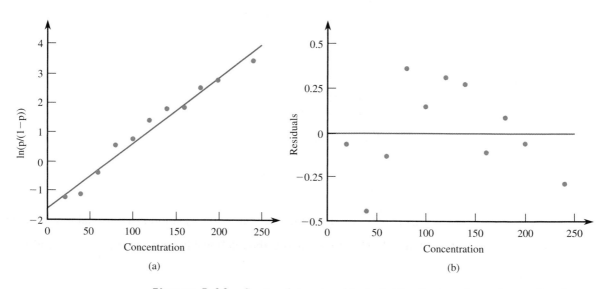

Figure 5.38 Scatterplot and residual plot for the transformed mortality data.

Example 5.21   The Call of the Wild Amazonian . . . Frog

Alan and Sandy Carey/Getty Images

● The Amazonian tree frog uses vocal communication to call for a mate. In a study of the relationship between calling behavior and the amount of rainfall ("How, When, and Where to Perform Visual Displays: The Case of the Amazonian Frog *Hyla parviceps*," *Herpetologica* [2004]: 420–429), the daily rainfall was recorded as well as observations of calling behavior by Amazonian frog males. Data consistent with the article are given in Table 5.8.

Table 5.8   Proportion Calling versus Daily Rainfall (mm)

Rainfall	0.2	0.3	0.4	0.5	0.7	0.8
Call rate	.17	.19	.20	.21	.27	.28
Rainfall	0.9	1.1	1.2	1.3	1.5	1.6
Call rate	.29	.34	.39	.41	.46	.49
Rainfall	1.7	2.0	2.2	2.4	2.6	2.8
Call rate	.53	.60	.67	.71	.75	.82
Rainfall	2.9	3.2	3.5	4.2	4.9	5.0
Call rate	.84	.88	.90	.97	.98	.98

Inspection of the scatterplot in Figure 5.39(a) reveals a pattern that is consistent with a logistic relation between the daily rainfall and the proportion of frogs

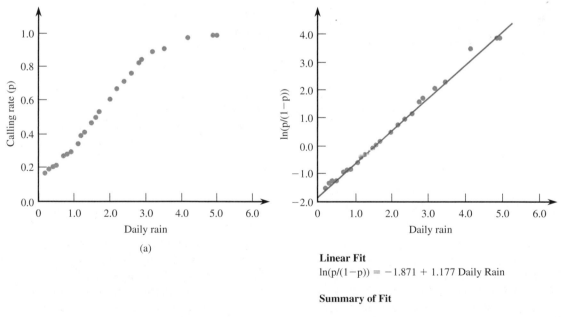

(a)

Linear Fit

$\ln(p/(1-p)) = -1.871 + 1.177$ Daily Rain

Summary of Fit

RSquare	0.996
RSquare Adj	0.996
s	0.103

(b)

Figure 5.39   Scatterplot of original and transformed data of Example 5.21.

● Data set available online

exhibiting calling behavior. The transformed data in Fig 5.39(b) show a clearly linear pattern. For these data the least-squares fit line is given by the equation $y' = -1.871 + 1.177$(Rainfall).

To predict calling proportion for a location with daily rainfall of 4.0 mm, we use the computed values of a and b in the logistic regression equation:

$$p = \frac{e^{-1.871 + 1.177R}}{1 + e^{-1.871 + 1.177R}} = \frac{e^{-1.871 + 1.177(4.0)}}{1 + e^{-1.871 + 1.177(4.0)}} = .945$$

■

Exercises 5.61–5.65

Turn to the Expanded Answers Section for answers not shown next to exercises.

5.61 ● Anabolic steroid abuse has been increasing despite increased press reports of adverse medical and psychiatric consequences. In a recent study, medical researchers studied the potential for addiction to testosterone in hamsters (*Neuroscience* [2004]: 971–981). Hamsters were allowed to self-administer testosterone over a period of days, resulting in the death of some of the animals. The data below show the proportion of hamsters surviving versus the peak self-administration of testosterone (μg). Fit a logistic regression equation and use the equation to predict the probability of survival for a hamster with a peak intake of 40μg. $\ln\left(\dfrac{p}{1-p}\right) = 4.589 - 0.0659x$; 0.876

Peak Intake (micrograms)	Survival Proportion (p)	$\dfrac{p}{1-p}$	$y' = \ln\left(\dfrac{p}{1-p}\right)$
10	0.980	49.0000	3.8918
30	0.900	9.0000	2.1972
50	0.880	7.3333	1.9924
70	0.500	1.0000	0.0000
90	0.170	0.2048	−1.5856

5.62 ● The paper "The Shelf Life of Bird Eggs: Testing Egg Viability Using a Tropical Climate Gradient" (*Ecology* [2005]: 2164–2175) investigated the effect of altitude and length of exposure on the hatch rate of thrasher eggs. Data consistent with the estimated probabilities of hatching after a number of days of exposure given in the paper are shown here.

Probability of Hatching

Exposure (days)	1	2	3	4	5	6	7	8
Proportion (lowland)	0.81	0.83	0.68	0.42	0.13	0.07	0.04	0.02
Proportion (mid-elevation)	0.73	0.49	0.24	0.14	0.037	0.040	0.024	0.030
Proportion (cloud forest)	0.75	0.67	0.36	0.31	0.14	0.09	0.06	0.07

a. Plot the data for the low- and mid-elevation experimental treatments versus exposure. Are the plots generally the shape you would expect from "logistic" plots?

b. Using the techniques introduced in this section, calculate $y' = \ln\left(\dfrac{p}{1-p}\right)$ for each of the exposure times in the cloud forest and fit the line $y' = a + b$(Days). What is the significance of a negative slope to this line?

c. Using your best-fit line from Part (b), what would you estimate the proportion of eggs that would, on average, hatch if they were exposed to cloud forest conditions for 3 days? 5 days?   0.438, 0.194

d. At what point in time does the estimated proportion of hatching for cloud forest conditions seem to cross from greater than 0.5 to less than 0.5?   Somewhere between two days ($p = .584$) and three days ($p = .438$)

5.63 ● As part of a study of the effects of timber management strategies (*Ecological Applications* [2003]: 1110–

1123) investigators used satellite imagery to study abundance of the lichen *Lobaria oregano* at different elevations. Abundance of a species was classified as "common" if there were more than 10 individuals in a plot of land. In the table below, approximate proportions of plots in which *Lobaria oregano* were common are given.

Proportions of Plots Where *Lobaria oregano* Are Common

Elevation (m)	400	600	800	1000	1200	1400	1600
Prop. of plots with Lichen (>10/plot)	0.99	0.96	0.75	0.29	0.077	0.035	0.01

a. As elevation increases, does *Lobaria oregano* become more common or less common? What aspect(s) of the table support your answer?   Less common

b. Using the techniques introduced in this section, calculate $y' = \ln\left(\dfrac{p}{1-p}\right)$ for each of the elevations and fit the line $y' = a + b$(Elevation). What is the equation of the best-fit line?

c. Using the best-fit line from Part (b), estimate the proportion of plots of land on which *Lobaria oregano* are classified as "common" at an elevation of 900 m.   0.632

5.64 ● The hypothetical data below are from a toxicity study designed to measure the effectiveness of different doses of a pesticide on mosquitoes. The table below summarizes the concentration of the pesticide, the sample sizes, and the number of critters dispatched.

Concentration (g/oc)	0.10	0.15	0.20	0.30	0.50	0.70	0.95
Number of mosquitoes	48	52	56	51	47	53	51
Number killed	10	13	25	31	39	51	49

a. Make a scatterplot of the proportions of mosquitoes killed versus the pesticide concentration.

b. Using the techniques introduced in this section, calculate $y' = \ln\left(\dfrac{p}{1-p}\right)$ for each of the concentrations and fit the line $y' = a + b$(Concentration). What is the significance of a positive slope for this line?

c. The point at which the dose kills 50% of the pests is sometimes called LD50, for "Lethal dose 50%." What would you estimate to be LD50 for this pesticide and for mosquitoes?   Approximately 0.23 g/cc

5.65 ● In the study of textiles and fabrics, the strength of a fabric is a very important consideration. Suppose that a significant number of swatches of a certain fabric are subjected to different "loads" or forces applied to the fabric. The data from such an experiment might look as follows:

Hypothetical Data on Fabric Strength

Load (lb/sq in.)	5	15	35	50	70	80	90
Proportion failing	0.02	0.04	0.20	0.23	0.32	0.34	0.43

a. Make a scatterplot of the proportion failing versus the load on the fabric.

b. Using the techniques introduced in this section, calculate $y' = \ln\left(\dfrac{p}{1-p}\right)$ for each of the loads and fit the line $y' = a + b$(Load). What is the significance of a positive slope for this line?

c. What proportion of the time would you estimate this fabric would fail if a load of 60 lb/sq in. were applied?

d. In order to avoid a "wardrobe malfunction," one would like to use fabric that has less than a 5% chance of failing. Suppose that this fabric is our choice for a new shirt. To have less than a 5% chance of failing, what would you estimate to be the maximum "safe" load in lb/sq in.?   15.5

5.6 Interpreting and Communicating the Results of Statistical Analyses

Using either a least-squares line to summarize a linear relationship or a correlation coefficient to describe the strength of a linear relationship is common in investigations that focus on more than a single variable. In fact, the methods described in this chapter are among the most widely used of all statistical tools. When numerical bivariate data are analyzed in journal articles and other published sources, it is common to find a scatterplot of the data and a least-squares line or a correlation coefficient.

■ Communicating the Results of Statistical Analyses

When reporting the results of a data analysis involving bivariate numerical data, it is important to include graphical displays as well as numerical summaries. Including a scatterplot and providing a description of what the plot reveals about the form of the relationship between the two variables under study establish the context in which numerical summary measures, such as the correlation coefficient or the equation of the least-squares line, can be interpreted.

In general, the goal of an analysis of bivariate data is to give a quantitative description of the relationship, if any, between the two variables. If there is a relationship, you can describe how strong or weak the relationship is or model the relationship in a way that allows various conclusions to be drawn. If the goal of the study is to describe the strength of the relationship and the scatterplot shows a linear pattern, you can report the value of the correlation coefficient or the coefficient of determination as a measure of the strength of the linear relationship. When you interpret the value of the correlation coefficient, it is a good idea to relate the interpretation to the pattern observed in the scatterplot. This is especially important before making a statement of no relationship between two variables, because a correlation coefficient near 0 does not necessarily imply that there is no relationship of any form. Similarly, a correlation coefficient near 1 or −1, by itself, does not guarantee that the relationship is linear. A curved pattern, such as the one we saw in Figure 5.15, can produce a correlation coefficient that is near 1.

If the goal of a study is prediction, then, when you report the results of the study, you should not only give a scatterplot and the equation of the least-squares line but also address how well the linear prediction model fits the data. At a minimum, you should include both the values of s_e (the standard deviation about the regression line) and r^2 (the coefficient of determination). Including a residual plot can also provide support for the appropriateness of the linear model for describing the relationship between the two variables.

■ What to Look for in Published Data

Here are a few things to consider when you read an article that includes an analysis of bivariate data:

- What two variables are being studied? Are they both numerical? Is a distinction made between a dependent variable and an independent variable?
- Does the article include a scatterplot of the data? If so, does there appear to be a relationship between the two variables? Can the relationship be described as linear, or is some type of nonlinear relationship a more appropriate description?

- Does the relationship between the two variables appear to be weak or strong? Is the value of a correlation coefficient reported?
- If the least-squares line is used to summarize the relationship between the dependent and independent variables, is any measure of goodness of fit reported, such as r^2 or s_e? How are these values interpreted, and what do they imply about the usefulness of the least-squares line?
- If a correlation coefficient is reported, is it interpreted properly? Be cautious of interpretations that claim a causal relationship.

The article "Rubbish Regression and the Census Undercount" (*Chance* [1992]: 33) describes work done by the Garbage Project at the University of Arizona. Project researchers had analyzed different categories of garbage for a number of households. They were asked by the Census Bureau to see whether any of the garbage data variables were related to household size. They reported that "the weight data for different categories of garbage were plotted on graphs against data on household size, dwelling by dwelling, and the resulting scatterplots were analyzed to see in which categories the weight showed a steady, monotonic rise relative to household size."

The researchers determined that the strongest linear relationship appeared to be that between the amount of plastic discarded and household size. The line used to summarize this relationship was stated to be $\hat{y} = 0.2815x$, where y = household size and x = weight (in pounds) of plastic during a 5-week collection. Note that this line has an intercept of 0. Scientists at the Census Bureau believed that this relationship would extend to entire neighborhoods, and so the amount of plastic discarded by a neighborhood could be measured (rather than having to measure house to house) and then used to approximate the neighborhood size.

An example of the use of r^2 and the correlation coefficient is found in the article "Affective Variables Related to Mathematics Achievement Among High-Risk College Freshmen" (*Psychological Reports* [1991]: 399–403). The researchers wrote that "an examination of the Pearson correlations indicated that the two scales, Confidence in Learning Mathematics and Mathematics Anxiety, were highly related, with values ranging from .87 to .91 for the men, women, and for the total group. Such a high relationship indicates that the two scales are measuring essentially the same concept, so for further analysis, one of the two scales could be eliminated."

The reported correlation coefficients (from .87 to .91) indicate a strong positive linear relationship between scores on the two scales. Students who scored high on the Confidence in Learning Mathematics scale tended also to score high on the Mathematics Anxiety scale. (The researchers explain that a high score on the Mathematics Anxiety scale corresponds to low anxiety toward mathematics.) We should be cautious of the statement that the two scales are measuring essentially the same concept. Although that interpretation may make sense in this context, given what the researchers know about these two scales, it does not follow solely from the fact that the two are highly correlated. (For example, there might be a correlation between number of years of education and income, but it would be hard to argue that these two variables are measuring the same thing!)

The same article also states that Mathematics Anxiety score accounted for 27% of the variability in first-quarter math grades. The reported value of 27% is the value of r^2 converted to a percentage. Although a value of 27% ($r^2 = .27$) may not seem particularly large, it is in fact surprising that the variable *Math Anxiety score* would be able to explain more than one-fourth of the student-to-student variability in first-quarter math grades.

■ A Word to the Wise: Cautions and Limitations ....

There are a number of ways to get into trouble when analyzing bivariate numerical data! Here are some of the things you need to keep in mind when conducting your own analyses or when reading reports of such analyses:

1. Correlation does not imply causation. A common media blunder is to infer a cause-and-effect relationship between two variables simply because there is a strong correlation between them. Don't fall into this trap! A strong correlation implies only that the two variables tend to vary together in a predictable way, but there are many possible explanations for why this is occurring besides one variable causing changes in the other.

    For example, the article "Ban Cell Phones? You May as Well Ban Talking Instead" (*USA Today*, April 27, 2000) gave data that showed a strong negative correlation between the number of cell phone subscribers and traffic fatality rates. During the years from 1985 to 1998, the number of cell phone subscribers increased from 200,000 to 60,800,000, and the number of traffic deaths per 100 million miles traveled decreased from 2.5 to 1.6 over the same period. However, based on this correlation alone, the conclusion that cell phone use improves road safety is not reasonable!

    Similarly, the *Calgary Herald* (April 16, 2002) reported that heavy and moderate drinkers earn more than light drinkers or those who do not drink. Based on the correlation between number of drinks consumed and income, the author of the study concluded that moderate drinking "causes" higher earnings. This is obviously a misleading statement, but at least the article goes on to state that "there are many possible reasons for the correlation. It could be because better-off men simply choose to buy more alcohol. Or it might have something to do with stress: Maybe those with stressful jobs tend to earn more after controlling for age, occupation, etc., and maybe they also drink more in order to deal with the stress."

2. A correlation coefficient near 0 does not necessarily imply that there is no relationship between two variables. Before such an interpretation can be given, it is important to examine a scatterplot of the data carefully. Although it may be true that the variables are unrelated, there may in fact be a strong but nonlinear relationship.

3. The least-squares line for predicting y from x is not the same line as the least-squares line for predicting x from y. The least-squares line is, by definition, the line that has the smallest possible sum of squared deviations of points from the line *in the y direction* (it minimizes $\sum(y - \hat{y})^2$). The line that minimizes the sum of squared deviations in the y direction is not generally the same as the line that minimizes the sum of the squared deviations in the x direction. So, for example, it is not appropriate to fit a line to data using y = house price and x = house size and then use the resulting least-squares line Price = $a + b$(Size) to predict the size of a house by substituting in a price and then solving for size. Make sure that the dependent and independent variables are clearly identified and that the appropriate line is fit.

4. Beware of extrapolation. It is dangerous to assume that a linear model fit to data is valid over a wider range of x values. Using the least-squares line to make predictions outside the range of x values in the data set often leads to poor predictions.

5. Be careful in interpreting the value of the slope and intercept in the least-squares line. In particular, in many instances interpreting the intercept as the value of y that would be predicted when $x = 0$ is equivalent to extrapolating way beyond the range of the x values in the data set, and this should be avoided unless $x = 0$ is within the range of the data.

6. Remember that the least-squares line may be the "best" line (in that it has a smaller sum of squared deviations than any other line), but that doesn't necessarily mean that the line will produce good predictions. Be cautious of predictions based on a least-squares line without any information about the adequacy of the linear model, such as s_e and r^2.

7. It is not enough to look at just r^2 or just s_e when assessing a linear model. These two measures address different aspects of the fit of the line. In general, we would like to have a small value for s_e (which indicates that deviations from the line tend to be small) and a large value for r^2 (which indicates that the linear relationship explains a large proportion of the variability in the y values). It is possible to have a small s_e combined with a small r^2 or a large r^2 combined with a large s_e. Remember to consider both values.

8. The value of the correlation coefficient as well as the values for the intercept and slope of the least-squares line can be sensitive to influential observations in the data set, particularly if the sample size is small. Because potentially influential observations are those whose x values are far away from most of the x values in the data set, it is important to look for such observations when examining the scatterplot. (Another good reason for *always* starting with a plot of the data!)

9. If the relationship between two variables is nonlinear, it is preferable to model the relationship using a nonlinear model rather than fitting a line to the data. A plot of the residuals from a linear fit is particularly useful in determining whether a nonlinear model would be a more appropriate choice.

Activity 5.1   Exploring Correlation and Regression Technology Activity (Applets)

CENGAGENOW

Explore this applet activity

Open the applet (available in CengageNOW at www.cengage.com/login) called CorrelationPoints. You should see a screen like the one shown below.

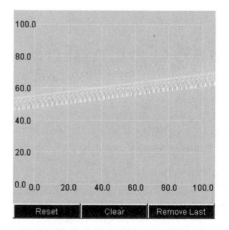

1. Using the mouse, you can click to add points to form a scatterplot. The value of the correlation coefficient for the data in the plot at any given time (once you have two or more points in the plot) will be displayed at the top of the screen. Try to add points to obtain a correlation coefficient that is close to $-.8$. Briefly describe the factors that influenced where you placed the points.

2. Reset the plot (by clicking on the Reset bar in the lower left corner of the screen) and add points trying to produce a data set with a correlation that is close to $+.4$. Briefly describe the factors that influenced where you placed the points. Now open the applet called RegDecomp. You should see a screen that looks like the one shown here:

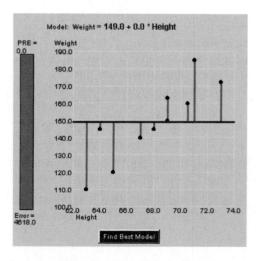

The black points in the plot represent the data set, and the blue line represents a possible line that might be used to describe the relationship between the two variables shown. The pink lines in the graph represent the deviations of points from the line, and the pink bar on the left-hand side of the display depicts the sum of squared errors for the line shown.

3. Using your mouse, you can move the line and see how the deviations from the line and the sum of squared errors change as the line changes. Try to move the line into a position that you think is close to the least-squares regression line. When you think you are close to the line that minimizes the sum of squared errors, you can click on the bar that says "Find Best Model" to see how close you were to the actual least-squares line.

Activity **5.2**   Age and Flexibility

Materials needed: Yardsticks.

In this activity you will investigate the relationship between age and a measure of flexibility. Flexibility will be measured by asking a person to bend at the waist as far as possible, extending his or her arms toward the floor. Using a yardstick, measure the distance from the floor to the fingertip closest to the floor.

1. Age and the measure of flexibility just described will be measured for a group of individuals. Our goal is to determine whether there is a relationship between age and this measure of flexibility. What are two reasons why it would not be a good idea to use just the students in your class as the subjects for your study?

2. Working as a class, decide on a reasonable way to collect data on the two variables of interest.

3. After your class has collected appropriate data, use them to construct a scatterplot. Comment on the interesting features of the plot. Does it look like there is a relationship between age and flexibility?

4. If there appears to be a relationship between age and flexibility, fit a model that is appropriate for describing the relationship.

5. *In the context of this activity*, write a brief description of the danger of extrapolation.

Summary of Key Concepts and Formulas

Term or Formula	Comment
Scatterplot	A picture of bivariate numerical data in which each observation (x, y) is represented as a point located with respect to a horizontal x axis and a vertical y axis.
Pearson's sample correlation coefficient $$r = \frac{\Sigma z_x z_y}{n - 1}$$	A measure of the extent to which sample x and y values are linearly related; $-1 \leq r \leq 1$, so values close to 1 or -1 indicate a strong linear relationship.
Principle of least squares	The method used to select a line that summarizes an approximate linear relationship between x and y. The least-squares line is the line that minimizes the sum of the squared vertical deviations for the points in the scatterplot.

Term or Formula	**Comment**
$b = \dfrac{\sum(x - \bar{x})(y - \bar{y})}{\sum(x - \bar{x})^2} = \dfrac{\sum xy - \dfrac{(\sum x)(\sum y)}{n}}{\sum x^2 - \dfrac{(\sum x)^2}{n}}$	The slope of the least-squares line.
$a = \bar{y} - b\bar{x}$	The intercept of the least-squares line.
Predicted (fitted) values $\hat{y}_1, \hat{y}_2, \ldots, \hat{y}_n$	Obtained by substituting the x value for each observation into the least-squares line; $\hat{y}_1 = a + bx_1, \ldots, \hat{y}_n = a + bx_n$
Residuals	Obtained by subtracting each predicted value from the corresponding observed y value: $y_1 - \hat{y}_1, \ldots, y_n - \hat{y}_n$. These are the vertical deviations from the least-squares line.
Residual plot	Scatterplot of the $(x, \text{residual})$ pairs. Isolated points or a pattern of points in a residual plot are indicative of potential problems.
Residual (error) sum of squares $\text{SSResid} = \sum(y - \hat{y})^2$	The sum of the squared residuals is a measure of y variation that cannot be attributed to an approximate linear relationship (unexplained variation).
Total sum of squares $\text{SSTo} = \sum(y - \bar{y})^2$	The sum of squared deviations from the sample mean is a measure of total variation in the observed y values.
Coefficient of determination $r^2 = 1 - \dfrac{\text{SSResid}}{\text{SSTo}}$	The proportion of variation in observed y's that can be attributed to an approximate linear relationship.
Standard deviation about the least-squares line $s_e = \sqrt{\dfrac{\text{SSResid}}{n - 2}}$	The size of a "typical" deviation from the least-squares line.
Transformation	A simple function of the x and/or y variable, which is then used in a regression.
Power transformation	An exponent, or power, p, is first specified, and then new (transformed) data values are calculated as *transformed value* = (original value)^p. A logarithmic transformation is identified with $p = 0$. When the scatterplot of original data exhibits curvature, a power transformation of x and/or y will often result in a scatterplot that has a linear appearance.
Logistic regression equation $p = \dfrac{e^{a+bx}}{1 + e^{a+bx}}$	The graph of this equation is an S-shaped curve. The logistic regression equation is used to describe the relationship between the probability of success and a numerical predictor variable.

Chapter Review Exercises 5.66–5.78

Turn to the Expanded Answers Section for answers not shown next to exercises.

CENGAGENOW Know exactly what to study! Take a pre-test and receive your Personalized Learning Plan.

5.66 ● The accompanying data represent x = the amount of catalyst added to accelerate a chemical reaction and y = the resulting reaction time:

x	1	2	3	4	5
y	49	46	41	34	25

a. Calculate r. Does the value of r suggest a strong linear relationship?
b. Construct a scatterplot. From the plot, does the word *linear* really provide the most effective description of the relationship between x and y? Explain.

5.67 ● The article "The Epiphytic Lichen *Hypogymnia physodes* as a Bioindicator of Atmospheric Nitrogen and Sulphur Deposition in Norway" (*Environmental Monitoring and Assessment* [1993]: 27–47) gives the following data (read from a graph in the paper) on x = NO_3 wet deposition (in grams per cubic meter) and y = lichen (% dry weight):

x	0.05	0.10	0.11	0.12	0.31
y	0.48	0.55	0.48	0.50	0.58
x	0.37	0.42	0.58	0.68	0.68
y	0.52	1.02	0.86	0.86	1.00
x	0.73	0.85	0.92		
y	0.88	1.04	1.70		

a. What is the equation of the least-squares regression line? $\hat{y} = 0.3651 + 0.9668x$
b. Predict lichen dry weight percentage for an NO_3 deposition of 0.5 g/m^3. 0.8485

5.68 ● Athletes competing in a triathlon participated in a study described in the paper "Myoglobinemia and Endurance Exercise" (*American Journal of Sports Medicine* [1984]: 113–118). The following data on finishing time x (in hours) and myoglobin level y (in nanograms per milliliter) were read from a scatterplot in the paper:

x	4.90	4.70	5.35	5.22	5.20
y	1590	1550	1360	895	865
x	5.40	5.70	6.00	6.20	6.10
y	905	895	910	700	675
x	5.60	5.35	5.75	5.35	6.00
y	540	540	440	380	300

a. Obtain the equation of the least-squares line.
b. Interpret the value of b.
c. What happens if the line in Part (a) is used to predict the myoglobin level for a finishing time of 8 hr? Is this reasonable? Explain.

5.69 The paper "Root Dentine Transparency: Age Determination of Human Teeth Using Computerized Densitometric Analysis" (*American Journal of Physical Anthropology* [1991]: 25–30) reported on an investigation of methods for age determination based on tooth characteristics. With y = age (in years) and x = percentage of root with transparent dentine, a regression analysis for premolars gave n = 36, SSResid = 5987.16, and SSTo = 17,409.60. Calculate and interpret the values of r^2 and s_e.

5.70 ● The paper cited in Exercise 5.69 gave a scatterplot in which x values were for anterior teeth. Consider the following representative subset of the data:

x	15	19	31	39	41
y	23	52	65	55	32
x	44	47	48	55	65
y	60	78	59	61	60

$$\sum x = 404 \quad \sum x^2 = 18{,}448 \quad \sum y = 545$$
$$\sum y^2 = 31{,}993 \quad \sum xy = 23{,}198$$
$$a = 32.080888 \quad b = 0.554929$$

a. Calculate the predicted values and residuals.
b. Use the results of Part (a) to obtain SSResid and r^2.
c. Does the least-squares line appear to give accurate predictions? Explain your reasoning.

5.71 ● The following data on the relationship between degree of exposure to ^{242}Cm alpha radiation particles (x) and the percentage of exposed cells without aberrations (y) appeared in the paper "Chromosome Aberrations Induced in Human Lymphocytes by DT Neutrons" (*Radiation Research* [1984]: 561–573):

x	0.106	0.193	0.511	0.527
y	98	95	87	85
x	1.08	1.62	1.73	2.36
y	75	72	64	55
x	2.72	3.12	3.88	4.18
y	44	41	37	40

Summary quantities are

$$n = 12 \qquad \sum x = 22.207 \qquad \sum y = 793$$
$$\sum x^2 = 62.600235 \qquad \sum xy = 1114.5 \qquad \sum y^2 = 57{,}939$$

a. Obtain the equation of the least-squares line.
b. Calculate SSResid and SSTo.
c. What percentage of observed variation in y can be explained by the approximate linear relationship between the two variables?   94.84%
d. Calculate and interpret the value of s_e.
e. Using just the results of Parts (a) and (c), what is the value of Pearson's sample correlation coefficient? $-.974$

5.72 ● The article "Reduction in Soluble Protein and Chlorophyll Contents in a Few Plants as Indicators of Automobile Exhaust Pollution" (*International Journal of Environmental Studies* [1983]: 239–244) reported the following data on $x =$ distance from a highway (in meters) and $y =$ lead content of soil at that distance (in parts per million):

x	0.3	1	5	10	15	20
y	62.75	37.51	29.70	20.71	17.65	15.41

x	25	30	40	50	75	100
y	14.15	13.50	12.11	11.40	10.85	10.85

a. Use a statistical computer package to construct scatterplots of y versus x, y versus $\log(x)$, $\log(y)$ versus $\log(x)$, and $\dfrac{1}{y}$ versus $\dfrac{1}{x}$.
b. Which transformation considered in Part (a) does the best job of producing an approximately linear relationship? Use the selected transformation to predict lead content when distance is 25 m.

5.73 The sample correlation coefficient between annual raises and teaching evaluations for a sample of $n = 353$ college faculty was found to be $r = .11$ ("Determination of Faculty Pay: An Agency Theory Perspective," *Academy of Management Journal* [1992]: 921–955).
a. Interpret this value.   Weak linear relationship
b. If a straight line were fit to the data using least squares, what proportion of variation in raises could be attributed to the approximate linear relationship between raises and evaluations?   .0121

5.74 ● An accurate assessment of oxygen consumption provides important information for determining energy ex-

penditure requirements for physically demanding tasks. The paper "Oxygen Consumption During Fire Suppression: Error of Heart Rate Estimation" (*Ergonomics* [1991]: 1469–1474) reported on a study in which $x =$ oxygen consumption (in milliliters per kilogram per minute) during a treadmill test was determined for a sample of 10 firefighters. Then $y =$ oxygen consumption at a comparable heart rate was measured for each of the 10 individuals while they performed a fire-suppression simulation. This resulted in the following data and scatterplot:

Firefighter	1	2	3	4	5
x	51.3	34.1	41.1	36.3	36.5
y	49.3	29.5	30.6	28.2	28.0

Firefighter	6	7	8	9	10
x	35.4	35.4	38.6	40.6	39.5
y	26.3	33.9	29.4	23.5	31.6

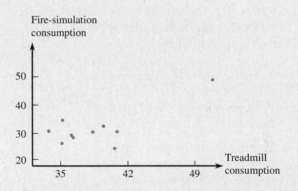

a. Does the scatterplot suggest an approximate linear relationship?
b. The investigators fit a least-squares line. The resulting MINITAB output is given in the following:

The regression equation is
firecon = −11.4 + 1.09 treadcon

Predictor	Coef	Stdev	t-ratio	p
Constant	−11.37	12.46	−0.91	0.388
treadcon	1.0906	0.3181	3.43	0.009

s = 4.70   R-sq = 59.5%   R-sq(adj) = 54.4%

Predict fire-simulation consumption when treadmill consumption is 40.   32.254
c. How effectively does a straight line summarize the relationship?
d. Delete the first observation, (51.3, 49.3), and calculate the new equation of the least-squares line and the value

of r^2. What do you conclude? (Hint: For the original data, $\sum x = 388.8$, $\sum y = 310.3$, $\sum x^2 = 15,338.54$, $\sum xy = 12,306.58$, and $\sum y^2 = 10,072.41$.)

5.75 ● The relationship between the depth of flooding and the amount of flood damage was examined in the paper "Significance of Location in Computing Flood Damage" (*Journal of Water Resources Planning and Management* [1985]: 65–81). The following data on x = depth of flooding (feet above first-floor level) and y = flood damage (as a percentage of structure value) were obtained using a sample of flood insurance claims:

x	1	2	3	4	5	6	7
y	10	14	26	28	29	41	43

x	8	9	10	11	12	13
y	44	45	46	47	48	49

a. Obtain the equation of the least-squares line.
b. Construct a scatterplot, and draw the least-squares line on the plot. Does it look as though a straight line provides an adequate description of the relationship between y and x? Explain.
c. Predict flood damage for a structure subjected to 6.5 ft of flooding.   34.5687
d. Would you use the least-squares line to predict flood damage when depth of flooding is 18 ft? Explain.

5.76 Consider the four (x, y) pairs $(0, 0)$, $(1, 1)$, $(1, -1)$, and $(2, 0)$.
a. What is the value of the sample correlation coefficient r? $r = 0$
b. If a fifth observation is made at the value $x = 6$, find a value of y for which $r > .5$.
c. If a fifth observation is made at the value $x = 6$, find a value of y for which $r < .5$.

5.77 ● The paper "Biomechanical Characteristics of the Final Approach Step, Hurdle, and Take-Off of Elite Amer-

ican Springboard Divers" (*Journal of Human Movement Studies* [1984]: 189–212) gave the following data on y = judge's score and x = length of final step (in meters) for a sample of seven divers performing a forward pike with a single somersault:

y	7.40	9.10	7.20	7.00	7.30	7.30	7.90
x	1.17	1.17	0.93	0.89	0.68	0.74	0.95

a. Construct a scatterplot.
b. Calculate the slope and intercept of the least-squares line. Draw this line on your scatterplot.
c. Calculate and interpret the value of Pearson's sample correlation coefficient. $r = 5815$; positive linear relationship

5.78 ● The following data on y = concentration of penicillin-G in pig's blood plasma (units per milliliter) and x = time (in minutes) from administration of a dose of penicillin (22 mg/kg body weight) appeared in the paper "Calculation of Dosage Regimens of Antimicrobial Drugs for Surgical Prophylaxis" (*Journal of the American Veterinary Medicine Association* [1984]: 1083–1087):

x	5	15	45	90	180
y	32.6	43.3	23.1	16.7	5.7
x	240	360	480	1440	
y	6.4	9.2	0.4	0.2	

a. Construct a scatterplot for these data.
b. Using the ladder of transformations of Section 5.4 (see Figure 5.31 and Table 5.5), suggest a transformation that might straighten the plot. Give reasons for your choice of transformation.

Personal Tutor
Do you need a live tutor
for homework problems?

CENGAGENOW
Are you ready? Take your
exam-prep post-test now.

Bold exercises answered in back   ● Data set available online but not required   ▼ Video solution available

Graphing Calculator Explorations

Exploration 5.1 Linear Regression

The scatterplot is a very versatile graph, not limited to its use in fitting a model to data. Thus, your calculator will most likely separate the scatterplot and the regression procedures, and you may have to perform two distinct procedures as you perform a regression and plot the results. The construction of a scatterplot was considered in a previous Graphing Calculator Exploration. Here we will point out some of the things you will encounter as you construct a best-fit line.

In general, you will need to perform three tasks to get the best-fit line:

1. Prepare for the plotting of the scatterplot, as discussed previously.
2. Navigate the menu to select linear regression.
3. Transfer the best-fit equation information back to the graphing screen.

Navigating your calculator's menu system means yet another session with your calculator manual. The procedure should end with a choice that looks like "Linear Regression," or possibly "LinReg" for short. When you arrive at that menu option you should be able to just press the appropriate key, and—voilà!—the best-fit line will be calculated for you. When you type the final keystroke or select the final menu item, you see

LinReg

What your calculator sees, however, is,

LinReg, List1, List2

which it interprets as "*find the best-fit line assuming that the x values are in List1, and the y values are in List2*." If you have set up your data in different lists, possibly because you wish to save the data from a previous analysis, you will need to key in the "complete" LinReg command to override the default lists, possibly something like this:

LinReg, List3, List4

Again, you will need to check your manual for the exact syntax, especially to see which variable is assigned the first list in the command. Figure 5.40 shows the output from a linear regression.

Now that we have calculated the best-fit line, we want to show it on the scatterplot. There are two ways to do this—one easy and one even easier. The easy way is to simply enter the equation as you would any other function. Then, when the graph is made, the line will appear along with the scatterplot. This is because the "data plotting screen" and the "function plotting screen" share the same memory and the same coordinate system. The easier way will differ from calculator to calculator, but will in general be one of these three possibilities:

1. There may be a "draw" menu option on the screen with the scatterplot. Pressing this will cause the best-fit line to be drawn.
2. You may be able to "copy and paste" the equation stored by the best-fit procedure. This would be similar to the copy-and-paste operation on computers.
3. On many calculators this copy-and-paste operation can be entered directly on the LinReg line when you specify the regression. For example, the LinReg line might look like:

LinReg List1, List2, Y1

which your calculator would interpret as "*find the best-fit line assuming that the x values are in List1, and the y values are in List2. Then, paste that best-fit line into function Y1 for graphing purposes*."

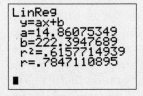

Figure 5.40 Linear regression output.

```
LinReg
 y=ax+b
 a=14.86075349
 b=222.3947689
 r²=.6157714939
 r=.7847110895
■
```

Exploration 5.2 The Calculator as Communicator

As many aficionados of both movies and television are aware, when movies are later shown on television there are some compromises made because of the space limitations of TV. *King Kong* simply doesn't look the same on television! There are similar compromises made when the designers of calculators create displays on calculator screens.

When we do "mathematical" work on paper, our instructors usually have certain requirements, which will vary from instructor to instructor. Those requirements usually assume access to 8.5-by-11-inch paper. It can safely be said that not very many calculators will have a screen that large, and this creates problems for calculator manufacturers. What is the best way to present information on a calculator screen? Calculator screens have limited space for effective display of text characters, and limited space for graphics. Although the designers of the calculators are very knowledgeable about how mathematics is written, they cannot work miracles and convert from 8.5-by-11-inch paper to a much smaller area. And in any case they cannot anticipate what each instructor will want shown for responses to different questions.

In general, it is our belief that how you solve a problem and report your method of solution should closely mirror what you would write if you were not using a calculator. We believe that when answering a question your fundamental task is not just to provide an answer, but also to communicate your method. The calculator, because of the compromises discussed above, is singularly unable to communicate methods effectively. We would like to use the context of regression to discuss some of the differences between what the calculator gives you and what you should give to your instructor. (You should check with your instructor to make sure what his or her requirements are—we will merely point out that there may well be a mismatch between instructors' requirements and a calculator's screen.) When performing linear regression on your calculator, you will probably see something like the accompanying figures for text and graphic output.

Consider first the regression equation in Figure 5.41. Notice all those digits? You probably don't want to report all of them! Also, notice that the equation is reported very generically. To the designer of the calculator screen this makes perfect sense. With a generic presentation the amount of screen area needed remains constant, irrespective of what variables you are using. However, in your work you should indicate what the variables are. We recommend presenting the regression information as follows:

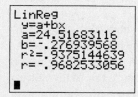

Figure 5.41 Linear regression output.

$$\text{Strength} = 24.52 - 0.277 \cdot \text{Depth} \quad \text{or}$$
$$\hat{S} = 24.52 - 0.277D$$

where S and D have been previously defined. Also, report $r = -0.97$, or $r = -0.968$, a more reasonable choice of significant digits.

Suppose you were asked to predict the strength of a core taken from a depth of $D = 25.0$ mm. Here is a possible template with our notes about what each line is saying to your instructor:

$$\hat{S} = 24.52 - 0.277D \quad \longleftarrow \text{("Here's the formula I'm using.")}$$
$$= 24.52 - 0.277(25.0) \quad \longleftarrow \text{("Here are numbers I'm substituting.")}$$
$$\hat{S} = 17.60 \text{ Mpa} \quad \longleftarrow \text{("Here's my result.")}$$

The advantages of this presentation are:

1. If you err in your keystrokes, your instructor can still evaluate your method and assess your understanding.
2. If your answer differs from that of your fellow statistics students, you can easily tell whether you made a conceptual error or a keystroke error without having to take time to duplicate your work.
3. Most important from our perspective, you are communicating in the language of statistics, where reliable communication demands a standard symbolism.

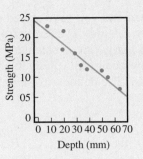

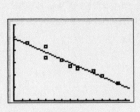

Figure 5.42 Scatter-
plot of strength versus
depth.

Figure 5.43 Best-fit
line and scatterplot.

Now consider the scatter plot as it appears on the calculator screen in Figure 5.42. The most serious omission is the
scale on the two axes. From looking at the graph, one cannot tell how large or small the numbers are. Also, there is no
indication of what the variables are. This is certainly understandable on a calculator screen, where all this information
would take up precious space. However, this is not how to
sketch a plot on a homework assignment. We recommend
that every graph have a title, indicate units, and show any
other relevant information, as in Figure 5.43.

$$\text{Strength} = 24.52 - 0.277\text{Depth} \quad \text{or}$$
$$r = -0.97$$

Our message to the student using a graphing calculator is really very simple: You must
realize not only the power of the calculator to save you time and effort, but also its limitations of communication.

Exploration 5.3 In the Matter of Residuals

Different calculators handle residuals in different ways. Calculator A may store the
residuals but not let you work with them; calculator B may not even store the residuals. In this Graphing Calculator Exploration we will offer several "workarounds" related to residuals, since analyzing the residuals is an essential part of regression analysis! We assume that your calculator is list-based, and that you can perform calculations
using those lists.

First, suppose that your calculator does not store the residuals. Let's see how we
might get around this problem. We will suppose that the x-variable data have been stored
in List1, and the y variable data in List2, and that you have successfully performed the
appropriate regression calculations, giving you a best-fit line: $\hat{y} = a + bx$.

Once you have the best-fit line you can calculate the residuals and put them in,
say, List3 as follows:

$$\text{List2} - (a + b * (\text{List1})) \rightarrow \text{List3}$$

As you have probably already noticed, we are simply using the definition of the residual for these calculations. This "List algebra" corresponds to our usual algebra, which
would be

$$y_i - \hat{y}_i \rightarrow \text{residual list} \quad \text{or} \quad y_i - (a + bx_i) \rightarrow \text{residual list}$$

Once the residuals are stored in this list we can plot them and use them in other ways
as well.

A second problem we would like to address is the calculation of the standard
deviation about the least-squares line, $s_e = \sqrt{\text{SSResid}/(n - 2)}$. You can obtain s_e directly from information at hand: the number of points, n, the sample standard deviation
of y, s_y, and the sample correlation coefficient, r. We will skip the straightforward but
slightly tedious algebra and cut directly to the chase:

$$s_e = \sqrt{\frac{n - 1}{n - 2}} \left(\sqrt{1 - r^2} \right) s_y$$

If your calculator is list-based and you have stored the residuals in a list as described above, you may be able to calculate s_e directly from the formula. Suppose that you have copied the residuals to List1. (Some calculators will not let you work with the residual list, and you will be forced to copy them to another list. In any case, making a copy preserves the originals.) The defining formula,

$$s_e = \sqrt{\frac{\text{SSResid}}{n-2}}$$

translates to

$$s_e = \sqrt{\frac{\sum \text{List1}^2}{n-2}}$$

This is an easy calculation on most calculators and might look something like this:

$$\sqrt{\sum \text{List1}^2/(n-2)}$$

For example, if we had $n = 10$ pairs of data points, we would calculate

$$\sqrt{\sum \text{List1}^2/8}$$

With these two workarounds, you should be able to circumvent a slightly statistically challenged calculator and work with residuals just as if you had one of those fancy statistical calculators!

Exploration 5.4 How Does Your Calculator Perform Nonlinear Regression?

Scientific graphing calculators come with built-in statistical capability, including linear regression. Some calculators will also have options for nonlinear regression, such as exponential regression ($\hat{y} = ab^x$), power regression ($\hat{y} = ax^b$), and logarithmic regression ($\hat{y} = a + \ln b$), as well as polynomial regression. The process of transforming data, fitting a line to the transformed data, and then "undoing" the transformation is used to get a good—though not "best" in the least-squares sense—fit of a function to the data.

Your calculator may transform variable(s) and fit a line to the transformed data, or it may use more complicated methods to estimate the least-squares solution. It is also possible that your calculator may produce unusual residual plots. This is not necessarily a bad thing, but does place some responsibility on your shoulders. Although your calculator does the arithmetic, it is *you* who are responsible for understanding and interpreting the calculator output.

We will use exponential regression to illustrate how you might explore the calculator's regression output. We will compare two approaches to the regression: using linear regression on the transformed data, and using the "exp reg" button on the calculator. Since your calculator may differ from ours, we will continue to avoid specific keystrokes. You should be able to reproduce this analysis for different nonlinear regressions. If you choose to do reexpression "by hand" (transforming the data, fitting a line to the transformed data, and then back-transforming), you do not need to worry about what the calculator "button" does. This analysis needs to be done only if you wish to get quicker answers by using your calculator's built-in nonlinear regression capabilities.

The strategy we will use is to pick pairs of points with the same x values, and choose y values on either side of our model, $f(x) - k$ and $f(x) + k$. Choosing points in this manner will produce data that will force a least-squares solution through $f(x)$. Generating data using the function $f(x) = e^x$, x values from 2 to 5, and $k = 5$ results in the data shown in the x and y columns of the accompanying table. Because of the way the data were generated, the least-squares exponential function is $\hat{y} = e^x$.

x	y (formula)	y (decimal)	ln (y)	Residual (calculator)	Residual (formula)
2	$e^2 + 5$	12.38905610	2.516813510	.7442163052	6.502935110
2	$e^2 - 5$	2.38905610	.870898350	−.901698854	−3.4970648
3	$e^3 + 5$	25.08553692	3.22229146	.3552438748	7.50051369
3	$e^3 - 5$	15.08553692	2.71373646	−.153311121	−2.4994863
4	$e^4 + 5$	59.59815003	4.08762453	.1261265648	7.06218576
4	$e^4 - 5$	49.59815003	3.90395353	−.057544433	−2.9378142
5	$e^5 + 5$	153.4131591	5.03313466	−.022813682	−3.5401474
5	$e^5 - 5$	143.4131591	4.96572968	−.090218662	−13.540147

Let's transform the data, fit a line, back-transform, and then compare the result to the calculator output when an exponential regression is requested. If the relationship between x and y is exponential (as it is here because of the way in which the data were generated), a log transformation would be an appropriate transformation to create a linear relationship. Taking the natural logarithm of y results in the transformed y values in the ln(y) column of the table. Using the calculator to fit a line to the $(x, \ln(y))$ data gives

$$\ln y = -0.416303559 + 1.094450382x$$

with $r^2 = 0.885784757$. Back-transforming gives the nonlinear relationship

$$y = 0.659480049(2.987540226)^x.$$

Notice this is different from the least-squares solution of $y = e^x$, but it still provides a reasonable fit to the data.

Figure 5.44a shows a scatterplot of ln(y) versus x and Figure 5.44b a plot of the residuals for the linear fit to the transformed data. The residual plot certainly does not look "random," but for our present purposes this is not a problem. The calculator

Figure 5.44 Plots using transformed data: (a) plot of ln(y) versus x; (b) Plot of ln(y) residuals versus ln(y) predicted values.

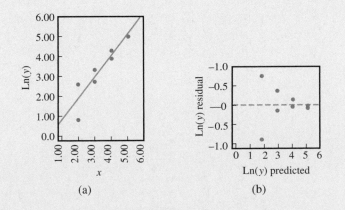

reports the residuals from this regression and we have reproduced these in the Residual (calculator) column in the previous table.

Now try the "exp reg" function on your calculator. You should find that the reported regression is $y = 0.659480049 \cdot 2.987540226^x$ with $r^2 = 0.886$, the same values that were obtained by transforming and then fitting a linear function to the transformed data. Note that the calculator did not produce the least-squares fit of $y = e^x$. If you now request a scatterplot and a residual plot, you will see something like the plots in Figures 5.45a and 5.45b.

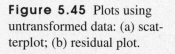

Figure 5.45 Plots using untransformed data: (a) scatterplot; (b) residual plot.

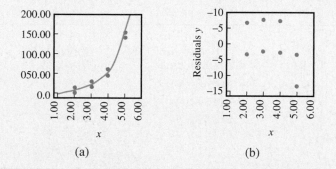

(a)   (b)

The scatterplots look different, which would be expected because the second plot is of the untransformed data. However, the pattern of residuals is very equation to get a predicted y and then computing residual $= y -$ predicted y rather than using the residuals from the linear fit to the transformed data, $\ln(y) -$ predicted $\ln(y)$. Another fact to note is that the r^2 reported by the exp reg button analysis is the r^2 for the *transformed* fit, even though the residuals reported are in the *untransformed* scale. The calculator is doing the transformations for us, which seems reasonable. However, the residuals from the linear regression are not the ones presented; moreover, the residuals that are reported have a very different pattern than those from the linear regression of the transformed data.

The lesson from all these calculations is a simple one: If you wish to use the built-in capabilities of the calculator, you need to understand what it is actually doing. The book that comes with the calculator may not be particularly clear, and you should perform an example such as the one above to be sure that you understand what is happening under the buttons! Remember, *you* are ultimately responsible for your analyses. Your calculator can only perform really quick arithmetic—you, the analyst, must provide the careful thinking and clear understanding of the results of that arithmetic.

TEACHING TIPS

Intent of this chapter:

After completing this chapter, students should be able to (1) identify the sample space for an experiment, (2) interpret probabilities as long-run relative frequencies, (3) use probability properties to calculate the likelihood of a simple event or a compound event, (4) calculate conditional probabilities, and (5) use simulations to estimate probabilities.

A PowerPoint® Lecture is available on the *Instructor's Resource Binder* CD.

■ **Before you begin . . .**

Students may or may not be familiar with counting rules. This will depend on the individual curriculum at your school. If your students have not seen counting rules before, you may want to do a short review. The counting principle is helpful in determining how many outcomes are in a sample space. For example, suppose you will flip four coins and want to determine the probability of tossing exactly three heads. Since each coin has two ways that it can land (heads and tails), to find the total number of outcomes multiply $2 \times 2 \times 2 \times 2$ to obtain 16 total outcomes. Combinations are sometimes useful in determining how many ways an event can occur. For example, suppose you wanted to know how many of the 16 outcomes would have exactly two heads. There are six different ways that flipping four coins will land with exactly two heads, $_4C_2 = 6$.

■ **Section 6.1**

This section introduces the concepts of sample spaces and events. Almost every student has some prior knowledge of probability; however, students are typically not familiar with the vocabulary presented in this section. Students should know that the probabilities of complementary events add to one. Also, some students are not familiar with the set notation: $\cup$ for union and $\cap$ for intersection. In this section students are introduced to four techniques to aid in the identification of the sample space: lists, tree diagrams, tables, and Venn diagrams. Although most students are familiar with Venn diagrams, some students may need more practice with Venn diagrams that represent two or more events. (Remember, there is a broad range of abilities in the students that enroll in an introductory statistics course, especially in critical thinking skills.)

Suggested Assignment: Exercises 6.3, 6.5, 6.8, 6.9, 6.11, 6.12

Activity: Bonus Activity 6.5—Efron's Dice (*Activities Manual*, p. 63)★ and "Chapter 6— Probability" (*Instructor's Resource Binder* & CD, Chapter 6 Activities Worksheets)

■ **Section 6.2**

Section 6.2 discusses three approaches to probability: the classical approach, the relative frequency approach, and the subjective approach. The classical approach, stated on page 289, is typically the approach with which students have past experience. It should be stressed that this approach is only applicable for equally likely events. The insurance scenario on page 290 is an excellent example of events that are not equally likely to occur.

The relative frequency approach to probability interprets probabilities as long-run relative frequencies. In this approach, probabilities are often estimated empirically based on the **observed** outcomes of a large number of repetitions of a chance experiment (see the definition on page 293). Figures 6.5 and 6.6 demonstrate the difference between short-run and long-run behavior. Most students understand that, in the short-run (a small number of trials), the observed relative frequency of heads may or may not be equal to 0.5. However, in the long run, the observed relative frequency of heads will get close and stay close to 0.5. A discussion of short-run versus long-run behavior, along with the examples in Figures 6.5 and 6.6, aid in students' understanding of the Law of Large Numbers. (See page 293.)

★ Kathy's personal favorite

The subjective approach is based on personal judgments. This approach has limited use in discussions of elementary probability since different individuals may assign different probabilities to the same outcomes.

■ **Section 6.3**

This section discusses four basic properties of probability: only values between 0 and 1, inclusive, are legitimate values for a probability; the sum of the probabilities of all of the simple events making up the sample space is 1; the probability of the union of two (or more) disjoint events equal the sum of the probabilities of each event; and the probabilities of complementary events sum to 1. The first property might seem to be a "no-brainer," but as amazing as it might seem, students will provide answers to probability questions that exceed 1! On past exams, students have given answers like the following: an event has a probability of 150%, or the P-value $= 3.45$. (Recall that the P-value is a probability [see Chapter 10].) Therefore, this rule should be stressed. In the event that a student calculates a probability that is not within the range of legitimate values, "red lights and sirens" should sound a warning in students' brains that a mistake has been made.

The second property, that the sum of the probabilities of all events in the sample space is equal to 1, can be extremely useful in calculating probabilities. Consider the following example. Suppose you will toss a coin two times and are interested in the probability of tossing *at least one* head. There are four (2×2) possible outcomes: HH, HT, TH, or TT. So the probability is ¾. Easy! However, what if you tossed the coin four times, or ten times? It quickly becomes very cumbersome to list all the possible outcomes. Notice, though, the probability of *at least one* head includes all the outcomes *except the one outcome that has no heads* (all tails). Therefore, the probability of seeing at least one head when flipping four coins can be calculated by $1 -$ probability that none of the four coins land heads; $1 - 1/32 = 0.9375$, since there are 32 possible outcomes in the sample space.

The third property introduces the addition rule for disjoint events. Be sure to stress that this rule applies *only* to disjoint events; if the events are *not* disjoint, a different method must be used to calculate the probability of the union. Venn diagrams are extremely useful in providing a visual representation of disjoint or mutually exclusive events. Further discussion of mutually exclusive events occurs in Section 6.6.

The last property involves complementary events. Like the second property, this property is very useful in calculating probabilities. Since probabilities of complementary events have a sum of 1, subtracting the probability of an event from 1 will result in the probability of the complement of the event. Often this rule is used in conjunction with the addition rule. (See Example 6.10 on page 299.)

In Exercise 6.25 on page 302, part (a) states that there are 10 possible outcomes. Students can make a list of possible outcomes or use a combination to find how many outcomes are possible. The number of ways to select two students from the five departments is $_5C_2 = 10$.

Suggested Assignment: Exercises 6.13, 6.18, 6.20, 6.21, 6.23, 6.26, 6.27
Activity: Activity 6.2: A Crisis for European Sports Fans? (p. 347 and *Activities Manual*, p. 55)★
Assessment: Chapter 6 Quiz 1 (*Test Bank* and *Instructor's Resource Binder* CD)

■ **Section 6.4**

Conditional probability is a new topic for statistics students. Students are therefore introduced to new notation, $P(E|F)$. To teach conditional probability more effectively, it is helpful to use an example like the one from the PowerPoint Lecture about a study

★ Kathy's personal favorite

("Motion Sickness in Public Road Transport: The Effect of Driver, Route and Vehicle" (*Ergonomics* (1999): 1646–1664)) performed to look at the relationship between motion sickness and seat position in a bus. The following table summarizes the data. The probability that a randomly selected individual in the study develops nausea is $P(N) = \dfrac{417}{3256}$.

	Seat Position in Bus			
	Front	Middle	Back	Total
Nausea	58	166	193	417
No Nausea	870	1163	806	2839
Total	928	1329	999	3256

The probability that a randomly selected individual in the study is seated in the front of the bus is $P(F) = \dfrac{928}{3256}$.

	Seat Position in Bus			
	Front	Middle	Back	Total
Nausea	58	166	193	417
No Nausea	870	1163	806	2839
Total	928	1329	999	3256

The probability that a randomly selected individual in the study is seated in the front of the bus **and** develops nausea is $P(F \cap N) = \dfrac{58}{3256}$. Notice that the numbers inside the table (outlined in blue) are the intersections ("and") of the various events.

	Seat Position in Bus			
	Front	Middle	Back	Total
Nausea	58	166	193	417
No Nausea	870	1163	806	2839
Total	928	1329	999	3256

Now consider a conditional probability. If a randomly selected individual in the study is seated in the front of the bus, what is the probability that the individual develops nausea? The condition narrows our focus so that we are interested *only* in the column for "Front of the bus," shaded green. Out of the 928 individuals seated in the front of the bus, 58 of them develop nausea.

Therefore, our probability is $P(N|F) = \dfrac{58}{928}$. Notice that this is $P(N|F) = \dfrac{P(N \cap F)}{P(F)}$ or $P(N|F) = \dfrac{P(and)}{P(given)}$. We have just *developed* the equation for conditional probability on page 305.

	Seat Position in Bus			
	Front	**Middle**	**Back**	**Total**
Nausea	**58**	166	193	417
No Nausea	870	1163	806	2839
Total	**928**	1329	999	3256

If a randomly selected individual in the study develops nausea, what is the probability that the individual is seated in the front of the bus? Our answer is $P(N\,|\,F) = \dfrac{58}{417}$.

	Seat Position in Bus			
	Front	**Middle**	**Back**	**Total**
Nausea	**58**	166	193	**417**
No Nausea	870	1163	806	2839
Total	928	1329	999	3256

It is very helpful for students to think of conditional probability in the form of $P(N\,|\,F) = \dfrac{P(and)}{P(given)}$. Instruct students to identify the condition, or "given," first and record that probability in the denominator. The numerator is then the probability of the intersection, or "and," of the two events. A Venn diagram like the one displayed with Example 6.12 on page 305 is also very helpful in aiding students' understanding of conditional probabilities.

Suggested Assignment: Exercises 6.29, 6.30, 6.32, 6.33, 6.37, 6.39, 6.40

Activity: "Chapter 6—Venn Diagrams and Conditional Probability" (*Instructor's Resource Binder* & CD, Chapter 6 Activities Worksheets)

■ **Section 6.5**

Students have a naive understanding of independence. In earlier math courses, students probably worked problems in which cards were drawn from a standard deck either with replacement (independent) or without replacement (dependent). This may be the extent of students' knowledge about independent events.

The last paragraph on page 314 is an important definition for independent events. Putting this into a context can aid understanding. For example, suppose a student is selected at random from your school and suppose that your school has both girls' and boys' soccer teams. Let $P(E)$ be the probability that the student plays soccer. Let $P(F)$ be the probability that the student is female. Does $P(E\,|\,F) = P(E)$? If it is known that the student is female, is the probability that the selected student plays soccer the same as it was without the knowledge that student was female? If the answer is "yes," then the two events E (student plays soccer) and F (student is female) are independent. Now, let $P(E)$ be the probability that the student plays football. Does $P(E\,|\,F) = P(E)$? If it is known that the student is female, is the probability that this student plays football the same as it was without the knowledge that student was female? If the answer is "no," then the two events E (playing football) and F (being female) are dependent events. Also note that if two events are independent, their complements are also independent. See the top of page 315.

The Multiplication Rule for Two Independent Events can be derived from the formula for conditional probability and the definition of independence (see page 315). Students have experience with this rule and have no difficulty calculating probabilities using this rule.

The last part of this section explores whether successive selections are dependent or independent when samples are selected with or without replacement. An important concept is stated in the blue box on page 319.

Suggested Assignment: Exercises 6.41, 6.42, 6.43, 6.45, 6.49, 6.51, 6.54, 6.55

Activity: "Chapter 6—Independence" (*Instructor's Resource Binder* & CD, Chapter 6 Activities Worksheets)

■ **Section 6.6**

In this section students are presented with two general probability rules: one involving addition and the other multiplication. Section 6.3 gives an addition rule *for two (or more) disjoint* events. The general rule in this section is for *any* two events: $P(E \cup F) = P(E) + P(F) - P(E \cap F)$. To clarify the difference between the general rule and the rule for disjoint events, Venn diagrams are very helpful. Students quickly see that in a Venn diagram (similar to the one on page 323) that, if you add $P(E)$ and $P(F)$, the center part, $P(E \cap F)$, is included twice. Similarly, if the two events E and F are disjoint, then in the Venn diagram, there is no overlap, so $P(E \cap F) = 0$.

In Section 6.5, students were given a multiplication rule for two independent events. The general multiplication rule, $P(E \cap F) = P(E|F)P(F)$, is for two events without the independence stipulation. Students typically feel some anxiety over this rule. Let's rewrite this rule using the commutative property of multiplication, $P(E \cap F) = P(F) \cdot P(E|F)$, and examine a problem in context. Suppose you select two red cards at random from a standard deck without replacement. Let F be the event the first card selected is red and E be the event the second card selected is red. The probability of *E and F* would be $P(F) \times P(E|F)$. The probability that the first card selected is red, $P(F)$, equals 0.5. The probability that the second card is red *given that a red card* was selected the first time, $P(E|F)$, equals 25/51. (A lightbulb has just switched on in the students' heads! This context they know—it was only the notation that was unfamiliar to them!)

So $P(E \cap F) = \left(\dfrac{1}{2}\right)\left(\dfrac{25}{51}\right) = \dfrac{25}{102}$.

At this point, students tend to get the concepts mutually exclusive and independence confused. These are two very different ideas and they need to be kept separate. Mutually exclusive events have no outcomes in common and $P(E|F) = 0$. We are interested in whether the events are mutually exclusive if, and *only* if, we are working with the union (*or*) of two (or more) events. Independent events are events for which the knowledge that one will occur (or has occurred) does not affect the probability of the other event. We are interested in whether the events are independent if, and *only* if, we are working with the intersection (*and*) of two (or more) events. A typical question that students ask is whether events that are mutually exclusive can also be independent. This is impossible. Remember, mutually exclusive events *do not* occur at the same time. Therefore, if we know that one event occurs, then we know the other event cannot occur. So if two events are mutually exclusive, then they *must* be dependent events.

Here is a thought process that helps students with these rules. Suppose the problem asks students to find $P(E \cap F)$ or $P(E \text{ and } F)$. The students should ask themselves, "Are these events independent?" If "yes," just multiply probabilities. If "no," multiply the probability of F times the conditional probability of E given F.

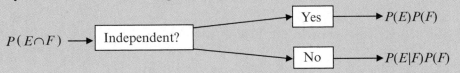

Suppose the problem asks students to find $P(E \cup F)$ or $P(E \text{ or } F)$. The students should ask themselves, "Are these events mutually exclusive?" If "yes," just add probabilities. If "no," be sure to add the probabilities and subtract the intersection. When students subtract the intersection (*and*), then they should use the process above.

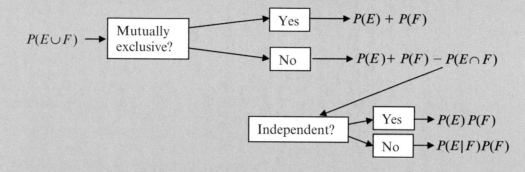

Students should be able to use Venn diagrams and probability trees, as well as formulas, to calculate probabilities. It is helpful to work problems in as many different ways as possible in order to provide students with a choice of methods. For different students, different representations may be more helpful for particular probability problems.

The Law of Total Probability and Bayes' Rule are not elementary topics. (For example, they are not in the Advanced Placement Statistics course description.) However, it is helpful for students to have some familiarity with these rules. Students can answer questions that are similar to Example 6.28 without the explicit use of Bayes' Rule. Two very different approaches can be utilized instead.

The first approach is to create a probability tree. Students have trouble deciding which event—S (store ordered from) or L (arrival time [late or not late])—should be placed on the first generation of branches. Whichever event encompasses the population is the one that should be placed on the first branches. Since the "store ordered from" is the event that encompasses the population, that is what should be placed on the first branches. We want to compare the following probabilities: $P(S_1 | L)$ and $P(S_2 | L)$.

$$P(L) = P(S_1 \cap L) \cup P(S_2 \cap L) = .3(.1) + .7(.08) = .086$$

$$P(S_1 | L) = \frac{P(S_1 \cap L)}{P(L)} = \frac{.3(.1)}{.086} = .349$$

$$P(S_2 | L) = \frac{P(S_2 \cap L)}{P(L)} = \frac{.7(.08)}{.086} = .651$$

Another approach is to create a table from a make-believe population of items, say of size 1000. (Why 1000? Because it makes calculations easy!) We will represent the events "store ordered from" as S_1 and S_2, and "arrival time" as L and L^C. The nice thing about the table approach is that it doesn't matter which event is presented in the rows! "Store ordered from" could go along the top or along the side. Let's put it on top.

	S_1	S_2	Total
L			
L^C			
Total			1000

What do we know about the population? We know 30% were ordered from Store 1. Next, what do we know about the arrival time of the 300 packages from Store 1? We know that 10% of them arrive late. So 270 packages arrive on time. Similarly, we know that 8% of the packages from Store 2 will be late and 644 will arrive on time. Then find the totals for L and L^C.

	S_1	S_2	Total
L	30	56	86
L^c	270	644	914
Total	300	700	1000

$$P(S_1|L) = \frac{P(S_1 \cap L)}{P(L)} = \frac{30}{86} = .349 \text{ and } P(S_2|L) = \frac{P(S_2 \cap L)}{P(L)} = \frac{56}{86} = .651$$

Most students find this approach easier to follow.

Suggested Assignment: Exercises 6.59, 6.61, 6.66, 6.67, 6.68, 6.69, 6.71, 6.73

Activity: Bonus Activity 6.4—The Monty Hall Problem (*Activities Manual*, p. 60)★

Assessment: Chapter 6 Quiz 2 (*Test Bank* and *Instructor's Resource Binder* CD)

■ **Section 6.7**

Simulation is a very useful tool for estimating probabilities in some circumstances where it is either difficult or impossible to calculate the probabilities exactly. Simulations employ some random mechanism (e.g., toss of a coin, roll of a die, numbers pulled from a hat, or random number table or generator) to represent outcomes of an experiment, taking care to match the probabilities of the simulation outcomes and those of the original experiment. The tan box on page 339 gives a sequence of steps to use in performing a simulation, and the text offers excellent examples for carrying out simulations.

On past AP Statistics exams, students have been asked to create histograms using outcomes from a simulation and then use the histograms to answer questions about long-run behaviors.

Suggested Assignment: Exercises 6.75, 6.78, 6.79, 6.81

Activity: Activity 6.1: Kisses (p. 347 and *Activities Manual*, p. 54)★ and Activity 6.3: The "Hot Hand" in Basketball (p. 347 and *Activities Manual*, p. 57)★

Assessment: Chapter 6 Quiz 3 (*Test Bank* and *Instructor's Resource Binder* CD) and/or Chapter 6 Concept Quiz (*Test Bank* and *Instructor's Resource Binder* CD)

Suggested Review Assignment: Exercises 6.84, 6.85, 6.86, 6.88, 6.89, 6.92, 6.96, 6.97

Assessment: Chapter 6 Test (*Test Bank* and *Instructor's Resource Binder* CD)

★ Kathy's personal favorite

Chapter 6

Probability

© Doug Menuez/Photodisc/Getty Images

We make decisions based on uncertainty every day. Should you buy an extended warranty for your new iPod? It depends on the likelihood that it will fail during the warranty period. Should you allow 45 min to get to your 8 A.M. class, or is 35 min enough? From experience, you may know that most mornings you can drive to school and park in 25 min or less. Most of the time, the walk from your parking space to class is 5 min or less. But how often will the drive to school or the walk to class take longer than you expect? When it takes longer than usual to drive to campus, is it more likely that it will also take longer to walk to class? less likely? Or are the driving and walking times unrelated? Some questions involving uncertainty are more serious: If an artificial heart has four key parts, how likely is each one to fail? How likely is it that at least one will fail? We can answer questions like these using the ideas and methods of probability, the systematic study of uncertainty.

6.1 Chance Experiments and Events

The basic ideas and terminology of probability are most easily introduced in situations that are both familiar and reasonably simple. Thus some of our initial examples involve such elementary activities as tossing a coin, selecting cards from a deck, and rolling a die. However, after considering some of these simplistic examples, we will move on to more interesting and realistic situations.

■ Chance Experiments

When a single coin is tossed, it can land with its heads side up or its tails side up. The selection of a single card from a well-mixed standard deck may result in the ace of spades, the five of diamonds, or any one of the other 50 possibilities. Consider rolling both a red die and a green die. One possible outcome is the red die lands with four dots facing up and the green die shows one dot on its upturned face. Another outcome is the red die lands with three dots facing up and the green die also shows three dots on its upturned face. There are 36 possible outcomes in all, and we do not know in advance what the result of a particular roll will be. Situations such as these are referred to as *chance experiments*.

> **D E F I N I T I O N**
>
> A **chance experiment** is any activity or situation in which there is uncertainty about which of two or more possible outcomes will result.

When the term *chance experiment* is used in a probability setting, we mean something different from what was meant by the term *experiment* in Chapter 2 (where experiments investigated the effect of two or more treatments on a certain response). For example, in an opinion poll or survey, there is uncertainty about whether an individual selected at random from the population of interest supports a school bond, and when a die is rolled, there is uncertainty about which face will land upturned. Both of these situations fit the definition of a chance experiment.

Consider a chance experiment designed to see whether men and women have different shopping preferences when buying a CD at a music store. Four types of music are sold at the store: classical, rock, country, and "other" (all other types of music). For this chance experiment, a randomly selected shopper will be asked what type of CD has been purchased, and the investigator will note the shopper's gender. Before we interact with the shopper, the outcome of this chance experiment is unknown to us. We do know, however, what the possible outcomes are. This set of possible outcomes is called the *sample space*.

> **D E F I N I T I O N**
>
> The collection of all possible outcomes of a chance experiment is the **sample space** for the experiment.

The sample space of a chance experiment can be represented in many ways. An obvious representation is a simple list of all the possible outcomes. For the CD purchase chance experiment, the possible outcomes are:

1. A male buying classical
2. A female buying classical
3. A male buying rock

4. A female buying rock
5. A male buying country
6. A female buying country
7. A male buying "other"
8. A female buying "other"

For brevity, we can use set notation and ordered pairs. A male purchasing a classical CD is represented as (male, classical). The sample space is then

$$\text{sample space} = \left\{ \begin{array}{l} (\text{male, classical}), (\text{female, classical}), (\text{male, rock}), (\text{female, rock}) \\ (\text{male, country}), (\text{female, country}), (\text{male, other}), (\text{female, other}) \end{array} \right\}$$

Another representation of the sample space is a tree diagram. A tree diagram (shown in Figure 6.1) for the outcomes of the CD shopper chance experiment has two sets of branches corresponding to the two pieces of information that were gathered. To identify any particular element of the sample space, you traverse the tree by first selecting a branch corresponding to gender and then a branch identified with a type of CD.

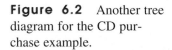

Figure 6.1 Tree diagram for the CD purchase example.

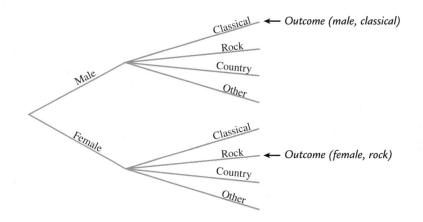

In the tree diagram of Figure 6.1, there is no particular reason for having the gender as the first generation branch and the type of CD as the second generation branch. Some chance experiments involve making observations in a particular order, in which case the order of the branches in the tree does matter. In this example, however, it would be acceptable to represent the sample space with the tree diagram shown in Figure 6.2.

Figure 6.2 Another tree diagram for the CD purchase example.

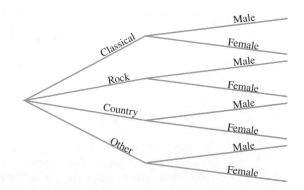

As we have seen, the sample space can be represented in several ways, but the representations all have one thing in common: *Every* element of the sample space is included in the representation.

■ Events

In the survey of CD purchasers, we might be interested in which particular outcome will result. Or we might focus on a group of outcomes that involve the purchase of classical music—the group consisting of (male, classical) and (female, classical). When we combine one or more individual outcomes in a collection, we are creating what is known as an *event*.

> **DEFINITION**
>
> An **event** is any collection of outcomes from the sample space of a chance experiment.
>
> A **simple event** is an event consisting of exactly one outcome.

We usually represent an event by an uppercase letter, such as A, B, C, and so on. On some occasions, the same letter with different numerical subscripts, such as E_1, E_2, E_3, . . . may be used for this purpose.

Example 6.1 Music Preferences

Reconsider the situation in which shoppers were categorized by gender (M or F) and type of music purchased (C = classical, R = rock, K = country, O = other). Using this notation, one possible representation of the sample space is

$$\text{sample space} = \{\text{MC, FC, MR, FR, MK, FK, MO, FO}\}$$

Because there are eight outcomes, there are eight simple events:

$$E_1 = \text{MC} \qquad E_2 = \text{FC} \qquad E_3 = \text{MR} \qquad E_4 = \text{FR}$$
$$E_5 = \text{MK} \qquad E_6 = \text{FK} \qquad E_7 = \text{MO} \qquad E_8 = \text{FO}$$

One event of interest might be the event consisting of all outcomes for which a classical CD was purchased. A symbolic description of the event *classical* is

$$classical = \{\text{MC, FC}\}$$

Another event is the event that the purchaser is male,

$$male = \{\text{MC, MR, MK, MO}\}$$

■

Example 6.2 Losing at Golf

Suppose you believe that after losing a game to an opponent, a golfer is more likely to lose the next game. You conduct a chance experiment that consists of watching two consecutive games for a particular player and observing whether the player won, tied, or lost each of the two games. In this case (using W, T, and L to represent win, tie, and loss, respectively), the sample space can be represented as

$$\text{sample space} = \{WW, WT, WL, TW, TT, TL, LW, LT, LL\}$$

The event *lose exactly one of the two games*, denoted by L_1, could then be defined as

$$L_1 = \{WL, TL, LW, LT\}$$

■

Only one outcome—and thus one simple event—occurs when a chance experiment is performed. We say that a given event occurs whenever one of the outcomes making up the event occurs. If the outcome in Example 6.1 is MC, then the simple event *male purchasing classical music* has occurred, and so has the nonsimple event *classical music purchased*.

■ **Forming New Events**   Once some events have been specified, they can be manipulated in several useful ways to create new events. Some new events are presented in the following definition.

Complementary events add to 1.

> ### DEFINITION
>
> Let A and B denote two events.
>
> 1. The event **not A** consists of all experimental outcomes that are not in event A. *Not A* is sometimes called the *complement* of A and is usually denoted by A^C, A', or $\overline{A}$.
> 2. The event **A or B** consists of all experimental outcomes that are in at least one of the two events, that is, in A or in B or in both of these. *A or B* is called the *union* of the two events and is denoted by $A \cup B$.
> 3. The event **A and B** consists of all experimental outcomes that are in both of the events A and B. *A and B* is called the *intersection* of the two events and is denoted by $A \cap B$.

Example 6.3   Turning Directions

An observer will stand at the bottom of a freeway off-ramp and record the turning direction (L = left or R = right) of each of three successive vehicles. The sample space contains eight outcomes:

all 3 cars turn left   *the 1st car turns left and the next 2 turn right*

   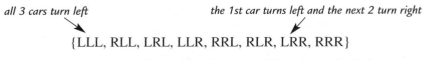

$$\{LLL, RLL, LRL, LLR, RRL, RLR, LRR, RRR\}$$

Each of these outcomes determines a simple event. Other events include

A = event that exactly one of the cars turns right = {RLL, LRL, LLR}
B = event that at most one of the cars turns right = {LLL, RLL, LRL, LLR}
C = event that all cars turn in the same direction = {LLL, RRR}

Some other events that can be formed from those just defined are

$$not\ C = C^C = \text{event that not all cars turn in the same direction}$$
$$= \{RLL, LRL, LLR, RRL, RLR, LRR\}$$
$$A\ or\ C = A \cup C = \text{event that exactly one of the cars turns}$$
$$\text{right or all cars turn in the same direction}$$
$$= \{RLL, LRL, LLR, LLL, RRR\}$$
$$B\ and\ C = B \cap C = \text{event that at most one car turns right and all cars}$$
$$\text{turn in the same direction}$$
$$= \{LLL\}$$

■

Example 6.4   More Losing at Golf

In Example 6.2, in addition to the event that a golfer loses exactly one of the two games, $L_1 = \{WL, TL, LW, LT\}$, we could also define events corresponding to $L_0 =$ neither game is lost and $L_2 =$ both games are lost. Then $L_0 = \{WW, WT, TW, TT\}$ and $L_2 = \{LL\}$.

Some other events that can be formed from those just defined include

$$L_2^C = \text{event that at most one game was lost}$$
$$= \{WW, WT, WL, TW, TT, TL, LW, LT\}$$
$$L_1 \cup L_2 = \text{event that at least one game was lost}$$
$$= \{WL, TL, LW, LT, LL\}$$
$$L_1 \cap L_2 = \text{event that exactly one game was lost } and \text{ two games were lost}$$
$$= \text{the empty set}$$

■

It frequently happens that two events have no common outcomes, as was the case for events L_1 and L_2 in the golf example. Such situations are described by special terminology.

D E F I N I T I O N

Two events that have no common outcomes are said to be **disjoint** or **mutually exclusive**.

Two events are disjoint if their intersection is the empty set, so any two *simple* events are disjoint.

It is sometimes useful to draw an informal picture of events to visualize relationships. In a **Venn diagram**, the collection of all possible outcomes is typically shown as the interior of a rectangle. Other events are then identified by specified regions inside this rectangle. Figure 6.3 illustrates several Venn diagrams.

Figure 6.3 Venn diagrams: (a) Gold region = *not A*; (b) gold region = *A or B*; (c) green region = *A and B;* (d) two disjoint events.

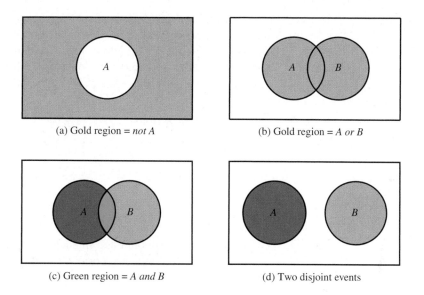

(a) Gold region = *not A*

(b) Gold region = *A or B*

(c) Green region = *A and B*

(d) Two disjoint events

The use of the *or* and *and* operations can be extended to form new events from more than two initially specified events.

> **DEFINITION**
>
> Let $A_1, A_2, \ldots, A_k$ denote k events.
>
> 1. The event A_1 *or* A_2 *or* ... *or* A_k consists of all outcomes in at least one of the individual events $A_1, A_2, \ldots, A_k$.
> 2. The event A_1 *and* A_2 *and* ... *and* A_k consists of all outcomes that are simultaneously in every one of the individual events $A_1, A_2, \ldots, A_k$.
>
> These k events are disjoint if no two of them have any common outcomes.

Figure 6.4 Venn diagrams: (a) A_1 *or* A_2 *or* A_3; (b) A_1 *and* A_2 *and* A_3; (c) three disjoint events.

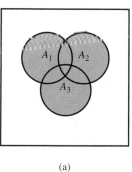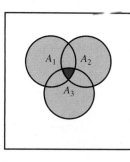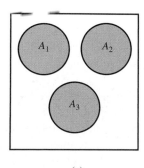

(a)

(b)

(c)

Example 6.5   Asking Questions

The instructor in a seminar class consisting of four students has an unusual way of asking questions. Four slips of paper numbered 1, 2, 3, and 4 are placed in a box. The instructor determines the student to whom any particular question is to be addressed by selecting one of these four slips. Suppose that one question is to be posed during each of the next two class meetings. One possible outcome could be represented as (3, 1)—the first question is addressed to Student 3 and the second question to Student 1. There are 15 other possibilities. Consider the following events:

the event that the same student is asked both questions ⟶ $A = \{(1,1), (2,2), (3,3), (4,4)\}$

the event that student 1 is asked at least one of the two questions ⟶ $B = \{(1,1), (1,2), (1,3), (1,4), (2,1), (3,1), (4,1)\}$

$C = \{(3,1), (2,2), (1,3)\}$

$D = \{(3,3), (3,4), (4,3)\}$

$E = \{(1,1), (1,3), (2,2), (3,1), (4,2), (3,3), (2,4), (4,4)\}$

$F = \{(1,1), (1,2), (2,1)\}$

Then

$$A \text{ or } C \text{ or } D = \{(1,1), (2,2), (3,3), (4,4), (3,1), (1,3), (3,4), (4,3)\}$$

The outcome (3,1) is contained in each of the events B, C, and E, as is the outcome (1,3). These are the only two common outcomes, so

$$B \text{ and } C \text{ and } E = \{(3,1), (1,3)\}$$

The events C, D, and F are disjoint because no outcome in any one of these events is contained in either of the other two events.

■

Exercises 6.1–6.12

Turn to the Expanded Answers Section for answers not shown next to exercises.

6.1 Define the term *chance experiment*, and give an example of a chance experiment with four possible outcomes.

6.2 Define the term *sample space*, and then give the sample space for the chance experiment you described in Exercise 6.1.

6.3 Consider the chance experiment in which the type of transmission—automatic (A) or manual (M)—is recorded for each of the next two cars purchased from a certain dealer.
a. What is the set of all possible outcomes (the sample space)? Sample space = {AA, AM, MA, MM}
b. Display the possible outcomes in a tree diagram.
c. List the outcomes in each of the following events. Which of these events are simple events?
  i. B the event that at least one car has an automatic transmission
  ii. C the event that exactly one car has an automatic transmission
  iii. D the event that neither car has an automatic transmission

d. What outcomes are in the event B and C? In the event B or C? $B \text{ and } C = \{(A,M), (M,A)\} = C$, $B \text{ or } C = \{(A,A), (A,M), (M,A)\} = B$

6.4 A tennis shop sells five different brands of rackets, each of which comes in either a midsize version or an oversize version. Consider the chance experiment in which brand and size are noted for the next racket purchased. One possible outcome is Head midsize, and another is Prince oversize. Possible outcomes correspond to cells in the following table:

	Head	Prince	Slazenger	Wimbledon	Wilson
Midsize					
Oversize					

a. Let A denote the event that an oversize racket is purchased. List the outcomes in A.
b. Let B denote the event that the name of the brand purchased begins with a W. List the outcomes in B.
c. List the outcomes in the event *not B*.
d. Head, Prince, and Wilson are U.S. companies. Let C

denote the event that the racket purchased is made by a U.S. company. List the outcomes in the event *B or C*.
e. List outcomes in *B and C*.
f. Display the possible outcomes on a tree diagram, with a first-generation branch for each brand.

6.5 ▼ Consider the chance experiment in which an automobile is selected and both the number of defective headlights (0, 1, or 2) and the number of defective tires (0, 1, 2, 3, or 4) are determined.
a. Display possible outcomes using a tree diagram.
b. Let *A* be the event that at most one headlight is defective and *B* be the event that at most one tire is defective. What outcomes are in A^C? in $A \cup B$? in $A \cap B$?
c. Let *C* denote the event that all four tires are defective. Are *A* and *C* disjoint events? Are *B* and *C* disjoint?   *A and C are not disjoint events. A and B are disjoint events.*

6.6 A college library has four copies of a certain book; the copies are numbered 1, 2, 3, and 4. Two of these are selected at random. The first selected book is placed on 2-hr reserve, and the second book can be checked out overnight.
a. Construct a tree diagram to display the 12 outcomes in the sample space.
b. Let *A* denote the event that at least one of the books selected is an even-numbered copy. What outcomes are in *A*?
c. Suppose that copies 1 and 2 are first printings, whereas copies 3 and 4 are second printings. Let *B* denote the event that exactly one of the copies selected is a first printing. What outcomes are contained in *B*?

6.7 A library has five copies of a certain textbook on reserve of which two copies (1 and 2) are first printings and the other three (3, 4, and 5) are second printings. A student examines these books in random order, stopping only when a second printing has been selected.
a. Display the possible outcomes in a tree diagram.
b. What outcomes are contained in the event *A*, that exactly one book is examined before the chance experiment terminates? $A = \{3, 4, 5\}$
c. What outcomes are contained in the event *C*, that the chance experiment terminates with the examination of book 5? $C = \{125, 15, 215, 25, 5\}$

6.8 Suppose that, starting at a certain time, batteries coming off an assembly line are examined one by one to see

whether they are defective (let D = defective and N = not defective). The chance experiment terminates as soon as a nondefective battery is obtained.
a. Give five possible experimental outcomes.
b. What can be said about the number of outcomes in the sample space?
c. What outcomes are in the event *E*, that the number of batteries examined is an even number?

6.9 Refer to Exercise 6.8, and now suppose that the chance experiment terminates only when two nondefective batteries have been obtained.
a. Let *A* denote the event that at most three batteries must be examined to terminate the chance experiment. What outcomes are contained in *A*?   A = {NN, DNN, NDN}
b. Let *B* be the event that exactly four batteries must be examined before the chance experiment terminates. What outcomes are in *B*?   B = {DDNN, NDDN, DNDN}
c. What can be said about the number of possible outcomes for this chance experiment?   *There are an infinite number of possible outcomes.*

6.10 A family consisting of three people—P_1, P_2, and P_3—belongs to a medical clinic that always has a physician at each of stations 1, 2, and 3. During a certain week, each member of the family visits the clinic exactly once and is randomly assigned to a station. One experimental outcome is (1, 2, 1), which means that P_1 is assigned to station 1, P_2 to station 2, and P_3 to station 1.
a. List the 27 possible outcomes. (Hint: First list the nine outcomes in which P_1 goes to station 1, then the nine in which P_1 goes to station 2, and finally the nine in which P_1 goes to station 3; a tree diagram might help.)
b. List all outcomes in the event *A*, that all three people go to the same station. $A = \{(1,1,1), (2,2,2), (3,3,3)\}$
c. List all outcomes in the event *B*, that all three people go to different stations.
d. List all outcomes in the event *C*, that no one goes to station 2.
e. Identify outcomes in each of the following events: B^C, C^C, $A \cup B$, $A \cap B$, $A \cap C$.

6.11 An engineering construction firm is currently working on power plants at three different sites. Define events E_1, E_2, and E_3 as follows:

E_1 = the plant at Site 1 is completed by the contract date
E_2 = the plant at Site 2 is completed by the contract date
E_3 = the plant at Site 3 is completed by the contract date

Bold exercises answered in back   ● Data set available online but not required   ▼ Video solution available

The following Venn diagram pictures the relationships among these events:

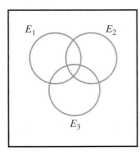

Shade the region in the Venn diagram corresponding to each of the following events (redraw the Venn diagram for each part of the problem):
a. At least one plant is completed by the contract date.
b. All plants are completed by the contract date.

c. None of the plants are completed by the contract date.
d. Only the plant at Site 1 is completed by the contract date.
e. Exactly one of the three plants is completed by the contract date.
f. Either the plant at Site 1 or both of the other two plants are completed by the contract date.

6.12 Consider a Venn diagram picturing two events A and B that are not disjoint.
a. Shade the event $(A \cup B)^C$. On a separate Venn diagram shade the event $A^C \cap B^C$. How are these two events related?
b. Shade the event $(A \cap B)^C$. On a separate Venn diagram shade the event $A^C \cup B^C$. How are these two events related? (Note: These two relationships together are called DeMorgan's laws.)

Bold exercises answered in back   ● Data set available online but not required   ▼ Video solution available

 Definition of Probability

Probabilistic reasoning of some sort may well be as old as civilization itself. Archeologists have found evidence that Egyptians used a small bone in mammals, the astragalus, as a sort of four-sided die as early as 3500 B.C. Games of chance were common in Greek and Roman times and during the Renaissance of Western Europe. Girolamo Cardano, a character of some ill-repute in early mathematical circles, wrote about such games and did some calculations of odds in the 16th century; Galileo mentioned dice in his writings in the early 17th century.

Probability is traditionally thought of as beginning with the correspondence between Blaise Pascal (1623–62) and Pierre de Fermat (1601–65) in 1654. Their letters discussed some problems related to gambling that were posed by the French nobleman Chevalier de Mère. The methods and solutions that resulted from this exchange, especially techniques of combinatorics (advanced counting), greatly enhanced the study of probability. From these beginnings to the later writings of the mathematician Pierre-Simon Laplace (1749–1827), the techniques, methods, and theory now known as classical probability grew to be a major contribution to the quantification of uncertainty.

From this early work, new interpretations of probability have evolved, including an empirical approach preferred by many statisticians today. We begin our discussion of probability with a look at some of the different approaches to probability: the classical, relative frequency, and subjective approaches.

■ Classical Approach to Probability

Early mathematicians' development of the theory of probability was primarily in the context of gambling and reflected the peculiarities of games of chance. A characteristic common to most games of chance is a physical device used to generate different outcomes. In children's games, a "spinner" might be used to create chance outcomes. In other games, the sequence of events is determined by playing cards. In the 17th century, the most popular gaming devices were dice. Dice were constructed so that the physical characteristics of each face were alike (except, of course, for the number of dots), ensuring that the different outcomes for an individual die were very close to equally likely. Therefore, it seemed quite natural for mathematicians of the time to assume, for example, that the probability of getting a five when a single six-sided die was rolled was one-sixth (1 chance in 6). In general, if there are *N equally likely* outcomes in a game of chance, the probability of each of the outcomes would be 1/*N*. The early probabilists often referred to honest dice or fair coins as physical devices that seemed to obey this assumption of equal probabilities. These early efforts led to what is known as the classical definition of probability.

Classical Approach to Probability for Equally Likely Outcomes

When the outcomes in the sample space of a chance experiment are equally likely, the **probability of an event *E***, denoted by **P(E)**, is the ratio of the number of outcomes favorable to *E* to the total number of outcomes in the sample space:

$$P(E) = \frac{\text{number of outcomes favorable to } E}{\text{number of outcomes in the sample space}}$$

According to this definition, the calculation of a probability consists of counting the number of outcomes that make up an event, counting the number of outcomes in the sample space, and then dividing.

Chance experiments that involve tossing fair coins, rolling fair dice, or selecting cards from a well-mixed deck have equally likely outcomes. For example, if a fair die is rolled once, each outcome (simple event) has probability 1/6. With *E* denoting the event that the number rolled is even, *P(E)* = 3/6. This is just the number of outcomes in *E* divided by the total number of possible outcomes.

Example 6.6 Calling the Toss

On some football teams, the honor of calling the toss at the beginning of a football game is determined by random selection. Suppose that this week a member of the offensive team will call the toss. There are 5 interior linemen on the 11-player offensive team. If we define the event *L* as the event that a lineman is selected to call the toss, 5 of the 11 possible outcomes are included in *L*. The probability that a lineman will be selected is then

$$P(L) = \frac{5}{11}$$

Example 6.7 Math Contest

Four students (Adam, Betina, Carlos, and Debra) submitted correct solutions to a math contest with two prizes. The contest rules specify that if more than two correct responses are submitted, the winners will be selected at random from those submitting correct responses. In this case, the set of possible outcomes for the chance experiment that consists of selecting the two winners from the four correct responses is

$$\{(A, B), (A, C), (A, D), (B, C), (B, D), (C, D)\}$$

Because the winners are selected at random, the six outcomes are equally likely and the probability of each individual outcome is 1/6.

Let E be the event that both selected winners are the same gender. Then

$$E = \{(A, C), (B, D)\}$$

Because E contains two outcomes, $P(E) = 2/6 = .333$. If F denotes the event that at least one of the selected winners is female, then F consists of all outcomes except (A, C) and $P(F) = 5/6 = .833$.

■

■ **Limitations of the Classical Approach to Probability**   The classical approach to probability works well with games of chance or other situations with a finite set of outcomes that can be regarded as equally likely. However, some situations that are clearly probabilistic in nature do not fit the classical model. Consider the accident rates of young adults driving cars. Information of this kind is used by insurance companies to set car insurance rates. Suppose that we have two individuals, one 18 years old and one 28 years old. A person purchasing a standard accident insurance policy at age 28 should have a lower annual premium than a person purchasing the same policy at age 18, because the 28-year-old has a much lower chance of an accident, based on prior experience. However, the classical probabilist, playing by the classical rules, would have to consider that there are two outcomes for a policy year—having one or more accidents or not having an accident—*and would consider those outcomes equally likely for both the 28-year-old and the 18-year-old.* This certainly seems to defy reason and experience.

Another example of a situation that cannot be handled using the classical approach has a geometric flavor. Suppose that you have a square that is 8 in. on each side and that a circle with a diameter of 8 in. is drawn inside the square, touching the square on all four sides, as shown.

This drawing will be used as a target in a dart game. A dart will be thrown at the target and any dart that completely misses the square is a "do-over" and will be thrown again. It seems reasonable to think about the probability of a dart landing in the circle as a ratio of the area of the circle to the area of the square. That ratio is

$$\frac{\text{area of circle}}{\text{area of square}} = \frac{\pi r^2}{s^2} = \frac{\pi(4)^2}{8^2} = \frac{\pi}{4} = .785$$

A practiced dart player who is aiming for the center of the target might have an even greater probability of hitting the inside of the circle. In any case, it is not reasonable to think that the two outcomes *inside the circle* and *not inside the circle* are equally likely, so the classical approach to probability will not help here.

From the standpoint of statistics, a major limitation of the classical approach to probability is that there seems to be no way to allow past experience to inform our expectation of the future. To return to the car insurance example, suppose that in the experience of the insurance companies, 0.5% of 18-year-olds have accidents during their 19th year. It seems reasonable that, all things being equal, we could expect this percentage to be stable enough to set the prices for an insurance policy. Although the classical approach's assumption of equal probabilities is based on and consistent with

Important concept

the experience with dice, it cannot accommodate using an observed proportion in real life to estimate a probability. For a statistician, this is a problem. Without an understanding of probability that allows such a generalization, statistical inference would be impossible.

■ Relative Frequency Approach to Probability ....

After the revolutionary writings of Sir Isaac Newton (1642–1727), modern science became solidly based on observation and the recording of data. Early scientific investigators were aware that chance experiments and observations do not always give the same results when repeated and that even in the most carefully replicated chance experiment, there is variation. Early on, it was noticed that the unpredictability of an individual chance experiment gave rise to a dependably stable regularity when a chance experiment was repeated many times. This became the basis for fair wagering in games of chance. For example, suppose that two friends, Chris and Jay, meet to play a game. A fair coin is flipped. If it lands heads up, Chris pays Jay $1; otherwise, Jay pays the same amount to Chris. After many repetitions, the proportion of the time that Chris wins will be close to one-half, and the two friends will have simply enjoyed the pleasure of each other's company for a few hours. This happy circumstance occurs because *in the long run* (i.e., over many repetitions) the proportion of heads is .5; half the time Chris wins, and half the time Jay wins.

When any given chance experiment is performed, some events are relatively likely to occur, whereas others are not so likely to occur. For a specified event E, we want to assign a probability that gives a precise indication of how likely it is that E will occur. In the relative frequency approach to probability, the probability describes how frequently E occurs when the chance experiment is performed repeatedly.

Example 6.8   Tossing a Coin

Frequently, we hear a coin described as fair, or we are told that there is a 50% chance of a coin landing heads up. What might be the meaning of the expression *fair*? Such a description cannot refer to the result of a single toss, because a single toss cannot result in both a head and a tail. Might "fairness" and "50%" refer to 10 successive tosses yielding exactly 5 heads and 5 tails? Not really, because it is easy to imagine a fair coin landing heads up on only 3 or 4 of the 10 tosses.

Consider the chance experiment of tossing a coin just once. Define the simple events

H = event that the coin lands with its heads side facing up
T = event that the coin lands with its tails side facing up.

Now suppose that we take a fair coin and begin to toss it over and over. After each toss, we compute the relative frequency of heads observed so far. This calculation gives the value of the ratio

$$\frac{\text{number of times event } H \text{ occurs}}{\text{number of tosses}}$$

The results of the first ten tosses might be as follows:

Toss	1	2	3	4	5	6	7	8	9	10
Cumulative number of H's	0	1	2	3	3	3	4	5	5	5
Relative frequency of H's	0	.5	.667	.75	.6	.5	.571	.625	.556	.5

Figure 6.5 illustrates how the relative frequency of heads fluctuates during a sample sequence of 50 tosses. *As the number of tosses increases, the relative frequency of heads does not continue to fluctuate wildly but instead stabilizes and approaches some fixed number (the limiting value).* This stabilization is illustrated for a sequence of 1000 tosses in Figure 6.6.

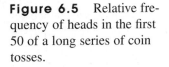

Figure 6.5 Relative frequency of heads in the first 50 of a long series of coin tosses.

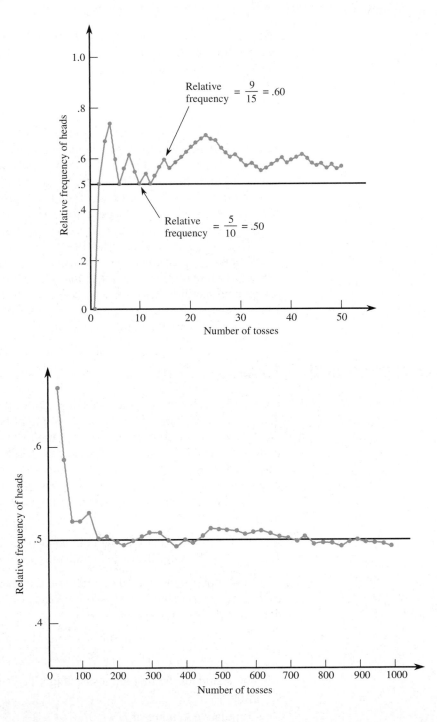

Figure 6.6 Stabilization of the relative frequency of heads in coin tossing.

It may seem natural to you that the proportion of H's would get closer and closer to the "real" probability of .5. That an empirically observed proportion could behave like this in real life also seemed reasonable to James Bernoulli in the early 18th century. He focused his considerable mathematical power on this topic and was able to prove mathematically what we now know as the law of large numbers.*

A Law of Large Numbers

As the number of repetitions of a chance experiment increases, the chance that the relative frequency of occurrence for an event will differ from the true probability of the event by more than any small number approaches 0.

Note that, as the number of repetitions of the chance experiment increases, the proportion of H's gets closer and closer to the real probability of H occurring in a single chance experiment *even if the value of this probability is not known.* This means that we can observe the outcomes of repetitions of a chance experiment and then use the observed outcomes to estimate probabilities.

Relative Frequency Approach to Probability

The **probability of an event E**, denoted by $P(E)$, is defined to be the value approached by the relative frequency of occurrence of E in a very long series of trials of a chance experiment. Thus, if the number of trials is quite large,

$$P(E) \approx \frac{\text{number of times } E \text{ occurs}}{\text{number of trials}}$$

The relative frequency definition of probability depends on being able to repeat a chance experiment under identical conditions. Suppose that we perform a chance experiment that consists of flipping a cap from a 20-oz bottle of soda and noting whether the cap lands with the open side up or down. The crucial difference from the previous coin-tossing chance experiment is that there is no particular reason to believe the cap is equally likely to land top up or top down. If we assume that the chance experiment can be repeated under similar conditions (which seems reasonable), then we can flip the cap a number of times and compute the relative frequency of the event *top up* observed so far:

$$\frac{\text{number of times the event } top\ up \text{ occurs}}{\text{number of tosses}}$$

---

*Technically, this should be referred to more specifically as the weak law of large numbers for Bernoulli trials. After Bernoulli's proof, mathematical statisticians proved more general laws of large numbers.

The results of the first ten flips, with U indicating top up and D indicating top down, might be:

Flip	1	2	3	4	5	6	7	8	9	10
Outcome	U	U	D	U	D	D	U	D	U	U
Cumulative number of *ups*	1	2	2	3	3	3	4	4	5	6
Relative frequency of *ups*	1.0	1.0	.67	.75	.6	.5	.57	.5	.56	.6

Figure 6.7 illustrates how the relative frequency of the event *top up* fluctuates during a sample sequence of 100 flips. Based on these results and faith in the law of large numbers, it is reasonable to think that the probability of the cap landing top up is about .7.

Figure 6.7 Stabilization of the relative frequency of a bottle cap landing top up.

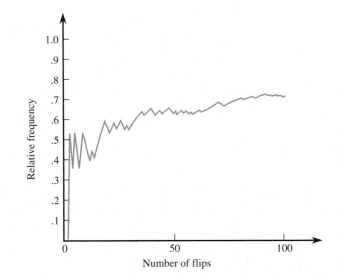

The relative frequency approach to probability is based on observation. From repeated observation, we can obtain stable relative frequencies that will, in the long run, provide good estimates of the probabilities of different events. The relative frequency interpretation of probability is intuitive, widely used, and most relevant for the inferential procedures introduced in later chapters.

■ Subjective Approach to Probability

A third distinct approach to probability is based on subjective judgments. In this view, probability can be interpreted as a personal measure of the strength of belief that a particular outcome will occur. A probability of 1 represents a belief that the outcome will certainly occur. A probability of 0 represents a belief that the outcome will certainly not occur—that it is impossible. Other probabilities are placed somewhere between 0 and 1, based on the strength of one's beliefs. For example, an airline passenger might report her subjective assessment of the probability of being denied a seat as a result of overbooking on a particular flight as .01. Because this probability is close to 0, she believes it is very unlikely that she will be denied a seat.

The subjective interpretation of probability presents some difficulties. For one thing, different people may assign different probabilities to the same outcome, because each could have a different subjective belief. Although subjective probabilities are useful in studying and analyzing decision making, they are of limited use because they are personal and not generally replicable by others.

■ **How Are Probabilities Determined?** ....

Probability statements are frequently encountered in newspapers and magazines (and, of course, in this textbook). What is the basis for these probability statements? In most cases, research allows reasonable estimates based on observation and analysis. In general, if a probability is stated, it is based on one of the following approaches:

1. *The classical approach:* This approach is appropriate only for modeling chance experiments with equally likely outcomes.
2. *The subjective approach:* In this case, the probability represents an individual's judgment based on facts combined with personal evaluation of other information.
3. *The relative frequency approach:* An estimate is based on an accumulation of experimental results. This estimate, usually derived empirically, presumes a replicable chance experiment.

When you see probabilities in this book and in other published sources, you should consider how the probabilities were produced. Be particularly cautious in the case of subjective probabilities.

6.3 Basic Properties of Probability

The view of probabilities as long-run relative frequencies is intuitively pleasing but has some serious limitations when it comes to calculating probabilities in real life. The most obvious problem is time. If computing probabilities even for simple chance experiments were to require hundreds or thousands of repeated trials, even the most ardent statistics students would avoid this task! An alternative approach that simplifies things in some situations is to study the axioms (fundamental properties) of probability. After presenting these basic propertics, we will develop some propositions that follow from them and that can be used to find probabilities of complex events. These fundamental properties are given in the accompanying box, and then each property is addressed in the discussion following the box.

Basic Properties of Probability

1. For any event E, $0 \leq P(E) \leq 1$.
2. If S is the sample space for an experiment, $P(S) = 1$.
3. If two events E and F are disjoint, then $P(E \text{ or } F) = P(E) + P(F)$.
4. For any event E, $P(E) + P(not\ E) = 1$. Therefore

$$P(not\ E) = 1 - P(E)$$

and

$$P(E) = 1 - P(not\ E).$$

■ **Property 1: For any event E, $0 \leq P(E) \leq 1$.**   To understand the first property of probability, recall the previous bottle cap chance experiment. You may remember that we were keeping track of the number of flips landing with the inside of the bottle cap facing up. Suppose that, after flipping the bottle cap N times, we have observed x occurrences of top up. What are the possible values of x? The fewest that could have been counted is 0, and the most that could have been counted is N. Therefore, the relative frequency falls between two numbers:

$$\frac{0}{N} \leq \text{relative frequency} \leq \frac{N}{N}$$

Thus, $0 \leq$ relative frequency ≤ 1. As N increases, the long-run value of the relative frequency, which defines the probability, must also lie between 0 and 1.

■ **Property 2: If S is the sample space for a chance experiment, $P(S) = 1$.**   Because the probability of any event is the proportion of time an outcome in the event will occur in the long run and because the sample space consists of all possible outcomes for a chance experiment, in the long run an outcome in S must occur 100% of the time. Thus $P(S) = 1$.

■ **Property 3: If two events E and F are disjoint, then $P(E \text{ or } F) = P(E) + P(F)$.**   This is one of the most important properties of probability because it provides a method for computing probabilities if the number of possible outcomes (simple events) in the sample space is finite. Any nonsimple event is just a collection of outcomes from the sample space—some outcomes are in the event and others are not. Because simple events are disjoint, any event can be viewed as a union of disjoint events. For example, suppose that a chance experiment has a sample space that consists of six outcomes, $O_1, O_2, \ldots, O_6$. Then $P(O_1)$ can be interpreted as the long run proportion of times that the outcome O_1 will occur, and the probabilities of the other outcomes can be interpreted in a similar way. Suppose that an event E is made up of outcomes O_1, O_2, and O_4. Because E will occur whenever O_1, O_2, or O_4 occurs and because O_1, O_2, and O_4 are disjoint, the long-run proportion of time that E will occur is just the sum of the proportion of time that each of the outcomes O_1, O_2, and O_4 will occur. Because it is always possible to express any event as a collection of disjoint simple events in this way, finding the probability of an event can be reduced to finding the probabilities of the simple events (outcomes) that make up the event and then adding. If the probabilities of all the simple events are known, it is then easy to compute the probability of more complex events constructed from the simple events.

■ **Property 4: For any event E, $P(E) + P(\text{not } E) = 1$. Therefore $P(\text{not } E) = 1 - P(E)$ and $P(E) = 1 - P(\text{not } E)$.**   Property 4 follows from Properties 2 and 3. Property 2 tells us that $P(E)$ is the sum of the probabilities in E and that $P(\text{not } E)$ is the sum of the probabilities for simple events corresponding to outcomes in *not E*. Every outcome in the sample space is in either E or *not E*, and we know from Property 3 that the sum of the probabilities of all the simple events is 1. It follows that $P(E) + P(\text{not } E)$ must equal 1. The implication of Property 4, namely, that $P(E) = 1 - P(\text{not } E)$ is surprisingly useful. There are many situations in which calculation of $P(\text{not } E)$ is much easier than direct determination of $P(E)$.

Example 6.9   Cash or Credit?

Customers at a certain department store pay for purchases with either cash or one of four types of credit card. Store records, kept for a long period of time, indicate that 30% of all purchases involve cash, 25% are made with the store's own credit card, 18% with MasterCard (MC), 15% with Visa (V), and the remaining 12% with American Express (AE). The following table displays the probabilities of the simple events for the chance experiment in which the mode of payment for a randomly selected transaction is observed:

Simple event	O_1 (Cash)	O_2 (Store)	O_3 (MC)	O_4 (V)	O_5 (AE)
Probability	.30	.25	.18	.15	.12

Let's create event E, the event *a randomly selected purchase is made with a nationally distributed credit card*. This event consists of the outcomes MC, V, and AE. Therefore

$$P(E) = P(O_3 \cup O_4 \cup O_5) = P(O_3) + P(O_4) + P(O_5) = .18 + .15 + .12 = .45$$

That is, in the long run, 45% of all purchases are made using one of the three national cards. In addition,

$$P(not\ E) = 1 - P(E) = 1 - .45 = .55$$

which could also have been obtained by noting that *not E* consists of outcomes corresponding to the simple events O_1 and O_2.

■

The discussion of these properties has been nonmathematical; instead, it appeals to your intuition. Students interested in a more formal mathematical treatment of these properties in particular and of probability in general are encouraged to consult a more advanced textbook (see the references by Devore or Mostellar, Rourke and Thomas listed in the back of the book).

Having established the fundamental properties of probability, we now present some practical probability rules that can be used to evaluate probabilities in some situations. One important aspect of each rule is its associated assumptions. A common error for beginning statistics students is to believe that the rules can be used in *any* probability calculation. This is *not* the case. You must be careful to use the rules only after verifying that the assumptions are met.

■ Equally Likely Outcomes

The first probability rule that we consider applies only when events are equally likely; it is *not* always true for all events in all situations. Recall that chance experiments involving tossing fair coins, rolling fair dice, or selecting cards from a well-mixed deck have equally likely outcomes. For example, if a fair die is rolled once, each outcome has probability 1/6. With E denoting the event that the outcome is an even number,

$$P(E) = \frac{1}{6} + \frac{1}{6} + \frac{1}{6} = \frac{3}{6}$$

This is just the ratio of the number of outcomes in E to the total number of possible outcomes. The following box presents the generalization of this result.

Calculating Probabilities When Outcomes Are Equally Likely

Consider an experiment that can result in any one of N possible outcomes. Denote the corresponding simple events by $O_1, O_2, \ldots, O_N$. If these simple events are equally likely to occur, then

1. $P(O_1) = \dfrac{1}{N}, P(O_2) = \dfrac{1}{N}, \ldots, P(O_N) = \dfrac{1}{N}$

2. For *any* event E,

$$P(E) = \frac{\text{number of outcomes in } E}{N}$$

■ Addition Rule for Disjoint Events ....

We have seen previously that the probability of an event can be calculated by adding together probabilities of the simple events that correspond to the outcomes making up the event. This addition process is also legitimate when calculating the probability of the union of two events that are disjoint but not necessarily simple.

The Addition Rule for Disjoint Events

Let E and F be two disjoint events. One of the basic properties (axioms) of probability is

$$P(E \text{ or } F) = P(E \cup F) = P(E) + P(F)$$

This property of probability is known as the addition rule for disjoint events.

More generally, if events $E_1, E_2, \ldots, E_k$ are disjoint, then

$$P(E_1 \text{ or } E_2 \text{ or } \ldots \text{ or } E_k) = P(E_1 \cup E_2 \cup \cdots \cup E_k) = P(E_1) + P(E_2) + \cdots + P(E_k)$$

In words, the probability that any of these k events occurs is the sum of the probabilities of the individual events.

Consider the chance experiment that consists of rolling a pair of fair dice. There are 36 possible outcomes for this chance experiment, such as (1, 1), (1, 2), and so on. Because these outcomes are equally likely, we can compute the probability of observing a total of 5 on the two dice by counting the number of outcomes that result in a total of 5. There are 4 such outcomes—(1, 4), (2, 3), (3, 2), and (4, 1)—giving

$$P(\text{total of 5}) = \frac{4}{36}$$

The probabilities for the other totals can be computed in a similar fashion; they are shown in the following table:

Total	Probability	Total	Probability
2	1/36	8	5/36
3	2/36	9	4/36
4	3/36	10	3/36
5	4/36	11	2/36
6	5/36	12	1/36
7	6/36		

What is the probability of getting a total of 3 or 5? Consider the two events E = total is 3 and F = total is 5. Clearly E and F are disjoint because the sum cannot simultaneously be both 3 and 5. Notice also that neither of these events is simple, because neither consists of only a single outcome from the sample space. We can apply the addition rule for disjoint events as follows:

$$P(E \text{ or } F) = P(E \cup F) = \frac{2}{36} + \frac{4}{36} = \frac{6}{36}$$

Example 6.10 Car Choices

A large auto center sells cars made by a number of different manufacturers. Three of these manufacturers are Japanese: Honda, Nissan, and Toyota. Consider the manufacturer and model of the next car purchased at this auto center, and define events E_1, E_2, and E_3 by E_1 = Honda, E_2 = Nissan, and E_3 = Toyota. Notice that E_1 is not a simple event because there is more than one model Honda (for example, Civics, Accords, and Insights).

Suppose that $P(E_1) = .25$, $P(E_2) = .18$, and $P(E_3) = .14$. Because E_1, E_2, and E_3 are disjoint, the addition rule gives

$$P(\text{Honda } or \text{ Nissan } or \text{ Toyota}) = P(E_1 \cup E_2 \cup E_3)$$
$$= P(E_1) + P(E_2) + P(E_3)$$
$$= .25 + .18 + .14 = .57$$

The probability that the next car purchased is *not* one of these three types is

$$P(not(E_1 \text{ or } E_2 \text{ or } E_3)) = P((E_1 \cup E_2 \cup E_3)^C) = 1 - .57 = .43$$

■

In Section 6.6, we will show how $P(E \cup F)$ can be calculated when the two events are not known to be disjoint.

■ **Exercises 6.13–6.27** .... Turn to the Expanded Answers Section for answers not shown next to exercises. ....

6.13 ▼ Insurance status—covered (C) or not covered (N) —is determined for each individual arriving for treatment at a hospital's emergency room. Consider the chance experiment in which this determination is made for two randomly selected patients. The simple events are $O_1 = (C, C)$, $O_2 = (C, N)$, $O_3 = (N, C)$, and $O_4 = (N, N)$. Suppose that probabilities are $P(O_1) = .81$, $P(O_2) = .09$, $P(O_3) = .09$, and $P(O_4) = .01$.
a. What outcomes are contained in A, the event that at most one patient is covered, and what is $P(A)$?
b. What outcomes are contained in B, the event that the two patients have the same status with respect to coverage, and what is $P(B)$? $B = \{(C,C), (N,N)\}; P(B) = 0.82$

6.14 "N.Y. Lottery Numbers Come Up 9-1-1 on 9/11" was the headline of an article that appeared in the *San Francisco Chronicle* (September 13, 2002). More than 5600 people had selected the sequence 9-1-1 on that date, many more than is typical for that sequence. A professor at the University of Buffalo is quoted as saying, "I'm a bit surprised, but I wouldn't characterize it as bizarre. It's randomness. Every number has the same chance of coming up."
a. The New York state lottery uses balls numbered 0–9 circulating in three separate bins. To select the winning sequence, one ball is chosen at random from each bin. What is the probability that the sequence 9-1-1 is the sequence selected on any particular day? (Hint: It may be helpful to think about the chosen sequence as a three-digit number.)   0.001
b. What approach (classical, relative frequency, or subjective) did you use to obtain the probability in Part (a)? Explain.   Classical or relative frequency

6.15 An article in the *New York Times* (March 2, 1994) reported that people who suffer cardiac arrest in New York City have only a 1 in 100 chance of survival. Using probability notation, an equivalent statement would be P(survival) $= .01$ for people who suffer a cardiac arrest in New York City. (The article attributed this poor survival rate to factors common in large cities: traffic congestion and the difficulty of finding victims in large buildings.)
a. Give a relative frequency interpretation of the given probability.
b. The research that was the basis for the *New York Times* article was a study of 2329 consecutive cardiac arrests in New York City. To justify the "1 in 100 chance of survival" statement, how many of the 2329 cardiac arrest sufferers do you think survived? Explain.   23 or 24, 1% of 2329 is 23.29

6.16 The article "Anxiety Increases for Airline Passengers After Plane Crash" (*San Luis Obispo Tribune*, November 13, 2001) reported that air passengers have a 1 in 11 million chance of dying in an airplane crash. This probability was then interpreted as "You could fly every day for 26,000 years before your number was up." Comment on why this probability interpretation is misleading.

6.17 Refer to the following information on births in the United States over a given period of time:

Type of Birth	Number of Births
Single birth	41,500,000
Twins	500,000
Triplets	5000
Quadruplets	100

Use this information to approximate the probability that a randomly selected pregnant woman who reaches full term
a. Delivers twins   .0119
b. Delivers quadruplets   .00000238
c. Gives birth to more than a single child   .012

6.18 Consider the chance experiment in which both tennis racket head size and grip size are noted for a randomly selected customer at a particular store. The six possible outcomes (simple events) and their probabilities are displayed in the following table:

Head size	Grip Size		
	$4\frac{3}{8}$ in.	$4\frac{1}{2}$ in.	$4\frac{5}{8}$ in.
Midsize	$O_1 (.10)$	$O_2 (.20)$	$O_3 (.15)$
Oversize	$O_4 (.20)$	$O_5 (.15)$	$O_6 (.20)$

a. The probability that grip size is $4\frac{1}{2}$ in. (event A) is

$$P(A) = P(O_2 \text{ or } O_5) = .20 + .15 = .35$$

How would you interpret this probability?

b. Use the result of Part (a) to calculate the probability that grip size is not $4\frac{1}{2}$ in. 0.65

c. What is the probability that the racket purchased has an oversize head (event B), and how would you interpret this probability? 0.55

d. What is the probability that grip size is *at least* $4\frac{1}{2}$ in.? 0.70

6.19 ▼ A mutual fund company offers its customers several different funds: a money market fund, three different bond funds, two stock funds, and a balanced fund. Among customers who own shares in just one fund, the percentages of customers in the different funds are as follows:

Money market	20%
Short-term bond	15%
Intermediate-term bond	10%
Long-term bond	5%
High-risk stock	18%
Moderate-risk stock	25%
Balanced fund	7%

A customer who owns shares in just one fund is to be selected at random.

a. What is the probability that the selected individual owns shares in the balanced fund? 0.07

b. What is the probability that the individual owns shares in a bond fund? 0.30

c. What is the probability that the selected individual does not own shares in a stock fund? 0.57

6.20 A radio station that plays classical music has a "by request" program each Saturday evening. The percentages of requests for composers on a particular night are as follows:

Bach	5%
Beethoven	26%
Brahms	9%
Dvorak	2%
Mendelssohn	3%
Mozart	21%
Schubert	12%
Schumann	7%
Tchaikovsky	14%
Wagner	1%

Suppose that one of these requests is to be selected at random.

a. What is the probability that the request is for one of the three B's? 0.40

b. What is the probability that the request is not for one of the two S's? 0.81

c. Neither Bach nor Wagner wrote any symphonies. What is the probability that the request is for a composer who wrote at least one symphony? 0.94

6.21 Refer to Exercise 6.18. Adding probabilities in the first row of the given table yields P(midsize) = .45, whereas from the first column, $P(4\frac{3}{8}$ in. grip) = .30. Is the following true?

$$P(\text{midsize } or \text{ } 4\tfrac{3}{8} \text{ in. grip}) = .45 + .30 = .75$$

Explain.

6.22 A deck of 52 cards is mixed well, and 5 cards are dealt.

a. It can be shown that (disregarding the order in which the cards are dealt) there are 2,598,960 possible hands, of which only 1287 are hands consisting entirely of spades. What is the probability that a hand will consist entirely of spades? What is the probability that a hand will consist entirely of a single suit? 0.000495198, 0.001980792

b. It can be shown that exactly 63,206 hands contain only spades and clubs, with both suits represented. What is the probability that a hand consists entirely of spades and clubs with both suits represented? 0.0243197

c. Using the result of Part (b), what is the probability that a hand contains cards from exactly two suits? 0.1459182

6.23 After all students have left the classroom, a statistics professor notices that four copies of the text were left under desks. At the beginning of the next lecture, the professor distributes the four books at random to the four students (1, 2, 3, and 4) who claim to have left books. One possible outcome is that 1 receives 2's book, 2 receives 4's book, 3 receives his or her own book, and 4 receives 1's book. This outcome can be abbreviated (2, 4, 3, 1).

a. List the 23 other possible outcomes.

b. Which outcomes are contained in the event that exactly two of the books are returned to their correct owners? Assuming equally likely outcomes, what is the probability of this event?

c. What is the probability that exactly one of the four students receives his or her own book? 8/24 = 0.3333

d. What is the probability that exactly three receive their own books? 0

e. What is the probability that at least two of the four students receive their own books? 7/24 = 0.2917

a senior. Does this information change our assessment of the likelihood that the selected student is an athlete? Because 150 of the 425 seniors participate in athletics, this suggests that

$$P(\text{athlete}\,|\,\text{senior}) = \frac{\text{number of senior athletes}}{\text{total number of seniors}} = \frac{150}{425} = .3529$$

The probability is calculated in this way because we know that the selected student is 1 of 425 seniors, each of whom is equally likely to have been the one selected. The interpretation of this conditional probability is that if we were to repeat the chance experiment of selecting a student at random, about 35.29% of the trials that resulted in a senior being selected would also result in the selection of someone who participates in athletics.

■

Example 6.12 GFI Switches

A GFI (ground fault interrupt) switch turns off power to a system in the event of an electrical malfunction. A spa manufacturer currently has 25 spas in stock, each equipped with a single GFI switch. Two different companies supply the switches, and some of the switches are defective, as summarized in the following table:

	Nondefective	Defective	Total
Company 1	10	5	15
Company 2	8	2	10
Total	18	7	

A spa is randomly selected for testing. Let

E = event that GFI switch in selected spa is from Company 1
F = event that GFI switch in selected spa is defective

The tabulated information implies that

$$P(E) = \frac{15}{25} = .60 \qquad P(F) = \frac{7}{25} = .28 \qquad P(E \text{ and } F) = P(E \cap F) = \frac{5}{25} = .20$$

Now suppose that testing reveals a defective switch. (Thus the chosen spa is one of the seven in the "defective" column.) How likely is it that the switch came from the first company? Because five of the seven defective switches are from Company 1,

$$P(E\,|\,F) = P(\text{company 1}\,|\,\text{defective}) = \frac{5}{7} = .714$$

Notice that this is larger than the unconditional probability $P(E)$. This is because Company 1 has a much higher defective rate than does Company 2.

An alternative expression for the conditional probability is

$$P(E|F) = \frac{5}{7} = \frac{5/25}{7/25} = \frac{P(E \text{ and } F)}{P(F)} = \frac{P(E \cap F)}{P(F)}$$

That is, $P(E|F)$ is a ratio of two previously specified probabilities: the probability that both events occur divided by the probability of the "conditioning event" F. Additional insight comes from the Venn diagram of Figure 6.8. Once it is known that the outcome lies in F, the likelihood of E (also) occurring is the "size" of (E and F) relative to the size of F.

Figure 6.8 Venn diagram for Example 6.12 (each dot represents one GFI switch).

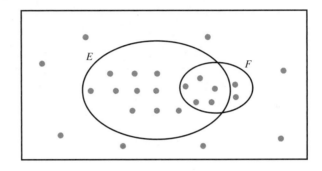

The results of the previous example lead us to a general definition of conditional probability.

> **D E F I N I T I O N**
>
> Let E and F be two events with $P(F) > 0$. The **conditional probability of the event E given that the event F has occurred**, denoted by $P(E|F)$, is
>
> $$P(E|F) = \frac{P(E \cap F)}{P(F)}$$

Notice the requirement that $P(F) > 0$. In addition to the standard warning about division by 0, there is another reason for requiring $P(F)$ to be positive. If the probability of F were 0, the event F would never occur; therefore, it would be unreasonable to calculate the probability of another event conditional on F having occurred.

Example 6.13 Internet Chat Room "Flaming"

One aspect of using the Internet is that diverse individuals from all over the world can form an electronic chat room and exchange opinions on various topics of interest. A side effect of such conversation is "flaming," which is negative criticism of others' contributions to the conversation. M. Dsilva and her colleagues were interested in the effect that personal criticism has on an individual ("Criticism on the

Internet: An Analysis of Participant Reactions," *Communication Research Reports* [1998]: 180–187). Would being criticized make one more likely to criticize others? Data from this study are reproduced here:

	Have Been Personally Criticized	Have Not Been Personally Criticized	Total
Have Criticized Others	19	8	**27**
Have Not Criticized Others	23	143	**166**
Total	**42**	**151**	**193**

We will assume that the table from the article is indicative of the larger group of chat room users and that the frequencies in the table are indicative of the long-run behavior of events in this situation. Suppose that a chat room user is randomly selected. Let

C = event that the individual has criticized others
O = event that the individual has been personally criticized by others

We can use the tabulated information to calculate

$$P(C) = \frac{27}{193} = .1399$$

$$P(O) = \frac{42}{193} = .2176$$

$$P(C \cap O) = \frac{19}{193} = .0984$$

Now suppose that it is known that the selection resulted in an individual that has been criticized. (Thus the individual is 1 of the 42 in the column labeled "Have been personally criticized.") How likely is it that he or she has criticized others? Using the probabilities previously computed, we get

$$P(C|O) = \frac{P(C \cap O)}{P(O)} = \frac{19/193}{42/193} = .4524$$

Notice again that this is larger than the original probability $P(C)$. We conclude that there is a higher probability of criticizing others among those who have been personally criticized.

It is also possible to compute $P(O|C)$, the probability that an individual has been criticized by others given that he or she has been critical of someone else:

$$P(O|C) = \frac{P(O \cap C)}{P(C)} = \frac{P(C \cap O)}{P(C)} = \frac{19/193}{27/193} = \frac{19}{27} = .7037$$

Let's look carefully at the interpretation of each of these probabilities:

1. $P(C) = .1399$ is interpreted as the proportion of chat room users who have been critical of others. Approximately 14% of chat room users have criticized others.

2. $P(O) = .2176$ gives the proportion of chat room users who have been criticized.
3. $P(C \cap O) = .0984$ is the proportion of chat room users who have been critical of others *and* who have also been criticized.
4. $P(C|O) = .4524$ is the proportion *of those who have been criticized* who have also been critical of others.
5. $P(O|C) = .7037$ is the proportion *of those who have been critical of others* who have themselves been criticized.

In particular, notice the difference between the probabilities in Interpretations 1– 3 and the conditional probabilities in Interpretations 4 and 5. The reference point for the unconditional probabilities is the entire group of interest (all chat room users), whereas the conditional probabilities are interpreted in a more restricted context defined by the "given" event.

∎

Example 6.14 demonstrates the calculation of conditional probabilities and also makes the point that we must be careful when translating real-world problems—especially probability problems—into mathematical form. Not only are probability problems sometimes difficult to formulate precisely, but the answers are frequently nonintuitive and occasionally counterintuitive.

Example 6.14 Two-Kid Families

Consider the population of all families with two children. Representing the gender of each child using G for girl and B for boy results in four possibilities: BB, BG, GB, GG. The gender information is sequential, with the first letter indicating the gender of the older sibling. Thus, a family having a girl first and then a boy is denoted GB. If we assume that a child is equally likely to be male or female, each of the four possibilities in the sample space for the chance experiment that selects at random from families with two children is equally likely. Consider the following two questions:

1. What is the probability of obtaining a family with two girls, given that the family has at least one girl?
2. What is the probability of obtaining a family with two girls, given that the older sibling is a girl?

To many people, these questions *appear* to be identical. However, by computing the appropriate probabilities, we can see that they are indeed different.

For Question 1 we obtain

$$P(\text{family with two girls} \mid \text{family has at least one girl})$$
$$= \frac{P(\text{family with two girls and family with at least one girl})}{P(\text{family with at least one girl})}$$
$$= \frac{P(GG)}{P(GG \text{ or } BG \text{ or } GB)}$$
$$= \frac{1/4}{3/4} = \frac{.25}{.75} = .3333$$

For question 2 we calculate

$$P(\text{family with two girls} \,|\, \text{family with older sibling a girl})$$

$$= \frac{P(\text{family with two girls } and \text{ family with older sibling a girl})}{P(\text{family with older sibling a girl})}$$

$$= \frac{P(GG)}{P(GB \text{ or } GG)}$$

$$= \frac{1/4}{1/2} = \frac{.25}{.50} = .50$$

The moral of this story is: Correct solutions to probability problems, especially conditional probability problems, are much more the result of careful consideration of the sample spaces and the probabilities than they are a product of what your intuition tells you!

■

As we mentioned in the opening paragraphs of this section, one of the most important practical uses of conditional probability is in making diagnoses. Your mechanic diagnoses your car by hooking it up to a machine and reading the pressures and speeds of the various components. A meteorologist diagnoses the weather by looking at temperatures, isobars, and wind speeds. And medical doctors diagnose the state of a person's health by performing various tests and gathering information, such as weight and blood pressure.

Doctors observe characteristics of their patients in an attempt to ascertain whether or not their patients have a certain disease. Many diseases are not actually observable—or at least not easily so—and the doctor must make a probabilistic judgment based on partial information. It is instructive to focus our attention on the probabilistic analysis that underpins diagnosis. As we will see, conditional probability plays a large role in evaluating diagnostic techniques.

We begin with the commonsense notion that a randomly selected individual either has or does not have a particular disease, for example, toxoplasmosis. Toxoplasmosis is a parasitic disease that usually does not result in any symptoms, but it may pose a significant risk to the developing fetus of a pregnant woman. The problem here is that the patient feels no symptoms and the doctor observes no symptoms in a routine checkup; the disease must be detected by prenatal screening.

Normally, a criterion or standard exists by which it would be unequivocally decided whether a person has a disease; this criterion is colloquially known as the "gold standard." The gold standard might be an invasive surgical procedure, expensive and dangerous. The gold standard might be a test that takes a long time to perform in the lab or that is costly. In some cases, definitive detection may be possible only postmortem. For a diagnostic test to improve on the gold standard, it would need to be faster, less expensive, or less invasive and yet still produce results that agree with the gold standard. In this context, *agreement* means (1) that the test generally comes out positive when the patient has the disease and (2) that the test generally comes out negative when the patient does not have the disease. These statements lead us to consideration of conditional probabilities.

Example 6.15 Diagnosing Tuberculosis

To illustrate the calculations involved in evaluating a diagnostic test, we consider the case of tuberculosis (TB), an infectious disease that typically attacks lung tissue. Before 1998, culturing was the existing gold standard for diagnosing TB. This method took 10 to 15 days to yield a positive or negative result. In 1998, investigators evaluated a DNA technique that turned out to be much faster ("LCx: A Diagnostic Alternative for the Early Detection of *Mycobacterium tuberculosis* Complex," *Diagnostic Microbiology and Infectious Diseases* [1998]: 259–264). The DNA technique for detecting tuberculosis was evaluated by comparing results from the test to the existing gold standard, with the following results for 207 patients exhibiting symptoms:

	Has Tuberculosis (gold standard)	Does Not Have Tuberculosis (gold standard)
DNA Positive Indication	14	0
DNA Negative Indication	12	181

Converting these data to proportions and inserting the column and row totals into the table, we get the following information:

	Has TB	Does Not Have TB	Total
DNA+	.0676	.0000	.0676
DNA−	.0580	.8744	.9324
Total	.1256	.8744	1.0000

A cursory look at the table indicates that the DNA technique seems to be working in a manner consistent with what we expect from a diagnostic test. Samples that tested positive with the technique agreed with the gold standard in every case. Samples that tested negative were generally in agreement with the gold standard, but the table also indicates some false-negative results. Consider a randomly selected individual who is tested for TB. Define the following events:

T = event that the individual has tuberculosis
N = event that the DNA test is negative

Let $P(T|N)$ denote the probability of the event T given that the event N has occurred. We calculate this probability as follows:

$$P(T|N) = P(\text{tuberculosis}|\text{negative DNA test})$$
$$= \frac{P(\text{tuberculosis} \cap \text{negative DNA test})}{P(\text{negative DNA test})}$$
$$= \frac{.0580}{.9324}$$
$$= .0622$$

Notice that .1256 of those tested had tuberculosis. The added information provided by the diagnostic test has altered the probability—and provided some measure of relief for the patients who test negative. Once it is known that the test result is negative, the estimated likelihood of the disease is cut in half. If the diagnostic test did not significantly alter the probability, the test would not be very useful to the doctor or patient.

■

Exercises 6.28–6.40

Turn to the Expanded Answers Section for answers not shown next to exercises.

6.28 Two different airlines have a flight from Los Angeles to New York that departs each weekday morning at a certain time. Let E denote the event that the first airline's flight is fully booked on a particular day, and let F denote the event that the second airline's flight is fully booked on that same day. Suppose that $P(E) = .7$, $P(F) = .6$, and $P(E \cap F) = .54$.
a. Calculate $P(E|F)$ the probability that the first airline's flight is fully booked given that the second airline's flight is fully booked.   0.9
b. Calculate $P(F|E)$.   0.771

6.29 Of the 60 movies reviewed last year by two critics on their joint television show, Critic 1 gave a "thumbs-up" rating to 15, Critic 2 gave this rating to 20, and 10 of the movies were rated thumbs-up by both critics. Suppose that 1 of these 60 movies is randomly selected.
a. Given that the movie was rated thumbs-up by Critic 1, what is the probability that it also received this rating from Critic 2?   0.6667
b. If the movie did not receive a thumbs-up rating from Critic 2, what is the probability that it also did not receive a thumbs up rating from Critic 1? (Hint: Construct a table with two rows for the first critic [for "up" and "not up"] and two columns for the second critic; then enter the relevant probabilities.)   0.875

6.30 The article "Chances Are You Know Someone with a Tattoo, and He's Not a Sailor" (Associated Press, June 11, 2006) included results from a survey of adults aged 18 to 50. The accompanying data are consistent with summary values given in the article.

	At Least One Tattoo	No Tattoo
Age 18–29	18	32
Age 30–50	6	44

Assuming these data are representative of adults in the United States and that a U.S. adult is selected at random, use the given information to estimate the following probabilities.
a. $P(\text{tattoo})$ 0.24
b. $P(\text{tattoo}|\text{age 18–29})$ .36
c. $P(\text{tattoo}|\text{age 30–50})$ .12
d. $P(\text{age 18–29}|\text{tattoo})$ 0.75

6.31 The newspaper article "Folic Acid Might Reduce Risk of Down Syndrome" (*USA Today*, September 29, 1999) makes the following statement: "Older women are at a greater risk of giving birth to a baby with Down Syndrome than are younger women. But younger women are more fertile, so most children with Down Syndrome are born to mothers under 30." Let D = event that a randomly selected baby is born with Down Syndrome and Y = event that a randomly selected baby is born to a young mother (under age 30). For each of the following probability statements, indicate whether the statement is consistent with the quote from the article, and if not, explain why not.
a. $P(D|Y) = .001$, $P(D|Y^C) = .004$, $P(Y) = .7$
b. $P(D|Y) = .001$, $P(D|Y^C) = .001$, $P(Y) = .7$
c. $P(D|Y) = .004$, $P(D|Y^C) = .004$, $P(Y) = .7$
d. $P(D|Y) = .001$, $P(D|Y^C) = .004$, $P(Y) = .4$
e. $P(D|Y) = .001$, $P(D|Y^C) = .001$, $P(Y) = .4$
f. $P(D|Y) = .004$, $P(D|Y^C) = .004$, $P(Y) = .4$

6.32 Suppose that an individual is randomly selected from the population of all adult males living in the United States. Let A be the event that the selected individual is over 6 ft in height, and let B be the event that the selected individual is a professional basketball player. Which do you think is larger, $P(A|B)$ or $P(B|A)$? Why?

6.33 ▼ Is ultrasound a reliable method for determining the gender of an unborn baby? The accompanying data on

1000 births are consistent with summary values that appeared in the online version of the *Journal of Statistics Education* ("New Approaches to Learning Probability in the First Statistics Course" [2001]).

	Ultrasound Predicted Female	Ultrasound Predicted Male
Actual Gender Is Female	432	48
Actual Gender Is Male	130	390

a. Use the given information to estimate the probability that a newborn baby is female, given that the ultrasound predicted the baby would be female.   0.769
b. Use the given information to estimate the probability that a newborn baby is male, given that the ultrasound predicted the baby would be male.   0.890
c. Based on your answers to Parts (a) and (b), do you think ultrasound is equally reliable for predicting gender for boys and for girls? Explain.   Ultrasound appears to be more reliable when the gender is male.
6.34 The article "Doctors Misdiagnose More Women, Blacks" (*San Luis Obispo Tribune*, April 20, 2000) gave the following information, which is based on a large study of more than 10,000 patients treated in emergency rooms in the eastern and midwestern United States:
1. Doctors misdiagnosed heart attacks in 2.1% of all patients $P(M) = 0.021$
2. Doctors misdiagnosed heart attacks in 4.3% of black patients. $P(M|B) = 0.043$
3. Doctors misdiagnosed heart attacks in 7% of women under 55 years old. $P(M|W) = 0.07$

Use the following event definitions: M = event that a heart attack is misdiagnosed, B = event that a patient is black, and W = event that a patient is a woman under 55 years old. Translate each of the three statements into probability notation.

6.35 The *Cedar Rapids Gazette* (November 20, 1999) reported the following information on compliance with child restraint laws for cities in Iowa:

City	Number of Children Observed	Number Properly Restrained
Cedar Falls	210	173
Cedar Rapids	231	206
Dubuque	182	135
Iowa City (city)	175	140
Iowa City (interstate)	63	47

a. Use the information provided to estimate the following probabilities:
i. The probability that a randomly selected child is properly restrained given that the child is observed in Dubuque.   0.7418
ii. The probability that a randomly selected child is properly restrained given that the child is observed in a city that has "Cedar" in its name.   0.8594
b. Suppose that you are observing children in the Iowa City area. Use a tree diagram to illustrate the possible outcomes of an observation that considers both the location of the observation (city or interstate) and whether the child observed was properly restrained.

6.36 The following table summarizes data on smoking status and perceived risk of smoking and is consistent with summary quantities obtained in a Gallup Poll conducted in November 2002:

Smoking Status	Perceived Risk			
	Very Harmful	Somewhat Harmful	Not Too Harmful	Not at All Harmful
Current Smoker	60	30	5	1
Former Smoker	78	16	3	2
Never Smoked	86	10	2	1

Assume that it is reasonable to consider these data as representative of the U.S. adult population.

a. What is the probability that a randomly selected U.S. adult is a former smoker? 0.3367

b. What is the probability that a randomly selected U.S. adult views smoking as very harmful? 0.7619

c. What is the probability that a randomly selected U.S. adult views smoking as very harmful given that the selected individual is a current smoker? 0.625

d. What is the probability that a randomly selected U.S. adult views smoking as very harmful given that the selected individual is a former smoker? 0.7879

e. What is the probability that a randomly selected U.S. adult views smoking as very harmful given that the selected individual never smoked? 0.8687

f. How do the probabilities computed in Parts (c), (d), and (e) compare? Does this surprise you? Explain.

6.37 *USA Today* (June 6, 2000) gave information on seat belt usage by gender. The proportions in the following table are based on a survey of a large number of adult men and women in the United States:

	Male	Female
Uses Seat Belts Regularly	.10	.175
Does Not Use Seat Belts Regularly	.40	.325

Assume that these proportions are representative of adults in the United States and that a U.S. adult is selected at random.

a. What is the probability that the selected adult regularly uses a seat belt? 0.275

b. What is the probability that the selected adult regularly uses a seat belt given that the individual selected is male?

c. What is the probability that the selected adult does not use a seat belt regularly given that the selected individual is female? 0.65

d. What is the probability that the selected individual is female given that the selected individual does not use a seat belt regularly? 0.448

e. Are the probabilities from Parts (c) and (d) equal? Write a couple of sentences explaining why this is so.

6.38 The *USA Today* article referenced in Exercise 6.37 also gave information on seat belt usage by age, which is summarized in the following table of counts:

Age	Does Not Use Seat Belt Regularly	Uses Seat Belt Regularly
18–24	59	41
25–34	73	27
35–44	74	26
45–54	70	30
55–64	70	30
65 and older	82	18

Consider the following events: S = event that a randomly selected individual uses a seat belt regularly, A_1 = event that a randomly selected individual is in age group 18–24, and A_6 = event that a randomly selected individual is in age group 65 and older.

a. Convert the counts to proportions and then use them to compute the following probabilities:
  i. $P(A_1)$ ii. $P(A_1 \cap S)$ iii. $P(A_1|S)$
  iv. $P(not\ A_1)$ v. $P(S|A_1)$ vi. $P(S|A_6)$

b. Using the probabilities $P(S|A_1)$ and $P(S|A_6)$ computed in Part (a), comment on how 18–24-year-olds and seniors differ with respect to seat belt usage.

6.39 The paper "Good for Women, Good for Men, Bad for People: Simpson's Paradox and the Importance of Sex-Specific Analysis in Observational Studies" (*Journal of Women's Health and Gender-Based Medicine* [2001]: 867–872) described the results of a medical study in which one treatment was shown to be better for men and better for women than a competing treatment. However, if the data for men and women are combined, it appears as though the competing treatment is better. To see how this can happen, consider the accompanying data tables constructed from information in the paper. Subjects in the study were given either Treatment A or Treatment B, and survival was noted. Let S be the event that a patient selected at random survives, A be the event that a patient selected at random received Treatment A, and B be the event that a patient selected at random received Treatment B.

a. The following table summarizes data for men and women combined:

	Survived	Died	Total
Treatment A	215	85	**300**
Treatment B	241	59	**300**
Total	**456**	**144**	

i. Find $P(S)$

ii. Find $P(S|A)$.

iii. Find $P(S|B)$.

iv. Which treatment appears to be better?

b. Now consider the summary data for the men who participated in the study:

	Survived	Died	Total
Treatment A	120	80	**200**
Treatment B	20	20	**40**
Total	**140**	**100**	

i. Find $P(S)$.

ii. Find $P(S|A)$.

iii. Find $P(S|B)$.

iv. Which treatment appears to be better?

c. Now consider the summary data for the women who participated in the study:

	Survived	Died	Total
Treatment A	95	5	**100**
Treatment B	221	39	**260**
Total	**316**	**144**	

i. Find $P(S)$.

ii. Find $P(S|A)$.

Bold exercises answered in back   ● Data set available online but not required   ▼ Video solution available

iii. Find $P(S|B)$.

iv. Which treatment appears to be better?

d. You should have noticed from Parts (b) and (c) that for both men and women, Treatment A appears to be better. But in Part (a), when the data for men and women are combined, it looks like Treatment B is better. This is an example of what is called Simpson's paradox. Write a brief explanation of why this apparent inconsistency occurs for this data set. (Hint: Do men and women respond similarly to the two treatments?)

6.40 According to a study conducted by a risk assessment firm (Associated Press, December 8, 2005), drivers residing within one mile of a restaurant are 30% more likely to be in an accident in a given policy year. Consider the following two events:

A = event that a driver has an accident during a policy year

R = event that a driver lives within one mile of a restaurant

Which of the following four probability statements is consistent with the findings of this survey? Justify your choice.

i. $P(A|R) = .3$ **iii.** $\dfrac{P(A|R)}{P(A|R^C)} = .3$

ii. $P(A|R^C) = .3$ **iv.** $\dfrac{P(A|R) - P(A|R^C)}{P(A|R^C)} = .3$

6.5 Independence

In Section 6.4, we saw that knowledge of the occurrence of one event can alter our assessment of the likelihood that some other event has occurred. We also saw how information about conditional probabilities could be used in medical diagnosis to revise assessments of patients in light of the outcome of a diagnostic procedure. However, it is also possible that knowledge that one event has occurred will not change our assessment of the probability of occurrence of a second event.

Example 6.16 Mortgage Choices

A large lending institution issues both adjustable-rate and fixed-rate mortgage loans on residential property, which it classifies into three categories: single-family houses, condominiums, and multifamily dwellings. The following table, sometimes called a *joint probability table*, displays probabilities based on the bank's long-run lending behavior:

	Single-Family	Condo	Multifamily	Total
Adjustable	.40	.21	.09	**.70**
Fixed	.10	.09	.11	**.30**
Total	**.50**	**.30**	**.20**	

From the table we see that 70% of all mortgages are adjustable rate, 50% of all mortgages are for single-family properties, 40% of all mortgages are adjustable rate for single-family properties (both adjustable-rate *and* single-family), and so on. Define the events E and F by

$$E = \text{event that a mortgage is adjustable rate}$$
$$F = \text{event that a mortgage is for a single-family property}$$

Then

$$P(E|F) = \frac{P(E \text{ and } F)}{P(F)} = \frac{.40}{.50} = .80$$

That is, 80% of loans made for single-family properties are adjustable-rate loans. Notice that $P(E|F)$ is larger than the original (unconditional) probability $P(E) = .70$. Also,

$$P(F|E) = \frac{P(E \text{ and } F)}{P(E)} = \frac{.40}{.70} = .571 > .5 = P(F)$$

If we define another event C by

$$C = \text{event that a mortgage is for a condominium}$$

we have

$$P(E|C) = \frac{P(E \text{ and } C)}{P(C)} = \frac{.21}{.30} = .70$$

Notice that $P(E|C) = P(E)$ so if we are told that a mortgage is for a condominium, the probability that it is adjustable remains unchanged.

■

Important definition

When two events E and F are such that $P(E|F) = P(E)$, the likelihood that event E has occurred is the same after we learn that F has occurred as it was before we had information about F's occurrence. We then say that E and F are independent of one another.

> **DEFINITION**
>
> Two events E and F are said to be **independent** if
>
> $$P(E|F) = P(E)$$
>
> If E and F are not independent, they are said to be **dependent** events.
>
> If $P(E|F) = P(E)$, it is also true that $P(F|E) = P(F)$, and vice versa.

In other words, independence of events E and F also implies the following additional three relationships:

$$P(not\ E|F) = P(not\ E)$$
$$P(E|not\ F) = P(E)$$
$$P(not\ E|not\ F) = P(not\ E)$$

That is, to say that E and F are independent implies that nothing we learn about F will change the likelihood of E or of *not E*.

The formula for conditional probability

$$P(E|F) = \frac{P(E \cap F)}{P(F)}$$

can be rearranged to give

$$P(E \cap F) = P(E|F)P(F)$$

When E and F are independent, $P(E|F) = P(E)$, so

$$P(E \cap F) = P(E|F)P(F) = P(E)P(F)$$

This result is called the multiplication rule for two independent events.

Multiplication Rule for Two Independent Events

The events E and F are independent if and only if

$$P(E \cap F) = P(E)P(F)$$

■

Example 6.17 Hitchhiker's Thumb

In humans, there is a gene that controls a characteristic known as hitchhiker's thumb, the ability to bend the last joint of the thumb back at an angle of 60° or more. Whether an offspring has hitchhiker's thumb is determined by two random events: which of two alleles is contributed by the father and which of two alleles is contributed by the mother. (You can think of these alleles as a parental vote of yes or no on the hitchhiker's thumb gene. If the votes by the two parents disagree, the dominant allele wins.) These two random events, the results of cell division in two different biological parents, are independent of each other. Suppose that there is a .10 probability that a parent contributes a positive hitchhiker's thumb allele. Because the events are independent, the probability that each parent contributes a positive hitchhiker's thumb allele, H+, to the offspring is

$$P(\text{mother contributes H+} \cap \text{father contributes H+})$$
$$= P(\text{mother contributes H+})P(\text{father contributes H+})$$
$$= (.10)(.10)$$
$$= .01$$

■

Example 6.18 Curious Guppies

© Maximilian Weinzier/Alamy

Let's look at another example, this time from the field of animal behavior, to illustrate how an investigator could judge whether two events are independent. In a number of fish species, including guppies, a phenomenon known as predator inspection has been reported. It is thought that predator inspection allows a guppy to assess the risk posed by a potential predator. In a typical inspection a guppy moves toward a predator, presumably to acquire information and then (hopefully) depart to inspect another day.

Investigators have observed that guppies sometimes approach and inspect a predator in pairs. Suppose that it is not known whether these predator inspections are independent or whether the guppies are operating as a team. Denote the probability that an individual guppy will inspect a predator by p. Let event E_i be the event that guppy i will approach and inspect a predator. Then the probability of two guppies, i and j, approaching the predator by chance if they are acting *independently* is

$$P(E_i \cap E_j) = P(E_i)P(E_j) = p \cdot p = p^2$$

Based on our analysis, if the inspections are in fact independent, we would expect the proportion of times that two guppies happen to simultaneously inspect a predator to be equal to the square of the proportion of times that a single fish does so. For example, if the probability a single guppy inspects a predator is .3, we would expect the proportion of the time that two guppies would inspect a predator simultaneously to be $(.3)^2 = .09$. Based on observations of the inspection behavior of a large number of guppies, scientists have concluded that the inspection behavior of guppies does not appear to be independent.

■

The concept of independence extends to more than two events. Consider three events, E_1, E_2, and E_3. Then independence means not only that

$$P(E_1|E_2) = P(E_1)$$
$$P(E_3|E_2) = P(E_3)$$

and so on but also that

$$P(E_1|E_2 \text{ and } E_3) = P(E_1)$$
$$P(E_1 \text{ and } E_3|E_2) = P(E_1 \text{ and } E_3)$$

and so on. There is also a multiplication rule for more than two independent events.

Multiplication Rule for *k* Independent Events

Events $E_1, E_2, \ldots, E_k$ are **independent** if knowledge that some number of the events have occurred does not change the probabilities that any particular one or more of the other events has occurred.

Independence implies that

$$P(E_1 \cap E_2 \cap \cdots \cap E_k) = P(E_1)P(E_2) \cdots P(E_k)$$

Thus when events are independent, the probability that all occur together is the product of the individual probabilities. Furthermore, this relationship remains valid if one or more E_i is replaced by the event E_i^C.

The independence of more than two events is an important concept in studying complex systems with many components. If these components are critical to the operation of a machine, an examination of the probability of the machine's failure is undertaken by analyzing the failure probabilities of the components. In Example 6.19 we take a rather simplified view of a desktop computer to illustrate the use of the multiplication rule.

Example 6.19 Computer Configurations

Suppose that a desktop computer system consists of a monitor, a mouse, a keyboard, the computer processor itself, and storage devices such as a disk drive. It is conventional wisdom that if a computer is going to fail, it will fail soon, a phenomenon known as infant mortality. Purchasers of new computer systems are advised to turn their computers on as soon as they are purchased and then to let them run for a few hours to see if any problems crop up.

Let

E_1 = event that a newly purchased monitor operates properly
E_2 = event that a newly purchased mouse operates properly
E_3 = event that a newly purchased disk drive operates properly
E_4 = event that a newly purchased computer processor operates properly

Suppose the four events are independent, with

$$P(E_1) = P(E_2) = .98 \qquad P(E_3) = .95 \qquad P(E_4) = .99$$

The probability that all these components operate properly is then

$$P(E_1 \cap E_2 \cap E_3 \cap E_4) = P(E_1)P(E_2)P(E_3)P(E_4)$$
$$= (.98)(.98)(.94)(.99)$$
$$= .89$$

We interpret this probability as follows: In the long run, 89% of such systems will run properly when tested shortly after purchase. (In reality, the reliability of these components is much higher than the numbers used in this example!) The probability that all components except the monitor will run properly is

$$P(E_1^C \cap E_2 \cap E_3 \cap E_4) = P(E_1^C)P(E_2)P(E_3)P(E_4)$$
$$= (1 - P(E_1))P(E_2)P(E_3)P(E_4)$$
$$= (.02)(.98)(.94)(.99)$$
$$= .018$$

■

■ Sampling With and Without Replacement

One area of statistics where the rules of probability are important is sampling. As we saw in Chapter 2, a well-designed sampling plan allows investigators to make inferences about a population based on information from a sample. Sampling methods can be classified into two categories: sampling with replacement and sampling without replacement. Most inferential methods presented in an introductory statistics course are based on the assumption of sampling *with* replacement, but when sampling from real populations, we almost always sample *without* replacement. This seemingly contradictory practice can be a source of confusion. Fortunately, in many circumstances, the distinction between sampling with and without replacement is unimportant for *practical* purposes.

DEFINITION

Sampling is **with replacement** if, once selected, an individual or object is put back into the population before the next selection. Sampling is **without replacement** if, once selected, an individual or object is *not* returned to the population prior to subsequent selections.

Example 6.20 Selecting Cards

Consider the process of selecting three cards from a standard deck. This selection can be made in two ways. One method is to deal three cards off the top of the deck. This method would constitute sampling without replacement. A second method, rarely seen in real games, is to select a card at random, note which card is observed, replace it in the deck, and shuffle before selecting the next card. This method is sampling with replacement. From the standpoint of probability, these procedures are analyzed differently. To see this, consider these events:

$$H_1 = \text{event that the first card is a heart}$$
$$H_2 = \text{event that the second card is a heart}$$
$$H_3 = \text{event that the third card is a heart}$$

For sampling with replacement, the probability of H_3 is .25, regardless of whether either H_1 or H_2 occurs, because replacing selected cards gives the same deck for the third selection as for the first two selections. Whether either of the first two cards is a heart has no bearing on the third card selected, so the three events, H_1, H_2, and H_3, are independent.

When sampling is without replacement, the chance of getting a heart on the third draw does depend on the results of the first two draws. If both H_1 and H_2 occur, only 11 of the 50 remaining cards are hearts. Because any 1 of these 50 has the same chance of being selected, the probability of H_3 in this case is

$$P(H_3 | H_1 \text{ and } H_2) = \frac{11}{50} = .22$$

Alternatively, if neither of the first 2 cards is a heart, then all 13 hearts remain in the deck for the third draw, so

$$P(H_3 | \text{ not } H_1 \text{ and not } H_2) = \frac{13}{50} = .26$$

Information about the occurrence of H_1 and H_2 affects the chance that H_3 has occurred; for sampling without replacement, the events are not independent.

■

In opinion polls and other types of surveys, sampling is virtually always done without replacement. For this method of sampling, the results of successive selections are not independent of one another. However, Example 6.21 suggests that, *under certain circumstances*, the fact that selections in sampling without replacement are not independent is not a cause for concern.

Example 6.21   Independence and Sampling Without Replacement

A lot of 10,000 industrial components consists of 2500 manufactured by one firm and 7500 manufactured by a second firm, all mixed together. Three of the components will be selected at random without replacement. Let

E_1 = event that first component selected was manufactured by Firm 1
E_2 = event that second component selected was manufactured by Firm 1
E_3 = event that third component selected was manufactured by Firm 1

Following the reasoning we used in Example 6.20, we get

$$P(E_3|E_1 \text{ and } E_2) = \frac{2498}{9998} = .24985$$

$$P(E_3|\text{not } E_1 \text{ and not } E_2) = \frac{2500}{9998} = .25005$$

Although these two probabilities differ slightly, when rounded to three decimal places they are both .250. We conclude that the occurrence or nonoccurrence of E_1 or E_2 has virtually no effect on the chance that E_3 will occur. *For practical purposes*, the three events can be considered independent.

▪

The essential difference between the situations of Example 6.20 and Example 6.21 is the size of the sample relative to the size of the population. In Example 6.20 a relatively large proportion of the population was sampled (3 out of 52), whereas in Example 6.21 the proportion of the population sampled was quite small (only 3 out of 10,000).

It turns out that the sample is typically small compared to the size of the population, whether we are sampling cards, people, light bulbs, or concentrations of metals in rivers and streams. The theory of sampling with replacement required for many inferential methods can coexist with the practice of sampling without replacement because of the following principle:

> If a random sample of size n is taken from a population of size N, the theoretical probabilities of successive selections calculated on the basis of sampling with replacement and on the basis of sampling without replacement differ by insignificant amounts when n is small compared to N. In practice, independence can be assumed for the purpose of calculating probabilities as long as n is not larger than 5% of N.

This principle justifies the assumption of independence in many statistical problems. The phrase *assumption of independence* does not signify that the investigators are in some sense fooling themselves; they are recognizing that, for all practical purposes, the results will not differ from the "right" answers.

In some sampling situations, the sample size might be a significant fraction of the population. For example, a school newspaper editor might easily sample more than 5% of the student body. In that instance, the editors would be wise to consult a statistician before proceeding. Does this mean that an investigator should not sample more than 5% of a population? Certainly not! Except for the time and resources used in sampling a population, it is almost always the case that a larger sample results in better inferences about the population. The only disadvantage of sampling more than 5% of the population is that the analysis of the resulting data is slightly more complicated.

Exercises 6.41–6.58

Turn to the Expanded Answers Section for answers not shown next to exercises.

6.41 Many fire stations handle emergency calls for medical assistance as well as calls requesting firefighting equipment. A particular station says that the probability that an incoming call is for medical assistance is .85. This can be expressed as P(call is for medical assistance) = .85.
a. Give a relative frequency interpretation of the given probability.
b. What is the probability that a call is not for medical assistance?   0.15
c. Assuming that successive calls are independent of one another, calculate the probability that two successive calls will both be for medical assistance.   0.7225
d. Still assuming independence, calculate the probability that for two successive calls, the first is for medical assistance and the second is not for medical assistance.
e. Still assuming independence, calculate the probability that exactly one of the next two calls will be for medical assistance. (Hint: There are two different possibilities. The one call for medical assistance might be the first call, or it might be the second call.)   0.255
f. Do you think that it is reasonable to assume that the requests made in successive calls are independent? Explain.

6.42 The article "SUVs Score Low in New Federal Rollover Ratings" (*San Luis Obispo Tribune*, January 6, 2001) gave information on death rates for various kinds of accidents by vehicle type for accidents reported to the police. Suppose that we randomly select an accident reported to the police and consider the following events: R = event that the selected accident is a single-vehicle rollover, F = event that the selected accident is a frontal collision, and D = event that the selected accident results in a death. Information in the article indicates that the following probability estimates are reasonable: $P(R) = .06$, $P(F) = .60$, $P(R|D) = .30$, $P(F|D) = .54$.

a. Interpret the value of $P(R|D)$.
b. Interpret the value of $P(F|D)$.
c. Are the events R and D independent? Justify your answer. $P(R|D) \neq P(R)$, so no, not independent.
d. Is $P(F \cap D) = P(F)P(D)$? Explain why or why not.
e. Is $F = R^C$? Explain how you can tell.

6.43 A Gallup survey of 2002 adults found that 46% of women and 37% of men experience pain daily (*San Luis Obispo Tribune*, April 6, 2000). Suppose that this information is representative of U.S. adults. If a U.S. adult is selected at random, are the events *selected adult is male* and *selected adult experiences pain daily* independent or dependent? Explain.

6.44 The Associated Press (*San Luis Obispo Telegram-Tribune*, August 23, 1995) reported on the results of mass screening of schoolchildren for tuberculosis (TB). For Santa Clara County, California, the proportion of all tested kindergartners who were found to have TB was .0006. The corresponding proportion for recent immigrants (thought to be a high-risk group) was .0075. Suppose that a Santa Clara County kindergartner is selected at random. Are the events *selected student is a recent immigrant* and *selected student has TB* independent or dependent events? Justify your answer using the given information.

6.45 The article "Men, Women at Odds on Gun Control" (*Cedar Rapids Gazette*, September 8, 1999) included the following statement: "The survey found that 56 percent of American adults favored stricter gun control laws. Sixty-six percent of women favored the tougher laws, compared with 45 percent of men." These figures are based on a large telephone survey conducted by Associated Press Polls. If an adult is selected at random, are the events *selected adult is female* and *selected adult favors stricter gun control* independent or dependent events? Explain.

6.46 The Australian newspaper *The Mercury* (May 30, 1995) reported that, based on a survey of 600 reformed and current smokers, 11.3% of those who had attempted to quit smoking in the previous 2 years had used a nicotine aid (such as a nicotine patch). It also reported that 62% of those who quit smoking without a nicotine aid began smoking again within 2 weeks and 60% of those who used a nicotine aid began smoking again within 2 weeks. If a smoker who is trying to quit smoking is selected at random, are the events *selected smoker who is trying to quit uses a nicotine aid* and *selected smoker who has attempted to quit begins smoking again within 2 weeks* independent or dependent events? Justify your answer using the given information.

6.47 In a small city, approximately 15% of those eligible are called for jury duty in any one calendar year. People are selected for jury duty at random from those eligible, and the same individual cannot be called more than once in the same year. What is the probability that a particular eligible person in this city is selected two years in a row? three years in a row?   0.225, 0.003375

6.48 Jeanie is a bit forgetful, and if she doesn't make a "to do" list, the probability that she forgets something she is supposed to do is .1. Tomorrow she intends to run three errands, and she fails to write them on her list.
a. What is the probability that Jeanie forgets all three errands? What assumptions did you make to calculate this probability?   0.001
b. What is the probability that Jeanie remembers at least one of the three errands?   0.999
c. What is the probability that Jeanie remembers the first errand but not the second or third?   0.009

6.49 ▼ Approximately 30% of the calls to an airline reservation phone line result in a reservation being made.
a. Suppose that an operator handles 10 calls. What is the probability that none of the 10 calls result in a reservation? $(.7)^{10} = 0.028$
b. What assumption did you make to calculate the probability in Part (a)?   Calls are independent.
c. What is the probability that at least one call results in a reservation being made?   0.972

6.50 Of the 10,000 students at a certain university, 7000 have Visa cards, 6000 have MasterCards, and 5000 have both. Suppose that a student is randomly selected.

a. What is the probability that the selected student has a Visa card?   0.7
b. What is the probability that the selected student has both cards?   0.5
c. Suppose you learn that the selected individual has a Visa card (was one of the 7000 with such a card). Now what is the probability that this student has both cards?
d. Are the events *has a Visa card* and *has a MasterCard* independent? Explain.
e. Answer the question posed in Part (d) if only 4200 of the students have both cards.   They are independent.

6.51 Consider a system consisting of four components, as pictured in the following diagram:

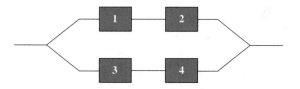

Components 1 and 2 form a series subsystem, as do Components 3 and 4. The two subsystems are connected in parallel. Suppose that $P(1 \text{ works}) = .9$, $P(2 \text{ works}) = .9$, $P(3 \text{ works}) = .9$, and $P(4 \text{ works}) = .9$ and that the four components work independently of one another.
a. The 1–2 subsystem works only if both components work. What is the probability of this happening?   0.81
b. What is the probability that the 1–2 subsystem doesn't work? that the 3–4 subsystem doesn't work?   0.19, 0.19
c. The system won't work if the 1–2 subsystem doesn't work and if the 3–4 subsystem also doesn't work. What is the probability that the system won't work? that it will work?   0.0361, 0.9639
d. How would the probability of the system working change if a 5–6 subsystem were added in parallel with the other two subsystems?   0.9931, it increases.
e. How would the probability that the system works change if there were three components in series in each of the two subsystems?   0.9266, it decreases.

6.52 Information from a poll of registered voters in Cedar Rapids, Iowa, to assess voter support for a new school tax was the basis for the following statements (*Cedar Rapids Gazette*, August 28, 1999):

> The poll showed 51 percent of the respondents in the Cedar Rapids school district are in favor of the tax. The approval rating rises to 56 percent for those with children in public schools. It falls to 45 percent for

those with no children in public schools. The older the respondent, the less favorable the view of the proposed tax: 36 percent of those over age 56 said they would vote for the tax compared with 72 percent of 18- to 25-year-olds.

Suppose that a registered voter from Cedar Rapids is selected at random, and define the following events: F = event that the selected individual favors the school tax, C = event that the selected individual has children in the public schools, O = event that the selected individual is over 56 years old, and Y = event that the selected individual is 18–25 years old.
a. Use the given information to estimate the values of the following probabilities:
  i. $P(F)$ 0.51
  ii. $P(F|C)$ 0.56
  iii. $P(F|C^C)$ 0.45
  iv. $P(F|O)$ 0.36
  v. $P(F|Y)$ 0.72
b. Are F and C independent? Justify your answer.
c. Are F and O independent? Justify your answer.

6.53 The following case study was reported in the article "Parking Tickets and Missing Women," which appeared in an early edition of the book *Statistics: A Guide to the Unknown*. In a Swedish trial on a charge of overtime parking, a police officer testified that he had noted the position of the two air valves on the tires of a parked car: To the closest hour, one was at the one o'clock position and the other was at the six o'clock position. After the allowable time for parking in that zone had passed, the policeman returned, noted that the valves were in the same position, and ticketed the car. The owner of the car claimed that he had left the parking place in time and had returned later. The valves just happened by chance to be in the same positions. An "expert" witness computed the probability of this occurring as $(1/12)(1/12) = 1/144$.
a. What reasoning did the expert use to arrive at the probability of 1/144?
b. Can you spot the error in the reasoning that leads to the stated probability of 1/144? What effect does this error have on the probability of occurrence? Do you think that 1/144 is larger or smaller than the correct probability of occurrence?

6.54 Three friends (A, B, and C) will participate in a round-robin tournament in which each one plays both of the others. Suppose that $P(A$ beats $B) = .7$, $P(A$ beats $C) = .8$,

$P(B$ beats $C) = .6$, and that the outcomes of the three matches are independent of one another.
a. What is the probability that A wins both her matches and that B beats C?   0.336
b. What is the probability that A wins both her matches?
c. What is the probability that A loses both her matches?
d. What is the probability that each person wins one match? (Hint: There are two different ways for this to happen.)   0.18

6.55 A shipment of 5000 printed circuit boards contains 40 that are defective. Two boards will be chosen at random, without replacement. Consider the two events E_1 = event that the first board selected is defective and E_2 = event that the second board selected is defective.
a. Are E_1 and E_2 dependent events? Explain in words.
b. Let *not* E_1 be the event that the first board selected is not defective (the event E_1^C). What is $P(not\ E_1)$?   0.992
c. How do the two probabilities $P(E_2|E_1)$ and $P(E_2|not\ E_1)$ compare?
d. Based on your answer to Part (c), would it be reasonable to view E_1 and E_2 as approximately independent?   Yes

6.56 A store sells two different brands of dishwasher soap, and each brand comes in three different sizes: small (S), medium (M), and large (L). The proportions of the two brands and of the three sizes purchased are displayed as marginal totals in the following table.

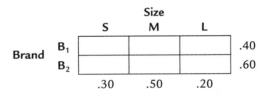

Suppose that any event involving brand is independent of any event involving size. What is the probability of the event that a randomly selected purchaser buys the small size of Brand B_1 (the event $B_1 \cap S$)? What are the probabilities of the other brand–size combinations?

6.57 The National Public Radio show *Car Talk* has a feature called "The Puzzler." Listeners are asked to send in answers to some puzzling questions—usually about cars but sometimes about probability (which, of course, must account for the incredible popularity of the program!). Suppose that for a car question, 800 answers are submitted, of which 50 are correct. Suppose also that the hosts randomly select two answers from those submitted *with replacement*.

a. Calculate the probability that both selected answers are correct. (For purposes of this problem, keep at least five digits to the right of the decimal.)

b. Suppose now that the hosts select the answers at random but *without replacement*. Use conditional probability to evaluate the probability that both answers selected are correct. How does this probability compare to the one computed in Part (a)?

6.58 Refer to Exercise 6.57. Suppose now that for a probability question, 100 answers are submitted, of which 50 are correct. Calculate the probabilities in Parts (a) and (b) of Exercise 6.57 for a probability question.   0.25, 0.24747

Bold exercises answered in back   ● Data set available online but not required   ▼ Video solution available

---

6.6   Some General Probability Rules

In previous sections, we saw how the probability of $P(E \cup F)$ could be easily computed when E and F are disjoint (mutually exclusive) and how $P(E \cap F)$ could be computed when E and F are independent. In this section, we develop more general rules: an addition rule that can be used when events are not necessarily disjoint and a multiplication rule that can be used when events are not necessarily independent.

■ General Addition Rule

The computation of $P(E \cup F)$ when the two events are not disjoint is a bit more complicated than in the case of disjoint events. Consider Figure 6.9, in which E and F overlap. The area of the colored region $(E \cup F)$ is not the sum of the area of E and the area of F, because when the two individual areas are added, the area of the intersection $(E \cap F)$ is counted twice. Similarly, $P(E) + P(F)$ includes $P(E \cap F)$ twice, so this intersection probability must then be subtracted to obtain $P(E \cup F)$. This reasoning leads to the general addition rule.

Figure 6.9 The shaded region is $P(E \cup F)$, and $P(E \cup F) \neq P(E) + P(F)$.

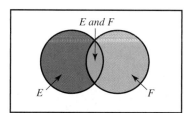

E and F

E   F

General Addition Rule for Two Events

For any two events E and F,

$$P(E \cup F) = P(E) + P(F) - P(E \cap F)$$

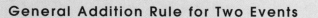

When E and F are disjoint, the general addition rule simplifies to the previous rule for disjoint events. This is because when $E \cap F$ contains no outcomes, $P(E \cap F) = 0$. The general addition rule can be used to determine any one of the four probabilities $P(E)$, $P(F)$, $P(E \cap F)$, or $P(E \cup F)$ provided that the other three probabilities are known.

Example 6.22 Insurance Policies

Suppose that 60% of all customers of a large insurance agency have automobile policies with the agency, 40% have homeowner's policies, and 25% have both types of policies. If a customer is selected at random, what is the probability that he or she has at least one of these two types of policies with the agency? Let

$$E = \text{event that a selected customer has auto insurance}$$
$$F = \text{event that a selected customer has homeowner's insurance}$$

The given information implies that

$$P(E) = .60 \qquad P(F) = .40 \qquad P(E \cap F) = .25$$

from which we obtain

$$P(\text{customer has at least one of the two types of policy})$$
$$= P(E \cup F)$$
$$= P(E) + P(F) - P(E \cap F)$$
$$= .60 + .40 - .25$$
$$= .75$$

The event that the customer has neither type of policy is $(E \cup F)^C$, so

$$P(\text{customer has neither types of policy}) = 1 - P(E \cup F) = .25$$

Now let's determine the probability that the selected customer has exactly one type of policy. Referring to the Venn diagram in Figure 6.10, we see that the event *at least one* can be thought of as consisting of two disjoint parts: *exactly one* and *both*. Thus

$$P(E \cup F) = P(\text{at least one})$$
$$= P(\text{exactly one } \cup \text{ both})$$
$$= P(\text{exactly one}) + P(\text{both})$$
$$= P(\text{exactly one}) + P(E \cap F)$$

Then

$$P(\text{exactly one}) = P(E \cup F) - P(E \cap F)$$
$$= .75 - .25$$
$$= .50$$

Figure 6.10 Representing $P(E \cup F)$ as the union of two disjoint events.

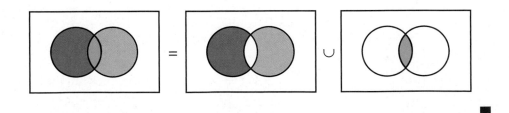

The general addition rule for more than two nondisjoint events is rather complicated. For example, in the case of three events,

$$P(E \cup F \cup G) = P(E) + P(F) + P(G) - P(E \cap F) - P(E \cap G)$$
$$- P(F \cap G) + P(E \cap F \cap G)$$

A more advanced treatment of probability can be consulted for examples and extensions of these methods.

■ General Multiplication Rule

Recall the definition of conditional probability: When $P(F) > 0$,

$$P(E|F) = \frac{P(E \cap F)}{P(F)}$$

Up to this point, we have used this formula to compute conditional probabilities when $P(E \cap F)$ is known. Sometimes, however, conditional probabilities are known or can be estimated. When this is the case, they can be used to calculate the probability of the intersection of two events. Multiplying both sides of the conditional probability formula by $P(F)$ gives a useful expression for the probability that both events will occur.

General Multiplication Rule for Two Events

For any two events E and F,

$$P(E \cap F) = P(E|F)P(F)$$

Example 6.23 Traffic School

Suppose that 20% of all teenage drivers in a certain county received a citation for a moving violation within the past year. Assume in addition that 80% of those receiving such a citation attended traffic school so that the citation would not appear on their permanent driving record. If a teenage driver from this county is randomly selected, what is the probability that he or she received a citation and attended traffic school?

Let's define two events E and F as follows:

$$E = \text{selected driver attended traffic school}$$
$$F = \text{selected driver received such a citation}$$

The question posed can then be answered by calculating $P(E \cap F)$. The percentages given in the problem imply that $P(F) = .20$ and $P(E|F) = .80$. Notice the difference between $P(E)$, which is the proportion in the entire population who attended traffic school (not given), and $P(E|F)$ which is the proportion of those receiving a citation that attended traffic school. Using the multiplication rule, we calculate

$$\begin{aligned} P(E \text{ and } F) &= P(E|F)P(F) \\ &= (.80)(.20) \\ &= .16 \end{aligned}$$

Thus 16% of all teenage drivers in this county received a citation and attended traffic school.

Example 6.24 DVD Player Warranties

The following table gives information on DVD players sold by a certain electronics store:

	Percentage of Customers Purchasing	Of Those Who Purchase, Percentage Who Purchase Extended Warranty
Brand 1	70	20
Brand 2	30	40

A purchaser is randomly selected from among all those who bought a DVD player from the store. What is the probability that the selected customer purchased a Brand 1 model and an extended warranty?

To answer this question, we first define the following events:

$$B_1 = \text{event that Brand 1 is purchased}$$
$$B_2 = \text{event that Brand 2 is purchased}$$
$$E = \text{event that an extended warranty is purchased}$$

The tabulated information implies that

$$P(\text{Brand 1 purchased}) = P(B_1) = .70$$
$$P(\text{extended warranty}|\text{Brand 1 purchased}) = P(E|B_1) = .20$$

Notice that the 20% is identified with a *conditional* probability; *among purchasers of Brand* 1, this is the percentage opting for an extended warranty. Substituting these numbers into the general multiplication rule yields

$$P(B_1 \text{ and } E) = P(E|B_1)P(B_1)$$
$$= (.20)(.70)$$
$$= .14$$

The tree diagram of Figure 6.11 gives a nice visual display of how the general multiplication rule is used here. The two first-generation branches are labeled with events B_1 and B_2 along with their probabilities. Two second-generation branches extend from each first-generation branch; these correspond to the two events E and *not* E. The *conditional* probabilities $P(E|B_1)$, $P(\text{not } E|B_1)$, $P(E|B_2)$, and $P(\text{not } E|B_2)$, appear on these branches. Application of the multiplication rule then consists of multiplying probabilities across the branches of the tree diagram. For example,

$$P(\text{Brand 2 and warranty purchased}) = P(B_2 \text{ and } E)$$
$$= P(B_2 \cap E)$$
$$= P(E|B_2)P(B_2)$$
$$= (.4)(.3)$$
$$= .12$$

and this probability is displayed to the right of the E branch that comes from the B_2 branch.

Figure 6.11 A tree diagram for the probability calculations of Example 6.24.

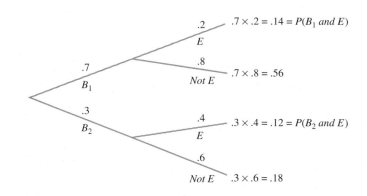

We can now easily calculate $P(E)$, the probability that an extended warranty is purchased. The event E can occur in two different ways: Buy Brand 1 *and* warranty, or buy Brand 2 *and* warranty. Symbolically, these events are $B_1 \cap E$ and $B_2 \cap E$. Furthermore, if each customer purchased a single DVD player, he or she could not have simultaneously purchased both Brand 1 and Brand 2, so the two events $B_1 \cap E$ and $B_2 \cap E$ are disjoint.

Thus

$$
\begin{aligned}
P(E) &= P(B_1 \cap E) + (B_2 \cap E) \\
&= P(E|B_1)P(B_1) + P(E|B_2)P(B_2) \\
&= (.2)(.7) + (.4)(.3) \\
&= .14 + .12 \\
&= .26
\end{aligned}
$$

This probability is the sum of two of the probabilities shown on the right-hand side of the tree diagram. Thus 26% of all DVD player purchasers selected an extended warranty.

■

The general multiplication rule can be extended to give an expression for the probability that several events occur together. In the case of three events E, F, and G, we have

$$P(E \cap F \cap G) = P(E|F \cap G)P(F|G)P(G)$$

When the events are all independent, $P(E|F \cap G) = P(E)$ and $P(F|G) = P(F)$ so the right-hand side of the equation for $P(E \cap F \cap G) = P(E)$ is simply the product of the three unconditional probabilities.

Example 6.25 Lost Luggage

Twenty percent of all passengers who fly from Los Angeles (LA) to New York (NY) do so on Airline G. This airline misplaces luggage for 10% of its passengers, and 90% of this lost luggage is subsequently recovered. If a passenger who has flown from LA to NY is randomly selected, what is the probability that the selected individual flew on Airline G (event G), had luggage misplaced (event F), and subsequently recovered the misplaced luggage (event E)? The given information implies that

$$P(G) = .20 \qquad P(F|G) = .10 \qquad P(E|F \cap G) = .90$$

Then

$$P(E \cap F \cap G) = P(E|F \cap G)P(F|G)P(G) = (.90)(.10)(.20) = .018$$

That is, about 1.8% of passengers flying from LA to NY fly on Airline G, have their luggage misplaced, and subsequently recover the lost luggage.

■

▪ Law of Total Probability

Let's reconsider the information on DVD player sales from Example 6.24. In this example, the following events were defined:

B_1 = event that Brand 1 is purchased
B_2 = event that Brand 2 is purchased
E = event that an extended warranty is purchased

Based on the information given in Example 6.24, the following probabilities are known:

$$P(B_1) = .7 \qquad P(B_2) = .3 \qquad P(E|B_1) = .2 \qquad P(E|B_2) = .4$$

Notice that the conditional probabilities $P(E|B_1)$ and $P(E|B_2)$ are known but that the unconditional probability $P(E)$ is not known.

To find $P(E)$ we note that the event E can occur in two ways: (1) A customer purchases an extended warranty *and* buys Brand 1 ($E \cap B_1$), or (2) a customer purchases an extended warranty *and* buys Brand 2 ($E \cap B_2$). Because these are the only ways in which E can occur, we can write the event E as

$$E = (E \cap B_1) \cup (E \cap B_2)$$

Again, the two events ($E \cap B_1$) and ($E \cap B_2$) are disjoint (since B_1 and B_2 are disjoint), so using the addition rule for disjoint events gives

$$\begin{aligned} P(E) &= P((E \cap B_1) \cup (E \cap B_2)) \\ &= P(E \cap B_1) + P(E \cap B_2) \end{aligned}$$

Finally, using the general multiplication rule to evaluate $P(E \cap B_1)$ and $P(E \cap B_2)$ results in

$$\begin{aligned} P(E) &= P(E \cap B_1) + P(E \cap B_2) \\ &= P(E|B_1)P(B_1) + P(E|B_2)P(B_2) \end{aligned}$$

Substituting in the known probabilities gives

$$\begin{aligned} P(E) &= P(E|B_1)P(B_1) + P(E|B_2)P(B_2) \\ &= (.2)(.7) + (.4)(.3) \\ &= .26 \end{aligned}$$

We would then conclude that 26% of the DVD player customers purchased an extended warranty.

As we have just illustrated, when conditional probabilities are known, they can sometimes be used to compute unconditional probabilities. The **law of total probability** formalizes this use of conditional probabilities.

The Law of Total Probability

If B_1 and B_2 are disjoint events with $P(B_1) + P(B_2) = 1$, then for any event E

$$P(E) = P(E \cap B_1) + P(E \cap B_2)$$
$$= P(E|B_1)P(B_1) + P(E|B_2)P(B_2)$$

More generally, if $B_1, B_2, \ldots, B_k$ are disjoint events with $P(B_1) + P(B_2) + \cdots + P(B_k) = 1$ then for any event E

$$P(E) = P(E \cap B_1) + P(E \cap B_2) + \cdots + P(E \cap B_k)$$
$$= P(E|B_1)P(B_1) + P(E|B_2)P(B_2) + \cdots + P(E|B_k)P(B_k)$$

Example 6.26 Bicycle Helmets

© Royalty-Free/Getty Images

The article "Association Between Bicycle Helmet Legislation, Bicycle Safety Education, and Use of Bicycle Helmets in Children" (*Archives of Pediatric and Adolescent Medicine* [1994]: 255–259) gave information on bicycle helmet usage in some Cleveland suburbs. In Beachwood, Ohio, a safety education program and a helmet law were in place, whereas Moreland Hills had neither a helmet law nor a safety education program. The article reported that 68% of elementary school students from Beachwood said that they always wear a helmet when bicycling, but only 21% of the elementary school students from Moreland Hills reported that they always wear a helmet.

For the purposes of this example, suppose that these two communities have the same number of elementary school students. A student from one of these two communities is selected at random. Let's define

B = selected student is from Beachwood
M = selected student is from Moreland Hills
H = selected student reports that he or she always wears a helmet when bicycling

From the information given, we can reason that $P(B) = P(M) = .5$ (because there are the same number of elementary school students in each community) and that

$$P(H|B) = .68$$
$$P(H|M) = .21$$

What proportion of elementary school students in these two communities always wear helmets? We can use the law of total probability to answer this question. Substituting B and M for B_1 and B_2, respectively, in the formula for the law of total probability, we get

$$P(H) = P(H|B)P(B) + P(H|M)P(M)$$
$$= (.68)(.5) + (.21)(.5)$$
$$= .34 + .105$$
$$= .445$$

That is, about 44.5% of the elementary school children in these two communities always wear a helmet when cycling.

■ Bayes' Rule

We conclude our discussion of probability rules by considering a formula discovered by the Reverend Thomas Bayes (1702–1761), an English Presbyterian minister. He discovered what is now known as Bayes' rule (or Bayes' theorem). Bayes' rule is a solution to what Bayes called the converse problem. To see what he meant by this, we return to the field of medical diagnosis.

Example 6.27 Lyme Disease

Lyme disease is the leading tick-borne disease in the United States and Europe. Diagnosis of the disease is difficult and is aided by a test that detects particular antibodies in the blood. The article "Laboratory Considerations in the Diagnosis and Management of Lyme Borreliosis" (*American Journal of Clinical Pathology* [1993]: 168–174) used the following notation:

$+$ represents a positive result on the blood test
$-$ represents a negative result on the blood test
L represents the event that the patient actually has Lyme disease
L^C represents the event that the patient actually does not have Lyme disease

The following probabilities were reported in the article:

$P(L) = .00207$ The prevalence of Lyme disease in the population; .207% of the population actually has Lyme disease.

$P(L^C) = .99793$ 99.793% of the population does not have Lyme disease.

$P(+|L) = .937$ 93.7% of those with Lyme disease test positive.
$P(-|L) = .063$ 6.3% of those with Lyme disease test negative.
$P(+|L^C) = .03$ 3% of those who do not have Lyme disease test positive.

$P(-|L^C) = .97$ 97% of those who do not have Lyme disease test negative.

Important concept

Notice the form of the known conditional probabilities; for example, $P(+|L)$ is the probability of a positive test given that a person selected at random from the population actually has Lyme disease. Bayes' converse problem poses a question of a different form: Given that a person tests positive for the disease, what is the probability that he or she actually has Lyme disease? This converse problem is the one that is of primary interest in medical diagnosis problems.

Bayes reasoned as follows to obtain the answer to the converse problem of finding $P(L|+)$. We know from the definition of conditional probability that

$$P(L|+) = \frac{P(L \cap +)}{P(+)}$$

Because $P(L \cap +) = P(+ \cap L)$, we can use the general multiplication rule to get

$$P(L \cap +) = P(+ \cap L) = P(+|L)P(L)$$

This helps, because both $P(+|L)$ and $P(L)$ are known. We now have

$$P(L|+) = \frac{P(+|L)P(L)}{P(+)}$$

The denominator $P(+)$ can be evaluated using the law of total probability, because L and L^C are disjoint with $P(L) + P(L^C) = 1$. Applying the law of total probability to the denominator, we obtain

$$P(+) = P(+ \cap L) + P(+ \cap L^C)$$
$$= P(+|L)P(L) + P(+|L^C)P(L^C)$$

We now have all we need to answer the converse problem:

$$P(L|+) = \frac{P(+|L)P(L)}{P(+|L)P(L) + P(+|L^C)P(L^C)}$$

$$= \frac{(.937)(.00207)}{(.937)(.00207) + (.03)(.99793)} = \frac{.0019}{.0319} = .0596$$

The probability $P(L|+)$ is a conditional probability and can be interpreted as such. $P(L|+) = .0596$ means that, in the long run, only 5.96% of those who test positive actually have the disease. Note the difference between $P(L|+)$ and the previously reported conditional probability $P(+|L) = .937$, which means that 93.7% of those with Lyme disease test positive.

■

The accompanying box formalizes this reasoning in the statement of Bayes' rule.

Bayes' Rule

If B_1 and B_2 are disjoint events with $P(B_1) + P(B_2) = 1$, then for any event E

$$P(B_1|E) = \frac{P(E|B_1)P(B_1)}{P(E|B_1)P(B_1) + P(E|B_2)P(B_2)}$$

More generally, if $B_1, B_2, \ldots, B_k$ are disjoint events with $P(B_1) + P(B_2) + \cdots + P(B_k) = 1$ then for any event E,

$$P(B_i|E) = \frac{P(E|B_i)P(B_i)}{P(E|B_1)P(B_1) + P(E|B_2)P(B_2) + \cdots + P(E|B_k)P(B_k)}$$

■

Example 6.28 Late Packages

Two shipping services offer overnight delivery of parcels, and both promise delivery before 10 A.M. A mail-order catalog company ships 30% of its overnight packages using Service 1 and 70% using Service 2. Service 1 fails to meet the 10 A.M. delivery promise 10% of the time, whereas Service 2 fails to deliver by 10 A.M. 8% of the time. Suppose that you made a purchase from this company and were expecting your package by 10 A.M., but it is late. Which shipping service is more likely to have been used?

Let's define the following events:

S_1 = event that package was shipped using Service 1
S_2 = event that package was shipped using Service 2
L = event that the package is late

The following probabilities are known:

$$P(S_1) = .3 \qquad P(S_2) = .7 \qquad P(L|S_1) = .1 \qquad P(L|S_2) = .08$$

Because you know that your package is late, you can use Bayes' rule to evaluate $P(S_1|L)$ and $P(S_2|L)$ as follows:

$$P(S_1|L) = \frac{P(L|S_1)P(S_1)}{P(L|S_1)P(S_1) + P(L|S_2)P(S_2)}$$
$$= \frac{(.1)(.3)}{(.1)(.3) + (.08)(.7)} = \frac{.03}{.086} = .3488$$

$$P(S_2|L) = \frac{P(L|S_2)P(S_2)}{P(L|S_1)P(S_1) + P(L|S_2)P(S_2)}$$
$$= \frac{(.08)(.7)}{(.1)(.3) + (.08)(.7)} = \frac{.056}{.086} = .6512$$

So you should call Service 2. Even though Service 2 has a lower percentage of late packages, it is more likely that a late package was shipped using Service 2. This is because many more packages (70%) are sent using Service 2.

■

Exercises 6.59–6.74

Turn to the Expanded Answers Section for answers not shown next to exercises.

6.59 ▼ A certain university has 10 vehicles available for use by faculty and staff. Six of these are vans and four are cars. On a particular day, only two requests for vehicles have been made. Suppose that the two vehicles to be assigned are chosen in a completely random fashion from among the 10.
a. Let E denote the event that the first vehicle assigned is a van. What is $P(E)$?   6/10 = 0.6
b. Let F denote the probability that the second vehicle assigned is a van. What is $P(F|E)$?   5/9 = 0.556
c. Use the results of Parts (a) and (b) to calculate $P(E \text{ and } F)$ (Hint: Use the definition of $P(F|E)$.)
1/3 = 0.333
6.60 A construction firm bids on two different contracts. Let E_1 be the event that the bid on the first contract is successful, and define E_2 analogously for the second contract. Suppose that $P(E_1) = .4$ and $P(E_2) = .3$ and that E_1 and E_2 are independent events.

a. Calculate the probability that both bids are successful (the probability of the event $E_1 \text{ and } E_2$).   0.12
b. Calculate the probability that neither bid is successful (the probability of the event $(not\ E_1) \text{ and } (not\ E_2)$).   0.42
c. What is the probability that the firm is successful in at least one of the two bids?   0.58

6.61 There are two traffic lights on the route used by a certain individual to go from home to work. Let E denote the event that the individual must stop at the first light, and define the event F in a similar manner for the second light. Suppose that $P(E) = .4$, $P(F) = .3$ and $P(E \cap F) = .15$.
a. What is the probability that the individual must stop at at least one light; that is, what is the probability of the event $E \cup F$?   0.55
b. What is the probability that the individual needn't stop at either light?   0.45

Bold exercises answered in back   ● Data set available online but not required   ▼ Video solution available

c. What is the probability that the individual must stop at exactly one of the two lights?   0.40

d. What is the probability that the individual must stop just at the first light? (Hint: How is the probability of this event related to $P(E)$ and $P(E \cap F)$? A Venn diagram might help.)   0.25

6.62 Let F denote the event that a randomly selected registered voter in a certain city has signed a petition to recall the mayor. Also, let E denote the event that a randomly selected registered voter actually votes in the recall election. Describe the event $E \cap F$ in words. If $P(F) = .10$ and $P(E|F) = .80$, determine $P(E \cap F)$.

6.63 Suppose that we define the following events: $C =$ event that a randomly selected driver is observed to be using a cell phone, $A =$ event that a randomly selected driver is observed driving a passenger automobile, $V =$ event that a randomly selected driver is observed driving a van or SUV, and $T =$ event that a randomly selected driver is observed driving a pickup truck. Based on the article "Three Percent of Drivers on Hand-Held Cell Phones at Any Given Time" (*San Luis Obispo Tribune*, July 24, 2001), the following probability estimates are reasonable: $P(C) = .03$, $P(C|A) = .026$, $P(C|V) = .048$, and $P(C|T) = .019$. Explain why $P(C)$ is not just the average of the three given conditional probabilities.

6.64 The article "Checks Halt over 200,000 Gun Sales" (*San Luis Obispo Tribune*, June 5, 2000) reported that required background checks blocked 204,000 gun sales in 1999. The article also indicated that state and local police reject a higher percentage of would-be gun buyers than does the FBI, stating, "The FBI performed 4.5 million of the 8.6 million checks, compared with 4.1 million by state and local agencies. The rejection rate among state and local agencies was 3 percent, compared with 1.8 percent for the FBI."

a. Use the given information to estimate $P(F)$, $P(S)$, $P(R|F)$, and $P(R|S)$, where $F =$ event that a randomly selected gun purchase background check is performed by the FBI, $S =$ event that a randomly selected gun purchase background check is performed by a state or local agency, and $R =$ event that a randomly selected gun purchase background check results in a blocked sale.

b. Use the probabilities from Part (a) to evaluate $P(S|R)$, and write a sentence interpreting this value in the context of this problem.

6.65 In an article that appears on the web site of the American Statistical Association (www.amstat.org), Carlton Gunn, a public defender in Seattle, Washington, wrote about how he uses statistics in his work as an attorney. He states:

> I personally have used statistics in trying to challenge the reliability of drug testing results. Suppose the chance of a mistake in the taking and processing of a urine sample for a drug test is just 1 in 100. And your client has a "dirty" (i.e., positive) test result. Only a 1 in 100 chance that it could be wrong? Not necessarily. If the vast majority of all tests given—say 99 in 100 —are truly clean, then you get one false dirty and one true dirty in every 100 tests, so that half of the dirty tests are false.

Define the following events as $TD =$ event that the test result is dirty, $TC =$ event that the test result is clean, $D =$ event that the person tested is actually dirty, and $C =$ event that the person tested is actually clean.

a. Using the information in the quote, what are the values of

  i. $P(TD|D)$ 0.99   **iii.** $P(C)$ 0.99

  ii. $P(TD|C)$ 0.01   **iv.** $P(D)$ 0.01

b. Use the law of total probability to find $P(TD)$.   0.02

c. Use Bayes' rule to evaluate $P(C|TD)$. Is this value consistent with the argument given in the quote? Explain.

0.5, it is consistent with the quote.

6.66 Consider the following information about travelers on vacation: 40% check work email, 30% use a cell phone to stay connected to work, 25% bring a laptop with them on vacation, 23% both check work email and use a cell phone to stay connected, and 51% neither check work email nor use a cell phone to stay connected nor bring a laptop. In addition 88% of those who bring a laptop also check work email and 70% of those who use a cell phone to stay connected also bring a laptop. With $E =$ event that a traveler on vacation checks work email, $C =$ event that a traveler on vacation uses a cell phone to stay connected, and $L =$ event that a traveler on vacation brought a laptop, use the given information to determine the following probabilities. A Venn diagram may help.

a. $P(E)$ 0.4

b. $P(C)$ 0.3

c. $P(L)$ 0.25

d. $P(E\ and\ C)$ 0.23

e. $P(E^C\ and\ C^C\ and\ L^C)$ 0.51

f. $P(E\ or\ C\ or\ L)$ 0.49

Bold exercises answered in back   ● Data set available online but not required   ▼ Video solution available

g. $P(E|L)$ 0.88   j. $P(E \text{ and } L)$ 0.22
h. $P(L|C)$ 0.70   k. $P(C \text{ and } L)$ 0.21
i. $P(E \text{ and } C \text{ and } L)$ 0.20   l. $P(C|E \text{ and } L)$ 0.91

6.67 The article "Birth Beats Long Odds for Leap Year Mom, Baby" (*San Luis Obispo Tribune*, March 2, 1996) reported that a leap year baby (someone born on February 29) became a leap year mom when she gave birth to a baby on February 29, 1996. The article stated that a hospital spokesperson said that the probability of a leap year baby giving birth on her birthday was one in 2.1 million (approximately .00000047).
a. In computing the given probability, the hospital spokesperson used the fact that a leap day occurs only once in 1461 days. Write a few sentences explaining how the hospital spokesperson computed the stated probability.
b. To compute the stated probability, the hospital spokesperson had to assume that the birth was equally likely to occur on any of the 1461 days in a four-year period. Do you think that this is a reasonable assumption? Explain.
c. Based on your answer to Part (b), do you think that the probability given by the hospital spokesperson is too small, about right, or too large? Explain.

6.68 Suppose that a new Internet company Mumble.com requires all employees to take a drug test. Mumble.com can afford only the inexpensive drug test—the one with a 5% false-positive rate and a 10% false-negative rate. (That means that 5% of those who are not using drugs will incorrectly test positive and that 10% of those who are actually using drugs will test negative.) Suppose that 10% of those who work for Mumble.com are using the drugs for which Mumble is checking. (Hint: It may be helpful to draw a tree diagram to answer the questions that follow.)
a. If one employee is chosen at random, what is the probability that the employee both uses drugs and tests positive? 0.09
b. If one employee is chosen at random, what is the probability that the employee does not use drugs but tests positive anyway? 0.045
c. If one employee is chosen at random, what is the probability that the employee tests positive? 0.135
d. If we know that a randomly chosen employee has tested positive, what is the probability that he or she uses drugs? 0.67

6.69 Refer to Exercise 6.68. Suppose that because of the high rate of false-positives for the drug test, Mumble.com

has instituted a mandatory second test for those who test positive on the first test.
a. If one employee is selected at random, what is the probability that the selected employee uses drugs and tests positive twice? 0.081
b. If one employee is selected at random, what is the probability that the employee tests positive twice?
c. If we know that the randomly chosen employee has tested positive twice, what is the probability that he or she uses drugs? 0.97297
d. What is the chance that an individual who does use drugs doesn't test positive twice (either this employee tests negative on the first round and doesn't need a retest, or this employee tests positive the first time and that result is followed by a negative result on the retest)? 0.19
e. Discuss the benefits and drawbacks of using a retest scheme such as the one proposed in Part (d).

6.70 According to a study released by the federal Substance Abuse and Mental Health Services Administration (*Knight Ridder Tribune*, September 9, 1999), approximately 8% of all adult full-time workers are drug users and approximately 70% of adult drug users are employed full-time.
a. Is it possible for both of the reported percentages to be correct? Explain.
b. Define the events D and E as D = event that a randomly selected adult is a drug user and E = event that a randomly selected adult is employed full-time. What are the estimated values of $P(D|E)$ and $P(E|D)$? 0.08, 0.70
c. Is it possible to determine $P(D)$, the probability that a randomly selected adult is a drug user, from the information given? If not, what additional information would be needed?

6.71 A company that manufactures video cameras produces a basic model and a deluxe model. Over the past year, 40% of the cameras sold have been the basic model. Of those buying the basic model, 30% purchase an extended warranty, whereas 50% of all purchasers of the deluxe model buy an extended warranty. If you learn that a randomly selected purchaser bought an extended warranty, what is the probability that he or she has a basic model? 0.286

6.72 At a large university, the Statistics Department has tried a different text during each of the last three quarters. During the fall quarter, 500 students used a book by Professor Mean; during the winter quarter, 300 stu-

dents used a book by Professor Median; and during the spring quarter, 200 students used a book by Professor Mode. A survey at the end of each quarter showed that 200 students were satisfied with the text in the fall quarter, 150 in the winter quarter, and 160 in the spring quarter.
a. If a student who took statistics during one of these three quarters is selected at random, what is the probability that the student was satisfied with the textbook?   0.51
b. If a randomly selected student reports being satisfied with the book, is the student most likely to have used the book by Mean, Median, or Mode? Who is the least likely author? (Hint: Use Bayes' rule to compute three probabilities.)   Most likely is Mean, least likely is Median.

6.73 ▼ A friend who works in a big city owns two cars, one small and one large. Three-quarters of the time he drives the small car to work, and one-quarter of the time he takes the large car. If he takes the small car, he usually has little trouble parking and so is at work on time with probability .9. If he takes the large car, he is on time to work with probability .6. Given that he was at work on

time on a particular morning, what is the probability that he drove the small car?   0.818

6.74 Only 0.1% of the individuals in a certain population have a particular disease (an incidence rate of .001). Of those who have the disease, 95% test positive when a certain diagnostic test is applied. Of those who do not have the disease, 90% test negative when the test is applied. Suppose that an individual from this population is randomly selected and given the test.
a. Construct a tree diagram having two first-generation branches, for *has disease* and *doesn't have disease*, and two second-generation branches leading out from each of these, for *positive test* and *negative test*. Then enter appropriate probabilities on the four branches.
b. Use the general multiplication rule to calculate P(has disease *and* positive test).   0.00095
c. Calculate P(positive test).   0.10085
d. Calculate P(has disease|positive test). Does the result surprise you? Give an intuitive explanation for the size of this probability.

Bold exercises answered in back   ● Data set available online but not required   ▼ Video solution available

6.7 Estimating Probabilities Empirically Using Simulation

In the examples presented so far, reaching conclusions required knowledge of the probabilities of various outcomes. In some cases, this is reasonable, and we know the true long-run proportion of the time that each outcome will occur. In other situations, these probabilities are not known and must be determined. Sometimes probabilities can be determined analytically, by using mathematical rules and probability properties, including the basic ones introduced in this chapter. When an analytical approach is impossible, impractical, or just beyond the limited probability tools of the introductory course, we can *estimate* probabilities empirically through observation or by simulation.

■ Estimating Probabilities Empirically

It is fairly common practice to use observed long-run proportions to estimate probabilities. The process of estimating probabilities is simple:

1. Observe a very large number of chance outcomes under controlled circumstances.
2. Estimate the probability of an event by using the observed proportion of occurrence and by appealing to the interpretation of probability as a long-run relative frequency and to the law of large numbers.

This process is illustrated in Examples 6.29 and 6.30.

Example 6.29 Fair Hiring Practices

The Biology Department at a university plans to recruit a new faculty member and intends to advertise for someone with a Ph.D. in biology and at least 10 years of college-level teaching experience. A member of the department expresses the belief that the experience requirement will exclude many potential applicants and will exclude far more female applicants than male applicants. The Biology Department would like to determine the probability that an applicant with a Ph.D. in biology would be eliminated from consideration because of the experience requirement.

A similar university just completed a search in which there was no requirement for prior teaching experience, but the information about prior teaching experience was recorded. The 410 applications yielded the following data:

| | Number of Applicants | | |
	Less Than 10 Years of Experience	10 Years of Experience or More	Total
Male	178	112	290
Female	99	21	120
Total	277	133	410

Let's assume that the populations of applicants for the two positions can be regarded as the same. We can use the available information to approximate the probability that an applicant will fall into each of the four gender–experience combinations.

Table 6.1 Estimated Probabilities for Example 6.29

	Less Than 10 Years of Experience	10 Years of Experience or More
Male	.4341	.2732
Female	.2415	.0512

The estimated probabilities (obtained by dividing the number of applicants for each gender–experience combination by 410) are given in Table 6.1. From Table 6.1, the estimate of P(candidate excluded because of the experience requirement) $= .4341 + .2415 = .6756$.

We can also assess the impact of the experience requirement separately for male and for female applicants. From the given information, the proportion of male applicants who have less than 10 years of experience is $178/290 = .6138$, whereas the corresponding proportion for females is $99/120 = .8250$. Therefore, approximately 61% of the male applicants would be eliminated by the experience requirement, and about 83% of the female applicants would be eliminated.

These subgroup proportions—.6138 for males and .8250 for females—are examples of *conditional probabilities*, which show how the original probability changes in light of new information. In this example, the probability that a potential candidate has less than 10 years of experience is .6756, but this probability changes to .8250 if we know that a candidate is female. These probabilities can be expressed as

P(less than 10 years of experience) $= .6756$ (an unconditional probability)

and

P(less than 10 years of experience|female) $= .8250$ (a conditional probability)

Example 6.30 Who Has the Upper Hand?

Men and women frequently express intimacy through the simple act of holding hands. Some researchers have suggested that hand-holding is not only an expression of intimacy but also communicates status differences. For two people to hold hands, one must assume an overhand grip and one an underhand grip. Research in this area has shown that it is predominantly the male who assumes the overhand grip. In the view of some investigators, the overhand grip is seen to imply status or superiority. The authors of the paper "Men and Women Holding Hands: Whose Hand Is Uppermost?" (*Perceptual and Motor Skills* [1999]: 537–549) investigated an alternative explanation—perhaps the positioning of hands is a function of the heights of the individuals? Because men, on average, tend to be taller than women, maybe comfort, not status, dictates the positioning. Investigators at two separate universities observed hand-holding male–female pairs and recorded the following data:

Number of Hand-Holding Couples

	Sex of Person with Uppermost Hand		
	Male	**Female**	**Total**
Man Taller	2149	299	**2,448**
Equal Height	780	246	**1,026**
Woman Taller	241	205	**446**
Total	**3,170**	**750**	**3,920**

Assuming that these hand-holding couples are representative of hand-holding couples in general, we can use the available information to estimate various probabilities. For example, if a hand-holding couple is selected at random, then

$$\text{estimate of } P(\text{man's hand uppermost}) = \frac{3170}{3920} = 0.809$$

For a randomly selected hand-holding couple, if the man is taller, then the probability that the male has the uppermost hand is

$$2149/2448 = 0.878.$$

On the other hand—so to speak—if the woman is taller, the probability that the female has the uppermost hand is

$$205/446 = 0.460.$$

Notice that these last two estimates are estimates of the conditional probabilities $P(\text{male uppermost}|\text{male taller})$ and $P(\text{female uppermost}|\text{female taller})$, respectively. Also, because $P(\text{male uppermost}|\text{male taller})$ is not equal to $P(\text{male uppermost})$, the events *male uppermost* and *male taller* are not independent events. But, even when the female is taller, the male is still more likely to have the upper hand!

■

■ Estimating Probabilities Using Simulation ....

Simulation provides a means of estimating probabilities when we are unable (or do not have the time or resources) to determine probabilities analytically and when it is impractical to estimate them empirically by observation. Simulation is a method that generates "observations" by performing a chance experiment that is as similar as possible in structure to the real situation of interest.

To illustrate the idea of simulation, consider the situation in which a professor wishes to estimate the probabilities of different possible scores on a 20-question true–false quiz when students are merely guessing at the answers. Because each question is a true–false question, a person who is guessing should be equally likely to answer correctly or incorrectly on any given question. Rather than asking a student to select true or false and then comparing the choice to the correct answer, an equivalent process would be to pick a ball at random from a box that contains half red balls and half blue balls, with a blue ball representing a correct answer. Making 20 selections from the box (with replacement) and then counting the number of correct choices (the number of times a blue ball is selected) is a physical substitute for an observation from a student who has guessed at the answers to 20 true–false questions. Any particular number of blue balls in 20 selections should have the same probability as the same number of correct responses to the quiz when a student is guessing.

For example, 20 selections of balls might yield the following results:

Selection	1	2	3	4	5	6	7	8	9	10
	R	R	B	R	B	B	R	R	R	B

Selection	11	12	13	14	15	16	17	18	19	20
	R	R	B	R	R	B	B	R	R	B

This would correspond to a quiz with eight correct responses, and it would provide us with one observation for estimating the probabilities of interest. This process could then be repeated a large number of times to generate additional observations. For example, we might find the following:

Repetition	Number of "Correct" Responses
1	8
2	11
3	10
4	12
⋮	⋮
1000	11

The 1000 simulated quiz scores could then be used to construct a table of estimated probabilities.

Taking this many balls out of a box and writing down the results would be cumbersome and tedious. The process can be simplified by using random digits to substitute for drawing balls from the box. For example, a single digit could be selected at random from the 10 digits 0, 1, 2, 3, 4, 5, 6, 7, 8, 9. When using random digits, each of the 10 possibilities is equally likely to occur, so we can use the even digits (including 0) to indicate a correct response and the odd digits to indicate an incorrect response. This would maintain the important property that a correct response and an incorrect response are equally likely, because correct and incorrect are each represented by 5 of the 10 digits.

To aid in carrying out such a simulation, tables of random digits (such as Appendix A Table 1) or computer-generated random digits can be used. The numbers in Appendix A Table 1 were generated using a computer's random number generator. You can think of the table as being produced by repeatedly drawing a chip from a box containing 10 chips numbered 0, 1, . . . , 9. After each selection, the result is recorded, the chip returned to the box, and the chips mixed. Thus, any of the digits is equally likely to occur on any of the selections.

To see how a table of random numbers can be used to carry out a simulation, let's reconsider the quiz example. We use a random digit to represent the guess on a single question, with an even digit representing a correct response. A series of 20 digits represents the answers to the 20 quiz questions. We pick an arbitrary starting point in Appendix A Table 1. Suppose that we start at row 10 and take the 20 digits in a row to represent one quiz. The first five "quizzes" and the corresponding number correct (number of even digits) are:

Quiz	Random Digits	Number Correct
1	9 4 6 0 6 9 7 8 8 2 5 2 9 6 0 1 4 6 0 5	13
2	6 6 9 5 7 4 4 6 3 2 0 6 0 8 9 1 3 6 1 8	12
3	0 7 1 7 7 7 2 9 7 8 7 5 8 8 6 9 8 4 1 0	9
4	6 1 3 0 9 7 3 3 6 6 0 4 1 8 3 2 6 7 6 8	11
5	2 2 3 6 2 1 3 0 2 2 6 6 9 7 0 2 1 2 5 8	13

This process would be repeated to generate a large number of observations, which would then be used to construct a table of estimated probabilities.

The method for generating observations must preserve the important characteristics of the actual process being considered if simulation is to be successful. For example, it would be easy to adapt the simulation procedure for the true–false quiz to one for a multiple-choice quiz. Suppose that each of the 20 questions on the quiz has five possible responses, only one of which is correct. For any particular question, we would expect a student to be able to guess the correct answer only one-fifth of the time in the long run. To simulate this situation, we could select at random from a box that contained four red balls and only one blue ball (or, more generally, four times as many red balls as blue balls). If we are using random digits for the simulation, we could use 0 and 1 to represent a correct response and 2, 3, . . . , 9 to represent an incorrect response.

Using Simulation to Approximate a Probability

1. Design a method that uses a random mechanism (such as a random number generator or table, the selection of a ball from a box, the toss of a coin, etc.) to represent an observation. Be sure that the important characteristics of the actual process are preserved.
2. Generate an observation using the method from Step 1, and determine whether the outcome of interest has occurred.
3. Repeat Step 2 a large number of times.
4. Calculate the estimated probability by dividing the number of observations for which the outcome of interest occurred by the total number of observations generated.

The simulation process is illustrated in Examples 6.31–6.33.

Example 6.31 Building Permits

Many California cities limit the number of building permits that are issued each year. Because of limited water resources, one such city plans to issue permits for only 10 dwelling units in the upcoming year. The city will decide who is to receive permits by holding a lottery. Suppose that you are one of 39 individuals who apply for permits. Thirty of these individuals are requesting permits for a single-family home, eight are requesting permits for a duplex (which counts as two dwelling units), and one person is requesting a permit for a small apartment building with eight units (which counts as eight dwelling units). Each request will be entered into the lottery. Requests will be selected at random one at a time, and if there are enough permits remaining, the request will be granted. This process will continue until all 10 permits have been issued. If your request is for a single-family home, what are your chances of receiving a permit? Let's use simulation to estimate this probability. (It is not easy to determine analytically.)

To carry out the simulation, we can view the requests as being numbered from 1 to 39 as follows:

01–30	Requests for single-family homes
31–38	Requests for duplexes
39	Request for 8-unit apartment

For ease of discussion, let's assume that your request is number 1.

One method for simulating the permit lottery consists of these three steps:

1. Choose a random number between 1 and 39 to indicate which permit request is selected first, and grant this request.
2. Select another random number between 1 and 39 to indicate which permit request is considered next. Determine the number of dwelling units for the selected request. Grant the request only if there are enough permits remaining to satisfy the request.
3. Repeat Step 2 until permits for 10 dwelling units have been granted.

We used MINITAB to generate random numbers between 1 and 39 to imitate the lottery drawing. (The random number table in Appendix A Table 1 could also be used by selecting two digits and ignoring 00 and any value over 39). For example, the first sequence generated by MINITAB is

Random Number	Type of Request	Total Number of Units So Far
25	Single-family home	1
07	Single-family home	2
38	Duplex	4
31	Duplex	6
26	Single-family home	7
12	Single-family home	8
33	Duplex	10

We would stop at this point, because permits for 10 units would have been issued. In this simulated lottery, Request 1 was not selected, so you would not have received a permit.

The next simulated lottery (using MINITAB to generate the selections) is as follows:

Random Number	Type of Request	Total Number of Units So Far
38	Duplex	2
16	Single-family home	3
30	Single-family home	4
39	Apartment—not granted, since there are not 8 permits remaining	4
14	Single-family home	5
26	Single-family home	6
36	Duplex	8
13	Single-family home	9
15	Single-family home	10

Again, Request 1 was not selected, so you would not have received a permit in this simulated lottery.

Now that a strategy for simulating a lottery has been devised, the tedious part of the simulation begins. We would now have to simulate a large number of lottery drawings, determining for each whether Request 1 was granted. We simulated 500 such drawings and found that Request 1 was selected in 85 of the lotteries. Thus,

$$\text{estimated probability of receiving a building permit} = \frac{85}{500} = .17$$

■

....

Example 6.32 One-Boy Family Planning

Suppose that couples who wanted children were to continue having children until a boy is born. Assuming that each newborn child is equally likely to be a boy or a girl, would this behavior change the proportion of boys in the population? This question was posed in an article that appeared in *The American Statistician* ("What Some Puzzling Problems Teach About the Theory of Simulation and the Use of Re-sampling" [1994]: 290–293), and many people answered the question incorrectly. We will use simulation to estimate the long-run proportion of boys in the population if families were to continue to have children until they have a boy. This proportion is an estimate of the probability that a randomly selected child from this population is a boy. Note that every sibling group would have exactly one boy.

Step-by-step technology instructions available online

We use a single-digit random number to represent a child. The odd digits (1, 3, 5, 7, 9) represent a male birth, and the even digits represent a female birth. An observation is constructed by selecting a sequence of random digits. If the first random number obtained is odd (a boy), the observation is complete. If the first selected number is even (a girl), another digit is chosen. We continue in this way until an odd digit is obtained. For example, reading across row 15 of the random number table (Appendix A Table 1), the first 10 digits are

$$0\ 7\ 1\ 7\ 4\ 2\ 0\ 0\ 0\ 1$$

Using these numbers to simulate sibling groups, we get

Sibling group 1	0 7	girl, boy
Sibling group 2	1	boy
Sibling group 3	7	boy
Sibling group 4	4 2 0 0 0 1	girl, girl, girl, girl, girl, boy

Continuing along row 15 of the random number table,

Sibling group 5	3	boy
Sibling group 6	1	boy
Sibling group 7	2 0 4 7	girl, girl, girl, boy
Sibling group 8	8 4 1	girl, girl, boy

After simulating eight sibling groups, we have 8 boys among 19 children. The proportion of boys is 8/19, which is close to .5. Continuing the simulation to obtain a large number of observations suggests that the long-run proportion of boys in the population would still be .5, which is indeed the case.

■

Example 6.33 ESP?

Can a close friend read your mind? Try the following chance experiment. Write the word *blue* on one piece of paper and the word *red* on another, and place the two slips of paper in a box. Select one slip of paper from the box, look at the word written on it, and then try to convey the word by sending a mental message to a friend who is seated in the same room. Ask your friend to select either red or blue, and record whether the response is correct. Repeat this 10 times and total the number of correct responses. How did your friend do? Is your friend receiving your mental messages or just guessing?

Let's investigate this issue by using simulation to get the approximate probabilities of the various possible numbers of correct responses for someone who is guessing. Someone who is guessing should have an equal chance of responding correctly or incorrectly. We can use a random digit to represent a response, with an even digit representing a correct response (C) and an odd digit representing an incorrect response (X). A sequence of 10 digits can be used to simulate one execution of the chance experiment.

For example, using the last 10 digits in row 25 of the random number table (Appendix A Table 1) gives

5	2	8	3	4	3	0	7	3	5
X	C	C	X	C	X	C	X	X	X

which is a simulated chance experiment resulting in four correct responses. We used MINITAB to generate 150 sequences of 10 random digits and obtained the following results:

Sequence Number	Digits	Number Correct
1	3996285890	5
2	1690555784	3
3	9133190550	2
⋮	⋮	⋮
149	3083994450	5
150	9202078546	7

Table 6.2 summarizes the results of our simulation.

Table 6.2 Estimated Probabilities for Example 6.33

Number Correct	Number of Sequences	Estimated Probability
0	0	.0000
1	1	.0067
2	8	.0533
3	16	.1067
4	30	.2000
5	36	.2400
6	35	.2333
7	17	.1133
8	7	.0467
9	0	.0000
10	0	.0000
Total	**150**	**1.0000**

The estimated probabilities in Table 6.2 are based on the assumption that a correct and an incorrect response are equally likely (guessing). Evaluate your friend's performance in light of the information in Table 6.2. Is it likely that someone who is guessing would have been able to get as many correct as your friend did? Do you think your friend was receiving your mental messages? How are the estimated probabilities in Table 6.2 used to support your answer?

■ **Exercises 6.75–6.83** .... Turn to the Expanded Answers Section for answers not shown next to exercises. ....

6.75 The *Los Angeles Times* (June 14, 1995) reported that the U.S. Postal Service is getting speedier, with higher overnight on-time delivery rates than in the past. The Price Waterhouse accounting firm conducted an independent audit by seeding the mail with letters and recording on-time delivery rates for these letters. Suppose that the results were as follows (these numbers are fictitious but are compatible with summary values given in the article):

	Number of Letters Mailed	Number of Letters Arriving on Time
Los Angeles	500	425
New York	500	415
Washington, D.C.	500	405
Nationwide	6000	5220

Use the given information to estimate the following probabilities:
a. The probability of an on-time delivery in Los Angeles
b. The probability of late delivery in Washington, D.C.
c. The probability that two letters mailed in New York are both delivered on time   0.689
d. The probability of on-time delivery nationwide   0.87

6.76 Five hundred first-year students at a state university were classified according to both high school GPA and whether they were on academic probation at the end of their first semester. The data are

High School GPA

Probation	2.5 to <3.0	3.0 to <3.5	3.5 and Above	Total
Yes	50	55	30	**135**
No	45	135	185	**365**
Total	**95**	**190**	**215**	**500**

a. Construct a table of the estimated probabilities for each GPA–probation combination.
b. Use the table constructed in Part (a) to approximate the probability that a randomly selected first-year student at this university will be on academic probation at the end of the first semester.   0.27
c. What is the estimated probability that a randomly selected first-year student at this university had a high school GPA of 3.5 or above?   0.43
d. Are the two outcomes *selected student has a high school GPA of 3.5 or above* and *selected student is on academic probation at the end of the first semester* independent outcomes? How can you tell?
e. Estimate the proportion of first-year students with high school GPAs between 2.5 and 3.0 who are on academic probation at the end of the first semester   0.5263
f. Estimate the proportion of those first-year students with high school GPAs 3.5 and above who are on academic probation at the end of the first semester.   0.1395

6.77 ▼ The table below describes (approximately) the distribution of students by gender and college at a mid-sized public university in the West. If we were to randomly select one student from this university:
a. What is the probability that the selected student is a male?   0.622
b. What is the probability that the selected student is in the College of Agriculture?   0.167
c. What is the probability that the selected student is a male in the College of Agriculture?   0.117
d. What is the probability that the selected student is a male who is not from Agriculture?   0.506

6.78 On April 1, 2000, the Bureau of the Census in the United States attempted to count every U.S. citizen and every resident. Suppose that the counts in the table at the top of the next page are obtained for four counties in one region:
a. If one person is selected at random from this region, what is the probability that the selected person is from Ventura County?   0.4247

Table for Exercise 6.77

Gender				College			
	Education	Engineering	Liberal Arts	Science and Math	Agriculture	Business	Architecture
Male	200	3200	2500	1500	2100	1500	200
Female	300	800	1500	1500	900	1500	300

Bold exercises answered in back   ● Data set available online but not required   ▼ Video solution available

Table for Exercise 6.78

| County | Race/Ethnicity | | | | |
	Caucasian	Hispanic	Black	Asian	American Indian
Monterey	163,000	139,000	24,000	39,000	4,000
San Luis Obispo	180,000	37,000	7,000	9,000	3,000
Santa Barbara	230,000	121,000	12,000	24,000	5,000
Ventura	430,000	231,000	18,000	50,000	7,000

b. If one person is selected at random from Ventura County, what is the probability that the selected person is Hispanic?   0.3139

c. If one Hispanic person is selected at random from this region, what is the probability that the selected individual is from Ventura?   0.4675

d. If one person is selected at random from this region, what is the probability that the selected person is an Asian from San Luis Obispo County?   0.0052

e. If one person is selected at random from this region, what is the probability that the person is either Asian or from San Luis Obispo County?   0.2014

f. If one person is selected at random from this region, what is the probability that the person is Asian or from San Luis Obispo County but not both?   0.1962

g. If two people are selected at random from this region, what is the probability that both are Caucasians?   0.335

h. If two people are selected at random from this region, what is the probability that neither is Caucasian?   0.1774

i. If two people are selected at random from this region, what is the probability that exactly one is a Caucasian?

j. If two people are selected at random from this region, what is the probability that both are residents of the same county?   0.2954

k. If two people are selected at random from this region, what is the probability that both are from different racial/ethnic groups?   0.5659

6.79 A medical research team wishes to evaluate two different treatments for a disease. Subjects are selected two at a time, and then one of the pair is assigned to each of the two treatments. The treatments are applied, and each is either a success (S) or a failure (F). The researchers keep track of the total number of successes for each treatment. They plan to continue the chance experiment until the number of successes for one treatment exceeds the number of successes for the other treatment by 2. For example, they might observe the results in the table below. The chance experiment would stop after the sixth pair, because Treatment 1 has 2 more successes than Treatment 2. The researchers would conclude that Treatment 1 is preferable to Treatment 2.

Suppose that Treatment 1 has a success rate of .7 (i.e., P(success) = .7 for Treatment 1) and that Treatment 2 has a success rate of .4. Use simulation to estimate the probabilities in Parts (a) and (b). (Hint: Use a pair of random digits to simulate one pair of subjects. Let the first digit represent Treatment 1 and use 1–7 as an indication of a

Table for Exercise 6.79

Pair	Treatment 1	Treatment 2	Total Number of Successes for Treatment 1	Total Number of Successes for Treatment 2
1	S	F	1	0
2	S	S	2	1
3	F	F	2	1
4	S	S	3	2
5	F	F	3	2
6	S	F	4	2

Bold exercises answered in back   ● Data set available online but not required   ▼ Video solution available

success and 8, 9, and 0 to indicate a failure. Let the second digit represent Treatment 2, with 1–4 representing a success. For example, if the two digits selected to represent a pair were 8 and 3, you would record failure for Treatment 1 and success for Treatment 2. Continue to select pairs, keeping track of the total number of successes for each treatment. Stop the trial as soon as the number of successes for one treatment exceeds that for the other by 2. This would complete one trial. Now repeat this whole process until you have results for at least 20 trials [more is better]. Finally, use the simulation results to estimate the desired probabilities.)

a. Estimate the probability that more than five pairs must be treated before a conclusion can be reached. (Hint: P(more than 5) = $1 - P$(5 or fewer).)

b. Estimate the probability that the researchers will incorrectly conclude that Treatment 2 is the better treatment.

6.80 Many cities regulate the number of taxi licenses, and there is a great deal of competition for both new and existing licenses. Suppose that a city has decided to sell 10 new licenses for $25,000 each. A lottery will be held to determine who gets the licenses, and no one may request more than three licenses. Twenty individuals and taxi companies have entered the lottery. Six of the 20 entries are requests for 3 licenses, 9 are requests for 2 licenses, and the rest are requests for a single license. The city will select requests at random, filling as many of the requests as possible. For example, the city might fill requests for 2, 3, 1, and 3 licenses and then select a request for 3. Because there is only one license left, the last request selected would receive a license, but only one.

a. An individual has put in a request for a single license. Use simulation to approximate the probability that the request will be granted. Perform *at least* 20 simulated lotteries (more is better!).   Answers will vary

b. Do you think that this is a fair way of distributing licenses? Can you propose an alternative procedure for distribution?   Answers will vary

6.81 Four students must work together on a group project. They decide that each will take responsibility for a particular part of the project, as follows:

Person	Maria	Alex	Juan	Jacob
Task	Survey design	Data collection	Analysis	Report writing

Because of the way the tasks have been divided, one student must finish before the next student can begin work. To ensure that the project is completed on time, a schedule is established, with a deadline for each team member. If any one of the team members is late, the timely completion of the project is jeopardized. Assume the following probabilities:

1. The probability that Maria completes her part on time is .8.
2. If Maria completes her part on time, the probability that Alex completes on time is .9, but if Maria is late, the probability that Alex completes on time is only .6.
3. If Alex completes his part on time, the probability that Juan completes on time is .8, but if Alex is late, the probability that Juan completes on time is only .5.
4. If Juan completes his part on time, the probability that Jacob completes on time is .9, but if Juan is late, the probability that Jacob completes on time is only .7.

Use simulation (with at least 20 trials) to estimate the probability that the project is completed on time. Think carefully about this one. For example, you might use a random digit to represent each part of the project (four in all). For the first digit (Maria's part), 1–8 could represent *on time* and 9 and 0 could represent *late*. Depending on what happened with Maria (late or on time), you would then look at the digit representing Alex's part. If Maria was on time, 1–9 would represent *on time* for Alex, but if Maria was late, only 1–6 would represent *on time*. The parts for Juan and Jacob could be handled similarly.

6.82 In Exercise 6.81, the probability that Maria completes her part on time was .8. Suppose that this probability is really only .6. Use simulation (with at least 20 trials) to estimate the probability that the project is completed on time.

6.83 Refer to Exercises 6.81 and 6.82. Suppose that the probabilities of timely completion are as in Exercise 6.81 for Maria, Alex, and Juan, but that Jacob has a probability of completing on time of .7 if Juan is on time and .5 if Juan is late.

a. Use simulation (with at least 20 trials) to estimate the probability that the project is completed on time.

b. Compare the probability from Part (a) to the one computed in Exercise 6.82. Which decrease in the probability of on-time completion (Maria's or Jacob's) made the bigger change in the probability that the project is completed on time?

Bold exercises answered in back   ● Data set available online but not required   ▼ Video solution available

Activity 6.1   Kisses

Background: The paper "What Is the Probability of a Kiss? (It's Not What You Think)" (*Journal of Statistics Education* (online) [2002]) posed the following question: What is the probability that a Hershey's Kiss will land on its base (as opposed to its side) if it is flipped onto a table? Unlike flipping a coin, there is no reason to believe that this probability would be .5.

Working as a class, develop a plan that would enable you to estimate this probability empirically.

Once you have an acceptable plan, carry it out and use the resulting data to produce an estimate of the desired probability. Do you think that a kiss is equally likely to land on its base or on its side? Explain.

Activity 6.2   A Crisis for European Sports Fans?

Background: The *New Scientist* (January 4, 2002) reported on a controversy surrounding the Euro coins that have been introduced as a common currency across Europe. Each country mints its own coins, but these coins are accepted in any of the countries that have adopted the Euro as their currency.

A group in Poland claims that the Belgium-minted Euro does not have an equal chance of landing heads or tails. This claim was based on 250 tosses of the Belgium-minted Euro, of which 140 (56%) came up heads. Should this be cause for alarm for European sports fans, who know that "important" decisions are made by the flip of a coin?

In this activity, we will investigate whether this should be cause for alarm by examining whether observing 140 heads out of 250 tosses is an unusual outcome if the coin is fair.

1. For this first step, you can either (a) flip a U.S. penny 250 times, keeping a tally of the number of heads and tails

observed (this won't take as long as you think), or (b) simulate 250 coin tosses by using your calculator or a statistics software package to generate random numbers (if you choose this option, give a brief description of how you carried out the simulation).

2. For your sequence of 250 tosses, calculate the proportion of heads observed.

3. Form a data set that consists of the values for proportion of heads observed in 250 tosses of a fair coin for the entire class. Summarize this data set by constructing a graphical display.

4. Working with a partner, write a paragraph explaining why European sports fans should or should not be worried by the results of the Polish experiment. Your explanation should be based on the observed proportion of heads from the Polish experiment and the graphical display constructed in Step 3.

Activity 6.3   The "Hot Hand" in Basketball

Background: Consider a mediocre basketball player who has consistently made only 50% of his free throws over several seasons. If we were to examine his free throw record over the last 50 free throw attempts, is it likely that we would see a streak of 5 in a row where he is successful in making the free throw? In this activity, we will investigate this question. We will assume that the outcomes of successive free throw attempts are independent and that the probability that the player is successful on any particular attempt is .5.

1. Begin by simulating a sequence of 50 free throws for this player. Because this player has probability of success of .5 for each attempt and the attempts are independent, we can model a free throw by tossing a coin. Using heads

to represent a successful free throw and tails to represent a missed free throw, simulate 50 free throws by tossing a coin 50 times, recording the outcome of each toss.

2. For your sequence of 50 tosses, identify the longest streak by looking for the longest string of heads in your sequence. Determine the length of this longest streak.

3. Combine your longest streak value with those from the rest of the class and construct a histogram or dotplot of these longest streak values.

4. Based on the graph from Step 3, does it appear likely that a player of this skill level would have a streak of 5 or more successes sometime during a sequence of 50 free throw attempts? Justify your answer based on the graph from Step 3.

5. Use the combined class data to estimate the probability that a player of this skill level has a streak of at least 5 somewhere in a sequence of 50 free throw attempts.

6. Using basic probability rules, we can calculate the probability that a player of this skill level is successful on the *next* 5 free throw attempts:

$$P(SSSSS) = \left(\frac{1}{2}\right)\left(\frac{1}{2}\right)\left(\frac{1}{2}\right)\left(\frac{1}{2}\right)\left(\frac{1}{2}\right) = \left(\frac{1}{2}\right)^5 = .031$$

which is relatively small. At first this might seem inconsistent with your answer in Step 5, but the estimated probability from Step 5 and the computed probability of .031 are really considering different situations. Explain why it is plausible that both probabilities could be correct.

7. Do you think that the assumption that the outcomes of successive free throws are independent is reasonable? Explain. (This is a hotly debated topic among both sports fans and statisticians!)

Summary of Key Concepts and Formulas

Term or Formula	Comment
Chance experiment	Any experiment for which there is uncertainty concerning the resulting outcome.
Sample space	The collection of all possible outcomes from a chance experiment.
Event	Any collection of possible outcomes from a chance experiment.
Simple event	Any event that consists of a single outcome.
Events 1. *not A*, A^C 2. *A or B*, $A \cup B$ 3. *A and B*, $A \cap B$	1. The event consisting of all outcomes not in A. 2. The event consisting of all outcomes in at least one of the two events. 3. The event consisting of outcomes common to both events.
Disjoint (mutually exclusive) events	Events that have no outcomes in common.
Basic properties of probability	Basic properties of probability 1. The probability of any event must be a number between 0 and 1. 2. If S is the sample space for a chance experiment, $P(S) = 1$ 3. If E and F are disjoint events, $P(E \cup F) = P(E) + P(F)$ 4. $P(E) + P(E^C) = 1$
$P(E) = \dfrac{\text{number of outcomes in } E}{N}$	$P(E)$ when the outcomes are equally likely and where N is the number of outcomes in the sample space.
$P(E \cup F) = P(E) + P(F)$ $P(E_1 \cup \cdots \cup E_k) = P(E_1) + \cdots + P(E_k)$	Addition rules when events are disjoint.
$P(E \mid F) = \dfrac{P(E \cap F)}{P(F)}$	The conditional probability of the event E given that the event F has occurred.

Term or Formula	Comment
Independence of events E and F $P(E\|F) = P(E)$	Events E and F are independent if the probability that E has occurred given F is the same as the probability that E will occur with no knowledge of F.
$P(E \cap F) = P(E)P(F)$ $P(E_1 \cap \cdots \cap E_k) = P(E_1)P(E_2) \cdots P(E_k)$	Multiplication rules for *independent* events.
$P(E \cup F) = P(E) + P(F) - P(E \cap F)$	The general addition rule for two events.
$P(E \cap F) = P(E\|F)P(F)$	The general multiplication rule for two events.
$P(E) = P(E\|B_1)P(B_1) + P(E\|B_2)P(B_2)$ $+ \cdots + P(E\|B_k)P(B_k)$	The law of total probability, where $B_1, B_2, \ldots, B_k$ are disjoint events with $P(B_1) + P(B_2) + \cdots + P(B_k) = 1$
$P(B_i\|E) = \dfrac{P(E\|B_i)P(B_i)}{P(E\|B_1)P(B_1) + P(E\|B_2)P(B_2) + \cdots + P(E\|B_k)P(B_k)}$	Bayes' rule, where $B_1, B_2, \ldots, B_k$ are disjoint events with $P(B_1) + P(B_2) + \cdots + P(B_k) = 1$

Chapter Review Exercises 6.84–6.98

Turn to the Expanded Answers Section for answers not shown next to exercises.

CENGAGENOW Know exactly what to study! Take a pre-test and receive your Personalized Learning Plan.

6.84 A company uses three different assembly lines—A_1, A_2, and A_3—to manufacture a particular component. Of those manufactured by A_1, 5% need rework to remedy a defect, whereas 8% of A_2's components and 10% of A_3's components need rework. Suppose that 50% of all components are produced by A_1, whereas 30% are produced by A_2 and 20% come from A_3.
a. Construct a tree diagram with first-generation branches corresponding to the three lines. Leading from each branch, draw one branch for rework (R) and another for no rework (N). Then enter appropriate probabilities on the branches.
b. What is the probability that a randomly selected component came from A_1 and needed rework?   0.025
c. What is the probability that a randomly selected component needed rework?   0.069

6.85 A certain company sends 40% of its overnight mail parcels by means of express mail service A_1. Of these parcels, 2% arrive after the guaranteed delivery time (use L to denote the event *late delivery*). If a record of an overnight mailing is randomly selected from the company's files, what is the probability that the parcel went by means of A_1 and was late?   0.008

6.86 Return to Exercise 6.85, and suppose that 50% of the overnight parcels are sent by means of express mail service A_2 and the remaining 10% are sent by means of A_3. Of those sent by means of A_2, only 1% arrived late, whereas 5% of the parcels handled by A_3 arrived late.
a. What is the probability that a randomly selected parcel arrived late? (Hint: A tree diagram should help.)   0.018
b. Suppose that the selected record shows that the parcel arrived late, but the name of the service does not appear on the record. What is the probability that the parcel was handled by A_1? That is, what is the probability of A_1 given L, denoted $P(A_1\|L)$? What is $P(A_2\|L)$? $P(A_3\|L)$?   0.44444, 0.27778, 0.27778

6.87 Two individuals, A and B, are finalists for a chess championship. They will play a sequence of games, each of which can result in a win for A, a win for B, or a draw. Suppose that the outcomes of successive games are independent, with $P(\text{A wins game}) = .3$, $P(\text{B wins game}) = .2$, and $P(\text{draw}) = .5$. Each time a player wins a game, he earns 1 point and his opponent earns no points. The first player to win 5 points wins the championship. For the sake of simplicity, assume that the championship will end in a draw if both players obtain 5 points at the same time.

Bold exercises answered in back   ● Data set available online but not required   ▼ Video solution available

mimics the behavior of a coin flip. To do this we need to generate random numbers on the calculator. Graphing calculator Exploration 2.2 illustrated the process of generating a random number between 0 and 1 using a sequence of keystrokes defined as *rand*. We will use this process now to flip some virtual coins. Define the probability of success, S, for the event of interest to be equal to p.

The calculator generates uniform random numbers on the interval from 0 to 1. One characteristic of this process is that for any number, p, the probability of generating a number between 0 and p is equal to p. Using this fact we can perform the chance experiment as follows: Generate a uniform random number between 0 and 1 using your *rand* keystroke sequence. If the number generated is less than or equal to p, count that as a "success." If the number generated is greater than p, count that as a "failure." For a fair coin, p is equal to .5, and therefore any number generated by *rand* that is less than or equal to .5 will count as a success. The following table shows the beginning of such a simulation:

Toss	rand	S/F	Cumulative Successes	Relative Frequency of Successes
1	.9351587791	F	0	0/1 = .0000
2	.1080114624	S	1	1/2 = .5000
3	.0062633066	S	2	2/3 = .6667
4	.5489861799	F	2	2/4 = .5000
5	.8555803143	F	2	2/5 = .4000

To see how the relative frequency of success behaves, plot toss number on the horizontal axis and relative frequency of success on the vertical axis of a scatterplot. A plot from the table above is shown in Figure 6.12.

To see that the relative frequency settles down and gets close to the true probability of success, we would need to simulate a very large number of tosses. This is a relatively slow process if you are flipping coins by hand, but lots faster on a calculator. Your calculator can perform this experiment many times and store the results, using some of its list commands. These commands will, of course, be different from calculator to calculator but the logic that underlies the process will be very similar. You will need to look up the correct syntax for the list commands in your calculator manual. (You may wish to refer to Graphing Calculator Exploration 3.1.) Here is a typical sequence of instructions to make this happen:

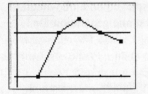

Figure 6.12 First five trials of the simulation.

Step	Generic Keystrokes	Description
1	Sequence(x, x, 1, 100) → List1	Generate a sequence of numbers from 1 to 100, and store this sequence in List1.
2	Sequence(*rand*, x, 1, 100) → List 2	Create a sequence of random numbers and store them in List2.
3	Int(p − List2) +1 → List 3	Convert the random numbers from List2 into 0's and 1's in List3.
4	CumSum(List3) → List 4	Store the cumulative frequency from List3 to List4.
5	List4/List1 → List 5	Convert the cumulative frequency to a relative frequency "so far" and store in List5.

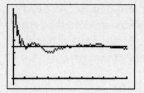

Figure 6.13
Continuation of the
simulation.

Now you can make a scatter plot using the numbers in List1 and the relative frequencies in List5. Figure 6.13 shows you what a typical screen might look like after performing the procedure. We used a $p = .5$ for our coin so we added the horizontal line corresponding to *relative frequency* $= .5$ for reference.

Exploration 6.2 Simulating Events That Are Independent

As noted earlier your calculator has a sequence of keystrokes that we have collectively referred to as "*rand.*" The numbers generated by *rand* can be used to construct simulations that involve independent events.

The *rand* function automatically generates "independent" random numbers. This means that, for example, the probability of *rand* returning a number between .5 and 1.0 is unaffected by the value of the previously returned random number. The practical effect of this is that simply pressing the *rand* button repeatedly generates a sequence of independently generated numbers. Let's see how we can use this capability in a simulation. We will consider a simplified version of an important problem from meteorology: In a world where a photon of light may be absorbed or deflected in its path, how far will it travel in its original direction? The safety of drivers in fog depends on enough light reaching an oncoming car to be aware of the presence of another car.

For purposes of our simulation we will assume that (1) an absorbed photon of light is no longer visible, (2) a photon once deflected does *not* return to its original path, and (3) there are constant probabilities of deflection and absorption. Furthermore the probability of the fate of the photon one second from now is independent of its status now. It is this real-life independence that requires us to generate "independent" random numbers to faithfully mirror the situation being modeled by our simulation.

Our model of a photon's travel will be a point traveling from left to right in the grid below, starting at square one. (*Note*: The probabilities we will use are for illustrative purposes and do not represent real probabilities in actual fog.)

1	2	3	4	5	6	7	8	9	10

At each position, the photon has three possible fates:

1. From square x it will go on to square $x + 1$ with probability .90.
2. From square x it will be absorbed with probability 0.05.
3. From square x it may be deflected "up" or "down," each with probability .025.

We arbitrarily define square 10 to be the position at which we would like an oncoming driver to be aware of our presence. We wish to find the probability a photon will

Photon Fate	Probability	*rand* Interval
Continue on path	.9	$0 \leq rand < .9$
Be absorbed	.05	$.9 \leq rand < .95$
Be deflected	.05	$.95 \leq rand < .1.0$

get all the way to square 10. To estimate this probability via simulation, we need to link the probabilities of the fates listed above with random numbers generated by *rand*.

Here are two rows of five random numbers generated by our *rand* procedure. A trial proceeds by reading across the first row, then across the second row.

| .812788 | .710859 | .704518 | .146311 | .956130 |
| .050035 | .126108 | .330590 | .390195 | .114339 |

In this trial, the photon of light would go straight for 4 intervals and then—according to the interpretation of the random number .956130—be deflected.

Here is a second trial:

| .576809 | .715678 | .404556 | .635070 | .811570 |
| .062257 | .490527 | .285129 | .758026 | .231308 |

In this second trial, the simulated photon makes it all the way to the oncoming car; all the random numbers are less than .90.

Forty trials were initiated with the following results, where Success is defined as a photon making it to square 10.

FSSFFFSFFFSSSFFFFSSSFFFFSFFSFFFSFFFFSSFF

Based on these 40 simulated trials, we estimate the probability of a photon arriving uninterrupted at square 10 to be approximately 14/40, or 0.35.

Here we have considered only whether the photon arrives successfully at square 10. This is not the only simulation we could consider, nor is it the most interesting. We could, for example, consider the typical distance a photon travels before it is absorbed or deflected. We might ask the question, At least how far do 50% of the photons travel on average? To address this question we would want to know the median of the distribution of photon travel distances. In this situation the outcome of a trial would not be success or failure, but the actual number of squares traveled by the photon before being absorbed or deflected. In the long run, the distribution of simulation results should look more and more like the "true" distribution of photon distances traveled, and the median of our simulation results should get closer and closer to the true median of photon travel distance.

Exploration 6.3 Simulating Events That Are Not Independent

To simulate nonindependent events we must use the definition of conditional probability. That means we must consider the outcome of one event before we can simulate the second event. We can simulate the outcome of an event by generating a random number; simulating two dependent events will require the generating of at least two random numbers, depending on the complexity of the simulation.

To illustrate, consider the following example. Burglaries occur about once a month in a neighborhood troubled by crime. Suppose the neighborhood decided to get a watchdog, whose function would be to bark loudly and scare away burglars. They have rescued an elderly (but vicious looking!) German shepherd from the local animal shelter. Unfortunately his bark is very much worse than his bite. Much of his time is spent sleeping, and there is some concern about his hearing. But he *does* have a relatively ferocious bark, and they got him cheap. From actual trial runs using local neighbors disguised as burglars (i.e., a simulation!), it was discovered that Herr Rover noticed the simulated burglars 90% of the time. When he noticed the simulated burglars he emitted a ferocious bark 75% of the time. The question the neighbors now have is, When can they expect a payoff from Rover? On average, how many months will go by before a burglar is noticed by Rover *and* frightened by Rover's ferocious bark?

Of course, basic probability rules could be used to calculate the probability that Rover notices a burglar *and* barks, but it isn't obvious how to, using probability rules, determine the number of months that will go by before this event occurs. A simulation can be used to provide some insight.

A particular burglar event can be modeled using the definition of conditional probability and two random numbers generated by *rand*:

1. If the first random number is less than or equal to .90, this will count as detected the burglar.
2. If the burglar is detected, we will use the second random number to determine whether Rover barks. A value less than or equal to .75 will indicate a "ferocious bark."
3. The first two steps represent the burglar event for one month. We will repeat steps 1 and 2 until we get a "ferocious bark," thus completing a trial of the simulation. We can note the number of months required to get an occurrence of "detected *and* ferocious bark."
4. Many repetitions (trials) would generate a distribution of times until a ferocious burglar-scaring bark occurs, and we could examine this distribution to find a "typical" number of months between successful burglar scarings.

Here is a record of 5 trials of this experiment, using random numbers generated by the calculator:

First Random Number	Detection?	Second Random Number	Ferocious Bark?
.8767844333	Yes	.7900592279	No
.9546136582	No	.5215775123	(not detected)
.5329329133	Yes	.2805102664	Yes—a 3-month wait
.1320755632	Yes	.7338039727	Yes—a 1-month wait
.6958584068	Yes	.5688014123	Yes—a 1-month wait
.9961435487	No	.6002686874	(not detected)
.7286825076	Yes	.3887292922	Yes—a 2-month wait
.4319389989	Yes	.0143592629	Yes—a 1-month wait

With only 5 trials we cannot reach any reliable conclusions, but after many repetitions we would be able to estimate the probability that a burglar would be detected and scared away after 1, 2, 3, or any other number of months.

You should notice that we were led to simulation because our basic probability rules alone were not adequate to solve the problem of interest. Of course, an analytic solution is preferable—if one can be found!—to one based on simulation because it gives a correct answer quickly. Simulation and an appeal to the law of large numbers can only give an approximate answer, and may take considerable time and bookkeeping to execute, but it DOES work!

Exploration 6.4 Simulations and the seq Function

Our examples of simulations have been fairly small so far. Usually simulations are undertaken with a large number of trials. Just as more data give us a better idea of a distribution of data, more trials in a simulation will yield a better idea about the distribution of outcomes.

If a simulation is not too complicated it is possible to use an interesting calculator function called a "Sequence" function, or "seq" for short. This function can be used with the list capability of your calculator to perform simulations and analyze the results graphically. The sequence command will allow a sequence of operations to be performed repetitiously, with results then stored in a list. If 1 random number is desired, you can just use *rand*; if 200 random numbers in a list are desired, the seq command can be used to fill the list. For example, the command

$$\text{seq}(rand, x, 1, 200) \rightarrow \text{List1}$$

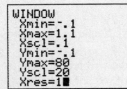

Figure 6.14 Window settings.

stores 200 random numbers in List1. As always, you should check your calculator manual to see what the syntax of the corresponding command is on your calculator.

Now we will perform a simulation that illustrates this command. Here's a question about random numbers generated using *rand*. Will the distribution of the average of random numbers look like the distribution of random numbers used to calculate the average? Let's see what we can find out. We will use the seq command to put 200 numbers generated by *rand*, in each of 5 lists.

$$\text{Seq}(rand, x, 1, 200) \rightarrow \text{List1}$$
$$\text{Seq}(rand, x, 1, 200) \rightarrow \text{List2}$$
$$\text{Seq}(rand, x, 1, 200) \rightarrow \text{List3}$$
$$\text{Seq}(rand, x, 1, 200) \rightarrow \text{List4}$$
$$\text{Seq}(rand, x, 1, 200) \rightarrow \text{List5}$$

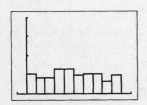

Figure 6.15 Histogram of data in List2.

Now we'll add the results, store them in List6, divide their sum by 5 to get the averages, and store in List6:

$$(\text{List1} + \text{List2} + \text{List3} + \text{List4} + \text{List5})/5 \rightarrow \text{List6}$$

The settings in the window shown in Figure 6.14 were used to generate the histograms below. The histogram in Figure 6.15 is of the data in List2, and is representative of the histograms in Lists 1–5. The histogram in Figure 6.16 is a histogram of List6, the sum of the first 5 lists, divided by 5 (the average of five numbers). Do they have the same shape? It doesn't look like it.

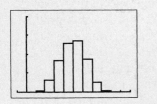

Figure 6.16 Histogram of data in List6.

We can see here the effect of averaging numbers. As you continue the study of statistics you will find that the behavior of these distributions governed by the Central Limit Theorem, an important result in statistics. Our point here is that a calculator, together with data analysis skills, can be used to perform nontrivial simulations and analyze the results statistically.

TEACHING TIPS

Intent of this chapter:

After completing this chapter, students should be able to (1) distinguish discrete and continuous numerical random variables, (2) use discrete probability distributions to compute probabilities, (3) use simple continuous probability distributions to compute probabilities, (4) calculate the mean and standard deviation of discrete probability distributions, (5) compute mean and standard deviation using linear transformation rules and linear combination rules, (6) compute probabilities using binomial and geometric distributions, (7) compute probabilities and percentiles for normal distributions, (8) assess normality of a data set, and (9) use normal distributions to approximate a binomial distribution.

A PowerPoint® Lecture is available on the *Instructor's Resource Binder* CD.

Chapter 7 covers many topics. It might be helpful to split this chapter into two parts: Section 7.1–Section 7.5 and Section 7.6–Section 7.8.

■ **Section 7.1**

This section reviews the concepts of discrete and continuous, and applies them to random variables (from Section 1.4). Remember, discrete random variables are typically associated with counts of something, whereas continuous random variables are associated with measurements.

Suggested Assignment: Exercises 7.1, 7.5, 7.7

■ **Section 7.2**

Section 7.2 introduces the probability distribution of a discrete random variable. For a discrete random variable that has a finite number of possible values, all the possible values of the random variable are listed and the sum of the probabilities of those values equals 1. As students work through examples, stress the importance of defining the random variable; for example, let x = the number of electric hot tubs purchased by the four customers (from Example 7.5). After defining the random variable, students seem to have less difficulty identifying the possible values for the discrete random variable. For relatively simple settings, students should be able to calculate the probability associated with each of the x-values. Using the probability properties and rules in Chapter 6 will aid in the calculation of these probabilities.

    Students should be able to create a probability histogram. Each value of x should have a bar centered over the value, having a height equal to the probability of the x-value. To use the TI calculator to create a probability histogram, input the values for the random variable into L1 and the corresponding probabilities into L2. In the Stat Plot window, use the following: xlist: L1 and Freq: L2. Students must set the window (using Zoom Stat does not work). The minimum x-value is -0.5, the maximum x-value is $(0.5 + \text{max})$, and the x-scale is 1. The y-values (being probabilities) are between 0 and 1. See Graphing Calculator Exploration 7.1 on page 434.

    Another important distinction for discrete probability distributions (as opposed to continuous distributions) is that the $P(x < 2)$ is *not* equal to $P(x \leq 2)$. In other words, whether the inequality is inclusive or exclusive changes the probability (see Exercise 7.9).

Suggested Assignment: Exercises 7.9, 7.11, 7.12, 7.15, 7.17, 7.19
Activity: Activity 7.2: Rotten Eggs? (p. 429 and *Activities Manual,* p. 65)★

■ **Section 7.3**

This section approximates the probability distribution of a discrete random variable with a smooth curve. The smooth curve describing the shape of the probability histogram is called a density curve. Density curves have two properties: (1) they are always on or above

★ Kathy's personal favorite

the horizontal axis and (2) the area under the curve is always 1. A density curve might be bell-shaped, like a normal curve; evenly or uniformly distributed, like a rectangle; or any other shape, like the triangle in Exercise 7.24. To calculate probabilities using a density curve, calculate the area under the curve for the interval of interest. For rectangular or other shapes, use basic geometric formulas to calculate the area. For bell-shaped curves, calculus must be used—but don't worry. Calculus has already been used to compute the needed areas, and the values appear in Appendix Table 2.

As a reminder, there is one important contrast in calculating probabilities for discrete probability distributions and continuous probability distributions. Remember, in a discrete distribution, $P(x < 2)$ and $P(x \leq 2)$ are not equal if 2 is a possible value. However, in a continuous distribution, $P(x < 2)$ and $P(x \leq 2)$ *are* equal. Look at Example 7.7 on page 369. Suppose we wanted to calculate $P(x = 5)$, denoted by the blue line on the graph below. What is the area of the line segment shown? Because line segments have no area, this value is 0. Therefore, in continuous distributions, the probability that the random variable is exactly equal to a value is always zero. So $P(x < 5)$ and $P(x \leq 5)$ would have the same probability since $P(x = 5)$ equals 0. A word of caution: $P(x < 5)$ includes all the values up to but not including 5. Students often want to use a value like 4.9 to calculate the probability. When doing so, students do not include values between 4.9 and 5, making their calculated probability too small.

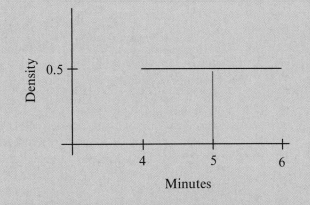

Suggested Assignment: Exercises 7.20, 7.22, 7.23, 7.24, 7.26

Assessment: Chapter 7 Quiz 1 (*Test Bank* and *Instructor's Resource Binder* CD)

■ **Section 7.4**

The mean and standard deviation of a random variable provide information about where the distribution is centered and about the spread of the distribution around the center. Since probability distributions identify all possible values for the random variable, probability distributions are similar to "populations." Therefore, Greek letters (parameters) are used to denote the mean (μ_x) and the standard deviations (σ_x) of a random variable x. In Example 7.8 on page 375, students might be asked, "How many attempts would we *expect* it to take before someone passed the licensing exam?" This idea of an expected value is the *mean* of the probability distribution. Also, students should be able to interpret the mean and the standard deviation of random variables in the context of the problem (see page 376, first paragraph).

Students should be familiar with the formulas for the mean and variance (σ^2) and should perform calculations using the formulas before learning the shortcut on the calculator. This ensures that the students understand what the calculator is doing for them. Using the TI calculator, students can easily obtain the values for mean and standard deviation of a random variable. As with graphing the probability histogram, input the x-values into L1 and the probabilities into L2. Use the following command to calculate the mean and standard deviation: "1-Var Stats L1,L2". (Note: This calculator is so "smart" it only provides the population standard deviation, leaving the sample standard deviation blank!) Although this calculator shortcut is a wonderful aid in making the calculations, students must still show what this shortcut accomplishes. It is not sufficient to just write "1-Var Stats L1,L2" as justification of work. Students must use the formulas to demonstrate how their values for mean and standard deviation were obtained.

The mean and standard deviation of a continuous random variable can be computed using calculus. However, students are not expected to carry out these computations. Most exercises provide the mean and standard deviation for continuous distributions. Example 7.13 on page 378 provides the basis for an excellent class discussion on the usefulness of the mean and standard deviation. The diagram in Figure 7.15 (found on the *Instructor's Resource Binder* CD) can be displayed. Ask students which Supplier, 1 or 2, would be the best choice and why.

The remaining part of this section discusses rules for linear transformation of random variables and rules for linear combination of random variables. The following examples will aid students in developing and understanding these rules.

Linear Transformations Suppose a company, named Stat Geeks, Inc. (goofy name is important!), has ten employees. The monthly salary for each employee is as follows: $1200, $1300, $1300, $1500, $1500, $2000, $2400, $2800, $3000, $5000. Calculate the mean and standard deviation (don't forget to use σ_x) for the monthly salary for these ten employees and create a boxplot for the data. Now suppose that the company has an excellent year (and of course it should if it deals with statistics). The owner decides to increase every employee's salary by $500. So, add $500 to each of the salaries above. Recalculate the mean and standard deviation. Notice that the mean salary increased by $500, but the standard deviation for the salaries stayed the same. Why? Create parallel boxplots of the original salaries and the increased salaries. What do you notice? *The distributions for both have the same shape and spread. The distribution for the increased salaries is just shifted to the right and centered over a different value.* (Compare this to transformations in algebra or precalculus. $f(x) + 3$ just *shifts* the function to the right 3 units.)

Next, suppose the owner decides to increase every employee's salary by 30% instead. To do this, multiply each salary by 1.3 (some students will require an explanation for why we use 1.3). Recalculate the mean and standard deviation. Notice that the mean salary increased by 30%, but the standard deviation also increased by 30%. Why? Create parallel boxplots of the original salaries and the increased salaries. What do you notice? *While the boxplots for the original salaries and the first increase of salaries have the same shape and spread, the increased salaries are* **shifted**. *When the salaries are multiplied by 1.3, the entire distribution has been* **stretched**. (Compare this to transformations in algebra or precalculus. $3(f(x))$ stretches the function by a factor of 3.) Whenever students encounter a problem that requires a linear transformation and they forget what to do, just say "Remember Stat Geeks Inc.," which helps to trigger the memory of these rules.

Linear Combinations Suppose we play a game, Stats Land, which requires each player to roll two standard dice, one red die and one blue die. What is the mean value and standard deviation of values for each die? (Notice that the mean and standard deviation are the same for both the red and blue die.) One version of this game allows the player to move forward the number of spaces equal to the sum of the two dice. List all 36 possible sums and calculate their mean and standard deviation (use σ_x). Notice that the mean of the sums equals the sum of the two means. Does the standard deviation of the sums equal the sum of the two standard deviations? No! Students typically do not know what has happened to the standard deviations, so you may have to say, "Let's think about this and return to it in a moment." Next, another version of this game requires a player to move the difference of the red die and the blue die (red − blue). What would happen if we had a negative difference? *Move backwards.* List all 36 possible differences and calculate their mean and standard deviation. Notice that the mean of the differences equals the difference of the two means. What do you notice about the standard deviation of the differences? *It equals the standard deviation of the sums!* Typically, some students will see the relationship and be able to give you the rule: the standard deviation of the sums (or differences) equals the square root of the sum of the variances for the two dice. This rule for the standard deviation for the sum or difference of two or more random variables *only* applies when the random variables are independent!

Suggested Assignment: Exercises 7.27, 7.29, 7.30, 7.33, 7.34, 7.36, 7.38, 7.39, 7.43

■ **Section 7.5**

Two special discrete probability distributions, binomial and geometric, are discussed in this section. Students should know the characteristics that lead to a binomial random variable and also those that lead to a geometric random variable. These characteristics are similar except that experiments that result in a geometric random variable do not have a fixed number of trials. The binomial random variable is defined as the *number of successes* out of a fixed number of trials and its possible values range from 0 to n, where n is the fixed number of trials. The geometric random variable is defined as the number of trials until the *first success* is observed and its possible values range from 1 to infinity.

Remember, in a discrete distribution, $P(x < 2)$ and $P(x \le 2)$ are not equal. This is true for binomial and geometric distributions since they are discrete distributions. Students can either use the binomial tables in the back of the textbook to calculate binomial probabilities or use a graphing calculator or computer software to perform these calculations. See Graphing Calculator Exploration 7.2 on page 435 for instructions. For a geometric distribution, most graphing calculators will compute these probabilities. See Graphing Calculator Exploration 7.3 on page 438.

Figure 7.16 (on page 391) displays a binomial probability histogram. It is helpful to have students display the binomial probability histograms for various values of π, such as 0.1, 0.2, . . . , 0.9. These histograms may be created using a graphing calculator. Input the x-values in L1. In L2, use the command binomialpdf(n,π,L1) (the cursor must be on top of L2) to calculate the binomial probabilities for each x-value. Have students display their binomial probability histograms on the board. Ask students why the binomial probability histogram is strongly skewed right for small values of π. *If the probability of success is small, then we do not* **expect** *very many successes out of n.* Similarly, ask students why the binomial probability histogram is strongly skewed left for large values of π. *If the probability of success is large, then we do* **expect** *many successes out of n.* Students can also compute the mean and standard deviation for each value of π. What happens to the means as the probability of success (π) increases? Why? *The means increase because, as the probability of success increases, we expect more success out of n.* What happens to the standard deviations as the probability of success (π) increases? *The standard deviations increase until $\pi = 0.5$, then decrease. They are symmetrical around $\pi = 0.5$.* For which value of π

is the standard deviation the largest? *0.5.* This last question is very important. In Chapter 9, this fact will come in handy when we need to determine the sample size required to achieve a specified margin of error when estimating a population proportion.

Similarly, students can create geometric probability histograms. As the probability of success (π) increases, these histograms will become more strongly skewed right. Why? *As the probability of success (π) increases, it is more likely the first success will occur on an earlier trial.*

Suggested Assignment: Exercises 7.46, 7.52, 7.54, 7.55, 7.56, 7.60, 7.61, 7.62

Activity: Bonus Activity 7.3—Pass the Message (*Activities Manual*, p. 68)★ and "Chapter 7—The Binomial Distribution: binomialpdf & binomialcdf" (*Instructor's Resource Binder* & CD, Chapter 7 Activities Worksheets)

Assessment: Chapter 7 Quiz 2 (*Test Bank* and *Instructor's Resource Binder* CD)

■ **Section 7.6**

This section discusses one of the most important continuous probability distributions, the normal distribution. Normal distributions (and the density curves that define them) are distinguished by a specified mean and standard deviation. A graphing calculator demonstration helps students to understand what happens to the normal curve as means and standard deviations change. In the "Y=" window, use the command normalpdf(x,μ,σ) to graph normal curves. (Note: This command is used for graphing *only*! Why? Remember, $P(x = 2) = 0$ in a continuous distribution. Thus, the normalpdf function should never be used to calculate probabilities.) The transition (inflection) points are located one standard deviation from the mean (see Figure 7.19, page 398).

All normal distributions can be standardized into the *standard normal distribution* using $z = \dfrac{x - \mu}{\sigma}$. The standard normal distribution has a mean of 0 and standard deviation of 1. The tables on the back inside cover of the textbook give the areas under the standard normal curve for specified z-values. The standard normal curve can be used to calculate probabilities for any normal distribution by first transforming to z-scores and then using the table to determine the appropriate probability. Remember, in a continuous distribution, $P(x < 2)$ and $P(x \le 2)$ are equal. It is helpful for students to practice calculating probabilities from a normal distribution using the standard normal tables; however, the normalcdf function on a graphing calculator can be used to compute the probabilities. See Graphing Calculator Exploration 7.4 on page 440 for instructions.

Students should be able to convert a z-score back into an x-value. A common mistake students make is finding an incorrect z-score from the table. Students should find the given probability *in the body* of the table and follow the lines *out* to find the appropriate z-score. In other words, use a reverse procedure. The common mistake is for students to *not* use this reverse procedure to find the z-score. The invNorm function on a graphing calculator will display the z-score when a probability is specified.

Finally, provide problems that incorporate the linear transformation and linear combinations rules (from Section 7.4) with normal distributions. For example, the heights of male teachers at a school are normally distributed with mean of 71 inches and standard deviation of 2.6 inches. The heights of female teachers are normally distributed with mean of 66 inches and standard deviation of 2.3 inches. Suppose a male teacher and a female teacher are randomly selected from the respective populations and the difference (male − female) in their heights is calculated, and suppose this procedure is repeated a

★ Kathy's personal favorite

large number of times. How would we describe the distribution of the heights difference? *Difference in height would be normally distributed with mean of 5 inches and standard deviation of 3.4713 inches.* We use the linear combination rules to calculate the mean and standard deviation. Since the two original distributions were normal, then the distribution of the difference will also be normal. What is the probability that the difference in heights of a randomly selected male teacher and female teacher (male − female) is negative? Using a graphing calculator, $P((x_M - x_F) < 0) = 0.0749$.

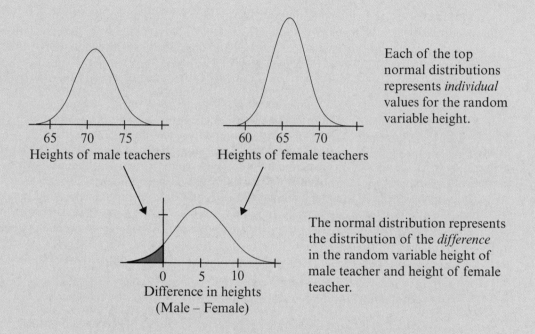

Each of the top normal distributions represents *individual* values for the random variable height.

The normal distribution represents the distribution of the *difference* in the random variable height of male teacher and height of female teacher.

Here is an example that uses the linear transformation rules. Suppose that the distribution of scores on a statistics test is approximately normal with a mean of 60 and standard deviation of 5. The teacher decides to compute "curved" scores by using the following: *Curve* = 1.2(*score*) + 3. Since the distribution of test scores is approximately normal, the distribution of curved scores will also be approximately normal with a mean of 75 and standard deviation of 6. What is the probability that a randomly selected student will have a curved score greater than 80? $P(curve) > 80) = .2023$.

Note: It is helpful to have students identify what variable's distribution is being described by each normal curve. A specific normal curve might describe the behavior of individual values for a random variable. In the examples above, normal distributions describe the behavior of the difference in random variables $(x_M - x_F)$ and the transformed score (x_C). In Chapter 8, we will see that under certain conditions, normal distributions are also used to describe the behavior of sample means ($\bar{x}$).

Suggested Assignment: Exercises 7.65, 7.67, 7.69, 7.71, 7.73, 7.77, 7.79, 7.80

Activity: Bonus Activity 7.2—The Sound of the Normal Distribution (*Activities Manual*, p. 67) and "Chapter 7—The Normal Distribution: normalcdf & invNorm" (*Instructor's Resource Binder* & CD, Chapter 7 Activities Worksheets)

Section 7.7

Many of the inferential procedures that are introduced in later chapters (Chapters 9–15) require that the population from which the sample is selected is approximately normally distributed. One way to assess whether it is reasonable to think the sample came from a normal population is to construct a normal probability plot. A normal probability plot is a scatterplot of the x-values graphed against a set of scores known to be normally distributed. Different sample sizes have different sets of normal scores. There are many different techniques for calculating normal scores. It is not necessary for students to know how to find normal scores—we will rely on technology to compute the normal scores for us.

To create the normal probability plot, the smallest observation from the data set is paired with the smallest score from the set of normal scores; the second smallest observation from the data set is paired with the second smallest score from the set of normal scores, and so on. We know the normal scores are normally distributed. What should happen if the data are from a population that is normally distributed? *The points should fall (approximately) on a straight line.* On the TI calculator, the normal probability plot is the last option under type of graph.

If a distribution is not normal (perhaps skewed), then transformations, similar to those used in Section 5.4, can be used to normalize the distribution. If the distribution is skewed right, either a logarithmic or square root transformation will typically normalize the distribution.

Suggested Assignment: Exercises 7.81, 7.82, 7.83, 7.85, 7.88, 7.92

Activity: "Chapter 7—Assessing Normality" (*Instructor's Resource Binder* & CD, Chapter 7 Activities Worksheets)

Assessment: Chapter 7 Quiz 3 (*Test Bank* and *Instructor's Resource Binder* CD)

Section 7.8

Prior to beginning this section, a brief discussion about the history of statistics enables students to appreciate the importance of the normal approximation to the binomial distribution. Before graphing calculators or computer software for statistics were common, the calculation of binomial probabilities was tedious. Binomial probabilities were computed either by using the tables provided in many books or, when appropriate, by using the normal distribution to approximate the binomial distribution. The inference procedures for proportions (Chapters 9 and 10) are based on this use of normal distributions to approximate binomial probabilities.

So when is it appropriate to use the normal distribution to approximate a binomial distribution? Remind students that in Section 7.5 we graphed binomial probability histograms for a given fixed number of trials (n) and a given probability of success (π). When the probability of success was small (or large), the probability histogram was strongly skewed. Let's repeat this activity with a large fixed number of trials and small probability of success. We will use the values for the number of trials and probability of success in Example 7.34, $n = 250$ and $\pi = 0.1$. Using a graphing calculator, input the x-values into L1 (since the calculator has a limited display window, use the values 0 to 45). Use the command binomialpdf(250,.1,L1) to calculate the probabilities in L2. Set the window of the calculator with an x-minimum $= -0.5$, x-maximum $= 45.5$, x-scale $= 1$, y-minimum $= 0$, y-maximum $= 0.2$, and y-scale $= 1$. Using the Stat Plot function, select a histogram with $x$list: L1 and Freq: L2. Before graphing, ask students what shape they predict for this binomial distribution. (*Most of them will probably say it will be skewed right.*) Surprise! Notice the histogram appears approximately normal. The mean of the binomial distribution is 25 ($n\pi$) and the standard deviation is 4.7434($\sqrt{n\pi(1 - \pi)}$). In the "Y1="

window, input normalpdf(x,25,4.7434). The normal curve fits very closely to the probability histogram. Through this exploration, students have a much better understanding of why and when it is appropriate to approximate a binomial distribution with a normal distribution (see the blue box on page 427).

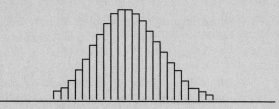

Suggested Assignment: Exercises 7.95, 7.96, 7.97, 7.101

Assessment: Chapter 7 Concept Quiz (*Test Bank* and *Instructor's Resource Binder* CD)

Suggested Review Assignment: Exercises 7.103, 7.104, 7.105, 7.107, 7.109, 7.112, 7.115, 7.117, 7.121, 7.123

Assessment: Chapter 7 Test (*Test Bank* and *Instructor's Resource Binder* CD)

Chapter 7

Random Variables and Probability Distributions

© Walter Bibikow/Getty Images

This chapter is the first of two chapters that together link the basic ideas of probability explored in Chapter 6 with the techniques of statistical inference. Chapter 6 used probability to describe the long-run relative frequency of occurrence of various types of outcomes. In this chapter we introduce probability models that can be used to describe the distribution of values of a variable. In Chapter 8, we will see how these same probability models can be used to describe the behavior of sample statistics. Such models are essential if we are to reach conclusions based on a sample from the population of interest.

In a chance experiment, we often focus on some numerical aspect of the outcome. An environmental scientist who obtains an air sample from a specified location might be especially concerned with the concentration of ozone (a major constituent of smog). A quality control inspector who must decide whether to accept a large shipment of components may base the decision on the number of defective components in a group of 20 components randomly selected from the shipment.

Before selection of the air sample, the value of the ozone concentration is uncertain. Similarly, the number of defective components among the 20 selected might be any whole number between 0 and 20. Because the value of a variable quantity such as ozone concentration or number of defective components is subject to uncertainty, such variables are called *random variables*.

In this chapter we begin by distinguishing between discrete and continuous numerical variables. We show how variation in both discrete and continuous numerical variables can be described by a probability distribution; this distribution can then be used to make probability statements about values of the random variable. Special emphasis is given to three commonly encountered probability distributions: the binomial, geometric, and normal distributions.

7.1 Random Variables

In most chance experiments, an investigator focuses attention on one or more variable quantities. For example, consider a management consultant who is studying the operation of a supermarket. The chance experiment might involve randomly selecting a customer leaving the store. One interesting numerical variable might be the number of items x purchased by the customer. Possible values of this variable are 0 (a frustrated customer), 1, 2, 3, and so on. Until a customer is selected and the number of items counted, the value of x is uncertain. Another variable of potential interest might be the time y (minutes) spent in a checkout line. One possible value of y is 3.0 min and another is 4.0 min, but *any* other number between 3.0 and 4.0 is also a possibility. Whereas possible values of x are isolated points on the number line, possible y values form an entire interval (a continuum) on the number line.

> **DEFINITION**
>
> A numerical variable whose value depends on the outcome of a chance experiment is called a **random variable**. A random variable associates a numerical value with each outcome of a chance experiment.
>
> A random variable is **discrete** if its set of possible values is a collection of isolated points on the number line. The variable is **continuous** if its set of possible values includes an entire interval on the number line.

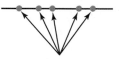

Possible values of a
discrete random variable

Possible values of a
continuous random variable

Figure 7.1 Two different types of random variables.

We use lowercase letters, such as x and y, to represent random variables.*

Figure 7.1 shows a set of possible values for each type of random variable. In practice, a discrete random variable almost always arises in connection with counting (e.g., the number of items purchased, the number of gas pumps in use, or the number of broken eggs in a carton). A continuous random variable is one whose value is typically obtained by measurement (temperature in a freezer compartment, weight of a pineapple, amount of time spent in the store, etc.). Because there is a limit to the accuracy of any measuring instrument, such as a watch or a scale, it may seem that any variable should be regarded as discrete. However, when there is a large number of closely spaced values, the variable's behavior is most easily studied by conceptualizing it as continuous. (Doing so allows the use of calculus to solve some types of probability problems.)

*In some books, uppercase letters are used to name random variables, with lowercase letters representing a particular value that the variable might assume. We have opted to use a simpler and less formal notation.

Example 7.1   Car Sales

Consider an experiment in which the type of car, new (N) or used (U), chosen by each of three successive customers at a discount car dealership is noted. Define a random variable x by

x = number of customers purchasing a new car

The experimental outcome in which the first and third customers purchase a new car and the second customer purchases a used car can be abbreviated NUN. The associated x value is 2, because two of the three customers selected a new car. Similarly, the x value for the outcome NNN (all three purchase a new car) is 3. We display each of the eight possible experimental outcomes and the corresponding value of x in the following table:

Outcome	UUU	NUU	UNU	UUN	NNU	NUN	UNN	NNN
x value	0	1	1	1	2	2	2	3

There are only four possible x values—0, 1, 2, and 3—and these are isolated points on the number line. Thus, x is a discrete random variable.

■

In some situations, the random variable of interest is discrete, but the number of possible values is not finite. This is illustrated in Example 7.2.

Example 7.2   This Could Be a Long Game . . .

Two friends agree to play a game that consists of a sequence of trials. The game continues until one player wins two trials in a row. One random variable of interest might be

x = number of trials required to complete the game

Let A denote a win for Player 1 and B denote a win for Player 2. The simplest possible experimental outcomes are AA (the case in which Player 1 wins the first two trials and the game ends) and BB (the case in which Player 2 wins the first two trials). With either of these two outcomes, $x = 2$. There are also two outcomes for which $x = 3$: ABB and BAA. Some other possible outcomes and associated x values are

Outcomes	x value
AA, BB	2
BAA, ABB	3
ABAA, BABB	4
ABABB, BABAA	5
⋮	⋮
ABABABABAA, BABABABABB	10

and so on.

Any positive integer that is at least 2 is a possible value. Because the values 2, 3, 4, . . . are isolated points on the number line (x is determined by counting), x is a discrete random variable even though there is no upper limit to the number of possible values.

■

Example 7.3   Stress

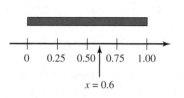

Figure 7.2 The bar for Example 7.3 and the outcome $x = 0.6$.

In an engineering stress test, pressure is applied to a thin 1-ft-long bar until the bar snaps. The precise location where the bar will snap is uncertain. Let x be the distance from the left end of the bar to the break. Then $x = 0.25$ is one possibility, $x = 0.9$ is another, and in fact any number between 0 and 1 is a possible value of x. (Figure 7.2 shows the case of the outcome $x = 0.6$.) This set of possible values is an entire interval on the number line, so x is a continuous random variable.

Even though in practice we may be able to measure the distance only to the nearest tenth of an inch or hundredth of an inch, the *actual* distance could be any number between 0 and 1. So, even though the recorded values might be rounded because of the accuracy of the measuring instrument, the variable is still continuous.

■

In data analysis, random variables often arise in the context of summarizing sample data when a sample is selected from some population. This is illustrated in Example 7.4.

Example 7.4   College Plans

Suppose that a counselor plans to select a random sample of 50 seniors at a large high school and to ask each student in the sample whether he or she plans to attend college after graduation. The process of sampling is a chance experiment. The sample space for this experiment consists of all the different possible random samples of size 50 that might result (there is a very large number of these), and for simple random sampling each of these outcomes is equally likely. Let

$$x = \text{number of successes in the sample}$$

where a success in this instance is defined as a student who plans to attend college. Then x is a random variable, because it associates a numerical value with each of the possible outcomes (random samples) that might occur. Possible values of x are 0, 1, 2, . . . , 50, and x is a discrete random variable.

■

■ Exercises 7.1–7.7

Turn to the Expanded Answers Section for answers not shown next to exercises.

7.1 State whether each of the following random variables is discrete or continuous:

a. The number of defective tires on a car   Discrete

b. The body temperature of a hospital patient   Continuous

c. The number of pages in a book   Discrete

d. The number of draws (with replacement) from a deck of cards until a heart is selected   Discrete

e. The lifetime of a lightbulb   Continuous

Bold exercises answered in back   ● Data set available online but not required   ▼ Video solution available

7.2 Classify each of the following random variables as either discrete or continuous:
a. The fuel efficiency (mpg) of an automobile   Continuous
b. The amount of rainfall at a particular location during the next year   Continuous
c. The distance that a person throws a baseball   Continuous
d. The number of questions asked during a 1-hr lecture
e. The tension (in pounds per square inch) at which a tennis racket is strung   Continuous
f. The amount of water used by a household during a given month   Continuous
g. The number of traffic citations issued by the highway patrol in a particular county on a given day   Discrete

7.3 Starting at a particular time, each car entering an intersection is observed to see whether it turns left (L) or right (R) or goes straight ahead (S). The experiment terminates as soon as a car is observed to go straight. Let y denote the number of cars observed. What are possible y values? List five different outcomes and their associated y values.   Positive integers.

7.4 A point is randomly selected from the interior of a square, as pictured:

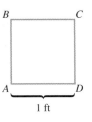

Let x denote the distance from the lower left-hand corner A of the square to the selected point. What are possible values of x? Is x a discrete or a continuous variable?

7.5 A point is randomly selected on the surface of a lake that has a maximum depth of 100 ft. Let y be the depth of the lake at the randomly chosen point. What are possible values of y? Is y discrete or continuous?   Any value between 0 and 100, continuous

7.6 A person stands at the corner marked A of the square pictured in Exercise 7.4 and tosses a coin. If it lands heads up, the person moves one corner clockwise, to B. If the coin lands tails up, the person moves one corner counterclockwise, to D. This process is then repeated until the person arrives back at A. Let y denote the number of coin tosses. What are possible values of y? Is y discrete or continuous?   All positive even integers, discrete

7.7 A box contains four slips of paper marked 1, 2, 3, and 4. Two slips are selected without replacement. List the possible values for each of the following random variables:
a. x = sum of the two numbers   3, 4, 5, 6, 7
b. y = difference between the first and second numbers
c. z = number of slips selected that show an even number
d. w = number of slips selected that show a 4   0, 1

7.2 Probability Distributions for Discrete Random Variables

The probability distribution for a random variable is a model that describes the long-run behavior of the variable. For example, suppose that the Department of Animal Regulation in a particular county is interested in studying the variable x = number of licensed dogs or cats for a household. County regulations prohibit more than five dogs or cats per household. If we consider the chance experiment of randomly selecting a household in this county, then x is a discrete random variable because it associates a numerical value (0, 1, 2, 3, 4, or 5) with each of the possible outcomes (households) in the sample space. Although we know what the possible values for x are, it would also be useful to know how this variable behaves in repeated observation. What would be the most common value? What proportion of the time would x = 5 be observed? x = 3? A probability distribution provides this type of information about the long-run behavior of a random variable.

> **DEFINITION**
>
> The **probability distribution of a discrete random variable** x gives the probability associated with each possible x value. Each probability is the limiting relative frequency of occurrence of the corresponding x value when the chance experiment is repeatedly performed.
>
> Common ways to display a probability distribution for a discrete random variable are a table, a probability histogram, or a formula.

If one possible value of x is 2, we often write $p(2)$ in place of $P(x = 2)$. Similarly, $p(5)$ denotes the probability that $x = 5$, and so on.

Example 7.5   Hot Tub Models

Suppose that each of four randomly selected customers purchasing a hot tub at a certain store chooses either an electric (E) or a gas (G) model. Assume that these customers make their choices independently of one another and that 40% of all customers select an electric model. This implies that for any particular one of the four customers, $P(E) = .4$ and $P(G) = .6$. One possible experimental outcome is EGGE, where the first and fourth customers select electric models and the other two choose gas models. Because the customers make their choices independently, the multiplication rule for independent events implies that

$$P(\text{EGGE}) = P(\text{1st chooses E } and \text{ 2nd chooses G } and \text{ 3rd chooses G } and \text{ 4th chooses E})$$
$$= P(E)P(G)P(G)P(E)$$
$$= (.4)(.6)(.6)(.4)$$
$$= .0576$$

Similarly,

$$P(\text{EGEG}) = P(E)P(G)P(E)P(G)$$
$$= (.4)(.6)(.4)(.6)$$
$$= .0576 \quad \text{(identical to } P(\text{EGGE}))$$

and

$$P(\text{GGGE}) = (.6)(.6)(.6)(.4) = .0864$$

The number among the four customers who purchase an electric hot tub is a random variable. Let

> $x =$ the number of electric hot tubs purchased by the four customers

Important to define random variable

Table 7.1 displays the 16 possible experimental outcomes, the probability of each outcome, and the value of the random variable x that is associated with each outcome.

The probability distribution of x is easily obtained from this information. Consider the smallest possible x value, 0. The only outcome for which $x = 0$ is GGGG, so

$$p(0) = P(x = 0) = P(\text{GGGG}) = .1296$$

Table 7.1 Outcomes and Probabilities for Example 7.5

Outcome	Probability	x Value	Outcome	Probability	x Value
GGGG	.1296	0	GEEG	.0576	2
EGGG	.0864	1	GEGE	.0576	2
GEGG	.0864	1	GGEE	.0576	2
GGEG	.0864	1	GEEE	.0384	3
GGGE	.0864	1	EGEE	.0384	3
EEGG	.0576	2	EEGE	.0384	3
EGEG	.0576	2	EEEG	.0384	3
EGGE	.0576	2	EEEE	.0256	4

There are four different outcomes for which $x = 1$, so $p(1)$ results from summing the four corresponding probabilities:

$$
\begin{aligned}
p(1) = P(x = 1) &= P(\text{EGGG } or \text{ GEGG } or \text{ GGEG } or \text{ GGGE}) \\
&= P(\text{EGGG}) + P(\text{GEGG}) + P(\text{GGEG}) + P(\text{GGGE}) \\
&= .0864 + .0864 + .0864 + .0864 \\
&= 4(.0864) \\
&= .3456
\end{aligned}
$$

Similarly,

$$
\begin{aligned}
p(2) &= P(\text{EEGG}) + \cdots + P(\text{GGEE}) = 6(.0576) = .3456 \\
p(3) &= 4(.0384) = .1536 \\
p(4) &= .0256
\end{aligned}
$$

The probability distribution of x is summarized in the following table:

x Value	0	1	2	3	4
$p(x)$ = Probability of Value	.1296	.3456	.3456	.1536	.0256

To interpret $p(3) = .1536$, think of performing the chance experiment repeatedly, each time with a new group of four customers. In the long run, 15.36% of these groups will have exactly three customers purchasing an electric hot tub. The probability distribution can be used to determine probabilities of various events involving x. For example, the probability that at least two of the four customers choose electric models is

$$
\begin{aligned}
P(x \geq 2) &= P(x = 2 \text{ or } x = 3 \text{ or } x = 4) \\
&= p(2) + p(3) + p(4) \\
&= .5248
\end{aligned}
$$

Thus, in the long run, 52.48% of the time a group of four hot tub purchasers will include at least two who select electric models.

■

A probability distribution table for a discrete variable shows the possible x values and also $p(x)$ for each possible x value. Because $p(x)$ is a probability, it must be a number between 0 and 1, and because the probability distribution lists all possible x values, the sum of all the $p(x)$ values must equal 1. These properties of discrete probability distributions are summarized in the following box.

Properties of Discrete Probability Distributions

1. For every possible x value, $0 \leq p(x) \leq 1$.
2. $\displaystyle\sum_{\text{all } x \text{ values}} p(x) = 1$

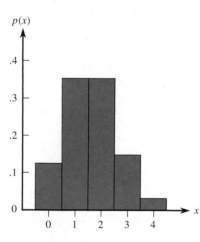

Figure 7.3 Probability histogram for the distribution of Example 7.5.

A pictorial representation of a discrete probability distribution is called a *probability histogram*. The picture has a rectangle centered above each possible value of x, and the area of each rectangle is the probability of the corresponding value. Figure 7.3 displays the probability histogram for the probability distribution of Example 7.5.

In Example 7.5, the probability distribution was derived by starting with a simple experimental situation and applying basic probability rules. When a derivation from fundamental probabilities is not possible because of the complexity of the experimental situation, an investigator often conjectures a probability distribution consistent with empirical evidence and prior knowledge. It must also be consistent with rules of probability. Specifically,

1. $p(x) \geq 0$ for every x value.
2. $\displaystyle\sum_{\text{all } x \text{ values}} p(x) = 1$

Example 7.6   Automobile Defects

A consumer organization that evaluates new automobiles customarily reports the number of major defects on each car examined. Let x denote the number of major defects on a randomly selected car of a certain type. A large number of automobiles were evaluated, and a probability distribution consistent with these observations is:

x	0	1	2	3	4	5	6	7	8	9	10
$p(x)$	.041	.010	.209	.223	.178	.114	.061	.028	.011	.004	.001

The corresponding probability histogram appears in Figure 7.4. The probabilities in this distribution reflect the organization's experience. For example, $p(3) = .223$ indicates that 22.3% of new automobiles had 3 major defects. The probability that the number of major defects is between 2 and 5 inclusive is

$$P(2 \leq x \leq 5) = p(2) + p(3) + p(4) + p(5) = .724$$

If car after car of this type were examined, in the long run, 72.4% would have 2, 3, 4, or 5 major defects.

Figure 7.4 Probability histogram for the distribution of the number of major defects on a randomly selected car.

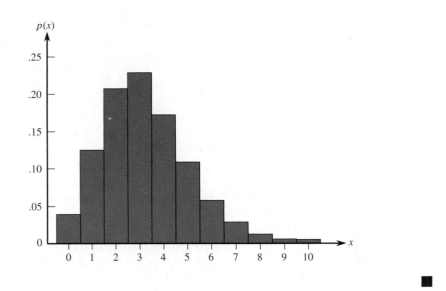

We have seen examples in which the probability distribution of a discrete random variable has been given as a table or as a probability (relative frequency) histogram. It is also possible to give a formula that allows calculation of the probability for each possible value of the random variable. Examples of this approach are given in Section 7.5.

Exercises 7.8–7.19

Turn to the Expanded Answers Section for answers not shown next to exercises.

7.8 Let x be the number of courses for which a randomly selected student at a certain university is registered. The probability distribution of x appears in the following table:

x	1	2	3	4	5	6	7
$p(x)$	.02	.03	.09	.25	.40	.16	.05

a. What is $P(x = 4)$?   0.25
b. What is $P(x \le 4)$?   0.39
c. What is the probability that the selected student is taking at most five courses?   0.79
d. What is the probability that the selected student is taking at least five courses? more than five courses?   0.61, 0.21
e. Calculate $P(3 \le x \le 6)$ and $P(3 < x < 6)$. Explain in words why these two probabilities are different.

7.9 ▼ Let y denote the number of broken eggs in a randomly selected carton of one dozen eggs. Suppose that the probability distribution of y is as follows:

y	0	1	2	3	4
$p(y)$	.65	.20	.10	.04	?

a. Only y values of 0, 1, 2, 3, and 4 have positive probabilities. What is $p(4)$?   0.01
b. How would you interpret $p(1) = .20$?
c. Calculate $P(y \le 2)$, the probability that the carton contains at most two broken eggs, and interpret this probability.   0.95
d. Calculate $P(y < 2)$, the probability that the carton contains *fewer than* two broken eggs. Why is this smaller than the probability in Part (c)?

Bold exercises answered in back   ● Data set available online but not required   ▼ Video solution available

e. What is the probability that the carton contains exactly 10 unbroken eggs? 0.1
f. What is the probability that at least 10 eggs are unbroken? 0.95

7.10 A restaurant has four bottles of a certain wine in stock. Unbeknownst to the wine steward, two of these bottles (Bottles 1 and 2) are bad. Suppose that two bottles are ordered, and let x be the number of good bottles among these two.
a. One possible experimental outcome is (1,2) (Bottles 1 and 2 are the ones selected) and another is (2,4). List all possible outcomes. (1,2), (1,3), (1,4), (2,3), (2,4), (3,4)
b. Assuming that the two bottles are randomly selected from among the four, what is the probability of each outcome in Part (a)? 1/6
c. The value of x for the (1,2) outcome is 0 (neither selected bottle is good), and $x = 1$ for the outcome (2,4). Determine the x value for each possible outcome. Then use the probabilities in Part (b) to determine the probability distribution of x.

7.11 Airlines sometimes overbook flights. Suppose that for a plane with 100 seats, an airline takes 110 reservations. Define the variable x as the number of people who actually show up for a sold-out flight. From past experience, the probability distribution of x is given in the following table:

x	95	96	97	98	99	100	101	102
$p(x)$	.05	.10	.12	.14	.24	.17	.06	.04

x	103	104	105	106	107	108	109	110
$p(x)$	.03	.02	.01	.005	.005	.005	.0037	.0013

a. What is the probability that the airline can accommodate everyone who shows up for the flight? 0.82
b. What is the probability that not all passengers can be accommodated? 0.18
c. If you are trying to get a seat on such a flight and you are number 1 on the standby list, what is the probability that you will be able to take the flight? What if you are number 3? 0.65, 0.27

7.12 Suppose that a computer manufacturer receives computer boards in lots of five. Two boards are selected from each lot for inspection. We can represent possible outcomes of the selection process by pairs. For example, the pair (1,2) represents the selection of Boards 1 and 2 for inspection.
a. List the 10 different possible outcomes.
b. Suppose that Boards 1 and 2 are the only defective boards in a lot of five. Two boards are to be chosen at ran-

dom. Define x to be the number of defective boards observed among those inspected. Find the probability distribution of x.

7.13 Simulate the chance experiment described in Exercise 7.12 using five slips of paper, with two marked *defective* and three marked *nondefective*. Place the slips in a box, mix them well, and draw out two. Record the number of defective boards. Replace the slips and repeat until you have 50 observations on the variable x. Construct a relative frequency distribution for the 50 observations, and compare this with the probability distribution obtained in Exercise 7.12.

7.14 Of all airline flight requests received by a certain discount ticket broker, 70% are for domestic travel (D) and 30% are for international flights (I). Let x be the number of requests among the next three requests received that are for domestic flights. Assuming independence of successive requests, determine the probability distribution of x. (Hint: One possible outcome is DID, with the probability $(.7)(.3)(.7) = .147$.)

7.15 Suppose that 20% of all homeowners in an earthquake-prone area of California are insured against earthquake damage. Four homeowners are selected at random; let x denote the number among the four who have earthquake insurance.
a. Find the probability distribution of x. (Hint: Let S denote a homeowner who has insurance and F one who does not. Then one possible outcome is SFSS, with probability $(.2)(.8)(.2)(.2)$ and associated x value of 3. There are 15 other outcomes.)
b. What is the most likely value of x? 0 and 1
c. What is the probability that at least two of the four selected homeowners have earthquake insurance? 0.1808

7.16 A box contains five slips of paper, marked $1, $1, $1, $10, and $25. The winner of a contest selects two slips of paper at random and then gets the larger of the dollar amounts on the two slips. Define a random variable w by w = amount awarded. Determine the probability distribution of w. (Hint: Think of the slips as numbered 1, 2, 3, 4, and 5, so that an outcome of the experiment consists of two of these numbers.)

7.17 Components coming off an assembly line are either free of defects (S, for success) or defective (F, for failure). Suppose that 70% of all such components are defect-free.

Components are independently selected and tested one by one. Let y denote the number of components that must be tested until a defect-free component is obtained.
a. What is the smallest possible y value, and what experimental outcome gives this y value? What is the second smallest y value, and what outcome gives rise to it?
b. What is the set of all possible y values?
c. Determine the probability of each of the five smallest y values. You should see a pattern that leads to a simple formula for $p(y)$, the probability distribution of y.

7.18 A contractor is required by a county planning department to submit anywhere from one to five forms (depending on the nature of the project) in applying for a building permit. Let y be the number of forms required of the next applicant. The probability that y forms are required is known to be proportional to y; that is, $p(y) = ky$ for $y = 1, \ldots, 5$.
a. What is the value of k? (Hint: $\Sigma\, p(y) = 1$.) $k = 1/15$
b. What is the probability that at most three forms are required?   0.4

c. What is the probability that between two and four forms (inclusive) are required?   0.6
d. Could $p(y) = y^2/50$ for $y = 1, 2, 3, 4, 5$ be the probability distribution of y? Explain.   No. The probabilities do not sum to 1.

7.19 A library subscribes to two different weekly news magazines, each of which is supposed to arrive in Wednesday's mail. In actuality, each one could arrive on Wednesday (W), Thursday (T), Friday (F), or Saturday (S). Suppose that the two magazines arrive independently of one another and that for each magazine $P(W) = .4$, $P(T) = .3$, $P(F) = .2$, and $P(S) = .1$. Define a random variable y by $y =$ the number of days beyond Wednesday that it takes for both magazines to arrive. For example, if the first magazine arrives on Friday and the second magazine arrives on Wednesday, then $y = 2$, whereas $y = 1$ if both magazines arrive on Thursday. Obtain the probability distribution of y. (Hint: Draw a tree diagram with two generations of branches, the first labeled with arrival days for Magazine 1 and the second for Magazine 2.)

Bold exercises answered in back   ● Data set available online but not required   ▼ Video solution available

---

7.3 Probability Distributions for Continuous Random Variables

A continuous random variable is one that has as its set of possible values an entire interval on the number line. An example is the weight x (in pounds) of a newborn child. Suppose for the moment that weight is recorded only to the nearest pound. Then possible x values are whole numbers, such as 4 or 9. The probability distribution can be pictured as a probability histogram in which the area of each rectangle is the probability of the corresponding weight value. The total area of all the rectangles is 1, and the probability that a weight (to the nearest pound) is between two values, such as 6 and 8, is the sum of the corresponding rectangular areas. Figure 7.5(a) illustrates this.

Excellent example

Now suppose that weight is measured to the nearest tenth of a pound. There are many more possible weight values than before, such as 5.0, 5.1, 5.7, 7.3, and 8.9. As shown in Figure 7.5(b), the rectangles in the probability histogram are much narrower, and this histogram has a much smoother appearance than the first one. Again, this histogram can be drawn so that the area of each rectangle equals the corresponding probability, and the total area of all the rectangles is 1.

Figure 7.5(c) shows what happens as weight is measured to a greater and greater degree of accuracy. The sequence of probability histograms approaches a smooth curve. The curve cannot go below the horizontal measurement scale, and the total area under the curve is 1 (because this is true of every probability histogram). The probability that

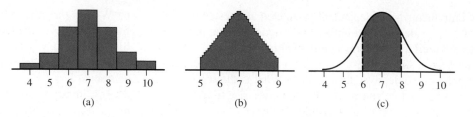

Figure 7.5   Probability distribution for birth weight: (a) weight measured to the nearest pound; (b) weight measured to the nearest tenth of a pound; (c) limiting curve as measurement accuracy increases; shaded area = $P(6 \leq \text{weight} \leq 8)$.

x falls in an interval such as $6 \leq x \leq 8$ is the area under the curve and above that interval.

> ### DEFINITION
>
> A **probability distribution for a continuous random variable x** is specified by a mathematical function denoted by $f(x)$ and called the **density function**. The graph of a density function is a smooth curve (the **density curve**). The following requirements must be met:
>
> 1. $f(x) \geq 0$ (so that the curve cannot dip below the horizontal axis).
> 2. The total area under the density curve is equal to 1.
>
> The probability that x falls in any particular interval is the area under the density curve and above the interval.

Many probability calculations for continuous random variables involve the following three events:

1. $a < x < b$, the event that the random variable x assumes a value between two given numbers, a and b
2. $x < a$, the event that the random variable x assumes a value less than a given number a
3. $b < x$, the event that the random variable x assumes a value greater than a given number b (this can also be written as $x > b$)

Figure 7.6 illustrates how the probabilities of these events are identified with areas under a density curve.

Figure 7.6   Probabilities as areas under a probability density curve.

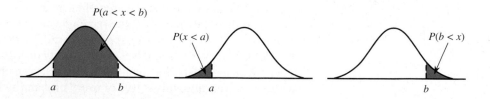

Example 7.7   Application Processing Times

Define a continuous random variable x by x = amount of time (in minutes) taken by a clerk to process a certain type of application form. Suppose that x has a probability distribution with density function

$$f(x) = \begin{cases} .5 & 4 < x < 6 \\ 0 & \text{otherwise} \end{cases}$$

The graph of $f(x)$, the density curve, is shown in Figure 7.7(a). It is especially easy to use this density curve to calculate probabilities, because it just requires finding the area of rectangles using the formula

$$\text{area} = (\text{base})(\text{height})$$

Figure 7.7   The uniform distribution for Example 7.7.

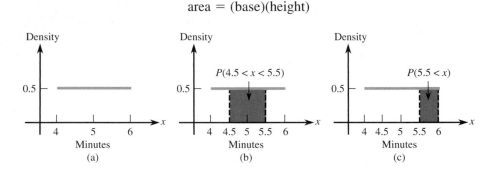

The curve has positive height, 0.5, only between $x = 4$ and $x = 6$. The total area under the curve is just the area of the rectangle with base extending from 4 to 6 and with height 0.5. This gives

$$\text{area} = (6 - 4)(0.5) = 1$$

as required.

When the density is constant over an interval (resulting in a horizontal density curve), the probability distribution is called a *uniform distribution*.

As illustrated in Figure 7.7(b), the probability that x is between 4.5 and 5.5 is

$$
\begin{aligned}
P(4.5 < x < 5.5) &= \text{area of shaded rectangle} \\
&= (\text{base width})(\text{height}) \\
&= (5.5 - 4.5)(.5) \\
&= .5
\end{aligned}
$$

Similarly (see Figure 7.7(c)), because in this context $x > 5.5$ is equivalent to $5.5 \leq x \leq 6$, we have

$$P(5.5 < x) = (6 - 5.5)(.5) = .25$$

According to this model, in the long run, 25% of all forms that are processed will have processing times that exceed 5.5 min.

■

The probability that a *discrete* random variable x lies in the interval between two limits a and b depends on whether either limit is included in the interval. Suppose, for example, that x is the number of major defects on a new automobile. Then

$$P(3 \leq x \leq 7) = p(3) + p(4) + p(5) + p(6) + p(7)$$

whereas

$$P(3 < x < 7) = p(4) + p(5) + p(6)$$

Important concept However, if x is a *continuous* random variable, such as task completion time, then

$$P(3 \leq x \leq 7) = P(3 < x < 7)$$

because the area under a density curve and above a single value such as 3 or 7 is 0. Geometrically, we can think of finding the area above a single point as finding the area of a rectangle with width = 0. The area above an interval of values therefore does not depend on whether either endpoint is included.

For any two numbers a and b with $a < b$,

$$P(a \leq x \leq b) = P(a < x \leq b) = P(a \leq x < b) = P(a < x < b)$$

when x is a continuous random variable.

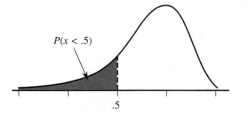

$P(x < .5)$

.5

Figure 7.8 A cumulative area under a density curve.

Probabilities for continuous random variables are often calculated using cumulative areas. A cumulative area is all of the area under the density curve to the left of a particular value. Figure 7.8 illustrates the cumulative area to the left of .5, which is $P(x < .5)$. The probability that x is in any particular interval, $P(a < x < b)$, is the difference between two cumulative areas.

The probability that a continuous random variable x lies between a lower limit a and an upper limit b is

$$P(a < x < b) = (\text{cumulative area to the left of } b) - (\text{cumulative area to the left of } a)$$
$$= P(x < b) - P(x < a)$$

The foregoing property is illustrated in Figure 7.9 for the case of $a = .25$ and $b = .75$. We will use this result extensively in Section 7.6 when we calculate probabilities using the normal distribution.

For some continuous distributions, cumulative areas can be calculated using methods from the branch of mathematics called integral calculus. However, because we are not assuming knowledge of calculus, we will rely on tables that have been constructed for the commonly encountered continuous probability distributions.

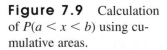

Figure 7.9 Calculation of $P(a < x < b)$ using cumulative areas.

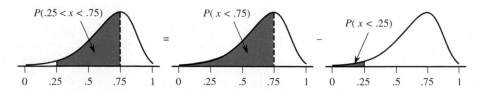

Exercises 7.20–7.26

7.20 Let x denote the lifetime (in thousands of hours) of a certain type of fan used in diesel engines. The density curve of x is as pictured:

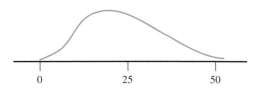

Shade the area under the curve corresponding to each of the following probabilities (draw a new curve for each part):

a. $P(10 < x < 25)$
b. $P(10 \le x \le 25)$
c. $P(x < 30)$
d. The probability that the lifetime is at least 25,000 hr
e. The probability that the lifetime exceeds 25,000 hr

7.21 A particular professor never dismisses class early. Let x denote the amount of time past the hour (minutes) that elapses before the professor dismisses class. Suppose that x has a uniform distribution on the interval from 0 to 10 min. The density curve is shown in the following figure:

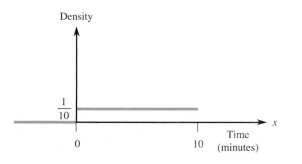

a. What is the probability that at most 5 min elapse before dismissal?   0.5
b. What is the probability that between 3 and 5 min elapse before dismissal?   0.2

7.22 Refer to the probability distribution given in Exercise 7.21. Put the following probabilities in order,

from smallest to largest:

$$P(2 < x < 3), P(2 \le x \le 3), P(x < 2), P(x > 7).$$

Explain your reasoning. $P(2 < x < 3) = P(2 \le x \le 3) < P(x < 2) < P(x > 7)$

7.23 The article "Modeling Sediment and Water Column Interactions for Hydrophobic Pollutants" (*Water Research* [1984]: 1169–1174) suggests the uniform distribution on the interval from 7.5 to 20 as a model for $x =$ depth (in centimeters) of the bioturbation layer in sediment for a certain region.

a. Draw the density curve for x.
b. What is the height of the density curve?   0.08
c. What is the probability that x is at most 12?   0.36
d. What is the probability that x is between 10 and 15? Between 12 and 17? Why are these two probabilities equal?   0.4, 0.4; they are equal because the associated areas are the same.

7.24 Let x denote the amount of gravel sold (in tons) during a randomly selected week at a particular sales facility. Suppose that the density curve has height $f(x)$ above the value x, where

$$f(x) = \begin{cases} 2(1 - x) & 0 \le x \le 1 \\ 0 & \text{otherwise} \end{cases}$$

The density curve (the graph of $f(x)$) is shown in the following figure:

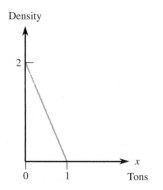

Use the fact that the area of a triangle $= \frac{1}{2}(\text{base})(\text{height})$ to calculate each of the following probabilities:

a. $P\left(x < \frac{1}{2}\right)$ 0.75

b. $P\left(x \leq \dfrac{1}{2}\right)$ 0.75

c. $P\left(x < \dfrac{1}{4}\right)$ 0.4375

d. $P\left(\dfrac{1}{4} < x < \dfrac{1}{2}\right)$ (Hint: Use the results of Parts (a)–(c).) 0.3125

e. The probability that gravel sold exceeds $\frac{1}{2}$ ton 0.25

f. The probability that gravel sold is at least $\frac{1}{4}$ ton 0.5625

7.25 Let x be the amount of time (in minutes) that a particular San Francisco commuter must wait for a BART train. Suppose that the density curve is as pictured (a uniform distribution):

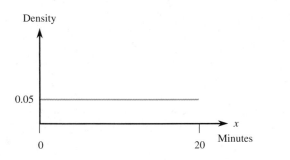

a. What is the probability that x is less than 10 min? more than 15 min? 0.5, 0.25

b. What is the probability that x is between 7 and 12 min?

c. Find the value c for which $P(x < c) = .9$.   18 mins

7.26 Referring to Exercise 7.25, let x and y be waiting times on two independently selected days. Define a new random variable w by $w = x + y$, the sum of the two waiting times. The set of possible values for w is the interval from 0 to 40 (because both x and y can range from 0 to 20). It can be shown that the density curve of w is as pictured (this curve is called a triangular distribution, for obvious reasons!):

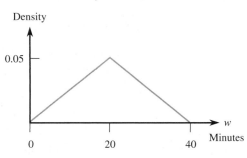

a. Verify that the total area under the density curve is equal to 1. (Hint: The area of a triangle is $= \frac{1}{2}$(base)(height).)

b. What is the probability that w is less than 20? less than 10? greater than 30?   0.5, 0.125, 0.125

c. What is the probability that w is between 10 and 30? (Hint: It might be easier first to find the probability that w is not between 10 and 30.)   0.75

Bold exercises answered in back   ● Data set available online but not required   ▼ Video solution available

---

7.4 # Mean and Standard Deviation of a Random Variable

We study a random variable x, such as the number of insurance claims made by a homeowner (a discrete variable) or the birth weight of a baby (a continuous variable), to learn something about how its values are distributed along the measurement scale. The sample mean $\bar{x}$ and sample standard deviation s summarize center and spread for the values in a sample. Similarly, the mean value and standard deviation of a random variable describe where the variable's probability distribution is centered and the extent to which it spreads out about the center.

> The **mean value of a random variable** x, denoted by $\boldsymbol{\mu}_x$ describes where the probability distribution of x is centered.
>
> The **standard deviation of a random variable** x, denoted by $\boldsymbol{\sigma}_x$ describes variability in the probability distribution. When σ_x is small, observed values of x will tend to be close to the mean value (little variability). When the value of σ_x is large, there will be more variability in observed x values.

Figure 7.10(a) shows two discrete probability distributions with the same standard deviation (spread) but different means (center). One distribution has a mean of $\mu_x = 6$ and the other has $\mu_x = 10$. Which is which? Figure 7.10(b) shows two continuous probability distributions that have the same mean but different standard deviations. Which distribution—(i) or (ii)—has the larger standard deviation? Finally, Figure 7.10(c) shows three continuous distributions with different means and standard deviations. Which of the three distributions has the largest mean? Which has a mean of about 5? Which distribution has the smallest standard deviation? (The correct answers to our questions are the following: Figure 7.10(a)(ii) has a mean of 6, and Figure 7.10(a)(i) has a mean of 10; Figure 7.10(b)(ii) has the larger standard deviation; Figure 7.10(c)(iii) has

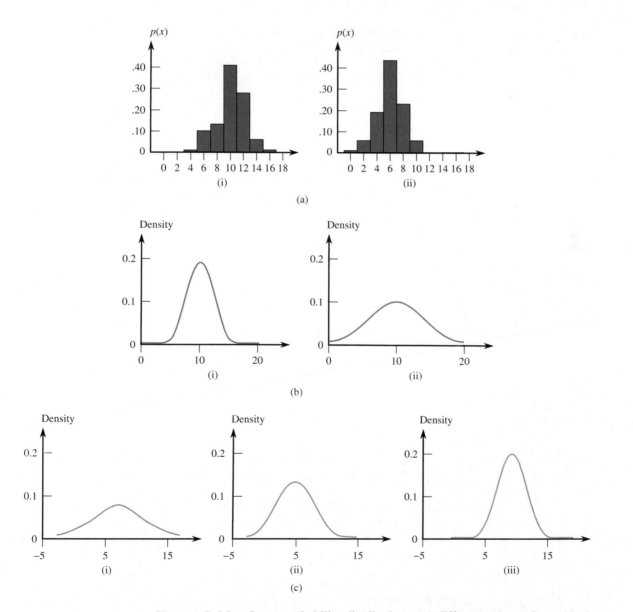

Figure 7.10 Some probability distributions: (a) different values of μ_x with the same value of σ_x; (b) different values of σ_x with the same value of μ_x; (c) different values of μ_x and σ_x.

the largest mean, Figure 7.10(c)(ii) has a mean of about 5, and Figure 7.10(c)(iii) has the smallest standard deviation.)

It is customary to use the terms *mean of the random variable x* and *mean of the probability distribution of x* interchangeably. Similarly, the standard deviation of the random variable *x* and the standard deviation of the probability distribution of *x* refer to the same thing. Although the mean and standard deviation are computed differently for discrete and continuous random variables, the interpretation is the same in both cases.

▪ Mean Value of a Discrete Random Variable ....

Consider an experiment consisting of the random selection of an automobile licensed in a particular state. Let the discrete random variable *x* be the number of low-beam headlights on the selected car that need adjustment. Possible *x* values are 0, 1, and 2, and the probability distribution of *x* might be as follows:

x value	0	1	2
Probability	.5	.3	.2

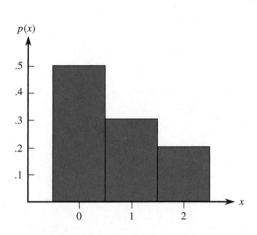

$p(x)$

Figure 7.11 Probability histogram for the distribution of the number of headlights needing adjustments.

The corresponding probability histogram appears in Figure 7.11. In a sample of 100 cars, the sample relative frequencies might differ somewhat from the given probabilities (which are the limiting relative frequencies). We might see:

x value	0	1	2
Frequency	46	33	21

The sample average value of *x* for these 100 observations is then the sum of 46 zeros, 33 ones, and 21 twos, all divided by 100:

$$\bar{x} = \frac{(46)(0) + (33)(1) + (21)(2)}{100}$$

$$= \left(\frac{46}{100}\right)(0) + \left(\frac{33}{100}\right)(1) + \left(\frac{21}{100}\right)(2)$$

$$= (\text{rel. freq. of } 0)(0) + (\text{rel. freq. of } 1)(1) + (\text{rel. freq. of } 2)(2)$$

$$= .75$$

As the sample size increases, each relative frequency approaches the corresponding probability. In a very long sequence of experiments, the value of $\bar{x}$ approaches

$$(\text{probability that } x = 0)(0) + (\text{probability that } x = 1)(1) + (\text{probability that } x = 2)(2)$$
$$= (.5)(0) + (.3)(1) + (.2)(2)$$
$$= .70$$
$$= \text{mean value of } x$$

Notice that the expression for $\bar{x}$ is a weighted average of possible *x* values; the weight of each value is the observed relative frequency. Similarly, the mean value of the random variable *x* is a weighted average, but now the weights are the probabilities from the probability distribution, as given in the definition in the following box.

Formula given on the
AP Exam

> ### DEFINITION
>
> The **mean value of a discrete random variable** x, denoted by μ_x, is computed by first multiplying each possible x value by the probability of observing that value and then adding the resulting quantities. Symbolically,
>
> $$\mu_x = \sum_{\text{all possible } x \text{ values}} x \cdot p(x)$$
>
> The term **expected value** is sometimes used in place of mean value, and $E(x)$ is alternative notation for μ_x.

....

Example 7.8 Exam Attempts

Individuals applying for a certain license are allowed up to four attempts to pass the licensing exam. Let x denote the number of attempts made by a randomly selected applicant. The probability distribution of x is as follows:

x	1	2	3	4
$p(x)$	.10	.20	.30	.40

Then x has mean value

$$
\begin{aligned}
\mu_x &= \sum_{x=1,2,3,4} x \cdot p(x) \\
&= (1)p(1) + (2)p(2) + (3)p(3) + (4)p(4) \\
&= (1)(.10) + (2)(.20) + (3)(.30) + (4)(.40) \\
&= .10 + .40 + .90 + 1.60 \\
&= 3.00
\end{aligned}
$$

■

Important concept

It is no accident that the symbol μ_x for the mean value is the same symbol used previously for a population mean. When the probability distribution describes how x values are distributed among the members of a population (and therefore the probabilities are population relative frequencies), the mean value of x is exactly the average value of x in the population.

....

Example 7.9 Apgar Scores

At 1 min after birth and again at 5 min, each newborn child is given a numerical rating called an Apgar score. Possible values of this score are 0, 1, 2, ..., 9, 10. A child's score is determined by five factors: muscle tone, skin color, respiratory effort, strength of heartbeat, and reflex, with a high score indicating a healthy infant. Let the random variable x denote the Apgar score (at 1 min) of a randomly selected newborn infant at a particular hospital, and suppose that x has the following probability distribution:

x	0	1	2	3	4	5	6	7	8	9	10
$p(x)$	.002	.001	.002	.005	.02	.04	.17	.38	.25	.12	.01

The mean value of x is

$$\mu_x = (0)p(0) + (1)p(1) + \cdots + (9)p(9) + (10)p(10)$$
$$= (0)(.002) + (1)(.001) + \cdots + (9)(.12) + (10)(.01)$$
$$= 7.16$$

The average Apgar score for a *sample* of newborn children born at this hospital may be $\bar{x} = 7.05, \bar{x} = 8.30$, or any one of a number of other possible values between 0 and 10. However, as child after child is born and rated, the average score will approach the value 7.16. This value can be interpreted as the mean Apgar score for the population of all babies born at this hospital.

■

■ Standard Deviation of a Discrete Random Variable ....

The mean value μ_x provides only a partial summary of a probability distribution. Two different distributions can have the same value of μ_x, yet a long sequence of sample values from one distribution might exhibit considerably more variability than a long sequence of values from the other distribution.

Example 7.10 Defective Components

A television manufacturer receives certain components in lots of four from two different suppliers. Let x and y denote the number of defective components in randomly selected lots from the first and second suppliers, respectively. The probability distributions for x and y are as follows:

x	0	1	2	3	4	y	0	1	2	3	4
$p(x)$	.4	.3	.2	.1	0	$p(y)$	.2	.6	.2	0	0

Probability histograms for x and y are given in Figure 7.12.

It is easy to verify that the mean values of both x and y are 1, so for either supplier the long-run average number of defective components per lot is 1. However, the two probability histograms show that the probability distribution for the second supplier is concentrated closer to the mean value than is the first supplier's distribution.

Figure 7.12 Probability distribution for the number of defective components in Example 7.10: (a) Supplier 1; (b) Supplier 2.

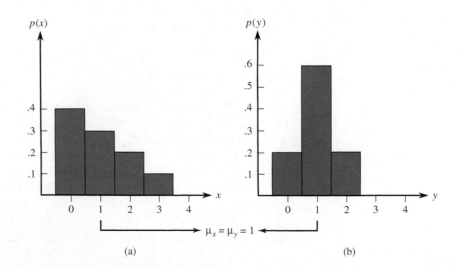

$\mu_x = \mu_y = 1$

(a)   (b)

The greater spread of the first distribution implies that there will be more variability in a long sequence of observed x values than in an observed sequence of y values. For example, the y sequence will contain no 3's, whereas in the long run, 10% of the observed x values will be 3.

■

As with s^2 and s, the variance and standard deviation of x involve squared deviations from the mean. A value far from the mean results in a large squared deviation. However, such a value contributes substantially to variability in x only if the probability associated with that value is not too small. For example, if $\mu_x = 1$ and $x = 25$ is a possible value, then the squared deviation is $(25 - 1)^2 = 576$. If, however, $P(x = 25) = .000001$, the value 25 will hardly ever be observed, so it won't contribute much to variability in a long sequence of observations. This is why each squared deviation is multiplied by the probability associated with the value to obtain a measure of variability.

Formula given on the AP Exam

> ### DEFINITION
>
> The **variance of a discrete random variable** x, denoted by σ_x^2, is computed by first subtracting the mean from each possible x value to obtain the deviations, then squaring each deviation and multiplying the result by the probability of the corresponding x value, and finally adding these quantities. Symbolically,
>
> $$\sigma_x^2 = \sum_{\text{all possible } x \text{ values}} (x - \mu)^2 p(x)$$
>
> The **standard deviation of** x, denoted by σ_x, is the square root of the variance.

When the probability distribution describes how x values are distributed among members of a population (so that the probabilities are population relative frequencies) σ_x^2 and σ_x are the population variance and standard deviation (of x), respectively.

Example 7.11 Defective Components Revised

For x = number of defective components in a lot from the first supplier in Example 7.10,

$$\begin{aligned}
\sigma_x^2 &= (0 - 1)^2 p(0) + (1 - 1)^2 p(1) + (2 - 1)^2 p(2) + (3 - 1)^2 p(3) \\
&= (1)(.4) + (0)(.3) + (1)(.2) + (4)(.1) \\
&= 1.0
\end{aligned}$$

Therefore $\sigma_x = 1.0$. For y = the number of defectives in a lot from the second supplier,

$$\sigma_y^2 = (0 - 1)^2(.2) + (1 - 1)^2(.6) + (2 - 1)^2(.2) = .4$$

Then $\sigma_y = \sqrt{.4} = .632$. The fact that $\sigma_x > \sigma_y$ confirms the impression conveyed by Figure 7.12 concerning the variability of x and y.

■

Example 7.12 More on Apgar scores

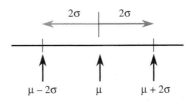

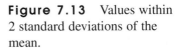

Figure 7.13 Values within 2 standard deviations of the mean.

Reconsider the distribution of Apgar scores for children born at a certain hospital, introduced in Example 7.9. What is the probability that a randomly selected child's score will be within 2 standard deviations of the mean score? As Figure 7.13 shows, values of x within 2 standard deviations of the mean are those for which $\mu - 2\sigma < x < \mu + 2\sigma$.

From Example 7.9 we already have $\mu_x = 7.16$. The variance is

$$\sigma^2 = \sum (x - \mu)^2 p(x) = \sum (x - 7.16)^2 p(x)$$
$$= (0 - 7.16)^2(.002) + (1 - 7.16)^2(.001) + \cdots + (10 - 7.16)^2(.01)$$
$$= 1.5684$$

and the standard deviation is

$$\sigma = \sqrt{1.5684} = 1.25$$

This gives (using the probabilities given in Example 7.9)

$$P(\mu - 2\sigma < x < \mu + 2\sigma) = P(7.16 - 2.50 < x < 7.16 + 2.50)$$
$$= P(4.66 < x < 9.66)$$
$$= p(5) + \cdots + p(9)$$
$$= .96$$

▪

▪ Mean and Standard Deviation When *x* Is Continuous

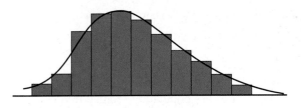

Figure 7.14 Approximating a density curve by a probability histogram.

Students should know these interpretations.

Figure 7.14 illustrates how the density curve for a continuous random variable can be approximated by a probability histogram of a discrete random variable. Computing the mean value and the standard deviation using this discrete distribution gives approximate values of μ_x and σ_x for the continuous random variable x. If an even more accurate approximating probability histogram is used (narrower rectangles), better approximations of μ_x and σ_x result.

In practice, such an approximation method is often unnecessary. Instead, μ_x and σ_x can be defined and computed using methods from calculus. The details need not concern us; what is important is that μ_x and σ_x play exactly the same role here as they did in the discrete case. The mean value μ_x locates the center of the continuous distribution and gives the approximate long-run average of many observed x values. The standard deviation σ_x measures the extent that the continuous distribution (density curve) spreads out about μ_x and gives information about the amount of variability that can be expected in a long sequence of observed x values.

Example 7.13 A "Concrete" Example

A company receives concrete of a certain type from two different suppliers. Define random variables x and y as follows:

x = compressive strength of a randomly selected batch from Supplier 1
y = compressive strength of a randomly selected batch from Supplier 2

Suppose that

$$\mu_x = 4650 \text{ lb/in.}^2 \qquad \sigma_x = 200 \text{ lb/in.}^2$$
$$\mu_y = 4500 \text{ lb/in.}^2 \qquad \sigma_y = 275 \text{ lb/in.}^2$$

The long-run average strength per batch for many, many batches from Supplier 1 will be roughly 4650 lb/in.^{2}. This is 150 lb/in.^{2} greater than the long-run average for batches from Supplier 2. In addition, a long sequence of batches from Supplier 1 will exhibit substantially less variability in compressive strength values than will a similar sequence from Supplier 2. The first supplier is preferred to the second both in terms of average value and variability. Figure 7.15 displays density curves that are consistent with this information.

Figure 7.15 Density curves for Example 7.13.

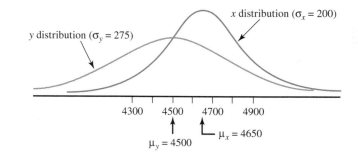

■

■ Mean and Variance of Linear Functions and Linear Combinations ....

We have seen how the mean and standard deviation of one or more random variables provide useful information about the variables' long-run behavior, but we might also be interested in the behavior of some function of these variables.

For example, consider the experiment in which a customer of a propane gas company is randomly selected. Suppose that the mean and standard deviation of the random variable

x = number of gallons required to fill a customer's propane tank

are known to be 318 gal and 42 gal, respectively. The company is considering two different pricing models:

Model 1: $3 per gal
Model 2: service charge of $50 + $2.80 per gal

The company is interested in the variable

y = amount billed

For each of the two models, y can be expressed as a function of the random variable x:

Model 1: $y_{model\ 1} = 3x$
Model 2: $y_{model\ 2} = 50 + 2.8x$

Both of these equations are examples of a linear function of x. The mean and standard deviation of a linear function of x can be computed from the mean and standard deviation of x, as described in the following box.

The Mean, Variance, and Standard Deviation of a Linear Function

If x is a random variable with mean μ_x and variance σ_x and a and b are numerical constants, the random variable y defined by

$$y = a + bx$$

is called a **linear function of the random variable x**.

The mean of $y = a + bx$ is

$$\mu_y = \mu_{a+bx} = a + b\mu_x$$

The variance of y is

$$\sigma_y^2 = \sigma_{a+bx}^2 = b^2\sigma_x^2$$

from which it follows that the standard deviation of y is

$$\sigma_y = \sigma_{a+bx} = |b|\sigma_x$$

We can use the results in the preceding box to compute the mean and standard deviation of the billing amount variable for the propane gas example, as follows:

For Model 1:

$$\mu_{model\,1} = \mu_{3x} = 3\mu_x = 3(318) = 954$$
$$\sigma_{model\,1}^2 = \sigma_{3x}^2 = 3^2\sigma_x^2 = 9(42)^2 = 15{,}876$$
$$\sigma_{model\,1} = \sqrt{15{,}876} = 126 = 3(42)$$

For Model 2:

$$\mu_{model\,2} = \mu_{50+2.8x} = 50 + 2.8\mu_x = 50 + 2.8(318) = 940.40$$
$$\sigma_{model\,2}^2 = \sigma_{50+2.8x}^2 = 2.8^2\sigma_x^2 = (2.8)^2(42)^2 = 13{,}829.76$$
$$\sigma_{model\,2} = \sqrt{13{,}829.76} = 117.60 = 2.8(42)$$

The mean billing amount for Model 1 is a bit higher than for Model 2, as is the variability in billing amounts. Model 2 results in slightly more consistency from bill to bill in the amount charged.

Now let's consider a different type of problem. Suppose that you have three tasks that you plan to complete on the way home from school: stop at the public library to return an overdue book for which you must pay a fine, deposit your most recent paycheck at the bank, and stop by the office supply store to purchase paper for your computer printer. Define the following variables:

$x_1 = $ time required to return book and pay fine
$x_2 = $ time required to deposit paycheck
$x_3 = $ time required to buy printer paper

We can then define a new variable, y, to represent the total amount of time to complete these tasks:

$$y = x_1 + x_2 + x_3$$

Defined in this way, y is an example of a linear combination of random variables.

If $x_1, x_2, \ldots, x_n$ are random variables and $a_1, a_2, \ldots, a_n$ are numerical constants, the random variable y defined as

$$y = a_2 x_1 + a_1 x_2 + \cdots + a_n x_n$$

is a **linear combination of the x_i's.**

For example, $y = 10x_1 - 5x_2 + 8x_3$ is a linear combination of x_1, x_2 and x_3 with $a_1 = 10$, $a_2 = -5$ and $a_3 = 8$. It is easy to compute the mean of a linear combination of x_i if the individual means $\mu_1, \mu_2, \ldots, \mu_n$ are known. The variance and standard deviation of a linear combination of the x_i are also easily computed *if the x_i are independent.* Two random variables x_i and x_j are independent if any event defined solely by x_i is independent of any event defined solely by x_j. When the x_i are not independent, computation of the variance and standard deviation of a linear combination of the x_i is more complicated; this case is not considered here.

Mean, Variance, and Standard Deviation for Linear Combinations

If $x_1, x_2, \ldots, x_n$ are random variables with means $\mu_1, \mu_2, \ldots, \mu_n$ and variances $\sigma_1^2, \sigma_2^2, \ldots, \sigma_n^2$, respectively, and

$$y = a_1 x_1 + a_2 x_2 + \cdots + a_n x_n$$

then

1. $\mu_y = \mu_{a_1 x_1 + a_2 x_2 + \cdots + a_n x_n} = a_1 \mu_1 + a_2 \mu_2 + \cdots + a_n \mu_n$

This result is true regardless of whether the x_i's are independent.

2. When $x_1, x_2, \ldots, x_n$ are independent random variables,

$$\sigma_y^2 = \sigma_{a_1 x_1 + a_2 x_2 + \cdots + a_n x_n}^2 = a_1^2 \sigma_1^2 + a_2^2 \sigma_2^2 + \cdots + a_n^2 \sigma_n^2$$

$$\sigma_y = \sigma_{a_1 x_1 + a_2 x_2 + \cdots + a_n x_n} = \sqrt{a_1^2 \sigma_1^2 + a_2^2 \sigma_2^2 + \cdots + a_n^2 \sigma_n^2}$$

This result is true only when the x_i's are independent.

Examples 7.14 –7.16 illustrate the use of these rules.

Example 7.14 Freeway Traffic

Three different roads feed into a particular freeway entrance. Suppose that during a fixed time period, the number of cars coming from each road onto the freeway is a random variable with mean values as follows:

Road	1	2	3
Mean	800	1000	600

With x_i representing the number of cars entering from road i, we can define $y = x_1 + x_2 + x_3$, the total number of cars entering the freeway. The mean value of y is

$$\begin{aligned}
\mu_y &= \mu_{x_1+x_2+x_3} \\
&= \mu_{x_1} + \mu_{x_2} + \mu_{x_3} \\
&= 800 + 1000 + 600 \\
&= 2400
\end{aligned}$$

■

Example 7.15 Combining Exam Subscores

A nationwide standardized exam consists of a multiple-choice section and a free response section. For each section, the mean and standard deviation are reported to be

	Mean	Standard Deviation
Multiple Choice	38	6
Free Response	30	7

Let's define x_1 and x_2 as the multiple-choice score and the free-response score, respectively, of a student selected at random from those taking this exam. We are also interested in the variable y = total score. Suppose that the total score is computed as $y = x_1 + 2x_2$. What are the mean and standard deviation of y?

Because $y = x_1 + 2x_2$ is a linear combination of x_1 and x_2, the mean of y is

$$\begin{aligned}
\mu_y &= \mu_{x_1+2x_2} \\
&= \mu_{x_1} + 2\mu_{x_2} \\
&= 38 + 2(30) \\
&= 98
\end{aligned}$$

What about the variance and standard deviation of y? To use Rule 2 in the preceding box, x_1 and x_2 must be independent. It is unlikely that the value of x_1 (a student's multiple-choice score) would be unrelated to the value of x_2 (the same student's free-response score), because it seems probable that students who score well on one section of the exam will also tend to score well on the other section. Therefore, it would not be appropriate to calculate the variance and standard deviation from the given information.

■

Example 7.16 Luggage Weights

A commuter airline flies small planes between San Luis Obispo and San Francisco. For small planes, the baggage weight is a concern, especially on foggy mornings, because the weight of the plane has an effect on how quickly the plane can ascend. Suppose that it is known that the variable x = weight of baggage checked by a randomly selected passenger has a mean and standard deviation of 42 and 16, respectively. Consider a flight on which 10 passengers, all traveling alone, are flying. If we use x_i to denote the baggage weight for passenger i (for i ranging from 1 to 10), the total weight of checked baggage, y, is then

$$y = x_1 + x_2 + \cdots + x_{10}$$

Note that y is a linear combination of the x_i. The mean value of y is

$$\mu_y = \mu_{x_1} + \mu_{x_2} + \cdots + \mu_{x_{10}}$$
$$= 42 + 42 + \cdots + 42$$
$$= 420$$

Since the ten passengers are all traveling alone, it is reasonable to think that the ten baggage weights are unrelated and that the x_i are independent. (This would not be a reasonable assumption if the 10 passengers were not traveling alone.) Then the variance of y is

$$\sigma_y^2 = \sigma_{x_1}^2 + \sigma_{x_2}^2 + \cdots + \sigma_{x_{10}}^2$$
$$= 16^2 + 16^2 + \cdots + 16^2$$
$$= 2560$$

and the standard deviation of y is

$$\sigma_y = \sqrt{2650} = 50.596$$

■

Exercises 7.27–7.44

Turn to the Expanded Answers Section for answers not shown next to exercises.

7.27 An express mail service charges a special rate for any package that weighs less than 1 lb. Let x denote the weight of a randomly selected parcel that qualifies for this special rate. The probability distribution of x is specified by the following density curve:

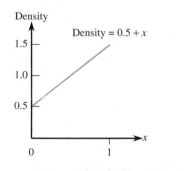

Use the fact that area of a trapezoid = (base)(average of two side lengths) to answer each of the following questions:
a. What is the probability that a randomly selected package of this type weighs at most 0.5 lb? between 0.25 and 0.5 lb? at least 0.75 lb?   0.375, 0.21875, 0.34375

b. It can be shown that $\mu_x = \frac{7}{12}$ and $\sigma_x^2 = \frac{11}{144}$. What is the probability that the value of x is more than 1 standard deviation from the mean value?   0.4012

7.28 The probability distribution of x, the number of defective tires on a randomly selected automobile checked at a certain inspection station, is given in the following table:

x	0	1	2	3	4
$p(x)$	.54	.16	.06	.04	.20

a. Calculate the mean value of x.   1.2
b. What is the probability that x exceeds its mean value?   0.30

7.29 Exercise 7.9 introduced the following probability distribution for y = the number of broken eggs in a carton:

y	0	1	2	3	4
$p(y)$	.65	.20	.10	.04	.01

a. Calculate and interpret μ_y.
b. In the long run, for what percentage of cartons is the number of broken eggs less than μ_y? Does this surprise you?

c. Why doesn't $\mu_y = (0 + 1 + 2 + 3 + 4)/5 = 2.0$. Explain.

7.30 Referring to Exercise 7.29, use the result of Part (a) along with the fact that a carton contains 12 eggs to determine the mean value of $z =$ the number of unbroken eggs. (Hint: z can be written as a linear function of x.)   11.44

7.31 The mean value of x, the number of defective tires, whose distribution appears in Exercise 7.28, is $\mu_x = 1.2$. Calculate σ_x^2 and σ_x. $\sigma_x^2 = 2.52, \sigma_x = 1.5875$

7.32 Exercise 7.8 gave the following probability distribution for $x =$ the number of courses for which a randomly selected student at a certain university is registered:

x	1	2	3	4	5	6	7
$p(x)$	.02	.03	.09	.25	.40	.16	.05

It can be easily verified that $\mu = 4.66$ and $\sigma = 1.20$.
a. Because $\mu - \sigma = 3.46$, the x values 1, 2, and 3 are more than 1 standard deviation below the mean. What is the probability that x is more than 1 standard deviation below its mean?   0.14
b. What x values are more than 2 standard deviations away from the mean value (i.e., either less than $\mu - 2\sigma$ or greater than $\mu + 2\sigma$)? What is the probability that x is more than 2 standard deviations away from its mean value?

7.33 Suppose that for a given computer salesperson, the probability distribution of $x =$ the number of systems sold in one month is given by the following table:

x	1	2	3	4	5	6	7	8
$p(x)$	.05	.10	.12	.30	.30	.11	.01	.01

a. Find the mean value of x (the mean number of systems sold).   4.12
b. Find the variance and standard deviation of x. How would you interpret these values?   1.9456, 1.3948
c. What is the probability that the number of systems sold is within 1 standard deviation of its mean value?   0.72
d. What is the probability that the number of systems sold is more than 2 standard deviations from the mean?   0.07

7.34 A local television station sells 15-sec, 30-sec, and 60-sec advertising spots. Let x denote the length of a randomly selected commercial appearing on this station, and suppose that the probability distribution of x is given by the following table:

x	15	30	60
$p(x)$	.1	.3	.6

a. Find the average length for commercials appearing on this station.   46.5
b. If a 15-sec spot sells for $500, a 30-sec spot for $800, and a 60-sec spot for $1000, find the average amount paid for commercials appearing on this station. (Hint: Consider a new variable, $y =$ cost, and then find the probability distribution and mean value of y.) $890

7.35 An author has written a book and submitted it to a publisher. The publisher offers to print the book and gives the author the choice between a flat payment of $10,000 and a royalty plan. Under the royalty plan the author would receive $1 for each copy of the book sold. The author thinks that the following table gives the probability distribution of the variable $x =$ the number of books that will be sold:

x	1000	5000	10,000	20,000
$p(x)$	.05	.30	.40	.25

Which payment plan should the author choose? Why?

7.36 A grocery store has an express line for customers purchasing at most five items. Let x be the number of items purchased by a randomly selected customer using this line. Give examples of two different assignments of probabilities such that the resulting distributions have the same mean but quite different standard deviations.

7.37 ▼ A gas station sells gasoline at the following prices (in cents per gallon, depending on the type of gas and service): 315.9, 318.9, 329.9, 339.9, 344.9, and 359.7. Let y denote the price per gallon paid by a randomly selected customer.
a. Is y a discrete random variable? Explain.
b. Suppose that the probability distribution of y is as follows:

y	315.9	318.9	329.9	339.9	344.9	359.7
$p(y)$	.36	.24	.10	.16	.08	.06

What is the probability that a randomly selected customer has paid more than $3.20 per gallon? Less than $3.40 per gallon?   0.4, 0.86
c. Refer to Part (b), and calculate the mean value and standard deviation of y. Interpret these values.   126.808, 13.45

7.38 A chemical supply company currently has in stock 100 lb of a certain chemical, which it sells to customers in 5-lb lots. Let $x =$ the number of lots ordered by a randomly chosen customer. The probability distribution of x is as follows:

x	1	2	3	4
$p(x)$	.2	.4	.3	.1

Bold exercises answered in back   ● Data set available online but not required   ▼ Video solution available

a. Calculate the mean value of x.   2.3
b. Calculate the variance and standard deviation of x.
0.81, 0.9
7.39 Return to Exercise 7.38, and let y denote the amount of material (in pounds) left after the next customer's order is shipped. Find the mean and variance of y. (Hint: y is a linear function of x.)   88.5, 20.25

7.40 An appliance dealer sells three different models of upright freezers having 13.5, 15.9, and 19.1 cubic feet of storage space. Let x = the amount of storage space purchased by the next customer to buy a freezer. Suppose that x has the following probability distribution:

x	13.5	15.9	19.1
p(x)	.2	.5	.3

a. Calculate the mean and standard deviation of x.
b. If the price of the freezer depends on the size of the storage space, x, such that Price = 25x − 8.5, what is the mean value of the variable *Price* paid by the next customer? $401.00
c. What is the standard deviation of the price paid?   49.96

7.41 ▼ To assemble a piece of furniture, a wood peg must be inserted into a predrilled hole. Suppose that the diameter of a randomly selected peg is a random variable with mean 0.25 in. and standard deviation 0.006 in. and that the diameter of a randomly selected hole is a random variable with mean 0.253 in. and standard deviation 0.002 in. Let x_1 = peg diameter, and let x_2 = denote hole diameter.
a. Why would the random variable y, defined as $y = x_2 − x_1$, be of interest to the furniture manufacturer?
b. What is the mean value of the random variable y?
c. Assuming that x_1 and x_2 are independent, what is the standard deviation of y?   0.006
d. Is it reasonable to think that x_1 and x_2 are independent? Explain.
e. Based on your answers to Parts (b) and (c), do you think that finding a peg that is too big to fit in the predrilled hole would be a relatively common or a relatively rare occurrence? Explain.

7.42 A multiple-choice exam consists of 50 questions. Each question has five choices, of which only one is correct. Suppose that the total score on the exam is computed as

$$y = x_1 − \frac{1}{4}x_2$$

where x_1 = number of correct responses and x_2 = number of incorrect responses. (Calculating a total score by

subtracting a term based on the number of incorrect responses is known as a correction for guessing and is designed to discourage test takers from choosing answers at random.)
a. It can be shown that if a totally unprepared student answers all 50 questions by just selecting one of the five answers at random, then μ_{x_1} = 10 and μ_{x_2} = 40. What is the mean value of the total score, y? Does this surprise you? Explain.
b. Explain why it is unreasonable to use the formulas given in this section to compute the variance or standard deviation of y.   It is not reasonable because x_1 and x_2 are not independent.

7.43 Consider a large ferry that can accommodate cars and buses. The toll for cars is $3, and the toll for buses is $10. Let x and y denote the number of cars and buses, respectively, carried on a single trip. Cars and buses are accommodated on different levels of the ferry, so the number of buses accommodated on any trip is independent of the number of cars on the trip. Suppose that x and y have the following probability distributions:

x	0	1	2	3	4	5
p(x)	.05	.10	.25	.30	.20	.10

y	0	1	2
p(y)	.50	.30	.20

a. Compute the mean and standard deviation of x.
b. Compute the mean and standard deviation of y.
c. Compute the mean and variance of the total amount of money collected in tolls from cars. $8.4, 14.94
d. Compute the mean and variance of the total amount of money collected in tolls from buses. $7, 61
e. Compute the mean and variance of z = total number of vehicles (cars and buses) on the ferry.   3.5, 2.27
f. Compute the mean and variance of w = total amount of money collected in tolls. $15.4, 75.94

7.44 Consider a game in which a red die and a blue die are rolled. Let x_R denote the value showing on the uppermost face of the red die, and define x_B similarly for the blue die.
a. The probability distribution of x_R is

x_R	1	2	3	4	5	6
$p(x_R)$	1/6	1/6	1/6	1/6	1/6	1/6

Find the mean, variance, and standard deviation of x_R.
b. What are the values of the mean, variance, and standard deviation of x_B? (You should be able to answer this question without doing any additional calculations.)

c. Suppose that you are offered a choice of the following two games:

Game 1: Costs $7 to play, and you win y_1 dollars, where
$$y_1 = x_R + x_B.$$

Game 2: Doesn't cost anything to play initially, but you "win" $3y_2$ dollars, where $y_2 = x_R - x_B$. If y_2 is negative, you must pay that amount; if it is positive, you receive that amount.

For Game 1, the net amount won in a game is $w_1 = y_1 - 7 = x_R + x_B - 7$. What are the mean and standard deviation of w_1?   0, 2.094

d. For Game 2, the net amount won in a game is $w_2 = 3y_2 = 3(x_R - x_B)$. What are the mean and standard deviation of w_2?   0, 5.124

e. Based on your answers to Parts (c) and (d), if you had to play, which game would you choose and why?

Bold exercises answered in back   ● Data set available online but not required   ▼ Video solution available

7.5 Binomial and Geometric Distributions

In this section we introduce two of the more commonly encountered discrete probability distributions: the binomial distribution and the geometric distribution. These distributions arise when the experiment of interest consists of making a sequence of dichotomous observations (two possible values for each observation). The process of making a single such observation is called a *trial*. For example, one characteristic of blood type is Rh factor, which can be either positive or negative. We can think of an experiment that consists of noting the Rh factor for each of 25 blood donors as a sequence of 25 dichotomous trials, where each trial consists of observing the Rh factor (positive or negative) of a single donor.

We could also conduct a different experiment that consists of observing the Rh factor of blood donors until a donor who is Rh-negative is encountered. This second experiment can also be viewed as a sequence of dichotomous trials, but the total number of trials in this experiment is not predetermined, as it was in the previous example, where we knew in advance that there would be 25 trials. Experiments of the two types just described are characteristic of those leading to the binomial and the geometric probability distributions, respectively.

■ Binomial Distributions

Suppose that we decide to record the gender of each of the next 25 newborn children at a particular hospital. What is the chance that at least 15 are female? What is the chance that between 10 and 15 are female? How many among the 25 can we expect to be female? These and other similar questions can be answered by studying the *binomial probability distribution*. This distribution arises when the experiment of interest is a *binomial experiment*, that is, an experiment having the characteristics listed in the following box.

Properties of a Binomial Experiment

A binomial experiment consists of a sequence of trials with the following conditions:

1. There are a fixed number of observations called trials.
2. Each trial can result in one of only two mutually exclusive outcomes labeled success (*S*) and failure (*F*).
3. Outcomes of different trials are independent.
4. The probability that a trial results in S is the same for each trial.

The **binomial random variable** x is defined as

x = number of successes observed when a binomial experiment is performed

The probability distribution of x is called the *binomial probability distribution.*

The term *success* here does not necessarily have any of its usual connotations. Which of the two possible outcomes is labeled "success" is determined by the random variable of interest. For example, if the variable counts the number of female births among the next 25 births at a particular hospital, then a female birth would be labeled a success (because this is what the variable counts). If male births were counted instead, a male birth would be labeled a success and a female birth a failure.

One illustration of a binomial probability distribution was given in Example 7.5. There, we considered x = number among four customers who selected an electric (as opposed to gas) hot tub. This is a binomial experiment with four trials and P(success) = $P(E)$ = .4. The 16 possible outcomes, along with their probabilities, were displayed in Table 7.1.

Consider now the case of five customers, a binomial experiment with five trials. Here the binomial distribution tells us the probability associated with each of the possible x values 0, 1, 2, 3, 4, and 5. There are 32 possible outcomes, and 5 of them yield $x = 1$: *SFFFF, FSFFF, FFSFF, FFFSF,* and *FFFFS.*

By independence, the first of these outcomes has probability

$$
\begin{aligned}
P(SFFFF) &= P(S)P(F)P(F)P(F)P(F) \\
&= (.4)(.6)(.6)(.6)(.6) \\
&= (.4)(.6)^4 \\
&= .05184
\end{aligned}
$$

The probability calculation will be the same for any outcome with only one success ($x = 1$). It does not matter where in the sequence the single success occurs. Thus

$$
\begin{aligned}
p(1) &= P(x = 1) \\
&= P(SFFFF \text{ or } FSFFF \text{ or } FFSFF \text{ or } FFFSF \text{ or } FFFFS) \\
&= .05184 + .05184 + .05184 + .05184 + .05184 \\
&= (5)(.05184) \\
&= .25920
\end{aligned}
$$

Similarly, there are ten outcomes for which $x = 2$, because there are 10 ways to select two from among the five trials to be the S's: *SSFFF, SFSFF,* . . . , and *FFFSS.* The probability of each results from multiplying together (.4) two times and (.6) three times. For example,

$$
\begin{aligned}
P(SSFFF) &= (.4)(.4)(.6)(.6)(.6) \\
&= (.4)^2(.6)^3 \\
&= .03456
\end{aligned}
$$

and so

$$
\begin{aligned}
p(2) &= P(x = 2) \\
&= P(SSFFF) + \cdots + P(FFFSS) \\
&= (10)(.4)^2(.6)^3 \\
&= .34560
\end{aligned}
$$

The general form of the distribution here is

$$p(x) = P(x\ S\text{'s among the five trials})$$
$$= (\text{no. of outcomes with } x\ S\text{'s}) \cdot (\text{probability of any particular outcome with } x\ S\text{'s})$$
$$= (\text{no. of outcomes with } x\ S\text{'s}) \cdot (.4)^x(.6)^{5-x}$$

This form was seen previously where $p(2) = 10(.4)^2(.6)^3$.

Let n denote the number of trials in the experiment. Then the number of outcomes with $x\ S$'s is the number of ways of selecting x from among the n trials to be the success trials. A simple expression for this quantity is

This is a combination, $_nC_x$.

$$\text{number of outcomes with } x \text{ successes} = \frac{n!}{x!(n-x)!}$$

where, for any positive whole number m, the symbol $m!$ (read "m factorial") is defined by

$$m! = m(m-1)(m-2)\cdots(2)(1)$$

and $0! = 1$.

The Binomial Distribution

Let

$n =$ number of independent trials in a binomial experiment

$\pi =$ constant probability that any particular trial results in a success*

Then

$$p(x) = P(x \text{ successes among } n \text{ trials})$$
$$= \frac{n!}{x!(n-x)!}\pi^x(1-\pi)^{n-x} \qquad x = 0, 1, 2, \ldots, n$$

The expressions $\binom{n}{x}$ or $_nC_x$ are sometimes used in place of $\frac{n!}{x!(n-x)!}$. Both are read as "n choose x" and represent the number of ways of choosing x items from a set of n. The binomial probability function can then be written as

$$p(x) = \binom{n}{x}\pi^x(1-\pi)^{n-x} \qquad x = 0, 1, 2, \ldots, n$$

or

$$p(x) = {}_nC_x\pi^x(1-\pi)^{n-x} \qquad x = 0, 1, 2, \ldots, n$$

*Some sources use p to represent the probability of success rather than π. We prefer the use of Greek letters for characteristics of a population or probability distribution, thus the use of π.

Notice that the probability distribution is specified using a formula that allows calculation of the various probabilities rather than by giving a table or a probability histogram.

Example 7.17 Computer Monitors

Sixty percent of all computer monitors sold by a large computer retailer have a flat panel display and 40% have a CRT display. The type of monitor purchased by each of the next 12 customers will be noted. Define a random variable x by

$$x = \text{number of monitors among these 12 that have a flat panel display}$$

Because x counts the number of flat panel displays, we use S to denote the sale of a flat panel monitor. Then x is a binomial random variable with $n = 12$ and $\pi = P(S) = .60$. The probability distribution of x is given by

$$p(x) = \frac{12!}{x!(12 - x)!}(.6)^x(.4)^{n-x} \qquad x = 0, 1, 2, \ldots, 12$$

The probability that exactly four monitors are flat panel displays is

$$
\begin{aligned}
p(4) &= P(x = 4)\\
&= \frac{12!}{4!8!}(.6)^4(.6)^8\\
&= (495)(.6)^4(.4)^8\\
&= .042
\end{aligned}
$$

If group after group of 12 purchases is examined, the long-run percentage of those with exactly four flat panel monitors will be 4.2%. According to this calculation, 495 of the possible outcomes (there are $2^{12} = 4096$) have $x = 4$.

The probability that between four and seven (inclusive) are flat panel displays is

$$P(4 \le x \le 7) = P(x = 4 \text{ or } x = 5 \text{ or } x = 6 \text{ or } x = 7)$$

Since these outcomes are disjoint, this is equal to

$$
\begin{aligned}
P(4 \le x \le 7) &= p(4) + p(5) + p(6) + p(7)\\
&= \frac{12!}{4!8!}(.6)^4(.4)^8 + \cdots + \frac{12!}{7!5!}(.6)^7(.4)^5\\
&= .042 + .101 + .177 + .227\\
&= .547
\end{aligned}
$$

Notice that

$$
\begin{aligned}
P(4 < x < 7) &= P(x = 5 \text{ or } x = 6)\\
&= p(5) + p(6)\\
&= .278
\end{aligned}
$$

so the probability depends on whether $<$ or $\le$ appears. (This is typical of *discrete* random variables.)

∎

The binomial distribution formula can be tedious to use unless n is small. Appendix Table 9 gives binomial probabilities for selected n in combination with various values of π. Appendix Table 9 should help you practice using the binomial distribution without getting bogged down in arithmetic.

Using Appendix Table 9

To find $p(x)$ for any particular value of x,

1. Locate the part of the table corresponding to your value of n (5, 10, 15, 20, or 25).
2. Move down to the row labeled with your value of x.
3. Go across to the column headed by the specified value of π.

The desired probability is at the intersection of the designated x row and π column. For example, when $n = 20$ and $\pi = .8$,

$$p(15) = P(x = 15) = \text{(entry at intersection of } n = 15 \text{ row and } \pi = .8 \text{ column)} = .175$$

Although $p(x)$ is positive for every possible x value, many probabilities are zero to three decimal places, so they appear as .000 in the table. More extensive binomial tables are available. Alternatively, most statistics software packages and graphing calculators are programmed to calculate these probabilities.

▪ **Sampling Without Replacement** Usually, sampling is carried out without replacement; that is, once an element has been selected for the sample, it is not a candidate for future selection. If the sampling was accomplished by selecting an element from the population, observing whether it is a success or a failure, and then returning it to the population before the next selection is made, the variable x = number of successes observed in the sample would fit all the requirements of a binomial random variable. When sampling is done without replacement, the trials (individual selections) are not independent. In this case, the number of successes observed in the sample does not have a binomial distribution but rather a different type of distribution called a *hypergeometric distribution*. The probability calculations for this distribution are even more tedious than for the binomial distribution. Fortunately, when the sample size n is much smaller than N, the population size, probabilities calculated using the binomial distribution and the hypergeometric distribution are very close in value. They are so close, in fact, that statisticians often ignore the difference and use the binomial probabilities in place of the hypergeometric probabilities. Most statisticians recommend the following guideline for determining whether the binomial probability distribution is appropriate when sampling without replacement.

Let x denote the number of S's in a sample of size n selected without replacement from a population consisting of N individuals or objects. If $(n/N) \leq 0.05$, i.e., at most 5% of the population is sampled, then the binomial distribution gives a good approximation to the probability distribution of x.

Example 7.18 Security Systems

A *Los Angeles Times* poll (November 10, 1991) reported that almost 20% of Southern California homeowners questioned had installed a home security system. Suppose that exactly 20% of all such homeowners have a system. Consider a random sample of $n = 20$ homeowners (much less than 5% of the population). Then x, the number of homeowners in the sample who have a security system, has (approximately) a binomial distribution with $n = 20$ and $\pi = .20$. The probability that five of those sampled have a system is

$$p(5) = P(x = 5)$$
$$= (\text{entry in } x \text{ row and } \pi = .20 \text{ column in Appendix Table 9 } (n = 20))$$
$$= .175$$

The probability that at least 40% of those in the sample—that is, eight or more—have a system is

$$P(x \geq 8) = P(x = 8, 9, 10, \ldots, 19, \text{ or } 20)$$
$$= p(8) + p(9) + \cdots + p(20)$$
$$= .022 + .007 + .002 + .000 + \cdots + .000$$
$$= .031$$

If, in fact, $\pi = .20$, only about 3% of all samples of size 20 would result in at least 8 homeowners having a security system. Because $P(x \geq 8)$ is so small when $\pi = .20$, if $x \geq 8$ were actually observed, we would have to wonder whether the reported value of $\pi = .20$ is correct. Although it is possible that we could observe $x \geq 8$ when $\pi = .20$ (this would happen about 3% of the time in the long run), it might also be the case that π is actually greater than .20. In Chapter 10, we show how hypothesis-testing methods can be used to decide which of two contradictory claims about a population (e.g., $\pi = .20$ or $\pi > .20$) is more plausible.

The binomial formula or tables can be used to compute each of the 21 probabilities $p(0), p(1), \ldots, p(20)$. Figure 7.16 shows the probability histogram for the binomial distribution with $n = 20$ and $\pi = .20$. Notice that the distribution is skewed to the right. (The binomial distribution is symmetric only when $\pi = .5$.)

Figure 7.16 The binomial probability histogram when $n = 20$ and $\pi = .20$.

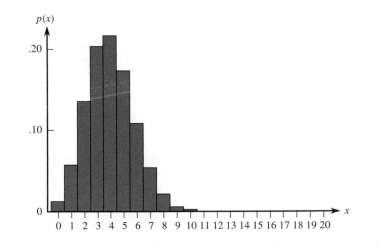

■ **Mean and Standard Deviation of a Binomial Random Variable**   A binomial random variable x based on n trials has possible values $0, 1, 2, \ldots, n$, so the mean value is

$$\mu_x = \sum x p(x) = (0)p(0) + (1)p(1) + \cdots + (n)p(n)$$

and the variance of x is

$$\sigma_x^2 = \sum (x - \mu_x)^2 \cdot p(x)$$
$$= (0 - \mu_x)^2 p(0) + (1 - \mu_x)^2 p(1) + \cdots + (n - \mu_x)^2 p(n)$$

These expressions appear to be very tedious to evaluate for any particular values of n and π. Fortunately, algebraic manipulation results in considerable simplification, making summation unnecessary.

The mean value and the standard deviation of a binomial random variable are, respectively,

$$\mu_x = n\pi \quad \text{and} \quad \sigma_x = \sqrt{n\pi(1 - \pi)}$$

Example 7.19 Credit Cards Paid in Full

Newsweek (December 2, 1991) reported that one-third of all credit card users pay their bills in full each month. This figure is, of course, an average across different cards and issuers. Suppose that 30% of all individuals holding Visa cards issued by a certain bank pay in full each month. A random sample of $n = 25$ cardholders is to be selected. The bank is interested in the variable $x =$ number in the sample who pay in full each month. Even though sampling is done without replacement, the sample size $n = 25$ is most likely very small compared to the total number of credit card holders, so we can approximate the probability distribution of x using a binomial distribution with $n = 25$ and $\pi = .3$. We have defined "paid in full" as a success because this is the outcome counted by the random variable x. The mean value of x is then

$$\mu_x = n\pi = 25(.30) = 7.5$$

and the standard deviation is

$$\sigma_x = \sqrt{n\pi(1 - \pi)} = \sqrt{25(.30)(.70)} = \sqrt{5.25} = 2.29$$

The probability that x is farther than 1 standard deviation from its mean value is

$$P(x < \mu_x - \sigma_x \text{ or } x > \mu_x + \sigma_x) = P(x < 5.21 \text{ or } x > 9.79)$$
$$= P(x \le 5) + P(x \ge 10)$$
$$= p(0) + \cdots + p(5) + p(10) + \cdots + p(25)$$
$$= .382 \quad \text{(using Appendix Table 9)}$$

The value of σ_x is 0 when $\pi = 0$ or $\pi = 1$. In these two cases, there is no uncertainty in x: We are sure to observe $x = 0$ when $\pi = 0$ and $x = n$ when $\pi = 1$. It is also easily verified that $\pi(1 - \pi)$ is largest when $\pi = .5$. Thus the binomial distribution spreads out the most when sampling from a 50–50 population. The farther π is from .5, the less spread out and the more skewed the distribution.

■ Geometric Distributions ....

A binomial random variable is defined as the number of successes in n independent trials, where each trial can result in either a success or a failure and the probability of success is the same for each trial. Suppose, however, that we are not interested in the number of successes in a fixed number of trials but rather in the number of trials that must be carried out before a success occurs. Two examples are counting the number of boxes of cereal that must be purchased before finding one with a rare toy and counting the number of games that a professional bowler must play before achieving a score over 250.

The variable

x = number of trials to first success

is called a *geometric random variable*, and the probability distribution that describes its behavior is called a *geometric probability distribution*.

Suppose an experiment consists of a sequence of trials with the following conditions:

1. The trials are independent.
2. Each trial can result in one of two possible outcomes, success and failure.
3. The probability of success is the same for all trials.

A **geometric random variable** is defined as

x = number of trials until the first success is observed (including the success trial)

The probability distribution of x is called the *geometric probability distribution*.

For example, suppose that 40% of the students who drive to campus at your university carry jumper cables. Your car has a dead battery and you don't have jumper cables, so you decide to stop students who are headed to the parking lot and ask them whether they have a pair of jumper cables. You might be interested in the number of students you would have to stop before finding one who has jumper cables. If we define success as a student with jumper cables, a trial would consist of asking an individual student for help. The random variable x = number of students who must be stopped before finding one with jumper cables is an example of a geometric random variable, because it can be viewed as the number of trials to the first success in a sequence of independent trials.

The probability distribution of a geometric random variable is easy to construct. We use π to denote the probability of success on any given trial. Possible outcomes can be denoted as follows:

Outcome	x = Number of Trials to First Success
S	1
FS	2
FFS	3
⋮	⋮
$FFFFFFS$	7
⋮	⋮

Each possible outcome consists of 0 or more failures followed by a single success. So,

$$p(x) = P(x \text{ trials to first success})$$
$$= P(FF \ldots FS)$$

↑

x − 1 failures followed by a success on trial x

Because the probability of success is π for each trial, the probability of failure for each trial is $1 - \pi$. Because the trials are independent,

$$p(x) = P(x \text{ trials to first success}) = P(FF \ldots FS)$$
$$= P(F)P(F) \cdots P(F)P(S)$$
$$= (1 - \pi)(1 - \pi) \cdots (1 - \pi)\pi$$
$$= (1 - \pi)^{x-1}\pi$$

This leads us to the formula for the geometric probability distribution.

Geometric Probability Distribution

If x is a geometric random variable with probability of success $= \pi$ for each trial, then

$$p(x) = (1 - \pi)^{x-1}\pi \qquad x = 1, 2, 3, \ldots$$

■

....

Example 7.20 Jumper Cables

© Stone +/Greg Ceo/Getty Images

Consider the jumper cable problem described previously. For this problem, $\pi = .4$, because 40% of the students who drive to campus carry jumper cables. The probability distribution of

$$x = \text{number of students who must be stopped before} $$
$$\text{finding a student with jumper cables}$$

is

$$p(x) = (.6)^{x-1}(.4) \qquad x = 1, 2, 3, \ldots$$

The probability distribution can now be used to compute various probabilities. For example, the probability that the first student stopped has jumper cables (i.e., $x = 1$) is

$$p(1) = (.6)^{1-1}(.4) = (.6)^0(.4) = .4$$

The probability that three or fewer students must be stopped is

$$P(x \leq 3) = p(1) + p(2) + p(3)$$
$$= (.6)^0(.4) + (.6)^1(.4) + (.6)^2(.4)$$
$$= .4 + .24 + .144$$
$$= .784$$

■

7.45 Consider two binomial experiments.
a. The first binomial experiment consists of six trials. How many outcomes have exactly one success, and what are these outcomes?
b. The second binomial experiment consists of 20 trials. How many outcomes have exactly 10 successes? exactly 15 successes? exactly 5 successes?

7.46 Suppose that in a certain metropolitan area, 9 out of 10 households have a VCR. Let x denote the number among four randomly selected households that have a VCR, so x is a binomial random variable with $n = 4$ and $\pi = .9$.
a. Calculate $p(2) = P(x = 2)$, and interpret this probability.
b. Calculate $p(4)$, the probability that all four selected households have a VCR. 0.6561
c. Determine $P(x \leq 3)$. 0.3439

7.47 ▼ The *Los Angeles Times* (December 13, 1992) reported that what airline passengers like to do most on long flights is rest or sleep; in a survey of 3697 passengers, almost 80% did so. Suppose that for a particular route the actual percentage is exactly 80%, and consider randomly selecting six passengers. Then x, the number among the selected six who rested or slept, is a binomial random variable with $n = 6$ and $\pi = .8$.
a. Calculate $p(4)$, and interpret this probability.
b. Calculate $p(6)$, the probability that all six selected passengers rested or slept. $p(6) = 0.262144$
c. Determine $P(x \geq 4)$. $p(x \geq 4) = 0.90112$

7.48 Refer to Exercise 7.47, and suppose that 10 rather than 6 passengers are selected ($n = 10$, $\pi = .8$), so that Appendix Table 9 can be used.
a. What is $p(8)$? 0.302
b. Calculate $P(x \leq 7)$. 0.323
c. Calculate the probability that more than half of the selected passengers rested or slept. 0.966

7.49 Twenty-five percent of the customers entering a grocery store between 5 P.M. and 7 P.M. use an express checkout. Consider five randomly selected customers, and let x denote the number among the five who use the express checkout.
a. What is $p(2)$, that is, $P(x = 2)$? 0.26367
b. What is $P(x \leq 1)$? 0.6328125
c. What is $P(2 \leq x)$? (Hint: Make use of your computation in Part (b).) 0.3671875
d. What is $P(x \neq 2)$? 0.73633

7.50 A breeder of show dogs is interested in the number of female puppies in a litter. If a birth is equally likely to result in a male or a female puppy, give the probability distribution of the variable x = number of female puppies in a litter of size 5.

7.51 The article "FBI Says Fewer than 25 Failed Polygraph Test" (*San Luis Obispo Tribune*, July 29, 2001) states that false-positives in polygraph tests (i.e., tests in which an individual fails even though he or she is telling the truth) are relatively common and occur about 15% of the time. Suppose that such a test is given to 10 trustworthy individuals.
a. What is the probability that all 10 pass? 0.1969
b. What is the probability that more than 2 fail, even though all are trustworthy? 0.1798
c. The article indicated that 500 FBI agents were required to take a polygraph test. Consider the random variable x = number of the 500 tested who fail. If all 500 agents tested are trustworthy, what are the mean and standard deviation of x? 75, 7.984
d. The headline indicates that fewer than 25 of the 500 agents tested failed the test. Is this a surprising result if all 500 are trustworthy? Answer based on the values of the mean and standard deviation from Part (c).

7.52 Industrial quality control programs often include inspection of incoming materials from suppliers. If parts are purchased in large lots, a typical plan might be to select 20 parts at random from a lot and inspect them. A lot might be judged acceptable if one or fewer defective parts are found among those inspected. Otherwise, the lot is rejected and returned to the supplier. Use Appendix Table 9 to find the probability of accepting lots that have each of the following (Hint: Identify success with a defective part):
a. 5% defective parts 0.735
b. 10% defective parts 0.392
c. 20% defective parts 0.070

7.53 An experiment was conducted to investigate whether a graphologist (a handwriting analyst) could distinguish a normal person's handwriting from that of a psychotic. A well-known expert was given 10 files, each containing handwriting samples from a normal person and from a person diagnosed as psychotic, and asked to identify the psychotic's handwriting. The graphologist made correct identifications in 6 of the 10 trials (data taken from *Statistics in the Real World*, by R. J. Larsen and D. F. Stroup [New York: Macmillan, 1976]). Does this evidence indi-

Bold exercises answered in back   ● Data set available online but not required   ▼ Video solution available

cate that the graphologist has an ability to distinguish the handwriting of psychotics? (Hint: What is the probability of correctly guessing 6 or more times out of 10? Your answer should depend on whether this probability is relatively small or relatively large.)

7.54 Suppose that the probability is .1 that any given citrus tree will show measurable damage when the temperature falls to 30°F. If the temperature does drop to 30°F, what is the expected number of citrus trees showing damage in orchards of 2000 trees? What is the standard deviation of the number of trees that show damage?   200, 13.4164

7.55 Thirty percent of all automobiles undergoing an emissions inspection at a certain inspection station fail the inspection.
a. Among 15 randomly selected cars, what is the probability that at most 5 fail the inspection?   0.722
b. Among 15 randomly selected cars, what is the probability that between 5 and 10 (inclusive) fail to pass inspection?
c. Among 25 randomly selected cars, what is the mean value of the number that pass inspection, and what is the standard deviation of the number that pass inspection?
d. What is the probability that among 25 randomly selected cars, the number that pass is within 1 standard deviation of the mean value?   0.618

7.56 You are to take a multiple-choice exam consisting of 100 questions with 5 possible responses to each question. Suppose that you have not studied and so must guess (select one of the five answers in a completely random fashion) on each question. Let x represent the number of correct responses on the test.
a. What kind of probability distribution does x have?
b. What is your expected score on the exam? (Hint: Your expected score is the mean value of the x distribution.)
c. Compute the variance and standard deviation of x.
d. Based on your answers to Parts (b) and (c), is it likely that you would score over 50 on this exam? Explain the reasoning behind your answer.

7.57 Suppose that 20% of the 10,000 signatures on a certain recall petition are invalid. Would the number of invalid signatures in a sample of size 1000 have (approximately) a binomial distribution? Explain.

7.58 A coin is spun 25 times. Let x be the number of spins that result in heads (H). Consider the following rule for deciding whether or not the coin is fair:

Judge the coin fair if $8 \le x \le 17$.
Judge the coin biased if either $x \le 7$ or $x \ge 18$.

a. What is the probability of judging the coin biased when it is actually fair?   0.044
b. What is the probability of judging the coin fair when $P(H) = .9$, so that there is a substantial bias? Repeat for $P(H) = .1$.   0.002, 0.002
c. What is the probability of judging the coin fair when $P(H) = .6$? when $P(H) = .4$? Why are the probabilities so large compared to the probabilities in Part (b)?
d. What happens to the "error probabilities" of Parts (a) and (b) if the decision rule is changed so that the coin is judged fair if $7 \le x \le 18$ and unfair otherwise? Is this a better rule than the one first proposed? Explain.

7.59 A city ordinance requires that a smoke detector be installed in all residential housing. There is concern that too many residences are still without detectors, so a costly inspection program is being contemplated. Let π be the proportion of all residences that have a detector. A random sample of 25 residences is selected. If the sample strongly suggests that $\pi < .80$ (less than 80% have detectors), as opposed to $\pi \ge .80$, the program will be implemented. Let x be the number of residences among the 25 that have a detector, and consider the following decision rule: Reject the claim that $\pi = .8$ and implement the program if $x \le 15$.
a. What is the probability that the program is implemented when $\pi = .80$?   0.017
b. What is the probability that the program is not implemented if $\pi = .70$? if $\pi = .60$?   0.811, 0.425
c. How do the "error probabilities" of Parts (a) and (b) change if the value 15 in the decision rule is changed to 14?

7.60 Suppose that 90% of all registered California voters favor banning the release of information from exit polls in presidential elections until after the polls in California close. A random sample of 25 California voters is to be selected.
a. What is the probability that more than 20 voters favor the ban?   0.902
b. What is the probability that at least 20 voters favor the ban?   0.967
c. What are the mean value and standard deviation of the number of voters who favor the ban?   22.5, 1.5
d. If fewer than 20 voters in the sample favor the ban, is this at odds with the assertion that (at least) 90% of the populace favors the ban? (Hint: Consider $P(x < 20)$ when $\pi = .9$.)

7.61 Sophie is a dog that loves to play catch. Unfortunately, she isn't very good, and the probability that she catches a ball is only .1. Let x be the number of tosses required until Sophie catches a ball.

a. Does x have a binomial or a geometric distribution?

b. What is the probability that it will take exactly two tosses for Sophie to catch a ball?   0.09

c. What is the probability that more than three tosses will be required?   0.729

7.62 Suppose that 5% of cereal boxes contain a prize and the other 95% contain the message, "Sorry, try again." Consider the random variable x, where x = number of boxes purchased until a prize is found.

a. What is the probability that at most two boxes must be purchased?   0.0975

b. What is the probability that exactly four boxes must be purchased?   0.0429

c. What is the probability that more than four boxes must be purchased?   0.8145

7.63 ▼ The article on polygraph testing of FBI agents referenced in Exercise 7.51 indicated that the probability of a false-positive (a trustworthy person who nonetheless fails the test) is .15. Let x be the number of trustworthy FBI agents tested until someone fails the test.

a. What is the probability distribution of x?   Geometric

b. What is the probability that the first false-positive will occur when the third person is tested?   0.1084

c. What is the probability that fewer than four are tested before the first false-positive occurs?   0.3859

d. What is the probability that more than three agents are tested before the first false-positive occurs?   0.6141

Bold exercises answered in back   ● Data set available online but not required   ▼ Video solution available

---

7.6   Normal Distributions

Normal distributions formalize the notion of mound-shaped histograms introduced in Chapter 4. Normal distributions are widely used for two reasons. First, they provide a reasonable approximation to the distribution of many different variables. They also play a central role in many of the inferential procedures that will be discussed in later chapters.

Figure 7.17   A normal distribution.

Normal distributions are continuous probability distributions that are bell shaped and symmetric, as shown in Figure 7.17. Normal distributions are sometimes referred to as *normal curves*.

There are many different normal distributions, and they are distinguished from one another by their mean μ and standard deviation σ. The mean μ of a normal distribution describes where the corresponding curve is centered, and the standard deviation σ describes how much the curve spreads out around that center. As with all continuous probability distributions, the total area under any normal curve is equal to 1. Three normal distributions are shown in Figure 7.18. Notice that the smaller the standard deviation, the taller and narrower the corresponding curve. Recall that areas under a continuous probability distribution curve represent probabilities, so when the standard deviation is small, a larger area is concentrated near the center of the curve and the chance of observing a value near the mean is much greater (because μ is at the center).

The value of μ is the number on the measurement axis lying directly below the top of the bell. The value of σ can also be ascertained from a picture of the curve. Consider the normal curve in Figure 7.19. Starting at the top of the bell (above $\mu = 100$) and moving to the right, the curve turns downward until it is above the value 110. After that point, it continues to decrease in height but is turning upward rather than downward. Similarly, to the left of $\mu = 100$, the curve turns downward until it reaches 90 and then begins to turn upward. The curve changes from turning downward to turning upward at a distance of 10 on either side of μ, so $\sigma = 10$. In general, σ is the distance to either side of μ at which a normal curve changes from turning downward to turning upward.

Figure 7.18 Three normal distributions.

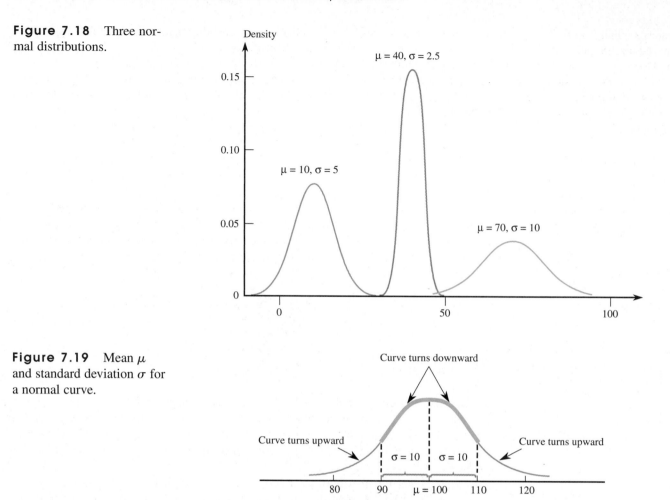

Figure 7.19 Mean μ and standard deviation σ for a normal curve.

If a particular normal distribution is to be used to describe the behavior of a random variable, a mean and a standard deviation must be specified. For example, a normal distribution with mean 7 and standard deviation 1 might be used as a model for the distribution of x = birth weight. If this model is a reasonable description of the probability distribution, we could use areas under the normal curve with $\mu = 7$ and $\sigma = 1$ to approximate various probabilities related to birth weight. The probability that a birth weight is over 8 lb (expressed symbolically as $P(x > 8)$) corresponds to the shaded area in Figure 7.20(a). The shaded area in Figure 7.20(b) is the (approximate) probability $P(6.5 < x < 8)$ of a birth weight falling between 6.5 and 8 lb.

Figure 7.20 Normal distribution for birth weight: (a) shaded area = $P(x > 8)$; (b) shaded area = $P(6.5 < x < 8)$.

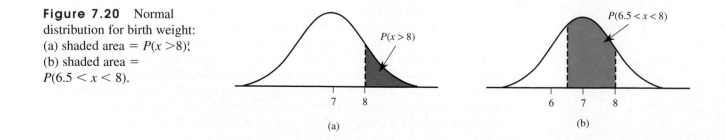

Unfortunately, direct computation of such probabilities (areas under a normal curve) is not simple. To overcome this difficulty, we rely on a table of areas for a reference normal distribution, called the *standard normal distribution*.

> ### DEFINITION
>
> The **standard normal distribution** is the normal distribution with
>
> $$\mu = 0 \quad \text{and} \quad \sigma = 1$$
>
> The corresponding density curve is called the *standard normal curve*. It is customary to use the letter *z* to represent a variable whose distribution is described by the standard normal curve. The term *z curve* is often used in place of *standard normal curve*.

Few naturally occurring variables have distributions that are well described by the standard normal distribution, but this distribution is important because it is also used in probability calculations for other normal distributions. When we are interested in finding a probability based on some other normal curve, we first translate our problem into an equivalent problem that involves finding an area under the standard normal curve. A table for the standard normal distribution is then used to find the desired area. To be able to do this, we must first learn to work with the standard normal distribution.

■ The Standard Normal Distribution ....

In working with normal distributions, we need two general skills:

1. We must be able to use the normal distribution to compute probabilities, which are areas under a normal curve and above given intervals.
2. We must be able to characterize extreme values in the distribution, such as the largest 5%, the smallest 1%, and the most extreme 5% (which would include the largest 2.5% and the smallest 2.5%).

Let's begin by looking at how to accomplish these tasks when the distribution of interest is the standard normal distribution.

The standard normal or *z* curve is shown in Figure 7.21(a). It is centered at $\mu = 0$, and the standard deviation, $\sigma = 1$, is a measure of the extent to which it spreads out

Figure 7.21   (a) A standard normal (*z*) curve and (b) a cumulative area.

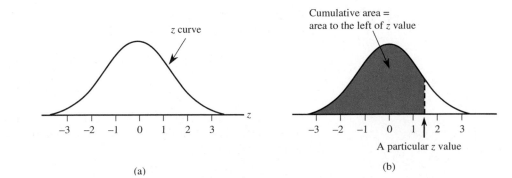

(a)

(b)

about its mean (in this case, 0). Note that this picture is consistent with the Empirical Rule of Chapter 4: About 95% of the area (probability) is associated with values that are within 2 standard deviations of the mean (between −2 and 2), and almost all of the area is associated with values that are within 3 standard deviations of the mean (between −3 and 3).

Appendix Table 2 tabulates cumulative z curve areas of the sort shown in Figure 7.21(b) for many different values of z. The smallest value for which the cumulative area is given is −3.89, a value far out in the lower tail of the z curve. The next smallest value for which the area appears is −3.88, then −3.87, then −3.86, and so on in increments of 0.01, terminating with the cumulative area to the left of 3.89.

Using the Table of Standard Normal Curve Areas

For any number z^* between −3.89 and 3.89 and rounded to two decimal places, Appendix Table 2 gives

(area under z curve to the left of z^*) = $P(z < z^*) = P(z \leq z^*)$

where the letter z is used to represent a random variable whose distribution is the standard normal distribution.

To find this probability, locate the following:

1. The row labeled with the sign of z^* and the digit to either side of the decimal point (for example, −1.7 or 0.5)
2. The column identified with the second digit to the right of the decimal point in z^* (for example, .06 if $z^* = -1.76$)

The number at the intersection of this row and column is the desired probability, $P(z < z^*)$.

A portion of the table of standard normal curve areas appears in Figure 7.22. To find the area under the z curve to the left of 1.42, look in the row labeled 1.4 and the column labeled .02 (the highlighted row and column in Figure 7.22). From the table, the corresponding cumulative area is .9222. So

z curve area to the left of 1.42 = .9222

We can also use the table to find the area to the right of a particular value. Because the total area under the z curve is 1, it follows that

$$(z \text{ curve area to the right of } 1.42) = 1 - (z \text{ curve area to the left of } 1.42)$$
$$= 1 - .9222$$
$$= .0778$$

These probabilities can be interpreted to mean that in a long sequence of observations, roughly 92.22% of the observed z values will be smaller than 1.42, and 7.78% will be larger than 1.42.

Figure 7.22   Portion of the table of standard normal curve areas.

z^*	.00	.01	.02	.03	.04	.05
0.0	.5000	.5040	.5080	.5120	.5160	.5199
0.1	.5398	.5438	.5478	.5517	.5557	.5596
0.2	.5793	.5832	.5871	.5910	.5948	.5987
0.3	.6179	.6217	.6255	.6293	.6331	.6368
0.4	.6554	.6591	.6628	.6664	.6700	.6736
0.5	.6915	.6950	.6985	.7019	.7054	.7088
0.6	.7257	.7291	.7324	.7357	.7389	.7422
0.7	.7580	.7611	.7642	.7673	.7704	.7734
0.8	.7881	.7910	.7939	.7967	.7995	.8023
0.9	.8159	.8186	.8212	.8238	.8264	.8289
1.0	.8413	.8438	.8461	.8485	.8508	.8531
1.1	.8643	.8665	.8686	.8708	.8729	.8749
1.2	.8849	.8869	.8888	.8907	.8925	.8944
1.3	.9032	.9049	.9066	.9082	.9099	.9115
1.4	.9192	.9207	.9222	.9236	.9251	.9265
1.5	.9332	.9345	.9357	.9370	.9382	.9394
1.6	.9452	.9463	.9474	.9484	.9495	.9505
1.7	.9554	.9564	.9573	.9582	.9591	.9599
1.8	.9641	.9649	.9656	.9664	.9671	.9678

$P(z < 1.42)$

Example 7.21   Finding Standard Normal Curve Areas

The probability $P(z < -1.76)$ is found at the intersection of the -1.7 row and the .06 column of the z table. The result is

$$P(z < -1.76) = .0392$$

as shown in the following figure:

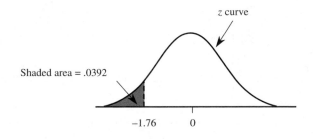

Shaded area = .0392

z curve

-1.76 0

Interpretation for probability of shaded area

In other words, in a long sequence of observations, roughly 3.9% of the observed z values will be smaller than -1.76. Similarly,

$$P(z \leq 0.58) = \text{entry in 0.5 row and .08 column of Table 2} = .7190$$

as shown in the following figure:

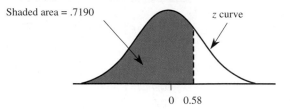

Now consider $P(z < -4.12)$. This probability does not appear in Appendix Table 2; there is no -4.1 row. However, it must be less than $P(z < -3.89)$, the smallest z value in the table, because -4.12 is farther out in the lower tail of the z curve. Since $P(z < -3.89) = .0000$ (that is, zero to four decimal places), it follows that

$$P(z < -4.12) \approx 0$$

Similarly,

$$P(z < 4.18) > P(z < 3.89) = 1.0000$$

from which we conclude that

$$P(z < 4.18) \approx 1$$

■

As illustrated in Example 7.21, we can use the cumulative areas tabulated in Appendix Table 2 to calculate other probabilities involving z. The probability that z is larger than a value c is

$P(z > c) =$ area under the z curve to the right of $c = 1 - P(z \leq c)$

In other words, the area to the right of a value (a right-tail area) is 1 minus the corresponding cumulative area. This is illustrated in Figure 7.23.

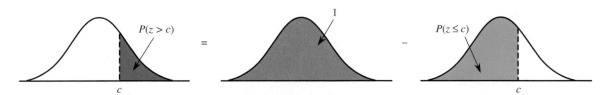

Figure 7.23 The relationship between an upper-tail area and a cumulative area.

Similarly, the probability that z falls in the interval between a lower limit a and an upper limit b is

$P(a < z < b) =$ area under the z curve and above the interval from a to b
$= P(z < b) - P(z < a)$

That is, $P(a < z < b)$ is the difference between two cumulative areas, as illustrated in Figure 7.24.

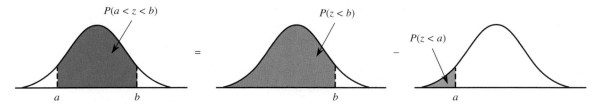

Figure 7.24 $P(a < z < b)$ as the difference between the two cumulative areas.

Example 7.22 More About Standard Normal Curve Areas

The probability that z is between -1.76 and 0.58 is

$$P(-1.76 < z < 0.58) = P(z < 0.58) - P(z < -1.76)$$
$$= .7190 - .0392$$
$$= .6798$$

as shown in the following figure:

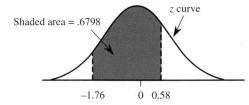

The probability that z is between -2 and $+2$ (within 2 standard deviations of its mean, since $\mu = 0$ and $\sigma = 1$) is

$$P(-2.00 < z < 2.00) = P(z < 2.00) - P(z < -2.00)$$
$$= .9772 - .0228$$
$$= .9544$$
$$\approx .95$$

as shown in the following figure:

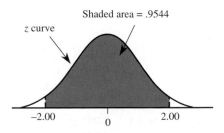

This last probability is the basis for one part of the Empirical Rule, which states that when a histogram is well approximated by a normal curve, roughly 95% of the values are within 2 standard deviations of the mean.

The probability that the value of z exceeds 1.96 is

$$P(z > 1.96) = 1 - P(z < 1.96)$$
$$= 1 - .9750$$
$$= .0250$$

as shown in the following figure:

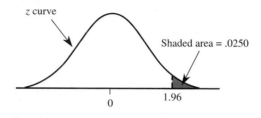

That is, 2.5% of the area under the z curve lies to the right of 1.96 in the upper tail. Similarly,

$$P(z > -1.28) = \text{area to the right of } -1.28$$
$$= 1 - P(z < -1.28)$$
$$= 1 - .1003$$
$$= .8997$$
$$\approx .90$$

■

■ Identifying Extreme Values ....

Suppose that we want to describe the values included in the smallest 2% of a distribution or the values making up the most extreme 5% (which includes the largest 2.5% and the smallest 2.5%). Let's see how we can identify extreme values in the distribution by working through Examples 7.23 and 7.24.

....

Example 7.23 Identifying Extreme Values

Suppose that we want to describe the values that make up the smallest 2% of the standard normal distribution. Symbolically, we are trying to find a value (call it z^*) such that

$$P(z < z^*) = .02$$

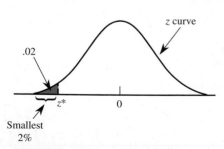

Figure 7.25 The smallest 2% of the standard normal distribution.

This is illustrated in Figure 7.25, which shows that the cumulative area for z^* is .02. Therefore we look for a cumulative area of .0200 in the body of Appendix Table 2. The closest cumulative area in the table is .0202, in the −2.0 row and .05 column; we will use $z^* = -2.05$, the best approximation from the table. Variable values less than −2.05 make up the smallest 2% of the standard normal distribution.

Now suppose that we had been interested in the largest 5% of all z values. We would then be trying to find a value of z^* for which

$$P(z > z^*) = .05$$

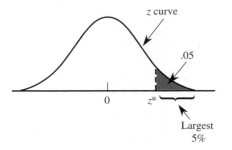

Figure 7.26 The largest 5% of the standard normal distribution.

as illustrated in Figure 7.26. Because Appendix Table 2 always works with cumulative area (area to the left), the first step is to determine

$$\text{area to the left of } z^* = 1 - .05 = .95$$

Looking for the cumulative area closest to .95 in Appendix Table 2, we find that .95 falls exactly halfway between .9495 (corresponding to a z value of 1.64) and .9505 (corresponding to a z value of 1.65). Because .9500 is exactly halfway between the two areas, we use a z value that is halfway between 1.64 and 1.65. (If one value had been closer to .9500 than the other, we would just use the z value corresponding to the closest area). This gives

$$z^* = \frac{1.64 + 1.65}{2} = 1.645$$

Values greater than 1.645 make up the largest 5% of the standard normal distribution. By symmetry, -1.645 separates the smallest 5% of all z values from the others.

■

Example 7.24 More Extremes

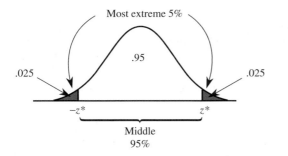

Figure 7.27 The most extreme 5% of the standard normal distribution.

Sometimes we are interested in identifying the most extreme (unusually large *or* small) values in a distribution. Consider describing the values that make up the most extreme 5% of the standard normal distribution. That is, we want to separate the middle 95% from the extreme 5%. This is illustrated in Figure 7.27.

Because the standard normal distribution is symmetric, the most extreme 5% is equally divided between the high side and the low side of the distribution, resulting in an area of .025 for each of the tails of the z curve. Symmetry about 0 implies that if z^* denotes the value that separates the largest 2.5%, the value that separates the smallest 2.5% is simply $-z^*$.

To find z^*, first determine the cumulative area for z^*, which is

$$\text{area to the left of } z^* = .95 + .025 = .975$$

The cumulative area .9750 appears in the 1.9 row and .06 column of Appendix Table 2, so $z^* = 1.96$. For the standard normal distribution, 95% of the variable values fall between -1.96 and 1.96; the most extreme 5% are those values that are either greater than 1.96 or less than -1.96.

■

■ Other Normal Distributions

We now show how z curve areas can be used to calculate probabilities and to describe values for any normal distribution. Remember that the letter z is reserved for those variables that have a standard normal distribution; the letter x is used more generally for any variable whose distribution is described by a normal curve with mean μ and standard deviation σ.

Suppose that we want to compute $P(a < x < b)$, the probability that the variable x lies in a particular range. This probability corresponds to an area under a normal curve and above the interval from a to b, as shown in Figure 7.28(a).

Figure 7.28 Equality of nonstandard and standard normal curve areas.

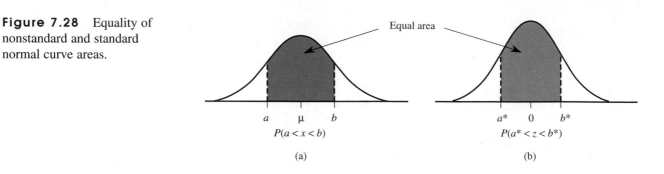

$P(a < x < b)$

(a)

$P(a^* < z < b^*)$

(b)

Our strategy for obtaining this probability is to find an equivalent problem involving the standard normal distribution. Finding an equivalent problem means determining an interval (a^*, b^*) that has the same probability for z (same area under the z curve) as does the interval (a, b) in our original normal distribution (Figure 7.28(b)). The asterisk is used to distinguish a and b, the values from the original normal distribution with mean μ and standard deviation σ, from a^* and b^*, the values from the z curve. To find a^* and b^*, we simply calculate z scores for the endpoints of the interval for which a probability is desired. This process is called **standardizing** the endpoints. For example, suppose that the variable x has a normal distribution with mean $\mu = 100$ and standard deviation $\sigma = 5$. To calculate

$$P(98 < x < 107)$$

we first translate this problem into an equivalent problem for the standard normal distribution. Recall from Chapter 4 that a z score, or standardized score, tells how many standard deviations away from the mean a value lies; the z score is calculated by first subtracting the mean and then dividing by the standard deviation. Converting the lower endpoint $a = 98$ to a z score gives

$$a^* = \frac{98 - 100}{5} = \frac{-2}{5} = -.40$$

and converting the upper endpoint yields

$$b^* = \frac{107 - 100}{5} = \frac{7}{5} = 1.40$$

Then

$$P(98 < x < 107) = P(-.40 < z < 1.40)$$

The probability $P(-.40 < z < 1.40)$ can now be evaluated using Appendix Table 2.

Finding Probabilities

To calculate probabilities for any normal distribution, standardize the relevant values and then use the table of z curve areas. More specifically, if x is a variable whose behavior is described by a normal distribution with mean μ and standard deviation σ, then

$$P(x < b) = P(z < b^*)$$
$$P(a < x) = P(a^* < z) \qquad \text{[Equivalently, } P(x > a) = P(z > a^*)]$$
$$P(a < x < b) = P(a^* < z < b^*)$$

(continued)

where z is a variable whose distribution is standard normal and

$$a^* = \frac{a - \mu}{\sigma} \qquad b^* = \frac{b - \mu}{\sigma}$$

Example 7.25 Children's Heights

Data from the article "The Osteological Paradox: Problems in Inferring Prehistoric Health from Skeletal Samples" (*Current Anthropology* [1992]: 343–370) suggest that a reasonable model for the probability distribution of the continuous numerical variable x = height of a randomly selected 5-year-old child is a normal distribution with a mean of $\mu = 100$ cm and standard deviation $\sigma = 6$ cm. What proportion of the heights is between 94 and 112 cm?

To answer this question, we must find

$$P(94 < x < 112)$$

First, we translate the interval endpoints to equivalent endpoints for the standard normal distribution:

$$a^* = \frac{a - \mu}{\sigma} = \frac{94 - 100}{6} = -1.00$$

$$b^* = \frac{b - \mu}{\sigma} = \frac{112 - 100}{6} = 2.00$$

Then

$$
\begin{aligned}
P(94 < x < 112) &= P(-1.00 < z < 2.00) \\
&= (z \text{ curve area to the left of } 2.00) \\
&\quad - (z \text{ curve area to the left of } -1.00) \\
&= .9772 - .1587 \\
&= .8185
\end{aligned}
$$

The probabilities for x and z are shown in Figure 7.29. If height were observed for many children from this population, about 82% of them would fall between 94 and 112 cm.

Figure 7.29
$P(94 < x < 112)$ and corresponding z curve area for the height problem of Example 7.25.

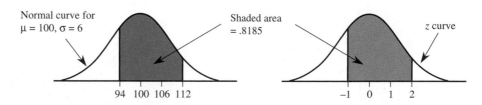

What is the probability that a randomly chosen child will be taller than 110 cm? To evaluate $P(x > 110)$, we first compute

$$a^* = \frac{a - \mu}{\sigma} = \frac{110 - 100}{6} = 1.67$$

Step-by-step technology instructions available online

Then (see Figure 7.30)

$$
\begin{aligned}
P(x > 110) &= P(z > 1.67) \\
&= z \text{ curve area to the right of } 1.67 \\
&= 1 - (z \text{ curve area to the left of } 1.67) \\
&= 1 - .9525 \\
&= .0475
\end{aligned}
$$

Figure 7.30 $P(x > 110)$ and corresponding z curve area for the height problem of Example 7.25.

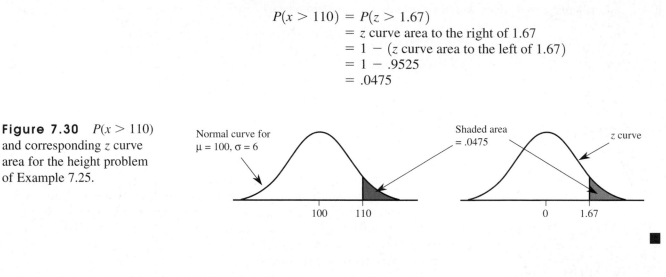

Normal curve for $\mu = 100$, $\sigma = 6$

Shaded area = .0475

z curve

100   110

0   1.67

■

Example 7.26 IQ Scores

Although there is some controversy regarding the appropriateness of IQ scores as a measure of intelligence, IQ scores are commonly used for a variety of purposes. One commonly used IQ scale has a mean of 100 and a standard deviation of 15, and IQ scores are approximately normally distributed. (IQ score is actually a discrete variable [because it is based on the number of correct responses on a test], but its population distribution closely resembles a normal curve.) If we define the random variable

$$x = \text{IQ score of a randomly selected individual}$$

then x has approximately a normal distribution with $\mu = 100$ and $\sigma = 15$.

One way to become eligible for membership in Mensa, an organization purportedly for those of high intelligence, is to have a Stanford–Binet IQ score above 130. What proportion of the population would qualify for Mensa membership? An answer to this question requires evaluating $P(x > 130)$. This probability is shown in Figure 7.31. With $a = 130$,

$$a^* = \frac{a - \mu}{\sigma} = \frac{130 - 100}{15} = 2.00$$

So (see Figure 7.32)

$$
\begin{aligned}
P(x > 130) &= P(z > 2.00) \\
&= z \text{ curve area to the right of } 2.00 \\
&= 1 - (z \text{ curve area to the left of } 2.00) \\
&= 1 - .9772 \\
&= .0228
\end{aligned}
$$

Only 2.28% of the population would qualify for Mensa membership.

Suppose that we are interested in the proportion of the population with IQ scores below 80—that is, $P(x < 80)$. With $b = 80$,

$$b^* = \frac{b - \mu}{\sigma} = \frac{80 - 100}{15} = -1.33$$

Figure 7.31 Normal distribution and desired proportion for Example 7.26.

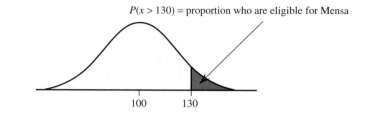

P(x > 130) = proportion who are eligible for Mensa

100   130

Figure 7.32 *P(x > 130)* and corresponding *z* curve area for the IQ problem of Example 7.26.

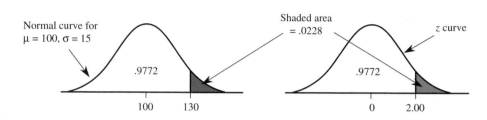

Normal curve for $\mu = 100$, $\sigma = 15$

.9772

100   130

Shaded area = .0228

z curve

.9772

0   2.00

So

$$P(x < 80) = P(z < -1.33)$$
$$= z \text{ curve area to the left of } -1.33$$
$$= .0918$$

as shown in Figure 7.33. This probability (.0918) tells us that just a little over 9% of the population has an IQ score below 80.

Figure 7.33 *P(x < 80)* and corresponding *z* curve area for the IQ problem of Example 7.26.

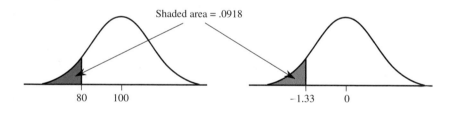

Shaded area = .0918

80   100

-1.33   0

Now consider the proportion of the population with IQs between 75 and 125. Using $a = 75$ and $b = 125$, we obtain

$$a^* = \frac{75 - 100}{15} = -1.67 \qquad b^* = \frac{125 - 100}{15} = 1.67$$

so

$$P(75 < x < 125) = P(-1.67 < z < 1.67)$$
$$= z \text{ curve area between } -1.67 \text{ and } 1.67$$
$$= (z \text{ curve area to the left of } 1.67)$$
$$\quad - (z \text{ curve area to the left of } -1.67)$$
$$= .9525 - .0475$$
$$= .9050$$

This is illustrated in Figure 7.34. The calculation tells us that 90.5% of the population has an IQ score between 75 and 125. Of the 9.5% whose IQ score is not between 75 and 125, half of them (4.75%) have scores over 125, and the other half have scores below 75.

Figure 7.34
$P(75 < x < 125)$ and corresponding z curve area for the IQ problem of Example 7.26.

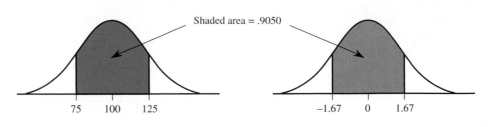

Shaded area = .9050

When we translate from a problem involving a normal distribution with mean μ and standard deviation σ to a problem involving the standard normal distribution, we convert to z scores:

$$z = \frac{x - \mu}{\sigma}$$

Because a z score can be interpreted as giving the distance of an x value from the mean in units of the standard deviation, a z score of 1.4 corresponds to an x value that is 1.4 standard deviations above the mean, and a z score of -2.1 corresponds to an x value that is 2.1 standard deviations below the mean.

Suppose that we are trying to evaluate $P(x < 60)$ for a variable whose distribution is normal with $\mu = 50$ and $\sigma = 5$. Converting the endpoint 60 to a z score gives

$$z = \frac{60 - 50}{5} = 2$$

which tells us that the value 60 is 2 standard deviations above the mean. We then have

$$P(x < 60) = P(z < 2)$$

where z is a standard normal variable. Notice that for the standard normal distribution, the value 2 is 2 standard deviations above the mean, because the mean is 0 and the standard deviation is 1. The value $z = 2$ is located the same distance (measured in standard deviations) from the mean of the standard normal distribution as is the value $x = 60$ from the mean in the normal distribution with $\mu = 50$ and $\sigma = 5$. This is why the translation using z scores results in an equivalent problem involving the standard normal distribution.

■ Describing Extreme Values in a Normal Distribution ....

To describe the extreme values for a normal distribution with mean μ and standard deviation σ, we first solve the corresponding problem for the standard normal distribution and then translate our answer into one for the normal distribution of interest. This process is illustrated in Example 7.27.

....

Example 7.27 Registration Times

Data on the length of time required to complete registration for classes using a telephone registration system suggest that the distribution of the variable

   $x =$ time to register

for students at a particular university can be well approximated by a normal distribution with mean $\mu = 12$ min and standard deviation $\sigma = 2$ min. (The normal distribution might not be an appropriate model for x = time to register at another university. Many factors influence the shape, center, and spread of such a distribution.) Because some students do not sign off properly, the university would like to disconnect students automatically after some amount of time has elapsed. It is decided to choose this time such that only 1% of the students are disconnected while they are still attempting to register. To determine the amount of time that should be allowed before disconnecting a student, we need to describe the largest 1% of the distribution of time to register. These are the individuals who will be mistakenly disconnected. This is illustrated in Figure 7.35(a). To determine the value of x^*, we first solve the analogous problem for the standard normal distribution, as shown in Figure 7.35(b).

Figure 7.35 Capturing the largest 1% in a normal distribution for the problem in Example 7.27.

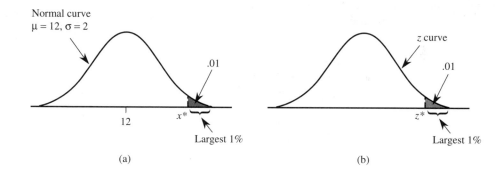

By looking at Appendix Table 2 for a cumulative area of .99, we find the closest entry (.9901) in the 2.3 row and the .03 column, from which $z^* = 2.33$. For the standard normal distribution, the largest 1% of the distribution is made up of those values greater than 2.33. An equivalent statement is that the largest 1% are those with z scores greater than 2.33. This implies that in the distribution of time to register x (or any other normal distribution), the largest 1% are those values with z scores greater than 2.33 or, equivalently, those x values more than 2.33 standard deviations above the mean. Here, the standard deviation is 2, so 2.33 standard deviations is 2.33(2), and it follows that

$$x^* = 12 + 2.33(2) = 12 + 4.66 = 16.66$$

The largest 1% of the distribution for time to register is made up of values that are greater than 16.66 min. If the university system was set to disconnect students after 16.66 min, only 1% of the students registering would be disconnected before completing their registration.

■

A general formula for converting a z score back to an x value results from solving $z^* = \dfrac{x^* - \mu}{\sigma}$ for x^*, as shown in the accompanying box.

To convert a z score z^* back to an x value, use

$$x = \mu + z^*\sigma$$

Example 7.28 Motor Vehicle Emissions

Data from the article "Determining Statistical Characteristics of a Vehicle Emissions Audit Procedure" (*Technometrics* [1980]: 483–493) suggest that the emissions of nitrogen oxides, which are major constituents of smog, can be plausibly modeled using a normal distribution. Let x denote the amount of this pollutant emitted by a randomly selected vehicle. The distribution of x can be described by a normal distribution with $\mu = 1.6$ and $\sigma = 0.4$.

Suppose that the EPA wants to offer some sort of incentive to get the worst polluters off the road. What emission levels constitute the worst 10% of the vehicles? The worst 10% would be the 10% with the highest emissions level, as shown in the illustration in the margin.

For the standard normal distribution, the largest 10% are those with z values greater than $z^* = 1.28$ (from Appendix Table 2, based on a cumulative area of .90). Then

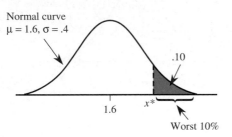

$$x^* = \mu + z^*\sigma$$
$$= 1.6 + 1.28(.4)$$
$$= 1.6 + .512$$
$$= 2.112$$

In the population of vehicles of the type considered, about 10% would have oxide emission levels greater than 2.112.

■

Exercises 7.64–7.80

Turn to the Expanded Answers Section for answers not shown next to exercises.

7.64 Determine the following standard normal (z) curve areas:

a. The area under the z curve to the left of 1.75   0.9599
b. The area under the z curve to the left of -0.68 0.2483
c. The area under the z curve to the right of 1.20   0.1151
d. The area under the z curve to the right of -2.82 0.9976
e. The area under the z curve between -2.22 and 0.53
f. The area under the z curve between -1 and 1   0.6826
g. The area under the z curve between -4 and 4   1

7.65 Determine each of the following areas under the standard normal (z) curve:
a. To the left of -1.28 0.1003
b. To the right of 1.28   0.1003

c. Between -1 and 2   0.8185
d. To the right of 0   0.5
e. To the right of -5 1
f. Between -1.6 and 2.5   0.9390
g. To the left of 0.23   0.5910

7.66 Let z denote a random variable that has a standard normal distribution. Determine each of the following probabilities:
a. $P(z < 2.36)$ 0.9909
b. $P(z \le 2.36)$ 0.9909
c. $P(z < -1.23)$ 0.1093
d. $P(1.14 < z < 3.35)$ 0.1267
e. $P(-0.77 \le z \le -0.55)$ 0.0706

Bold exercises answered in back ● Data set available online but not required ▼ Video solution available

f. $P(z > 2)$ 0.0228
g. $P(z \geq -3.38)$ 0.9996
h. $P(z < 4.98)$ Approximately 1

7.67 Let z denote a random variable having a normal distribution with $\mu = 0$ and $\sigma = 1$. Determine each of the following probabilities:
a. $P(z < 0.10)$ 0.5398
b. $P(z < -0.10)$ 0.4602
c. $P(0.40 < z < 0.85)$ 0.1469
d. $P(-0.85 < z < -0.40)$ 0.1469
e. $P(-0.40 < z < 0.85)$ 0.4577
f. $P(z > -1.25)$ 0.8944
g. $P(z < -1.50 \text{ or } z > 2.50)$ 0.0730

7.68 Let z denote a variable that has a standard normal distribution. Determine the value z^* to satisfy the following conditions:
a. $P(z < z^*) = .025$ −1.96
b. $P(z < z^*) = .01$ −2.33
c. $P(z < z^*) = .05$ −1.645
d. $P(z > z^*) = .02$ 2.05
e. $P(z > z^*) = .01$ 2.33
f. $P(z > z^* \text{ or } z < -z^*) = .20$ 1.28

7.69 Determine the value z^* that
a. Separates the largest 3% of all z values from the others
b. Separates the largest 1% of all z values from the others
c. Separates the smallest 4% of all z values from the others   −1.75
d. Separates the smallest 10% of all z values from the others   −1.28

7.70 Determine the value of z^* such that
a. $-z^*$ and z^* separate the middle 95% of all z values from the most extreme 5%   1.96
b. $-z^*$ and z^* separate the middle 90% of all z values from the most extreme 10%   1.645
c. $-z^*$ and z^* separate the middle 98% of all z values from the most extreme 2%   2.33
d. $-z^*$ and z^* separate the middle 92% of all z values from the most extreme 8%   1.75

7.71 Because $P(z < .44) = .67$, 67% of all z values are less than .44, and .44 is the 67th percentile of the standard normal distribution. Determine the value of each of the following percentiles for the standard normal distribution (Hint: If the cumulative area that you must look for does not appear in the z table, use the closest entry):
a. The 91st percentile (Hint: Look for area .9100.)   1.34

b. The 77th percentile   0.74
c. The 50th percentile   0
d. The 9th percentile   −1.34
e. What is the relationship between the 70th z percentile and the 30th z percentile?   They are negatives of one another.

7.72 Consider the population of all 1-gal cans of dusty rose paint manufactured by a particular paint company. Suppose that a normal distribution with mean $\mu = 5$ ml and standard deviation $\sigma = 0.2$ ml is a reasonable model for the distribution of the variable $x =$ amount of red dye in the paint mixture. Use the normal distribution model to calculate the following probabilities:
a. $P(x < 5.0)$ 0.5   d. $P(4.6 < x < 5.2)$ 0.8185
b. $P(x < 5.4)$ 0.9772   e. $P(x > 4.5)$ 0.9938
c. $P(x \leq 5.4)$ 0.9772   f. $P(x > 4.0)$ Approximately 1

7.73 Consider babies born in the "normal" range of 37–43 weeks gestational age. Extensive data support the assumption that for such babies born in the United States, birth weight is normally distributed with mean 3432 g and standard deviation 482 g ("Are Babies Normal?" *The American Statistician* [1999]: 298–302).
a. What is the probability that the birth weight of a randomly selected baby of this type exceeds 4000 g? is between 3000 and 4000 g?   0.119, 0.6969
b. What is the probability that the birth weight of a randomly selected baby of this type is either less than 2000 g or greater than 5000 g?   0.0021
c. What is the probability that the birth weight of a randomly selected baby of this type exceeds 7 lb? (Hint: 1 lb = 453.59 g.)   0.7019
d. How would you characterize the most extreme 0.1% of all birth weights?
e. If x is a random variable with a normal distribution and a is a numerical constant ($a \neq 0$), then $y = ax$ also has a normal distribution. Use this formula to determine the distribution of birth weight expressed in pounds (shape, mean, and standard deviation), and then recalculate the probability from Part (c). How does this compare to your previous answer?

7.74 A machine that cuts corks for wine bottles operates in such a way that the distribution of the diameter of the corks produced is well approximated by a normal distribution with mean 3 cm and standard deviation 0.1 cm. The specifications call for corks with diameters between 2.9 and 3.1 cm. A cork not meeting the specifications is considered defective. (A cork that is too small leaks and causes the wine to deteriorate; a cork that is too large doesn't fit in the bottle.) What proportion of corks produced by this machine are defective?   0.3174

7.75 Refer to Exercise 7.74. Suppose that there are two machines available for cutting corks. The machine described in the preceding problem produces corks with diameters that are approximately normally distributed with mean 3 cm and standard deviation 0.1 cm. The second machine produces corks with diameters that are approximately normally distributed with mean 3.05 cm and standard deviation 0.01 cm. Which machine would you recommend? (Hint: Which machine would produce fewer defective corks?)   The second machine

7.76 A gasoline tank for a certain car is designed to hold 15 gal of gas. Suppose that the variable x = actual capacity of a randomly selected tank has a distribution that is well approximated by a normal curve with mean 15.0 gal and standard deviation 0.1 gal.
a. What is the probability that a randomly selected tank will hold at most 14.8 gal?   0.0228
b. What is the probability that a randomly selected tank will hold between 14.7 and 15.1 gal?   0.8400
c. If two such tanks are independently selected, what is the probability that both hold at most 15 gal?   0.25

7.77 ▼ The time that it takes a randomly selected job applicant to perform a certain task has a distribution that can be approximated by a normal distribution with a mean value of 120 sec and a standard deviation of 20 sec. The fastest 10% are to be given advanced training. What task times qualify individuals for such training?   94.4 seconds or less

7.78 A machine that produces ball bearings has initially been set so that the true average diameter of the bearings it produces is 0.500 in. A bearing is acceptable if its diameter is within 0.004 in. of this target value. Suppose, however, that the setting has changed during the course of

production, so that the distribution of the diameters produced is well approximated by a normal distribution with mean 0.499 in. and standard deviation 0.002 in. What percentage of the bearings produced will not be acceptable?   0.0730

7.79 ▼ Suppose that the distribution of net typing rate in words per minute (wpm) for experienced typists can be approximated by a normal curve with mean 60 wpm and standard deviation 15 wpm ("Effects of Age and Skill in Typing", *Journal of Experimental Psychology* [1984]: 345–371).
a. What is the probability that a randomly selected typist's net rate is at most 60 wpm? less than 60 wpm?   0.5, 0.5
b. What is the probability that a randomly selected typist's net rate is between 45 and 90 wpm?   0.8185
c. Would you be surprised to find a typist in this population whose net rate exceeded 105 wpm? (Note: The largest net rate in a sample described in the paper is 104 wpm.)
d. Suppose that two typists are independently selected. What is the probability that both their typing rates exceed 75 wpm?   0.0252
e. Suppose that special training is to be made available to the slowest 20% of the typists. What typing speeds would qualify individuals for this training?   47.4 or less words per minute.

7.80 Consider the variable x = time required for a college student to complete a standardized exam. Suppose that for the population of students at a particular university, the distribution of x is well approximated by a normal curve with mean 45 min and standard deviation 5 min.
a. If 50 min is allowed for the exam, what proportion of students at this university would be unable to finish in the allotted time?   0.1587
b. How much time should be allowed for the exam if we wanted 90% of the students taking the test to be able to finish in the allotted time?   51.4 minutes
c. How much time is required for the fastest 25% of all students to complete the exam?   41.65 minutes

Bold exercises answered in back   ● Data set available online but not required   ▼ Video solution available

---

7.7 # Checking for Normality and Normalizing Transformations

Some of the most frequently used statistical methods are valid only when a sample $x_1, x_2, \ldots, x_n$ has come from a population distribution that is at least approximately normal. One way to see whether an assumption of population normality is plausible is to construct a **normal probability plot** of the data. One version of this plot uses quantities called **normal scores**. The values of the normal scores depend on the sample size n. For example, the normal scores when $n = 10$ are as follows:

$$-1.539 \quad -1.001 \quad -.656 \quad -.376 \quad -.123$$
$$.123 \quad\quad .376 \quad\quad .656 \quad\quad 1.001 \quad\quad 1.539$$

To interpret these numbers, think of selecting sample after sample from a standard normal distribution, each one consisting of $n = 10$ observations. Then -1.539 is the long-run average of the smallest observation from each sample, -1.001 is the long-run average of the second smallest observation from each sample, and so on. In other words, -1.539 is the mean value of the smallest observation in a sample of size 10 from the z distribution, -1.001 is the mean value of the second smallest observation, and so on.

Extensive tabulations of normal scores for many different sample sizes are available. Alternatively, many software packages (such as MINITAB and SAS) and some graphing calculators can compute these scores on request and then construct a normal probability plot. Not all calculators and software packages use the same algorithm to compute normal scores. However, this does not change the overall character of a normal probability plot, so either the tabulated values or those given by the computer or calculator can be used.

After the sample observations are ordered from smallest to largest, the smallest normal score is paired with the smallest observation, the second smallest normal score with the second smallest observation, and so on. The first number in a pair is the normal score, and the second number in the pair is the observed data value. A normal probability plot is just a scatterplot of the (normal score, observed value) pairs.

If the sample has been selected from a *standard* normal distribution, the second number in each pair should be reasonably close to the first number (ordered observation $\approx$ corresponding mean value). Then the n plotted points will fall near a line with slope equal to 1 (a 45° line) passing through $(0, 0)$. When the sample has been obtained from *some* normal population distribution (but not necessarily the standard normal distribution), the plotted points should be close to *some* straight line.

> ### DEFINITION
>
> A **normal probability plot** is a scatter plot of the (normal score, observed value) pairs. A substantial linear pattern in a normal probability plot suggests that population normality is plausible. On the other hand, a systematic departure from a straight-line pattern (such as curvature in the plot) casts doubt on the legitimacy of assuming a normal population distribution.

Example 7.29 Window Widths

The following 10 observations are widths of contact windows in integrated circuit chips:

$$3.21 \quad 2.49 \quad 2.94 \quad 4.38 \quad 4.02 \quad 3.62 \quad 3.30 \quad 2.85 \quad 3.34 \quad 3.81$$

The 10 pairs for the normal probability plot are then

$$(-1.539, 2.49) \quad\quad (0.123, 3.34)$$
$$(-1.001, 2.85) \quad\quad (0.376, 3.62)$$
$$(-0.656, 2.94) \quad\quad (0.656, 3.81)$$

$$(-0.376, 3.21) \qquad (1.001, 4.02)$$
$$(-0.123, 3.30) \qquad (1.539, 4.38)$$

The normal probability plot is shown in Figure 7.36. The linearity of the plot supports the assumption that the window width distribution from which these observations were drawn is normal.

Figure 7.36 A normal probability plot for Example 7.29.

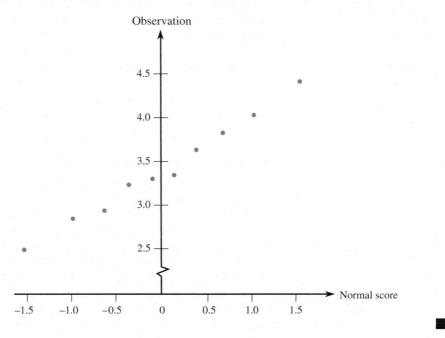

The decision as to whether a plot shows a substantial linear pattern is somewhat subjective. Particularly when n is small, normality should not be ruled out unless the departure from linearity is clear-cut. Figure 7.37 displays several plots that suggest a nonnormal population distribution.

Figure 7.37 Plots suggesting nonnormality: (a) indication that the population distribution is skewed; (b) indication that the population distribution has heavier tails than a normal curve; (c) presence of an outlier.

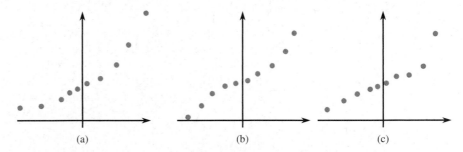

■ Using the Correlation Coefficient to Check Normality ....

The correlation coefficient r was introduced in Chapter 5 as a quantitative measure of the extent to which the points in a scatterplot fall close to a straight line. Consider the n (normal score, observed value) pairs:

(smallest normal score, smallest observation)
$$\vdots$$
(largest normal score, largest observation)

Then the correlation coefficient can be computed as discussed in Chapter 5. The normal probability plot always slopes upward (because it is based on values ordered from smallest to largest), so r will be a positive number. A value of r quite close to 1 indicates a very strong linear relationship in the normal probability plot. If r is too much smaller than 1, normality of the underlying distribution is questionable.

How far below 1 does r have to be before we begin to seriously doubt the plausibility of normality? The answer depends on the sample size n. If n is small, an r value somewhat below 1 is not surprising, even when the distribution is normal, but if n is large, only an r value very close to 1 supports the assumption of normality. For selected values of n, Table 7.2 gives critical values to which r can be compared to check for normality. If your sample size is in between two tabulated values of n, use the critical value for the larger sample size. (For example, if $n = 46$, use the value .966 for sample size 50.)

Table 7.2   Values to Which r Can Be Compared to Check for Normality *

n	5	10	15	20	25	30	40	50	60	75
Critical r	.832	.880	.911	.929	.941	.949	.960	.966	.971	.976

*Source: *MINITAB User's Manual*.

If

$r <$ critical r for corresponding n

considerable doubt is cast on the assumption of population normality.

How were the critical values in Table 7.2 obtained? Consider the critical value .941 for $n = 25$. Suppose that the underlying distribution is actually normal. Consider obtaining a large number of different samples, each one consisting of 25 observations, and computing the value of r for each one. Then it can be shown that only 1% of the samples result in an r value less than the critical value .941. That is, .941 was chosen to guarantee a 1% error rate: In only 1% of all cases will we judge normality implausible when the distribution is really normal. The other critical values are also chosen to yield a 1% error rate for the corresponding sample sizes.

It might have occurred to you that another type of error is possible: obtaining a large value of r and concluding that normality is a reasonable assumption when the distribution is actually nonnormal. This type of error is more difficult to control than the type mentioned previously, but the procedure we have described generally does a good job in both respects.

Example 7.30   Window Widths Continued

The sample size for the contact window width data of Example 7.29 is $n = 10$. The critical r, from Table 7.2 is then .880. The correlation coefficient calculated using the (normal score, observed value) pairs is $r = .995$. Because r is larger than the critical r for a sample of size 10, it is plausible that the population distribution of window widths from which this sample was drawn is approximately normal.

■

■ **Transforming Data to Obtain a Distribution That Is Approximately Normal** ....

Many of the most frequently used statistical methods are valid only when the sample is selected at random from a population whose distribution is at least approximately normal. When a sample histogram shows a distinctly nonnormal shape, it is common to use a transformation or reexpression of the data. By *transforming* data, we mean applying some specified mathematical function (such as the square root, logarithm, or reciprocal) to each data value to produce a set of transformed data. We can then study and summarize the distribution of these transformed values using methods that require normality. We saw in Chapter 5 that, with bivariate data, one or both of the variables can be transformed in an attempt to find two variables that are linearly related. With univariate data, a transformation is usually chosen to yield a distribution of transformed values that is more symmetric and more closely approximated by a normal curve than was the original distribution.

....

Example 7.31 Rainfall Data

● Data that have been used by several investigators to introduce the concept of transformation (e.g., "Exploratory Methods for Choosing Power Transformations," *Journal of the American Statistical Association* [1982]: 103–108) consist of values of March precipitation for Minneapolis–St. Paul over a period of 30 years. These values are given in Table 7.3, along with the square root of each value. Histograms of both the original and the transformed data appear in Figure 7.38. The distribution of the original data is clearly skewed, with a long upper tail. The square-root transformation results in a substantially more symmetric distribution, with a typical (i.e., central) value near the 1.25 boundary between the third and fourth class intervals.

Table 7.3   Original and Square-Root-Transformed Values of March Precipitation in Minneapolis–St. Paul over a 30-year Period

Year	Precipitation	$\sqrt{\text{Precipitation}}$	Year	Precipitation	$\sqrt{\text{Precipitation}}$
1	.77	.88	16	1.62	1.27
2	1.74	1.32	17	1.31	1.14
3	.81	.90	18	.32	.57
4	1.20	1.10	19	.59	.77
5	1.95	1.40	20	.81	.90
6	1.20	1.10	21	2.81	1.68
7	.47	.69	22	1.87	1.37
8	1.43	1.20	23	1.18	1.09
9	3.37	1.84	24	1.35	1.16
10	2.20	1.48	25	4.75	2.18
11	3.00	1.73	26	2.48	1.57
12	3.09	1.76	27	.96	.98
13	1.51	1.23	28	1.89	1.37
14	2.10	1.45	29	.90	.95
15	.52	.72	30	2.05	1.43

Bold exercises answered in back   ● Data set available online but not required   ▼ Video solution available

Figure 7.38 Histograms of the precipitation data used in Example 7.31: (a) untransformed data; (b) square-root transformed data.

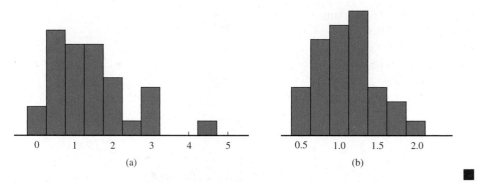

Logarithmic transformations are also common and, as with bivariate data, either the natural logarithm or the base 10 logarithm can be used. A logarithmic transformation is usually applied to data that are positively skewed (a long upper tail). This affects values in the upper tail substantially more than values in the lower tail, yielding a more symmetric—and often more nearly normal—distribution.

Example 7.32 Beryllium Exposure

● Exposure to beryllium is known to produce adverse effects on lungs as well as on other tissues and organs in both laboratory animals and humans. The article "Time Lapse Cinematographic Analysis of Beryllium: Lung Fibroblast Interactions" (*Environmental Research* [1983]: 34–43) reported the results of experiments designed to study the behavior of certain individual cells that had been exposed to beryllium. An important characteristic of such an individual cell is its interdivision time (IDT). IDTs were determined for a large number of cells under both exposed (treatment) and unexposed (control) conditions. The authors of the article stated, "The IDT distributions are seen to be skewed, but the natural logs do have an approximate normal distribution." The same property holds for $\log_{10}$ transformed data. We give representative IDT data in Table 7.4 and the resulting histograms in Figure 7.39, which are in agreement with the authors' statement.

Table 7.4 Original and $\log_{10}$(IDT) Values

IDT	$\log_{10}$(IDT)	IDT	$\log_{10}$(IDT)	IDT	$\log_{10}$(IDT)
28.1	1.45	31.2	1.49	13.7	1.14
46.0	1.66	25.8	1.41	16.8	1.23
34.8	1.54	62.3	1.79	28.0	1.45
17.9	1.25	19.5	1.29	21.1	1.32
31.9	1.50	28.9	1.46	60.1	1.78
23.7	1.37	18.6	1.27	21.4	1.33
26.6	1.42	26.2	1.42	32.0	1.51
43.5	1.64	17.4	1.24	38.8	1.59
30.6	1.49	55.6	1.75	25.5	1.41
52.1	1.72	21.0	1.32	22.3	1.35
15.5	1.19	36.3	1.56	19.1	1.28
38.4	1.58	72.8	1.86	48.9	1.69
21.4	1.33	20.7	1.32	57.3	1.76
40.9	1.61				

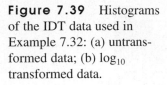

Figure 7.39 Histograms of the IDT data used in Example 7.32: (a) untransformed data; (b) $\log_{10}$ transformed data.

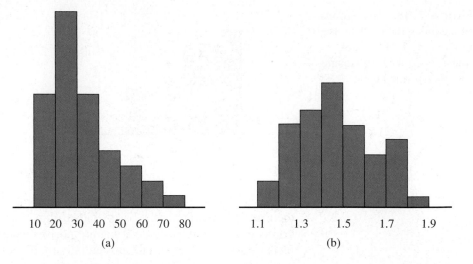

(a)   (b)

The sample size for the IDT data is $n = 40$. The correlation coefficient for the (normal score, original [untransformed] data) pairs is .950, which is less than the critical r for $n = 40$ (critical $r = .960$). The correlation coefficient using the transformed data is .998, which is much larger than the critical r, supporting the assertion that $\log_{10}(\text{IDT})$ has approximately a normal distribution. Figure 7.40 displays

Figure 7.40 MINITAB-generated normal probability plots for Example 7.32: (a) original IDT data; (b) log-transformed IDT.

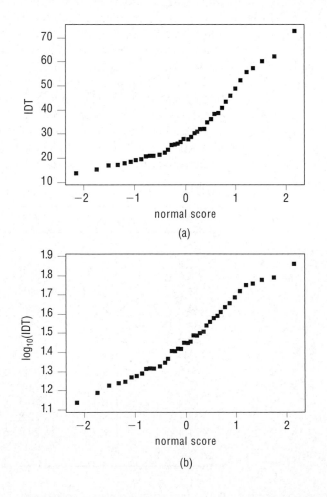

MINITAB normal probability plots for the original data and for the transformed data. The plot for the transformed data is clearly more linear in appearance than the plot for the original data.

■

■ Selecting a Transformation ....

Occasionally, a particular transformation can be dictated by some theoretical argument, but often this is not the case and you may wish to try several different transformations to find one that is satisfactory. Figure 7.41, from the article "Distribution of Sperm Counts in Suspected Infertile Men" (*Journal of Reproduction and Fertility* [1983]: 91–96), shows what can result from such a search. Other investigators in this field had previously used all three of the transformations illustrated.

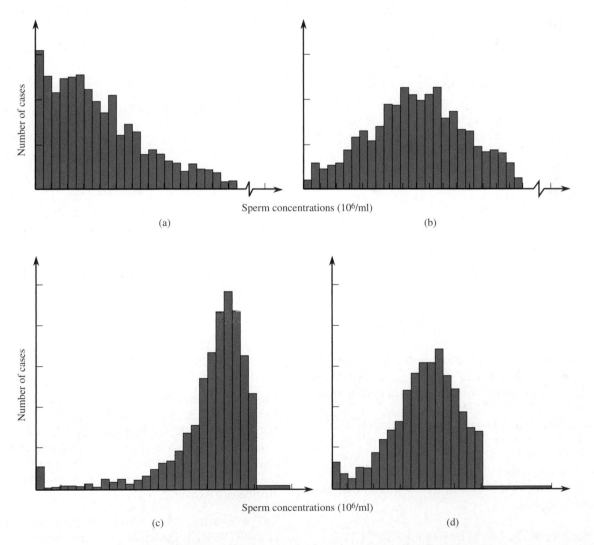

Figure 7.41 Histograms of sperm concentrations for 1711 suspected infertile men: (a) untransformed data (highly skewed); (b) log-transformed data (reasonably symmetric); (c) square-root-transformed data; (d) cube-root-transformed data.

Exercises 7.81–7.92

Turn to the Expanded Answers Section for answers not shown next to exercises.

7.81 Ten measurements of the steam rate (in pounds per hour) of a distillation tower were used to construct the following normal probability plot ("A Self-Descaling Distillation Tower," *Chemical Engineering Process* [1968]: 79–84). Based on the plot, do you think it is reasonable to assume that the normal distribution provides an adequate description of the steam rate distribution? Explain.

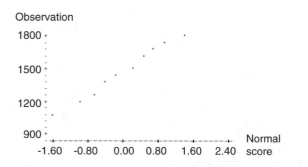

7.82 The following normal probability plot was constructed using part of the data appearing in the paper "Trace Metals in Sea Scallops" (*Environmental Concentration and Toxicology* 19: 1326–1334).

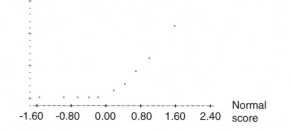

The variable under study was the amount of cadmium in North Atlantic scallops. Do the sample data suggest that the cadmium concentration distribution is not normal? Explain.

7.83 ● Consider the following 10 observations on the lifetime (in hours) for a certain type of component: 152.7, 172.0, 172.5, 173.3, 193.0, 204.7, 216.5, 234.9, 262.6, 422.6. Construct a normal probability plot, and comment on the plausibility of a normal distribution as a model for component lifetime.

7.84 The paper "The Load-Life Relationship for M50 Bearings with Silicon Nitride Ceramic Balls" (*Lubrication Engineering* [1984]: 153–159) reported the following data on bearing load life (in millions of revolutions); the corresponding normal scores are also given:

x	Normal Score	x	Normal Score
47.1	−1.867	240.0	0.062
68.1	−1.408	240.0	0.187
68.1	−1.131	278.0	0.315
90.8	−0.921	278.0	0.448
103.6	−0.745	289.0	0.590
106.0	−0.590	289.0	0.745
115.0	−0.448	367.0	0.921
126.0	−0.315	385.9	1.131
146.6	−0.187	392.0	1.408
229.0	−0.062	395.0	1.867

Construct a normal probability plot. Is normality plausible?

7.85 ● The following observations are DDT concentrations in the blood of 20 people:

24 26 30 35 35 38 39 40 40 41 42 52
56 58 61 75 79 88 102 42

Use the normal scores from Exercise 7.84 to construct a normal probability plot, and comment on the appropriateness of a normal probability model.

7.86 ● Consider the following sample of 25 observations on the diameter x (in centimeters) of a disk used in a certain system:

16.01	16.08	16.13	15.94	16.05	16.27	15.89
15.84	15.95	16.10	15.92	16.04	15.82	16.15
16.06	15.66	15.78	15.99	16.29	16.15	16.19
16.22	16.07	16.13	16.11			

The 13 largest normal scores for a sample of size 25 are 1.965, 1.524, 1.263, 1.067, 0.905, 0.764, 0.637, 0.519, 0.409, 0.303, 0.200, 0.100, and 0. The 12 smallest scores result from placing a negative sign in front of each of the given nonzero scores. Construct a normal probability plot. Does it appear plausible that disk diameter is normally distributed? Explain.

7.87 ● Example 7.31 examined rainfall data for Minneapolis–St. Paul. The square-root transformation was used to obtain a distribution of values that was more symmetric than the distribution of the original data. Another power transformation that has been suggested by meteorologists is the cube root: transformed value = (original value)$^{1/3}$. The original values and their cube roots (the transformed values) are given in the following table:

Original	Transformed	Original	Transformed
0.32	0.68	1.51	1.15
0.47	0.78	1.62	1.17
0.52	0.80	1.74	1.20
0.59	0.84	1.87	1.23
0.77	0.92	1.89	1.24
0.81	0.93	1.95	1.25
0.81	0.93	2.05	1.27
0.90	0.97	2.10	1.28
0.96	0.99	2.20	1.30
1.18	1.06	2.48	1.35
1.20	1.06	2.81	1.41
1.20	1.06	3.00	1.44
1.31	1.09	3.09	1.46
1.35	1.11	3.37	1.50
1.43	1.13	4.75	1.68

Construct a histogram of the transformed data. Compare your histogram to those given in Figure 7.38. Which of the cube-root and square-root transformations appear to result in the more symmetric histogram(s)?

7.88 ● The following data are a sample of survival times (days from diagnosis) for patients suffering from chronic leukemia of a certain type (*Statistical Methodology for Survival Time Studies* [Bethesda, MD: National Cancer Institute, 1986]):

7	47	58	74	177	232	273	285
317	429	440	445	455	468	495	497
532	571	579	581	650	702	715	779
881	900	930	968	1077	1109	1314	1334
1367	1534	1712	1784	1877	1886	2045	2056
2260	2429	2509					

a. Construct a relative frequency distribution for this data set, and draw the corresponding histogram.
b. Would you describe this histogram as having a positive or a negative skew?   Positive skew
c. Would you recommend transforming the data? Explain.

7.89 ● In a study of warp breakage during the weaving of fabric (*Technometrics* [1982]: 63), 100 pieces of yarn were tested. The number of cycles of strain to breakage was recorded for each yarn sample. The resulting data are given in the following table:

86	146	251	653	98	249	400	292	131	176
76	264	15	364	195	262	88	264	42	321
180	198	38	20	61	121	282	180	325	250
196	90	229	166	38	337	341	40	40	135
597	246	211	180	93	571	124	279	81	186
497	182	423	185	338	290	398	71	246	185
188	568	55	244	20	284	93	396	203	829
239	236	277	143	198	264	105	203	124	137
135	169	157	224	65	315	229	55	286	350
193	175	220	149	151	353	400	61	194	188

a. Construct a frequency distribution using the class intervals 0 to < 100, 100 to < 200, and so on.
b. Draw the histogram corresponding to the frequency distribution in Part (a). How would you describe the shape of this histogram?   The histogram is positively skewed.
c. Find a transformation for these data that results in a more symmetric histogram than what you obtained in Part (b).   A square root transformation will result in a more symmetrical histogram.

7.90 The article "The Distribution of Buying Frequency Rates" (*Journal of Marketing Research* [1980]: 210–216) reported the results of a $3\frac{1}{2}$-year study of dentifrice purchases. The investigators conducted their research using a national sample of 2071 households and recorded the number of toothpaste purchases for each household participating in the study. The results are given in the following frequency distribution:

Number of Purchases	Number of Households (Frequency)
10 to <20	904
20 to <30	500
30 to <40	258
40 to <50	167
50 to <60	94
60 to <70	56
70 to <80	26
80 to <90	20
90 to <100	13
100 to <110	9
110 to <120	7
120 to <130	6
130 to <140	6
140 to <150	3
150 to <160	0
160 to <170	2

Bold exercises answered in back   ● Data set available online but not required   ▼ Video solution available

a. Draw a histogram for this frequency distribution. Would you describe the histogram as positively or negatively skewed?

b. Does the square-root transformation result in a histogram that is more symmetric than that of the original data? (Be careful! This one is a bit tricky, because you don't have the raw data; transforming the endpoints of the class intervals will result in class intervals that are not necessarily of equal widths, so the histogram of the transformed values will have to be drawn with this in mind.)

7.91 ● The paper "Temperature and the Northern Distributions of Wintering Birds" (*Ecology* [1991]: 2274–2285) gave the following body masses (in grams) for 50 different bird species:

7.7	10.1	21.6	8.6	12.0	11.4	16.6	9.4
11.5	9.0	8.2	20.2	48.5	21.6	26.1	6.2
19.1	21.0	28.1	10.6	31.6	6.7	5.0	68.8
23.9	19.8	20.1	6.0	99.6	19.8	16.5	9.0
448.0	21.3	17.4	36.9	34.0	41.0	15.9	12.5
10.2	31.0	21.5	11.9	32.5	9.8	93.9	10.9
19.6	14.5						

a. Construct a stem-and-leaf display in which 448.0 is listed separately beside the display as an outlier on the high side, the stem of an observation is the tens digit, the leaf is the ones digit, and the tenths digit is suppressed (e.g., 21.5 has stem 2 and leaf 1). What do you perceive as the most prominent feature of the display?

b. Draw a histogram based on class intervals 5 to <10, 10 to <15, 15 to <20, 20 to <25, 25 to <30, 30 to <40, 40 to <50, 50 to <100, and 100 to <500. Is a transformation of the data desirable? Explain.

c. Use a calculator or statistical computer package to calculate logarithms of these observations, and construct a histogram. Is the log transformation successful in producing a more symmetric distribution?

d. Consider transformed value $= \dfrac{1}{\sqrt{\text{original value}}}$ and construct a histogram of the transformed data. Does it appear to resemble a normal curve?

7.92 The following figure appeared in the paper "EDTA-Extractable Copper, Zinc, and Manganese in Soils of the Canterbury Plains" (*New Zealand Journal of Agricultural Research* [1984]: 207–217). A large number of topsoil samples were analyzed for manganese (Mn), zinc (Zn), and copper (Cu), and the resulting data were summarized using histograms. The investigators transformed each data set using logarithms in an effort to obtain more symmetric distributions of values. Do you think the transformations were successful? Explain.

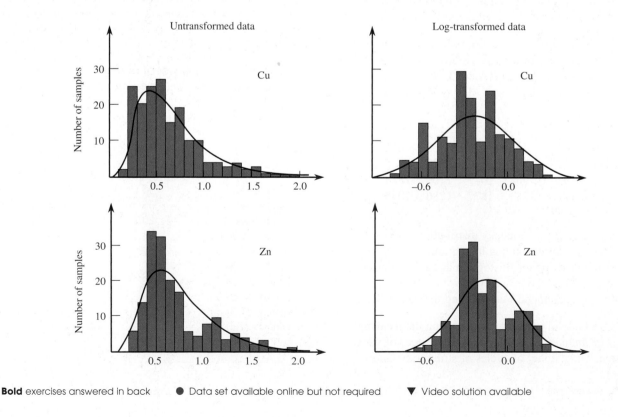

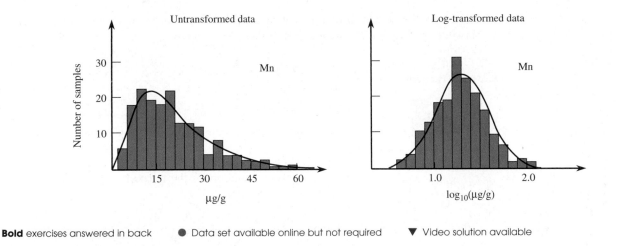

7.8 Using the Normal Distribution to Approximate a Discrete Distribution

The distribution of many random variables can be approximated by a carefully chosen normal distribution. In this section, we show how probabilities for some discrete random variables can be approximated using a normal curve. The most important case of this is the approximation of binomial probabilities.

■ The Normal Curve and Discrete Variables

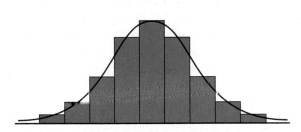

Figure 7.42 A normal curve approximation to a probability histogram.

The probability distribution of a discrete random variable x is represented pictorially by a probability histogram. The probability of a particular value is the area of the rectangle centered at that value. Possible values of x are isolated points on the number line, usually whole numbers. For example, if $x =$ the IQ of a randomly selected 8-year-old child, then x is a discrete random variable, because an IQ score must be a whole number.

Often a probability histogram can be well approximated by a normal curve, as illustrated in Figure 7.42. In such cases, it is customary to say that x has approximately a normal distribution. The normal distribution can then be used to calculate approximate probabilities of events involving x.

Example 7.33 Express Mail Packages

The number of express mail packages mailed at a certain post office on a randomly selected day is approximately normally distributed with mean 18 and standard deviation 6. Let's first calculate the approximate probability that $x = 20$. Figure 7.43(a) shows a portion of the probability histogram for x with the approximating normal curve superimposed. The area of the shaded rectangle is $P(x = 20)$. The left edge of this rectangle is at 19.5 on the horizontal scale, and the right edge is at 20.5.

Therefore, the desired probability is approximately the area under the normal curve between 19.5 and 20.5. Standardizing these limits gives

$$\frac{20.5 - 18}{6} = .42 \qquad \frac{19.5 - 18}{6} = .25$$

from which we get

$$P(x = 20) \approx P(.25 < z < .42) = .6628 - .5987 = .0641$$

In a similar fashion, Figure 7.43(b) shows that $P(x \leq 10)$ is approximately the area under the normal curve to the left of 10.5. Then

$$P(x \leq 10) \approx P\left(z \leq \frac{10.5 - 18}{6} \right) = P(z \leq -1.25)$$
$$= .1056$$

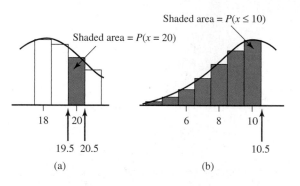

Shaded area = $P(x = 20)$

Shaded area = $P(x \leq 10)$

(a)   (b)

Figure 7.43   The normal approximation for Example 7.33.

■

The calculation of probabilities in Example 7.33 illustrates the use of what is known as a **continuity correction**. Because the rectangle for $x = 10$ extends to 10.5 on the right, we use the normal curve area to the left of 10.5 rather than 10. In general, if possible x values are consecutive whole numbers, then $P(a \leq x \leq b)$ will be approximately the normal curve area between limits $a - \frac{1}{2}$ and $b + \frac{1}{2}$.

■ Normal Approximation to a Binomial Distribution

Figure 7.44 shows the probability histograms for two binomial distributions, one with $n = 25$, $\pi = .4$, and the other with $n = 25$, $\pi = .1$. For each distribution, we computed $\mu = n\pi$ and $\sigma = \sqrt{n\pi(1 - \pi)}$ and then we superimposed a normal curve with this μ and σ on the corresponding probability histogram. A normal curve fits the probability histogram well in the first case (Figure 7.44(a)). When this happens, binomial probabilities can be accurately approximated by areas under the normal curve. Because of this, statisticians say that both x (the number of successes) and x/n (the proportion of successes) are approximately normally distributed. In the second case (Figure 7.44(b)), the normal curve does not give a good approximation because the probability histogram is skewed, whereas the normal curve is symmetric.

Figure 7.44   Normal approximations to binomial distributions.

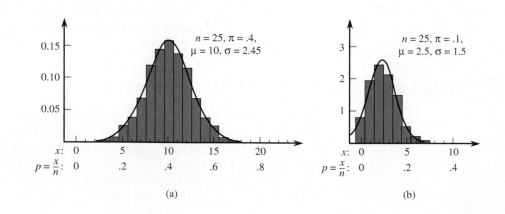

Let x be a binomial random variable based on n trials and success probability π, so that

$$\mu = n\pi \quad \text{and} \quad \sigma = \sqrt{n\pi(1 - \pi)}$$

If n and π are such that

$$n\pi \geq 10 \text{ and } n(1 - \pi) \geq 10$$

then x has approximately a normal distribution. Combining this result with the continuity correction implies that

$$P(a \leq x \leq b) = P\left(\frac{a - \frac{1}{2} - \mu}{\sigma} \leq z \leq \frac{b + \frac{1}{2} - \mu}{\sigma}\right)$$

That is, the probability that x is between a and b inclusive is approximately the area under the approximating normal curve between $a - \frac{1}{2}$ and $b + \frac{1}{2}$.

Similarly,

$$P(x \leq b) \approx P\left(z \leq \frac{b + \frac{1}{2} - \mu}{\sigma}\right) \qquad P(a \leq x) \approx P\left(\frac{a - \frac{1}{2} - \mu}{\sigma} \leq z\right)$$

Important concept

When either $n\pi < 10$ or $n(1 - \pi) < 10$, the binomial distribution is too skewed for the normal approximation to give accurate results.

Example 7.34 Premature Babies

Premature babies are those born more than 3 weeks early. *Newsweek* (May 16, 1988) reported that 10% of the live births in the United States are premature. Suppose that 250 live births are randomly selected and that the number x of "preemies" is determined. Because

$$n\pi = 250(.1) = 25 \geq 10$$
$$n(1 - \pi) = 250(.9) = 225 \geq 0$$

x has approximately a normal distribution, with

$$\mu = 250(.1) = 25$$
$$\sigma = \sqrt{250(.1)(.9)} = 4.743$$

The probability that x is between 15 and 30 (inclusive) is

$$P(15 \leq x \leq 30) = P\left(\frac{14.5 - 25}{4.743} \leq z \leq \frac{30.5 - 25}{4.743}\right)$$
$$= P(-2.21 \leq z \leq 1.16)$$
$$= .8770 - .0136$$
$$= .8634$$

as shown in the following figure:

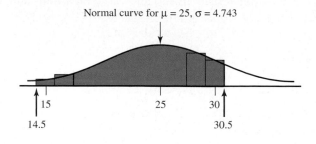

Normal curve for $\mu = 25$, $\sigma = 4.743$

Exercises 7.93–7.101

Turn to the Expanded Answers Section for answers not shown next to exercises.

7.93 Let x denote the IQ for an individual selected at random from a certain population. The value of x must be a whole number. Suppose that the distribution of x can be approximated by a normal distribution with mean value 100 and standard deviation 15. Approximate the following probabilities:
a. $P(x = 100)$ 0.0240
b. $P(x \le 110)$ 0.7580
c. $P(x < 110)$ (Hint: $x < 110$ is the same as $x \le 109$.)
d. $P(75 \le x \le 125)$ 0.9108

7.94 Suppose that the distribution of the number of items x produced by an assembly line during an 8-hr shift can be approximated by a normal distribution with mean value 150 and standard deviation 10.
a. What is the probability that the number of items produced is at most 120? 0.0016
b. What is the probability that at least 125 items are produced? 0.9946
c. What is the probability that between 135 and 160 (inclusive) items are produced? 0.7926

7.95 The number of vehicles leaving a turnpike at a certain exit during a particular time period has approximately a normal distribution with mean value 500 and standard deviation 75. What is the probability that the number of cars exiting during this period is
a. At least 650? 0.0233
b. Strictly between 400 and 550? (*Strictly* means that the values 400 and 550 are not included.) 0.6536
c. Between 400 and 550 (inclusive)? 0.6585

7.96 Let x have a binomial distribution with $n = 50$ and $\pi = .6$, so that $\mu = n\pi = 30$ and $\sigma = \sqrt{n\pi(1 - \pi)} = 3.4641$. Calculate the following probabilities using the normal approximation with the continuity correction:
a. $P(x = 30)$ 0.1147
b. $P(x = 25)$ 0.04079
c. $P(x \le 25)$ 0.09697
d. $P(25 \le x \le 40)$ 0.9426
e. $P(25 < x < 40)$ (Hint: $25 < x < 40$ is the same as $26 \le x \le 39$.) 0.89998

7.97 Seventy percent of the bicycles sold by a certain store are mountain bikes. Among 100 randomly selected bike purchases, what is the approximate probability that
a. At most 75 are mountain bikes? 0.8849
b. Between 60 and 75 (inclusive) are mountain bikes?
c. More than 80 are mountain bikes? 0.011
d. At most 30 are not mountain bikes? 0.5438

7.98 Suppose that 25% of the fire alarms in a large city are false alarms. Let x denote the number of false alarms in a random sample of 100 alarms. Give approximations to the following probabilities:
a. $P(20 \le x \le 30)$ 0.79598
b. $P(20 < x < 30)$ 0.7013
c. $P(35 \le x)$ 0.01412
d. The probability that x is farther than 2 standard deviations from its mean value 0.04964

7.99 Suppose that 65% of all registered voters in a certain area favor a 7-day waiting period before purchase of a

handgun. Among 225 randomly selected voters, what is the probability that
a. At least 150 favor such a waiting period?   0.3264
b. More than 150 favor such a waiting period?   0.2776
c. Fewer than 125 favor such a waiting period?   0.0013

7.100 Flash bulbs manufactured by a certain company are sometimes defective.
a. If 5% of all such bulbs are defective, could the techniques of this section be used to approximate the probability that at least 5 of the bulbs in a random sample of size 50 are defective? If so, calculate this probability; if not, explain why not.
b. Reconsider the question posed in Part (a) for the probability that at least 20 bulbs in a random sample of size 500 are defective.

7.101 A company that manufactures mufflers for cars offers a lifetime warranty on its products, provided that ownership of the car does not change. Suppose that only 20% of its mufflers are replaced under this warranty.
a. In a random sample of 400 purchases, what is the approximate probability that between 75 and 100 (inclusive) mufflers are replaced under warranty?   0.7497
b. Among 400 randomly selected purchases, what is the probability that at most 70 mufflers are ultimately replaced under warranty?   0.117
c. If you were told that fewer than 50 among 400 randomly selected purchases were ever replaced under warranty, would you question the 20% figure? Explain.
0.0001

Bold exercises answered in back   ● Data set available online but not required   ▼ Video solution available

Activity **7.1**   Rotten Eggs?

Background: *The Salt Lake Tribune* (October 11, 2002) printed the following account of an exchange between a restaurant manager and a health inspector:

> The recipe calls for four fresh eggs for each quiche. A Salt Lake County Health Department inspector paid a visit recently and pointed out that research by the Food and Drug Administration indicates that one in four eggs carries salmonella bacterium, so restaurants should never use more than three eggs when preparing quiche. The manager on duty wondered aloud if simply throwing out three eggs from each dozen and using the remaining nine in four-egg quiches would serve the same purpose.

1. Working in a group or as a class, discuss the folly of the above statement!
2. Suppose the following argument is made for three-egg quiches rather than four-egg quiches: Let $x \le$ number of eggs that carry salmonella. Then

$$p(0) = p(x = 0) = (0.75)^3 = .422$$

for three-egg quiches and

$$p(0) = p(x = 0) = (0.75)^4 = .316$$

for four-egg quiches. What assumption must be made to justify these probability calculations? Do you think this is reasonable or not? Explain.

3. Suppose that a carton of one dozen eggs does happen to have exactly three eggs that carry salmonella and that the manager does as he proposes: selects three eggs at random and throws them out, then uses the remaining nine eggs in four-egg quiches. Let x = number of eggs that carry salmonella among four eggs selected at random from the remaining nine.

Working with a partner, conduct a simulation to approximate the distribution of x by carrying out the following sequence of steps:
a. Take 12 identical slips of paper and write "Good" on 9 of them and "Bad" on the remaining 3. Place the slips of paper in a paper bag or some other container.
b. Mix the slips and then select three at random and remove them from the bag.
c. Mix the remaining slips and select four "eggs" from the bags.
d. Note the number of bad eggs among the four selected. (This is an observed x value.)
e. Replace all slips, so that the bag now contains all 12 "eggs."
f. Repeat Steps (b)–(d) at least 10 times, each time recording the observed x value.
4. Combine the observations from your group with those from the other groups. Use the resulting data to approximate the distribution of x. Comment on the resulting distribution in the context of the risk of salmonella exposure if the manager's proposed procedure is used.

Summary of Key Terms and Concepts

Term or Formula	Comment
Random variable: discrete or continuous	A numerical variable with a value determined by the outcome of a chance experiment. It is discrete if its possible values are isolated points along the number line and continuous if its possible values form an entire interval on the number line.
Probability distribution $p(x)$ of a discrete random variable x	A formula, table, or graph that gives the probability associated with each x value. Conditions on $p(x)$ are (1) $p(x) \geq 0$, and (2) $\sum p(x) = 1$, where the sum is over all possible x values.
Probability distribution of a continuous random variable x	Specified by a smooth (density) curve for which the total area under the curve is 1. The probability $P(a < x < b)$ is the area under the curve and above the interval from a to b; this is also $P(a \leq x \leq b)$.
μ_x and σ_x	The mean and standard deviation, respectively, of a random variable x. These quantities describe the center and extent of spread about the center of the variable's probability distribution.
$\mu_x = \sum xp(x)$	The mean value of a discrete random variable x; it locates the center of the variable's probability distribution.
$\sigma_x^2 = \sum(x - \mu_x)^2 p(x)$ $\sigma_x = \sqrt{\sigma_x^2}$	The variance and standard deviation, respectively, of a discrete random variable; these are measures of the extent to which the variable's distribution spreads out about the mean μ_x.
Binomial probability distribution $p(x) = \dfrac{n!}{x!(n - x)!}\, \pi^x(1 - \pi)^{n-x}$	This formula gives the probability of observing x successes $(x = 0, 1, \ldots, n)$ among n trials of a binomial experiment.
$\mu_x = n\pi$ $\sigma_x = \sqrt{n\pi(1 - \pi)}$	The mean and standard deviation of a binomial random variable.
Normal distribution	A continuous probability distribution that has a bell-shaped density curve. A particular normal distribution is determined by specifying values of μ and σ.
Standard normal distribution	This is the normal distribution with $\mu = 0$ and $\sigma = 1$. The density curve is called the z curve, and z is the letter commonly used to denote a variable having this distribution. Areas under the z curve to the left of various values are given in Appendix Table 2.
z critical value	A number on the z measurement scale that captures a specified tail area or central area.
$z = \dfrac{x - \mu}{\sigma}$	z is obtained by "standardizing": subtracting the mean and then dividing by the standard deviation. When x has a normal distribution, z has a standard normal distribution.

Term or Formula	Comment
	This fact implies that probabilities involving any normal random variable (any μ or σ) can be obtained from z curve areas.
Normal probability plot	A picture used to judge the plausibility of the assumption that a sample has been selected from a normal population distribution. If the plot is reasonably straight, this assumption is reasonable.
Normal approximation to the binomial distribution	When both $n\pi \geq 10$ and $n(1 - \pi) \geq 10$, binomial probabilities are well approximated by corresponding areas under a normal curve with $\mu = n\pi$ and $\sigma = \sqrt{n\pi(1 - \pi)}$.

Chapter Review Exercises 7.102–7.124

Turn to the Expanded Answers Section for answers not shown next to exercises.

CENGAGENOW™ Know exactly what to study! Take a pre-test and receive your Personalized Learning Plan.

7.102 An article in the *Los Angeles Times* (December 8, 1991) reported that there are 40,000 travel agencies nationwide, of which 11,000 are members of the American Society of Travel Agents (booking a tour through an ASTA member increases the likelihood of a refund in the event of cancellation).
a. If x is the number of ASTA members among 5000 randomly selected agencies, could you use the methods of Section 7.8 to approximate $P(1200 < x < 1400)$? Why or why not?
b. In a random sample of 100 agencies, what are the mean value and standard deviation of the number of ASTA members? 27.5, 4.46514
c. If the sample size in Part (b) is doubled, does the standard deviation double? Explain.

7.103 A soft-drink machine dispenses only regular Coke and Diet Coke. Sixty percent of all purchases from this machine are diet drinks. The machine currently has 10 cans of each type. If 15 customers want to purchase drinks before the machine is restocked, what is the probability that each of the 15 is able to purchase the type of drink desired? (Hint: Let x denote the number among the 15 who want a diet drink. For which possible values of x is everyone satisfied?) 0.784

7.104 A mail-order computer software business has six telephone lines. Let x denote the number of lines in use at a specified time. The probability distribution of x is as follows:

x	0	1	2	3	4	5	6
$p(x)$	.10	.15	.20	.25	.20	.06	.04

Write each of the following events in terms of x, and then calculate the probability of each one:
a. At most three lines are in use 0.70
b. Fewer than three lines are in use 0.45
c. At least three lines are in use 0.55
d. Between two and five lines (inclusive) are in use 0.71
e. Between two and four lines (inclusive) are not in use
f. At least four lines are not in use 0.45

7.105 Refer to the probability distribution of Exercise 7.104.
a. Calculate the mean value and standard deviation of x.
b. What is the probability that the number of lines in use is farther than 3 standard deviations from the mean value? 0

7.106 A new battery's voltage may be acceptable (A) or unacceptable (U). A certain flashlight requires two batteries, so batteries will be independently selected and tested until two acceptable ones have been found. Suppose that 80% of all batteries have acceptable voltages, and let y denote the number of batteries that must be tested.
a. What is $p(2)$, that is, $P(y = 2)$? 0.64
b. What is $p(3)$? (Hint: There are two different outcomes that result in $y = 3$.) 0.256
c. In order to have $y = 5$, what must be true of the fifth battery selected? List the four outcomes for which $y = 5$, and then determine $p(5)$.

Bold exercises answered in back ● Data set available online but not required ▼ Video solution available

d. Use the pattern in your answers for Parts (a)–(c) to obtain a general formula for $p(y)$.

7.107 A pizza company advertises that it puts 0.5 lb of real mozzarella cheese on its medium pizzas. In fact, the amount of cheese on a randomly selected medium pizza is normally distributed with a mean value of 0.5 lb and a standard deviation of 0.025 lb.
a. What is the probability that the amount of cheese on a medium pizza is between 0.525 and 0.550 lb?   0.1573
b. What is the probability that the amount of cheese on a medium pizza exceeds the mean value by more than 2 standard deviations?   0.02275
c. What is the probability that three randomly selected medium pizzas all have at least 0.475 lb of cheese?   0.59555

7.108 Suppose that fuel efficiency for a particular model car under specified conditions is normally distributed with a mean value of 30.0 mpg and a standard deviation of 1.2 mpg.
a. What is the probability that the fuel efficiency for a randomly selected car of this type is between 29 and 31 mpg?   0.59534
b. Would it surprise you to find that the efficiency of a randomly selected car of this model is less than 25 mpg?
c. If three cars of this model are randomly selected, what is the probability that all three have efficiencies exceeding 32 mpg?   0.0001091
d. Find a number c such that 95% of all cars of this model have efficiencies exceeding c (i.e., $P(x > c) = .95$).
28.026
7.109 The amount of time spent by a statistical consultant with a client at their first meeting is a random variable having a normal distribution with a mean value of 60 min and a standard deviation of 10 min.
a. What is the probability that more than 45 min is spent at the first meeting?   0.9332
b. What amount of time is exceeded by only 10% of all clients at a first meeting?   72.8 min
c. If the consultant assesses a fixed charge of $10 (for overhead) and then charges $50 per hour, what is the mean revenue from a client's first meeting? $60

7.110 The lifetime of a certain brand of battery is normally distributed with a mean value of 6 hr and a standard deviation of 0.8 hr when it is used in a particular cassette player. Suppose that two new batteries are independently selected and put into the player. The player ceases to function as soon as one of the batteries fails.
a. What is the probability that the player functions for at least 4 hr?   0.98762

b. What is the probability that the cassette player works for at most 7 hr?   0.9888
c. Find a number z^* such that only 5% of all cassette players will function without battery replacement for more than z^* hr.   6.608

7.111 A machine producing vitamin E capsules operates so that the actual amount of vitamin E in each capsule is normally distributed with a mean of 5 mg and a standard deviation of 0.05 mg. What is the probability that a randomly selected capsule contains less than 4.9 mg of vitamin E? at least 5.2 mg?   0.0228, ≈ 0

7.112 Accurate labeling of packaged meat is difficult because of weight decrease resulting from moisture loss (defined as a percentage of the package's original net weight). Suppose that moisture loss for a package of chicken breasts is normally distributed with mean value 4.0% and standard deviation 1.0%. (This model is suggested in the paper "Drained Weight Labeling for Meat and Poultry: An Economic Analysis of a Regulatory Proposal," *Journal of Consumer Affairs* [1980]: 307–325.) Let x denote the moisture loss for a randomly selected package.
a. What is the probability that x is between 3.0% and 5.0%?   0.6826
b. What is the probability that x is at most 4.0%?   0.5
c. What is the probability that x is at least 7.0%?   0.0013
d. Find a number z^* such that 90% of all packages have moisture losses below z^*%.   5.28
e. What is the probability that moisture loss differs from the mean value by at least 1%?   .3174

7.113 The *Wall Street Journal* (February 15, 1972) reported that General Electric was sued in Texas for sex discrimination over a minimum height requirement of 5 ft 7 in. The suit claimed that this restriction eliminated more than 94% of adult females from consideration. Let x represent the height of a randomly selected adult woman. Suppose that x is approximately normally distributed with mean 66 in. (5 ft 6 in.) and standard deviation 2 in.
a. Is the claim that 94% of all women are shorter than 5 ft 7 in. correct?   No.
b. What proportion of adult women would be excluded from employment as a result of the height restriction?

7.114 The longest "run" of S's in the sequence *SSFSSSSFFS* has length 4, corresponding to the S's on the fourth, fifth, sixth, and seventh trials. Consider a binomial experiment with $n = 4$, and let y be the length (number of trials) in the longest run of S's.

a. When $\pi = .5$, the 16 possible outcomes are equally likely. Determine the probability distribution of y in this case (first list all outcomes and the y value for each one). Then calculate μ_y.
b. Repeat Part (a) for the case $\pi = .6$.
c. Let z denote the longest run of either S's or F's. Determine the probability distribution of z when $\pi = .5$.

7.115 Two sisters, Allison and Teri, have agreed to meet between 1 and 6 P.M. on a particular day. In fact, Allison is equally likely to arrive at exactly 1 P.M., 2 P.M., 3 P.M., 4 P.M., 5 P.M., or 6 P.M. Teri is also equally likely to arrive at each of these six times, and Allison's and Teri's arrival times are independent of one another. Thus there are 36 equally likely (Allison, Teri) arrival-time pairs, for example, $(2, 3)$ or $(6, 1)$. Suppose that the first person to arrive waits until the second person arrives; let w be the amount of time the first person has to wait.
a. What is the probability distribution of w?
b. How much time do you expect to elapse between the two arrivals? 1.9444 hours

7.116 Four people—a, b, c, and d—are waiting to give blood. Of these four, a and b have type AB blood, whereas c and d do not. An emergency call has just come in for some type AB blood. If blood samples are taken one by one from the four people in random order for blood typing and x is the number of samples taken to obtain an AB individual (so possible x values are 1, 2, and 3), what is the probability distribution of x?

7.117 Bob and Lygia are going to play a series of Trivial Pursuit games. The first person to win four games will be declared the winner. Suppose that outcomes of successive games are independent and that the probability of Lygia winning any particular game is .6. Define a random variable x as the number of games played in the series.
a. What is $p(4)$? (Hint: Either Bob or Lygia could win four straight games.) 0.1552
b. What is $p(5)$? (Hint: For Lygia to win in exactly five games, what has to happen in the first four games and in Game 5?) 0.2688
c. Determine the probability distribution of x.
d. How many games can you expect the series to last? 5.69728
7.118 Refer to Exercise 7.117, and let y be the number of games won by the series loser. Determine the probability distribution of y.

7.119 A sporting goods store has a special sale on three brands of tennis balls—call them D, P, and W. Because the sale price is so low, only one can of balls will be sold to each customer. If 40% of all customers buy Brand W, 35% buy Brand P, and 25% buy Brand D and if x is the number among three randomly selected customers who buy Brand W, what is the probability distribution of x?

7.120 Suppose that your statistics professor tells you that the scores on a midterm exam were approximately normally distributed with a mean of 78 and a standard deviation of 7. The top 15% of all scores have been designated A's. Your score is 89. Did you receive an A? Explain.
Yes, $P(x > 89) = 0.0582$ so 89 is in the top 15%.
7.121 Suppose that the pH of soil samples taken from a certain geographic region is normally distributed with a mean pH of 6.00 and a standard deviation of 0.10. If the pH of a randomly selected soil sample from this region is determined, answer the following questions about it:
a. What is the probability that the resulting pH is between 5.90 and 6.15? 0.7745
b. What is the probability that the resulting pH exceeds 6.10? 0.1587
c. What is the probability that the resulting pH is at most 5.95? 0.3085
d. What value will be exceeded by only 5% of all such pH values? 6.1645

7.122 The lightbulbs used to provide exterior lighting for a large office building have an average lifetime of 700 hr. If length of life is approximately normally distributed with a standard deviation of 50 hr, how often should all the bulbs be replaced so that no more than 20% of the bulbs will have already burned out? 658 hours

7.123 Suppose that 16% of all drivers in a certain city are uninsured. Consider a random sample of 200 drivers.
a. What is the mean value of the number who are uninsured, and what is the standard deviation of the number who are uninsured? 32, 5.18459
b. What is the (approximate) probability that between 25 and 40 (inclusive) drivers in the sample were uninsured?
c. If you learned that more than 50 among the 200 drivers were uninsured, would you doubt the 16% figure? Explain.
0.0002
7.124 Let x denote the duration of a randomly selected pregnancy (the time elapsed between conception and birth). Accepted values for the mean value and standard deviation of x are 266 days and 16 days, respectively. Suppose that the probability distribution of x is (approximately) normal.
a. What is the probability that the duration of pregnancy is between 250 and 300 days? 0.8247

Bold exercises answered in back ● Data set available online but not required ▼ Video solution available

b. What is the probability that the duration of pregnancy is at most 240 days? 0.0516

c. What is the probability that the duration of pregnancy is within 16 days of the mean duration? 0.68826

d. A "Dear Abby" column dated January 20, 1973, contained a letter from a woman who stated that the duration of her pregnancy was exactly 310 days. (She wrote that the last visit with her husband, who was in the navy, occurred 310 days before birth.) What is the probability that the duration of pregnancy is at least 310 days? Does this probability make you a bit skeptical of the claim?

e. Some insurance companies will pay the medical expenses associated with childbirth only if the insurance has been in effect for more than 9 months (275 days). This restriction is designed to ensure that the insurance company pays benefits for only those pregnancies for which conception occurred during coverage. Suppose that conception occurred 2 weeks after coverage began. What is the probability that the insurance company will refuse to pay benefits because of the 275-day insurance requirement? 0.3783

Personal Tutor
Do you need a live tutor for homework problems?

CENGAGENOW
Are you ready? Take your exam-prep post-test now.

Bold exercises answered in back   ● Data set available online but not required   ▼ Video solution available

Graphing Calculator Explorations

Exploration 7.1 Discrete Probability Distributions

The calculator is at its finest when used with random variables, transforming minutes of mindless calculation into seconds of easy button-pushing. In our calculator presentation of random variables we will capitalize extensively on the list capabilities of your calculator. We will show you not only how to graph a discrete probability distribution, but also how to find the mean and standard deviation of a discrete random variable.

First recall that we have encountered a similar problem before when we considered the problem of graphing a relative frequency histogram of a frequency distribution. At that time frequencies were converted into relative frequencies for plotting; now, these relative frequencies have morphed into probabilities. You begin by entering the possible values of the random variable in your calculator's equivalent of List1 and the corresponding probabilities of these values in List2. For our example we will use a numerical rating of newborn children called an Apgar score. The Apgar score has eleven possible values, 0, 1, . . . , 10 based on factors such as muscle tone, skin color, etc. Suppose that the scores have the following probability distribution.

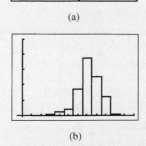

(a)

(b)

Figure 7.45 (a) Apgar score probability distribution; (b) probability histogram for Apgar score.

x	0	1	2	3	4	5	6	7	8	9	10
$p(x)$	.002	.001	.002	.005	.02	.04	.17	.38	.25	.12	.01

In Figure 7.45(a) a portion of the calculator screen after data entry is shown. After you enter the data you can graph the probability distribution by supplying the proper lists in the histogram command as was done for the relative frequency histogram. The graph for the Apgar probability distribution is shown in Figure 7.45(b). The window is set so that the horizontal and vertical axis would show in the screen to give a more informative display. The horizontal axis runs from −.5 to 12, and the vertical axis runs from −.01 to 0.5.

Now we turn our attention to calculating the mean and standard deviation of the random variable. The data are already entered and we begin by recalling the definition of the mean of a discrete random variable,

$$\mu_x = \sum_{\text{all possible } x \text{ values}} xp(x)$$

Because we have stored exactly what is needed in Lists 1 and 2, we can virtually duplicate this definition using the language of lists and list operations for our calculator:

$$\mu_x = \sum_{\text{all possible } x \text{ values}} \text{List1} * \text{List2}$$

The strategy for finding the mean of the Apgar random variable, translated from math symbols to English is the following: calculate the products of numbers in our Lists 1 and 2, store the results in List3, and then find the sum of all the numbers in List3. Multiplying to obtain the product is fairly easy as we appeal again to the language of lists:

List1 * List2 → List3

Now we need to find the sum of the numbers in List3. Exactly how this is done will vary from calculator to calculator. The most likely scenario is for you to calculate the "1-variable statistics" for List3. Calculators will usually report the sum—look for this symbol: $(\sum x)$. Be careful as you scan for the right choice in your calculator window! Don't be misled by the symbol for the *mean*; we want the *sum*. You should get the value 7.16 for the Apgar mean.

We use a similar strategy to compute the standard deviation. We will find the variance first, then take the square root. The formula for the variance,

$$\sigma_x^2 = \sum_{\text{all possible } x \text{ values}} (x - \mu_x)^2 p(x)$$

also easily translates into the language of lists:

$$\sigma_x^2 = \sum_{\text{all possible } x \text{ values}} (\text{List1} - 7.16)^2 * \text{List2}$$

The list language is only slightly more complicated than for the mean:

$(\text{List1} - 7.16)^2 * \text{List2} \rightarrow \text{List3}$

After this production the sum of the numbers in List3 is the variance of the random variable; the square root is the standard deviation. Performing these calculations, you should get a variance of 1.5684 (from $(\sum x)$ for List3; again, don't be misled and choose the σ_x or s_x. The standard deviation is then found by taking the square root of 1.5684, resulting in 1.2524.

Exploration 7.2 Binomial Probability Calculations

Most calculations having to do with random variables are of one of three types. They are (1) the probability the variable will assume a value between two given numbers, (2) the probability the variable will assume a value less than a given number, or (3) the probability the random variable will assume a value greater than a given number. Because these calculations are so common in statistics your calculator may have a built-in capability for finding these probabilities. In the case of a discrete random variable such as the binomial distribution there is a special case of (1) above, the probability that the random variable will actually assume a particular value.

For the binomial distribution, we will illustrate these calculations using an example. Suppose that 60% of all computer monitors have a flat panel display and 40% have a CRT display. Suppose further that the next 12 purchases monitored, and the random variable is defined as the number of flat panel monitors in the next 12 purchases. As an example, the probability exactly four monitors would be flat panel displays is

$$p(x) = {}_nC_x\pi^x(1 - \pi)^{n-x} = {}_{12}C_4(.6)^4(1 - .6)^{12-4} = .042$$

Even if your calculator does not have special binomial functions it is likely to have a key for the combinations, $({}_nC_r)$, possibly cleverly hidden in the "math" or "probability" menu. The calculator keystrokes might look like this (don't forget that to perform the ${}_nC_r$ calculation above you will have to press n, then the ${}_nC_r$ key, and then r):

12 ${}_nC_r$ 4 * .6^4 * .4^8

If your calculator has built-in binomial capabilities you will have fewer keystrokes. Let's consider these problems one at a time, starting with the function names. If your calculator has a built-in function for binomial calculations, it probably has two: a function for finding the probability that "x is equal to a given value," and a function for finding the probability that "x is less than or equal to a given value." The first function is known to statisticians as a "density" function, and is commonly abbreviated "pdf" for "probability density function." The second is known as a "cumulative distribution function" and is commonly labeled "cdf." These two functions on your calculator will in all likelihood mirror these abbreviations.

The second problem you will face is that to find the binomial probabilities the calculator will need more than just one number, and the order you enter the numbers *does* make a difference! Look in your calculator manual for something that looks like "binomial," especially with a "pdf" somewhere. The function could be obvious, like "binompdf," or it may be a little more cryptic, like "binpdf." Your manual will be very careful to specify both what the needed function parameters are, and the order you should enter them. As an example, one type of calculator has the following:

binompdf(*numtrials*, p, [x])

The manual informs that "*numtrials*" is the number of trials, "p" is the probability of success, and that "x" can be either an integer or a list of integers. This information collectively explains what is known as the "syntax" of the function. It is your responsibility to get the numbers right, and get them in the right order! The square brackets, "[ ]," are a standard notation in the calculator world. They indicate the bracketed quantity is either optional or defaults to a preselected option if you do not enter a number in that space. For our example, the number of trials is 12, and the probability of success is 0.6. Since the probability of exactly 4 flat panel monitors is desired, we enter

binompdf(12, .6, 4)

Our calculator gives us 0.042042, which is the correct answer. This is a good sign! Now try this on your calculator. Remember, you must navigate to the function in the manner presented in your manual, and you have to pay attention to the syntax. While you are learning how to use this function (or any calculator function), it is a very good idea to use examples with known answers and check the results.

Now suppose you wish to find the probability of getting 4 *or fewer* flat panel monitors out of 12. The appropriate function here is the cumulative density function, or "cdf":

binomcdf(*numtrials*, p, [x])

Does this look disturbingly familiar? Except for the "c" instead of the "p," they look exactly alike! The good news is that we already understand the syntax; the bad news is that if we aren't careful we might get the wrong function in haste. Be careful! The function **binomcdf**(12, .6, 4) gives the answer, .0573099213. If we are not convinced of our prowess with **binomcdf** we can use **binompdf** to check the result:

binompdf(12, .6, 0) + **binompdf**(12, .6, 1) + $\cdots$ + **binompdf**(12, .6, 4)
    = .000016777 + .0003019898 + $\cdots$ + .042042
    = .05731

Now let's move on to another of the common calculations with random variables: What is the probability that the random variable will assume a value between 4 and 7? One common source of confusion here is that the word *between* is disturbingly ambiguous. Do we mean to include 4 and 7, or do we mean only the probability of getting a 5 or 6? In this case we wish to be inclusive. To evaluate the probability desired, we will use the *cumulative* distribution function for the binomial, **binomcdf**. (Your calculator function may have a different name!) The logic is elementary, as Sherlock Holmes would say. The probability that the binomial random variable will assume a value between 4 and 7 (inclusive) is equal to the probability of assuming a value less than or equal to 7, minus the probability of assuming a value less than or equal to 3 (*not* 4!). In symbols,

$$p(4 \leq x \leq 7) = p(x \leq 7) - p(x \leq 3)$$

which is found as follows:

binomcdf(12, .6, 7) − **binomcdf**(12, .6, 3)

gives us 0.5465545.

Our last binomial random variable calculation problem is finding the probability of a value greater than a given value. What, for instance, is the probability of more than 7 monitors having flat panel displays? Using a fundamental property of probability, we know that:

$$p(7 < x) + p(x \leq 7) = 1$$

and thus

$$p(7 < x) = 1 - p(x \leq 7)$$

which we translate into:

1 − **binomcdf**(12, .6, 7)

giving us 0.438178222.

We have gone into some detail to explain how these binomial probability problems can be solved using the calculator. This detail is justified not only because of the importance of the binomial distribution, but also because these same calculator procedures will be used for finding probabilities involving the geometric and normal random variables, yet to come. Because the discussions in this exploration have been detailed, the discussions in those cases will be less so.

In Exploration 7.1 we discussed how to graph a discrete distribution. When we graphed the probability density function for the Apgar scores we manually entered the outcomes and their associated probabilities. Anticipating that you may wish to consider binomial chance experiments with many potential successes, we will streamline the data entry process using some commands and functions we have already discussed in previous calculator explorations.

Graphing a binomial distribution will involve three steps:

1. Construct the list of possible values in List1 using the seq command (or your calculator's equivalent).
2. Construct the probabilities in List2 using the binompdf function (or your calculator's equivalent).
3. Draw the graph (in the form of a histogram) of the probability distribution.

Consider the binomial probability distribution for $n = 20$ and $\pi = .20$. Carrying out the steps below puts the integers 0 to 20 in List1, and $p(x)$ for x values from 0 to 20 in List2.

1. seq($x, x, 0, 20$) → List1 puts a sequence of 21 integers into List1. (Remember to verify your calculator syntax and the order of the information to be entered for your calculator!)
2. binompdf(20, .2) → List2. (Remember to verify . . .)
3. Now graph the probability distribution, where List1 contains the possible data values and List2 contains the probabilities.

To check your work, partial calculator screen output for this problem is given in Figure 7.46(a) , and the graph of the distribution is shown in Figure 7.46(b).

Figure 7.46
(a) Binomial probabilities;
(b) histogram of binomial distribution.

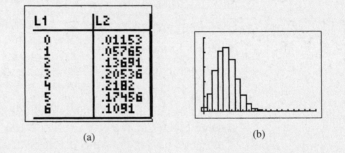

(a)   (b)

Exploration 7.3 Geometric Probability Calculations

Our calculator exploration of geometric random variables will be an echo of the binomial random variables we have already discussed in Exploration 7.2. We again consider (1) the probability the variable will assume a value between two given numbers, (2) the probability the variable will assume a value less than a given number, and (3) the probability the random variable will assume a value greater than a given number.

Our example here will be about jumper cables. Suppose that 40% of students who drive to campus carry jumper cables. If your car has a dead battery, and you aren't one of the forward thinking 40%, how many students will you have to ask before you find one with jumper cables?

Consider the first problem, the probability of a particular number. The probability the first student stopped has jumper cables is:

$$p(1) = (1 - \pi)^{1-1}\pi = (1 - .4)^{1-1}(.4) = .4$$

The corresponding keystrokes for finding this probability will be something like

$(1 - 0.4)^0 * 0.4.$

Now let's consider problems (2) and (3). If your calculator has density and cumulative density functions for the geometric distribution the functions are probably named something like geompdf and geomcdf, similar to the names for the binomial functions. The calculator syntax for the probability density function will probably look something like

geompdf(p, x)

where p is the probability of success and x is in this example the number of students you would ask until success. We want the probability of jumper cables on the very first stop. We enter **geompdf**(.4, 1), and the function returns .4.

Now suppose you wish to find the probability of jumper cables after 4 *or fewer* stops. Using the cumulative density function, "**geomcdf**" (which has the same parameters as the geompdf function), we enter **geomcdf**(.4, 4), which returns 0.8704. As with the binomial, we can check this by summing:

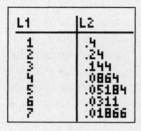

(a)

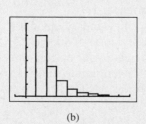

(b)

Figure 7.47 (a) Geometric probability distribution; (b) histogram of geometric probability distribution.

geompdf(.4, 1) + **geompdf**(.4, 2) + **geompdf**(.4, 4) + **geompdf**(.4, 4)
$$= 0.4 + 0.24 + 0.144 + 0.0864$$
$$= 0.87041$$

The probability that a geometric random variable will assume a value between 4 and 7 (inclusive) is equal to the probability of observing a value less than or equal to 7, minus the probability of observing a value less than or equal to 3 (*not* 4). In symbols,

$$P(4 \leq x \leq 7) = P(x \leq 7) - P(x \leq 3)$$

which is found using **geomcdf**(.4, 7) − **geomcdf**(.4, 3), giving 0.1880064.

What is the probability of more than 7 stops before we get jumper cables?

1 − **geomcdf**(.4, 7) gives us 0.0279936.

Graphing an entire geometric probability distribution is not possible, since there is an infinite number of possible values—1, 2, 3, ... Nevertheless, we can graph parts of the distribution. The method for graphing is similar to that for the binomial random variable. Use the **seq** function to create a list of integers in List1; then use the **geompdf** function to find the corresponding probabilities and store them in List2; and finally plot the distribution as you would a histogram, and as we have previously done with the binomial. These steps for the geometric distribution of Example 7.20 are summarized below:

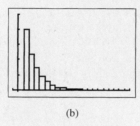

(a)

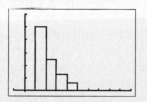

1. seq(x, x, 1, 20) → List1
2. geompdf(.4, List1) → List2
3. Graph a histogram with the domain List1, and probabilities in List2.

The data editing window is shown in Figure 7.47(a) and a graph of this geometric distribution appears in Figure 7.47(b).

We are really calculating probabilities for only part of the distribution, since the number of possible values is infinite. The graph should tail to the right in a gradual manner, not suddenly drop out of sight. You may notice a sudden plummeting in your graph but it could be that there are more significant probabilities to the right. As an example, suppose we consider the chance experiment of flipping a coin until a head appears. The distribution of x = number of tosses is geometric with success probability .5. If the distribution is plotted using the previous steps but only using a sequence of integers from 1 to 4, the results are shown in Figure 7.48 (a). Clearly, there are values with probabilities different from zero that are not represented in the graph. The solution is to construct the sequence of integers over a larger range of values, say 1 to 16.

(b)

Figure 7.48 (a) Incorrect display; (b) correct display.

At some point, of course, the geometric probabilities become very close to zero, as Figure 7.48 (b). If your graph looks similar to the one on the right, tailing off gradually, you can be fairly certain you have captured the essential behavior of the particular geometric distribution.

Exploration 7.4 Normal Curves and the Normal Probability Distribution

The normal distribution is arguably the most famous distribution in all of statistics. As we have learned "the" normal distribution is really a family of distributions with the same shape, but different means and standard deviations. The "standard" normal distribution is the normal probability distribution with $\mu = 0$, and $\sigma = 1.0$. From the calculator perspective working with the normal distribution is slightly different from the binomial and geometric distributions because the normal distribution is continuous. Consequently it will not be graphed as a histogram; normal curves are graphed just as any other function is graphed. The normal curve, however, is not particularly simple. Fortunately you calculator, if it has statistical functions, will have "normal" already in it somewhere. It might be something like this:

normalpdf(x, [μ, σ]).

If you are a glutton for punishment or your calculator does not have a built-in normalpdf function, here is the formula for a normal curve with mean μ and standard deviation σ:

$$y = \frac{1}{\sqrt{2\pi}}\, e^{[(x-\mu)^2/(2\sigma^2)]}$$

Here are the keystrokes for the formula:

y1 = 1/(sqr(2*π)*σ)*exp($-(x-\mu)$^2/(2*σ^2))

Assuming you are smiling because of your foresight in purchasing a calculator with a built-in normalpdf function, let's put it to good use. The syntax above for the **normalpdf** function might seem complicated but actual use is simple once you get used to it. You should check your calculator manual for two *very important* pieces of information. First, make sure you know the required order for the information you must provide. Second, look closely at the sigma, wherever it is in your calculator's syntax. Make sure you check whether you must enter (the standard deviation) or (the variance). Now let's tackle the notation. First of all, if your calculator's syntax has those square brackets—[μ, σ]—remember that they indicate numbers that are optional. If you leave them out, the normalpdf function will simply default to the standard normal curve, with mean 0 and standard deviation (or variance) 1.

Let's graph the three normal curves. The first has a mean and standard deviation of 10 and 5, respectively. The second has a mean and standard deviation of 40 and 2.5, and the third a mean and standard deviation of 70 and 10. Navigate your calculator's menu system to find the normal curve function, and paste this function into the function definition window where you usually define simpler functions. Using the syntax above, you should see your calculator's equivalent of the following:

y1 = **normalpdf**(x, 10, 5)
y2 = **normalpdf**(x, 40, 2.5)
y3 = **normalpdf**(x, 70, 10)

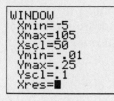

(a)

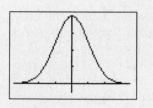

(b)

Figure 7.49 (a) Window settings; (b) normal curves.

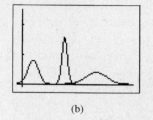

7.50 The standard normal distribution.

```
Normal C.D.

Lower  :0
Upper  :0
σ    :0
μ    :0
Execute
```

7.51 Setup for normal calculations.

Graphing these functions using the window setting in Figure 7.49(a), we see the graphs in Figure 7.49 (b).

Now let's graph the standard normal distribution. If your calculator syntax indicates that it defaults to a standard normal, you will only have to enter your calculator equivalent of

$y1 =$ **normalpdf**(x).

It is also possible that your calculator does *not* default to standard normal, in which case you would have to specify the mean and standard deviation as 0 and 1, something like

$y1 =$ **normalpdf**$(x, 0, 1)$.

Set your graphing window with x values running from about -3.5 to 3.5 and the y values from to 0.40. These values should be fine for the standard normal distribution. If you don't see a distribution filling the screen as in Figure 7.50, something is amiss and you need to verify your keystrokes and check your calculator's manual.

Since the normal probability distribution is a continuous distribution the probability that x would be equal to a specific value is, of course, 0. For continuous distributions we are usually interested in finding (1) the area under the curve between two specific values; and (2) the area in the extremes, or "tails," of the distribution. The function that we will use to find these values will be symbolized with the notation "**normalcdf**," which stands for the "normal cumulative distribution function." This actually is a misnomer, because the functions calling themselves "cdf" functions on many calculators actually calculate the probability that the standard normal variable is *between* two values. Calculators seem to get it right for the discrete probability density functions, but for some reason have elected to use similar names for very different kinds of calculation when they get to the continuous probability density functions —don't let this minor inconvenience confuse you!

Two strategies are used by calculator manufacturers for evaluating the probability that z is between two values, a and b. It is possible your calculator has a table for you to fill in the values as in Figure 7.51. For a calculator utilizing this strategy, you would have to fill in the lower bound, upper bound, and standard deviation (σ) and mean (μ).

Other calculators ask for the mean and standard deviation as parameters of the function. If your calculator uses this strategy, your built-in cumulative distribution function will have syntax something like this:

normalcdf(lower bound, upper bound [, μ, σ)

For a calculator using this syntax you would fill in the lower bound and upper bound with the appropriate values for z, and ignore the optional parameters, since z will have a standard normal distribution. You will specify other values for μ and σ when performing calculations that are not already in terms of z scores. After navigating your calculator's menus, you will enter something like this:

normalcdf(z-lower, z-upper).

Let's find the probability that z is between -1.76 and .58. We enter the function as **normalcdf**$(-1.76, 0.58)$ and the calculator will return a value of 0.6798388789. We would not suggest writing all those digits; rounding off to .6798 is perfectly fine (as you may have surmised from considering Appendix Table 2).

As you might guess from its name, the **normalcdf** function can also be used for calculation of the cumulative distribution function—that is, finding the probability that *z* will be below a specific value. Suppose we want to find the probability that *z* is less than -1.76. Remembering that the set of possible values for a standard normal random variable is the entire real line, you might think to enter the following:

normalcdf($-\infty$, -1.76).

If so, your thinking is right on target, except for one thing: there is no "$-\infty$" on your calculator. Some calculators will have a special symbol for "$-\infty$" which the calculator translates internally to its equivalent of a "very small number." You should check your manual for this number and how to find it. The representation will probably be something like "$-1E99$" or "$-1e999$" which is calculator-speak for -1 times 10 raised to the highest power the calculator can handle. In the case of the standard normal curve, it may be just as easy to enter a different but still very small number in place of the "$-\infty$," perhaps **normalcdf**(-10, -1.76). On our calculator .0392038577 is returned, which agrees with the tabled answer. (If you are squeamish about -10, use -50; using -50 we get .0392038577 also!)

Finding the probability that a *z* is greater than a particular value is also easy. For example, we find the area to the right of $z = 1.42$ as follows:

$1 -$ **normalcdf**("$-\infty$", 1.42)

Using -10 for the lower bound, we get 0.0778038883.

The last type of problem examined will be the identification of extreme values. The easiest way to do this is with a built-in function, typically called "InvNormal," which stands for "inverse normal." The "InvNormal" function—or whatever it is named on your calculator—will be the reverse of finding the probability that *z* is less than a specified value. Earlier in our discussion, we found the probability that *z* is less than -1.76 to be 0.0392038577. The InvNormal function returns a *z* value when given the probability. Thus, InvNormal(.0392038577) equals -1.76. Except for the difference in function name, the syntax for this function should be the same as for **normalcdf**:

InvNormal(cumulative probability, [, μ, σ])

On our calculator, **InvNormal**(0.0392038577) returns -1.760000538.

Exploration 7.5 The Normal Approximation to the Binomial Distribution

In an earlier Exploration we showed you how to use your calculator to find probabilities associated with the binomial and normal distributions, using built-in calculator functions. We generically used the terms *binompdf, binomcdf, normalpdf,* and *normalcdf* to refer to these functions. In this Exploration we would like to focus on the normal approximation to the binomial distribution. Whenever a continuous distribution is used to approximate a discrete distribution the question naturally occurs, "How good is the approximation?" The answer usually given by statistics instructors is, "it depends." In the case of the normal approximation to the binomial, the goodness of fit depends on the two quantities which define the binomial distribution: *n* and π. Most statisticians have a simple "rule of thumb" they apply for approximating the binomial with a normal distribution, such as:

When either $n\pi < 10$ or $n(1 - \pi) < 10$, the binomial distribution is too skewed for the normal approximation to give accurate results.

Different statisticians have different rules of thumb, some feeling comfortable with the accuracy provided by using 5 instead of 10 in the rule of thumb above. In days of yore—that is, the precalculator days—students would have to accept the rule of thumb as one of the mysteries of statistics. In more modern times a statistics student, armed with her calculator, can not only understand what the rules of thumb are all about, but evaluate the various rules of thumb for a particular n and π pair.

It might be argued that using the normal distribution to approximate a distribution that we can evaluate exactly seems a little foolish. There is something to this argument, but remember: we will not always be able to find exact probabilities in other situations in statistics, and must rely on approximations. Using an approximation involves a fundamental tradeoff between ease of calculation and exactness of answer. An understanding of this with the normal approximation to the binomial will give us a better understanding of the issues involved when we encounter similar tradeoffs in statistics courses yet to come. (At least you'll be more tolerant of those "rules of thumb!")

We shall reacquaint ourselves with some syntax and warm up with a distribution of the number of express mail packages mailed at a certain post office in a day. The number is approximately normally distributed with $\mu = 18$ and $\sigma = 6$. Suppose we wish to find the probability that 20 express mail packages are mailed in a given day. We calculate the probability that in a normal distribution with $\mu = 18$ and $\sigma = 6$, the event $x = 20$ would happen. Remembering the syntax from our earlier discussion,

normalcdf(lower bound, upper bound [, μ, σ])

we enter: normalcdf(19.5, 20.5, 18, 6), and our calculator returns 0.062832569.

We will now compare binomial calculations with the normal approximations. It is reported that 10% of live births in the United States are premature. Suppose we randomly select 250 live births and define the random variable x to be the number of these that are premature. We wish to calculate the probability that x is between 15 and 30 (inclusive). To find the binomial probability we recall that we must use the built-in function we called binomcdf. This function includes the rightmost interval indicated; therefore we subtract the probability of getting x less than or equal to 14 from the probability of getting x less than or equal to 30.

$$P(15 \leq x \leq 30) = \text{binomcdf}(250, .1, 30) - \text{binomcdf}(250, .1, 14)$$
$$= 0.8753286537 - 0.00931244187$$
$$= 0.8660162088.$$

To evaluate this probability using the normal curve approximation we will use the machine accuracy of the calculator with the mean $\mu = 25$ and $\sigma = 4.74341649$:

normalcdf (lower bound, upper bound [, μ, σ])
$$= \text{normalcdf}(14.5, 30.5 , 25, 4.74341649)$$
$$= 0.8634457937$$

The difference between the two probabilities to machine accuracy is 0.0025704151. This does not seem to be a large difference, but it *is* a difference. According to the rule of thumb this approximation meets the test, but the investigator in the context of his or her situation must evaluate the practical importance of the difference.

Now lets redo the calculations, not with a sample size of 250, but a sample size of only 50. Keeping the results proportionally the same by dividing by 5, we will consider approximating the probability of getting between 3 and 6 preemies (inclusive) from a random sample of 50 babies. In this case,

$$n\pi = 50(.10) = 5 < 10$$
$$n(1 - \pi) = 50(1 - .10) = 45 \geq 10$$

Since $n\pi < 10$ our rule of thumb would regard the binomial distribution too skewed for the normal curve approximation to give accurate results. Let's see what happens:

$$P(3 \le x \le 6) = \text{binomcdf}(50,.1,6) - \text{binomcdf}(50,.1,2)$$
$$= 0.7702268435 - 0.1117287563$$
$$= 0.6584980872$$

To evaluate this probability using the normal curve approximation we will use the machine accuracy of the calculator with $\mu = 50(.1) = 5$ and $\sigma = \sqrt{50(.1)(.9)} = 2.121320344$

normalcdf(lower bound, upper bound [, μ, σ])
$$= \text{normalcdf}(2.5, 6.5 , 5, 2.121320344)$$
$$= 0.6409535402$$

The difference between the binomial and the normal approximation in this case is 0.017544547. It is interesting to note that using a rule of thumb with 5 instead of 10 would call this difference "acceptable." We do not argue with this rule of thumb in principle, but once again point out that the individual judgment by the investigator on site must be used in evaluating the goodness of the approximation.

Finally, we will superimpose the appropriate normal distribution over the binomial distribution to get a visual sense of the approximation. It is entirely possible that a given approximation will do a better job for different choices of values of the end points of the interval, and the graphs may give us an overall sense of when a normal approximation might be acceptable.

Graphing a binomial distribution and a normal distribution at the same time involves skills we have seen in previous Explorations. (You may want to refer back to the calculator explorations about the binomial and normal distributions to refresh your memory.)

We will graph the binomial and normal distributions for four distributions, each with sample size 20, but with probabilities of success of .05, .1, .25, and .5 We will change the windows to make the graphs fill the windows, but this should not affect any interpretations of the goodness of fit to the binomial by the normal distribution. As a reminder, our binomial preparations for the first graph are

1. seq(x, x, 0, 20) → List1
2. binompdf(20, .05) → List2
3. Specify that we want a histogram with the values in List1, and the corresponding binomial probabilities in List2.

For the normal curve plot, define the graphing function by supplying the mean and standard deviation of the binomial as parameters for the normalpdf function:

Y1 = normalpdf(x, 1, 0.97468)

The four plots appear in Figure 7.52.

As can be seen from a comparison of the plots, the normal approximation gets "closer and closer" to the binomial as gets closer and closer to 0.5. For $\pi = .25$ the rule of thumb is satisfied for $n = 20$, and for $\pi = .5$, the rule of thumb is satisfied using $n = 10$. It is a bit difficult to judge whether or not the normal approximation to the binomial is "adequate" for a particular situation by just looking at the plots.

Modern technology makes it possible to do binomial calculations quickly, so the normal approximation to the binomial is not as widely used as it once was. However, there are other distributions in statistics that are "approximately" normal as long as certain conditions are satisfied. We hope that working with the approximation to the binomial has given you an appreciation for the uses of the normal distribution to approximate these other distributions.

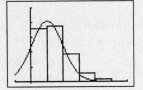

$n = 20; \pi = .05$

(a)

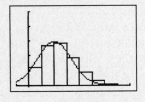

$n = 20; \pi = .10$

(b)

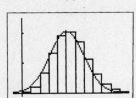

$n = 20; \pi = .25$

(c)

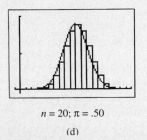

$n = 20; \pi = .50$

(d)

Figure 7.52 Binomial distributions: (a) $n = 20$, $\pi = .05$; (b) $n = 20$, $\pi = .10$; (c) $n = 20$, $\pi = .25$; (d) $n = 20$, $\pi = .50$.

TEACHING TIPS

Intent of this chapter:

After completing this chapter, students should be able to (1) define sampling variability and sampling distributions, (2) investigate the general properties of the sampling distribution of $\bar{x}$, (3) use the general properties of the sampling distribution of $\bar{x}$ to calculate probabilities, (4) investigate the general properties of the sampling distribution of p, and (5) use the general properties of the sampling distribution of p to calculate probabilities.

A PowerPoint® Lecture is available on the *Instructor's Resource Binder* CD.

■ **Section 8.1**

Students typically have an intuitive understanding that statistics calculated from different samples of the same size from the same population will vary, i.e., sampling variability. A sampling distribution is the distribution of values of a statistic for *all* possible samples of the same size from the same population. Remind students of the "Random Rectangles" activity in Chapter 2. The dotplots were approximate sampling distributions; that is, they were constructed using the values of the sample mean for some (but not all) samples of size five from the population of rectangles. *Sampling distributions are the foundation for all inference!*

Suggested Assignment: Exercises 8.1, 8.2, 8.3, 8.6, 8.9

Activity: "Chapter 8—Sampling Distributions" (*Instructor's Resource Binder* & CD, Chapter 8 Activities Worksheets)

■ **Section 8.2**

This section investigates the general properties of the sampling distribution of $\bar{x}$. Suppose that we have a fish pond in our school's courtyard that contains five goldfish with lengths (in cm) 4, 7, 10, 11, and 13. Find the mean and standard deviation for this small population ($\mu_x = 9$ and $\sigma_x = 3.162$). Find all possible samples of size 2 selected without replacement from this population of fish (there are 10 possible samples, $_5C_2 = 10$) and calculate the sample mean for each sample ($\bar{x} = 5.5, 7, 7.5, 8.5, 8.5, 9, 10, 10.5, 11.5, 12$). These are all the possibilities for the sample mean of samples of size 2 from the fish population, so we can use them to construct the sampling distribution of $\bar{x}$. For example, we might display this sampling distribution as a table that lists each of the possible $\bar{x}$ values. Find the mean and standard deviation for this sampling distribution of $\bar{x}$. ($\mu_{\bar{x}} = 9$ and $\sigma_{\bar{x}} = 1.936$.) (Note: Students should understand this notation.) What do you notice? *The mean of the sampling distribution equals the population mean (parameter). The standard deviation of the sampling distribution is smaller than the population standard deviation.*

Repeat the activity above with samples of size 3. The sample means are 7, 7.3333, 8, 8.3333, 9, 9.3333, 9.3333, 10, 10.3333, and 11.333. The mean and standard deviation of these sample means are $\mu_{\bar{x}} = 9$ and $\sigma_{\bar{x}} = 1.291$. What do you notice? *The mean of the sampling distribution equals the population mean (parameter). The standard deviation of the sampling distribution gets smaller as the sample size increases.*

Students "see" Rule 1: $\mu_{\bar{x}} = \mu$ (on page 455). Next we can consider Rule 2:

$$\sigma_{\bar{x}} = \frac{\sigma}{\sqrt{n}} = \frac{3.162}{\sqrt{2}} = 2.2359.$$ What happened? Why isn't this equal to the value of the standard deviation we calculated for samples of size 2 ($\sigma_{\bar{x}} = 1.936$)? In this example, the sample was selected without replacement. Lead students to think about how big the sample size is compared to the population size. We have exceeded the 10% caveat! However, we can adjust for this by using the finite population correction factor (students do

not need to know this factor). $\sigma_{\bar{x}} = \frac{\sigma}{\sqrt{n}} \times \sqrt{\frac{N-n}{N-1}}$, where N is the population size. So

$\sigma_{\bar{x}} = \dfrac{3.162}{\sqrt{2}} \times \sqrt{\dfrac{3}{4}} = 1.936$. Therefore, we have verified Rule 2 and have also stressed

the fact that the $\sigma_{\bar{x}} = \dfrac{\sigma}{\sqrt{n}}$ is only true when sampling with replacement or approximately

true when the sample is no more than 10% of the population.

To explore Rule 3 and Rule 4, one or more of the following activities may be used. In the first activity, students sample from three different populations: approximately normal, uniform, and skewed. Each population consists of 100 numbers, placed on squares of poster board, foam board, or poker chips. The table gives an example of these populations.

	Numbers	10	9,11	8,12	7,13	6,14	5,15	4,16	3,17	2,18	1,19	0,20
Approximately Normal	On this many squares	10	9	9	8	6	5	3	2	1	1	1
	Numbers	0	1	2	3	4	5	6	7	8	9	
Uniform	On this many squares	10	10	10	10	10	10	10	10	10	10	
	Numbers	0,1,2	3,4,5	6,7	8	9	10	11	12	13	14	15
Skewed	On this many squares	1	2	3	6	8	10	12	15	13	11	10

(Adapted from "Statistics: Decisions with Data," NSF/COMAP, Lexington MA: COMAP, 1992.)

Place the 100 squares into a plastic container or a gallon-size, self-closing plastic bag. Create a set of these populations for every group of four students. Each student randomly selects samples without replacement of size 3, size 5, and size 8 from each population and calculates the sample means. Next, have each student plot the sample means on the appropriate dotplot. (For smaller classes, have students repeat this procedure once or twice. Have students keep this data for use in Chapter 9.) These dotplots can be used to illustrate several important ideas. All of the approximate sampling distributions of $\bar{x}$ should be centered at the population mean (parameter). If the population is approximately normal to begin with, then the sampling distribution of $\bar{x}$ is also approximately normal. As the sample size increases, the sample means cluster closer to the population mean, producing a smaller standard deviation in the sample means. Finally, with populations that are not normal, the sampling distribution of $\bar{x}$ is still approximately normal when the sample size is sufficiently large, the Central Limit Theorem. (Since this skewed distribution is only slightly skewed, we see the sampling distribution of $\bar{x}$ start to approach a normal distribution with a sample size of 8. However, if the distribution was more strongly skewed, we would need a larger sample size before the sampling distribution of $\bar{x}$ becomes approximately normal.)

The applet listed below also allows students to investigate Rules 3 and 4. This applet simulates the activity just discussed. Students can quickly draw 10,000 samples from a normal population, a uniform population, and a skewed population. An added feature is that students also have the ability to create a custom population. By creating a population that is extremely non-normal, students can explore how large a sample is "sufficiently large" so that the sampling distribution of $\bar{x}$ will be approximately normal. Graphing Calculator Exploration 8.1 (on page 471) also explores the Central Limit Theorem.

Finally, these properties of the sampling distribution of $\bar{x}$ allow us to use normal distributions to calculate probabilities (see Examples 8.5 and 8.6).

Suggested Assignment: Exercises 8.10, 8.11, 8.13, 8.16, 8.19, 8.22

Activity: Activity 8.1: Do Students Who Take the SAT Multiple Times Have an Advantage in College Admissions? (p. 468 and *Activities Manual*, p. 71)★ or "Chapter 8—Sampling Distribution of $\bar{x}$" (*Instructor's Resource Binder* & CD, Chapter 8 Activities Worksheets)

Applet: http://www.ruf.rice.edu/~lane/stat_sim/sampling_dist/index.html

Assessment: Chapter 8 Quiz 1 (*Test Bank* and *Instructor's Resource Binder* CD)

■ **Section 8.3**

This section investigates the general properties of the sampling distribution of p. The general properties are stated on page 465. In order to explore these properties, suppose that a club at school consists of the following officers (gender is given in parenthesis): Alice (F), Bob (M), Chris (M), Debbie (F), Edward (M), and Frank (M). These officers will represent a small finite population. What is the population proportion (parameter) of female officers? ($\pi = 1/3$) Suppose we were interested in the distribution of the proportion of females if two officers were selected at random. Using a procedure similar to the one used in the activity in Section 8.2, find all possible samples of size 2 selected without replacement from this population and calculate the sample proportion of females (p). The table below provides each sample and its sample proportion of females.

pairs	A,B	A,C	A,D	A,E	A,F	B,C	B,D	B,E	B,F	C,D	C,E	C,F	D,E	D,F	E,F
p	.5	.5	1	.5	.5	0	.5	0	0	.5	0	0	.5	.5	0

Calculate the mean of the sample proportions and the standard deviation of the sample proportions. $\mu_p = 1/3$, $\sigma_p = .2981$. Notice that the population proportion (π) equals the mean of the sampling distribution of p (μ_p). This is Rule 1. Verify Rule 2:

$\sigma_p = \sqrt{\dfrac{\pi(1-\pi)}{n}}$. Using this formula, $\sigma_p = 1/3$. Why does this not match the standard

deviation we calculated from our samples? *Because we sampled without replacement and the sample size is more than 10% of our population.* Using the finite population correction

factor $\sqrt{\dfrac{N-n}{N-1}}$, we can demonstrate Rule 2.

Another helpful activity is to have students flip a coin 20 times and record the number of heads. Then ask the students to calculate a sample proportion (p) for the number of heads out of the 20 flips and plot the proportion on a dotplot. (M&Ms® could also be used for this activity.) Notice that the dotplot is approximately normal and is centered at 0.5. Be sure to stress that this normal curve consists of values for sample proportions (p). Discuss the sampling variability that occurs. Ask students what would happen if they flip the coin 50 times and compute the proportion of heads. (The sample proportions would cluster closer to the parameter. In other words, the standard deviation of the sampling distribution of p (σ_p) would decrease.) Graphing Calculator Exploration 8.2 (on page 473) also explores sampling distributions of p.

The properties of the sampling distribution of p allow us to use normal distributions to calculate probabilities (see Example 8.9).

Suggested Assignment: Exercises 8.23, 8.24, 8.25, 8.28, 8.31

Activity: Bonus Activity 8.2—Defective M&Ms (*Activities Manual*, p. 75)★ or "Chapter 8—Sampling Distribution of p" (*Instructor's Resource Binder* & CD, Chapter 8 Activities Worksheets)

★ Kathy's personal favorite

Assessment: Chapter 8 Quiz 2 (*Test Bank* and *Instructor's Resource Binder* CD) and/or Chapter 8 Concept Quiz (*Test Bank* and *Instructor's Resource Binder* CD)

Suggested Review Assignment: Exercises 8.32, 8.33, 8.34, 8.35, 8.36

Assessment: Chapter 8 Test (*Test Bank* and *Instructor's Resource Binder* CD)

Chapter 8

Sampling Variability and Sampling Distributions

© Christian Peterson/Getty Images

The inferential methods presented in Chapters 9–15 use information contained in a sample to reach conclusions about one or more characteristics of the population from which the sample was selected. For example, let μ denote the true mean fat content of quarter-pound hamburgers marketed by a national fast-food chain. To learn something about μ, we might obtain a sample of $n = 50$ hamburgers and determine the fat content for each one. The sample data might produce a mean of $\bar{x} = 28.4$ g. How close is this sample mean to the population mean, μ? If we selected another sample of 50 quarter-pound burgers and then determine the sample mean fat content, would this second value be near 28.4, or might it be quite different? These questions can be addressed by studying what is called the *sampling distribution* of $\bar{x}$. Just as the distribution of a numerical variable describes its long-run behavior, the sampling distribution of $\bar{x}$ provides information about the long-run behavior of $\bar{x}$ when sample after sample is selected.

In this chapter, we also consider the sampling distribution of a sample proportion (the fraction of individuals or objects in a sample that have some characteristic of interest). The sampling distribution of a sample proportion, p, provides information about the long-run behavior of the sample proportion that is necessary for making inferences about a population proportion.

8.1 Statistics and Sampling Variability

A quantity computed from the values in a sample is called a **statistic**. Values of statistics such as the sample mean $\bar{x}$, the sample median, the sample standard deviation s, or the proportion of individuals in a sample that possess a particular property p, are our primary sources of information about various population characteristics.

The usual way to obtain information regarding the value of a population characteristic is by selecting a sample from the population. For example, to gain insight about the mean credit card balance for students at a particular university, we might select a sample of 50 students at the university. Each student would be asked about his or her credit card balance to yield a value of x = current balance. We could construct a histogram of the 50 sample x values, and we could view this histogram as a rough approximation of the population distribution of x. In a similar way, we could view the sample mean $\bar{x}$ (the mean of a sample of n values) as an approximation of μ, the mean of the population distribution. It would be nice if the value of $\bar{x}$ were equal to the value of the population mean μ, but this would be an unusual occurrence. Moreover, not only will the value of $\bar{x}$ for a particular sample from a population usually differ from μ, but also the $\bar{x}$ values from different samples typically differ from one another. (For example, two different samples of 50 student credit card balances will usually result in different $\bar{x}$ values.) This sample-to-sample variability makes it challenging to generalize from a sample to the population from which it was selected. To meet this challenge, we must understand sample-to-sample variability.

> **DEFINITION**
>
> Any quantity computed from values in a sample is called a **statistic**.
>
> The observed value of a statistic depends on the particular sample selected from the population; typically, it varies from sample to sample. This variability is called **sampling variability**.

Example 8.1 Exploring Sampling Variability

Consider a small population consisting of the 20 students enrolled in an upper division class. The amount of money (in dollars) each of the 20 students spent on textbooks for the current semester is shown in the following table:

Student	Amount Spent on Books	Student	Amount Spent on Books	Student	Amount Spent on Books
1	367	8	370	15	433
2	358	9	378	16	284
3	442	10	268	17	331
4	361	11	419	18	259
5	375	12	363	19	330
6	395	13	365	20	423
7	322	14	362		

For this population,

$$\mu = \frac{367 + 358 + \cdots + 423}{20} = 360.25$$

Suppose we don't know the value of the population mean, so we decide to estimate μ by taking a random sample of five students and computing the sample mean amount spent on textbooks, $\bar{x}$. Is this a reasonable thing to do? Is the estimate that results likely to be close to the value of μ, the population mean? To answer these questions, consider a simple experiment that allows us to examine the behavior of the statistic $\bar{x}$ when random samples of size 5 are repeatedly selected. (Note that this scenario is not realistic. If a population consisted of only 20 individuals, we would probably conduct a census rather than select a sample. However, this small population size is easier to work with as we develop the idea of sampling variability.)

Let's first select a random sample of size 5 from this population. This can be done by writing the numbers from 1 to 20 on otherwise identical slips of paper, mixing them well, and then selecting 5 slips without replacement. The numbers on the slips selected identify which of the 20 students will be included in our sample. Alternatively, either a table of random digits or a random number generator can be used to determine which 5 students should be selected. We used MINITAB to obtain 5 numbers between 1 and 20, resulting in 17, 20, 7, 11, and 9, and the following sample of amounts spent on books:

331   423   322   419   378

For this sample,

$$\bar{x} = \frac{1873}{5} = 374.60$$

The sample mean is larger than the population mean of \$360.25 by about \$15. Is this difference typical, or is this particular sample mean unusually far away from μ? Taking additional samples will provide some additional insight.

Four more random samples (Samples 2–5) from this same population are shown here.

Sample 2		Sample 3		Sample 4		Sample 5	
Student	x	Student	x	Student	x	Student	x
4	361	15	433	20	423	18	259
15	433	12	363	16	284	8	370
12	363	3	442	19	330	9	378
1	367	7	322	1	367	7	322
18	259	18	259	8	370	14	362
$\bar{x}$	**356.60**	$\bar{x}$	**363.80**	$\bar{x}$	**354.80**	$\bar{x}$	**338.20**

Because $\mu = 360.25$, we can see the following:

1. The value of $\bar{x}$ varies from one random sample to another (sampling variability).
2. Some samples produced $\bar{x}$ values larger than μ (Samples 1 and 3), whereas others produced values smaller than μ (Samples 2, 4, and 5).

3. Samples 2, 3, and 4 produced $\bar{x}$ values that were fairly close to the population mean, but Sample 5 resulted in a value that was $22 below the population mean.

Continuing with the experiment, we selected 45 additional random samples (each of size $n = 5$). The resulting sample means are as follows:

Sample	$\bar{x}$	Sample	$\bar{x}$	Sample	$\bar{x}$
6	374.6	21	355.0	36	353.4
7	356.6	22	407.2	37	379.6
8	363.8	23	380.0	38	352.6
9	354.8	24	377.4	39	342.2
10	338.2	25	341.2	40	362.6
11	375.6	26	316.0	41	315.4
12	379.2	27	370.0	42	366.2
13	341.6	28	401.0	43	361.4
14	355.4	29	347.0	44	375
15	363.8	30	373.8	45	401.4
16	339.6	31	382.8	46	337
17	348.2	32	320.4	47	387.4
18	430.8	33	313.6	48	349.2
19	388.8	34	387.6	49	336.8
20	352.8	35	314.8	50	364.6

Figure 8.1, a density histogram of the 50 sample means, provides insight about the behavior of $\bar{x}$. Most samples resulted in $\bar{x}$ values that are reasonably near $\mu = 360.25$, falling between 335 and 395. A few samples, however, produced values that were far from μ. If we were to take a sample of size 5 from this population and use $\bar{x}$ as an estimate of the population mean μ, we should *not* necessarily expect $\bar{x}$ to be close to μ.

Figure 8.1 Density histogram of $\bar{x}$ values from 50 random samples for Example 8.1.

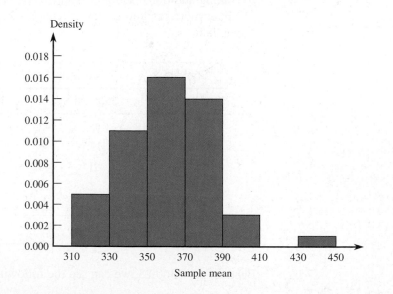

The density histogram in Figure 8.1 visually conveys information about the sampling variability in the statistic $\bar{x}$. It provides an approximation to the distribution of $\bar{x}$ values that would have been observed if we had considered every different possible sample of size 5 from this population.

■

In the example just considered, we obtained the approximate sampling distribution of the statistic $\bar{x}$ by considering just 50 different samples. The actual sampling distribution comes from considering *all* possible samples of size n.

Important concept

DEFINITION

The distribution that would be formed by considering the value of a sample statistic for every possible different sample of a given size from a population is called its **sampling distribution**.

The sampling distribution of a statistic, such as $\bar{x}$, provides important information about variation in the values of the statistic and how this variation relates to the values of various population characteristics. The density histogram of Figure 8.1 is an *approximation* of the sampling distribution of the statistic $\bar{x}$ for samples of size 5 from the population described in Example 8.1. We could have determined the true sampling distribution of $\bar{x}$ by considering every possible different sample of size 5 from the population of 20 students, computing the mean for each sample, and then constructing a density histogram of the $\bar{x}$ values, but this would have been a lot of work—there are 15,504 different possible samples of size 5. And, for more realistic situations with larger population and sample sizes, the situation becomes even worse because there are so many possible samples that must be considered. Fortunately, as we look at a few more examples in the sections that follow, patterns emerge that enable us to describe some important aspects of the sampling distributions for some statistics without actually having to look at all possible samples.

■ **Exercises 8.1–8.9** .... Turn to the Expanded Answer Section for answers not shown next to exercises. ....

8.1 Explain the difference between a population characteristic and a statistic.

8.2 What is the difference between $\bar{x}$ and μ? between s and σ?

8.3 ▼ For each of the following statements, identify the number that appears in boldface type as the value of either a population characteristic or a statistic:
a. A department store reports that **84%** of all customers who use the store's credit plan pay their bills on time.
b. A sample of 100 students at a large university had a mean age of **24.1** years.   Statistic

c. The Department of Motor Vehicles reports that **22%** of all vehicles registered in a particular state are imports.
d. A hospital reports that based on the 10 most recent cases, the mean length of stay for surgical patients is **6.4** days.   Statistic
e. A consumer group, after testing 100 batteries of a certain brand, reported an average life of **63** hr of use.
Statistic
8.4 Consider a population consisting of the following five values, which represent the number of video rentals during the academic year for each of five housemates:

8   14   16   10   11

a. Compute the mean of this population.   11.8
b. Select a random sample of size 2 by writing the numbers on slips of paper, mixing them, and then selecting 2. Compute the mean of your sample.
c. Repeatedly select samples of size 2, and compute the $\bar{x}$ value for each sample until you have the results of 25 samples.
d. Construct a density histogram using the 25 $\bar{x}$ values. Are most of the $\bar{x}$ values near the population mean? Do the $\bar{x}$ values differ a lot from sample to sample, or do they tend to be similar?   Answers will vary. See solutions manual for an example.
8.5 Select 10 additional random samples of size 5 from the population of 20 students given in Example 8.1, and compute the mean amount spent on books for each of the 10 samples. Are the $\bar{x}$ values consistent with the results of the sampling experiment summarized in Figure 8.1? Answers will vary. See solutions manual for an example.
8.6 Suppose that the sampling experiment described in Example 8.1 had used samples of size 10 rather than size 5. If 50 samples of size 10 were selected, the $\bar{x}$ value for each sample computed, and a density histogram constructed, how do you think this histogram would differ from the density histogram constructed for samples of size 5 (Figure 8.1)? In what way would it be similar?

8.7 ▼ Consider the following population: {1, 2, 3, 4}. Note that the population mean is

$$\mu = \frac{1 + 2 + 3 + 4}{4} = 2.5$$

a. Suppose that a random sample of size 2 is to be selected without replacement from this population. There are 12 possible samples (provided that the order in which observations are selected is taken into account):

1, 2	1, 3	1, 4	2, 1	2, 3	2, 4
3, 1	3, 2	3, 4	4, 1	4, 2	4, 3

Compute the sample mean for each of the 12 possible samples. Use this information to construct the sampling

distribution of $\bar{x}$. (Display the sampling distribution as a density histogram.)
b. Suppose that a random sample of size 2 is to be selected, but this time sampling will be done with replacement. Using a method similar to that of Part (a), construct the sampling distribution of $\bar{x}$. (Hint: There are 16 different possible samples in this case.)
c. In what ways are the two sampling distributions of Parts (a) and (b) similar? In what ways are they different?

8.8 Simulate sampling from the population of Exercise 8.7 by using four slips of paper individually marked 1, 2, 3, and 4. Select a sample of size 2 without replacement, and compute $\bar{x}$. Repeat this process 50 times, and construct a density histogram of the 50 $\bar{x}$ values. How does this sampling distribution compare to the sampling distribution of $\bar{x}$ derived in Exercise 8.7, Part (a)?   Answers will vary. See solutions manual for an example.
8.9 Consider the following population: {2, 3, 3, 4, 4}. The value of μ is 3.2, but suppose that this is not known to an investigator, who therefore wants to estimate μ from sample data. Three possible statistics for estimating μ are

Statistic 1: the sample mean, $\bar{x}$
Statistic 2: the sample median
Statistic 3: the average of the largest and the smallest values in the sample

A random sample of size 3 will be selected without replacement. Provided that we disregard the order in which the observations are selected, there are 10 possible samples that might result (writing 3 and 3*, 4 and 4* to distinguish the two 3's and the two 4's in the population):

2, 3, 3*	2, 3, 4	2, 3, 4*	2, 3*, 4	2, 3*, 4*
2, 4, 4*	3, 3*, 4	3, 3*, 4*	3, 4, 4*	3*, 4, 4*

For each of these 10 samples, compute Statistics 1, 2, and 3. Construct the sampling distribution of each of these statistics. Which statistic would you recommend for estimating μ and why?

8.2 The Sampling Distribution of a Sample Mean

When the objective of a statistical investigation is to make an inference about the population mean μ, it is natural to consider the sample mean $\bar{x}$ as an estimate of μ. To understand how inferential procedures based on $\bar{x}$ work, we must first study how sampling variability causes $\bar{x}$ to vary in value from one sample to another. The behavior of $\bar{x}$ is described by its sampling distribution. The sample size n and characteristics of the population—its shape, mean value μ, and standard deviation σ—are important in determining properties of the sampling distribution of $\bar{x}$.

It is helpful first to consider the results of some sampling experiments. In Examples 8.2 and 8.3, we start with a specified x population distribution, fix a sample size n, and select 500 different random samples of this size. We then compute $\bar{x}$ for each sample and construct a sample histogram of these 500 $\bar{x}$ values. Because 500 is reasonably large (a reasonably long sequence of samples), the histogram of the $\bar{x}$ values should closely resemble the true sampling distribution of $\bar{x}$ (which would be obtained from an unending sequence of $\bar{x}$ values). We repeat the experiment for several different values of n to see how the choice of sample size affects the sampling distribution. Careful examination of these histograms will aid in understanding the general properties to be stated shortly.

....

Example 8.2   Blood Platelet Size

The article "Changes in Platelet Size and Count in Unstable Angina Compared to Stable Angina or Non-Cardiac Chest Pain" (*European Heart Journal* [1998]: 80–84) presented data suggesting that the distribution of platelet size for patients with non-cardiac chest pain is approximately normal with mean $\mu = 8.25$ and standard deviation $\sigma = 0.75$. Figure 8.2 shows a normal curve centered at 8.25, the mean value of platelet size. The value of the population standard deviation, 0.75, determines the extent to which the x distribution spreads out about its mean value.

Figure 8.2 Normal distribution of platelet size x with $\mu = 8.25$ and $\sigma = 0.75$ for Example 8.2.

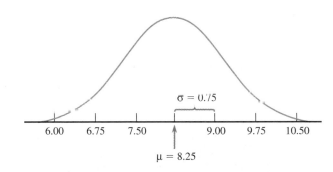

$\sigma = 0.75$

6.00   6.75   7.50   9.00   9.75   10.50

$\mu = 8.25$

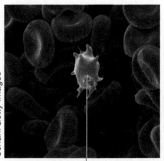

Activated platelet

We first used MINITAB to select 500 random samples from this normal distribution, with each sample consisting of $n = 5$ observations. A histogram of the resulting 500 $\bar{x}$ values appears in Figure 8.3(a). This procedure was repeated for samples of size $n = 10$, $n = 20$, and $n = 30$. The resulting sample histograms of the $\bar{x}$ values are displayed in Figures 8.3(b)–(d).

The first thing to notice about the histograms is their shape. To a reasonable approximation, each of the four histograms looks like a normal curve. The resemblance would be even more striking if each histogram had been based on many more than 500 $\bar{x}$ values. Second, notice that each histogram is centered approximately at 8.25, the mean of the population being sampled. Had the histograms been based on an unending sequence of values, their centers would have been exactly the population mean, 8.25.

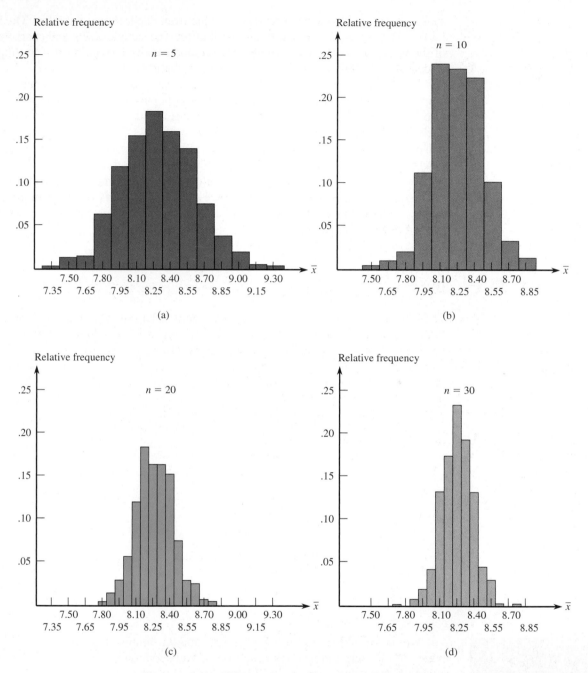

Figure 8.3 Histograms for $\bar{x}$ based on 500 samples, each consisting of n observations, for Example 8.2: (a) $n = 5$; (b) $n = 10$; (c) $n = 20$; (d) $n = 30$.

The final aspect of the histograms to note is their spread relative to one another. The smaller the value of n, the greater the extent to which the sampling distribution spreads out about the population mean value. This is why the histograms for $n = 20$ and $n = 30$ are based on narrower class intervals than those for the two smaller sample sizes. For the larger sample sizes, most of the $\bar{x}$ values are quite close to 8.25. This is the effect of averaging. When n is small, a single unusual x value can result in an $\bar{x}$ value far from the center. With a larger sample size, any unusual x

values, when averaged with the other sample values, still tend to yield an $\bar{x}$ value close to μ. Combining these insights yields a result that should appeal to your intuition: *$\bar{x}$ based on a large sample size will tend to be closer to μ than $\bar{x}$ will based on a small sample size.*

■

Example 8.3   Length of Overtime Period in Hockey

Now consider properties of the $\bar{x}$ distribution when the population is quite skewed (and thus very unlike a normal distribution). The Winter 1995 issue of *Chance* magazine gave data on the length of the overtime period for all 251 National Hockey League play-off games between 1970 and 1993 that went into overtime. In hockey, the overtime period ends as soon as one of the teams scores a goal. Figure 8.4 displays a histogram of the data. The histogram has a long upper tail, indicating that although most overtime periods lasted less than 20 min, a few games had long overtime periods.

If we think of the 251 values as a population, the histogram in Figure 8.4 shows the distribution of values in that population. The skewed shape makes identification of the mean value from the picture more difficult than for a normal distribution, but we computed the average of the 251 values to be $\mu = 9.841$. The median value for the population, 8.000, is less than μ, a consequence of the fact that the distribution is positively skewed.

For each of the sample sizes $n = 5$, 10, 20, and 30, we selected 500 random samples of size n. This was done with replacement to approximate more nearly the usual situation, in which the sample size n is only a small fraction of the population size. We then constructed a histogram of the 500 $\bar{x}$ values for each of the four sample sizes. These histograms are displayed in Figure 8.5.

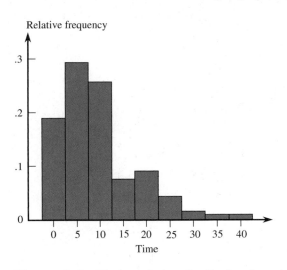

Figure 8.4 The population distribution for Example 8.3 ($\mu = 9.841$).

As with samples from a normal population, the averages of the 500 $\bar{x}$ values for the four different sample sizes are all close to the population mean $\mu = 9.841$. If each histogram had been based on an unending sequence of sample means rather than just 500 of them, each one would have been centered at exactly 9.841. Comparison of the four $\bar{x}$ histograms in Figure 8.5 also shows that as n increases, the histogram's spread about its center decreases. This was also true of increasing sample sizes from a normal population: $\bar{x}$ is less variable for a large sample size than it is for a small sample size.

One aspect of these histograms distinguishes them from the distribution of $\bar{x}$ based on a sample from a normal population. They are skewed and differ in shape more, but they become progressively more symmetric as the sample size increases. The four histograms in Figure 8.5 were drawn using the same horizontal scale, to make it easy to see the decrease in spread as n increases. However, this makes it harder to see how the histograms' shapes change with sample size. Figure 8.6 is a histogram based on narrower class intervals for the $\bar{x}$ values from samples of size 30. This figure shows that for $n = 30$, the histogram has a shape much like a normal curve. Again this is the effect of averaging. Even when n is large, one of the few large x values in the population appears infrequently in the sample. When one does

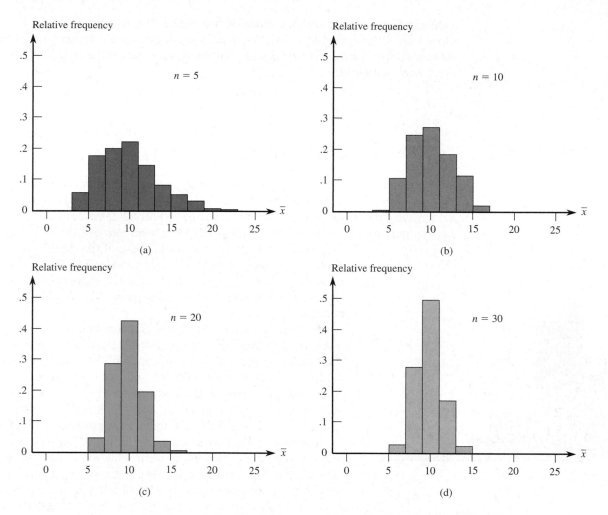

Figure 8.5 Four histograms of 500 $\bar{x}$ values for Example 8.3.

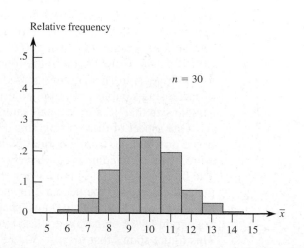

Figure 8.6 Histogram of 500 $\bar{x}$ values for Example 8.3 when $n = 30$.

appear, its contribution to $\bar{x}$ is swamped by the contributions of more typical sample values. The normal shape of the histogram for $n = 30$ is what is predicted by the Central Limit Theorem, which will be introduced shortly. According to this theorem, even if the population distribution bears no resemblance whatsoever to a normal curve, the $\bar{x}$ sampling distribution is approximately normal in shape when the sample size n is reasonably large.

■

■ General Properties of the Sampling Distribution of $\bar{x}$ ....

Examples 8.2 and 8.3 suggest that for any n, the center of the $\bar{x}$ distribution (the mean value of $\bar{x}$) coincides with the mean of the population being sampled and that the spread of the $\bar{x}$ distribution decreases as n increases, indicating that the standard deviation of $\bar{x}$ is smaller for large n than for small n. The sample histograms also suggest that in some cases, the $\bar{x}$ distribution is approximately normal in shape. These observations are stated more formally in the following general rules.

General Properties of the Sampling Distribution of $\bar{x}$

Let $\bar{x}$ denote the mean of the observations in a random sample of size n from a population having mean μ and standard deviation σ. Denote the mean value of the $\bar{x}$ distribution by $\mu_{\bar{x}}$ and the standard deviation of the $\bar{x}$ distribution by $\sigma_{\bar{x}}$. Then the following rules hold:

Rule 1. $\mu_{\bar{x}} = \mu$.

Rule 2. $\sigma_{\bar{x}} = \dfrac{\sigma}{\sqrt{n}}$. This rule is exact if the population is infinite, and is approximately correct if the population is finite and no more than 10% of the population is included in the sample.

Rule 3. When the population distribution is normal, the sampling distribution of $\bar{x}$ is also normal for any sample size n.

Rule 4. (Central Limit Theorem) When n is sufficiently large, the sampling distribution of $\bar{x}$ is well approximated by a normal curve, even when the population distribution is not itself normal.

■

Rule 1, $\mu_{\bar{x}} = \mu$, states that the sampling distribution of $\bar{x}$ is always centered at the mean of the population sampled. Rule 2, $\sigma_{\bar{x}} = \dfrac{\sigma}{\sqrt{n}}$, not only states that the spread of the sampling distribution of $\bar{x}$ decreases as n increases, but also gives a precise relationship between the standard deviation of the $\bar{x}$ distribution and the population standard deviation and sample size. When $n = 4$, for example,

$$\sigma_{\bar{x}} = \frac{\sigma}{\sqrt{n}} = \frac{\sigma}{\sqrt{4}} = \frac{\sigma}{2}$$

so the $\bar{x}$ distribution has a standard deviation only half as large as the population standard deviation. Rules 3 and 4 specify circumstances under which the $\bar{x}$ distribution is normal (when the population is normal) or approximately normal (when the sample

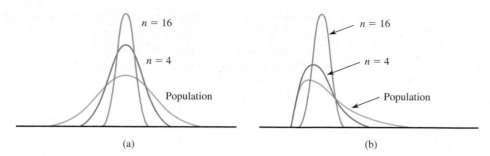

Figure 8.7 Population distribution and sampling distributions of $\bar{x}$: (a) symmetric population; (b) skewed population.

size is large). Figure 8.7 illustrates these rules by showing several $\bar{x}$ distributions superimposed over a graph of the population distribution.

The Central Limit Theorem of Rule 4 states that when n is sufficiently large, the $\bar{x}$ distribution is approximately normal, no matter what the population distribution looks like. This result has enabled statisticians to develop procedures for making inferences about a population mean μ using a large sample, even when the shape of the population distribution is unknown.

Recall that a variable is standardized by subtracting the mean value and then dividing by its standard deviation. Using Rules 1 and 2 to standardize $\bar{x}$ gives an important consequence of the last two rules.

If n is large or the population distribution is normal, the standardized variable

$$z = \frac{\bar{x} - \mu_{\bar{x}}}{\sigma_{\bar{x}}} = \frac{\bar{x} - \mu}{\dfrac{\sigma}{\sqrt{n}}}$$

has (at least approximately) a standard normal (z) distribution.

Application of the Central Limit Theorem in specific situations requires a rule of thumb for deciding whether n is indeed sufficiently large. Such a rule is not as easy to come by as one might think. Look back at Figure 8.5, which shows the approximate sampling distribution of $\bar{x}$ for $n = 5$, 10, 20, and 30 when the population distribution is quite skewed. Certainly the histogram for $n = 5$ is not well described by a normal curve, and this is still true of the histogram for $n = 10$, particularly in the tails of the histogram (far away from the mean value). Among the four histograms, only the histogram for $n = 30$ has a reasonably normal shape.

Important concept

On the other hand, when the population distribution is normal, the sampling distribution of $\bar{x}$ is normal for any n. If the population distribution is somewhat skewed but not to the extent of Figure 8.4, we might expect the $\bar{x}$ sampling distribution to be a bit skewed for $n = 5$ but quite well fit by a normal curve for n as small as 10 or 15. How large an n is needed for the $\bar{x}$ distribution to approximate a normal curve depends on how much the population distribution differs from a normal distribution. The closer the population distribution is to being normal, the smaller the value of n necessary for the Central Limit Theorem approximation to be accurate.

Many statisticians recommend the following conservative rule:

The Central Limit Theorem can safely be applied if *n* is greater than or equal to 30.

If the population distribution is believed to be reasonably close to a normal distribution, an *n* of 15 or 20 is often large enough for $\bar{x}$ to have approximately a normal distribution. At the other extreme, we can imagine a distribution with a much longer tail than that of Figure 8.4, in which case even *n* = 40 or 50 would not suffice for approximate normality of $\bar{x}$. In practice, however, few population distributions are likely to be this badly behaved.

····················

Example 8.4   Bluethroat Song Duration

© Jan Baks/Alamy

Male bluethroats have a complex song, which is thought to be used to attract female birds. Let *x* denote the duration of a randomly selected song (in seconds) for a male bluethroat. Suppose that the mean value of song duration is μ = 13.8 sec and that the standard deviation of song duration is σ = 11.8 sec. (These values are suggested in the paper "Response of Male Bluethroats *Luscinia svecica* to Song Playback," *Ornis Fennica* [2000]: 43–47.) The authors of the paper noted that the song length distribution is not normal (which could also be deduced from the relative values of the mean and standard deviation and the fact that song length cannot be negative). The sampling distribution of $\bar{x}$ based on a random sample of *n* = 25 song durations then also has mean value

$$\mu_{\bar{x}} = 13.8 \text{ sec}$$

That is, the sampling distribution of $\bar{x}$ is centered at 13.8. The standard deviation of $\bar{x}$ is

$$\sigma_{\bar{x}} = \frac{\sigma}{\sqrt{n}} = \frac{11.8}{\sqrt{25}} = 2.36$$

which is only one-fifth as large as the population standard deviation σ. Because the population distribution is not normal and because the sample size is not larger than 30, we cannot assume that the sampling distribution of $\bar{x}$ is normal in shape.

■

····················

Example 8.5   Soda Volumes

A soft-drink bottler claims that, on average, cans contain 12 oz of soda. Let $\bar{x}$ denote the actual volume of soda in a randomly selected can. Suppose that *x* is normally distributed with σ = 0.16 oz. Sixteen cans are to be selected, and the soda volume will be determined for each one; let $\bar{x}$ denote the resulting sample mean soda volume. Because the *x* distribution is normal, the sampling distribution of $\bar{x}$ is also normal. *If the bottler's claim is correct,* the sampling distribution of $\bar{x}$ has a mean value of $\mu_{\bar{x}} = \mu$ = 12 and a standard deviation of

$$\sigma_{\bar{x}} = \frac{\sigma}{\sqrt{n}} = \frac{.16}{\sqrt{16}} = .04$$

To calculate a probability involving $\bar{x}$, we standardize by subtracting the mean value, 12, and dividing by the standard deviation (of $\bar{x}$), which is 0.04. For example, the probability that the sample mean soda volume is between 11.96 oz and 12.08 oz is the area between 11.96 and 12.08 under the normal density curve with mean 12 and standard deviation 0.04, as shown in the following figure:

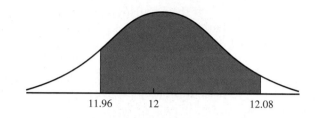

This area is calculated by first standardizing the interval limits:

$$\text{Lower limit: } a^* = \frac{11.96 - 12}{.04} = -1.0$$

$$\text{Upper limit: } b^* = \frac{12.08 - 12}{.04} = 2.0$$

Then

$$
\begin{aligned}
P(11.96 \leq \bar{x} \leq 12.08) &= \text{area under the } z \text{ curve between } -1.0 \text{ and } 2.0 \\
&= (\text{area to the left of } 2.0) - (\text{area to the left of } -1.0) \\
&= .9772 - .1587 \\
&= .8185
\end{aligned}
$$

The probability that the sample mean soda volume is at most 11.9 oz is

$$P(\bar{x} \leq 11.9) = P\left(z \leq \frac{11.9 - 12}{.04} = -2.5\right)$$
$$= (\text{area under the } z \text{ curve to the left of } -2.5) = .0062$$

If the x distribution is as described and the claim is correct, a sample mean soda volume based on 16 observations is less than 11.9 oz for less than 1% of all such samples. Thus, observation of an $\bar{x}$ value that is smaller than 11.9 oz would cast doubt on the bottler's claim that the average soda volume is 12 oz.

■

Example 8.6   Fat Content of Hot Dogs

A hot dog manufacturer asserts that one of its brands of hot dogs has an average fat content of $\mu = 18$ g per hot dog. Consumers of this brand would probably not be disturbed if the mean is less than 18 but would be unhappy if it exceeds 18. Let x denote the fat content of a randomly selected hot dog, and suppose that σ, the standard deviation of the x distribution, is 1.

An independent testing organization is asked to analyze a random sample of 36 hot dogs. Let $\bar{x}$ be the average fat content for this sample. The sample size, $n = 36$, is large enough to rely on the Central Limit Theorem and to regard the $\bar{x}$ distribution as approximately normal. The standard deviation of the $\bar{x}$ distribution is

$$\sigma_{\bar{x}} = \frac{\sigma}{\sqrt{n}} = \frac{1}{\sqrt{36}} = .1667$$

If the manufacturer's claim is correct, we know that $\mu_{\bar{x}} = \mu = 18$ g. Suppose that the sample resulted in a mean of $\bar{x} = 18.4$ g. Does this result suggest that the manufacturer's claim is incorrect?

We can answer this question by looking at the sampling distribution of $\bar{x}$. Because of sampling variability, even if $\mu = 18$, we know that $\bar{x}$ typically will not be exactly 18. But, is it likely that we would see a sample mean at least as large as 18.4 when the population mean is really 18? *If the company's claim is correct,*

$$P(\bar{x} \geq 18.4) \approx P\left(z \geq \frac{18.4 - 18}{.1667}\right)$$
$$= P(z \geq 2.4)$$
$$= \text{area under the } z \text{ curve to the right of } 2.4$$
$$= .0082$$

Values of $\bar{x}$ at least as large as 18.4 will be observed only approximately 0.82% of the time when a random sample of size 36 is taken from a population with mean 18 and standard deviation 1. The value $\bar{x} = 18.4$ exceeds 18 by enough to cast substantial doubt on the manufacturer's claim.

■

■ Other Cases

We now know a great deal about the sampling distribution of $\bar{x}$ in two cases: for a normal population distribution and for a large sample size. What happens when the population distribution is not normal and n is small? Although it is still true that $\mu_{\bar{x}} = \mu$ and $\sigma_{\bar{x}} = \frac{\sigma}{\sqrt{n}}$, unfortunately there is no general result about the shape of the distribution. When the objective is to make an inference about the center of such a population, one way to proceed is to replace the normality assumption with some other assumption about the shape of the distribution. Statisticians have proposed and studied a number of such models. Theoretical methods or simulation can be used to describe the $\bar{x}$ distribution corresponding to the assumed model. An alternative strategy is to use one of the transformations presented in Chapter 7 to create a data set that more closely resembles a sample from a normal population and then to base inferences on the transformed data. Yet another path is to use an inferential procedure based on a statistic other than $\bar{x}$.

Exercises 8.10–8.22

Turn to the Expanded Answer Section for answers not shown next to exercises.

8.10 A random sample is selected from a population with mean $\mu = 100$ and standard deviation $\sigma = 10$. Determine the mean and standard deviation of the $\bar{x}$ sampling distribution for each of the following sample sizes:

Bold exercises answered in back   ● Data set available online but not required   ▼ Video solution available

a. $n = 9$ 100, 3.333 **d.** $n = 50$ 100, 1.414
b. $n = 15$ 100, 2.582 **e.** $n = 100$ 100, 1.0
c. $n = 36$ 100, 1.667 **f.** $n = 400$ 100, 0.5

8.11 For which of the sample sizes given in Exercise 8.10 would it be reasonable to think that the $\bar{x}$ sampling distribution is approximately normal in shape? For $n = 36, 50, 100,$ and 400

8.12 Explain the difference between σ and $\sigma_{\bar{x}}$ and between μ and $\mu_{\bar{x}}$.

8.13 ▼ Suppose that a random sample of size 64 is to be selected from a population with mean 40 and standard deviation 5.
a. What are the mean and standard deviation of the $\bar{x}$ sampling distribution? Describe the shape of the $\bar{x}$ sampling distribution. $\mu_{\bar{x}} = 40, \sigma_{\bar{x}} = 0.625,$ approximately normal.
b. What is the approximate probability that $\bar{x}$ will be within 0.5 of the population mean μ? 0.5762
c. What is the approximate probability that $\bar{x}$ will differ from μ by more than 0.7? 0.2628

8.14 The time that a randomly selected individual waits for an elevator in an office building has a uniform distribution over the interval from 0 to 1 min. It can be shown that for this distribution $\mu = 0.5$ and $\sigma = 0.289$.
a. Let $\bar{x}$ be the sample average waiting time for a random sample of 16 individuals. What are the mean and standard deviation of the sampling distribution of $\bar{x}$? 0.5, 0.072
b. Answer Part (a) for a random sample of 50 individuals. In this case, sketch a picture of a good approximation to the actual $\bar{x}$ distribution. Approximately normal with mean 0.5 and standard deviation 0.041

8.15 Let x denote the time (in minutes) that it takes a fifth-grade student to read a certain passage. Suppose that the mean value and standard deviation of x are $\mu = 2$ min and $\sigma = 0.8$ min, respectively.
a. If $\bar{x}$ is the sample average time for a random sample of $n = 9$ students, where is the $\bar{x}$ distribution centered, and how much does it spread out about the center (as described by its standard deviation)? $\mu_{\bar{x}} = 2, \sigma_{\bar{x}} = 0.267$
b. Repeat Part (a) for a sample of size of $n = 20$ and again for a sample of size $n = 100$. How do the centers and spreads of the three $\bar{x}$ distributions compare to one another? Which sample size would be most likely to result in an $\bar{x}$ value close to μ, and why?

8.16 In the library on a university campus, there is a sign in the elevator that indicates a limit of 16 persons. Furthermore, there is a weight limit of 2500 lb. Assume that the average weight of students, faculty, and staff on campus is 150 lb, that the standard deviation is 27 lb, and that the distribution of weights of individuals on campus is approximately normal. If a random sample of 16 persons from the campus is to be taken:
a. What is the expected value of the distribution of the sample mean? 150
b. What is the standard deviation of the sampling distribution of the sample mean weight? 6.75
c. What average weights for a sample of 16 people will result in the total weight exceeding the weight limit of 2500 lb? $\bar{x} > 156.25$
d. What is the chance that a random sample of 16 persons on the elevator will exceed the weight limit? 0.1772

8.17 Suppose that the mean value of interpupillary distance (the distance between the pupils of the left and right eyes) for adult males is 65 mm and that the population standard deviation is 5 mm.
a. If the distribution of interpupillary distance is normal and a sample of $n = 25$ adult males is to be selected, what is the probability that the sample average distance $\bar{x}$ for these 25 will be between 64 and 67 mm? at least 68 mm?
b. Suppose that a sample of 100 adult males is to be obtained. Without assuming that interpupillary distance is normally distributed, what is the approximate probability that the sample average distance will be between 64 and 67 mm? at least 68 mm? 0.9772, 0

8.18 Suppose that a sample of size 100 is to be drawn from a population with standard deviation 10.
a. What is the probability that the sample mean will be within 2σ of the value of μ? 0.9544
b. For this example ($n = 100, \sigma = 10$), complete each of the following statements by computing the appropriate value:
  i. Approximately 95% of the time, $\bar{x}$ will be within _____ of μ. 2
  ii. Approximately 0.3% of the time, $\bar{x}$ will be farther than _____ from μ. 3

8.19 A manufacturing process is designed to produce bolts with a 0.5-in. diameter. Once each day, a random sample of 36 bolts is selected and the diameters recorded. If the resulting sample mean is less than 0.49 in. or greater than 0.51 in., the process is shut down for adjustment. The standard deviation for diameter is 0.02 in. What is the probability that the manufacturing line will be shut down unnecessarily? (Hint: Find the probability of observing an $\bar{x}$ in the shutdown range when the true process mean really is 0.5 in.)   0.0026

8.20 College students with a checking account typically write relatively few checks in any given month, whereas nonstudent residents typically write many more checks during a month. Suppose that 50% of a bank's accounts are held by students and that 50% are held by nonstudent residents. Let x denote the number of checks written in a given month by a randomly selected bank customer.
a. Give a sketch of what the probability distribution of x might look like.
b. Suppose that the mean value of x is 22.0 and that the standard deviation is 16.5. If a random sample of $n = 100$ customers is to be selected and $\bar{x}$ denotes the sample average number of checks written during a particular month, where is the sampling distribution of $\bar{x}$ centered, and what is the standard deviation of the $\bar{x}$ distribution? Sketch a rough picture of the sampling distribution.
c. Referring to Part (b), what is the approximate probability that $\bar{x}$ is at most 20? at least 25?   0.1127, 0.0345

8.21 ▼ An airplane with room for 100 passengers has a total baggage limit of 6000 lb. Suppose that the total weight of the baggage checked by an individual passenger is a random variable x with a mean value of 50 lb and a stan

dard deviation of 20 lb. If 100 passengers will board a flight, what is the approximate probability that the total weight of their baggage will exceed the limit? (Hint: With $n = 100$, the total weight exceeds the limit when the average weight $\bar{x}$ exceeds 6000/100.)   Approximately 0

8.22 The thickness (in millimeters) of the coating applied to disk drives is a characteristic that determines the usefulness of the product. When no unusual circumstances are present, the thickness (x) has a normal distribution with a mean of 3 mm and a standard deviation of 0.05 mm. Suppose that the process will be monitored by selecting a random sample of 16 drives from each shift's production and determining $\bar{x}$, the mean coating thickness for the sample.
a. Describe the sampling distribution of $\bar{x}$ (for a sample of size 16).
b. When no unusual circumstances are present, we expect $\bar{x}$ to be within $3\sigma_{\bar{x}}$ of 3 mm, the desired value. An $\bar{x}$ value farther from 3 than $3\sigma_{\bar{x}}$ is interpreted as an indication of a problem that needs attention. Compute $3 \pm 3\sigma_{\bar{x}}$. (A plot over time of $\bar{x}$ values with horizontal lines drawn at the limits $\mu \pm 3\sigma_{\bar{x}}$ is called a process control chart.)
c. Referring to Part (b), what is the probability that a sample mean will be outside $3 \pm 3\sigma_{\bar{x}}$ just by chance (i.e., when there are no unusual circumstances)?   0.0026
d. Suppose that a machine used to apply the coating is out of adjustment, resulting in a mean coating thickness of 3.05 mm. What is the probability that a problem will be detected when the next sample is taken? (Hint: This will occur if $\bar{x} > 3 + 3\sigma_{\bar{x}}$ or $\bar{x} < 3 - 3\sigma_{\bar{x}}$ when $\mu = 3.05$.)   0.8413

Bold exercises answered in back   ● Data set available online but not required   ▼ Video solution available

---

8.3 The Sampling Distribution of a Sample Proportion

The objective of many statistical investigations is to draw a conclusion about the proportion of individuals or objects in a population that possess a specified property—for example, Maytag washers that don't require service during the warranty period or coffee drinkers who regularly drink decaffeinated coffee. Traditionally, any individual or object that possesses the property of interest is labeled a success (S), and one that does not possess the property is termed a failure (F). The Greek letter π denotes the proportion of successes in the population. The value of π is a number between 0 and 1, and 100π is the percentage of successes in the population. If $\pi = .75$, 75% of the population members are successes, and if $\pi = .01$, the population contains only 1% successes and 99% failures.

The value of π is usually unknown to an investigator. When a random sample of size n is selected from this type of population, some of the individuals in the sample are successes, and the rest are failures. The statistic that provides a basis for making inferences about π is p, the **sample proportion of successes**:

$$p = \frac{\text{number of S's in the sample}}{n}$$

For example, if $n = 5$ and three successes result, then $p = 3/5 = .6$.

Just as making inferences about μ requires knowing something about the sampling distribution of the statistic $\bar{x}$, making inferences about π requires first learning about properties of the sampling distribution of the statistic p. For example, when $n = 5$, the six possible values of p are 0, .2 (from 1/5), .4, .6, .8, and 1. The sampling distribution of p gives the probability of each of these six possible values, the long-run proportion of the time that each value would occur if samples with $n = 5$ were selected over and over again.

As we did for the distribution of the sample mean, we will look at some simulation experiments to develop an intuitive understanding of the distribution of the sample proportion before stating general rules. In each example, 500 random samples (each of size n) are selected from a population having a specified value of π. We compute p for each sample and then construct a sample histogram of the 500 values.

Example 8.7   Gender of College Students

In the fall of 2004, there were 16,636 students enrolled at California Polytechnic State University, San Luis Obispo. Of these students, 6983 (42%) were female. To illustrate properties of the sampling distribution of a sample proportion, we will simulate sampling from this Cal Poly student population. With S denoting a female student and F a male student, the proportion of S's in the population is $\pi = .42$. A statistical software package was used to select 500 samples of size $n = 5$, then 500 samples of size $n = 10$, then 500 samples with $n = 25$, and finally 500 samples with $n = 50$. The sample histograms of the 500 values of p for each of the four sample sizes are displayed in Figure 8.8.

The most noticeable feature of the histogram shapes is the progression toward the shape of a normal curve as n increases. The histogram for $n = 10$ is more bell shaped than the histogram for samples of size 5, although it is still slightly skewed. The histograms for $n = 25$ and $n = 50$ look much like normal curves.

Although the skewness of the first two histograms makes it a bit difficult to locate their centers, all four histograms appear to be centered at roughly .42, the value of π for the population sampled. Had the histograms been based on an unending sequence of samples, each histogram would have been centered at exactly .42. Finally, as was the case with the sampling distribution of $\bar{x}$, the histograms spread out more for small sample sizes than for large sample sizes. Not surprisingly, the value of p based on a large sample size tends to be closer to π, the population proportion of successes, than does p from a small sample.

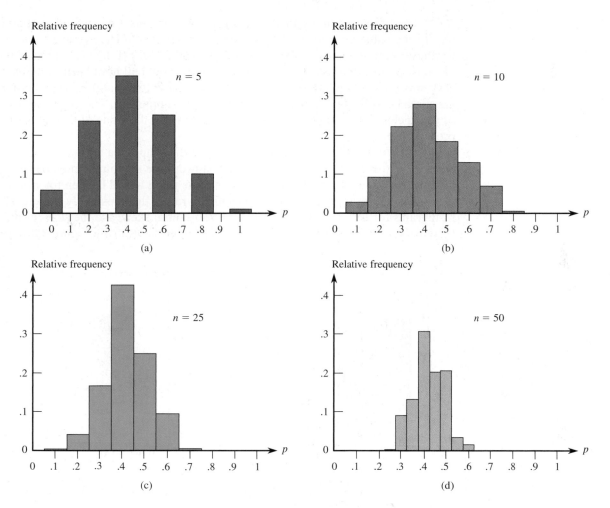

Figure 8.8   Histograms for 500 values of p ($\pi = .42$) for Example 8.7: (a) $n = 5$; (b) $n = 10$; (c) $n = 25$; (d) $n = 50$.

■

Example 8.8   Contracting Hepatitis from Blood Transfusion

The development of viral hepatitis subsequent to a blood transfusion can cause serious complications for a patient. The article "Lack of Awareness Results in Poor Autologous Blood Transfusion" (*Health Care Management*, May 15, 2003) reported that hepatitis occurs in 7% of patients who receive multiple blood transfusions during heart surgery. Here, we simulate sampling from the population of blood recipients, with S denoting a recipient who contracts hepatitis (not the sort of characteristic one usually thinks of as a success, but the S–F labeling is arbitrary), so $\pi = .07$. Figure 8.9 displays histograms of 500 values of p for the four sample sizes $n = 10$, 25, 50, and 100.

As was the case in Example 8.7, all four histograms are centered at approximately the value of π for the population being sampled. (The average values of p for these simulations are .0690, .0677, .0707, and .0694.) If the histograms had been based on an unending sequence of samples, they would all have been centered at exactly $\pi = .07$. Again, the spread of a histogram based on a large n is smaller than the spread of a histogram resulting from a small sample size. The larger the value of n, the closer the sample proportion p tends to be to the value of the population proportion π.

Furthermore, there is a progression toward the shape of a normal curve as n increases. However, the progression is much slower here than in the previous example, because the value of π is so extreme. (The same thing would happen for $\pi = .93$, except that the histograms would be negatively rather than positively skewed.) The histograms for $n = 10$ and $n = 25$ exhibit substantial skew, and the skew of the histogram for $n = 50$ is still moderate (compare Figure 8.9(c) with Figure 8.8(d)). Only

Figure 8.9 Histograms of 500 values of p ($\pi = .07$) for Example 8.8: (a) $n = 10$; (b) $n = 25$; (c) $n = 50$; (d) $n = 100$.

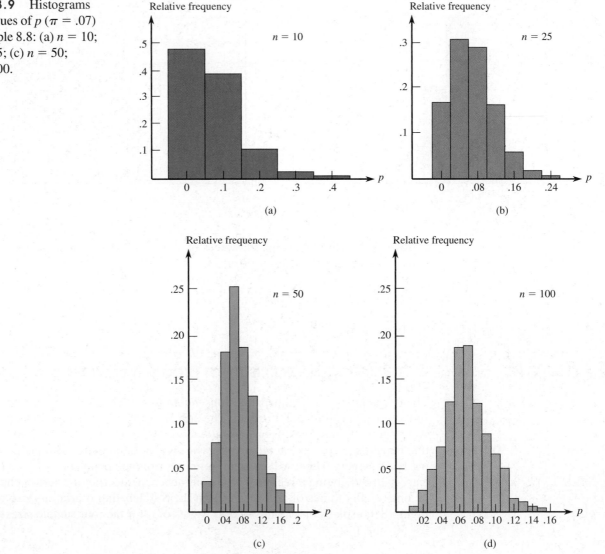

the histogram for $n = 100$ is reasonably well fit by a normal curve. It appears that whether a normal curve provides a good approximation to the sampling distribution of p depends on the values of both n and π. Knowing only that $n = 50$ is not enough to guarantee that the shape of the histogram is approximately normal.

■

■ General Properties of the Sampling Distribution of p ....

Examples 8.7 and 8.8 suggest that the sampling distribution of p depends on both n, the sample size, and π, the proportion of successes in the population. Key results are stated more formally in the following general rules.

General Properties of the Sampling Distribution of p

Let p be the proportion of successes in a random sample of size n from a population whose proportion of S's is π. Denote the mean value of p by μ_p and the standard deviation by σ_p. Then the following rules hold.

Rule 1. $\mu_p = \pi$.

Rule 2. $\sigma_p = \sqrt{\dfrac{\pi(1 - \pi)}{n}}$. This rule is exact if the population is infinite, and is approximately correct if the population is finite and no more than 10% of the population is included in the sample.

Rule 3. When n is large and π is not too near 0 or 1, the sampling distribution of p is approximately normal.

■

Thus, the sampling distribution of p is always centered at the value of the population success proportion π, and the extent to which the distribution spreads out about π decreases as the sample size n increases.

Examples 8.7 and 8.8 indicate that both π and n must be considered in judging whether the sampling distribution of p is approximately normal.

The farther the value of π is from .5, the larger n must be for a normal approximation to the sampling distribution of p to be accurate. A conservative rule of thumb is that if both $n\pi \geq 10$ and $n(1 - \pi) \geq 10$, then a normal distribution provides a reasonable approximation to the sampling distribution of p.

■

A sample size of $n = 100$ is not by itself sufficient to justify the use of a normal approximation. If $\pi = .01$, the distribution of p is extremely positively skewed, so a bell-shaped curve does not give a good approximation. Similarly, if $n = 100$ and $\pi = .99$ (so that $n(1 - \pi) = 1 < 10$), the distribution of p has a substantial negative skew. The conditions $n\pi \geq 10$ and $n(1 - \pi) \geq 10$ ensure that the sampling distribution of p is not too skewed. If $\pi = .5$, the normal approximation can be used for n as small as 20, whereas for $\pi = .05$ or .95, n should be at least 200.

Example 8.9   Blood Transfusions Continued

The proportion of all cardiac patients receiving blood transfusions who contract hepatitis was given as .07 in the article referenced in Example 8.8. Suppose that a new blood screening procedure is believed to reduce the incidence rate of hepatitis. Blood screened using this procedure is given to $n = 200$ blood recipients. Only 6 of the 200 patients contract hepatitis. This appears to be a favorable result, because $p = 6/200 = .03$. The question of interest to medical researchers is, Does this result indicate that the true (long-run) proportion of patients who contract hepatitis when the new screening procedure is used is less than .07, or could this result be plausibly attributed to sampling variability (i.e., to the fact that p typically differs from the population proportion, π)? *If the screening procedure is ineffective and $\pi = .07$,*

$$\mu_p = \pi = .07$$

$$\sigma_p = \sqrt{\frac{\pi(1 - \pi)}{200}} = \sqrt{\frac{(.07)(.93)}{200}} = .018$$

Furthermore, because

$$n\pi = 200(.07) = 14 \geq 10$$

and

$$n(1 - \pi) = 200(.93) = 186 \geq 10$$

the sampling distribution of p is approximately normal. Then, if the screening procedure is ineffective,

$$P(p < .03) = P\left(z < \frac{.03 - .07}{.018}\right)$$
$$= P(z < -2.22)$$
$$= .0132$$

Thus, it is unlikely that a sample proportion .03 or smaller would be observed if the screening procedure was ineffective. The new screening procedure appears to yield a smaller incidence rate for hepatitis.

■

   Exercises 8.23–8.31 ....   Turn to the Expanded Answer Section for answers not shown next to exercises.

8.23 A random sample is to be selected from a population that has a proportion of successes $\pi = .65$. Determine the mean and standard deviation of the sampling distribution of p for each of the following sample sizes:

a. $n = 10$ 0.65, 0.1508   **d.** $n = 50$ 0.65, 0.0675
b. $n = 20$ 0.65, 0.1067   **e.** $n = 100$ 0.65, 0.0477
c. $n = 30$ 0.65, 0.0871   **f.** $n = 200$ 0.65, 0.0337

Bold exercises answered in back   ● Data set available online but not required   ▼ Video solution available

8.24 For which of the sample sizes given in Exercise 8.23 would the sampling distribution of p be approximately normal if $\pi = .65$? if $\pi = .2$?

8.25 ▼ The article "Unmarried Couples More Likely to Be Interracial" (*San Luis Obispo Tribune*, March 13, 2002) reported that 7% of married couples in the United States are mixed racially or ethnically. Consider the population consisting of all married couples in the United States.
a. A random sample of $n = 100$ couples will be selected from this population and p, the proportion of couples that are mixed racially or ethnically, will be computed. What are the mean and standard deviation of the sampling distribution of p? $\mu_p = 0.07, \sigma_p = 0.0255$
b. Is it reasonable to assume that the sampling distribution of p is approximately normal for random samples of size $n = 100$? Explain.   No, because $n\pi < 10$.
c. Suppose that the sample size is $n = 200$ rather than $n = 100$, as in Part (b). Does the change in sample size change the mean and standard deviation of the sampling distribution of p? If so, what are the new values for the mean and standard deviation? If not, explain why not.
d. Is it reasonable to assume that the sampling distribution of p is approximately normal for random samples of size $n = 200$? Explain.
e. When $n = 200$, what is the probability that the proportion of couples in the sample who are racially or ethnically mixed will be greater than .10?   0.0485

8.26 The article referenced in Exercise 8.25 reported that for unmarried couples living together, the proportion that are racially or ethnically mixed is .15. Answer the questions posed in Parts (a)–(e) of Exercise 8.25 for the population of unmarried couples living together.

8.27 ▼ A certain chromosome defect occurs in only 1 out of 200 adult Caucasian males. A random sample of $n = 100$ adult Caucasian males is to be obtained.
a. What is the mean value of the sample proportion p, and what is the standard deviation of the sample proportion?
b. Does p have approximately a normal distribution in this case? Explain.
c. What is the smallest value of n for which the sampling distribution of p is approximately normal? $n \geq 2000$

8.28 The article "Should Pregnant Women Move? Linking Risks for Birth Defects with Proximity to Toxic Waste Sites" (*Chance* [1992]: 40–45) reported that in a large study carried out in the state of New York, approximately 30% of the study subjects lived within 1 mi of a hazardous waste site. Let π denote the proportion of all New York residents who live within 1 mi of such a site, and suppose that $\pi = .3$.
a. Would p based on a random sample of only 10 residents have approximately a normal distribution? Explain why or why not.   No, because $n\pi = 10(0.3) = 3 < 10$.
b. What are the mean value and standard deviation of p based on a random sample of size 400?
c. When $n = 400$, what is $P(.25 \leq p \leq .35)$?   0.9709
d. Is the probability calculated in Part (c) larger or smaller than would be the case if $n = 500$? Answer without actually calculating this probability.   Smaller, because as n increases, σ_p decreases.

8.29 The article "Thrillers" (*Newsweek*, April 22, 1985) stated, "Surveys tell us that more than half of America's college graduates are avid readers of mystery novels." Let π denote the actual proportion of college graduates who are avid readers of mystery novels. Consider a sample proportion p that is based on a random sample of 225 college graduates.
a. If $\pi = .5$, what are the mean value and standard deviation of p? Answer this question for $\pi = .6$. Does p have approximately a normal distribution in both cases? Explain.
b. Calculate $P(p \geq .6)$ for both $\pi = .5$ and $\pi = .6$.
c. Without doing any calculations, how do you think the probabilities in Part (b) would change if n were 400 rather than 225?

8.30 Suppose that a particular candidate for public office is in fact favored by 48% of all registered voters in the district. A polling organization will take a random sample of 500 voters and will use p, the sample proportion, to estimate π. What is the approximate probability that p will be greater than .5, causing the polling organization to incorrectly predict the result of the upcoming election?   0.1853

8.31 ▼ A manufacturer of computer printers purchases plastic ink cartridges from a vendor. When a large shipment is received, a random sample of 200 cartridges is selected, and each cartridge is inspected. If the sample proportion of defective cartridges is more than .02, the entire shipment is returned to the vendor.
a. What is the approximate probability that a shipment will be returned if the true proportion of defective cartridges in the shipment is .05?   0.9744
b. What is the approximate probability that a shipment will not be returned if the true proportion of defective cartridges in the shipment is .10?   Approximately 0

Activity **8.1**   Do Students Who Take the SATs Multiple Times Have an Advantage in College Admissions?

Technology activity: Requires use of a computer or a graphing calculator.

Background: The *Chronicle of Higher Education* (January 29, 2003) summarized an article that appeared on the *American Prospect* web site titled "College Try: Why Universities Should Stop Encouraging Applicants to Take the SATs Over and Over Again." This paper argued that current college admission policies that permit applicants to take the SAT exam multiple times and then use the highest score for consideration of admission favor students from families with higher incomes (who can afford to take the exam many times). The author proposed two alternatives that he believes would be fairer than using the highest score: (1) Use the average of all test scores, or (2) use only the most recent score.

In this activity, you will investigate the differences between the three possibilities by looking at the sampling distributions of three statistics for a test taker who takes the exam twice and for a test taker who takes the exam five times. The three statistics are

> Max = maximum score
> Mean = average score
> Recent = most recent score

An individual's score on the SAT exam fluctuates between test administrations. Suppose that a particular student's "true ability" is reflected by an SAT score of 1200 but, because of chance fluctuations, the test score on any particular administration of the exam can be considered a random variable that has a distribution that is approximately normal with mean 1200 and standard deviation 30. If we select a sample from this normal distribution, the resulting set of observations can be viewed as a collection of test scores that might have been obtained by this student.

Part 1: Begin by considering what happens if this student takes the exam twice. You will use simulation to generate samples of two test scores, Score1 and Score2, for this student. Then you will compute the values of Max, Mean, and Recent for each pair of scores. The resulting values of Max, Mean, and Recent will be used to construct approximations to the sampling distributions of the three statistics.

The instructions that follow assume the use of MINITAB. If you are using a different software package

or a graphing calculator, your instructor will provide alternative instructions.

a. Obtain 500 sets of 2 test scores by generating observations from a normal distribution with mean 1200 and standard deviation 30.

MINITAB: Calc → Random Data → Normal
   Enter 500 in the Generate box (to get 500 sets of scores)
   Enter C1-C2 in the Store in Columns box (to get two test scores in each set)
   Enter 1200 in the Mean box (because we want scores from a normal distribution with mean 1200)
   Enter 30 in the Standard Deviation box (because we want scores from a normal distribution with standard deviation 30)
   Click on OK

b. Looking at the MINITAB worksheet, you should now see 500 rows of values in each of the first two columns. The two values in any particular row can be regarded as the test scores that might be observed when the student takes the test twice. For each pair of test scores, we now calculate the values of Max, Mean, and Recent.

   i. Recent is just the last test score, so the values in C2 are the values of Recent. Name this column recent2 by typing the name into the gray box at the top of C2.

   ii. Compute the maximum test score (Max) for each pair of scores, and store the values in C3, as follows:

MINITAB: Calc → Row statistics
   Click the button for maximum
   Enter C1-C2 in the Input variables box Enter C3 in the Store Result In box.
   Click on OK

You should now see the maximum value for each pair in C3. Name this column max2.

   iii. Compute the average test score (Mean) for each pair of scores, and store the values in C4, as follows:

MINITAB: Calc → Row statistics
   Click the button for mean
   Enter C1-C2 in the Input Variables box
   Enter C4 in the Store Result In box.
   Click on OK

You should now see the average for each pair in C4. Name this column mean2.

c. Construct density histograms for each of the three statistics (these density histograms approximate the sampling distributions of the three statistics), as follows:

MINITAB: Graph → Histogram

    Enter max2, mean2, and recent2 into the first three rows of the Graph Variables box

    Click on the Options button. Select Density. Click on OK. (This will produce histograms that use the density scale rather than the frequency scale.)

    Click on the Frame drop-down menu, and select Multiple Graphs. Select Same X and Same Y. (This will cause MINITAB to use the same scales for all three histograms, so that they can be easily compared.)

    Click on OK.

Part 2: Now you will produce approximate sampling distributions for these same three statistics, but for the case of a student who takes the exam five times. Follow the same steps as in Part 1, with the following modifications:

a. Obtain 500 sets of 5 test scores, and store these values in columns C11– C15.

b. Recent will just be the values in C15; name this column recent5. Compute the Max and Mean values, and store them in columns C16 and C17. Name these columns max5 and mean5.

c. Construct density histograms for max5, mean5, and recent5.

Part 3: Now use the approximate sampling distributions constructed in Parts 1 and 2 to answer the following questions.

a. The statistic that is the average of the test scores is just a sample mean (for a sample of size 2 in Part 1 and for a sample of size 5 in Part 2). How do the sampling distributions of mean2 and mean5 compare to what is expected based on the general properties of the $\bar{x}$ distribution given in Section 8.2? Explain.

b. Based on the three distributions from Part 1, for a two-time test taker, describe the advantage of using the maximum score compared to using either the average score or the most recent score.

c. Now consider the approximate sampling distributions of the maximum score for two-time and for five-time test takers. How do these two distributions compare?

d. Does a student who takes the exam five times have a big advantage over a student of equal ability who takes the exam only twice if the maximum score is used for college admission decisions? Explain.

e. If you were writing admission procedures for a selective university, would you recommend using the maximum test score, the average test score, or the most recent test score in making admission decisions? Write a paragraph explaining your choice.

Summary of Key Concepts and Formulas

Term or Formula	Comment
Statistic	Any quantity whose value is computed from sample data.
Sampling distribution	The probability distribution of a statistic: The sampling distribution describes the long-run behavior of the statistic.
Sampling distribution of $\bar{x}$	The probability distribution of the sample mean $\bar{x}$ based on a random sample of size n. Properties of the $\bar{x}$ sampling distribution: $\mu_{\bar{x}} = \mu$ and $\sigma_{\bar{x}} = \dfrac{\sigma}{\sqrt{n}}$ (where μ and σ are the population mean and standard deviation, respectively). In addition, when the population distribution is normal or the sample size is large, the sampling distribution of $\bar{x}$ is (approximately) normal.
Central Limit Theorem	This important theorem states that when n is sufficiently large, the $\bar{x}$ distribution will be approximately normal. The standard rule of thumb is that the theorem can safely be applied when n exceeds 30.

Term or Formula	Comment
Sampling distribution of p	The probability distribution of the sample proportion p, based on a random sample of size n. When the sample size is sufficiently large, the sampling distribution of p is approximately normal, with $\mu_p = \pi$ and $\sigma_p = \sqrt{\dfrac{\pi(1 - \pi)}{n}}$ where π is the value of the population proportion.

Chapter Review Exercises 8.32–8.37

Turn to the Expanded Answer Section for answers not shown next to exercises.

CENGAGENOW™ Know exactly what to study! Take a pre-test and receive your Personalized Learning Plan.

8.32 The nicotine content in a single cigarette of a particular brand has a distribution with mean 0.8 mg and standard deviation 0.1 mg. If 100 of these cigarettes are analyzed, what is the probability that the resulting sample mean nicotine content will be less than 0.79? less than 0.77?   0.1587, 0.0013

8.33 Let $x_1, x_2, \ldots, x_{100}$ denote the actual net weights (in pounds) of 100 randomly selected bags of fertilizer. Suppose that the weight of a randomly selected bag has a distribution with mean 50 lb and variance 1 lb^2. Let $\bar{x}$ be the sample mean weight ($n = 100$).
a. Describe the sampling distribution of $\bar{x}$.
b. What is the probability that the sample mean is between 49.75 lb and 50.25 lb?   0.9876
c. What is the probability that the sample mean is less than 50 lb?   0.5

8.34 Suppose that 20% of the subscribers of a cable television company watch the shopping channel at least once a week. The cable company is trying to decide whether to replace this channel with a new local station. A survey of 100 subscribers will be undertaken. The cable company has decided to keep the shopping channel if the sample proportion is greater than .25. What is the approximate probability that the cable company will keep the shopping channel, even though the true proportion who watch it is only .20?   0.1056

8.35 Water permeability of concrete can be measured by letting water flow across the surface and determining the amount lost (in inches per hour). Suppose that the permeability index x for a randomly selected concrete specimen of a particular type is normally distributed with mean value 1000 and standard deviation 150.
a. How likely is it that a single randomly selected specimen will have a permeability index between 850 and 1300?   0.8185
b. If the permeability index is to be determined for each specimen in a random sample of size 10, how likely is it that the sample average permeability index will be between 950 and 1100? between 850 and 1300?   0.8357, 0.9992

8.36 *Newsweek* (November 23, 1992) reported that 40% of all U.S. employees participate in "self-insurance" health plans ($\pi = .40$).
a. In a random sample of 100 employees, what is the approximate probability that at least half of those in the sample participate in such a plan?   0.0207
b. Suppose you were told that at least 60 of the 100 employees in a sample from your state participated in such a plan. Would you think $\pi = .40$ for your state? Explain.

8.37 The amount of money spent by a customer at a discount store has a mean of $100 and a standard deviation of $30. What is the probability that a randomly selected group of 50 shoppers will spend a total of more than $5300? (Hint: The total will be more than $5300 when the sample average exceeds what value?)   0.0786

Personal Tutor
Do you need a live tutor for homework problems?

CENGAGENOW™
Are you ready? Take your exam-prep post-test now.

Bold exercises answered in back   ● Data set available online but not required   ▼ Video solution available

Graphing Calculator Explorations

Exploration 8.1 Sampling Distribution of the Mean

The characteristics of the sampling distribution of a statistic are important for drawing inferences. Because the sample mean is widely used, its sampling distribution is of particular interest. The Central Limit Theorem guarantees that as long as n is large enough, the sampling distribution of the sample mean is approximately normal even though the parent population may not be normal. We relied on the results of simulations to illustrate this property of the sampling distribution of $\bar{x}$. The graphing calculator can be used to simulate the process of sampling from a large population. We have already discussed most of the calculator commands needed to perform such a simulation.

Although simulations are generally done on a very large scale, we operate on a very small scale in this calculator exploration. Our methods are suitable for performing a small number of trials of a simulation and then combining the results over a whole statistics class. To go further into simulation of sampling distributions would require programming your calculator (more on this later) or using a computer with appropriate software.

The process of selecting a random sample from a very large population is generally modeled in the following manner:

1. Construct a mathematical description—a probability distribution—of the population distribution of interest.
2. Generate a random sample from the specified distribution.
3. Calculate the statistic of interest (here, the sample mean) for the simulated random sample.
4. Repeat Steps 2 and 3 a large number of times.
5. Construct a distribution of the resulting values of the sample statistic.

We illustrate how this might be done using a graphing calculator. First, we need a population to sample from. A convenient population, though not a common one, is the uniform distribution from 0 to 1. (This distribution is convenient for the calculator, as it is easy to use the *rand* function to generate the random sample.)

Now we need to sample from this population—that is, to generate random selections from a uniform population distribution. We do this by using the seq command and storing the results in a list, as discussed in Graphing Calculator Exploration 6.4. If you are following along with calculator in hand, you should note a couple of things. First, the number of random numbers you can generate depends on both your calculator's capabilities and the amount of memory your calculator has left to use. For this example we use $n = 500$. If your calculator gives you an error message, try a number smaller than 500 to check that you have the calculator syntax correct. If your calculator works for a small value of n but not a large value, the large value may exceed a maximum list size or you may have already stored many programs, pictures, and/or data in your calculator and may not have enough memory left to handle the requested computation. Second, the seq command *can* take a while to complete. Do not be worried if your calculator's attention seems to be wandering.

Here is a generic statement to generate 500 random numbers and store them in a list:

seq(*rand*, X, 1, 500) → List1

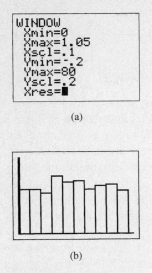

(a)

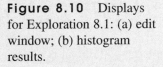

(b)

Figure 8.10 Displays for Exploration 8.1: (a) edit window; (b) histogram results.

Now that we have generated a sample of size 500, let's check our results by making a histogram. If we have been successful, we should see a distribution that mirrors the population—that is, a fairly uniform distribution of results. *It is important to check this by displaying your simulated sampling results!* Verifying these results is the equivalent of checking to see that you have a "right answer so far." Although you should expect some fluctuation from a perfect match—we are, after all, sampling—you should not get seriously distorted results with an *n* as large as 500. Our edit window and histogram results are displayed in Figure 8.10. Pay particular attention to the Xmin and Xscl values in Figure 8.10(a). (Recall from an earlier calculator exploration that these values determine the class size for a histogram.) We want to partition the interval from 0 to 1 into 10 equal intervals, and other choices for these values will give you a graph that is difficult to interpret. From Figure 8.10(b), the sample appears to be representative of the population from which we are sampling.

Now we calculate the sample mean of the data in List1. Our result is $\bar{x} =$ 0.50223. This sample mean represents *one trial* of our simulation. To do a respectable simulation, we would have to repeat the preceding commands a large number of times. This single mean from our one trial is what we meant earlier by a very small-scale simulation. Even on this small scale, however, combining your results with similar results from your classmates can generate good simulations. Obtaining 10 trials each from 25 students is easily accomplished in a class period, and 250 trials is nothing to sneeze at!

We should note that this small-scale restriction of generating sample means one at a time is easily overcome *if* you have someone in your class who knows how to do a little programming on the calculator. It is even possible that if you have *no* experience programming a calculator, you could do a simulation. Having spent many frustrating hours programming, we cannot in good conscience suggest that you learn to program your calculator just for simulations. However, if you want to explore simulations on a medium-size scale and if you have someone to help you set up the accompanying program, you could be doing some simulations quickly. As written, the following program performs 500 trials of the simulation we have been discussing, using a sample size of 50. Remember, syntax will vary among calculators, but something close to this program will perform Steps 2, 3, and 4 of the simulation and store the results in List2. Consult your friendly local calculator gurus for details.

```
For (X, 1, 500)
seq(rand, X, 1, 50) → List1
mean(List1) → List2(X)
End
```

In our example, we generated random numbers from a population uniformly distributed from 0 to 1. Sampling distributions of the mean from other populations can also be considered; the populations available are limited only by your random number generation capability. Some populations might be more appropriately modeled as normal or binomial. Some calculators have built-in functions to generate the random values you would get if you were sampling from a normal or binomial population. Here are some other possible sampling simulations:

From a population of integers, 1 to 10:

seq(int(*rand* * 10) +1, X, 1, 500) → List1

or, if there is a built-in function (keystrokes denoted here by randInt):

seq(randInt, X, 1, 500) → List1

From a normal population (assuming a built-in function, keystrokes denoted by randNormal):

seq(randNormal, X, 1, 500) → List1

Your calculator may have different built-in functions or different syntax, but your exploration of the sampling distribution of the mean can begin as soon as you check that manual!

Exploration 8.2 Sampling Distribution of the Sample Proportion

We saw in the previous exploration how to simulate sampling from a population to study the sampling distribution of the sample mean. In this exploration we consider a similar process for studying the sampling distribution of the sample proportion. To review a bit, the process of random sampling from a large population is modeled as follows:

1. Construct a mathematical description—a probability distribution—of the population.
2. Generate a random sample from this population density function.
3. Calculate the statistic of interest (here, the sample proportion) for the simulated random sample.
4. Repeat Steps 2 and 3 a large number of times.
5. Construct a distribution of the resulting values of the sample proportion.

To model the sampling distribution of a sample proportion, we conceive of a population consisting of a set of 0's and 1's, where the proportion of 1's in the population is equal to the given population proportion π. The process of simulating a random sample consists of generating a fixed number n of 0's and 1's and finding the proportion of 1's in the sample. This value is the simulated sample proportion. For example, suppose we are simulating sampling from a population with a population proportion $\pi = .75$ using a sample size of $n = 12$. One simulated sample might look like this:

1 1 1 0 1 0 0 1 1 1 1 0

We note that 8 of the 12 values are 1's, giving a sample proportion of $p = .67$. How might we accomplish this simulation on our calculator? The easiest method is to capitalize on a built-in function to generate outcomes that are from a binomial distribution. If your calculator has such a function, it is probably somewhere in a Math or Prob menu—look in your manual's index for something like RandBin or RandBinom. We use the generic name RandBin to refer to this built-in calculator function. This function needs two numbers to generate a binomial random variable: the sample size and the population proportion. For our example, you might enter RandBin(12, .75) or RandBin(.75, 12), depending on the function's syntax for your calculator.

On your calculator, find the built-in function and enter the appropriate syntax with the parameters $n = 12$ and $\pi = .75$. When you do this, the output might be 8.

We simulate the process of sampling and calculating a sample proportion by using the seq function, as we did earlier, but now we convert the outcomes to proportions:

seq(RandBin(12, .75)/12, X, 1, 500) → List1

The distribution of our results and summary values are shown in Figure 8.11. Note that the simulated results are close to the theoretical values for the mean (0.75) and standard deviation (0.125) for the sampling distribution of p.

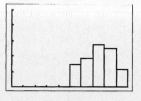

(a)

(b)

Figure 8.11 Displays for Exploration 8.2: (a) histogram; (b) summary values.

If your calculator does not have a RandBin built-in function, you can still simulate random sampling from a population by using a slightly modified program from the one used for the sampling distribution of the mean. Be forewarned; this slight modification involves some mathematics, but it is fairly easy to implement.

The use of the same program as before capitalizes on the fact that the sample proportion is actually the mean of the 0's and 1's in the sample. The programming goal is to generate 12 random numbers between 0 and 1 and then to convert them to 12 0's and 1's before calculating the mean. That's where the "slight modification" rears its ugly head! The hitch in the plan is to find a simple (for the calculator) function that translates random numbers on the interval from 0 to 1 into 0's and 1's with the right probability. We use two common functions to make this happen: the absolute value function, usually called abs on calculators, and the greatest integer function, usually named Int or possibly Floor. The mathematical function that will perform the translation is

$Y = \text{abs}(\text{int}(rand - \pi))$, where π is the population proportion

For our example, this would be

$Y = \text{abs}(\text{int}(rand - .75))$

To generate a random sample, substitute the expression for *rand* appropriate to your calculator in the seq function:

```
For (X, 1, 500)
seq(abs(int(rand .75)), X, 1, 12) → List1
mean(List1) → List2(X)
End
```

Before you use the program to study the sampling distribution of the sample proportion, be sure to make a couple of trial runs and check the results with the theoretical expectations!

TEACHING TIPS

Intent of this chapter: After completing this chapter, students should be able to (1) explain why unbiasedness and small standard error are desirable properties of statistics used to produce a point estimate, (2) interpret a confidence level, (3) compute and interpret a confidence interval for a population proportion (π), (4) compute and interpret a confidence interval for a population mean (μ), (5) calculate the sample size necessary to achieve a given bound on error of estimation for a specified confidence level, and (6) identify how changes in sample size or confidence level (and in the value of p when constructing a confidence interval for a proportion) affect the width of a confidence interval.

A PowerPoint® Lecture is available on the *Instructor's Resource Binder* CD.

■ **Section 9.1** Suppose we wanted to know the value of a population proportion, such as the proportion of registered voters that plan to vote in the next election, or of a population mean, such as the average number of hours teenagers "text" per week. These population parameters are almost always unknown. Therefore, we must estimate the value of these parameters. If a sample is randomly selected from the population, the sample mean or sample proportion could be computed. This single value is a *point estimate* for the population parameter. Although this is an easy approach, it has limitations. Think back to the sampling distributions in Chapter 8. Due to sampling variability, the calculated statistic may or may not be close to the true parameter. We typically don't know where our calculated statistic falls in the distribution of all possible values for that statistic. In Section 9.2, we will learn a more informative approach to estimating the population parameter, which is to compute a confidence interval.

When using a statistic (point estimate) to estimate a parameter, it is helpful to know if the statistic is biased or unbiased. Remind students of the "Random Rectangle" activity in Chapter 2. The first step was for students to use their judgment to select five rectangles that best represent the population of rectangles. When the sample means ($\bar{x}$) were plotted on a dotplot, the resulting distribution isn't usually centered at the true mean size of the rectangles (parameter)—it is usually centered at a value that was noticeably higher than the true mean. Using individual judgment to select rectangles biased the results. When other sampling methods (SRS, stratified random sampling, systematic sampling, and cluster sampling) were employed, the distributions of the sample means were centered at about the true mean of the rectangles. Refer also to the sampling distributions of $\bar{x}$ and sampling distributions of p from the "fish pond" and "club" activities in Chapter 8. Recall that the mean of all the sample means equaled the population mean. Likewise, the mean of all the sample proportions equaled the population proportion. Therefore, for random sampling the sample mean, $\bar{x}$, and sample proportion, p, are *unbiased* estimators respectively of a population mean and of a population proportion. (See the definition on page 478.)

Suggested Assignment: Exercises 9.1, 9.2, 9.4, 9.5, 9.6, 9.8, 9.9

■ **Section 9.2** Begin this section with the following activity. Ask students to rate their level of confidence (0% to 100%) in guessing your age within 10 years. Note: Within 10 years provides a 20-year interval. Ask students to rate their level of confidence (0% to 100%) in guessing your age within 5 years. Within 1 year. What happened to their confidence level as the width of the interval decreased? *The confidence level decreased.* (Thanks to Michael Legacy for this activity.) This relationship is important with confidence intervals. *Larger* confidence levels produce *wider* confidence intervals. This is sometimes counterintuitive for students.

The purpose of a confidence interval is to provide an *interval* of plausible values for a population parameter. The confidence level refers to the method used to produce the intervals. The confidence level is the proportion of *all* possible samples that will result in an interval that contains the parameter. For example, a 95% confidence level means that of all possible samples that might be used to compute a confidence interval using this method, 95% of the intervals will contain the population parameter. In Section 8.3, students flipped a coin 20 times and calculated the proportion of heads. Using this sample proportion, have students calculate 95% confidence intervals. Next, students should plot their confidence intervals to create a graph similar to Figure 9.4 on page 487. Do all the intervals contain the parameter 0.5? If not, why not? Also, Activity 9.2 will allow students to investigate the relationship between the confidence level and the proportion of intervals that contain the parameter, in the long run.

The discussion on pages 484 and 485 develops the formula used to compute a confidence interval for a population proportion (π). When computing a confidence interval, students should always *check the relevant assumptions* to be sure that it is appropriate to be using the procedure and students should always *give an interpretation* of the computed interval in context. The tan box on page 488 provides the assumptions that should be checked and the formula for the confidence interval for π. Graphing Calculator Exploration 8.1 (on page 521) provides instructions for using the TI calculator to compute the confidence interval.

The critical value (z^*) for the confidence level of 95% is the **upper** z-score in the standard normal curve so that the middle 95% of the area under the curve falls between $-z^*$ and z^*. This value can be found by locating the z-score with an area of 0.975 to its left. Notice the curve in Figure 9.2 on page 484. This value can also be found using the invNorm function on the TI calculator. Type "invNorm(.975)" to find the critical value for a 95% confidence level. (Note: The same area that is used on the chart is used in the calculator.)

If we want to calculate the standard deviation of the sampling distribution of p, we use the formula $\sqrt{\dfrac{\pi(1-\pi)}{n}}$. Typically, we do not know the population proportion, so we will use the sample proportion as an estimate for π, $\sqrt{\dfrac{p(1-p)}{n}}$. The estimate of the standard deviation of the statistic is known as the standard error. Also, the portion of the confidence interval equation, (critical value)(standard deviation of the statistic), is the *bound on the error of estimation* or *margin of error*. This portion of the confidence interval equation is used in the calculation of the sample size needed to achieve a given margin of error. (See Example 9.6.) Notice that π is unknown and since we haven't selected a sample yet, the sample proportion (p) is also unknown. What value should we use for π in this calculation? The value that will result in the largest standard deviation, and therefore the largest sample size, is $\pi = 0.5$. So the rule of thumb when determining sample size for estimating a proportion is if you have a good estimate of π (such as the value of p from a pilot study), use that value and if you don't know π and do not have a good estimate, substitute 0.5.

Suggested Assignment: Exercises 9.11, 9.12, 9.13, 9.14, 9.16, 9.17, 9.19, 9.21, 9.29

Activity: Activity 9.2: An Alternative Confidence Interval for a Population Proportion (p. 515 and *Activities Manual*, p. 80) ★, Activity 9.3: Verifying Signatures on a Recall Petition (p. 516 and *Activities Manual*, p. 82)★, "Chapter 9—Confidence Interval for π" (*Instructor's Resource Binder* & CD, Chapter 9 Activities Worksheets), "Chapter 9—1-PropZInt"

★ Kathy's personal favorite

(*Instructor's Resource Binder* & CD, Chapter 9 Activities Worksheets), and/or "Chapter 9—Sample Size for Estimating π" (*Instructor's Resource Binder* & CD, Chapter 9 Activities Worksheets)

Assessment: Chapter 9 Quiz 1 (*Test Bank* and *Instructor's Resource Binder* CD)

■ Section 9.3

The first part of this section discusses the z confidence interval for μ, used when the population standard deviation, σ, is known (which is rarely true). Even though in practice the z confidence interval for μ is rarely used, the z confidence interval provides a nice transition from confidence intervals for π to confidence intervals for μ. Refer students to the activity in Section 8.2 in which squares were randomly selected from a population. Using the sample means from this activity, students can compute 95% z confidence intervals for the population mean and plot the confidence intervals on a graph to reinforce the concept that different samples will produce different sample means and different confidence intervals.

Because the population standard deviation is not usually known, it is much more common to use the t confidence interval to estimate μ, where the critical value is based on a t-distribution. William Gosset developed t-distributions when he was working to maintain the quality of beer for Guinness Brewing Company. Gosset found that small sample sizes tend to have the distribution with more variation than the standard normal distribution. t-distributions are bell-shaped, centered at zero, and distinguishable from one another by a number called degrees of freedom. For the one sample t confidence interval, df $= n - 1$. A graphing calculator can be used to compare the standard normal distribution to different t-distributions. The first example below plots the curves for the standard normal distribution and a t-distribution with 2 df.

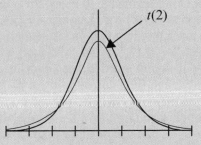

Notice that the t-distribution is centered at 0 but has more variability than the standard normal distribution. Next, plot the curves for the standard normal distribution, a t-distribution with 2 df, and a t-distribution with 5 df.

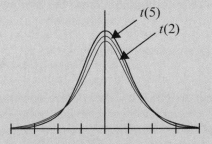

The lowest curve is the t-distribution with 2 df and the middle curve is the t-distribution with 5 df. What happens as the degrees of freedom increase? *The t-distributions approach*

the standard normal distribution. For what number of degrees of freedom would the *t*-distribution very closely approximate the standard normal distribution? (Students will guess many different amounts—caution them to think it through.) Where does the 30 df come from? *The justification is based on the Central Limit Theorem.*

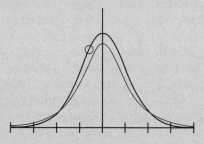

Notice that the *t*-distribution with 30 df closely approximates the standard normal distribution. (There appears to be only two curves, but "Y3=" is plotting a *t*-curve that is almost the same as the normal curve.) However, the *t*-distribution will always have a larger variance than the standard normal distribution. (See Figure 9.6 on page 498.)

The *t* confidence interval for μ is used when the population standard deviation (σ) is unknown. The critical value for a 95% *t* confidence interval with a sample size of 10 is the **upper** *t*-value in the *t*-distribution with 9 df so that the middle 95% of the area under the curve falls between –*t** and *t**. This critical value can be located in the table on the inside back cover of the textbook. Notice that not all degrees of freedom are listed. The table jumps from 30 df to 40 df, and so on. When using the table to find critical values, use the "Price is Right" rule—that is, use the closest df without going over. For example, if the df = 36, use the row for the 30 df. Some graphing calculators will also provide the *t* critical value using this command: inv*t*(area,df).

The tan box on page 500 states the assumptions that need to be checked to verify that the *t*-procedure is appropriate and provides the formula for the *t* confidence interval for μ. Graphing Calculator Exploration 9.2 (on page 522) provides instructions for the graphing calculator.

The last portion of this section explains how to calculate a sample size needed in order to achieve a specified margin of error with a specified level of confidence. Note that an estimate of the population standard deviation is needed to complete the sample size calculation. When computing sample size, be sure to always *round up* to the next whole number.

Suggested Assignment: Exercises 9.30, 9.31, 9.32, 9.33, 9.34, 9.36, 9.40, 9.41, 9.43, 9.48, 9.49

Activity: Activity 9.1: Getting a Feel for Confidence Level (p. 514 and *Activities Manual*, p. 76)★, Bonus Activity 9.5—Water Stains (*Activities Manual*, p. 84)★, "Chapter 9—Confidence Interval for μ" (*Instructor's Resource Binder* & CD, Chapter 9 Activities Worksheets), and/or "Chapter 9—TInterval" (*Instructor's Resource Binder* & CD, Chapter 9 Activities Worksheets)

Applet: http://lstat.kuleuven.be/java/index.htm

Assessment: Chapter 9 Quiz 2 (*Test Bank* and *Instructor's Resource Binder* CD)

★ Kathy's personal favorite

■ **Section 9.4**

This section provides excellent suggestions on communication and on the interpretation of confidence intervals. Students should also pay close attention to "A Word to the Wise: Cautions and Limitations."

Students often confuse the terms accuracy and precision. In statistics, these terms have very specific meanings. *Accuracy* refers to where the sampling distribution is centered. If a sampling distribution of a statistic is centered at the true parameter, the statistic is considered to be unbiased and is often described as accurate, meaning it doesn't consistently under- or overestimate. *Precision* refers to the variability of the sampling distribution. An unbiased statistic whose sampling distribution has a smaller standard error is more precise (and will have a narrower confidence interval) than an unbiased statistic with a larger standard error.

Suggested Review Assignment: Exercises 9.51, 9.52, 9.54, 9.59, 9.61, 9.65, 9.67, 9.73

Activity: Activity 9.4: A Meaningful Paragraph (p. 516 and *Activities Manual*, p. 83)

Assessment: Chapter 9 Concept Quiz (*Test Bank* and *Instructor's Resource Binder* CD) and Chapter 9 Test (*Test Bank* and *Instructor's Resource Binder* CD)

Chapter 9

© George Hall/Corbis

Estimation Using a Single Sample

Affirmative action in university admissions is a controversial topic. Some believe that affirmative action programs are no longer needed, whereas others argue that using race and ethnicity as factors in university admissions is necessary to achieve diverse student populations. To assess public opinion on this issue, investigators conducted a survey of 1013 randomly selected U.S. adults. The results are summarized in the article "Poll Finds Sharp Split on Affirmative Action" (*San Luis Obispo Tribune*, March 8, 2003). The investigators wanted to use the survey data to estimate the true proportion of U.S. adults who believed that programs that give "advantages and preferences to Blacks, Hispanics, and other minorities in hiring, promotions, and college admissions" should be continued. The methods introduced in this chapter will be used to produce the desired estimate. Because the estimate is based only on a sample rather than on a census of all U.S. adults, it is important that this estimate be constructed in a way that also conveys information about the anticipated accuracy.

The objective of inferential statistics is to use sample data to decrease our uncertainty about some characteristic of the corresponding population, such as a population mean μ or a population proportion π. One way to accomplish this uses the sample data to arrive at a single number that represents a plausible value for the characteristic of interest. Alternatively, an entire range of plausible values for the characteristic can be reported. These two estimation techniques, *point estimation* and *interval estimation*, are introduced in this chapter.

9.1 Point Estimation

The simplest approach to estimating a population characteristic involves using sample data to compute a single number that can be regarded as a plausible value of the characteristic. For example, sample data might suggest that 1000 hr is a plausible value for μ, the true mean lifetime for lightbulbs of a particular brand. In a different setting, a sample survey of students at a particular university might lead to the statement that .41 is a plausible value for π, the true proportion of students who favor a fee for recreational facilities.

> **DEFINITION**
>
> A **point estimate** of a population characteristic is a single number that is based on sample data and represents a plausible value of the characteristic.

In the examples just given, 1000 is a point estimate of μ and .41 is a point estimate of π. The adjective *point* reflects the fact that the estimate corresponds to a single point on the number line.

A point estimate is obtained by first selecting an appropriate statistic. The estimate is then the value of the statistic for the given sample. For example, the computed value of the sample mean provides a point estimate of a population mean μ.

Example 9.1   Support for Affirmative Action

One of the purposes of the survey on affirmative action described in the chapter introduction was to estimate the proportion of the U.S. population who believe that affirmative action programs should be continued. The article reported that 537 of the 1013 people surveyed believed that affirmative action programs should be continued. Let's use this information to estimate π, where π is the true proportion of all U.S. adults who favor continuing affirmative action programs. With success identified as a person who believes that affirmative action programs should continue, π is then just the population proportion of successes. The statistic

$$p = \frac{\text{number of successes in the sample}}{n}$$

which is the sample proportion of successes, is an obvious choice for obtaining a point estimate of π. Based on the reported information, the point estimate of π is

$$p = \frac{537}{1013} = .530$$

That is, based on this random sample, we estimate that 53% of the adults in the United States believe that affirmative action programs should be continued.

■

For purposes of estimating a population proportion π, there is no obvious alternative to the statistic p. In other situations, such as the one illustrated in Example 9.2, there may be several statistics that can be used to obtain an estimate.

Example 9.2   Internet Use by College Students

● The article "Online Extracurricular Activity" (*USA Today*, March 13, 2000) reported the results of a study of college students conducted by a polling organization called The Student Monitor. One aspect of computer use examined in this study was the number of hours per week spent on the Internet. Suppose that the following observations represent the number of Internet hours per week reported by 20 college students (these data are compatible with summary values given in the article):

| 4.00 | 5.00 | 5.00 | 5.25 | 5.50 | 6.25 | 6.25 | 6.50 | 6.50 | 7.00 |
| 7.25 | 7.75 | 8.00 | 8.00 | 8.00 | 8.25 | 8.50 | 8.50 | 9.50 | 10.50 |

A dotplot of the data is shown here:

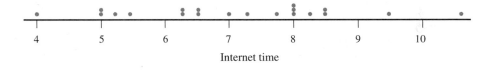

Internet time

If a point estimate of μ, the true mean Internet time per week for college students, is desired, an obvious choice of a statistic for estimating μ is the sample mean $\bar{x}$. However, there are other possibilities. We might consider using a trimmed mean or even the sample median, because the data set exhibits some symmetry. (If the corresponding population distribution is symmetric, the population mean μ and the population median are equal).

The three statistics and the resulting estimates of μ calculated from the data are

$$\text{sample mean} = \bar{x} = \frac{\sum x}{n} = \frac{141.50}{20} = 7.075$$

$$\text{sample median} = \frac{7.0 + 7.25}{2} = 7.125$$

$$10\% \text{ trimmed mean} = \left(\begin{array}{c} \text{average of middle} \\ \text{16 observations} \end{array} \right) = \frac{112.5}{16} = 7.031$$

The estimates of the mean Internet time per week for college students differ somewhat from one another. The choice from among them should depend on which statistic tends, on average, to produce an estimate closest to the true value of μ. The following subsection discusses criteria for choosing among competing statistics.

■

▪ Choosing a Statistic for Computing an Estimate ....

The point of Example 9.2 is that more than one statistic may be reasonable to use to obtain a point estimate of a specified population characteristic. Loosely speaking, the statistic used should be one that tends to yield an accurate estimate—that is, an estimate close to the value of the population characteristic. Information about the accuracy of estimation for a particular statistic is provided by the statistic's sampling distribution. Figure 9.1 displays the sampling distributions of three different statistics. The value of the population characteristic, which we refer to as the *true value*, is marked on the measurement axis.

The distribution in Figure 9.1(a) is that of a statistic unlikely to yield an estimate close to the true value. The distribution is centered to the right of the true value, making it very likely that an estimate (a value of the statistic for a particular sample) will be larger than the true value. If this statistic is used to compute an estimate based on a first sample, then another estimate based on a second sample, and another estimate based on a third sample, and so on, the long-run average value of these estimates will exceed the true value.

Figure 9.1 Sampling distributions of three different statistics for estimating a population characteristic.

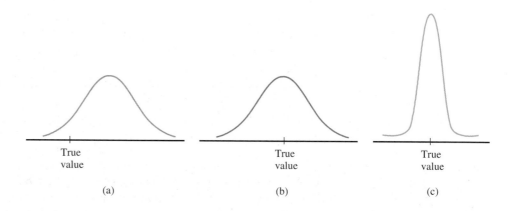

The sampling distribution of Figure 9.1(b) is centered at the true value. Thus, although one estimate may be smaller than the true value and another may be larger, when this statistic is used many times over with different samples, there will be no long-run tendency to over- or underestimate the true value. Note that even though the sampling distribution is correctly centered, it spreads out quite a bit about the true value. Because of this, some estimates resulting from the use of this statistic will be far above or far below the true value, even though there is no systematic tendency to underestimate or overestimate the true value.

In contrast, the mean value of the statistic with the distribution shown in Figure 9.1(c) is equal to the true value of the population characteristic (implying no systematic error in estimation), and the statistic's standard deviation is relatively small. Estimates based on this third statistic will almost always be quite close to the true value—certainly more often than estimates resulting from the statistic with the sampling distribution shown in Figure 9.1(b).

DEFINITION

A statistic whose mean value is equal to the value of the population characteristic being estimated is said to be an **unbiased statistic**. A statistic that is not unbiased is said to be **biased**.

As an example of a statistic that is biased, consider using the sample range as an estimate of the population range. Because the range of a population is defined as the difference between the largest value in the population and the smallest value, the range for a sample tends to underestimate the population range. This is true because the largest value in a sample must be less than or equal to the largest value in the population and the smallest sample value must be greater than or equal to the smallest value in the population. The sample range equals the population range *only* if the sample includes both the largest and the smallest values in the population; in all other instances, the sample range is smaller than the population range. Thus, $\mu_{\text{sample range}} <$ population range, implying bias.

Let $x_1, x_2, ..., x_n$ represent the values in a random sample. One of the general results concerning the sampling distribution of $\bar{x}$, the sample mean, is that $\mu_{\bar{x}} = \mu$. This result says that the $\bar{x}$ values from all possible random samples of size n center around μ, the population mean. For example, if $\mu = 100$, the distribution is centered at 100, whereas if $\mu = 5200$, then the distribution is centered at 5200. Therefore, $\bar{x}$ is an unbiased statistic for estimating μ. Similarly, because the sampling distribution of p is centered at π, it follows that p is an unbiased statistic for estimating a population proportion.

Using an unbiased statistic that also has a small standard deviation ensures that there will be no systematic tendency to under- or overestimate the value of the population characteristic *and* that estimates will almost always be relatively close to the true value.

Given a choice between several unbiased statistics that could be used for estimating a population characteristic, the best statistic to use is the one with the smallest standard deviation.

Consider the problem of estimating a population mean, μ. The obvious choice of statistic for obtaining a point estimate of μ is the sample mean, $\bar{x}$, an unbiased statistic for this purpose. However, when the population distribution is symmetric, $\bar{x}$ is not the only choice. Other unbiased statistics for estimating μ in this case include the sample median and any trimmed mean (with the same number of observations trimmed from each end of the ordered sample). Which statistic should be used? The following facts may be helpful in making a choice.

1. If the population distribution is normal, then $\bar{x}$ has a smaller standard deviation than any other unbiased statistic for estimating μ. However, in this case, a trimmed mean with a small trimming percentage (such as 10%) performs almost as well as $\bar{x}$.
2. When the population distribution is symmetric with heavy tails compared to the normal curve, a trimmed mean is a better statistic than $\bar{x}$ for estimating μ.

When the population distribution is unquestionably normal, the choice is clear: Use $\bar{x}$ to estimate μ. However, with a heavy-tailed distribution, a trimmed mean gives protection against one or two outliers in the sample that might otherwise have a large effect on the value of the estimate.

Now consider estimating another population characteristic, the population variance σ^2. The sample variance

$$s^2 = \frac{\sum(x - \bar{x})^2}{n - 1}$$

is a good choice for obtaining a point estimate of the population variance σ^2. It can be shown that s^2 is an unbiased statistic for estimating σ^2; that is, whatever the value of σ^2, the sampling distribution of s^2 is centered at that value. It is precisely for this reason—to obtain an unbiased statistic—that the divisor $(n - 1)$ is used. An alternative statistic is the average squared deviation

$$\frac{\sum(x - \bar{x})^2}{n}$$

which one might think has a more natural divisor than s^2. However, the average squared deviation is biased, with its values tending to be smaller, on average, than σ^2.

....

Example 9.3   Airborne Times for Flight 448

● The Bureau of Transportation Statistics provides data on U.S. airline flights. The airborne times (in minutes) for United Airlines flight 448 from Albuquerque to Denver on 10 randomly selected days between January 1, 2003, and March 31, 2003, are

$$57 \quad 54 \quad 55 \quad 51 \quad 56 \quad 48 \quad 52 \quad 51 \quad 59 \quad 59$$

For these data $\sum x = 542$, $\sum x^2 = 29{,}498$, $n = 10$, and

$$\sum(x - \bar{x})^2 = \sum x^2 - \frac{(\sum x)^2}{n}$$

$$= 29{,}498 - \frac{(542)^2}{10}$$

$$= 121.6$$

Let σ^2 denote the true variance in airborne time for flight 448. Using the sample variance s^2 to provide a point estimate of σ^2 yields

$$s^2 = \frac{\sum(x - \bar{x})^2}{n - 1} = \frac{121.6}{9} = 13.51$$

Using the average squared deviation (with divisor $n = 10$), the resulting point estimate is

$$\frac{\sum(x - \bar{x})^2}{n} = \frac{121.6}{10} = 12.16$$

Because s^2 is an unbiased statistic for estimating σ^2, most statisticians would recommend using the point estimate 13.51.

■

An obvious choice of a statistic for estimating the population standard deviation σ is the sample standard deviation s. For the data given in Example 9.3,

$$s = \sqrt{13.51} = 3.68$$

Unfortunately, the fact that s^2 is an unbiased statistic for estimating σ^2 does not imply that s is an unbiased statistic for estimating σ. The sample standard deviation tends to underestimate slightly the true value of σ. However, unbiasedness is not the only criterion by which a statistic can be judged, and there are other good reasons for using s to estimate σ. In what follows, whenever we need to estimate σ based on a single random sample, we use the statistic s to obtain a point estimate.

● Data set available online

Exercises 9.1–9.10

Turn to the Expanded Answer Section for answers not shown next to exercises.

9.1 ▼ Three different statistics are being considered for estimating a population characteristic. The sampling distributions of the three statistics are shown in the following illustration:

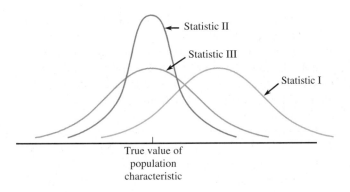

Which statistic would you recommend? Explain your choice.

9.2 Why is an unbiased statistic generally preferred over a biased statistic for estimating a population characteristic? Does unbiasedness alone guarantee that the estimate will be close to the true value? Explain. Under what circumstances might you choose a biased statistic over an unbiased statistic if two statistics are available for estimating a population characteristic?

9.3 Consumption of fast food is a topic of interest to researchers in the field of nutrition. The article "Effects of Fast-Food Consumption on Energy Intake and Diet Quality Among Children" (*Pediatrics* [2004]: 112–118) reported that 1720 of those in a random sample of 6212 U.S. children indicated that on a typical day they ate fast food. Estimate π, the proportion of children in the U.S. who eat fast food on a typical day.   0.2769

9.4 ● Data consistent with summary quantities in the article referenced in Exercise 9.3 on total calorie consumption on a particular day are given for a sample of children who did not eat fast food on that day and for a sample of children who did eat fast food on that day. Assume that it is reasonable to regard these samples as representative of the population of children in the United States.

No Fast Food

| 2331 | 1918 | 1009 | 1730 | 1469 | 2053 | 2143 | 1981 |
| 1852 | 1777 | 1765 | 1827 | 1648 | 1506 | 2669 | |

Fast Food

| 2523 | 1758 | 934 | 2328 | 2434 | 2267 | 2526 | 1195 |
| 890 | 1511 | 875 | 2207 | 1811 | 1250 | 2117 | |

a. Use the given information to estimate the mean calorie intake for children in the United States on a day when no fast food is consumed.   1845.2

b. Use the given information to estimate the mean calorie intake for children in the United States on a day when fast food is consumed.   1775.1

c. Use the given information to produce estimates of the standard deviations of calorie intake for days when no fast food is consumed and for days when fast food is consumed.   No fast food: $s = 386.346$, fast food: $s = 620.660$

9.5 Each person in a random sample of 20 students at a particular university was asked whether he or she is registered to vote. The responses (R = registered, N = not registered) are given here:

R R N R N N R R N R R R R N R R R N

Use these data to estimate π, the true proportion of all students at the university who are registered to vote.   0.70

9.6 A study reported in *Newsweek* (December 23, 1991) involved a sample of 935 smokers. Each individual received a nicotine patch, which delivers nicotine to the bloodstream but at a much slower rate than cigarettes do. Dosage was decreased to 0 over a 12-week period. Suppose that 245 of the subjects were still not smoking 6 months after treatment (this figure is consistent with information given in the article). Estimate the percentage of all smokers who, when given this treatment, would refrain from smoking for at least 6 months.   0.262

9.7 ● The article "Sensory and Mechanical Assessment of the Quality of Frankfurters" (*Journal of Texture Studies* [1990]: 395–409) reported the following salt content (percentage by weight) for 10 frankfurters:

2.26 2.11 1.64 1.17 1.64 2.36 1.70 2.10 2.19 2.40

a. Use the given data to produce a point estimate of μ, the true mean salt content for frankfurters.   1.957

b. Use the given data to produce a point estimate of σ^2, the variance of salt content for frankfurters. 0.15945

c. Use the given data to produce an estimate of σ, the standard deviation of salt content. Is the statistic you used to produce your estimate unbiased? 0.3993, the statistic used is not an unbiased estimator.

9.8 ● The following data on gross efficiency (ratio of work accomplished per minute to calorie expenditure per minute) for trained endurance cyclists were given in the article "Cycling Efficiency Is Related to the Percentage of Type I Muscle Fibers" (*Medicine and Science in Sports and Exercise* [1992]: 782–88):

18.3	18.9	19.0	20.9	21.4	20.5	20.1	20.1
20.8	20.5	19.9	20.5	20.6	22.1	21.9	21.2
20.5	22.6	22.6					

a. Assuming that the distribution of gross energy in the population of all endurance cyclists is normal, give a point estimate of μ, the population mean gross efficiency.

b. Making no assumptions about the shape of the population distribution, estimate the proportion of all such cyclists whose gross efficiency is at most 20. $p = 0.2105$

9.9 ● A random sample of $n = 12$ four-year-old red pine trees was selected, and the diameter (in inches) of each tree's main stem was measured. The resulting observations are as follows:

11.3	10.7	12.4	15.2	10.1	12.1	16.2	10.5
11.4	11.0	10.7	12.0				

a. Compute a point estimate of σ, the population standard deviation of main stem diameter. What statistic did you use to obtain your estimate? $s = 1.886$

b. Making no assumptions about the shape of the population distribution of diameters, give a point estimate for the population median diameter. What statistic did you use to obtain the estimate? Sample median = 11.35

c. Suppose that the population distribution of diameter is symmetric but with heavier tails than the normal distribution. Give a point estimate of the population mean diameter based on a statistic that gives some protection against the presence of outliers in the sample. What statistic did you use?

d. Suppose that the diameter distribution is normal. Then the 90th percentile of the diameter distribution is $\mu + 1.28\sigma$ (so 90% of all trees have diameters less than this value). Compute a point estimate for this percentile. (Hint: First compute an estimate of μ in this case; then use it along with your estimate of σ from Part (a).) $11.967 + (1.28)(1.886) = 14.31$

9.10 ● A random sample of 10 houses in a particular area, each of which is heated with natural gas, is selected, and the amount of gas (in therms) used during the month of January is determined for each house. The resulting observations are as follows:

103	156	118	89	125	147	122	109	138	99

a. Let μ_J denote the average gas usage during January by all houses in this area. Compute a point estimate of μ_J.

b. Suppose that 10,000 houses in this area use natural gas for heating. Let τ denote the total amount of gas used by all of these houses during January. Estimate τ using the data of Part (a). What statistic did you use in computing your estimate? $10,000(\bar{x}) = 1,206,000$

c. Use the data in Part (a) to estimate π, the proportion of all houses that used at least 100 therms. 0.8

d. Give a point estimate of the population median usage based on the sample of Part (a). Which statistic did you use? Sample median = 120

Bold exercises answered in back   ● Data set available online but not required   ▼ Video solution available

---

9.2 Large-Sample Confidence Interval for a Population Proportion

In Section 9.1 we saw how to use a statistic to produce a point estimate of a population characteristic. The value of a point estimate depends on which sample, out of all the possible samples, happens to be selected. Different samples usually yield different estimates as a result of chance differences from one sample to another. Because of sampling variability, rarely is the point estimate from a sample exactly equal to the true value of the population characteristic. We hope that the chosen statistic produces an estimate that is close, on average, to the true value. Although a point estimate may represent our best single-number guess for the value of the population characteristic, it is not the only plausible value. These considerations suggest the need to indicate in some way a range of plausible values for the population characteristic. A point estimate by itself does not provide this information.

As an alternative to a point estimate, we can report an interval of reasonable values for the population characteristic based on the sample data. For example, we might be confident that for all calls made from AT&T pay phones, the proportion π of calls that are billed to a credit card is in the interval from .53 to .57. The narrowness of this interval implies that we have rather precise information about the value of π. If, with the same high degree of confidence, we could only state that π was between .32 and .74, it would be clear that we had relatively imprecise knowledge of the value of π.

DEFINITION

A **confidence interval (CI)** for a population characteristic is an interval of plausible values for the characteristic. It is constructed so that, with a chosen degree of confidence, the value of the characteristic will be captured between the lower and upper endpoints of the interval.

Associated with each confidence interval is a *confidence level*. The confidence level provides information on how much "confidence" we can have in the *method* used to construct the interval estimate (*not* our confidence in any one particular interval). Usual choices for confidence levels are 90%, 95%, and 99%, although other levels are also possible. If we were to construct a 95% confidence interval using the technique to be described shortly, we would be using a method that is "successful" 95% of the time. That is, if this method was used to generate an interval estimate over and over again with different samples, in the long run 95% of the resulting intervals would capture the true value of the characteristic being estimated. Similarly, a 99% confidence interval is one that is constructed using a method that is, in the long run, successful in capturing the true value of the population characteristic 99% of the time.

DEFINITION

The **confidence level** associated with a confidence interval estimate is the success rate of the *method* used to construct the interval.

Many factors influence the choice of confidence level. These factors will be discussed after we develop the method for constructing confidence intervals. We first consider a large-sample confidence interval for a population proportion π.

Often an investigator wishes to make an inference about the proportion of individuals or objects in a population that possess a particular property of interest. For example, a university administrator might be interested in the proportion of students who prefer a new web-based computer registration system to the previous registration method. In a different setting, a quality control engineer might be concerned about the proportion of defective parts manufactured using a particular process.

Let π be the proportion of the population that possess the property of interest. Previously, we used the sample proportion

$$p = \frac{\text{number in the sample that possess the property of interest}}{n}$$

to calculate a point estimate of π. We can also use p to form a confidence interval for π.

Although a small-sample confidence interval for π can be obtained, our focus is on the large-sample case. The justification for the large-sample interval rests on properties of the sampling distribution of the statistic p:

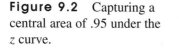

1. The sampling distribution of p is centered at π; that is, $\mu_p = \pi$. Therefore, p is an unbiased statistic for estimating π.

2. The standard deviation of p is $\sigma_p = \sqrt{\dfrac{\pi(1-\pi)}{n}}$

3. As long as n is large ($n\pi \geq 10$ and $n(1-\pi) \geq 10$) and the sample size is less than 10% of the population size, the sampling distribution of p is well approximated by a normal curve.

The accompanying box summarizes these properties.

When n is large, the statistic p has a sampling distribution that is approximately normal with mean π and standard deviation $\sqrt{\dfrac{\pi(1-\pi)}{n}}$.

The development of a confidence interval for π is easier to follow if we select a particular confidence level. For a confidence level of 95%, Appendix Table 2, the table of standard normal (z) curve areas, can be used to determine a value z^* such that a central area of .95 falls between $-z^*$ and z^*. In this case, the remaining area of .05 is divided equally between the two tails, as shown in Figure 9.2. The total area to the left of the desired z^* is .975 (.95 central area + .025 area below $-z^*$). By locating .9750 in the body of Appendix Table 2, we find that the corresponding z critical value is 1.96.

Figure 9.2 Capturing a central area of .95 under the z curve.

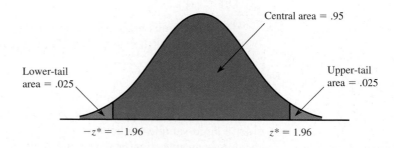

Generalizing this result to normal distributions other than the standard normal distribution tells us that for *any* normal distribution, about 95% of the values are within 1.96 standard deviations of the mean. Because (for large random samples) the sampling distribution of p is approximately normal with mean $\mu_p = \pi$ and standard deviation $\sigma_p = \sqrt{\dfrac{\pi(1-\pi)}{n}}$, we get the following result.

When n is large, approximately 95% of all samples of size n will result in a value of p that is within $1.96\sigma_p = 1.96\sqrt{\dfrac{\pi(1-\pi)}{n}}$ of the true population proportion π.

If p is within $1.96\sqrt{\dfrac{\pi(1-\pi)}{n}}$ of π, this means the interval

$$p - 1.96\sqrt{\frac{\pi(1-\pi)}{n}} \quad \text{to} \quad p + 1.96\sqrt{\frac{\pi(1-\pi)}{n}}$$

will capture π (and this will happen for 95% of all possible samples). However, if p is farther away from π than $1.96\sqrt{\dfrac{\pi(1-\pi)}{n}}$ (which will happen for about 5% of all possible samples), the interval will not include the true value of π. This is shown in Figure 9.3.

Figure 9.3 The population proportion π is captured in the interval from $p - 1.96\sqrt{\dfrac{\pi(1-\pi)}{n}}$ to $p + 1.96\sqrt{\dfrac{\pi(1-\pi)}{n}}$ when p is within $1.96\sqrt{\dfrac{\pi(1-\pi)}{n}}$ of π.

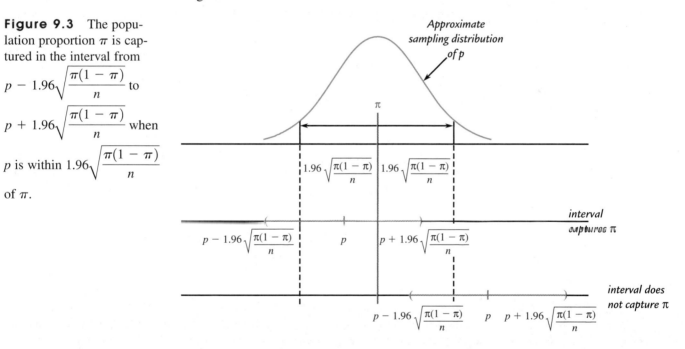

Because p is within $1.96\sigma_p$ of π 95% of the time, this implies that in repeated sampling, 95% of the time the interval

$$p - 1.96\sqrt{\frac{\pi(1-\pi)}{n}} \quad \text{to} \quad p + 1.96\sqrt{\frac{\pi(1-\pi)}{n}}$$

will contain π.

Since π is unknown, $\sqrt{\dfrac{\pi(1-\pi)}{n}}$ must be estimated. As long as the sample size

Important concept   is large, the value of $\sqrt{\dfrac{p(1-p)}{n}}$ can be used in place of $\sqrt{\dfrac{\pi(1-\pi)}{n}}$.

When n is large, a 95% confidence interval for π is

$$\left(p - 1.96\sqrt{\frac{p(1-p)}{n}},\, p + 1.96\sqrt{\frac{p(1-p)}{n}}\right)$$

An abbreviated formula for the interval is

$$p \pm 1.96\sqrt{\frac{p(1-p)}{n}}$$

where $p + 1.96\sqrt{\dfrac{p(1-p)}{n}}$ gives the upper endpoint of the interval and $p - 1.96\sqrt{\dfrac{p(1-p)}{n}}$

gives the lower endpoint of the interval.

The interval can be used as long as

1. $np \geq 10$ and $n(1-p) \geq 10$,
2. the sample size is less than 10% of the population size if sampling is without replacement,
3. the sample can be regarded as a random sample from the population of interest.

Example 9.4   Affirmative Action Continued

Let's return to the information from the survey on attitudes toward affirmative action (see the chapter introduction and Example 9.1):

> Total number of people surveyed: 1013
> Number who believe affirmative action programs should be continued: 537

A point estimate of π, the true proportion of U.S. adults who believe that affirmative action programs should be continued, is

$$p = \frac{537}{1013} = .530$$

Check of assumptions   Because $np = (1013)(.53) = 537 \geq 10$ and $n(1-p) = (1013)(.47) = 476 \geq 10$ and the adults in the sample were randomly selected from a large population, the large-sample interval can be used. For a 95% confidence level, a confidence interval for π is

$$p \pm 1.96\sqrt{\frac{p(1-p)}{n}} = 5.30 \pm 1.96\sqrt{\frac{(.530)(.470)}{1013}}$$
$$= .530 \pm (1.96)(.016)$$
$$= .530 \pm .031$$
$$= (.499, .561)$$

Interpretation of confidence interval in context   Based on this sample, we can be 95% confident that π, the true proportion who believe that affirmative action programs should continue, is between .499 and .561. We used a *method* to construct this estimate that in the long run will successfully capture the true value of π 95% of the time.

■

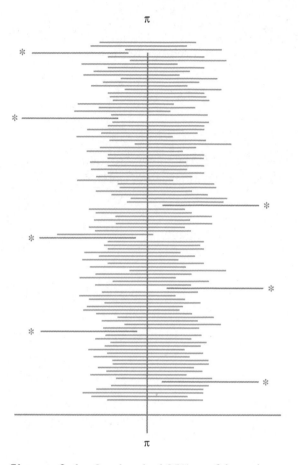

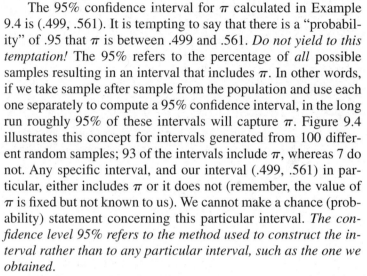

The 95% confidence interval for π calculated in Example 9.4 is (.499, .561). It is tempting to say that there is a "probability" of .95 that π is between .499 and .561. *Do not yield to this temptation!* The 95% refers to the percentage of *all* possible samples resulting in an interval that includes π. In other words, if we take sample after sample from the population and use each one separately to compute a 95% confidence interval, in the long run roughly 95% of these intervals will capture π. Figure 9.4 illustrates this concept for intervals generated from 100 different random samples; 93 of the intervals include π, whereas 7 do not. Any specific interval, and our interval (.499, .561) in particular, either includes π or it does not (remember, the value of π is fixed but not known to us). We cannot make a chance (probability) statement concerning this particular interval. *The confidence level 95% refers to the method used to construct the interval rather than to any particular interval, such as the one we obtained.*

The formula given for a 95% confidence interval can easily be adapted for other confidence levels. The choice of a 95% confidence level led to the use of the z value 1.96 (chosen to capture a central area of .95 under the standard normal curve) in the formula. Any other confidence level can be obtained by using an appropriate z critical value in place of 1.96. For example, suppose that we wanted to achieve a confidence level of 99%. To obtain a central area of .99, the approximate z critical value would have a cumulative area (area to the left) of .995, as illustrated in Figure 9.5. From Appendix Table 2, we find that the corresponding z critical value is 2.58. A 99% confidence interval for π is then obtained by using 2.58 in place of 1.96 in the formula for the 95% confidence interval.

Figure 9.4   One hundred 95% confidence intervals for π computed from 100 different random samples (asterisks identify intervals that do not include π).

Figure 9.5   Finding the z critical value for a 99% confidence level.

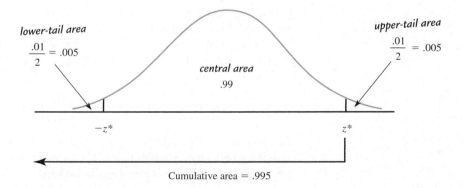

The Large-Sample Confidence Interval for π

The general formula for a confidence interval for a population proportion π when

1. p is the sample proportion from a **random sample**, and
2. the sample size **n is large** ($np \geq 10$ and $n(1 - p) \geq 10$), and
3. if the sample is selected without replacement, **the sample size is small relative to the population size** (n is at most 10% of the population size)

is

$$p \pm (z \text{ critical value})\sqrt{\frac{p(1 - p)}{n}}$$

The desired confidence level determines which z critical value is used. The three most commonly used confidence levels, 90%, 95%, and 99%, use z critical values 1.645, 1.96, and 2.58, respectively.

Note: This interval is not appropriate for small samples. It is possible to construct a confidence interval in the small-sample case, but this is beyond the scope of this textbook.

Why settle for 95% confidence when 99% confidence is possible? Because the higher confidence level comes with a price tag. The resulting interval is wider than the 95% interval. The width of the 95% interval is $2\left(1.96\sqrt{\frac{p(1 - p)}{n}}\right)$, whereas the 99% interval has width $2\left(2.58\sqrt{\frac{p(1 - p)}{n}}\right)$. The higher *reliability* of the 99% interval (where "reliability" is specified by the confidence level) entails a loss in precision (as indicated by the wider interval). In the opinion of many investigators, a 95% interval is a reasonable compromise between reliability and precision.

Example 9.5 Dangerous Driving

The article "Nine Out of Ten Drivers Admit in Survey to Having Done Something Dangerous" (*Knight Ridder Newspapers*, July 8, 2005) reported the results of a survey of 1100 drivers. Of those surveyed, 990 admitted to careless or aggressive driving during the previous six months. Assuming that it is reasonable to regard this sample of 1100 as representative of the population of drivers, we can use this information to construct an estimate of π, the true proportion of drivers who have engaged in careless or aggressive driving in the past six months.

For this sample

$$p = \frac{990}{1100} = .900$$

Because $np = 990$ and $n(1 - p) = 110$ are both greater than or equal to 10, the sample size is large enough to use the formula for a large-sample confidence interval. A 90% confidence interval for π is then

$$p \pm (z \text{ critical value})\sqrt{\frac{p(1-p)}{n}} = .900 \pm 1.645\sqrt{\frac{(.900)(.100)}{1100}}$$
$$= .900 \pm (1.645)(.009)$$
$$= .900 \pm .015$$
$$= (.885, .915)$$

Based on these sample data, we can be 90% confident that the true proportion of drivers who have engaged in careless or aggressive driving in the past six months is between .885 and .915. We have used a *method* to construct this interval estimate that has a 10% error rate.

■

The confidence level for the z confidence interval for a population proportion is only approximate. That is, when we report a 95% confidence interval for a population proportion, the 95% confidence level implies that we have used a method that produces an interval that includes the actual value of the population proportion 95% of the time in repeated sampling. In fact, because the normal distribution is only an approximation to the sampling distribution of p, the true confidence level may differ somewhat from the reported value. If the conditions (1) $np \geq 10$ and $n(1-p) \geq 10$ and (2) n is at most 10% of the population size if sampling without replacement are met, the normal approximation is reasonable and the actual confidence level is usually quite close to the reported level; this is why it is important to check these conditions before computing and reporting a z confidence interval for a population proportion.

What should you do if these conditions are not met? If the sample size is too small to satisfy the np and $n(1-p) \geq 10$ condition, an alternative procedure can be used. Consult a statistician or a more advanced textbook in this case. If the condition that the sample size is less than 10% of the population size when sampling without replacement is not satisfied, the z confidence interval tends to be conservative (i.e., it tends to be wider than is necessary to achieve the desired confidence level). In this case, a finite population correction factor can be used to obtain a more precise interval. Again, it would be wise to consult a statistician or a more advanced textbook.

■ An Alternative to the Large-Sample z Interval

Investigators have shown that in some instances, even when the sample size conditions of the large-sample z confidence interval for a population proportion are met, the actual confidence level associated with the method may be noticeably different from the reported confidence level. A modified interval that has an actual confidence level that is closer to the reported confidence level is based on a modified sample proportion, p_{mod}, the proportion of successes after adding two successes and two failures to the sample. Then p_{mod} is

$$p_{\text{mod}} = \frac{\text{number of successes} + 2}{n + 4}$$

p_{mod} is used in place of p in the usual confidence interval formula. Properties of this modified confidence interval are investigated in Activity 9.2 at the end of the chapter.

■ General Form of a Confidence Interval ....

Many confidence intervals have the same general form as the large-sample z interval for π just considered. We started with a statistic p, from which a point estimate for π was obtained. The standard deviation of this statistic is $\sqrt{\pi(1 - \pi)/n}$. This resulted in a confidence interval of the form

$$\left(\begin{array}{c} \text{point estimate using} \\ \text{a specified statistic} \end{array}\right) \pm (\text{critical value}) \left(\begin{array}{c} \text{standard deviation} \\ \text{of the statistic} \end{array}\right)$$

Because π was unknown, we estimated the standard deviation of the statistic by $\sqrt{p(1 - p)/n}$, which yielded the interval

$$\left(\begin{array}{c} \text{point estimate using} \\ \text{a specified statistic} \end{array}\right) \pm (\text{critical value}) \left(\begin{array}{c} \text{estimated} \\ \text{standard deviation} \\ \text{of the statistic} \end{array}\right)$$

For a population characteristic other than π, a statistic for estimating the characteristic is selected. Then (drawing on statistical theory) a formula for the standard deviation of the statistic is given. In practice, it is almost always necessary to estimate this standard deviation (using something analogous to $\sqrt{p(1 - p)/n}$ rather than $\sqrt{\pi(1 - \pi)/n}$, for example), so that the interval

$$\left(\begin{array}{c} \text{point estimate using} \\ \text{a specified statistic} \end{array}\right) \pm (\text{critical value}) \left(\begin{array}{c} \text{estimated} \\ \text{standard deviation} \\ \text{of the statistic} \end{array}\right)$$

is the prototype confidence interval. It is common practice to refer to both the standard deviation of a statistic and the *estimated* standard deviation of a statistic as the *standard error*. In this textbook, when we use the term *standard error*, we mean the estimated standard deviation of a statistic.

> ### DEFINITION
>
> The **standard error** of a statistic is the estimated standard deviation of the statistic.

The 95% confidence interval for π is based on the fact that, for approximately 95% of all random samples, p is within $1.96\sqrt{\dfrac{\pi(1 - \pi)}{n}}$ of π. The quantity $1.96\sqrt{\dfrac{\pi(1 - \pi)}{n}}$ is sometimes called the *bound on the error of estimation* associated with a 95% confidence level—we have 95% confidence that the point estimate p is no farther than this quantity from π.

> **DEFINITION**
>
> If the sampling distribution of a statistic is (at least approximately) normal, the **bound on error of estimation, B**, associated with a 95% confidence interval is $(1.96) \cdot$ (standard error of the statistic).

■ Choosing the Sample Size

Before collecting any data, an investigator may wish to determine a sample size for which a particular value of the bound on the error is achieved. For example, with π representing the true proportion of students at a university who purchase textbooks over the Internet, the objective of an investigation may be to estimate π to within .05 with 95% confidence. The value of n necessary to achieve this is obtained by equating

.05 to $1.96\sqrt{\dfrac{\pi(1 - \pi)}{n}}$ and solving for n.

In general, suppose that we wish to estimate π to within an amount B (the specified bound on the error of estimation) with 95% confidence. Finding the necessary sample size requires solving the equation

$$B = 1.96\sqrt{\frac{\pi(1 - \pi)}{n}}$$

Solving this equation for n results in

$$n = \pi(1 - \pi)\left(\frac{1.96}{B}\right)^2$$

Unfortunately, the use of this formula requires the value of π, which is unknown. One possible way to proceed is to carry out a preliminary study and use the resulting data to get a rough estimate of π. In other cases, prior knowledge may suggest a reasonable estimate of π. If there is no reasonable basis for estimating π and a preliminary study is not feasible, a conservative solution follows from the observation that $\pi(1 - \pi)$ is never larger than .25 (its value when $\pi = .5$). Replacing $\pi(1 - \pi)$ with .25, the maximum value, yields

$$n = .25\left(\frac{1.96}{B}\right)^2$$

Using this formula to obtain n gives us a sample size for which we can be 95% confident that p will be within B of π, no matter what the value of π.

The sample size required to estimate a population proportion π to within an amount B with 95% confidence is

$$n = \pi(1 - \pi)\left(\frac{1.96}{B}\right)^2$$

The value of π may be estimated using prior information. In the absence of any such information, using $\pi = .5$ in this formula gives a conservatively large value for the required sample size (this value of π gives a larger n than would any other value).

Example 9.6 Sniffing Out Cancer

Researchers have found biochemical markers of cancer in the exhaled breath of cancer patients, but chemical analysis of breath specimens has not yet proven effective in clinical diagnosis. The authors of the paper "Diagnostic Accuracy of Canine Scent Detection in Early- and Late-Stage Lung and Breast Cancers" (*Integrative Cancer Therapies* [2006]: 1–10) describe a study to investigate whether dogs can be trained to identify the presence or absence of cancer by sniffing breath specimens. Suppose we want to collect data that would allow us to estimate the long-run proportion of accurate identifications for a particular dog that has completed training. The dog has been trained to lie down when presented with a breath specimen from a cancer patient and to remain standing when presented with a specimen from a person who does not have cancer. How many different breath specimens should be used if we want to estimate the long-run proportion of correct identifications for this dog to within .05 with 95% confidence?

Using a conservative value of $\pi = .5$ in the formula for required sample size gives

$$n = \pi(1 - \pi)\left(\frac{1.96}{B}\right)^2 = .25\left(\frac{1.96}{.05}\right)^2 = 384.16$$

Always round answer *up* to the next whole number.

Thus, a sample of at least 385 breath specimens should be used. Note that in sample size calculations, we always round up.

■

■ Exercises 9.11–9.29

Turn to the Expanded Answer Section for answers not shown next to exercises.

9.11 ▼ For each of the following choices, explain which would result in a wider large-sample confidence interval for π:
a. 90% confidence level or 95% confidence level
b. $n = 100$ or $n = 400$ $n = 100$

9.12 The formula used to compute a large-sample confidence interval for π is

$$p \pm (z \text{ critical value})\sqrt{\frac{p(1 - p)}{n}}$$

What is the appropriate z critical value for each of the following confidence levels?
a. 95%   1.96
b. 90%   1.645
c. 99%   2.58
d. 80%   1.28
e. 85%   Approximately 1.44

9.13 The use of the interval

$$p \pm (z \text{ critical value})\sqrt{\frac{p(1 - p)}{n}}$$

requires a large sample. For each of the following combinations of n and p, indicate whether the given interval would be appropriate.
a. $n = 50$ and $p = .30$ Yes
b. $n = 50$ and $p = .05$ No
c. $n = 15$ and $p = .45$ No
d. $n = 100$ and $p = .01$ No
e. $n = 100$ and $p = .70$ Yes
f. $n = 40$ and $p = .25$ Yes
g. $n = 60$ and $p = .25$ Yes
h. $n = 80$ and $p = .10$ No

Bold exercises answered in back   ● Data set available online but not required   ▼ Video solution available

9.14 Discuss how each of the following factors affects the width of the confidence interval for π:
a. The confidence level
b. The sample size
c. The value of p

9.15 According to an AP-Ipsos poll (June 15, 2005), 42% of 1001 randomly selected adult Americans made plans in May 2005 based on a weather report that turned out to be wrong.
a. Construct and interpret a 99% confidence interval for the proportion of Americans who made plans in May 2005 based on an incorrect weather report.
b. Do you think it is reasonable to generalize this estimate to other months of the year? Explain. No, weather reports may be more or less reliable during other months.

9.16 The article "Students Increasingly Turn to Credit Cards" (*San Luis Obispo Tribune*, July 21, 2006) reported that 37% of college freshmen and 48% of college seniors carry a credit card balance from month to month. Suppose that the reported percentages were based on random samples of 1000 college freshmen and 1000 college seniors.
a. Construct a 90% confidence interval for the proportion of college freshmen who carry a credit card balance from month to month. (0.3449, 0.3951)
b. Construct a 90% confidence interval for the proportion of college seniors who carry a credit card balance from month to month. (0.45401, 0.50599)
c. Explain why the two 90% confidence intervals from Parts (a) and (b) are not the same width.

9.17 ▼ The article "CSI Effect Has Juries Wanting More Evidence" (*USA Today*, August 5, 2004) examines how the popularity of crime-scene investigation television shows is influencing jurors' expectations of what evidence should be produced at a trial. In a survey of 500 potential jurors, one study found that 350 were regular watchers of at least one crime-scene forensics television series.
a. Assuming that it is reasonable to regard this sample of 500 potential jurors as representative of potential jurors in the United States, use the given information to construct and interpret a 95% confidence interval for the true proportion of potential jurors who regularly watch at least one crime-scene investigation series. (0.6598, 0.7402)
b. Would a 99% confidence interval be wider or narrower than the 95% confidence interval from Part (a)? Wider

9.18 In a survey of 1000 randomly selected adults in the United States, participants were asked what their most favorite and what their least favorite subject was when they were in school (Associated Press, August 17, 2005). In what might seem like a contradiction, math was chosen more often than any other subject in both categories! Math was chosen by 230 of the 1000 as the favorite subject, and it was also chosen by 370 of the 1000 as the least favorite subject.
a. Construct a 95% confidence interval for the proportion of U.S. adults for whom math was the favorite subject in school. (0.2039, 0.2561)
b. Construct a 95% confidence interval for the proportion of U.S. adults for whom math was the least favorite subject. (0.3401, 0.3999)

9.19 The report "2005 Electronic Monitoring & Surveillance Survey: Many Companies Monitoring, Recording, Videotaping—and Firing—Employees" (American Management Association, 2005) summarized the results of a survey of 526 U.S. businesses. The report stated that 137 of the 526 businesses had fired workers for misuse of the Internet and 131 had fired workers for email misuse. For purposes of this exercise, assume that it is reasonable to regard this sample as representative of businesses in the United States.
a. Construct and interpret a 95% confidence interval for the proportion of U.S. businesses that have fired workers for misuse of the Internet. (0.2225, 0.2975)
b. What are two reasons why a 90% confidence interval for the proportion of U.S. businesses that have fired workers for misuse of email would be narrower than the 95% confidence interval computed in Part (a). The confidence level is lower and the estimated standard error is smaller.

9.20 In an AP-AOL sports poll (Associated Press, December 18, 2005), 394 of 1000 randomly selected U.S. adults indicated that they considered themselves to be baseball fans. Of the 394 baseball fans, 272 stated that they thought the designated hitter rule should either be expanded to both baseball leagues or eliminated.
a. Construct a 95% confidence interval for the proportion of U.S. adults that consider themselves to be baseball fans. (0.3637, 0.4243)
b. Construct a 95% confidence interval for the proportion of those who consider themselves to be baseball fans that think the designated hitter rule should be expanded to both leagues or eliminated. (0.6443, 0.7357)
c. Explain why the confidence intervals of Parts (a) and

(b) are not the same width even though they both have a confidence level of 95%. Because the sample sizes and the sample proportions are different

9.21 The article "Viewers Speak Out Against Reality TV" (Associated Press, September 12, 2005) included the following statement: "Few people believe there's much reality in reality TV: a total of 82 percent said the shows are either 'totally made up' or 'mostly distorted'." This statement was based on a survey of 1002 randomly selected adults. Compute and interpret a bound on the error of estimation for the reported percentage. Bound on error = 0.0238

9.22 One thousand randomly selected adult Americans participated in a survey conducted by the Associated Press (June, 2006). When asked "Do you think it is sometimes justified to lie or do you think lying is never justified?" 52% responded that lying was never justified. When asked about lying to avoid hurting someone's feelings, 650 responded that this was often or sometimes OK.
a. Construct a 90% confidence interval for the proportion of adult Americans who think lying is never justified.
b. Construct a 90% confidence interval for the proportion of adult American who think that it is often or sometimes OK to lie to avoid hurting someone's feelings.
c. Based on the confidence intervals from Parts (a) and (b), comment on the apparent inconsistency in the responses given by the individuals in this sample.

9.23 The article "Doctors Cite Burnout in Mistakes" (*San Luis Obispo Tribune*, March 5, 2002) reported that many doctors who are completing their residency have financial struggles that could interfere with training. In a sample of 115 residents, 38 reported that they worked moonlighting jobs and 22 reported a credit card debt of more than $3000. Suppose that it is reasonable to consider this sample of 115 as a random sample of all medical residents in the United States.
a. Construct and interpret a 95% confidence interval for the proportion of U.S. medical residents who work moonlighting jobs. (0.2441, 0.4159)
b. Construct and interpret a 90% confidence interval for the proportion of U.S. medical residents who have a credit card debt of more than $3000. (0.1307, 0.2513)
c. Give two reasons why the confidence interval in Part (a) is wider than the confidence interval in Part (b).

9.24 The National Geographic Society conducted a study that included 3000 respondents, age 18 to 24, in nine different countries (*San Luis Obispo Tribune*, November 21, 2002). The society found that 10% of the participants could not identify their own country on a blank world map.
a. Construct a 90% confidence interval for the proportion who can identify their own country on a blank world map.
b. What assumptions are necessary for the confidence interval in Part (a) to be valid?
c. To what population would it be reasonable to generalize the confidence interval estimate from Part (a)? All 18- to 24-year-olds in the nine different countries in the study.

9.25 ▼ "Tongue Piercing May Speed Tooth Loss, Researchers Say" is the headline of an article that appeared in the *San Luis Obispo Tribune* (June 5, 2002). The article describes a study of 52 young adults with pierced tongues. The researchers found receding gums, which can lead to tooth loss, in 18 of the participants. Construct a 95% confidence interval for the proportion of young adults with pierced tongues who have receding gums. What assumptions must be made for use of the z confidence interval to be appropriate?

9.26 *USA Today* (October 14, 2002) reported that 36% of adult drivers admit that they often or sometimes talk on a cell phone when driving. This estimate was based on data from a sample of 1004 adult drivers, and a bound on the error of estimation of 3.1% was reported. Explain how the given bound on the error can be justified.

9.27 The Gallup Organization conducts an annual survey on crime. It was reported that 25% of all households experienced some sort of crime during the past year. This estimate was based on a sample of 1002 randomly selected adults. The report states, "One can say with 95% confidence that the margin of sampling error is ± 3 percentage points." Explain how this statement can be justified.

9.28 In a study of 1710 schoolchildren in Australia (*Herald Sun*, October 27, 1994), 1060 children indicated that they normally watch TV before school in the morning. (Interestingly, only 35% of the parents said their children watched TV before school!) Construct a 95% confidence interval for the true proportion of Australian children who say they watch TV before school. What assumption about the sample must be true for the method used to construct the interval to be valid?

9.29 ▼ A consumer group is interested in estimating the proportion of packages of ground beef sold at a particular

Bold exercises answered in back   ● Data set available online but not required   ▼ Video solution available

store that have an actual fat content exceeding the fat content stated on the label. How many packages of ground beef should be tested to estimate this proportion to within .05 with 95% confidence?   385

9.3 Confidence Interval for a Population Mean

In this section, we consider how to use information from a random sample to construct a confidence interval estimate of a population mean, μ. We begin by considering the case in which (1) σ, **the population standard deviation, is known** (not realistic, but we will see shortly how to handle the more realistic situation where σ is unknown) and (2) **the sample size n is large** enough for the Central Limit Theorem to apply. In this case, the following three properties about the sampling distribution of $\bar{x}$ hold:

1. The sampling distribution of $\bar{x}$ is centered at μ, so $\bar{x}$ is an unbiased statistic for estimating μ $(\mu_{\bar{x}} = \mu)$.

2. The standard deviation of $\bar{x}$ is $\sigma_{\bar{x}} = \dfrac{\sigma}{\sqrt{n}}$.

3. As long as n is large (generally $n \geq 30$), the sampling distribution of $\bar{x}$ is approximately normal, even when the population distribution itself is not normal.

The same reasoning that was used to develop the large-sample confidence interval for a population proportion π can be used to obtain a confidence interval estimate for μ.

The One-Sample z Confidence Interval for μ

The general formula for a confidence interval for a population mean μ when

1. $\bar{x}$ is the sample mean from a **random sample**,
2. the **sample size n is large** (generally $n > 30$), and
3. σ, **the population standard deviation, is known**

is

$$\bar{x} \pm (z \text{ critical value})\left(\frac{\sigma}{\sqrt{n}}\right)$$

Example 9.7 Cosmic Radiation

Cosmic radiation levels rise with increasing altitude, prompting researchers to consider how pilots and flight crews might be affected by increased exposure to cosmic radiation. The paper "Estimated Cosmic Radiation Doses for Flight Personnel" (*Space Medicine and Medical Engineering* [2002]: 265–269) reported a mean annual cosmic radiation dose of 219 mrems for a sample of flight personnel of Xinjiang Airlines. Suppose that this mean was based on a random sample of 100 flight crew members.

Let μ denote the true mean annual cosmic radiation exposure for Xinjiang Airlines flight crew members. Although σ, the true population standard deviation, is not usually known, suppose for illustrative purposes that $\sigma = 35$ mrem is known. Because the sample size is large and σ is known, a 95% confidence interval for μ is

$$\bar{x} \pm (z \text{ critical value})\left(\frac{\sigma}{\sqrt{n}}\right) = 219 \pm (1.96)\left(\frac{35}{\sqrt{100}}\right)$$
$$= 219 \pm 6.86$$
$$= (212.14, \ 225.86)$$

Based on this sample, *plausible* values of μ, the true mean annual cosmic radiation exposure for Xinjaing Airlines flight crew members, are between 212.14 and 225.86 mrem. A 95% confidence level is associated with the method used to produce this interval estimate.

■

The confidence interval just introduced is appropriate when σ is known and n is large, and it can be used regardless of the shape of the population distribution. This is because this confidence interval is based on the Central Limit Theorem, which says that when n is sufficiently large, the sampling distribution of $\bar{x}$ is approximately normal for any population distribution. When n is small, the Central Limit Theorem cannot be used to justify the normality of the $\bar{x}$ sampling distribution, so the z confidence interval can not be used. One way to proceed in the small-sample case is to make a specific assumption about the shape of the population distribution and then to use a method of estimation that is valid only under this assumption.

The one instance where this is easy to do is when it is reasonable to believe that the population distribution is normal in shape. Recall that for a normal population distribution the sampling distribution of $\bar{x}$ is normal even for small sample sizes. So, if n is small but the population distribution is normal, the same confidence interval formula just introduced can still be used.

If n is small (generally $n < 30$) but it is reasonable to believe that the distribution of values in the population is normal, a confidence interval for μ (when σ is known) is

$$\bar{x} \pm (z \text{ critical value})\left(\frac{\sigma}{\sqrt{n}}\right)$$

■

Important concept

There are several ways that sample data can be used to assess the plausibility of normality. The two most common ways are to look at a normal probability plot of the sample data (looking for a plot that is reasonably straight) and to construct a boxplot of the data (looking for approximate symmetry and no outliers).

■ Confidence Interval for μ When σ Is Unknown

The confidence intervals just developed have an obvious drawback: To compute the interval endpoints, σ must be known. Unfortunately, this is rarely the case in practice. We now turn our attention to the situation when σ is unknown. The development of the confidence interval in this instance depends on the assumption that the population distribution is normal. This assumption is not critical if the sample size is large, but it is important when the sample size is small.

To understand the derivation of this confidence interval, it is instructive to begin by taking another look at the previous 95% confidence interval. We know that $\mu_{\bar{x}} = \mu$ and $\sigma_{\bar{x}} = \dfrac{\sigma}{\sqrt{n}}$. Also, when the population distribution is normal, the $\bar{x}$ distribution is normal. These facts imply that the standardized variable

$$z = \frac{\bar{x} - \mu}{\dfrac{\sigma}{\sqrt{n}}}$$

has approximately a standard normal distribution. Because the interval from -1.96 to 1.96 captures an area of .95 under the z curve, approximately 95% of all samples result in an $\bar{x}$ value that satisfies

$$-1.96 < \frac{\bar{x} - \mu}{\dfrac{\sigma}{\sqrt{n}}} < 1.96$$

Manipulating these inequalities to isolate μ in the middle results in the equivalent inequalities:

$$\bar{x} - 1.96\left(\frac{\sigma}{\sqrt{n}}\right) < \mu < \bar{x} + 1.96\left(\frac{\sigma}{\sqrt{n}}\right)$$

The term $\bar{x} - 1.96\left(\dfrac{\sigma}{\sqrt{n}}\right)$ is the lower endpoint of the 95% large-sample confidence interval for μ, and $\bar{x} + 1.96\left(\dfrac{\sigma}{\sqrt{n}}\right)$ is the upper endpoint.

If σ is unknown, we must use the sample data to estimate σ. If we use the sample standard deviation as our estimate, the result is a different standardized variable denoted by t:

$$t = \frac{\bar{x} - \mu}{\dfrac{s}{\sqrt{n}}}$$

The value of s may not be all that close to σ, especially when n is small. As a consequence, the use of s in place of σ introduces extra variability, so the distribution of t is more spread out than the standard normal (z) distribution. (The value of z varies from sample to sample, because different samples generally result in different $\bar{x}$ values. There is even more variability in t, because different samples may result in different values of both $\bar{x}$ and s.)

To develop an appropriate confidence interval, we must investigate the probability distribution of the standardized variable t for a sample from a normal population. This requires that we first learn about probability distributions called t *distributions*.

■ *t* Distributions

Just as there are many different normal distributions, there are also many different t distributions. Although normal distributions are distinguished from one another by their mean μ and standard deviation σ, t distributions are distinguished by a positive whole number called the number of *degrees of freedom* (df). There is a t distribution with 1 df, another with 2 df, and so on.

Important Properties of *t* Distributions

1. The t distribution corresponding to any fixed number of degrees of freedom is bell shaped and centered at zero (just like the standard normal (z) distribution).
2. Each t distribution is more spread out than the standard normal (z) distribution.
3. As the number of degrees of freedom increases, the spread of the corresponding t distribution decreases.
4. As the number of degrees of freedom increases, the corresponding sequence of t distributions approaches the standard normal (z) distribution.

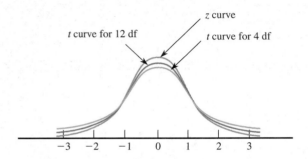

Figure 9.6 Comparison of the z curve and t curves for 12 df and 4 df.

The properties discussed in the preceding box are illustrated in Figure 9.6, which shows two t curves along with the z curve.

Appendix Table 3 gives selected critical values for various t distributions. The central areas for which values are tabulated are .80, .90, .95, .98, .99, .998, and .999. To find a particular critical value, go down the left margin of the table to the row labeled with the desired number of degrees of freedom. Then move over in that row to the column headed by the desired central area. For example, the value in the 12-df row under the column corresponding to central area .95 is 2.18, so 95% of the area under the t curve with 12 df lies between -2.18 and 2.18. Moving over two columns, we find the critical value for central area .99 (still with 12 df) to be 3.06 (see Figure 9.7). Moving down the .99 column to the 20-df row, we see the critical value is 2.85, so the area between -2.85 and 2.85 under the t curve with 20 df is .99.

Figure 9.7 t critical values illustrated.

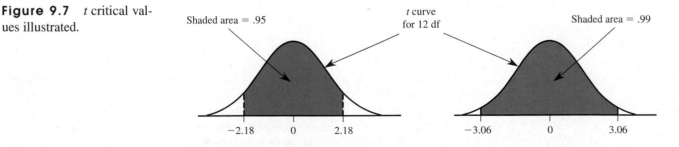

Notice that the critical values increase from left to right in each row of Appendix Table 3. This makes sense because as we move to the right, we capture larger central areas. In each column, the critical values decrease as we move downward, reflecting decreasing spread for t distributions with larger degrees of freedom.

The larger the number of degrees of freedom, the more closely the t curve resembles the z curve. To emphasize this, we have included the z critical values as the last row of the t table. Furthermore, once the number of degrees of freedom exceeds 30, the critical values change little as the number of degrees of freedom increases. For this reason, Appendix Table 3 jumps from 30 df to 40 df, then to 60 df, then to 120 df, and finally to the row of z critical values. If we need a critical value for a number of degrees of freedom between those tabulated, we just use the critical value for the closest df. For df > 120, we use the z critical values. Many graphing calculators calculate t critical values for any number of degrees of freedom; so, if you are using such a calculator, it is not necessary to approximate the t critical values as described.

■ One-Sample t Confidence Interval

The fact that the sampling distribution of $\dfrac{\bar{x} - \mu}{(\sigma/\sqrt{n})}$ is approximately the z (standard normal) distribution when n is large led to the z confidence interval when σ is known. In the same way, the following proposition provides the key to obtaining a confidence interval when the population distribution is normal but σ is unknown.

Let $x_1, x_2, ..., x_n$ constitute a random sample from a normal population distribution. Then the probability distribution of the standardized variable

$$t = \frac{\bar{x} - \mu}{\dfrac{s}{\sqrt{n}}}$$

is the t distribution with df $= n - 1$.

To see how this result leads to the desired confidence interval, consider the case $n = 25$. We use the t distribution with df $= 24$ $(n - 1)$. From Appendix Table 3, the interval between -2.06 and 2.06 captures a central area of $.95$ under the t curve with 24 df. This means that 95% of all samples (with $n = 25$) from a normal population result in values of $\bar{x}$ and s for which

$$-2.06 < \frac{\bar{x} - \mu}{\dfrac{s}{\sqrt{n}}} < 2.06$$

Algebraically manipulating these inequalities to isolate μ yields

$$\bar{x} - 2.06\left(\frac{s}{\sqrt{25}}\right) < \mu < \bar{x} + 2.06\left(\frac{s}{\sqrt{25}}\right)$$

The 95% confidence interval for μ in this situation extends from the lower endpoint $\bar{x} - 2.06\left(\dfrac{s}{\sqrt{25}}\right)$ to the upper endpoint $\bar{x} + 2.06\left(\dfrac{s}{\sqrt{25}}\right)$. This interval can be written

$$\bar{x} \pm 2.06\left(\frac{s}{\sqrt{25}}\right)$$

The differences between this interval and the interval when σ is known are the use of the t critical value 2.06 rather than the z critical value 1.96 and the use of the sample standard deviation as an estimate of σ. The extra uncertainty that results from estimating σ causes the t interval to be wider than the z interval.

Important concept

If the sample size is something other than 25 or if the desired confidence level is something other than 95%, a different t critical value (obtained from Appendix Table 3) is used in place of 2.06.

The One-Sample t Confidence Interval for μ

The general formula for a confidence interval for a population mean μ based on a sample of size n when

1. $\bar{x}$ is the sample mean from a **random sample**,
2. the **population distribution is normal**, *or* the **sample size n is large** (generally $n \geq 30$), and
3. σ, **the population standard deviation, is unknown**

is

$$\bar{x} \pm (t \text{ critical value}) \left(\frac{s}{\sqrt{n}} \right)$$

where the t critical value is based on df $= n - 1$. Appendix Table 3 gives critical values appropriate for each of the confidence levels 90%, 95%, and 99%, as well as several other less frequently used confidence levels.

If n is large (generally $n \geq 30$), the normality of the population distribution is not critical. *However, this confidence interval is appropriate for small n only when the population distribution is (at least approximately) normal.* If this is not the case, as might be suggested by a normal probability plot or boxplot, another estimation method should be used.

....

Example 9.8   Waiting for Surgery

The Cardiac Care Network in Ontario, Canada collected information on the time between the date a patient was recommended for heart surgery and the surgery date for cardiac patients in Ontario ("Wait Times Data Guide," Ministry of Health and Long-Term Care, Ontario, Canada, 2006). The reported mean waiting time (in days) for samples of patients for two cardiac procedures are given in the accompanying table. (The standard deviations in the table were estimated from information on wait-time variability included in the report.)

Surgical Procedure	Sample Size	Mean Wait Time	Standard Deviation
Bypass	539	19	10
Angiography	847	18	9

If we had access to the raw data (the $539 + 847 = 1386$ individual wait time observations), we might begin by looking at boxplots. Data consistent with the given summary quantities were used to generate the boxplots of Figure 9.8. The boxplots for the two surgical procedures are similar. There are outliers in both data sets, which might cause us to question the normality of the two wait-time distributions, but because the sample sizes are large, it is still appropriate to use the t confidence interval.

Figure 9.8 Boxplots for Example 9.8.

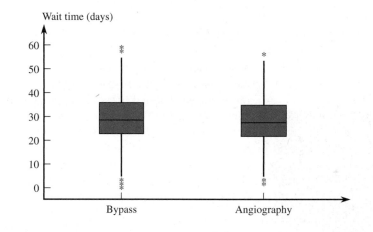

As a next step, we can use the confidence interval of this section to estimate the true mean wait time for each of the two procedures. Let's first focus on the sample of bypass patients. For this group,

$$\text{sample size} = n = 539$$
$$\text{sample mean wait time} = \bar{x} = 19$$
$$\text{sample standard deviation} = s = 10$$

The report referenced here indicated that it is reasonable to regard these data as representative of the Ontario population. So, with μ denoting the mean wait time for bypass surgery in Ontario, we can estimate μ using a 90% confidence interval.

From Appendix Table 3, we use t critical value $= 1.645$ (from the z critical value row because df $= n - 1 = 538 > 120$, the largest number of degrees of freedom in the table). The 90% confidence interval for μ is

$$\bar{x} \pm (t \text{ critical value})\left(\frac{s}{\sqrt{n}}\right) = 19 \pm (1.645)\left(\frac{10}{\sqrt{539}}\right)$$
$$= 19 \pm .709$$
$$= (18.291, 19.709)$$

Based on this sample, we are 90% confident that μ is between 18.291 days and 19.709 days. This interval is fairly narrow indicating that our information about the value of μ is relatively precise.

A graphing calculator or any of the commercially available statistical computing packages can produce t confidence intervals. Confidence interval output from MINITAB for the angiography data is shown here.

One-Sample T

N	Mean	StDev	SE Mean	90% CI
847	18.0000	9.0000	0.3092	(17.4908, 18.5092)

The 90% confidence interval for mean wait time for angiography extends from 17.4908 days to 18.5092 days. This interval is narrower than the 90% interval for bypass surgery wait time for two reasons: the sample size is larger (847 rather than 539) and the sample standard deviation is smaller (9 rather than 10).

■

Example 9.9   Selfish Chimps?

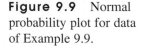

● The article "Chimps Aren't Charitable" (*Newsday*, November 2, 2005) summarized the results of a research study published in the journal *Nature*. In this study, chimpanzees learned to use an apparatus that dispensed food when either of two ropes was pulled. When one of the ropes was pulled, only the chimp controlling the apparatus received food. When the other rope was pulled, food was dispensed both to the chimp controlling the apparatus and also to a chimp in the adjoining cage. The accompanying data (approximated from a graph in the paper) represent the number of times out of 36 trials that each of seven chimps chose the option that would provide food to both chimps (the "charitable" response).

23   22   21   24   19   20   20

Figure 9.9 is a normal probability plot of these data. The plot is reasonably straight, so it seems plausible that the population distribution is approximately normal.

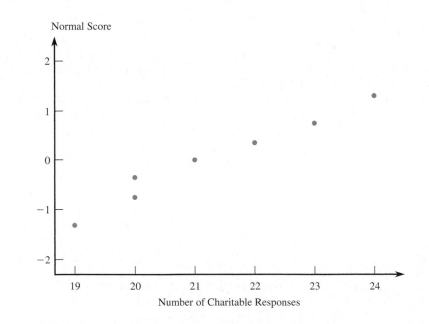

Figure 9.9 Normal probability plot for data of Example 9.9.

Calculation of a confidence interval for the population mean number of charitable responses requires $\bar{x}$ and s. From the given data, we compute

$$\bar{x} = 21.29 \qquad s = 1.80$$

The t critical value for a 99% confidence interval based on 6 df is 3.71. The interval is

$$\bar{x} \pm (t \text{ critical value})\left(\frac{s}{\sqrt{n}}\right) = 21.29 \pm (3.71)\left(\frac{1.80}{\sqrt{7}}\right)$$
$$= 21.29 \pm 2.52$$
$$= (18.77, 23.81)$$

A statistical software package could also have been used to compute the 99% confidence interval. The following is output from SPSS. The slight discrepancy between the hand-calculated interval and the one reported by SPSS occurs because SPSS uses more decimal accuracy in $\bar{x}$, s, and t critical values.

One-Sample Statistics

	N	Mean	Std. Deviation	Std. Error Mean
CharitableResponses	7	21.2857	1.79947	.68014

One-Sample

	99% Confidence Interval	
	Lower	Upper
CharitableResponses	18.7642	23.8073

With 99% confidence, we estimate the population mean number of charitable responses (out of 36 trials) to be between 18.77 and 23.81. Remember that the 99% confidence level implies that if the same formula is used to calculate intervals for sample after sample randomly selected from the population, in the long run 99% of these intervals will capture μ between the lower and upper confidence limits.

Notice that based on this interval, we would conclude that on average chimps choose the charitable option more than half the time (18 out of 36 trials). The *Newsday* headline "Chimps Aren't Charitable" was based on additional data from the study indicating that chimps' charitable behavior was no different when there was another chimp in the adjacent cage than when the adjacent cage was empty. We will revisit this study in Chapter 11 to investigate this further.

■

....

Example 9.10 Housework

How much time do school-age children spend helping with housework? The article "The Three Corners of Domestic Labor: Mothers', Fathers', and Children's Weekday and Weekend Housework" (*Journal of Marriage and the Family* [1994]: 657–668) gave information on the number of minutes per weekday spent on housework. The following mean and standard deviation are for a random sample of 26 girls in two-parent families where both parents work full-time:

$$n = 26 \qquad \bar{x} = 14.0 \qquad s = 8.6$$

The authors of the article analyzed these data using methods designed for population distributions that are approximately normal. This assumption appears a bit questionable based on the reported mean and standard deviation (it is impossible to spend less than 0 min per day on housework, so the smallest possible value, 0, is only 1.63 standard deviations below the mean). However, because the authors reported that there were no outliers in the data and because n is relatively close to 30, we use the t confidence interval formula of this section to compute a 95% confidence interval.

Because $n = 26$, $df = 25$, and the appropriate t critical value is 2.06, the confidence interval is then

$$\bar{x} \pm (t \text{ critical value})\left(\frac{s}{\sqrt{n}}\right) = 14.0 \pm (2.06)\left(\frac{8.6}{\sqrt{26}}\right)$$
$$= 14.0 \pm 3.5$$
$$= (10.5, 17.5)$$

Based on the sample data, we believe that the true mean time per weekday spent on housework is between 10.5 and 17.5 min for girls in two-parent families where both parents work. We used a method that has a 5% error rate to construct this interval. We should be somewhat cautious in interpreting this confidence interval because of the concern expressed about the normality of the population distribution.

◼

▪ Choosing the Sample Size ....

When estimating μ using a large sample or a small sample from a normal population with known σ, the bound B on the error of estimation associated with a 95% confidence interval is

$$B = 1.96\left(\frac{\sigma}{\sqrt{n}}\right)$$

Before collecting any data, an investigator may wish to determine a sample size for which a particular value of the bound is achieved. For example, with μ representing the average fuel efficiency (in miles per gallon, mpg) for all cars of a certain type, the objective of an investigation may be to estimate μ to within 1 mpg with 95% confidence. The value of n necessary to achieve this is obtained by setting $B = 1$ and then solving $1 = 1.96\left(\frac{\sigma}{\sqrt{n}}\right)$ for n.

In general, suppose that we wish to estimate μ to within an amount B (the specified bound on the error of estimation) with 95% confidence. Finding the necessary sample size requires solving the equation $B = 1.96\left(\frac{\sigma}{\sqrt{n}}\right)$ for n. The result is

$$n = \left(\frac{1.96\sigma}{B}\right)^2$$

Notice that, in general, a large value of σ forces n to be large, as does a small value of B.

Use of the sample-size formula requires that σ be known, but this is rarely the case in practice. One possible strategy for estimating σ is to carry out a preliminary study and use the resulting sample standard deviation (or a somewhat larger value, to be conservative) to determine n for the main part of the study. Another possibility is simply to make an educated guess about the value of σ and to use that value to calculate n. For a population distribution that is not too skewed, dividing the range (the difference between the largest and the smallest values) by 4 often gives a rough idea of the value of the standard deviation.

The sample size required to estimate a population mean μ to within an amount B with 95% confidence is

$$n = \left(\frac{1.96\sigma}{B}\right)^2$$

If σ is unknown, it may be estimated based on previous information or, for a population that is not too skewed, by using (range)/4.

If the desired confidence level is something other than 95%, 1.96 is replaced by the appropriate z critical value (e.g., 2.58 for 99% confidence).

Example 9.11 Cost of Textbooks

The financial aid office wishes to estimate the mean cost of textbooks per quarter for students at a particular university. For the estimate to be useful, it should be within $20 of the true population mean. How large a sample should be used to be 95% confident of achieving this level of accuracy?

To determine the required sample size, we must have a value for σ. The financial aid office is pretty sure that the amount spent on books varies widely, with most values between $50 and $450. A reasonable estimate of σ is then

$$\frac{range}{4} = \frac{450 - 50}{4} = \frac{400}{4} = 100$$

The required sample size is

$$n = \left(\frac{1.96\sigma}{B}\right)^2 = \left(\frac{(1.96)(100)}{20}\right)^2 = (9.8)^2 = 96.04$$

Rounding up, a sample size of 97 or larger is recommended.

▪

Exercises 9.30–9.50

Turn to the Expanded Answer Section for answers not shown next to exercises.

9.30 Given a variable that has a t distribution with the specified degrees of freedom, what percentage of the time will its value fall in the indicated region?
a. 10 df, between -1.81 and 1.81   90%
b. 10 df, between -2.23 and 2.23   95%
c. 24 df, between -2.06 and 2.06   95%
d. 24 df, between -2.80 and 2.80   99%
e. 24 df, outside the interval from -2.80 to 2.80   1%
f. 24 df, to the right of 2.80   0.5
g. 10 df, to the left of -1.81 5

9.31 The formula used to compute a confidence interval for the mean of a normal population when n is small is

$$\bar{x} \pm (t \text{ critical value})\frac{s}{\sqrt{n}}$$

What is the appropriate t critical value for each of the following confidence levels and sample sizes?
a. 95% confidence, $n = 17$ 2.12
b. 90% confidence, $n = 12$ 1.80
c. 99% confidence, $n = 24$ 2.81

Bold exercises answered in back   ● Data set available online but not required   ▼ Video solution available

d. 90% confidence, $n = 25$ 1.71
e. 90% confidence, $n = 13$ 1.78
f. 95% confidence, $n = 10$ 2.26

9.32 The two intervals (114.4, 115.6) and (114.1, 115.9) are confidence intervals for μ = true average resonance frequency (in hertz) for all tennis rackets of a certain type.
a. What is the value of the sample mean resonance frequency?   115.0
b. The confidence level for one of these intervals is 90% and for the other it is 99%. Which is which, and how can you tell?

9.33 ▼ Samples of two different types of automobiles were selected, and the actual speed for each car was determined when the speedometer registered 50 mph. The resulting 95% confidence intervals for true average actual speed were (51.3, 52.7) and (49.4, 50.6). Assuming that the two sample standard deviations are identical, which confidence interval is based on the larger sample size? Explain your reasoning.   (49.4, 50.6) is based on the larger sample size because the interval is narrower.

9.34 Suppose that a random sample of 50 bottles of a particular brand of cough medicine is selected and the alcohol content of each bottle is determined. Let μ denote the average alcohol content for the population of all bottles of the brand under study. Suppose that the sample of 50 results in a 95% confidence interval for μ of (7.8, 9.4).
a. Would a 90% confidence interval have been narrower or wider than the given interval? Explain your answer.
b. Consider the following statement: There is a 95% chance that μ is between 7.8 and 9.4. Is this statement correct? Why or why not?
c. Consider the following statement: If the process of selecting a sample of size 50 and then computing the corresponding 95% confidence interval is repeated 100 times, 95 of the resulting intervals will include μ. Is this statement correct? Why or why not?   Incorrect; can't be certain that exactly 95 of the 100 will contain μ.

9.35 ▼ Acrylic bone cement is sometimes used in hip and knee replacements to fix an artificial joint in place. The force required to break an acrylic bone cement bond was measured for six specimens under specified conditions, and the resulting mean and standard deviation were 306.09 Newtons and 41.97 Newtons, respectively. Assuming that it is reasonable to assume that breaking force under these conditions has a distribution that is approximately normal, estimate the true average breaking force for acrylic bone cement under the specified conditions.

9.36 The article "The Association Between Television Viewing and Irregular Sleep Schedules Among Children Less Than 3 Years of Age" (*Pediatrics* [2005]: 851–856) reported the accompanying 95% confidence intervals for average TV viewing time (in hours per day) for three different age groups.

Age Group	95% Confidence Interval
Less than 12 months	(0.8, 1.0)
12 to 23 months	(1.4, 1.8)
24 to 35 months	(2.1, 2.5)

a. Suppose that the sample sizes for each of the three age group samples were equal. Based on the given confidence intervals, which of the age group samples had the greatest variability in TV viewing time? Explain your choice.
b. Now suppose that the sample standard deviations for the three age group samples were equal, but that the three sample sizes might have been different. Which of the three age group samples had the largest sample size? Explain your choice.
c. The interval (.768, 1.302) is either a 90% confidence interval or a 99% confidence interval for the mean TV viewing time for children less than 12 months old. Is the confidence level for this interval 90% or 99%? Explain your choice.

9.37 Five hundred randomly selected working adults living in Calgary, Canada were asked how long, in minutes, their typical daily commute was (*Calgary Herald* Traffic Study, Ipsos, September 17, 2005). The resulting sample mean and standard deviation of commute time were 28.5 minutes and 24.2 minutes, respectively. Construct and interpret a 90% confidence interval for the mean commute time of working adult Calgary residents.

9.38 The article "Most Canadians Plan to Buy Treats, Many Will Buy Pumpkins, Decorations and/or Costumes" (Ipsos-Reid, October 24, 2005) summarized results from a survey of 1000 randomly selected Canadian residents. Each individual in the sample was asked how much he or she anticipated spending on Halloween during 2005. The resulting sample mean and standard deviation were $46.65 and $83.70 respectively.
a. Explain how it could be possible for the standard deviation of the anticipated Halloween expense to be larger than the mean anticipated expense.

b. Is it reasonable to think that the distribution of the variable *anticipated Halloween expense* is approximately normal? Explain why or why not.

c. Is it appropriate to use the *t* confidence interval to estimate the mean anticipated Halloween expense for Canadian residents? Explain why or why not.

d. If appropriate, construct and interpret a 99% confidence interval for the mean anticipated Halloween expense for Canadian residents.   (39.82, 53.48)

9.39 Because of safety considerations, in May 2003 the Federal Aviation Administration (FAA) changed its guidelines for how small commuter airlines must estimate passenger weights. Under the old rule, airlines used 180 lb as a typical passenger weight (including carry-on luggage) in warm months and 185 lb as a typical weight in cold months. The *Alaska Journal of Commerce* (May 25, 2003) reported that Frontier Airlines conducted a study to estimate average passenger plus carry-on weights. They found an average summer weight of 183 lb and a winter average of 190 lb. Suppose that each of these estimates was based on a random sample of 100 passengers and that the sample standard deviations were 20 lb for the summer weights and 23 lb for the winter weights.

a. Construct and interpret a 95% confidence interval for the mean summer weight (including carry-on luggage) of Frontier Airlines passengers.   (179.03, 186.97)

b. Construct and interpret a 95% confidence interval for the mean winter weight (including carry-on luggage) of Frontier Airlines passengers.   (185.44, 194.56)

c. The new FAA recommendations are 190 lb for summer and 195 lb for winter. Comment on these recommendations in light of the confidence interval estimates from Parts (a) and (b).

9.40 "Heinz Plays Catch-up After Under-Filling Ketchup Containers" is the headline of an article that appeared on CNN.com (November 30, 2000). The article stated that Heinz had agreed to put an extra 1% of ketchup into each ketchup container sold in California for a 1-year period. Suppose that you want to make sure that Heinz is in fact fulfilling its end of the agreement. You plan to take a sample of 20-oz bottles shipped to California, measure the amount of ketchup in each bottle, and then use the resulting data to estimate the mean amount of ketchup in each bottle. A small pilot study showed that the amount of ketchup in 20-oz bottles varied from 19.9 to 20.3 oz. How many bottles should be included in the sample if you want to estimate the true mean amount of ketchup to within 0.1 oz with 95% confidence?   4

9.41 ● ▼ Example 9.3 gave the following airborne times for United Airlines flight 448 from Albuquerque to Denver on 10 randomly selected days:

| 57 | 54 | 55 | 51 | 56 | 48 | 52 | 51 | 59 | 59 |

a. Compute and interpret a 90% confidence interval for the mean airborne time for flight 448.   (52.07, 56.33)

b. Give an interpretation of the 90% confidence level associated with the interval estimate in Part (a).

c. Based on your interval in Part (a), if flight 448 is scheduled to depart at 10 A.M., what would you recommend for the published arrival time? Explain.   10.57 a.m.

9.42 The authors of the paper "Short-Term Health and Economic Benefits of Smoking Cessation: Low Birth Weight" (*Pediatrics* [1999]: 1312–1320) investigated the medical cost associated with babies born to mothers who smoke. The paper included estimates of mean medical cost for low-birth-weight babies for different ethnic groups. For a sample of 654 Hispanic low-birth-weight babies, the mean medical cost was $55,007 and the standard error $(s/\sqrt{n})$ was $3011. For a sample of 13 Native American low-birth-weight babies, the mean and standard error were $73,418 and $29,577, respectively. Explain why the two standard errors are so different.   Standard error depends on sample size, and here the sample sizes are very different.

9.43 ● ▼ A study of the ability of individuals to walk in a straight line ("Can We Really Walk Straight?" *American Journal of Physical Anthropology* [1992]: 19–27) reported the following data on cadence (strides per second) for a sample of $n = 20$ randomly selected healthy men:

| 0.95 | 0.85 | 0.92 | 0.95 | 0.93 | 0.86 | 1.00 | 0.92 | 0.85 | 0.81 |
| 0.78 | 0.93 | 0.93 | 1.05 | 0.93 | 1.06 | 1.06 | 0.96 | 0.81 | 0.96 |

Construct and interpret a 99% confidence interval for the population mean cadence.   (0.874, 0.978)

9.44 ● Fat contents (in percentage) for 10 randomly selected hot dogs were given in the article "Sensory and Mechanical Assessment of the Quality of Frankfurters" (*Journal of Texture Studies* [1990]: 395–409). Use the following data to construct a 90% confidence interval for the true mean fat percentage of hot dogs:

| 25.2 | 21.3 | 22.8 | 17.0 | 29.8 | 21.0 | 25.5 | 16.0 | 20.9 | 19.5 |

(19.51, 24.29)

Bold exercises answered in back   ● Data set available online but not required   ▼ Video solution available

9.45 ● Five students visiting the student health center for a free dental examination during National Dental Hygiene Month were asked how many months had passed since their last visit to a dentist. Their responses were as follows:

| 6 | 17 | 11 | 22 | 29 |

Assuming that these five students can be considered a random sample of all students participating in the free checkup program, construct a 95% confidence interval for the mean number of months elapsed since the last visit to a dentist for the population of students participating in the program. (5.77, 28.23)

9.46 The article "First Year Academic Success: A Prediction Combining Cognitive and Psychosocial Variables for Caucasian and African American Students" (*Journal of College Student Development* [1999]: 599–610) reported that the sample mean and standard deviation for high school grade point average (GPA) for students enrolled at a large research university were 3.73 and 0.45, respectively. Suppose that the mean and standard deviation were based on a random sample of 900 students at the university.
a. Construct a 95% confidence interval for the mean high school GPA for students at this university. (3.70, 3.76)
b. Suppose that you wanted to make a statement about the range of GPAs for students at this university. Is it reasonable to say that 95% of the students at the university have GPAs in the interval you computed in Part (a)? Explain.

9.47 ● ▼ The following data are the calories per half-cup serving for 16 popular chocolate ice cream brands reviewed by *Consumer Reports* (July 1999):

| 270 | 150 | 170 | 140 | 160 | 160 | 160 | 290 |
| 190 | 190 | 160 | 170 | 150 | 110 | 180 | 170 |

Is it reasonable to use the t confidence interval to compute a confidence interval for μ, the true mean calories per half-cup serving of chocolate ice cream? Explain why or why not.

9.48 The Bureau of Alcohol, Tobacco, and Firearms (BATF) has been concerned about lead levels in California wines. In a previous testing of wine specimens, lead levels ranging from 50 to 700 parts per billion were recorded (*San Luis Obispo Telegram Tribune*, June 11, 1991). How many wine specimens should be tested if the BATF wishes to estimate the true mean lead level for California wines to within 10 parts per billion with 95% confidence? 1015

9.49 ▼ The article "*National Geographic*, the Doomsday Machine," which appeared in the March 1976 issue of the *Journal of Irreproducible Results* (yes, there really is a journal by that name—it's a spoof of technical journals!) predicted dire consequences resulting from a nationwide buildup of *National Geographic* magazines. The author's predictions are based on the observation that the number of subscriptions for *National Geographic* is on the rise and that no one ever throws away a copy of *National Geographic*. A key to the analysis presented in the article is the weight of an issue of the magazine. Suppose that you were assigned the task of estimating the average weight of an issue of *National Geographic*. How many issues should you sample to estimate the average weight to within 0.1 oz with 95% confidence? Assume that σ is known to be 1 oz. 385

9.50 The formula described in this section for determining sample size corresponds to a confidence level of 95%. What would be the appropriate formula for determining sample size when the desired confidence level is 90%? 98%?

Bold exercises answered in back ● Data set available online but not required ▼ Video solution available

---

9.4 Interpreting and Communicating the Results of Statistical Analyses

The purpose of most surveys and many research studies is to produce estimates of population characteristics. One way of providing such an estimate is to construct and report a confidence interval for the population characteristic of interest.

■ Communicating the Results of a Statistical Analysis ....

When using sample data to estimate a population characteristic, a point estimate or a confidence interval estimate might be used. Confidence intervals are generally preferred because a point estimate by itself does not convey any information about the accuracy of the estimate. For this reason, whenever you report the value of a point estimate, it is a good idea to also include an estimate of the bound on the error of estimation.

Reporting and interpreting a confidence interval estimate requires a bit of care. First, always report both the confidence interval and the confidence level associated with the method used to produce the interval. Then, remember that both the confidence interval and the confidence level should be interpreted. A good strategy is to begin with an interpretation of the confidence interval in the context of the problem and then to follow that with an interpretation of the confidence level. For example, if a 90% confidence interval for π, the proportion of students at a particular university who own a computer, is (.56, .78), we might say

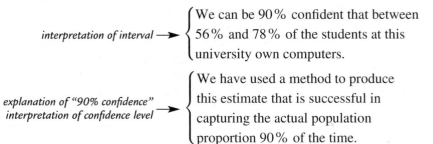

interpretation of interval → $\Big\{$ We can be 90% confident that between 56% and 78% of the students at this university own computers.

explanation of "90% confidence"
interpretation of confidence level → $\Big\{$ We have used a method to produce this estimate that is successful in capturing the actual population proportion 90% of the time.

When providing an interpretation of a confidence interval, remember that the interval is an estimate of a population characteristic and be careful *not* to say that the interval applies to individual values in the population or to the values of sample statistics. For example, if a 99% confidence interval for μ, the mean amount of ketchup in bottles labeled as 12 oz, is (11.94, 11.98), this does not tell us that 99% of 12-oz ketchup bottles contain between 11.94 and 11.98 oz of ketchup. Nor does it tell us that 99% of samples of the same size would have sample means in this particular range. The confidence interval is an estimate of the *mean* for all bottles in the *population* of interest.

■ Interpreting the Results of a Statistical Analysis ....

Unfortunately, there is no customary way of reporting the estimates of population characteristics in published sources. Possibilities include

confidence interval
estimate $\pm$ bound on error
estimate $\pm$ standard error

If the population characteristic being estimated is a population mean, then you may also see

sample mean $\pm$ sample standard deviation

If the interval reported is described as a confidence interval, a confidence level should accompany it. These intervals can be interpreted just as we have interpreted the confidence intervals in this chapter, and the confidence level specifies the long-run error rate associated with the method used to construct the interval (e.g., a 95% confidence level specifies a 5% long-run error rate).

A form particularly common in news articles is estimate ± bound on error, where the bound on error is also sometimes called the **margin of error**. The bound on error reported is usually 2 times the standard deviation of the estimate. This method of reporting is a little more informal than a confidence interval and, if the sample size is reasonably large, is roughly equivalent to reporting a 95% confidence interval. You can interpret these intervals as you would a confidence interval with approximate confidence level of 95%.

You must use care in interpreting intervals reported in the form of an estimate ± standard error. Recall from Section 9.2 that the general form of a confidence interval is

estimate ± (critical value)(standard deviation of the estimate)

In journal articles, the estimated standard deviation of the estimate is usually referred to as the *standard error*. The critical value in the confidence interval formula was determined by the form of the sampling distribution of the estimate and by the confidence level. Note that the reported form, estimate ± standard error, is equivalent to a confidence interval with the critical value set equal to 1. For a statistic whose sampling distribution is (approximately) normal (such as the mean of a large sample or a large-sample proportion), a critical value of 1 corresponds to an approximate confidence level of about 68%. Because a confidence level of 68% is rather low, you may want to use the given information and the confidence interval formula to convert to an interval with a higher confidence level.

When researchers are trying to estimate a population mean, they sometimes report sample mean ± sample standard deviation. Be particularly careful here. To convert this information into a useful interval estimate of the population mean, you must first convert the sample standard deviation to the standard error of the sample mean (by dividing by $\sqrt{n}$) and then use the standard error and an appropriate critical value to construct a confidence interval.

For example, suppose that a random sample of size 100 is used to estimate the population mean. If the sample resulted in a sample mean of 500 and a sample standard deviation of 20, you might find the published results summarized in any of the following ways:

95% confidence interval for the population mean: (496.08, 503.92)
mean ± bound on error: 500 ± 4
mean ± standard error: 500 ± 2
mean ± standard deviation: 500 ± 20

■ What to Look For in Published Data ....

Here are some questions to ask when you encounter interval estimates in research reports.

- ■ Is the reported interval a confidence interval, mean ± bound on error, mean ± standard error, or mean ± standard deviation? If the reported interval is not a confidence interval, you may want to construct a confidence interval from the given information.
- ■ What confidence level is associated with the given interval? Is the choice of confidence level reasonable? What does the confidence level say about the long-run error rate of the method used to construct the interval?

■ Is the reported interval relatively narrow or relatively wide? Has the population characteristic been estimated precisely?

For example, the article "Use of a Cast Compared with a Functional Ankle Brace After Operative Treatment of an Ankle Fracture" (*Journal of Bone and Joint Surgery* [2003]: 205–211) compared two different methods of immobilizing an ankle after surgery to repair damage from a fracture. The article includes the following statement:

> The mean duration (and standard deviation) between the operation and return to work was 63±13 days (median, sixty-three days; range, thirty three to ninety-eight days) for the cast group and 65±19 days (median, sixty-two days; range, eight to 131 days) for the brace group; the difference was not significant.

This is an example of a case where we must be careful—the reported intervals are of the form estimate ± standard deviation. We can use this information to construct a confidence interval for the mean time between surgery and return to work for each method of immobilization. One hundred patients participated in the study, with 50 wearing a cast after surgery and 50 wearing an ankle brace (random assignment was used to assign patients to treatment groups). Because the sample sizes are both large, we can use the *t* confidence interval formula

$$\text{mean} \pm (t \text{ critical value})\left(\frac{s}{\sqrt{n}}\right)$$

Each sample has df $= 50 - 1 = 49$. The closest df value in Appendix Table 3 is for df $= 40$, and the corresponding *t* critical value for a 95% confidence level is 2.02. The corresponding intervals are

$$\text{Cast: } 63 \pm 2.02\left(\frac{13}{\sqrt{50}}\right) = 63 \pm 3.71 = (59.29, 66.71)$$

$$\text{Brace: } 65 \pm 2.02\left(\frac{19}{\sqrt{50}}\right) = 65 \pm 5.43 = (59.57, 70.43)$$

The chosen confidence level of 95% implies that the method used to construct each of the intervals has a 5% long-run error rate. Assuming that it is reasonable to view these samples as representative of the patient population, we can interpret these intervals as follows: We can be 95% confident that the mean return-to-work time for those treated with a cast is between 59.29 and 66.71 days, and we can be 95% confident that the mean return-to-work time for those treated with an ankle brace is between 59.57 and 70.43 days. These intervals are relatively wide, indicating that the values of the treatment means have not been estimated as precisely as we might like. This is not surprising, given the sample sizes and the variability in each sample. Note that the two intervals overlap. This supports the statement that the difference between the two immobilization methods was not significant. Formal methods for directly comparing two groups, covered in Chapter 11, could be used to further investigate this issue.

■ A Word to the Wise: Cautions and Limitations ....

When working with point and confidence interval estimates, here are a few things you need to keep in mind;

1. In order for an estimate to be useful, we must know something about accuracy. You should beware of point estimates that are not accompanied by a bound on error or some other measure of accuracy.

2. A confidence interval estimate that is wide indicates that we don't have very precise information about the population characteristics being estimated. Don't be fooled by a high confidence level if the resulting interval is wide. High confidence, while desirable, is not the same thing as saying we have precise information about the value of a population characteristic.

   The width of a confidence interval is affected by the confidence level, the sample size, and the standard deviation of the statistic used (e.g. p or $\bar{x}$) as the basis for constructing the interval. The best strategy for decreasing the width of a confidence interval is to take a larger sample. It is far better to think about this before collecting data and to use the required sample size formulas to determine a sample size that will result in a confidence interval estimate that is narrow enough to provide useful information.

3. The accuracy of estimates depends on the sample size, not the population size. This may be counter to intuition, but as long as the sample size is small relative to the population size (n less than 10% of the population size), the bound on error for estimating a population proportion with 95% confidence is approximately $2\sqrt{\dfrac{p(1-p)}{n}}$ and for estimating a population mean with 95% confidence is approximately $2\dfrac{s}{\sqrt{n}}$.

   Note that each of these involves the sample size n, and both bounds decrease as the sample size increases. Neither approximate bound on error depends on the population size.

   The size of the population does need to be considered if sampling is without replacement and the sample size is more than 10% of the population size. In this case, a **finite population correction factor** $\sqrt{\dfrac{N-n}{N-1}}$ is used to adjust the bound on error (the given bound is multiplied by the correction factor). Since this correction factor is always less than 1, the adjusted bound on error is smaller.

4. Assumptions and "plausibility" conditions are important. The confidence interval procedures of this chapter require certain assumptions. If these assumptions are met, the confidence intervals provide us with a method for using sample data to estimate population characteristics with confidence. When the assumptions associated with a confidence interval procedure are in fact true, the confidence level specifies a correct success rate for the method. However, assumptions (such as the assumption of a normal population distribution) are rarely exactly met in practice. Fortunately, in most cases, as long as the assumptions are approximately met, the confidence interval procedures still work well.

   In general we can only determine if assumptions are "plausible" or approximately met, and that we are in the situation where we expect the inferential procedure to work reasonably well. This is usually confirmed by knowledge of the data collection process and by using the sample data to check certain "plausibility conditions".

   The formal assumptions for the z confidence interval for a population proportion are

   1. The sample is a random sample from the population of interest.
   2. The sample size is large enough for the sampling distribution of p to be approximately normal.
   3. Sampling is without replacement.

Whether the random sample assumption is plausible will depend on how the sample was selected and the intended population. Plausibility conditions for the other two assumptions are the following:

$np \geq 10$ and $n(1 - p) \geq 10$ (so the sampling distribution of p is approximately normal), and

n is less than 10% of the population size (so sampling with replacement approximates sampling without replacement).

The formal assumptions for the t confidence interval for a population mean are

1. The sample is a random sample from the population of interest.

2. The population distribution is normal, so that the distribution of $t = \dfrac{\bar{x} - \mu}{s/\sqrt{n}}$

   has a t distribution.

The plausibility of the random sample assumption, as was the case for proportions, will depend on how the sample was selected and the population of interest. The plausibility conditions for the normal population distribution assumption are the following:

A normal probability plot of the data is reasonably straight (indicating that the population distribution is approximately normal), or

The data distribution is approximately symmetric and there are no outliers. This may be confirmed by looking at a dotplot, boxplot, stem-and-leaf display, or histogram of the data. (This would indicate that the population is approximately normal.)

Alternatively, if n is large ($n \geq 30$), the sampling distribution of $\bar{x}$ will be approximately normal even for nonnormal population distributions. This implies that use of the t interval is appropriate even if population normality is not plausible.

In the end, you must decide that the assumptions are met or that they are "plausible" and that the inferential method used will provide reasonable results. This is also true for the inferential methods introduced in the chapters that follow.

5. Watch out for the "$\pm$" when reading published reports. Don't fall into the trap of thinking confidence interval every time you see a $\pm$ in an expression. As was discussed earlier in this section, published reports are not consistent, and in addition to confidence intervals, it is common to see estimate $\pm$ standard error and estimate $\pm$ sample standard deviation reported.

A c t i v i t y 9.1   Getting a Feel for Confidence Level

Technology Activity (Applet): Open the applet (available in CengageNOW at www.cengage.com/login) called ConfidenceIntervals. You should see a screen like the one shown.

Simulating Confidence Intervals

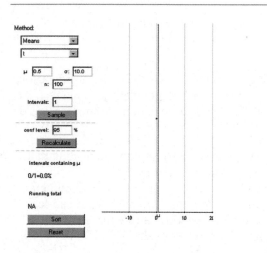

Getting Started: If the "Method" box does not say "Means," use the drop-down menu to select Means. In the box just below, select "t" from the drop-down menu. This applet will select a random sample from a specified normal population distribution and then use the sample to construct a confidence interval for the population mean. The interval is then plotted on the display at the right, and we can see if the resulting interval contains the actual value of the population mean.

For purposes of this activity, we will sample from a normal population with mean 100 and standard deviation 5. We will begin with a sample size of $n = 10$. In the applet window, set $\mu = 100$, $\sigma = 5$ and $n = 10$. Leave the conf-level box set at 95%. Click the "Recalculate" button to rescale the picture on the right. Now click on the sample button. You should see a confidence interval appear on the display on the right hand side. If the interval contains the actual mean of 100, the interval is drawn in green; if 100 is not in the confidence interval, the interval is shown in red. Your screen should look something like the following.

Simulating Confidence Intervals

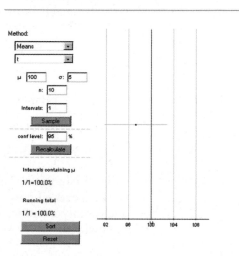

Part 1: Click on the "Sample" button several more times, and notice how the confidence interval estimate changes from sample to sample. Also notice that at the bottom of the left-hand side of the display, the applet is keeping track of the proportion of all the intervals calculated so far that include the actual value of μ. If we were to construct a large number of intervals, this proportion should closely approximate the capture rate for the confidence interval method.

To look at more than 1 interval at a time, change the "Intervals" box from 1 to 100, and then click the sample button. You should see a screen similar to the one at the top of page 515, with 100 intervals in the display on the right-hand side. Again, intervals containing 100 (the value of μ in this case) will be green and those that do not contain 100 will be red. Also note that the capture proportion on the left-hand side has also been updated to reflect what happened with the 100 newly generated intervals.

Simulating Confidence Intervals

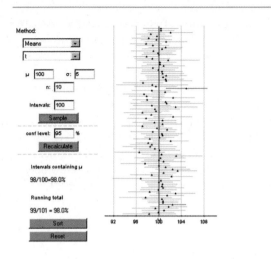

Continue generating intervals until you have seen at least 1000 intervals, and then answer the following question:

a. How does the proportion of intervals constructed that contain $\mu = 100$ compare to the stated confidence level of 95%? On how many intervals was your proportion based? (Note—if you followed the instructions, this should be at least 1000.)

Experiment with three other confidence levels of your choice, and then answer the following question:

b. In general, is the proportion of computed t confidence intervals that contain $\mu = 100$ close to the stated confidence level?

Part 2: When the population is normal but σ is unknown, we construct a confidence interval for a population mean using a t critical value rather than a z critical value. How important is this distinction?

Let's investigate. Use the drop-down menu to change the box just below the method box that's says "Means" from "t" to "z with s." The applet will now construct intervals using the sample standard deviation, but will use a z critical value rather than the t critical value.

Use the applet to construct at least 1000 95% intervals, and then answer the following question:

c. Comment on how the proportion of the computed intervals that include the actual value of the population mean compares to the stated confidence level of 95%. Is this surprising? Explain why or why not.

Now experiment with some different samples sizes. What happens when $n = 20$? $n = 50$? $n = 100$? Use what you have learned to write a paragraph explaining what these simulations tell you about the advisability of using a z critical value in the construction of a confidence interval for μ when σ is unknown.

Activity 9.2   An Alternative Confidence Interval for a Population Proportion

CENGAGENOW™

Explore this applet activity

Technology Activity (Applet): This activity presumes that you have already worked through Activity 9.1.

Background: In Section 9.2, it was suggested that a confidence interval of the form

$$P_{mod} \pm (z \text{ critical value}) \sqrt{\frac{P_{mod}(1 - P_{mod})}{n}}$$

where $P_{mod} = \dfrac{\text{successes} + 2}{n + 4}$ is an alternative to the usual large-sample z confidence interval. This alternative interval is preferred by many statisticians because, in repeated sampling, the proportion of intervals constructed that include the actual value of the population proportion, π, tends to be closer to the stated confidence level. In this activity, we will explore how the "capture rates" for the two different interval estimation methods compare.

Open the applet (available in CengageNOW at www.cengage.com/login) called ConfidenceIntervals. You should see a screen like the one shown.

Simulating Confidence Intervals

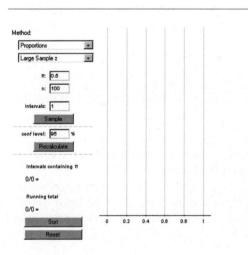

Select "Proportion" from the Method box drop-down menu, and then select "Large Sample z" from the drop-down menu of the second box. We will consider sampling from a population with $\pi = .3$ using a sample size of 40. In the applet window, enter $\pi = .3$ and $n = 40$. Note that $n = 40$ is large enough to satisfy $n\pi \geq 10$ and $n(1 - \pi) \geq 10$.

Set the "Intervals" box to 100, and then use the applet to construct a large number (at least 1000) of 95% confidence intervals.

1. How does the proportion of intervals constructed that include $\pi = .3$, the population proportion, compare to 95%? Does this surprise you? Explain.

Now use the drop-down menu to change "Large Sample z" to "Modified." Now the applet will construct the alternative confidence interval that is based on P_{mod}. Use the applet to construct a large number (at least 1000) of 95% confidence intervals.

2. How does the proportion of intervals constructed that include $\pi = .3$, the population proportion, compare to 95%? Is this proportion closer to 95% than was the case for the large-sample z interval?

3. Experiment with different combinations of values of sample size and population proportion π. Can you find a combination for which the large sample z interval has a capture rate that is close to 95%? Can you find a combination for which it has a capture rate that is even farther from 95% than it was for $n = 40$ and $\pi = .3$? How does the modified interval perform in each of these cases?

Activity 9.3   Verifying Signatures on a Recall Petition

Background: In 2003, petitions were submitted to the California Secretary of State calling for the recall of Governor Gray Davis. Each of California's 58 counties then had to report the number of valid signatures on the petitions from that county so that the State could determine whether there were enough valid signatures to certify the recall and set a date for the recall election. The following paragraph appeared in the *San Luis Obispo Tribune* (July 23, 2003):

> In the campaign to recall Gov. Gray Davis, the secretary of state is reporting 16,000 verified signatures from San Luis Obispo County. In all, the County Clerk's Office received 18,866 signatures on recall petitions and was instructed by the state to check a random sample of 567. Out of those, 84.48% were good. The verification process includes checking whether the signer is a registered voter and whether the address and signature on the recall petition match the voter registration.

1. Use the data from the random sample of 567 San Luis Obispo County signatures to construct a 95% confidence interval for the proportion of petition signatures that are valid.

2. How do you think that the reported figure of 16,000 verified signature for San Luis Obispo County was obtained?

3. Based on your confidence interval from Step 1, explain why you think that the reported figure of 16,000 verified signatures is or is not reasonable.

Activity 9.4   A Meaningful Paragraph

Write a meaningful paragraph that includes the following six terms: **sample, population, confidence level, estimate, mean, margin of error**.

A "meaningful paragraph" is a coherent piece writing in an appropriate context that uses all of the listed words. The paragraph should show that you understand the meaning of the terms and their relationship to one another. A sequence of sentences that just define the terms is *not* a meaningful paragraph. When choosing a context, think carefully about the terms you need to use. Choosing a good context will make writing a meaningful paragraph easier.

Summary of Key Concepts and Formulas

Term or Formula	**Comment**
Point estimate	A single number, based on sample data, that represents a plausible value of a population characteristic.
Unbiased statistic	A statistic that has a sampling distribution with a mean equal to the value of the population characteristic to be estimated.
Confidence interval	An interval that is computed from sample data and provides a range of plausible values for a population characteristic.
Confidence level	A number that provides information on how much "confidence" we can have in the method used to construct a confidence interval estimate. The confidence level specifies the percentage of all possible samples that will produce an interval containing the true value of the population characteristic.
$p \pm (z \text{ critical value}) \sqrt{\dfrac{p(1-p)}{n}}$	A formula used to construct a confidence interval for π when the sample size is large.
$n = \pi(1-\pi)\left(\dfrac{1.96}{B}\right)^2$	A formula used to compute the sample size necessary for estimating π to within an amount B with 95% confidence. (For other confidence levels, replace 1.96 with an appropriate z critical value.)
$\bar{x} \pm (z \text{ critical value}) \dfrac{\sigma}{\sqrt{n}}$	A formula used to construct a confidence interval for μ when σ is known and either the sample size is large or the population distribution is normal.
$\bar{x} \pm (t \text{ critical value}) \dfrac{s}{\sqrt{n}}$	A formula used to construct a confidence interval for μ when σ is unknown and either the sample size is large or the population distribution is normal.
$n = \left(\dfrac{1.96\sigma}{B}\right)^2$	A formula used to compute the sample size necessary for estimating μ to within an amount B with 95% confidence. (For other confidence levels, replace 1.96 with an appropriate z critical value.)

Chapter Review Exercises 9.51-9.73

Turn to the Expanded Answer Section for answers not shown next to exercises.

CENGAGENOW· Know exactly what to study! Take a pre-test and receive your Personalized Learning Plan.

9.51 Despite protests from civil libertarians and gay rights activists, many people favor mandatory AIDS testing of certain at-risk groups, and some people even believe that all citizens should be tested. What proportion of the adults in the United States favor mandatory testing for all citizens? To assess public opinion on this issue, researchers conducted a survey of 1014 randomly selected adult U.S. citizens ("Large Majorities Continue to Back AIDS Testing," *Gallup Poll Monthly* [1991]: 25–28). The article reported that 466 of the 1014 people surveyed believed that all citizens should be tested. Use this information to estimate π, the true proportion of all U.S. adults who favor AIDS testing of all citizens.   0.4596

9.52 The article "Consumers Show Increased Liking for Diesel Autos" (*USA Today*, January 29, 2003) reported that 27% of U.S. consumers would opt for a diesel car if it ran as cleanly and performed as well as a car with a gas engine. Suppose that you suspect that the proportion might be different in your area and that you want to conduct a survey to estimate this proportion for the adult residents of your city. What is the required sample size if you want to estimate this proportion to within .05 with 95% confidence? Compute the required sample size first using .27 as a preliminary estimate of π and then using the conservative value of .5. How do the two sample sizes compare? What sample size would you recommend for this study?

9.53 In the article "Fluoridation Brushed Off by Utah" (Associated Press, August 24, 1998), it was reported that a small but vocal minority in Utah has been successful in keeping fluoride out of Utah water supplies despite evidence that fluoridation reduces tooth decay and despite the fact that a clear majority of Utah residents favor fluoridation. To support this statement, the article included the result of a survey of Utah residents that found 65% to be in favor of fluoridation. Suppose that this result was based on a random sample of 150 Utah residents. Construct and interpret a 90% confidence interval for π, the true proportion of Utah residents who favor fluoridation. Is this interval consistent with the statement that fluoridation is favored by a clear majority of residents?   (0.586, 0.714) Yes, because the lower bound is greater than 0.5.

9.54 Seventy-seven students at the University of Virginia were asked to keep a diary of a conversation with their mothers, recording any lies they told during these conversations (*San Luis Obispo Telegram-Tribune*, August 16, 1995). It was reported that the mean number of lies per conversation was 0.5. Suppose that the standard deviation (which was not reported) was 0.4.
a. Suppose that this group of 77 is a random sample from the population of students at this university. Construct a 95% confidence interval for the mean number of lies per conversation for this population.   (0.409, 0.591)
b. The interval in Part (a) does not include 0. Does this imply that all students lie to their mothers? Explain.

9.55 The article "Selected Characteristics of High-Risk Students and Their Enrollment Persistence" (*Journal of College Student Development* [1994]: 54–60) examined factors that affect whether students stay in college. The following summary statistics are based on data from a sample of 44 students who did not return to college after the first quarter (the nonpersisters) and a sample of 257 students who did return (the persisters):

Number of Hours Worked per Week During the First Quarter

	Mean	Standard Deviation
Nonpersisters	25.62	14.41
Persisters	18.10	15.31

a. Consider the 44 nonpersisters as a random sample from the population of all nonpersisters at the university where the data were collected. Compute a 98% confidence interval for the mean number of hours worked per week for nonpersisters.   (20.371, 30.869)
b. Consider the 257 persisters as a random sample from the population of all persisters at the university where the data were collected. Compute a 98% confidence interval for the mean number of hours worked per week for persisters.   (15.864, 20.336)

Bold exercises answered in back   ● Data set available online but not required   ▼ Video solution available

c. The 98% confidence interval for persisters is narrower than the corresponding interval for nonpersisters, even though the standard deviation for persisters is larger than that for nonpersisters. Explain why this happened.

d. Based on the interval in Part (a), do you think that the mean number of hours worked per week for nonpersisters is greater than 20? Explain.   Yes, the lower end of the confidence interval is greater than 20.

9.56 An Associated Press article on potential violent behavior reported the results of a survey of 750 workers who were employed full time (*San Luis Obispo Tribune*, September 7, 1999). Of those surveyed, 125 indicated that they were so angered by a coworker during the past year that they felt like hitting the coworker (but didn't). Assuming that it is reasonable to regard this sample of 750 as a random sample from the population of full-time workers, use this information to construct and interpret a 90% confidence interval estimate of π, the true proportion of full-time workers so angered in the last year that they wanted to hit a colleague.   (0.1446, 0.1894)

9.57 The 1991 publication of the book *Final Exit*, which includes chapters on doctor-assisted suicide, caused a great deal of controversy in the medical community. The Society for the Right to Die and the American Medical Association quoted very different figures regarding the proportion of primary-care physicians who have participated in some form of doctor-assisted suicide for terminally ill patients (*USA Today*, July 1991). Suppose that a survey of physicians is to be designed to estimate this proportion to within .05 with 95% confidence. How many primary-care physicians should be included in a random sample?   385

9.58 Retailers report that the use of cents-off coupons is increasing. The Scripps Howard News Service (July 9, 1991) reported the proportion of all households that use coupons as .77. Suppose that this estimate was based on a random sample of 800 households (i.e., $n = 800$ and $p = .77$). Construct a 95% confidence interval for π, the true proportion of all households that use coupons.   (0.741, 0.799)

9.59 A manufacturer of small appliances purchases plastic handles for coffeepots from an outside vendor. If a handle is cracked, it is considered defective and must be discarded. A large shipment of plastic handles is received.

The proportion of defective handles π is of interest. How many handles from the shipment should be inspected to estimate π to within 0.1 with 95% confidence?   97

9.60 An article in the *Chicago Tribune* (August 29, 1999) reported that in a poll of residents of the Chicago suburbs, 43% felt that their financial situation had improved during the past year. The following statement is from the article: "The findings of this Tribune poll are based on interviews with 930 randomly selected suburban residents. The sample included suburban Cook County plus DuPage, Kane, Lake, McHenry, and Will Counties. In a sample of this size, one can say with 95% certainty that results will differ by no more than 3 percent from results obtained if all residents had been included in the poll."

Comment on this statement. Give a statistical argument to justify the claim that the estimate of 43% is within 3% of the true proportion of residents who feel that their financial situation has improved.   The point estimate is .43 and the bound on error is approximately .03.

9.61 The McClatchy News Service (*San Luis Obispo Telegram-Tribune*, June 13, 1991) reported on a study of violence on television during primetime hours. The following table summarizes the information reported for four networks:

Network	Mean Number of Violent Acts per Hour
ABC	15.6
CBS	11.9
FOX	11.7
NBC	11.0

Suppose that each of these sample means was computed on the basis of viewing $n = 50$ randomly selected prime-time hours and that the population standard deviation for each of the four networks is known to be $\sigma = 5$.

a. Compute a 95% confidence interval for the true mean number of violent acts per prime-time hour for ABC.

b. Compute 95% confidence intervals for the mean number of violent acts per prime-time hour for each of the other three networks.

c. The National Coalition on Television Violence claims that shows on ABC are more violent than those on the other networks. Based on the confidence intervals from

Parts (a) and (b), do you agree with this conclusion? Explain.   Yes

9.62 The *Chronicle of Higher Education* (January 13, 1993) reported that 72.1% of those responding to a national survey of college freshmen were attending the college of their first choice. Suppose that $n = 500$ students responded to the survey (the actual sample size was much larger).
a. Using the sample size $n = 500$, calculate a 99% confidence interval for the proportion of college students who are attending their first choice of college.   (0.0669, 0.773)
b. Compute and interpret a 95% confidence interval for the proportion of students who are *not* attending their first choice of college.   (0.240, 0.138)
c. The actual sample size for this survey was much larger than 500. Would a confidence interval based on the actual sample size have been narrower or wider than the one computed in Part (a)?   Narrower

9.63 Increases in worker injuries and disability claims have prompted renewed interest in workplace design and regulation. As one particular aspect of this, employees required to do regular lifting should not have to handle unsafe loads. The article "Anthropometric, Muscle Strength, and Spinal Mobility Characteristics as Predictors of the Rating of Acceptable Loads in Parcel Sorting" (*Ergonomics* [1992]: 1033–1044) reported on a study involving a random sample of $n = 18$ male postal workers. The sample mean rating of acceptable load attained with a work-simulating test was found to be $\bar{x} = 9.7$ kg. and the sample standard deviation was $s = 4.3$ kg. Suppose that in the population of all male postal workers, the distribution of rating of acceptable load can be modeled approximately using a normal distribution with mean value μ. Construct and interpret a 95% confidence interval for μ.   (7.56, 11.84)

9.64 The Gallup Organization conducted a telephone survey on attitudes toward AIDS (*Gallup Monthly*, 1991). A total of 1014 individuals were contacted. Each individual was asked whether they agreed with the following statement: "Landlords should have the right to evict a tenant from an apartment because that person has AIDS." One hundred one individuals in the sample agreed with this statement. Use these data to construct a 90% confidence interval for the proportion who are in agreement with this statement. Give an interpretation of your interval.   (0.084, 0.115)

9.65 A manufacturer of college textbooks is interested in estimating the strength of the bindings produced by a particular binding machine. Strength can be measured by recording the force required to pull the pages from the binding. If this force is measured in pounds, how many books should be tested to estimate with 95% confidence to within 0.1 lb, the average force required to break the binding? Assume that σ is known to be 0.8 lb.   246

9.66 Recent high-profile legal cases have many people reevaluating the jury system. Many believe that juries in criminal trials should be able to convict on less than a unanimous vote. To assess support for this idea, investigators asked each individual in a random sample of Californians whether they favored allowing conviction by a 10–2 verdict in criminal cases not involving the death penalty. The Associated Press (*San Luis Obispo Telegram-Tribune*, September 13, 1995) reported that 71% supported the 10–2 verdict. Suppose that the sample size for this survey was $n = 900$. Compute and interpret a 99% confidence interval for the proportion of Californians who favor the 10–2 verdict.   (0.671, 0.749)

9.67 The Center for Urban Transportation Research released a report stating that the average commuting distance in the United States is 10.9 mi (*USA Today*, August 13, 1991). Suppose that this average is actually the mean of a random sample of 300 commuters and that the sample standard deviation is 6.2 mi. Estimate the true mean commuting distance using a 99% confidence interval.   (9.976, 11.824)

9.68 In 1991, California imposed a "snack tax" (a sales tax on snack food) in an attempt to help balance the state budget. A proposed alternative tax was a 12¢-per-pack increase in the cigarette tax. In a poll of 602 randomly selected California registered voters, 445 responded that they would have preferred the cigarette tax increase to the snack tax (*Reno Gazette-Journal*, August 26, 1991). Estimate the true proportion of California registered voters who preferred the cigarette tax increase; use a 95% confidence interval.   (0.704, 0.774)

9.69 The confidence intervals presented in this chapter give both lower and upper bounds on plausible values for the population characteristic being estimated. In some instances, only an upper bound or only a lower bound is appropriate. Using the same reasoning that gave the large sample interval in Section 9.3, we can say that when n is large, 99% of all samples have

$$\mu < \bar{x} + 2.33\frac{s}{\sqrt{n}}$$

(because the area under the z curve to the left of 2.33 is .99). Thus, $\bar{x} + 2.33\frac{s}{\sqrt{n}}$ is a 99% upper confidence bound for μ. Use the data of Example 9.8 to calculate the 99% upper confidence bound for the true wait time for bypass patients in Ontario.   20.004

9.70 The Associated Press (December 16, 1991) reported that in a random sample of 507 people, only 142 correctly described the Bill of Rights as the first 10 amendments to the U.S. Constitution. Calculate a 95% confidence interval for the proportion of the entire population that could give a correct description.   (0.241, 0.319)

9.71 When n is large, the statistic s is approximately unbiased for estimating σ and has approximately a normal distribution. The standard deviation of this statistic when the population distribution is normal is $\sigma_s \approx \dfrac{\sigma}{\sqrt{2n}}$ which can be estimated by $\dfrac{s}{\sqrt{2n}}$. A large-sample confidence interval for the population standard deviation σ is then

$$s \pm (z \text{ critical value})\frac{s}{\sqrt{2n}}$$

Use the data of Exercise 9.67 to obtain a 95% confidence interval for the true standard deviation of commuting distance.   (5.704, 6.696)

9.72 The interval from -2.33 to 1.75 captures an area of .95 under the z curve. This implies that another large-sample 95% confidence interval for μ has lower limit $\bar{x} - 2.33\dfrac{\sigma}{\sqrt{n}}$ and upper limit $\bar{x} + 1.75\dfrac{\sigma}{\sqrt{n}}$. Would you recommend using this 95% interval over the 95% interval $\bar{x} \pm 1.96\dfrac{\sigma}{\sqrt{n}}$ discussed in the text? Explain. (Hint: Look at the width of each interval.)   No, because this interval is wider (less precise).

9.73 The eating habits of 12 bats were examined in the article "Foraging Behavior of the Indian False Vampire Bat" (*Biotropica* [1991]: 63–67). These bats consume insects and frogs. For these 12 bats, the mean time to consume a frog was $\bar{x} = 21.9$ min. Suppose that the standard deviation was $s = 7.7$ min. Construct and interpret a 90% confidence interval for the mean suppertime of a vampire bat whose meal consists of a frog. What assumptions must be reasonable for the one-sample t interval to be appropriate?   (17.9, 25.9)

Personal Tutor

Do you need a live tutor
for homework problems?

CENGAGENOW™

Are you ready? Take your
exam-prep post-test now.

Bold exercises answered in back   ● Data set available online but not required   ▼ Video solution available

Graphing Calculator Explorations

Exploration 9.1 The Confidence Interval for a
Population Proportion

Because confidence intervals are widely used, it will come as no surprise to you that your calculator may have a built-in capability to determine a confidence interval for a population proportion. Once again you will need to navigate your calculator's menu system, looking for key words such as INTR for "interval," or possibly TESTS. (Confidence intervals are frequently associated with "hypothesis tests," a topic to come later in Chapter 10.) Once you find the right menu, look for words like "1" and "prop" and "z"—together these key words should indicate a one-sample z confidence interval for a proportion.

```
1-Prop Zinterval
C-Level   :
x      :
n      :
Execute
```

```
1-Prop Zinterval
   Left =0.42889
   Right=0.49024
   p̂ =0.46
   n=1014
```

```
1-PropZInt
x  :
n  :
C-Level:
Calculate
```

```
1-PropZInt
(.42889, .49024)
p̂=.459566075
n=1014
```

Figure 9.10 Common screen presentations for confidence interval data entry.

Figure 9.11 Calculator screens for confidence intervals.

Once you select the correct choice, you will be presented with a screen for providing the information needed to calculate a confidence interval: the number of successes, the sample size, and the confidence level. Two representative screens are given in Figure 9.10.

From Example 9.4, supply the following information: $x = 466$, $n = 1014$, and the C-Level = .95. Move your cursor down to Execute or Calculate, and press the Enter, Execute, or Calculate button, depending on your calculator. The confidence interval should appear immediately. Again, two representative screens are given in Figure 9.11.

Notice that the formatting in these screens is slightly different. Which of these is the "right" format? Probably neither one! Recall a previous graphing calculator exploration in Chapter 3, where we discussed the differences between a calculator's presentation of information and the appropriate way to communicate this information. *Check with your instructor to ascertain her or his preferences about what format information is required.* You will almost certainly want to report more information than what is presented on these calculator screens. Specifically, you will notice that the calculator does not verify the appropriateness of the large-sample assumption! Some confidence intervals may stray outside the interval from 0 to 1, which makes sense to the calculator, but not to the statistician. Thus, you may have to modify the interval returned by your calculator. Also, don't forget that your calculator does not know the context of the problem; you will have to provide that contextual information irrespective of the form of the calculator presentation. Remember: The calculator only does the calculation; you must do the thinking.

Exploration 9.2 A Confidence Interval for a Population Mean

Finding the confidence interval for a single mean on your calculator will require you to navigate the menu system much as you did to find the confidence interval for a proportion. The confidence interval for the mean will have some added challenges, however:

1. You will need to decide whether to base the interval on the z or t distribution, and
2. You may use previous calculations of the sample mean and standard deviation, or the calculator will evaluate these statistics from data contained in a List.

We will use the data from Exercise 9.43 to construct a 99% confidence interval. We have entered the data in List1 and are ready to proceed. Since the population standard deviation is not known, we will use the t confidence interval. For the t confidence interval, the normality of the population becomes an issue. The original data are at hand, and we can assess the plausibility of a normal population via a normal probability plot. Figure 9.12 shows the normal probability plot.

After verifying the plausibility of the normality of the population, we can now construct the confidence interval. In this case, we have a choice of entering the sample

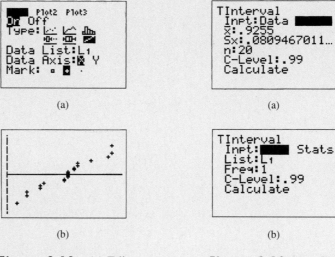

(a)

(b)

Figure 9.12 (a) Edit window for normal probability plot; (b) normal probability plot.

Figure 9.13 (a) Confidence interval input screen for raw data; (b) confidence interval input screen for summary statistics.

Figure 9.14 Calculated confidence interval.

calculations or letting the calculator evaluate the sample mean and standard deviation. Based on that choice, we see one of the screens shown in Figure 9.13.

Figure 9.14 shows one calculator's version of the confidence interval. Once again, we caution you that the calculator only calculates—you must still do the thinking and present the solution in the context of the particular problem at hand.

TEACHING TIPS

Intent of this chapter:

After completing this chapter, students should be able to (1) write appropriate hypotheses, (2) identify Type I and Type II errors in context, (3) test hypotheses about a population proportion, (4) test hypotheses about a population mean, and (5) understand the relationships between α, β, and power.

A PowerPoint® Lecture is available on the *Instructor's Resource Binder* CD.

■ **Section 10.1**

In general, students are not necessarily familiar with the concept of hypotheses. Using the analogy of a murder trial is very helpful throughout this chapter. A defendant is brought to trial because we suspect he/she is guilty of committing murder. The purpose of a trial is to examine the facts to see if there is a sufficient amount of evidence to find the defendant guilty. In the United States justice system, the trial begins with the supposition that the defendant is "innocent."

In writing hypotheses, we might follow this murder trial analogy. We begin with the assumption of innocence; for example, that the company's claim is true. This is our null hypothesis, denoted as H_0. The null hypothesis should be an equality statement: H_0: $\mu = 1000$. Our alternative hypothesis, H_a, is what the prosecutor is trying to prove, i.e., a guilty verdict. (In a scientific study, of course, we are not so much trying to "prove" something, as we are searching for reasonable and credible hypotheses about the real world. Our subsequent credible hypothesis might be connected with a null or an alternative hypothesis, depending on the study.) The alternative hypothesis may be a one-sided (or one-tailed) statement such as H_a: $\mu < 1000$ or H_a: $\mu > 1000$. The alternative hypothesis may also be a two-sided (or two-tailed) statement such as H_a: $\mu \neq 1000$. (See the blue box on page 528 and the tan box on page 543.) One should never wait to look at the data to determine which alternative hypothesis to use. The context of the problem and the question that we are trying to answer should determine the form of the alternative hypothesis. If no specific direction ($<$ or $>$) is implied by the question posed, then students should use the "not equal" alternative as the default. Hypotheses should *always* be written as statements about the population; hypotheses are statements about parameters and not statistics, so the symbols used should reflect this.

Like the murder trial that has two possible verdicts, "guilty" or "not guilty," there are two possible conclusions in a test of hypotheses, "reject H_0" or "fail to reject H_0." The decision is always worded in terms of the null hypothesis. If we reject H_0, we are saying that there is enough evidence to suggest that our beginning assumption that "the null hypothesis is true" is not believable. (This is our guilty verdict.) However, if we fail to reject H_0, we are saying that there is insufficient evidence to convince us that our initial assumption that "the null hypothesis is true" is incorrect. (This is our not guilty verdict.) Interestingly, in a murder trial, we *never* decide that the defendant is innocent (our initial assumption)! Likewise, in hypotheses testing, we will *never accept* the null hypothesis as being true after reviewing the evidence. We never have strong evidence that the null is true, but we might be convinced to "change our mind" and decide that the alternative hypothesis is more credible given what we have observed in the sample.

Suggested Assignment: Exercises 10.2, 10.4, 10.5, 10.7, 10.8, 10.9, 10.10

■ **Section 10.2**

Because there are two possible conclusions in a hypothesis test, there are two possible errors that might result. See the tan box on page 531 for the definitions of Type I and Type II errors. A table is often helpful in aiding students' understanding of error types.

Students can see that if the null hypothesis is true and we fail to reject it, we are making a correct decision. Likewise, if the null hypothesis is false and we reject it, that is also a correct decision. Additionally, students see that rejecting a true null hypothesis is a Type I error. Let's return to the murder trial scenario. Many recent news reports have included stories about individuals convicted of murder being released from jail because DNA testing (unavailable at the time of trial) has established their innocence. What type of error was made in the initial trial? (*Type 1*)

| | Reality | |
	H_0 is true	H_0 is not true
Reject H_0	Type I error α	Correct decision
Fail to reject H_0	Correct decision	Type II error β

Another graphical approach uses normal curves to demonstrate this. Suppose we were testing the hypotheses H_0: $\mu = 10$ versus H_a: $\mu > 10$. We would have one curve (representing the sampling distribution of the sample mean) centered at $\mu_0 = 10$, the value specified by the null hypothesis. If the null hypothesis is true and we use $\alpha = .05$ (or a 5% significance level, discussed in more detail below), then 5% of the time our test procedure would reject the null hypothesis even though the null hypothesis is true, which is a Type I error. If our sample mean fell far enough away from the hypothesized value μ_0, past the line that corresponds to $\alpha = .05$, then we would reject the null hypothesis.

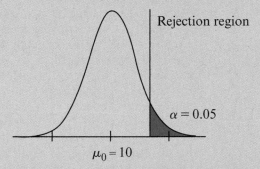

If we reject H_0, we conclude that the mean is greater than 10. In this case, the sampling distribution is centered at a value greater than 10. Imagine the curve centered at 10 shifting to the right. In order to see the two curves without them overlapping, let's drop down the second curve and extend the line that denotes the rejection region downward (see figure on the following page). Notice that if the null hypothesis is false and the alternative hypothesis is true, then the "correct" curve is the one centered at a value greater than 10. If we fail to reject a false H_0, then we have made a Type II error. Notice the orange shaded part, labeled β, in the alternative curve. Type I (significance level) and Type II errors are on opposite sides of the line that denotes the boundary of the rejection region. Imagine shifting the boundary. A left-shift increases α, but decreases β. A right-shift decreases α, but increases β. The probability of a Type I error, α, and the probability of a Type II error, β, are inversely related.

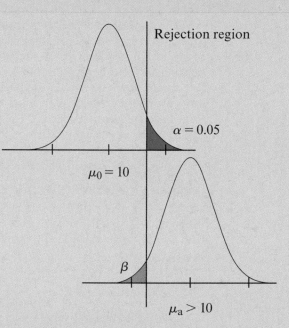

The tan box (on page 532) states that the probability of a Type I error is denoted by α, which is also the level of significance, and that the probability of a Type II error is denoted by β. The level of significance determines how extreme the sample results must be before we are willing to reject the null hypothesis. Before performing this next activity, you must first "rig" a deck of cards. Purchase two decks of cards that are identical. Carefully open the outside plastic wrapping at the *bottom of each box.* Remove all the cards. Replace all the red cards into one box and all the black cards into the other box. Reseal the bottom plastic wrapping so that it appears that the deck of cards is brand new! In class, dramatically open the sealed deck of cards. Shuffle the deck a few times, but be careful not to show the cards to the students! Offer a piece of candy to each student that selects a black card from the all-red deck (or a red card from the all-black deck). Fan out the cards and have students, one at a time, randomly select a card. Show the card to the class and replace the card into the deck. Surprisingly, it takes only three to five attempts for students to begin to doubt that the deck is a "fair" deck! Be sure to summarize: as you were opening the deck, what did students assume was true? *The students assumed that the deck was a standard deck with half red and half black cards.* They begin to doubt this assumption after about 3 students in a row lose, and usually they become "convinced" that the deck is not half red and half black after 4 or 5 cards have been selected. A quick probability calculation confirms that the evidence against the initial assumption was convincing when it would have occurred by chance less than $\left(\dfrac{1}{2}\right)^4$ to $\left(\dfrac{1}{2}\right)^5$ of the time. This is around 0.05 and can be used to motivate the usual choices for significance level. Of course you may want to give the class a piece of candy anyway!

Students should be able to describe Type I and Type II errors *in context* and should also be able to discuss the possible consequence of making each type of error. The consequence of an error is what happens as a result of making a wrong decision. An analysis of the severity of the consequences for Type I and Type II errors informs the decision of what level of significance to use. If the consequences of a Type I error are deemed more serious than those of a Type II error, then choosing a smaller value for α ensures that the probability of making a Type I error is small. (But remember, it also increases the probability of making a Type II error, other things being equal.) Similarly, if the consequences of a Type II error are deemed more serious than those of a Type I error, then choosing a larger value of α would decrease the probability of making a Type II error. Researchers have control over the significance level used. The probability of a Type II error, β, is dependent on the choice of α, the standard deviation of the sampling distribution, and the actual value for the population parameter when the null hypothesis is false. Typically, hypotheses and the level of significance are selected at the beginning of a study, prior to data collection.

Suggested Assignment: Exercises 10.12, 10.13, 10.15, 10.17, 10.20, 10.22

Activity: "Chapter 10—Hypothesis Test Basics" (*Instructor's Resource Binder* & CD, Chapter 10 Activities Worksheets), "Chapter 10—Type I and Type II Errors" (*Instructor's Resource Binder* & CD, and/or Chapter 10 Activities Worksheets)

Assessment: Chapter 10 Quiz 1 (*Test Bank* and *Instructor's Resource Binder* CD)

■ **Section 10.3**

This section introduces a large-sample hypothesis test for a population proportion. Like the confidence interval for π, this hypothesis test is based on what we know about the sampling distribution of p, which is approximately normal with $\mu_p = \pi$ and $\sigma_p = \sqrt{\dfrac{\pi(1 - \pi)}{n}}$ when n is large. Since both the hypothesis test and the confidence interval for π are based on the same sampling distribution, the assumptions are the same. (See the tan box on page 545.)

Once the assumptions have been checked and the hypotheses have been stated, students can proceed to calculate the test statistic and P-value. The generic formula for a test statistic is test statistic $= \dfrac{\text{statistic} - \text{parameter}}{\text{standard deviation of the statistic}}$. In the case of proportions, the test statistic is $z = \dfrac{p - \pi}{\sqrt{\dfrac{\pi(1 - \pi)}{n}}}$. The test statistic essentially tells us how many standard deviations the sample proportion is from the hypothesized value of π in the sampling distribution of p. The P-value is the probability of observing a test statistic this extreme or more extreme, assuming the null hypothesis is true. Remember that in continuous distributions, like normal curves, the probability of "exactly equal to" is zero. Therefore, we must define the P-value as "this extreme or more extreme." Why does the definition for the P-value include "assuming the null is true"? *In Section 10.1, we said that we begin the hypothesis test assuming the null is true.* So the P-value is based on that initial assumption. On the graphing calculator, students can use the normalcdf function to calculate the P-value. With one-sided tests (the alternative hypothesis is either $<$ or $>$), the P-value is the area under the standard normal curve and in the appropriate tail. For two-sided tests, the P-value is the combined area in both tails. (See the tan box on page 543.) Graphing Calculator Exploration 10.1 (on page 580) provides instructions for the graphing calculator.

We use the P-value to make a decision about whether to reject or fail to reject the null hypothesis by comparing the P-value to the level of significance (α). (See the blue box on page 541.) Our conclusion should contain two parts: (1) the decision about the null and *why* we make that decision, and (2) what this decision means in context. Be sure to stress once more that the decision is always worded in terms of the null hypothesis. Here is an example for part (1): "I fail to reject the null hypothesis because the P-value of 0.132 is greater than α of 0.05." The statement in part (2) should be in context and in terms of the *alternative* hypothesis. Warn students to be careful. A common mistake, especially when the decision is "fail to reject," is to write part (2) in terms of the null hypothesis. Thus, we end by making a statement "accepting" the null hypothesis. We *never accept* the null hypothesis! Here is an example for part (2): "There is insufficient evidence to suggest that the true proportion of adults who believe in ghosts exceeds 40%." (In this scenario, $H_a: \pi > .4$)

Suggested Assignment: Exercises 10.25, 10.26, 10.27, 10.33, 10.34, 10.37, 10.39, 10.43

Activity: "Chapter 10—Testing Hypotheses about π" (*Instructor's Resource Binder* & CD, Chapter 10 Activities Worksheets) and/or "Chapter 10—1PropZTest" (*Instructor's Resource Binder* & CD, Chapter 10 Activities Worksheets)

■ **Section 10.4**   This section introduces hypothesis tests for a population mean. Like the confidence interval for μ, these hypothesis tests are based on the sampling distribution of $\bar{x}$, which is approximately normal when n is large (or judged to be approximately normal by graphical displays) with $\mu_{\bar{x}} = \mu$ and $\sigma_{\bar{x}} = \dfrac{\sigma}{\sqrt{n}}$. Since both the hypothesis test and the confidence interval for μ are based on the same sampling distribution, the assumptions to be checked are the same. (See tan box on page 553.)

For the hypothesis test for μ, if we knew the population standard deviation (σ), we would be able to use the formula $z = \dfrac{\bar{x} - \mu}{\sigma/\sqrt{n}}$ to compute the value of the test statistic. Of course, the population standard deviation is almost always unknown. We therefore substitute the sample standard deviation (s) as a reasonable estimate of σ. When we do this, more uncertainty—in the form of more variability—has been introduced in the sampling distribution of the test statistic. Also, the sampling distribution of the test statistic is no longer normal. We now must use a t-procedure based upon $n - 1$ df. In a t-test, the test statistic is computed using the formula $t = \dfrac{\bar{x} - \mu}{s/\sqrt{n}}$. The P-value is computed using the same procedure as in Section 10.3, but using the appropriate t distribution instead of the standard normal distribution. For a t-test statistic, students with a graphing calculator can use the tcdf function to calculate P-values. Graphing Calculator Exploration 10.2 (on page 581) provides instructions.

The fundamental ideas of a hypothesis test remain the same no matter which test students are performing: Students should be sure to report both the test statistic and the P-value. For any t-procedure, students should also report the appropriate degrees of freedom. In the following chapters, students will learn how to compute test statistics for testing different types of hypotheses. However, the process for stating hypotheses, for calculating

P-values, and for making decisions remains the same. For any particular hypothesis test, we select the test statistic and check the associated assumptions appropriate to the problem at hand.

How are confidence intervals and hypothesis tests about means related? Confidence intervals estimate an unknown parameter. Hypothesis tests assess the plausibility of a claim about a population parameter. Look at Example 10.12 on page 546. There we tested H_0: $\pi = .71$ versus H_a: $\pi \neq .71$. At $\alpha = .01$, we reject the null hypothesis. Notice that when we are performing a two-tail test, an area of $\alpha/2$ goes into each tail. Thus area under the curve in the center part is $1 - \alpha$ (for this example, the central area is 0.99). If we construct a 99% confidence interval using the data from Example 10.12, we would get the interval (.819, .861). Notice that the hypothesized value for π, .71, is not contained in this confidence interval.

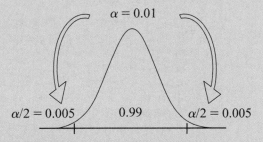

$\alpha = 0.01$

$\alpha/2 = 0.005$ 0.99 $\alpha/2 = 0.005$

Now look at Example 10.14 on page 555. We tested H_0: $\mu = 120$ versus H_a: $\mu < 120$. At $\alpha = .05$, we fail to reject the null hypothesis. This is a one-sided test; the area specified by the significance level all goes into the lower tail. Remember, though, that the confidence intervals of this chapter are two-sided. Since the z and the t distributions are symmetrical, if we put an area of .05 in the lower tail and also .05 in the upper tail, the central area would be .90. A 90% confidence interval for the data of Example 10.14 is (111.32, 122.28). Notice that the hypothesized value for μ, 120, is contained in the interval.

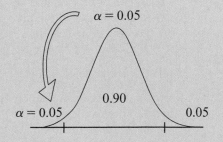

$\alpha = 0.05$

$\alpha = 0.05$ 0.90   0.05

In summary, hypothesis tests and confidence intervals about means are related. If the hypothesis test rejects the null hypothesis, then the appropriate confidence interval will *not* contain the hypothesized value. However, if the hypothesis test fails to reject the null hypothesis, then the appropriate confidence interval will contain the hypothesized value. Students must be very careful to match the correct confidence level to the significance level, especially for one-tailed tests. (Note that the hypothesis test and confidence interval for a proportion occasionally give different results because the standard error used in the hypothesis test and the standard error used in the confidence interval are not identical—the confidence interval uses p and the hypothesis test uses the hypothesized value from the null hypothesis in the calculation.)

Suggested Assignment: Exercises 10.47, 10.49, 10.53, 10.56, 10.59, 10.61, 10.63

Activity: Activity 10.1: Comparing the t and z Distributions (p. 574 and *Activities Manual*, p. 86) ★, Bonus Activity 10.3—Elapsed Time (*Activities Manual*, p. 91), "Chapter 10—Computing the P-Value" (*Instructor's Resource Binder* & CD, Chapter 10 Activities Worksheets), "Chapter 10—Testing Hypotheses about μ" (*Instructor's Resource Binder* & CD, Chapter 10 Activities Worksheets), and/or "Chapter 10—TTest" (*Instructor's Resource Binder* & CD, Chapter 10 Activities Worksheets)

Assessment: Chapter 10 Quiz 2 (*Test Bank* and *Instructor's Resource Binder* CD)

■ **Section 10.5**

The *power* of a test is the probability of rejecting the null hypothesis. When the null hypothesis is false, this is the probability of making a correct decision to reject the null hypothesis. Because we are usually interested in determining whether there is convincing evidence supporting the alternative hypothesis, we would like for the power to be high when the null hypothesis is false, since that would mean that we would be more likely to detect that the null hypothesis is false. First, let's explore the definition of power. Next, we will see how to calculate the probability of power (optional). Last, we will investigate what affects the power of a test.

	Reality	
	H_0 **is true**	H_0 **is not true**
Reject H_0	Type I error α	Correct decision **Power**
Fail to reject H_0	Correct decision	Type II error β

Refer to Section 10.2 for the definition of Type I and Type II errors. Looking at the table above, we see that there are two combinations that result in a correct decision. In the case where the null hypothesis is not true, the probability of a correct decision is the power.

Consider the curves on the next page. The part of the lower curve shaded yellow represents the probability associated with correctly rejecting the false null hypothesis. What do you notice about power and β, the probability of a Type II error? *Power and β are complements (they total 1).* You might not want to have students compute these probabilities by hand—the calculations are tedious. However, if you are so inclined, having students perform these calculations on two or three homework problems can add to students' understanding of the relationships between α, β, and power.

★ Kathy's personal favorite

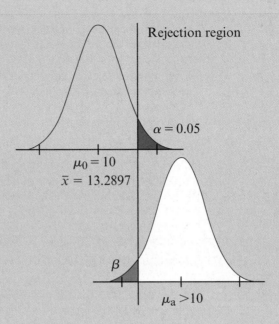

Suppose that the null hypothesis is false and that $\mu_a = 14$. How would the value for power change? Imagine the lower curve shifting left one unit. The value of power decreases. So, the difference between the hypothesized and alternative means $|\mu_0 - \mu_a|$ is one thing that affects power. **The greater the difference $|\mu_0 - \mu_a|$, the larger the power of the test.**

Suppose we increase the significance level, α (shifting the line to the left in the diagram above). How does the power change? The value of power increases. So, **increasing the significance level (α) increases the power of a test.**

Suppose our sample size increased. How would the curves in the diagram above change? *The curves would become taller and thinner (as shown in the figure on the following page) because $\sigma_{\bar{x}}$ would decrease when n increases.* Notice that the significance level (α) remains .05, so the rejection line must shift left. Due to the movement of the rejection line and the smaller $\sigma_{\bar{x}}$, β has decreased. So, **increasing the sample size will increase the power of the test.**

Students need to understand how changing the difference $|\mu_0 - \mu_a|$, changing the significance level (α), and changing the sample size (n) affect the power of the test. The applet listed on the next page allows students to investigate these relationships. Students can change various values and see how power is affected.

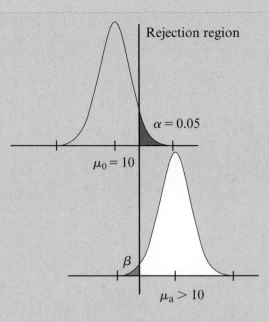

Suggested Assignment: Exercises 10.66, 10.67, 10.69

Activity: Activity 10.2: A Meaningful Paragraph (p. 575 and *Activities Manual*, p. 90)

Applet: http://wise.cgu.edu/power/powerapplet1.html

Assessment: Chapter 10 Concept Quiz (*Test Bank* and *Instructor's Resource Binder* CD)

■ **Section 10.6**   This section provides excellent suggestions on communication and interpretation in a hypothesis testing setting. Students should also pay close attention to "A Word to the Wise: Cautions and Limitations."

Suggested Review Assignment: Exercises 10.72, 10.75, 10.79, 10.80, 10.81, 10.89, 10.91

Assessment: Chapter 10 Test (*Test Bank* and *Instructor's Resource Binder* CD)

© Royalty-Free/Corbis

Chapter 10

Hypothesis Testing Using a Single Sample

In Chapter 9, we considered situations in which the primary goal was to estimate the unknown value of some population characteristic. Sample data can also be used to decide whether some claim or *hypothesis* about a population characteristic is plausible. For example, cross-border purchasing of prescription drugs is a controversial topic of current interest, and there has been a great deal of media coverage on the practice of importing prescription medications from Canada or Mexico. But is this really a common practice? The article "Much Ado About Cross-Border Prescription Purchasing" (Ipsos, February 19, 2004) summarized the results of a poll of 1000 randomly selected adult Americans and reported that only 15 of the 750 adults in the sample who had purchased prescription drugs in the past year had made a purchase from a pharmacy in Canada or Mexico. Let π denote the proportion of all American adults who have made a prescription drug purchase in the last year from a Canadian or Mexican pharmacy. The hypothesis testing methods presented in this chapter can be used to decide whether the sample data from this survey provide strong support for the hypothesis that this proportion is small, for example $\pi < .05$.

As another example, a report released by the National Association of Colleges and Employers stated that the average starting salary for students graduating in 2006 with a degree in accounting was $45,656 ("Starting Salary Offers to New College Grads Continue to Climb," July 12, 2006, available at www.naceweb.org/press). Suppose that you are interested in investigating whether the mean starting salary for students graduating with an accounting degree from your university this year is greater than the 2006 average of $45,656. You select a random sample of $n = 40$ accounting graduates

Improve your understanding and save time! Visit www.cengage.com/login where you will find:
- Step-by-step instructions for MINITAB, Excel, TI-83, and JMP
- Video solutions to selected exercises
- Data sets available for selected examples and exercises
- Exam-prep pre-tests that build a Personalized Learning Plan based on your results so that you know exactly what to study
- Help from a live statistics tutor 24 hours a day

from the current graduating class of your university and determine the starting salary of each one. If this sample produced a mean starting salary of $45,958 and a standard deviation of $1214, is it reasonable to conclude that μ the current mean starting salary for all accounting graduates in the current graduating class at your university, is greater than $45,656 (i.e., $\mu = 45,656$)? We will see in this chapter how these sample data can be analyzed to decide whether $\mu > 45,656$ is a reasonable conclusion.

10.1 Hypotheses and Test Procedures

A hypothesis is a claim or statement about the value of a single population characteristic or the values of several population characteristics. The following are examples of legitimate hypotheses:

$\mu = 1000$, where μ is the mean number of characters in an email message
$\pi < .01$, where π is the proportion of email messages that are undeliverable

In contrast, the statements $\bar{x} = 1000$ and $p = .01$ are *not* hypotheses, because $\bar{x}$ and p are *sample* characteristics.

A **test of hypotheses** or **test procedure** is a method that uses sample data to decide between two competing claims (hypotheses) about a population characteristic. One hypothesis might be $\mu = 1000$ and the other $\mu \neq 1000$ or one hypothesis might be $\pi = .01$ and the other $\pi < .01$. If it were possible to carry out a census of the entire population, we would know which of the two hypotheses is correct, but usually we must decide between them using information from a sample.

A criminal trial is a familiar situation in which a choice between two contradictory claims must be made. The person accused of the crime must be judged either guilty or not guilty. Under the U.S. system of justice, the individual on trial is initially presumed not guilty. Only strong evidence to the contrary causes the not guilty claim to be rejected in favor of a guilty verdict. The burden is thus put on the prosecution to prove the guilty claim. The French perspective in criminal proceedings is the opposite of ours. There, once enough evidence has been presented to justify bringing an individual to trial, the initial assumption is that the accused is guilty. The burden of proof then falls on the accused to establish otherwise.

As in a judicial proceeding, we initially assume that a particular hypothesis, called the *null hypothesis,* is the correct one. We then consider the evidence (the sample data) and reject the null hypothesis in favor of the competing hypothesis, called the *alternative hypothesis,* only if there is *convincing* evidence against the null hypothesis.

> **DEFINITION**
>
> The **null hypothesis**, denoted by H_0, is a claim about a population characteristic that is initially assumed to be true.
>
> The **alternative hypothesis**, denoted by H_a, is the competing claim.
>
> In carrying out a test of H_0 versus H_a, the hypothesis H_0 will be rejected in favor of H_a only if sample evidence strongly suggests that H_0 is false. If the sample does not provide such evidence, H_0 will not be rejected. The two possible conclusions are then *reject H_0* or *fail to reject H_0.*

Example 10.1 Tennis Ball Diameters

Because of variation in the manufacturing process, tennis balls produced by a particular machine do not have identical diameters. Let μ denote the true average diameter for tennis balls currently being produced. Suppose that the machine was initially calibrated to achieve the design specification $\mu = 3$ in. However, the manufacturer is now concerned that the diameters no longer conform to this specification. That is, $\mu \neq 3$ in. must now be considered a possibility. If sample evidence suggests that $\mu \neq 3$ in., the production process will have to be halted while the machine is recalibrated. Because stopping production is costly, the manufacturer wants to be quite sure that $\mu \neq 3$ in. before undertaking recalibration. Under these circumstances, a sensible choice of hypotheses is

H_0: $\mu = 3$ (the specification is being met, so recalibration is unnecessary)
H_a: $\mu \neq 3$ (the specification is not being met, so recalibration is necessary)

Only compelling sample evidence would then result in H_0 being rejected in favor of H_a.

■

Example 10.2 Lightbulb Lifetimes

Kmart brand 60W lightbulbs state on the package "Avg. Life 1000 Hr." Let μ denote the true mean life of Kmart 60-W lightbulbs. Then the advertised claim is $\mu = 1000$ hr. People who purchase this brand would be unhappy if μ is actually less than the advertised value. Suppose that a sample of Kmart lightbulbs is selected and the lifetime for each bulb in the sample is recorded. The sample results can then be used to test the hypothesis $\mu = 1000$ hr against the hypothesis $\mu < 1000$ hr. The accusation that the company is overstating the mean lifetime is a serious one, and it is reasonable to require compelling evidence from the sample before concluding that $\mu < 1000$. This suggests that the claim $\mu = 1000$ should be selected as the null hypothesis and that $\mu < 1000$ should be selected as the alternative hypothesis. Then

$$H_0: \mu = 1000$$

would be rejected in favor of

$$H_a: \mu < 1000$$

only if sample evidence strongly suggests that the initial assumption, $\mu = 1000$ hr, is not plausible.

■

Because the alternative hypothesis in Example 10.2 asserted that $\mu < 1000$ (true average lifetime is less than the advertised value), it might have seemed sensible to state H_0 as the inequality $\mu \geq 1000$. The assertion $\mu \geq 1000$ is in fact the *implicit* null hypothesis, but we will state H_0 explicitly as a claim of equality. There are several reasons for this. First of all, the development of a decision rule is most easily understood if there is only a single hypothesized value of μ (or π or whatever other population characteristic is under consideration). Second, suppose that the sample data provided

Important concept

compelling evidence that H_0: $\mu = 1000$ should be rejected in favor of H_a: $\mu < 1000$. This means that we were convinced by the sample data that the true mean was smaller than 1000. It follows that we would have also been convinced that the true mean could not have been 1001 or 1010 or any other value that was larger than 1000. As a consequence, the conclusion when testing H_0: $\mu = 1000$ versus H_a: $\mu < 1000$ is always the same as the conclusion for a test where the null hypothesis is H_0: $\mu \geq 1000$. For these reasons it is customary to state the null hypothesis H_0 as a claim of equality.

The form of a null hypothesis is

H_0: population characteristic = hypothesized value

where the hypothesized value is a specific number determined by the problem context.

The alternative hypothesis will have one of the following three forms:

H_a: population characteristic > hypothesized value
H_a: population characteristic < hypothesized value
H_a: population characteristic ≠ hypothesized value

Thus, we might test H_0: $\pi = .1$ versus H_a: $\pi < .1$; but we won't test H_0: $\mu = 50$ versus H_a: $\mu > 100$. The number appearing in the alternative hypothesis must be identical to the hypothesized value in H_0.

Example 10.3 illustrates how the selection of H_0 (the claim initially believed true) and H_a depend on the objectives of a study.

Example 10.3 Evaluating a New Medical Treatment

A medical research team has been given the task of evaluating a new laser treatment for certain types of tumors. Consider the following two scenarios:

Scenario 1: The current standard treatment is considered reasonable and safe by the medical community, has no major side effects, and has a known success rate of 0.85 (85%).

Scenario 2: The current standard treatment sometimes has serious side effects, is costly, and has a known success rate of 0.30 (30%).

In the first scenario, research efforts would probably be directed toward determining whether the new treatment has a higher success rate than the standard treatment. Unless convincing evidence of this is presented, it is unlikely that current medical practice would be changed. With π representing the true proportion of successes for the laser treatment, the following hypotheses would be tested:

$$H_0: \ \pi = .85 \text{ versus } H_a: \ \pi > .85$$

In this case, rejection of the null hypothesis is indicative of compelling evidence that the success rate is higher for the new treatment.

In the second scenario, the current standard treatment does not have much to recommend it. The new laser treatment may be considered preferable because of cost or because it has fewer or less serious side effects, as long as the success rate for the new procedure is no worse than that of the standard treatment. Here, researchers might decide to test the hypothesis

$$H_0: \pi = .30 \text{ versus } H_a: \pi < .30$$

If the null hypothesis is rejected, the new treatment will not be put forward as an alternative to the standard treatment, because there is strong evidence that the laser method has a lower success rate.

If the null hypothesis is not rejected, we are able to conclude only that there is not convincing evidence that the success rate for the laser treatment is lower than that for the standard. This is *not* the same as saying that we have evidence that the laser treatment is as good as the standard treatment. If medical practice were to embrace the new procedure, it would not be because it has a higher success rate but rather because it costs less or has fewer side effects, and there is not strong evidence that it has a lower success rate than the standard treatment.

■

Important concept

You should be careful in setting up the hypotheses for a test. *Remember that a statistical hypothesis test is only capable of demonstrating strong support for the alternative hypothesis (by rejection of the null hypothesis). When the null hypothesis is not rejected, it does not mean strong support for H_0—only lack of strong evidence against it.* In the lightbulb scenario of Example 10.2, if $H_0: \mu = 1000$ is rejected in favor of $H_a: \mu < 1000$, it is because we have strong evidence for believing that true average lifetime is less than the advertised value. However, nonrejection of H_0 does not necessarily provide strong support for the advertised claim. If the objective is to demonstrate that the average lifetime is greater than 1000 hr, the hypotheses to be tested are $H_0: \mu = 1000$ versus $H_a: \mu > 1000$. Now rejection of H_0 indicates strong evidence that $\mu > 1000$. When deciding which alternative hypothesis to use, *keep the research objectives in mind.*

■ Exercises 10.1–10.11

Turn to the Expanded Answers Section for aswers not shown next to exercises.

10.1 Explain why the statement $\bar{x} = 50$ is not a legitimate hypothesis.

10.2 For the following pairs, indicate which do not comply with the rules for setting up hypotheses, and explain why:
a. $H_0: \mu = 15, H_a: \mu = 15$
b. $H_0: \pi = .4, H_a: \pi > .6$
c. $H_0: \mu = 123, H_a: \mu < 123$ Does comply
d. $H_0: \mu = 123, H_a: \mu = 125$
e. $H_0: p = .1, H_a: p \neq .1$ Does not comply, p is not a population characteristic.

10.3 To determine whether the pipe welds in a nuclear power plant meet specifications, a random sample of welds is selected and tests are conducted on each weld in the sample. Weld strength is measured as the force required to break the weld. Suppose that the specifications state that the mean strength of welds should exceed 100 lb/in.^{2}. The inspection team decides to test $H_0: \mu = 100$ versus $H_a: \mu > 100$. Explain why this alternative hypothesis was chosen rather than $\mu < 100$.

10.4 Do state laws that allow private citizens to carry concealed weapons result in a reduced crime rate? The author of a study carried out by the Brookings Institution is reported as saying, "The strongest thing I could say is that I don't see any strong evidence that they are reducing crime" (*San Luis Obispo Tribune,* January 23, 2003).

Bold exercises answered in back   ● Data set available online but not required   ▼ Video solution available

a. Is this conclusion consistent with testing

H_0: concealed weapons laws reduce crime

versus

H_a: concealed weapons laws do not reduce crime

or with testing

H_0: concealed weapons laws do not reduce crime

versus

H_a: concealed weapons laws reduce crime

Explain.

b. Does the stated conclusion indicate that the null hypothesis was rejected or not rejected? Explain. The null hypothesis was not rejected.

10.5 Consider the following quote from the article "Review Finds No Link Between Vaccine and Autism" (*San Luis Obispo Tribune,* October 19, 2005): "'We found no evidence that giving MMR causes Crohn's disease and/or autism in the children that get the MMR,' said Tom Jefferson, one of the authors of *The Cochrane Review.* 'That does not mean it doesn't cause it. It means we could find no evidence of it.'" (MMR is a measles-mumps-rubella vaccine.) In the context of a hypothesis test with the null hypothesis being that MMR does not cause autism, explain why the author could not just conclude that the MMR vaccine does not cause autism.

10.6 A certain university has decided to introduce the use of plus and minus with letter grades, as long as there is evidence that more than 60% of the faculty favor the change. A random sample of faculty will be selected, and the resulting data will be used to test the relevant hypotheses. If π represents the true proportion of all faculty that favor a change to plus–minus grading, which of the following pair of hypotheses should the administration test:

H_0: $\pi = .6$ versus H_a: $\pi < .6$

or

H_0: $\pi = .6$ versus H_a: $\pi > .6$

Explain your choice.

10.7 ▼ A certain television station has been providing live coverage of a particularly sensational criminal trial. The station's program director wishes to know whether more than half the potential viewers prefer a return to regular daytime programming. A survey of randomly selected viewers is conducted. Let π represent the true proportion of viewers who prefer regular daytime programming. What hypotheses should the program director test to answer the question of interest? H_0: $\pi = 0.5$ versus H_a: $\pi > 0.5$

10.8 Researchers have postulated that because of differences in diet, Japanese children have a lower mean blood cholesterol level than U.S. children do. Suppose that the mean level for U.S. children is known to be 170. Let μ represent the true mean blood cholesterol level for Japanese children. What hypotheses should the researchers test? H_0: $\mu = 170$ versus H_a: $\mu < 170$

10.9 A county commissioner must vote on a resolution that would commit substantial resources to the construction of a sewer in an outlying residential area. Her fiscal decisions have been criticized in the past, so she decides to take a survey of constituents to find out whether they favor spending money for a sewer system. She will vote to appropriate funds only if she can be fairly certain that a majority of the people in her district favor the measure. What hypotheses should she test? H_0: $\pi = 0.5$ versus H_a: $\pi > 0.5$

10.10 The mean length of long-distance telephone calls placed with a particular phone company was known to be 7.3 min under an old rate structure. In an attempt to be more competitive with other long-distance carriers, the phone company lowered long-distance rates, thinking that its customers would be encouraged to make longer calls and thus that there would not be a big loss in revenue. Let μ denote the true mean length of long-distance calls after the rate reduction. What hypotheses should the phone company test to determine whether the mean length of long-distance calls increased with the lower rates? H_0: $\mu = 7.3$ versus H_a: $\mu > 7.3$

10.11 ▼ Many older homes have electrical systems that use fuses rather than circuit breakers. A manufacturer of 40-amp fuses wants to make sure that the mean amperage at which its fuses burn out is in fact 40. If the mean amperage is lower than 40, customers will complain because the fuses require replacement too often. If the mean amperage is higher than 40, the manufacturer might be liable for damage to an electrical system as a result of fuse malfunction. To verify the mean amperage of the fuses, a sample of fuses is selected and tested. If a hypothesis test is performed using the resulting data, what null and alternative hypotheses would be of interest to the manufacturer? H_0: $\mu = 40$ versus H_a: $\mu \neq 40$

10.2 Errors in Hypothesis Testing

Once hypotheses have been formulated, we employ a method called a **test procedure** to use sample data to determine whether H_0 should be rejected. Just as a jury may reach the wrong verdict in a trial, there is some chance that using a test procedure with sample data may lead us to the wrong conclusion about a population characteristic. In this section, we discuss the kinds of errors that can occur and consider how the choice of a test procedure influences the chances of these errors.

One erroneous conclusion in a criminal trial is for a jury to convict an innocent person, and another is for a guilty person to be set free. Similarly, there are two different types of errors that might be made when making a decision in a hypothesis testing problem. One type of error involves rejecting H_0 even though the null hypothesis is true. The second type of error results from failing to reject H_0 when it is false. These errors are known as Type I and Type II errors, respectively.

D E F I N I T I O N

Type I error: the error of rejecting H_0 when H_0 is true
Type II error: the error of failing to reject H_0 when H_0 is false

---

Important concept

The only way to guarantee that neither type of error occurs is to base the decision on a census of the entire population. The risk of error is the price researchers pay for basing an inference on a sample. With any reasonable sample-based procedure, there is some chance that a Type I error will be made and some chance that a Type II error will result.

Example 10.4 On-Time Arrivals

The U.S. Bureau of Transportation Statistics reports that for 2005, 78.6% of all domestic passenger flights arrived on time (meaning within 15 min of the scheduled arrival). Suppose that an airline with a poor on-time record decides to offer its employees a bonus if, in an upcoming month, the airline's proportion of on-time flights exceeds the overall industry rate of 0.786. Let π be the true proportion of the airline's flights that are on time during the month of interest. A random sample of flights might be selected and used as a basis for choosing between

$$H_0: \pi = .786 \text{ and } H_a: \pi > .786$$

In this context, a Type I error (rejecting a true H_0) results in the airline rewarding its employees when in fact their true proportion of on-time flights did not exceed .786. A Type II error (not rejecting a false H_0) results in the airline employees *not* receiving a reward that in fact they deserved.

■

....

Example 10.5 Slowing the Growth of Tumors

In 2004, Vertex Pharmaceuticals, a biotechnology company, issued a press release announcing that it had filed an application with the Food and Drug Administration to begin clinical trials of an experimental drug VX-680 that had been found to reduce the growth rate of pancreatic and colon cancer tumors in animal studies (*New York Times,* February 24, 2004).

Let μ denote the true mean growth rate of tumors for patients receiving the experimental drug. Data resulting from the planned clinical trials can be used to test

H_0: μ = mean growth rate of tumors for patients not taking the experimental drug

versus

H_a: μ < mean growth rate of tumors for patients not taking the experimental drug

The null hypothesis states that the experimental drug is not effective—that the mean growth rate of tumors for patients receiving the experimental drug is the same as for patients who do not take the experimental drug. The alternative hypothesis states that the experimental drug is effective in reducing the mean growth rate of tumors. In this context, a Type I error consists of incorrectly concluding that the experimental drug is effective in slowing the growth rate of tumors. A Type II error consists of concluding that the experimental drug is ineffective when in fact the mean growth rate of tumors is reduced.

■

Examples 10.4 and 10.5 illustrate the two different types of error that might occur when testing hypotheses. Type I and Type II errors—and the associated consequences of making such errors—are quite different. The accompanying box introduces the terminology and notation used to describe error probabilities.

> ### DEFINITION
>
> The **probability of a Type I error** is denoted by α and is called the **level of significance** of the test. Thus, a test with $\alpha = .01$ is said to have a level of significance of .01 or to be a level .01 test.
>
> The **probability of a Type II error** is denoted by β.

....

Example 10.6 Blood Test for Ovarian Cancer

Women with ovarian cancer usually are not diagnosed until the disease is in an advanced stage, when it is most difficult to treat. A blood test has been developed that appears to be able to identify ovarian cancer at its earliest stages. In a report issued by the National Cancer Institute and the Food and Drug Administration (February 8, 2002), the following information from a preliminary evaluation of the blood test was given:

■ The test was given to 50 women known to have ovarian cancer, and it correctly identified all of them as having cancer.

■ The test was given to 66 women known not to have ovarian cancer, and it correctly identified 63 of these 66 as being cancer free.

We can think of using this blood test to choose between two hypotheses:

H_0: woman has ovarian cancer
H_a: woman does not have ovarian cancer

Note that although these are not "statistical hypotheses" (statements about a population characteristic), the possible decision errors are analogous to Type I and Type II errors.

In this situation, believing that a woman with ovarian cancer is cancer free would be a Type I error—rejecting the hypothesis of ovarian cancer when it is in fact true. Believing that a woman who is actually cancer free does have ovarian cancer is a Type II error—not rejecting the null hypothesis when it is in fact false. Based on the preliminary study results, we can estimate the error probabilities. The probability of a Type I error, α, is approximately $0/50 = 0$. The probability of a Type II error, β, is approximately $3/66 = .046$.

■

The ideal test procedure would result in both $\alpha = 0$ and $\beta = 0$. However, if we must base our decision on incomplete information—a sample rather than a census —it is impossible to achieve this ideal. The standard test procedures allow us to control α, but they provide no direct control over β. Because α represents the probability of rejecting a true null hypothesis, selecting a significance level $\alpha = .05$ results in a test procedure that, used over and over with different samples, rejects a *true* H_0 about 5 times in 100. Selecting $\alpha = .01$ results in a test procedure with a Type I error rate of 1% in long-term repeated use. Choosing a small value for α implies that the user wants to use a procedure for which the risk of a Type I error is quite small.

One question arises naturally at this point: If we can select α, the probability of making a Type I error, why would we ever select $\alpha = .05$ rather than $\alpha = .01$? Why not always select a very small value for α? To achieve a small probability of making a Type I error, we would need the corresponding test procedure to require the evidence against H_0 to be very strong before the null hypothesis can be rejected. Although this makes a Type I error unlikely, it increases the risk of a Type II error (*not* rejecting H_0 when it should have been rejected). Frequently the investigator must balance the consequences of Type I and Type II errors. If a Type II error has serious consequences, it may be a good idea to select a somewhat larger value for α.

In general, there is a compromise between small α and small β, leading to the following widely accepted principle for specifying a test procedure.

After assessing the consequences of Type I and Type II errors, identify the largest α that is tolerable for the problem. Then employ a test procedure that uses this maximum acceptable value —rather than anything smaller—as the level of significance (because using a smaller α increases β). In other words, don't make α smaller than it needs to be.

■

Example 10.7 Lead in Tap Water

In 1991, the Environmental Protection Agency (EPA) adopted what is known as the Lead and Copper Rule, which defines drinking water as unsafe if the concentration of lead is 15 parts per billion (ppb) or greater or if the concentration of copper is 1.3 parts per million (ppm) or greater. The "2003 National Public Water Systems Compliance Report" (EPA, September 2005) indicates that 6% of public water systems reported violation of a health-based drinking water standard in 2003 and that 5% of these were violation of the Lead and Copper Rule. With μ denoting the mean concentration of lead, a water system monitoring lead levels might use lead level measurements from a sample of water specimens to test

$$H_0: \mu = 15 \text{ versus } H_a: \mu < 15$$

The null hypothesis (which is equivalent to the assertion $\mu \geq 15$) states that the mean lead concentration is excessive by EPA standards. The alternative hypothesis states that the mean lead concentration is at an acceptable level and that the water system meets EPA standards for lead.

In this context, a Type I error leads to the conclusion that a water source meets EPA standards for lead when in fact it does not. Possible consequences of this type of error include health risks associated with excessive lead consumption (e.g., increased blood pressure, hearing loss, and, in severe cases, anemia and kidney damage). A Type II error is to conclude that the water does not meet EPA standards for lead when in fact it actually does. Possible consequences of a Type II error include elimination of a community water source. Because a Type I error might result in potentially serious public health risks, a small value of α (Type I error probability), such as $\alpha = .01$, could be selected. Of course, selecting a small value for α increases the risk of a Type II error. If the community has only one water source, a Type II error could also have very serious consequences for the community, and we might want to rethink our choice of α.

■

▪ Exercises 10.12–10.22

Turn to the Expanded Answers Section for aswers not shown next to exercises.

10.12 Researchers at the University of Washington and Harvard University analyzed records of breast cancer screening and diagnostic evaluations ("Mammogram Cancer Scares More Frequent than Thought," *USA Today*, April 16, 1998). Discussing the benefits and downsides of the screening process, the article states that, although the rate of false-positives is higher than previously thought, if radiologists were less aggressive in following up on suspicious tests, the rate of false-positives would fall but the rate of missed cancers would rise. Suppose that such a screening test is used to decide between a null hypothesis of H_0: no cancer is present and an alternative hypothesis of H_a: cancer is present. (Although these are not hypotheses about a population characteristic, this exercise illustrates the definitions of Type I and Type II errors.)

a. Would a false-positive (thinking that cancer is present when in fact it is not) be a Type I error or a Type II error?

b. Describe a Type I error in the context of this problem, and discuss the consequences of making a Type I error.

c. Describe a Type II error in the context of this problem, and discuss the consequences of making a Type II error.

d. What aspect of the relationship between the probability of Type I and Type II errors is being described by the statement in the article that if radiologists were less aggressive in following up on suspicious tests, the rate of false-positives would fall but the rate of missed cancers would rise? Lowering the risk of a Type I error will increase the risk of a Type II error.

10.13 Medical personnel are required to report suspected cases of child abuse. Because some diseases have symptoms that mimic those of child abuse, doctors who see a child with these symptoms must decide between two competing hypotheses:

> H_0: symptoms are due to child abuse
> H_a: symptoms are due to disease

(Although these are not hypotheses about a population characteristic, this exercise illustrates the definitions of Type I and Type II errors.) The article "Blurred Line Between Illness, Abuse Creates Problem for Authorities" (*Macon Telegraph,* February 28, 2000) included the following quote from a doctor in Atlanta regarding the consequences of making an incorrect decision: "If it's disease, the worst you have is an angry family. If it is abuse, the other kids (in the family) are in deadly danger."

a. For the given hypotheses, describe Type I and Type II errors.

b. Based on the quote regarding consequences of the two kinds of error, which type of error does the doctor quoted consider more serious? Explain. Based on the doctor's quote, a Type I error is more serious.

10.14 Ann Landers, in her advice column of October 24, 1994 (*San Luis Obispo Telegram-Tribune*), described the reliability of DNA paternity testing as follows: "To get a completely accurate result, you would have to be tested, and so would (the man) and your mother. The test is 100 percent accurate if the man is *not* the father and 99.9 percent accurate if he is."

a. Consider using the results of DNA paternity testing to decide between the following two hypotheses:

> H_0: a particular man is the father
> H_a: a particular man is not the father

In the context of this problem, describe Type I and Type II errors. (Although these are not hypotheses about a population characteristic, this exercise illustrates the definitions of Type I and Type II errors.)

b. Based on the information given, what are the values of α, the probability of Type I error, and β, the probability of Type II error? $\alpha = 0.001, \beta = 0$

c. Ann Landers also stated, "If the mother is not tested, there is a 0.8 percent chance of a false positive." For the hypotheses given in Part (a), what are the values of α and

β if the decision is based on DNA testing in which the mother is not tested? $\alpha = 0.001, \beta = 0.008$

10.15 ▼ Pizza Hut, after test-marketing a new product called the Bigfoot Pizza, concluded that introduction of the Bigfoot nationwide would increase its sales by more than 14% (*USA Today,* April 2, 1993). This conclusion was based on recording sales information for a random sample of Pizza Hut restaurants selected for the marketing trial. With μ denoting the mean percentage increase in sales for all Pizza Hut restaurants, consider using the sample data to decide between H_0: $\mu = 14$ and H_a: $\mu > 14$.

a. Is Pizza Hut's conclusion consistent with a decision to reject H_0 or to fail to reject H_0? Reject H_0

b. If Pizza Hut is incorrect in its conclusion, is the company making a Type I or a Type II error? Type I error

10.16 A television manufacturer claims that (at least) 90% of its TV sets will need no service during the first 3 years of operation. A consumer agency wishes to check this claim, so it obtains a random sample of $n = 100$ purchasers and asks each whether the set purchased needed repair during the first 3 years after purchase. Let p be the sample proportion of responses indicating no repair (so that no repair is identified with a success). Let π denote the true proportion of successes for all sets made by this manufacturer. The agency does not want to claim false advertising unless sample evidence strongly suggests that $\pi < .9$. The appropriate hypotheses are then H_0: $\pi = .9$ versus H_a: $\pi < .9$.

a. In the context of this problem, describe Type I and Type II errors, and discuss the possible consequences of each.

b. Would you recommend a test procedure that uses $\alpha = .10$ or one that uses $\alpha = .01$? Explain.

10.17 A manufacturer of hand-held calculators receives large shipments of printed circuits from a supplier. It is too costly and time-consuming to inspect all incoming circuits, so when each shipment arrives, a sample is selected for inspection. Information from the sample is then used to test H_0: $\pi = .05$ versus H_a: $\pi > .05$, where π is the true proportion of defective circuits in the shipment. If the null hypothesis is not rejected, the shipment is accepted, and the circuits are used in the production of calculators. If the null hypothesis is rejected, the entire shipment is returned to the supplier because of inferior quality. (A shipment is defined to be of inferior quality if it contains more than 5% defective circuits.)

a. In this context, define Type I and Type II errors.
b. From the calculator manufacturer's point of view, which type of error is considered more serious?
c. From the printed circuit supplier's point of view, which type of error is considered more serious?

10.18 Water samples are taken from water used for cooling as it is being discharged from a power plant into a river. It has been determined that as long as the mean temperature of the discharged water is at most 150°F, there will be no negative effects on the river's ecosystem. To investigate whether the plant is in compliance with regulations that prohibit a mean discharge water temperature above 150°F, researchers will take 50 water samples at randomly selected times and record the temperature of each sample. The resulting data will be used to test the hypotheses $H_0: \mu = 150°F$ versus $H_a: \mu > 150°F$. In the context of this example, describe Type I and Type II errors. Which type of error would you consider more serious? Explain.

10.19 ▼ Occasionally, warning flares of the type contained in most automobile emergency kits fail to ignite. A consumer advocacy group wants to investigate a claim against a manufacturer of flares brought by a person who claims that the proportion of defective flares is much higher than the value of .1 claimed by the manufacturer. A large number of flares will be tested, and the results will be used to decide between $H_0: \pi = .1$ and $H_a: \pi > .1$, where π represents the true proportion of defective flares made by this manufacturer. If H_0 is rejected, charges of false advertising will be filed against the manufacturer.
a. Explain why the alternative hypothesis was chosen to be $H_a: \pi > .1$.
b. In this context, describe Type I and Type II errors, and discuss the consequences of each.

10.20 Suppose that you are an inspector for the Fish and Game Department and that you are given the task of determining whether to prohibit fishing along part of the Oregon coast. You will close an area to fishing if it is determined that fish in that region have an unacceptably high mercury content.
a. Assuming that a mercury concentration of 5 ppm is considered the maximum safe concentration, which of the following pairs of hypotheses would you test:

$$H_0: \mu = 5 \text{ versus } H_a: \mu > 5$$

or

$$H_0: \mu = 5 \text{ versus } H_a: \mu < 5$$

Give the reasons for your choice.
b. Would you prefer a significance level of .1 or .01 for your test? Explain. .01, because the consequences of a Type I error are more serious than those of a Type II error.

10.21 The National Cancer Institute conducted a 2-year study to determine whether cancer death rates for areas near nuclear power plants are higher than for areas without nuclear facilities (*San Luis Obispo Telegram-Tribune,* September 17, 1990). A spokesperson for the Cancer Institute said, "From the data at hand, there was no convincing evidence of any increased risk of death from any of the cancers surveyed due to living near nuclear facilities. However, no study can prove the absence of an effect."
a. Let π denote the true proportion of the population in areas near nuclear power plants who die of cancer during a given year. The researchers at the Cancer Institute might have considered the two rival hypotheses of the form

$$H_0: \pi = \text{value for areas without nuclear facilities}$$
$$H_a: \pi > \text{value for areas without nuclear facilities}$$

Did the researchers reject H_0 or fail to reject H_0?
b. If the Cancer Institute researchers were incorrect in their conclusion that there is no increased cancer risk associated with living near a nuclear power plant, are they making a Type I or a Type II error? Explain. Type II error
c. Comment on the spokesperson's last statement that no study can *prove* the absence of an effect. Do you agree with this statement?

10.22 An automobile manufacturer is considering using robots for part of its assembly process. Converting to robots is an expensive process, so it will be undertaken only if there is strong evidence that the proportion of defective installations is lower for the robots than for human assemblers. Let π denote the true proportion of defective installations for the robots. It is known that human assemblers have a defect proportion of .02.
a. Which of the following pairs of hypotheses should the manufacturer test:

$$H_0: \pi = .02 \text{ versus } H_a: \pi < .02$$

or

$$H_0: \pi = .02 \text{ versus } H_a: \pi > .02$$

Explain your answer. $H_0: \pi = 0.02 \text{ versus } H_a: \pi < 0.02$

b. In the context of this exercise, describe Type I and Type II errors.

c. Would you prefer a test with $\alpha = .01$ or $\alpha = .1$? Explain your reasoning.

Bold exercises answered in back   ● Data set available online but not required   ▼ Video solution available

10.3 Large-Sample Hypothesis Tests for a Population Proportion

Now that some general concepts of hypothesis testing have been introduced, we are ready to turn our attention to the development of procedures for using sample information to decide between a null and an alternative hypothesis. There are two possible conclusions: We either reject H_0 or else fail to reject H_0. The fundamental idea behind hypothesis-testing procedures is this: *We reject the null hypothesis if the observed sample is very unlikely to have occurred when H_0 is true.* In this section, we consider testing hypotheses about a population proportion when the sample size n is large.

Let π denote the proportion of individuals or objects in a specified population that possess a certain property. A random sample of n individuals or objects is selected from the population. The sample proportion

$$p = \frac{\text{number in the sample that possess property}}{n}$$

is the natural statistic for making inferences about π.

The large-sample test procedure is based on the same properties of the sampling distribution of p that were used previously to obtain a confidence interval for π, namely:

Based on sampling distribution of p

1. $\mu_p = \pi$

2. $\sigma_p = \sqrt{\dfrac{\pi(1 - \pi)}{n}}$

3. When n is large, the sampling distribution of p is approximately normal.

These three results imply that the standardized variable

$$z = \frac{p - \pi}{\sqrt{\dfrac{\pi(1 - \pi)}{n}}}$$

has approximately a standard normal distribution when n is large. Example 10.8 shows how this information allows us to make a decision.

Example 10.8   Impact of Food Labels

In June of 2006, an Associated Press survey was conducted to investigate how people use the nutritional information provided on food package labels. Interviews were conducted with 1003 randomly selected adult Americans, and each participant was asked a series of questions, including the following two:

Question 1: When purchasing packaged food, how often to you check the nutrition labeling on the package?

Question 2: How often do you purchase foods that are bad for you, even after you've checked the nutrition labels?

It was reported that 582 responded "frequently" to the question about checking labels and 441 responded very often or somewhat often to the question about purchasing "bad" foods even after checking the label.

Let's start by looking at the responses to the first question. Based on these data, is it reasonable to conclude that a majority of adult Americans frequently check the nutritional labels when purchasing packaged foods? We can answer this question by testing hypotheses, where

π = true proportion of adult Americans who frequently check nutritional labels

H_0: $\pi = .5$
H_a: $\pi > .5$ (The proportion of adult Americans who frequently check nutritional labels is greater than .5. That is, more than half (a majority) frequently check nutritional labels.)

Recall that in a hypothesis test, the null hypothesis is rejected only if there is convincing evidence against it—in this case, convincing evidence that $\pi > .5$. If H_0 is rejected, there is strong support for the claim that a majority of adult Americans frequently check nutritional labels when purchasing packaged foods.

For this sample,

$$p = \frac{582}{1003} = .58$$

The observed sample proportion is certainly greater than .5, but this could just be due to sampling variability. That is, when $\pi = .5$ (meaning H_0 is true), the sample proportion p usually differs somewhat from .5 simply because of chance variation from one sample to another. Is it plausible that a sample proportion of $p = .58$ occurred as a result of this chance variation, or is it unusual to observe a sample proportion this large when $\pi = .5$?

To answer this question, we form a *test statistic,* the quantity used as a basis for making a decision between H_0 and H_a. Creating a test statistic involves replacing π with the hypothesized value in the z variable $z = \dfrac{p - \pi}{\sqrt{\pi(1 - \pi)/n}}$ to obtain

$$z = \frac{p - .5}{\sqrt{\dfrac{(.5)(.5)}{n}}}$$

If the null hypothesis is true, this statistic should have approximately a standard normal distribution, because when the sample size is large and H_0 is true,

1. $\mu_p = .5,$

2. $\sigma_p = \sqrt{\dfrac{(.5)(.5)}{n}}$

3. p has approximately a normal distribution.

Important concept

The calculated value of z expresses the distance between p and the hypothesized value as a number of standard deviations. If, for example, $z = 3$, then the value of

p that came from the sample is 3 standard deviations (of p) greater than what we would have expected if the null hypothesis were true. How likely is it that a z value at least this contradictory to H_0 would be observed if in fact H_0 is true? The test statistic z is constructed using the hypothesized value from the null hypothesis; if H_0 is true, the test statistic has (approximately) a standard normal distribution. Therefore

$$P(z \geq 3 \text{ when } H_0 \text{ is true}) = \text{area under the } z \text{ curve to the right of } 3.00 = .0013$$

That is, if H_0 is true, very few samples (much less than 1% of all samples) produce a value of z at least as contradictory to H_0 as $z = 3$. Because this z value is in the most extreme 1%, it is sensible to reject H_0.

For our data,

$$z = \frac{p - .5}{\sqrt{\dfrac{(.5)(.5)}{n}}} = \frac{.58 - .5}{\sqrt{\dfrac{(.5)(.5)}{1003}}} = \frac{.08}{.016} = 5.00$$

That is, $p = .58$ is 5 standard deviations greater than what we would expect it to be if the null hypothesis H_0: $\pi = .5$ were true. The sample data appear to be much more consistent with the alternative hypothesis, H_a: $\pi > .5$. In particular,

$$P(\text{value of } z \text{ is at least as contradictory to } H_0 \text{ as } 5.00 \text{ when } H_0 \text{ is true})$$
$$= P(z \geq 5.00 \text{ when } H_0 \text{ is true})$$
$$= \text{area under the } z \text{ curve to the right of } 5.00$$
$$\approx 0$$

There is virtually no chance of seeing a sample proportion and corresponding z value this extreme as a result of chance variation alone when H_0 is true. If p is 5 standard deviations or more away from .5, how can we believe that $\pi = .5$? The evidence for rejecting H_0 in favor of H_a is very compelling.

Interestingly, in spite of the fact that there is strong evidence that a majority of adult Americans frequently check nutritional labels, the data on responses to the second question suggest that the percentage of people who then ignore the information on the label and purchase "bad" foods anyway is not small—the sample proportion who responded very often or somewhat often was .44.

■

Fundamental concept of hypothesis tests

The preceding example illustrates the rationale behind large-sample procedures for testing hypotheses about π (and other test procedures as well). We begin by assuming that the null hypothesis is correct. The sample is then examined in light of this assumption. If the observed sample proportion would not be unusual when H_0 is true, then chance variability from one sample to another is a plausible explanation for what has been observed, and H_0 should not be rejected. On the other hand, if the observed sample proportion would have been quite unlikely when H_0 is true, then we would take the sample as convincing evidence against the null hypothesis and we should reject H_0. We base a decision to reject or to fail to reject the null hypothesis on an assessment of how extreme or unlikely the observed sample is if H_0 is true.

The assessment of how contradictory the observed data are to H_0 is based on first computing the value of the test statistic

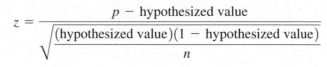

$$z = \frac{p - \text{hypothesized value}}{\sqrt{\dfrac{(\text{hypothesized value})(1 - \text{hypothesized value})}{n}}}$$

We then calculate the *P-value,* the probability, assuming that H_0 is true, of obtaining a z value at least as contradictory to H_0 as what was actually observed.

DEFINITION

A **test statistic** is the function of sample data on which a conclusion to reject or fail to reject H_0 is based.

The *P*-**value** (also sometimes called the **observed significance level**) is a measure of inconsistency between the hypothesized value for a population characteristic and the observed sample. It is the probability, assuming that H_0 is true, of obtaining a test statistic value at least as inconsistent with H_0 as what actually resulted.

Example 10.9 Detecting Plagiarism

Plagiarism is a growing concern among college and university faculty members, and many universities are now using software tools to detect student work that is not original. Researchers at the University of Luton conducted a survey of 321 faculty members at a variety of academic institutions ("Technical Review of Plagiarism Detection Software," University of Luton, 2001). Included in the survey were questions about strategies used to uncover instances of plagiarism. It was reported that 36% of those surveyed said they occasionally used online searches with key words from student work to check for plagiarism.

Assuming it is reasonable to regard this sample as representative of university faculty members, does the sample provide convincing evidence that more than one-third of faculty members occasionally use key word searches to check student work?
With

π = true proportion faculty members who use key word searches to check student work

the relevant hypotheses are

$$H_0: \pi = \frac{1}{3} = .33$$
$$H_a: \pi > .33$$

The sample proportion was reported to be $p = .36$. Does the value of p exceed one-third by enough to cast substantial doubt on H_0?

Because the sample size is large, the statistic

$$z = \frac{p - .33}{\sqrt{\dfrac{(.33)(1 - .33)}{n}}}$$

has approximately a standard normal distribution when H_0 is true. The calculated value of the test statistic is

$$z = \frac{.36 - .33}{\sqrt{\dfrac{(.33)(1 - .33)}{321}}} = \frac{.03}{.026} = 1.15$$

The probability that a z value at least this inconsistent with H_0 would be observed if in fact H_0 is true is

$$P\text{-value} = P(z \geq 1.15 \text{ when } H_0 \text{ is true})$$
$$= \text{area under the } z \text{ curve to the right of } 1.15$$
$$= 1 - .8749$$
$$= .1251$$

This probability indicates that when $\pi = .33$, it would not be all that unusual to observe a sample proportion as large as .36. When H_0 is true, roughly 12.5% of all samples would have a sample proportion larger than .36, so a sample proportion of .36 is reasonably consistent with the null hypothesis. Although .36 is larger than the hypothesized value of $\pi = .33$, chance variation from sample to sample is a plausible explanation for what was observed. There is not strong evidence that the proportion of faculty members who use key word searches to check student work for plagiarism is greater than one-third.

■

As illustrated by Examples 10.8 and 10.9, small P-values indicate that sample results are inconsistent with H_0, whereas larger P-values are interpreted as meaning that the data are consistent with H_0 and that sampling variability alone is a plausible explanation for what was observed in the sample. As you probably noticed, the two cases examined (P-value ≈ 0 and P-value $= .1251$) were such that a decision between rejecting or not rejecting H_0 was clear-cut. A decision in other cases might not be so obvious. For example, what if the sample had resulted in a P-value of .04? Is this unusual enough to warrant rejection of H_0? How small must the P-value be before H_0 should be rejected?

The answer depends on the significance level α (the probability of a Type I error) selected for the test. For example, suppose that we set $\alpha = .05$. This implies that the probability of rejecting a true null hypothesis is .05. To obtain a test procedure with this probability of Type I error, we would reject the null hypothesis if the sample result is among the most unusual 5% of all samples when H_0 is true. That is, H_0 is rejected if the computed P-value $\leq .05$. If we had selected $\alpha = .01$, H_0 would be rejected only if we observed a sample result so extreme that it would be among the most unusual 1% if H_0 is true (i.e., if P-value $\leq .01$).

A decision as to whether H_0 should be rejected results from comparing the P-value to the chosen α:

H_0 should be rejected if P-value $\leq \alpha$.
H_0 should not be rejected if P-value $> \alpha$.

■

Suppose, for example, that the P-value $= .0352$ and that a significance level of .05 is chosen. Then, because

$$P\text{-value} = .0352 \leq .05 = \alpha$$

H_0 would be rejected. This would not be the case, though, for $\alpha = .01$, because then P-value $> \alpha$.

■ Computing a P-Value for a Large-Sample Test Concerning π ....

The computation of the P-value depends on the form of the inequality in the alternative hypothesis, H_a. Suppose, for example, that we wish to test

$$H_0: \pi = .6 \qquad \text{versus} \qquad H_a: \pi > .6$$

based on a large sample. The appropriate test statistic is

$$z = \frac{p - .6}{\sqrt{\dfrac{(.6)(1 - .6)}{n}}}$$

Values of p contradictory to H_0 and much more consistent with H_a are those much *larger* than .6 (because $\pi = .6$ when H_0 is true and $\pi > .6$ when H_0 is false and H_a is true). Such values of p correspond to z values considerably greater than 0. If $n = 400$ and $p = .679$, then

$$z = \frac{.679 - .6}{\sqrt{\dfrac{(.6)(1 - .6)}{400}}} = \frac{.079}{.025} = 3.16$$

The value $p = .679$ is more than 3 standard deviations larger than what we would have expected if H_0 were true. Thus

$$\begin{aligned}
P\text{-value} &= P(z \text{ at least as contradictory to } H_0 \text{ as 3.16 when } H_0 \text{ is true}) \\
&= P(z \geq 3.16 \text{ when } H_0 \text{ is true}) \\
&= \text{area under the } z \text{ curve to the right of 3.16} \\
&= 1 - .9992 \\
&= .0008
\end{aligned}$$

This P-value is illustrated in Figure 10.1. If H_0 is true, in the long run only 8 out of 10,000 samples would result in a z value as or more extreme than what actually resulted; most of us would consider such a z quite unusual. Using a significance level of .01, we reject the null hypothesis because P-value $= .0008 \leq .01 = \alpha$.

Now consider testing $H_0: \pi = .3$ versus $H_a: \pi \neq .3$. A value of p either *much* greater than .3 *or* much less than .3 is inconsistent with H_0 and provides support for H_a. Such a p corresponds to a z value far out in *either* tail of the z curve. If

$$z = \frac{p - .3}{\sqrt{\dfrac{(.3)(1 - .3)}{n}}} = 1.75$$

then (as shown in Figure 10.2)

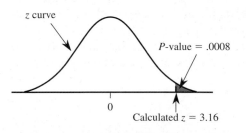

z curve

P-value = .0008

0

Calculated z = 3.16

Figure 10.1 Calculating a P-value.

$$\begin{aligned}
P\text{-value} &= P(z \text{ value at least as inconsistent with } H_0 \text{ as 1.75 when } H_0 \text{ is true}) \\
&= P(z \geq 1.75 \text{ or } z \leq -1.75 \text{ when } H_0 \text{ is true}) \\
&= (z \text{ curve area to the right of 1.75}) + (z \text{ curve area to the left of } -1.75) \\
&= (1 - .9599) + .0401 \\
&= .0802
\end{aligned}$$

The P-value in this situation is also .0802 if $z = -1.75$, because 1.75 and -1.75 are equally inconsistent with H_0.

Figure 10.2 *P*-value as the sum of two tail areas.

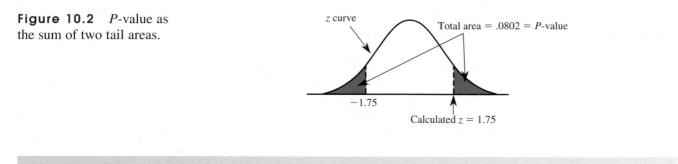

Determination of the *P*-Value When the Test Statistic Is *z*

1. **Upper-tailed test:**

   H_a: $\pi >$ hypothesized value

   P-value computed as illustrated:

   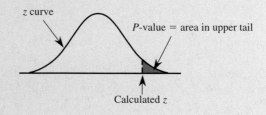

2. **Lower-tailed test:**

   H_a: $\pi <$ hypothesized value

   P-value computed as illustrated:

   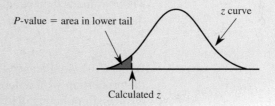

3. **Two-tailed test:**

   H_a: $\pi \neq$ hypothesized value

   P-value computed as illustrated:

   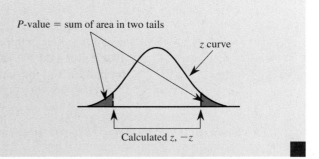

The symmetry of the *z* curve implies that when the test is two-tailed (the "not equal" alternative), it is not necessary to add two curve areas. Instead,

If *z* is positive, *P*-value = 2(area to the right of *z*).
If *z* is negative, *P*-value = 2(area to the left of *z*).

Example 10.10 Water Conservation

In December 2005 a countywide water conservation campaign was conducted in a particular county. In January 2006 a random sample of 500 homes was selected, and water usage was recorded for each home in the sample. The county supervisors wanted to know whether their data supported the claim that fewer than half the households in the county reduced water consumption. The relevant hypotheses are

$$H_0: \pi = .5 \qquad \text{versus} \qquad H_a: \pi < .5$$

where π is the true proportion of households in the county with reduced water usage.

Suppose that the sample results were $n = 500$ and $p = .440$. Because the sample size is large and this is a lower-tailed test, we can compute the P-value by first calculating the value of the z test statistic

$$z = \frac{p - .5}{\sqrt{\dfrac{(.5)(1 - .5)}{n}}}$$

and then finding the area under the z curve to the left of this z.

Based on the observed sample data,

$$z = \frac{.440 - .5}{\sqrt{\dfrac{(.5)(1 - .5)}{500}}} = \frac{-.060}{.0224} = -2.68$$

The P-value is then equal to the area under the z curve and to the left of -2.68. From the entry in the -2.6 row and $.08$ column of Appendix Table 2, we find that

$$P\text{-value} = .0037$$

Using a .01 significance level, we reject H_0 (because $.0037 \leq .01$), suggesting that the proportion with reduced water usage was less than .5. Notice that rejection of H_0 would not be justified if a *very* small significance level, such as .001, had been selected.

■

Example 10.10 illustrates the calculation of a P-value for a lower-tailed test. The use of P-values in upper-tailed and two-tailed tests is illustrated in Examples 10.11 and 10.12. But first we summarize large-sample tests of hypotheses about a population proportion and introduce a step-by-step procedure for carrying out a hypothesis test.

Summary of Large-Sample z Test for π

Null hypothesis: $H_0: \pi = $ hypothesized value

Test statistic: $z = \dfrac{p - \text{hypothesized value}}{\sqrt{\dfrac{(\text{hypothesized value})(1 - \text{hypothesized})}{n}}}$

(*continued*)

Alternative Hypothesis:

H_a: $\pi >$ hypothesized value

H_a: $\pi <$ hypothesized value

H_a: $\pi \neq$ hypothesized value

P-Value:

Area under z curve to right of calculated z

Area under z curve to left of calculated z

(1) 2(area to right of z) if z is positive, or

(2) 2(area to left of z) if z is negative

Assumptions:
1. p is the sample proportion from a *random sample*.
2. The *sample size is large*. This test can be used if n satisfies both n(hypothesized value) ≥ 10 and $n(1 -$ hypothesized value) ≥ 10.
3. If sampling is without replacement, the sample size is no more than 10% of the population size.

We recommend that the following sequence of steps be used when carrying out a hypothesis test.

Steps in a Hypothesis-Testing Analysis

1. Describe the population characteristic about which hypotheses are to be tested.
2. State the null hypothesis H_0.
3. State the alternative hypothesis H_a.
4. Select the significance level α for the test.
5. Display the test statistic to be used, with substitution of the hypothesized value identified in Step 2 but without any computation at this point.
6. Check to make sure that any assumptions required for the test are reasonable.
7. Compute all quantities appearing in the test statistic and then the value of the test statistic itself.
8. Determine the *P*-value associated with the observed value of the test statistic.
9. State the conclusion (which is to reject H_0 if *P*-value $\leq \alpha$ and not to reject H_0 otherwise). The conclusion should then be stated in the context of the problem, and the level of significance should be included.

Steps 1–4 constitute a statement of the problem, Steps 5–8 give the analysis that leads to a decision, and Step 9 provides the conclusion.

Example 10.11 Unfit Teens

The article "7 Million U.S. Teens Would Flunk Treadmill Tests" (Associated Press, December 11, 2005) summarized the results of a study in which 2205 adolescents aged 12 to 19 took a cardiovascular treadmill test. The researchers conducting the study indicated that the sample was selected in such a way that it could be regarded as representative of adolescents nationwide. Of the 2205 adolescents tested, 750 showed a poor level of cardiovascular fitness. Does this sample provide support for the claim that more than 30% of adolescents have a low level of cardiovascular fit-

Step-by-step technology instructions available online

ness? We answer this question by carrying out a hypothesis test using a .05 significance level.

1. Population characteristic of interest:

    π = true proportion of adolescents who have a low level of cardiovascular fitness

2. Null hypothesis: $H_0: \pi = .3$
3. Alternative hypothesis: $H_a: \pi > .3$ (the percentage of adolescents with a low fitness level is greater than 30%)
4. Significance level: $\alpha = .05$
5. Test statistic:

$$z = \frac{p - \text{hypothesized value}}{\sqrt{\dfrac{(\text{hypothesized value})(1 - \text{hypothesized value})}{n}}} = \frac{p - .3}{\sqrt{\dfrac{(.3)(1 - .3)}{n}}}$$

6. Assumptions: This test requires a random sample and a large sample size. The given sample was considered to be representative of adolescents nationwide, and if this is the case it is reasonable to regard the sample as if it were a random sample. The sample size was $n = 2205$. Since $2205(.3) \geq 10$ and $2205(1 - .3) \geq 10$, the large-sample test is appropriate. The sample size is small compared to the population (adolescents) size.
7. Computations: $n = 2205$ and $p = 750/2205 = .34$, so

$$z = \frac{.34 - .3}{\sqrt{\dfrac{(.3)(1 - .3)}{2205}}} = \frac{.04}{.010} = 4.00$$

8. *P*-value: This is an upper-tailed test (the inequality in H_a is >), so the *P*-value is the area to the right of the computed z value. Since $z = 4.00$ is so far out in the upper tail of the standard normal distribution, the area to its right is negligible. Thus,

$$P\text{-value} \approx 0$$

9. Conclusion: Since *P*-value $\leq \alpha$ ($0 \leq .05$), H_0 is rejected at the .05 level of significance. We conclude that the proportion of adolescents who have a low level of cardiovascular fitness is greater than .3. That is, the sample provides convincing evidence to support the claim that more than 30% of adolescents have a low fitness level.

■

Example 10.12 Single-Family Homes

The Public Policy Institute of California reported that 71% of people nationwide prefer to live in a single-family home. To determine whether the preferences of Californians are consistent with this nationwide figure, a random sample of 2002 Californians were interviewed. Of those interviewed, 1682 said that they consider a single-family home the ideal (Associated Press, November 13, 2001). Can we reasonably conclude that the proportion of Californians who prefer a single-family home is different from the national figure? We answer this question by carrying out a hypothesis test with $\alpha = .01$.

1. π = proportion of all Californians who prefer a single-family home.
2. H_0: $\pi = .71$.
3. H_a: $\pi \neq .71$ (differs from the national proportion).
4. Significance level: $\alpha = .01$.
5. Test statistic:

$$z = \frac{p - \text{hypothesized value}}{\sqrt{\dfrac{(\text{hypothesized value})(1 - \text{hypothesized value})}{n}}} = \frac{p - .71}{\sqrt{\dfrac{(.71)(.29)}{n}}}$$

6. Assumptions: This test requires a random sample and a large sample size. The given sample was a random sample, the population size is much larger than the sample size, and the sample size was $n = 2002$. Because $2002(.71) \geq 10$ and $2002(.29) \geq 10$, the large-sample test is appropriate.
7. Computations: $p = 1682/2002 = .84$, from which

$$z = \frac{.84 - .71}{\sqrt{\dfrac{(.71)(.29)}{2002}}} = \frac{.13}{.0101} = 12.87$$

8. *P*-value: The area under the z curve to the right of 12.87 is approximately 0, so *P*-value $\approx 2(0) = 0$.
9. Conclusion: At significance level .01, we reject H_0 because *P*-value $\approx 0 < .01 = \alpha$. The data provide convincing evidence that the proportion in California who prefer a single-family home differs from the nationwide proportion.

■

Most statistical computer packages and graphing calculators can calculate and report *P*-values for a variety of hypothesis-testing situations, including the large sample test for a proportion. MINITAB was used to carry out the test of Example 10.10, and the resulting computer output follows (MINITAB uses *p* instead of π to denote the population proportion):

Test and Confidence Interval for One Proportion

Test of p = 0.5 vs p < 0.5

Sample	X	N	Sample p	95.0 % CI	Z-Value	P-Value
1	220	500	0.440000	(0.396491, 0.483509)	−2.68	0.004

From the MINITAB output, $z = -2.68$, and the associated *P*-value is .004. The small difference in the *P*-value is the result of rounding.

It is also possible to compute the value of the z test statistic and then use a statistical computer package or graphing calculator to determine the corresponding *P*-value as an area under the standard normal curve. For example, the user can specify a value and MINITAB will determine the area to the left of this value for any particular normal distribution. Because of this, the computer can be used in place of Appendix Table 2. In Example 10.10 the computed z was -2.68. Using MINITAB gives the following output:

Normal with mean = 0 and standard deviation = 1.00000

x	P(X ≤ x)
−2.6800	0.0037

Thus we learn that the area to the left of $-2.68 = .0037$, which agrees with the value obtained by using the tables.

▪ Exercises 10.23–10.44

Turn to the Expanded Answers Section for aswers not shown next to exercises.

10.23 Use the definition of the P-value to explain the following:
a. Why H_0 would certainly be rejected if P-value $= .0003$
b. Why H_0 would definitely not be rejected if P-value $= .350$

10.24 For which of the following P-values will the null hypothesis be rejected when performing a level .05 test:
a. .001   d. .047
b. .021   e. .148   0.001, 0.021, 0.047
c. .078

10.25 Pairs of P-values and significance levels, α, are given. For each pair, state whether the observed P-value leads to rejection of H_0 at the given significance level.
a. P-value $= .084$, $\alpha = .05$ Fail to reject
b. P-value $= .003$, $\alpha = .001$ Fail to reject
c. P-value $= .498$, $\alpha = .05$ Fail to reject
d. P-value $= .084$, $\alpha = .10$ Reject
e. P-value $= .039$, $\alpha = .01$ Fail to reject
f. P-value $= .218$, $\alpha = .10$ Fail to reject

10.26 Let π denote the proportion of grocery store customers that use the store's club card. For a large sample z test of H_0: $\pi = .5$ versus H_a: $\pi > .5$, find the P-value associated with each of the given values of the test statistic:
a. 1.40   0.0808   d. 2.45   0.0071
b. 0.93   0.1762   e. −0.17   0.5675
c. 1.96   0.0250

10.27 Assuming a random sample from a large population, for which of the following null hypotheses and sample sizes n is the large-sample z test appropriate:
a. H_0: $\pi = .2$, $n = 25$ Not appropriate
b. H_0: $\pi = .6$, $n = 210$ Appropriate
c. H_0: $\pi = .9$, $n = 100$ Appropriate
d. H_0: $\pi = .05$, $n = 75$ Not appropriate

10.28 The article "Poll Finds Most Oppose Return to Draft, Wouldn't Encourage Children to Enlist" (Associated Press, December 18, 2005) reports that in a random sample of 1000 American adults, 700 indicated that they oppose the reinstatement of a military draft. Is there convincing evidence that the proportion of American adults who oppose reinstatement of the draft is greater than two-thirds? Use a significance level of .05. $z = 2.21$, P-value $= 0.0136$, reject H_0

10.29 The poll referenced in the previous exercise ("Military Draft Study," AP-Ipsos, June 2005) also included the following question: "If the military draft were reinstated, would you favor or oppose drafting women as well as men?" Forty-three percent of the 1000 people responding said that they would favor drafting women if the draft were reinstated. Using a .05 significance level, carry out a test to determine if there is convincing evidence that fewer than half of adult Americans would favor the drafting of women. $z = -4.43$, P-value ≈ 0, reject H_0

10.30 The article "Irritated by Spam? Get Ready for Spit" (*USA Today*, November 10, 2004) predicts that "spit," spam that is delivered via Internet phone lines and cell phones, will be a growing problem as more people turn to web-based phone services. In a 2004 poll of 5500 cell phone users conducted by the Yankee Group, 20% indicated that they had received commercial messages and ads on their cell phones. Is there sufficient evidence that the proportion of cell phone users who have received commercial messages or ads in 2004 was greater than the proportion of .13 reported for the previous year? $z = 15.44$, P-value ≈ 0, reject H_0

10.31 In a survey conducted by Yahoo Small Business, 1432 of 1813 adults surveyed said that they would alter their shopping habits if gas prices remain high (Associated Press, November 30, 2005). The article did not say how the sample was selected, but for purposes of this exercise, assume that it is reasonable to regard this sample as representative of adult Americans. Based on these survey data, is it reasonable to conclude that more than three-quarters of adult Americans plan to alter their shopping habits if gas prices remain high? $z = 3.93$, P-value ≈ 0, reject H_0

10.32 According to a *Washington Post-ABC News* poll, 331 of 502 randomly selected U.S. adults interviewed said they would not be bothered if the National Security

Agency collected records of personal telephone calls they had made. Is there sufficient evidence to conclude that a majority of U.S. adults feel this way? Test the appropriate hypotheses using a .01 significance level. $z = 7.17$, P-value ≈ 0, reject H_0

10.33 According to a survey of 1000 adult Americans conducted by Opinion Research Corporation, 210 of those surveyed said playing the lottery would be the most practical way for them to accumulate $200,000 in net wealth in their lifetime ("One in Five Believe Path to Riches Is the Lottery," *San Luis Obispo Tribune,* January 11, 2006). Although the article does not describe how the sample was selected, for purposes of this exercise, assume that the sample can be regarded as a random sample of adult Americans. Is there convincing evidence that more than 20% of adult Americans believe that playing the lottery is the best strategy for accumulating $200,000 in net wealth? $z = 0.79$, P-value $= 0.2148$, fail to reject H_0

10.34 The article "Theaters Losing Out to Living Rooms" (*San Luis Obispo Tribune,* June 17, 2005) states that movie attendance declined in 2005. The Associated Press found that 730 of 1000 randomly selected adult Americans preferred to watch movies at home rather than at a movie theater. Is there convincing evidence that the majority of adult Americans prefer to watch movies at home? Test the relevant hypotheses using a .05 significance level. $z = 14.55$, P-value ≈ 0, reject H_0

10.35 The article referenced in Exercise 10.34 also reported that 470 of 1000 randomly selected adult Americans thought that the quality of movies being produced was getting worse.
a. Is there convincing evidence that fewer than half of adult Americans believe that movie quality is getting worse? Use a significance level of .05.
b. Suppose that the sample size had been 100 instead of 1000, and that 47 thought that the movie quality was getting worse (so that the sample proportion is still .47). Based on this sample of 100, is there convincing evidence that fewer than half of adult Americans believe that movie quality is getting worse? Use a significance level of .05.
c. Write a few sentences explaining why different conclusions were reached in the hypothesis tests of Parts (a) and (b).

10.36 The report "2005 Electronic Monitoring & Surveillance Survey: Many Companies Monitoring, Recording, Videotaping—and Firing—Employees" (American Management Association, 2005) summarized the results of a survey of 526 U.S. businesses. Four hundred of these com-

panies indicated that they monitor employees' web site visits. For purposes of this exercise, assume that it is reasonable to regard this sample as representative of businesses in the United States.
a. Is there sufficient evidence to conclude that more than 75% of U.S. businesses monitor employees' web site visits? Test the appropriate hypotheses using a significance level of .01. $z = 0.53$, P-value $= 0.2981$, fail to reject H_0
b. Is there sufficient evidence to conclude that a majority of U.S. businesses monitor employees' web site visits? Test the appropriate hypotheses using a significance level of .01. $z = 11.93$, P-value ≈ 0, reject H_0

10.37 In an AP-AOL sports poll (Associated Press, December 18, 2005), 272 of 394 randomly selected baseball fans stated that they thought the designated hitter rule should either be expanded to both baseball leagues or eliminated. Based on the given information, is there sufficient evidence to conclude that a majority of baseball fans feel this way? $z = 7.54$, P-value ≈ 0, reject H_0

10.38 In a representative sample of 1000 adult Americans, only 430 could name at least one justice who is currently serving on the U.S. Supreme Court (Ipsos, January 10, 2006). Using a significance level of .01, carry out a hypothesis test to determine if there is convincing evidence to support the claim that fewer than half of adult Americans can name at least one justice currently serving on the Supreme Court. $z = -4.43$, P-value ≈ 0, reject H_0

10.39 ▼ In a national survey of 2013 adults, 1590 responded that lack of respect and courtesy in American society is a serious problem, and 1283 indicated that they believe that rudeness is a more serious problem than in past years (Associated Press, April 3, 2002). Is there convincing evidence that more than three-quarters of U.S. adults believe that rudeness is a worsening problem? Test the relevant hypotheses using a significance level of .05. $z = -11.7$, P-value $= 1$, fail to reject H_0

10.40 The success of the U.S. census depends on people filling out and returning census forms. Despite extensive advertising, many Americans are skeptical about claims that the Census Bureau will guard the information it collects from other government agencies. In a *USA Today* poll (March 13, 2000), only 432 of 1004 adults surveyed said that they believe the Census Bureau when it says the information you give about yourself is kept confidential. Is there convincing evidence that, despite the advertising campaign, fewer than half of U.S. adults believe the Cen-

"Smoking Abstinence Impairs Time Estimation Accuracy in Cigarette Smokers" (*Psychopharmacology Bulletin* [2003]: 90–95). After a 24-hr smoking abstinence, 20 smokers were asked to estimate how much time had passed during a 45-sec period. Suppose the resulting data on perceived elapsed time (in seconds) were as shown (these data are artificial but are consistent with summary quantities given in the paper):

| 69 | 65 | 72 | 73 | 59 | 55 | 39 | 52 | 67 | 57 |
| 56 | 50 | 70 | 47 | 56 | 45 | 70 | 64 | 67 | 53 |

From these data, we obtain

$$n = 20 \qquad \bar{x} = 59.30 \qquad s = 9.84$$

The researchers wanted to determine whether smoking abstinence had a negative impact on time perception, causing elapsed time to be overestimated. We can answer this question by testing

H_0: $\mu = 45$ (no consistent tendency to overestimate the time elapsed)

versus

H_a: $\mu > 45$ (tendency for elapsed time to be overestimated)

The null hypothesis is rejected only if there is convincing evidence that $\mu > 45$. The observed value, 59.30, is certainly larger than 45, but can a sample mean as large as this be plausibly explained by chance variation from one sample to another when $\mu = 45$? To answer this question, we carry out a hypothesis test with a significance level of .05 using the step-by-step procedure described in Section 10.3.

1. Population characteristic of interest:

   μ = mean perceived elapsed time for smokers who have abstained from smoking for 24 hours

2. Null hypothesis: H_0: $\mu = 45$
3. Alternative hypothesis: H_a: $\mu > 45$
4. Significance level: $\alpha = .05$
5. Test statistic: $t = \dfrac{\bar{x} - \text{hypothesized value}}{\dfrac{s}{\sqrt{n}}} = \dfrac{\bar{x} - 45}{\dfrac{s}{\sqrt{n}}}$

6. Assumptions: This test requires a random sample and either a large sample size or a normal population distribution. The authors of the paper believed that it was reasonable to consider this sample as representative of smokers in general, so we regard it as if it were a random sample. Because the sample size is only 20, for the t test to be appropriate, we must be willing to presume that the population distribution of perceived elapsed times is at least approximately normal. Is this reasonable? The following graph gives a boxplot of the data:

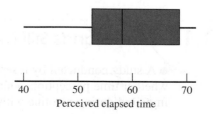

Perceived elapsed time

Although the boxplot is not perfectly symmetric, it is not too skewed and there are no outliers, so we judge the use of the t test to be reasonable.

7. Computations: $n = 20$, $\bar{x} = 59.30$, and $s = 9.84$, so

$$t = \frac{59.30 - 45}{\frac{9.84}{\sqrt{20}}} = \frac{14.30}{2.20} = 6.50$$

8. P-value: This is an upper-tailed test (the inequality in H_a is "greater than"), so the P-value is the area to the right of the computed t value. Because df $= 20 - 1 = 19$, we can use the df $= 19$ column of Appendix Table 4 to find the P-value. With $t = 6.50$, we obtain P-value $=$ area to the right of $6.50 \approx 0$ (because 6.50 is greater than 4.0, the largest tabulated value).

9. Conclusion: Because P-value $\leq \alpha$, we reject H_0 at the .05 level of significance. There is virtually no chance of seeing a sample mean (and hence a t value) this extreme as a result of just chance variation when H_0 is true. There is convincing evidence that the mean perceived time elapsed is greater than the actual time elapsed of 45 sec.

This paper also looked at perception of elapsed time for a sample of nonsmokers and for a sample of smokers who had not abstained from smoking. The investigators found that the null hypothesis of $\mu = 45$ could not be rejected for either of these groups.

■

Example 10.14 Goofing Off at Work

● A growing concern of employers is time spent in activities like surfing the Internet and emailing friends during work hours. The *San Luis Obispo Tribune* summarized the findings from a survey of a large sample of workers in an article that ran under the headline "Who Goofs Off 2 Hours a Day? Most Workers, Survey Says" (August 3, 2006). Suppose that the CEO of a large company wants to determine whether the average amount of wasted time during an eight-hour work day for employees of her company is less than the reported 120 minutes. Each person in a random sample of 10 employees was contacted and asked about daily wasted time at work. (Participants would probably have to be guaranteed anonymity to obtain truthful responses!) The resulting data are the following:

| 108 | 112 | 117 | 130 | 111 | 131 | 113 | 113 | 105 | 128 |

Summary quantities are $n = 10$, $\bar{x} = 116.80$, and $s = 9.45$.

Do these data provide evidence that the mean wasted time for this company is less than 120 min? To answer this question, let's carry out a hypothesis test with $\alpha = .05$.

1. μ = mean daily wasted time for employees of this company
2. H_0: $\mu = 120$
3. H_a: $\mu < 120$
4. $\alpha = .05$

5. $t = \dfrac{\bar{x} - \text{hypothesized value}}{\dfrac{s}{\sqrt{n}}} = \dfrac{\bar{x} - 120}{\dfrac{s}{\sqrt{n}}}$

6. This test requires a random sample and either a large sample or a normal population distribution. The given sample was a random sample of employees. Because the sample size is small, we must be willing to assume that the population distribution of times is at least approximately normal. The following normal probability plot appears to be reasonably straight, and although the normal probability plot and the boxplot reveal some skewness in the sample, there are no outliers:

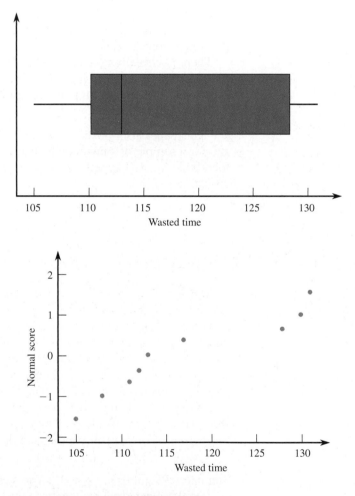

Correlations (Pearson)
Correlation of Time and Normal Score = 0.943

Also, the correlation between the expected normal scores and the observed data for this sample is .943, which is well above the critical *r* value for *n* = 10 of .880 (see Chapter 5 for critical *r* values). Based on these observations, it is plausible that the population distribution is approximately normal, so we proceed with the *t* test.

7. Test statistic: $t = \dfrac{116.80 - 120}{\dfrac{9.45}{\sqrt{10}}} = -1.07$

8. From the df = 9 column of Appendix Table 4 and by rounding the test statistic value to −1.1, we get

$$P\text{-value} = \text{area to the left of} -1.1 = \text{area to the right of } 1.1 = .150$$

as shown:

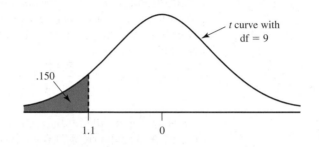

.150

t curve with
df = 9

1.1   0

9. Because the P-value > α, we fail to reject H_0. There is not sufficient evidence to conclude that the mean wasted time per eight-hour work day for employees at this company is less than 120 minutes.

MINITAB could also have been used to carry out the test, as shown in the accompanying output:

One-Sample T: Wasted Time
Test of mu = 120 vs < 120

Variable	N	Mean	StDev	SE Mean	95% Upper Bound	T	P
Wasted Time	10	116.800	9.449	2.988	122.278	−1.07	0.156

Although we had to round the computed t value to −1.1 to use Appendix Table 4, MINITAB was able to compute the P-value corresponding to the actual value of the test statistic.

■

Example 10.15 Cricket Love

© Dynamic Graphics/Creatas/Alamy

The article "Well-Fed Crickets Bowl Maidens Over" (*Nature Science Update*, February 11, 1999) reported that female field crickets are attracted to males that have high chirp rates and hypothesized that chirp rate is related to nutritional status. The usual chirp rate for male field crickets was reported to vary around a mean of 60 chirps per second. To investigate whether chirp rate was related to nutritional status, investigators fed male crickets a high protein diet for 8 days, after which chirp rate was measured. The mean chirp rate for the crickets on the high protein diet was reported to be 109 chirps per second. Is this convincing evidence that the mean chirp rate for crickets on a high protein diet is greater than 60 (which would then imply an advantage in attracting the ladies)? Suppose that the sample size and sample standard deviation are $n = 32$ and $s = 40$. We test the relevant hypotheses with $\alpha = .01$.

1. μ = mean chirp rate for crickets on a high protein diet
2. H_0: $\mu = 60$
3. H_a: $\mu > 60$
4. $\alpha = .01$

5. $t = \dfrac{\bar{x} - \text{hypothesized value}}{\dfrac{s}{\sqrt{n}}} = \dfrac{\bar{x} - 60}{\dfrac{s}{\sqrt{n}}}$

6. This test requires a random sample and either a large sample or a normal population distribution. Because the sample size is large ($n = 32$), it is reasonable to proceed with the t test as long as we are willing to consider the 32 male field crickets in this study as if they were a random sample from the population of male field crickets.

7. Test statistic: $t = \dfrac{109 - 60}{\dfrac{40}{\sqrt{32}}} = \dfrac{49}{7.07} = 6.93$

8. This is an upper-tailed test, so the P-value is the area under the t curve with df = 31 and to the right of 6.93. From Appendix Table 4, P-value ≈ 0.

9. Because P-value ≈ 0, which is less than the significance level, α, we reject H_0. There is convincing evidence that the mean chirp rate is higher for male field crickets that eat a high protein diet.

■

■ Statistical Versus Practical Significance ....

Important concept

Carrying out a hypothesis test amounts to deciding whether the value obtained for the test statistic could plausibly have resulted when H_0 is true. When the value of the test statistic leads to rejection of H_0, it is customary to say that the result is **statistically significant** at the chosen level α. The finding of statistical significance means that, in the investigator's opinion, the observed deviation from what was expected under H_0 cannot plausibly be attributed only to chance variation. However, statistical significance cannot be equated with the conclusion that the true situation differs from what H_0 states in any practical sense. That is, even after H_0 has been rejected, the data may suggest that there is no *practical* difference between the true value of the population characteristic and what the null hypothesis states that value to be. This is illustrated in Example 10.16.

....

Example 10.16 "Significant" but Unimpressive Test Score Improvement

Let μ denote the true average score on a standardized test for children in a certain region of the United States. The average score for all children in the United States is 100. Regional education authorities are interested in testing $H_0: \mu = 100$ versus $H_a: \mu > 100$ using a significance level of .001. A sample of 2500 children resulted in the values $n = 2500$, $\bar{x} = 101.0$, and $s = 15.0$. Then

$$t = \dfrac{101.0 - 100}{\dfrac{15}{\sqrt{2500}}} = 3.3$$

This is an upper-tailed test, so (using the z column of Appendix Table 4 because df = 2499) P-value = area to the right of 3.33 $\approx$ .000. Because P-value < .001, we reject H_0. The true mean score for this region does appear to exceed 100.

However, with $n = 2500$, the point estimate $\bar{x} = 101.0$ is almost surely very close to the true value of μ. Therefore it looks as though H_0 was rejected because $\mu \approx 101$ rather than 100. And, from a practical point of view, a 1-point difference is most likely of no practical importance. A statistically significant result does not necessarily mean that there are any practical consequences.

■

10.45 Newly purchased automobile tires of a certain type are supposed to be filled to a pressure of 30 psi. Let μ denote the true average pressure. Find the P-value associated with each of the following given z statistic values for testing H_0: $\mu = 30$ versus H_a: $\mu \neq 30$ when σ is known:

a. 2.10   0.0358
b. −1.75   0.0802
c. 0.58   0.562
d. 1.44   0.1498
e. −5.00   0

10.46 The desired percentage of silicon dioxide in a certain type of cement is 5.0%. A random sample of $n = 36$ specimens gave a sample average percentage of $\bar{x} = 5.21$. Let μ be the true average percentage of silicon dioxide in this type of cement, and suppose that σ is known to be 0.38. Test H_0: $\mu = 5$ versus H_a: $\mu \neq 5$ using a significance level of .01. $z = 3.32$ P-value $= 0.0010$, reject H_0

10.47 Give as much information as you can about the P-value of a t test in each of the following situations:

a. Upper-tailed test, df = 8, $t = 2.0$ 0.040
b. Upper-tailed test, $n = 14$, $t = 3.2$ 0.003
c. Lower-tailed test, df = 10, $t = -2.4$ 0.019
d. Lower-tailed test, $n = 22$, $t = -4.2$ 0
e. Two-tailed test, df = 15, $t = -1.6$ 0.13
f. Two-tailed test, $n = 16$, $t = 1.6$ 0.13
g. Two-tailed test, $n = 16$, $t = 6.3$ 0

10.48 Give as much information as you can about the P-value of a t test in each of the following situations:

a. Two-tailed test, df = 9, $t = 0.73$
b. Upper-tailed test, df = 10, $t = -0.5$
c. Lower-tailed test, $n = 20$, $t = -2.1$
d. Lower-tailed test, $n = 20$, $t = -5.1$
e. Two-tailed test, $n = 40$, $t = 1.7$

10.49 Paint used to paint lines on roads must reflect enough light to be clearly visible at night. Let μ denote the true average reflectometer reading for a new type of paint under consideration. A test of H_0: $\mu = 20$ versus H_a: $\mu > 20$ based on a sample of 15 observations gave $t = 3.2$. What conclusion is appropriate at each of the following significance levels?

a. $\alpha = .05$ Reject H_0
b. $\alpha = .01$ Reject H_0
c. $\alpha = .001$ Fail to reject H_0

10.50 A certain pen has been designed so that true average writing lifetime under controlled conditions (involving the use of a writing machine) is at least 10 hr. A random sample of 18 pens is selected, the writing lifetime of each is determined, and a normal probability plot of the resulting data support the use of a one-sample t test. The relevant hypotheses are H_0: $\mu = 10$ versus H_a: $\mu < 10$.

a. If $t = -2.3$ and $\alpha = .05$ is selected, what conclusion is appropriate?   Reject
b. If $t = -1.83$ and $\alpha = .01$ is selected, what conclusion is appropriate?   Fail to reject
c. If $t = 0.47$, what conclusion is appropriate?   Fail to reject

10.51 The true average diameter of ball bearings of a certain type is supposed to be 0.5 in. What conclusion is appropriate when testing H_0: $\mu = 0.5$ versus H_a: $\mu \neq 0.5$ in each of the following situations:

a. $n = 13$, $t = 1.6$, $\alpha = .05$ Fail to reject H_0
b. $n = 13$, $t = -1.6$, $\alpha = .05$ Fail to reject H_0
c. $n = 25$, $t = -2.6$, $\alpha = .01$ Fail to reject H_0
d. $n = 25$, $t = -3.6$ H_0 would be rejected for any $\alpha > 0.002$

10.52 A credit bureau analysis of undergraduate students credit records found that the average number of credit cards in an undergraduate's wallet was 4.09 ("Undergraduate Students and Credit Cards in 2004," Nellie Mae, May 2005). It was also reported that in a random sample of 132 undergraduates, the sample mean number of credit cards carried was 2.6. The sample standard deviation was not reported, but for purposes of this exercise, suppose that it was 1.2. Is there convincing evidence that the mean number of credit cards that undergraduates report carrying is less than the credit bureau's figure of 4.09? $t = 14.3$, P-value ≈ 0, reject H_0

10.53 ● Medical research has shown that repeated wrist extension beyond 20 degrees increases the risk of wrist and hand injuries. Each of 24 students at Cornell University used a proposed new mouse design, and while using the mouse their wrist extension was recorded for each one. Data consistent with summary values given in the paper "Comparative Study of Two Computer Mouse Designs" (Cornell Human Factors Laboratory Technical Report RP7992) are given. Use these data to test the hypothesis that the mean wrist extension for people using this new mouse design is greater than 20 degrees. Are any assumptions required in order for it to be appropriate to generalize the results of your test to the population of Cornell students? To the population of all university students? (data on next page)

| 27 | 28 | 24 | 26 | 27 | 25 | 25 | 24 | 24 | 24 | 25 | 28 |
| 22 | 25 | 24 | 28 | 27 | 26 | 31 | 25 | 28 | 27 | 27 | 25 |

10.54 The international polling organization Ipsos reported data from a survey of 2000 randomly selected Canadians who carry debit cards (Canadian Account Habits Survey, July 24, 2006). Participants in this survey were asked what they considered the minimum purchase amount for which it would be acceptable to use a debit card. Suppose that the sample mean and standard deviation were $9.15 and $7.60 respectively. (These values are consistent with a histogram of the sample data that appears in the report.) Do these data provide convincing evidence that the mean minimum purchase amount for which Canadians consider the use of a debit card to be appropriate is less than $10? Carry out a hypothesis test with a significance level of .01. $t = -5.0$, P-value ≈ 0, reject H_0

10.55 A comprehensive study conducted by the National Institute of Child Health and Human Development tracked more than 1000 children from an early age through elementary school (*New York Times*, November 1, 2005). The study concluded that children who spent more than 30 hours a week in child care before entering school tended to score higher in math and reading when they were in the third grade. The researchers cautioned that the findings should not be a cause for alarm because the effects of child care were found to be small. Explain how the difference between the mean math score for third graders who spent long hours in child care and the overall mean for third-graders could be small but the researchers could still reach the conclusion that the mean for the child care group is significantly higher than the overall mean for third-graders.

10.56 In a study of computer use, 1000 randomly selected Canadian Internet users were asked how much time they spend using the Internet in a typical week (Ipsos Reid, August 9, 2005). The mean of the 1000 resulting observations was 12.7 hours.
a. The sample standard deviation was not reported, but suppose that it was 5 hours. Carry out a hypothesis test with a significance level of .05 to decide if there is convincing evidence that the mean time spent using the Internet by Canadians is greater than 12.5 hours.
b. Now suppose that the sample standard deviation was 2 hours. Carry out a hypothesis test with a significance level of .05 to decide if there is convincing evidence that the mean time spent using the Internet by Canadians is greater than 12.5 hours. $t = 3.16$, P-value ≈ 0.0008, reject H_0

c. Explain why the null hypothesis was rejected in the test of Part (b) but not in the test of Part (a).

10.57 A survey of teenagers and parents in Canada conducted by the polling organization Ipsos ("Untangling the Web: The Facts About Kids and the Internet," January 25, 2006) included questions about Internet use. It was reported that for a sample of 534 randomly selected teens, the mean number of hours per week spent online was 14.6 and the standard deviation was 11.6.
a. What does the large standard deviation, 11.6 hours, tell you about the distribution of online times for this sample of teens? There is a lot of variability in online time.
b. Do the sample data provide convincing evidence that the mean number of hours that teens spend online is greater than 10 hours per week? $t = 9.16$, P-value ≈ 0, reject H_0

10.58 The same survey referenced in the previous exercise reported that for a random sample of 676 parents of Canadian teens, the mean number of hours parents thought their teens spent online was 6.5 and the sample standard deviation was 8.6.
a. Do the sample data provide convincing evidence that the mean number of hours that parents think their teens spend online is less than 10 hours per week?
b. Write a few sentences commenting on the results of the test in Part (a) and of the test in Part (b) of the previous exercise.

10.59 The paper titled "Music for Pain Relief" (*The Cochrane Database of Systematic Reviews*, April 19, 2006) concluded, based on a review of 51 studies of the effect of music on pain intensity, that "Listening to music reduces pain intensity levels . . . However, the magnitude of these positive effects is small, the clinical relevance of music for pain relief in clinical practice is unclear." Are the authors of this paper claiming that the pain reduction attributable to listening to music is not statistically significant, not practically significant, or neither statistically nor practically significant? Explain.

10.60 Typically, only very brave students are willing to speak out in a college classroom. Student participation may be especially difficult if the individual is from a different culture or country. The article "An Assessment of Class Participation by International Graduate Students" (*Journal of College Student Development* [1995]: 132–140) considered a numerical "speaking-up" scale, with possible values from 3 to 15 (a low value means that a

student rarely speaks). For a random sample of 64 males from Asian countries where English is not the official language, the sample mean and sample standard deviation were 8.75 and 2.57, respectively. Suppose that the mean for the population of all males having English as their native language is 10.0 (suggested by data in the article). Does it appear that the population mean for males from non-English-speaking Asian countries is smaller than 10.0?
$t = -3.89$, P-value ≈ 0, reject H_0

10.61 ● ▼ A well-designed and safe workplace can contribute greatly to increasing productivity. It is especially important that workers not be asked to perform tasks, such as lifting, that exceed their capabilities. The following data on maximum weight of lift (MWOL, in kilograms) for a frequency of 4 lifts per minute were reported in the article "The Effects of Speed, Frequency, and Load on Measured Hand Forces for a Floor-to-Knuckle Lifting Task" (*Ergonomics* [1992]: 833–843):

25.8   36.6   26.3   21.8   27.2

Suppose that it is reasonable to regard the sample as a random sample from the population of healthy males, age 18–30. Do the data suggest that the population mean MWOL exceeds 25? Carry out a test of the relevant hypotheses using a .05 significance level. $t = 1.0$, P-value $=$ 0.187, fail to reject H_0

10.62 An article titled "Teen Boys Forget Whatever It Was" appeared in the Australian newspaper *The Mercury* (April 21, 1997). It described a study of academic performance and attention span and reported that the mean time to distraction for teenage boys working on an independent task was 4 min. Although the sample size was not given in the article, suppose that this mean was based on a random sample of 50 teenage Australian boys and that the sample standard deviation was 1.4 min. Is there convincing evidence that the average attention span for teenage boys is less than 5 min? Test the relevant hypotheses using $\alpha = .01$. $t = -5.05$, P-value ≈ 0, reject H_0

10.63 ● ▼ Many consumers pay careful attention to stated nutritional contents on packaged foods when making purchases. It is therefore important that the information on packages be accurate. A random sample of $n = 12$ frozen dinners of a certain type was selected from production during a particular period, and the calorie content of each one was determined. (This determination entails destroying the product, so a census would certainly not be desirable!) Here are the resulting observations, along with a boxplot and normal probability plot:

255   244   239   242   265   245   259   248
225   226   251   233

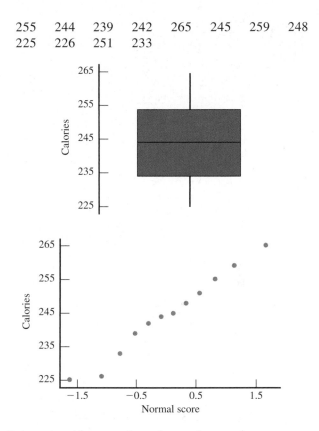

a. Is it reasonable to test hypotheses about true average calorie content μ by using a t test?
b. The stated calorie content is 240. Does the boxplot suggest that true average content differs from the stated value? Explain your reasoning.
c. Carry out a formal test of the hypotheses suggested in Part (b). $t = 1.21$, P-value ≈ 0.256, fail to reject H_0

10.64 ● Much concern has been expressed in recent years regarding the practice of using nitrates as meat preservatives. In one study involving possible effects of these chemicals, bacteria cultures were grown in a medium containing nitrates. The rate of uptake of radio-labeled amino acid was then determined for each culture, yielding the following observations:

7251   6871   9632   6866   9094   5849   8957   7978
7064   7494   7883   8178   7523   8724   7468

Suppose that it is known that the true average uptake for cultures without nitrates is 8000. Do the data suggest that the addition of nitrates results in a decrease in the true average uptake? Test the appropriate hypotheses using a significance level of .10. $t = -0.816$, P-value ≈ 0.214, fail to reject H_0

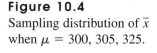

10.5 Power and Probability of Type II Error

In this chapter, we have introduced test procedures for testing hypotheses about population characteristics, such as μ and π. What characterizes a "good" test procedure? It makes sense to think that a good test procedure is one that has both a small probability of rejecting H_0 when it is true (a Type I error) and a high probability of rejecting H_0 when it is false. The test procedures presented in this chapter allow us to directly control the probability of rejecting a true H_0 by our choice of the significance level α. But what about the probability of rejecting H_0 when it is false? As we will see, several factors influence this probability. Let's begin by considering an example.

Suppose that the student body president at a university is interested in studying the amount of money that students spend on textbooks each semester. The director of the financial aid office believes that the average amount spent on books is $300 per semester and uses this figure to determine the amount of financial aid for which a student is eligible. The student body president plans to ask each individual in a random sample of students how much he or she spent on books this semester and has decided to use the resulting data to test

$$H_0\text{: }\mu = 300 \quad \text{versus} \quad H_a\text{: }\mu > 300$$

using a significance level of .05. If the true mean is 300 (or less than 300), the correct decision is to fail to reject the null hypothesis (incorrectly rejecting the null hypothesis is a Type I error). On the other hand, if the true mean is 325 or 310 or even 301, the correct decision is to reject the null hypothesis (not rejecting the null hypothesis is a Type II error). How likely is it that the null hypothesis will in fact be rejected?

If the true mean is 301, the probability that we reject H_0: $\mu = 300$ is not very great. This is because when we carry out the test, we are essentially looking at the sample mean and asking, Does this look like what we would expect to see if the population mean were 300? As illustrated in Figure 10.4, if the true mean is greater than but very close to 300, chances are that the sample mean will look pretty much like what we would expect to see if the population mean were 300, and we will be unconvinced that the null hypothesis should be rejected. If the true mean is 325, it is less likely that the sample will be mistaken for a sample from a population with mean 300; sample means will tend to cluster around 325, and so it is more likely that we will correctly reject H_0. If the true mean is 350, rejection of H_0 is even more likely.

Figure 10.4
Sampling distribution of $\bar{x}$
when $\mu = 300, 305, 325$.

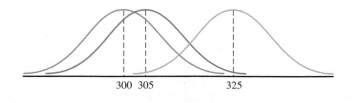

300 305   325

When we consider the probability of rejecting the null hypothesis, we are looking at what statisticians refer to as the **power** of the test.

The **power of a test** is the probability of rejecting the null hypothesis.

From the previous discussion, it should be apparent that the power of the test when a hypothesis about a population mean is being tested depends on the true value of the population mean, μ. Because the true value of μ is unknown (if we knew the value of μ we wouldn't be doing the hypothesis test!), we cannot know what the power is for the actual true value of μ. It is possible, however, to gain some insight into the power of a test by looking at a number of "what if" scenarios. For example, we might ask, What is the power if the true mean is 325? or What is the power if the true mean is 310? and so on. That is, we can determine the power at $\mu = 325$, the power at $\mu = 310$, and the power at any other value of interest. Although it is technically possible to consider power when the null hypothesis is true, an investigator is usually concerned about the power only at values for which the null hypothesis is false.

In general, when testing a hypothesis about a population characteristic, there are three factors that influence the power of the test:

1. The size of the difference between the true value of the population characteristic and the hypothesized value (the value that appears in the null hypothesis)
2. The choice of significance level, α, for the test
3. The sample size

Effect of Various Factors on the Power of a Test

1. The larger the size of the discrepancy between the hypothesized value and the true value of the population characteristic, the higher the power.
2. The larger the significance level, α, the higher the power of the test.
3. The larger the sample size, the higher the power of the test.

Let's consider each of these three statements. The first statement has already been discussed in the context of the textbook example. Because power is the probability of rejecting the null hypothesis, it makes sense that the power will be higher when the true value of a population characteristic is quite different from the hypothesized value than when it is close to that value.

The effect of significance level on power is not quite as obvious. To understand the relationship between power and significance level, it helps to see the relationship between power and β, the probability of a Type II error.

When H_0 is false, power $= 1 - \beta$.

This relationship follows from the definitions of power and Type II error. A Type II error results from *not* rejecting a false H_0. Because power is the probability of rejecting H_0, it follows that *when H_0 is false*

$$\begin{aligned}
\text{power} &= \text{probability of rejecting a false } H_0 \\
&= 1 - \text{probability of not rejecting a false } H_0 \\
&= 1 - \beta
\end{aligned}$$

Recall from Section 10.2 that the choice of α, the Type I error probability, affects the value of β, the Type II error probability. Choosing a larger value for α results in a smaller value for β (and thus a larger value for $1 - \beta$). In terms of power, this means that choosing a larger value for α results in a larger value for the power of the test. That is, the larger the Type I error probability we are willing to tolerate, the more likely it is that the test will be able to detect any particular departure from H_0.

The third factor that affects the power of a test is the sample size. When H_0 is false, the power of a test is the probability that we will in fact "detect" that H_0 is false and, based on the observed sample, reject H_0. Intuition suggests that we will be more likely to detect a departure from H_0 with a large sample than with a small sample. This is in fact the case—the larger the sample size, the higher the power.

Consider testing the hypotheses presented previously:

$$H_0: \mu = 300 \text{ versus } H_a: \mu > 300$$

The observations about power imply the following, for example:

1. For any value of μ exceeding 300, the power of a test based on a sample of size 100 is higher than the power of a test based on a sample of size 75 (assuming the same significance level).
2. For any value of μ exceeding 300, the power of a test using a significance level of .05 is higher than the power of a test using a significance level of .01 (assuming the same sample size).
3. For any value of μ exceeding 300, the power of the test is greater if the true mean is 350 than if the true mean is 325 (assuming the same sample size and significance level).

As was mentioned previously in this section, it is impossible to calculate the *actual* power of a test because in practice we do not know the true value of population characteristics. However, we can evaluate the power at a selected alternative value if we want to know whether the power would be high or low if this alternative value is the true value.

The following optional subsection shows how Type II error probabilities and power can be evaluated for selected tests.

■ Calculating Power and Type II Error Probabilities for Selected Tests (Optional)

The test procedures presented in this chapter are designed to control the probability of a Type I error (rejecting H_0 when H_0 is true) at the desired level α. However, little has been said so far about calculating the value of β, the probability of a Type II error (not rejecting H_0 when H_0 is false). Here we consider the determination of β and power for the hypothesis tests previously introduced.

When we carry out a hypothesis test, we specify the desired value of α, the probability of a Type I error. The probability of a Type II error, β, is the probability of not rejecting H_0 even though it is false. Suppose that we are testing

$$H_0: \mu = 1.5 \text{ versus } H_a: \mu > 1.5$$

Because we do not know the true value of μ, we cannot calculate the actual value of β. However, the vulnerability of the test to Type II error can be investigated by calculating β for several different potential values of μ, such as $\mu = 1.55$, $\mu = 1.6$, and $\mu = 1.7$. Once the value of β has been determined, the power of the test at the corresponding alternative value is just $1 - \beta$.

Example 10.17 Calculating Power

A cigarette manufacturer claims that the mean nicotine content of its cigarettes is 1.5 mg. We might investigate this claim by testing

$$H_0: \mu = 1.5 \text{ versus } H_a: \mu > 1.5$$

where μ is the true mean nicotine content. A random sample of $n = 36$ cigarettes is to be selected, and the resulting data will be used to reach a conclusion. Suppose that the standard deviation of nicotine content (σ) is known to be 0.20 mg and that a significance level of .01 is to be used. Our test statistic (because $\sigma = 0.20$) is

$$z = \frac{\bar{x} - 1.5}{\dfrac{.20}{\sqrt{n}}} = \frac{\bar{x} - 1.5}{\dfrac{.20}{\sqrt{36}}} = \frac{\bar{x} - 1.5}{.0333}$$

The inequality in H_a implies that

$$P\text{-value} = \text{area under } z \text{ curve to the right of calculated } z$$

From Appendix Table 2, it is easily verified that the z critical value 2.33 captures an upper-tail z curve area of .01. Thus, P-value $\leq .01$ if and only if $z \geq 2.33$. This is equivalent to the decision rule

$$\text{reject } H_0 \text{ if calculated } z \geq 2.33$$

which becomes

$$\text{reject } H_0 \text{ if } \frac{\bar{x} - 1.5}{.0333} \geq 2.33$$

Solving this inequality for $\bar{x}$ we get

$$\bar{x} \geq 1.5 + 2.33(.0333)$$

or

$$\bar{x} \geq 1.578$$

So if $\bar{x} \geq 1.578$, we will reject H_0, and if $\bar{x} < 1.578$, we will fail to reject H_0 This decision rule corresponds to $\alpha = .01$.

Suppose now that $\mu = 1.6$ (so that H_0 is false). A Type II error will then occur if $\bar{x} < 1.578$. What is the probability that this occurs? If $\mu = 1.6$, the sampling distribution of $\bar{x}$ is approximately normal, centered at 1.6, and has a standard deviation of .0333. The probability of observing an $\bar{x}$ value less than 1.578 can then be determined by finding an area under a normal curve with mean 1.6 and standard deviation .0333, as illustrated in Figure 10.5.

Figure 10.5 β when $\mu = 1.6$ in Example 10.17.

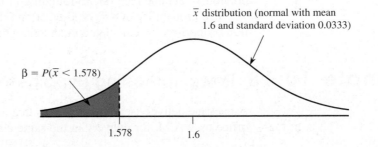

Because the curve in Figure 10.5 is not the standard normal (z) curve, we must first convert to a z score before using Appendix Table 2 to find the area. Here,

$$z \text{ score for } 1.578 = \frac{1.578 - \mu_{\bar{x}}}{\sigma_{\bar{x}}} = \frac{1.578 - 1.6}{.0333} = -.66$$

and

$$\text{area under } z \text{ curve to left of } -0.66 = .2546$$

So, if $\mu = 1.6$, $\beta = .2546$. This means that if μ is 1.6, about 25% of all samples would still result in $\bar{x}$ values less than 1.578 and failure to reject H_0.

The power of the test at $\mu = 1.6$ is then

$$\begin{align}(\text{power at } \mu = 1.6) &= 1 - (\beta \text{ when } \mu \text{ is } 1.6) \\ &= 1 - .2546 \\ &= .7454\end{align}$$

Thus, if the true mean is 1.6, the probability of rejecting H_0: $\mu = 1.5$ in favor of H_a: $\mu > 1.5$ is .7454. That is, if μ is 1.6 and the test is used repeatedly with random samples selected from the population, in the long run about 75% of the samples will result in the correct conclusion to reject H_0.

Now consider β and power when $\mu = 1.65$. The normal curve in Figure 10.5 would then be centered at 1.65. Because β is the area to the left of 1.578 and the curve has shifted to the right, β decreases. Converting 1.578 to a z score and using Appendix Table 2 gives $\beta = .0154$. Also,

$$(\text{power at } \mu = 1.65) = 1 - .0154 = .9846$$

As expected, the power at $\mu = 1.65$ is higher than the power at $\mu = 1.6$ because 1.65 is farther from the hypothesized value of 1.5.

■

MINITAB can calculate the power for specified values of σ, α, n, and the difference between the true and hypothesized values of μ. The following output shows power calculations corresponding to those in Example 10.17:

1-Sample Z Test
Testing mean = null (versus > null)
Alpha = 0.01   Sigma = 0.2   Sample Size = 36

Difference	Power
0.10	0.7497
0.15	0.9851

The slight differences between the power values computed by MINITAB and those previously obtained are due to rounding in Example 10.17.

The probability of a Type II error and the power for z tests concerning a population proportion are calculated in an analogous manner.

....

Example 10.18 Power for Testing Hypotheses About Proportions

A package delivery service advertises that at least 90% of all packages brought to its office by 9 A.M. for delivery in the same city are delivered by noon that day. Let π

denote the proportion of all such packages actually delivered by noon. The hypotheses of interest are

$$H_0:\ \pi = .9 \quad \text{versus} \quad H_a:\ \pi < .9$$

where the alternative hypothesis states that the company's claim is untrue. The value $\pi = .8$ represents a substantial departure from the company's claim. If the hypotheses are tested at level .01 using a sample of $n = 225$ packages, what is the probability that the departure from H_0 represented by this alternative value will go undetected?

At significance level .01, H_0 is rejected if P-value $\le .01$. For the case of a lower-tailed test, this is the same as rejecting H_0 if

$$z = \frac{p - \mu_p}{\sigma_p} = \frac{p - .9}{\sqrt{\dfrac{(.9)(.1)}{225}}} = \frac{p - .9}{.02} \le -2.33$$

(Because -2.33 captures a lower-tail z curve area of .01, the smallest 1% of all z values satisfy $z \le -2.33$.) This inequality is equivalent to $p \le .853$, so H_0 is *not* rejected if $p > .853$. When $\pi = .8$, p has approximately a normal distribution with

$$\mu_p = .8 \qquad \sigma_p = \sqrt{\frac{(.8)(.2)}{225}} = .0267$$

Then β is the probability of obtaining a sample proportion greater than .853, as illustrated in Figure 10.6.

Figure 10.6 β when $\pi = .8$ in Example 10.18.

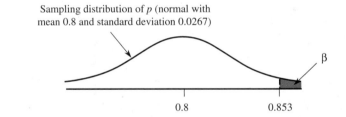

Sampling distribution of p (normal with mean 0.8 and standard deviation 0.0267)

β

0.8   0.853

Converting to a z score results in

$$z = \frac{.853 - .8}{.0267} = 1.99$$

and Appendix Table 2 gives

$$\beta = 1 - .9767 = .0233$$

When $\pi = .8$ and a level .01 test is used, less than 3% of all samples of size $n = 225$ will result in a Type II error. The power of the test at $\pi = .8$ is $1 - .0233 = .9767$. This means that the probability of rejecting $H_0:\ \pi = .9$ in favor of $H_a:\ \pi < .9$ when π is really .8 is .9767, which is quite high.

■

■ β and Power for the *t* Test (Optional)

The power and β values for t tests can be determined by using a set of curves specially constructed for this purpose or by utilizing appropriate software. As with the z test, the value of β depends not only on the true value of μ but also on the selected significance level α; β increases as α is made smaller. In addition, β depends on the number of degrees of freedom, $n - 1$. For any fixed level α, it should be easier for the test to detect a specific departure from H_0 when n is large than when n is small. This is indeed the case; for a fixed alternative value, β decreases as $n - 1$ increases.

Unfortunately, there is one other quantity on which β depends: the population standard deviation σ. As σ increases, so does $\sigma_{\bar{x}}$. This in turn makes it more likely that a value far from μ will be observed, resulting in an erroneous conclusion. Once α is specified and n is fixed, the determination of β at a particular alternative value of μ requires that a value of σ be chosen, because each different value of σ yields a different value of β. (This did not present a problem with the z test because when using a z test, the value of σ is known.) If the investigator can specify a range of plausible values for σ, then using the largest such value will give a pessimistic β (one on the high side) and a pessimistic value of power (one on the low side).

Figure 10.7 shows three different β curves for a one-tailed t test (appropriate for H_a: $\mu >$ hypothesized value or for H_a: $\mu <$ hypothesized value). A more complete set of curves for both one- and two-tailed tests when $\alpha = .05$ and when $\alpha = .01$ appears in Appendix Table 5. To determine β, first compute the quantity

$$d = \frac{|\text{alternative value} - \text{hypothesized value}|}{\sigma}$$

Then locate d on the horizontal axis, move directly up to the curve for $n - 1$ df, and move over to the vertical axis to read β.

Figure 10.7 β curves for the one-tailed t test.

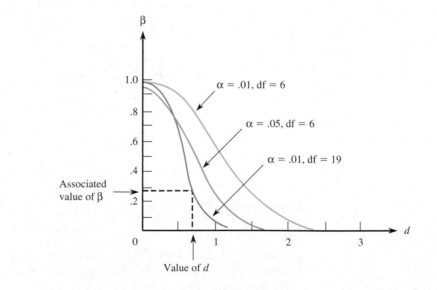

Example 10.19 β and Power for t Tests

Consider testing

$$H_0: \mu = 100 \text{ versus } H_a: \mu > 100$$

and focus on the alternative value $\mu = 110$. Suppose that $\sigma = 10$, the sample size is $n = 7$, and a significance level of .01 has been selected. For $\sigma = 10$,

$$d = \frac{|110 - 100|}{10} = \frac{10}{10} = 1$$

Figure 10.7 (using df $= 7 - 1 = 6$) gives $\beta \approx .6$. The interpretation is that if $\sigma = 10$ and a level .01 test based on $n = 7$ is used when $\mu = 110$ (and thus H_0 is false), roughly 60% of all samples result in erroneously not rejecting H_0! Equivalently, the power of the test at $\mu = 110$ is only $1 - .6 = .4$. The probability of rejecting H_0 when $\mu = 110$ is not very large. If a level .05 test is used instead, then $\beta \approx .3$, which is still rather large. Using a level .01 test with $n = 20$ (df $= 19$) yields, from Figure 10.7, $\beta \approx .05$. At the alternative value $\mu = 110$, for $\sigma = 10$ the level .01 test based on $n = 20$ has smaller β than the level .05 test with $n = 7$. Substantially increasing n counterbalances using the smaller α.

Now consider the alternative $\mu = 105$, again with $\sigma = 10$, so that

$$d = \frac{|105 - 100|}{10} = \frac{5}{10} = .5$$

Then, from Figure 10.7, $\beta = .95$ when $\alpha = .01$, $n = 7$; $\beta = .7$ when $\alpha = .05$, $n = 7$; and $\beta = .65$ when $\alpha = .01$, $n = 20$. These values of β are all quite large; with $\sigma = 10$, $\mu = 105$ is too close to the hypothesized value of 100 for any of these three tests to have a good chance of detecting such a departure from H_0. A substantial decrease in β necessitates using a much larger sample size. For example, from Appendix Table 5, $\beta = .08$ when $\alpha = .05$ and $n = 40$.

The curves in Figure 10.7 also give β when testing H_0: $\mu = 100$ versus H_a: $\mu < 100$. If the alternative value $\mu = 90$ is of interest and $\sigma = 10$,

$$d = \frac{|90 - 100|}{10} = \frac{10}{10} = 1$$

and values of β are the same as those given in the first paragraph of this example.

■

Because curves for only selected degrees of freedom appear in Appendix Table 5, other degrees of freedom require a visual approximation. For example, the 27-df curve (for $n = 28$) lies between the 19-df and 29-df curves, which do appear, and it is closer to the 29-df curve. This type of approximation is adequate because it is the general magnitude of β—large, small, or moderate—that is of primary concern.

MINITAB can also evaluate power for the t test. For example, the following output shows MINITAB calculations for power at $\mu = 110$ for samples of size 7 and 20 when $\alpha = .01$. The corresponding approximate values from Appendix Table 5 found in Example 10.19 are fairly close to the MINITAB values.

1-Sample t Test
Testing mean = null (versus > null)
Calculating power for mean = null + 10
Alpha = 0.01   Sigma = 10

Sample Size	Power
7	0.3968
20	0.9653

The β curves in Appendix Table 5 are those for t tests. When the alternative value in H_a corresponds to a value of d relatively close to 0, β for a t test may be rather large. One might ask whether there is another type of test that has the same level of significance α as does the t test and smaller values of β. The following result provides the answer to this question.

When the population distribution is normal, the t test for testing hypotheses about μ has smaller β than does any other test procedure that has the same level of significance α.

Stated another way, among all tests with level of significance α, the t test makes β as small as it can possibly be when the population distribution is normal. In this sense, the t test is a best test. Statisticians have also shown that when the population distribution is not too far from a normal distribution, no test procedure can improve on the t test by very much (i.e., no test procedure can have the same α and substantially smaller β). However, when the population distribution is believed to be strongly non-normal (heavy-tailed, highly skewed, or multimodal), the t test should not be used. Then it's time to consult your friendly neighborhood statistician, who can provide you with alternative methods of analysis.

Exercises 10.65–10.71
Turn to the Expanded Answers Section for aswers not shown next to exercises.

10.65 The power of a test is influenced by the sample size and the choice of significance level.
a. Explain how increasing the sample size affects the power (when significance level is held fixed).
b. Explain how increasing the significance level affects the power (when sample size is held fixed).

10.66 Water samples are taken from water used for cooling as it is being discharged from a power plant into a river. It has been determined that as long as the mean temperature of the discharged water is at most 150°F, there will be no negative effects on the river ecosystem. To investigate whether the plant is in compliance with regulations that prohibit a mean discharge water temperature above 150°F, a scientist will take 50 water samples at randomly selected times and will record the water temperature of each sample. She will then use a z statistic

$$z = \frac{\bar{x} - 150}{\dfrac{\sigma}{\sqrt{n}}}$$

to decide between the hypotheses H_0: $\mu = 150$ and H_a: $\mu > 150$, where μ is the true mean temperature of discharged water. Assume that σ is known to be 10.
a. Explain why use of the z statistic is appropriate in this setting.   Because σ is known and the sample size is large
b. Describe Type I and Type II errors in this context.
c. The rejection of H_0 when $z \geq 1.8$ corresponds to what value of α? (That is, what is the area under the z curve to the right of 1.8?)
d. Suppose that the true value for μ is 153 and that H_0 is to be rejected if $z \geq 1.8$. Draw a sketch (similar to that of Figure 10.5) of the sampling distribution of $\bar{x}$, and shade the region that would represent β, the probability of making a Type II error.
e. For the hypotheses and test procedure described, compute the value of β when $\mu = 153$.   0.3745
f. For the hypotheses and test procedure described, what is the value of β if $\mu = 160$?   Approximately 0
g. If H_0 is rejected when $z \geq 1.8$ and $\bar{x} = 152.4$, what is the appropriate conclusion? What type of error might have been made in reaching this conclusion?   Fail to reject, a Type II error

10.67 ▼ Let μ denote the true average lifetime for a certain type of pen under controlled laboratory conditions. A test of H_0: $\mu = 10$ versus H_a: $\mu < 10$ will be based on a sample of size 36. Suppose that σ is known to be 0.6, from which $\sigma_{\bar{x}} = 0.1$. The appropriate test statistic is then

$$z = \frac{\bar{x} - 10}{0.1}$$

a. What is α for the test procedure that rejects H_0 if $z \leq -1.28$?   0.1003

b. If the test procedure of Part (a) is used, calculate β when $\mu = 9.8$, and interpret this error probability.   0.2358

c. Without doing any calculation, explain how β when $\mu = 9.5$ compares to β when $\mu = 9.8$. Then check your assertion by computing β when $\mu = 9.5$.

d. What is the power of the test when $\mu = 9.8$? when $\mu = 9.5$?   0.7642, 1

10.68 The city council in a large city has become concerned about the trend toward exclusion of renters with children in apartments within the city. The housing coordinator has decided to select a random sample of 125 apartments and determine for each whether children are permitted. Let π be the true proportion of apartments that prohibit children. If π exceeds .75, the city council will consider appropriate legislation.

a. If 102 of the 125 sampled apartments exclude renters with children, would a level .05 test lead you to the conclusion that more than 75% of all apartments exclude children?   Yes, $z = 1.71$, P-value $= 0.0436$, reject H_0

b. What is the power of the test when $\pi = .8$ and $\alpha = .05$?   0.6480

10.69 The amount of shaft wear after a fixed mileage was determined for each of 7 randomly selected internal combustion engines, resulting in a mean of 0.0372 in. and a standard deviation of 0.0125 in.

a. Assuming that the distribution of shaft wear is normal, test at level .05 the hypotheses H_0: $\mu = .035$ versus H_a: $\mu > .035$. $t = 0.47$, P-value ≈ 0.317, fail to reject H_0.

b. Using $\sigma = 0.0125$, $\alpha = .05$, and Appendix Table 5, what is the approximate value of β, the probability of a Type II error, when $\mu = .04$? $\beta \approx 0.75$

c. What is the approximate power of the test when $\mu = .04$ and $\alpha = .05$?   Power ≈ 0.25

10.70 Optical fibers are used in telecommunications to transmit light. Current technology allows production of fibers that transmit light about 50 km (*Research at Rensselaer*, 1984). Researchers are trying to develop a new type of glass fiber that will increase this distance. In evaluating a new fiber, it is of interest to test H_0: $\mu = 50$ versus H_a: $\mu > 50$, with μ denoting the true average transmission distance for the new optical fiber.

a. Assuming $\sigma = 10$ and $n = 10$, use Appendix Table 5 to find β, the probability of a Type II error, for each of the given alternative values of μ when a level .05 test is employed:

i. 52   ii. 55   iii. 60   iv. 70

b. What happens to β in each of the cases in Part (a) if σ is actually larger than 10? Explain your reasoning.   As σ increases, d decreases in value and the value of β increases.

10.71 Let μ denote the true average diameter for bearings of a certain type. A test of H_0: $\mu = 0.5$ versus H_a: $\mu \neq 0.5$ will be based on a sample of n bearings. The diameter distribution is believed to be normal. Determine the value of β in each of the following cases:

a. $n = 15$, $\alpha = .05$, $\sigma = 0.02$, $\mu = 0.52$ $\beta \approx 0.06$

b. $n = 15$, $\alpha = .05$, $\sigma = 0.02$, $\mu = 0.48$ $\beta \approx 0.06$

c. $n = 15$, $\alpha = .01$, $\sigma = 0.02$, $\mu = 0.52$ $\beta \approx 0.21$

d. $n = 15$, $\alpha = .05$, $\sigma = 0.02$, $\mu = 0.54$ $\beta \approx 0$

e. $n = 15$, $\alpha = .05$, $\sigma = 0.04$, $\mu = 0.54$ $\beta \approx 0.06$

f. $n = 20$, $\alpha = .05$, $\sigma = 0.04$, $\mu = 0.54$ $\beta \approx 0.01$

g. Is the way in which β changes as n, α, σ, and μ vary consistent with your intuition? Explain.   Answers will vary.

Bold exercises answered in back   ● Data set available online but not required   ▼ Video solution available

10.6 Interpreting and Communicating the Results of Statistical Analyses

The step-by-step procedure that we have proposed for testing hypotheses provides a systematic approach for carrying out a complete test. However, you rarely see the results of a hypothesis test reported in publications in such a complete way.

■ Communicating the Results of Statistical Analyses ....

When summarizing the results of a hypothesis test, it is important that you include several things in the summary to have all the relevant information. These are

1. *Hypotheses.* Whether specified in symbols or described in words, it is important that both the null and the alternative hypotheses be clearly stated. If you are using symbols to define the hypotheses, be sure to describe them in the context of the problem at hand (e.g., μ = population mean calorie intake).
2. *Test procedure.* You should be clear about what test procedure was used (e.g., large-sample z test for proportions) and why you think it was reasonable to use this procedure. The plausibility of any required assumptions should be satisfactorily addressed.
3. *Test statistic.* Be sure to include the value of the test statistic and the P-value. Including the P-value allows a reader who may have chosen a different significance level to see whether she would have reached the same or a different conclusion.
4. *Conclusion in context.* Never end the report of a hypothesis test with the statement "I rejected (or did not reject) H_0." Always provide a conclusion that is in the context of the problem and that answers the original research question which the hypothesis test was designed to answer. Be sure also to indicate the level of significance used as a basis for the decision.

■ Interpreting the Results of Statistical Analyses ....

When the results of a hypothesis test are reported in a journal article or other published source, it is common to find only the value of the test statistic and the associated P-value accompanying the discussion of conclusions drawn from the data. Sometimes, even the exact P-value doesn't appear, but instead "coded" information is given. For example, * = significant (P-value $< .05$), ** = very significant (P-value $< .01$), and *** = highly significant (P-value $< .001$). Often, especially in newspaper articles, only sample summary statistics are given, with the conclusion immediately following. You may have to fill in some of the intermediate steps for yourself to see whether or not the conclusion is justified.

For example, the article "Physicians' Knowledge of Herbal Toxicities and Adverse Herb-Drug Interactions" (*European Journal of Emergency Medicine*, August 2004) summarizes the results of a study to assess doctors' familiarity with adverse effects of herbal remedies as follows: "A total of 142 surveys and quizzes were completed by 59 attending physicians, 57 resident physicians, and 26 medical students. The mean subject score on the quiz was only slightly higher than would have occurred from random guessing." The quiz consisted of 16 multiple-choice questions. If each question had four possible choices, the statement that the mean quiz score was only slightly higher than would have occurred from random guessing suggests that the researchers considered the hypotheses $H_0 : \mu = 4$ and $H_a : \mu > 4$, where μ represents the true mean score for the population of physicians and medical students and the null hypothesis corresponds to the expected number of correct choices for someone who is guessing. Assuming that it is reasonable to regard this sample as representative of the population of interest, the data from the sample could be used to carry out a test of these hypotheses.

■ What to Look For in Published Data ....

Here are some questions to consider when you are reading a report that contains the results of a hypothesis test:

- What hypotheses are being tested? Are the hypotheses about a population mean, a population proportion, or some other population characteristic?
- Was the appropriate test used? Does the validity of the test depend on any assumptions about the population from which the sample was selected? If so, are the assumptions reasonable?
- What is the *P*-value associated with the test? Was a significance level selected for the test (as opposed to simply reporting the *P*-value)? Is the chosen significance level reasonable?
- Are the conclusions drawn consistent with the results of the hypothesis test?

For example, consider the following statement from the paper "Didgeridoo Playing as Alternative Treatment for Obstructive Sleep Apnoea Syndrome" (*British Medical Journal* [2006]: 266–270): "We found that four months of training of the upper airways by didgeridoo playing reduces daytime sleepiness in people with snoring and obstructive apnoea syndrome." This statement was supported by data on a measure of daytime sleepiness called the Epworth scale. For the 14 participants in the study, the mean improvement in Epworth scale was 4.4 and the standard deviation was 3.7. The paper does not indicate what test was performed or what the value of the test statistic was. It appears that the hypotheses of interest are H_0: $\mu = 0$ (no improvement) versus H_a: $\mu > 0$, where μ represents the true mean improvement in Epworth score after four months of didgeridoo playing for people with snoring and obstructive sleep apnoea. Because the sample size is not large, the one-sample *t* test would be appropriate if the sample can be considered a random sample and the distribution of Epworth scale improvement scores is approximately normal. If these assumptions are reasonable (something that was not addressed in the paper), the *t* test results in $t = 4.45$ and an associated *P*-value of .000. Because the reported *P*-value is so small H_0 would be rejected, supporting the conclusion in the paper that didgeridoo playing is an effective treatment. (In case you are wondering, a didgeridoo is an Australian Aboriginal woodwind instrument.)

■ A Word to the Wise: Cautions and Limitations...

There are several things you should watch for when conducting a hypothesis test or when evaluating a written summary of such a test.

1. The result of a hypothesis test can never show strong support for the null hypothesis. Make sure that you don't confuse "There is no reason to believe the null hypothesis is not true" with the statement "There is convincing evidence that the null hypothesis is true." These are very different statements!
2. If you have complete information for the population, don't carry out a hypothesis test! It should be obvious that no test is needed to answer questions about a population if you have complete information and don't need to generalize from a sample, but people sometimes forget this fact. For example, in an article on growth in the number of prisoners by state, the *San Luis Obispo Tribune* (August 13, 2001) reported "California's numbers showed a statistically insignificant change, with 66 fewer prisoners at the end of 2000." The use of the term "statistically insignificant" implies some sort of statistical inference, which is not appropriate when a complete accounting of the entire prison population is known. Perhaps the author confused statistical and practical significance. Which brings us to . . .
3. Don't confuse statistical significance with practical significance. When statistical significance has been declared, be sure to step back and evaluate the result in light of its practical importance. For example, we may be convinced that the proportion

who respond favorably to a proposed medical treatment is greater than .4, the known proportion that responds favorably for the currently recommended treatments. But if our estimate of this proportion for the proposed treatment is .405, is this of any practical interest? It might be if the proposed treatment is less costly or has fewer side effects, but in other cases it may not be of any real interest. Results must always be interpreted in context.

Activity **10.1**   Comparing the *t* and *z* Distributions

Technology Activity: Requires use of a computer or a graphing calculator.

The instructions that follow assume the use of MINITAB. If you are using a different software package or a graphing calculator, your instructor will provide alternative instructions.

Background: Suppose a random sample will be selected from a population that is known to have a normal distribution. Then the statistic

$$z = \frac{\bar{x} - \mu}{\dfrac{\sigma}{\sqrt{n}}}$$

has a standard normal (z) distribution. Since it is rarely the case that σ is known, inferences for population means are usually based on the statistic $t = \dfrac{\bar{x} - \mu}{(s/\sqrt{n})}$, which has a t distribution rather than a z distribution. The informal justification for this was that the use of s to estimate σ introduces additional variability, resulting in a statistic whose distribution is more spread out than is the z distribution.

In this activity, you will use simulation to sample from a known normal population, and then investigate how the behavior of $t = \dfrac{\bar{x} - \mu}{s/\sqrt{n}}$ compares to $z = \dfrac{\bar{x} - \mu}{\sigma/(\sqrt{n})}$.

1. Generate 200 random samples of size 5 from a normal population with mean 100 and standard deviation 10.

Using MINTAB, go to the Calc Menu. Then

Calc → Random Data → Normal
In the "Generate" box, enter 200
In the "Store in columns" box, enter c1-c5
In the mean box, enter 100
In the standard deviation box, enter 10
Click on OK

You should now see 200 rows of data in each of the first 5 columns of the MINITAB worksheet.

2. Each row contains five values that have been randomly selected from a normal population with mean 100 and standard deviation 10. Viewing each row as a sample of size 5 from this population, calculate the mean and standard deviation for each of the 200 samples (the 200 rows) by using MINITAB's row statistics functions, which can also be found under the Calc menu:

Calc → Row statistics
Choose the "Mean" button
In the "Input Variables" box, enter c1-c5
In the "Store result in" box, enter c7
Click on OK

You should now see the 200 sample means in column 7 of the MINITAB worksheet. Name this column "x-bar", by typing the name in the gray box at the top of c7.

Now follow a similar process to compute the 200 sample standard deviations, and store them in c8. Name c8 "s."

3. Next, calculate the value of the z statistic for each of the 200 samples. We can calculate z in this example because we know that the samples were selected from a population for which $\sigma = 10$. Use the calculator function of MINITAB to compute $z = \dfrac{\bar{x} - \mu}{(\sigma/\sqrt{n})} = \dfrac{\bar{x} - 100}{(10/\sqrt{5})}$ as follows:

Calc → Calculator
In the "Store results in" box, enter c10
In the "Expression box" type in the following: (c7-100)/(10/sqrt(5))
Click on OK

You should now see the z values for the 200 samples in c10. Name c10 "z".

4. Now calculate the value of the t statistic for each of the 200 samples. Use the calculator function of MINITAB to compute $t = \dfrac{\bar{x} - \mu}{(s/\sqrt{n})} = \dfrac{\bar{x} - 100}{(s/\sqrt{5})}$ as follows:

Calc → Calculator
In the "Store results in" box, enter c11
In the "Expression box" type in the following: (c7-100)/(c8/sqrt(5))
Click on OK

You should now see the t values for the 200 samples in c10. Name c10 "t."

5. Graphs, at last! Now construct histograms of the 200 z values and the 200 t values. These two graphical displays will provide insight about how each of these two statistics behaves in repeated sampling. Use the same scale for the two histograms so that it will be easier to compare the two distributions.

Graph → Histogram
In the "Graph variables" box, enter c10 for graph 1 and c11 for graph 2
Click the Frame dropdown menu and select multiple graphs.
Then under the scale choices, select "Same X and same Y."

6. Now use the histograms from Step 5 to answer the following questions:
a. Write a brief description of the shape, center and spread for the histogram of the z values. Is what you see in the histogram consistent with what you would have expected to see? Explain. (Hint: In theory, what is the distribution of the z statistic?)

b. How does the histogram of the t values compare to the z histogram? Be sure to comment on center, shape, and spread.
c. Is your answer to Part (b) consistent with what would be expected for a statistic that has a t distribution? Explain.
d. The z and t histograms are based on only 200 samples, and they only approximate the corresponding sampling distributions. The 5th percentile for the standard normal distribution is -1.645 and the 95th percentile is $+1.645$. For a t distribution with df $= 5 - 1 = 4$, the 5th and 95th percentiles are -2.13 and $+2.13$, respectively. How do these percentiles compare to those of the distributions displayed in the histograms? (Hint: Sort the 200 z values— in MINITAB, choose "Sort" from the Manip menu. Once the values are sorted, percentiles from the histogram can be found by counting in 10 [which is 5% of 200] values from either end of the sorted list. Then repeat this with the t values.)
e. Are the results of your simulation and analysis consistent with the statement that the statistic $z = \dfrac{\bar{x} - \mu}{(\sigma/\sqrt{n})}$ has a standard normal (z) distribution and the statistic $t = \dfrac{\bar{x} - \mu}{(s/\sqrt{n})}$ has a t distribution? Explain.

Activity **10.2**   A Meaningful Paragraph

Write a meaningful paragraph that includes the following six terms: **hypotheses, P-value, reject H_0, Type I error, statistical significance, practical significance**.

A "meaningful paragraph" is a coherent piece of writing in an appropriate context that uses all of the listed words. The paragraph should show that you understand the meaning of the terms and their relationship to one another. A sequence of sentences that just define the terms is *not* a meaningful paragraph. When choosing a context, think carefully about the terms you need to use. Choosing a good context will make writing a meaningful paragraph easier.

Summary of Key Concepts and Formulas

Term or Formula	Comment
Hypothesis	A claim about the value of a population characteristic.
Null hypothesis, H_0	The hypothesis initially assumed to be true. It has the form H_0: population characteristic = hypothesized value.
Alternative hypothesis, H_a	A hypothesis that specifies a claim that is contradictory to H_0 and is judged the more plausible claim when H_0 is rejected.

Term or Formula	**Comment**
Type I error	Rejection of H_0 when H_0 is true; the probability of a Type I error is denoted by α and is referred to as the significance level for the test.
Type II error	Nonrejection of H_0 when H_0 is false; the probability of a Type II error is denoted by β.
Test statistic	The quantity computed from sample data and used to make a decision between H_0 and H_a.
P-value	The probability, computed assuming H_0 to be true, of obtaining a value of the test statistic at least as contradictory to H_0 as what actually resulted. H_0 is rejected if P-value $\leq \alpha$ and not rejected if P-value $> \alpha$, where α is the chosen significance level.
$z = \dfrac{p - \text{hypothesized value}}{\sqrt{\dfrac{(\text{hyp. val})(1 - \text{hyp. val})}{n}}}$	A test statistic for testing H_0: π = hypothesized value when the sample size is large. The P-value is determined from the z curve.
$z = \dfrac{\bar{x} - \text{hypothesized value}}{\dfrac{\sigma}{\sqrt{n}}}$	A test statistic for testing H_0: μ = hypothesized value when σ is known and either the population distribution is normal or the sample size is large. The P-value is determined from the z curve.
$t = \dfrac{\bar{x} - \text{hypothesized value}}{\dfrac{s}{\sqrt{n}}}$	A test statistic for testing H_0: μ = hypothesized value when σ is unknown and either the population distribution is normal or the sample size is large. The P-value is determined from the t curve with df $= n - 1$.
Power	The power of a test is the probability of rejecting the null hypothesis. Power is affected by the size of the difference between the hypothesized value and the true value, the sample size, and the significance level.

Chapter Review Exercises 10.72–10.95

Turn to the Expanded Answers Section for aswers not shown next to exercises.

CENGAGENOW Know exactly what to study! Take a pre-test and receive your Personalized Learning Plan.

10.72 The authors of the article "Perceived Risks of Heart Disease and Cancer Among Cigarette Smokers" (*Journal of the American Medical Association* [1999]: 1019–1021) expressed the concern that a majority of smokers do not view themselves as being at increased risk of heart disease or cancer. A study of 737 current smokers selected at random from U.S. households with telephones found that of 737 smokers surveyed, 295 indicated that they believed they have a higher than average risk of cancer. Do these data suggest that π, the true proportion of smokers who view themselves as being at increased risk of cancer is in fact less than .5, as claimed by the authors of the paper? Test the relevant hypotheses using $\alpha = .05$. $z = -5.43$, P-value ≈ 0, reject H_0

10.73 A number of initiatives on the topic of legalized gambling have appeared on state ballots. Suppose that a political candidate has decided to support legalization of casino gambling if he is convinced that more than two-thirds of U.S. adults approve of casino gambling. *USA Today* (June 17, 1999) reported the results of a Gallup poll in which 1523 adults (selected at random from households with telephones) were asked whether they approved of casino gambling. The number in the sample who approved was 1035. Does the sample provide convincing evidence that more than two-thirds approve? $z = 1.08$, *P*-value = 0.1401, fail to reject H_0

10.74 The article "Credit Cards and College Students: Who Pays, Who Benefits?" (*Journal of College Student Development* [1998]: 50–56) described a study of credit card payment practices of college students. According to the authors of the article, the credit card industry asserts that at most 50% of college students carry a credit card balance from month to month. However, the authors of the article report that, in a random sample of 310 college students, 217 carried a balance each month. Does this sample provide sufficient evidence to reject the industry claim? $z = 7.04$, *P*-value ≈ 0, reject H_0

10.75 Although arsenic is known to be a poison, it also has some beneficial medicinal uses. In one study of the use of arsenic to treat acute promyelocytic leukemia (APL), a rare type of blood cell cancer, APL patients were given an arsenic compound as part of their treatment. Of those receiving arsenic, 42% were in remission and showed no signs of leukemia in a subsequent examination (*Washington Post*, November 5, 1998). It is known that 15% of APL patients go into remission after the conventional treatment. Suppose that the study had included 100 randomly selected patients (the actual number in the study was much smaller). Is there sufficient evidence to conclude that the proportion in remission for the arsenic treatment is greater than .15, the remission proportion for the conventional treatment? Test the relevant hypotheses using a .01 significance level. $z = 7.56$, *P*-value ≈ 0, reject H_0

10.76 According to the article "Which Adults Do Underage Youth Ask for Cigarettes?" (*American Journal of Public Health* [1999]: 1561–1564), 43.6% of the 149 18- to 19-year-olds in a random sample have been asked to buy cigarettes for an underage smoker.
a. Is there convincing evidence that fewer than half of 18- to 19-year-olds have been approached to buy cigarettes by an underage smoker?
b. The article went on to state that of the 110 nonsmoking 18- to 19-year-olds, only 38.2% had been approached to buy cigarettes for an underage smoker. Is there evidence

that less than half of nonsmoking 18- to 19-year-olds have been approached to buy cigarettes? $z = -2.48$ *P*-value = 0.0066, reject H_0 (at $\alpha = 0.05$ and $\alpha = 0.10$)

10.77 Many people have misconceptions about how profitable small, consistent investments can be. In a survey of 1010 randomly selected U.S. adults (Associated Press, October 29, 1999), only 374 responded that they thought that an investment of $25 per week over 40 years with a 7% annual return would result in a sum of over $100,000 (the correct amount is $286,640). Is there sufficient evidence to conclude that less than 40% of U.S. adults are aware that such an investment would result in a sum of over $100,000? Test the relevant hypotheses using $\alpha = .05$. $z = -1.93$, *P*-value = 0.0268, reject H_0

10.78 The same survey described in Exercise 10.77 also asked the individuals in the sample what they thought was their best chance to obtain more than $500,000 in their lifetime. Twenty-eight percent responded "win a lottery or sweepstakes." Does this provide convincing evidence that more than one-fourth of U.S. adults see a lottery or sweepstakes win as their best chance of accumulating $500,000? Carry out a test using a significance level of .01. $z = 2.20$ *P*-value = 0.0139, fail to reject H_0

10.79 The state of Georgia's HOPE scholarship program guarantees fully paid tuition to Georgia public universities for Georgia high school seniors who have a B average in academic requirements as long as they maintain a B average in college. Of 137 randomly selected students enrolling in the Ivan Allen College at the Georgia Institute of Technology (social science and humanities majors) in 1996 who had a B average going into college, 53.2% had a GPA below 3.0 at the end of their first year ("Who Loses HOPE? Attrition from Georgia's College Scholarship Program," *Southern Economic Journal* [1999]: 379–390). Do these data provide convincing evidence that a majority of students at Ivan Allen College who enroll with a HOPE scholarship lose their scholarship? $z = 0.7491$, *P*-value = 02269, fail to reject H_0

10.80 Speed, size, and strength are thought to be important factors in football performance. The article "Physical and Performance Characteristics of NCAA Division I Football Players" (*Research Quarterly for Exercise and Sport* [1990]: 395–401) reported on physical characteristics of Division I starting football players in the 1988 football season. Information for teams ranked in the top 20 was easily obtained, and it was reported that the mean weight of starters on top-20 teams was 105 kg. A random sample of 33 starting players (various positions were represented) from Division I teams that were not ranked in the top 20 resulted in a sample mean weight of 103.3 kg and a sample standard deviation of 16.3 kg. Is there sufficient evidence to conclude that the mean weight for non-top-20 starters is less than 105, the known value for top-20 teams? $t = -0.60$, P-value ≈ 0.277, fail to reject H_0

10.81 Are young women delaying marriage and marrying at a later age? This question was addressed in a report issued by the Census Bureau (Associated Press, June 8, 1991). The report stated that in 1970 (based on census results) the mean age of brides marrying for the first time was 20.8 years. In 1990 (based on a sample, because census results were not yet available), the mean was 23.9. Suppose that the 1990 sample mean had been based on a random sample of size 100 and that the sample standard deviation was 6.4. Is there sufficient evidence to support the claim that in 1990 women were marrying later in life than in 1970? Test the relevant hypotheses using $\alpha = .01$. (Note: It is probably not reasonable to think that the distribution of age at first marriage is normal in shape.) $t = 4.84$, P-value ≈ 0, reject H_0

10.82 According to the article "Workaholism in Organizations: Gender Differences" (*Sex Roles* [1999]: 333–346), the following data were reported on 1996 income for random samples of male and female MBA graduates from a certain Canadian business school:

	N	$\bar{x}$	s
Males	258	$133,442	$131,090
Females	233	$105,156	$98,525

Note: These salary figures are in Canadian dollars.

a. Test the hypothesis that the mean salary of male MBA graduates from this school was in excess of $100,000 in 1996. $t = 4.098$, P-value ≈ 0, reject H_0
b. Is there convincing evidence that the mean salary for all female MBA graduates is above $100,000? Test using $\alpha = .10$. $t = 0.799$, P-value ≈ 0.2122, fail to reject H_0
c. If a significance level of .05 or .01 were used instead of .10 in the test of Part (b), would you still reach the same conclusion? Explain.

10.83 Duck hunting in populated areas faces opposition on the basis of safety and environmental issues. The *San Luis Obispo Telegram-Tribune* (June 18, 1991) reported the results of a survey to assess public opinion regarding duck hunting on Morro Bay (located along the central coast of California). A random sample of 750 local residents included 560 who strongly opposed hunting on the bay. Does this sample provide sufficient evidence to conclude that the majority of local residents oppose hunting on Morro Bay? Test the relevant hypotheses using $\alpha = .01$. $z = 13.51$, P-value ≈ 0, reject H_0

10.84 Seat belts help prevent injuries in automobile accidents, but they certainly don't offer complete protection in extreme situations. A random sample of 319 front-seat occupants involved in head-on collisions in a certain region resulted in 95 people who sustained no injuries ("Influencing Factors on the Injury Severity of Restrained Front Seat Occupants in Car-to-Car Head-on Collisions," *Accident Analysis and Prevention* [1995]: 143–150). Does this suggest that the true (population) proportion of uninjured occupants exceeds .25? State and test the relevant hypotheses using a significance level of .05. $z = 1.97$, P-value $= 0.244$, reject H_0

10.85 White remains the most popular car color in the United States, but its popularity appears to be slipping. According to an annual survey by DuPont (*Los Angeles Times*, February 22, 1994), white was the color of 20% of the vehicles purchased during 1993, a decline of 4% from the previous year. (According to a DuPont spokesperson, white represents "innocence, purity, honesty, and cleanliness.") A random sample of 400 cars purchased during this period in a certain metropolitan area resulted in 100 cars that were white. Does the proportion of all cars purchased in this area that are white appear to differ from the national percentage? Test the relevant hypotheses using $\alpha = .05$. Does your conclusion change if $\alpha = .01$ is used?

10.86 When a published article reports the results of many hypothesis tests, the P-values are not usually given. Instead, the following type of coding scheme is frequently used: $*p \le .05$, $**p \le .01$, $***p \le .001$, $****p \le .0001$. Which of the symbols would be used to code for each of the following P-values?
a. .037 *
b. .0026 **
c. .072 None
d. .0003 ***

10.87 A random sample of $n = 44$ individuals with a B.S. degree in accounting who started with a Big Eight accounting firm and subsequently changed jobs resulted in a sample mean time to change of 35.02 months and a sample standard deviation of 18.94 months ("The Debate over Post-Baccalaureate Education: One University's Experience," *Issues in Accounting Education* [1992]: 18–36). Can it be concluded that the true average time to change exceeds 2 years? Test the appropriate hypotheses using a significance level of .01. $t = 3.86$, P-value ≈ 0, reject H_0

10.88 What motivates companies to offer stock ownership plans to their employees? In a random sample of 87 companies having such plans, 54 said that the primary rationale was tax related ("The Advantages and Disadvantages of ESOPs: A Long-Range Analysis," *Journal of Small Business Management* [1991]: 15–21). Does this information provide strong support for concluding that more than half of all such firms feel this way? $z = 2.25$, P-value $= 0.0122$, reject H_0

10.89 The article "Caffeine Knowledge, Attitudes, and Consumption in Adult Women" (*Journal of Nutrition Education* [1992]: 179–184) reported the following summary statistics on daily caffeine consumption for a random sample of adult women: $n = 47$, $\bar{x} = 215$ mg, $s = 235$ mg, and the data values ranged from 5 to 1176.
a. Does it appear plausible that the population distribution of daily caffeine consumption is normal? Is it necessary to assume a normal population distribution to test hypotheses about the value of the population mean consumption? Explain your reasoning.
b. Suppose that it had previously been believed that mean consumption was at most 200 mg. Does the given information contradict this prior belief? Test the appropriate hypotheses at significance level .10. $t = 0.44$, P-value ≈ 0.346, fail to reject H_0

10.90 Past experience has indicated that the true response rate is 40% when individuals are approached with a request to fill out and return a particular questionnaire in a stamped and addressed envelope. An investigator believes that if the person distributing the questionnaire is stigmatized in some obvious way, potential respondents would feel sorry for the distributor and thus tend to respond at a rate higher than 40%. To investigate this theory, a distributor is fitted with an eye patch. Of the 200 questionnaires distributed by this individual, 109 were returned. Does this strongly suggest that the response rate in this situation exceeds the rate in the past? State and test the appropriate hypotheses at significance level .05. $z = 4.19$, P-value ≈ 0, reject H_0

10.91 ● An automobile manufacturer who wishes to advertise that one of its models achieves 30 mpg (miles per gallon) decides to carry out a fuel efficiency test. Six nonprofessional drivers are selected, and each one drives a car from Phoenix to Los Angeles. The resulting fuel efficiencies (in miles per gallon) are:

27.2   29.3   31.2   28.4   30.3   29.6

Assuming that fuel efficiency is normally distributed under these circumstances, do the data contradict the claim that true average fuel efficiency is (at least) 30 mpg? $t = -1.16$, P-value $= 0.15$, at $\alpha = 0.01$: fail to reject H_0

10.92 A student organization uses the proceeds from a particular soft-drink dispensing machine to finance its activities. The price per can had been $0.75 for a long time, and the average daily revenue during that period had been $75.00. The price was recently increased to $1.00 per can. A random sample of $n = 20$ days after the price increase yielded a sample average daily revenue and sample standard deviation of $70.00 and $4.20, respectively. Does this information suggest that the true average daily revenue has decreased from its value before the price increase? Test the appropriate hypotheses using $\alpha = .05$. $t = -5.32$, P-value ≈ 0, reject H_0

10.93 A hot tub manufacturer advertises that with its heating equipment, a temperature of 100°F can be achieved in at most 15 min. A random sample of 25 tubs is selected, and the time necessary to achieve a 100°F temperature is determined for each tub. The sample average time and sample standard deviation are 17.5 min and 2.2 min, respectively. Does this information cast doubt on the company's claim? Carry out a test of hypotheses using significance level .05. $t = 5.68$, P-value $= 0$, reject H_0

10.94 Let π denote the proportion of voters in a certain state who favor a particular proposed constitutional amendment. Consider testing H_0: $\pi = .5$ versus H_a: $\pi > .5$ at significance level .05 based on a sample of size $n = 50$.
a. Suppose that H_0 is in fact true. Use Appendix Table 1 (our table of random numbers) to simulate selecting a sample, and use the resulting data to carry out the test.
b. If you repeated Part (a) a total of 100 times (a simulation consisting of 100 replications), how many times would you expect H_0 to be rejected?   About 5 times
c. Now suppose that $\pi = .6$, which implies that H_0 is false. Again, use Appendix Table 1 to simulate selecting a sample, and carry out the test. If you repeated this a total of 100 times, would you expect H_0 to be rejected more frequently than when H_0 is true?   Answers will vary based on simulation results.

10.95 A type of lie detector that measures brain waves was developed by a professor of neurobiology at Northwestern University (Associated Press, July 7, 1988). He said, "It would probably not falsely accuse any innocent people and it would probably pick up 70% to 90% of guilty people." Suppose that the result of this lie detector test is allowed as evidence in a criminal trial as the sole basis of a decision between two rival hypotheses: accused is innocent versus accused is guilty. Although these are not "statistical hypotheses" (statements about a population characteristic), the possible decision errors are analogous to Type I and Type II errors.

In this situation, a Type I error is finding an innocent person guilty—rejecting the null hypothesis of innocence when it is in fact true. A Type II error is finding a guilty person innocent—not rejecting the null hypothesis of in-

nocence when it is in fact false. If the developer of the lie detector is correct in his statements, what is the probability of a Type I error, α? What can you say about the probability of a Type II error, β? $\alpha = 0, 0.1 < \beta < 0.3$

Personal Tutor
Do you need a live tutor for homework problems?

CENGAGENOW
Are you ready? Take your exam-prep post-test now.

Bold exercises answered in back   ● Data set available online but not required   ▼ Video solution available

Graphing Calculator Explorations

Exploration 10.1   Hypothesis Test for a Population Proportion

Using your calculator's hypothesis-testing capability begins, as usual, by navigating your calculator's menu system, this time looking for key words such as "hypothesis" and "tests." As before with confidence intervals, look for words such as "1" and "prop." and "z."

Once you select the correct choice, you will be presented with a screen for providing information. The information you must provide is exactly what you would need to test the hypotheses on paper: the sample number of successes, the sample size, the level of significance, and so on. Figure 10.8 shows two representative screens, with information filled in from Example 10.10 in the text (the screen in Figure 10.8(b) has the "less than" alternative hypothesis selected by shading, although the shading doesn't show up in the figure).

Move your cursor down to Execute or Calculate, and press the Enter, Execute, or Calculate button, depending on your calculator. The results should appear immediately. Again, we show two representative screens in Figure 10.9.

Notice the slight difference between the text and calculator answers that results from rounding in the hand calculations.

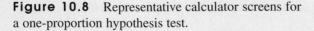

1-Prop ZTest
Prop : $<p_0$
p_0 :.5
x :220
n :500
Execute

(a)

1-PropZTest
p_0:.5
x:220
n:500
Prop $\neq p_0$ $<p_0$ $>p_0$
Calculate Draw

(b)

Figure 10.8   Representative calculator screens for a one-proportion hypothesis test.

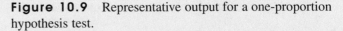

1-Prop ZTest
Prop<0.5
Z $= -2.683281573$
p $= .003645226$
$\hat{p} = .44$
n $= 500$

(a)

1-PropZInt
prop<0.5
z $= -2.683281573$
p $= .003645226$
$\hat{p} = .44$
n $= 500$

(b)

Figure 10.9   Representative output for a one-proportion hypothesis test.

The calculator does the work for only a few of the steps in a hypothesis test. Remember, there is still work for you to do in completing the other steps!

Exploration 10.2   Hypothesis Test for a Population Mean

Testing a hypothesis for a single mean on your calculator requires you to navigate the menu system once again. As was true with the confidence interval for the mean,

1. you must decide whether to base the test on the z or a t distribution, and
2. you can use previous calculations of the sample mean and standard deviation, or the calculator will evaluate these statistics from data contained in a list.

We follow Example 10.14, and use the t test after entering the data in List1. The normality of the population must be assessed as before. A calculator boxplot and normal probability plot are shown in Figure 10.10.

Figure 10.10 Plots for data of Example 10.14: (a) boxpolt; (b) normal probability plot.

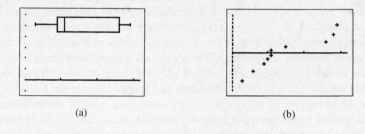

(a)   (b)

After assessing the plausibility of the normality of the population, we test the hypotheses. Again we have the choice of entering the sample calculations or providing data in a list and letting the calculator evaluate the sample mean and standard deviation. Based on the choice, we see one of the screens shown in Figure 10.11.

Figure 10.11 Representative calculator screens for a single-sample t test.

(a)   (b)

Figure 10.12 shows calculator output for the hypothesis test. Remember, just writing the calculator output is *not* a complete response to a hypothesis testing task —there are necessary steps in the hypothesis-testing procedure that you must write yourself.

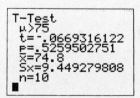

Figure 10.12 Calculator output for the single-sample t test.

mean $(\mu_{\bar{x}_f})$ of 66 inches and standard deviation $(\sigma_{\bar{x}_f})$ of $\dfrac{2.3}{\sqrt{25}}$ inches. The difference in the sample means, $(\bar{x}_m - \bar{x}_f)$, is just a linear combination of the two sample means. Using the linear combination rules in Section 7.4 and the results about the two individual sampling distributions (from Section 8.2), the distribution of the difference in sample means is approximately normal with mean $(\mu_{\bar{x}_m - \bar{x}_f}) = 6$ and standard deviation

$$\sigma_{\bar{x}_m - \bar{x}_f} = \sqrt{\left(\frac{2.6}{\sqrt{25}}\right)^2 + \left(\frac{2.3}{\sqrt{25}}\right)^2} = \sqrt{\frac{2.6^2}{25} + \frac{2.3^2}{25}}.$$ (See the following figure.) It's

amazing that students can develop this standard deviation on their own.

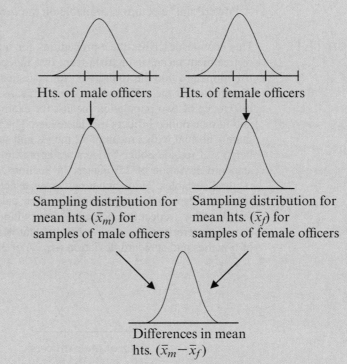

Hts. of male officers   Hts. of female officers

Sampling distribution for
mean hts. $(\bar{x}_m)$ for
samples of male officers

Sampling distribution for
mean hts. $(\bar{x}_f)$ for
samples of female officers

Differences in mean
hts. $(\bar{x}_m - \bar{x}_f)$

This activity can be simulated by using computer software or a graphing calculator to generate random samples from a normal distribution. On a graphing calculator, students could use the following command in L1: randNorm(μ,σ,n). They can simulate random samples from each population, calculate the sample means, find the difference in sample means, and plot the differences in sample means on a dotplot.

A discussion like the one above gives students a clear picture of the sampling distribution of $\bar{x}_1 - \bar{x}_2$, the foundation for inference procedures about the difference in population or treatment means based on independent samples. The properties for the sampling distribution of $\bar{x}_1 - \bar{x}_2$ are stated in the tan box on page 585.

Remind students that hypotheses must be written about the populations. The purpose of two-sample tests is to compare two populations (or two treatment groups). So, the null hypothesis is often written as a difference, such as $\mu_1 - \mu_2$. Often we are interested in whether or not there is a difference in the populations; in this case the hypothesized value in the null hypothesis would be zero (i.e., there is no difference in the means of the populations). The graphing calculator is programmed to test the null hypothesis that states there is no difference in the populations (i.e., H_0: $\mu_1 - \mu_2 = 0$). Sometimes, though, we are interested in whether the difference is a specific value, such as H_0: $\mu_1 - \mu_2 = 3$. Be sure to do a few examples where the hypothesized value of the difference is not zero.

For both the hypothesis test and confidence interval procedures for means, statisticians use t-procedures when the population standard deviations are unknown. The blue box on page 586 gives the formula for the approximate degree of freedom for t-procedures based on two independent samples. Graphing calculators automatically use this formula to calculate the degree of freedom and report the df along with the test statistic and the P-value.

It is helpful to use the generic formula for the test statistic,

$$\left(\text{test statistic} = \frac{\text{statistic} - \text{parameter}}{\text{standard deviation of the statistic}} \right),$$ and ask students to fill in each part

for a specific procedure. For the two-sample t-test, what is the statistic? (Answer is $\bar{x}_1 - \bar{x}_2$.) What is the parameter? (Answer is $\mu_1 - \mu_2$.) What is the standard deviation (standard error) of the statistic? (Answer is $\sqrt{\dfrac{s_1^2}{n_1} + \dfrac{s_2^2}{n_2}}$.) The P-value is computed by using the same procedure as for a one-sample t-test.

The tan box on page 596 provides the formula for the two-sample t confidence interval. As above, it is helpful to use the generic formula for a confidence interval, statistic $\pm$ (critical value)(standard deviation of the statistic), and have students fill in each part. Notice that the 95% confidence interval computed in Example 11.4 does not contain zero. This suggests that there is a difference in the mean diastolic blood pressure for the two treatments. Graphing Calculator Exploration 11.1 (on page 641) provides instructions for performing two-sample t-procedures using the graphing calculator.

As with all hypothesis tests and confidence intervals, students need to be sure to check the necessary assumptions and to write the conclusions in context. It is also helpful to ask questions about Type I and II errors and power in the two-sample scenarios.

Suggested Assignment: Exercises 11.1, 11.2, 11.4, 11.6, 11.7, 11.9, 11.12, 11.17, 11.22, 11.25

Activities: Bonus Activity 11.5—Which Weighs More, Coke or Diet Coke? (*Activities Manual*, p. 100)★, "Chapter 11—Comparing Population Means: Independent Samples CI" (*Instructor's Resource Binder* & CD, Chapter 11 Activities Worksheets), and "Chapter 11—Comparing Population Means: Independent Samples Test" (*Instructor's Resource Binder* & CD, Chapter 11 Activities Worksheets)

Assessment: Chapter 11 Quiz 1 (*Test Bank* and *Instructor's Resource Binder* CD)

■ **Section 11.2**

One experimental design introduced in Chapter 2 is the matched-pairs design. It is helpful to briefly review the matched-pairs design in which the assignment of experimental units to treatments results in **paired** samples. When performing inference procedures with dependent samples, we are interested in the difference between the two population means, but we actually analyze the sample differences. Therefore, the first step is to compute the differences for each pair, creating a sample of differences. Now we work just with the differences, using the one-sample inference procedures. That is, we use the one-sample t-interval or t-test from Chapters 9 and 10. Typically, in a paired-t test the null hypothesis is that there is no difference in the population or treatment means: H_0: $\mu_d = 0$, where μ_d is the true mean of the difference population. Refer to the tan box on page 609 for additional

★ Kathy's personal favorite

information. Graphing Calculator Exploration 11.2 (on page 643) provides instructions for the graphing calculator.

Suggested Assignment: Exercises 11.29, 11.31, 11.32, 11.33, 11.35, 11.38, 11.40, 11.41, 11.43

Activity: Activity 11.1: Helium-Filled Footballs (p. 632 and *Activities Manual*, p. 93)★, Activity 11.2: Thinking about Data Collection (p. 633 and *Activities Manual*, p. 95), "Chapter 11—Independent or Dependent?" (*Instructor's Resource Binder* & CD, Chapter 11 Activities Worksheets), "Chapter 11—Comparing Population Means: Dependent Samples CI" (*Instructor's Resource Binder* & CD, Chapter 11 Activities Worksheets), and/ or "Chapter 11—Comparing Population Means: Dependent Samples Test" (*Instructor's Resource Binder* & CD, Chapter 11 Activities Worksheets)

Assessment: Chapter 11 Quiz 2 (*Test Bank* and *Instructor's Resource Binder* CD)

■ **Section 11.3**

This section describes the inference procedures for the difference in two population or treatment proportions, $\pi_1 - \pi_2$. Let's investigate the sampling distribution of the difference in two sample proportions, $p_1 - p_2$. Ask, "What is the probability that a spinning coin will land heads?" *It's not necessarily 0.5.* Studies have shown that the probability of heads when a coin is spun is approximately 0.4 for newer pennies. (Coins minted in different years have different probabilities of heads when spun; see *Activity-Based Statistics* by Richard Scheaffer et al., p. 129.) Pair the students and have one student flip a penny 25 times while the other spins a penny 25 times.

To flip: Students should toss the coin in the air so that it flips end over end, and catch it in midair. Always start with heads facing up.

To spin: Students should spin the penny by standing it on its edge and flicking it with a finger. Always start with heads facing the student spinning the penny. Students should then compute the proportion of heads.

Since the presumed probability of heads when flipping pennies is .50, we will use $\pi_f = 0.5$. We suppose the probability of heads when spinning pennies is .40, so $\pi_s = 0.4$. The sampling distribution of p for the proportion of heads when flipping is approximately normal (because $n\pi = 25(.5) = 12.5$ and $n(1 - \pi) = (25)(.5) = 12.5$ are both greater than 10) with $\mu_{p_f} = .5$ and $\sigma_{p_f} = \sqrt{\dfrac{.5(.5)}{25}}$. The sampling distribution of p for the proportion of heads when spinning is approximately normal (because $n\pi = 25(.4) = 10$ and $n(1 - \pi) = (25)(.6) = 15$ are both greater than or equal to 10) with $\mu_{p_s} = .4$ and

$$\sigma_{p_s} = \sqrt{\frac{.4(.6)}{25}}.$$

Each pair of students should compute the difference in their sample proportions, $p_1 - p_2$, and plot the difference on a dotplot. Note that some of these differences will be negative. The dotplot will display a partial sampling distribution of the difference in sample proportions and should (assuming that your class is large enough to have a reasonable number of differences plotted) be centered at approximately $\mu_{p_f - p_s} = .5 - .4 = .1$ with an approximate standard deviation of $\mu_{p_f - p_s} = \sqrt{\dfrac{.5(.5)}{25} + \dfrac{.4(.6)}{25}}$. Again, students generally have little difficulty developing this standard deviation.

★ Kathy's personal favorite

This activity gives students a clear picture of the sampling distribution of $p_1 - p_2$ which is the foundation for inference procedures for two population proportions. The properties of the sampling distribution of $p_1 - p_2$ are stated in the tan box on page 620.

The formula for the confidence interval for $\pi_1 - \pi_2$ is

$$(p_1 - p_2) \pm z^* \sqrt{\frac{p_1(1 - p_1)}{n_1} + \frac{p_2(1 - p_2)}{n_2}}.$$ (See the tan box on page 625.) As in Section 11.1, it is helpful to give the generic formula for a confidence interval and ask students to fill in the parts.

Hypotheses about the difference between two population proportions are similar to the hypotheses about a difference in means. The null hypothesis typically states that there is no difference in the two population proportions: $H_0: \pi_1 - \pi_2 = 0$.

However, the standard error used in the computation of the confidence interval is *not* used in the hypothesis test! Under the assumption that the null hypothesis is true, the respective variances of the sampling distributions of the two population proportions are the same (even though we don't know this value). So we estimate the true population proportion by combining information from both samples. This combined estimate is used in the

calculation of the standard error of the test statistic, $z = \dfrac{p_1 - p_2}{\sqrt{\dfrac{p_c(1 - p_c)}{n_1} + \dfrac{p_c(1 - p_c)}{n_2}}}$.

(See both tan boxes on page 621.)

Finally, students need to be sure to check the necessary assumptions and to write the conclusions in context. Be sure to ask questions about Type I and II errors and power in the two-sample scenarios.

Suggested Assignment: Exercises 11.44, 11.45, 11.46, 11.47, 11.51, 11.56, 11.57, 11.59

Activity: Bonus Activity 11.4—More on Defective M&Ms (*Activities Manual*, p. 98)★, "Chapter 11—Estimating the Difference Between Two Proportions" (*Instructor's Resource Binder* & CD, Chapter 11 Activities Worksheets) and "Chapter 11—Testing for a Difference Between Two Proportions" (*Instructor's Resource Binder* & CD, Chapter 11 Activities Worksheets)

Assessment: Chapter 11 Quiz 3 (*Test Bank* and *Instructor's Resource Binder* CD)

■ Section 11.4

This section provides excellent suggestions on communicating and interpreting confidence intervals. Students should also pay close attention to "A Word to the Wise: Cautions and Limitations."

Suggested Review Assignment: Exercises 11.64, 11.66, 11.67, 11.68, 11.70, 11.71, 11.72, 11.74, 11.77, 11.81, 11.82, 11.88, 11.89

Activity: Activity 11.3: A Meaningful Paragraph (p. 633 and *Activities Manual*, p. 97)

Assessment: Chapter 11 Concept Quiz (*Test Bank* and *Instructor's Resource Binder* CD) and Chapter 11 Test (*Test Bank* and *Instructor's Resource Binder* CD)

© Randy Wells/Getty Images

Chapter 11

Comparing Two Populations or Treatments

Many investigations are carried out for the purpose of comparing two populations or treatments. For example, the article "Learn More, Earn More?" (*ETS Policy Notes* [1999]: 1–12) described how the job market treats high school graduates who do not go to college. By comparing data from a random sample of students who had high grades in high school with data from a random sample of students who had low grades, the authors of the article were able to investigate whether the proportion employed for those with high grades was higher than the proportion employed for those with low grades. The authors were also interested in whether the mean monthly salary was higher for those with good grades than for those who did not earn good grades. To answer these questions, hypothesis tests that compare the proportions or means for two different populations were used. In this chapter we introduce hypothesis tests and confidence intervals that can be used to compare two populations or treatments.

11.1 Inferences Concerning the Difference Between Two Population or Treatment Means Using Independent Samples

In this section, we consider using sample data to compare two population means or two treatment means. An investigator may wish to estimate the difference between two population means or to test hypotheses about this difference. For example, a university

Improve your understanding and save time! Visit www.cengage.com/login where you will find:

- Step-by-step instructions for MINITAB, Excel, TI-83, SPSS, and JMP
- Video solutions to selected exercises
- Data sets available for selected examples and exercises

- Exam-prep pre-tests that build a Personalized Learning Plan based on your results so that you know exactly what to study
- Help from a live statistics tutor 24 hours a day

financial aid director may want to determine whether the mean cost of textbooks is different for students enrolled in the engineering college than for students enrolled in the liberal arts college. Here, two populations (one consisting of all students enrolled in the engineering college and the other consisting of all students enrolled in the liberal arts college) are to be compared on the basis of their respective mean textbook costs. Information from two random samples, one from each population, could be the basis for making such a comparison.

In other cases, an experiment might be carried out to compare two different treatments or to compare the effect of a treatment with the effect of no treatment (treatment versus control). For example, an agricultural experimenter might wish to compare weight gains for animals placed on two different diets (each diet is a treatment), or an educational researcher might wish to compare online instruction to traditional classroom instruction by studying the difference in mean scores on a common final exam (each type of instruction is a treatment).

In previous chapters, the symbol μ was used to denote the mean of a single population under study. When comparing two populations or treatments, we must use notation that distinguishes between the characteristics of the first and those of the second. This is accomplished by using subscripts on quantities such as μ and σ^2. Similarly, subscripts on sample statistics, such as $\bar{x}$, indicate to which sample these quantities refer.

Notation

	Mean	Variance	Standard Deviation
Population or Treatment 1	μ_1	σ_1^2	σ_1
Population or Treatment 2	μ_2	σ_2^2	σ_2

	Sample Size	Mean	Variance	Standard Deviation
Sample from Population or Treatment 1	n_1	$\bar{x}_1$	s_1^2	s_1
Sample from Population or Treatment 2	n_2	$\bar{x}_2$	s_2^2	s_2

A comparison of means focuses on the difference, $\mu_1 - \mu_2$. When $\mu_1 - \mu_2 = 0$, the two population or treatment means are identical. That is,

$\mu_1 - \mu_2 = 0$ is equivalent to $\mu_1 = \mu_2$

Similarly,

$\mu_1 - \mu_2 > 0$ is equivalent to $\mu_1 > \mu_2$

and

$\mu_1 - \mu_2 < 0$ is equivalent to $\mu_1 < \mu_2$

Before developing inferential procedures concerning $\mu_1 - \mu_2$, we must consider how the two samples, one from each population, are selected. Two samples are said to

Important concept

be **independent** samples if the selection of the individuals or objects that make up one sample does not influence the selection of individuals or objects in the other sample. However, when observations from the first sample are paired in some meaningful way with observations in the second sample, the samples are said to be **paired**. For example, to study the effectiveness of a speed-reading course, the reading speed of subjects could be measured before they take the class and again after they complete the course. This gives rise to two related samples—one from the population of individuals who have not taken this particular course (the "before" measurements) and one from the population of individuals who have had such a course (the "after" measurements). These samples are paired. The two samples are not independently chosen, because the selection of individuals from the first (before) population completely determines which individuals make up the sample from the second (after) population. In this section, we consider procedures based on independent samples. Methods for analyzing data resulting from paired samples are presented in Section 11.2.

Because $\bar{x}_1$ provides an estimate of μ_1 and $\bar{x}_2$ gives an estimate of μ_2, it is natural to use $\bar{x}_1 - \bar{x}_2$ as a point estimate of $\mu_1 - \mu_2$. The value of $\bar{x}_1$ varies from sample to sample (it is a *statistic*), as does the value of $\bar{x}_2$. Since the difference $\bar{x}_1 - \bar{x}_2$ is calculated from sample values, it is also a statistic and therefore has a sampling distribution.

Properties of the Sampling Distribution of $\bar{x}_1 - \bar{x}_2$

If the random samples on which $\bar{x}_1$ and $\bar{x}_2$ are based are selected independently of one another, then

1. $\mu_{\bar{x}_1 - \bar{x}_2} = \left(\begin{matrix} \text{mean value} \\ \text{of } \bar{x}_1 - \bar{x}_2 \end{matrix} \right) = \mu_{\bar{x}_1} - \mu_{\bar{x}_2} = \mu_1 - \mu_2$

   Thus, the sampling distribution of $\bar{x}_1 - \bar{x}_2$ is always centered at the value of $\mu_1 - \mu_2$, so $\bar{x}_1 - \bar{x}_2$ is an unbiased statistic for estimating $\mu_1 - \mu_2$.

2. $\sigma^2_{\bar{x}_1 - \bar{x}_2} = \left(\begin{matrix} \text{variance of} \\ \bar{x}_1 - \bar{x}_2 \end{matrix} \right) = \sigma^2_{\bar{x}_1} + \sigma^2_{\bar{x}_2} = \dfrac{\sigma^2_1}{n_1} + \dfrac{\sigma^2_2}{n_2}$

   and

   $\sigma_{\bar{x}_1 - \bar{x}_2} = \left(\begin{matrix} \text{standard deviation} \\ \text{of } \bar{x}_1 - \bar{x}_2 \end{matrix} \right) = \sqrt{\dfrac{\sigma^2_1}{n_1} + \dfrac{\sigma^2_2}{n_2}}$

3. If n_1 and n_2 are both large or the population distributions are (at least approximately) normal, $\bar{x}_1$ and $\bar{x}_2$ each have (at least approximately) a normal distribution. This implies that the sampling distribution of $\bar{x}_1 - \bar{x}_2$ is also normal or approximately normal.

Properties 1 and 2 follow from the following general results:

These are the properties of the sampling distribution of $\bar{x}_1 - \bar{x}_2$.

1. The mean value of a difference in means is the difference of the two individual mean values.
2. The variance of a difference of *independent* quantities is the *sum* of the two individual variances.

When the sample sizes are large or when the population distributions are approximately normal, the properties of the sampling distribution of $\bar{x}_1 - \bar{x}_2$ imply that $\bar{x}_1 - \bar{x}_2$ can be standardized to obtain a variable with a sampling distribution that is approximately the standard normal (z) distribution. This gives the following result.

When n_1 and n_2 are both large or the population distributions are (at least approximately) normal, the distribution of

$$z = \frac{\bar{x}_1 - \bar{x}_2 - (\mu_1 - \mu_2)}{\sqrt{\dfrac{\sigma_1^2}{n_1} + \dfrac{\sigma_2^2}{n_2}}}$$

is described (at least approximately) by the standard normal (z) distribution.

Although it is possible to base a test procedure and confidence interval on this result, the values of σ_1^2 and σ_2^2 are rarely known. As a result, the applicability of z is limited. When σ_1^2 and σ_2^2 are unknown, we must estimate them using the corresponding sample variances, s_1^2 and s_2^2. The result on which both a test procedure and confidence interval are based is given in the accompanying box.

When two random samples are independently selected and when n_1 and n_2 are both large or if the population distributions are normal, the standardized variable

$$t = \frac{\bar{x}_1 - \bar{x}_2 - (\mu_1 - \mu_2)}{\sqrt{\dfrac{s_1^2}{n_1} + \dfrac{s_2^2}{n_2}}}$$

has approximately a t distribution with

$$df = \frac{(V_1 + V_2)^2}{\dfrac{V_1^2}{n_1 - 1} + \dfrac{V_2^2}{n_2 - 1}} \quad \text{where } V_1 = \frac{s_1^2}{n_1} \text{ and } V_2 = \frac{s_2^2}{n_2}$$

The computed value of df should be truncated (rounded down) to obtain an integer value of df.

If one or both sample sizes are small, we must consider the shape of the population distributions. We can use normal probability plots or boxplots to evaluate whether it is reasonable to proceed as if the population distributions are normal.

■ Test Procedures

In a test designed to compare two population means, the null hypothesis is of the form

$$H_0: \mu_1 - \mu_2 = \text{hypothesized value}$$

Often the hypothesized value is 0, indicating that there is no difference between the population means. The alternative hypothesis involves the same hypothesized value but uses one of three inequalities (less than, greater than, or not equal to), depending on the research question of interest. As an example, let μ_1 and μ_2 denote the average fuel efficiencies (in miles per gallon, mpg) for two models of a certain type of car equipped with 4-cylinder and 6-cylinder engines, respectively. The hypotheses under consideration might be

$$H_0:\ \mu_1 - \mu_2 = 5 \quad \text{versus} \quad H_a:\ \mu_1 - \mu_2 > 5$$

The null hypothesis is equivalent to the claim that average efficiency for the 4-cylinder engine exceeds the average efficiency for the 6-cylinder engine by 5 mpg. The alternative hypothesis states that the difference between the true average efficiencies is more than 5 mpg.

A test statistic is obtained by replacing $\mu_1 - \mu_2$ in the standardized t variable (given in the previous box) with the hypothesized value that appears in H_0. Thus, the t statistic for testing $H_0:\ \mu_1 - \mu_2 = 5$ is

$$t = \frac{\bar{x}_1 - \bar{x}_2 - 5}{\sqrt{\dfrac{s_1^2}{n_1} + \dfrac{s_2^2}{n_2}}}$$

When H_0 is true and the sample sizes are large or when the population distributions are normal, the sampling distribution of the test statistic is approximately a t distribution. The P-value for the test is obtained by first computing the appropriate number of degrees of freedom and then using Appendix Table 4. The following box gives a general description of the test procedure.

Summary of the Two-Sample t Test for Comparing Two Population Means

Null hypothesis: $H_0:\ \mu_1 - \mu_2 =$ hypothesized value

Test statistic: $t = \dfrac{\bar{x}_1 - \bar{x}_2 - \text{hypothesized value}}{\sqrt{\dfrac{s_1^2}{n_1} + \dfrac{s_2^2}{n_2}}}$

The appropriate df for the two-sample t test is

$$\text{df} = \frac{(V_1 + V_2)^2}{\dfrac{V_1^2}{n_1 - 1} + \dfrac{V_2^2}{n_2 - 1}} \quad \text{where } V_1 = \frac{s_1^2}{n_1} \text{ and } V_2 = \frac{s_2^2}{n_2}$$

The computed number of degrees of freedom should be truncated (rounded down) to an integer.

Alternative Hypothesis:	**_P_-Value:**
$H_a:\ \mu_1 - \mu_2 >$ hypothesized value	Area under appropriate t curve to the right of the computed t
$H_a:\ \mu_1 - \mu_2 <$ hypothesized value	Area under appropriate t curve to the left of the computed t
$H_a:\ \mu_1 - \mu_2 \neq$ hypothesized value	(1) 2(area to the right of the computed t) if t is positive or (2) 2(area to the left of the computed t) if t is negative

Assumptions: 1. The two samples are *independently selected random samples*.
2. The *sample sizes are large* (generally 30 or larger) or *the population distributions are (at least approximately) normal.*

Example 11.1 Brain Size

Do children diagnosed with attention deficit/hyperactivity disorder (ADHD) have smaller brains than children without this condition? This question was the topic of a research study described in the paper "Developmental Trajectories of Brain Volume Abnormalities in Children and Adolescents with Attention Deficit/Hyperactivity Disorder" (*Journal of the American Medical Association* [2002]: 1740–1747). Brain scans were completed for 152 children with ADHD and 139 children of similar age without ADHD. Summary values for total cerebral volume (in milliliters) are given in the following table:

	n	$\overline{x}$	s
Children with ADHD	152	1059.4	117.5
Children without ADHD	139	1104.5	111.3

Do these data provide evidence that the mean brain volume of children with ADHD is smaller than the mean for children without ADHD? Let's test the relevant hypotheses using a .05 level of significance.

1. μ_1 = true mean brain volume for children with ADHD
   μ_2 = true mean brain volume for children without ADHD
   $\mu_1 - \mu_2$ = difference in mean brain volume
2. $H_0: \mu_1 - \mu_2 = 0$
3. $H_a: \mu_1 - \mu_2 < 0$
4. Significance level: $\alpha = .05$
5. Test statistic: $t = \dfrac{\overline{x}_1 - \overline{x}_2 - \text{hypothesized value}}{\sqrt{\dfrac{s_1^2}{n_1} + \dfrac{s_2^2}{n_2}}} = \dfrac{\overline{x}_1 - \overline{x}_2 - 0}{\sqrt{\dfrac{s_1^2}{n_1} + \dfrac{s_2^2}{n_2}}}$
6. Assumptions: The paper states that the study controlled for age and that the participants were "recruited from the local community." This is not equivalent to random sampling, but the authors of the paper (five of whom were doctors at well-known medical institutions) believed that it was reasonable to regard these samples as representative of the two groups under study. Both sample sizes are large, so it is reasonable to proceed with the two-sample t test.
7. Calculation:

$$t = \frac{(1059.4 - 1104.5) - 0}{\sqrt{\dfrac{(117.5)^2}{152} + \dfrac{(111.3)^2}{139}}} = \frac{-45.10}{\sqrt{90.831 + 89.120}} = \frac{-45.10}{13.415} = -3.36$$

8. *P*-value: We first compute the df for the two-sample t test:

$$V_1 = \frac{s_1^2}{n_1} = 90.831 \qquad V_2 = \frac{s_2^2}{n_2} = 89.120$$

$$\text{df} = \frac{(V_1 + V_2)^2}{\dfrac{V_1^2}{n_1 - 1} + \dfrac{V_2^2}{n_2 - 1}} = \frac{(90.831 + 89.120)^2}{\dfrac{(90.831)^2}{151} + \dfrac{(89.120)^2}{138}} = \frac{32,382.362}{112.191} = 288.636$$

We truncate the number of degrees of freedom to 288. Appendix Table 4 shows that the area under the t curve with 288 df (using the z critical value column because 288 is larger than 120 df) to the left of -3.36 is approximately 0. Therefore

$$P\text{-value} \approx 0$$

9. Conclusion: Because P-value $\approx 0 \leq .05$, we reject H_0. There is convincing evidence that the mean brain volume for children with ADHD is smaller than the mean for children without ADHD.

■

Example 11.2 Oral Contraceptive Use and Bone Mineral Density

● To assess the impact of oral contraceptive use on bone mineral density (BMD), researchers in Canada carried out a study comparing BMD for women who had used oral contraceptives for at least 3 months to BMD for women who had never used oral contraceptives ("Oral Contraceptive Use and Bone Mineral Density in Premenopausal Women," *Canadian Medical Association Journal* [2001]: 1023–1029). Data on BMD (in grams per centimeter) consistent with summary quantities given in the paper appear in the following table (the actual sample sizes for the study were much larger):

Never used oral contraceptives	0.82	0.94	0.96	1.31	0.94	1.21	1.26	1.09	1.13	1.14
Used oral contraceptives	0.94	1.09	0.97	0.98	1.14	0.85	1.30	0.89	0.87	1.01

The authors of the paper believed that it was reasonable to view the samples used in the study as representative of the two populations of interest—women who used oral contraceptives for at least 3 months and women who never used oral contraceptives. For the purposes of this example, we will assume that it is also justifiable to consider the two samples given here as representative of the populations. We use the given information and a significance level of .05 to determine whether there is evidence that women who use oral contraceptives have a lower mean BMD than women who have never used oral contraceptives.

1. μ_1 = true mean bone mineral density for women who never used oral contraceptives
   μ_2 = true mean bone mineral density for women who used oral contraceptives
   $\mu_1 - \mu_2$ = difference in mean bone mineral density
2. $H_0: \mu_1 - \mu_2 = 0$
3. $H_a: \mu_1 - \mu_2 > 0$
4. Significance level: $\alpha = .05$
5. Test statistic: $t = \dfrac{\bar{x}_1 - \bar{x}_2 - \text{hypothesized value}}{\sqrt{\dfrac{s_1^2}{n_1} + \dfrac{s_2^2}{n_2}}} = \dfrac{\bar{x}_1 - \bar{x}_2 - 0}{\sqrt{\dfrac{s_1^2}{n_1} + \dfrac{s_2^2}{n_2}}}$
6. Assumptions: For the two-sample t test to be appropriate, we must be willing to assume that the two samples can be viewed as independently selected random samples from the two populations of interest. As previously noted, we assume that

Step-by-step technology instructions available online   ● Data set available online

this is reasonable. Because both of the sample sizes are small, it is also necessary to assume that the BMD distribution is approximately normal for each of these two populations. Boxplots constructed using the sample data are shown here:

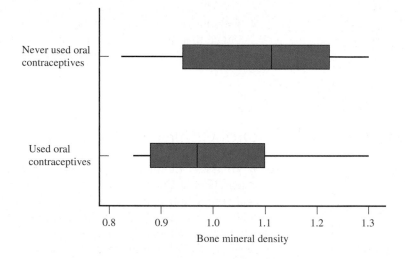

Because the boxplots are reasonably symmetric and because there are no outliers, the assumption of normality is plausible.

7. Calculation: For the given data: $\bar{x}_1 = 1.08$ $s_1 = .16$ $\bar{x}_2 = 1.00$ $s_2 = .14$, and

$$t = \frac{(1.08 - 1.00) - 0}{\sqrt{\dfrac{(.16)^2}{10} + \dfrac{(.14)^2}{10}}} = \frac{.08}{\sqrt{.003 + .002}} = \frac{.08}{.071} = 1.13$$

8. *P*-value: We first compute the df for the two-sample *t* test:

$$V_1 = \frac{s_1^2}{n_1} = .003 \qquad V_2 = \frac{s_2^2}{n_2} = .002$$

$$\text{df} = \frac{(V_1 + V_2)^2}{\dfrac{V_1^2}{n_1 - 1} + \dfrac{V_2^2}{n_2 - 1}} = \frac{(.003 + .002)^2}{\dfrac{(.003)^2}{9} + \dfrac{(.002)^2}{9}} = \frac{.000025}{.0000014} = 17.86$$

We truncate df to 17. Appendix Table 4 shows that the area under the *t* curve with 17 df to the right of 1.1 is .143, so *P*-value = .143.

9. Conclusion: Because *P*-value = .143 > .05, we fail to reject H_0. There is not convincing evidence to support the claim that mean bone mineral density is lower for women who used oral contraceptives.

■

Many statistical computer packages can perform the calculations for the two sample *t* test. The accompanying partial SPSS output shows summary statistics for the two groups ("Use = no" refers to the group that never used contraceptives and "Use = yes" refers to the group that used contraceptives). The second part of the output gives the number of degrees of freedom, the test statistic value, and a two-sided *P*-value. Since the test in Example 11.2 is a one-sided test, we need to divide the two-sided

P-value in half to obtain the correct value for our test, which is $\frac{.272}{2} = .136$. This *P*-value differs from the value in Example 11.2 because Appendix Table 4 gives only tail areas for *t* values to one decimal place and so we rounded the test statistic to 1.1. As a consequence, the *P*-value given in the example is only approximate; the *P*-value from SPSS is more accurate.

Group Statistics					
	Use	N	Mean	Std. Deviation	Std. Error Mean
BMD	**no**	10	1.0800	.15972	.05051
	yes	10	1.0040	.13906	.04397

	t	df	Sig. (2-tailed)	Mean Difference	Std. Error Difference	95% Confidence Interval of the Difference	
						Lower	Upper
Equal Variances Not Assumed	1.135	17.665	.272	.07600	.06697	−.06489	.21689

■ Comparing Treatments ....

When an experiment is carried out to compare two treatments (or to compare a single treatment with a control), the investigator is interested in the effect of the treatments on some response variable. The treatments are "applied" to individuals (as in an experiment to compare two different medications for decreasing blood pressure) or objects (as in an experiment to compare two different baking temperatures on the density of bread), and the value of some response variable (e.g., blood pressure, density) is recorded. Based on the resulting data, the investigator might wish to determine whether there is a difference in the mean response for the two treatments.

Important concept

In many actual experimental situations, the individuals or objects to which the treatments will be applied are not selected at random from some larger population. A consequence of this is that it is not possible to generalize the results of the experiment to some larger population. However, *if the experimental design provides for random assignment of treatments to the individuals or objects used in the experiment (or for random assignment of the individuals or objects to treatments), it is possible to test hypotheses about treatment differences.*

It is common practice to use the two-sample *t* test statistic previously described if the experiment employs random assignment and if either the sample sizes are large or it is reasonable to think that the treatment response distributions (the distributions of response values that would result if the treatments were applied to a *very* large number of individuals or objects) are approximately normal.

Two-Sample *t* Test for Comparing Two Treatments

When

1. *treatments are randomly assigned* to individuals or objects (or vice versa; that is, individuals or objects are randomly assigned to treatments), and
2. the *sample sizes are large* (generally 30 or larger)
   or the *treatment response distributions are approximately normal,*

the two-sample *t* test can be used to test H_0: $\mu_1 - \mu_2 = $ *hypothesized value*, where μ_1 and μ_2 represent the mean response for treatments 1 and 2, respectively.

In this case, these two conditions replace the assumptions previously stated for comparing two population means. The validity of the assumption of normality of the treatment response distributions can be assessed by constructing a normal probability plot or a boxplot of the response values in each sample.

When the two-sample *t* test is used to compare two treatments when the individuals or objects used in the experiment are not randomly selected from some population, it is only an approximate test (the reported *P*-values are only approximate). However, this is still the most common way to analyze such data.

Example 11.3 Reading Emotions

© Royalty-Free/Getty Images

The paper "How Happy Was I, Anyway? A Retrospective Impact Bias" (*Social Cognition* [2003]: 421–446) reported on an experiment designed to assess the extent to which people rationalize poor performance. In this study, 246 college undergraduates were assigned at random to one of two groups—a negative feedback group or a positive feedback group. Each participant took a test in which they were asked to guess the emotions displayed in photographs of faces. At the end of the test, those in the negative feedback group were told that they had correctly answered 21 of the 40 items and were assigned a "grade" of D. Those in the positive feedback group were told that they had answered 35 of 40 correctly and were assigned an A grade. After a brief time, participants were asked to answer two sets of questions. One set of questions asked about the validity of the test and the other set of questions asked about the importance of being able to read faces. The researchers hypothesized that those in the negative feedback group would tend to rationalize their poor performance by rating both the validity of the test and the importance of being a good face reader lower than those in the positive feedback group. Do the data from this experiment support the researchers' hypotheses?

Group	Sample Size	Test Validity Rating		Face Reading Importance Rating	
		Mean	Standard Deviation	Mean	Standard Deviation
Negative feedback	123	5.51	.79	5.36	1.00
Positive feedback	123	6.95	1.09	6.62	1.19

We will test the relevant hypotheses using a significance level of .01, beginning with the hypotheses about the test validity rating.

1. Let μ_1 denote the mean test validity score for the negative feedback group and define μ_2 analogously for the positive feedback group. Then $\mu_1 - \mu_2$ is the difference between the mean test validity scores for the two treatment groups.
2. H_0: $\mu_1 - \mu_2 = 0$
3. H_a: $\mu_1 - \mu_2 < 0$
4. Significance level: $\alpha = .01$
5. Test statistic: $t = \dfrac{\bar{x}_1 - \bar{x}_2 - \text{hypothesized value}}{\sqrt{\dfrac{s_1^2}{n_1} + \dfrac{s_2^2}{n_2}}} = \dfrac{\bar{x}_1 - \bar{x}_2 - 0}{\sqrt{\dfrac{s_1^2}{n_1} + \dfrac{s_2^2}{n_2}}}$
6. Assumptions: Subjects were randomly assigned to the treatment groups, and both sample sizes are large, so use of the two-sample t test is reasonable.
7. Calculation: $t = \dfrac{(5.51 - 6.95) - 0}{\sqrt{\dfrac{(.79)^2}{123} + \dfrac{(1.09)^2}{123}}} = \dfrac{-1.44}{0.1214} = -11.86$
8. P-value: We first compute the df for the two-sample t test:

$$V_1 = \frac{s_1^2}{n_1} = .0051 \qquad V_2 = \frac{s_2^2}{n_2} = .0097$$

$$\text{df} = \frac{(V_1 + V_2)^2}{\dfrac{V_1^2}{n_1 - 1} + \dfrac{V_2^2}{n_2 - 1}} = \frac{(.0051 + .0097)^2}{\dfrac{(.0051)^2}{122} + \dfrac{(.0097)^2}{122}} = \frac{.000219}{.000001} = 219$$

This is a lower-tailed test, so the P-value is the area under the t curve with df $= 219$ and to the left of -11.86. Since -11.86 is so far out in the lower tail of this t curve, P-value ≈ 0.

9. Conclusion: Since P-value $\leq \alpha$, H_0 is rejected. There is evidence that the mean validity rating score for the positive feedback group is higher. The data support the conclusion that those who received negative feedback did not rate the validity of the test, on average, as highly as those who thought they had done well on the test.

We will use MINITAB to test the researchers' hypothesis that those in the negative feedback group would also not rate the importance of being able to read faces as highly as those in the positive group.

1. Let μ_1 denote the mean face reading importance rating for the negative feedback group and define μ_2 analogously for the positive feedback group. Then $\mu_1 - \mu_2$ is the difference between the mean face reading ratings for the two treatment groups.
2. H_0: $\mu_1 - \mu_2 = 0$
3. H_a: $\mu_1 - \mu_2 < 0$
4. Significance level: $\alpha = .01$
5. Test statistic: $t = \dfrac{\bar{x}_1 - \bar{x}_2 - \text{hypothesized value}}{\sqrt{\dfrac{s_1^2}{n_1} + \dfrac{s_2^2}{n_2}}} = \dfrac{\bar{x}_1 - \bar{x}_2 - 0}{\sqrt{\dfrac{s_1^2}{n_1} + \dfrac{s_2^2}{n_2}}}$
6. Assumptions: Subjects were randomly assigned to the treatment groups, and both sample sizes are large, so use of the two-sample t test is reasonable.

7. Calculation: MINITAB output is shown here. From the output, $t = -8.99$.

Two-Sample T-Test and CI

Sample	N	Mean	StDev	SE Mean
1	123	5.36	1.00	0.090
2	123	6.62	1.19	0.11

Difference = mu (1) − mu (2)
Estimate for difference: −1.26000
95% upper bound for difference: −1.02856
T-Test of difference = 0 (vs <): T-Value = −8.99   P-Value = 0.000   DF = 236

8. *P*-value: From the MINITAB output, *P*-value = 0.000.
9. Conclusion: Since *P*-value $\leq \alpha$, H_0 is rejected. There is evidence that the mean face reading importance rating for the positive feedback group is higher.

■

You have probably noticed that evaluating the formula for number of degrees of freedom for the two-sample *t* test involves quite a bit of arithmetic. An alternative approach is to compute a conservative estimate of the *P*-value—one that is close to but larger than the actual *P*-value. If H_0 is rejected using this conservative estimate, then it will also be rejected if the actual *P*-value is used. *A conservative estimate of the P-value for the two-sample t test can be found by using the t curve with the number of degrees of freedom equal to the smaller of $(n_1 - 1)$ and $(n_2 - 1)$.*

■ The Pooled *t* Test

The two-sample *t* test procedure just described is appropriate when it is reasonable to assume that the population distributions are approximately normal. If it is also known that the variances of the two populations are equal ($\sigma_1^2 = \sigma_2^2$), an alternative procedure known as the *pooled t test* can be used. This test procedure combines information from both samples to obtain a "pooled" estimate of the common variance and then uses this pooled estimate of the variance in place of s_1^2 and s_2^2 in the *t* test statistic. This test procedure was widely used in the past, but it has fallen into some disfavor because it is quite sensitive to departures from the assumption of equal population variances. If the population variances are equal, the pooled *t* procedure has a slightly better chance of detecting departures from H_0 than does the two-sample *t* test of this section. However, *P*-values based on the pooled *t* procedure can be seriously in error if the population variances are not equal, so, in general, the two-sample *t* procedure is a better choice than the pooled *t* test.

■ Comparisons and Causation

If the assignment of treatments to the individuals or objects used in a comparison of treatments is not made by the investigators, the study is observational. As an example, the article "Lead and Cadmium Absorption Among Children near a Nonferrous Metal Plant" (*Environmental Research* [1978]: 290–308) reported data on blood lead concentrations for two different samples of children. The first sample was drawn from a population residing within 1 km of a lead smelter, whereas those in the second sample were selected from a rural area much farther from the smelter. It was the parents of the children, rather than the investigators, who determined whether the children would be in the close-to-smelter group or the far-from-smelter group. As a second example, a letter in the *Journal of the American Medical Association* (May 19, 1978) reported on a comparison of doctors' longevity after medical school graduation for those with an academic affiliation and those in private practice. (The letter writer's stated objective was to see whether "publish or perish" really meant "publish *and* perish.") Here again, an investigator did not start out with a group of doctors, assigning some to academic and others to nonacademic careers. The doctors themselves selected their groups.

The difficulty with drawing conclusions based on an observational study is that a statistically significant difference may be due to some underlying factors that have not been controlled rather than to conditions that define the groups. Does the type of medical practice itself have an effect on longevity, or is the observed difference in lifetimes caused by other factors, which themselves led graduates to choose academic or nonacademic careers? Similarly, is the observed difference in blood lead concentration levels due to proximity to the smelter? Perhaps other physical and socioeconomic factors are related both to choice of living area and to concentration.

In general, rejection of H_0: $\mu_1 - \mu_2 = 0$ in favor of H_a: $\mu_1 - \mu_2 > 0$ suggests that, on average, higher values of the variable are *associated* with individuals in the first population or receiving the first treatment than with those in the second population or receiving the second treatment. But *association does not imply causation.* Strong statistical evidence for a causal relationship can be built up over time through many different comparative studies that point to the same conclusions (as in the many investigations linking smoking to lung cancer). A **randomized controlled experiment**, in which investigators assign subjects at random to the treatments or conditions being compared, is particularly effective in suggesting causality. With such random assignment, the investigator and other interested parties can have more confidence in the conclusion that an observed difference is caused by the difference in treatments or conditions. (Recall the discussion in Chapter 2 of the role of randomization in designing an experiment.)

■ A Confidence Interval

A confidence interval for $\mu_1 - \mu_2$ is easily obtained from the basic t variable of this section. Both the derivation of and the formula for the interval are similar to those of the one-sample t interval discussed in Chapter 9.

The Two-Sample *t* Confidence Interval for the Difference Between Two Population or Treatment Means

The general formula for a confidence interval for $\mu_1 - \mu_2$ when

1. the two samples are *independently chosen random samples,* and
2. the *sample sizes are both large* (generally $n_1 \geq 30$ and $n_2 \geq 30$)
   OR
   the *population distributions are approximately normal*

(continued)

is

$$\bar{x}_1 - \bar{x}_2 \pm (t \text{ critical value}) \sqrt{\frac{s_1^2}{n_1} + \frac{s_2^2}{n_2}}$$

The t critical value is based on

$$df = \frac{(V_1 + V_2)^2}{\dfrac{V_1^2}{n_1 - 1} + \dfrac{V_2^2}{n_2 - 1}} \quad \text{where } V_1 = \frac{s_1^2}{n_1} \text{ and } V_2 = \frac{s_2^2}{n_2}$$

df should be truncated (rounded down) to an integer. The t critical values for the usual confidence levels are given in Appendix Table 3.

For a comparison of two treatments, when

1. *treatments are randomly assigned* to individuals or objects (or vice versa), and
2. the *sample sizes are large* (generally 30 or larger) or the *treatment response distributions are approximately normal*,

the two-sample t confidence interval formula can be used to estimate $\mu_1 - \mu_2$.

··

Example 11.4 Effect of Talking on Blood Pressure

● Does talking elevate blood pressure, contributing to the tendency for blood pressure to be higher when measured in a doctor's office than when measured in a less stressful environment? (This well-documented effect is called the "white coat effect.") The article "The Talking Effect and 'White Coat' Effect in Hypertensive Patients: Physical Effort or Emotional Content" (*Behavioral Medicine* [2001]: 149–157) described a study in which patients with high blood pressure were randomly assigned to one of two groups. Those in the first group (the talking group) were asked questions about their medical history and about the sources of stress in their lives in the minutes before their blood pressure was measured. Those in the second group (the counting group) were asked to count aloud from 1 to 100 four times before their blood pressure was measured. The following data values for diastolic blood pressure (in millimeters of Hg) are consistent with summary quantities appearing in the paper:

Talking	104	110	107	112	108	103	108	118
		$n_1 = 8$		$\bar{x}_1 = 108.75$		$s_1 = 4.74$		
Counting	110	96	103	98	100	109	97	105
		$n_2 = 8$		$\bar{x}_2 = 102.25$		$s_2 = 5.39$		

Subjects were randomly assigned to the two treatments. Because both sample sizes are small, we must first investigate whether it is reasonable to assume that the diastolic blood pressure distributions are approximately normal for the two treatments. There are no outliers in either data set, and the boxplots are reasonably symmetric, suggesting that the assumption of approximate normality is reasonable.

▌▎ Step-by-step technology instructions available   ● Data set available online

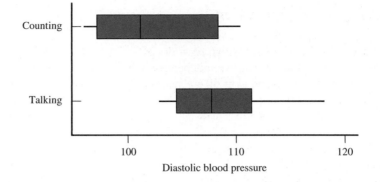

To estimate $\mu_1 - \mu_2$, the difference in mean diastolic blood pressure for the two treatments, we will calculate a 95% confidence interval.

$$V_1 = \frac{s_1^2}{n_1} = \frac{(4.74)^2}{8} = 2.81 \qquad V_2 = \frac{s_2^2}{n_2} = \frac{(5.39)^2}{8} = 3.63$$

$$\text{df} = \frac{(V_1 + V_2)^2}{\dfrac{V_1^2}{n_1 - 1} + \dfrac{V_2^2}{n_2 - 1}} = \frac{(2.81 + 3.63)^2}{\dfrac{(2.81)^2}{7} + \dfrac{(3.63)^2}{7}} = \frac{41.47}{3.01} = 13.78$$

Truncating to an integer gives df = 13. In the 13-df row of Appendix Table 3, the t critical value for a 95% confidence level is 2.16. The interval is then

$$\bar{x}_1 - \bar{x}_2 \pm (t \text{ critical value})\sqrt{\frac{s_1^2}{n_1} + \frac{s_2^2}{n_2}}$$

$$= (108.75 - 102.25) \pm (2.16)\sqrt{\frac{(4.74)^2}{8} + \frac{(5.39)^2}{8}}$$

$$= 6.5 \pm (2.16)(2.54)$$

$$= 6.5 \pm 5.49$$

$$= (1.01, 11.99)$$

This interval is rather wide, because the two sample variances are large and the sample sizes are small. Notice that the interval does not include 0, so 0 is not one of the plausible values for $\mu_1 - \mu_2$. Based on this computed interval, we estimate that the mean diastolic blood pressure when talking is higher than the mean when counting by somewhere between 1.01 and 11.99 mm Hg. This result supports the existence of a talking effect over and above the white coat effect. The 95% confidence level means that we used a method to produce this estimate that correctly captures the true value of $\mu_1 - \mu_2$ 95% of the time in repeated sampling.

Most statistical computer packages can compute the two-sample t confidence interval. MINITAB was used to construct a 95% confidence interval using the data of this example; the resulting output is shown here:

Two-Sample T-Test and CI: Talking, Counting

Two-sample T for Talking vs Counting

	N	Mean	StDev	SE Mean
Talking	8	108.75	4.74	1.7
Counting	8	102.25	5.39	1.9

95% CI for difference: (1.01, 11.99)

Exercises 11.1–11.28

Turn to the Explanded Answers Section for answers not shown next to exercises.

11.1 Consider two populations for which $\mu_1 = 30$, $\sigma_1 = 2$, $\mu_2 = 25$, and $\sigma_2 = 3$. Suppose that two independent random samples of sizes $n_1 = 40$ and $n_2 = 50$ are selected. Describe the approximate sampling distribution of $\bar{x}_1 - \bar{x}_2$ (center, spread, and shape). *Approximately normal with mean 5 and standard deviation 0.529*

11.2 An individual can take either a scenic route to work or a nonscenic route. She decides that use of the non-scenic route can be justified only if it reduces true average travel time by more than 10 min.
a. If μ_1 refers to the scenic route and μ_2 to the nonscenic route, what hypotheses should be tested?
b. If μ_1 refers to the nonscenic route and μ_2 to the scenic route, what hypotheses should be tested? *H_0: $\mu_1 - \mu_2 = -10$ H_a: $\mu_1 - \mu_2 < -10$*

11.3 ▼ Reduced heart rate variability (HRV) is known to be a predictor of mortality after a heart attack. One measure of HRV is the average of normal-to-normal beat interval (in milliseconds) for a 24-hour time period. Twenty-two heart attack patients who were dog owners and 80 heart attack patients who did not own a dog participated in a study of the effect of pet ownership on HRV, resulting in the summary statistics shown in the accompanying table ("Relationship Between Pet Ownership and Heart Rate Variability in Patients with Healed Myocardial Infarcts," *The American Journal of Cardiology* [2003]: 718–721).

	Measure of HRV (average normal-to-normal beat interval)	
	Mean	Standard Deviation
Owns Dog	873	136
Does Not Own Dog	800	134

a. The authors of this paper used a two-sample t test to test H_0: $\mu_1 - \mu_2 = 0$ versus H_a: $\mu_1 - \mu_2 \neq 0$. What assumptions must be reasonable in order for this to be an appropriate method of analysis?
b. The paper indicates that the null hypothesis from Part (a) was rejected and reports that the P-value is less than .05. Carry out the two-sample t test. Is your conclusion consistent with that of the paper?

11.4 Each person in a random sample of 228 male teenagers and a random sample of 306 female teenagers was asked how many hours he or she spent online in a typical week (Ipsos, January 25, 2006). The sample mean and standard deviation were 15.1 hours and 11.4 hours for males and 14.1 and 11.8 for females.
a. The standard deviation for each of the samples is large, indicating a lot of variability in the responses to the question. Explain why it is not reasonable to think that the distribution of responses would be approximately normal for either the population of male teenagers or the population of female teenagers. Hint: The number of hours spent online in a typical week cannot be negative.
b. Given your response to Part (a), would it be appropriate to use the two-sample t test to test the null hypothesis that there is no difference in the mean number of hours spent online in a typical week for male teenagers and female teenagers? Explain why or why not.
c. If appropriate, carry out a test to determine if there is convincing evidence that the mean number of hours spent online in a typical week is greater for male teenagers than for female teenagers. Use a .05 significance level. *$t = 0.988$, df = 505, P-value ≈ 0.159, fail to reject H_0*

11.5 Each person in random samples of 247 male and 253 female working adults living in Calgary, Canada was asked how long, in minutes, his or her typical daily commute was ("Calgary Herald Traffic Study," Ipsos, September 17, 2005). Use the accompanying summary statistics and an appropriate hypothesis test to determine if there is convincing evidence that the mean commute times for male and female working Calgary residents differ. Use a significance level of .05. *$t = 1.06$, df = 497, P-value ≈ 0.2876, fail to reject H_0*

Males			Females		
Sample Size	$\bar{x}$	s	Sample Size	$\bar{x}$	s
247	29.6	24.3	253	27.3	24.0

11.6 The paper "Effects of Fast-Food Consumption on Energy Intake and Diet Quality Among Children in a National Household Survey" (*Pediatrics* [2004]: 112–118) investigated the effect of fast-food consumption on other dietary variables. For a sample of 663 teens who reported that they did not eat fast food during a typical day, the mean daily calorie intake was 2258 and the sample standard deviation was 1519. For a sample of 413 teens who reported that they did eat fast food on a typical day, the mean calorie intake was 2637 and the standard deviation was 1138.

a. What assumptions about the two samples must be reasonable in order for the use of the two-sample t confidence interval to be appropriate?

b. Use the given information to estimate the difference in mean daily calorie intake for teens who do eat fast food on a typical day and those who do not.   For a 95% confidence level, $(-538.42, -219.58)$.

11.7 In a study of malpractice claims where a settlement had been reached, two random samples were selected: a random sample of 515 closed malpractice claims that were found not to involve medical errors and a random sample of 889 claims that were found to involve errors (*New England Journal of Medicine* [2006]: 2024–2033). The following statement appeared in the referenced paper: "When claims not involving errors were compensated, payments were significantly lower on average than were payments for claims involving errors ($313,205 vs. $521,560, $P = 0.004$)."

a. What hypotheses must the researchers have tested in order to reach the stated conclusion?

b. Which of the following could have been the value of the test statistic for the hypothesis test? Explain your reasoning.

    i. $t = 5.00$ ii. $t = 2.65$ iii. $t = 2.33$ iv. $t = 1.47$

11.8 In a study of the effect of college student employment on academic performance, the following summary statistics for GPA were reported for a sample of students who worked and for a sample of students who did not work (*University of Central Florida Undergraduate Research Journal,* Spring 2005):

	Sample Size	Mean GPA	Standard Deviation
Students Who Are Employed	184	3.12	.485
Students Who Are Not Employed	114	3.23	.524

The samples were selected at random from working and nonworking students at the University of Central Florida. Does this information support the hypothesis that for students at this university, those who are not employed have a higher mean GPA than those who are employed? $t = -1.81$, df $= 227$, P-value ≈ 0.036, reject H_0

11.9 ● Acrylic bone cement is commonly used in total joint replacement to secure the artificial joint. Data on the force (measured in Newtons, N) required to break a cement bond under two different temperature conditions and in two different mediums appear in the accompanying table. (These data are consistent with summary quantities appearing in the paper "Validation of the Small-Punch Test as a Technique for Characterizing the Mechanical Properties of Acrylic Bone Cement" (*Journal of Engineering in Medicine* [2006]: 11–21).)

Temperature	Medium	Data on Breaking Force
22 degrees	Dry	100.8, 141.9, 194.8, 118.4, 176.1, 213.1
37 degrees	Dry	302.1, 339.2, 288.8, 306.8, 305.2, 327.5
22 degrees	Wet	385.3, 368.3, 322.6, 307.4, 357.9, 321.6
37 degrees	Wet	363.5, 377.7, 327.7, 331.9, 338.1, 394.6

a. Estimate the difference between the mean breaking force in a dry medium at 37 degrees and the mean breaking force at the same temperature in a wet medium using a 90% confidence interval. $(-69, -19)$

b. Is there sufficient evidence to conclude that the mean breaking force in a dry medium at the higher temperature is greater than the mean breaking force at the lower temperature by more than 100 N? Test the relevant hypotheses using a significance level of .10. $t = 2.76$, df $= 6$, P-value ≈ 0.016, reject H_0

11.10 ● The article "Genetic Tweak Turns Promiscuous Animals Into Loyal Mates" (*Los Angeles Times,* June 17, 2004) summarizes the results of a research study that appeared in the June 2004 issue of *Nature.* In this study, 11 male meadow voles who had a single gene introduced into a specific part of the brain were compared to 20 male meadow voles who did not undergo this genetic manipulation. All of the voles were paired with a receptive female partner for 24 hours. At the end of the 24 hour period, the male was placed in a situation where he could choose either the partner from the previous 24 hours or a different female. The percentage of the time during a three-hour trial that the male spent with his previous partner was recorded. The accompanying data are approximate values read from a graph that appeared in the *Nature* article. Do these data support the researchers' hypothesis that the mean percentage of the time spent with the previous partner is significantly greater for genetically altered voles than for voles that did not have the gene introduced? Test the relevant hypotheses using $\alpha = .05$. $t = 3.34$, df $= 28$, P-value ≈ 0.001, reject H_0

	Percent of Time Spent with Previous Partner							
Genetically Altered	59	62	73	80	84	85	89	92
	92	93	100					
Not Genetically Altered	2	5	13	28	34	40	48	50
	51	54	60	67	70	76	81	84
	85	92	97	99				

11.11 A newspaper story headline reads "Gender Plays Part in Monkeys' Toy Choices, Research Finds—Like Humans, Male Monkeys Choose Balls and Cars, While Females Prefer Dolls and Pots" (*Knight Ridder Newspapers*, December 8, 2005). The article goes on to summarize findings published in the paper "Sex Differences in Response to Children's Toys in Nonhuman Primates" (*Evolution and Human Behavior* [2002] 467–479). Forty-four male monkeys and 44 female monkeys were each given a variety of toys, and the time spent playing with each toy was recorded. The accompanying table gives means and standard deviations (approximate values read from graphs in the paper) for the percentage of the time that a monkey spent playing with a particular toy. Assume that it is reasonable to regard these two samples of 44 monkeys as representative of the population of male and female mon-

keys. Use a .05 significance level for any hypothesis tests that you carry out when answering the various parts of this exercise.
a. The police car was considered a "masculine toy." Do these data provide convincing evidence that the mean percentage of the time spent playing with the police car is greater for male monkeys than for female monkeys?
b. The doll was considered a "feminine toy." Do these data provide convincing evidence that the mean percentage of the time spent playing with the doll is greater for female monkeys than for male monkeys?
c. The furry dog was considered a "neutral toy." Do these data provide convincing evidence that the mean percentage of the time spent playing with the furry dog is not the same for male and female monkeys?
d. Based on the conclusions from the hypothesis tests of Parts (a)–(c), is the quoted newspaper story headline a reasonable summary of the findings? Explain.
e. Explain why it would be inappropriate to use the two-sample *t* test to decide if there was evidence that the mean percentage of the time spent playing with the police car and the mean percentage of the time spent playing with the doll is not the same for female monkeys.

11.12 ● The paper "The Observed Effects of Teenage Passengers on the Risky Driving Behavior of Teenage Drivers" (*Accident Analysis and Prevention* [2005]: 973–982) investigated the driving behavior of teenagers by observing their vehicles as they left a high school parking lot and then again at a site approximately $\frac{1}{2}$ mile from the school. Assume that it is reasonable to regard the teen drivers in this study as representative of the population of teen drivers. Use a .01 level of significance for any hypothesis tests.

Table for Exercise 11.11

		Percent of Time					
		Female Monkeys			Male Monkeys		
		n	Sample Mean	Sample Standard Deviation	*n*	Sample Mean	Sample Standard Deviation
Toy	Police Car	44	8	4	44	18	5
	Doll	44	20	4	44	9	2
	Furry Dog	44	20	5	44	25	5

a. Data consistent with summary quantities appearing in the paper are given in the accompanying table. The measurements represent the difference between the observed vehicle speed and the posted speed limit (in miles per hour) for a sample of male teenage drivers and a sample of female teenage drivers. Do these data provide convincing support for the claim that, on average, male teenage drivers exceed the speed limit by more than do female teenage drivers? $t = 2.99$, df $= 14$, P-value ≈ 0.004, reject H_0

Amount by Which Speed Limit Was Exceeded

Male Driver	Female Driver
1.3	−0.2
1.3	0.5
0.9	1.1
2.1	0.7
0.7	1.1
1.3	1.2
3	0.1
1.3	0.9
0.6	0.5
2.1	0.5

b. Consider the average miles per hour over the speed limit for teenage drivers with passengers shown in the accompanying table. For purposes of this exercises, suppose that each driver-passenger combination mean is based on a sample of size $n = 40$ and that all sample standard deviations are equal to .8.

	Male Passenger	Female Passenger
Male Driver	5.2	.3
Female Driver	2.3	.6

i. Is there sufficient evidence to conclude that the average number of miles per hour over the speed limit is greater for male drivers with male passengers than it is for male drivers with female passengers?

ii. Is there sufficient evidence to conclude that the average number of miles per hour over the speed limit is greater for female drivers with male passengers than it is for female drivers with female passengers?

iii. Is there sufficient evidence to conclude that the average number of miles per hour over the speed limit is smaller for male drivers with female passengers than it is for female drivers with male passengers?

c. Write a few sentences commenting on the effect of a passenger on a teen driver's speed.

11.13 Many people take ginkgo supplements advertised to improve memory. Are these over-the-counter supplements effective? In a study reported in the paper "Ginkgo for Memory Enhancement" (*Journal of the American Medical Association* [2002]: 835–840), elderly adults were assigned at random to either a treatment group or a control group. The 104 participants who were assigned to the treatment group took 40 mg of ginkgo three times a day for six weeks. The 115 participants assigned to the control group took a placebo pill three times a day for six weeks. At the end of six weeks, the Wechsler Memory Scale (a test of short-term memory) was administered. Higher scores indicate better memory function. Summary values are given in the following table.

	n	$\bar{x}$	s
Ginkgo	104	5.6	.6
Placebo	115	5.5	.6

Based on these results, is there evidence that taking 40 mg of ginkgo three times a day is effective in increasing mean performance on the Wechsler Memory Scale? Test the relevant hypotheses using $\alpha = .05$. $t = 1.23$, df $= 214$, P-value ≈ 0.110, fail to reject H_0

11.14 Fumonisins are environmental toxins produced by a type of mold and have been found in corn and in products made from raw corn. The Center for Food Safety and Applied Nutrition provided recommendations on allowable fumonisin levels in human food and in animal feed based on a study of corn meal. The study compared corn meal made from partially degermed corn (corn that has had the germ, the part of the kernel located at the bottom center of the kernel that is used to produce corn oil, partially removed) and corn meal made from corn that has not been degermed. Specimens of corn meal were analyzed and the total fumonisin level (ppm) was determined for each specimen. Summary statistics for total fumonisin level from the U.S. Food and Drug Administration's web site are given here.

	$\bar{x}$	s
Partially Degermed	.59	1.01
Not Degermed	1.21	1.71

a. If the given means and standard deviations had been based on a random sample of 10 partially degermed specimens and a random sample of 10 specimens made from corn that was not degermed, explain why it would not be appropriate to carry out a two-sample t test to determine if there is a significant difference in the mean fumonisin level for the two types of corn meal.

b. Suppose instead that each of the samples had included 50 corn meal specimens. Explain why it would now be reasonable to carry out a two-sample t test.

c. Assuming that each sample size was 50, carry out a test to determine if there is a significant difference in mean fumonisin level for the two types of corn meal. Use a significance level of .01. $t = -2.21$, df = 79, P-value ≈ 0.032, fail to reject H_0

11.15 Do faculty and students have similar perceptions of what types of behavior are inappropriate in the classroom? This question was examined by the author of the article "Faculty and Student Perceptions of Classroom Etiquette" (*Journal of College Student Development* (1998): 515–516). Each individual in a random sample of 173 students in general education classes at a large public university was asked to judge various behaviors on a scale from 1 (totally inappropriate) to 5 (totally appropriate). Individuals in a random sample of 98 faculty members also rated the same behaviors. The mean rating for two of the behaviors studied are shown here (the means are consistent with data provided by the author of the article). The sample standard deviations were not given, but for purposes of this exercise, assume that they are all equal to 1.0.

Student Behavior	Student Mean Rating	Faculty Mean Rating
Wearing hats in the classroom	2.80	3.63
Addressing instructor by first name	2.90	2.11
Talking on a cell phone	1.11	1.10

a. Is there sufficient evidence to conclude that the mean "appropriateness" score assigned to wearing a hat in class differs for students and faculty?

b. Is there sufficient evidence to conclude that the mean "appropriateness" score assigned to addressing an instructor by his or her first name is higher for students than for faculty? $t = -6.25$, df = 201, P-value ≈ 0, reject H_0

c. Is there sufficient evidence to conclude that the mean "appropriateness" score assigned to talking on a cell phone differs for students and faculty? Does the result of your test imply that students and faculty consider it acceptable to talk on a cell phone during class? $t = 0.07$, df = 201, P-value ≈ 0.92, fail to reject H_0

11.16 Does clear-cutting of trees in an area cause local extinction of the tailed frog? We don't really know, but an article that begins to address that issue attempts to quantify aspects of the habitat and microhabitat of streams with and without tailed frogs. The following data are from "Distribution and Habitat of *Ascaphus truei* in Streams on Managed, Young Growth Forests in North Coastal California" (*Journal of Herpetology* [1999]: 71–79).

	Sites With Tailed Frogs		
	n	$\bar{x}$	s
Habitat Characteristics			
Stream gradient (%)	18	9.1	6.00
Water temp. (°C)	18	12.2	1.71
Microhabitat Characteristics			
Depth (cm)	82	5.32	2.27

	Sites Without Tailed Frogs		
	n	$\bar{x}$	s
Habitat Characteristics			
Stream gradient (%)	31	5.9	6.29
Water temp. (°C)	31	12.8	1.33
Microhabitat Characteristics			
Depth (cm)	267	8.46	5.95

Assume that the habitats and microhabitats examined were selected independently and at random from those that have tailed frogs and those that do not have tailed frogs. Use a significance level of .01 for any hypothesis tests required to answer the following questions.

a. Is there evidence of a difference in mean stream gradient between streams with and streams without tailed frogs? (*Note*: Assume that it is reasonable to think that the distribution of stream gradients is approximately normal for both types of streams. It is possible for gradient percentage to be either positive or negative, so the fact that the mean -2 standard deviations is negative is not, by itself, an indication of nonnormality in this case.)

b. Is there evidence of a difference in the mean water temperature between streams with and streams without tailed frogs?

c. The article reported that the two depth distributions are both quite skewed. Does this imply that it would be unreasonable to use the independent samples t test to compare the mean depth for the two types of streams? If not, carry out a test to determine whether the sample data support the claim that the mean depth of streams without tailed frogs is greater than the mean depth of streams with tailed frogs. $t = -7.1$, df $= 332$, P-value ≈ 0, reject H_0

11.17 Are girls less inclined to enroll in science courses than boys? One study ("Intentions of Young Students to Enroll in Science Courses in the Future: An Examination of Gender Differences" *Science Education* [1999]: 55–76) asked randomly selected fourth-, fifth-, and sixth-graders how many science courses they intend to take. The following data were obtained:

	n	Mean	Standard Deviation
Males	203	3.42	1.49
Females	224	2.42	1.35

Calculate a 99% confidence interval for the difference between males and females in mean number of science courses planned. Interpret your interval. Based on your interval, how would you answer the question posed at the beginning of the exercise?

11.18 ● The article "Movement and Habitat Use by Lake Whitefish During Spawning in a Boreal Lake: Integrating Acoustic Telemetry and Geographic Information Systems" (*Transactions of the American Fisheries Society*, [1999]: 939–952) included the accompanying data on weights of 10 fish caught in 1995 and 10 caught in 1996.

1995	776	580	539	648	538	891	673	783	571	627
1996	571	627	727	727	867	1042	804	832	764	727

Is it appropriate to use the independent samples t test to compare the mean weight of fish for the 2 years? Explain why or why not.

11.19 ▼ Techniques for processing poultry were examined in the article "Texture Profiles of Canned Boned Chicken as Affected by Chilling-Aging Times" (*Poultry Science* [1994]: 1475–1478). Whole chickens were chilled 0, 2, 8, or 24 hours before being cooked and canned. To determine whether the chilling time affected the texture of the canned chicken, samples were evaluated by trained tasters. One characteristic of interest was hardness. The accompanying summary quantities were obtained. Each mean is based on 36 ratings.

	Chilling Time			
	0 hr	2 hr	8 hr	24 hr
Mean Hardness	7.52	6.55	5.70	5.65
Standard Deviation	.96	1.74	1.32	1.50

a. Do the data suggest that there is a difference in mean hardness for chicken chilled 0 hours before cooking and chicken chilled 2 hours before cooking? Use $\alpha = .05$.

b. Do the data suggest that there is a difference in mean hardness for chicken chilled 8 hours before cooking and chicken chilled 24 hours before cooking? Use $\alpha = .05$.

c. Use a 90% confidence interval to estimate the difference in mean hardness for chicken chilled 2 hours before cooking and chicken chilled 8 hours before cooking. (0.242, 1.458)

11.20 The Rape Myth Acceptance Scale (RMAS) was administered to 333 male students at a public university. Of these students, 155 were randomly selected from students who did not belong to a fraternity, and 178 were randomly selected from students who belonged to a fraternity. The higher the score on the RMAS, the greater the acceptance of rape myths, so lower scores are considered desirable. The accompanying data appeared in the article "Rape Supportive Attitudes Among Greek Students Before and After a Date Rape Prevention Program" (*Journal of College Student Development* [1994]: 450–455). Do the data support the researchers' claim that the mean RMA score is lower for fraternity members? Use $\alpha = .05$. $t = -2.77$, df $= 330$, P-value ≈ 0.002, reject H_0

Sample	n	Mean RMAS Score	sd
Fraternity members	178	25.63	6.16
Not fraternity members	155	27.40	5.51

Bold exercises answered in back   ● Data set available online but not required   ▼ Video solution available

Table for Exercise 11.24

	Data								n	$\bar{x}$	s
Advanced	44.70	26.31	55.75	28.54	46.99	39.46			6	40.3	11.3
Intermediate	15.58	19.16	24.13	10.56	32.88	21.47	14.32	33.09	8	21.4	8.3

11.21 The discharge of industrial wastewater into rivers affects water quality. To assess the effect of a particular power plant on water quality, 24 water specimens were taken 16 km upstream and 4 km downstream of the plant. Alkalinity (mg/L) was determined for each specimen, resulting in the summary quantities in the accompanying table. Do the data suggest that the true mean alkalinity is higher downstream than upstream by more than 50 mg/L? Use a .05 significance level. $t = 113.17$, df = 45, P-value ≈ 0, reject H_0

Location	n	Mean	Standard Deviation
Upstream	24	75.9	1.83
Downstream	24	183.6	1.70

11.22 According to the Associated Press (*San Luis Obispo Telegram-Tribune,* June 23, 1995), a study by Italian researchers indicated that low cholesterol and depression were linked. The researchers found that among 331 randomly selected patients hospitalized because they had attempted suicide, the mean cholesterol level was 198. The mean cholesterol level of 331 randomly selected patients admitted to the hospital for other reasons was 217. The sample standard deviations were not reported, but suppose that they were 20 for the group who had attempted suicide and 24 for the other group. Do these data provide sufficient evidence to conclude that the mean cholesterol level is lower for those who have attempted suicide? Test the relevant hypotheses using $\alpha = .05$. $t = -11.06$, df = 639, P-value ≈ 0, reject H_0

11.23 The article "The Sorority Rush Process: Self-Selection, Acceptance Criteria, and the Effect of Rejection" (*Journal of College Student Development* [1994]: 346–353) reported on a study of factors associated with the decision to rush a sorority. Fifty-four women who rushed a sorority and 51 women who did not were asked how often they drank alcoholic beverages. For the sorority rush group, the mean was 2.72 drinks per week and the standard deviation .86. For the group who did not rush, the mean was 2.11 and the standard deviation 1.02. Is there evidence to support the claim that those who rush a sorority drink more than those who do not rush? Test the relevant hypotheses using $\alpha = .01$. What assumptions are required in order for the two-sample t test to be appropriate? $t = 3.30$, df = 98, P-value ≈ 0.001, reject H_0

11.24 ● Tennis elbow is thought to be aggravated by the impact experienced when hitting the ball. The article "Forces on the Hand in the Tennis One-Handed Backhand" (*International Journal of Sport Biomechanics* [1991]: 282–292) reported the force (*N*) on the hand just after impact on a one-handed backhand drive for six advanced players and for eight intermediate players. Summary statistics from the article, as well as data consistent with these summary quantities, appear in the table at the top of the page.

The authors of the article assumed in their analysis of the data that both force distributions (advanced and intermediate) were normal. Use the given information to determine whether the mean force after impact is greater for advanced tennis players than it is for intermediate players. $t = 3.46$, df = 8, P-value ≈ 0.004, reject H_0

11.25 ● British health officials have expressed concern about problems associated with vitamin D deficiency among certain immigrants. Doctors have conjectured that such a deficiency could be related to the amount of fiber in a person's diet. An experiment was designed to compare the vitamin D plasma half-life for two groups of healthy individuals. One group was placed on a normal diet, whereas the second group was placed on a high-fiber diet. The accompanying table gives the resulting data (from "Reduced Plasma Half-Lives of Radio-Labeled 25(OH)D3 in Subjects Receiving a High-Fibre Diet," *British Journal of Nutrition* [1993]: 213–216).

Normal diet	19.1	24.0	28.6	29.7	30.0	34.8	
High-fiber diet	12.0	13.0	13.6	20.5	22.7	23.7	24.8

Use the following MINITAB output to determine whether the data indicate that the mean half-life is higher for those on a normal diet than those on a high-fiber diet. Assume that treatments were assigned at random and the two plasma half-life distributions are normal. Test the appropriate hypotheses using $\alpha = .01$. *t = 2.97, df = 10, P-value ≈ 0.007, reject H_0*

Two-sample T for normal vs high

	N	Mean	StDev	SE Mean
Normal	6	27.70	5.44	2.2
High	7	18.61	5.55	2.1

95% C.I. for mu normal – mu high: (2.3, 15.9)
T-Test mu normal = mu high(vs >):T = 2.97 P = 0.0070 DF = 10

11.26 A researcher at the Medical College of Virginia conducted a study of 60 randomly selected male soccer players and concluded that frequently "heading" the ball in soccer lowers players' IQs (*USA Today*, August 14, 1995). The soccer players were divided into two groups, based on whether they averaged 10 or more headers per game. Mean IQs were reported in the article, but the sample sizes and standard deviations were not given. Suppose that these values were as given in the accompanying table.

	n	Sample Mean	Sample sd
Fewer Than 10 Headers	35	112	10
10 or More Headers	25	103	8

Do these data support the researcher's conclusion? Test the relevant hypotheses using $\alpha = .05$. Can you conclude that heading the ball *causes* lower IQ? *t = 3.87, df = 57, P-value = 0.0001, reject H_0*

11.27 The effect of loneliness among college students was examined in the article "The Importance of Perceived Duration: Loneliness and Its Relationship to Self- Esteem and Academic Performance" (*Journal College Student Development* [1994]: 456–460). Based on reported frequency and duration of loneliness, subjects were divided into two groups. The first group ($n_1 = 72$) was the short-duration loneliness group, and the second ($n_2 = 17$) was the long-duration loneliness group. A self-esteem inventory was administered to students in both groups. For the short-duration group, the reported mean self-esteem score was 76.78 and the standard deviation was 17.80. For the long-

duration group, the mean and standard deviation were 64.00 and 15.68, respectively. Do the data support the researcher's claim that mean self-esteem is lower for students classified as having long-duration loneliness? Test the relevant hypotheses using $\alpha = .01$. Be sure to state any assumptions that are necessary for your test to be valid. *t = 2.94, df = 26, P-value ≈ 0.003, reject H_0*

11.28 Do certain behaviors result in a severe drain on energy resources because a great deal of energy is expended in comparison to energy intake? The article "The Energetic Cost of Courtship and Aggression in a Plethodontid Salamander" (*Ecology* [1983]: 979–983) reported on one of the few studies concerned with behavior and energy expenditure. The accompanying table gives oxygen consumption (mL/g/hr) for male-female salamander pairs. (The determination of consumption values is rather complicated. It is partly for this reason that so few studies of this type have been carried out.)

Behavior	Sample Size	Sample Mean	Sample sd
Noncourting	11	.072	.0066
Courting	15	.099	.0071

a. The pooled *t* test is a test procedure for testing H_0: $\mu_1 - \mu_2 = $ *hypothesized value* when it is reasonable to assume that the two population distributions are normal with equal standard deviations ($\sigma_1 = \sigma_2$). The test statistic for the pooled *t* test is obtained by replacing both s_1 and s_2 in the two-sample *t* test statistic with s_p, where

$$s_p = \sqrt{\frac{(n_1 - 1)s_1^2 + (n_2 - 1)s_2^2}{n_1 + n_2 - 2}}$$

When the population distributions are normal with equal standard deviations and H_0 is true, the resulting pooled *t* statistic has a *t* distribution with df = $n_1 + n_2 - 2$. For the reported data, the two sample standard deviations are similar. Use the pooled *t* test with $\alpha = .05$ to determine whether the mean oxygen consumption for courting pairs is higher than the mean oxygen consumption for noncourting pairs. *pooled t =9.86, df = 24, P-value ≈ 0, reject H_0*

b. Would the conclusion in Part (a) have been different if the two-sample *t* test had been used rather than the pooled *t* test?

11.2 Inferences Concerning the Difference Between Two Population or Treatment Means Using Paired Samples

Two samples are said to be *independent* if the selection of the individuals or objects that make up one of the samples has no bearing on the selection of individuals or objects in the other sample. In some situations, an experiment with independent samples is not the best way to obtain information about a possible difference between the populations. For example, suppose that an investigator wants to determine whether regular aerobic exercise affects blood pressure. A random sample of people who jog regularly and a second random sample of people who do not exercise regularly are selected independently of one another. The researcher then uses the two-sample *t* test to conclude that a significant difference exists between the average blood pressures for joggers and nonjoggers. But is it reasonable to think that the difference in mean blood pressure is attributable to jogging? It is known that blood pressure is related to both diet and body weight. Might it not be the case that joggers in the sample tend to be leaner and adhere to a healthier diet than the nonjoggers and that *this* might account for the observed difference? On the basis of this study, the researcher would not be able to rule out the possibility that the observed difference in blood pressure is explained by weight differences between the people in the two samples and that aerobic exercise itself has no effect.

One way to avoid this difficulty is to match subjects by weight. The researcher would find pairs of subjects so that the jogger and nonjogger in each pair were similar in weight (although weights for different pairs might vary widely). The factor *weight* could then be ruled out as a possible explanation for an observed difference in average blood pressure between the two groups. Matching the subjects by weight results in two samples for which each observation in the first sample is coupled in a meaningful way with a particular observation in the second sample. Such samples are said to be **paired**.

Studies can be designed to yield paired data in a number of different ways. Some studies involve using the same group of individuals with measurements recorded both before and after some intervening treatment. Other experiments use naturally occurring pairs, such as twins or husbands and wives, and some investigations construct pairs by matching on factors with effects that might otherwise obscure differences (or the lack of them) between the two populations of interest (as might weight in the jogging example). Paired samples often provide more information than independent samples because extraneous effects are screened out.

Example 11.5 Benefits of Ultrasound

● Ultrasound is often used in the treatment of soft tissue injuries. In an experiment to investigate the effect of an ultrasound and stretch therapy on knee extension, range of motion was measured both before and after treatment for a sample of physical therapy patients. A subset of the data appearing in the paper "Location of Ultrasound Does Not Enhance Range of Motion Benefits of Ultrasound and Stretch Treatment"

● Data set available online

(University of Virginia Thesis, Trae Tashiro, 2003) is given in the accompanying table.

	Range of Motion						
Subject	**1**	**2**	**3**	**4**	**5**	**6**	**7**
Pre-treatment	31	53	45	57	50	43	32
Post-treatment	32	59	46	64	49	45	40

We can regard the data as consisting of two samples—a sample of knee range of motion measurements for physical therapy patients prior to treatment and a sample of physical therapy patients after ultrasound and stretch treatment. The samples are paired rather than independent because both samples are composed of observations on the same seven patients.

Is there evidence that the ultrasound and stretch treatment increases range of motion? Let μ_1 denote the mean range of motion for the population of all physical therapy patients prior to treatment. Similarly, let μ_2 denote the mean range of motion for physical therapy patients after ultrasound and stretch treatment. Hypotheses of interest might be

$$H_0:\ \mu_1 - \mu_2 = 0 \qquad \text{versus} \qquad H_a:\ \mu_1 - \mu_2 < 0$$

with the null hypothesis indicating that the mean range of motion before treatment and the mean after treatment are equal and the alternative hypothesis stating that the mean range of motion after treatment exceeds the mean before treatment. Notice that in six of the seven data pairs, the range of motion measurement is higher after treatment than before treatment. Intuitively, this suggests that the population means may not be equal.

Disregarding the paired nature of the samples results in a loss of information. Both the pre-treatment and post-treatment range of motion measurements vary from one patient to another. It is this variability that may obscure the difference when the two-sample t test is used. If we were to (incorrectly) use the two-sample t test for independent samples on the given data, the resulting t test statistic value would be $-.61$. This value would not allow for rejection of the hypothesis even at level of significance .10. This result might surprise you at first, but remember that this test procedure ignores the information about how the samples are paired. Two plots of the data are given in Figure 11.1. The first plot (Figure 11.1(a)) ignores the pairing, and the

Figure 11.1 Two plots of the paired data from Example 11.5: (a) pairing ignored; (b) pairs identified.

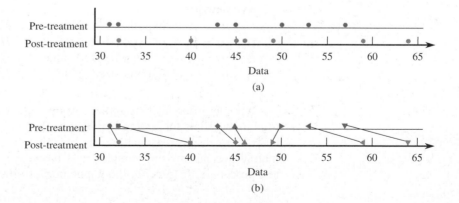

two samples look quite similar. The plot in which pairs are identified (Figure 11.1(b)) does suggest a difference, because for six of the seven pairs the actual post-treatment observation exceeds the pre-treatment observation.

■

Example 11.5 suggests that the methods of inference developed for independent samples are not adequate for dealing with paired samples. When sample observations from the first population are paired in some meaningful way with sample observations from the second population, inferences can be based on the differences between the two observations within each sample pair. The n sample differences can then be regarded as having been selected from a large population of differences. Thus, in Example 11.5, we can think of the seven (pre-treatment − post-treatment) differences as having been selected from an entire population of differences.

Let

Important concept

$$\mu_d = \text{mean value of the difference population}$$

and

$$\sigma_d = \text{standard deviation of the difference population}$$

The relationship between the two individual population means and the mean difference is

$$\mu_d = \mu_1 - \mu_2$$

Therefore, when the samples are paired, inferences about $\mu_1 - \mu_2$ are equivalent to inferences about μ_d. Since inferences about μ_d can be based on the n observed sample differences, the original two-sample problem becomes a familiar one-sample problem.

■ Paired *t* Test

To compare two population or treatment means when the samples are paired, we first translate the hypothesis of interest from one about the value of $\mu_1 - \mu_2$ to an equivalent one involving μ_d:

Hypothesis	Equivalent Hypothesis When Samples Are Paired
H_0: $\mu_1 - \mu_2 =$ hypothesized value	H_0: $\mu_d =$ hypothesized value
H_a: $\mu_1 - \mu_2 = >$ hypothesized value	H_a: $\mu_d >$ hypothesized value
H_a: $\mu_1 - \mu_2 = <$ hypothesized value	H_a: $\mu_d <$ hypothesized value
H_a: $\mu_1 - \mu_2 \neq$ hypothesized value	H_a: $\mu_d \neq$ hypothesized value

Sample differences (Sample 1 value − Sample 2 value) are then computed and used as the basis for testing hypotheses about μ_d. When the number of differences is large or when it is reasonable to assume that the population of differences is approximately normal, the one-sample t test based on the differences is the recommended test procedure. In general, the population of differences is normal if each of the two individual populations is normal. A normal probability plot or boxplot of the differences can be used to support this assumption.

Summary of the Paired *t* Test for Comparing Two Population or Treatment Means

Null hypothesis: H_0: μ_d = hypothesized value

Test statistic: $t = \dfrac{\bar{x}_d - \text{hypothesized value}}{\dfrac{s_d}{\sqrt{n}}}$

where n is the number of sample differences and $\bar{x}_d$ and s_d are the sample mean and standard deviation of the differences. This test is based on df = $n - 1$.

Alternative Hypothesis:	**P-Value:**
H_a: $\mu_d >$ hypothesized value	Area under the appropriate t curve to the right of the calculated t
H_a: $\mu_d <$ hypothesized value	Area under the appropriate t curve to the left of the calculated t
H_a: $\mu_d \neq$ hypothesized value	(1) 2(area to the right of t) if t is positive, or (2) 2(area to the left of t) if t is negative

Assumptions:
1. The samples are *paired*.
2. The n sample differences can be viewed as a *random sample* from a population of differences.
3. The *number of sample differences is large* (generally at least 30) or the *population distribution of differences is approximately normal*.

Example 11.6   Improve Memory by Playing Chess?

© Royalty-Free/Getty Images

∎ Can taking chess lessons and playing chess daily improve memory? The online article "The USA Junior Chess Olympics Research: Developing Memory and Verbal Reasoning" (*New Horizons for Learning,* April 2001; available at www.newhorizons .org) described a study in which sixth-grade students who had not previously played chess participated in a program where they took chess lessons and played chess daily for 9 months. Each student took a memory test (the Test of Cognitive Skills) before starting the chess program and again at the end of the 9-month period. Data (read from a graph in the article) and computed differences are given in the accompanying table.

The author of the article proposed using these data to test the theory that students who participated in the chess program tend to achieve higher memory scores after completion of the program. We can consider the pre-test scores as a sample of scores from the population of sixth-grade students who have not participated in the chess program and the post-test scores as a sample of scores from the population of sixth-grade students who have completed the chess training program. The samples

∎ Data set available online

were not independently chosen, because each sample is composed of the same 12 students.

	Memory Test Score		
Student	Pre-test	Post-test	Difference
1	510	850	-340
2	610	790	-180
3	640	850	-210
4	675	775	-100
5	600	700	-100
6	550	775	-225
7	610	700	-90
8	625	850	-225
9	450	690	-240
10	720	775	-55
11	575	540	35
12	675	680	-5

Let

μ_1 = mean memory score for sixth-graders with no chess training
μ_2 = mean memory score for sixth-graders after chess training

and

$\mu_d = \mu_1 - \mu_2$ = mean memory score difference between students with no chess training and students who have completed chess training

The question of interest can be answered by testing the hypothesis

$$H_0: \mu_d = 0 \qquad \text{versus} \qquad H_a: \mu_d < 0$$

Using the 12 differences, we compute

$$\sum diff = -1735 \qquad \sum (diff)^2 = 383{,}325$$

$$\bar{x}_d = \frac{\sum diff}{n} = \frac{-1735}{12} = -144.58$$

$$s_d^2 = \frac{\sum (diff)^2 - \frac{(\sum diff)^2}{n}}{n-1} = \frac{383{,}325 - \frac{(-1735)^2}{12}}{11} = 12{,}042.99$$

$$s_d = \sqrt{s_d^2} = 109.74$$

We now use the paired t test with a significance level of .05 to carry out the hypothesis test.

1. μ_d = mean memory score difference between students with no chess training and students with chess training
2. $H_0: \mu_d = 0$

3. H_a: $\mu_d < 0$
4. Significance level: $\alpha = .05$
5. Test statistic: $t = \dfrac{\bar{x}_d - \text{hypothesized value}}{\dfrac{s_d}{\sqrt{n}}}$
6. Assumptions: Although the sample of 12 sixth-graders was not a random sample, the author believed that it was reasonable to view the 12 sample differences as a random sample of all such differences. A boxplot of the differences is approximately symmetric and does not show any outliers, so the assumption of normality is not unreasonable and we will proceed with the paired t test.
7. Calculation: $t = \dfrac{-144.6 - 0}{\dfrac{109.74}{\sqrt{12}}} = -4.56$
8. P-value: This is a lower-tailed test, so the P-value is the area to the left of the computed t value. The appropriate df for this test is df $= 12 - 1 = 11$. From the 11-df column of Appendix Table 4, we find that P-value $< .001$ because the area to the left of -4.0 is $.001$ and the test statistic (-4.56) is even farther out in the lower tail.
9. Conclusion: Because P-value $\le \alpha$, we reject H_0. The data support the theory that the mean memory score is higher for sixth-graders after completion of the chess training than the mean score before training.

■

Using the two-sample t test (for independent samples) for the data in Example 11.6 would have been incorrect, because the samples are not independent. Inappropriate use of the two-sample t test would have resulted in a computed test statistic value of -4.25. The conclusion would still be to reject the hypothesis of equal mean memory scores in this particular example, but this is not always the case

Example 11.7 Charitable Chimps

● The authors of the paper "Chimpanzees Are Indifferent to the Welfare of Unrelated Group Members" (*Nature* [2005]: 1357–1359) concluded that "chimpanzees do not take advantage of opportunities to deliver benefits to individuals at no cost to themselves." This conclusion was based on data from an experiment in which a sample of chimpanzees was trained to use an apparatus that would deliver food just to the subject chimpanzee when one lever was pushed and would deliver food both to the subject chimpanzee and another chimpanzee in an adjoining cage when another lever was pushed. After training, the chimps were observed when there was no chimp in the adjoining cage and when there was another chimp in the adjoining cage.

The researchers hypothesized that if chimpanzees were motivated by the welfare of others, they would choose the option that provided food to both chimpanzees more often when there was a chimpanzee in the adjoining cage. Data on the number

● Data set available online

of times the "feed both" option was chosen out of 36 opportunities (approximate values read from a graph in the paper) are given in the accompanying table.

Chimp	Number of Times "Feed Both" Option Was Chosen	
	No Chimp in Adjoining Cage	Chimp in Adjoining Cage
1	21	23
2	22	22
3	23	21
4	21	23
5	18	19
6	16	19
7	19	19

Most statistical software packages will perform a paired t-test, and we will use MINITAB to carry out a test to determine if there is convincing evidence that the mean number of times the "feed both" option is selected is higher when another chimpanzee is present in the adjoining cage than when the subject chimpanzee is alone.

1. μ_d = difference between mean number of "feed both" selections for chimpanzees who are alone and for chimpanzees who have company in the adjoining cage
2. H_0: $\mu_d = 0$
3. H_a: $\mu_d < 0$
4. Significance level: $\alpha = .05$
5. Test statistic: $t = \dfrac{\bar{x}_d - \text{hypothesized value}}{\dfrac{s_d}{\sqrt{n}}}$

6. Assumptions: Although the chimpanzees in this study were not randomly selected, the authors considered them to be representative of the population of chimpanzees. A boxplot of the differences is approximately symmetric and does not show any outliers, so the assumption of normality is not unreasonable and we will proceed with the paired t test.
7. Calculation: From the given MINITAB output, $t = -1.35$.

Paired T-Test and CI: Alone, Companion

Paired T for Alone - Companion

	N	Mean	StDev	SE Mean
Alone	7	20.0000	2.4495	0.9258
Companion	7	20.8571	1.8645	0.7047
Difference	7	−0.857143	1.676163	0.633530

95% CI for mean difference: (−2.407335, 0.693050)
T-Test of mean difference = 0 (vs not = 0): T-Value = −1.35   P-Value = 0.225

8. P-value: From the MINITAB output, P-value $= .225$.
9. Conclusion: Since P-value $> \alpha$, H_0 is not rejected. The data do not provide evidence that the mean number of times that the "feed both" option is chosen is

greater when there is a chimpanzee in the adjoining cage. This is the basis for the statement quoted at the beginning of this example.

■

Important concept   Notice that the numerators $\bar{x}_d$ and $\bar{x}_1 - \bar{x}_2$ of the paired t and the two-sample t test statistics are always equal. The difference lies in the denominator. The variability in differences is usually smaller than the variability in each sample separately (because measurements in a pair tend to be similar). As a result, the value of the paired t statistic is usually larger in magnitude than the value of the two-sample t statistic. Pairing typically reduces variability that might otherwise obscure small but nevertheless significant differences.

■ A Confidence Interval

The one-sample t confidence interval for μ given in Chapter 9 is easily adapted to obtain an interval estimate for μ_d.

Paired t Confidence Interval for μ_d

When

1. the samples are *paired*,
2. the n sample differences can be viewed as a *random sample* from a population of differences, and
3. the *number of sample differences is large* (generally at least 30) or the *population distribution of differences is approximately normal,*

the paired t confidence interval for is

$$\bar{x}_d \pm (t\ \text{critical value}) \cdot \frac{s_d}{\sqrt{n}}$$

For a specified confidence level, the $(n - 1)$ df row of Appendix Table 3 gives the appropriate t critical values.

■

Example 11.8 Benefits of Ultrasound Revisited

● Let's use the data from Example 11.5 to estimate the difference in mean range of motion prior to treatment and the mean range of motion after ultrasound and stretch treatment for physical therapy patients. The data and the computed differences are shown in the accompanying table.

| Subject | Range of Motion | | | | | | |
	1	2	3	4	5	6	7
Pre-treatment	31	53	45	57	50	43	32
Post-treatment	32	59	46	64	49	45	40
Difference	**−1**	**−6**	**−1**	**−7**	**1**	**−2**	**−8**

Step-by-step technology instructions available online   ● Data set available online

We will use these data to estimate the mean change in range of motion using a 95% confidence interval, assuming that the 7 patients participating in this study can be considered as representative of physical therapy patients. The following boxplot of the 7 sample differences is not inconsistent with a difference population that is approximately normal, so the paired t confidence interval is appropriate:

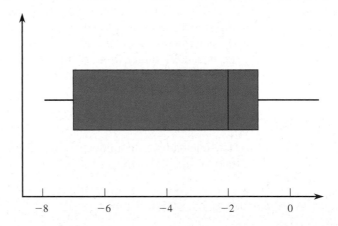

The mean and standard deviation computed using the seven sample differences are -3.43 and 3.51, respectively. The t critical value for df $= 6$ and a 95% confidence level is 2.45, and therefore the confidence interval is

$$\bar{x}_d \pm (t \text{ critical value}) \cdot \frac{s_d}{\sqrt{n}} = -3.43 \pm (2.45) \cdot \frac{3.51}{\sqrt{7}}$$
$$= -3.43 \pm 3.25$$
$$= (-6.68, -0.18)$$

Based on the sample data, we can be 95% confident that the difference in mean range of motion is between -6.68 and -0.18. That is, we are 95% confident that the mean increase in range of motion after ultrasound and stretch therapy is somewhere between 0.18 and 6.68.

MINITAB output is also shown. MINITAB carries a bit more decimal accuracy, and reports a 95% confidence interval of $(-6.67025, -0.18690)$.

Paired T-Test and CI: Pre, Post

Paired T for Pre − Post

	N	Mean	StDev	SE Mean
Pre	7	44.4286	9.9976	3.7787
Post	7	47.8571	10.8847	4.1140
Difference	7	−3.42857	3.50510	1.32480

95% CI for mean difference: (−6.67025, −0.18690)
T-Test of mean difference = 0 (vs not = 0): T-Value = −2.59   P-Value = 0.041

When two populations must be compared to draw a conclusion on the basis of sample data, a researcher might choose to use independent samples or paired samples. In many situations, paired data provide a more effective comparison by screening out the effects of extraneous variables that might obscure differences between the two populations or that might suggest a difference when none exists.

Exercises 11.29–11.43

Turn to the Explanded Answers Section for answers not shown next to exercises.

11.29 Suppose that you were interested in investigating the effect of a drug that is to be used in the treatment of patients who have glaucoma in both eyes. A comparison between the mean reduction in eye pressure for this drug and for a standard treatment is desired. Both treatments are applied directly to the eye.
a. Describe how you would go about collecting data for your investigation.
b. Does your method result in paired data?
c. Can you think of a reasonable method of collecting data that would result in independent samples? Would such an experiment be as informative as a paired experiment? Comment.

11.30 Two different underground pipe coatings for preventing corrosion are to be compared. The effect of a coating (as measured by maximum depth of corrosion penetration on a piece of pipe) may vary with depth, orientation, soil type, pipe composition, etc. Describe how an experiment that filters out the effects of these extraneous factors could be carried out.

11.31 ● ▼ To determine if chocolate milk was as effective as other carbohydrate replacement drinks, nine male cyclists performed an intense workout followed by a drink and a rest period. At the end of the rest period, each cyclist performed an endurance trial where he exercised until exhausted and time to exhaustion was measured. Each cyclist completed the entire regimen on two different days. On one day the drink provided was chocolate milk and on the other day the drink provided was a carbohydrate replacement drink. Data consistent with summary quantities appearing in the paper "The Efficacy of Chocolate Milk as a Recovery Aid" (*Medicine and Science in Sports and Exercise* [2004], S126) appear in the table below. Is there evidence that the mean time to exhaustion is greater after chocolate milk than after carbohydrate replacement drink? Use a significance level of .05. $t = 4.46$, df $= 8$, P-value $\approx$ 0, reject H_0

11.32 ● The humorous paper "Will Humans Swim Faster or Slower in Syrup?" (*American Institute of Chemical Engineers Journal* [2004]: 2646–2647) investigates the fluid mechanics of swimming. Twenty swimmers each swam a specified distance in a water-filled pool and in a pool where the water was thickened with food grade guar gum to create a syrup-like consistency. Velocity, in meters per second, was recorded. Values estimated from a graph that appeared in the paper are given. The authors of the paper concluded that swimming in guar syrup does not change swimming speed. Are the given data consistent with this conclusion? Carry out a hypothesis test using a .01 significance level. $t = -0.52$, df $= 19$, P-value $\approx$ 0622, fail to reject H_0

| | Velocity (m/s) | |
Swimmer	Water	Guar Syrup
1	0.90	0.92
2	0.92	0.96
3	1.00	0.95
4	1.10	1.13
5	1.20	1.22
6	1.25	1.20
7	1.25	1.26
8	1.30	1.30
9	1.35	1.34
10	1.40	1.41
11	1.40	1.44
12	1.50	1.52
13	1.65	1.58
14	1.70	1.70
15	1.75	1.80
16	1.80	1.76
17	1.80	1.84
18	1.85	1.89
19	1.90	1.88
20	1.95	1.95

Table for Exercise 11.31

| | Time to Exhaustion (minutes) | | | | | | | | |
Cyclist	1	2	3	4	5	6	7	8	9
Chocolate Milk	24.85	50.09	38.30	26.11	36.54	26.14	36.13	47.35	35.08
Carbohydrate Replacement	10.02	29.96	37.40	15.52	9.11	21.58	31.23	22.04	17.02

Bold exercises answered in back   ● Data set available online but not required   ▼ Video solution available

11.33 ● Breast feeding sometimes results in a temporary loss of bone mass as calcium is depleted in the mother's body to provide for milk production. The paper "Bone Mass Is Recovered from Lactation to Postweaning in Adolescent Mothers with Low Calcium Intakes" (*American Journal of Clinical Nutrition* [2004]: 1322–1326) gave the following data on total body bone mineral content (g) for a sample of mothers both during breast feeding (B) and in the postweaning period (P). Do the data suggest that true average total body bone mineral content during postweaning exceeds that during breast feeding by more than 25 g? State and test the appropriate hypotheses using a significance level of .05. $t = 2.46$, df = 9, P-value ≈ 0.017, reject H_0

Subject	1	2	3	4	5	6
B	1928	2549	2825	1924	1628	2175
P	2126	2885	2895	1942	1750	2184

Subject	7	8	9	10
B	2114	2621	1843	2541
P	2164	2626	2006	2627

11.34 ● A deficiency of the trace element selenium in the diet can negatively impact growth, immunity, muscle and neuromuscular function, and fertility. The introduction of selenium supplements to dairy cows is justified when pastures have low selenium levels. Authors of the paper "Effects of Short-Term Supplementation with Selenised Yeast on Milk Production and Composition of Lactating Cows" (*Australian Journal of Dairy Technology,* [2004]: 199–203) supplied the following data on milk selenium concentration (mg/L) for a sample of cows given a selenium supplement (the treatment group) and a control sample given no supplement, both initially and after a 9-day period.

Initial Measurement		After 9 Days	
Treatment	Control	Treatment	Control
11.4	9.1	138.3	9.3
9.6	8.7	104.0	8.8
10.1	9.7	96.4	8.8
8.5	10.8	89.0	10.1
10.3	10.9	88.0	9.6
10.6	10.6	103.8	8.6
11.8	10.1	147.3	10.4
9.8	12.3	97.1	12.4
10.9	8.8	172.6	9.3
10.3	10.4	146.3	9.5

Initial Measurement		After 9 Days	
Treatment	Control	Treatment	Control
10.2	10.9	99.0	8.4
11.4	10.4	122.3	8.7
9.2	11.6	103.0	12.5
10.6	10.9	117.8	9.1
10.8		121.5	
8.2		93.0	

a. Use the given data for the treatment group to determine if there is there sufficient evidence to conclude that the mean selenium concentration is greater after 9 days of the selenium supplement.

b. Are the data for the cows in the control group (no selenium supplement) consistent with the hypothesis of no significant change in mean selenium concentration over the 9-day period?

c. Would you use the paired t test of this section to determine if there was a significant difference in the initial mean selenium concentration for the control group and the treatment group? Explain why or why not.   No, the two samples are not paired.

11.35 ● The paper "Quantitative Assessment of Glenohumeral Translation in Baseball Players" (*The American Journal of Sports Medicine* [2004]: 1711–1715) considered various aspects of shoulder motion for a sample of pitchers and another sample of position players. The authors kindly supplied the following data (see p. 617) on anteroposterior translation (mm), a measure of the extent of anterior and posterior motion, both for the dominant arm and the nondominant arm.

a. Estimate the true average difference in translation between dominant and nondominant arms for pitchers using a 95% confidence interval.   (2.032, 6.10)

b. Estimate the true average difference in translation between dominant and nondominant arms for position players using a 95% confidence interval.   (−0.54, 1.005)

c. The authors asserted that pitchers have greater difference in mean anteroposterior translation of their shoulders than do position players. Do you agree? Explain.

11.36 ● The report of a European commission on radiation protection titled "Cosmic Radiation Exposure of Aircraft Crew" (2004) measured the exposure to radiation on eight international flights from Madrid using several different methods for measuring radiation. Data for two of the methods are given in the accompanying table. Use these data to

Data for Exercise 11.35

Player	Position Player Dominant Arm	Position Player Nondominant Arm	Pitcher	Pitcher Dominant Arm	Pitcher Nondominant Arm
1	30.31	32.54	1	27.63	24.33
2	44.86	40.95	2	30.57	26.36
3	22.09	23.48	3	32.62	30.62
4	31.26	31.11	4	39.79	33.74
5	28.07	28.75	5	28.50	29.84
6	31.93	29.32	6	26.70	26.71
7	34.68	34.79	7	30.34	26.45
8	29.10	28.87	8	28.69	21.49
9	25.51	27.59	9	31.19	20.82
10	22.49	21.01	10	36.00	21.75
11	28.74	30.31	11	31.58	28.32
12	27.89	27.92	12	32.55	27.22
13	28.48	27.85	13	29.56	28.86
14	25.60	24.95	14	28.64	28.58
15	20.21	21.59	15	28.58	27.15
16	33.77	32.48	16	31.99	29.46
17	32.59	32.48	17	27.16	21.26
18	32.60	31.61			
19	29.30	27.46			

test the hypothesis that there is no significant difference in mean radiation measurement for the two methods. $t = -0.59$, df $= 7$, P-value ≈ 0.568, fail to reject H_0

Flight	Method 1	Method 2
1	27.5	34.4
2	41.3	38.6
3	3.5	3.5
4	24.3	21.9
5	27	24.4
6	17.7	21.4
7	12	11.8
8	20.9	24.1

11.37 Two proposed computer mouse designs were compared by recording wrist extension in degrees for 24 people who each used both mouse types ("Comparative Study of Two Computer Mouse Designs," Cornell Human Factors Laboratory Technical Report RP7992). The difference in wrist extension was computed by subtracting extension for mouse type B from the wrist extension for mouse type A for each student. The mean difference was reported to be 8.82 degrees. Assume that it is reasonable to regard this sample of 24 people as representative of the population of computer users.

a. Suppose that the standard deviation of the differences was 10 degrees. Is there convincing evidence that the mean wrist extension for mouse type A is greater than for mouse type B? Use a .05 significance level.

b. Suppose that the standard deviation of the differences was 26 degrees. Is there convincing evidence that the mean wrist extension for mouse type A is greater than for mouse type B? Use a .05 significance level.

c. Briefly explain why a different conclusion was reached in the hypothesis tests of Parts (a) and (b).

11.38 ● The article "More Students Taking AP Tests" (*San Luis Obispo Tribune*, January 10, 2003) provided the following information on the percentage of students in grades 11 and 12 taking one or more AP exams and the percentage of exams that earned credit in 1997 and 2002 for seven high schools on the central coast of California.

School	Percentage of Students Taking One or More AP Exams		Percentage of Exams That Earned College Credit	
	1997	2002	1997	2002
1	13.6	18.4	61.4	52.8
2	20.7	25.9	65.3	74.5
3	8.9	13.7	65.1	72.4
4	17.2	22.4	65.9	61.9
5	18.3	43.5	42.3	62.7
6	9.8	11.4	60.4	53.5
7	15.7	17.2	42.9	62.2

a. Assuming it is reasonable to regard these seven schools as a random sample of high schools located on the central coast of California, carry out an appropriate test to determine if there is convincing evidence that the mean percentage of exams earning college credit at central coast high schools declined between 1997 and 2002.

b. Do you think it is reasonable to generalize the conclusion of the test in Part (a) to all California high schools? Explain.

c. Would it be reasonable to use the paired t test with the data on percentage of students taking one or more AP classes? Explain. No, there is an outlier in the sample distribution of the differences.

11.39 ● The Oregon Department of Health web site provides information on the cost-to-charge ratio (the percentage of billed charges that are actual costs to the hospital). The cost-to-charge ratios for both inpatient and outpatient care in 2002 for a sample of six hospitals in Oregon follow.

Hospital	2002 Inpatient Ratio	2002 Outpatient Ratio
1	68	54
2	100	75
3	71	53
4	74	56
5	100	74
6	83	71

Is there evidence that the mean cost-to-charge ratio for Oregon hospitals is lower for outpatient care than for inpatient care? Use a significance level of .05. $t = 8.13$, df = 5, P-value ≈ 0, reject H_0

11.40 Babies born extremely prematurely run the risk of various neurological problems and tend to have lower IQ and verbal ability scores than babies that are not premature. The article "Premature Babies May Recover Intelligence, Study Says" (*San Luis Obispo Tribune*, February 12, 2003) summarized the results of medical research that suggests that the deficit observed at an early age may decrease as children age. Children who were born prematurely were given a test of verbal ability at age 3 and again at age 8. The test is scaled so that a score of 100 would be average for a normal-birth-weight child. Data that are consistent with summary quantities given in the paper for 50 children who were born prematurely were used to generate the accompanying MINITAB output, where Age3 represents the verbal ability score at age 3 and Age8 represents the verbal ability score at age 8. Use the MINITAB output to carry out a test to determine if there is evidence that the mean verbal ability score for children born prematurely increases between age 3 and age 8. You may assume that it is reasonable to regard the sample of 50 children as a random sample from the population of all children born prematurely. P-value = 0.001, reject H_0

Paired T-Test and CI: Age8, Age3

Paired T for Age8 − Age3

	N	Mean	StDev	SE Mean
Age8	50	97.21	16.97	2.40
Age3	50	87.30	13.84	1.96
Difference	50	9.91	22.11	3.13

T-Test of mean difference = 0 (vs > 0): T-Value = 3.17   P-Value = 0.001

11.41 The article "Report: Mixed Progress in Math" (*USA Today*, August 3, 2001) gave data for all 50 states on the percentage of public school students who were at or above the proficient level in mathematics testing in 1996 and 2000. Explain why it is not necessary to use an inference procedure such as the paired t test if you want to know if the mean percentage proficient for the 50 states increased from 1996 to 2002.

11.42 Do girls think they don't need to take as many science classes as boys? The article "Intentions of Young Students to Enroll in Science Courses in the Future: An Examination of Gender Differences" (*Science Education* [1999]: 55–76) gives information from a survey of children in grades 4, 5, and 6. The 224 girls participating in the survey each indicated the number of science courses they intended to take in the future, and they also indicated the number of science courses they thought boys their age should take in the future. For each girl, the authors calculated the difference between the number of science classes she intends to take and the number she thinks boys should take.

a. Explain why these data are paired.

b. The mean of the differences was $-.83$ (indicating girls intended, on average, to take fewer classes than they thought boys should take), and the standard deviation was 1.51. Construct and interpret a 95% confidence interval for the mean difference. $(-1.03, -0.63)$

11.43 ● The effect of exercise on the amount of lactic acid in the blood was examined in the article "A Descriptive Analysis of Elite-Level Racquetball" (*Research Quarterly for Exercise and Sport* [1991]: 109–114). Eight men and seven women who were attending a weeklong training camp participated in the experiment. Blood lactate levels were measured before and after playing three games of racquetball, as shown in the accompanying table.

Men			Women		
Player	Before	After	Player	Before	After
1	13	18	1	11	21
2	20	37	2	16	26
3	17	40	3	13	19
4	13	35	4	18	21
5	13	30	5	14	14
6	16	20	6	11	31
7	15	33	7	13	20
8	16	19			

a. Estimate the mean change in blood lactate level for male racquetball players using a 95% confidence interval.

b. Estimate the mean change for female players using a 95% confidence interval. $(2.071, 13.929)$

c. Based on the intervals from Parts (a) and (b), do you think the mean change in blood lactate level is the same for men as it is for women? Explain.

Bold exercises answered in back   ● Data set available online but not required   ▼ Video solution available

....

11.3 Large Sample Inferences Concerning a Difference Between Two Population or Treatment Proportions

Large-sample methods for estimating and testing hypotheses about a single population proportion were presented in Chapters 9 and 10. The symbol π was used to represent the proportion of individuals in the population who possess some characteristic (the successes). Inferences about the value of π were based on p, the corresponding sample proportion of successes.

Many investigations are carried out to compare the proportion of successes in one population (or resulting from one treatment) to the proportion of successes in a second population (or from a second treatment). As was the case for means, we use the subscripts 1 and 2 to distinguish between the two population proportions, sample sizes, and sample proportions.

Notation

Population or Treatment 1:   Proportion of "successes" $= \pi_1$
Population or Treatment 2:   Proportion of "successes" $= \pi_2$

	Sample Size	Proportion of Successes
Sample from Population or Treatment 1	n_1	p_1
Sample from Population or Treatment 2	n_2	p_2

When comparing two populations or treatments on the basis of "success" proportions, it is common to focus on the quantity $\pi_1 - \pi_2$, the difference between the two proportions. Because p_1 provides an estimate of π_1 and p_2 provides an estimate of π_2, the obvious choice for an estimate of $\pi_1 - \pi_2$ is $p_1 - p_2$.

Because p_1 and p_2 each vary in value from sample to sample, so will the difference $p_1 - p_2$. For example, a first sample from each of two populations might yield

$$p_1 = .69 \qquad p_2 = .70 \qquad p_1 - p_2 = .01$$

A second sample from each might result in

$$p_1 = .79 \qquad p_2 = .67 \qquad p_1 - p_2 = .12$$

and so on. Because the statistic $p_1 - p_2$ is the basis for drawing inferences about $\pi_1 - \pi_2$, we need to know something about its behavior.

Properties of the Sampling Distribution of $p_1 - p_2$

If two random samples are selected independently of one another, the following properties hold:

1. $\mu_{p_1-p_2} = \pi_1 - \pi_2$

   This says that the sampling distribution of $p_1 - p_2$ is centered at $\pi_1 - \pi_2$, so $p_1 - p_2$ is an unbiased statistic for estimating $\pi_1 - \pi_2$.

2. $\sigma^2_{p_1-p_2} = \sigma^2_{p_1} + \sigma^2_{p_2} = \dfrac{\pi_1(1 - \pi_1)}{n_1} + \dfrac{\pi_2(1 - \pi_2)}{n_2}$

   and

   $$\sigma_{p_1-p_2} = \sqrt{\frac{\pi_1(1 - \pi_1)}{n_1} + \frac{\pi_2(1 - \pi_2)}{n_2}}$$

3. If both n_1 and n_2 are large (that is, if $n_1\pi_1 \geq 10$, $n_1(1 - \pi_1) \geq 10$, $n_2\pi_2 \geq 10$, and $n_2(1 - \pi_2) \geq 10$), then each have a sampling distribution that is approximately normal, and their difference $p_1 - p_2$ also has a sampling distribution that is approximately normal.

The properties in the box imply that when the samples are independently selected and when both sample sizes are large, the distribution of the standardized variable

$$z = \frac{p_1 - p_2 - (\pi_1 - \pi_2)}{\sqrt{\dfrac{\pi_1(1 - \pi_1)}{n_1} + \dfrac{\pi_2(1 - \pi_2)}{n_2}}}$$

is described approximately by the standard normal (z) curve.

■ A Large-Sample Test Procedure ....

Comparisons of π_1 and π_2 are often based on large, independently selected samples, and we restrict ourselves to this case. The most general null hypothesis of interest has the form

$$H_0: \pi_1 - \pi_2 = \text{hypothesized value}$$

However, when the hypothesized value is something other than 0, the appropriate test statistic differs somewhat from the test statistic used for H_0: $\pi_1 - \pi_2 = 0$. Because this H_0 is almost always the relevant one in applied problems, we focus exclusively on it.

Our basic testing principle has been to use a procedure that controls the probability of a Type I error at the desired level α. This requires using a test statistic with a sampling distribution that is known when H_0 is true. That is, the test statistic should be developed under the assumption that $\pi_1 = \pi_2$ (as specified by the null hypothesis $\pi_1 - \pi_2 = 0$). In this case, π can be used to denote the common value of the two population proportions. The z variable obtained by standardizing $p_1 - p_2$ then simplifies to

$$z = \frac{p_1 - p_2}{\sqrt{\dfrac{\pi(1 - \pi)}{n_1} + \dfrac{\pi(1 - \pi)}{n_2}}}$$

Unfortunately, this cannot serve as a test statistic, because the denominator cannot be computed: H_0 says that there is a common value π, but it does not specify what that value is. A test statistic can be obtained, though, by first *estimating* π from the sample data and then using this estimate in the denominator of z.

When $\pi_1 = \pi_2$, both p_1 and p_2 are estimates of the common proportion π. However, a better estimate than either p_1 or p_2 is a weighted average of the two, in which more weight is given to the sample proportion based on the larger sample.

DEFINITION

The **combined estimate of the common population proportion** is

$$p_c = \frac{n_1 p_1 + n_2 p_2}{n_1 + n_2} = \frac{\text{total number of } S\text{'s in the two samples}}{\text{total of the two sample sizes}}$$

The test statistic for testing H_0: $\pi_1 - \pi_2 = 0$ results from using p_c, the combined estimate, in place of π in the standardized variable z given previously. This z statistic has approximately a standard normal distribution when H_0 is true, so a test that has the desired significance level α can be obtained by calculating a P value using the z table.

Summary of Large-Sample z Tests for $\pi_1 - \pi_2 = 0$

Null hypothesis: H_0: $\pi_1 - \pi_2 = 0$

Test statistic: $z = \dfrac{p_1 - p_2}{\sqrt{\dfrac{p_c(1 - p_c)}{n_1} + \dfrac{p_c(1 - p_c)}{n_2}}}$

Alternative Hypothesis:	**P-Value:**
H_a: $\pi_1 - \pi_2 > 0$	Area under the z curve to the right of the computed z
H_a: $\pi_1 - \pi_2 < 0$	Area under the z curve to the left of the computed z
H_a: $\pi_1 - \pi_2 \neq 0$	(1) 2(area to the right of z) if z is positive
	or
	(2) 2(area to the left of z) is z is negative

(continued)

Assumptions:
1. The samples are *independently chosen random samples*, or *treatments were assigned at random to individuals or objects* (or subjects were assigned at random to treatments).
2. Both *sample sizes are large*:

$$n_1p_1 \geq 10 \qquad n_1(1 - p_1) \geq 10 \qquad n_2p_2 \geq 10 \qquad n_2(1 - p_2) \geq 10$$

....

Example 11.9 Duct Tape to Remove Warts?

Some people seem to believe that you can fix anything with duct tape. Even so, many were skeptical when researchers announced that duct tape may be a more effective and less painful alternative to liquid nitrogen, which doctors routinely use to freeze warts. The article "What a Fix-It: Duct Tape Can Remove Warts" (*San Luis Obispo Tribune*, October 15, 2002) described a study conducted at Madigan Army Medical Center. Patients with warts were randomly assigned to either the duct tape treatment or the more traditional freezing treatment. Those in the duct tape group wore duct tape over the wart for 6 days, then removed the tape, soaked the area in water, and used an emery board to scrape the area. This process was repeated for a maximum of 2 months or until the wart was gone. Data consistent with values in the article are summarized in the following table:

Treatment	n	Number with Wart Successfully Removed
Liquid nitrogen freezing	100	60
Duct tape	104	88

Do the data suggest that freezing is less successful than duct tape in removing warts? Let π_1 represent the true proportion of warts that would be successfully removed by freezing, and let π_2 represent the true proportion of warts that would be successfully removed with the duct tape treatment. We test the relevant hypotheses

$$H_0: \pi_1 - \pi_2 = 0 \qquad \text{versus} \qquad H_a: \pi_1 - \pi_2 < 0$$

using $\alpha = .01$. For these data,

$$p_1 = \frac{60}{100} = .60$$

$$p_2 = \frac{88}{104} = .85$$

Suppose that $\pi_1 = \pi_2$; let π denote the common value. Then the combined estimate of π is

$$p_c = \frac{n_1p_1 + n_2p_2}{n_1 + n_2} = \frac{100(.60) + 104(.85)}{100 + 104} = .73$$

The nine-step procedure can now be used to perform the hypothesis test:

1. $\pi_1 - \pi_2$ is the difference between the true proportions of warts removed by freezing and by the duct tape treatment.
2. H_0: $\pi_1 - \pi_2 = 0$ ($\pi_1 = \pi_2$)
3. H_a: $\pi_1 - \pi_2 < 0$ ($\pi_1 < \pi_2$, in which case the proportion of warts removed by freezing is lower than the proportion by duct tape.)
4. Significance level: $\alpha = .01$
5. Test statistic: $z = \dfrac{p_1 - p_2}{\sqrt{\dfrac{p_c(1 - p_c)}{n_1} + \dfrac{p_c(1 - p_c)}{n_2}}}$
6. Assumptions: The subjects were assigned randomly to the two treatments. Checking to make sure that the sample sizes are large enough, we have

$$
\begin{aligned}
n_1 p_1 &= 100(.60) = 60 \geq 10 \\
n_1(1 - p_1) &= 100(.40) = 40 \geq 10 \\
n_2 p_2 &= 104(.85) = 88.4 \geq 10 \\
n_2(1 - p_2) &= 104(.15) = 15.6 \geq 10
\end{aligned}
$$

7. Calculations:

$$
n_1 = 100 \qquad n_2 = 104 \qquad p_1 = .60 \qquad p_2 = .85 \qquad p_c = .73
$$

and so

$$
z = \frac{.60 - .85}{\sqrt{\dfrac{(.73)(.27)}{100} + \dfrac{(.73)(.27)}{104}}} = \frac{-.25}{.062} = -4.03
$$

8. *P*-value: This is a lower-tailed test, so the *P*-value is the area under the z curve and to the left of the computed $z = -4.03$. From Appendix Table 2, *P*-value ≈ 0.
9. Conclusion: Since *P*-value $\leq \alpha$, the null hypothesis is rejected at level .01. There is convincing evidence that the proportion of warts successfully removed is lower for freezing than for the duct tape treatment.

■

MINITAB can also be used to carry out a two-sample z test to compare two population proportions, as illustrated in the following example.

Example 11.10 AIDS and Housing Availability

The authors of the article "Accommodating Persons with AIDS: Acceptance and Rejection in Rental Situations" (*Journal of Applied Social Psychology* [1999]: 261–270) stated that, even though landlords participating in a telephone survey indicated that they would generally be willing to rent to persons with AIDS, they wondered whether this was true in actual practice. To investigate, the researchers independently selected two random samples of 80 advertisements for rooms for rent from newspaper advertisements in three large cities. An adult male caller responded to each ad in the first sample of 80 and inquired about the availability of the room and was told that the room was still available in 61 of these calls. The same caller also responded to each ad in the second sample. In these calls, the caller indicated that he was currently receiving some treatment for AIDS and was about to be released from the hospital and would require a place to live. The caller was told that a room was available in 32 of these calls. Based on this information, the authors concluded that "reference to AIDS substantially decreased the likelihood of a room being described as available." Do the data support this conclusion? Let's carry out a hypothesis test with $\alpha = .01$.

No AIDS Reference	$n_1 = 80$	$p_1 = 61/80 = .763$
AIDS Reference	$n_2 = 80$	$p_2 = 32/80 = .400$

1. π_1 = proportion of rooms available when there is no AIDS reference
   π_2 = proportion of rooms available with AIDS reference
2. H_0: $\pi_1 - \pi_2 = 0$
3. H_a: $\pi_1 - \pi_2 > 0$
4. Significance level: $\alpha = .01$
5. Test statistic: $z = \dfrac{p_1 - p_2}{\sqrt{\dfrac{p_c(1 - p_c)}{n_1} + \dfrac{p_c(1 - p_c)}{n_2}}}$
6. Assumptions: The two samples are independently chosen random samples. Checking to make sure that the sample sizes are large enough by using $n_1 = 80$, $p_1 = .763$, $n_2 = 80$, and $p_2 = .400$, we have

$$n_1 p_1 = 61.04 \geq 10$$
$$n_1(1 - p_1) = 18.96 \geq 10$$
$$n_2 p_2 = 32.00 \geq 10$$
$$n_2(1 - p_2) = 48.00 \geq 10$$

7. Calculations: MINITAB output is shown below. From the output, $z = 4.65$.

Test for Two Proportions

Sample	X	N	Sample p
1	61	80	0.762500
2	32	80	0.400000

Difference = p (1) − p (2)
Estimate for difference: 0.3625
Test for difference = 0 (vs > 0): Z = 4.65 P-Value = 0.000

8. *P*-value: From the computer output, *P*-value = 0.000
9. Conclusion: Because *P*-value $\leq \alpha$, the null hypothesis is rejected at level .01. There is strong evidence that the proportion of rooms reported as available is smaller with the AIDS reference than without it. This supports the claim made by the authors of the article.

■

■ A Confidence Interval

A large-sample confidence interval for $\pi_1 - \pi_2$ is a special case of the general z interval formula

point estimate $\pm$ (z critical value)(estimated standard deviation)

The statistic $p_1 - p_2$ gives a point estimate of $\pi_1 - \pi_2$, and the standard deviation of this statistic is

$$\sigma_{p_1 - p_2} = \sqrt{\frac{\pi_1(1 - \pi_1)}{n_1} + \frac{\pi_2(1 - \pi_2)}{n_2}}$$

An estimated standard deviation is obtained by using the sample proportions p_1 and p_2 in place of π_1 and π_2, respectively, under the square-root symbol. Notice that this estimated standard deviation differs from the one used previously in the test statistic. When constructing a confidence interval, there isn't a null hypothesis that claims $\pi_1 = \pi_2$, so there is no assumed common value of π to estimate.

A Large-Sample Confidence Interval for $\pi_1 - \pi_2$

When

1. the samples are *independently selected random samples* or *treatments were assigned at random to individuals or objects* (or vice versa), and
2. both *sample sizes are large*:

$$n_1 p_1 \geq 10 \quad n_1(1 - p_1) \geq 10 \quad n_2 p_2 \geq 10 \quad n_2(1 - p_2) \geq 10$$

a large-sample confidence interval for $\pi_1 - \pi_2$ is

$$(p_1 - p_2) \pm (z \text{ critical value})\sqrt{\frac{p_1(1 - p_1)}{n_1} + \frac{p_2(1 - p_2)}{n_2}}$$

Example 11.11 Opinions on Freedom of Speech

The article "Freedom of What?" (*Associated Press*, February 1, 2005) described a study in which high school students and high school teachers were asked whether they agreed with the following statement: "Students should be allowed to report controversial issues in their student newspapers without the approval of school authorities." It was reported that 58% of the students surveyed and 39% of the teachers surveyed agreed with the statement. The two samples 10,000 high school students and 8,000 high school teachers were selected from 544 different schools across the country.

We will use the given information to estimate the difference between the proportion of high school students who agree that students should be allowed to report controversial issues in their student newspapers without the approval of school authorities, π_1, and the proportion of high school teachers who agree with the statement, π_2.

The sample sizes are large enough for the large sample interval to be valid ($n_1 p_1 = 10,000(.58) \geq 10$, $n_1(1 - p_1) = 10,000(.42) \geq 10$, etc.). A 90% confidence interval for $\pi_1 - \pi_2$ is

$$(p_1 - p_2) \pm (z \text{ critical value})\sqrt{\frac{p_1(1 - p_1)}{n_1} + \frac{p_2(1 - p_2)}{n_2}}$$

$$= (.58 - .39) \pm (1.645)\sqrt{\frac{(.58)(.42)}{10,000} + \frac{(.39)(.61)}{8000}}$$

$$= .19 \pm (1.645)(.0074)$$

$$= .19 \pm .012$$

$$(.178, .202)$$

Statistical software or a graphing calculator could also have been used to compute the endpoints of the confidence interval. MINITAB output is shown here.

Test and CI for Two Proportions

Sample	X	N	Sample p
1	5800	10000	0.580000
2	3120	8000	0.390000

Difference = p (1) − p (2)
Estimate for difference: 0.19
90% CI for difference: (0.177902, 0.202098)
Test for difference = 0 (vs not = 0): Z = 25.33 P-Value = 0.000

Assuming that it is reasonable to regard these two samples as being independently selected and also that they are representative of the two populations of interest, we can say that we believe that the proportion of high school students who agree that students should be allowed to report controversial issues in their student newspapers without the approval of school authorities exceeds that for teachers by somewhere between .178 and .202. We used a method to construct this estimate that captures the true difference in proportions 90% of the time in repeated sampling. ■

■ **Exercises 11.44–11.60** .... Turn to the Explanded Answers Section for answers not shown next to exercises. ....

11.44 A university is interested in evaluating registration processes. Students can register for classes by using either a telephone registration system or an online system that is accessed through the university's web site. Independent random samples of 80 students who registered by phone and 60 students who registered online were selected. Of those who registered by phone, 57 reported that they were satisfied with the registration process. Of those who registered online, 50 reported that they were satisfied. Based on these data, is it reasonable to conclude that the proportion who are satisfied is higher for those who register online? Test the appropriate hypotheses using $\alpha = .05$. $z = -1.67$, P-value $= 0.0475$, reject H_0

11.45 Some commercial airplanes recirculate approximately 50% of the cabin air in order to increase fuel efficiency. The authors of the paper "Aircraft Cabin Air Recirculation and Symptoms of the Common Cold" (*Journal of the American Medical Association* [2002]: 483–486) studied 1100 airline passengers who flew from San Francisco to Denver between January and April 1999. Some passengers traveled on airplanes that recirculated air and others traveled on planes that did not recirculate air. Of the 517 passengers who flew on planes that did not recirculate air, 108

reported post-flight respiratory symptoms, while 111 of the 583 passengers on planes that did recirculate air reported such symptoms. Is there sufficient evidence to conclude that the proportion of passengers with post-flight respiratory symptoms differs for planes that do and do not recirculate air? Test the appropriate hypotheses using $\alpha = .05$. You may assume that it is reasonable to regard these two samples as being independently selected and as representative of the two populations of interest. $z = 0.77$, P-value $= 0.4412$, fail to reject H_0

11.46 "Doctors Praise Device That Aids Ailing Hearts" (*Associated Press*, November 9, 2004) is the headline of an article that describes the results of a study of the effectiveness of a fabric device that acts like a support stocking for a weak or damaged heart. In the study, 107 people who consented to treatment were assigned at random to either a standard treatment consisting of drugs or the experimental treatment that consisted of drugs plus surgery to install the stocking. After two years, 38% of the 57 patients receiving the stocking had improved and 27% of the patients receiving the standard treatment had improved. Do these data provide convincing evidence that the proportion of patients who improve is higher for the experimental treatment than

---

for the standard treatment? Test the relevant hypotheses using a significance level of .05. $z = 1.21$, P-value $= 0.1131$, fail to reject H_0

11.47 The article "Portable MP3 Player Ownership Reaches New High" (*Ipsos Insight*, June 29, 2006) reported that in 2006, 20% of those in a random sample of 1112 Americans age 12 and older indicated that they owned an MP3 player. In a similar survey conducted in 2005, only 15% reported owning an MP3 player. Suppose that the 2005 figure was also based on a random sample of size 1112. Estimate the difference in the proportion of Americans age 12 and older who owned an MP3 player in 2006 and the corresponding proportion for 2005 using a 95% confidence interval. Is zero included in the interval? What does this tell you about the change in this proportion from 2005 to 2006?

11.48 The article referenced in the previous exercise also reported that 24% of the males and 16% of the females in the 2006 sample reported owning an MP3 player. Suppose that there were the same number of males and females in the sample of 1112. Do these data provide convincing evidence that the proportion of females that owned an MP3 player in 2006 is smaller than the corresponding proportion of males? Carry out a test using a significance level of .01. $z = 3.33$, P-value $= 0.0004$, reject H_0

11.49 Olestra is a fat substitute approved by the FDA for use in snack foods. To investigate reports of gastrointestinal problems associated with olestra consumption, an experiment was carried out to compare olestra potato chips with regular potato chips ("Gastrointestinal Symptoms Following Consumption of Olestra or Regular Triglyceride Potato Chips," *Journal of the American Medical Association* [1998]: 150–152). Subjects were assigned at random to either the olestra chip group or the regular chip group. Of the 529 individuals in the regular chip group, 93 experienced gastrointestinal symptoms, whereas 90 of the 563 individuals in the olestra chip group experienced symptoms. Carry out a hypothesis test at the 5% significance level to decide whether the proportion of individuals who experience symptoms after consuming olestra chips differs from the proportion who experience symptoms after consuming regular chips. $z = -0.71$, P-value $= 0.4778$, fail to reject H_0

11.50 Public Agenda conducted a survey of 1379 parents and 1342 students in grades 6–12 regarding the importance of science and mathematics in the school curriculum (Associated Press, February 15, 2006). It was reported that 50% of students thought that understanding science and having strong math skills are essential for them to succeed in life after school, whereas 62% of the parents thought it was crucial for today's students to learn science and higher-level math. The two samples—parents and students—were selected independently of one another. Is there sufficient evidence to conclude that the proportion of parents who regard science and mathematics as crucial is different than the corresponding proportion for students in grades 6–12? Test the relevant hypotheses using a significance level of .05. $z = -6.32$, P-value $= 0$, reject H_0

11.51 The authors of the paper "Inadequate Physician Knowledge of the Effects of Diet on Blood Lipids and Lipoproteins" (*Nutrition Journal* [2003]: 19–26) summarize the responses to a questionnaire on basic knowledge of nutrition that was mailed to 6000 physicians selected at random from a list of physicians licensed in the United States. Sixteen percent of those who received the questionnaire completed and returned it. The authors report that 26 of 120 cardiologists and 222 of 419 internists did not know that carbohydrate was the diet component most likely to raise triglycerides.

a. Estimate the difference between the proportion of cardiologists and the proportion of internists who did not know that carbohydrate was the diet component most likely to raise triglycerides using a 95% confidence interval.

b. What potential source of bias might limit your ability to generalize the estimate from Part (a) to the populations of all cardiologists and all internists? The sample consisted only of volunteers.

11.52 The article "Spray Flu Vaccine May Work Better Than Injections for Tots" (*San Luis Obispo Tribune*, May 2, 2006) described a study that compared flu vaccine administered by injection and flu vaccine administered as a nasal spray. Each of the 8000 children under the age of 5 that participated in the study received both a nasal spray and an injection, but only one was the real vaccine and the other was salt water. At the end of the flu season, it was determined that 3.9% of the 4000 children receiving the real vaccine by nasal spray got sick with the flu and 8.6% of the 4000 receiving the real vaccine by injection got sick with the flu.

a. Why would the researchers give every child both a nasal spray and an injection?

b. Use the given data to estimate the difference in the proportion of children who get sick with the flu after being vaccinated with an injection and the proportion of children who get sick with the flu after being vaccinated with the nasal spray using a 99% confidence interval. Based on the confidence interval, would you conclude that the proportion of children who get the flu is different for the two vaccination methods? $(-0.061, -0.033)$ Yes, because 0 is not included in the interval.

11.53 "Smartest People Often Dumbest About Sunburns" is the headline of an article that appeared in the *San Luis Obispo Tribune* (July 19, 2006). The article states that "those with a college degree reported a higher incidence of sunburn that those without a high school degree— 43 percent versus 25 percent." For purposes of this exercise, suppose that these percentages were based on random samples of size 200 from each of the two groups of interest (college graduates and those without a high school degree). Is there convincing evidence that the proportion experiencing a sunburn is higher for college graduates than it is for those without a high school degree? Answer based on a test with a .05 significance level. $z = 3.86$, P-value ≈ 0, reject H_0

11.54 The following quote is from the article "Canadians Are Healthier Than We Are" (*Associated Press*, May 31, 2006): "The Americans also reported more heart disease and major depression, but those differences were too small to be statistically significant." This statement was based on the responses of a sample of 5183 Americans and a sample of 3505 Canadians. The proportion of Canadians who reported major depression was given as .082.

a. Assuming that the researchers used a significance level of .05, could the sample proportion of Americans reporting major depression have been as large as .09? Explain why or why not.

b. Assuming that the researchers used a significance level of .05, could the sample proportion of Americans reporting major depression have been as large as .10? Explain why or why not.

11.55 Researchers at the National Cancer Institute released the results of a study that examined the effect of weed-killing herbicides on house pets (Associated Press, September 4, 1991). Dogs, some of whom were from homes where the herbicide was used on a regular basis, were examined for the presence of malignant lymphoma. The following data are compatible with summary values given in the report:

Group	Sample Size	Number with Lymphoma	p
Exposed	827	473	.572
Unexposed	130	19	.146

Estimate the difference between the proportion of exposed dogs that develop lymphoma and the proportion of unexposed dogs that develop lymphoma using a 95% confidence interval. $(.357, .495)$

11.56 The article "A 'White' Name Found to Help in Job Search" (*Associated Press*, January 15, 2003) described an experiment to investigate if it helps to have a "white-sounding" first name when looking for a job. Researchers sent 5000 resumes in response to ads that appeared in the *Boston Globe* and *Chicago Tribune*. The resumes were identical except that 2500 of them had "white-sounding" first names, such as Brett and Emily, whereas the other 2500 had "black-sounding" names such as Tamika and Rasheed. Resumes of the first type elicited 250 responses and resumes of the second type only 167 responses. Do these data support the theory that the proportion receiving positive responses is higher for those resumes with "white-sounding first" names? $z = 4.25$, P-value ≈ 0, reject H_0

11.57 "Mountain Biking May Reduce Fertility in Men, Study Says" was the headline of an article appearing in the *San Luis Obispo Tribune* (December 3, 2002). This conclusion was based on an Austrian study that compared sperm counts of avid mountain bikers (those who ride at least 12 hours per week) and nonbikers. Ninety percent of the avid mountain bikers studied had low sperm counts, as compared to 26% of the nonbikers. Suppose that these percentages were based on independent samples of 100 avid mountain bikers and 100 nonbikers and that it is reasonable to view these samples as representative of Austrian avid mountain bikers and nonbikers.

a. Do these data provide convincing evidence that the proportion of Austrian avid mountain bikers with low sperm count is higher than the proportion of Austrian nonbikers? $z = 9.17$, P-value $= 0$, reject H_0

b. Based on the outcome of the test in Part (a), is it reasonable to conclude that mountain biking 12 hours per week or more causes low sperm count? Explain.   No, this was an observational study.

11.58 In a study of a proposed approach for diabetes prevention, 339 people under the age of 20 who were thought to be at high risk of developing type I diabetes were assigned at random to two groups. One group received twice-daily injections of a low dose of insulin. The other group (the control) did not receive any insulin, but was closely monitored. Summary data (from the article "Diabetes Theory Fails Test," *USA Today*, June 25, 2001) follow.

Group	n	Number Developing Diabetes
Insulin	169	25
Control	170	24

a. Use the given data to construct a 90% confidence interval for the difference in the proportion that develop diabetes for the control group and the insulin group.
b. Give an interpretation of the confidence interval and the associated confidence level.
c. Based on your interval from Part (a), write a few sentences commenting on the effectiveness of the proposed prevention treatment.

11.59 Women diagnosed with breast cancer whose tumors have not spread may be faced with a decision between two surgical treatments—mastectomy (removal of the breast) or lumpectomy (only the tumor is removed). In a long-term study of the effectiveness of these two treatments, 701 women with breast cancer were randomly assigned to one of two treatment groups. One group received mastec-

tomies and the other group received lumpectomies and radiation. Both groups were followed for 20 years after surgery. It was reported that there was no statistically significant difference in the proportion surviving for 20 years for the two treatments (*Associated Press*, October 17, 2002). What hypotheses do you think the researchers tested in order to reach the given conclusion? Did the researchers reject or fail to reject the null hypothesis? $H_0: \pi_1 - \pi_2 = 0$ vs. $H_a: \pi_1 - \pi_2 \neq 0$, fail to reject

11.60 In December 2001, the Department of Veterans Affairs announced that it would begin paying benefits to soldiers suffering from Lou Gehrig's disease who had served in the Gulf War (*The New York Times*, December 11, 2001). This decision was based on an analysis in which the Lou Gehrig's disease incidence rate (the proportion developing the disease) for the approximately 700,000 soldiers sent to the Gulf between August 1990 and July 1991 was compared to the incidence rate for the approximately 1.8 million other soldiers who were not in the Gulf during this time period. Based on these data, explain why it is not appropriate to perform a formal inference procedure (such as the two-sample z test) and yet it is still reasonable to conclude that the incidence rate is higher for Gulf War veterans than for those who did not serve in the Gulf War.

Bold exercises answered in back   ● Data set available online but not required   ▼ Video solution available

11.4 Interpreting and Communicating the Results of Statistical Analyses

Many different types of research involve comparing two populations or treatments. It is easy to find examples of the two-sample hypothesis tests introduced in this chapter in published sources in a wide variety of disciplines.

■ Communicating the Results of Statistical Analyses

As was the case with one-sample hypothesis tests, it is important to include a description of the hypotheses, the test procedure used, the value of the test statistic and the *P*-value, and a conclusion in context when summarizing the results of a two-sample test.

Correctly interpreting confidence intervals in the two-sample case is more difficult than in the one-sample case, so take particular care when providing a two-sample confidence interval interpretation. Because the two-sample confidence intervals of this chapter estimate a difference ($\mu_1 - \mu_2$ or $\pi_1 - \pi_2$), the most important thing to note is whether or not the interval includes 0. If both endpoints of the interval are positive, then it is correct to say that, based on the interval, you believe that μ_1 is greater than μ_2 (or that π_1 is greater than π_2 if you are working with proportions) and then the interval provides an estimate of how much greater. Similarly, if both interval endpoints are negative, you would say that μ_1 is less than μ_2 (or that π_1 is less than π_2), with the interval providing an estimate of the size of the difference. If 0 is included in the interval, it is plausible that μ_1 and μ_2 (or π_1 and π_2) are equal.

▪ Interpreting the Results of Statistical Analyses ....

As with one-sample tests, it is common to find only the value of the test statistic and the associated *P*-value (or sometimes only the *P*-value) in published reports. You may have to think carefully about the missing steps to determine whether or not the conclusions are justified.

▪ What to Look For in Published Data ....

Here are some questions to consider when you are reading a report that contains the result of a two-sample hypothesis test or confidence interval:

- ▪ Are only two groups being compared? If more than two groups are being compared two at a time, then a different type of analysis is preferable (see Chapter 15).
- ▪ Were the samples selected independently, or were the samples paired? If the samples were paired, was the performed analysis appropriate for paired samples?
- ▪ If a confidence interval is reported, is it correctly interpreted as an estimate of a population or treatment difference in means or proportions?
- ▪ What hypotheses are being tested? Is the test one- or two-tailed?
- ▪ Does the validity of the test performed depend on any assumptions about the sampled populations (such as normality)? If so, do the assumptions appear to be reasonable?
- ▪ What is the *P*-value associated with the test? Does the *P*-value lead to rejection of the null hypothesis?
- ▪ Are the conclusions consistent with the results of the hypothesis test? In particular, if H_0 was rejected, does this indicate practical significance or only statistical significance?

For example, the paper "Ginkgo for Memory Enhancement" (*Journal of the American Medical Association* [2003]: 835–840) included the following statement in the summary of conclusions from an experiment where participants were randomly assigned to receive ginkgo or a placebo:

> Figure 2 shows the 95% confidence intervals (CIs) for differences (treatment group minus control) for performance on each test in the modified intent-to-treat analysis. Each interval contains a zero, indicating that none of the differences are statistically significant.

Because participants were assigned at random to the two treatments and the sample sizes were large (115 in each sample), use of the two-sample t confidence interval was appropriate. The 95% confidence intervals included in the paper (for example, $(-1.71, 0.65)$ and $(-2.25, 0.20)$ for two different measures of logical memory) did all include 0 and were interpreted correctly in the quoted conclusion.

As another example, we consider a study reported in the article "The Relationship Between Distress and Delight in Males' and Females' Reactions to Frightening Films" (*Human Communication Research* [1991]: 625–637).The investigators measured emotional responses of 50 males and 60 females after the subjects viewed a segment from a horror film. The article included the following statement: "Females were much more likely to express distress than were males. While males did express higher levels of delight than females, the difference was not statistically significant." The following summary information was also contained in the article:

Gender	Distress Index Mean	Delight Index Mean
Males	31.2	12.02
Females	40.4	9.09
	P-value < .001	Not significant (P-value > .05)

The P-values are the only evidence of the hypothesis tests that support the given conclusions. The P-value < .001 for the distress index means that the hypothesis $H_0: \mu_F - \mu_M = 0$ was rejected in favor of $H_a: \mu_F - \mu_M > 0$, where μ_F and μ_M are the true mean distress indexes for females and males, respectively.

The nonsignificant P-value (P-value > .05) reported for the delight index means that the hypothesis $H_0: \mu_F - \mu_M = 0$ (where μ_F and μ_M now refer to mean delight index for females and males, respectively) could not be rejected. Chance sample-to-sample variability is a plausible explanation for the observed difference in sample means (12.02 − 9.09). Thus we would not want to put much emphasis on the author's statement that males express higher levels of delight than females, because it is based only on the fact that 12.02 > 9.09, which could plausibly be due entirely to chance.

The article describes the samples as consisting of undergraduates selected from the student body of a large Midwestern university. The authors extrapolate their results to American men and women in general. If this type of generalization is considered unreasonable, we could be more conservative and view the sampled populations as male and female university students or male and female Midwestern university students or even male and female students at this particular university.

The comparison of males and females was based on two independently selected groups (not paired). Because the sample sizes were large, the two-sample t test for means could reasonably have been used, and this would have required no specific assumptions about the two underlying populations.

In a newspaper article, you may find even less information than in a journal article. For example, the article "Prayer Is Little Help to Some Heart Patients, Study Shows" (*Chicago Tribune*, March 31, 2006) included the following paragraphs:

> Bypass patients who consented to take part in the experiment were divided randomly into three groups. Some patients received prayers but were not informed of that. In the second group the patients got no prayers, and also were not informed one way or the other. The third group got prayers and were told so.
>
> There was virtually no difference in complication rates between the patients in the first two groups. But the third group, in which patients knew they were receiving prayers, had a complication rate of 59 percent—significantly more than the rate of 52 percent in the no-prayer group.

Earlier in the article, the total number of participants in the experiment was given as 1800. The author of this article has done a good job of describing the important aspects of the experiment. The final comparison in the quoted paragraph was probably based on a two-sample z test for proportions, comparing the sample proportion with complications for the 600 patients in the no-prayer group with the sample proportion with complications for the 600 participants who knew that someone was praying for them. For the reported sample sizes and sample proportions, the test statistic for testing $H_0: \pi_1 - \pi_2 = 0$ versus $H_a: \pi_1 - \pi_2 < 0$ (where π_1 represents the complication proportion for patients who did not receive prayers and π_2 represents the complication proportion for patients who knew they were receiving prayers) is $z = -2.10$. The associated P-value is .036, supporting the conclusion stated in the article.

■ A Word to the Wise: Cautions and Limitations ....

The three cautions that appeared at the end of Chapter 10 apply here as well. They were (see Chapter 10 for more detail):

1. Remember that the result of a hypothesis test can never show strong support for the null hypothesis. In two-sample situations, this means that we shouldn't be *convinced* that there is no difference between population means or proportions based on the outcome of a hypothesis test.
2. If you have complete information (a census) of both populations, there is no need to carry out a hypothesis test or to construct a confidence interval—in fact, it would be inappropriate to do so.
3. Don't confuse statistical significance and practical significance. In the two-sample setting, it is possible to be convinced that two population means or proportions are not equal even in situations where the actual difference between them is small enough that it is of no practical interest. After rejecting a null hypothesis of no difference (statistical significance), it is useful to look at a confidence interval estimate of the difference to get a sense of practical significance.

And here's one new caution to keep in mind for two-sample tests:

4. Be sure to think carefully about how the data were collected, and make sure that an appropriate test procedure or confidence interval is used. A common mistake is to overlook pairing and to analyze paired samples as if they were independent. The question, Are the samples paired? is usually easy to answer—you just have to remember to ask!

Activity 11.1 Helium-Filled Footballs?

Technology activity: Requires Internet access.

Background: Do you think that a football filled with helium will travel farther than a football filled with air? Two researchers at the Ohio State University investigated this question by performing an experiment in which 39 people each kicked a helium-filled football and an air-filled football. Half were assigned to kick the air-filled football first and then the helium-filled ball, whereas the other half kicked the helium-filled ball first followed by the air-filled ball. Distance (in yards) was measured for each kick.

In this activity, you will use the Internet to obtain the data from this experiment and then carry out a hypothesis test to determine whether the mean distance is greater for helium-filled footballs than for air-filled footballs.

1. Do you think that helium-filled balls will tend to travel farther than air-filled balls when kicked? Before looking at the data, write a few sentences indicating what you think the outcome of this experiment was and describing the reasoning that supports your prediction.
2. The data from this experiment can be found in the Data and Story Library at the following web site:

http://lib.stat.cme.edu/DASL/Datafiles/heliumfootball.html

Go to this web site and print out the data for the 39 trials.
3. There are two samples in this data set. One consists of distances traveled for the 39 kicks of the air-filled football, and the other consists of the 39 distances for the helium-filled football. Are these samples independent or paired? Explain.
4. Carry out an appropriate hypothesis test to determine whether there is convincing evidence that the mean

distance traveled is greater for a helium-filled football than for an air-filled football.
5. Is the conclusion in the test of Step 4 consistent with your initial prediction of the outcome of this experiment? Explain.
6. Write a paragraph for the sports section of your school newspaper describing this experiment and the conclusions that can be drawn from it.

Activity 11.2   Thinking About Data Collection

Background: In this activity you will design two experiments that would allow you to investigate whether people tend to have quicker reflexes when reacting with their dominant hand than with their nondominant hand.
1. Working in a group, design an experiment to investigate the given research question that would result in independent samples. Be sure to describe how you plan to measure quickness of reflexes, what extraneous variables will be directly controlled, and the role that randomization plays in your design.
2. How would you modify the design from Step 1 so that the resulting data are paired? Is the way in which random-

ization is incorporated into the new design different from the way it is incorporated in the design from Step 1? Explain.
3. Which of the two proposed designs would you recommend, and why?
4. If assigned to do so by your instructor, carry out one of your experiments and analyze the resulting data. Write a brief report that describes the experimental design, includes both graphical and numerical summaries of the resulting data, and communicates the conclusions that follow from your data analysis.

Activity 11.3   A Meaningful Paragraph

Write a meaningful paragraph that includes the following six terms: **paired samples**, **significantly different**, *P*-value, **sample**, **population**, **alternative hypothesis**.
   A "meaningful paragraph" is a coherent piece of writing in an appropriate context that uses all of the listed words. The paragraph should show that you understand the

meaning of the terms and their relationship to one another. A sequence of sentences that just define the terms is *not* a meaningful paragraph. When choosing a context, think carefully about the terms you need to use. Choosing a good context will make writing a meaningful paragraph easier,

Summary of Key Concepts and Formulas

Term or Formula	Comment
Independent samples	Two samples where the individuals or objects in the first sample are selected independently from those in the second sample.
Paired samples	Two samples for which each observation in one sample is paired in a meaningful way with a particular observation in a second sample.
$t = \dfrac{(\bar{x}_1 - \bar{x}_2) - \text{hypothesized value}}{\sqrt{\dfrac{s_1^2}{n_1} + \dfrac{s_2^2}{n_2}}}$	The test statistic for testing $H_0: \mu_1 - \mu_2 =$ hypothesized value when the samples are independently selected and the sample sizes are large or it is reasonable to assume that both population distributions are normal.

Term or Formula	**Comment**
$(\bar{x}_1 - \bar{x}_2) \pm (t \text{ critical value})\sqrt{\dfrac{s_1^2}{n_1} + \dfrac{s_2^2}{n_2}}$	A formula for constructing a confidence interval for $\mu_1 - \mu_2$ when the samples are independently selected and the sample sizes are large or it is reasonable to assume that the population distributions are normal.
$\text{df} = \dfrac{(V_1 + V_2)^2}{\dfrac{V_1^2}{n_1 - 1} + \dfrac{V_2^2}{n_2 - 1}}$ where $V_1 = \dfrac{s_1^2}{n_1}$ and $V_2 = \dfrac{s_2^2}{n_2}$	The formula for determining df for the two-sample t test and confidence interval.
$\bar{x}_d$	The sample mean difference.
s_d	The standard deviation of the sample differences.
μ_d	The mean value for the population of differences.
σ_d	The standard deviation for the population of differences.
$t = \dfrac{\bar{x}_d - \text{hypothesized value}}{\dfrac{s_d}{\sqrt{n}}}$	The paired t test statistic for testing $H_0: \mu_d = \text{hypothesized value.}$
$\bar{x}_d \pm (t \text{ critical value})\dfrac{s_d}{\sqrt{n}}$	The paired t confidence interval formula.
$p_c = \dfrac{n_1 p_1 + n_2 p_2}{n_1 + n_2}$	p_c is the statistic for estimating the common population proportion when $\pi_1 = \pi_2$.
$z = \dfrac{p_1 - p_2}{\sqrt{\dfrac{p_c(1 - p_c)}{n_1} + \dfrac{p_c(1 - p_c)}{n_2}}}$	The test statistic for testing $H_0: \pi_1 - \pi_2 = 0$ when both sample sizes are large.
$(p_1 - p_2) \pm (z \text{ critical value})\sqrt{\dfrac{p_1(1 - p_1)}{n_1} + \dfrac{p_2(1 - p_2)}{n_2}}$	A formula for constructing a confidence interval for $\pi_1 - \pi_2$ when both sample sizes are large.

Chapter Review Exercises 11.61–11.90

Turn to the Explanded Answers Section for answers not shown next to exercises.

CENGAGENOW Know exactly what to study! Take a pre-test and receive your Personalized Learning Plan.

11.61 When a surgeon repairs injuries, sutures (stitched knots) are used to hold together and stabilize the injured area. If these knots elongate and loosen through use, the injury may not heal properly because the tissues would not be optimally positioned. Researchers at the University of California, San Francisco, tied a series of different types of knots with two types of suture material, Maxon and Ticron.

Suppose that 112 tissue specimens were available and that for each specimen the type of knot and suture material were randomly assigned. The investigators tested the knots to see how much the loops elongated; the elongations (in mm) were measured and the resulting data are summarized here. For purposes of this exercise, assume it is reasonable to regard the elongation distributions as approximately normal.

Bold exercises answered in back   ● Data set available online but not required   ▼ Video solution available

Types of knot	Maxon		
	n	$\bar{x}$	sd
Square (control)	10	10.0	.1
Duncan Loop	15	11.0	.3
Overhand	15	11.0	.9
Roeder	10	13.5	.1
Snyder	10	13.5	2.0

Types of knot	Ticron		
	n	$\bar{x}$	sd
Square (control)	10	2.5	.06
Duncan Loop	11	10.9	.40
Overhand	11	8.1	1.00
Roeder	10	5.0	.04
Snyder	10	8.1	.06

a. Is there a significant difference in elongation between the square knot and the Duncan loop for Maxon thread?
b. Is there a significant difference between the square knot and the Duncan loop for Ticron thread?
c. For the Duncan loop data, is there a significant difference between the elongations of Maxon versus Ticron threads? $t = 0.698$, df $= 17$, P-value ≈ 0.247, fail to reject H_0

11.62 Snake experts believe that venomous snakes inject different amounts of venom when killing their prey. Researchers at the University of Wyoming tested this hypothesis to determine whether young prairie rattlesnakes use more venom to kill larger mice and less venom for smaller mice ("Venom Metering by Juvenile Prairie Rattlesnakes, Crotalus v. Viridis: Effects of Prey Size and Experience," *Animal Behavior* [1995]: 33–40). In the first trial, the researchers used three groups of seven randomly selected snakes that were inexperienced hunters. In the second trial, three different groups of seven randomly selected "experienced" snakes were used. The amount (mg) of venom each snake injected was recorded and categorized according to the size of the mouse.

	Small Mouse		Medium Mouse		Large Mouse	
	$\bar{x}$	s	$\bar{x}$	s	$\bar{x}$	s
Inexperienced	3.1	1.0	3.4	.4	1.8	.3
Experienced	2.6	.3	2.9	.6	4.7	.3

For the small prey, is there a significant difference between the inexperienced and experienced snakes? For the medium prey? For the large prey?

11.63 The coloration of male guppies may affect the mating preference of the female guppy. To test this hypothesis, scientists first identified two types of guppies, Yarra and Paria, that display different colorations ("Evolutionary Mismatch of Mating Preferences and Male Colour Patterns in Guppies," *Animal Behaviour* [1997]: 343–51). The relative area of orange was calculated for fish of each type. A random sample of 30 Yarra guppies resulted in a mean relative area of .106 and a standard deviation of .055. A random sample of 30 Paria guppies resulted in a mean relative area of .178 and a standard deviation .058. Is there evidence of a difference in coloration? Test the relevant hypotheses to determine whether the mean area of orange is different for the two types of guppies. $t = -4.94$, df $= 57$, P-value ≈ 0, reject H_0

11.64 The article "Trial Lawyers and Testosterone: Blue-Collar Talent in a White-Collar World" (*Journal of Applied Social Psychology* [1998]: 84–94) compared trial lawyers and nontrial lawyers on the basis of mean testosterone level. Random samples of 35 male trial lawyers, 31 male nontrial lawyers, 13 female trial lawyers, and 18 female nontrial lawyers were selected for study. The article includes the following statement: "Trial lawyers had higher testosterone levels than did nontrial lawyers. This was true for men, $t(64) = 3.75$, $p < .001$, and for women, $t(29) = 2.26$, $p < .05$."
a. Based on the information given, is there a significant difference in the mean testosterone level for male trial and nontrial lawyers? Yes
b. Based on the information given, is there a significant difference in the mean testosterone level for female trial and nontrial lawyers? Yes
c. Do you have enough information to carry out a test to determine whether there is a significant difference in the mean testosterone levels of male and female trial lawyers? If so, carry out such a test. If not, what additional information would you need to be able to conduct the test?

11.65 The article "So Close, Yet So Far: Predictors of Attrition in College Seniors" (*Journal of College Student Development* [1999]: 343–354) attempts to describe differences between college seniors who disenroll before graduating and those who do graduate. Researchers randomly selected 42 nonreturning and 48 returning seniors, none of whom were transfer students. These 90 students rated themselves on personal contact and campus involvement. The resulting data are summarized here:

	Returning (n = 48)		Nonreturning (n = 42)	
	Mean	Standard Deviation	Mean	Standard Deviation
Personal Contact	3.22	.93	2.41	1.03
Campus Involvement	3.21	1.01	3.31	1.03

a. Construct and interpret a 95% confidence interval for the difference in mean campus involvement rating for returning and nonreturning students. Does your interval support the statement that students who do not return are less involved, on average, than those who do? Explain.
b. Do students who don't return have a lower mean personal contact rating than those who do return? Test the relevant hypotheses using a significance level of .01.

11.66 The article "Workaholism in Organizations: Gender Differences" (*Sex Roles* [1999]: 333–346) gave the following data on 1996 income (in Canadian dollars) for random samples of male and female MBA graduates from a particular Canadian business school:

	n	$\bar{x}$	s
Males	258	$133,442	$131,090
Females	233	$105,156	$98,525

a. For what significance levels would you conclude that the mean salary of female MBA graduates of this business school is above $100,000? For any value of $\alpha > 0.212$
b. Is there convincing evidence that the mean salary for female MBA graduates of this business school is lower than the mean salary for the male graduates? $t = 2.72$, df = 473, P-value ≈ 0.003, reject H_0
11.67 The article "The Relationship of Task and Ego Orientation to Sportsmanship Attitudes and the Perceived Legitimacy of Injurious Acts" (*Research Quarterly for Exercise and Sport* [1991]: 79–87) examined the extent of approval of unsporting play and cheating. High school basketball players completed a questionnaire that was

used to arrive at an approval score, with higher scores indicating greater approval. A random sample of 56 male players resulted in a mean approval rating for unsportsmanlike play of 2.76, whereas the mean for a random sample of 67 female players was 2.02. Suppose that the two sample standard deviations were .44 for males and .41 for females. Is it reasonable to conclude that the mean approval rating is higher for male players than for female players by more than .5? Use $\alpha = .05$. $t = 3.1$, df = 113, P-value ≈ 0.001, reject H_0

11.68 ● In a study of memory recall, eight students from a large psychology class were selected at random and given 10 min to memorize a list of 20 nonsense words. Each was asked to list as many of the words as he or she could remember both 1 hr and 24 hr later, as shown in the accompanying table. Is there evidence to suggest that the mean number of words recalled after 1 hr exceeds the mean recall after 24 hr by more than 3? Use a level .01 test. $t = 0.86$, df = 7, P-value ≈ 0.21, fail to reject H_0

Subject	1	2	3	4	5	6	7	8
1 hr later	14	12	18	7	11	9	16	15
24 hr later	10	4	14	6	9	6	12	12

11.69 As part of a study to determine the effects of allowing the use of credit cards for alcohol purchases in Canada (see "Changes in Alcohol Consumption Patterns Following the Introduction of Credit Cards in Ontario Liquor Stores," *Journal of Studies on Alcohol* [1999]: 378–382), randomly selected individuals were given a questionnaire asking them (among other things) how many drinks they had consumed during the previous week. A year later (after liquor stores started accepting credit cards for purchases), these same individuals were again asked how many drinks they had consumed in the previous week. The data shown are consistent with summary statistics presented in the article.

	n	1994 Mean	1995 Mean	$\bar{d}$	s_d
Credit-Card Shoppers	96	6.72	6.34	.38	5.52
Non-Credit Card Shoppers	850	4.09	3.97	.12	4.58

a. The standard deviation of the difference was quite large. Explain how this could be the case.
b. Calculate a 95% confidence interval for the mean difference in drink consumption for credit card shoppers

between 1994 and 1995. Is there evidence that the mean number of drinks decreased? $(-0.738, 1.498)$

c. Test the hypothesis that there was no change in the mean number of drinks between 1994 and 1995 for the non-credit card shoppers. Be sure to calculate and interpret the P-value for this test. $t = 0.764$, df = 849, P-value = 0.445, fail to reject H_0

11.70 ● Several methods of estimating the number of seeds in soil samples have been developed by ecologists. An article in the *Journal of Ecology* ("A Comparison of Methods for Estimating Seed Numbers in the Soil" [1990]: 1079–1093) considered three such methods. The accompanying data give number of seeds detected by the direct method and by the stratified method for 27 soil specimens.

Specimen	Direct	Stratified	Specimen	Direct	Stratified
1	24	8	2	32	36
3	0	8	4	60	56
5	20	52	6	64	64
7	40	28	8	8	8
9	12	8	10	92	100
11	4	0	12	68	56
13	76	68	14	24	52
15	32	28	16	0	0
17	36	36	18	16	12
19	92	92	20	4	12
21	40	48	22	24	24
23	0	0	24	8	12
25	12	40	26	16	12
27	40	76			

Do the data provide sufficient evidence to conclude that the mean number of seeds detected differs for the two methods? Test the relevant hypotheses using $\alpha = .05$.
$t = -1.34$, df = 26, P-value = 0.192, fail to reject H_0

11.71 Many people who quit smoking complain of weight gain. The results of an investigation of the relationship between smoking cessation and weight gain are given in the article "Does Smoking Cessation Lead to Weight Gain?" (*American Journal of Public Health* [1983]: 1303–1305). Three hundred twenty-two subjects, selected at random from those who successfully participated in a program to quit smoking, were weighed at the beginning of the program and again 1 year later. The mean change in weight was 5.15 lb, and the standard deviation of the weight

changes was 11.45 lb. Is there sufficient evidence to conclude that the true mean change in weight is positive? Use $\alpha = .05$. $z = 8.07$, P-value ≈ 0, reject H_0

11.72 Do teachers find their work rewarding and satisfying? The article "Work-Related Attitudes" (*Psychological Reports* [1991]: 443–450) reported the results of a survey of random samples of 395 elementary school teachers and 266 high school teachers. Of the elementary school teachers, 224 said they were very satisfied with their jobs, whereas 126 of the high school teachers were very satisfied with their work. Based on these data, is it reasonable to conclude that the proportion very satisfied is different for elementary school teachers than it is for high school teachers? Test the appropriate hypotheses using a .05 significance level. $z = 2.36$, P-value = 0.0182, reject H_0

11.73 The article "Foraging Behavior of the Indian False Vampire Bat" (*Biotropica* [1991]: 63–67) reported that 36 of 193 female bats in flight spent more than 5 min in the air before locating food. For male bats, 64 of 168 spent more than 5 min in the air. Is there sufficient evidence to conclude that the proportion of flights longer than 5 min in length differs for males and females? Test the relevant hypotheses using $\alpha = .01$. $z = -4.12$, P-value ≈ 0, reject H_0

11.74 Gender differences in student needs and fears were examined in the article "A Survey of Counseling Needs of Male and Female College Students" (*Journal of College Student Development* [1998]: 205–208). Random samples of male and female students were selected from those attending a particular university. Of 234 males surveyed, 27.5% said that they were concerned about the possibility of getting AIDS. Of 568 female students surveyed, 42.7% reported being concerned about the possibility of getting AIDS. Is there sufficient evidence to conclude that the proportion of female students concerned about the possibility of getting AIDS is greater than the corresponding proportion for males? $z = -4.09$, P-value ≈ 0, reject H_0

11.75 The state of Georgia's HOPE scholarship program guarantees fully paid tuition to Georgia public universities for Georgia high school seniors who have a B average in academic requirements as long as they maintain a B average in college. (See "Who Loses HOPE? Attrition from Georgia's College Scholarship Program" (*Southern Economic Journal* [1999]: 379–390).) It was reported that 53.2% of a random sample of 137 students entering Ivan Allen College at Georgia Tech (social science and human-

Bold exercises answered in back ● Data set available online but not required ▼ Video solution available

ities) with a HOPE scholarship lost the scholarship at the end of the first year because they had a GPA of less than 3.0. It was also reported that 72 of a random sample of 111 students entering the College of Computing with a B average had lost their HOPE scholarship by the end of the first year. Is there evidence that the proportion who lose HOPE scholarships is different for the College of Computing than for the Ivan Allen College? $z = -1.84$, P-value $= 0.066$, fail to reject H_0

11.76 An Associated Press article (*San Luis Obispo Telegram-Tribune*, September 23, 1995) examined the changing attitudes of Catholic priests. National surveys of priests aged 26 to 35 were conducted in 1985 and again in 1993. The priests surveyed were asked whether they agreed with the following statement: Celibacy should be a matter of personal choice for priests. In 1985, 69% of those surveyed agreed; in 1993, 38% agreed. Suppose that the samples were randomly selected and that the sample sizes were both 200. Is there evidence that the proportion of priests who agreed that celibacy should be a matter of personal choice declined from 1985 to 1993? Use $\alpha = .05$. $z = 6.21$, P-value ≈ 0, reject H_0

11.77 Are college students who take a freshman orientation course more or less likely to stay in college than those who do not take such a course? The article "A Longitudinal Study of the Retention and Academic Performance of Participants in Freshmen Orientation Courses" (*Journal of College Student Development* [1994]: 444–449) reported that 50 of 94 randomly selected students who did not participate in an orientation course returned for a second year. Of 94 randomly selected students who did take the orientation course, 56 returned for a second year. Construct a 95% confidence interval for $\pi_1 - \pi_2$, the difference in the proportion returning for students who do not take an orientation course and those who do. Give an interpretation of this interval. ($-0.2053, 0.0777$)

11.78 The article "Truth and DARE: Tracking Drug Education to Graduation" (*Social Problems* [1994]: 448–456) compared the drug use of 238 randomly selected high school seniors exposed to a drug education program (DARE) and 335 randomly selected high school seniors who were not exposed to such a program. Data for marijuana use are given in the accompanying table. Is there evidence that the proportion using marijuana is lower for students exposed to the DARE program? Use $\alpha = .05$. $z = -1.263$, P-value $= 0.1033$, fail to reject H_0

	n	Number Who Use Marijuana
Exposed to DARE	288	141
Not Exposed to DARE	335	181

11.79 The article "Softball Sliding Injuries" (*American Journal of Diseases of Children* [1988]: 715–716) provided a comparison of breakaway bases (designed to reduce injuries) and stationary bases. Consider the accompanying data (which agree with summary values given in the paper).

	Number of Games Played	Number of Games Where a Player Suffered a Sliding Injury
Stationary Bases	1250	90
Breakaway Bases	1250	20

Does the use of breakaway bases reduce the proportion of games with a player suffering a sliding injury? Answer by performing a level .01 test. What assumptions are necessary in order for your test to be valid? Do you think they are reasonable in this case? $z = 6.83$, P-value ≈ 0, reject H_0

11.80 The positive effect of water fluoridation on dental health is well documented. One study that validates this is described in the article "Impact of Water Fluoridation on Children's Dental Health: A Controlled Study of Two Pennsylvania Communities" (*American Statistical Association Proceedings of the Social Statistics Section* [1981]: 262–265). Two communities were compared. One had adopted fluoridation in 1966, whereas the other had no such program. Of 143 randomly selected children from the town without fluoridated water, 106 had decayed teeth, and 67 of 119 randomly selected children from the town with fluoridated water had decayed teeth. Let π_1 denote the true proportion of children drinking fluoridated water who have decayed teeth, and let π_2 denote the analogous proportion for children drinking unfluoridated water. Estimate $\pi_1 - \pi_2$ using a 90% confidence interval. Does the interval contain 0? Interpret the interval.

Bold exercises answered in back ● Data set available online but not required ▼ Video solution available

11.81 Wayne Gretzky was one of ice hockey's most prolific scorers when he played for the Edmonton Oilers. During his last season with the Oilers, Gretzky played in 41 games and missed 17 games due to injury. The article "The Great Gretzky" (*Chance* [1991]: 16–21) looked at the number of goals scored by the Oilers in games with and without Gretzky, as shown in the accompanying table. If we view the 41 games with Gretzky as a random sample of all Oiler games in which Gretzky played and the 17 games without Gretzky as a random sample of all Oiler games in which Gretzky did not play, is there evidence that the mean number of goals scored by the Oilers is higher for games in which Gretzky played? Use $\alpha = .01$.
$t = 2.43$, df $= 32$, P-value ≈ 0.0105, fail to reject H_0

	n	Sample Mean	Sample sd
Games with Gretzky	41	4.73	1.29
Games without Gretzky	17	3.88	1.18

11.82 Here's one to sink your teeth into: The authors of the article "Analysis of Food Crushing Sounds During Mastication: Total Sound Level Studies" (*Journal of Texture Studies* [1990]: 165–178) studied the nature of sounds generated during eating. Peak loudness (in decibels at 20 cm away) was measured for both open-mouth and closed-mouth chewing of potato chips and of tortilla chips. Forty subjects participated, with ten assigned at random to each combination of conditions (such as closed-mouth, potato chip, and so on). We are not making this up! Summary values taken from plots given in the article appear in the accompanying table.

	n	$\overline{x}$	s
Potato Chip			
Open mouth	10	63	13
Closed mouth	10	54	16
Tortilla Chip			
Open mouth	10	60	15
Closed mouth	10	53	16

a. Construct a 95% confidence interval for the difference in mean peak loudness between open-mouth and closed-mouth chewing of potato chips. Interpret the resulting interval. $(-4.75, 22.75)$

b. For closed-mouth chewing (the recommended method!), is there sufficient evidence to indicate that there is a difference between potato chips and tortilla chips with respect to mean peak loudness? Test the relevant hypotheses using $\alpha = .01$.
c. The means and standard deviations given here were actually for stale chips. When ten measurements of peak loudness were recorded for closed-mouth chewing of fresh tortilla chips, the resulting mean and standard deviation were 56 and 14, respectively. Is there sufficient evidence to conclude that fresh tortilla chips are louder than stale chips? Use $\alpha = .05$. $t = 0.4462$, df $= 17$, P-value ≈ 0.331, fail to reject H_0

11.83 Are very young infants more likely to imitate actions that are modeled by a person or simulated by an object? This question was the basis of a research study summarized in the article "The Role of Person and Object in Eliciting Early Imitation" (*Journal of Experimental Child Psychology* [1991]: 423–433). One action examined was mouth opening. This action was modeled repeatedly by either a person or a doll, and the number of times that the infant imitated the behavior was recorded. Twenty-seven infants participated, with 12 exposed to a human model and 15 exposed to the doll. Summary values are given here. Is there sufficient evidence to conclude that the mean number of imitations is higher for infants who watch a human model than for infants who watch a doll? Test the relevant hypotheses using a .01 significance level.
$t = 2.94$, df $= 21$, P-value ≈ 0.0039, reject H_0

	Person Model	Doll Model
$\overline{x}$	5.14	3.46
s	1.60	1.30

11.84 Dentists make many people nervous (even more so than statisticians!). To see whether such nervousness elevates blood pressure, the blood pressure and pulse rates of 60 subjects were measured in a dental setting and in a medical setting ("The Effect of the Dental Setting on Blood Pressure Measurement," *American Journal of Public Health* [1983]: 1210–1214). For each subject, the difference (dental-setting blood pressure minus medical-setting blood pressure) was calculated. The analogous differences were also calculated for pulse rates. Summary data follows.

	Mean Difference	Standard Deviation of Differences
Systolic Blood Pressure	4.47	8.77
Pulse (beats/min)	−1.33	8.84

a. Do the data strongly suggest that true mean blood pressure is higher in a dental setting than in a medical setting? Use a level .01 test. $t = 3.95$, df $= 59$, P-value ≈ 0, reject H_0
b. Is there sufficient evidence to indicate that true mean pulse rate in a dental setting differs from the true mean pulse rate in a medical setting? Use a significance level of .05. $t = −1.165$, df $= 59$, P-value ≈ 0.248, fail to reject H_0

11.85 Do teenage boys worry more than teenage girls? This is one of the questions addressed by the authors of the article "The Relationship of Self-Esteem and Attributional Style to Young People's Worries" (*Journal of Psychology* [1987]: 207–215). A scale called the Worries Scale was administered to a group of teenagers, and the results are summarized in the accompanying table.

Gender	n	Sample Mean Score	Sample sd
Girls	108	62.05	9.5
Boys	78	67.59	9.7

Is there sufficient evidence to conclude that teenage boys score higher on the Worries Scale than teenage girls? Use a significance level of $\alpha = .05$. $t = 3.88$, df $= 163$, P-value ≈ 0, reject H_0

11.86 Data on self-esteem, leadership ability, and GPA were used to compare college students who were hired as resident assistants at Mississippi State University and those who were not hired ("Wellness as a Factor in University Housing," *Journal of College Student Development* [1994]: 248–254), as shown in the accompanying table.

	Self-Esteem		Leadership		GPA	
	$\bar{x}$	s	$\bar{x}$	s	$\bar{x}$	s
Hired ($n = 69$)	83.28	12.21	62.51	3.05	2.94	.61
Not Hired ($n = 47$)	81.96	12.78	62.43	3.36	2.60	.79

a. Do the data suggest that the true mean self-esteem score for students hired by the university as resident assistants differs from the mean for students who are not hired? Test the relevant hypotheses using $\alpha = .05$.
b. Do the data suggest that the true mean leadership score for students hired by the university as resident assistants differs from the mean for students who are not hired? Test the relevant hypotheses using $\alpha = .05$.
c. Do the data suggest that the true mean GPA for students hired by the university as resident assistants differs from the mean for students who are not hired? Test the relevant hypotheses using $\alpha = .05$. $t = 2.4882$, df $= 811$, P-value ≈ 0.0148, reject H_0

11.87 The article "Religion and Well-Being Among Canadian University Students: The Role of Faith Groups on Campus" (*Journal for the Scientific Study of Religion* [1994]: 62–73) compared the self-esteem of students who belonged to Christian clubs and students who did not belong to such groups. Each student in a random sample of $n = 169$ members of Christian groups (the affiliated group) completed a questionnaire designed to measure self-esteem. The same questionnaire was also completed by each student in a random sample of $n = 124$ students who did not belong to a religious club (the unaffiliated group). The mean self-esteem score for the affiliated group was 25.08, and the mean for the unaffiliated group was 24.55. The sample standard deviations weren't given in the article, but suppose that they were 10 for the affiliated group and 8 for the unaffiliated group. Is there evidence that the true mean self-esteem score differs for affiliated and unaffiliated students? Test the relevant hypotheses using a significance level of .01. $t = 0.504$, df $= 288$, P-value ≈ 0.6150, fail to reject H_0

11.88 Key terms in survey questions too often are not well understood, and such ambiguity can affect responses. As an example, the article "How Unclear Terms Affect Survey Data" (*Public Opinion Quarterly* [1992]: 218–231) described a survey in which each individual in a sample was asked, "Do you exercise or play sports regularly?" But what constitutes exercise? The following revised question was then asked of each individual in the same sample: "Do you do any sports or hobbies involving physical activities, or any exercise, including walking, on a regular basis?" The resulting data are shown in the accompanying table.

	Yes	No
Initial Question	48	52
Revised Question	60	40

Is there any difference between the true proportions of yes responses to these questions? Can a procedure from this chapter be used to answer the question posed? If yes, use it; if not, explain why not.

11.89 An electronic implant that stimulates the auditory nerve has been used to restore partial hearing to a number of deaf people. In a study of implant acceptability (*Los Angeles Times*, January 29, 1985), 250 adults born deaf and 250 adults who went deaf after learning to speak were followed for a period of time after receiving an implant. Of those deaf from birth, 75 had removed the implant, whereas only 25 of those who went deaf after learning to speak had done so. Does this suggest that the true proportion who remove the implants differs for those that were born deaf and those that went deaf after learning to speak? Test the relevant hypotheses using a .01 significance level. $z = 5.59$, P-value ≈ 0, reject H_0

11.90 ● Samples of both surface soil and subsoil were taken from eight randomly selected agricultural locations

in a particular county. The soil samples were analyzed to determine both surface pH and subsoil pH, with the results shown in the accompanying table.

Location	1	2	3	4	5	6	7	8
Surface pH	6.55	5.98	5.59	6.17	5.92	6.18	6.43	5.68
Subsoil pH	6.78	6.14	5.80	5.91	6.10	6.01	6.18	5.88

a. Compute a 90% confidence interval for the true average difference between surface and subsoil pH for agricultural land in this county. ($-.1857$, $.1107$)
b. What assumptions are necessary to validate the interval in Part (a)? Need to assume that the pH distributions are normal.

Personal Tutor

Do you need a live tutor for homework problems?

CENGAGENOW™

Are you ready? Take your exam-prep post-test now.

Bold exercises answered in back ● Data set available online but not required ▼ Video solution available

Graphing Calculator Explorations

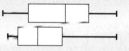

Figure 11.2 Boxplots of the two samples.

Exploration 11.1 Testing Hypotheses

Using your calculator to construct confidence intervals and test hypotheses about differences between two population means involves actions similar to working with a single mean. You must navigate the menu system of your calculator to the point where these procedures can be selected. Of course, by now you are an old hand at such navigation! We will use Example 11.2 to illustrate an independent *t* hypothesis test of equal population means.

Enter the data in your calculator, using the lists of your choice. (We will use List1 and List2.) We will check the assumption of approximate normality of the populations by displaying box plots of the two data sets, as shown in Figure 11.2. (Be sure to choose the modified box plot option!)

After verifying the plausibility of the normality of the populations, navigate your calculator's menu system to choose a procedure. For the two-sample *t* test, the common calculator screens are not large enough to show all relevant data, so in Figure 11.3 we again show "larger-than-life" representations rather than capture the actual screens.

Notice from Figure 11.3 that you have choices similar to the hypothesis testing for a single population mean, with one exception: pooling. Pooling with two samples has fallen into disfavor among statisticians, so your choice here should be "Off" or "No." The subsequent output also requires more than one calculator screen, and we therefore show a "big screen" representation in Figure 11.4.

Figure 11.3 Windows from two different calculators for the two-sample *t* test.

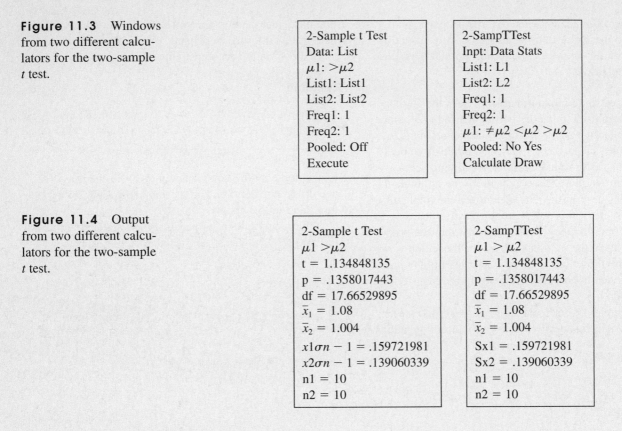

2-Sample t Test
Data: List
$\mu 1: >\mu 2$
List1: List1
List2: List2
Freq1: 1
Freq2: 1
Pooled: Off
Execute

2-SampTTest
Inpt: Data Stats
List1: L1
List2: L2
Freq1: 1
Freq2: 1
$\mu 1: \neq \mu 2 <\mu 2 >\mu 2$
Pooled: No Yes
Calculate Draw

Figure 11.4 Output from two different calculators for the two-sample *t* test.

2-Sample t Test
$\mu 1 >\mu 2$
t = 1.134848135
p = .1358017443
df = 17.66529895
$\bar{x}_1 = 1.08$
$\bar{x}_2 = 1.004$
$x1\sigma n - 1 = .159721981$
$x2\sigma n - 1 = .139060339$
n1 = 10
n2 = 10

2-SampTTest
$\mu 1 > \mu 2$
t = 1.134848135
p = .1358017443
df = 17.66529895
$\bar{x}_1 = 1.08$
$\bar{x}_2 = 1.004$
Sx1 = .159721981
Sx2 = .139060339
n1 = 10
n2 = 10

Once again we will remind you that although this information is "all there" in the sense that the results of the hypothesis test are completely displayed, it is *not* in a form that should be used for communicating your results to others. You should still complete and report all of the appropriate steps for a hypothesis test.

▪ Confidence Intervals

2-Sample t Interval
Data: List
C-Level: .95
List1: List1
List2: List2
Freq1: 1
Freq2: 1
Pooled: Off
Execute

2-SampTInt
Inpt: Data Stats
List1: L1
List2: L2
Freq1: 1
Freq2: 1
C-Level: .95
Pooled: No Yes
Calculate

Figure 11.5 Windows from two different calculators for the two-sample *t* confidence interval.

We will now turn our attention to confidence intervals for the difference between two population means, using data from Example 11.4.

Enter the data in your calculator, using the lists of your choice. (We again use List1 and List2.) After verifying the plausibility of the normality of the populations, navigate the calculator's menu system to choose a procedure. For the two-sample *t* confidence interval the data entry will be the same as for a hypothesis test, and there should be an additional piece of information needed: the confidence level. Two potential displays are shown in Figure 11.5. Remember, your calculator may display only some of this information, and you will thus have to scroll to see it all.

We can see from the screen representations in Figure 11.5 that the data are contained in a list and that the confidence levels are 95%. Once again, choose *not* to pool. The results from the calculations are displayed in Figure 11.6. These boxes represent two common displays calculators use to report confidence intervals.

Figure 11.6 Output from two different calculators for the two-sample *t* confidence interval.

2-Sample tInterval
Left = 1.0461
Right = 11.954
df = 13.7763
$\bar{x}_1$ = 108.75
$\bar{x}_2$ = 102.25
$x1\sigma n - 1$ = 4.7434
$x2\sigma n - 1$ = 5.3918
n1 = 8
n2 = 8

2-SampTInt
(1.0461, 11.954)
df = 13.77631629
$\bar{x}_1$ = 108.75
$\bar{x}_2$ = 102.25
Sx1 = 4.74341649
Sx2 = 5.3917927
n1 = 8
n2 = 8

Exploration 11.2 Inferences About Differences in Means Using Paired Samples

Procedures for making inferences about the difference between two population means using paired samples is similar to inference using a single sample. To test hypotheses or to construct confidence intervals using these differences, you must enter the observations in your calculator, and create the differences by subtraction. We will use Example 11.6 to illustrate how this is done on the calculator. Recall that the data are from the study of memory improvement from taking chess lessons and playing daily.

If you are following along with your calculator, enter the pretest data into List1 and the posttest data into List2. After entering the data, use list commands to subtract the posttest scores from the pretest scores (i.e., pre-test minus post-test) and store the result in List3:

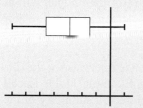

Figure 11.7 Edit data screen.

List1 − List2 → List3

At this point, you should see something like Figure 11.7 in your data edit screen. The list of interest for our paired *t* procedures is List3. We proceed from here as we did with inference for a single sample. Because the sample size in this example is small, the approximate normality of the differences is an important consideration—we must assess the plausibility of a normal population *of differences*.

The boxplot for the differences (List3) shown in Figure 11.8 does not indicate any difficulties with the assumption of approximate normality, so we proceed to the hypothesis test and construction of a confidence interval as discussed in Graphing Calculator Exploration 11.1. The results are shown in Figure 11.9. The alternative hypothesis is that the mean difference would be less than 0, a one-tailed test. In the case of the confidence interval we used a 95% confidence level. You should compare the results shown in Figure 11.9 with your calculator results to check that you have correctly implemented the paired *t* calculator process. Remember, the calculator will typically give answers with more decimal accuracy than is shown in the text, so slight differences should not be cause for worry.

Figure 11.8 Boxplot of differences.

Exploration 11.3 Inference about Differences in Proportions

Using your calculator for making inferences about the difference in population proportions is similar to the procedure for making inferences about the difference in population means using independent samples. Once again, navigate your calculator's menu system, this time looking for something like hypothesis tests and confidence intervals for "2 sample proportions."

Figure 11.9 Calculator output for the *t*-test and confidence interval windows.

T-Test
$\mu < 0$
$t = -4.563958109$
$p = 4.0568244E\text{-}4$
$\overline{x} = -144.5833333$
$Sx = 109.7405687$
$n = 12$

Tinterval
$(-214.3, -74.86)$
$\overline{x} = -144.5833333$
$Sx = 109.7405687$
$n = 12$

■ **Testing Hypotheses** Data entry for a test of the null hypothesis of equal population proportions is very simple. Once you correctly identify the number of "successes" in the problem at hand, the most difficult part is remembering to check the large sample assumptions! We will use Example 11.9 for illustrative purposes. Recall that this example considered the efficacy of duct tape in removing warts. The data on successful wart removal are summarized in the accompanying table.

	Liquid Nitrogen Treatment	Duct Tape Treatment
Sample Size	$n_1 = 100$	$n_2 = 104$
Number of Successes	60	88
Sample Proportion	$p_1 = 60/100 = .6$	$p_2 = 88/104 = .846$

2-PropZTest
x1: 60
n1: 100
x2: 88
n2: 104
p1: $\neq p2 < p2 > p2$
Calculate Draw

Figure 11.10 Setup window for a two-proportion *z* test.

2-PropZTest
p1 < p2
$z = -3.938342737$
$p = 4.1040568E\text{-}5$
$\hat{p}1 = .6$
$\hat{p}2 = .8461538462$
$\hat{p} = .7254901961$
$n1 = 100$
$n2 = 104$

Figure 11.11 Output screen for a two-proportion *z* test.

After transferring this data to the calculator screen, you should have something similar to the Figure 11.10.

Notice that on this calculator the notation for the hypothesized population proportions is *p* rather than π. Notice also that you must choose your alternative hypothesis from among the three listed—in Example 11.9, the alternative hypothesis is H_a: $\pi_1 - \pi_2 < 0$, equivalent to $\pi_1 < \pi_2$. (On this calculator, the corresponding alternative hypothesis would be selected by moving the cursor to "<p2" and pressing Enter.)

When Enter is pressed, we are again presented with more information than will fit on one screen, so Figure 11.11 is another "larger than life" screen. This calculator screen presents enough information to reconstruct the hypothesis test of the example.

■ **Confidence Intervals** We now turn our attention to confidence intervals for the difference between two population proportions, using data from Example 11.11. Recall that this example looked at opinions on freedom of speech. Here is a summary of the data:

Group	Sample Size	Number in Agreement	P
Students	10,000	5,800	.58
Teachers	8,000	3,120	.39

After navigating your calculator's menu system, you should find a screen for data entry. The entry of the data is similar to that for the hypothesis test, but here you

```
2-PropZInt
x1:5800
n1:10000
x2:3120
n2:8000
C-Level: .90
Calculate
```

Figure 11.12 Setup window for a two-proportion z interval.

```
2-PropZInt
(.1779, .2021)
p̂1 .5800
p̂2 .3900
n1=10000.0000
n2=8000.0000
```

Figure 11.13 Output screen for a two-proportion z interval.

must specify the confidence level, which was unnecessary for the hypothesis test. Figure 11.12 shows a representative screen.

After entering the appropriate values and pressing Enter, you should see something like Figure 11.13.

Notice that there is no actual point estimate of the difference in population proportions on this screen. If you want to express your confidence interval in point estimate $\pm$ error bound form, you must find the point estimate by subtracting the sample proportions. The values to add and subtract from the point estimate can be found by computing one-half the interval length given in the screen. These calculations are illustrated here:

point estimate $= .5800 - .3900 = .1900$
error bound $= (.2021 - .1779)/2 = .0121$.

The 90% confidence interval is then $.1900 \pm .0121$. Except for the effects of rounding, these values are in agreement with Example 11.11.

TEACHING TIPS

Intent of this chapter:

After completing this chapter, students should be able to (1) test hypotheses about the distribution of a categorical variable using a goodness-of-fit chi-square test, (2) test the hypothesis that the distribution of a categorical variable is the same for two or more populations using a chi-square test of homogeneity, and (3) test hypotheses about the association between two categorical variables using a chi-square test of independence.

A PowerPoint® Lecture is available on the *Instructor's Resource Binder* CD.

■ Section 12.1

Show the students a die and ask them what it means for the die to be "fair." (An unusually shaped die or a rigged die makes this activity more interesting. These types of die are available for purchase online.) Students usually understand that a "fair" die means that, in the long run, the proportion of rolls that land as one is equal to the proportion of rolls that land as two, etc. In symbols, $\pi_1 = \pi_2 = \pi_3 = \pi_4 = \pi_5 = \pi_6$. A chi-square goodness-of-fit test can be performed to test the null hypothesis that the die is fair. A short discussion of the chi-square distribution is helpful. The chi-square curve is a continuous, unimodal curve described by its degrees of freedom. The graphing calculator can be used to investigate the chi-square curves (see page 652). By graphing one curve at a time, students can see how the chi-square curve changes as the degrees of freedom increase.

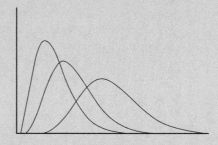

An important point to stress with students is that the chi-square test statistic *must* be computed using counts (the number observed in each category), not proportions (the subject of the hypotheses.) The test statistic formula, $X^2 = \sum \dfrac{(\text{observed count} - \text{expected count})^2}{\text{expected count}}$, and the associated assumptions are the same for all three chi-square tests. Students should verify that the assumptions are plausible before performing the test. The assumptions are as follows: random sample (independent random samples for the test of homogeneity) and large sample size(s) (which is verified by making sure all *expected* counts are at least 5). While the assumptions and formula are the same for all three chi-square tests, the hypotheses, the way in which expected counts are obtained, and the way that degrees of freedom are computed differ.

There is a discussion about the X^2 value on the top of page 652. A word of caution: Just because the X^2 value is small (say 6 or 7) doesn't mean that the test results will not be significant. A X^2 value of 6 or 7 could be significant if the number of degrees of freedom is small.

Section 12.1 describes the goodness-of-fit chi-square test. The goodness-of-fit test is used with univariate, categorical data from a single population. The observed counts are typically displayed in a one-way frequency table. Be sure to write the hypotheses in terms of the population, using the parameters π_i. For the goodness-of-fit test, the expected count for each category is obtained by multiplying the sample size (n) times the hypothesized proportion for that category (π_i). However, the number of degrees of freedom is computed as the number of categories (k) minus 1 (df $= k - 1$). Note: Although this may look familiar, this is a different procedure than what we used for t-distributions. (See the tan box on page 653.)

Refer back to the die example. Have each student roll the die to be tested ten times, recording the numbers rolled. Combine the data from the class. Perform a goodness-of-fit test to see if there is evidence that the die is not fair.

A graphing calculator can be used to perform the goodness-of-fit test. Input the observed counts into L1 and the expected counts into L2. See Graphing Calculator Exploration 12.1 on page 685 for instructions. The P-value can be computed using the following command: χ^2cdf(lower boundary, upper boundary, df).

Suggested Assignment: Exercises 12.1, 12.3, 12.6, 12.9, 12.11

Activity: Activity 12.1: Pick a Number Any Number . . . (p. 680 and *Activities Manual*, p. 102)★, "Chapter 12—Goodness-of-Fit Test" (*Instructor's Resource Binder* & CD, Chapter 12 Activities Worksheets) **(Note: The percentage for the color distribution of M&Ms listed in this activity is out of date; the company has changed their distribution! See the M&M website for the most current color distribution.)**

Assessment: Chapter 12 Quiz 1 (*Test Bank* and *Instructor's Resource Binder* CD)

■ **Section 12.2**

This section discusses both the chi-square test of homogeneity and the chi-square test for independence. The computational procedures for these two tests are identical *but* the hypotheses are different, and therefore written differently for the two tests.

For both tests, observed counts are summarized in a two-way frequency table. The formula for calculating the expected counts is as follows:

$$\text{expected count} = \frac{(\text{row total})(\text{column total})}{\text{grand total}}$$

The expected counts can be displayed with the corresponding observed counts in the two-way table by placing the parentheses around the expected counts. The number of degrees of freedom is obtained by using the formula df $=$ (number of rows $-$ 1)(number of columns $-$ 1) or by covering up one row and one column and counting the remaining cells. (Note: The cells in the total row or column are not used in computing the df.)

The following two scenarios help to clarify the difference in chi-square tests of homogeneity and independence for students.

Suppose we survey a random sample of boys from our school about which type of movie they prefer to see: action, drama, or comedy. We survey a random sample of girls from our school, asking them the same question. (Key point: There were two independently selected random samples.) We record only the movie preference for each student. We are interested in determining if the movie preferences are different for males and females. These are univariate, categorical data (movie preference) from two independent samples (boys and girls). A chi-square test of *homogeneity* is used in this situation. (See the tan box on page 664.)

Now suppose we take a single random sample of *students* from our school and ask each student which type of movie he or she prefers to see, but we record gender as well as preference for each student. We are interested in determining if there is an association between gender and movie preferences. These are bivariate, categorical data (gender and movie preference) from a single sample (students). A chi-square test of *independence* is used in this situation. (See the tan box on page 668.)

★ Kathy's personal favorite

See Graphing Calculator Exploration 12.2 on page 686 for instructions for the graphing calculator.

Suggested Assignment: Exercises 12.15, 12.16, 12.20, 12.23, 12.29, 12.36

Activity: Activity 12.2: Color and Perceived Taste (p. 680 and *Activities Manual*, p. 104)★, Bonus Activity 12.3—Peanut and Plain M&M Defects (*Activities Manual*, p. 106), "Chapter 12—Test of Homogeneity" (*Instructor's Resource Binder* & CD, Chapter 12 Activities Worksheets), and/or "Chapter 12—Test of Independence" (*Instructor's Resource Binder* & CD, Chapter 12 Activities Worksheets)

Assessment: Chapter 12 Quiz 2 (*Test Bank* and *Instructor's Resource Binder* CD)

■ **Section 12.3**

This section provides excellent suggestions for communicating and interpreting results of chi-square tests. Students should also pay close attention to "A Word to the Wise: Cautions and Limitations."

Suggested Review Assignment: Exercises 12.38, 12.39, 12.40, 12.42, 12.45, 12.47

Assessment: Chapter 12 Concept Quiz (*Test Bank* and *Instructor's Resource Binder* CD) and Chapter 12 Test (*Test Bank* and *Instructor's Resource Binder* CD)

★ Kathy's personal favorite

Chapter 12

The Analysis of Categorical Data and Goodness-of-Fit Tests

© Greg Fiume/NewSport/Corbis

It is often the case that information is collected on categorical variables, such as political affiliation, gender, or college major. As with numerical data, categorical data sets can be univariate (consisting of observations on a single categorical variable), bivariate (observations on two categorical variables), or even multivariate. In this chapter, we will first consider inferential methods for analyzing univariate categorical data sets and then turn to techniques appropriate for use with bivariate categorical data.

12.1 Chi-Square Tests for Univariate Data

Univariate categorical data sets arise in a variety of settings. If each student in a sample of 100 is classified according to whether he or she is enrolled full-time or part-time, data on a categorical variable with two categories result. Each airline passenger in a sample of 50 might be classified into one of three categories based on type of ticket—coach, business class, or first class. Each registered voter in a sample of 100 selected from those registered in a particular city might be asked which of the five city council members he or she favors for mayor. This would yield observations on a categorical variable with five categories.

Univariate categorical data are most conveniently summarized in a **one-way frequency table**. For example, the article "Fees Keeping American Taxpayers From Using Credit Cards to Make Tax Payments" (*IPSOS Insight*, March 24, 2006) surveyed American taxpayers regarding their intent to pay taxes with a credit card. Suppose that 100 randomly selected taxpayers participated in such a survey, with possible responses being definitely will use a credit card to pay taxes next year, probably will use a credit card, probably won't use a credit card, and definitely won't use a credit card. The first few observations might be

Probably will	Definitely will not	Probably will not
Probably will not	Definitely will	Definitely will not

Counting the number of observations of each type might then result in the following one-way table:

	Outcome			
	Definitely Will	Probably Will	Probably Will Not	Definitely Will Not
Frequency	14	12	24	50

For a categorical variable with k possible values (k different levels or categories), sample data are summarized in a one-way frequency table consisting of k cells, which may be displayed either horizontally or vertically.

In this section, we consider testing hypotheses about the proportion of the population that falls into each of the possible categories. For example, the manager of a tax preparation company might be interested in determining whether the four possible responses to the tax credit card question occur equally often. If this is indeed the case, the long-run proportion of responses falling into each of the four categories is 1/4, or .25. The test procedure to be presented shortly would allow the manager to decide whether the hypothesis that all four category proportions are equal to .25 is plausible.

Notation

k = number of categories of a categorical variable
π_1 = true proportion for Category 1
π_2 = true proportion for Category 2
$\vdots$
π_k = true proportion for Category k
      (Note: $\pi_1 + \pi_2 + \cdots + \pi_k = 1$)

The hypotheses to be tested have the form

H_0: π_1 = hypothesized proportion for Category 1
      π_2 = hypothesized proportion for Category 2
$\vdots$
      π_k = hypothesized proportion for Category k
H_a: H_0 is not true, so at least one of the true category proportions differs from the corresponding hypothesized value.

For the example involving responses to the tax survey, let

π_1 = the proportion of all taxpayers who will definitely pay by credit card
π_2 = the proportion of all taxpayers who will probably pay by credit card
π_3 = the proportion of all taxpayers who will probably not pay by credit card

and

π_4 = the proportion of all taxpayers who will definitely not pay by credit card

The null hypothesis of interest is then

$$H_0: \ \pi_1 = .25, \ \pi_2 = .25, \ \pi_3 = .25, \ \pi_4 = .25$$

A null hypothesis of the type just described can be tested by first selecting a random sample of size n and then classifying each sample response into one of the k possible categories. To decide whether the sample data are compatible with the null hypothesis, we compare the observed cell counts (frequencies) to the cell counts that would have been expected when the null hypothesis is true. The expected cell counts are

Expected cell count for Category 1 = $n\pi_1$
Expected cell count for Category 2 = $n\pi_2$

and so on. The expected cell counts when H_0 is true result from substituting the corresponding hypothesized proportion for each π_i.

Example 12.1 Births and the Lunar Cycle

● A common urban legend is that more babies than expected are born during certain phases of the lunar cycle, especially near the full moon. The paper "The Effect of the Lunar Cycle on Frequency of Births and Birth Complications" (*American Journal of Obstetrics and Gynecology* [2005]: 1462–1464) classified births according to the lunar cycle. Data for a sample of randomly selected births occurring during 24 lunar cycles consistent with summary quantities appearing in the paper are given in the accompanying table.

Lunar Phase	Number of Days	Number of Births
New moon	24	7680
Waxing crescent	152	48442
First quarter	24	7579
Waxing gibbous	149	47814
Full moon	24	7711
Waning gibbous	150	47595
Last quarter	24	7733
Waning crescent	152	48230

● Data set available online

Let's define lunar phase category proportions as follows:

π_1 = proportion of births that occur during the new moon
π_2 = proportion of births that occur during the waxing crescent moon
π_3 = proportion of births that occur during the first quarter moon
π_4 = proportion of births that occur during the waxing gibbous moon
π_5 = proportion of births that occur during the full moon
π_6 = proportion of births that occur during the waning gibbous moon
π_7 = proportion of births that occur during the last quarter moon
π_8 = proportion of births that occur during the waning crescent moon

If there is no relationship between number of births and the lunar cycle, then the number of births in each lunar cycle category should be proportional to the number of days included in that category. Since there are a total of 699 days in the 24 lunar cycles considered and 24 of those days are in the new moon category, if there is no relationship between number of births and lunar cycle,

$$\pi_1 = \frac{24}{699} = .0343$$

Similarly, in the absence of any relationship,

$$\pi_2 = \frac{152}{699} = .2175 \qquad \pi_3 = \frac{24}{699} = .0343$$

$$\pi_4 = \frac{149}{699} = .2132 \qquad \pi_5 = \frac{24}{699} = .0343$$

$$\pi_6 = \frac{150}{699} = .2146 \qquad \pi_7 = \frac{24}{699} = .0343$$

$$\pi_8 = \frac{152}{699} = .2175$$

The hypotheses of interest are then

H_0: $\pi_1 = .0343$, $\pi_2 = .2175$, $\pi_3 = .0343$, $\pi_4 = .2132$, $\pi_5 = .0343$, $\pi_6 = .2146$, $\pi_7 = .0343$, $\pi_8 = .2175$
H_a: H_0 is not true.

There were a total of 222,784 births in the sample, so if H_0 is true, the expected counts for the first two categories are

$$\begin{pmatrix} \text{expected count} \\ \text{for new moon} \end{pmatrix} = n \begin{pmatrix} \text{hypothesized proportion} \\ \text{for new moon} \end{pmatrix}$$
$$= 222{,}784(.0343) = 7641.49$$

$$\begin{pmatrix} \text{expected count} \\ \text{for waxing crescent} \end{pmatrix} = n \begin{pmatrix} \text{hypothesized proportion} \\ \text{for waxing crescent} \end{pmatrix}$$
$$= 222{,}784(.2175) = 48{,}455.52$$

Expected counts for the other six categories are computed in a similar fashion, and observed and expected cell counts are given in the following table.

Lunar Phase	Observed Number of Births	Expected Number of Births
New moon	7680	7641.49
Waxing crescent	48442	48455.52
First quarter	7579	7641.49
Waxing gibbous	47814	47497.55
Full moon	7711	7641.49
Waning gibbous	47595	47809.45
Last quarter	7733	7641.49
Waning crescent	48230	48455.52

Important concept

Because the observed counts are based on a *sample* of births, it would be somewhat surprising to see *exactly* 3.43% of the sample falling in the first category, exactly 21.75% in the second, and so on, even when H_0 is true. If the differences between the observed and expected cell counts can reasonably be attributed to sampling variation, the data are considered compatible with H_0. On the other hand, if the discrepancy between the observed and the expected cell counts is too large to be attributed solely to chance differences from one sample to another, H_0 should be rejected in favor of H_a. Thus, we need an assessment of how different the observed and expected counts are.

■

The goodness-of-fit statistic, denoted by X^2, is a quantitative measure of the extent to which the observed counts differ from those expected when H_0 is true. (The Greek letter χ is often used in place of X. The symbol X^2 is referred to as the chi-square [χ^2] statistic. In using X^2 rather than χ^2, we are adhering to the convention of denoting sample quantities by Roman letters.)

The **goodness-of-fit statistic**, X^2, results from first computing the quantity

$$\frac{(\text{observed cell count} - \text{expected cell count})^2}{\text{expected cell count}}$$

for each cell, where, for a sample of size n,

$$\left(\begin{array}{c}\text{expected cell} \\ \text{count}\end{array}\right) = n\left(\begin{array}{c}\text{hypothesized value of corresponding} \\ \text{population proportion}\end{array}\right)$$

The X^2 statistic is the sum of these quantities for all k cells:

$$X^2 = \sum_{\text{all cells}} \frac{(\text{observed cell count} - \text{expected cell count})^2}{\text{expected cell count}}$$

Important concept

The value of the X^2 statistic reflects the magnitude of the discrepancies between observed and expected cell counts. When the differences are sizable, the value of X^2 tends to be large. Therefore, large values of X^2 suggest rejection of H_0. A small value of X^2 (it can never be negative) occurs when the observed cell counts are quite similar to those expected when H_0 is true and so would be consistent with H_0.

As with previous test procedures, a conclusion is reached by comparing a P-value to the significance level for the test. The P-value is computed as the probability of observing a value of X^2 at least as large as the observed value when H_0 is true. This requires information about the sampling distribution of X^2 when H_0 is true.

When the null hypothesis is correct and the sample size is sufficiently large, the behavior of X^2 is described approximately by a **chi-square distribution**. A chi-square curve has no area associated with negative values and is asymmetric, with a longer tail on the right. There are actually many chi-square distributions, each one identified with a different number of degrees of freedom. Curves corresponding to several chi-square distributions are shown in Figure 12.1.

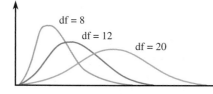

Figure 12.1 Chi-square curves.

For a test procedure based on the X^2 statistic, the associated P-value is the area under the appropriate chi-square curve and to the right of the computed X^2 value. Appendix Table 8 gives upper-tail areas for chi-square distributions with up to 20 df. Our chi-square table has a different appearance from the t table used in previous chapters. In the t table, there is a single "value" column on the far left and then a column of P-values (tail areas) for each different number of degrees of freedom. A single column of t values works for the t table because all t curves are centered at 0, and the t curves approach the z curve as the number of degrees of freedom increases. However, because the chi-square curves move farther and farther to the right and spread out more as the number of degrees of freedom increases, a single "value" column is impractical in this situation.

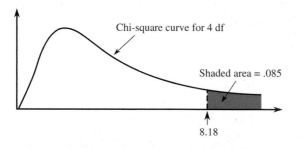

Figure 12.2 A chi-square upper-tail area.

To find the area to the right for a particular X^2 value, locate the appropriate df column in Appendix Table 8. Determine which listed value is closest to the X^2 value of interest, and read the right-tail area corresponding to this value from the left-hand column of the table. For example, for a chi-square distribution with df = 4, the area to the right of $X^2 = 8.18$ is .085, as shown in Figure 12.2. For this same chi-square distribution (df = 4), the area to the right of 9.70 is approximately .045 (the area to the right of 9.74, the closest entry in the table for df = 4).

It is also possible to use computer software or a graphing calculator to compute areas under a chi-square curve.

■ Goodness-of-Fit Tests ....

When H_0 is true, the X^2 goodness-of-fit statistic has approximately a chi-square distribution with df = $(k - 1)$, as long as none of the expected cell counts are too small. *It is generally agreed that use of the chi-square distribution is appropriate when the sample size is large enough for every expected cell count to be at least 5.* If any of the expected cell frequencies are less than 5, categories can be combined in a sensible way to create acceptable expected cell counts. Just remember to compute the number of degrees of freedom based on the reduced number of categories.

Goodness-of-Fit Test Procedure

Hypotheses: H_0: π_1 = hypothesized proportion for Category 1
$$\vdots$$
π_k = hypothesized proportion for Category k
H_a: H_0 is not true

Test statistic: $X^2 = \sum_{\text{all cells}} \dfrac{(\text{observed cell count} - \text{expected cell count})^2}{\text{expected cell count}}$

P-values: When H_0 is true and all expected counts are at least 5, X^2 has approximately a chi-square distribution with df $= k - 1$. Therefore, the P-value associated with the computed test statistic value is the area to the right of X^2 under the df $= k - 1$ chi-square curve. Upper-tail areas for chi-square distributions are found in Appendix Table 8.

Assumptions: 1. Observed cell counts are based on a *random sample*.
2. The *sample size is large*. The sample size is large enough for the chi-square test to be appropriate as long as every expected cell count is at least 5.

Example 12.2   Births and the Lunar Cycle Revisited

We use the births data of Example 12.1 to test the hypothesis that number of births is unrelated to lunar cycle. Let's use a .05 level of significance and the nine-step hypothesis-testing procedure illustrated in previous chapters.

1. Let $\pi_1, \pi_2, \pi_3, \pi_4, \pi_5, \pi_6, \pi_7$, and π_8 denote the proportions of all births falling in the eight lunar cycle categories as defined in Example 12.1.
2. H_0: $\pi_1 = .0343, \pi_2 = .2175, \pi_3 = .0343, \pi_4 = .2132, \pi_5 = .0343, \pi_6 = .2146, \pi_7 = .0343, \pi_8 = .2175$
3. H_a: H_0 is not true.
4. Significance level: $\alpha = .05$.
5. Test statistic: $X^2 = \sum_{\text{all cells}} \dfrac{(\text{observed cell count} - \text{expected cell count})^2}{\text{expected cell count}}$
6. Assumptions: The expected cell counts (from Example 12.1) are all greater than 5. The births represent a random sample of births occurring during the lunar cycles considered.
7. Calculation:
$$X^2 = \frac{(7680 - 7641.49)^2}{7641.49} + \frac{(48442 - 48455.52)^2}{48455.52} + \cdots + \frac{(48230 - 48455.52)^2}{48455.52}$$
$$= .194 + .004 + .511 + 2.108 + .632 + .962 + 1.096 + 1.050$$
$$= 6.557$$
8. P-value: The P-value is based on a chi-square distribution with df $= 8 - 1 = 7$. The computed value of X^2 is smaller than 12.01 (the smallest entry in the df $= 7$ column of Appendix Table 8), so P-value $> .10$.
9. Conclusion: Because P-value $> \alpha$, H_0 cannot be rejected. There is not sufficient evidence to conclude that number of births and lunar cycle are related. This is

consistent with the conclusion in the paper: "We found no statistical evidence that deliveries occurred in a predictable pattern across the phases of the lunar cycle."

MINITAB will perform a chi-square goodness-of-fit test. MINITAB output for the data and hypothesized proportions of this example is shown here.

Chi-Square Goodness-of-Fit Test for Observed Counts in Variable: Number of Births
Using category names in Lunar Phase

Category	Observed	Test Proportion	Expected	Contribution to Chi-Sq
First Quarter	7579	0.0343	7641.5	0.51105
Full Moon	7711	0.0343	7641.5	0.63227
Last Quarter	7733	0.0343	7641.5	1.09584
New Moon	7680	0.0343	7641.5	0.19406
Waning Crescent	48230	0.2175	48455.5	1.04961
Waning Gibbous	47595	0.2146	47809.4	0.96189
Waxing Crescent	48442	0.2175	48455.5	0.00377
Waxing Gibbous	47814	0.2132	47497.5	2.10835

N	DF	Chi-Sq	P-Value
222784	7	6.55683	0.476

Note that MINITAB has reordered the categories from smallest to largest based on the observed count. MINITAB also carried a bit more decimal accuracy in the computation of the chi-square statistic, reporting $X^2 = 6.55683$ and an associated P-value of .476. The computed P-value $= .476$ is consistent with the statement P-value $> .10$ from step 8 of the hypothesis test.

■

Example 12.3 Hybrid Car Purchases

● *USA Today* ("Hybrid Car Sales Rose 81% Last Year," April 25, 2005) reported the top five states for sales of hybrid cars in 2004 as California, Virginia, Washington, Florida, and Maryland. Suppose that each car in a sample of 2004 hybrid car sales is classified by state where the sale took place. Sales from states other than the top five were excluded from the sample, resulting in the accompanying table.

State	Observed Frequency
California	250
Virginia	56
Washington	34
Florida	33
Maryland	33
Total	**406**

● Data set available online

(The given observed counts are artificial, but they are consistent with hybrid sales figures given in the article.)

We will use the X^2 goodness-of-fit test and a significance level of $\alpha = .01$ to test the hypothesis that hybrid sales for these five states are proportional to the 2004 population for these states. 2004 population estimates from the Census Bureau web site are given in the following table. The population proportion for each state was computed by dividing each state population by the total population for all five states.

State	2004 Population	Population Proportion
California	35842038	0.495
Virginia	7481332	0.103
Washington	6207046	0.085
Florida	17385430	0.240
Maryland	5561332	0.077
Total	**72477178**	

If these same population proportions hold for hybrid car sales, the expected counts are

$$\text{Expected count for California} = 406(.495) = 200.970$$
$$\text{Expected count for Virginia} = 406(.103) = 41.818$$
$$\text{Expected count for Washington} = 406(.085) = 34.510$$
$$\text{Expected count for Florida} = 406(.240) = 97.440$$
$$\text{Expected count for Maryland} = 406(.077) = 31.362$$

These expected counts have been entered in Table 12.1.

Table 12.1   Observed and Expected Counts for Example 12.3

State	Observed Counts	Expected Counts
California	250	200.970
Virginia	56	41.818
Washington	34	34.510
Florida	33	97.440
Maryland	33	31.262

1. Let $\pi_1, \pi_2, \ldots, \pi_5$ denote the true proportion of hybrid car sales for the five states in the following order: California, Virginia, Washington, Florida, and Maryland.
2. H_0: $\pi_1 = .495$, $\pi_2 = .103$, $\pi_3 = .085$, $\pi_4 = .240$, $\pi_5 = .077$
3. H_a: H_0 is not true.
4. Significance level: $\alpha = .01$
5. Test statistic: $X^2 = \displaystyle\sum_{\text{all cells}} \dfrac{(\text{observed cell count} - \text{expected cell count})^2}{\text{expected cell count}}$

6. Assumptions: The sample was a random sample of hybrid car sales. All expected counts are greater than 5, so the sample size is large enough to use the chi-square test.

7. Calculation: From MINITAB

Chi-Square Goodness-of-Fit Test for Observed Counts in Variable: Hybrid Sales

Using category names in State

Category	Observed	Test Proportion	Expected	Contribution to Chi-Sq
California	250	0.495	200.970	11.9617
Florida	33	0.240	97.440	42.6161
Maryland	33	0.077	31.262	0.0966
Virginia	56	0.103	41.818	4.8096
Washington	34	0.085	34.510	0.0075

N	DF	Chi-Sq	P-Value
406	4	59.4916	0.000

8. P-value: All expected counts exceed 5, so the P-value can be based on a chi-square distribution with df $= 5 - 1 = 4$. From MINITAB, the P-value is 0.000.

9. Conclusion: Since P-value $\leq \alpha$, H_0 is rejected. There is convincing evidence that hybrid sales are not proportional to population size for at least one of the five states.

Based on the hybrid sales data, we have determined that there is convincing evidence that at least one of these five states has hybrid sales that are not proportional to population size. Looking back at the MINITAB output, notice that there is a column labeled "Contribution to Chi-Sq." This column shows the individual values of

$$\frac{(\text{observed cell count } - \text{ expected cell count})^2}{\text{expected cell count}}$$, which are summed to produce the

value of the chi-square statistic. Notice that the two states with the largest contribution to the chi-square statistic are Florida and California. For Florida, observed hybrid sales were smaller than expected (observed $= 33$, expected $= 97.44$), whereas for California observed sales were higher than expected (observed $= 250$, expected $= 200.970$).

■

Exercises 12.1–12.14

Turn to the Expanded Answers Section for answers not shown next to exercises.

12.1 From the given information in each case below, state what you know about the P-value for a chi-square test and give the conclusion for a significance level of $\alpha = .01$.

a. $X^2 = 7.5$, df $= 2$
b. $X^2 = 13.0$, df $= 6$
c. $X^2 = 18.0$, df $= 9$
d. $X^2 = 21.3$, df $= 4$
e. $X^2 = 5.0$, df $= 3$

12.2 A particular paperback book is published in a choice of four different covers. A certain bookstore keeps copies of each cover on its racks. To test the hypothesis that sales are equally divided among the four choices, a random sample of 100 purchases is identified.

a. If the resulting X^2 value were 6.4, what conclusion would you reach when using a test with significance level .05?   Fail to reject H_0
b. What conclusion would be appropriate at significance level .01 if $X^2 = 15.3$?   Reject H_0
c. If there were six different covers rather than just four, what would you conclude if $X^2 = 13.7$ and a test with $\alpha = .05$ was used?   Reject H_0

12.3 Packages of mixed nuts made by a certain company contain four types of nuts. The percentages of nuts of Types 1, 2, 3, and 4 are supposed to be 40%, 30%, 20%, and 10%, respectively. A random sample of nuts is selected, and each one is categorized by type.
a. If the sample size is 200 and the resulting test statistic value is $X^2 = 19.0$, what conclusion would be appropriate for a significance level of .001?
b. If the random sample had consisted of only 40 nuts, would you use the chi-square test here? Explain your reasoning.

12.4 ● The paper "Cigarette Tar Yields in Relation to Mortality from Lung Cancer in the Cancer Prevention Study II Prospective Cohort" (*British Medical Journal* [2004]: 72–79) included the accompanying data on the tar level of cigarettes smoked for a sample of male smokers who subsequently died of lung cancer.

Tar Level	Frequency
0–7 mg	103
8–14 mg	378
15–21 mg	563
>22 mg	150

Assume it is reasonable to regard the sample as representative of male smokers who die of lung cancer. Is there convincing evidence that the proportion of male smoker lung cancer deaths is not the same for the four given tar level categories? $X^2 = 457.46$, P-value < 0.001, reject H_0

12.5 ● The paper referenced in the previous exercise also gave the accompanying data on the age at which smoking started for a sample of 1031 men who smoked low-tar cigarettes.

Age	Frequency
<16	237
16–17	258
18–20	320
≥21	216

a. Use a chi-square goodness-of-fit test to test the null hypothesis H_0: $\pi_1 = .25$, $\pi_2 = .2$, $\pi_3 = .3$, $\pi_4 = .25$, where π_1 = proportion of male low-tar cigarette smokers who started smoking before age 16, and π_2, π_3, and π_4 are defined in a similar way for the other three age groups.
b. The null hypothesis from Part (a) specifies that half of male smokers of low-tar cigarettes began smoking between the ages of 16 and 20. Explain why $\pi_2 = .2$ and $\pi_3 = .3$ is consistent with the ages between 16 and 20 being equally likely to be when smoking started.

12.6 ● The report "Fatality Facts 2004: Bicycles" (Insurance Institute, 2004) included the following table classifying 715 fatal bicycle accidents according to time of day the accident occurred.

Time of Day	Number of Accidents
Midnight to 3 A.M.	38
3 A.M. to 6 A.M.	29
6 A.M. to 9 A.M.	66
9 A.M. to Noon	77
Noon to 3 P.M.	99
3 P.M. to 6 P.M.	127
6 P.M. to 9 P.M.	166
9 P.M. to Midnight	113

a. Assume it is reasonable to regard the 715 bicycle accidents summarized in the table as a random sample of fatal bicycle accidents in 2004. Do these data support the hypothesis that fatal bicycle accidents are not equally likely to occur in each of the 3-hour time periods used to construct the table? Test the relevant hypotheses using a significance level of .05. $X^2 = 166.94$, P-value < 0.001, reject H_0
b. Suppose a safety office proposes that bicycle fatalities are twice as likely to occur between noon and midnight as during midnight to noon and suggests the following

hypothesis: H_0: $\pi_1 = 1/3$, $\pi_2 = 2/3$, where π_1 is the proportion of accidents occurring between midnight and noon and π_2 is the proportion occurring between noon and midnight. Do the given data provide evidence against this hypothesis, or are the data consistent with it? Justify your answer with an appropriate test. (Hint: Use the data to construct a one-way table with just two time categories.) $X^2 = 20.66$, P-value < 0.001, reject H_0

12.7 ● The report referenced in the previous exercise ("Fatality Facts 2004: Bicycles") also classified 719 fatal bicycle accidents according to the month in which the accident occurred, resulting in the accompanying table.

Month	Number of Accidents
January	38
February	32
March	43
April	59
May	78
June	74
July	98
August	85
September	64
October	66
November	42
December	40

a. Use the given data to test the null hypothesis H_0: $\pi_1 = 1/12$, $\pi_2 = 1/12, \ldots, \pi_{12} = 1/12$, where π_1 is the proportion of fatal bicycle accidents that occur in January, π_2 is the proportion for February, and so on. Use a significance level of .01. $X^2 = 82.2$, P-value < 0.001, reject H_0
b. The null hypothesis in Part (a) specifies that fatal accidents were equally likely to occur in any of the 12 months. But not all months have the same number of days. What null and alternative hypotheses would you test to determine if some months are riskier than others if you wanted to take differing month lengths into account? (Hint: 2004 was a leap year, with 366 days.)
c. Test the hypotheses proposed in Part (b) using a .05 significance level. $X^2 = 78.04$, P-value < 0.001, reject H_0

12.8 ● Each observation in a random sample of 100 bicycle accidents resulting in death was classified according to the day of the week on which the accident occurred. Data consistent with information given on the web site www.highwaysafety.com are given in the following table

Day of Week	Frequency
Sunday	14
Monday	13
Tuesday	12
Wednesday	15
Thursday	14
Friday	17
Saturday	15

Based on these data, is it reasonable to conclude that the proportion of accidents is not the same for all days of the week? Use $\alpha = .05$. $X^2 = 1.08$, P-value > 0.100, fail to reject H_0

12.9 ● ▼ The color vision of birds plays a role in their foraging behavior: Birds use color to select and avoid certain types of food. The authors of the article "Colour Avoidance in Northern Bobwhites: Effects of Age, Sex, and Previous Experience" (*Animal Behaviour* [1995]: 519–526) studied the pecking behavior of 1-day-old bobwhites. In an area painted white, they inserted four pins with different colored heads. The color of the pin chosen on the bird's first peck was noted for each of 33 bobwhites, resulting in the accompanying table.

Color	First Peck Frequency
Blue	16
Green	8
Yellow	6
Red	3

Do the data provide evidence of a color preference? Test using $\alpha = .01$. $X^2 = 11.242$, $0.010 < P$-value < 0.015, fail to reject H_0

12.10 An article about the California lottery that appeared in the *San Luis Obispo Tribune* (Dec. 15, 1999) gave the following information on the age distribution of adults in California: 35% are between 18 and 34 years old, 51% are between 35 and 64 years old, and 14% are 65 years old or older. The article also gave information on the age distribution of those who purchase lottery tickets. The following table is consistent with the values given in the article:

Age of Purchaser	Frequency
18–34	36
35–64	130
65 and over	34

Suppose that the data resulted from a random sample of 200 lottery ticket purchasers. Based on these sample data, is it reasonable to conclude that one or more of these three age groups buys a disproportionate share of lottery tickets? Use a chi-square goodness-of-fit test with $\alpha = .05$. $X^2 = 25.48$, P-value < 0.001, reject H_0

12.11 When public opinion surveys are conducted by mail, a cover letter explaining the purpose of the survey is usually included. To determine whether the wording of the cover letter influences the response rate, three different cover letters were used in a survey of students at a Midwestern university ("The Effectiveness of Cover-Letter Appeals," *Journal of Social Psychology* [1984]: 85–91). Suppose that each of the three cover letters accompanied questionnaires sent to an equal number of randomly selected students. Returned questionnaires were then classified according to the type of cover letter (I, II, or III). Use the accompanying data to test the hypothesis that $\pi_1 = 1/3$, $\pi_2 = 1/3$, and $\pi_3 = 1/3$, where π_1, π_2, and π_3 are the true proportions of all returned questionnaires accompanied by cover letters I, II, and III, respectively. Use a .05 significance level.

	Cover-letter Type		
	I	II	III
Frequency	48	44	39

12.12 A certain genetic characteristic of a particular plant can appear in one of three forms (phenotypes). A researcher has developed a theory, according to which the hypothesized proportions are $\pi_1 = .25$, $\pi_2 = .50$, and $\pi_3 = .25$. A random sample of 200 plants yields $X^2 = 4.63$.

a. Carry out a test of the null hypothesis that the theory is correct, using level of significance $\alpha = .05$.

b. Suppose that a random sample of 300 plants had resulted in the same value of X^2. How would your analysis and conclusion differ from those in Part (a)?   The analysis would not change.

12.13 ▼ The article "Linkage Studies of the Tomato" (*Transactions of the Royal Canadian Institute* [1931]: 1–19) reported the accompanying data on phenotypes resulting from crossing tall cut-leaf tomatoes with dwarf potato-leaf tomatoes. There are four possible phenotypes: (1) tall cut-leaf, (2) tall potato-leaf, (3) dwarf cut-leaf, and (4) dwarf potato-leaf.

	Phenotype			
	1	2	3	4
Frequency	926	288	293	104

Mendel's laws of inheritance imply that $\pi_1 = 9/16$, $\pi_2 = 3/16$, $\pi_3 = 3/16$, and $\pi_4 = 1/16$. Are the data from this experiment consistent with Mendel's laws? Use a .01 significance level. $X^2 = 1.47$, P-value > 0.100, fail to reject H_0

12.14 ● It is hypothesized that when homing pigeons are disoriented in a certain manner, they will exhibit no preference for any direction of flight after takeoff. To test this, 120 pigeons are disoriented and released, and the direction of flight of each is recorded. The resulting data are given in the accompanying table.

Direction	Frequency
0° to < 45°	12
45° to < 90°	16
90° to < 135°	17
135° to < 180°	15
180° to < 225°	13
225° to < 270°	20
270° to < 315°	17
315° to < 360°	10

Use the goodness-of-fit test with significance level .10 to determine whether the data are consistent with this hypothesis. $X^2 = 4.8$, P-value > 0.100, fail to reject H_0

Bold exercises answered in back   ● Data set available online but not required   ▼ Video solution available

12.2 Tests for Homogeneity and Independence in a Two-way Table

Data resulting from observations made on two different categorical variables can also be summarized using a tabular format. As an example, suppose that residents of a particular city can watch national news on affiliate stations of ABC, CBS, NBC, or PBS. A researcher wishes to know whether there is any relationship between political philosophy (liberal, moderate, or conservative) and preferred news program among those residents who regularly watch the national news. Let x denote the variable *political philosophy* and y the variable *preferred network*. A random sample of 300 regular watchers is selected, and each individual is asked for his or her x and y values. The data set is bivariate and might initially be displayed as follows:

Observation	x Value	y Value
1	Liberal	CBS
2	Conservative	ABC
3	Conservative	PBS
⋮	⋮	⋮
299	Moderate	NBC
300	Liberal	PBS

Bivariate categorical data of this sort can most easily be summarized by constructing a **two-way frequency table**, or **contingency table**. This is a rectangular table that consists of a row for each possible value of x (each category specified by this variable) and a column for each possible value of y. There is then a cell in the table for each possible (x, y) combination. Once such a table has been constructed, the number of times each particular (x, y) combination occurs in the data set is determined, and these numbers (frequencies) are entered in the corresponding cells of the table. The resulting numbers are called **observed cell counts**. The table for the example relating political philosophy to preferred network contains 3 rows and 4 columns (because x and y have 3 and 4 possible values, respectively). Table 12.2 is one possible table.

Table 12.2 An Example of a 3 × 4 Frequency Table

	ABC	CBS	NBC	PBS	Row Marginal Total
Liberal	20	20	25	15	**80**
Moderate	45	35	50	20	**150**
Conservative	15	40	10	5	**70**
Column Marginal Total	**80**	**95**	**85**	**40**	**300**

Marginal totals are obtained by adding the observed cell counts in each row and also in each column of the table. The row and column marginal totals, along with the total of all observed cell counts in the table—the **grand total**—have been included in Table 12.2. The marginal totals provide information on the distribution of observed values for each variable separately. In this example, the row marginal totals reveal that the sample consisted of 80 liberals, 150 moderates, and 70 conservatives. Similarly, column marginal totals indicate how often each of the preferred program categories occurred: 80 preferred ABC news, 95 preferred CBS, and so on. The grand total, 300, is the number of observations in the bivariate data set.

Two-way frequency tables are often characterized by the number of rows and columns in the table (specified in that order: rows first, then columns). Table 12.2 is called a 3×4 table. The smallest two-way frequency table is a 2×2 table, which has only two rows and two columns and thus four cells.

Two-way tables arise naturally in two different types of investigations. A researcher may be interested in comparing two or more populations or treatments on the basis of a single categorical variable and so may obtain independent samples from each population or treatment. For example, data could be collected at a university to compare students, faculty, and staff on the basis of primary mode of transportation to campus (car, bicycle, motorcycle, bus, or on foot). One random sample of 200 students, another of 100 faculty members, and a third of 150 staff members might be chosen, and the selected individuals could be interviewed to obtain the necessary transportation information. Data from such a study could be summarized in a 3×5 two-way frequency table with row categories of student, faculty, and staff and column categories corresponding to the five possible modes of transportation. The observed cell counts could then be used to gain insight into differences and similarities among the three groups with respect to the means of transportation. This type of bivariate categorical data set is characterized by having one set of marginal totals predetermined (the sample sizes from the different groups). In the 3×5 situation just discussed, the row totals would be fixed at 200, 100, and 150.

A two-way table also arises when the values of two different categorical variables are observed for all individuals or items in a single sample. For example, a sample of 500 registered voters might be selected. Each voter could then be asked both if he or she favored a particular property tax initiative and if he or she was a registered Democrat, Republican, or Independent. This would result in a bivariate data set with x representing the variable *political affiliation* (with categories Democrat, Republican, and Independent) and y representing the variable *response* (favors initiative or opposes initiative). The corresponding 3×2 frequency table could then be used to investigate any association between position on the tax initiative and political affiliation. This type of bivariate categorical data set is characterized by having only the grand total predetermined (by the sample size).

■ Comparing Two or More Populations or Treatments: A Test of Homogeneity ....

When the value of a categorical variable is recorded for members of independent random samples obtained from each population or treatment under study, the central issue is whether the category proportions are the same for all the populations or treatments. As in Section 12.1, the test procedure uses a chi-square statistic that compares the observed counts to those that would be expected if there were no differences.

Example 12.4 Risky Soccer??

● The paper "No Evidence of Impaired Neurocognitive Performance in Collegiate Soccer Players" (*American Journal of Sports Medicine* [2002]:157–162) compared collegiate soccer players, athletes in sports other than soccer, and a group of students who were not involved in collegiate sports with respect to history of head injuries. Table 12.3, a 3 × 4 two-way frequency table, is the result of classifying each student in independently selected random samples of 91 soccer players, 96 non-soccer athletes, and 53 non-athletes according to the number of previous concussions the student reported on a medical history questionnaire.

Table 12.3   Observed Counts for Example 12.4

	Number of Concussions				
	0 Concussions	1 Concussion	2 Concussions	3 or More Concussions	Row Marginal Total
Soccer Players	45	25	11	10	91
Non-Soccer Athletes	68	15	8	5	96
Non-Athletes	45	5	3	0	53
Column Marginal Total	158	45	22	15	240

© Royalty-Free/Getty Images

Estimates of expected cell counts can be thought of in the following manner: There were 240 responses on number of concussions, of which 158 were "0 concussions." The proportion of the total responding "0 concussions" is then

$$\frac{158}{240} = .658$$

If there were no difference in response for the different groups, we would then expect about 65.8% of the soccer players to have responded "0 concussions," 65.8% of the non-soccer athletes to have responded "0 concussions," and so on. Therefore the estimated expected cell counts for the three cells in the "0 concussions" column are

Expected count for soccer player *and* 0 concussions cell = .658(91) = 59.9
Expected count for non-soccer athlete *and* 0 concussions cell = .658(96) = 63.2
Expected count for non-athlete *and* 0 concussions cell = .658(53) = 34.9

Note that the expected cell counts need not be whole numbers. The expected cell counts for the remaining cells can be computed in a similar manner. For example,

$$\frac{45}{240} = .188$$

of all responses were in the "1 concussion" category, so

Expected count for soccer player *and* 1 concussion cell = .188(91) = 17.1
Expected count for non-soccer athlete *and* 1 concussion cell = .188(96) = 18.0
Expected count for non-athlete *and* 1 concussion cell = .188(53) = 10.0

It is common practice to display the observed cell counts and the corresponding expected cell counts in the same table, with the expected cell counts enclosed in parentheses. Expected cell counts for the remaining cells have been computed and entered into Table 12.4. Except for small differences resulting from rounding, each marginal total for the expected cell counts is identical to that of the corresponding observed counts.

Table 12.4   Observed and Expected Counts for Example 12.4

	Number of Concussions				
	0 Concussions	1 Concussion	2 Concussions	3 or More Concussions	Row Marginal Total
Soccer players	45 (59.9)	25 (17.1)	11 (8.3)	10 (5.7)	91
Non-Soccer Athletes	68 (63.2)	15 (18.0)	8 (8.8)	5 (6.0)	96
Non-Athletes	45 (34.9)	5 (10.0)	3 (4.9)	0 (3.3)	53
Column Marginal Total	158	45	22	15	240

A quick comparison of the observed and expected cell counts in Table 12.4 reveals some large discrepancies, suggesting that the proportions falling into the concussion categories may not be the same for all three groups. This will be explored further in Example 12.5.

■

In Example 12.4, the expected count for a cell corresponding to a particular group–response combination was computed in two steps. First, the response *marginal proportion* was computed (e.g., 158/240 for the "0 concussions" response). Then this proportion was multiplied by a marginal group total (e.g., 91(158/240) for the soccer player group). Algebraically, this is equivalent to first multiplying the row and column marginal totals and then dividing by the grand total:

$$\frac{(91)(158)}{240}$$

To compare two or more populations or treatments on the basis of a categorical variable, calculate an **expected cell count** for each cell by selecting the corresponding row and column marginal totals and then computing

$$\text{expected cell count} = \frac{(\text{row marginal total})(\text{column marginal total})}{\text{grand total}}$$

These quantities represent what would be expected when there is no difference between the groups under study.

■

The X^2 statistic, introduced in Section 12.1, can now be used to compare the observed cell counts to the expected cell counts. A large value of X^2 results when there are substantial discrepancies between the observed and expected counts and suggests that the hypothesis of no differences between the populations should be rejected. A formal test procedure is described in the accompanying box.

Comparing Two or More Populations or Treatments Using the X^2 Statistic

Null hypothesis: H_0: The true category proportions are the same for all the populations (homogeneity of populations).

Alternative hypothesis: H_a: The true category proportions are not all the same for all of the populations.

Test Statistic: $$X^2 = \sum_{\text{all cells}} \frac{(\text{observed cell count} - \text{expected cell count})^2}{\text{expected cell count}}$$

The expected cell counts are estimated from the sample data (assuming that H_0 is true) using the formula

$$\text{expected cell count} = \frac{(\text{row marginal total})(\text{column marginal total})}{\text{grand total}}$$

P-values: When H_0 is true, X^2 has approximately a chi-square distribution with df = (number of rows − 1)(number of columns − 1) The *P*-value associated with the computed test statistic value is the area to the right of X^2 under the chi-square curve with the appropriate df. Upper-tail areas for chi-square distributions are found in Appendix Table 8.

Assumptions: 1. The data consists of *independently chosen random samples or subjects were assigned at random to treatment groups.*
   2. *The sample size is large*: all expected counts are at least 5. If some expected counts are less than 5, rows or columns of the table may be combined to achieve a table with satisfactory expected counts.

....

Example 12.5 Risky Soccer Revisited

The following table of observed and expected cell counts appeared in Example 12.4:

	Number of Concussions				
	0 Concussions	**1 Concussion**	**2 Concussions**	**3 or more Concussions**	**Row Marginal Total**
Soccer Players	45 (59.9)	25 (17.1)	11 (8.3)	10 (5.7)	**91**
Non-Soccer Athletes	68 (63.2)	15 (18.0)	8 (8.8)	5 (6.0)	**96**
Non-Athletes	45 (34.9)	5 (10.0)	3 (4.9)	0 (3.3)	**53**
Column Marginal Total	**158**	**45**	**22**	**15**	**240**

Hypotheses: H_0: Proportions in each response (number of concussions) category are the same for all three groups

  H_a: H_0 is not true.

Significance level: A significance level of $\alpha = .05$ will be used.

Test statistic: $X^2 = \sum\limits_{\text{all cells}} \dfrac{(\text{observed cell count} - \text{expected cell count})^2}{\text{expected cell count}}$

Combine categories if expected counts are too small.

Assumptions: The random samples were independently chosen, so use of the test is appropriate if the sample size is large enough. One of the expected cell counts (in the 3 or more concussions column) is less than 5, so we will combine the last two columns of the table prior to carrying out the chi-square test. The table we will work with is then

	Number of Concussions			
	0 Concussions	1 Concussion	2 or More Concussions	Row Marginal Total
Soccer Players	45 (59.9)	25 (17.1)	21 (14.0)	91
Non-Soccer Athletes	68 (63.2)	15 (18.0)	13 (14.8)	96
Non-Athletes	45 (34.9)	5 (10.0)	3 (8.2)	53
Column Marginal Total	158	45	22	240

Calculation:

$$X^2 = \frac{(45 - 59.9)^2}{59.9} + \cdots + \frac{(3 - 8.2)^2}{8.2} = 20.6$$

P-value: The two-way table for this example has 3 rows and 3 columns, so the appropriate df is $(3 - 1)(3 - 1) = 4$. Since 20.6 is greater than 18.46, the largest entry in the 4-df column of Appendix Table 8,

$$P\text{-value} < .001$$

Conclusion: P-value $\leq \alpha$, so H_0 is rejected. There is strong evidence to support the claim that the proportions in the number of concussions categories are not the same for the three groups compared. The largest differences between the observed frequencies and those that would be expected if there were no group differences occur in the response categories for soccer players and for non-athletes, with soccer players having higher than expected proportions in the 1 and 2 or more concussion categories and non-athletes having a higher than expected proportion in the 0 concussion category.

◼

Most statistical computer packages can calculate expected cell counts, the value of the X^2 statistic, and the associated P-value. This is illustrated in the following example.

....

Example 12.6 Keeping the Weight Off

● The article "Daily Weigh-ins Can Help You Keep Off Lost Pounds, Experts Say" (*Associated Press*, October 17, 2005) describes an experiment in which 291 people who had lost at least 10% of their body weight in a medical weight loss program were assigned at random to one of three groups for follow-up. One group met monthly in person, one group "met" online monthly in a chat room, and one group received a monthly newsletter by mail. After 18 months, participants in each group were classified according to whether or not they had regained more than 5 pounds, resulting in the data given in Table 12.5.

Table 12.5   Observed and Expected Counts for Example 12.6

	Amount of Weight Gained		
	Regained 5 Lb or Less	Regained More Than 5 Lb	Row Marginal Total
In-Person	52 (41.0)	45 (56.0)	97
Online	44 (41.0)	53 (56.0)	97
Newsletter	27 (41.0)	70 (56.0)	97

Does there appear to be a difference in the weight regained proportions for the three follow-up methods? The relevant hypotheses are

H_0: True proportions for the two weight regained categories are the same for the three follow-up methods.

H_a: H_0 is not true.

Significance level: $\alpha = .01$

Test statistic: $X^2 = \displaystyle\sum_{\text{all cells}} \frac{(\text{observed cell count } - \text{ expected cell count})^2}{\text{expected cell count}}$

Assumptions: Table 12.5 contains the computed expected counts, all of which are greater than 5. The subjects in this experiment were assigned at random to the treatment groups.

Calculation: MINITAB output follows. From the output, $X^2 = 13.773$.

```
Chi-Square Test
Expected counts are printed below observed counts
Chi-Square contributions are printed below expected counts

                   <=5      >5     Total
In-person           52      45       97
                 41.00   56.00
                 2.951   2.161

Online              44      53       97
                 41.00   56.00
                 0.220   0.161

Newsletter          27      70       97
                 41.00   56.00
                 4.780   3.500

Total              123     168      291

Chi-Sq = 13.773,   DF = 2,   P-Value = 0.001
```

P-value: From the MINITAB output, *P*-value $= .001$.

Conclusion: Since *P*-value $\leq \alpha$, H_0 is rejected. The data indicate that the proportions who have regained more than five pounds are not the same for the three follow-up methods. Comparing the observed and expected cell counts, we can see that the observed number in the newsletter group who had regained more than five pounds was higher than would have been expected and the observed number in the in-person group who had regained five or more pounds was lower than would have been expected if there were no difference in the three follow-up methods.

■

■ Testing for Independence of Two Categorical Variables ....

The X^2 test statistic and test procedure can also be used to investigate association between two categorical variables in a single population. As an example, television viewers in a particular city might be categorized with respect to both preferred network (ABC, CBS, NBC, or PBS) and favorite type of programming (comedy, drama, or information and news). The question of interest is often whether knowledge of one variable's value provides any information about the value of the other variable—that is, are the two variables independent?

Continuing the example, suppose that those who favor ABC prefer the three types of programming in proportions .4, .5, and .1 and that these proportions are also correct for individuals favoring any of the other three networks. Then, learning an individual's preferred network provides no added information about that individual's favorite type of programming. The categorical variables *preferred network* and *favorite program type* would be independent.

To see how expected counts are obtained in this situation, recall from Chapter 6 that if two outcomes *A* and *B* are independent, then

$$P(A \text{ and } B) = P(A)P(B)$$

so the proportion of time that the two outcomes occur together in the long run is the product of the two individual long-run relative frequencies. Similarly, two categorical variables are independent in a population if, for each particular category of the first variable and each particular category of the second variable,

$$\begin{pmatrix} \text{proportion of individuals} \\ \text{in a particular category} \\ \text{combination} \end{pmatrix} = \begin{pmatrix} \text{proportion in} \\ \text{specified category} \\ \text{of first variable} \end{pmatrix} \cdot \begin{pmatrix} \text{proportion in} \\ \text{specified category} \\ \text{of second variable} \end{pmatrix}$$

Thus, if 30% of all viewers prefer ABC and the proportions of program type preferences are as previously given, then, assuming that the two variables are independent, the proportion of individuals who both favor ABC and prefer comedy is $(.3)(.4) = .12$ (or 12%).

Multiplying the right-hand side of this expression by the sample size gives us the expected number of individuals in the sample who are in both specified categories of the two variables when the variables are independent. However, these expected counts cannot be calculated, because the individual population proportions are not known.

The solution to this dilemma is to estimate each population proportion using the corresponding sample proportion:

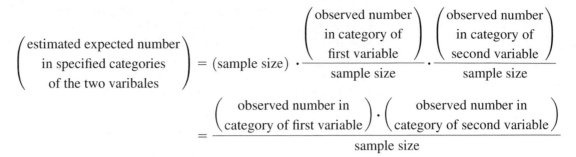

$$\begin{pmatrix} \text{estimated expected number} \\ \text{in specified categories} \\ \text{of the two varibales} \end{pmatrix} = (\text{sample size}) \cdot \dfrac{\begin{pmatrix} \text{observed number} \\ \text{in category of} \\ \text{first variable} \end{pmatrix}}{\text{sample size}} \cdot \dfrac{\begin{pmatrix} \text{observed number} \\ \text{in category of} \\ \text{second variable} \end{pmatrix}}{\text{sample size}}$$

$$= \dfrac{\begin{pmatrix} \text{observed number in} \\ \text{category of first variable} \end{pmatrix} \cdot \begin{pmatrix} \text{observed number in} \\ \text{category of second variable} \end{pmatrix}}{\text{sample size}}$$

Suppose that the observed counts are displayed in a rectangular table in which rows correspond to the categories of the first variable and columns to the categories of the second variable. Then the numerator in the preceding expression for expected counts is just the product of the row and column marginal totals. This is exactly how expected counts were computed in the test for homogeneity of several populations, even though the reasoning used to arrive at the formula is different.

X^2 Test for Independence

Null hypothesis: H_0: The two variables are independent.

Alternative hypothesis: H_a: The two variables are not independent.

Test statistic: $X^2 = \sum\limits_{\text{all cells}} \dfrac{(\text{observed cell count} - \text{expected cell count})^2}{\text{expected cell count}}$

The expected cell counts are estimated (assuming H_0 is true) by the formula

$$\text{expected cell count} = \dfrac{(\text{row marginal total})(\text{column marginal total})}{\text{grand total}}$$

P-values: When H_0 is true and the assumptions of the X^2 test are satisfied, X^2 has approximately a chi-square distribution with

$$df = (\text{number of rows} - 1)(\text{number of columns} - 1)$$

The *P*-value associated with the computed test statistic value is the area to the right of X^2 under the chi-square curve with the appropriate df. Upper-tail areas for chi-square distributions are found in Appendix Table 8.

Assumptions: 1. The observed counts are from a *random sample*.
2. The *sample size is large*: All expected counts are at least 5. If some expected counts are less than 5, rows or columns of the table should be combined to achieve a table with satisfactory expected counts.

Example 12.7 A Pained Expression

● The paper "Facial Expression of Pain in Elderly Adults with Dementia" (*Journal of Undergraduate Research* [2006]) examined the relationship between a nurse's assessment of a patient's facial expression and his or her self-reported level of pain. Data for 89 patients are summarized in Table 12.6.

The authors were interested in determining if there is evidence of a relationship between a facial expression that reflects pain and self-reported pain because patients with dementia do not always give a verbal indication that they are in pain.

Using a .05 significance level, we will test

H_0: Facial expression and self-reported pain are independent.
H_a: Facial expression and self-reported pain are not independent.

Significance level: $\alpha = .05$

Test statistic: $X^2 = \sum_{\text{all cells}} \frac{(\text{observed cell count} - \text{expected cell count})^2}{\text{expected cell count}}$

Table 12.6 Observed Counts for Example 12.7

Facial Expression	Self-Report	
	No Pain	Pain
No Pain	17	40
Pain	3	29

Assumptions: Before we can check the assumptions we must first compute the expected cell counts.

Cell		Expected Cell Count
Row	Column	
1	1	$\frac{(57)(20)}{89} = 12.81$
1	2	$\frac{(57)(69)}{89} = 44.19$
2	1	$\frac{(32)(20)}{89} = 7.19$
2	2	$\frac{(32)(69)}{89} = 24.81$

All expected cell counts are greater than 5. Although the participants in the study were not randomly selected, they were thought to be representative of the population of nursing home patients with dementia. The observed and expected counts are given together in Table 12.7.

Table 12.7 Observed and Expected Counts for Example 12.7

Facial Expression	Self-Report	
	No Pain	Pain
No Pain	17 (12.81)	40 (44.19)
Pain	3 (7.19)	29 (24.81)

Step-by-step technology instructions available online   ● Data set available online

Calculation: $X^2 = \dfrac{(17 - 12.81)^2}{12.81} + \cdots + \dfrac{(29 - 24.81)^2}{24.81} = 4.92$

P-value: The table has 2 rows and 2 columns, so df $= (2 - 1)(2 - 1) = 1$. The entry closest to 4.92 in the 1-df column of Appendix Table 8 is 5.02, so the approximate P-value for this test is

$$P\text{-value} \approx .025$$

Conclusion: Since P-value $\leq \alpha$, we reject H_0 and conclude that there is an association between a nurse's assessment of facial expression and self-reported pain.

■

······

Example 12.8 Stroke Mortality and Education

● Table 12.8 was constructed using data from the article "Influence of Socioeconomic Status on Mortality After Stroke" (*Stroke* [2005]: 310–314). One of the questions of interest to the author was whether there was an association between survival after a stroke and level of education. Medical records for a sample of 2333 residents of Vienna, Austria who had suffered a stroke were used to classify each individual according to two variables—survival (survived, died) and level of education (no basic education, secondary school graduation, technical training/apprenticed, higher secondary school degree, university graduate). Expected cell counts (computed under the assumption of no association between survival and level of education) appear in parentheses in the table.

Table 12.8 Observed and Expected Counts for Example 12.8

	No Basic Education	Secondary School Graduation	Technical Training/ Apprenticed	Higher Secondary School Degree	University Graduate
Died	13 (17.40)	91 (77.18)	196 (182.68)	33 (41.91)	36 (49.82)
Survived	97 (92.60)	397 (410.82)	959 (972.32)	232 (223.09)	279 (265.18)

The X^2 test with a significance level of .01 will be used to test the relevant hypotheses:

H_0: Survival and level of education are independent.
H_a: Survival and level of education are not independent.

Significance level: $\alpha = .01$

Test statistic: $X^2 = \displaystyle\sum_{\text{all cells}} \dfrac{(\text{observed cell count} - \text{expected cell count})^2}{\text{expected cell count}}$

Assumptions: All expected cell counts are at least 5. Assuming that the data can be viewed as a random sample of adults, the X^2 test can be used.

● Data set available online

Calculation: MINITAB output is shown. From the MINITAB output, $X^2 = 12.219$.

Chi-Square Test
Expected counts are printed below observed counts
Chi-Square contributions are printed below expected counts

	1	2	3	4	5	Total
1	13	91	196	33	36	369
	17.40	77.18	182.68	41.91	49.82	
	1.112	2.473	0.971	1.896	3.835	
2	97	397	959	232	279	1964
	92.60	410.82	972.32	223.09	265.18	
	0.209	0.465	0.182	0.356	0.720	
Total	110	488	1155	265	315	2333

Chi-Sq = 12.219,   DF = 4,   P-Value = 0.016

P-value: From the MINITAB output, P-value $= .016$.

Conclusion: Since P-value $> \alpha$, H_0 is not rejected. There is not sufficient evidence to conclude that an association exists between level of education and survival.

■

In some investigations, values of more than two categorical variables are recorded for each individual in the sample. For example, in addition to the variable *survival* and *level of education*, the researchers in the study referenced in Example 12.8 also collected information on occupation. A number of interesting questions could then be explored: Are all three variables independent of one another? Is it possible that occupation and survival are dependent but that the relationship between them does not depend on level of education? For a particular education level group, is there an association between survival and occupation? The X^2 test procedure described in this section for analysis of bivariate categorical data can be extended for use with *multivariate categorical data*. Appropriate hypothesis tests can then be used to provide insight into the relationships between variables. However, the computations required to calculate expected cell counts and to compute the value of X^2 are quite tedious, so they are seldom done without the aid of a computer. Most statistical computer packages can perform this type of analysis. Consult the references by Agresti and Findlay, Everitt, or Mosteller and Rourke in the back of the book for further information on the analysis of categorical data.

■ **Exercises 12.15–12.36** .... Turn to the Expanded Answers Section for answers not shown next to exercises.

12.15 A particular state university system has six campuses. On each campus, a random sample of students will be selected, and each student will be categorized with respect to political philosophy as liberal, moderate, or conservative. The null hypothesis of interest is that the proportion of students falling in these three categories is the same at all six campuses.
a. On how many degrees of freedom will the resulting X^2 test be based?   10

b. How does your answer in Part (a) change if there are seven campuses rather than six?   12

c. How does your answer in Part (a) change if there are four rather than three categories for political philosophy?   15

12.16 A random sample of 1000 registered voters in a certain county is selected, and each voter is categorized with respect to both educational level (four categories) and preferred candidate in an upcoming election for county supervisor (five possibilities). The hypothesis of interest is that educational level and preferred candidate are independent factors.

a. If $X^2 = 7.2$, what would you conclude at significance level .10?   *P*-value > 0.10, fail to reject H_0

b. If there were only four candidates vying for election, what would you conclude if $X^2 = 14.5$ and $\alpha = .05$?   *P*-value > 0.10 (or *P*-value = 0.106), fail to reject H_0

12.17 ● ▼ The polling organization Ipsos conducted telephone surveys in March of 2004, 2005, and 2006. In each year, 1001 people age 18 or older were asked about whether they planned to use a credit card to pay federal income taxes that year. The data given in the accompanying table are from the report "Fees Keeping Taxpayers from Using Credit Cards to Make Tax Payments" (*IPSOS Insight*, March 24, 2006). Is there evidence that the proportion falling in the three credit card response categories is not the same for all three years? Test the relevant hypotheses using a .05 significance level.

Intent to Pay Taxes with a Credit Card

	2004	2005	2006
Definitely/Probably Will	40	50	40
Might/Might Not/Probably Not	180	190	160
Definitely Will Not	781	761	801

12.18 ● In November 2005, an international study to assess public opinion on the treatment of suspected terrorists was conducted ("Most in U.S., Britain, S. Korea and France Say Torture Is OK in at Least Rare Instances," *Associated Press*, December 7, 2005). Each individual in random samples of 1000 adults from each of nine different countries was asked the following question: "Do you feel the use of torture against suspected terrorists to obtain information about terrorism activities is justified?"

Responses consistent with percentages given in the article for the samples from Italy, Spain, France, the United States, and South Korea are summarized in the accompanying table. Based on these data, is it reasonable to conclude that the response proportions are not the same for all five countries? Use a .01 significance level to test the appropriate hypotheses. $X^2 = 881.326$, *P*-value < 0.001, reject H_0

Country	Response				
	Never	Rarely	Some- times	Often	Not Sure
Italy	600	140	140	90	30
Spain	540	160	140	70	90
France	400	250	200	120	30
United States	360	230	270	110	30
South Korea	100	330	470	60	40

12.19 ● In a study to determine if hormone therapy increases risk of venous thrombosis in menopausal women, each person in a sample of 579 women who had been diagnosed with venous thrombosis was classified according to hormone use. Each woman in a sample of 2243 women who had not been diagnosed with venous thrombosis was also classified according to hormone use. Data from the study are given in the accompanying table (*Journal of the American Medical Association* [2004]: 1581–1587). The women in each of the two samples were selected at random from the patients at a large HMO in the state of Washington.

a. Is there convincing evidence that the proportions falling in to each of the hormone use categories is not the same for women who have been diagnosed with venous thrombosis and those who have not?

b. To what population would it be reasonable to generalize the conclusions of Part (a)? Explain.

	Current Hormone Use		
	None	Esterified Estrogen	Conjugated Equine Estrogen
Venous Thrombosis	372	86	121
No Venous Thrombosis	1439	515	289

12.20 ● The Harvard University Institute of Politics surveys undergraduates across the United States annually. Responses to the question "When it comes to voting, do you consider yourself to be affiliated with the Democratic Party, the Republican Party, or are you Independent or unaffiliated with a major party?" for the survey conducted in 2003, 2004, and 2005 are summarized in the given table. The samples for each year were independently selected and are considered to be representative of the population of undergraduate students in the year the survey was conducted. Is there evidence that the distribution of political affiliation is not the same for all three years for which data are given? $X^2 = 26.175$, P-value < 0.001, reject H_0

| | Year | | |
Political Affiliation	2005	2004	2003
Democrat	397	409	325
Republican	301	349	373
Independent/unaffiliated	458	397	457
Other	60	48	48

12.21 ● The survey described in the previous exercise also asked the following question: "Please tell me whether you trust the President to do the right thing all of the time, most of the time, some of the time, or never. Use the data in the given table and an appropriate hypothesis test to determine if there is evidence that trust in the President was not the same in 2005 as it was in 2002. $X^2 = 95.921$, P-value < 0.001, reject H_0

| | Year | |
Response	2005	2002
All of the time	132	180
Most of the time	337	528
Some of the time	554	396
Never	169	96

12.22 ● The report "Undergraduate Students and Credit Cards in 2004" (Nellie Mae, May 2005) included information collected from individuals in a random sample of undergraduate students in the United States. Students were

classified according to region of residence and whether or not they have one or more credit cards, resulting in the accompanying two-way table. Carry out a test to determine if there is evidence that region of residence and having a credit card are not independent. Use $\alpha = .05$.
$X^2 = 15.106$, $0.001 < P$-value < 0.005, reject H_0

| | Credit Card? | |
Region	At Least One Credit Card	No Credit Cards
Northeast	401	164
Midwest	162	36
South	408	115
West	104	23

12.23 ● The report described in the previous exercise also classified students according to region of residence and whether or not they had a credit card with a balance of more than $7000. Do these data support the conclusion that there is an association between region of residence and whether or not the student has a balance exceeding $7000? Test the relevant hypotheses using a .01 significance level. $X^2 = 339.99$, P-value < 0.001, reject H_0

| | Balance Over $7000? | |
Region	No	Yes
Northeast	28	537
Midwest	162	182
South	42	481
West	9	118

12.24 ● The paper "Overweight Among Low-Income Preschool Children Associated with the Consumption of Sweet Drinks" (*Pediatrics* [2005]: 223–229) described a study of children who were underweight or normal weight at age 2. Children in the sample were classified according to the number of sweet drinks consumed per day and whether or not the child was overweight one year after the study began. Is there evidence of an association between whether or not children are overweight after one

Bold exercises answered in back ● Data set available online but not required ▼ Video solution available

year and the number of sweet drinks consumed? Assume that it is reasonable to regard the sample of children in this study as representative of 2- to 3-year-old children and then test the appropriate hypotheses using a .05 significance level. $X^2 = 3.03$, P-value > 0.100, fail to reject H_0

Number of Sweet Drinks Consumed per Day	Overweight?	
	Yes	No
0	22	930
1	73	2074
2	56	1681
3 or More	102	3390

12.25 ● Data consistent with summary quantities given in a paper that appeared in the *British Medical Journal* ("Cigarette Tar Yields in Relation to Mortality from Lung Cancer in the Cancer Prevention Study II Prospective Cohort" [2004]: 72–79) are shown in the accompanying table. Suppose that the table was constructed by classifying individuals in a random sample of smokers according to gender and the age at which they began smoking. Do these data support the hypothesis that there is an association between gender and the age at which smoking began? $X^2 = 12.091$, $0.005 < P$-value < 0.010, reject H_0

Age When Smoking Began	Gender	
	Male	Female
<16	25	10
16–17	24	17
18–20	28	32
≥21	19	34

12.26 ● The 2006 Expedia Vacation Deprivation Survey (*Ipsos Insight*, May 18, 2006) described the results of a poll of working adults in Canada. Each person in a random sample was classified according to gender and the number of vacation days he or she usually took each year. The resulting data are summarized in the given table. Is it reasonable to conclude that there is an association between gender and the number of vacation days taken? To

what population would it be reasonable to generalize this conclusion? $X^2 = 9.858$, P-value > 0.100, fail to reject H_0

Days of Vacation	Gender	
	Male	Female
None	51	42
1–5	21	25
6–10	67	79
11–15	111	94
16–20	71	70
21–25	82	58
More than 25	118	79

12.27 ● The report "Health Insurance Coverage of the Near Elderly" (Urban Institute, July 2000) includes information from a study of American adults. The sample used in the study was a random sample of those who participated in the National Survey of America's Families, a national survey of more than 100,000 people, and was considered to be representative of American adults. Using data consistent with summary quantities appearing in the report, the following table classifies individuals in the sample according to age group and whether or not the individual was considered to be in good health. Use the given data to test the hypothesis that age group and whether or not an individual is in good health are independent. Use $\alpha = .01$. $X^2 = 68.953$, P-value < 0.001, reject H_0

	Health Status	
Age	Good Health	Poor Health
18–34	920	80
35–54	860	140
55–64	790	210

12.28 ● A survey was conducted in the San Francisco Bay area in which each participating individual was classified according to the type of vehicle used most often and city of residence. A subset of the resulting data are given in the accompanying table (*The Relationship of Vehicle Type Choice to Personality, Lifestyle, Attitudinal and Demographic Variables*, Technical Report UCD-ITS-RR02-06, DaimlerCrysler Corp., 2002).

Vehicle Type	City		
	Concord	Pleasant Hills	North San Francisco
Small	68	83	221
Compact	63	68	106
Midsize	88	123	142
Large	24	18	11

Do the data provide convincing evidence of an association between city of residence and vehicle type? Use a significance level of .05. You may assume that it is reasonable to regard the sample as a random sample of Bay area residents. $X^2 = 49.81$, P-value < 0.001, reject H_0

12.29 ● ▼ Do women have different patterns of work behavior than men? The article "Workaholism in Organizations: Gender Differences" (*Sex Roles: A Journal of Research* [1999]: 333–346) attempts to answer this question. Each person in a random sample of 423 graduates of a business school in Canada were polled and classified by gender and workaholism type, resulting in the accompanying table:

Workaholism Types	Gender	
	Female	Male
Work Enthusiasts	20	41
Workaholics	32	37
Enthusiastic Workaholics	34	46
Unengaged Workers	43	52
Relaxed Workers	24	27
Disenchanted Workers	37	30

a. Test the hypothesis that gender and workaholism type are independent. $X^2 = 6.852$, P-value > 0.100, fail to reject H_0
b. The author writes "women and men fell into each of the six workaholism types to a similar degree." Does the outcome of the test you performed in Part (a) support this conclusion? Explain.

12.30 ● The article "Motion Sickness in Public Road Transport: The Effect of Driver, Route and Vehicle" (*Ergonomics* [1999]: 1646–1664) reported that seat position within a bus may have some effect on whether one experiences motion sickness. The accompanying table classifies each person in a random sample of bus riders by the location of his or her seat and whether nausea was reported. $X^2 = 73.789$, P-value < 0.001, reject H_0

	Nausea	No Nausea
Front	58	870
Middle	166	1163
Rear	193	806

Based on these data, can you conclude that there is an association between seat location and nausea? Test the relevant hypotheses using $\alpha = .05$.

12.31 ● The article "Mortality Among Recent Purchasers of Handguns" (*New England Journal of Medicine* [1999]: 1583–1589) examined the relationship between handgun purchases and cause of death. Suppose that a random sample of 4000 death records was examined and cause of death was noted for each individual in the sample. Gun registration records were also examined to determine which of these 4000 individuals had purchased a handgun within the previous year. Data consistent with summary values given in the article are shown in the accompanying table.

	Suicide	Not Suicide
Handgun Purchased	4	12
No Handgun Purchased	63	3921

Is it reasonable to use the chi-square test to determine whether there is an association between handgun purchase within the year prior to death and whether the death was a suicide? Justify your answer.

12.32 ● A story describing a date rape was read by 352 high school students. To investigate the effect of the victim's clothing on subject's judgment of the situation described, the story was accompanied by either a photograph of the victim dressed provocatively, a photo of the victim dressed conservatively, or no picture. Each student was asked whether the situation described in the story was one of rape. Data from the article "The Influence of

Victim's Attire on Adolescent Judgments of Date Rape" (*Adolescence* [1995]: 319–323) are given in the accompanying table. Is there evidence that the proportion who believe that the story described a rape differs for the three different photo groups? Test the relevant hypotheses using $\alpha = .01$. $X^2 = 28.81$, P-value < 0.001, reject H_0

| | Picture | | |
Response	Provocative	Conservative	No Picture
Rape	80	104	92
Not Rape	47	12	17

12.33 ● Can people tell the difference between a female nose and a male nose? This important (?) research question was examined in the article "You Can Tell by the Nose: Judging Sex from an Isolated Facial Feature" (*Perception* [1995]: 969–973). Eight Caucasian males and eight Caucasian females posed for nose photos. The article states that none of the volunteers wore nose studs or had prominent nasal hair. Each person placed a black Lycra tube over his or her head in such a way that only the nose protruded through a hole in the material. Photos were then taken from three different angles: front view, three-quarter view, and profile. These photos were shown to a sample of undergraduate students. Each student in the sample was shown one of the nose photos and asked whether it was a photo of a male or a female; and the response was classified as either correct or incorrect. The accompanying table was constructed using summary values reported in the article. Is there evidence that the proportion of correct sex identifications differs for the three different nose views? $X^2 = 1.978$, P-value > 0.10, fail to reject H_0

| | View | | |
Sex ID	Front	Profile	Three-Quarter
Correct	23	26	29
Incorrect	17	14	11

12.34 ● The accompanying table appeared in the article "Europe's Receptivity to New Religious Movements: Round Two" (*Journal for the Scientific Study of Religion*

[1993]: 389–397). Suppose that the percentages in the table were based on random samples of size 200 from each of the six countries.

Country	Percentage Who Believe in Fortune-Tellers
Great Britain	42
West Germany	32
East Germany	22
Slovenia	55
Ireland	27
Northern Ireland	33

a. Use the given percentages to construct a two-way table with rows corresponding to the six countries and columns corresponding to the categories "believe in fortune-tellers" and "don't believe in fortune-tellers." Enter the observed counts in the table. (*Hint*: The observed counts for Great Britain are $(.42)(200) = 84$ for the "believe" column and $(.58)(200) = 116$ for the "don't believe" column.)
b. Is there evidence that the proportion who believe in fortune-tellers is not the same for all six countries? Test the relevant hypotheses using $\alpha = .01$. $X^2 = 60.95$, P-value < 0.001, reject H_0

12.35 ● Job satisfaction of professionals was examined in the article "Psychology of the Scientist: Work-Related Attitudes of U.S. Scientists" (*Psychological Reports* [1991]: 443–450). Each person in a random sample of 778 teachers was classified according to a job satisfaction variable and also by teaching level, resulting in the accompanying two-way table. Can we conclude that there is an association between job satisfaction and teaching level? Test the relevant hypotheses using $\alpha = .05$.

| | Job Satisfaction | |
Teaching Level	Satisfied	Unsatisfied
College	74	43
High School	224	171
Elementary	126	140

12.36 ● The article "Victim's Race Affects Killer's Sentence, Study Finds" (*New York Times*, April 20, 2001) included the following information:

Defendant's Race	Victim's Race	Death Penalty	No Death Penalty
Not white	White	33	251
White	White	33	508
Not white	Not white	29	587
White	Not white	4	76

The table was based on a study of *all* homicide cases in North Carolina for the period 1993 to 1997 where it was possible that the person convicted of murder could receive the death penalty. Explain why it would not be necessary (or appropriate) to use a chi-square test to determine if there was an association between the defendant–victim race combination and whether or not the convicted murderer receive a death sentence for convicted murders in North Carolina during 1993 to 1997.

Bold exercises answered in back　　● Data set available online but not required　　▼ Video solution available

12.3　Interpreting and Communicating the Results of Statistical Analyses

Many studies, particularly those in the social sciences, result in categorical data. The questions of interest in such studies often lead to an analysis that involves using a chi-square test.

■ Communicating the Results of Statistical Analyses

Three different chi-square tests were introduced in this chapter—the goodness-of-fit test, the test for homogeneity, and the test for independence. They are used in different settings and to answer different questions. When summarizing the results of a chi-square test, be sure to indicate which chi-square test was performed. One way to do this is to be clear about how the data were collected and the nature of the hypotheses being tested.

It is also a good idea to include a table of observed and expected counts in addition to reporting the computed value of the test statistic and the *P*-value. And finally, make sure to give a conclusion in context, and make sure that the conclusion is worded appropriately for the type of test conducted. For example, don't use terms such as *independence* and *association* to describe the conclusion if the test performed was a test for homogeneity.

■ Interpreting the Results of Statistical Analyses

As with the other hypothesis tests considered, it is common to find the result of a chi-square test summarized by giving the value of the chi-square test statistic and an associated *P*-value. Because categorical data can be summarized compactly in frequency tables, the data often are given in the article (unlike data for numerical variables, which are rarely given).

■ What to Look For in Published Data

Here are some questions to consider when you are reading an article that contains the results of a chi-square test:

- Are the variables of interest categorical rather than numerical?
- Are the data given in the article in the form of a frequency table?

- If a two-way frequency table is involved, is the question of interest one of homogeneity or one of independence?
- What null hypothesis is being tested? Are the results of the analysis reported in the correct context (homogeneity, etc.)?
- Is the sample size large enough to make use of a chi-square test reasonable? (Are all expected counts at least 5?)
- What is the value of the test statistic? Is the associated P-value given? Should the null hypothesis be rejected?
- Are the conclusions drawn by the authors consistent with the results of the test?
- How different are the observed and expected counts? Does the result have practical significance as well as statistical significance?

The authors of the article "Predicting Professional Sports Game Outcomes from Intermediate Game Scores" (*Chance* [1992]: 18–22) used a chi-square test to determine whether there was any merit to the idea that basketball games are not settled until the last quarter, whereas baseball games are over by the seventh inning. They also considered football and hockey. Data were collected for 189 basketball games, 92 baseball games, 80 hockey games, and 93 football games. The analyzed games were sampled randomly from all games played during the 1990 season for baseball and football and for the 1990–1991 season for basketball and hockey. For each game, the late-game leader was determined, and then it was noted whether the late-game leader actually ended up winning the game. The resulting data are summarized in the following table:

Sport	Late-Game Leader Wins	Late-Game Leader Loses
Basketball	150	39
Baseball	86	6
Hockey	65	15
Football	72	21

The authors stated that the "*late-game leader* is defined as the team that is ahead after three quarters in basketball and football, two periods in hockey, and seven innings in baseball. The chi-square value (with three degrees of freedom) is 10.52 ($P < .015$)." They also concluded that "the sports of basketball, hockey, and football have remarkably similar percentages of late-game reversals, ranging from 18.8% to 22.6%. The sport that is an anomaly is baseball. Only 6.5% of baseball games resulted in late reversals. . . . [The chi-square test] is statistically significant due almost entirely to baseball."

In this particular analysis, the authors are comparing four populations (games from each of the four sports) on the basis of a categorical variable with two categories (late-game leader wins and late-game leader loses). The appropriate null hypothesis is then

H_0: The true proportion in each category (leader wins, leader loses) is the same for all four sports.

Based on the reported value of the chi-square statistic and the associated P-value, this null hypothesis is rejected, leading to the conclusion that the category proportions are not the same for all four sports.

The validity of the chi-square test requires that the sample sizes be large enough so that no expected counts are less than 5. Is this reasonable here? The following MINITAB output shows the expected cell counts and the computation of the X^2 statistic:

Chi-Square Test

Expected counts are printed below observed counts

	Leader W	Leader L	Total
1	150	39	189
	155.28	33.72	
2	86	6	92
	75.59	16.41	
3	65	15	80
	65.73	14.27	
4	72	21	93
	76.41	16.59	
Total	373	81	454

Chi-Sq = 0.180 + 0.827 +
1.435 + 6.607 +
0.008 + 0.037 +
0.254 + 1.171 = 10.518
DF = 3, P-Value = 0.015

The smallest expected count is 14.27, so the sample sizes are large enough to justify the use of the X^2 test. Note also that the two cells in the table that correspond to baseball contribute a total of $1.435 + 6.607 = 8.042$ to the value of the X^2 statistic of 10.518. This is due to the large discrepancies between the observed and expected counts for these two cells. There is reasonable agreement between the observed and the expected counts in the other cells. This is probably the basis for the authors' conclusion that baseball is the only anomaly and that the other sports were similar.

■ A Word to the Wise: Cautions and Limitations ....,. ....

Be sure to keep the following in mind when analyzing categorical data using one of the chi-square tests presented in this chapter:

1. Don't confuse tests for homogeneity with tests for independence. The hypotheses and conclusions are different for the two types of test. Tests for homogeneity are used when the individuals in each of two or more independent samples are classified according to a single categorical variable. Tests for independence are used when individuals in a *single* sample are classified according to two categorical variables.
2. As was the case for the hypothesis tests of earlier chapters, remember that we can never say we have strong support for the null hypothesis. For example, if we do not reject the null hypothesis in a chi-square test for independence, we cannot conclude that there is convincing evidence that the variables are independent. We can only say that we were not convinced that there is an association between the variables.
3. Be sure that the assumptions for the chi-square test are reasonable. P-values based on the chi-square distribution are only approximate, and if the large sample

conditions are not met, the true *P*-value may be quite different from the approximate one based on the chi-square distribution. This can sometimes lead to erroneous conclusions. Also, for the chi-square test of homogeneity, the assumption of *independent* samples is particularly important.

4. Don't jump to conclusions about causation. Just as a strong correlation between two numerical variables does not mean that there is a cause-and-effect relationship between them, an association between two categorical variables does not imply a causal relationship.

Activity **12.1**   Pick a Number, Any Number . . .

Background: There is evidence to suggest that human beings are not very good random number generators. In this activity, you will investigate this phenomenon by collecting and analyzing a set of human-generated "random" digits.

For this activity, work in a group with four or five other students.

1. Each member of the group should complete this step individually. Ask 25 different people to pick a digit from 0 to 9 at random. Record the responses.

2. Combine the responses you collected with those of the other members of your group to form a single sample. Summarize the resulting data in a one-way frequency table.

3. If people are adept at picking digits at random, what would you expect for the proportion of the responses in the sample that were 0? that were 1?

4. State a null hypothesis and an alternative hypothesis that could be tested to determine whether there is evidence that the 10 digits from 0 to 9 are not selected an equal proportion of the time when people are asked to pick a digit at random.

5. Carry out the appropriate hypothesis test, and write a few sentences indicating whether or not the data support the theory that people are not good random number generators.

Activity **12.2**   Color and Perceived Taste

Background: Does the color of a food or beverage affect the way people perceive its taste? In this activity you will conduct an experiment to investigate this question and analyze the resulting data using a chi-square test.

You will need to recruit at least 30 subjects for this experiment, so it is advisable to work in a large group (perhaps even the entire class) to complete this activity.

Subjects for the experiment will be assigned at random to one of two groups. Each subject will be asked to taste a sample of gelatin (e.g., Jell-O) and rate the taste as not very good, acceptable, or very good. Subjects assigned to the first group will be asked to taste and rate a cube of lemon-flavored gelatin. Subjects in the second group will be asked to taste and rate a cube of lemon-flavored gelatin that has been colored an unappealing color by adding food coloring to the gelatin mix before the gelatin sets.

Note: You may choose to use something other than gelatin, such as lemonade. Any food or beverage whose color can be altered using food coloring can be used. You can experiment with the food colors to obtain a color that you think is particularly unappealing!

1. As a class, develop a plan for collecting the data. How will subjects be recruited? How will they be assigned to one of the two treatment groups (unaltered color, altered color)? What extraneous variables will be directly controlled, and how will you control them?

2. After the class is satisfied with the data collection plan, assign members of the class to prepare the gelatin to be used in the experiment.

3. Carry out the experiment, and summarize the resulting data in a two-way table like the one shown:

	Taste Rating		
Treatment	Not Very Good	Acceptable	Very Good
Unaltered Color			
Altered Color			

4. The two-way table summarizes data from two independent samples (as long as subjects were assigned *at random* to the two treatments, the samples are independent). Carry out an appropriate test to determine whether the proportion for each of the three taste rating categories is the same when the color is altered as for when the color is not altered.

Summary of Key Concepts and Formulas

Term or Formula	Comment
One-way frequency table	A compact way of summarizing data on a categorical variable; it gives the number of times each of the possible categories in the data set occurs (the frequencies).
Goodness-of-fit statistic, $$X^2 = \sum_{\text{all cells}} \frac{(\text{observed cell count} - \text{expected cell count})^2}{\text{expected cell count}}$$	A statistic used to provide a comparison between observed counts and those expected when a given hypothesis is true. When none of the expected counts are too small, X^2 has approximately a chi-square distribution.
X^2 goodness-of-fit test	A hypothesis test performed to determine whether the true category proportions are different from those specified by the given null hypothesis.
Two-way frequency table (contingency table)	A rectangular table used to summarize a bivariate categorical data set; two-way tables are used to compare several populations on the basis of a categorical variable or to identify whether an association exists between two categorical variables.
X^2 test for homogeneity	The hypothesis test performed to determine whether the true category proportions are the same for two or more populations or treatments.
X^2 test for independence	The hypothesis test performed to determine whether an association exists between two categorical variables.

Chapter Review Exercises 12.37–12.47

Turn to the Expanded Answers Section for answers not shown next to exercises.

CENGAGENOW™　Know exactly what to study! Take a pre-test and receive your Personalized Learning Plan.

12.37 ● According to Census Bureau data, in 1998 the California population consisted of 50.7% whites, 6.6% blacks, 30.6% Hispanics, 10.8% Asians, and 1.3% other ethnic groups. Suppose that a random sample of 1000 students graduating from California colleges and universities in 1998 resulted in the accompanying data on ethnic group. These data are consistent with summary statistics contained in the article titled "Crumbling Public School System a Threat to California's Future (*Investor's Business Daily*, November 12, 1999).

Bold exercises answered in back　　● Data set available online but not required　　▼ Video solution available

Ethnic Group	Number in Sample
White	679
Black	51
Hispanic	77
Asian	190
Other	3

Do the data provide evidence that the proportion of students graduating from colleges and universities in California for these ethnic group categories differs from the respective proportions in the population for California? Test the appropriate hypotheses using $\alpha = .01$. $X^2 = 303.09$, P-value < 0.001, reject H_0

12.38 Criminologists have long debated whether there is a relationship between weather and violent crime. The author of the article "Is There a Season for Homicide?" (*Criminology* [1988]: 287–296) classified 1361 homicides according to season, resulting in the accompanying data. Do these data support the theory that the homicide rate is not the same over the four seasons? Test the relevant hypotheses using a significance level of .05.

	Season		
Winter	Spring	Summer	Fall
328	334	372	327

12.39 ● The following table is based on data reported in "Light-to-Moderate Alcohol Consumption and Risk of Stroke Among U.S. Male Physicians" (*New England Journal of Medicine* [1999]: 1557–1564). The table is based on 21,870 male physicians age 40–84 who are participating in the Physician's Health Study.
a. Calculate row percentages by dividing each observed count by the corresponding row total. Are the proportions falling in each of the alcohol consumption categories similar for the three smoking categories?

b. Test the hypothesis that smoking status and alcohol consumption are independent.
c. Is the result of your test in Part (b) consistent with what you expected based on your answer to Part (a)? If not, explain why your initial impression based on the percentages may not have been accurate. What aspect of the data set factors into your explanation?

12.40 ● Each boy in a sample of Mexican American males, age 10–18, was classified according to smoking status and response to a question asking whether he likes to do risky things. The following table is based on data given in the article "The Association Between Smoking and Unhealthy Behaviors Among a National Sample of Mexican-American Adolescents" (*Journal of School Health* [1998]: 376–379):

	Smoking Status	
	Smoker	Nonsmoker
Likes Risky Things	45	46
Doesn't Like Risky Things	36	153

Assume that it is reasonable to regard the sample as a random sample of Mexican-American male adolescents.
a. Is there sufficient evidence to conclude that there is an association between smoking status and desire to do risky things? Test the relevant hypotheses using $\alpha = .05$.
b. Based on your conclusion in Part (a), is it reasonable to conclude that smoking *causes* an increase in the desire to do risky things? Explain. No, this is an observational study.

12.41 ● The article "Cooperative Hunting in Lions: The Role of the Individual" (*Behavioral Ecology and Sociobiology* [1992]: 445–454) discusses the different roles taken by lionesses as they attack and capture prey. The authors were interested in the effect of the position in line as stalking occurs; an individual lioness may be in the center

Table for Exercise 12.39

	Alcohol Consumption (No. of Drinks)				
	<1/Wk	1/Wk	2–4/Wk	5–6/Wk	1/Day
Never Smoked	3577	1711	2430	1211	1910
Smoked in the Past	1595	1068	1999	1264	2683
Currently Smokes	524	289	470	296	849

of the line or on the wing (end of the line) as they advance toward their prey. In addition to position, the role of the lioness was also considered. A lioness could initiate a chase (be the first one to charge the prey), or she could participate and join the chase after it has been initiated. Data from the article are summarized in the accompanying table.

| | Role | |
| | Initiate | Participate |
Position	Chase	in Chase
Center	28	48
Wing	66	41

Is there evidence of an association between position and role? Test the relevant hypotheses using $\alpha = .01$. What assumptions about how the data were collected must be true for the chi-square test to be an appropriate way to analyze these data? $X^2 = 10.97, P\text{-value} < 0.001$, reject H_0

12.42 ● The authors of the article "A Survey of Parent Attitudes and Practices Regarding Underage Drinking" (*Journal of Youth and Adolescence* [1995]: 315–334) conducted a telephone survey of parents with preteen and teenage children. One of the questions asked was "How effective do you think you are in talking to your children about drinking?" Responses are summarized in the accompanying 3×2 table. Using a significance level of .05, carry out a test to determine whether there is an association between age of children and parental response.

| | Age of Children | |
Response	Preteen	Teen
Very Effective	126	149
Somewhat Effective	44	41
Not at All Effective or Don't Know	51	26

12.43 ● The article "Regional Differences in Attitudes Toward Corporal Punishment" (*Journal of Marriage and Family* [1994]: 314–324) presents data resulting from a random sample of 978 adults. Each individual in the sample was asked whether he or she agreed with the following statement: "Sometimes it is necessary to discipline a child with a good, hard spanking." Respondents were also classified according to the region of the United States in which they lived. The resulting data are summarized in the accompanying table. Is there an association between response (agree, disagree) and region of residence? Use $\alpha = .01$. $X^2 = 22.855, P\text{-value} < 0.0001$, reject H_0

| | Response | |
Region	Agree	Disagree
Northeast	130	59
West	146	42
Midwest	211	52
South	291	47

12.44 When charities solicit donations, they must know what approach will encourage people to donate. Would a picture of a needy child tug at the heartstrings and convince someone to contribute to the charity? As reported in the article "The Effect of Photographs of the Handicapped on Donation" (*Journal of Applied Social Psychology* [1979]: 426–431), experimenters went door-to-door with four types of displays: a picture of a smiling child, a picture of an unsmiling child, a verbal message, and an identification of the charity only. At each door, they would randomly select which display to present. Suppose 18 of 30 subjects contributed when shown the picture of the smiling child, 14 of 30 contributed when shown the picture of an unsmiling child, 16 of 26 contributed when shown the verbal message, and 18 of 22 contributed when shown only the identification of the charity. What conclusions can be drawn from these data? Explain your procedure and reasoning. $X^2 = 6.621, P\text{-value} > 0.05$, fail to reject H_0

12.45 ● Jail inmates can be classified into one of the following four categories according to the type of crime committed: violent crime, crime against property, drug offenses, and public-order offenses. Suppose that random samples of 500 male inmates and 500 female inmates are selected, and each inmate is classified according to type of offense. The data in the accompanying table are based on summary values given in the article "Profile of Jail Inmates" (*USA Today*, April 25, 1991). We would like to know whether male and female inmates differ with respect to type of offense.

	Gender	
Type of Crime	Male	Female
Violent	117	66
Property	150	160
Drug	109	168
Public-Order	124	106

a. Is this a test of homogeneity or a test of independence?
b. Test the relevant hypotheses using a significance level of .05. $X^2 = 28.51$, P-value < 0.001, reject H_0

12.46 ● Drivers born under the astrological sign of Capricorn are the worst drivers in Australia, according to an article that appeared in the Australian newspaper *The Mercury* (October 26, 1998). This statement was based on a study of insurance claims that resulted in the following data for male policyholders of a large insurance company.

Astrological Sign	Number of Policyholders
Aquarius	35,666
Aries	37,926
Cancer	38,126
Capricorn	54,906
Gemini	37,179
Leo	37,354
Libra	37,910
Pisces	36,677
Sagittarius	34,175
Scorpio	35,352
Taurus	37,179
Virgo	37,718

a. Assuming that it is reasonable to treat the policyholders of this particular insurance company as a random sample of insured drivers in Australia, are the observed data consistent with the hypothesis that the proportion of policyholders is the same for each of the 12 astrological signs?
b. Why do you think that the proportion of Capricorn policyholders is so much higher than would be expected if the proportions are the same for all astrological signs?
c. Suppose that a random sample of 1000 accident claims submitted to this insurance company is selected and each claim classified according to the astrological sign of the driver. (The accompanying table is consistent with accident rates given in the article.)

Astrological Sign	Observed Number in Sample
Aquarius	85
Aries	83
Cancer	82
Capricorn	88
Gemini	83
Leo	83
Libra	83
Pisces	82
Sagittarius	81
Scorpio	85
Taurus	84
Virgo	81

Test the null hypothesis that the proportion of accident claims submitted by drivers of each astrological sign is consistent with the proportion of policyholders of each sign. Use the information on the distribution of policyholders from Part (a) to compute expected frequencies and then carry out an appropriate test.

12.47 ● The article "Identification of Cola Beverages" (*Journal of Applied Psychology* [1962]: 356–360) reported on an experiment in which each of 79 subjects was presented with glasses of cola in pairs and asked to identify which glass contained a specific brand of cola. The accompanying data appeared in the article. Do these data suggest that individuals' abilities to make correct identification differ for the different brands of cola?

	Number of Correct Identifications			
Cola	0	1	2	3 or 4
Coca-Cola	13	23	24	19
Pepsi Cola	12	20	26	21
Royal Crown	18	28	19	14

Personal Tutor
Do you need a live tutor for homework problems?

CENGAGENOW
Are you ready? Take your exam-prep post-test now.

Graphing Calculator Explorations

Exploration 12.1 The Goodness-of-Fit Test

As computers have seemingly taken over the world, many strange new words have been added to our collective vocabulary. One of these "words" is WYSIWYG, an acronym for "What you see is what you get." Goodness-of-fit tests on your calculator are an example of this idea, in that the table of observed and expected counts corresponds very closely to the lists on your calculator. Because of this match, the calculator commands will also appear very similar to the chi-square formula in a goodness-of-fit test, making the calculator procedure very easy to remember.

We will use Example 12.3, the hybrid car data. The goodness-of-fit procedure, as it develops, will have two parts. First we will calculate the value of X^2; then we will evaluate its statistical significance. We begin by entering the observed counts for the categories in List1. For your reference, observed counts follow.

State	Observed Count
California	250
Virginia	56
Washington	34
Florida	33
Maryland	33

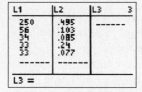

Figure 12.3 Calculator screen with observed counts and hypothesized proportions.

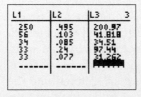

Figure 12.4 Expected counts added in List3.

Now we need to compute the expected counts. It might seem natural to calculate the expected counts as was done in the text, and then enter them in List2. However, a different approach actually turns out to be easier. Since expected counts usually follow from the hypothesized category proportions, you can enter the hypothesized population proportions in List2 and use them to calculate the expected values for each cell. After entering the proportions in List2 you should see something like Figure 12.3.

The expected values are calculated by multiplying the proportions in List2 by the total number of observations. You can find the total number of observations by using the statistics capabilities of the calculator to find the sum of the numbers in List1. You can get lots of statistics, such as the mean, sum, and standard deviation, by pressing the "Stat" button. On some calculators it is possible to get the sum alone by keying in something like "Sum(List1)." If your calculator has a "Sum" list function, you may be able to key in something like:

Sum(List1)

If not, you can find the sum (406 in this example) and then key in

406 * List2 → List3.

You should now see three lists, similar to the ones shown in Figure 12.4, on your calculator screen.

Now we are in a position to actually calculate X^2. The formula for X^2 is a sum of quantities of the form

$$\frac{(\text{observed cell count} - \text{expected cell count})^2}{(\text{expected cell count})}$$

On your calculator, these quantities can be calculated all at once and then stored in List4:

(List1 − List3)^2/List3 → List4.

The value of the X^2 statistic is then found by finding the sum of the numbers in List4. The calculator we are using has a list function called Sum, so we enter "Sum(List4)" and the calculator returns: 59.4916.

Now that you have calculated the value of the X^2 statistic, you can proceed to assess its significance. Since the goodness-of-fit test is a one-tailed test, we must find the area to the right of $X^2 = 59.4916$ under the chi-square curve with 4 degrees of freedom. At this point it is possible to be misled by the menu system of your calculator. There may be a "Chi square test" on your calculator, but it will *not* be the test for goodness-of-fit. (The "Chi square test" on the calculator is the test for association or homogeneity in a two-way table and will be the topic of the next Graphing Calculator Exploration.)

To find the *P*-value on your calculator, you must navigate your menu system to a function that returns cumulative probabilities for X^2 values—most likely called a "Chisqcdf" or "χ^2 cdf." The two parameters necessary to find the *P*-value are the X^2 statistic value and the number of degrees of freedom. Some calculators, as we saw with the normal distribution calculator exploration, use the abbreviation "cdf" loosely, actually finding the probability of a statistic being *between* two values. In the case of the chi-square distribution this is not too much of a problem since the X^2 statistic cannot be less than zero. If your calculator is one of those that uses the abbreviation "cdf" loosely, you should be able to find the cumulative area by keying in (using the current problem numbers) something like

χ^2 cdf(0, 59.4916, 4)

The calculator should return a number suspiciously close to 1. (Our calculator returns 1.000000000.) The actual *P*-value, remember, will be the area to the right of X^2, 1 − .1.000000000 = .00000000, which agrees with the computer output.

In the case of the chi-square calculations, your calculator may differ in its syntax from what we have written here. Your calculator manual will once again come to the rescue if there are syntax differences.

Exploration 12.2 Homogeneity, Independence, and the Chi-Square Statistic

In Exploration 12.1, we discussed the use of the chi-square statistic to assess the "goodness of fit" of data to a model. Using the chi-square statistic for the assessment of homogeneity of proportions and the independence of two categorical variables requires a bit more discussion. The calculator procedures for homogeneity and independence are more complicated than the goodness-of-fit procedure because there is more than one row of data to consider. However, once the data are entered into the calculator, the calculation of the expected values is an automatic feature built into the calculator—a great time saver!

There are three steps in performing a chi-square test for independence or homogeneity:

1. Prepare to enter the data.
2. Enter the data.
3. Perform the chi-square calculations.

Matrix
Mat A : 2x3
Mat B : 2x2
Mat C : None

Names Math Edit
1: [A] 2x3
2: [B] 2x2
3: [C]
4: [D]
5: [E]
6: [F]
7: [G]

Figure 12.5 Common calculator displays for matrix operations.

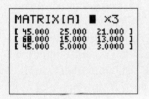

Figure 12.6 Previous numbers in Matrix A.

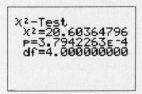

Figure 12.7 Numbers from Example 12.4.

```
X²-Test
  X²=20.60364796
  P=3.7942263E-4
  df=4.000000000
```

Figure 12.8 Calculator display after calculation.

Preparing to enter the data sounds fairly easy, possibly something like (a) pull up a chair, (b) sharpen your pencil, (c) adjust the lighting. Actually, we're not referring to the preparation of your work environment, but preparation of the *calculator*! In past Explorations you have entered data at the cursor into individual places on the screen, and sets of data into lists. The chi-square data entry will be a little different, because you will enter data into a "matrix." Be very careful here: your calculator may have another possibility for data entry, that being "tables." Look in your manual under *matrix*. Matrices (plural of matrix) are different from both lists and tables.

The calculator details for working with matrices will differ from calculator to calculator, but generally you must (a) create or redefine a matrix so that it has the same number of rows and columns as the table you are analyzing, and (b) enter numbers into that matrix. Your calculator may already have some predefined matrices, usually labeled "A," "B," etc. Some calculators may have a "matrix" or "matr" key, or you may have to navigate your menu system to get to a menu item with these terms. If you are following along with your calculator, your goal is to navigate to a matrix screen and set up the matrices for data entry. Two common methods for indicating what matrices are already defined are shown in Figure 12.5.

The screens in Figure 12.5 show that two matrices are defined. The first one, Matrix "A," has two rows and three columns; the second, Matrix "B," has two rows and two columns. To change, or edit, a matrix you must move your cursor to that matrix and press Enter. Editing matrices is another place where calculators will differ, and we refer you to your manual for edification. On the calculator we are using, we move the cursor over to "Edit" and press Enter. The calculator then shows the screen illustrated in Figure 12.6.

We will use Example 12.4, the data on concussions. Those data, you may notice, use a 3 × 3 table. Our plan is to use Matrix A for the observed values, and happily our existing Matrix A is a 3 × 3. (If it were not, we would consult our calculator manual and change the dimensions!) The matrix after data entry appears as shown in Figure 12.7.

We also need a matrix for the expected values for the chi-square procedure. (Some calculators may have a "dedicated" matrix for this purpose. If so, it will be named "Expected" or "MatAns" or something similar.) The calculator we are using requires us to specify a matrix to hold the soon-to-be calculated expected values, and we will use Matrix B. (Your calculator will probably automatically resize the matrix of expected values during the chi-square calculation—If so, you don't need to manually resize Matrix D.) After resizing the necessary matrices and entering the data from Example 12.4, you are now ready to proceed with the actual chi-square calculations. The chi-square test is a statistical hypothesis test, so your navigation through your calculator's menus should be a familiar one. When you get to the menu item, "χ^2-test," press Enter or Execute as appropriate. You will be asked to indicate which matrix contains the observed values, Matrix A in our case. If your calculator does not have a dedicated matrix for the expected values, you must specify that matrix also, in our case Matrix B. Then move your cursor to "Calculate" or "Execute" as appropriate.

Your calculator should return a screen that looks something like Figure 12.8. The value of the chi-square statistic, the P-value, and the number of degrees of freedom are all reported. (Be really careful here—the calculator has switched to scientific notation. That P-value is 0.0003794!) Once again, be sure to compare your numbers with the "correct" answer. If your results differ significantly, enough that round-off error seems unlikely, review your methods and see where you might have made an error. Especially remember to check the original data—incorrect data entry is a common source of errors when working with calculators.

TEACHING TIPS

Intent of this chapter: After completing this chapter, students should be able to (1) know the assumptions necessary for inference in a regression setting and understand when use of the linear regression model is appropriate; (2) compute and interpret confidence intervals for the slope of the regression line; (3) perform a hypothesis test for the slope of the regression line; (4) verify that assumptions needed for inference procedures for the slope of the regression line are plausible; (5) compute and interpret confidence intervals for the mean y-value for a given x-value (optional); (6) compute and interpret prediction intervals for an individual value of y for a given x-value (optional); and (7) test hypotheses about the population correlation coefficient (optional).

A PowerPoint® Lecture is available on the *Instructor's Resource Binder* CD.

■ **Section 13.1**

This section discusses the additive probabilistic model for linear regression (the simple linear regression model): $y = \alpha + \beta x + e$, where α is the population intercept, β is the population slope, and e is the random deviation of an observed point from the line.

The basic assumptions for the simple linear regression model deal with the random deviation e (see page 691). Refer back to the example in the Teaching Tips for Section 5.3 about the height and weight of a population of adult women. Explaining these assumptions in the height/weight scenario aids students' understanding. For a given height, there is a distribution of weights that are normally distributed with the true regression line going through the mean of the distribution (see figure below). The distribution of the random deviation, e, represents the distribution of the deviation of an individual weight from the mean weight for a given height in this example. Thus, e is normally distributed with a mean of zero. For each value of x, the distribution of e is assumed to have the same standard deviation, σ. Because weights of different women of the same height are independent, corresponding values of e are also independent.

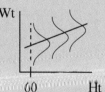

At the top of page 697, the difference between a point estimate of the mean of y and the prediction of an individual y-value is developed. Although the values (and computation) of the point estimate of the mean and the prediction of an individual value are the same, their interpretations are very different. The point estimate of the mean y-value is the estimate of the mean of y at a given x—e.g., the mean weight of all women of a particular height. One of the assumptions is that the true regression line runs through the mean of y. Since the least squares regression line, $\hat{y} = a + bx$, is an estimate for the true regression line, then $\hat{y}$ is a point estimate for the mean of y. The least squares regression line is also used to predict the value of a particular y for a given x-value. Thus, $\hat{y}$ can also be interpreted as the predicted y-value for a given x-value. Section 13.4 (optional) introduces a confidence interval for the mean of y and a prediction interval for an individual y-value.

The last part of Section 13.1 describes the computation of the standard error of the random deviation e (s_e). Because we do not know the true standard deviation about the regression line (σ), we must use the estimate s_e. Notice that the degrees of freedom is $n - 2$ because the calculation of the residuals used to estimate the standard deviation of the errors requires estimation of two unknown parameters (α and β).

Suggested Assignment: Exercises 13.1, 13.3 13.4, 13.5, 13.8, 13.11

Activity: "Chapter 13—Simple Linear Regression Model" (*Instructor's Resource Binder* & CD, Chapter 13 Activities Worksheets)

Assessment: Chapter 13 Quiz 1 (*Test Bank* and *Instructor's Resource Binder* CD)

■ Section 13.2

Section 13.2 describes how to compute and interpret a confidence interval for the slope of the population regression line, as well as how to test hypotheses about the slope. Because most inferential procedures are based on knowledge of a sampling distribution, let's begin by examining the sampling distribution of b.

Suppose heights and weights are known for a population of adult women. A random sample of women is selected and the least-squares line is computed for predicting weight using height. The line, $\hat{y} = a + bx$, is then an estimate of the population regression line (the line that would have been obtained if the entire population was used to fit the least squares line), and b is an estimate for β, the slope of the population line. Suppose that a large number of random samples are selected from this population and that the least squares line is computed for each sample. A dotplot of the sample slopes, b, would be an empirical estimate of the sampling distribution of b. If the assumptions of the regression model are satisfied (from Section 13.1), then the sampling distribution of b will have the properties listed in the tan box on page 703.

The applet listed on the next page can be used to simulate the sampling distributions of b. Students can input the population (true) slope and intercept, the mean and standard deviation of the x-values, and the standard deviation of e (σ). The applet will select samples of a given size and then display the sample lines that are computed and a dotplot for the sample slopes (b) from these lines (see the following figure). Students can see that the mean of the sample slopes is approximately equal to the true slope β (1.0 in the case of the figure below). If we could find all possible samples of the same size from this population, then the mean of bs from those samples would equal β, ($\mu_b = \beta$).

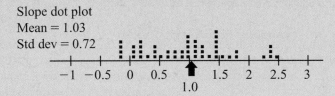

Slope dot plot
Mean = 1.03
Std dev = 0.72

Be sure to stress the difference between σ and σ_b. Students confuse these two standard deviations. σ is the standard deviation of the random deviation about the true regression line for a given x-value. It is the standard deviation of the normal curves for each x-value in the regression model. σ_b, on the other hand, is the standard deviation of the sampling distribution of b. It describes how much the slope differs from sample to sample—the standard deviation of the dotplot shown in the figure above.

The t-distribution with $n - 2$ degrees of freedom is used to compute the confidence interval for the population slope. Remind students how to use the general formula for confidence intervals to find the formula for the confidence interval for the slope of the population regression line:

statistic ± critical value(standard deviation of the statistic) and $b \pm t * (s_b)$

The tan boxes on pages 706–707 provide a summary of the hypothesis test for β. Note that the assumptions deal with e (the random deviation). Section 13.3 discusses how to verify that these assumptions are plausible.

We can test whether the true slope of the regression line is equal to *any* hypothesized value, but we typically test to see if $\beta = 0$ (the model utility test). Draw a horizontal line on a scatterplot, similar to Figure 13.10(a) on page 707. Ask students if this line is helpful in predicting *y*-values from *x*-values. *(They should discover that it's not very useful!)* That line would predict the same value of *y* irrespective of the value of *x*. The model utility test is a procedure that determines if the model is "useful" (i.e., better than the horizontal line) for making predictions by testing H_0: $\beta = 0$. Graphing Calculator Exploration 13.1 on page 746 provides instructions for performing the model utility test.

Also, be sure that students can read computer-generated output for regression analysis, like that at the top of page 706.

Suggested Assignment: Exercises 13.13, 13.15, 13.18, 13.23, 13.24

Activity: Activity 13.1: Are Tall Women from "Big" Families? (p. 739 and *Activities Manual*, p. 108), Bonus Activity 13.2—Golden Rectangles (*Activities Manual*, p. 111)★, and/ or "Chapter 13—Inferences about Slope" (*Instructor's Resource Binder* & CD, Chapter 13 Activities Worksheets)

Applet: http://www.rossmanchance.com/applets/regcoeff/regcoeff.html

■ **Section 13.3**

This section discusses how to verify that the assumptions about *e*, the random deviation of an observed *y*-value from the population regression line $\alpha + \beta x$, are plausible. The key assumptions are as follows: (1) for any *x*-value, the distribution of *e* is normal, and (2) for all *x*-values, the standard deviation of *e* (σ) remains constant (see the bottom of page 713).

To verify the second assumption, the standard deviation of *e* (σ) remains constant for all *x*-values, we examine a residual plot (or a standardized residual plot if you have access to software). The residual plot should resemble Figure 13.5(a) on page 717. The scatter does not appear to change for different *x*-values, which indicates that the equal standard deviation assumption is plausible.

For verification of the first assumption, refer to the Teaching Tips for Section 5.3. For each *x*-value, there is a distribution of *y*-values with the true regression line, $\alpha + \beta x$, passing through the mean of these distributions. We need to verify that the random deviation of observed *y*-values from the population line (*e*) credibly represents deviations from a normal distribution of errors. In samples from our population, we typically get only one or two *y*-values for any particular *x*-value. So we cannot plot the *y*-values for each *x*-value to see if the distribution is credibly normal. But imagine taking each distribution of *e* for a particular *x*-value and sliding (translating) the distributions along the population line so that the distributions of *e* are on top of each other, as shown in the figure below. We have already verified that the standard deviations of *e* are the same. By looking at the residuals (or standardized residuals), we can see how far from the population line each *y*-value is at its given *x*-value. Using a normal probability plot (or histogram or boxplot), plot the residuals to confirm that, as a group, the residuals are at least approximately normal. Thus, if we separate the distributions of *e*, there would be no reason to doubt the normality of each individual distribution. (See plots on pages 720 and 721.)

★ Kathy's personal favorite

Suggested Assignment: Exercises 13.27, 13.28, 13.31

Activity: "Chapter 13—Model Adequacy" (*Instructor's Resource Binder* & CD, Chapter 13 Activities Worksheets)

Assessment: Chapter 13 Quiz 2 (*Test Bank* and *Instructor's Resource Binder* CD)

■ Section 13.4 (Optional)

Section 13.4 introduces the confidence interval for the mean y-value at a given x and the prediction interval for a single y-value.

The tan box on page 727 provides the formula for the confidence interval for the mean y-value. The standard error, s_{a+bx^*} is larger the farther x^* is from $\bar{x}$. Therefore, the confidence interval widens as x^* moves away from $\bar{x}$. This interval estimates the true value of $\alpha + \beta x^*$.

The prediction interval is the interval for a single y value, y^*. This interval estimates the true value of y^*. Notice that the standard error for the prediction interval, $\sqrt{s_e^2 + s_{a+bx^*}^2}$, is larger than the standard error used in the confidence interval for the mean of y. This makes the prediction interval wider than the confidence interval. Look at Figure 13.22 on page 731. This figure shows the regression line with a 90% confidence interval for the mean y-value. Notice that the confidence interval is wider at the ends than in the center. A 90% prediction interval is also shown. This interval is much wider than the confidence interval, yet both intervals are centered at the regression line.

Suggested Assignment: Exercises 13.34, 13.35, 13.36, 13.40, 13.41, 13.48

■ Section 13.5 (Optional)

This section presents the hypothesis test for the population correlation coefficient ρ (pronounced "row" as in rowing a boat). A bivariate normal distribution is one where the y-values for each x are normally distributed *and* the x-values for each y are also normally distributed. Height and weight of adult males is an example of a pair of variables that might have a bivariate normal distribution. In a bivariate normal distribution, $\rho = 0$ is equivalent to x and y being independent. For a bivariate normal population, testing the null hypothesis $\rho = 0$ is a test for independence. The test is a t test with $n - 2$ degrees of freedom. (See the tan box on page 735.) Notice that the model utility test can also be used to test for independence in a bivariate normal population.

Suggested Assignment: Exercises 13.50, 13.51, 13.53

Activity: Bonus Activity 13.3—Name Lengths (*Activities Manual*, p.114)★

Assessment: Chapter 13 Quiz 3 (*Test Bank* and *Instructor's Resource Binder* CD)

■ Section 13.6

Section 13.6 provides excellent suggestions for communicating and interpreting results from hypothesis tests and confidence intervals. Students should also pay close attention to "A Word to the Wise: Cautions and Limitations."

Suggested Review Assignment: Exercises 13.61, 13.62, 13.63, 13.64, 13.65, 13.69, 13.73

Assessment: Chapter 13 Concept Quiz (*Test Bank* and *Instructor's Resource Binder* CD) and Chapter 13 Test (*Test Bank* and *Instructor's Resource Binder* CD)

★ Kathy's personal favorite

Chapter 13

Simple Linear Regression and Correlation: Inferential Methods

egression and correlation were introduced in Chapter 5 as techniques for describing and summarizing bivariate data consisting of (x, y) pairs. For example, consider a regression of the dependent variable y = percentage of courses taught by teachers with inappropriate or no license on the independent variable x = spending per pupil based on a sample of Missouri public school districts ("Is Teacher Pay Adequate?," *Research Working Papers Series*, Kennedy School of Government, Harvard University, October 2005). A scatterplot of the data shows a surprising linear pattern; the sample correlation coefficient is $r = .27$; and the equation of the least-squares line had a positive slope, indicating that school districts with higher expenditures per student also tended to have a higher percentage of courses taught by teachers with an inappropriate license or no license. Could the pattern observed in the scatterplot be plausibly explained by chance, or does the sample provide convincing evidence of a linear relationship between these two variables for school districts in Missouri? If there is evidence of a meaningful relationship between these two variables, the regression line could be used as the basis for predicting the percentage of teachers with inappropriate or no license for a school district with a specified expenditure per student or for estimating the average percentage of teachers with inappropriate or no license for all school districts with a specified expenditure per student. In this chapter, we develop inferential methods for bivariate numerical data, including a confidence interval (interval estimate) for a mean y value, a prediction interval for a single y value, and a test of hypotheses regarding the extent of correlation in the entire population of (x, y) pairs.

13.1 Simple Linear Regression Model

A *deterministic relationship* is one in which the value of y is completely determined by the value of an independent variable x. Such a relationship can be described using traditional mathematical notation, such as $y = f(x)$ where $f(x)$ is a specified function of x. For example, we might have

$$y = f(x) = 10 + 2x$$

or

$$y = f(x) = 4 - (10)^{2x}$$

However, in most situations, the variables of interest are not deterministically related. For example, the value of $y =$ first-year college grade point average is certainly not determined solely by $x =$ high school grade point average, and $y =$ crop yield is determined partly by factors other than $x =$ amount of fertilizer used.

A description of the relation between two variables x and y that are not deterministically related can be given by specifying a **probabilistic model**. The general form of an **additive probabilistic model** allows y to be larger or smaller than $f(x)$ by a random amount e. The **model equation** is of the form

$$y = \text{deterministic function of } x + \text{random deviation}$$
$$= f(x) + e$$

Let x^* denote some particular value of x, and suppose that an observation on y is made when $x = x^*$. Then

$$y > f(x^*) \quad \text{if } e > 0$$
$$y < f(x^*) \quad \text{if } e < 0$$
$$y = f(x^*) \quad \text{if } e = 0$$

Thinking geometrically, if $e > 0$, the observed point (x^*, y) will lie above the graph of $y = f(x)$; $e < 0$ implies that this point will fall below the graph. This is illustrated in Figure 13.1. If $f(x)$ is a function used in a probabilistic model relating y to x and if observations on y are made for various values of x, the resulting (x, y) points will be distributed about the graph of $f(x)$, some falling above it and some falling below it.

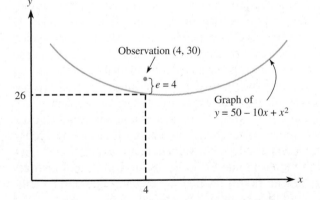

Figure 13.1   A deviation from the deterministic part of a probabilistic model.

■ Simple Linear Regression

The simple linear regression model is a special case of the general probabilistic model in which the deterministic function $f(x)$ is linear (so its graph is a straight line).

DEFINITION

The **simple linear regression model** assumes that there is a line with vertical or y intercept α and slope β, called the **true** or **population regression line**. When a value of the independent variable x is fixed and an observation on the dependent variable y is made,

$$y = \alpha + \beta x + e$$

Without the random deviation e, all observed (x, y) points would fall exactly on the population regression line. The inclusion of e in the model equation recognizes that points will deviate from the line by a random amount.

Figure 13.2 shows several observations in relation to the population regression line.

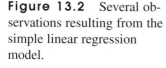

Figure 13.2 Several observations resulting from the simple linear regression model.

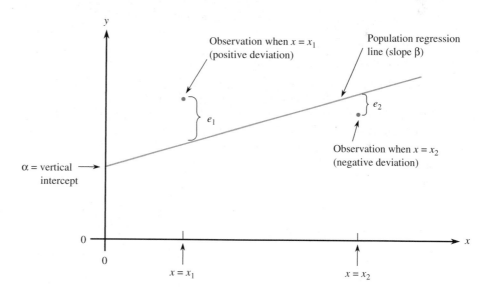

Before we make an observation on y for any particular value of x, we are uncertain about the value of e. It could be negative, positive, or even 0. Also, it might be quite large in magnitude (a point far from the population regression line) or quite small (a point very close to the line). In this chapter, we make some assumptions about the distribution of e in repeated sampling at any particular x value.

Basic Assumptions of the Simple Linear Regression Model

1. The distribution of e at any particular x value has mean value 0. That is, $\mu_e = 0$.
2. The standard deviation of e (which describes the spread of its distribution) is the same for any particular value of x. This standard deviation is denoted by σ.
3. The distribution of e at any particular x value is normal.
4. The random deviations $e_1, e_2, \ldots, e_n$ associated with different observations are independent of one another.

Randomness in e implies that y itself is subject to uncertainty. The foregoing assumptions about the distribution of e imply that the distribution of y values in repeated sampling satisfies certain properties. Consider y when x has some fixed value x^*, so that

$$y = a + \beta x^* + e$$

Because α and β are fixed numbers, $\alpha + \beta x^*$ is also a fixed number. The sum of a fixed number and a normally distributed variable is again a normally distributed variable (the bell-shaped curve is simply relocated), so y itself has a normal distribution. Furthermore, $\mu_e = 0$ implies that the mean value of y is just $\alpha + \beta x^*$, the height of the population regression line above the value x^*. Finally, because there is no variability in the fixed number $\alpha + \beta x^*$, the standard deviation of y is the same as that of e. These properties are summarized in the following box.

For any fixed x value, y itself has a normal distribution, with

$$\begin{pmatrix} \text{mean } y \text{ value} \\ \text{for fixed } x \end{pmatrix} = \begin{pmatrix} \text{height of the population} \\ \text{regression line above } x \end{pmatrix} = \alpha + \beta x$$

and

$$(\text{standard deviation of } y \text{ for a fixed } x) = \sigma$$

The slope β of the population regression line is the *average* change in y associated with a 1-unit increase in x. The y intercept α is the height of the population line when $x = 0$. The value of σ determines the extent to which (x, y) observations deviate from the population line; when σ is small, most observations will be quite close to the line, but with large σ, there are likely to be some substantial deviations.

The key features of the model are illustrated in Figures 13.3 and 13.4. Notice that the three normal curves in Figure 13.3 have identical spreads. This is a consequence of $\sigma_e = \sigma$, which implies that the variability does not depend on the value of x.

Figure 13.3 Illustration of the simple linear regression model.

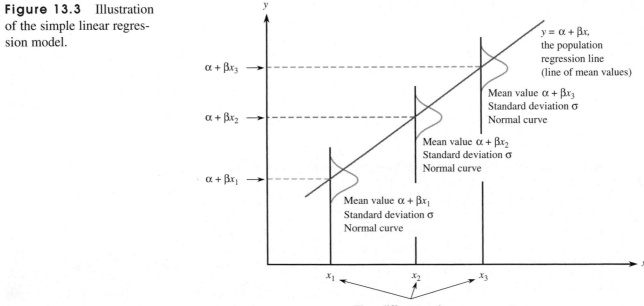

Figure 13.4 Data from the simple linear regression model: (a) small σ; (b) large σ.

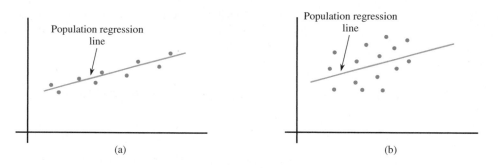

(a)   (b)

Example 13.1 Stand on Your Head to Lose Weight?

© ImageState Royalty-Free/Alamy

The authors of the article "On Weight Loss by Wrestlers Who Have Been Standing on Their Heads" (paper presented at the Sixth International Conference on Statistics, Combinatorics, and Related Areas, Forum for Interdisciplinary Mathematics, 1999, with the data also appearing in *A Quick Course in Statistical Process Control*, Mick Norton, Pearson Prentice Hall, 2005) stated that "amateur wrestlers who are overweight near the end of the weight certification period, but just barely so, have been known to stand on their heads for a minute or two, get on their feet, step back on the scale, and establish that they are in the desired weight class. Using a headstand as the method of last resort has become a fairly common practice in amateur wrestling."

Does this really work? Data were collected in an experiment where weight loss was recorded for each wrestler after exercising for 15 min and then doing a headstand for 1 min 45 sec. Based on these data, the authors of the article concluded that there was in fact a demonstrable weight loss that was greater than that for a control group that exercised for 15 min but did not do the headstand. (The authors give a plausible explanation for why this might be the case based on the way blood and other body fluids collect in the head during the headstand and the effect of weighing while these fluids are draining immediately after standing.) The authors also concluded that a simple linear regression model was a reasonable way to describe the relationship between the variables

$$y = \text{weight loss (in pounds)}$$

and

$$x = \text{body weight prior to exercise and headstand (in pounds)}$$

Suppose that the actual model equation has $\alpha = 0$, $\beta = 0.001$, and $\sigma = 0.09$ (these values are consistent with the findings in the article). The population regression line is shown in Figure 13.5.

Figure 13.5 The population regression line for Example 13.1.

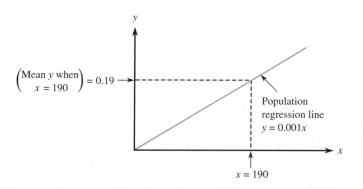

If the distribution of the random errors at any fixed weight (x value) is normal, then the variable y = weight loss is normally distributed with

$$\mu_y = 0 + 0.001x$$
$$\sigma_y = .09$$

For example, when $x = 190$ (corresponding to a 190-lb wrestler), weight loss has mean value

$$\mu_y = 0 + .001(190) = .19$$

Because the standard deviation of y is $\sigma = 0.09$, the interval $0.19 \pm 2(0.09) = (0.01, 0.37)$ includes y values that are within 2 standard deviations of the mean value for y when $x = 190$. Roughly 95% of the weight loss observations made for 190-lb wrestlers will be in this range.

The slope $\beta = 0.001$ is the change in average weight associated with each additional pound of body weight.

■

More insight into model properties can be gained by thinking of the population of all (x, y) pairs as consisting of many smaller populations. Each one of these smaller populations contains pairs for which x has a fixed value. Suppose, for example, that the variables

x = grade point average in major courses

and

y = starting salary after graduation

are related according to the simple linear regression model. Then there is the population of all pairs with $x = 3.20$, the population of all pairs having $x = 2.75$, and so on. The model assumes that for each such population, y is normally distributed with the same standard deviation, and the *mean y value* (rather than y itself) is linearly related to x.

Check of the assumptions

In practice, the judgment of whether the simple linear regression model is appropriate must be based on how the data were collected and on a scatterplot of the data. The sample observations should be independent of one another. In addition, the scatterplot should show a linear rather than a curved pattern, and the vertical spread of points should be relatively homogeneous throughout the range of x values. Figure 13.6 shows plots with three different patterns; only the first is consistent with the model assumptions.

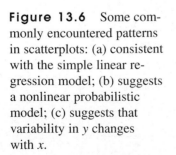

Figure 13.6 Some commonly encountered patterns in scatterplots: (a) consistent with the simple linear regression model; (b) suggests a nonlinear probabilistic model; (c) suggests that variability in y changes with x.

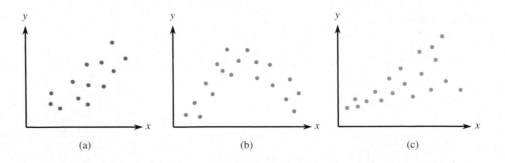

▪ **Estimating the Population Regression Line** ....

The values of α and β (y intercept and slope of the population regression line) will almost never be known to an investigator. Instead, these values must first be estimated from the sample data $(x_1, y_1), \ldots, (x_n, y_n)$. We now assume that these n (x, y) pairs were obtained independently of one another and that each observed y is related to the corresponding x by means of the model equation for simple linear regression.

Let a and b denote point estimates of α and β, respectively. These estimates come from applying the method of least squares introduced in Chapter 5; the sum of squared vertical deviations of points in the scatterplot from the least-squares line is smaller than for any other line.

The point estimates of β, the slope, and α, the y intercept of the population regression line, are the slope and y intercept, respectively, of the least-squares line. That is,

$$b = \text{point estimate of } \beta = \frac{S_{xy}}{S_{xx}}$$
$$a = \text{point estimate of } \alpha = \bar{y} - b\bar{x}$$

where

$$S_{xy} = \sum xy - \frac{(\sum x)(\sum y)}{n} \quad \text{and} \quad S_{xx} = \sum x^2 - \frac{(\sum x)^2}{n}$$

The estimated regression line is then just the least-squares line

$$\hat{y} = a + bx$$

Let x^* denote a specified value of the predictor variable x. Then $a + bx^*$ has two different interpretations:

1. It is a point estimate of the mean y value when $x = x^*$.
2. It is a point prediction of an individual y value to be observed when $x = x^*$.

Example 13.2 Mother's Age and Baby's Birth Weight

● Medical researchers have noted that adolescent females are much more likely to deliver low-birth-weight babies than are adult females. Because low-birth-weight babies have higher mortality rates, a number of studies have examined the relationship between birth weight and mother's age for babies born to young mothers.

One such study is described in the article "The Risk of Teen Mothers Having Low Birth Weight Babies: Implications of Recent Medical Research for School Health Personnel" (*Journal of School Health* [1998]: 271–274). The following data on

$$x = \text{maternal age (in years)}$$

and

$$y = \text{birth weight of baby (in grams)}$$

are consistent with summary values given in the referenced article and also with data published by the National Center for Health Statistics.

	Observation									
	1	**2**	**3**	**4**	**5**	**6**	**7**	**8**	**9**	**10**
x	15	17	18	15	16	19	17	16	18	19
y	2289	3393	3271	2648	2897	3327	2970	2535	3138	3573

A scatterplot (Figure 13.7) supports the appropriateness of the simple linear regression model.

Figure 13.7 Scatterplot of the data from Example 13.2.

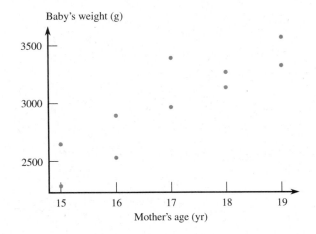

The summary statistics are

$$n = 10 \qquad \sum x = 170 \qquad \sum y = 30{,}041$$
$$\sum x^2 = 2910 \qquad \sum xy = 515{,}600 \qquad \sum y^2 = 91{,}785{,}351$$

from which

$$S_{xy} = \sum xy - \frac{(\sum x)(\sum y)}{n} = 515{,}600 - \frac{(170)(30{,}041)}{10} = 4903.0$$

$$S_{xx} = \sum x^2 - \frac{(\sum x)^2}{n} = 2910 - \frac{(170)^2}{10} = 20.0$$

$$\bar{x} = \frac{170}{10} = 17.0 \qquad \bar{y} = \frac{30041}{10} = 3004.1$$

This gives

$$b = \frac{S_{xy}}{S_{xx}} = \frac{4903.0}{20.0} = 245.15$$
$$a = \bar{y} - b\bar{x} = 3004.1 - (245.1)(17.0) = -1163.45$$

The equation of the estimated regression line is then

$$\hat{y} = a + bx = -1163.45 + 245.15x$$

A point estimate of the average birth weight of babies born to 18-year-old mothers results from substituting $x = 18$ into the estimated equation:

$$\text{(estimated average } y \text{ when } x = 18) = a + bx$$
$$= -1163.45 + 245.15(18)$$
$$= 3249.25 \text{grams}$$

Similarly, we would predict the birth weight of a baby to be born to a particular 18-year-old mother to be

$$\text{(predicted } y \text{ value when } x = 18) = a + b(18) = 3249.25 \text{ grams}$$

Important concept

The point estimate and the point prediction are identical, because the same x value was used in each calculation. However, the interpretation of each is different. One represents our prediction of the weight of a single baby whose mother is 18, whereas the other represents our estimate of the average weight of *all* babies born to 18-year-old mothers. This distinction will become important in Section 13.4, when we consider interval estimates and predictions.

The least-squares line could have also been fit using a graphing calculator or a statistical software package. MINITAB output resulting from the computation of the regression line for the data of this example is shown here. Note that MINITAB has rounded the values of the estimated coefficients in the equation of the regression line, which would result in small differences in predictions based on the line.

Regression Analysis: Birth Weight versus Maternal Age

The regression equation is
Birth Weight = −1163 + 245 Maternal Age

Predictor	Coef	SE Coef	T	P
Constant	−1163.4	783.1	−1.49	0.176
Maternal Age	245.15	45.91	5.34	0.001

S = 205.308   R-Sq = 78.1%   R-Sq(adj) = 75.4%

In Example 13.2, the x values in the sample ranged from 15 to 19. An estimate or prediction should not be attempted for any x value much outside this range. Without sample data for such values, there is no evidence that the estimated linear relationship can be extrapolated very far. Statisticians refer to this potential pitfall as the **danger of extrapolation**.

■ Estimating σ^2 and σ ....

The value of σ determines the extent to which observed points (x, y) tend to fall close to or far away from the population regression line. A point estimate of σ is based on

$$\text{SSResid} = \sum (y - \hat{y})^2$$

where $\hat{y}_1 = a + bx_1, \ldots, \hat{y}_n = a + bx_n$ are the fitted or predicted y values and the residuals are $y_1 - \hat{y}_1, \ldots, y_n - \hat{y}_n$. SSResid is a measure of the extent to which the sample data spread out about the estimated regression line.

> **DEFINITION**
>
> The statistic for estimating the variance σ^2 is
>
> $$s_e^2 = \frac{\text{SSResid}}{n - 2}$$
>
> where
>
> $$\text{SSResid} = \sum(y - \hat{y})^2 = \sum y^2 - a\sum y - b\sum xy$$
>
> The subscript e in s_e^2 reminds us that we are estimating the variance of the "errors" or residuals.
>
> The estimate of σ is the **estimated standard deviation**
>
> $$s_e = \sqrt{s_e^2}$$
>
> The number of degrees of freedom associated with estimating σ^2 or σ in simple linear regression is $n - 2$.

The estimates and number of degrees of freedom here have analogs in our previous work involving a single sample $x_1, x_2, \ldots, x_n$. The sample variance s^2 had a numerator of $\sum(x - \bar{x})^2$, a sum of squared deviations (residuals), and denominator $n - 1$, the number of degrees of freedom associated with s^2 and s. The use of $\bar{x}$ as an estimate of μ in the formula for s^2 reduces the number of degrees of freedom by 1, from n to $n - 1$. In simple linear regression, estimation of α and β results in a loss of 2 degrees of freedom, leaving $n - 2$ as the number of degrees of freedom for SSResid, s_e^2, and s_e.

The coefficient of determination was defined previously (see Chapter 5) as

$$r^2 = 1 - \frac{\text{SSResid}}{\text{SSTo}}$$

where

$$\text{SSTo} = \sum(y - \bar{y})^2 = \sum y^2 - \frac{(\sum y)^2}{n} = S_{yy}$$

The value of r^2 can now be interpreted as the proportion of observed y variation that can be explained by (or attributed to) the model relationship. The estimate s_e also gives another assessment of model performance. Roughly speaking, the value of σ represents the magnitude of a typical deviation of a point (x, y) in the population from the population regression line. Similarly, in a rough sense, s_e is the magnitude of a typical sample deviation (residual) from the least-squares line. The smaller the value of s_e, the closer the points in the sample fall to the line and the better the line does in predicting y from x.

....

Example 13.3 Predicting Election Outcomes

● The authors of the paper "Inferences of Competence from Faces Predict Election Outcomes" (*Science* [2005]: 1623–1626) found that they could successfully predict the outcome of a U.S. congressional election substantially more than half the time

● Data set available online

based on the facial appearance of the candidates. In the study described in the paper, participants were shown photos of two candidates for a U.S. Senate or House of Representatives election. Each participant was asked to look at the photos and then indicate which candidate he or she thought was more competent. The two candidates were labeled A and B. If a participant recognized either candidate, data from that participant were not used in the analysis. The proportion of participants who chose candidate A as the more competent was computed. After the election, the difference in votes (candidate A − candidate B) expressed as a proportion of the total votes cast in the election was also computed. This difference falls between +1 and −1. It is 0 for an election where both candidates receive the same number of votes, positive for an election where candidate A received more votes than candidate B (with +1 indicating that candidate A received all of the votes), and negative for an election where candidate A received fewer votes than candidate B.

This process was carried out for a large number of congressional races. A subset of the resulting data (approximate values read from a graph that appears in the paper) is given in the accompanying table, which also includes the predicted values and residuals for the least-squares line fit to these data.

Competent Proportion	Difference in Vote Proportion	Predicted y Value	Residual
0.20	−0.70	−0.389	−0.311
0.23	−0.40	−0.347	−0.053
0.40	−0.35	−0.109	−0.241
0.35	0.18	−0.179	0.359
0.40	0.38	−0.109	0.489
0.45	−0.10	−0.040	−0.060
0.50	0.20	0.030	0.170
0.55	−0.30	0.100	−0.400
0.60	0.30	0.170	0.130
0.68	0.18	0.281	−0.101
0.70	0.50	0.309	0.191
0.76	0.22	0.393	−0.173

The scatterplot (Figure 13.8) gives evidence of a positive linear relationship between

x = proportion of participants who judged candidate A as the more competent

and

y = difference in vote proportion.

The summary statistics are

$$n = 12 \qquad \sum x = 5.82 \qquad \sum y = 0.11$$
$$\sum x^2 = 3.1804 \qquad \sum xy = 0.5526 \qquad \sum y^2 = 1.5101$$

from which we calculate

$$b = 1.3957 \qquad a = -0.6678$$
$$\text{SSResid} = .81228 \qquad \text{SSTo} = 1.50909$$

Figure 13.8 MINITAB
scatterplot for Example 13.3.

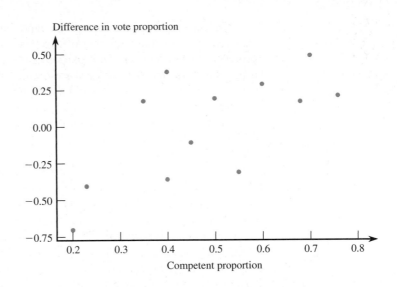

Thus,

$$r^2 = 1 - \frac{\text{SSResid}}{\text{SSTo}} = 1 - \frac{0.81228}{1.50909} = 1 - .538 = .462$$

$$s_e^2 = \frac{\text{SSResid}}{n-2} = \frac{0.81228}{10} = .081$$

and

$$s_e = \sqrt{.081} = .285$$

Approximately 46.2% of the observed variation in the difference in vote proportion y can be attributed to the probabilistic linear relationship with proportion of participants who judged the candidate to be more competent based on facial appearance. The magnitude of a typical sample deviation from the least-squares line is about .285, which is reasonably small in comparison to the y values themselves. The model appears to be useful for estimation and prediction; in Section 13.2, we show how a model utility test can be used to judge whether this is indeed the case.

■

A key assumption of the simple linear regression model is that the random deviation e in the model equation is normally distributed. In Section 13.3, we will indicate how the residuals can be used to determine whether this is plausible.

■ **Exercises 13.1–13.11** .... Turn to the Expanded Answers Section for answers not shown next to exercises. ....

13.1 Let x be the size of a house (sq ft) and y be the amount of natural gas used (therms) during a specified period. Suppose that for a particular community, x and y are related according to the simple linear regression model with

β = slope of population regression line = .017
α = y intercept of population regression line = -5.0

a. What is the equation of the population regression line?
b. Graph the population regression line by first finding the point on the line corresponding to $x = 1000$ and then the point corresponding to $x = 2000$, and drawing a line through these points.
c. What is the mean value of gas usage for houses with 2100 sq ft of space? 30.7

Bold exercises answered in back ● Data set available online but not required ▼ Video solution available

d. What is the average change in usage associated with a 1-sq-ft increase in size?   0.017

e. What is the average change in usage associated with a 100-sq-ft increase in size?   1.7

f. Would you use the model to predict mean usage for a 500-sq-ft house? Why or why not? (*Note*: There are no small houses in the community in which this model is valid.)   No, the model should not be used to predict outside the range of the data.

13.2 The flow rate in a device used for air quality measurement depends on the pressure drop x (inches of water) across the device's filter. Suppose that for x values between 5 and 20, these two variables are related according to the simple linear regression model with true regression line $y = -0.12 + 0.095x$.

a. What is the true average flow rate for a pressure drop of 10 in.? A drop of 15 in.?   0.83, 1.305

b. What is the true average change in flow rate associated with a 1-in. increase in pressure drop? Explain.   0.095

c. What is the average change in flow rate when pressure drop decreases by 5 in.? -0.475

13.3 Data presented in the article "Manganese Intake and Serum Manganese Concentration of Human Milk-Fed and Formula-Fed Infants" (*American Journal of Clinical Nutrition* [1984]: 872–878) suggest that a simple linear regression model is reasonable for describing the relationship between y = serum manganese (Mn) and x = Mn intake (mg/kg/day). Suppose that the true regression line is $y = -2 + 1.4x$ and that $\sigma = 1.2$. Then for a fixed x value, y has a normal distribution with mean $-2 + 1.4x$ and standard deviation 1.2.

a. What is the mean value of serum Mn when Mn intake is 4.0? When Mn intake is 4.5?   3.6, 4.3

b. What is the probability that an infant whose Mn intake is 4.0 will have serum Mn greater than 5?   0.121

c. Approximately what proportion of infants whose Mn intake is 5 will have a serum Mn greater than 5? Less than 3.8?   0.5, 0.1587

13.4 A sample of small cars was selected, and the values of x = horsepower and y = fuel efficiency (mpg) were determined for each car. Fitting the simple linear regression model gave the estimated regression equation $\hat{y} = 44.0 - .150x$.

a. How would you interpret $b = -.150$?

b. Substituting $x = 100$ gives $\hat{y} = 29.0$. Give two different interpretations of this number.

c. What happens if you predict efficiency for a car with a 300-horsepower engine? Why do you think this has occurred?

d. Interpret r^2 in the context of this problem.

e. Interpret s_e in the context of this problem.

13.5 Suppose that a simple linear regression model is appropriate for describing the relationship between y = house price and x = house size (sq ft) for houses in a large city. The true regression line is $y = 23,000 + 47x$ and $\sigma = 5000$.

a. What is the average change in price associated with one extra sq ft of space? With an additional 100 sq ft of space?   47, 4700

b. What proportion of 1800-sq-ft homes would be priced over \$110,000? Under \$100,000?   0.3156, 0.0643

13.6 a. Explain the difference between the line $y = \alpha + \beta x$ and the line $\hat{y} = a + bx$.

b. Explain the difference between β and b.

c. Let x^* denote a particular value of the independent variable. Explain the difference between $\alpha + \beta x^*$ and $a + bx^*$.

d. Explain the difference between σ and s_e.

13.7 ▼ Legumes, such as peas and beans, are important crops whose production is greatly affected by pests. The article "Influence of Wind Speed on Residence Time of *Uroleucon ambrosiae alatae* on Bean Plants" (*Environmental Entomology* [1991]: 1375–1380) reported on a study in which aphids were placed on a bean plant, and the elapsed time until half of the aphids had departed was observed. Data on x = wind speed (m/sec) and y = residence half time were given and used to produce the following information.

$a = 0.0119$ $b = 3.4307$ $n = 13$
SSTo = 73.937   SSResid = 27.890

a. What percentage of observed variation in residence half time can be attributed to the simple linear regression model? $r^2 = 0.6228$

b. Give a point estimate of σ and interpret the estimate.

c. Estimate the mean change in residence half time associated with a 1-m/sec increase in wind speed. $b = 3.4307$

d. Calculate a point estimate of true average residence half time when wind speed is 1 m/sec.   3.4426

13.8 ● The accompanying data on x = treadmill run time to exhaustion (min) and y = 20-km ski time (min) were

Bold exercises answered in back   ● Data set available online but not required   ▼ Video solution available

taken from the article "Physiological Characteristics and Performance of Top U.S. Biathletes" (*Medicine and Science in Sports and Exercise* [1995]: 1302–1310):

x	7.7	8.4	8.7	9.0	9.6	9.6
y	71.0	71.4	65.0	68.7	64.4	69.4

x	10.0	10.2	10.4	11.0	11.7
y	63.0	64.6	66.9	62.6	61.7

$$\sum x = 106.3 \qquad \sum x^2 = 1040.95$$
$$\sum y = 728.70 \qquad \sum xy = 7009.91 \qquad \sum y^2 = 48390.79$$

a. Does a scatterplot suggest that the simple linear regression model is appropriate?
b. Determine the equation of the estimated regression line, and draw the line on your scatterplot.
c. What is your estimate of the average change in ski time associated with a 1-min increase in treadmill time?
d. What would you predict ski time to be for an individual whose treadmill time is 10 min?   65.4606
e. Should the model be used as a basis for predicting ski time when treadmill time is 15 min? Explain.
f. Calculate and interpret the value of r^2.
g. Calculate and interpret the value of s_e.

13.9 The accompanying summary quantities resulted from a study in which x was the number of photocopy machines serviced during a routine service call and y was the total service time (min):

$$n = 16 \quad \sum(y - \bar{y})^2 = 22{,}398.05 \quad \sum(y - \hat{y})^2 = 2620.57$$

a. What proportion of observed variation in total service time can be explained by a linear probabilistic relationship between total service time and the number of machines serviced? $r^2 = 0.883$

b. Calculate the value of the estimated standard deviation s_e. What is the number of degrees of freedom associated with this estimate? $s_e = 13.682$, df = 14

13.10 Exercise 5.48 described a regression situation in which y = hardness of molded plastic and x = amount of time elapsed since termination of the molding process. Summary quantities included $n = 15$, SSResid = 1235.470, and SSTo = 25,321.368.
a. Calculate a point estimate of σ. On how many degrees of freedom is the estimate based? $s_e = 9.749$, df = 13
b. What percentage of observed variation in hardness can be explained by the simple linear regression model relationship between hardness and elapsed time? $r^2 = 0.951$

13.11 ● The accompanying data on x = advertising share and y = market share for a particular brand of cigarettes during 10 randomly selected years are from the article "Testing Alternative Econometric Models on the Existence of Advertising Threshold Effect" (*Journal of Marketing Research* [1984]: 298–308).

x	.103	.072	.071	.077	.086	.047	.060	.050	.070	.052
y	.135	.125	.120	.086	.079	.076	.065	.059	.051	.039

a. Construct a scatterplot for these data. Do you think the simple linear regression model would be appropriate for describing the relationship between x and y?
b. Calculate the equation of the estimated regression line and use it to obtain the predicted market share when the advertising share is .09.
c. Compute r^2. How would you interpret this value?
d. Calculate a point estimate of σ. On how many degrees of freedom is your estimate based? $s_e = 0.0263$, df = 8

Bold exercises answered in back   ● Data set available online but not required   ▼ Video solution available

---

13.2 Inferences About the Slope of the Population Regression Line

The slope coefficient β in the simple linear regression model represents the average or expected change in the dependent variable y that is associated with a 1-unit increase in the value of the independent variable x. For example, consider x = the size of a house (in square feet) and y = selling price of the house. If we assume that the simple linear regression model is appropriate for the population of houses in a particular city, β would be the average increase in selling price associated with a 1-ft^2 increase in size. As another example, if x = amount of time per week a computer system is used and y = the resulting annual maintenance expense, then β would be the expected change in expense associated with using the computer system one additional hour per week.

Important concept

Because the value of β is almost always unknown, it has to be estimated from the n independently selected observations $(x_1, y_1), \ldots, (x_n, y_n)$. The slope b of the least-squares line gives a point estimate. As with any point estimate, though, it is desirable to have some indication of how accurately b estimates β. In some situations, the value of the statistic b may vary greatly from sample to sample, so b computed from a single sample may well be rather different from the true slope β. In other situations, almost all possible samples yield b values that are quite close to β, so the error of estimation is almost sure to be small. To proceed further, we need some facts about the sampling distribution of b: information about the shape of the sampling distribution curve, where the curve is centered relative to β, and how much the curve spreads out about its center.

Properties of the Sampling Distribution of b

When the four basic assumptions of the simple linear regression model are satisfied, the following conditions are met:

1. The mean value of b is β. That is, $\mu_b = \beta$, so the sampling distribution of b is always centered at the value of β. Thus, b is an unbiased statistic for estimating β.
2. The standard deviation of the statistic b is

$$\sigma_b = \frac{\sigma}{\sqrt{S_{xx}}}$$

3. The statistic b has a normal distribution (a consequence of the model assumption that the random deviation e is normally distributed).

The fact that b is unbiased means only that the sampling distribution is centered at the right place; it gives no information about dispersion. If σ_b is large, then the sampling distribution of b will be quite spread out around β and an estimate far from β may well result. For σ_b to be small, the numerator σ should be small (little variability about the population line) and/or the denominator $\sqrt{S_{xx}}$ or, equivalently, $S_{xx} = \Sigma(x - \bar{x})^2$ itself should be large. Because $\Sigma(x - \bar{x})^2$ is a measure of how much the observed x values spread out, β tends to be more precisely estimated when the x values in the sample are spread out rather than when they are close together.

The normality of the sampling distribution of b implies that the standardized variable

$$z = \frac{b - \beta}{\sigma_b}$$

has a standard normal distribution. However, inferential methods cannot be based on this variable, because the value of σ_b is not available (since the unknown σ appears in the numerator of σ_b). The obvious way out of this dilemma is to estimate σ with s_e, yielding an estimated standard deviation.

The **estimated standard deviation of the statistic b** is

$$S_b = \frac{s_e}{\sqrt{S_{xx}}}$$

When the four basic assumptions of the simple linear regression model are satisfied, the probability distribution of the standardized variable

$$t = \frac{b - \beta}{s_b}$$

is the t distribution with df $= (n - 2)$.

In the same way that $t = \dfrac{\bar{x} - \mu}{\dfrac{s}{\sqrt{n}}}$ was used in Chapter 9 to develop a confidence interval for μ, the t variable in the preceding box can be used to obtain a confidence interval (interval estimate) for β.

Confidence Interval for β

When the four basic assumptions of the simple linear regression model are satisfied, a **confidence interval for β**, the slope of the population regression line, has the form

$$b \pm (t \text{ critical value}) \cdot s_b$$

where the t critical value is based on df $= n - 2$. Appendix Table 3 gives critical values corresponding to the most frequently used confidence levels.

The interval estimate of β is centered at b and extends out from the center by an amount that depends on the sampling variability of b. When s_b is small, the interval is narrow, implying that the investigator has relatively precise knowledge of β.

....

Example 13.4 Athletic Performance and Cardiovascular Fitness

● Is cardiovascular fitness (as measured by time to exhaustion from running on a treadmill) related to an athlete's performance in a 20-km ski race? The following data on

$$x = \text{treadmill time to exhaustion (in minutes)}$$

and

$$y = \text{20-km ski time (in minutes)}$$

Step-by-step technology instructions available online   ● Data set available online

were taken from the article "Physiological Characteristics and Performance of Top U.S. Biathletes" (*Medicine and Science in Sports and Exercise* [1995]: 1302–1310):

x	7.7	8.4	8.7	9.0	9.6	9.6	10.0	10.2	10.4	11.0	11.7
y	71.0	71.4	65.0	68.7	64.4	69.4	63.0	64.6	66.9	62.6	61.7

A scatterplot of the data appears in the following figure:

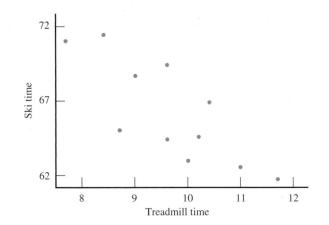

The plot shows a linear pattern, and the vertical spread of points does not appear to be changing over the range of x values in the sample. If we assume that the distribution of errors at any given x value is approximately normal, then the simple linear regression model seems appropriate.

The slope β in this context is the average change in ski time associated with a 1-min increase in treadmill time. The scatterplot shows a negative linear relationship, so the point estimate of β is negative.

Straightforward calculation gives

$$n = 11 \qquad \sum x = 106.3 \qquad \sum y = 728.70$$
$$\sum x^2 = 1040.95 \qquad \sum xy = 7009.91 \qquad \sum y^2 = 48{,}390.79$$

from which

$$b = -2.3335 \qquad a = 88.796$$
$$\text{SSResid} = 43.097 \qquad \text{SSTo} = 117.727$$
$$r^2 = .634 \quad (63.4\% \text{ of the observed variation in ski time can be explained}$$
$$\text{by the simple linear regression model})$$
$$s_e^2 = 4.789 \qquad s_e = 2.188$$
$$s_b = \frac{s_e}{\sqrt{S_{xx}}} = \frac{2.188}{3.702} = .591$$

Calculation of the 95% confidence interval for β requires a t critical value based on df $= n - 2 = 11 - 2 = 9$, which (from Appendix Table 3) is 2.26. The resulting interval is then

$$b \pm (t \text{ critical value}) \cdot s_b = -2.3335 \pm (2.26)(.591)$$
$$= -2.3335 \pm 1.336$$
$$= (-3.671, -.999)$$

We interpret this interval as follows: Based on the sample data, we are 95% confident that the true average decrease in ski time associated with a 1-minute increase in treadmill time is between 1 and 3.7 minutes.

Figure 13.9   Partial
MINITAB output for the
data of Example 13.4.

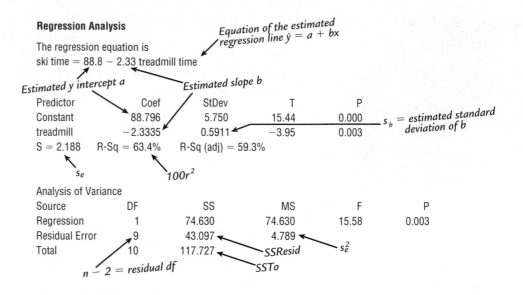

Output from any of the standard statistical computer packages routinely includes the computed values of a, b, SSResid, SSTo, and s_b. Figure 13.9 displays partial MINITAB output for the data of Example 13.4. The format from other packages is similar. Rounding occasionally leads to small discrepancies between hand-calculated and computer-calculated values, but there are no such discrepancies in this example.

■ Hypothesis Tests Concerning β ....

Hypotheses about β can be tested using a t test similar to the t tests discussed in Chapters 10 and 11. The null hypothesis states that β has a specified hypothesized value. The t statistic results from standardizing b, the point estimate of β, under the assumption that H_0 is true. When H_0 is true, the sampling distribution of this statistic is the t distribution with df $= n - 2$.

Summary of Hypothesis Tests Concerning β

Null hypothesis: H_0: $\beta = $ hypothesized value

Test statistic: $t = \dfrac{b - \text{hypothesized value}}{s_b}$

The test is based on df $= n - 2$.

Alternative Hypothesis:	P-Value:
H_a: $\beta > $ hypothesized value	Area to the right of the computed t under the appropriate t curve
H_a: $\beta < $ hypothesized value	Area to the left of the computed t under the appropriate t curve
H_a: $\beta \neq $ hypothesized value	(1) 2(area to the right of t) if t is positive or (2) 2(area to the left of the t) if t is negative

(continued)

Assumptions: For this test to be appropriate, the four basic assumptions of the simple linear regression model must be met:

1. The distribution of e at any particular x value has mean value 0 (that is $\mu_e = 0$).
2. The standard deviation of e is σ, which does not depend on x.
3. The distribution of e at any particular x value is normal.
4. The random deviations $e_1, e_2, \ldots, e_n$ associated with different observations are independent of one another.

Frequently, the null hypothesis of interest is $\beta = 0$. When this is the case, the population regression line is a horizontal line, and the value of y in the simple linear regression model does not depend on x. That is,

$$y = \alpha + 0 \cdot x + e$$

or equivalently,

$$y = \alpha + e$$

In this situation, knowledge of x is of no use in predicting y. On the other hand, if $\beta \neq 0$, there is a useful linear relationship between x and y, and knowledge of x is useful for predicting y. This is illustrated by the scatterplots in Figure 13.10.

Figure 13.10   (a) $\beta = 0$; (b) $\beta \neq 0$.

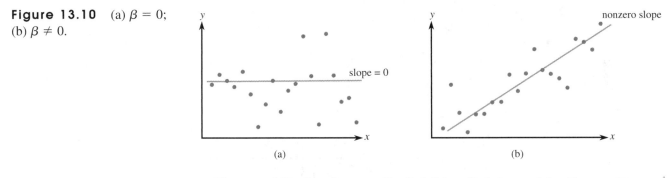

The test of H_0: $\beta = 0$ versus H_a: $\beta \neq 0$ is called the *model utility test for simple linear regression*.

The Model Utility Test for Simple Linear Regression

The **model utility test for simple linear regression** is the test of

$$H_0: \beta = 0$$

versus

$$H_a: \beta \neq 0$$

The null hypothesis specifies that there is *no* useful linear relationship between x and y, whereas the alternative hypothesis specifies that there *is* a useful linear relationship between x and y. If H_0 is rejected, we conclude that the simple linear regression model is useful for predicting y. The test procedure in the previous box (with hypothesized value = 0) is used to carry out the model utility test; in particular, the test statistic is the *t* **ratio** $t = b/s_b$.

If a scatterplot and the r^2 value do not provide convincing evidence for a useful linear relationship, we recommend that the model utility test be carried out before using the regression line to make inferences.

Example 13.5 University Graduation Rates

● The accompanying data on six-year graduation rate (%), student-related expenditure per full-time student, and median SAT score for a random sample of the primarily undergraduate public universities and colleges in the United States with enrollments between 10,000 and 20,000 were taken from College Results Online, The Education Trust.

Median SAT	Expenditure	Graduation Rate
1065	7970	49
950	6401	33
1045	6285	37
990	6792	49
950	4541	22
970	7186	38
980	7736	39
1080	6382	52
1035	7323	53
1010	6531	41
1010	6216	38
930	7375	37
1005	7874	45
1090	6355	57
1085	6261	48

Let's first investigate the relationship between graduation rate and median SAT score. With y = graduation rate and x = median SAT score, the summary statistics necessary for a simple linear regression analysis are as follows:

$$n = 15 \qquad \sum x = 15{,}195 \qquad \sum y = 638$$
$$\sum x^2 = 15{,}430{,}725 \qquad \sum xy = 651{,}340 \qquad \sum y^2 = 28{,}294$$

from which

$$b = 0.132 \qquad a = -91.31 \qquad \text{SSResid} = 491.01$$
$$s_e = 6.146 \qquad r^2 = .576 \qquad S_{xx} = 38190$$

Because $r^2 = .576$, about 57.6% of observed variation in graduation rates can be explained by the simple linear regression model. It appears from this that there is a useful linear relation between the two variables, but a confirmation requires a formal model utility test. We will use a significance level of .05 to carry out this test.

1. β = the true average change in graduation rate associated with an increase of 1 in median SAT score

2. $H_0: \beta = 0$
3. $H_a: \beta \neq 0$
4. $\beta = .05$
5. Test statistic: $t = \dfrac{b - \text{hypothesized value}}{s_b} = \dfrac{b - 0}{s_b} = \dfrac{b}{s_b}$
6. Assumptions: The accompanying scatterplot of the data shows a linear pattern and the variability of points does not appear to be changing with x:

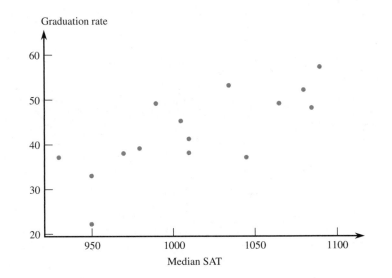

Assuming that the distribution of errors at any given x value is approximately normal, the assumptions of the simple linear regression model are appropriate.
7. Calculation: The calculation of t requires

$$s_b = \frac{s_e}{\sqrt{S_{xx}}} = \frac{6.146}{195.423} = .031$$

yielding

$$t = \frac{0.132 - 0}{.031} = 4.26$$

8. P-value: Appendix Table 4 shows that for a t test based on 13 df, $P(t > 4.26) <$ .001. The inequality in H_a requires a two-tailed test, so

$$P\text{-value} < 2(.001) = .002.$$

9. Conclusion: Since P-value $< .002$ is smaller than the significance level .05, H_0 is rejected. We conclude that there is a useful linear relationship between graduation rate and median SAT score.

Figure 13.11 shows partial MINITAB output from a simple linear regression analysis. The Coef column gives $b = 0.13213$; $s_b = 0.03145$ is in the SE Coef column; the T column (for t ratio) contains the value of the test statistic for testing $H_0: \beta = 0$; and the P-value for the model utility test is given in the last column as 0.001 (slightly different from the ones given in Step 8 because of rounding and the use of the table that produces only approximate P-values). Other commonly used statistical packages also include this information in their output.

Figure 13.11 MINITAB output for the data of Example 13.5.

Regression Analysis: Graduation Rate versus Median SAT

The regression equation is
Graduation Rate = -91.3 + 0.132 Median SAT

Predictor	Coef	SE Coef	T	P
Constant	-91.31	31.90	-2.86	0.013
Median SAT	0.13213	0.03145	4.20	0.001

S = 6.14574   R-Sq = 57.6%   R-Sq(adj) = 54.3%

Analysis of Variance

Source	DF	SS	MS	F	P
Regression	1	666.72	666.72	17.65	0.001
Residual Error	13	491.01	37.77		
Total	14	1157.73			

Let's next consider the relationship between graduation rate and expenditure per full-time student. Figure 13.12 shows partial MINITAB output from a simple linear regression with expenditure as the predictor variable.

Figure 13.12 MINITAB output using expenditure as the predictor.

The regression equation is
Graduation Rate = 10.9 + 0.00468 Expenditure

Predictor	Coef	SE Coef	T	P
Constant	10.95	17.51	0.63	0.543
Expenditure	0.004680	0.002574	1.82	0.092

S = 8.42608   R-Sq = 20.3%   R-Sq(adj) = 14.1%

The value of the test statistic for the model utility test in this case is $t = 1.82$ and the associated P-value is .092. For a .05 level of significance, we would not reject the null hypothesis of $H_0: \beta = 0$. There is not convincing evidence of a linear relationship between graduation rate and expenditure per full-time student.

▪

When $H_0: \beta = 0$ cannot be rejected by the model utility test at a reasonably small significance level, the search for a useful model must continue. One possibility is to relate y to x using a nonlinear model—an appropriate strategy if the scatterplot shows curvature. Alternatively, a multiple regression model using more than one predictor variable can be employed. We introduce such models in Chapter 14.

 Exercises 13.12–13.26   Turn to the Expanded Answers Section for answers not shown next to exercises.

13.12 What is the difference between σ and σ_b? What is the difference between σ_b and s_b?

13.13 Suppose that a single y observation is made at each of the x values 5, 10, 15, 20, and 25.

a. If $\sigma = 4$, what is the standard deviation of the statistic b? $\sigma_b = 0.253$

b. Now suppose that a second observation is made at every x value listed in Part (a) (for a total of 10 observa-

Bold exercises answered in back   ● Data set available online but not required   ▼ Video solution available

tions). Is the resulting value of σ_b half of what it was in Part (a)? $\sigma_b = 0.179$, no

c. How many observations at each x value in Part (a) are required to yield a σ_b value that is half the value calculated in Part (a)? Verify your conjecture.   It would require four observations at each x value.

13.14 Refer back to Example 13.3 in which the simple linear regression model was fit to data on x = proportion who judged candidate A as more competent and y = vote difference proportion. For the purpose of estimating β as accurately as possible, would it have been preferable to have observations with x values .05, .1, .2, .3, .4, .5, .6, .7, .8, .9, .95 and .98? Explain your reasoning.

13.15 Exercise 13.10 presented information from a study in which y was the hardness of molded plastic and x was the time elapsed since termination of the molding process. Summary quantities included

$$n = 15 \qquad b = 2.50 \qquad \text{SSResid} = 1235.470$$
$$\sum(x - \bar{x})^2 = 4024.20$$

a. Calculate the estimated standard deviation of the statistic b. $s_e = 9.7486,\ s_b = 0.1537$

b. Obtain a 95% confidence interval for β, the slope of the true regression line.   (2.17, 2.83)

c. Does the interval in Part (b) suggest that β has been precisely estimated? Explain.   Yes, the interval is relatively narrow.

13.16 A study was carried out to relate sales revenue y (in thousands of dollars) to advertising expenditure x (also in thousands of dollars) for fast-food outlets during a 3-month period. A sample of 15 outlets yielded the accompanying summary quantities.

$$\sum x = 14.10 \qquad \sum y = 1438.50 \qquad \sum x^2 = 13.92$$
$$\sum y^2 = 140,354 \qquad \sum xy = 1387.20$$
$$\sum(y - \bar{y})^2 = 2401.85 \qquad \sum(y - \hat{y})^2 = 561.46$$

a. What proportion of observed variation in sales revenue can be attributed to the linear relationship between revenue and advertising expenditure?   0.766

b. Calculate s_e and s_b. $s_e = 6.572,\ s_b = 8.053$

c. Obtain a 90% confidence interval for β, the average change in revenue associated with a $1000 (that is, 1-unit) increase in advertising expenditure.   (38.313, 66.821)

13.17 ▼ An experiment to study the relationship between x = time spent exercising (min) and y = amount of oxygen consumed during the exercise period resulted in the following summary statistics.

$$n = 20 \qquad \sum x = 50 \qquad \sum y = 16,705 \qquad \sum x^2 = 150$$
$$\sum y^2 = 14,194,231 \qquad \sum xy = 44,194$$

a. Estimate the slope and y intercept of the population regression line. $a = 592.1,\ b = 97.26$

b. One sample observation on oxygen usage was 757 for a 2-min exercise period. What amount of oxygen consumption would you predict for this exercise period, and what is the corresponding residual?

c. Compute a 99% confidence interval for the true average change in oxygen consumption associated with a 1-min increase in exercise time.   (87.76, 106.76).

13.18 Are workers less likely to quit their jobs when wages are high than when they are low? The paper "Investigating the Causal Relationship Between Quits and Wages: An Exercise in Comparative Dynamics" (*Economic* Inquiry [1986]: 61–83) gave data on x = average hourly wage and y = quit rate for a sample of industries. These data were used to produce the accompanying MINITAB output

The regression equation is
quit rate = 4.86 − 0.347 wage

Predictor	Coef	Stdev	t-ratio	p
Constant	4.8615	0.5201	9.35	0.000
wage	0.34655	0.05866	5.91	0.000

$s = 0.4862$ R-sq = 72.9%   R-sq(adj) = 70.8%

Analysis of Variance

Source	DF	SS	MS	F	p
Regression	1	8.2507	8.2507	34.90	0.000
Error	13	3.0733	0.2364		
Total	14	11.3240			

a. Based on the given P-value, does there appear to be a useful linear relationship between average wage and quit rate? Explain your reasoning.

b. Calculate an estimate of the average change in quit rate associated with a $1 increase in average hourly wage, and do so in a way that conveys information about the precision and reliability of the estimate.

13.19 The article "Cost-Effectiveness in Public Education" (*Chance* [1995]: 38–41) reported that, for a sample of $n = 44$ New Jersey school districts, a regression of y = average SAT score on x = expenditure per pupil (thousands of dollars) gave $b = 15.0$ and $s_b = 5.3$.

a. Does the simple linear regression model specify a useful relationship between x and y?

b. Calculate and interpret a confidence interval for β based on a 95% confidence level.

13.20 ● The article "Root Dentine Transparency: Age Determination of Human Teeth Using Computerized Densitometric Analysis" (*American Journal of Physical Anthropology* [1991]: 25–30) described a study in which the objective was to predict age (y) from percentage of a tooth's root with transparent dentine. The accompanying data are for anterior teeth.

x	15	19	31	39	41	44	47	48	55	65
y	23	52	65	55	32	60	78	59	61	60

Use the accompanying MINITAB output to decide whether the simple linear regression model is useful.

The regression equation is
Age = 32.1 + 0.555percent

Predictor	Coef	Stdev	t-ratio	p
Constant	32.08	13.32	2.41	0.043
percent	0.5549	0.3101	1.79	0.111

s = 14.30   R-sq = 28.6%   R-sq(adj) = 19.7%

Analysis of Variance

Source	DF	SS	MS	F	p
Regression	1	654.8	654.8	3.20	0.111
Error	8	1635.7	204.5		
Total	9	2290.5			

13.21 ● The accompanying data were read from a plot (and are a subset of the complete data set) given in the article "Cognitive Slowing in Closed-Head Injury" (*Brain and Cognition* [1996]: 429–440). The data represent the mean response times for a group of individuals with closed-head injury (CHI) and a matched control group without head injury on 10 different tasks. Each observation was based on a different study, and used different subjects, so it is reasonable to assume that the observations are independent.

Mean Response Time

Study	Control	CHI
1	250	303
2	360	491
3	475	659
4	525	683
5	610	922
6	740	1044

Mean Response Time

Study	Control	CHI
7	880	1421
8	920	1329
9	1010	1481
10	1200	1815

a. Fit a linear regression model that would allow you to predict the mean response time for those suffering a closed-head injury from the mean response time on the same task for individuals with no head injury.
b. Do the sample data support the hypothesis that there is a useful linear relationship between the mean response time for individuals with no head injury and the mean response time for individuals with CHI? Test the appropriate hypotheses using $\alpha = .05$.
c. It is also possible to test hypotheses about the y intercept in a linear regression model. For these data, the null hypothesis H_0: $\alpha = 0$ cannot be rejected at the .05 significance level, suggesting that a model with a y intercept of 0 might be an appropriate model. Fitting such a model results in an estimated regression equation of

CHI = 1.48(Control)

Interpret the estimated slope of 1.48.

13.22 ● The article "Effects of Enhanced UV-B Radiation on Ribulose-1,5-Biphosphate, Carboxylase in Pea and Soybean" (*Environmental and Experimental Botany* [1984]: 131–143) included the accompanying data on pea plants, with y = sunburn index and x = distance (cm) from an ultraviolet light source.

x	18	21	25	26	30	32	36	40
y	4.0	3.7	3.0	2.9	2.6	2.5	2.2	2.0

x	40	50	51	54	61	62	63
y	2.1	1.5	1.5	1.5	1.3	1.2	1.1

$$\sum x = 609 \qquad \sum y = 33.1 \qquad \sum x^2 = 28{,}037$$
$$\sum y^2 = 84.45 \qquad \sum xy = 1156.8$$

Estimate the mean change in the sunburn index associated with an increase of 1 cm in distance in a way that includes information about the precision of estimation.

13.23 Exercise 13.16 described a regression analysis in which y = sales revenue and x = advertising expenditure. Summary quantities given there yield

$$n = 15 \qquad b = 52.27 \qquad s_b = 8.05$$

a. Test the hypothesis $H_0: \beta = 0$ versus $H_a: \beta \neq 0$ using a significance level of .05. What does your conclusion say about the nature of the relationship between x and y?

b. Consider the hypothesis $H_0: \beta = 40$ versus $H_a: \beta > 40$. The null hypothesis states that the average change in sales revenue associated with a 1-unit increase in advertising expenditure is (at most) \$40,000. Carry out a test using significance level .01. $t = 1.56$, P-value $= 0.079$, fail to reject H_0.

13.24 ● The article "Technology, Productivity, and Industry Structure" (*Technological Forecasting and Social Change* [1983]: 1–13) included the accompanying data on x = research and development expenditure and y = growth rate for eight different industries.

x	2024	5038	905	3572	1157	327	378	191
y	1.90	3.96	2.44	0.88	0.37	−0.90	0.49	1.01

a. Would a simple linear regression model provide useful information for predicting growth rate from research and development expenditure? Use a .05 level of significance.

b. Use a 90% confidence interval to estimate the average change in growth rate associated with a 1-unit increase in expenditure. Interpret the resulting interval.   (0.000086, 0.001064)

13.25 ● ▼ The article "Effect of Temperature on the pH of Skim Milk" (*Journal of Dairy Research* [1988]: 277–280) reported on a study involving x = temperature (°C) under specified experimental conditions and y = milk pH. The accompanying data (read from a graph) are a representative subset of that which appeared in the article:

x	4	4	24	24	25	38	38	40
y	6.85	6.79	6.63	6.65	6.72	6.62	6.57	6.52

x	45	50	55	56	60	67	70	78
y	6.50	6.48	6.42	6.41	6.38	6.34	6.32	6.34

$$\sum x = 678 \qquad \sum y = 104.54 \qquad \sum x^2 = 36{,}056$$
$$\sum y^2 = 683.4470 \qquad \sum xy = 4376.36$$

Do these data strongly suggest that there is a negative linear relationship between temperature and pH? State and test the relevant hypotheses using a significance level of .01. $t = -17.57$, P-value ≈ 0, reject H_0.

13.26 ● In anthropological studies, an important characteristic of fossils is cranial capacity. Frequently skulls are at least partially decomposed, so it is necessary to use other characteristics to obtain information about capacity. One such measure that has been used is the length of the lambda-opisthion chord. The article "Vertesszollos and the Presapiens Theory" (*American Journal of Physical Anthropology* [1971]) reported the accompanying data for $n = 7$ *Homo erectus* fossils.

x (chord length in mm)	78	75	78	81	84	86	87
y (capacity in cm^3)	850	775	750	975	915	1015	1030

Suppose that from previous evidence, anthropologists had believed that for each 1-mm increase in chord length, cranial capacity would be expected to increase by 20 cm^3. Do these new experimental data strongly contradict prior belief?

Bold exercises answered in back   ● Data set available online but not required   ▼ Video solution available

13.3 Checking Model Adequacy

The simple linear regression model equation is

$$y = \alpha + \beta x + e$$

where e represents the random deviation of an observed y value from the population regression line $\alpha + \beta x$. The inferential methods presented in Section 13.2 required some assumptions about e. These assumptions include:

1. At any particular x value, the distribution of e is a normal distribution.
2. At any particular x value, the standard deviation of e is σ, which is constant over all values of x (i.e., σ does not depend on x).

Inferences based on the simple linear regression model continue to be reliable when model assumptions are slightly violated (e.g., mild nonnormality of the random deviation distribution). However, using an estimated model in the face of grossly violated assumptions can result in misleading conclusions. Therefore, it is desirable to have easily applied methods available for identifying such serious violations and for suggesting how a satisfactory model can be obtained.

■ Residual Analysis

If the deviations $e_1, e_2, \ldots, e_n$ from the population line were available, they could be examined for any inconsistencies with model assumptions. For example, a normal probability plot would suggest whether or not the normality assumption was tenable. But, because

$$e_1 = y_1 - (\alpha + \beta x_1)$$
$$\vdots$$
$$e_n = y_n - (\alpha + \beta x_n)$$

these deviations can be calculated only if the equation of the population line is known. In practice, this will never be the case. Instead, diagnostic checks must be based on the residuals

$$y_1 - \hat{y}_1 = y_1 - (a + b x_1)$$
$$\vdots$$
$$y_n - \hat{y}_n = y_n - (a + b x_n)$$

which are the deviations from the *estimated* line.

The values of the residuals will vary from sample to sample. When all model assumptions are met, the mean value of the residuals at any particular x value is 0.

Any observation that gives a large positive or negative residual should be examined carefully for any anomalous circumstances, such as a recording error or exceptional experimental conditions. Identifying residuals with unusually large magnitudes is made easier by inspecting **standardized residuals**.

Recall that a quantity is standardized by subtracting its mean value (0 in this case) and dividing by its true or estimated standard deviation. Thus

$$\text{standardized residual} = \frac{\text{residual}}{\text{estimated standard deviation of residual}}$$

The value of a standardized residual tells how many standard deviations the corresponding residual lies from its expected value, 0.

Because residuals at different x values have different standard deviations (depending on the value of x for that observation),* computing the standardized residuals can be tedious. Fortunately, many computer regression programs provide standardized residuals as part of the output.

In Chapter 7, a normal probability plot was introduced as a technique for deciding whether the n observations in a random sample could plausibly have come from a normal population distribution. To assess whether the assumption that $e_1, e_2, \ldots, e_n$ all come from the same normal distribution is reasonable, we recommend a normal probability plot of the standardized residuals.

_____

*The estimated standard deviation of the ith residual, $y_i - \hat{y}_i$ is $s_e\sqrt{1 - \dfrac{1}{n} - \dfrac{(x_i - \bar{x})^2}{S_{xx}}}$

Example 13.6   Political Faces

Example 13.3 introduced data on

> x = proportion who judged candidate A as the more competent of two candidates based on facial appearance

and

> y = vote difference (candidate A − candidate B) expressed as a proportion of the total number of votes cast

for a sample of 12 congressional elections. (See Example 13.3 for a more detailed description of the study.)

The scatterplot in Figure 13.13 is consistent with the assumptions of the simple linear regression model.

Figure 13.13
MINITAB output for the data of Example 13.6.

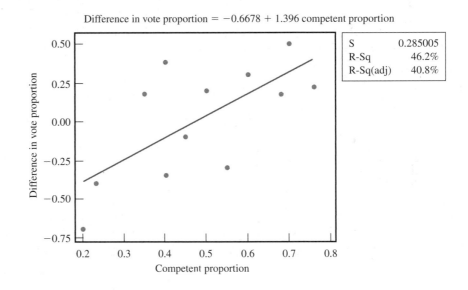

Difference in vote proportion = −0.6678 + 1.396 competent proportion

S	0.285005
R-Sq	46.2%
R-Sq(adj)	40.8%

The residuals, their standard deviations, and the standardized residuals are given in Table 13.1. For the residual with the largest magnitude, 0.49, the standardized residual is 1.81. That is, the residual is approximately 1.8 standard deviations above its expected value of 0, which is not particularly unusual in a sample of this size. On the standardized scale, no residual here is surprisingly large.

Figure 13.14 displays a normal probability plot of the standardized residuals and also one of the residuals. Notice that in this case the plots are nearly identical; it is usually the case that the two plots are similar. Although it is preferable to work with the standardized residuals, if you do not have access to a computer package or calculator that will produce standardized residuals, a plot of the unstandardized residuals should suffice. Few plots are straighter than these! The plots would not cause us to question the assumption of normality.

! Step-by-step technology instructions available online

Table 13.1   Data, Residuals, and Standardized Residuals for Example 13.6

Observation	Competent Proportion x	Difference in Vote Proportion y	$\hat{y}$	Residual	Estimated Standard Deviation of Residual	Standardized Residual
1	0.20	−0.70	−0.39	−0.31	0.24	−1.32
2	0.23	−0.40	−0.35	−0.05	0.24	−0.22
3	0.40	−0.35	−0.11	−0.24	0.27	−0.89
4	0.35	0.18	−0.18	0.36	0.27	1.35
5	0.40	0.38	−0.11	0.49	0.27	1.81
6	0.45	−0.10	−0.04	−0.06	0.27	−0.22
7	0.50	0.20	0.03	0.17	0.27	0.62
8	0.55	−0.30	0.10	−0.40	0.27	−1.48
9	0.60	0.30	0.17	0.13	0.27	0.49
10	0.68	0.18	0.28	−0.10	0.26	−0.39
11	0.70	0.50	0.31	0.19	0.25	0.75
12	0.76	0.22	0.39	−0.17	0.24	−0.72

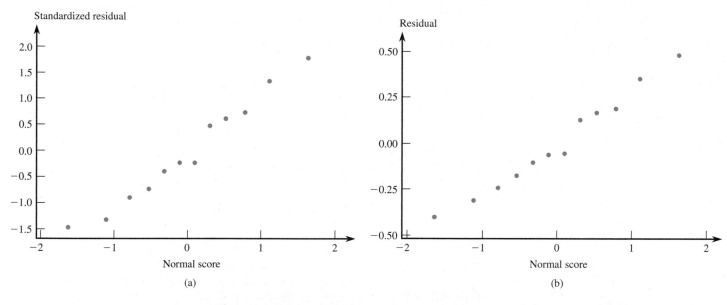

Figure 13.14   Normal probability plots for Example 13.6 (from MINITAB): (a) standardized residuals; (b) residuals.

■ **Plotting the Residuals**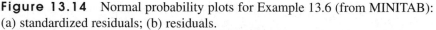

A plot of the (x, residual) pairs is called a **residual plot**, and a plot of the (x, standardized residual) pairs is a **standardized residual plot**. Residual and standardized residual plots typically exhibit the same general shapes. If you are using a computer package or graphing calculator that calculates standardized residuals, we recommend using the standardized residual plot. If not, it is acceptable to use the residual plot instead.

A standardized residual plot or a residual plot is often helpful in identifying unusual or highly influential observations and in checking for violations of model assumptions. A desirable plot is one that exhibits no particular pattern (such as curvature or a much greater spread in one part of the plot than in another) or one that has no point that is far removed from all the others. A point falling far above or far below the horizontal line at height 0 corresponds to a large standardized residual, which can indicate some kind of unusual behavior, such as a recording error, a nonstandard experimental condition, or an atypical experimental subject. A point that has an *x* value that differs greatly from others in the data set could have exerted excessive influence in determining the fitted line.

A standardized residual plot, such as the one pictured in Figure 13.15(a) is desirable, because no point lies much outside the horizontal band between −2 and 2 (so there is no unusually large residual corresponding to an outlying observation); there is no

Figure 13.15 Examples of residual plots: (a) satisfactory plot; (b) plot suggesting that a curvilinear regression model is needed; (c) plot indicating nonconstant variance; (d) plot showing a large residual; (e) plot showing a potentially influential observation.

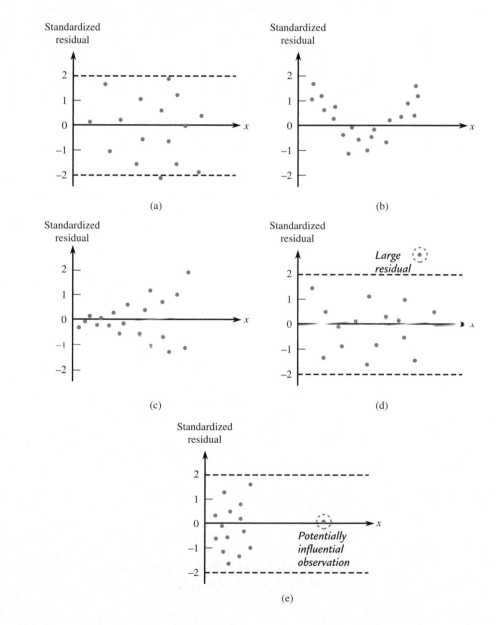

point far to the left or right of the others (thus no observation that might greatly influence the fit), and there is no pattern to indicate that the model should somehow be modified. When the plot has the appearance of Figure 13.15(b), the fitted model should be changed to incorporate curvature (a nonlinear model).

The increasing spread from left to right in Figure 13.15(c) suggests that the variance of y is not the same at each x value but rather increases with x. A straight-line model may still be appropriate, but the best-fit line should be selected by using *weighted least squares* rather than ordinary least squares. This involves giving more weight to observations in the region exhibiting low variability and less weight to observations in the region exhibiting high variability. A specialized regression analysis textbook or a statistician should be consulted for more information on using weighted least squares.

The standardized residual plots of Figures 13.15(d) and 13.15(e) show an extreme outlier and a potentially influential observation, respectively. Consider deleting the observation corresponding to such a point from the data set and refitting the same model. Substantial changes in estimates and various other quantities warn of instability in the data. The investigator should certainly carry out a more careful analysis and perhaps collect more data before drawing any firm conclusions. Improved computing power has allowed statisticians to develop and implement a variety of diagnostic tests for identifying unusual observations in a regression data set.

Example 13.7 Political Faces Revisited

Figure 13.16 displays a standardized residual plot and a residual plot for the data of Example 13.6 on perceived competence based on facial appearance and election outcome. The first observation was at $x_1 = 0.20$, and the corresponding standardized residual was -1.32, so the first plotted point in the standardized residual plot is

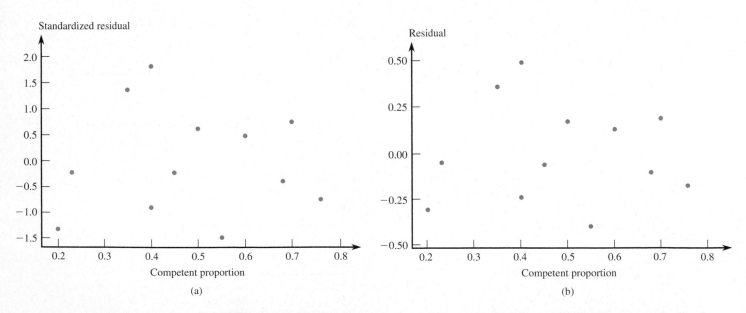

Figure 13.16 Plots for the data of Example 13.6 (from MINITAB): (a) standardized residual plot; (b) residual plot.

(0.20, −1.32). Other points are similarly obtained and plotted. The standardized residual plot shows no unusual behavior that might call for model modifications or further analysis. Note that the general pattern in the residual plot is similar to that of the standardized residual plot.

■

....

Example 13.8 Snow Cover and Temperature

● The article "Snow Cover and Temperature Relationships in North America and Eurasia" (*Journal of Climate and Applied Meteorology* [1983]: 460–469) explored the relationship between October–November continental snow cover (x, in millions of square kilometers) and December–February temperature (y, in °C). The following data refer to Eurasia during the $n = 13$ time periods (1969–1970, 1970–1971, . . . , 1981–1982):

x	y	Standardized Residual	x	y	Standardized Residual
13.00	−13.5	−0.11	22.40	−18.9	−1.54
12.75	−15.7	−2.19	16.20	−14.8	0.04
16.70	−15.5	−0.36	16.70	−13.6	1.25
18.85	−14.7	1.23	13.65	−14.0	−0.28
16.60	−16.1	−0.91	13.90	−12.0	−1.54
15.35	−14.6	−0.12	14.75	−13.5	0.58
13.90	−13.4	0.34			

A simple linear regression analysis done by the authors yielded $r^2 = .52, r = .72$, suggesting a significant linear relationship. This is confirmed by a model utility test. The scatterplot and standardized residual plot are displayed in Figure 13.17. There are no unusual patterns, although one standardized residual, −2.19, is a bit on the large side. The most interesting feature is the observation (22.40, −18.9), corresponding to a point far to the right of the others in these plots. This observation may have had a substantial influence on all aspects of the fit. The estimated slope when all 13 observations are included is $b = -0.459$, and $s_b = 0.133$. When the potentially influential observation is deleted, the estimate of β based on the remaining 12 observations is $b = -0.228$. Thus

$$\begin{aligned} \text{change in slope} &= \text{original } b - \text{new } b \\ &= -.459 - (-.288) \\ &= -.231 \end{aligned}$$

The change expressed in standard deviations is $-.231/.133 = -1.74$. Because b has changed by substantially more than 1 standard deviation, the observation under consideration appears to be highly influential.

● Data set available online

Figure 13.17 Plots for the data of Example 13.8 (from MINITAB): (a) scatterplot; (b) standardized residual plot.

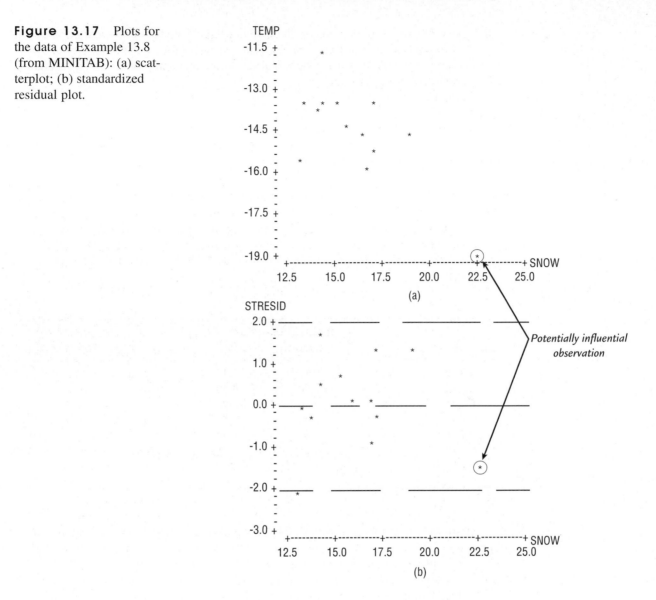

In addition, r^2 based just on the 12 observations is only .13, and the t ratio for testing $\beta = 0$ is not significant. Evidence for a linear relationship is much less conclusive in light of this analysis. The investigators should seek a climatological explanation for the influential observation and collect more data, which can be used to find an effective relationship.

Example 13.9 Treadmill Time and Ski Time Revisited

Example 13.4 presented data on $x =$ treadmill time and $y =$ ski time. A simple linear regression model was fit to the data, and a confidence interval for β, the average change in ski time associated with a 1-min increase in treadmill time, was constructed. The validity of the confidence interval depends on the assumptions that the distribution of the residuals from the population regression line at any fixed x is approximately normal and that the variance of this distribution does not depend on x. Constructing a normal probability plot of the standardized residuals and a standardized residual plot will provide insight into whether these assumptions are in fact reasonable.

MINITAB was used to fit the simple linear regression model and compute the standardized residuals, resulting in the values shown in Table 13.2.

Table 13.2 Data, Residuals, and Standardized Residuals for Example 13.9

Observation	Treadmill	Ski Time	Residual	Standardized Residual
1	7.7	71.0	0.172	0.10
2	8.4	71.4	2.206	1.13
3	8.7	65.0	3.494	1.74
4	9.0	68.7	0.906	0.44
5	9.6	64.4	1.994	0.96
6	9.6	69.4	3.006	1.44
7	10.0	63.0	2.461	1.18
8	10.2	64.6	0.394	0.19
9	10.4	66.9	2.373	1.16
10	11.0	62.6	0.527	0.27
11	11.7	61.7	0.206	0.12

Figure 13.18 shows a normal probability plot of the standardized residuals and a standardized residual plot. The normal probability plot is quite straight, and the standardized residual plot does not show evidence of any patterns or of increasing spread. These observations support the use of the confidence interval in Example 13.4.

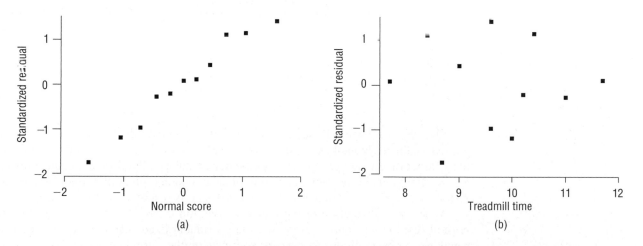

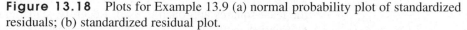

Figure 13.18 Plots for Example 13.9 (a) normal probability plot of standardized residuals; (b) standardized residual plot.

Example 13.10 More on University Graduation Rates

Use of the model utility test in Example 13.5 led to the conclusion that there was a useful linear relationship between y = graduation rate and x = median SAT score for public undergraduate universities. The validity of this test requires that the distribution of random errors be normal and that the variability of the error distribution not change with x. Let's construct a normal probability plot of the standardized residuals and a standardized residual plot to see whether these assumptions are reasonable. Table 13.3 gives the residuals and standardized residuals that are needed to construct the plots shown in Figure 13.19 (see page 723).

Table 13.3 Data, Residuals, and Standardized Residuals for Example 13.10

Observation	Median SAT	Graduation Rate	Residual	Standardized Residual
1	1065	49	−0.404	−0.071
2	950	33	−1.209	−0.216
3	1045	37	−9.762	−1.668
4	990	49	9.506	1.613
5	950	22	−12.209	−2.181
6	970	38	1.148	0.199
7	980	39	0.827	0.141
8	1080	52	0.614	0.111
9	1035	53	7.560	1.282
10	1010	41	−1.137	−0.192
11	1010	38	−4.137	−0.697
12	930	37	5.433	1.019
13	1005	45	3.524	0.594
14	1090	57	4.293	0.792
15	1085	48	−4.047	−0.737

The normal probability plot is quite straight, and the standardized residual plot does not show evidence of any pattern or increasing spread. This supports the use of the model utility test in Example 13.5.

■

Occasionally, you will see a residual plot or a standardized residual plot with $\hat{y}$ plotted on the horizontal axis rather than x. Because $\hat{y}$ is just a linear function of x, using $\hat{y}$ rather than x changes the scale of the horizontal axis but does not change the pattern of the points in the plot. As a consequence, residual plots that use $\hat{y}$ on the horizontal axis can be interpreted in the same manner as residual plots that use x.

When the distribution of the random deviation e has heavier tails than does the normal distribution, observations with large standardized residuals are not that unusual. Such observations can have great effects on the estimated regression line when the least-squares approach is used. Recently, statisticians have proposed a number of alternative methods—called **robust**, or **resistant**, methods—for fitting a line. Such

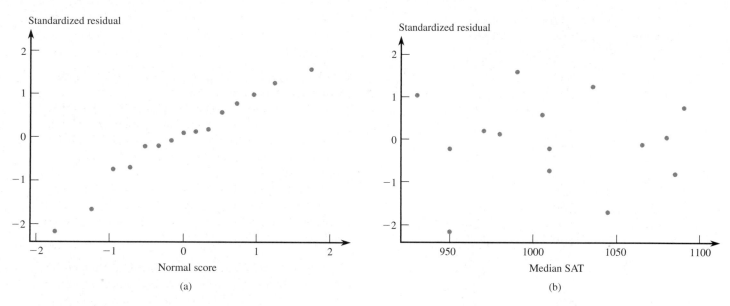

Figure 13.19 Plots for Example 13.10: (a) normal probability plot of standardized residuals; (b) standardized residual plot.

methods give less weight to outlying observations than does the least-squares method without deleting the outliers from the data set. The most widely used robust procedures require a substantial amount of computation, so a good computer program is necessary. Associated confidence interval and hypothesis testing procedures are still in the developmental stage.

Exercises 13.27–13.33

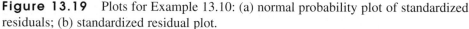

Turn to the Expanded Answers Section for answers not shown next to exercises.

13.27 ● ▼ Exercise 13.8 gave data on x = treadmill run time to exhaustion and y = 20-km ski time. The x values and corresponding standardized residuals from a simple linear regression are as follows.

x	7.7	8.4	8.7	9.0	9.6	9.6
St. resid.	0.10	1.13	−1.74	0.44	−0.96	1.44

x	10.0	10.2	10.4	11.0	11.7
St. resid.	−1.18	−0.19	1.16	−0.27	0.12

Construct a standardized residual plot. Does the plot exhibit any unusual features?

13.28 ● The article "Vital Dimensions in Volume Perception: Can the Eye Fool the Stomach?" (*Journal of Marketing Research* [1999]: 313–326) gave the accompanying data on the dimensions of 27 representative food products

(Gerber baby food, Cheez Whiz, Skippy Peanut Butter, and Ahmed's tandoori paste, to name a few).

Product	Maximum Width	Minimum Width
1	2.50	1.80
2	2.90	2.70
3	2.15	2.00
4	2.90	2.60
5	3.20	3.15
6	2.00	1.80
7	1.60	1.50
8	4.80	3.80
9	5.90	5.00
10	5.80	4.75

(*continued*)

Bold exercises answered in back ● Data set available online but not required ▼ Video solution available

Product	Maximum Width	Minimum Width
11	2.90	2.80
12	2.45	2.10
13	2.60	2.20
14	2.60	2.60
15	2.70	2.60
16	3.10	2.90
17	5.10	5.10
18	10.20	10.20
19	3.50	3.50
20	2.70	1.20
21	3.00	1.70
22	2.70	1.75
23	2.50	1.70
24	2.40	1.20
25	4.40	1.20
26	7.50	7.50
27	4.25	4.25

a. Fit the simple linear regression model that would allow prediction of the maximum width (in cm) of a food container based on its minimum width (in cm).
b. Calculate the standardized residuals (or just the residuals if you don't have access to a computer program that gives standardized residuals), and plot them to determine whether there are any outliers. Would eliminating these outliers increase the accuracy of predictions using this regression? If so, eliminate the outlying data point (a 1-liter Coke bottle) and refit the regression.
c. Interpret the estimated slope and, if appropriate, the intercept.
d. Do you think that the assumptions of the simple linear regression model are reasonable? Give statistical evidence for your answer.

13.29 ● ▼ The authors of the article "Age, Spacing and Growth Rate of *Tamarix* as an Indication of Lake Boundary Fluctuations at Sebkhet Kelbia, Tunisia" (*Journal of Arid Environments* [1982]: 43–51) used a simple linear regression model to describe the relationship between $y =$ vigor (average width in centimeters of the last two annual rings) and $x =$ stem density (stems/m^2). The estimated model was based on the following data. Also given are the standardized residuals.

x	4	5	6	9	14
y	0.75	1.20	0.55	0.60	0.65
St. resid.	−0.28	1.92	−0.90	−0.28	0.54

x	15	15	19	21	22
y	0.55	0.00	0.35	0.45	0.40
St. resid.	0.24	−2.05	−0.12	0.60	0.52

a. What assumptions are required for the simple linear regression model to be appropriate?
b. Construct a normal probability plot of the standardized residuals. Does the assumption that the random deviation distribution is normal appear to be reasonable? Explain.
c. Construct a standardized residual plot. Are there any unusually large residuals?
d. Is there anything about the standardized residual plot that would cause you to question the use of the simple linear regression model to describe the relationship between x and y?

13.30 ● The article "Effects of Gamma Radiation on Juvenile and Mature Cuttings of Quaking Aspen" (*Forest Science* [1967]: 240–245) reported the following data on $x =$ exposure to radiation (kR/6 hours) and $y =$ dry weight of roots (mg):

x	0	2	4	6	8
y	110	123	119	86	62

a. Construct a scatterplot for these data. Does the plot suggest that the simple linear regression model might be appropriate?
b. The estimated regression line for these data is $\hat{y} = 127 - 6.65x$ and the standardized residuals are as given.

x	0	2	4	6	8
St. resid.	−1.55	0.68	1.25	−0.05	−1.06

Construct a standardized residual plot. What does the plot suggest about the adequacy of the simple linear regression model?

13.31 ● ▼ Carbon aerosols have been identified as a contributing factor in a number of air quality problems. In a chemical analysis of diesel engine exhaust, $x =$ mass (μg/cm^2) and $y =$ elemental carbon (μg/cm^2) were recorded ("Comparison of Solvent Extraction and Thermal Optical Carbon Analysis Methods: Application to Diesel Vehicle Exhaust Aerosol" *Environmental Science Technology* [1984]: 231–234). The estimated regression line for this data set is $\hat{y} = 31 + .737x$. The accompanying table gives the observed x and y values and the corresponding standardized residuals.

x	164.2	156.9	109.8	111.4	87.0
y	181	156	115	132	96
St. resid.	2.52	0.82	0.27	1.64	0.08

Bold exercises answered in back ● Data set available online but not required ▼ Video solution available

x	161.8	230.9	106.5	97.6	79.7
y	170	193	110	94	77
St. resid.	1.72	−0.73	0.05	−0.77	−1.11

x	118.7	248.8	102.4	64.2	89.4
y	106	204	98	76	89
St. resid.	−1.07	−0.95	−0.73	−0.20	−0.68

x	108.1	89.4	76.4	131.7	100.8
y	102	91	97	128	88
St. resid.	−0.75	−0.51	0.85	0.00	−1.49

x	78.9	387.8	135.0	82.9	117.9
y	86	310	141	90	130
St. resid.	−0.27	−0.89	0.91	−0.18	1.05

a. Construct a standardized residual plot. Are there any unusually large residuals? Do you think that there are any influential observations?

b. Is there any pattern in the standardized residual plot that would indicate that the simple linear regression model is not appropriate?

c. Based on your plot in Part (a), do you think that it is reasonable to assume that the variance of y is the same at each x value? Explain.

13.32 ● An investigation of the relationship between traffic flow x (thousands of cars per 24 hr) and lead content y of bark on trees near the highway (mg/g dry weight) yielded the accompanying data. A simple linear regression model was fit, and the resulting estimated regression line was $\hat{y} = 28.7 + 33.3x$. Both residuals and standardized residuals are also given.

x	8.3	8.3	12.1	12.1	17.0
y	227	312	362	521	640
Residual	−78.1	6.9	−69.6	89.4	45.3
St. resid.	−0.99	0.09	−0.81	1.04	0.51

x	17.0	17.0	24.3	24.3	24.3
y	539	728	945	738	759
Residual	−55.7	133.3	107.2	−99.8	−78.8
St. resid.	−0.63	1.51	1.35	−1.25	−0.99

a. Plot the (x, residual) pairs. Does the resulting plot suggest that a simple linear regression model is an appropriate choice? Explain your reasoning.

b. Construct a standardized residual plot. Does the plot differ significantly in general appearance from the plot in Part (a)?

13.33 ● The accompanying data on $x =$ U.S. population (millions) and $y =$ crime index (millions) appeared in the article "The Normal Distribution of Crime" (*Journal of Police Science and Administration* [1975]: 312–318). The author comments that "The simple linear regression analysis remains one of the most useful tools for crime prediction." When observations are made sequentially in time, the residuals or standardized residuals should be plotted in time order (that is, first the one for time $t = 1$ (1963 here), then the one for time $t = 2$, and so on). Notice that here x increases with time, so an equivalent plot is of residuals or standardized residuals versus x. Using $\hat{y} = 47.26 + .260x$, calculate the residuals and plot the (x, residual) pairs. Does the plot exhibit a pattern that casts doubt on the appropriateness of the simple linear regression model? Explain.

Year	1963	1964	1965	1966	1967	1968
x	188.5	191.3	193.8	195.9	197.9	199.9
y	2.26	2.60	2.78	3.24	3.80	4.47

Year	1969	1970	1971	1972	1973
x	201.9	203.2	206.3	208.2	209.9
y	4.99	5.57	6.00	5.89	8.64

Bold exercises answered in back ● Data set available online but not required ▼ Video solution available

13.4 Inferences Based on the Estimated Regression Line (Optional)

The number obtained by substituting a particular x value, x^*, into the equation of the estimated regression line has two different interpretations: It is a point estimate of the average y value when $x = x^*$, and it is also a point prediction of a single y value to be observed when $x = x^*$. How precise is this estimate or prediction? That is, how close is $a + bx^*$ to the actual mean value $\alpha + \beta x^*$ or to a particular y observation? Because both a and b vary in value from sample to sample (each one is a statistic), the statistic $a + bx^*$ also has different values for different samples. The way in which this statistic varies in value with different samples is summarized by its sampling distribution. Properties of the sampling distribution are used to obtain both a confidence interval formula for $\alpha + \beta x^*$ and a prediction interval formula for a particular y observation. The width of the corresponding interval conveys information about the precision of the estimate or prediction.

Properties of the Sampling Distribution of $a + bx$ for a Fixed x Value

Let x^* denote a particular value of the independent variable x. When the four basic assumptions of the simple linear regression model are satisfied, the sampling distribution of the statistic $a + bx^*$ has the following properties:

1. The mean value of $a + bx^*$ is $\alpha + \beta x^*$, so $a + bx^*$ is an unbiased statistic for estimating the average y value when $x = x^*$.
2. The standard deviation of the statistic $a + bx^*$, denoted by σ_{a+bx^*}, is given by

$$\sigma_{a+bx^*} = \sigma\sqrt{\frac{1}{n} + \frac{(x^* - \bar{x})^2}{S_{xx}}}$$

3. The distribution of $a + bx^*$ is normal.

As you can see from the formula for σ_{a+bx^*} the standard deviation of $a + bx^*$ is larger when $(x^* - \bar{x})^2$ is large than when $(x^* - \bar{x})^2$ is small; that is, $a + bx^*$ tends to be a more precise estimate of $\alpha + \beta x^*$ when x^* is close to the center of the x values at which observations were made than when x^* is far from the center.

The standard deviation σ_{a+bx^*} cannot be calculated from the sample data, because the value of σ is unknown. However, σ_{a+bx^*} can be estimated by using s_e in place of σ. Using the mean and estimated standard deviation to standardize $a + bx^*$ gives a variable with a t distribution.

The **estimated standard deviation of the statistic $a + bx^*$**, denoted by s_{a+bx^*}, is given by

$$s_{a+bx^*} = s_e\sqrt{\frac{1}{n} + \frac{(x^* - \bar{x})^2}{S_{xx}}}$$

When the four basic assumptions of the simple linear regression model are satisfied, the probability distribution of the standardized variable

$$t = \frac{a + bx^* - (\alpha + \beta x^*)}{s_{a+bx^*}}$$

is the t distribution with df $= n - 2$.

■ Inferences About the Mean y Value $\alpha + \beta x^*$ ....

In previous chapters, standardized variables were manipulated algebraically to give confidence intervals of the form

(point estimate) $\pm$ (critical value)(estimated standard deviation)

A parallel argument leads to the following interval.

Confidence Interval for a Mean y Value

When the basic assumptions of the simple linear regression model are met, a **confidence interval for $\alpha + \beta x^*$**, the average y value when x has value x^*, is

$$a + bx^* \pm (t \text{ critical value}) \cdot s_{a+bx^*}$$

where the t critical value is based on df $= n - 2$. Appendix Table 3 gives critical values corresponding to the most frequently used confidence levels.

Important concept | Because s_{a+bx^*} is larger the farther x^* is from $\bar{x}$, the confidence interval widens as x^* moves away from the center of the data.

Example 13.11 Shark Length and Jaw Width

● Physical characteristics of sharks are of interest to surfers and scuba divers as well as to marine researchers. The following data on $x =$ length (in feet) and $y =$ jaw width (in inches) for 44 sharks were found in various articles appearing in the magazines *Skin Diver* and *Scuba News*:

x	18.7	12.3	18.6	16.4	15.7	18.3	14.6	15.8	14.9	17.6	12.1
y	17.5	12.3	21.8	17.2	16.2	19.9	13.9	14.7	15.1	18.5	12.0
x	16.4	16.7	17.8	16.2	12.6	17.8	13.8	12.2	15.2	14.7	12.4
y	13.8	15.2	18.2	16.7	11.6	17.4	14.2	14.8	15.9	15.3	11.9
x	13.2	15.8	14.3	16.6	9.4	18.2	13.2	13.6	15.3	16.1	13.5
y	11.6	14.3	13.3	15.8	10.2	19.0	16.8	14.2	16.9	16.0	15.9
x	19.1	16.2	22.8	16.8	13.6	13.2	15.7	19.7	18.7	13.2	16.8
y	17.9	15.7	21.2	16.3	13.0	13.3	14.3	21.3	20.8	12.2	16.9

Because it is difficult to measure jaw width in living sharks, researchers would like to determine whether it is possible to estimate jaw width from body length, which is more easily measured. A scatterplot of the data (Figure 13.20) shows a linear pattern and is consistent with use of the simple linear regression model.

Figure 13.20 A scatterplot for the data of Example 13.11.

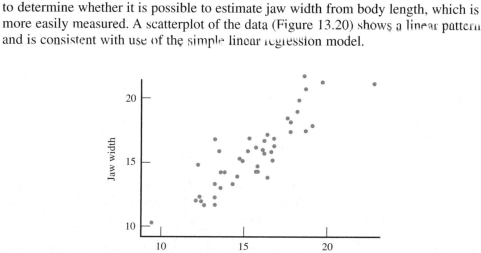

From the accompanying MINITAB output, it is easily verified that

$$a = .688 \qquad b = .96345 \qquad \text{SSResid} = 79.49$$
$$\text{SSTo} = 339.02 \qquad s_e = 1.376 \qquad r^2 = .766$$

Regression Analysis

The regression equation is

Jaw Width = 0.69 + 0.963 Length

Predictor	Coef	StDev	T	P
Constant	0.688	1.299	0.53	0.599
Length	0.96345	0.08228	11.71	0.000

S = 1.376   R-Sq = 76.6%   R-Sq(adj) = 76.0%

Analysis of Variance

Source	DF	SS	MS	F	P
Regression	1	259.53	259.53	137.12	0.000
Residual Error	42	79.49	1.89		
Total	43	339.02			

From the data, we can also compute $S_{xx} = 279.8718$. Because $r^2 = .766$, the simple linear regression model explains 76.6% of the variability in jaw width. The model utility test also confirms the usefulness of this model (P-value = .000).

Let's use the data to compute a 90% confidence interval for the mean jaw width for 15-ft-long sharks. The mean jaw width when length is 15 ft is $\alpha + \beta(15)$. The point estimate is

$$a + b(15) = .688 + .96345(15) = 15.140 \text{ in.}$$

Since

$$\bar{x} = \frac{\Sigma x}{n} = \frac{685.80}{44} = 15.586$$

the estimated standard deviation of $a + b(15)$ is

$$s_{a+b(15)} = s_e \sqrt{\frac{1}{n} + \frac{(15 - \bar{x})^2}{S_{xx}}}$$
$$= (1.376)\sqrt{\frac{1}{44} + \frac{(15 - 15.586)^2}{279.8718}}$$
$$= .213$$

The t critical value for df = 42 is 1.68 (using the tabulated value for df = 40 from Appendix Table 3). We now have all the relevant quantities needed to compute a 90% confidence interval:

$$a + b(15) \pm (t \text{ critical value}) \cdot s_{a+bx^*} = 15.140 \pm (1.68)(.213)$$
$$= 15.140 \pm .358$$
$$= (14.782, 15.498)$$

Based on these sample data, we can be 90% confident that the mean jaw width for sharks whose length is 15 ft is between 14.782 and 15.498 in. As with all confidence intervals, the 90% confidence level means that we have used a *method* that has a 10% error rate to construct this interval estimate.

■

We have just considered estimation of the mean y value at a fixed $x = x^*$. When the basic assumptions of the simple linear regression model are met, the true value of this mean is $\alpha + \beta x^*$. The reason that our point estimate $a + bx^*$ is not exactly equal to $\alpha + \beta x^*$ is that the values of α and β are not known, so they have been estimated from sample data. As a result, the estimate $a + bx^*$ is subject to sampling variability and the extent to which the estimated line might differ from the true population line is reflected in the width of the confidence interval.

■ Prediction Interval for a Single *y* ....

We now turn our attention to the problem of predicting a single y value at a particular $x = x^*$ (rather than estimating the mean y value when $x = x^*$). This problem is equivalent to trying to predict the y value of an individual point in a scatterplot of the population. If we use the estimated regression line to obtain a point prediction $a + bx^*$, this prediction will probably not be exactly equal to the true y value for two reasons. First, as was the case when estimating a mean y value, the estimated line is not going to be exactly equal to the true population regression line. But, in the case of predicting a single y value, there is an additional source of error: e, the deviation from the line. Even if we knew the true population line, individual points would not fall exactly on the population line. This implies that there is more uncertainty associated with predicting a single y value at a particular x^* than with estimating the mean y value at x^*. This extra uncertainty is reflected in the width of the corresponding intervals.

An interval for a single y value, y^*, is called a **prediction interval** (to distinguish it from the confidence interval for a mean y value). The interpretation of a prediction interval is similar to the interpretation of a confidence interval. A 95% prediction interval for y^* is constructed using a method for which 95% of all possible samples would yield interval limits capturing y^*; only 5% of all samples would give an interval that did not include y^*.

Manipulation of a standardized variable similar to the one from which a confidence interval was obtained gives the following prediction interval.

Prediction Interval for a Single *y* Value

When the four basic assumptions of the simple linear regression model are met, a **prediction interval for y^***, a single y observation made when $x = x^*$, has the form

$$a + bx^* \pm (t \text{ critical value}) \cdot \sqrt{s_e^2 + s_{a+bx^*}^2}$$

■

The prediction interval and the confidence interval are centered at exactly the same place, $a + bx^*$. The addition of s_e^2 under the square-root symbol makes the prediction interval wider—often substantially so—than the confidence interval.

Example 13.12 Jaws II

In Example 13.11, we computed a 90% confidence interval for the mean jaw width of sharks whose length is 15 ft. Suppose that we are interested in predicting the jaw width of a single shark of length 15 ft. The required calculations for a 90% prediction interval for y^* are

$$a + b(15) = .688 + .96245(15) = 15.140$$
$$s_e^2 = (1.376)^2 = 1.8934$$
$$s_{a+b(15)}^2 = (.213)^2 = .0454$$

The t critical value for df = 42 and a 90% prediction level is 1.68 (using the tabled value for df = 40). Substitution into the prediction interval formula then gives

$$a + b(15) \pm (t \text{ critical value})\sqrt{s_e^2 + s_{a+b(15)}^2} = 15.140 \pm (1.68)\sqrt{1.9388}$$
$$= 15.140 \pm 2.339$$
$$= (12.801, 17.479)$$

We can be 90% confident that an individual shark with length 15 ft will have a jaw width between 12.801 and 17.479 in. Notice that, as expected, this 90% prediction interval is much wider than the 90% confidence interval when $x^* = 15$ from Example 13.11.

Figure 13.21 gives MINITAB output that includes a 95% confidence interval and a 95% prediction interval when $x^* = 15$ and when $x^* = 20$. The intervals for $x^* = 20$ are wider than the corresponding intervals for $x^* = 15$ because 20 is farther from $\bar{x}$ (the center of the sample x values) than is 15. Each prediction interval is wider than the corresponding confidence interval. Figure 13.22 is a MINITAB plot that shows the estimated regression line as well as 90% confidence limits and prediction limits.

Figure 13.21 MINITAB output for the data of Example 13.12.

Regression Analysis

The regression equation is
Jaw Width = 0.69 + 0.963 Length

Predictor	Coef	StDev	T	P
Constant	0.688	1.299	0.53	0.599
Length	0.96345	0.08228	11.71	0.000

S = 1.376   R-Sq = 76.6%   R-Sq(adj) = 76.0%

Analysis of Variance

Source	DF	SS	MS	F	P
Regression	1	259.53	259.53	137.12	0.000
Residual Error	42	79.49	1.89		
Total	43	399.02			

Predicted Values

	Fit	StDev Fit	95.0% CI	95.0% PI
$x^* = 15 \rightarrow$	15.140	0.213	(14.710, 15.569)	(12.330, 17.949)
$x^* = 20 \rightarrow$	19.957	0.418	(19.113, 20.801)	(17.055, 22.859)

Figure 13.22 MINITAB plot showing estimated regression line and 90% confidence and prediction limits for the data of Example 13.12.

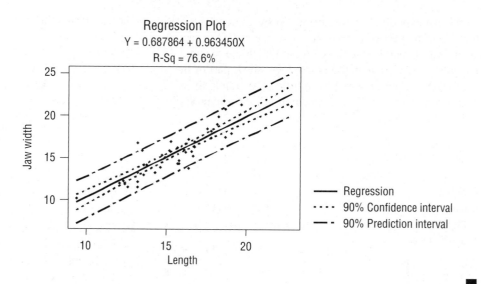

Regression Plot

Y = 0.687864 + 0.963450X

R-Sq = 76.6%

— Regression

···· 90% Confidence interval

-·- 90% Prediction interval

Exercises 13.34–13.48   Turn to the Expanded Answers Section for answers not shown next to exercises.

13.34 Explain the difference between a confidence interval and a prediction interval. How can a prediction level of 95% be interpreted?

13.35 Suppose that a regression data set is given and you are asked to obtain a confidence interval. How would you tell from the phrasing of the request whether the interval is for β or for $\alpha + \beta x^*$?

13.36 In Exercise 13.17, we considered a regression of y = oxygen consumption on x = time spent exercising. Summary quantities given there yield

$$n = 20 \qquad \bar{x} = 2.50 \qquad S_{xx} = 25$$
$$b = 97.26 \qquad a = 592.10 \qquad s_e = 16.486$$

a. Calculate $s_{a+b(2.0)}$ the estimated standard deviation of the statistic $a + b(2.0)$.   4.038

b. Without any further calculation, what is $s_{a+b(3.0)}$ and what reasoning did you use to obtain it?

c. Calculate the estimated standard deviation of the statistic $a + b(2.8)$.   3.817

d. For what value x^* is the estimated standard deviation of $a + bx^*$ smallest, and why? $x^* = \bar{x} = 2.5$

13.37 Example 13.3 gave data on x = proportion who judged candidate A as more competent and y = vote difference proportion. Calculate a confidence interval

using a 95% confidence level for the mean vote difference proportion for congressional races where 60% judge candidate A as more competent. $(-0.459, 0.798)$

13.38 The data of Exercise 13.25, in which x = milk temperature and y = milk pH, yield

$$n = 16 \qquad \bar{x} = 43.375 \qquad S_{xx} = 7325.75$$
$$b = -.00730608 \qquad a = 6.843345 \qquad s_e = .0356$$

a. Obtain a 95% confidence interval for $\alpha + \beta(40)$, the true average milk pH when the milk temperature is 40°C.

b. Calculate a 99% confidence interval for the true average milk pH when the milk temperature is 35°C.

c. Would you recommend using the data to calculate a 95% confidence interval for the true average pH when the temperature is 90°C? Why or why not?   No, 90 is outside the range of x values in the data set.

13.39 Return to the regression of y = milk pH on x = milk temperature described in Exercise 13.38.

a. Obtain a 95% prediction interval for a single pH observation to be made when milk temperature = 40°C.

b. Calculate a 99% prediction interval for a single pH observation when milk temperature = 35°C. $(6.4779, 6.6973)$

c. When the milk temperature is 60°C, would a 99% prediction interval be wider than the intervals of Parts (a) and (b)? Answer without calculating the interval.   Yes, because $x^* = 60$ is farther from $\bar{x}$.

Bold exercises answered in back   ● Data set available online but not required   ▼ Video solution available

13.40 An experiment was carried out by geologists to see how the time necessary to drill a distance of 5 ft in rock (y, in min) depended on the depth at which the drilling began (x, in ft, between 0 and 400). We show part of the MINITAB output obtained from fitting the simple linear regression model ("Mining Information," *American Statistician* [1991]: 4–9).

The regression equation is
Time = 4.79 + 0.0144depth

Predictor	Coef	Stdev	t-ratio	p
Constant	4.7896	0.6663	7.19	0.000
depth	0.014388	0.002847	5.05	0.000

s = 1.432   R-sq = 63.0%   R-sq(adj) = 60.5%

Analysis of Variance

Source	DF	SS	MS	F	p
Regression	1	52.378	52.378	25.54	0.000
Error	15	30.768	2.051		
Total	16	83.146			

a. What proportion of observed variation in time can be explained by the simple linear regression model?
b. Does the simple linear regression model appear to be useful?
c. MINITAB reported that $s_{a+b(200)}$ = .347. Calculate a confidence interval at the 95% confidence level for the true average time when depth = 200 ft.   (6.928, 8.406)
d. A single observation on time is to be made when drilling starts at a depth of 200 ft. Use a 95% prediction interval to predict the resulting value of time.
e. MINITAB gave (8.147, 10.065) as a 95% confidence interval for true average time when depth = 300. Calculate a 99% confidence interval for this average.   (7.778, 10.434)

13.41 ● According to "Reproductive Biology of the Aquatic Salamander *Amphiuma tridactylum* in Louisiana" (*Journal of Herpetology* [1999]: 100–105), the size of a female salamander's snout is correlated with the number of eggs in her clutch. The following data are consistent with summary quantities reported in the article. MINITAB output is also included.

Snout-Vent Length	32	53	53	53	54
Clutch Size	45	215	160	170	190

Snout-Vent Length	57	57	58	58	59
Clutch Size	200	270	175	245	215

Snout-Vent Length	63	63	64	67
Clutch Size	170	240	245	280

The regression equation is
Y = −133 + 5.92x

Predictor	Coef	StDev	T	P
Constant	133.02	64.30	2.07	0.061
X	5.919	1.127	5.25	0.000

S = 33.90   R-Sq = 69.7%   R-Sq(adj) = 67.2%

Additional summary statistics are

$$n = 14 \quad \bar{x} = 56.5 \quad \bar{y} = 201.4$$
$$\sum x^2 = 45,958 \quad \sum y^2 = 613,550 \quad \sum xy = 164,969$$

a. What is the equation of the regression line for predicting clutch size based on snout-vent length?
b. Calculate the standard deviation of b. $s_b = 1.127$
c. Is there sufficient evidence to conclude that the slope of the population line is positive.
d. Predict the clutch size for a salamander with a snout-vent length of 65 using a 95% interval.   251.715
e. Predict the clutch size for a salamander with snout-vent length of 105.

13.42 The article first introduced in Exercise 13.28 of Section 13.3 gave data on the dimensions of 27 representative food products.
a. Use the information given there to test the hypothesis that there is a positive linear relationship between the minimum width and the maximum width of an object.
b. Calculate and interpret s_e. $s_e = 0.6725$
c. Calculate a 95% confidence interval for the mean maximum width of products with a minimum width of 6 cm.
d. Calculate a 95% prediction interval for the maximum width of a food package with a minimum width of 6 cm.   (4.715, 7.641)

13.43 ● The shelf life of packaged food depends on many factors. Dry cereal is considered to be a moisture-sensitive product (no one likes soggy cereal!) with the shelf life determined primarily by moisture content. In a study of the shelf life of one particular brand of cereal, x = time on shelf (stored at 73°F and 50% relative humidity) and y = moisture content were recorded. The resulting data are from "Computer Simulation Speeds Shelf Life Assessments" (*Package Engineering* [1983]: 72–73).

x	0	3	6	8	10	13	16
y	2.8	3.0	3.1	3.2	3.4	3.4	3.5

x	20	24	27	30	34	37	41
y	3.1	3.8	4.0	4.1	4.3	4.4	4.9

a. Summary quantities are

$\sum x = 269$ $\sum y = 51$ $\sum xy = 1081.5$
$\sum y^2 = 7745$ $\sum x^2 = 190.78$

Find the equation of the estimated regression line for predicting moisture content from time on the shelf.
b. Does the simple linear regression model provide useful information for predicting moisture content from knowledge of shelf time?
c. Find a 95% interval for the moisture content of an individual box of cereal that has been on the shelf 30 days.
d. According to the article, taste tests indicate that this brand of cereal is unacceptably soggy when the moisture content exceeds 4.1. Based on your interval in Part (c), do you think that a box of cereal that has been on the shelf 30 days will be acceptable? Explain.

13.44 For the cereal data of Exercise 13.43, the average x value is 19.21. Would a 95% confidence interval with $x^* = 20$ or $x^* = 17$ be wider? Explain. Answer the same question for a prediction interval.

13.45 High blood-lead levels are associated with a number of different health problems. The article "A Study of the Relationship between Blood Lead Levels and Occupational Lead Levels" (*American Statistician* [1983]: 471) gave data on $x = $ air-lead level ($\mu g/m^3$) and $y = $ blood-lead level ($\mu g/dL$). Summary quantities (based on a subset of the data given in a plot in the article) are

$n = 15$ $\sum x = 1350$ $\sum y = 600$
$\sum x^2 = 155,400$ $\sum y^2 = 24,869.33$ $\sum xy = 57,760$

a. Find the equation of the estimated regression line.
b. Estimate the mean blood-lead level for people who work where the air-lead level is 100 $\mu g/m^3$ using a 90% interval.   (38.354, 43.864)
c. Construct a 90% prediction interval for the blood-lead level of a particular person who works where the air-lead level is 100 $\mu g/m^3$.   (30.311, 51.907)
d. Explain the difference in interpretation of the intervals computed in Parts (b) and (c).

13.46 A regression of $y = $ sunburn index for a pea plant on $x = $ distance from an ultraviolet light source was considered in Exercise 13.22. The data and summary statistics presented there give

$n = 15$ $\bar{x} = 40.60$ $\sum (x - \bar{x})^2 = 3311.60$
$b = -.0565$ $a = 4.500$ SSResid $= .8430$

a. Calculate a 95% confidence interval for the true average sunburn index when the distance from the light source is 35 cm.   (2.371, 2.674)
b. When two 95% confidence intervals are computed, it can be shown that the simultaneous confidence level is at least $[100 - 2(5)]\% = 90\%$. That is, if both intervals are computed for a first sample, for a second sample, yet again for a third, and so on, in the long run at least 90% of the samples will result in intervals both of which capture the values of the corresponding population characteristics. Calculate confidence intervals for the true mean sunburn index when the distance is 35 cm and when the distance is 45 cm in such a way that the simultaneous confidence level is at least 90%.
c. If two 99% intervals were computed, what do you think could be said about the simultaneous confidence level?
d. If a 95% confidence interval were computed for the true mean index when $x = 35$, another 95% confidence interval were computed when $x = 40$, and yet another one when $x = 45$, what do you think would be the simultaneous confidence level for the three resulting intervals?
e. Return to Part (d) and answer the question posed there if the individual confidence level for each interval were 99%. The simultaneous confidence level would be at least 97%.

13.47 By analogy with the discussion in Exercise 13.46, when two different prediction intervals are computed, each at the 95% prediction level, the *simultaneous prediction level* is at least $[100 - 2(5)]\% = 90\%$.
a. Return to Exercise 13.46 and obtain prediction intervals for sunburn index both when distance is 35 cm and when distance is 45 cm, so that the simultaneous prediction level is at least 90%.
b. If three different 99% prediction intervals are calculated for distances (cm) of 35, 40, and 45, respectively, what can be said about the simultaneous prediction level? The simultaneous prediction level would be at least 97%.

13.48 ● The article "Performance Test Conducted for a Gas Air-Conditioning System" (*American Society of Heating, Refrigerating, and Air Conditioning Engineering* [1969]: 54) reported the following data on maximum outdoor temperature (x) and hours of chiller operation per day (y) for a 3-ton residential gas air-conditioning system:

x	72	78	80	86	88	92
y	4.8	7.2	9.5	14.5	15.7	17.9

Suppose that the system is actually a prototype model, and the manufacturer does not wish to produce this model unless the data strongly indicate that when maximum outdoor temperature is 82°F, the true average number of hours of chiller operation is less than 12. The appropriate hypothesis is then

$$H_0:\ \alpha + \beta(82) = 12 \quad \text{versus} \quad H_a:\ \alpha + \beta(82) < 12$$

Use the statistic

$$t = \frac{a + b(82) - 12}{s_{a+b(82)}}$$

which has a t distribution based on $(n - 2)$ df when H_0 is true, to test the hypotheses at significance level .01. $t = -2.80$, P-value ≈ 0.024, fail to reject H_0.

Bold exercises answered in back   ● Data set available online but not required   ▼ Video solution available

13.5 Inferences About the Population Correlation Coefficient (Optional)

The sample correlation coefficient r, defined in Chapter 5, measures how strongly the x and y values in a *sample* of pairs are linearly related to one another. There is an analogous measure of how strongly x and y are related in the entire *population* of pairs from which the sample $(x_1, y_1), \ldots, (x_n, y_n)$ was obtained. It is called the **population correlation coefficient** and is denoted by ρ. As with r, ρ must be between -1 and 1, and it assesses the extent of any *linear* association in the population. To have $\rho = 1$ or $\rho = -1$, all (x, y) pairs in the population must lie exactly on a straight line. The value of ρ is a population characteristic and is generally unknown. The sample correlation coefficient r can be used as the basis for making inferences about ρ.

■ Test for Independence ($\rho = 0$)

Investigators are often interested in detecting not just linear association but also association of *any* kind. When there is no association of any type between the x and y values, statisticians say that the two variables are *independent*. In general, $\rho = 0$ is not equivalent to the independence of x and y. However, there is one special—yet frequently occurring—situation in which the two conditions ($\rho = 0$ and independence) are identical. This is when the pairs in the population have what is called a **bivariate normal distribution**. The essential feature of such a distribution is that for *any* fixed x value, the distribution of associated y values is normal, *and* for any fixed y value, the distribution of x values is normal.

As an example, suppose that height x and weight y have a bivariate normal distribution in the American adult male population. (There is good empirical evidence for this.) Then, when $x = 68$ in., weight y has a normal distribution; when $x = 72$ in., weight is normally distributed; when $y = 160$ lb, height x has a normal distribution; when $y = 175$ lb, height has a normal distribution; and so on. In this example, of course, x and y are not independent, because large height values tend to be paired with large weight values and small height values tend to be paired with small weight values.

Check of the assumptions   There is no easy way to check the assumption of bivariate normality, especially when the sample size n is small. A partial check can be based on the following property: If (x, y) has a bivariate normal distribution, then x alone has a normal distribution and so does y. This suggests constructing a normal probability plot of $x_1, x_2, \ldots, x_n$ and a separate normal probability plot of $y_1, y_2, \ldots, y_n$. If either plot shows a substantial departure from a straight line, then bivariate normality is a questionable assumption. If both plots are reasonably straight, then bivariate normality is plausible, although no guarantee can be given.

For a bivariate normal population, the test of independence (correlation = 0) is a *t* test. The formula for the test statistic essentially involves standardizing the estimate *r* under the assumption that the null hypothesis H_0: $\rho = 0$ is true.

A Test for Independence in a Bivariate Normal Population

Null hypothesis: H_0: $\rho = 0$

Test statistic: $t = \dfrac{r}{\sqrt{\dfrac{1 - r^2}{n - 2}}}$

The test is based on df $= n - 2$.

Alternative Hypothesis:	**P-Value:**
H_a: $\rho > 0$ (positive dependence)	Area under the appropriate *t* curve to the right of the computed *t*
H_a: $\rho < 0$ (negative dependence)	Area under the appropriate *t* curve to the left of the computed *t*
H_a: $\rho \neq 0$ (dependence)	(1) 2(area to the right of *t*) if *t* is positive or (2) 2(area to the left of *t*) if *t* is negative

Assumption: *r* is the correlation coefficient for a *random sample* from a *bivariate normal population.*

Example 13.13 Sleepless Nights

The relationship between sleep duration and the level of the hormone leptin (a hormone related to energy intake and energy expenditure) in the blood was investigated in the paper "Short Sleep Duration is Associated with Reduced Leptin, Elevated Ghrelin, and Increased Body Mass Index" (*Public Library of Science Medicine*, December 2004, pages 210-217). Average nightly sleep (*x*, in hours) and blood leptin level (*y*) were recorded for each person in a sample of 716 participants in the Wisconsin Sleep Cohort Study. The sample correlation coefficient was *r* = 0.11. Does this support the claim in the title of the paper that short sleep duration is associated with reduced leptin? Let's carry out a test using a significance level of .01.

1. ρ = the correlation between average nightly sleep and blood leptin level for the population of adult Americans.
2. H_0: $\rho = 0$
3. H_a: $\rho > 0$ (small values of *y* = average nightly sleep tend to be associated with small values of *x* = blood leptin level—that is, short sleep duration tends to be paired with lower leptin levels)
4. $\alpha = .01$
5. Test statistic: $t = \dfrac{r}{\sqrt{\dfrac{1 - r^2}{n - 2}}}$

6. Assumptions: Without the actual data, it is not possible to assess whether the assumption of bivariate normality is reasonable. For purposes of this example, we will assume that it is reasonable and proceed with the test. Had the data been available, we would have looked at normal probability plots of the x values and of the y values. We will also assume (as did the authors of the paper) that it is reasonable to regard the sample of participants in the study as representative of the larger population of adult Americans.

7. Calculation: $t = \dfrac{.11}{\sqrt{\dfrac{1 - (.11)^2}{714}}} = \dfrac{.11}{\sqrt{.00138}} = 2.96$

8. P-value: The t curve with 714 df is essentially indistinguishable from the z curve, and Appendix Table 4 shows that the area under this curve and to the right of 2.96 is .0015. That is, P-value $= .0015$.

9. Conclusion: P-value $\le .01$. We reject H_0 as did the authors of the article, and confirm the conclusion that there is a positive association between sleep duration and blood leptin level. Notice, though, that according to the guidelines described in Chapter 5, $r = .11$ suggests only a weak linear relationship. Since $r^2 = .0121$, fitting the simple linear regression model to the data would result in only about 1% of observed variation in average nightly sleep being explained.

■

In the context of regression analysis, the hypothesis of no linear relationship (H_0: $\beta = 0$) was tested using the t ratio $t = \dfrac{b}{s_b}$. Some algebraic manipulation shows that

$$\frac{r}{\sqrt{\dfrac{1 - r^2}{n - 2}}} = \frac{b}{s_b}$$

Therefore the model utility test can also be used to test for independence in a bivariate normal population. The reason for using the formula for t that involves r is that when we are interested only in correlation, the extra effort involved in computing the regression quantities b, a, SSResid, s_e, and s_b need not be expended.

Other inferential procedures for drawing conclusions about ρ—a confidence interval or a test of hypothesis with nonzero hypothesized value—are somewhat complicated. The reference by Neter, Wasserman, and Kutner in the back of the book can be consulted for details.

 Exercises 13.49–13.57 .... Turn to the Expanded Answers Section for answers not shown next to exercises. ....

13.49 Discuss the difference between r and ρ.

13.50 If the sample correlation coefficient is equal to 1, is it necessarily true that $\rho = 1$? If $\rho = 1$, is it necessarily true that $r = 1$?

13.51 A sample of $n = 353$ college faculty members was obtained, and the values of $x =$ teaching evaluation index and $y =$ annual raise were determined ("Determination of Faculty Pay: An Agency Theory Perspective," *Academy of Management Journal* [1992]: 921–955). The resulting value of r was .11. Does there appear to be a linear association between these variables in the population from which the sample was selected? Carry out a test of hypothesis using a significance level of .05. Does the conclusion surprise you? Explain.

13.52 It seems plausible that higher rent for retail space could be justified only by a higher level of sales. A random sample of $n = 53$ specialty stores in a chain was selected, and the values of x = annual dollar rent per square foot and y = annual dollar sales per square foot were determined, resulting in $r = .37$ ("Association of Shopping Center Anchors with Performance of a Nonanchor Specialty Chain Store," *Journal of Retailing* [1985]: 61–74). Carry out a test at significance level .05 to see whether there is in fact a positive linear association between x and y in the population of all such stores.

13.53 Television is regarded by many as a prime culprit for the difficulty many students have in performing well in school. The article "The Impact of Athletics, Part-Time Employment, and Other Activities on Academic Achievement" (*Journal of College Student Development* [1992]: 447–453) reported that for a random sample of $n = 528$ college students, the sample correlation coefficient between time spent watching television (x) and grade point average (y) was $r = -.26$.
a. Does this suggest that there is a negative correlation between these two variables in the population from which the 528 students were selected? Use a test with significance level .01.
b. If y were regressed on x, would the regression explain a substantial percentage of the observed variation in grade point average? Explain your reasoning.

13.54 The accompanying summary quantities for x = particulate pollution ($\mu g/m^3$) and y = luminance (.01 cd/m^2) were calculated from a representative sample of data that appeared in the article "Luminance and Polarization of the Sky Light at Seville (Spain) Measured in White Light" (*Atmospheric Environment* [1988]: 595–599).

$$n = 15 \qquad \sum x = 860 \qquad \sum y = 348$$
$$\sum x^2 = 56,700 \qquad \sum y^2 = 8954 \qquad \sum xy = 22,265$$

a. Test to see whether there is a positive correlation between particulate pollution and luminance in the population from which the data were selected.
b. What proportion of observed variation in luminance can be attributed to the approximate linear relationship between luminance and particulate pollution?

13.55 ● In a study of bacterial concentration in surface and subsurface water ("Pb and Bacteria in a Surface Microlayer" *Journal of Marine Research* [1982]: 1200–1206), the accompanying data were obtained.

Concentration ($\times 10^6$ /mL)

Surface	48.6	24.3	15.9	8.29	5.75
Subsurface	5.46	6.89	3.38	3.72	3.12
Surface	10.8	4.71	8.26	9.41	
Subsurface	3.39	4.17	4.06	5.16	

Summary quantities are

$$\sum x = 136.02 \qquad \sum y = 39.35$$
$$\sum x^2 = 3602.65 \qquad \sum y^2 = 184.27 \qquad \sum xy = 673.65$$

Using a significance level of .05, determine whether the data support the hypothesis of a linear relationship between surface and subsurface concentration.

13.56 A sample of $n = 500$ (x, y) pairs was collected and a test of $H_0: \rho = 0$ versus $H_a: \rho \neq 0$ was carried out. The resulting P-value was computed to be .00032.
a. What conclusion would be appropriate at level of significance .001?
b. Does this small P-value indicate that there is a very strong linear relationship between x and y (a value of ρ that differs considerably from zero)? Explain.

13.57 A sample of $n = 10,000$ (x, y) pairs resulted in $r = .022$. Test $H_0: \rho = 0$ versus $H_a: \rho \neq 0$ at significance level .05. Is the result statistically significant? Comment on the practical significance of your analysis.

Bold exercises answered in back   ● Data set available online but not required   ▼ Video solution available

---

13.6 Interpreting and Communicating the Results of Statistical Analyses

Although regression analysis can be used as a tool for summarizing bivariate data, it is also widely used to enable researchers to make inferences about the way in which two variables are related.

■ What to Look For in Published Data ....

Here are some things to consider when you evaluate research that involves fitting a simple linear regression model:

- Which variable is the dependent variable? Is it a numerical (rather than a qualitative) variable?
- If sample data have been used to estimate the coefficients in a simple linear regression model, is it reasonable to think that the basic assumptions required for inference are met?
- Does the model appear to be useful? Are the results of a model utility test reported? What is the *P*-value associated with the test?
- Has the model been used in an appropriate way? Has the regression equation been used to predict *y* for values of the independent variable that are outside the range of the data?
- If a correlation coefficient is reported, is it accompanied by a test of significance? Are the results of the test interpreted properly?

Both correlation and linear regression methods were used by the authors of the paper "Obesity, Cigarette Smoking, and Telomere Length in Women" (*Lancet* [2005]: 6632–664) in the analysis of factors related to aging. The telomere is a region at the end of a chromosome, and because telomeres can erode each time a chromosome replicates, telomere length is thought to decrease with age. The paper states

> Telomere length decreased steadily with age at a mean rate of 27 bp per year and a highly significant negative correlation was detected. The proportion of the variance in telomere length accounted for by age was 20.6%. Squared and cubed terms were also added to the model and had no significant effect on telomere length (p = 0.92 and p = 0.98, respectively) suggesting a linear relation between [telomere length] and age."

A correlation coefficient of -0.455 and an associated *P*-value of .0001 was reported to support the statement that there was a significant negative correlation. This also implies that a model utility test would have also indicated a useful linear relationship between $y = $ telomere length and $x = $ age. A scatterplot of telomere length versus age was given in the paper, and it was consistent with the basic assumptions required for the validity of the model utility test. From the quoted passage, $r^2 = .206$ (from 20.6% of variability in telomere length explained by age). Although the authors did not report the equation of the least-squares regression line, we can tell from the quoted passage that the slope of the line is -27 (the mean change in telomere length for a change of 1 in age).

The effects of caffeine were examined in the article "Withdrawal Syndrome After the Double-Blind Cessation of Caffeine Consumption" (*New England Journal of Medicine* [1992]: 1109–1113). The authors found that the dose of caffeine was significantly correlated with a measure of insomnia and also with latency on a test of reaction time. They reported $r = .26$ (*P*-value $= .042$) for insomnia and $r = .31$ (*P*-value $= .014$) for latency. Because both *P*-values are small, the authors concluded that the population correlation coefficients differ from 0. We should also note, however, that the reported correlation coefficients are not particularly large and do not indicate a strong linear relationship.

■ A Word to the Wise: Cautions and Limitations ....

In addition to the cautions and limitations described in Chapter 5 (which also apply here), here are a few additional things to keep in mind:

1. It doesn't make sense to use a regression line as the basis for making inferences about a population if there is no convincing evidence of a useful linear relationship between the two variables under study. It is particularly important to remember this when you are working with small samples. If the sample size is small, it is not uncommon for a weak linear pattern in the scatterplot to be due to chance rather than to a meaningful relationship in the population of interest.

2. As with all inferential procedures, it makes sense to carry out a model utility test or to construct confidence or prediction intervals only if the linear regression is based on a random sample from some larger population. It is not unusual to see a least-squares line used as a descriptive tool in situations where the data used to construct the line cannot reasonably be viewed as a sample. In this case, the inferential methods of this chapter are not appropriate.

3. Don't forget to check assumptions. If you are used to checking assumptions before doing much in the way of computation, it is sometimes easy to forget to check them in this setting because the equation of the least-squares line and then the residuals must be computed before a residual plot or a normal probability plot of the standardized residuals can be constructed. Be sure to step back and think about whether the linear regression model is appropriate and reasonable before using the model to draw inferences about a population.

Activity 13.1   Are Tall Women from "Big" Families?

In this activity, you should work with a partner (or in a small group).

Consider the following data on height (in inches) and number of siblings for a random sample of 10 female students at a large university.

1. Construct a scatterplot of the given data. Does there appear to be a linear relationship between y = height and x = number of siblings?

Height (y)	Number of Siblings (x)	Height (y)	Number of Siblings (x)
64.2	2	65.5	1
65.4	0	67.2	2
64.6	2	66.4	2
66.1	6	63.3	0
65.1	3	61.7	1

2. Compute the value of the correlation coefficient. Is the value of the correlation coefficient consistent with your answer from Step 1? Explain.

3. What is the equation of the least-squares line for these data?

4. Is the slope of the least-squares regression line from Step 3 equal to 0? Does this necessarily mean that there is a meaningful relationship between height and number of siblings in the population of female students at this university? Discuss this with your partner, and then write a few sentences of explanation.

5. For the population of all female students at the university, do you think it is reasonable to assume that the distribution of heights at each particular x value is approximately normal and that the standard deviation of the height distribution at each particular x value is the same? That is, do you think it is reasonable to assume that the distribution of heights for female students with zero siblings is approximately normal and that the distribution of heights for female students with one sibling is approximately normal with the same standard deviation as for female students with no siblings, and so on? Discuss this with your partner, and then write a few sentences of explanation.

6. Carry out the model utility test ($H_0: \beta = 0$). Explain why the conclusion from this test is consistent with your explanation in Step 4.

7. Would you recommend using the least-squares regression line as a way of predicting heights for women at this university? Explain.

8. After consulting with your partner, write a paragraph explaining why it is a good idea to include a model utility test (H_0: $\beta = 0$) as part of a regression analysis.

Summary of Key Concepts and Formulas

Term or Formula	Comment
Simple linear regression model, $y = \alpha + \beta x + e$	This model assumes that there is a line with slope β and y intercept α, called the population (true) regression line, such that an observation deviates from the line by a random amount e. The random deviation is assumed to have a normal distribution with mean zero and standard deviation σ, and random deviation for different observations are assumed to be independent of one another.
Estimated regression line, $\hat{y} = a + bx$	The least-squares line introduced in Chapter 5.
$s_e = \sqrt{\dfrac{\text{SSResid}}{n - 2}}$	The point estimate of the standard deviation σ, with associated degrees of freedom $n - 2$.
$s_b = \dfrac{s_e}{\sqrt{S_{xx}}}$	The estimated standard deviation of the statistic b.
$b \pm (t \text{ critical value})s_b$	A confidence interval for the slope β of the population regression line, where the t critical value is based on $(n - 2)$ df.
$t = \dfrac{b - \text{hypothesized value}}{s_b}$	The test statistic for testing hypotheses about β. The test is based on $(n - 2)$ df.
Model utility test, with test statistic $t = \dfrac{b}{s_b}$	A test of H_0: $\beta = 0$, which asserts that there is no useful linear relationship between x and y, versus H_a: $\beta \neq 0$, the claim that there is a useful linear relationship.
Residual analysis	Methods based on the residuals or standardized residuals for checking the assumptions of a regression model.
Standardized residual	A residual divided by its standard deviation.
Standardized residual plot	A plot of the (x, standardized residual) pairs. A pattern in this plot suggests a problem with the simple linear regression model.
$s_{a+bx^*} = s_e\sqrt{\dfrac{1}{n} + \dfrac{(x^* - \bar{x})^2}{S_{xx}}}$	The estimated standard deviation of the statistic $a + bx^*$, where x^* denotes a particular value of x.
$a + bx^* \pm (t \text{ critical value})s_{a+bx^*}$	A confidence interval for $\alpha + \beta x^*$, the average value of y when $x = x^*$.
$a + bx^* \pm (t \text{ critical value})\sqrt{s_e^2 + s_{a+bx^*}^2}$	A prediction interval for a single y value to be observed when $x = x^*$.

Term or Formula	Comment
Population correlation coefficient ρ	A measure of the extent to which the x and y values in an entire population are linearly related.
$$t = \frac{r}{\sqrt{\dfrac{1 - r^2}{n - 2}}}$$	The test statistic for testing H_0: $\rho = 0$, according to which (assuming a bivariate normal population distribution) x and y are independent of one another.

Chapter Review Exercises 13.58–13.75

Turn to the Expanded Answers Section for answers not shown next to exercises.

CENGAGENOW™ Know exactly what to study! Take a pre-test and receive your Personalized Learning Plan.

13.58 The effects of grazing animals on grasslands have been the focus of numerous investigations by ecologists. One such study, reported in "The Ecology of Plants, Large Mammalian Herbivores, and Drought in Yellowstone National Park" (*Ecology* [1992]: 2043–2058), proposed using the simple linear regression model to relate y = green biomass concentration (g/cm^3) to x = elapsed time since snowmelt (days).
a. The estimated regression equation was given as $\hat{y} = 106.3 - .640x$. What is the estimate of average change in biomass concentration associated with a 1-day increase in elapsed time?
b. What value of biomass concentration would you predict when elapsed time is 40 days?
c. The sample size was $n = 58$, and the reported value of the coefficient of determination was .470. Does this suggest that there is a useful linear relationship between the two variables? Carry out an appropriate test.

13.59 A random sample of $n = 347$ students was selected, and each one was asked to complete several questionnaires, from which a Coping Humor Scale value x and a Depression Scale value y were determined ("Depression and Sense of Humor" (*Psychological Reports* [1994]: 1473–1474). The resulting value of the sample correlation coefficient was $-.18$.
a. The investigators reported that P-value $< .05$. Do you agree?
b. Is the sign of r consistent with your intuition? Explain. (Higher scale values correspond to more developed sense of humor and greater extent of depression.)
c. Would the simple linear regression model give accurate predictions? Why or why not?

13.60 Data on x = depth of flooding and y = flood damage were given in Exercise 5.75. Summary quantities are

$$n = 13 \qquad \sum x = 91 \qquad \sum x^2 = 819$$
$$\sum y = 470 \qquad \sum y^2 = 19{,}118 \qquad \sum xy = 3867$$

a. Do the data suggest the existence of a positive linear relationship (one in which an increase in y tends to be associated with an increase in x)? Test using a .05 significance level.
b. Predict flood damage resulting from a claim made when depth of flooding is 3.5 ft, and do so in a way that conveys information about the precision of the prediction.

13.61 Exercise 13.8 gave data on x = treadmill run time to exhaustion and y = 20-km ski time for a sample of 11 biathletes. Use the accompanying MINITAB output to answer the following questions

The regression equation is
ski = −88.8 − 2.33tread

Predictor	Coef	Stdev	t-ratio	p
Constant	88.796	5.750	15.44	0.000
tread	2.3335	0.5911	3.95	0.003

s = 2.188   R-sq = 63.4%   R-sq(adj) = 59.3%
Analysis of Variance

Source	DF	SS	MS	F	p
Regression	1	74.630	74.630	15.58	0.003
Error	9	43.097	4.789		
Total	10	117.727			

a. Carry out a test at significance level .01 to decide whether the simple linear regression model is useful.

Bold exercises answered in back   ● Data set available online but not required   ▼ Video solution available

b. Estimate the average change in ski time associated with a 1-minute increase in treadmill time, and do so in a way that conveys information about the precision of estimation.

c. MINITAB reported that $s_{a+b(10)} = .689$. Predict ski time for a single biathlete whose treadmill time is 10 min, and do so in a way that conveys information about the precision of prediction.

d. MINITAB also reported that $s_{a+b(11)} = 1.029$. Why is this larger than $s_{a+b(10)}$?

13.62 Exercise 5.46 presented data on x = squawfish length and y = maximum size of salmonid consumed, both in mm. Use the accompanying MINITAB output along with the values $\bar{x} = 343.27$ and $S_{xx} = 69,112.18$ to answer the following questions.

The regression equation is
Size = −89.1 = 0.729length

Predictor	Coef	Stdev	t-ratio	p
Constant	89.09	16.83	5.29	0.000
length	0.72907	0.04778	15.26	0.000

s = 12.56 R-sq = 96.3% R-sq(adj) = 95.9% Analysis of Variance

Source	DF	SS	MS	F	p
Regression	1	36736	36736	232.87	0.000
Error	9	1420	158		
Total	10	38156			

a. Does there appear to be a useful linear relationship between length and size?

b. Does it appear that the average change in maximum size associated with a 1-mm increase in length is less than .8 mm? State and test the appropriate hypotheses.

c. Estimate average maximum size when length is 325 mm in a way that conveys information about the precision of estimation.

d. How would the estimate when length is 250 mm compare to the estimate of Part (c)? Answer without actually calculating the new estimate.

13.63 A sample of $n = 61$ penguin burrows was selected, and values of both y = trail length (m) and x = soil hardness (force required to penetrate the substrate to a depth of 12 cm with a certain gauge, in kg) were determined for each one ("Effects of Substrate on the Distribution of Magellanic Penguin Burrows," *The Auk* [1991]: 923–933). The equation of the least-squares line was $\hat{y} = 11.607 - 1.4187x$, and $r^2 = .386$.

a. Does the relationship between soil hardness and trail length appear to be linear, with shorter trails associated with harder soil (as the article asserted)? Carry out an appropriate test of hypotheses.

b. Using $s_e = 2.35$, $\bar{x} = 4.5$, and $\sum(x - \bar{x})^2 = 250$, predict trail length when soil hardness is 6.0 in a way that conveys information about the reliability and precision of the prediction.

c. Would you use the simple linear regression model to predict trail length when hardness is 10.0? Explain your reasoning.

13.64 ● The article "Photocharge Effects in Dye Sensitized Ag[Br,I] Emulsions at Millisecond Range Exposures" (*Photographic Science and Engineering* [1981]: 138–144) gave the accompanying data on x = % light absorption and y = peak photovoltage.

x	4.0	8.7	12.7	19.1	21.4	24.6	28.9	29.8	30.5
y	0.12	0.28	0.55	0.68	0.85	1.02	1.15	1.34	1.29

$$\sum x = 179.7 \quad \sum x^2 = 4334.41$$
$$\sum y = 7.28 \quad \sum y^2 = 7.4028 \quad \sum xy = 178.683$$

a. Construct a scatterplot of the data. What does it suggest?
b. Assuming that the simple linear regression model is appropriate, obtain the equation of the estimated regression line.
c. How much of the observed variation in peak photovoltage can be explained by the model relationship?
d. Predict peak photovoltage when percent absorption is 19.1, and compute the value of the corresponding residual.
e. The authors claimed that there is a useful linear relationship between the two variables. Do you agree? Carry out a formal test.
f. Give an estimate of the average change in peak photovoltage associated with a 1% increase in light absorption. Your estimate should convey information about the precision of estimation.
g. Give an estimate of true average peak photovoltage when percentage of light absorption is 20, and do so in a way that conveys information about precision.

13.65 ● Reduced visual performance with increasing age has been a much-studied phenomenon in recent years. This decline is due partly to changes in optical properties of the eye itself and partly to neural degeneration throughout the visual system. As one aspect of this problem, the article "Morphometry of Nerve Fiber Bundle Pores in the Optic Nerve Head of the Human" (*Experimental Eye Research* [1988]: 559–568) presented the accompanying data on x = age and y = percentage of the cribriform area of the lamina scleralis occupied by pores.

x	22	25	27	39	42	43	44	46	46
y	75	62	50	49	54	49	59	47	54

x	48	50	57	58	63	63	74	74
y	52	58	49	52	49	31	42	41

a. Suppose that the researchers had believed a priori that the average decrease in percentage area associated with a 1-year age increase was .5%. Do the data contradict this prior belief? State and test the appropriate hypotheses using a .10 significance level.
b. Estimate true average percentage area covered by pores for all 50-year-olds in the population in a way that conveys information about the precision of estimation.

13.66 ● Occasionally an investigator may wish to compute a confidence interval for α, the y intercept of the true regression line, or test hypotheses about α. The estimated y intercept is simply the height of the estimated line when $x = 0$, since $a + b(0) = a$. This implies that s_a the estimated standard deviation of the statistic a, results from substituting $x^* = 0$ in the formula for s_{a+bx^*}. The desired confidence interval is then

$$a \pm (t \text{ critical value})s_a$$

and a test statistic is

$$t = \frac{a - \text{hypothesized value}}{s_a}$$

a. The article "Comparison of Winter-Nocturnal Geostationary Satellite Infrared-Surface Temperature with Shelter-Height Temperature in Florida" (*Remote Sensing of the Environment* [1983]: 313–327) used the simple linear regression model to relate surface temperature as measured by a satellite (y) to actual air temperature (x) as determined from a thermocouple placed on a traversing vehicle. Selected data are given (read from a scatterplot in the article).

x	−2	−1	0	1	2	3	4
y	−3.9	−2.1	−2.0	−1.2	0.0	1.9	0.6

x	5	6	7
y	2.1	1.2	3.0

Estimate the true regression line.

b. Compute the estimated standard deviation s_a. Carry out a test at level of significance .05 to see whether the y intercept of the true regression line differs from zero.
c. Compute a 95% confidence interval for α. Does the result indicate that $\alpha = 0$ is plausible? Explain.

13.67 ● In some studies, an investigator has n (x, y) pairs sampled from one population and m (x, y) pairs from a second population. Let β and β' denote the slopes of the first and second population lines, respectively, and let b and b' denote the estimated slopes calculated from the first and second samples, respectively. The investigator may then wish to test the null hypothesis H_0: $\beta - \beta' = 0$ (that is, $\beta = \beta'$) against an appropriate alternative hypothesis. Suppose that σ^2, the variance about the population line, is the same for both populations. Then this common variance can be estimated by

$$s^2 = \frac{\text{SSResid} + \text{SSResid}'}{(n - 2) + (m - 2)}$$

where SSResid and SSResid' are the residual sums of squares for the first and second samples, respectively. With S_{xx} and S'_{xx} denoting the quantity $\sum(x - \bar{x})^2$ for the first and second samples, respectively, the test statistic is

$$t = \frac{b - b'}{\sqrt{\dfrac{s^2}{S_{xx}} + \dfrac{s^2}{S'_{xx}}}}$$

When H_0 is true, this statistic has a t distribution based on $(n + m - 4)$ df.

The given data are a subset of the data in the article "Diet and Foraging Model of *Bufa marinus* and *Leptodactylus ocellatus*" (*Journal of Herpetology* [1984]: 138–146). The independent variable x is body length (cm) and the dependent variable y is mouth width (cm), with $n = 9$ observations for one type of nocturnal frog and $m = 8$ observations for a second type. Test at level .05 to determine if the slopes of the true regression lines for the two different frog populations are equal. (Summary statistics are given in the table.)

Leptodactylus ocellatus

x	3.8	4.0	4.9	7.1	8.1	8.5	8.9	9.1	9.8
y	1.0	1.2	1.7	2.0	2.7	2.5	2.4	2.9	3.2

Bufa marinus

x	3.8	4.3	6.2	6.3	7.8	8.5	9.0	10.0
y	1.6	1.7	2.3	2.5	3.2	3.0	3.5	3.8

	Leptodactylus	*Bufa*
Sample size:	9	8
$\sum x$	64.2	55.9
$\sum x^2$	500.78	425.15
$\sum y$	19.6	21.6
$\sum y^2$	47.28	62.92
$\sum xy$	153.36	163.36

13.68 ● Consider the following four (x, y) data sets: the first three have the same x values, so these values are listed only once (from "Graphs in Statistical Analysis" *American Statistician* [1973]: 17–21).

Data Set	1–3	1	2	3	4	4
Variable	*x*	*y*	*y*	*y*	*x*	*y*
	10.0	8.04	9.14	7.46	8.0	6.58
	8.0	6.95	8.14	6.77	8.0	5.76
	13.0	7.58	8.74	12.74	8.0	7.71
	9.0	8.81	8.77	7.11	8.0	8.84
	11.0	8.33	9.26	7.81	8.0	8.47
	14.0	9.96	8.10	8.84	8.0	7.04
	6.0	7.24	6.13	6.08	8.0	5.25
	4.0	4.26	3.10	5.39	19.0	12.50
	12.0	10.84	9.13	8.15	8.0	5.56
	7.0	4.82	7.26	6.42	8.0	7.91
	5.0	5.68	4.74	5.73	8.0	6.89

For each of these data sets, the values of the summary quantities $\bar{x}$, $\bar{y}$, $\sum(x - \bar{x})^2$, and $\sum(x - \bar{x})(y - \bar{y})$ are identical, so all quantities computed from these will be identical for the four sets: the estimated regression line, SSResid, s_e, r^2, and so on. The summary quantities provide no way of distinguishing among the four data sets.

Based on a scatterplot for each set, comment on the appropriateness or inappropriateness of fitting the simple linear regression model in each case.

13.69 The accompanying scatterplot, based on 34 sediment samples with x = sediment depth (cm) and y = oil and grease content (mg/kg), appeared in the article "Mined Land Reclamation Using Polluted Urban Navigable Waterway Sediments" (*Journal of Environmental Quality* [1984]: 415– 422). Discuss the effect that the observation (20, 33,000) will have on the estimated regression line. If this point were omitted, what can you say about the slope of the estimated regression line? What do you think will

happen to the slope if this observation is included in the computations?

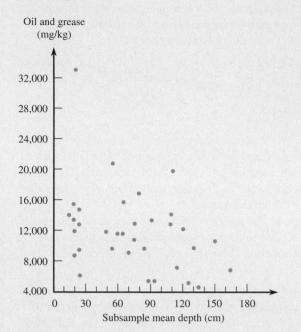

13.70 ● The article "Improving Fermentation Productivity with Reverse Osmosis" (*Food Technology* [1984]: 92–96) gave the following data (read from a scatterplot) on y = glucose concentration (g/L) and x = fermentation time (days) for a blend of malt liquor.

x	1	2	3	4	5	6	7	8
y	74	54	52	51	52	53	58	71

a. Use the data to calculate the estimated regression line.
b. Do the data indicate a linear relationship between y and x? Test using a .10 significance level.
c. Using the estimated regression line of Part (a), compute the residuals and construct a plot of the residuals versus x (that is, of the $(x, \text{residual})$ pairs).
d. Based on the plot in Part (c), do you think that a linear model is appropriate for describing the relationship between y and x? Explain.

13.71 The employee relations manager of a large company was concerned that raises given to employees during a recent period might not have been based strictly on objective performance criteria. A sample of $n = 20$ employees was selected, and the values of x, a quantitative measure of productivity, and y, the percentage salary increase, were determined for each one. A computer package was used to fit the simple linear regression model, and the resulting output gave the *P*-value = .0076 for the model utility test. Does the percentage raise appear to be linearly related to productivity? Explain.

13.72 ● The article "Statistical Comparison of Heavy Metal Concentrations in Various Louisiana Sediments" (*Environmental Monitoring and Assessment* [1984]: 163–170) gave the accompanying data on depth (m), zinc concentration (ppm), and iron concentration (%) for 17 core samples.

Core	Depth	Zinc	Iron
1	0.2	86	3.4
2	2.0	77	2.9
3	5.8	91	3.1
4	6.5	86	3.4
5	7.6	81	3.2
6	12.2	87	2.9
7	16.4	94	3.2
8	20.8	92	3.4
9	22.5	90	3.1
10	29.0	108	4.0
11	31.7	112	3.4
12	38.0	101	3.6
13	41.5	88	3.7
14	60.0	99	3.5
15	61.5	90	3.4
16	72.0	98	3.5
17	104.0	70	4.8

a. Using a .05 significance level, test appropriate hypotheses to determine whether a correlation exists between depth and zinc concentration.
b. Using a .05 significance level, do the data strongly suggest a correlation between depth and iron concentration?
c. Calculate the slope and intercept of the estimated regression line relating y = iron concentration and x = depth.
d. Use the estimated regression equation to construct a 95% prediction interval for the iron concentration of a single core sample taken at a depth of 50 m.
e. Compute and interpret a 95% interval estimate for the true average iron concentration of core samples taken at 70 m.

13.73 The accompanying figure is from the article "Root and Shoot Competition Intensity Along a Soil Depth

Gradient" (*Ecology* [1995]: 673–682). It shows the relationship between above-ground biomass and soil depth within the experimental plots. The relationship is described by the linear equation: biomass = −9.85 + 25.29(soil depth) and r^2 = .65; $P < 0.001$; $n = 55$. Do you think the simple linear regression model is appropriate here? Explain. What would you expect to see in a plot of the standardized residuals versus x?

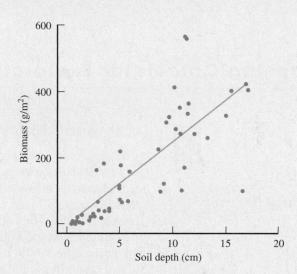

13.74 Give a brief answer, comment, or explanation for each of the following.
a. What is the difference between $e_1, e_2, \ldots, e_n$ and the n residuals?
b. The simple linear regression model states that $y = \alpha + \beta x$.
c. Does it make sense to test hypotheses about b?
d. SSResid is always positive.
e. A student reported that a data set consisting of $n = 6$ observations yielded residuals 2, 0, 5, 3, 0, and 1 from the least-squares line.
f. A research report included the following summary quantities obtained from a simple linear regression analysis:

$$\sum(y - \bar{y})^2 = 615 \qquad \sum(y - \hat{y})^2 = 731$$

13.75 Some straightforward but slightly tedious algebra shows that

$$\text{SSResid} = (1 - r^2)\sum(y - \bar{y})^2$$

from which it follows that

$$s_e = \sqrt{\frac{n-1}{n-2}}(\sqrt{1 - r^2})s_y$$

Unless n is quite small, $(n - 1)/(n - 2) \approx 1$, so

$$s_e \approx (\sqrt{1 - r^2})s_y$$

a. For what value of r is s_e as large as s_y? What is the equation of the least-squares line in this case?

b. For what values of r will s_e be much smaller than s_y?

Personal Tutor
Do you need a live tutor for homework problems?

CENGAGENOW™
Are you ready? Take your exam-prep post-test now.

Bold exercises answered in back ● Data set available online but not required ▼ Video solution available

Graphing Calculator Exploration

Exploration 13.1 Inference for the Regression Slope

By this time you should be an old hand at performing hypothesis tests. You can navigate your calculator menu with aplomb, and punch keys with the best of them! With this background, inferences concerning the slope of a population regression line using your calculator will hold few surprises.

Enter the data from Example 13.4, the treadmill data, into your calculator. Enter the treadmill run time in List1 and the ski time in List2. Navigating the menu system on the calculator should lead you to a screen similar to the one in Figure 13.23.

In Figure 13.23, L1 and L2 are the default values chosen by the calculator; if you entered the data in a different list, you will have to change these lists. The line $\beta \& \rho: \neq 0 < 0 > 0$ is where the alternative hypothesis is selected. For this example, we have selected the alternative hypothesis $\beta \neq 0$.

Placing the calculator's cursor on Calculate and pressing Enter produces the screen shown in Figure 13.24, another larger-than-life presentation. Comparing the screen in Figure 13.24 with the quantities calculated in the text, you will find all the information needed to complete the hypothesis test. The notation is standard, except for the equation s = 2.1883. This s, after checking with the quantities in the discussion of Example 13.4, turns out to be s_e, the estimated standard deviation of the random deviations. It is *not* the estimated standard deviation of the sample slope, s_b. This is unfortunate, since to construct a confidence interval for b we need to calculate $b \pm (t \text{ critical value}) \cdot s_b$.

Worry not! You can construct the confidence interval for b with the aid of algebra and the statistics capabilities of your calculator. The standard error for b can be calculated from the information given and the sample variance of the treadmill times:

```
LinRegTTest
Xlist: L1
Ylist: L2
Freq: 1
β & ρ: ≠0 <0 >0:
RegEq:
Calculate
```

Figure 13.23 Setup window.

```
LinRegTTest
Y = a+bx
β ≠ 0 and ρ ≠ 0
t = −3.9478
p = .0034
df = 9.0000
a = 88.7956
b = −2.3335
s = 2.1883
r² = .6339
r = .7962
```

Figure 13.24 Output screen.

$$s_b = \frac{s_e}{\sqrt{S_{xx}}}$$

$$= \frac{s_e}{\sqrt{(x - \bar{x})^2}}$$

$$= \frac{s_e}{\sqrt{(n - 1)s_x^2}}$$

$$= \frac{s_e}{s_x\sqrt{(n - 1)}}$$

There are 11 data points; $s_x = 1.1707030$ (from the statistics menu for List1). Substituting gives

$$s_b = \frac{s_e}{s_x\sqrt{(n-1)}}$$

$$= \frac{2.188286324}{1.1707030\sqrt{10}}$$

$$= .5910952$$

With the information now available, you can calculate the 95% confidence interval for the population slope:

$b \pm (t \text{ critical value}) \cdot s_b$
$-2.3335 \pm (2.26) \cdot .5910952$
$(-3.6694, -.9976)$

TEACHING TIPS

Intent of this chapter: After completing this chapter, students should be able to (1) recognize multiple regression model equations for the general additive model, including polynomial models and models with interaction terms; (2) distinguish between and utilize both quantitative and qualitative predictor variables; (3) fit a multiple regression model and assess its utility; and (4) be able to read and comprehend the standard computer multiple regression output.

A PowerPoint® Lecture is available on the *Instructor's Resource Binder* CD.

■ **Section 14.1**

Section 14.1 discusses the additive probabilistic model for linear regression (the multiple linear regression model): $y = \alpha + \beta_1 x_1 + \beta_2 x_2 + \cdots + e$, where α is the population intercept, β_i s are the population regression coefficients, and e is the random deviation of an observed point from the model. This section extends the regression methodology and concepts developed in Chapters 5 and 13 to include multiple predictor variables. Although Sections 14.1 and 14.2 are short introductions to multiple regression, a number of concepts that are presented will lay the foundation for future study in statistics. The multiple linear regression model is a direct extension of the simple linear model. The concepts of regression coefficients and normally distributed errors will be familiar to the students. The "multiple" part of the multiple regression allows some interesting variations to be presented: polynomial models, models with interaction terms, and qualitative variables.

The ideas of interaction and the incorporation of qualitative variables are difficult for students to grasp algebraically. Because of this difficulty, the use of graphic representations in the context of models with two predictor variables aids in the understanding of these concepts. Graphs, such as those in Figure 14.3 on page 754 and Figure 14.5 on page 759, will be very helpful in explaining both interaction and qualitative variables in multiple regression.

Suggested Assignment: Exercises 14.1, 14.2, 14.7, 14.9

Activity: Activity 14.1: Exploring the Relationship Between Number of Predictors and Sample Size (p. 780 and *Activities Manual*, p. 117), "Chapter 14—General Additive Multiple Regression" (*Instructor's Resource Binder* & CD, Chapter 14 Activities Worksheets)

■ **Section 14.2**

This section extends the ideas of model fitting, the model utility test, and model assessment that were developed for simple regression models in Chapter 13 to the multiple regression setting. As you teach this section it will be very helpful to students to point out the similarities and common ideas between fitting a model with a single predictor variable and fitting a model with multiple predictors. Concepts such as the principle of least squares, the idea of using sample data to estimate the beta values and to obtain the estimated regression function, predicted values and residuals, and of course the coefficient of determination have already been discussed for models with a single predictor. The computations involved in fitting a multiple regression model make statistical software a necessity, and it will be essential for students to see and interpret computer output.

The F distribution and the F test for model utility in a multiple regression setting will be new for students. By now students have seen many sampling distributions, so the introduction of a new one should not be particularly troublesome. Instructors will vary with respect to how much they require students to work with the F table; however, even when computer output is used exclusively, it will be helpful to students to have seen a generic F distribution such as in Figure 14.7 on page 768. This graphic assist will be reminiscent of representing P-values using other sampling distributions and will help students connect their prior experience of inference to inference in a multiple regression setting.

Suggested Assignment: Exercises 14.15, 14.21, 14.25

Activity: "Chapter 14—Model Utility and Utility of Individual Predictors" (*Instructor's Resource Binder* & CD, Chapter 14 Activities Worksheets), "Chapter 14—Inferences Based on an Estimated Regression Function" (*Instructor's Resource Binder* & CD, Chapter 14 Activities Worksheets)

Assessment: Chapter 14 Concept Quiz (*Test Bank* and *Instructor's Resource Binder* CD), Chapter 14 Quiz 1 (*Test Bank* and *Instructor's Resource Binder* CD), and/or Chapter 14 Test (*Test Bank* and *Instructor's Resource Binder* CD)

© Andrew Michael/Getty Images

Chapter 14

Multiple Regression Analysis

The general objective of regression analysis is to model the relationship between a dependent variable y and one or more independent (i.e., predictor or explanatory) variables. The simple linear regression model $y = \alpha + \beta x + e$, discussed in Chapter 13, has been used successfully by many investigators in a wide variety of disciplines to relate y to a single predictor variable x. In many situations, however, the relationship between y and any single predictor variable is not strong, but knowing the values of several independent variables may considerably reduce uncertainty about the associated y value. For example, some variation in house prices in a large city can certainly be attributed to house size, but knowledge of size by itself would not usually enable a bank appraiser to accurately predict a home's value. Price is also determined to some extent by other variables, such as age, lot size, number of bedrooms and bathrooms, and distance from schools.

In this chapter, we extend the regression methodology developed in the previous chapter to *multiple regression models*, which include at least two predictor variables. Fortunately, many of the concepts developed in the context of simple linear regression carry over to multiple regression with little or no modification. The calculations required to fit such a model and make further inferences are *much* more tedious than those for simple linear regression, so a computer is an indispensable tool for multiple regression analysis. The computer's ability to perform a huge number of computations in a short time has spurred the development of new methods for analyzing large data sets with many predictor variables. These include techniques for fitting numerous alternative models and choosing between them, tools for identifying influential observations, and both algebraic and graphical diagnostics designed to reveal potential violations of model assumptions. A single chapter can do little more than scratch the surface of this important subject area.

14.1   Multiple Regression Models

The relationship between a dependent or response variable y and two or more independent or predictor variables is deterministic if the value of y is completely determined, with no uncertainty, once values of the independent variables have been specified. Consider, for example, a school district in which teachers with no prior teaching experience and no college credits beyond a bachelor's degree start at an annual salary of $38,000. Suppose that for each year of teaching experience up to 20 years, a teacher receives an additional $800 per year and that each unit of postcollege coursework up to 75 units results in an extra $60 per year. Define three variables:

y = salary of a teacher who has at most 20 years teaching experience and at most 75 postcollege units

x_1 = number of years teaching experience

x_2 = number of postcollege units

Previously, x_1 and x_2 denoted the first two observations on the single variable x. In the usual notation for multiple regression, however, x_1 and x_2 represent two different variables.

The value of y is entirely determined by values of x_1 and x_2 through the equation

$$y = 38,000 + 800x_1 + 60x_2$$

Thus, if $x_1 = 10$ and $x_2 = 30$,

$$\begin{aligned} y &= 38,000 + 800(10) + 60(30) \\ &= 38,000 + 8000 + 1800 \\ &= 47,800 \end{aligned}$$

If two different teachers both have the same x_1 values and the same x_2 values, they will also have identical y values.

Only rarely is y deterministically related to predictors $x_1, \ldots, x_k$; a probabilistic model is more realistic in most situations. A probabilistic model results from adding a random deviation e to a deterministic function of the x_i's.

Invite students to compare this model with the simple linear regression model on page 691.

DEFINITION

A **general additive multiple regression model**, which relates a dependent variable y to k predictor variables $x_1, x_2, \ldots, x_k$, is given by the model equation

$$y = \alpha + \beta_1 x_1 + \beta_2 x_2 + \cdots + \beta_k x_k + e$$

The random deviation e is assumed to be normally distributed with mean value 0 and standard deviation σ for any particular values of $x_1, \ldots x_k$. This

(continued)

implies that for fixed $x_1, x_2, \ldots, x_k$ values, y has a normal distribution with standard deviation σ and

$$\begin{pmatrix} \text{mean } y \text{ value for fixed} \\ x_1, \ldots, x_k \text{ values} \end{pmatrix} = \alpha + \beta_1 x_1 + \beta_2 x_2 + \cdots + \beta_k x_k$$

The β_i's are called **population regression coefficients**; each β_i can be interpreted as the true average change in y when the predictor x_i increases by 1 unit *and* the values of all the other predictors remain fixed.

The deterministic portion $\alpha + \beta_1 x_1 + \beta_2 x_2 + \cdots + \beta_k x_k$ is called the **population regression function**.

As in simple linear regression, if σ (the standard deviation of the random error distribution) is quite close to 0, any particular observed y will tend to be quite near its mean value. When σ is large, many of the y observations may deviate substantially from their mean y values.

Example 14.1 Sophomore Success

What factors contribute to the academic success of college sophomores? Data collected in a survey of approximately 1000 second-year college students suggest that GPA at the end of the second year is related to the student's level of interaction with faculty and staff and to the student's commitment to his or her major ("An Exploration of the Factors that Affect the Academic Success of College Sophomores," *College Student Journal* [2005] 367–376). Consider the variables

y = GPA at the end of the sophomore year
x_1 = level of faculty and staff interaction (measured on a scale from 1 to 5)
x_2 = level of commitment to major (measured on a scale from 1 to 5)

One possible population model might be

$$y = 1.4 + .33x_1 + .16x_2 + e$$

with

$$\sigma = 0.15$$

The population regression function is

$$(\text{mean } y \text{ value for fixed } x_1, x_2) = 1.4 + .33x_1 + .16x_2$$

For sophomore students whose level of interaction with faculty and staff is rated at 4.2 and whose level of commitment to their major was rated as 2.1,

$$(\text{mean value of GPA}) = 1.4 + .33(4.2) + .16(2.1) = 3.12$$

With $2\sigma = 2(.15) = .30$, it is quite likely that an actual y value will be within .30 of the mean value (i.e., in the interval from 2.82 to 3.42 when $x_1 = 4.2$ and $x_2 = 2.1$).

■ A Special Case: Polynomial Regression ....

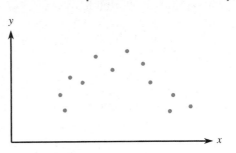

Figure 14.1 A scatterplot that suggests the appropriateness of a quadratic probabilistic model.

Consider again the case of a single independent variable x, and suppose that a scatterplot of the n sample (x, y) pairs has the appearance of Figure 14.1. The simple linear regression model is clearly not appropriate, but it does look as though a parabola (quadratic function) with equation $y = \alpha + \beta_1 x + \beta_2 x^2$ would provide a very good fit to the data for appropriately chosen values of α, β_1, and β_2. Just as the inclusion of the random deviation e in simple linear regression allowed an observation to deviate from the population regression line by a random amount, adding e to this quadratic function yields a probabilistic model in which an observation is allowed to fall above or below the parabola. The model equation is

$$y = \alpha + \beta_1 x + \beta_2 x^2 + e$$

Let's rewrite the model equation by using x_1 to denote x and x_2 to denote x^2. The model equation then becomes

$$y = \alpha + \beta_1 x_1 + \beta_2 x_2 + e$$

This is a special case of the general multiple regression model with $k = 2$. You may wonder about the legitimacy of allowing one predictor variable to be a mathematical function of another predictor—here, $x_2 = (x_1)^2$. However, there is absolutely nothing in the general multiple regression model that prevents this. *In the model $y = \alpha + \beta_1 x_1 + \beta_2 x_2 + \cdots + \beta_k x_k + e$ it is permissible to have several predictors that are mathematical functions of other predictors.* For example, starting with the two independent variables x_1 and x_2, we could create a model with $k = 4$ predictors in which x_1 and x_2 themselves are the first two predictor variables and $x_3 = (x_1)^2$, $x_4 = x_1 x_2$. (We will soon discuss the consequences of using a predictor such as x_4.) In particular, the general polynomial regression model begins with a single independent variable x and creates predictors $x_1 = x$, $x_2 = x^2$, $x_3 = x^3$, ..., $x_k = x^k$ for some specified value of k.

> **D E F I N I T I O N**
>
> The *k*th-degree polynomial regression model
>
> $$y = \alpha + \beta_1 x + \beta_2 x^2 + \cdots + \beta_k x^k + e$$
>
> is a special case of the general multiple regression model with
>
> $$x_1 = x, x_2 = x^2, x_3 = x^3, \ldots, x_k = x^k.$$
>
> The **population regression function** (mean value of y for fixed values of the predictors) is
>
> $$\alpha + \beta_1 x + \beta_2 x^2 + \cdots + \beta_k x^k$$
>
> The most important special case other than simple linear regression ($k = 1$) is the **quadratic regression model**
>
> $$y = \alpha + \beta_1 x + \beta_2 x^2 + e$$
>
> This model replaces the line of mean values $\alpha + \beta x$ in simple linear regression with a parabolic curve of mean values $\alpha + \beta_1 x + \beta_2 x^2$. If $\beta_2 > 0$, the curve opens upward, whereas if $\beta_2 < 0$, the curve opens downward. A less frequently encountered special case is that of cubic regression, in which $k = 3$. (See Figure 14.2.)

This quadratic regression model was introduced early in Section 5.4 in the discussion of nonlinear models.

Figure 14.2 Polynomial regression: (a) quadratic regression model with $\beta_2 < 0$; (b) quadratic regression model with $\beta_2 > 0$; (c) cubic regression model with $\beta_3 > 0$.

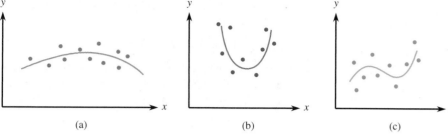

(a)   (b)   (c)

Example 14.2 Increased Risk of Heart Attack

Many researchers have examined a number of factors that are believed to contribute to the risk of heart attack. The authors of the paper "Obesity and the Risk of Myocardial Infarction in 27,000 Participants from 53 Countries: A Case-Control Study" (*The Lancet* [2005]: 1640–1649) found that hip-to-waist ratio was a better predictor of heart attacks than body-mass index. A plot that appeared in the paper of a measure of the risk of heart attack (y) versus hip-to-waist ratio (x) exhibited a curved relationship. Larger values of y indicate a higher risk of heart attack. A model consistent with summary values given in the paper is

$$y = 1.023 + 0.024x + 0.060x^2 + e$$

Then the population regression function is

$$\begin{pmatrix} \text{mean value of the} \\ \text{risk of heart attack measure} \end{pmatrix} = 1.023 + 0.024x + 0.060x^2$$

For example, if $x = 1.3$

$$\begin{pmatrix} \text{mean value of the} \\ \text{risk of heart attack measure} \end{pmatrix} = 1.023 + 0.024(1.3) + 0.060(1.3)^2 = 1.16$$

If $\sigma = .25$, then it is quite likely that the risk of heart attack measure for a person with a hip-to-waist ratio of 1.3 would be between .66 and 1.66.

■

The interpretation of β_i previously given for the general multiple regression model cannot be applied in polynomial regression. This is because all predictors are functions of the single variable x, so $x_i = x^i$ cannot be increased by 1 unit without changing the values of all the other predictor variables as well. *In general, the interpretation of regression coefficients requires extra care when some predictor variables are mathematical functions of other variables.*

■ Interaction Between Variables

Suppose that an industrial chemist is interested in the relationship between product yield (y) from a certain chemical reaction and two independent variables, x_1 = reaction temperature and x_2 = pressure at which the reaction is carried out. The chemist initially suggests that for temperature values between 80 and 110 in combination with pressure values ranging from 50 to 70, the relationship can be well described by the probabilistic model

$$y = 1200 + 15x_1 - 35x_2 + e$$

The regression function, which gives the mean y value for any specified values of x_1 and x_2 is then $1200 + 15x_1 - 35x_2$. Consider this mean y value for three different particular temperature values:

$x_1 = 90$:   mean y value $= 1200 + 15(90) - 35x_2 = 2550 - 35x_2$
$x_1 = 95$:   mean y value $= 2625 - 35x_2$
$x_1 = 100$:   mean y value $= 2700 - 35x_2$

Graphs of these three mean value functions (each a function only of pressure x_2, because the temperature value has been specified) are shown in Figure 14.3(a). Each graph is a straight line, and the three lines are parallel, each one having slope -35. Because of this, the average change in yield when pressure x_2 is increased by 1 unit is -35, regardless of the fixed temperature value.

Figure 14.3 Graphs of mean y value for two different models: (a) $1200 + 15x_1 - 35x_2$; (b) $-4500 + 75x_1 + 60x_2 - x_1x_2$.

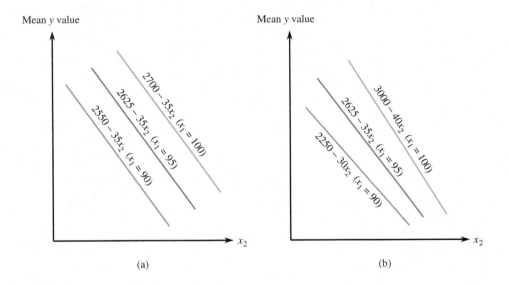

(a)   (b)

Because chemical theory suggests that the decline in average yield when pressure x_2 increases should be more rapid for a high temperature than for a low temperature, the chemist now has reason to doubt the appropriateness of the proposed model. Rather than the lines being parallel, the line for a temperature of 100 should be steeper than the line for a temperature of 95, and that line in turn should be steeper than the one for $x_1 = 90$. A model that has this property includes, in addition to predictors x_1 and x_2 separately, a third predictor variable $x_3 = x_1x_2$. One such model is

$$y = -4500 + 75x_1 + 60x_2 - x_1x_2 + e$$

which has regression function $-4500 + 75x_1 + 60x_2 - x_1x_2$. Then

(mean y when $x_1 = 100$) $= -4500 + 75(100) + 60x_2 - 100x_2$
                                        $= 3000 - 40x_2$

whereas

(mean y when $x_1 = 95$) $= 2625 - 35x_2$
(mean y when $x_1 = 90$) $= 2250 - 30x_2$

These functions are graphed in Figure 14.3(b), where it is clear that the three slopes are different. In fact, each different value of x_1 yields a different slope, so the average change in yield associated with a 1-unit increase in x_2 depends on the value of x_1. When this is the case, the two variables are said to *interact*.

DEFINITION

If the change in the mean y value associated with a 1-unit increase in one independent variable depends on the value of a second independent variable, there is **interaction** between these two variables. When the variables are denoted by x_1 and x_2, such interaction can be modeled by including x_1x_2, the product of the variables that interact, as a predictor variable.

The general equation for a multiple regression model based on two independent variables x_1 and x_2 that also includes an interaction predictor is

$$y = \alpha + \beta_1 x_1 + \beta_2 x_2 + \beta_3 x_1 x_2 + e$$

When x_1 and x_2 do interact, this model usually gives a much better fit to the resulting sample data—and thus explains more variation in y—than does the no interaction model. Failure to consider a model with interaction often leads an investigator to conclude incorrectly that there is no strong relationship between y and a set of independent variables.

More than one interaction predictor can be included in the model when more than two independent variables are available. If, for example, there are three independent variables x_1, x_2, and x_3, one possible model is

$$y = \alpha + \beta_1 x_1 + \beta_2 x_2 + \beta_3 x_3 + \beta_4 x_4 + \beta_5 x_5 + \beta_6 x_6 + e$$

where

$$x_4 = x_1 x_2 \qquad x_5 = x_1 x_3 \qquad x_6 = x_2 x_3$$

One could even include a three-way interaction predictor $x_7 = x_1 x_2 x_3$ (the product of all three independent variables), although in practice this is rarely done.

In applied work, quadratic terms, such as x_1^2 and x_2^2 are often included to model a curved relationship between y and several independent variables. For example, a frequently used model involving just two independent variables x_1 and x_2 but $k = 5$ predictors is the *full quadratic* or **complete second-order model**

$$y = \alpha + \beta_1 x_1 + \beta_2 x_2 + \beta_3 x_1 x_2 + \beta_4 x_1^2 + \beta_5 x_2^2 + e$$

This model replaces the straight lines of Figure 14.3 with parabolas (each one is the graph of the regression function for different values of x_2 when x_1 has a fixed value). With four independent variables, one could examine a model containing four quadratic predictors and six two-way interaction predictor variables. Clearly, with just a few independent variables, one could examine a great many different multiple regression models. In Section 14.5 we briefly discuss methods for selecting one model from a number of competing models.

When developing a multiple regression model, scatterplots of y with each potential predictor can be informative. This is illustrated in Example 14.3, which describes a model that includes a term that is a function of one of the predictor variables and also an interaction term.

Example 14.3 Wind Chill Factor

● The wind chill index, often included in winter weather reports, combines information on air temperature and wind speed to describe how cold it really feels. In 2001, the National Weather Service announced that it would begin using a new wind chill formula beginning in the fall of that year (*USA Today*, August 13, 2001). The following table gives the wind chill index for various combinations of air temperature and wind speed.

Wind (mph)	Temperature (°F)														
	35	**30**	**25**	**20**	**15**	**10**	**5**	**0**	**−5**	**−10**	**−15**	**−20**	**−25**	**−30**	**−35**
5	31	25	19	13	7	1	−5	−11	−16	−22	−28	−34	−40	−46	−52
10	27	21	15	9	3	−4	−10	−16	−22	−28	−35	−41	−47	−53	−59
15	25	19	13	6	0	−7	−13	−19	−26	−32	−39	−45	−51	−58	−64
20	24	17	11	4	−2	−9	−15	−22	−29	−35	−42	−48	−55	−61	−68
25	23	16	9	3	−4	−11	−17	−24	−31	−37	−44	−51	−58	−64	−71
30	22	15	8	1	−5	−12	−19	−26	−33	−39	−46	−53	−60	−67	−73
35	21	14	7	0	−7	−14	−21	−27	−34	−41	−48	−55	−62	−69	−76
40	20	13	6	−1	−8	−15	−22	−29	−36	−43	−50	−57	−64	−71	−78
45	19	12	5	−2	−9	−16	−23	−30	−37	−44	−51	−58	−65	−72	−79

Figure 14.4(a) shows a scatterplot of wind chill index versus air temperature with the different wind speeds denoted by different colors in the plot. It appears that the wind chill index increases linearly with air temperature at each of the wind speeds, but the linear patterns for the different wind speeds are not quite parallel. This suggests that to model the relationship between y = wind chill index and the two variables x_1 = air temperature and x_2 = wind speed, we should include both x_1 and an interaction term that involves x_2. Figure 14.4(b) shows a scatterplot of wind chill index versus wind speed with the different temperatures denoted by different colors. This plot reveals that the relationship between wind chill index and wind speed is nonlinear at each of the different temperatures, and because the pattern is more markedly curved at some temperatures than at others, an interaction is suggested.

These observations are consistent with the new model used by the National Weather Service for relating wind chill index to air temperature and wind speed. The model used is

$$\text{mean } y = 35.74 + 0.621x_1 - 35.75(x_2') + 0.4275x_1x_2'$$

where

$$x_2' = x_2^{0.16}$$

which incorporates a transformed x_2 (to model the nonlinear relationship between wind chill index and wind speed) and an interaction term.

● Data set available online

Figure 14.4 Scatter-plots of wind chill index data of Example 14.3: (a) wind chill index versus air temperature; (b) wind chill index versus wind speed.

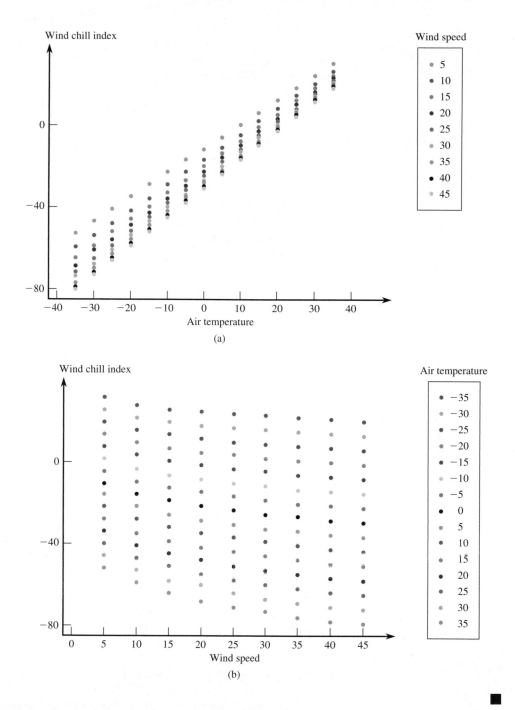

Qualitative Predictor Variables

Up to this point, we have explicitly considered the inclusion only of quantitative (numerical) predictor variables in a multiple regression model. Using simple numerical coding, qualitative (categorical) variables can also be incorporated into a model. Let's focus first on a dichotomous variable, one with just two possible categories: male or female, U.S. or foreign manufacture, a house with or without a view, and so on. With any such variable, we associate a numerical variable x whose possible values are 0 and 1, where 0 is identified with one category (e.g., married) and 1 is identified with the other possible category (e.g., not married). This 0–1 variable is often called a **dummy variable** or **indicator variable**.

....

Example 14.4 Predictors of Writing Competence

The article "Grade Level and Gender Differences in Writing Self-Beliefs of Middle School Students" (*Contemporary Educational Psychology* [1999]: 390–405) considered relating writing competence score to a number of predictor variables, including perceived value of writing and gender. Both writing competence and perceived value of writing were represented by a numerically scaled variable, but gender was a qualitative predictor.

Let

$$y = \text{writing competence score}$$
$$x_1 = \begin{cases} 0 & \text{if male} \\ 1 & \text{if female} \end{cases}$$
$$x_2 = \text{perceived value of writing}$$

One possible multiple regression model is

$$y = \alpha + \beta_1 x_1 + \beta_2 x_2 + e$$

Considering the mean y value first when $x_1 = 0$ and then when $x_1 = 1$ yields

$$\text{average score} = \alpha + \beta_2 x_2 \qquad \text{when } x_1 = 0 \text{ (males)}$$
$$\text{average score} = \alpha + \beta_1 + \beta_2 x_2 \qquad \text{when } x_1 = 1 \text{ (females)}$$

The coefficient β_1 is the difference in average writing competence score between males and females when perceived value of writing is held fixed.

A second possibility is a model with an interaction term:

$$y = \alpha + \beta_1 x_1 + \beta_2 x_2 + \beta_3 x_1 x_2 + e$$

The regression function for this model is $\alpha + \beta_1 x_1 + \beta_2 x_2 + \beta_3 x_3$ where $x_3 = x_1 x_2$. Now the two cases $x_1 = 0$ and $x_1 = 1$ give

$$\text{average score} = \alpha + \beta_2 x_2 \qquad \text{when } x_1 = 0 \text{ (males)}$$
$$\text{average score} = \alpha + \beta_1 + (\beta_2 + \beta_3) x_2 \qquad \text{when } x_1 = 1 \text{ (females)}$$

For each model, the graph of the average writing competence score, when regarded as a function of perceived value of writing, is a line for either gender (Figure 14.5).

In the no-interaction model, the coefficient on x_2 is β_2 both when $x_1 = 0$ and when $x_1 = 1$, so the two lines are parallel, although their intercepts are different (unless $\beta_1 = 0$). With interaction, the lines not only have different intercepts but also have different slopes (unless $\beta_3 = 0$). For this model, the change in average writing competence score when perceived value of writing increases by 1 unit depends on gender—the two variables *perceived value* and *gender* interact.

Figure 14.5 Regression functions for models with one qualitative variable (x_1) and one quantitative variable (x_2): (a) no interaction; (b) interaction.

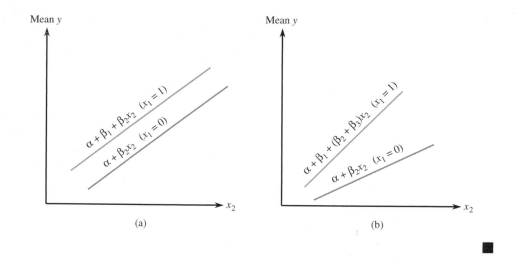

(a)

(b)

You might think that the way to handle a three-category situation is to define a single numerical variable with coded values such as 0, 1, and 2 corresponding to the three categories. This is incorrect, because it imposes an ordering on the categories that is not necessarily implied by the problem. The correct approach to modeling a categorical variable with three categories is to define *two* different $0 - 1$ variables, as illustrated in Example 14.5.

··

Example 14.5 Location, Location, Location

One of the factors that has an effect on the price of a house is location. We might want to incorporate location, as well as numerical predictors such as size and age, into a multiple regression model for predicting house price. Suppose that in a California beach community houses can be classified by location into three categories—ocean-view and beachfront, ocean-view but not beachfront, and no ocean view. Let

$$x_1 = \begin{cases} 1 & \text{if the house is ocean-view and beachfront} \\ 0 & \text{otherwise} \end{cases}$$

$$x_2 = \begin{cases} 1 & \text{if the house has an ocean-view but is not beachfront} \\ 0 & \text{otherwise} \end{cases}$$

$x_3 =$ house size

$x_4 =$ house age

Thus, $x_1 = 1$, $x_2 = 0$ indicates a beachfront ocean-view house; $x_1 = 0$, $x_2 = 1$ indicates a house with an ocean view but not beachfront; and $x_1 = x_2 = 0$ indicates a house that does not have an ocean view. ($x_1 = x_2 = 1$ is not possible.) We could then consider a multiple regression model of the form

$$y = \alpha + \beta_1 x_1 + \beta_2 x_2 + \beta_3 x_3 + \beta_4 x_4 + e$$

This model allows individual adjustments to the predicted price for a house with no ocean view for the other two location categories. For example, β_1 is the amount that would be added to the predicted price for a home with no ocean view to adjust for an oceanfront location (assuming that age and size were the same).

In general, incorporating a categorical variable with c possible categories into a regression model requires the use of $c - 1$ indicator variables. Thus, even one such categorical variable can add many predictors to a model.

■ Nonlinear Multiple Regression Models

Many nonlinear relationships can be put into the form $y = \alpha + \beta_1 x_1 + \cdots + \beta_k x_k + e$ by transforming one or more of the variables. An appropriate transformation could be suggested by theory or by various plots of the data (e.g., residual plots after fitting a particular model). There are also relationships that cannot be linearized by transformations, in which case more complicated methods of analysis must be used. A general discussion of nonlinear regression is beyond the scope of this textbook; you can learn more by consulting the book by Neter, Wasserman, and Kutner listed in the references in the back of the book.

■ Exercises 14.1–14.14

Turn to the Expanded Answers Section for answers not shown next to exercises.

14.1 Explain the difference between a deterministic and a probabilistic model. Give an example of a dependent variable y and two or more independent variables that might be related to y deterministically. Give an example of a dependent variable y and two or more independent variables that might be related to y in a probabilistic fashion.

14.2 A number of investigations have focused on the problem of assessing loads that can be manually handled in a safe manner. The article "Anthropometric, Muscle Strength, and Spinal Mobility Characteristics as Predictors in the Rating of Acceptable Loads in Parcel Sorting" (*Ergonomics* [1992]: 1033–1044) proposed using a regression model to relate the dependent variable

y = individual's rating of acceptable load (kg)

to $k = 3$ independent (predictor) variables:

x_1 = extent of left lateral bending (cm)
x_2 = dynamic hand grip endurance (sec)
x_3 = trunk extension ratio (N/kg)

Suppose that the model equation is

$y = 30 + .90x_1 + .08x_2 - 4.50x_3 + e$

and that $\sigma = 5$.
a. What is the population regression function?
b. What are the values of the population regression coefficients? $\beta_0 = 30, \beta_1 = 0.9, \beta_2 = 0.08, \beta_3 = -4.50$
c. Interpret the value of β_1.
d. Interpret the value of β_3.

e. What is the mean value of rating of acceptable load when extent of left lateral bending is 25 cm, dynamic hand grip endurance is 200 sec, and trunk extension ratio is 10 N/kg? 23.5
f. If repeated observations on rating are made on different individuals, all of whom have the values of x_1, x_2, and x_3 specified in Part (e), in the long run approximately what percentage of ratings will be between 13.5 kg and 33.5 kg? 0.9544

14.3 The following statement appeared in the article "Dimensions of Adjustment Among College Women" (*Journal of College Student Development* [1998]: 364):

> Regression analyses indicated that academic adjustment and race made independent contributions to academic achievement, as measured by current GPA.

Suppose

y = current GPA
x_1 = academic adjustment score
x_2 = race (with white = 0, other = 1)

What multiple regression model is suggested by the statement? Did you include an interaction term in the model? Why or why not?

14.4 According to "Assessing the Validity of the Post-Materialism Index" (*American Political Science Review* [1999]: 649–664), one may be able to predict an individual's level of support for ecology based on demographic and ideological characteristics. The multiple regression model proposed by the authors was

$$y = 3.60 - .01x_1 + .01x_2 - .07x_3 + .12x_4 + .02x_5 \\ - .04x_6 - .01x_7 - .04x_8 - .02x_9 + e$$

where the variables are defined as follows

y = ecology score (higher values indicate a greater concern for ecology)
x_1 = age times 10
x_2 = income (in thousands of dollars)
x_3 = gender (1 = male, 0 = female)
x_4 = race (1 = white, 0 = nonwhite)
x_5 = education (in years)
x_6 = ideology (4 = conservative, 3 = right of center, 2 = middle of the road, 1 = left of center, and 0 = liberal)
x_7 = social class (4 = upper, 3 = upper middle, 2 = middle, 1 = lower middle, 0 = lower)
x_8 = postmaterialist (1 if postmaterialist, 0 otherwise)
x_9 = materialist (1 if materialist, 0 otherwise)

a. Suppose you knew a person with the following characteristics: a 25-year-old, white female with a college degree (16 years of education), who has a $32,000-per-year job, is from the upper middle class and considers herself left of center, but who is neither a materialist nor a postmaterialist. Predict her ecology score.   1.15
b. If the woman described in Part (a) were Hispanic rather than white, how would the prediction change?
c. Given that the other variables are the same, what is the estimated mean difference in ecology score for men and women?   −0.07
d. How would you interpret the coefficient of x_2?
e. Comment on the numerical coding of the ideology and social class variables. Can you suggest a better way of incorporating these two variables into the model?

14.5 ▼ The article "The Influence of Temperature and Sunshine on the Alpha-Acid Contents of Hops" (*Agricultural Meteorology* [1974]: 375–382) used a multiple regression model to relate y = yield of hops to x_1 = mean temperature (°C) between date of coming into hop and date of picking and x_2 = mean percentage of sunshine during the same period. The model equation proposed is

$$y = 415.11 - 6.60x_1 - 4.50x_2 + e$$

a. Suppose that this equation does indeed describe the true relationship. What mean yield corresponds to a temperature of 20 and a sunshine percentage of 40?   103.11

b. What is the mean yield when the mean temperature and percentage of sunshine are 18.9 and 43, respectively?
c. Interpret the values of the population regression coefficients.

14.6 The article "Readability of Liquid Crystal Displays: A Response Surface" (*Human Factors* [1983]: 185–190) used a multiple regression model with four independent variables, where

y = error percentage for subjects reading a four-digit liquid crystal display
x_1 = level of backlight (from 0 to 122 cd/m)
x_2 = character subtense (from .025° to 1.34°)
x_3 = viewing angle (from 0° to 60°)
x_4 = level of ambient light (from 20 to 1500 lx)

The model equation suggested in the article is

$$y = 1.52 + .02x_1 - 1.40x_2 + .02x_3 - .0006x_4 + e$$

a. Assume that this is the correct equation. What is the mean value of y when $x_1 = 10$, $x_2 = .5$, $x_3 = 50$, and $x_4 = 100$?   1.96
b. What mean error percentage is associated with a backlight level of 20, character subtense of .5, viewing angle of 10, and ambient light level of 30?   1.402
c. Interpret the values of β_2 and β_3.

14.7 The article "Pulp Brightness Reversion: Influence of Residual Lignin on the Brightness Reversion of Bleached Sulfite and Kraft Pulps" (*TAPPI* [1964]: 653–662) proposed a quadratic regression model to describe the relationship between x = degree of delignification during the processing of wood pulp for paper and y = total chlorine content. Suppose that the actual model is

$$y = 220 + 75x - 4x^2 + e$$

a. Graph the regression function $220 + 75x - 4x^2$ over x values between 2 and 12. (Substitute $x = 2, 4, 6, 8, 10$, and 12 to find points on the graph, and connect them with a smooth curve.)
b. Would mean chlorine content be higher for a degree of delignification value of 8 or 10?
c. What is the change in mean chlorine content when the degree of delignification increases from 8 to 9? From 9 to 10?

14.8 The relationship between yield of maize, date of planting, and planting density was investigated in the article "Development of a Model for Use in Maize Replant Decisions" (*Agronomy Journal* [1980]: 459–464). Let

y = percent maize yield
x_1 = planting date (days after April 20)
x_2 = planting density (10,000 plants/ha)

The regression model with both quadratic terms ($y = \alpha + \beta_1 x_1 + \beta_2 x_2 + \beta_3 x_3 + \beta_4 x_4 + e$ where $x_3 = x_1^2$ and $x_4 = x_2^2$) provides a good description of the relationship between y and the independent variables.

a. If $\alpha = 21.09$, $\beta_1 = .653$, $\beta_2 = .0022$, $\beta_3 = -.0206$, and $\beta_4 = 0.4$, what is the population regression function?

b. Use the regression function in Part (a) to determine the mean yield for a plot planted on May 6 with a density of 41,180 plants/ha.

c. Would the mean yield be higher for a planting date of May 6 or May 22 (for the same density)?   It is smaller.

d. Is it legitimate to interpret $\beta_1 = .653$ as the true average change in yield when planting date increases by one day and the values of the other three predictors are held fixed? Why or why not?   No, because $x_3 = x_1^2$. So if x_1 increased by 1, x_3 cannot remain fixed.

14.9 Suppose that the variables y, x_1, and x_2 are related by the regression model

$$y = 1.8 + .1x_1 + .8x_2 + e$$

a. Construct a graph (similar to that of Figure 14.5) showing the relationship between mean y and x_2 for fixed values 10, 20, and 30 of x_1.

b. Construct a graph depicting the relationship between mean y and x_1 for fixed values 50, 55, and 60 of x_2.

c. What aspect of the graphs in Parts (a) and (b) can be attributed to the lack of an interaction between x_1 and x_2?

d. Suppose the interaction term $.03x_3$ where $x_3 = x_1 x_2$ is added to the regression model equation. Using this new model, construct the graphs described in Parts (a) and (b). How do they differ from those obtained in Parts (a) and (b)?

14.10 A manufacturer of wood stoves collected data on y = particulate matter concentration and x_1 = flue temperature for three different air intake settings (low, medium, and high).

a. Write a model equation that includes dummy variables to incorporate intake setting, and interpret all the β coefficients.

b. What additional predictors would be needed to incorporate interaction between temperature and intake setting?

14.11 Consider a regression analysis with three independent variables x_1, x_2, and x_3. Give the equation for the following regression models:

a. The model that includes as predictors all independent variables but no quadratic or interaction terms

b. The model that includes as predictors all independent variables and all quadratic terms

c. All models that include as predictors all independent variables, no quadratic terms, and exactly one interaction term

d. The model that includes as predictors all independent variables, all quadratic terms, and all interaction terms (the full quadratic model)

14.12 The article "The Value and the Limitations of High-Speed Turbo-Exhausters for the Removal of Tar-Fog from Carburetted Water-Gas" (*Society of Chemical Industry Journal* [1946]: 166–168) presented data on y = tar content (grains/100 ft^3) of a gas stream as a function of x_1 = rotor speed (rev/min) and x_2 = gas inlet temperature (°F). A regression model using x_1, x_2, $x_3 = x_2^2$ and $x_4 = x_1 x_2$ was suggested:

$$\text{mean } y \text{ value} = 86.8 - .123x_1 + 5.09x_2 - .0709x_3 + .001x_4$$

a. According to this model, what is the mean y value if $x_1 = 3200$ and $x_2 = 57$?

b. For this particular model, does it make sense to interpret the value of any individual β_i (β_1, β_2, β_3, or β_4) in the way we have previously suggested? Explain.

14.13 ▼ Consider the dependent variable y = fuel efficiency of a car (mpg).

a. Suppose that you want to incorporate size class of car, with four categories (subcompact, compact, midsize, and large), into a regression model that also includes x_1 = age of car and x_2 = engine size. Define the necessary dummy variables, and write out the complete model equation.

b. Suppose that you want to incorporate interaction between age and size class. What additional predictors would be needed to accomplish this?

14.14 If we knew the width and height of cylindrical tin cans of food, could we predict the volume of these cans with precision and accuracy?

a. Give the equation that would allow us to make such predictions.   volume = π(height)(width)^2

b. Is the relationship between volume and its predictors, height and width, a linear one?   No

c. Should we use an additive multiple regression model to predict a volume of a can from its height and width? Explain.

d. If you were to take logarithms of each side of the equation in Part (a), would the relationship be linear?

14.2 Fitting a Model and Assessing Its Utility

In Section 14.1, we introduced multiple regression models containing several different types of predictors. Let's now suppose that a particular set of k predictor variables $x_1, x_2, \ldots, x_k$ has been selected for inclusion in the model

$$y = \alpha + \beta_1 x_1 + \beta_2 x_2 + \cdots + \beta_k x_k + e$$

It is then necessary to estimate the model coefficients $\alpha, \beta_1, \ldots, \beta_k$ and the regression function $\alpha + \beta_1 x_1 + \cdots + \beta_k x_k$ (the mean y value for specified values of the predictors), assess the model's utility, and perhaps use the estimated model to make further inferences. All this, of course, requires sample data. As before, n denotes the number of observations in the sample. With just one predictor variable, the sample consisted of n (x, y) pairs. Now each observation consists of $k + 1$ numbers: a value of x_1, a value of $x_2, \ldots$, a value of x_k, and the associated value of y. The n observations are assumed to have been selected independently of one another.

Example 14.6 Graduation Rates at Small Colleges

● One way colleges measure success is by graduation rates. The Education Trust publishes six-year graduation rates along with other college characteristics on its web site (www.collegeresults.org). We will consider the following variables:

y = six-year graduation rate
x_1 = median SAT score of students accepted to the college
x_2 = student-related expense per full-time student (in dollars)
$x_3 = \begin{cases} 1 & \text{if college has only female students or only male students} \\ 0 & \text{if college has both male and female students} \end{cases}$

The following data represent a random sample of 22 colleges selected from the 1037 colleges in the United States with enrollments under 5000 students. The data consist of 22 observations on each of these four variables.

College	y	x_1	x_2	x_3
Cornerstone University	0.391	1065	9482	0
Barry University	0.389	950	13149	0
Wilkes University	0.532	1090	9418	0
Colgate University	0.893	1350	26969	0
Lourdes College	0.313	930	8489	0
Concordia University at Austin	0.315	985	8329	0
Carleton College	0.896	1390	29605	0
Letourneau University	0.545	1170	13154	0
Ohio Valley College	0.288	950	10887	0
Chadron State College	0.469	990	6046	0
Meredith College	0.679	1035	14889	1
Tougaloo College	0.495	845	11694	0
Hawaii Pacific University	0.41	1000	9911	0

(continued)

College	y	x_1	x_2	x_3
University Of Michigan-Dearborn	0.497	1065	9371	0
Whittier College	0.553	1065	14051	0
Wheaton College	0.845	1325	18420	0
Southampton College Of Long Island	0.465	1035	13302	0
Keene State College	0.541	1005	8098	0
Mount St Mary's College	0.579	918	12999	1
Wellesley College	0.912	1370	35393	1
Fort Lewis College	0.298	970	5518	0
Bowdoin College	0.891	1375	35669	0

One possible model that could be considered to model the relationship between y and these three predictor variables is

$$y = \alpha + \beta_1 x_1 + \beta_2 x_2 + \beta_3 x_3 + e$$

We will return to this example after we see how sample data are used to estimate model coefficients.

■

As in simple linear regression, the principle of least squares is used to estimate the coefficients $\alpha, \beta_1, \ldots, \beta_k$. For specified estimates $a, b_1, \ldots, b_k$

$$y - (a + b_1 x_1 + b_2 x_2 + \cdots + b_k x_k)$$

is the deviation between the observed y value for a particular observation and the predicted value using the estimated regression function $a + b_1 x_1 + \cdots + b_k x_k$. For example, the first observation in the data set of Example 14.6 is

$$(x_1, x_2, x_3, y) = (1065, 9482, 0, 0.391).$$

The resulting deviation between observed and predicted y values is

$$0.391 - [a + b_1(1065) + b_2(9482) + b_3(0)]$$

Deviations corresponding to other observations are expressed in a similar manner. The principle of least squares then says to use as estimates of α, β_1, β_2, and β_3 the values of a, b_1, b_2, and b_3 that minimize the sum of these squared deviations.

D EFINITION

According to the principle of least squares, the fit of a particular estimated regression function $a + b_1 x_1 + \cdots + b_k x_k$ to the observed data is measured by the sum of squared deviations between the observed y values and the y values predicted by the estimated function:

$$\sum [y - (a + b_1 x_1 + \cdots + b_k x_k)]^2$$

The **least-squares estimates** of $\alpha, \beta_1, \ldots, \beta_k$ are those values of $a, b_1, \ldots, b_k$ that make this sum of squared deviations as small as possible.

The least-squares estimates for a given data set are obtained by solving a system of $k + 1$ equations in the $k + 1$ unknowns $a, b_1, \ldots, b_k$ (called the *normal equations*). In the case $k = 1$ (simple linear regression), there are only two equations, and we gave their general solution—the expressions for b and a—in Chapter 5. For $k \geq 2$, it is not as easy to write general expressions for the estimates without using advanced mathematical notation. Fortunately, the computer saves us! Formulas for the estimates have been programmed into all the commonly used statistical software packages.

Example 14.7   More on Graduation Rates at Small Colleges

Figure 14.6 displays MINITAB output from a regression command requesting that the model $y = \alpha + \beta_1 x_1 + \beta_2 x_2 + \beta_3 x_3 + e$ be fit to the small college data of Example 14.6. Focus on the column labeled Coef (for coefficient) in the table near the top of the figure. The four numbers in this column are the estimated model coefficients:

$$a = -0.3906 \text{ (the estimate of the constant term } \alpha)$$
$$b_1 = 0.0007602 \text{ (the estimate of the coefficient } \beta_1)$$
$$b_2 = 0.00000697 \text{ (the estimate of the coefficient } \beta_2)$$
$$b_3 = 0.12495 \text{ (the estimate of the coefficient } \beta_3)$$

Thus, we estimate that the average change in six-year graduation rate associated with a 1 dollar increase in expenditure per full-time student while type of institution (same sex or coed) and median SAT score remains fixed is 0.000007. A similar interpretation applies to b_1. The variable x_3 is an indicator variable that takes on a value of 1 for colleges that have either all female students or all male students. We would interpret the estimated value of $b_3 = 0.125$ as the "correction" that we would make to the predicted six-year graduation rate of a coed college with the same

Figure 14.6 MINITAB output for the regression analysis of Example 14.7.

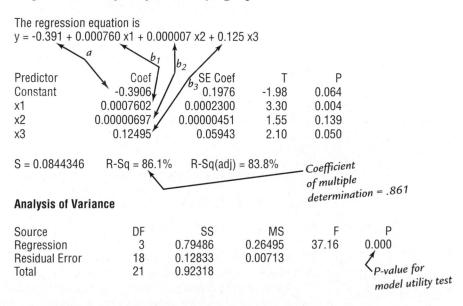

Regression Analysis: y versus x₁, x₂, x₃

The regression equation is
y = -0.391 + 0.000760 x1 + 0.000007 x2 + 0.125 x3

Predictor	Coef	SE Coef	T	P
Constant	-0.3906	0.1976	-1.98	0.064
x1	0.0007602	0.0002300	3.30	0.004
x2	0.00000697	0.00000451	1.55	0.139
x3	0.12495	0.05943	2.10	0.050

S = 0.0844346   R-Sq = 86.1%   R-Sq(adj) = 83.8%

Coefficient of multiple determination = .861

Analysis of Variance

Source	DF	SS	MS	F	P
Regression	3	0.79486	0.26495	37.16	0.000
Residual Error	18	0.12833	0.00713		
Total	21	0.92318			

P-value for model utility test

median SAT and expenditure per full-time student to incorporate the difference associated with having only female or only male students. The estimated regression function is

$$\begin{pmatrix} \text{estimated mean value of } y \\ \text{for specified } x_1, x_2, \text{ and } x_3 \text{ values} \end{pmatrix} = -0.3906 + 0.0007602x_1$$
$$+ 0.00000697x_2 + 0.12495x_3$$

Substituting $x_1 = 1000$, $x_2 = 11{,}000$, and $x_3 = 0$ gives

$$-0.3906 + 0.0007602(1000) + 0.00000697(11{,}000) + 0.12495(0) = .4462$$

which can be interpreted either as a point estimate for the mean six-year graduation rate of coed colleges with a median SAT of 1000 and an expenditure per full-time student of \$11,000 or as a point prediction for a single college with these same characteristics.

■

■ Is the Model Useful?

The utility of an estimated model can be assessed by examining the extent to which predicted y values based on the estimated regression function are close to the y values actually observed.

> ### DEFINITION
>
> The first predicted value $\hat{y}_1$ is obtained by taking the values of the predictor variables $x_1, x_2, \ldots, x_k$ for the first sample observation and substituting these values into the estimated regression function. Doing this successively for the remaining observations yields the **predicted values** $\hat{y}_2, \ldots, \hat{y}_n$. The **residuals** are then the differences $y_1 - \hat{y}_1, y_2 - \hat{y}_2, \ldots, y_n - \hat{y}_n$ between the observed and predicted y values.

The predicted values and residuals are defined here exactly as they were in simple linear regression, but computation of the values is more tedious because there is more than one predictor. Fortunately, the $\hat{y}$'s and $(y - \hat{y})$'s are automatically computed and displayed in the output of all good statistical software packages. Consider again the college data discussed in Examples 14.6 and 14.7. Because the first y observation, $y_1 = 0.391$, was made with $x_1 = 1065$, $x_2 = 9482$ and $x_3 = 0$, the first predicted value is

$$\hat{y} = -0.3906 + 0.0007602(1065) + 0.00000697(9482) + 0.12495(0) = 0.485$$

The first residual is then

$$y_1 - \hat{y}_1 = 0.391 - 0.485 = -0.094$$

The other predicted values and residuals are computed in a similar fashion. The sum of residuals from a least-squares fit should, except for rounding effects, be 0.

As in simple linear regression, the sum of squared residuals is the basis for several important summary quantities that are indicative of a model's utility.

DEFINITION

The **residual (or error) sum of squares, SSResid**, and **total sum of squares, SSTo**, are given by

$$\text{SSResid} = \sum(y - \hat{y})^2 \qquad \text{SSTo} = \sum(y - \bar{y})^2$$

where $\bar{y}$ is the mean of the y observations in the sample.

The number of degrees of freedom associated with SSResid is $n - (k + 1)$, because $k + 1$ df are lost in estimating the $k + 1$ coefficients $\alpha, \beta_1, \ldots, \beta_k$.

An estimate of the random deviation variance σ^2 is given by

$$s_e^2 = \frac{\text{SSResid}}{n - (k + 1)}$$

and $s_e = \sqrt{s_e^2}$ is the estimate of σ.

The **coefficient of multiple determination**, R^2, interpreted as the proportion of variation in observed y values that is explained by the fitted model, is

$$R^2 = 1 - \frac{\text{SSResid}}{\text{SSTo}}$$

In order to help students see that simple linear regression is a special case of multiple regression, refer to the formulas for s_e^2 and R^2 on page 698.

Example 14.8 Small Colleges Revisited

Looking again at Figure 14.6, which contains MINITAB output for the college data fit by a three-predictor model, residual sum of squares is found in the Residual Error row and SS column of the table headed Analysis of Variance: SSResid = 0.12833. The associated number of degrees of freedom is $n - (k + 1) = 22 - (3 + 1) = 18$, which appears in the DF column just to the left of SSResid. The sample average y value is $\bar{y} = .5544$, and SSTo $= \sum(y - .5544)^2 = 0.92318$ appears in the Total row and SS column of the Analysis of Variance table just under the value of SSResid. The values of s_e, s_e^2, and R^2 are then

$$s_e^2 = \frac{\text{SSResid}}{n - (k + 1)} = \frac{0.12833}{18} = 0.007$$

(in the MS column of the MINITAB output)

$$s_e = \sqrt{s_e^2} = \sqrt{.007} = 0.084$$

(which appears just above the Analysis of Variance table)

$$R^2 = 1 - \frac{\text{SSResid}}{\text{SSTo}} = 1 - \frac{0.12833}{0.92318} = 1 - .139 = .861$$

Thus, the percentage of variation explained is $100R^2 = 86.1\%$, which appears on the output as R-Sq = 86.1%. The values of R^2 and s_e suggest that the chosen model has been very successful in relating y to the predictors.

■

In general, a desirable model is one that results in both a large R^2 value and a small s_e value. However, there is a catch. These two conditions can be achieved by fitting a model that contains a large number of predictors. Such a model might be successful in explaining y variation in the data in our sample, but it almost always specifies a relationship that cannot be generalized to the population and that may be unrealistic and difficult to interpret. What we really want is a simple model—that is, a model that has relatively few predictors whose roles are easily interpreted and that also explains much of the variation in y.

All statistical software packages include R^2 and s_e in their output, and most also give SSResid. In addition, some packages compute the quantity called the **adjusted R^2**:

$$\text{adjusted } R^2 = 1 - \left[\frac{n-1}{n-(k+1)}\right]\left(\frac{\text{SSResid}}{\text{SSTo}}\right)$$

Because the quantity in square brackets exceeds 1, the number subtracted from 1 is larger than SSResid/SSTo, so the adjusted R^2 is smaller than R^2. The value of R^2 must be between 0 and 1, but the adjusted R^2 can, on rare occasions, be negative. If a large R^2 has been achieved by using just a few model predictors, the adjusted R^2 will differ little from R^2. However, the adjustment can be substantial when a great many predictors (relative to the number of observations) have been used or when R^2 itself is small to moderate (which could happen even when there is no relationship between y and the predictors). In Example 14.7, the adjusted $R^2 = .838$, which is not much less than R^2 because the model included only two predictor variables and the sample size was 22.

■ *F* Distributions

The model utility test in simple linear regression was based on the fact that when $H_0\colon \beta = 0$ is true, the test statistic has a t distribution. The model utility test for multiple regression uses a type of probability distribution called an *F distribution*. We digress briefly to describe some general properties of *F* distributions. An *F* distribution always arises in connection with a ratio in which the numerator involves one sum of squares and the denominator involves a second sum of squares. Each sum of squares has associated with it a specified number of degrees of freedom, so a particular *F* distribution is determined by fixing values of df_1 = numerator degrees of freedom and df_2 = denominator degrees of freedom. There is a different *F* distribution for each different df_1 and df_2 combination. For example, there is an *F* distribution based on 4 numerator degrees of freedom and 12 denominator degrees of freedom, another *F* distribution based on 3 numerator degrees of freedom and 20 denominator degrees of freedom, and so on. A typical *F* curve for fixed numerator and denominator degrees of freedom appears in Figure 14.7. All *F* tests presented in this book are upper-tailed. Recall that for an upper-tailed t test, the *P*-value is the area under the associated t curve to the right of the calculated t. Similarly, the *P*-value for an upper-tailed *F* test is the area under the associated *F* curve to the right of the calculated *F*. Figure 14.7 illustrates this for a test based on $df_1 = 4$ and $df_2 = 6$.

Unfortunately, tabulation of these upper-tail areas is much more cumbersome than for t distributions, because here 2 df are involved. For each of a number of different *F* distributions, our *F* table (Appendix

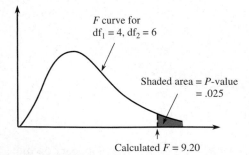

Figure 14.7 A *P*-value for an upper-tailed *F* test.

Table 6) tabulates only four numbers: the values that capture tail areas .10, .05, .01, and .001. Different columns correspond to different values of df_1, and each different group of rows is for a different value of df_2. Figure 14.8 shows how this table is used to obtain P-value information.

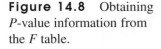

Figure 14.8 Obtaining P-value information from the F table.

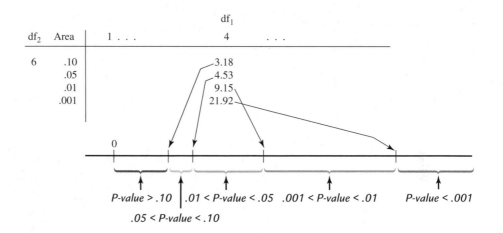

For example, for a test with $df_1 = 4$ and $df_2 = 6$,

calculated $F = 5.70 \rightarrow .01 < P\text{-value} < .05$
calculated $F = 2.16 \rightarrow P\text{-value} > .10$
calculated $F = 25.03 \rightarrow P\text{-value} < .001$

Only if calculated F equals a tabulated value do we obtain an exact P-value (e.g., if calculated $F = 4.53$, then $P\text{-value} = .05$). If $.01 < P\text{-value} < .05$, we should reject the null hypothesis at a significance level of .05 but not at a level of .01. When P-value $< .001$, H_0 would be rejected at any reasonable significance level.

■ The F Test for Model Utility

In the simple linear model with regression function $\alpha + \beta x$, if $\beta = 0$, there is no useful linear relationship between y and the single predictor variable x. Similarly, if all k coefficients $\beta_1, \beta_2, \ldots, \beta_k$ are 0 in the general k-predictor multiple regression model, there is no useful linear relationship between y and *any* of the predictor variables $x_1, x_2, \ldots, x_k$ included in the model. Before using an estimated model to make further inferences (e.g., predictions and estimates of mean values), you should confirm the model's utility through a formal test procedure.

Recall that SSTo is a measure of total variation in the observed y values and that SSResid measures the amount of total variation that has not been explained by the fitted model. The difference between total and error sums of squares is itself a sum of squares, called the **regression sum of squares**, which is denoted by SSRegr:

$$SSRegr = SSTo - SSResid$$

SSRegr is interpreted as the amount of total variation that *has* been explained by the model. Intuitively, the model should be judged useful if SSRegr is large relative to SSResid and the model uses a small number of predictors relative to the sample size.

The number of degrees of freedom associated with SSRegr is k, the number of model predictors, and the number of degrees of freedom for SSResid is $n - (k + 1)$ The model utility F test is based on the following distributional result.

When all k β_i's are 0 in the model $y = \alpha + \beta_1 x_1 + \beta_2 x_2 + \cdots + \beta_k x_k + e$ and when the distribution of e is normal with mean 0 and variance σ^2 for any particular values of $x_1, x_2, \ldots, x_k$, the statistic

$$F = \frac{\text{SSRegr}/k}{\text{SSResid}/(n - (k + 1))}$$

has an F probability distribution based on numerator df $= k$ and denominator df $= n - (k + 1)$.

The value of F tends to be larger when at least one β_i is not 0 than when all the β_i's are 0, because more variation is typically explained by the model in the former case than in the latter case. An F statistic value far out in the upper tail of the associated F distribution can be more plausibly attributed to at least one nonzero β_i than to something extremely unusual having occurred when all the β_i's are 0. This is why the F test for model utility is upper-tailed.

The *F* Test for Utility of the Model $y = \alpha + \beta_1 x_1 + \beta_2 x_2 + \cdots + \beta_k x_k + e$

Null hypothesis: $H_0: \beta_1 = \beta_2 = \cdots = \beta_k = 0$
(There is no useful linear relationship between y and *any* of the predictors.)

Alternative hypothesis: H_a: At least one among $\beta_1, \ldots, \beta_k$ is not zero.
(There is a useful linear relationship between y and *at least one* of the predictors.)

Test statistic: $F = \dfrac{\text{SSRegr}/k}{\text{SSResid}/(n - (k + 1))}$

where SSRegr = SSTo − SSResid.
An equivalent formula is

$$F = \frac{R^2/k}{(1 - R^2)/(n - (k + 1))}$$

The test is upper-tailed, and the information in Appendix Table 6 is used to obtain a bound or bounds on the P-value using numerator df $= k$ and denominator df $= n - (k + 1)$.

Assumptions: For any particular combination of predictor variable values, the distribution of e, the random deviation, is *normal* with mean 0 and *constant variance*.

For the model utility test, the null hypothesis is the claim that the model is not useful. Unless H_0 can be rejected at a small level of significance, the model has not demonstrated its utility, in which case the investigator must search further for a model that can be judged useful. The alternative formula for F allows the test to be carried out when only R^2, k, and n are available, as is frequently the case in published articles.

Example 14.9   Small Colleges One Last Time

The model fit to the college data introduced in Example 14.6 involved $k = 3$ predictors. The MINITAB output in Figure 14.6 contains the relevant information for carrying out the model utility test.

1. The model is $y = \alpha + \beta_1 x_1 + \beta_2 x_2 + \beta_3 x_3 + e$ where y = six-year graduation rate, x_1 = median SAT score, x_2 = expenditure per full-time student, and x_3 is an indicator variable that is equal to 1 for a college that has only female or only male students and is equal to 0 if the college is coed.
2. H_0: $\beta_1 = \beta_2 = \beta_3 = 0$
3. H_a: At least one of the three β_i's is not zero.
4. Significance level: $\alpha = .05$
5. Test statistic: $F = \dfrac{\text{SSRegr}/k}{\text{SSResid}/[n - (k + 1)]}$
6. Assumptions: The accompanying table gives the residuals and standardized residuals (from MINITAB) for the model under consideration.

Obs.	y	x1	x2	x3	Residual	Standardized Residual
1	0.391	1065	9482	0	−0.094	−1.166
2	0.389	950	13149	0	−0.034	−0.442
3	0.532	1090	9418	0	0.028	0.358
4	0.893	1350	26969	0	0.069	0.908
5	0.313	930	8489	0	−0.062	−0.779
6	0.315	985	8329	0	−0.101	−1.244
7	0.896	1390	29605	0	0.024	0.319
8	0.545	1170	13154	0	−0.045	−0.575
9	0.288	950	10887	0	−0.119	−1.497
10	0.469	990	6046	0	0.065	0.812
11	0.679	1035	14889	1	0.054	0.806
12	0.495	845	11694	0	0.162	2.388
13	0.410	1000	9911	0	−0.029	−0.350
14	0.497	1065	9371	0	0.013	0.158
15	0.553	1065	14051	0	0.036	0.441
16	0.845	1325	18420	0	0.100	1.381
17	0.465	1035	13302	0	−0.024	−0.292
18	0.541	1005	8098	0	0.111	1.371
19	0.579	918	12999	1	0.056	0.857
20	0.912	1370	35393	1	−0.110	−1.793
21	0.298	970	5518	0	−0.087	−1.091
22	0.891	1375	35669	0	−0.012	−0.189

A normal probability plot of the standardized residuals is shown here; the plot is quite straight, indicating that the assumption of normality of the random deviation distribution is reasonable:

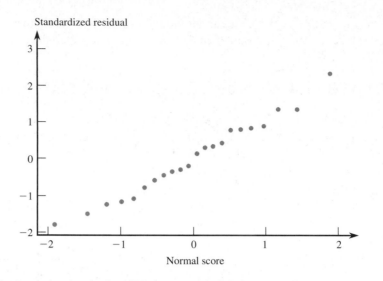

7. Directly from the Analysis of Variance table in Figure 14.6, the SS column gives SSRegr = 0.79486 and SSResid = 0.12833. Thus,

$$F = \frac{0.79486/3}{0.12833/18} = \frac{0.26495}{0.00713} = 37.16 \quad \text{(in the column labeled F in Figure 14.6)}$$

8. Appendix Table 6 shows that for a test based on $df_1 = k = 3$ and $df_2 = n - (k + 1) = 22 - (3 + 1) = 18$, the value 8.49 captures upper-tail F curve area .001. Since calculated $F = 37.16 > 8.49$, it follows that P-value $< .001$. In fact, Figure 14.6 shows that to three decimal places, P-value $= 0$.

9. Because P-value $< .001 \le .05 = \alpha$, H_0 should be rejected. The conclusion would be the same using $\alpha = .01$ or $\alpha = .001$. The utility of the model is resoundingly confirmed.

■

Example 14.10 School Board Politics

A multiple regression analysis presented in the article "The Politics of Bureaucratic Discretion: Educational Access as an Urban Service" (*American Journal of Political Science* [1991]: 155–177) considered a model in which the dependent variable was

y = percentage of school board members in a school district who are black

and the predictors were

x_1 = black-to-white income ratio in the district
x_2 = percentage of whites in the district below the poverty line
x_3 = indicator for whether district was in the South
x_4 = percentage of blacks in the district with a high-school education
x_5 = black population percentage in the district

Summary quantities included $n = 140$ and $R^2 = .749$.

1. The fitted model was $y = \alpha + \beta_1 x_1 + \beta_2 x_2 + \cdots + \beta_5 x_5 + e$
2. H_0: $\beta_1 = \beta_2 = \beta_3 = \beta_4 = \beta_5 = 0$
3. H_a: at least one of the β_i's is not zero
4. Significance level: $\alpha = .01$

5. Test statistic: $F = \dfrac{R^2/k}{(1 - R^2)/(n - (k + 1))}$

6. Assumptions: The raw data were not given in this article, so we are unable to compute standardized residuals or construct a normal probability plot. For this test to be valid, we must be willing to assume that the random deviation distribution is normal.

7. $F = \dfrac{.749/5}{.251/(140 - (5 + 1))} = \dfrac{.1498}{.001873} = 80.0$

8. The test is based on $df_1 = k = 5$ and $df_2 = n - (k + 1) = 134$. This latter df is not included in the F table. However, the .001 cutoff value for $df_2 = 120$ is 4.42, and for $df_2 = 240$ it is 4.25; so for $df_2 = 134$, the cutoff value is roughly 4.4. Clearly, 80.0 greatly exceeds this value, implying that P-value $< .001$.

9. Since P-value $< .001 \le .01 = \alpha$, H_0 is rejected at significance level .01. There appears to be a useful linear relationship between y and at least one of the five predictors.

■

In the next section, we presume that a model has been judged useful after performing an F test and then show how the estimated coefficients and regression function can be used to draw further conclusions. However, you should realize that in many applications, more than one model's utility could be confirmed by the F test. Also, just because the model utility test indicates that the multiple regression model is useful does not necessarily mean that all the predictors included in the model contribute to the usefulness of the model. This is illustrated in Example 14.11, and strategies for selecting a model are briefly considered later in Section 14.4.

Example 14.11 The Cost of Energy Bars

● What factors contribute to the price of energy bars promoted to provide endurance and increase muscle power? The article "Energy Bars, Unwrapped" (*Consumer Reports*, June 2003, 19–21) included the following data on price, calorie content, protein content (in grams), and fat content (in grams) for a sample of 19 energy bars:

Price	Calories	Protein	Fat
1.40	180	12	3.0
1.28	200	14	6.0
1.31	210	16	7.0
1.10	220	13	6.0
2.29	220	17	11.0
1.15	230	14	4.5
2.24	240	24	10.0
1.99	270	24	5.0
2.57	320	31	9.0
0.94	110	5	30.0

(continued)

● Data set available online

Price	Calories	Protein	Fat
1.40	180	10	4.5
0.53	200	7	6.0
1.02	220	8	5.0
1.13	230	9	6.0
1.29	230	10	2.0
1.28	240	10	4.0
1.44	260	6	5.0
1.27	260	7	5.0
1.47	290	13	6.0

Figure 14.9 displays MINITAB output from a regression for the model

$$y = \alpha + \beta_1 x_1 + \beta_2 x_2 + \beta_3 x_3 + e$$

where

$y = $ price $x_1 = $ calorie content $x_2 = $ protein content $x_3 = $ fat content

Figure 14.9 MINITAB output for the energy bar data of Example 14.11.

The regression equation is
Price = 0.252 + 0.00125 Calories + 0.0485 Protein + 0.0444 Fat

Predictor	Coef	SE Coef	T	P
Constant	0.2511	0.3524	0.71	0.487
Calories	0.001254	0.001724	0.73	0.478
Protein	0.04849	0.01353	3.58	0.003
Fat	0.04445	0.03648	1.22	0.242

S = 0.2789   R-Sq = 74.7%   R-Sq(adj) = 69.6%

Analysis of Variance

Source	DF	SS	MS	F	P
Regression	3	3.4453	1.1484	14.76	0.000
Residual Error	15	1.1670	0.0778		
Total	18	4.6122			

From the MINITAB output, $F = 14.76$, with an associated P-value of 0.000, indicating that the null hypothesis in the model utility test, $H_0\!: \beta_1 = \beta_2 = \beta_3 = 0$, would be rejected. We would conclude that there is a useful linear relationship between y and x_1, x_2, and x_3. However, consider the MINITAB output shown in Figure 14.10, which resulted from fitting a model that uses only $x_2 = $ protein content as a predictor. Notice that the F test would also indicate the utility of this model and also that the R^2 and adjusted R^2 values of 71.1% and 69.4% are quite similar to those of the model that included all three predictors (74.7% and 69.6% from the MINITAB output of Figure 14.9). This suggests that protein content alone explains about the same amount of the variability in price as all three variables together, and so the simpler model with just one predictor may be preferred over the more complicated model with three predictor variables.

Figure 14.10 MINITAB output for the energy bar data of Example 14.11 when only $x_2 =$ protein content is included as a predictor.

The regression equation is
Price = 0.607 + 0.0623 Protein

Predictor	Coef	SE Coef	T	P
Constant	0.6072	0.1419	4.28	0.001
Protein	0.062256	0.009618	6.47	0.000

S = 0.2798   R-Sq = 71.1%   R-Sq(adj) = 69.4%

Analysis of Variance

Source	DF	SS	MS	F	P
Regression	1	3.2809	3.2809	41.90	0.000
Residual Error	17	1.3313	0.0783		
Total	18	4.6122			

Exercises 14.15–14.33

Turn to the Expanded Answers Section for answers not shown next to exercises.

14.15 ▼ When coastal power stations take in large quantities of cooling water, it is inevitable that a number of fish are drawn in with the water. Various methods have been designed to screen out the fish. The article "Multiple Regression Analysis for Forecasting Critical Fish Influxes at Power Station Intakes" (*Journal of Applied Ecology* [1983]: 33–42) examined intake fish catch at an English power plant and several other variables thought to affect fish intake:

$y =$ fish intake (number of fish)
$x_1 =$ water temperature (°C)
$x_2 =$ number of pumps running
$x_3 =$ sea state (values 0, 1, 2, or 3)
$x_4 =$ speed (knots)

Part of the data given in the article were used to obtain the estimated regression equation

$$\hat{y} = 92 - 2.18x_1 - 19.20x_2 - 9.38x_3 + 2.32x_4$$

(based on $n = 26$). SSRegr = 1486.9 and SSResid = 2230.2 were also calculated.
a. Interpret the values of b_1 and b_4.
b. What proportion of observed variation in fish intake can be explained by the model relationship? $R^2 = 0.40$
c. Estimate the value of σ. $s_e = 10.305$
d. Calculate adjusted R^2. How does it compare to R^2 itself? Adjusted $R^2 = 0.2857$; adjusted R^2 is smaller than R^2.

14.16 Obtain as much information as you can about the P-value for an upper-tailed F test in each of the following situations:

a. $df_1 = 3$, $df_2 = 15$, calculated $F = 4.23$
b. $df_1 = 4$, $df_2 = 18$, calculated $F = 1.95$
c. $df_1 = 5$, $df_2 = 20$, calculated $F = 4.10$
d. $df_1 = 4$, $df_2 = 35$, calculated $F = 4.58$

14.17 Obtain as much information as you can about the P-value for the F test for model utility in each of the following situations:
a. $k = 2$, $n = 21$, calculated $F = 2.47$ P-value > 0.10
b. $k = 8$, $n = 25$, calculated $F = 5.98$
c. $k = 5$, $n = 26$, calculated $F = 3.00$
d. The full quadratic model based on x_1 and x_2 is fit, $n = 20$, and calculated $F = 8.25$. P-value < 0.001
e. $k = 5$, $n = 100$, calculated $F = 2.33$ $0.01 < P$-value < 0.05

14.18 The ability of ecologists to identify regions of greatest species richness could have an impact on the preservation of genetic diversity, a major objective of the World Conservation Strategy. The article "Prediction of Rarities from Habitat Variables: Coastal Plain Plants on Nova Scotian Lakeshores" (*Ecology* [1992]: 1852–1859) used a sample of $n = 37$ lakes to obtain the estimated regression equation

$$\hat{y} = 3.89 + .033x_1 + .024x_2 + .023x_3 + .008x_4 - .13x_5 - .72x_6$$

where $y =$ species richness, $x_1 =$ watershed area, $x_2 =$ shore width, $x_3 =$ drainage (%), $x_4 =$ water color (total color units), $x_5 =$ sand (%), and $x_6 =$ alkalinity. The coefficient of multiple determination was reported as $R^2 = .83$.

Bold exercises answered in back   ● Data set available online but not required   ▼ Video solution available

Use a test with significance level .01 to decide whether the chosen model is useful.

14.19 The article "Impacts of On-Campus and Off-Campus Work on First-Year Cognitive Outcomes" (*Journal of College Student Development* [1994]: 364–370) reported on a study in which y = spring math comprehension score was regressed against x_1 = previous fall test score, x_2 = previous fall academic motivation, x_3 = age, x_4 = number of credit hours, x_5 = residence (1 if on campus, 0 otherwise), x_6 = hours worked on campus, and x_7 = hours worked off campus. The sample size was $n = 210$, and $R^2 = .543$. Test to see whether there is a useful linear relationship between y and at least one of the predictors.

14.20 Is the model fit in Exercise 14.15 useful? Carry out a test using a significance level of .10.

14.21 The accompanying MINITAB output results from fitting the model described in Exercise 14.12 to data.

Predictor	Coef	Stdev	t-ratio
Constant	86.85	85.39	1.02
X1	−0.12297	0.03276	−3.75
X2	5.090	1.969	2.58
X3	−0.07092	0.01799	−3.94
X4	0.0015380	0.0005560	2.77

S = 4.784   R-sq = 90.8%   R-sq(adj) = 89.4%

Analysis of Variance

	DF	SS	MS
Regression	4	5896.6	1474.2
Error	26	595.1	22.9
Total	30	6491.7

a. What is the estimated regression equation?
b. Using a .01 significance level, perform the model utility test.
c. Interpret the values of R^2 and s_e given in the output.

14.22 For the multiple regression model in Exercise 14.4, the value of R^2 was .06 and the adjusted R^2 was .06. The model was based on a data set with 1136 observations. Perform a model utility test for this regression. $F = 7.986$, P-value < 0.001, reject H_0, and conclude that the model is useful.
14.23 ● *This exercise requires the use of a computer package.* The article "Movement and Habitat Use by Lake Whitefish During Spawning in a Boreal Lake: Integrating Acoustic Telemetry and Geographic Information Systems" (*Transactions of the American Fisheries Society* [1999]:

939–952) included the accompanying data on 17 fish caught in two consecutive years.

Year	Fish Number	Weight (g)	Length (mm)	Age (years)
Year 1	1	776	410	9
	2	580	368	11
	3	539	357	15
	4	648	373	12
	5	538	361	9
	6	891	385	9
	7	673	380	10
	8	783	400	12
Year 2	9	571	407	12
	10	627	410	13
	11	727	421	12
	12	867	446	19
	13	1042	478	19
	14	804	441	18
	15	832	454	12
	16	764	440	12
	17	727	427	12

a. Fit a multiple regression model to describe the relationship between weight and the predictors *length* and *age*. $\hat{y} = -511 + 3.06 \text{ length} - 1.11 \text{ age}$
b. Carry out the model utility test to determine whether the predictors *length* and *age*, together, are useful for predicting weight.

14.24 ● *This exercise requires the use of a computer package.* The authors of the article "Absolute Versus per Unit Body Length Speed of Prey as an Estimator of Vulnerability to Predation" (*Animal Behaviour* [1999]: 347–352) found that the speed of a prey (twips/s) and the length of a prey (twips × 100) are good predictors of the time (s) required to catch the prey. (A twip is a measure of distance used by programmers.) Data were collected in an experiment where subjects were asked to "catch" an animal of prey moving across his or her computer screen by clicking on it with the mouse. The investigators varied the length of the prey and the speed with which the prey moved across the screen. The following data are consistent with summary values and a graph given in the article. Each value represents the average catch time over all subjects. The order of the various speed–length combinations was randomized for each subject.

Prey Length	Prey Speed	Catch Time
7	20	1.10
6	20	1.20
5	20	1.23
4	20	1.40
3	20	1.50
3	40	1.40
4	40	1.36
6	40	1.30
7	40	1.28
7	80	1.40
6	60	1.38
5	80	1.40
7	100	1.43
6	100	1.43
7	120	1.70
5	80	1.50
3	80	1.40
6	100	1.50
3	120	1.90

a. Fit a multiple regression model for predicting catch time using prey length and speed as predictors.
b. Predict the catch time for an animal of prey whose length is 6 and whose speed is 50.   1.3245
c. Is the multiple regression model useful for predicting catch time? Test the relevant hypotheses using $\alpha = .05$.
d. The authors of the article suggest that a simple linear regression model with the single predictor

$$x = \frac{\text{length}}{\text{speed}}$$

might be a better model for predicting catch time. Calculate the x values and use them to fit this linear regression model.
e. Which of the two models considered (the multiple regression model from Part (a) or the simple linear regression model from Part (d)) would you recommend for predicting catch time? Justify your choice.

14.25 ● *This exercise requires the use of a computer package.* The article "Vital Dimensions in Volume Perception: Can the Eye Fool the Stomach?" (*Journal of Marketing Research* [1999]: 313–326) gave the data at

the top of page 778 on dimensions of 27 representative food products.
a. Fit a multiple regression model for predicting the volume (in ml) of a package based on its minimum width, maximum width, and elongation score.
b. Why should we consider adjusted R^2 instead of R^2 when attempting to determine the quality of fit of the data to our model?
c. Perform a model utility test.

14.26 The article "The Caseload Controversy and the Study of Criminal Courts" (*Journal of Criminal Law and Criminology* [1979]: 89–101) used a multiple regression analysis to help assess the impact of judicial caseload on the processing of criminal court cases. Data were collected in the Chicago criminal courts on the following variables:

> y = number of indictments
> x_1 = number of cases on the docket
> x_2 = number of cases pending in criminal court trial system

The estimated regression equation (based on $n = 367$ observations) was

$$\hat{y} = 28 - .05x_1 - .003x_2 + .00002x_3$$

where $x_3 = x_1x_2$.
a. The reported value of R^2 was .16. Conduct the model utility test. Use a .05 significance level.
b. Given the results of the test in Part (a), does it surprise you that the R^2 value is so low? Can you think of a possible explanation for this?
c. How does adjusted R^2 compare to R^2?

14.27 ● ▼ The article "The Undrained Strength of Some Thawed Permafrost Soils" (*Canadian Geotechnical Journal* [1979]: 420–427) contained the accompanying data (see page 778) on y = shear strength of sandy soil (kPa), x_1 = depth (m), and x_2 = water content (%). The predicted values and residuals were computed using the estimated regression equation

$$\hat{y} = -151.36 - 16.22x_1 + 13.48x_2 + .094x_3 - .253x_4 + .492x_5$$

where $x_3 = x_1^2$, $x_4 = x_2^2$, and $x_5 = x_1x_2$.

Data for
Exercise 14.25

Product	Material	Height	Maximum Width	Minimum Width	Elongation	Volume
1	glass	7.7	2.50	1.80	1.50	125
2	glass	6.2	2.90	2.70	1.07	135
3	glass	8.5	2.15	2.00	1.98	175
4	glass	10.4	2.90	2.60	1.79	285
5	plastic	8.0	3.20	3.15	1.25	330
6	glass	8.7	2.00	1.80	2.17	90
7	glass	10.2	1.60	1.50	3.19	120
8	plastic	10.5	4.80	3.80	1.09	520
9	plastic	3.4	5.90	5.00	0.29	330
10	plastic	6.9	5.80	4.75	0.59	570
11	tin	10.9	2.90	2.80	1.88	340
12	plastic	9.7	2.45	2.10	1.98	175
13	glass	10.1	2.60	2.20	1.94	240
14	glass	13.0	2.60	2.60	2.50	240
15	glass	13.0	2.70	2.60	2.41	360
16	glass	11.0	3.10	2.90	1.77	310
17	cardboard	8.7	5.10	5.10	0.85	635
18	cardboard	17.1	10.20	10.20	0.84	1250
19	glass	16.5	3.50	3.50	2.36	650
20	glass	16.5	2.70	1.20	3.06	305
21	glass	9.7	3.00	1.70	1.62	315
22	glass	17.8	2.70	1.75	3.30	305
23	glass	14.0	2.50	1.70	2.80	245
24	glass	13.6	2.40	1.20	2.83	200
25	plastic	27.9	4.40	1.20	3.17	1205
26	tin	19.5	7.50	7.50	1.30	2330
27	tin	13.8	4.25	4.25	1.62	730

Data for Exercise 14.27

y	x_1	x_2	Predicted y	Residual
14.7	8.9	31.5	23.35	−8.65
48.0	36.6	27.0	46.38	1.62
25.6	36.8	25.9	27.13	−1.53
10.0	6.1	39.1	10.99	−0.99
16.0	6.9	39.2	14.10	1.90
16.8	6.9	38.3	16.54	0.26
20.7	7.3	33.9	23.34	−2.64
38.8	8.4	33.8	25.43	13.37
16.9	6.5	27.9	15.63	1.27
27.0	8.0	33.1	24.29	2.71
16.0	4.5	26.3	15.36	0.64
24.9	9.9	37.8	29.61	−4.71
7.3	2.9	34.6	15.38	−8.08
12.8	2.0	36.4	7.96	4.84

a. Use the given information to compute SSResid, SSTo, and SSRegr.

b. Calculate R^2 for this regression model. How would you interpret this value?

c. Use the value of R^2 from Part (b) and a .05 level of significance to conduct the appropriate model utility test.

14.28 The article "Readability of Liquid Crystal Displays: A Response Surface" (*Human Factors* [1983]: 185–190) used the estimated regression equation to describe the relationship between y = error percentage for subjects reading a four-digit liquid crystal display and the independent variables x_1 = level of backlight, x_2 = character subtense, x_3 = viewing angle, and x_4 = level of ambient light. From a table given in the article, SSRegr = 19.2, SSResid = 20.0, and n = 30.

a. Does the estimated regression equation specify a useful relationship between y and the independent variables? Use the model utility test with a .05 significance level.

b. Calculate R^2 and s_e for this model. Interpret these values.

c. Do you think that the estimated regression equation would provide reasonably accurate predictions of error rate? Explain.

14.29 The article "Effect of Manual Defoliation on Pole Bean Yield" (*Journal of Economic Entomology* [1984]: 1019–1023) used a quadratic regression model to describe the relationship between y = yield (kg/plot) and x = defoliation level (a proportion between 0 and 1). The estimated regression equation based on $n = 24$ was $\hat{y} = 12.39 + 6.67x_1 - 15.25x_2$ where $x_1 = x$ and $x_2 = x^2$. The article also reported that R^2 for this model was .902. Does the quadratic model specify a useful relationship between y and x? Carry out the appropriate test using a .01 level of significance.

14.30 Suppose that a multiple regression data set consists of $n = 15$ observations. For what values of k, the number of model predictors, would the corresponding model with $R^2 = .90$ be judged useful at significance level .05? Does such a large R^2 value necessarily imply a useful model? Explain.

14.31 *This exercise requires the use of a computer package.* Use the data given in Exercise 14.27 to verify that the true regression function

$$\text{mean } y \text{ value} = \alpha + \beta_1 x_1 + \beta_2 x_2 + \beta_3 x_3 + \beta_4 x_4 + \beta_5 x_5$$

is estimated by

$$\hat{y} = -151.36 - 16.22x_1 + 13.48x_2 + .094x_3 - .253x_4 + .492x_5$$

14.32 ● *This exercise requires the use of a computer package.* The accompanying data resulted from a study of the relationship between y = brightness of finished paper and the independent variables x_1 = hydrogen peroxide (% by weight), x_2 = sodium hydroxide (% by weight), x_3 = silicate (% by weight), and x_4 = process temperature ("Advantages of CE-HDP Bleaching for High Brightness Kraft Pulp Production," *TAPPI* [1964]: 107A–173A).

x_1	x_2	x_3	x_4	y
.2	.2	1.5	145	83.9
.4	.2	1.5	145	84.9
.2	.4	1.5	145	83.4
.4	.4	1.5	145	84.2
.2	.2	3.5	145	83.8
.4	.2	3.5	145	84.7
.2	.4	3.5	145	84.0
.4	.4	3.5	145	84.8
.2	.2	1.5	175	84.5
.4	.2	1.5	175	86.0
.2	.4	1.5	175	82.6
.4	.4	1.5	175	85.1
.2	.2	3.5	175	84.5
.4	.2	3.5	175	86.0
.2	.4	3.5	175	84.0
.4	.4	3.5	175	85.4
.1	.3	2.5	160	82.9
.5	.3	2.5	160	85.5
.3	.1	2.5	160	85.2
.3	.5	2.5	160	84.5
.3	.3	0.5	160	84.7
.3	.3	4.5	160	85.0
.3	.3	2.5	130	84.9
.3	.3	2.5	190	84.0
.3	.3	2.5	160	84.5
.3	.3	2.5	160	84.7
.3	.3	2.5	160	84.6
.3	.3	2.5	160	84.9
.3	.3	2.5	160	84.9
.3	.3	2.5	160	84.5
.3	.3	2.5	160	84.6

a. Find the estimated regression equation for the model that includes all independent variables, all quadratic terms, and all interaction terms.

b. Using a .05 significance level, perform the model utility test.

c. Interpret the values of the following quantities: SSResid, R^2, s_e.

14.33 ● *This exercise requires the use of a computer package.* The cotton aphid poses a threat to cotton crops in Iraq. The accompanying data on

y = infestation rate (aphids/100 leaves)

x_1 = mean temperature (°C)

x_2 = mean relative humidity

appeared in the article "Estimation of the Economic Threshold of Infestation for Cotton Aphid" (*Mesopotamia Journal of Agriculture* [1982]: 71–75). Use the data to find the estimated regression equation and assess the utility of the multiple regression model

$$y = \alpha + \beta_1 x_1 + \beta_2 x_2 + e$$

y	x_1	x_2	y	x_1	x_2
61	21.0	57.0	77	24.8	48.0
87	28.3	41.5	93	26.0	56.0
98	27.5	58.0	100	27.1	31.0
104	26.8	36.5	118	29.0	41.0
102	28.3	40.0	74	34.0	25.0
63	30.5	34.0	43	28.3	13.0
27	30.8	37.0	19	31.0	19.0

y	x_1	x_2	y	x_1	x_2
14	33.6	20.0	23	31.8	17.0
30	31.3	21.0	25	33.5	18.5
67	33.0	24.5	40	34.5	16.0
6	34.3	6.0	21	34.3	26.0
18	33.0	21.0	23	26.5	26.0
42	32.0	28.0	56	27.3	24.5
60	27.8	39.0	59	25.8	29.0
82	25.0	41.0	89	18.5	53.5
77	26.0	51.0	102	19.0	48.0
108	18.0	70.0	97	16.3	79.5

Personal Tutor

Do you need a live tutor for homework problems?

CENGAGENOW

Are you ready? Take your exam-prep post-test now.

Bold exercises answered in back   ● Data set available online but not required   ▼ Video solution available

---

Activity 14.1   Exploring the Relationship Between Number of Predictors and Sample Size

This activity requires the use of a statistical computer package capable of fitting multiple regression models.

Background: The given data on y, x_1, x_2, x_3, and x_4 were generated using a computer package capable of producing random observations from any specified normal distribution. Because the data were generated at random, there is no reason to believe that y is related to any of the proposed predictor variables x_1, x_2, x_3, and x_4.

y	x_1	x_2	x_3	x_4
20.5	18.6	22.0	17.1	18.5
20.1	23.9	19.1	21.1	21.3
20.0	20.9	20.7	19.4	20.6
21.7	18.7	18.1	20.9	18.1
20.7	21.1	21.7	23.7	17.0

1. Construct four scatterplots—one of y versus each of x_1, x_2, x_3, and x_4. Do the scatterplots look the way you

expected based on the way the data were generated? Explain.

2. Fit each of the following regression models:
   i. y with x_1
   ii. y with x_1 and x_2
   iii. y with x_1 and x_2 and x_3
   iv. y with x_1 and x_2 and x_3 and x_4

3. Make a table that gives the R^2, the adjusted R^2, and s_e values for each of the models fit in Step 2. Write a few sentences describing what happens to each of these three quantities as additional variables are added to the multiple regression model.

4. Given the manner in which these data were generated, what is the implication of what you observed in Step 3? What does this suggest about the relationship between number of predictors and sample size?

Summary of Key Concepts and Formulas

Term or Formula	**Comment**
Additive multiple regression model, $y = \alpha + \beta_1 x_1 + \beta_2 x_2 + \cdots + \beta_k x_k + e$	This equation specifies a general probabilistic relationship between y and k predictor variables $x_1, x_2, \ldots, x_k$, where $\alpha, \beta_1, \ldots, \beta_k$ are population regression coefficients and $\alpha + \beta_1 x_1 + \beta_2 x_2 + \cdots + \beta_k x_k$ is the population regression function (the mean value of y for fixed values of $x_1, x_2, \ldots, x_k$).
Estimated regression function, $\hat{y} = a + b_1 x_1 + b_2 x_2 + \cdots + b_k x_k$	The estimates $a, b_1, \ldots, b_k$ of $\alpha, \beta_1, \ldots, \beta_k$ result from applying the principle of least squares.
Coefficient of multiple determination, $R^2 = 1 - \dfrac{\text{SSResid}}{\text{SSTo}}$	The proportion of observed y variation that can be explained by the model relationship, where SSResid is defined as it is in simple linear regression but is now based on $n - (k + 1)$ degrees of freedom.
Adjusted R^2	A downward adjustment of R^2 that depends on the number of predictors k relative to the sample size n.
F distribution	A type of probability distribution used in many different inferential procedures. A particular F distribution results from specifying both numerator df and denominator df.
$F = \dfrac{R^2/k}{(1 - R^2)/(n - (k + 1))}$ or $F = \dfrac{\text{SSRegr}/k}{\text{SSResid}/(n - (k + 1))}$	The test statistic for testing $H_0: \beta_1 = \beta_2 = \cdots = \beta_k = 0$, which asserts that there is no useful linear relationship between y and any of the model predictors. The F test is upper-tailed and is based on numerator df $= k$ and denominator df $= n - (k + 1)$.

TEACHING TIPS

Intent of this chapter:

After completing this chapter, students should (1) understand single-factor ANOVA to be a comparison of several means, (2) be able to identify the assumptions necessary for single-factor ANOVA, (3) carry out a test of the hypothesis that several population means are equal, and (4) be able to interpret the results of a single-factor ANOVA as presented in the standard ANOVA table.

A PowerPoint® Lecture is available on the *Instructor's Resource Binder* CD.

■ Before you begin . . .

With the ANOVA calculations comes a significant increase in notational complexity. Graphic representations with measures of center and spread—for example, comparative boxplots such as those on page 792—may be very helpful in portraying the contrasting ideas of variability due to differences in population means and variability due to the natural fluctuation in any population.

■ Section 15.1

The fundamental notions and terminology of ANOVA are introduced in this section. Conceptually, the single-factor ANOVA is an easy extension of the t procedures that have been previously introduced. However, there are some statistical differences between the t procedures and single-factor ANOVA procedures. First, there is a greater pedagogical focus on partitioning the total variability using the fundamental identity for single-factor ANOVA, anticipating the future study of more complex ANOVA procedures. Second, the test statistic is the F statistic, not the t statistic from the previous test where we compared two populations using data from independent samples. Finally, unlike the t procedures, the single-factor ANOVA procedure involves pooling, and the hypothesis test is not as robust with respect to a violation of unequal variances in the populations, especially with small and unequal sample sizes. This is why the assumptions for single-factor ANOVA include one about equal variances.

These statistical differences are of some importance in practice; however, the conceptual similarity between inference using ANOVA and previous inferential methods is perhaps more important. We still have a test statistic with a sampling distribution, assumptions to check, and P-values. We are conceptually pursuing the same inferential strategy as with previous hypothesis testing—slightly different trees, but the same forest!

Suggested Assignment: Exercises 15.1, 15.3, 15.5, 15.6, 15.14–15.17

Activity: Activity 15.1: Exploring Single-Factor ANOVA (p. 808 and *Activities Manual*, p. 119)

■ Section 15.2

This section presents the Tukey-Kramer multiple-comparisons procedure, one of many strategies for assessing differences between pairs of population means after an initial rejection of the null hypothesis in a single-factor ANOVA. Once again we see new formulas, tables, and procedures within the same inferential framework.

Students, when confronted with yet another new procedure, may wonder why their old friend the t procedure cannot be used to test the equality of population means two at a time. The distinction between the Type I error rate for a single comparison of two of the p means, for example, and the accumulated Type I error rate when $\dfrac{p(p-1)}{2}$ such pairs are tested is not obvious! Depending on the sophistication of the students and their recollection of random variables from earlier chapters, your approach might range from "Trust me" to an explanation of the mathematics. A very readable and accessible treatment of the error rate among a "family" of comparisons is given in R. O. Kuehl (2000) *Design of*

Experiments: Statistical Principles of Research Design and Analysis (2nd edition) Brooks/Cole: Pacific Grove, CA.

Another pedagogical decision revolves around whether to depend on software or to use a table of critical values for the Studentized range calculations. Opinions of statistics teachers fall somewhere between "Students need to know the general skill of reading statistical tables, and every new table helps" and "For goodness sake, the computer will do all this in real life, so what's the gain?" We will certainly not attempt to arbitrate this dispute, except to note that by Chapter 15 one can be sympathetic to a certain amount of "table fatigue" on the part of students in what is, after all, an introductory course!

Suggested Assignment: Exercises 15.23, 15.26, 15.31, 15.34

Activity: "Chapter 15—Single Factor ANOVA" (*Instructor's Resource Binder* & CD, Chapter 15 Activities Worksheets)

Assessment: Chapter 15 Concept Quiz (*Test Bank* and *Instructor's Resource Binder* CD), Chapter 15 Quiz 1 (*Test Bank* and *Instructor's Resource Binder* CD), and/or Chapter 15 Test (*Test Bank* and *Instructor's Resource Binder* CD)

Chapter 75

Analysis of Variance

In Chapter 11 we discussed methods for testing H_0: $\mu_1 - \mu_2 = 0$ (i.e., $\mu_1 = \mu_2$), where μ_1 and μ_2 are the means of two different populations or the true mean responses when two different treatments are applied. Many investigations involve a comparison of more than two population or treatment means. For example, an investigation was carried out to study possible consequences of the high incidence of head injuries among soccer players ("No Evidence of Impaired Neurocognitive Performance in Collegiate Soccer Players," *The American Journal of Sports Medicine* [2002]: 157–162). Three groups of college students (soccer athletes, nonsoccer athletes, and a control group consisting of students who did not participate in intercollegiate sports) were considered in the study, and the following information on scores from the Hopkins Verbal Learning Test (which measures immediate memory recall) was given in the paper:

Group	Soccer Athletes	Nonsoccer Athletes	Control
Sample Size	86	95	53
Sample Mean Score	29.90	30.94	29.32
Sample Standard Deviation	3.73	5.14	3.78

Let μ_1, μ_2, and μ_3 denote the true average (i.e., population mean) scores on the Hopkins test for soccer athletes, nonsoccer athletes, and students who do not participate in collegiate athletics, respectively. Do the data support the claim that $\mu_1 = \mu_2 = \mu_3$, or does it appear that at least two of the μ's are different from one another? This is an example of a **single-factor analysis of variance (ANOVA)** problem, in which the objective is to decide whether the means for more than two populations or treatments are identical. The first two sections of this chapter discuss various aspects of single-factor ANOVA. In Sections 15.3 and 15.4 (which can be found online), we consider more complex ANOVA situations and methods.

15.1 Single-Factor ANOVA and the *F* Test

When two or more populations or treatments are being compared, the characteristic that distinguishes the populations or treatments from one another is called the **factor** under investigation. For example, an experiment might be carried out to compare three different methods for teaching reading (three different treatments), in which case the factor of interest would be *teaching method*, a qualitative factor. If the growth of fish raised in waters having different salinity levels—0%, 10%, 20%, and 30%—is of interest, the factor *salinity level* is quantitative.

A **single-factor analysis of variance (ANOVA)** problem involves a comparison of k population or treatment means $\mu_1, \mu_2, \ldots, \mu_k$. The objective is to test

$$H_0: \mu_1 = \mu_2 = \cdots = \mu_k$$

against

$$H_a: \text{At least two of the } \mu\text{'s are different}$$

When comparing populations, the analysis is based on independently selected random samples, one from each population. When comparing treatment means, the data typically result from an experiment and the analysis assumes random assignment of the experimental units (subjects or objects) to treatments. If, in addition, the experimental units are chosen at random from a population of interest, it is also possible to generalize the results of the analysis to this population. (See Chapter 2 for a more detailed discussion of conclusions that can reasonably be drawn based on data from an experiment.)

Whether the null hypothesis of a single-factor ANOVA should be rejected depends on how substantially the samples from the different populations or treatments differ from one another. Figure 15.1 displays two possible data sets that might result when observations are selected from each of three populations under study. Each data set

Figure 15.1 Two possible ANOVA data sets when three populations are under investigation: green circle = observation from Population 1; orange circle = observation from Population 2; blue circle = observation from Population 3.

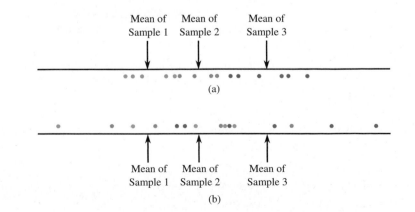

consists of five observations from the first population, four observations from the second population, and six observations from the third population. For both data sets, the three sample means are located by arrows. The means of the two samples from Population 1 are identical, and a similar statement holds for the two samples from Population 2 and those from Population 3.

After looking at the data set in Figure 15.1(a), almost anyone would readily agree that the claim $\mu_1 = \mu_2 = \mu_3$ appears to be false. Not only are the three sample means different, but also the three samples are clearly separated. In other words, differences between the three sample means are quite large relative to the variability within each sample. (If all data sets gave such obvious messages, statisticians would not be in such great demand!)

The situation pictured in Figure 15.1(b) is much less clear-cut. The sample means are as different as they were in the first data set, but now there is considerable overlap among the three samples. The separation between sample means here can plausibly be attributed to substantial variability in the populations (and therefore the samples) rather than to differences between μ_1, μ_2, and μ_3. The phrase *analysis of variance* comes from the idea of analyzing variability in the data to see how much can be attributed to differences in the μ's and how much is due to variability in the individual populations. In Figure 15.1(a), there is little within-sample variability relative to the amount of between-sample variability, whereas in Figure 15.1(b), a great deal more of the total variability is due to variation within each sample. If differences between the sample means can be explained by within sample variability, there is no compelling reason to reject H_0.

▪ Notations and Assumptions

Notation in single-factor ANOVA is a natural extension of the notation used in Chapter 11 for comparing two population or treatment means.

ANOVA Notation

k = number of populations or treatments being compared

Population or treatment	1	2		k
Population or treatment mean	μ_1	μ_2	...	μ_k
Population or treatment variance	σ_1^2	σ_2^2	...	σ_k^2
Sample size	n_1	n_2	...	n_k
Sample mean	$\bar{x}_1$	$\bar{x}_2$	...	$\bar{x}_k$
Sample variance	s_1^2	s_2^2	...	s_k^2

$N = n_1 + n_2 + \cdots + n_k$ (the total number of observations in the data set)

T = grand total = sum of all N observations = $n_1\bar{x}_1 + n_2\bar{x}_2 + \cdots + n_k\bar{x}_k$

$\bar{\bar{x}}$ = grand mean = $\dfrac{T}{N}$

A decision between H_0 and H_a is based on examining the $\bar{x}$ values to see whether observed discrepancies are small enough to be attributable simply to sampling variability or whether an alternative explanation for the differences is more plausible.

Example 15.1 An Indicator of Heart Attack Risk

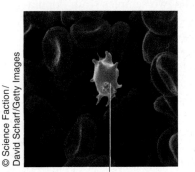

Activated platelet

The article "Could Mean Platelet Volume Be a Predictive Marker for Acute Myocardial Infarction?" (*Medical Science Monitor* [2005]: 387–392) described an experiment in which four groups of patients seeking treatment for chest pain were compared with respect to mean platelet volume (MPV, measured in fL). The four groups considered were based on the clinical diagnosis and were (1) noncardiac chest pain, (2) stable angina pectoris, (3) unstable angina pectoris, and (4) myocardial infarction (heart attack). The purpose of the study was to determine if the mean MPV differed for the four groups, and in particular if the mean MPV was different for the heart attack group, because then MPV could be used as an indicator of heart attack risk and an antiplatelet treatment could be administered in a timely fashion, potentially reducing the risk of heart attack.

To carry out this study, patients seen for chest pain were divided into groups according to diagnosis. The researchers then selected a random sample of 35 from each of the resulting $k = 4$ groups. The researchers believed that this sampling process would result in samples that were representative of the four populations of interest and that could be regarded as if they were random samples from these four populations. Table 15.1 presents summary values given in the paper.

Table 15.1 Summary Values for MPV Data of Example 15.1

Group Number	Group Description	Sample Size	Sample Mean	Sample Standard Deviation
1	Noncardiac chest pain	35	10.89	0.69
2	Stable angina pectoris	35	11.25	0.74
3	Unstable angina pectoris	35	11.37	0.91
4	Myocardial infarction (heart attack)	35	11.75	1.07

With μ_i denoting the true mean MPV for group i ($i = 1, 2, 3, 4$), let's consider the null hypothesis $H_0: \mu_1 = \mu_2 = \mu_3 = \mu_4$. Figure 15.2 shows a comparative boxplot for the four samples (based on data consistent with summary values given in the

Figure 15.2 Boxplots for Example 15.1.

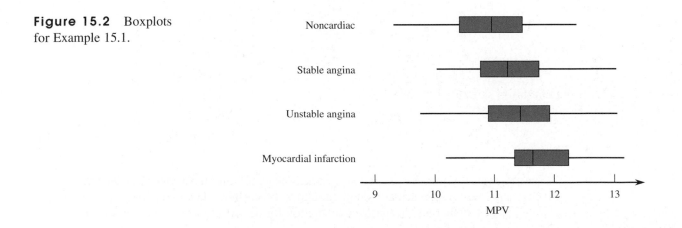

paper). The mean MPV for the heart attack sample is larger than for the other three samples and the boxplot for the heart attack sample appears to be shifted a bit higher than the boxplots for the other three samples. However, because the four boxplots show substantial overlap, it is not obvious whether H_0 is true or false. In situations such as this, we need a formal test procedure.

■

As with the inferential methods of previous chapters, the validity of the ANOVA test for H_0: $\mu_1 = \mu_2 = \cdots = \mu_k$ requires some assumptions.

Assumptions for ANOVA

1. Each of the k population or treatment response distributions is normal.
2. $\sigma_1 = \sigma_2 = \cdots = \sigma_k$ (The k normal distributions have identical standard deviations.)
3. The observations in the sample from any particular one of the k populations or treatments are independent of one another.
4. When comparing population means, k random samples are selected independently of one another. When comparing treatment means, treatments are assigned at random to subjects or objects (or subjects are assigned at random to treatments).

In practice, the test based on these assumptions works well as long as the assumptions are not too badly violated. If the sample sizes are reasonably large, normal probability plots or boxplots of the data in each sample are helpful in checking the assumption of normality. Often, however, sample sizes are so small that a separate normal probability plot or boxplot for each sample is of little value in checking normality. In this case, a single combined plot can be constructed by first subtracting $\bar{x}_1$ from each observation in the first sample, $\bar{x}_2$ from each value in the second sample, and so on and then constructing a normal probability or boxplot of all N deviations from their respective means. The plot should be reasonably straight. Figure 15.3 shows such a normal probability plot for the data of Example 15.1.

There is a formal procedure for testing the equality of population standard deviations. Unfortunately, it is quite sensitive to even a small departure from the normality

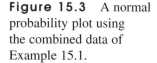

Figure 15.3 A normal probability plot using the combined data of Example 15.1.

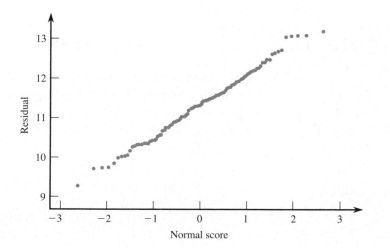

assumption, so we do not recommend its use. Instead, we suggest that the ANOVA F test (to be described subsequently) can safely be used if the largest of the sample standard deviations is at most twice the smallest one. The largest standard deviation in Example 15.1 is $s_4 = 1.07$, which is only about 1.5 times the smallest standard deviation ($s_1 = 0.69$). The book *Beyond ANOVA: The Basics of Applied Statistics* by Rupert (see the references in the back of the book) is a good source for alternative methods of analysis if there appears to be a violation of assumptions.

The test procedure is based on the following measures of variation in the data.

DEFINITION

A measure of disparity among the sample means is the **treatment sum of squares**, denoted by **SSTr** and given by

$$\text{SSTr} = n_1(\bar{x}_1 - \bar{\bar{x}})^2 + n_2(\bar{x}_2 - \bar{\bar{x}})^2 + \cdots + n_k(\bar{x}_k - \bar{\bar{x}})^2$$

A measure of variation within the k samples, called **error sum of squares** and denoted by **SSE**, is

$$\text{SSE} = (n_1 - 1)s_1^2 + (n_2 - 1)s_2^2 + \cdots + (n_k - 1)s_k^2$$

Each sum of squares has an associated df:

$$\text{treatment df} = k - 1 \qquad \text{error df} = N - k$$

A **mean square** is a sum of squares divided by its df. In particular,

$$\text{mean square for treatments} = \text{MSTr} = \frac{\text{SSTr}}{k - 1}$$

$$\text{mean square for error} = \text{MSE} = \frac{\text{SSE}}{N - k}$$

The number of error degrees of freedom comes from adding the number of degrees of freedom associated with each of the sample variances:

$$(n_1 - 1) + (n_2 - 1) + \cdots + (n_k - 1) = n_1 + n_2 + \cdots + n_k - 1 - 1 - \cdots - 1$$
$$= N - k$$

....

Example 15.2 Heart Attack Calculations

Let's return to the mean platelet volume (MPV) data of Example 15.1. The grand mean $\bar{\bar{x}}$ was computed to be 11.315. Notice that because the sample sizes are all equal, the grand mean is just the average of the four sample means (this will not usually be the case when the sample sizes are unequal). With $\bar{x}_1 = 10.89$, $\bar{x}_2 = 11.25$, $\bar{x}_3 = 11.37$, $\bar{x}_4 = 11.75$, and $n_1 = n_2 = n_3 = n_4 = 35$,

$$\begin{aligned}
\text{SSTr} &= n_1(\bar{x}_1 - \bar{\bar{x}})^2 + n_2(\bar{x}_2 - \bar{\bar{x}})^2 + \cdots + n_k(\bar{x}_k - \bar{\bar{x}})^2 \\
&= 35(10.89 - 11.315)^2 + 35(11.25 - 11.315)^2 + 35(11.37 - 11.315)^2 \\
&\quad + 35(11.75 - 11.315)^2 \\
&= 6.322 + 0.148 + 0.106 + 6.623 \\
&= 13.199
\end{aligned}$$

Because $s_1 = 0.69$, $s_2 = 0.74$, $s_3 = 0.91$, and $s_4 = 1.07$

$$
\begin{aligned}
\text{SSE} &= (n_1 - 1)s_1^2 + (n_2 - 1)s_2^2 + \cdots + (n_k - 1)s_k^2 \\
&= (35 - 1)(0.69)^2 + (35 - 1)(0.74)^2 + (35 - 1)(0.91)^2 + (35 - 1)(1.07)^2 \\
&= 101.888
\end{aligned}
$$

The numbers of degrees of freedom are

treatment df $= k - 1 = 3$ error df $= N - k = 35 + 35 + 35 + 35 - 4 = 136$

from which

$$
\text{MSTr} = \frac{\text{SSTr}}{k - 1} = \frac{13.199}{3} = 4.400
$$

$$
\text{MSE} = \frac{\text{SSE}}{N - k} = \frac{101.888}{136} = 0.749
$$

    Both MSTr and MSE are quantities whose values can be calculated once sample data are available; that is, they are statistics. Each of these statistics varies in value from data set to data set. Both statistics MSTr and MSE have sampling distributions, and these sampling distributions have mean values. The following box describes the key relationship between MSTr and MSE and the mean values of these two statistics.

When H_0 is true ($\mu_1 = \mu_2 = \cdots = \mu_k$),

$\mu_{\text{MSTr}} = \mu_{\text{MSE}}$

However, when H_0 is false,

$\mu_{\text{MSTr}} > \mu_{\text{MSE}}$

and the greater the differences among the μ's, the larger μ_{MSTr} will be relative to μ_{MSE}.

    According to this result, when H_0 is true, we expect the two mean squares to be close to one another, whereas we expect MSTr to substantially exceed MSE when some μ's differ greatly from others. Thus, a calculated MSTr that is much larger than MSE casts doubt on H_0. In Example 15.2, MSTr $= 4.400$ and MSE $= 0.749$, so MSTr is about 6 times as large as MSE. Can this difference be attributed solely to sampling variability, or is the ratio MSTr/MSE of sufficient magnitude to suggest that H_0 is false? Before we can describe a formal test procedure, it is necessary to revisit F distributions, first introduced in multiple regression analysis (Chapter 14).

    Many ANOVA test procedures are based on a family of probability distributions called F distributions. As explained in Chapter 14, an F distribution always arises in connection with a ratio. A particular F distribution is obtained by specifying both numerator degrees of freedom (df_1) and denominator degrees of freedom (df_2). Figure 15.4 shows an F curve for a particular choice of df_1 and df_2. All F tests in this book are upper-tailed, so the P-value is the area under the F curve to the right of the calculated F.

Figure 15.4 An F curve and P-value for an upper-tailed test.

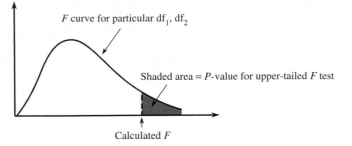

Tabulation of these upper-tail areas is cumbersome, because there are two degrees of freedom rather than just one (as in the case of t distributions). For selected (df_1, df_2) pairs, our F table (Appendix Table 6) gives only the four numbers that capture tail areas .10, .05, .01, and .001, respectively. Here are the four numbers for $df_1 = 4$, $df_2 = 10$ along with the statements that can be made about the P-value:

Tail area	.10	.05	.01	.001
Value	2.61	3.48	5.99	11.28
	↑	↑	↑	↑	↑
	a	b	c	d	e

a. $F < 2.61 \rightarrow$ tail area $= P$-value $> .10$
b. $2.61 < F < 3.48 \rightarrow .05 < P$-value $< .10$
c. $3.48 < F < 5.99 \rightarrow .01 < P$-value $< .05$
d. $5.99 < F < 11.28 \rightarrow .001 < P$-value $< .01$
e. $F > 11.28 \rightarrow P$-value $< .001$

Thus, if $F = 7.12$, then $.001 < P$-value $< .01$. If a test with $\alpha < .05$ is used, H_0 should be rejected, because P-value $\leq \alpha$. The most frequently used statistical computer packages can provide exact P-values for F tests.

The Single-Factor ANOVA F Test

Null hypothesis: $H_0: \mu_1 = \mu_2 = \cdots = \mu_k$

Test statistic: $F = \dfrac{\text{MSTr}}{\text{MSE}}$

When H_0 is true and the ANOVA assumptions are reasonable, F has an F distribution with $df_1 = k - 1$ and $df_2 = N - k$.

Values of F more contradictory to H_0 than what was calculated are values even farther out in the upper tail, so the P-value is the area captured in the upper tail of the corresponding F curve.

Example 15.3 Heart Attacks Revisited

The two mean squares for the MPV data given in Example 15.1 were calculated in Example 15.2 as

$$\text{MSTr} = 4.400 \qquad \text{MSE} = 0.749$$

The value of the *F* statistic is then

$$F = \frac{\text{MSTr}}{\text{MSE}} = \frac{4.400}{0.749} = 5.87$$

with $df_1 = k - 1 = 3$ and $df_2 = N - k = 140 - 4 = 136$. Using $df_1 = 3$ and $df_2 = 120$ (the closest value to 136 that appears in the table), Appendix Table 6 shows that 5.78 captures tail area .001. Since 5.87 > 5.78, it follows that *P*-value = captured tail area < .001. The *P*-value is smaller than any reasonable α, so there is compelling evidence for rejecting $H_0: \mu_1 = \mu_2 = \cdots = \mu_4$. We can conclude that the true mean MPV is not the same for all four patient populations. Techniques for determining which means differ are introduced in Section 15.2.

■

Example 15.4 Hormones and Body Fat

● The article "Growth Hormone and Sex Steroid Administration in Healthy Aged Women and Men" (*Journal of the American Medical Association* [2002]: 2282–2292) described an experiment to investigate the effect of four treatments on various body characteristics. In this double-blind experiment, each of 57 female subjects who were over age 65 was assigned at random to one of the following four treatments: (1) placebo "growth hormone" and placebo "steroid" (denoted by P + P), (2) placebo "growth hormone" and the steroid estradiol (denoted by P + S), (3) growth hormone and placebo "steroid" (denoted by G + P), and (4) growth hormone and the steroid estradiol (denoted by G + S).

The following table lists data on change in body fat mass over the 26-week period following the treatments that are consistent with summary quantities given in the article

Treatment	Change In Body Fat Mass (kg)			
	P + P	P + S	G + P	G + S
	0.1	−0.1	−1.6	−3.1
	0.6	0.2	−0.4	−3.2
	2.2	0.0	0.4	−2.0
	0.7	−0.4	−2.0	−2.0
	−2.0	−0.9	−3.4	−3.3
	0.7	−1.1	−2.8	−0.5
	0.0	1.2	−2.2	−4.5
	−2.6	0.1	−1.8	−0.7
	−1.4	0.7	−3.3	−1.8
	1.5	−2.0	−2.1	−2.3
	2.8	−0.9	−3.6	−1.3
	0.3	−3.0	−0.4	−1.0

(*continued*)

	Change in Body Fat Mass (kg)			
Treatment	**P + P**	**P + S**	**G + P**	**G + S**
	−1.0	1.0	−3.1	−5.6
	−1.0	1.2		−2.9
				−1.6
				−0.2
n	14	14	13	16
$\bar{x}$	0.064	−0.286	−2.023	−2.250
s	1.545	1.218	1.264	1.468
s^2	2.387	1.484	1.598	2.155

Also, $N = 57$, grand total $= -65.4$, and $\bar{\bar{x}} = \dfrac{-65.4}{57} = -1.15$.

Let's carry out an F test to see whether true mean change in body fat mass differs for the four treatments.

1. Let μ_1, μ_2, μ_3, and μ_4 denote the true mean change in body fat for treatments P + P, P + S, G + P, and G + S, respectively.
2. H_0: $\mu_1 = \mu_2 = \mu_3 = \mu_4$
3. H_a: At least two among μ_1, μ_2, μ_3, and μ_4 are different.
4. Significance level: $\alpha = .01$
5. Test statistic: $F = \dfrac{\text{MSTr}}{\text{MSE}}$
6. Assumptions: Figure 15.5 shows boxplots for the four samples. The boxplots are roughly symmetric, and there are no outliers. The largest standard deviation ($s_1 = 1.545$) is not more than twice as big as the smallest ($s_2 = 1.264$). The subjects were randomly assigned to treatments. The assumptions of ANOVA are reasonable.
7. Computation:

$$\begin{aligned} \text{SSTr} &= n_1(\bar{x}_1 - \bar{\bar{x}})^2 + n_2(\bar{x}_2 - \bar{\bar{x}})^2 + \cdots + n_k(\bar{x}_k - \bar{\bar{x}})^2 \\ &= 14(0.064 - (-1.15))^2 + 14(-0.286 - (-1.15))^2 \\ &\quad + 13(-2.023 - (-1.15))^2 + 16(-2.250 - (-1.15))^2 \\ &= 60.37 \end{aligned}$$

Figure 15.5 Boxplots for the data of Example 15.4.

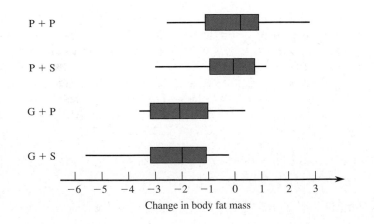

Change in body fat mass

treatment df $= k - 1 = 3$

$$SSE = (n_1 - 1)s_1^2 + (n_2 - 1)s_2^2 + \cdots + (n_k - 1)s_k^2$$
$$= 13(2.387) + 13(1.484) + 12(1.598) + 15(2.155)$$
$$= 101.81$$

error df $= N - k = 57 - 4 = 53$

Thus,

$$F = \frac{MSTr}{MSE} = \frac{SSTr/\text{treatment df}}{SSE/\text{error df}} = \frac{60.37/3}{101.81/53} = \frac{20.12}{1.92} = 10.48$$

8. *P*-value: Appendix Table 6 shows that for $df_1 = 3$ and $df_2 = 60$ (the closest tabled df to df $= 53$), the value 6.17 captures upper-tail area .001. Because $F = 10.48 > 6.17$, it follows *P*-value $< .001$.

9. Conclusion: Since *P*-value $\leq \alpha$, we reject H_0. The mean change in body fat mass is not the same for all four treatments.

■

■ Summarizing an ANOVA ....

ANOVA calculations are often summarized in a tabular format called an ANOVA table. To understand such a table, we must define one more sum of squares.

Total sum of squares, denoted by **SSTo**, is given by

$$SSTo = \sum_{\text{all } N \text{ obs.}} (x - \bar{\bar{x}})^2$$

with associated df $= N - 1$.

The relationship between the three sums of squares is

$$SSTo = SSTr + SSE$$

which is often called the *fundamental identity for single-factor ANOVA.*

■

The quantity SSTo, the sum of squared deviations about the grand mean, is a measure of total variability in the data set consisting of all k samples. The quantity SSE results from measuring variability separately within each sample and then combining as indicated in the formula for SSE. Such within-sample variability is present regardless of whether or not H_0 is true. The magnitude of SSTr, on the other hand, has much to do with the status of H_0 (whether it is true or false). The more the μ's differ from one another, the larger SSTr will tend to be. Thus, SSTr represents variation that can (at least to some extent) be explained by any differences between means. An informal paraphrase of the fundamental identity for single-factor ANOVA is

total variation = explained variation + unexplained variation

Once any two of the sums of squares have been calculated, the remaining one is easily obtained from the fundamental identity. Often SSTo and SSTr are calculated first (using computational formulas given in the online appendix to this chapter), and then SSE is obtained by subtraction: SSE = SSTo − SSTr. All the degrees of freedom, sums of squares, and mean squares are entered in an ANOVA table, as displayed in Table 15.2. The *P*-value usually appears to the right of *F* when the analysis is done by a statistical software package.

Table 15.2   General Format for a Single-Factor ANOVA Table

Source of Variation	df	Sum of Squares	Mean Square	F
Treatments	$k - 1$	SSTr	$MSTr = \dfrac{SSTr}{k - 1}$	$F = \dfrac{MSTr}{MSE}$
Error	$N - k$	SSE	$MSE = \dfrac{SSE}{N - k}$	
Total	$N - 1$	SSTo		

An ANOVA table from MINITAB for the change in body fat mass data of Example 15.4 is shown in Table 15.3. The reported *P*-value is .000, consistent with our previous conclusion that *P*-value < .001.

Table 15.3   An ANOVA Table from MINITAB for the Data of Example 15.4

One-way ANOVA

Source	DF	SS	MS	F	P
Factor	3	60.37	20.12	10.48	0.000
Error	53	101.81	1.92		
Total	56	162.18			

  Exercises 15.1–15.22   ....   Turn to the Expanded Answers Section for answers not shown next to exercises.

15.1 Give as much information as you can about the *P*-value for an upper-tailed *F* test in each of the following situations.
a. $df_1 = 4$, $df_2 = 15$, $F = 5.37$ $0.001 < P\text{-value} < 0.01$
b. $df_1 = 4$, $df_2 = 15$, $F = 1.90$ $P\text{-value} > 0.10$
c. $df_1 = 4$, $df_2 = 15$, $F = 4.89$ $P\text{-value} = 0.01$
d. $df_1 = 3$, $df_2 = 20$, $F = 14.48$ $P\text{-value} < 0.001$
e. $df_1 = 3$, $df_2 = 20$, $F = 2.69$ $0.05 < P\text{-value} < 0.10$
f. $df_1 = 4$, $df_2 = 50$, $F = 3.24$ $0.01 < P\text{-value} < 0.05$
(using $df_1 = 4$ and $df_2 = 60$)

15.2 Give as much information as you can about the *P*-value of the single-factor ANOVA *F* test in each of the following situations.
a. $k = 5$, $n_1 = n_2 = n_3 = n_4 = n_5 = 4$, $F = 5.37$
b. $k = 5$, $n_1 = n_2 = n_3 = 5$, $n_4 = n_5 = 4$, $F = 2.83$
c. $k = 3$, $n_1 = 4$, $n_2 = 5$, $n_3 = 6$, $F = 5.02$
d. $k = 3$, $n_1 = n_2 = 4$, $n_3 = 6$, $F = 15.90$
e. $k = 4$, $n_1 = n_2 = 15$, $n_3 = 12$, $n_4 = 10$, $F = 1.75$
$P\text{-value} > 0.10$

Bold exercises answered in back   ● Data set available online but not required   ▼ Video solution available

15.3 Employees of a certain state university system can choose from among four different health plans. Each plan differs somewhat from the others in terms of hospitalization coverage. Four samples of recently hospitalized individuals were selected, each sample consisting of people covered by a different health plan. The length of the hospital stay (number of days) was determined for each individual selected.

a. What hypotheses would you test to decide whether average length of stay was related to health plan? (*Note*: Carefully define the population characteristics of interest.)

b. If each sample consisted of eight individuals and the value of the ANOVA *F* statistic was $F = 4.37$, what conclusion would be appropriate for a test with $\alpha = .01$?

c. Answer the question posed in Part (b) if the *F* value given there resulted from sample sizes $n_1 = 9$, $n_2 = 8$, $n_3 = 7$, and $n_4 = 8$.   0.01 < *P*-value < 0.05, fail to reject H_0

15.4 ● The experiment described in Example 15.4 also gave data on change in body fat mass for men ("Growth Hormone and Sex Steroid Administration in Healthy Aged Women and Men," *Journal of the American Medical Association* [2002]: 2282–2292). Each of 74 male subjects who were over age 65 was assigned at random to one of the following four treatments: (1) placebo "growth hormone" and placebo "steroid" (denoted by P + P), (2) placebo "growth hormone" and the steroid testosterone (denoted by P + S), (3) growth hormone and placebo "steroid" (denoted by G + P), and (4) growth hormone and the steroid testosterone (denoted by G + S). The accompanying table lists data on change in body fat mass over the 26-week period following the treatment that are consistent with summary quantities given in the article

Change in Body Fat Mass (kg)

Treatment	P + P	P + S	G + P	G + S
	0.3	−3.7	−3.8	−5.0
	0.4	−1.0	−3.2	−5.0
	−1.7	0.2	−4.9	−3.0
	−0.5	−2.3	−5.2	−2.6
	−2.1	1.5	−2.2	−6.2
	1.3	−1.4	−3.5	−7.0
	0.8	1.2	−4.4	−4.5
	1.5	−2.5	−0.8	−4.2
	−1.2	−3.3	−1.8	−5.2
	−0.2	0.2	−4.0	−6.2
	1.7	0.6	−1.9	−4.0
	1.2	−0.7	−3.0	−3.9

Change in Body Fat Mass (kg)

Treatment	P + P	P + S	G + P	G + S
	0.6	−0.1	−1.8	−3.3
	0.4	−3.1	−2.9	−5.7
	−1.3	0.3	−2.9	−4.5
	−0.2	−0.5	−2.9	−4.3
	0.7	−0.8	−3.7	−4.0
		−0.7		−4.2
		−0.9		−4.7
		−2.0		
		−0.6		
n	17	21	17	19
$\bar{x}$	0.100	−0.933	−3.112	−4.605
s	1.139	1.443	1.178	1.122
s^2	1.297	2.082	1.388	1.259

Also, $N = 74$, grand total $= -158.3$, and $\bar{\bar{x}} = \dfrac{-158.3}{74} = -2.139$. Carry out an *F* test to see whether true mean change in body fat mass differs for the four treatments. $F = 53.8$, *P*-value < 0.001, reject H_0

15.5 ● The article "Compression of Single-Wall Corrugated Shipping Containers Using Fixed and Floating Text Platens" (*Journal of Testing and Evaluation* [1992]: 318–320) described an experiment in which several different types of boxes were compared with respect to compression strength (in pounds). The data at the top of page 796 resulted from a single-factor experiment involving $k = 4$ types of boxes (the sample means and standard deviations are in close agreement with values given in the paper). Do these data provide evidence to support the claim that the mean compression strength is not the same for all four box types? Test the relevant hypothesis using a significance level of .01. $F = 25.09$, *P*-value < 0.001, reject H_0

15.6 ● The article "The Soundtrack of Recklessness: Musical Preferences and Reckless Behavior Among Adolescents" (*Journal of Adolescent Research* [1992]: 313–331) described a study whose purpose was to determine whether adolescents who preferred certain types of music reported higher rates of reckless behaviors, such as speeding, drug use, shoplifting, and unprotected sex. Independently chosen random samples were selected from each of four groups of students with different musical preferences at a large high school: (1) acoustic/pop, (2) mainstream rock,

Table for Exercise 15.5

Type of Box	Compression Strength (lb)						Sample Mean	Sample SD
1	655.5	788.3	734.3	721.4	679.1	699.4	713.00	46.55
2	789.2	772.5	786.9	686.1	732.1	774.8	756.93	40.34
3	737.1	639.0	696.3	671.7	717.2	727.1	698.07	37.20
4	535.1	628.7	542.4	559.0	586.9	520.0	562.02	39.87
							$\overline{\overline{x}} = 682.50$	

(3) hard rock, and (4) heavy metal. Each student in these samples was asked how many times he or she had engaged in various reckless activities during the last year. The following table lists data and summary quantities on driving over 80 mph that is consistent with summary quantities given in the article (the sample sizes in the article were much larger, but for the purposes of this exercise, we use $n_1 = n_2 = n_3 = n_4 = 20$):

Number of Times Driving Over 80 mph by Musical Preference

	Acoustic/ Pop	Main-stream Rock	Hard Rock	Heavy Metal
	2	3	3	4
	3	2	4	3
	4	1	3	4
	1	2	1	3
	3	3	2	3
	3	4	1	3
	3	3	4	3
	3	2	2	3
	2	4	2	2
	2	4	2	4
	1	4	3	4
	3	4	3	5
	2	2	4	4
	2	3	3	5
	2	2	3	3
	3	2	2	4
	2	2	3	5
	2	3	4	4
	3	1	2	2
	4	3	4	3
n	20	20	20	20
$\overline{x}$	2.50	2.70	2.75	3.55
s	.827	.979	.967	.887
s^2	.6839	.9584	.9351	.7868

Also, $N = 80$, grand total $= 230.0$, and $\overline{\overline{x}} = 230.0/80 = 2.875$. Carry out an F test to determine if these data provide convincing evidence that the true mean number of times driving over 80 mph varies with musical preference. $F = 5.09, 0.001 < P\text{-value} < 0.01$, reject H_0

15.7 ● Suppose that a random sample of size $n = 5$ was selected from the vineyard properties for sale in Sonoma County, California, in each of three years. The following data are consistent with summary information on price per acre (in dollars, rounded to the nearest thousand) for disease-resistant grape vineyards in Sonoma County (*Wines and Vines*, November 1999).

1996	30,000	34,000	36,000	38,000	40,000
1997	30,000	35,000	37,000	38,000	40,000
1998	40,000	41,000	43,000	44,000	50,000

a. Construct boxplots for each of the three years on a common axis, and label each by year. Comment on the similarities and differences.

b. Carry out an ANOVA to determine whether there is evidence to support the claim that the mean price per acre for vineyard land in Sonoma County was not the same for the three years considered. Use a significance level of .05 for your test. $F = 6.83, 0.01 < P\text{-value} < 0.05$, reject H_0

15.8 The article "An Analysis of Job Sharing, Full-Time, and Part-Time Arrangements" (*American Business Review* [1989]: 34–40) reported on a study of hospital employees from three different groups. Each employee reported a level of satisfaction with his or her work schedule (1 = very dissatisfied to 7 = very satisfied), as shown in the accompanying table.

Group	Sample Size	Sample Mean
1. Job sharers	24	6.60
2. Full-time employees	24	5.37
3. Part-time employees	20	5.20

a. What are numerator and denominator df's for the *F* test?
b. The test statistic value is $F = 6.62$. Use a test with significance level .05 to decide whether it is plausible that true average satisfaction levels are identical for the three groups. $F = 6.62, 0.001 < P\text{-value} < 0.01$, reject H_0
c. What is the value of MSE?   MSE = 2.028

15.9 ▼ High productivity and carbohydrate storage ability of the Jerusalem artichoke make it a promising agricultural crop. The article "Leaf Gas Exchange and Tuber Yield in Jerusalem Artichoke Cultivars" (*Field Crops Research* [1991]: 241–252) reported on various plant characteristics. Consider the accompanying data on chlorophyll concentration (gm/m^2) for four varieties of Jerusalem artichoke:

Variety	BI	RO	WA	TO
Sample mean	.30	.24	.41	.33

Suppose that the sample sizes were 5, 5, 4, and 6, respectively, and also that MSE = .0130. Do the data suggest that true average chlorophyll concentration depends on the variety? State and test the appropriate hypotheses using a significance level of .05. $F = 1.71, P\text{-value} > 0.10$, fail to reject H_0

15.10 It has been reported that varying work schedules can lead to a variety of health problems for workers. The article "Nutrient Intake in Day Workers and Shift Workers" (*Work and Stress* [1994]: 332–342) reported on blood glucose levels (mmol/L) for day-shift workers and workers on two different types of rotating shifts. The sample sizes were $n_1 = 37$ for the day shift, $n_2 = 34$ for the second shift, and $n_3 = 25$ for the third shift. A single-factor ANOVA resulted in $F = 3.834$. At a significance level of .05, does true average blood glucose level appear to depend on the type of shift? $0.01 < P\text{-value} < 0.05$, reject H_0

15.11 In the introduction to this chapter, we considered a study comparing three groups of college students (soccer athletes, nonsoccer athletes, and a control group consisting of students who did not participate in intercollegiate sports). The following information on scores from the Hopkins Verbal Learning Test (which measures immediate memory recall) was

Group	Soccer Athletes	Nonsoccer Athletes	Control
Sample size	86	95	53
Sample mean score	29.90	30.94	29.32
Sample standard deviation	3.73	5.14	3.78

In addition, $\bar{\bar{x}} = 30.19$. Suppose that it is reasonable to regard these three samples as random samples from the three student populations of interest. Is there sufficient evidence to conclude that the mean Hopkins score is not the same for the three student populations? Use $\alpha = .05$.
$F = 2.62, 0.05 < P\text{-value} < 0.10$, fail to reject H_0

15.12 The article "Utilizing Feedback and Goal Setting to Increase Performance Skills of Managers" (*Academy of Management Journal* [1979]: 516–526) reported the results of an experiment to compare three different interviewing techniques for employee evaluations. One method allowed the employee being evaluated to discuss previous evaluations, the second involved setting goals for the employee, and the third did not allow either feedback or goal setting. After the interviews were concluded, the evaluated employee was asked to indicate how satisfied he or she was with the interview. (A numerical scale was used to quantify level of satisfaction.) The authors used ANOVA to compare the three interview techniques. An *F* statistic value of 4.12 was reported.
a. Suppose that a total of 33 subjects were used, with each technique applied to 11 of them. Use this information to conduct a level .05 test of the null hypothesis of no difference in mean satisfaction level for the three interview techniques. $F = 4.12, 0.01 < P\text{-value} < 0.05$, reject H_0
b. The actual number of subjects on which each technique was used was 45. After studying the *F* table, explain why the conclusion in Part (a) still holds.

15.13 An investigation carried out to study the toxic effects of mercury was described in the article "Comparative Responses of the Action of Different Mercury Compounds on Barley" (*International Journal of Environmental Studies* [1983]: 323–327). Ten different concentrations of mercury (0, 1, 5, 10, 50, 100, 200, 300, 400, and 500 mg/L) were compared with respect to their effects on average dry weight (per 100 seven-day-old seedlings). The basic experiment was replicated four times for a total of 40 dry-weight observations (four for each treatment level). The article reported an ANOVA *F* statistic value of 1.895. Using a significance level of .05, test the null hypothesis that the true mean dry weight is the same for all 10 concentration levels. $F = 1.895, .05 < P\text{-value} < 0.10$, fail to reject H_0

15.14 ● The accompanying data on calcium content of wheat are consistent with summary quantities that appeared in the article "Mineral Contents of Cereal Grains as Affected by Storage and Insect Infestation" (*Journal of*

Stored Products Research [1992]: 147–151). Four different storage times were considered. Partial output from the SAS computer package is also shown.

Storage Period	Observations					
0 months	58.75	57.94	58.91	56.85	55.21	57.30
1 month	58.87	56.43	56.51	57.67	59.75	58.48
2 months	59.13	60.38	58.01	59.95	59.51	60.34
4 months	62.32	58.76	60.03	59.36	59.61	61.95

Dependent Variable: CALCIUM

Source	DF	Sum of Squares	Mean Square	F Value	Pr>F
Model	3	32.13815000	10.71271667	6.51	0.0030
Error	20	32.90103333	1.64505167		
Corrected Total	23	65.03918333			

R-Square	C.V.	Root MSE	CALCIUM Mean
0.494135	2.180018	1.282596	58.8341667

a. Verify that the sums of squares and df's are as given in the ANOVA table.

b. Is there sufficient evidence to conclude that the mean calcium content is not the same for the four different storage times? Use the value of F from the ANOVA table to test the appropriate hypotheses at significance level .05.
$F = 6.51$, P-value $= 0.003$, reject H_0

15.15 In an experiment to investigate the performance of four different brands of spark plugs intended for use on a 125-cc motorcycle, five plugs of each brand were tested, and the number of miles (at a constant speed) until failure was observed. A partially completed ANOVA table is given. Fill in the missing entries, and test the relevant hypotheses using a .05 level of significance.

Source of Variation	df	Sum of Squares	Mean Square	F
Treatments				
Error		235,419.04		
Total		310,500.76		

15.16 The partially completed ANOVA table given in this problem is taken from the article "Perception of Spatial Incongruity" (*Journal of Nervous and Mental Disease* [1961]: 222), in which the abilities of three different groups to identify a perceptual incongruity were assessed and compared. All individuals in the experiment had been hospitalized to undergo psychiatric treatment. There were 21 individuals in the depressive group, 32 individuals in the functional "other" group, and 21 individuals in the brain-damaged group. Complete the ANOVA table. Carry out the appropriate test of hypothesis (use $\alpha = .01$), and interpret your results.

Source of Variation	df	Sum of Squares	Mean Square	F
Treatments			76.09	
Error				
Total		1123.14		

15.17 Research carried out to investigate the relationship between smoking status of workers and short-term absenteeism rate (hr/mo) yielded the accompanying summary information ("Work-Related Consequences of Smoking Cessation," *Academy of Management Journal* [1989]: 606–621). In addition, $F = 2.56$. Construct an ANOVA table, and then state and test the appropriate hypotheses using a .01 significance level.

Status	Sample Size	Sample Mean
Continuous smoker	96	2.15
Recent ex-smoker	34	2.21
Long-term ex-smoker	86	1.47
Never smoked	206	1.69

15.18 An investigation carried out to study purchasers of luxury automobiles reported data on a number of different attributes that might affect purchase decisions, including comfort, safety, styling, durability, and reliability ("Measuring Values Can Sharpen Segmentation in the Luxury Car Market," *Journal of Advertising Research* [1995]: 9–22). Here is summary information on the level of importance of speed, rated on a seven-point scale:

Type of Car	American	German	Japanese
Sample size	58	38	59
Sample mean rating	3.05	2.87	2.67

In addition, SSE = 459.04. Carry out a hypothesis test to determine if there is sufficient evidence to conclude that the mean importance rating of speed is not the same for owners of these three types of cars. $F = 0.70$, *P*-value > 0.10, fail to reject H_0

15.19 ● The Gunning Fog index is a measure of reading difficulty based on the average number of words per sentence and the percentage of words with three or more syllables. High values of the Gunning Fog index are associated with difficult reading levels. Independent random samples of six advertisements were taken from three different magazines, and Gunning Fog indices were computed to obtain the data given in the accompanying table ("Readability Levels of Magazine Advertisements," *Journal of Advertising Research* [1981]: 45–50). Construct an ANOVA table, and then use a significance level of .01 to test the null hypothesis of no difference between the mean Gunning Fog index levels for advertisements appearing in the three magazines. $F = 6.97$, $0.001 < $ *P*-value $ < 0.01$, reject H_0

Scientific American	15.75	11.55	11.16	9.92	9.23	8.20
Fortune	12.63	11.46	10.77	9.93	9.87	9.42
New Yorker	9.27	8.28	8.15	6.37	6.37	5.66

15.20 ● Some investigators think that the concentration (mg/mL) of a particular antigen in supernatant fluids could be related to onset of meningitis in infants. The accompanying data are typical of that given in plots in the article "Type-Specific Capsular Antigen Is Associated with Virulence in Late-Onset Group B Streptococcal Type III Disease" (*Infection and Immunity* [1984]: 124–129). Construct an ANOVA table, and use it to test the null hypothesis of no difference in mean antigen concentrations for the three groups.

Asymptomatic infants	1 56	1.06	0.87	1.39	0.71	0.87	
Infants with late onset sepsis	1.51	1.78	1.45	1.13	1.87	1.89	1.07 1.72
Infants with late onset meningitis	1.21	1.34	1.95	2.27	0.88	1.67	2.57

15.21 ● Parents are frequently concerned when their child seems slow to begin walking (although when the child finally walks, the resulting havoc sometimes has the parents wishing they could turn back the clock!). The article "Walking in the Newborn" (*Science*, 176 [1972]: 314–315)

reported on an experiment in which the effects of several different treatments on the age at which a child first walks were compared. Children in the first group were given special walking exercises for 12 min per day beginning at age 1 week and lasting 7 weeks. The second group of children received daily exercises but not the walking exercises administered to the first group. The third and fourth groups were control groups: They received no special treatment and differed only in that the third group's progress was checked weekly, whereas the fourth group's progress was checked just once at the end of the study. Observations on age (in months) when the children first walked are shown in the accompanying table. Also given is the ANOVA table, obtained from the SPSS computer package.

	Age			*n*	Total
Treatment 1	9.00	9.50	9.75	6	60.75
	10.00	13.00	9.50		
Treatment 2	11.00	10.00	10.00	6	68.25
	11.75	10.50	15.00		
Treatment 3	11.50	12.00	9.00	6	70.25
	11.50	13.25	13.00		
Treatment 4	13.25	11.50	12.00	5	61.75
	13.50	11.50			

Analysis of Variance

Source	df	Sum of sq.	Mean Sq.	F Ratio	F Prob
Between Groups	3	14.778	4.926	2.142	.129
Within Groups	19	43.690	2.299		
Total	22	58.467			

a. Verify the entries in the ANOVA table.
b. State and test the relevant hypotheses using a significance level of .05. $F = 2.142$, *P*-value > 0.10, fail to reject H_0

15.22 The article "Heavy Drinking and Problems Among Wine Drinkers" (*Journal of Studies on Alcohol* [1999]: 467–471) analyzed drinking problems among Canadians. For each of several different groups of drinkers, the mean and standard deviation of "highest number of drinks consumed" were calculated:

	$\bar{x}$	*s*	*n*
Beer only	7.52	6.41	1256
Wine only	2.69	2.66	1107
Spirits only	5.51	6.44	759

(*continued*)

	$\bar{x}$	s	n
Beer and wine	5.39	4.07	1334
Beer and spirits	9.16	7.38	1039
Wine and spirits	4.03	3.03	1057
Beer, wine, and spirits	6.75	5.49	2151

Assume that each of the seven samples studied can be viewed as a random sample for the respective group. Is there sufficient evidence to conclude that the mean value of highest number of drinks consumed is not the same for all seven groups? $F = 186.524$, P-value < 0.001, reject H_0

Bold exercises answered in back   ● Data set available online but not required   ▼ Video solution available

15.2 Multiple Comparisons

When $H_0: \mu_1 = \mu_2 = \cdots = \mu_k$ is rejected by the F test, we believe that there are differences among the k population or treatment means. A natural question to ask at this point is, Which means differ? For example, with $k = 4$, it might be the case that $\mu_1 = \mu_2 = \mu_4$, with μ_3 different from the other three means. Another possibility is that $\mu_1 = \mu_4$ and $\mu_2 = \mu_3$. Still another possibility is that all four means are different from one another. A **multiple comparisons procedure** is a method for identifying differences among the μ's once the hypothesis of overall equality has been rejected. We present one such method, the **Tukey–Kramer** (T–K) multiple comparisons procedure.

The T–K procedure is based on computing confidence intervals for the difference between each possible pair of μ's. For $k = 3$, there are three differences to consider:

$$\mu_1 - \mu_2 \qquad \mu_1 - \mu_3 \qquad \mu_2 - \mu_3$$

(The difference $\mu_2 - \mu_1$ is not considered, because the interval for $\mu_1 - \mu_2$ provides the same information. Similarly, intervals for $\mu_3 - \mu_1$ and $\mu_3 - \mu_2$ are not necessary.) Once all confidence intervals have been computed, each is examined to determine whether the interval includes 0. If a particular interval does not include 0, the two means are declared "significantly different" from one another. An interval that does include 0 supports the conclusion that there is no significant difference between the means involved.

Suppose, for example, that $k = 3$ and that the three confidence intervals are

Difference	T-K Interval
$\mu_1 - \mu_2$	$(-.9, 3.5)$
$\mu_1 - \mu_3$	$(2.6, 7.0)$
$\mu_2 - \mu_3$	$(1.2, 5.7)$

Generally, "studentized" refers to standardizing a statistic by dividing that statistic by an estimate of the population standard deviation. Though "studentized" is named after Student of Student's t, the Studentized range is not the t procedure confidence interval.

Because the interval for $\mu_1 - \mu_2$ includes 0, we judge that μ_1 and μ_2 do not differ significantly. The other two intervals do not include 0, so we conclude that $\mu_1 \neq \mu_3$ and $\mu_2 \neq \mu_3$.

The T–K intervals are based on critical values for a probability distribution called the *Studentized range distribution*. These critical values appear in Appendix Table 7. To find a critical value, enter the table at the column corresponding to the number of populations or treatments being compared, move down to the rows corresponding to the number of error degrees of freedom, and select either the value for a 95% confidence level or the one for a 99% level.

The Tukey–Kramer Multiple Comparison Procedure

When there are k populations or treatments being compared, $\dfrac{k(k-1)}{2}$ confidence intervals must be computed. Denoting the relevant Studentized range critical value by q, the intervals are as follows:

$$\text{For } \mu_i - \mu_j: \quad (\bar{x}_i - \bar{x}_j) \pm q\sqrt{\dfrac{\text{MSE}}{2}\left(\dfrac{1}{n_i} + \dfrac{1}{n_j}\right)}$$

Two means are judged to differ significantly if the corresponding interval does not include zero.

If the sample sizes are all the same, let n denote the common value of $n_1, \ldots, n_k$. In this case, the $\pm$ factor for each interval is the same quantity

$$q\sqrt{\dfrac{\text{MSE}}{n}}$$

Example 15.5 Hormones and Body Fat Revisited

● Example 15.4 introduced the accompanying data on change in body fat mass resulting from a double-blind experiment designed to compare the following four treatments: (1) placebo "growth hormone" and placebo "steroid" (denoted by P + P), (2) placebo "growth hormone" and the steroid estradiol (denoted by P + S), (3) growth hormone and placebo "steroid" (denoted by G + P), and (4) growth hormone and the steroid estradiol (denoted by G + S).

Treatment	Change in Body Fat Mass (kg)			
	P + P	P + S	G + P	G + S
	0.1	−0.1	−1.6	−3.1
	0.6	0.2	−0.4	−3.2
	2.2	0.0	0.4	−2.0
	0.7	−0.4	−2.0	−2.0
	−2.0	−0.9	−3.4	−3.3
	0.7	−1.1	−2.8	−0.5
	0.0	1.2	−2.2	−4.5
	−2.6	0.1	−1.8	−0.7
	−1.4	0.7	−3.3	−1.8
	1.5	−2.0	−2.1	−2.3
	2.8	−0.9	−3.6	−1.3
	0.3	−3.0	−0.4	−1.0

(continued)

● Data set available online

Treatment	Change in Body Fat Mass (kg)			
	P + P	**P + S**	**G + P**	**G + S**
	−1.0	1.0	−3.1	−5.6
	−1.0	1.2		−2.9
				−1.6
				−0.2
n	14	14	13	16
$\overline{x}$	0.064	−0.286	−2.023	−2.250
s	1.545	1.218	1.264	1.468
s^2	2.387	1.484	1.598	2.155

From Example 15.4, MSTr = 20.12, MSE = 1.92, and F = 10.48 with an associated P-value $<$ .001. We concluded that the mean change in body fat mass is not the same for all four treatments.

Appendix Table 7 gives the 95% Studentized range critical value q = 3.74 (using error df = 60, the closest tabled value to df = $N - k$ = 53). The first two T–K intervals are

$$\mu_1 - \mu_2: (0.064 - (-0.286)) \pm 3.74 \sqrt{\left(\frac{1.92}{2}\right)\left(\frac{1}{14} + \frac{1}{14}\right)}$$

$$= 0.35 \pm 1.39$$
$$= (-1.04, 1.74) \leftarrow \textit{Includes 0}$$

$$\mu_1 - \mu_3: (0.064 - (-2.023)) \pm 3.74 \sqrt{\left(\frac{1.92}{2}\right)\left(\frac{1}{14} + \frac{1}{13}\right)}$$

$$= 2.09 \pm 1.41$$
$$= (0.68, 3.50) \leftarrow \textit{Does not include 0}$$

The remaining intervals are

$\mu_1 - \mu_4$	(0.97, 3.66)	← *Does not include 0*
$\mu_2 - \mu_3$	(0.32, 3.15)	← *Does not include 0*
$\mu_2 - \mu_4$	(0.62, 3.31)	← *Does not include 0*
$\mu_3 - \mu_4$	(−1.145, 1.60)	← *Includes 0*

This numeric presentation of intervals should be followed by the graphic procedures such as the MINITAB output on page 803 or the Summary of Results on page 804.

We would conclude that μ_1 is not significantly different from μ_2 and that μ_3 is not significantly different from μ_4. We would also conclude that μ_1 and μ_2 are significantly different than μ_3 and μ_4. Note that Treatments 1 and 2 were treatments that administered a placebo in place of the growth hormone and Treatments 3 and 4 were treatments that included the growth hormone. This analysis was the basis of the researchers' conclusion that growth hormone, with or without sex steroids, decreased body fat mass.

■

MINITAB can be used to construct T–K intervals if raw data are available. Typical output (based on Example 15.5) is shown in Figure 15.6. From the output, we see

Tukey 95% Simultaneous Confidence Intervals
All Pairwise Comparisons

Individual confidence level = 98.95%

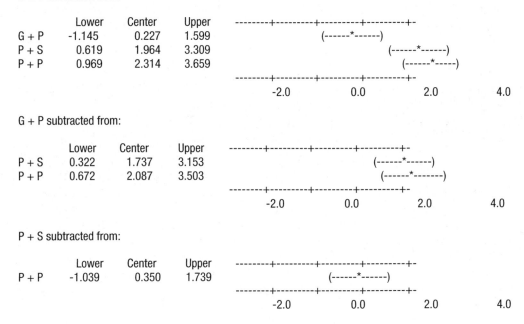

G + S subtracted from:

	Lower	Center	Upper
G + P	-1.145	0.227	1.599
P + S	0.619	1.964	3.309
P + P	0.969	2.314	3.659

G + P subtracted from:

	Lower	Center	Upper
P + S	0.322	1.737	3.153
P + P	0.672	2.087	3.503

P + S subtracted from:

	Lower	Center	Upper
P + P	-1.039	0.350	1.739

Figure 15.6 The T–K intervals for Example 15.5 (from MINITAB).

that the confidence interval for μ_1 (P + P) − μ_2 (P + S) is (−1.039, 1.739), that for μ_2 (P + S) − μ_4 (G + S) is (0.619, 3.309), and so on.

Why calculate the T–K intervals rather than use the t confidence interval for a difference between μ's from Chapter 11? The answer is that the T–K intervals control the **simultaneous confidence level** at approximately 95% (or 99%). That is, if the procedure is used repeatedly on many different data sets, in the long run only about 5% (or 1%) of the time would at least one of the intervals not include the value of what the interval is estimating. Consider using separate 95% t intervals, each one having a 5% error rate. Then the chance that at least one interval would make an incorrect statement about a difference in μ's increases dramatically with the number of intervals calculated. The MINITAB output in Figure 15.6 shows that to achieve a simultaneous confidence level of about 95% (experimentwise or "family" error rate of 5%) when $k = 4$ and error df = 76, the individual confidence level must be 98.95% (individual error rate 1.05%).

An effective display for summarizing the results of any multiple comparisons procedure involves listing the $\bar{x}$'s and underscoring pairs judged to be not significantly different. The process for constructing such a display is described in the accompanying box.

Summarizing the Results of the Tukey–Kramer Procedure

1. List the sample means in increasing order, identifying the corresponding population just above the value of each $\bar{x}$.

2. Use the T–K intervals to determine the group of means that do not differ significantly from the first in the list. Draw a horizontal line extending from the smallest mean to the last mean in the group identified. For example, if there are five means, arranged in order,

Population	3	2	1	4	5
Sample mean	$\bar{x}_3$	$\bar{x}_2$	$\bar{x}_1$	$\bar{x}_4$	$\bar{x}_5$

   and μ_3 is judged to be not significantly different from μ_2 or μ_1, but is judged to be significantly different from μ_4 and μ_5, draw the following line:

Population	3	2	1	4	5
Sample mean	$\bar{x}_3$	$\bar{x}_2$	$\bar{x}_1$	$\bar{x}_4$	$\bar{x}_5$

3. Use the T–K intervals to determine the group of means that are not significantly different from the second smallest. (You need consider only means that appear to the right of the mean under consideration.) If there is already a line connecting the second smallest mean with all means in the new group identified, no new line need be drawn. If this entire group of means is not underscored with a single line, draw a line extending from the second smallest to the last mean in the new group. Continuing with our example, if μ_2 is not significantly different from μ_1 but is significantly different from μ_4 and μ_5, no new line need be drawn. However, if μ_2 is not significantly different from either μ_1 or μ_4 but is judged to be different from μ_5, a second line is drawn as shown:

Population	3	2	1	4	5
Sample mean	$\bar{x}_3$	$\bar{x}_2$	$\bar{x}_1$	$\bar{x}_4$	$\bar{x}_5$

4. Continue considering the means in the order listed, adding new lines as needed.

To illustrate this summary procedure, suppose that four samples with $\bar{x}_1 = 19$, $\bar{x}_2 = 27$, $\bar{x}_3 = 24$, and $\bar{x}_4 = 10$ are used to test $H_0: \mu_1 = \mu_2 = \mu_3 = \mu_4$ and that this hypothesis is rejected. Suppose the T–K confidence intervals indicate that μ_2 is significantly different from both μ_1 and μ_4, and that there are no other significant differences. The resulting summary display would then be

Population	4	1	3	2
Sample mean	10	19	24	27

....

Example 15.6 Sleep Time

● A biologist wished to study the effects of ethanol on sleep time. A sample of 20 rats, matched for age and other characteristics, was selected, and each rat was given an oral injection having a particular concentration of ethanol per body weight.

! Step-by-step technology instructions available online　　　● Data set available online

The rapid eye movement (REM) sleep time for each rat was then recorded for a 24-hr period, with the results shown in the following table:

Treatment	Observations					$\bar{x}$
1. 0 (control)	88.6	73.2	91.4	68.0	75.2	79.28
2. 1 g/kg	63.0	53.9	69.2	50.1	71.5	61.54
3. 2 g/kg	44.9	59.5	40.2	56.3	38.7	47.92
4. 4 g/kg	31.0	39.6	45.3	25.2	22.7	32.76

Table 15.4 (an ANOVA table from SAS) leads to the conclusion that true average REM sleep time depends on the treatment used; the P-value for the F test is .0001.

Table 15.4 SAS ANOVA Table for Example 15.6

Analysis of Variance Procedure
Dependent Variable: TIME

Source	DF	Sum of Squares	Mean Square	F Value	Pr > F
Model	3	5882.35750	1960.78583	21.09	0.0001
Error	16	1487.40000	92.96250		
Total	**19**	**7369.75750**			

The T–K intervals are

Difference	Interval	Includes 0?
$\mu_1 - \mu_2$	17.74 ± 17.446	no
$\mu_1 - \mu_3$	31.36 ± 17.446	no
$\mu_1 - \mu_4$	46.24 ± 17.446	no
$\mu_2 - \mu_3$	13.08 ± 17.446	yes
$\mu_2 - \mu_4$	28.78 ± 17.446	no
$\mu_3 - \mu_4$	15.16 ± 17.446	yes

The only T–K intervals that include zero are those for $\mu_2 - \mu_3$ and $\mu_3 - \mu_4$. The corresponding underscoring pattern is

$\bar{x}_4$	$\bar{x}_3$	$\bar{x}_2$	$\bar{x}_1$
32.76	47.92	61.54	79.28

Figure 15.7 displays the SAS output that agrees with our underscoring; letters are used to indicate groupings in place of the underscoring.

Alpha = 0.05 df = 16 MSE = 92.9625
Critical Value of Studentized Range = 4.046
Minimum Significant Difference = 17.446
Means with the same letter are not significantly different.

Tukey Grouping		Mean	N	Treatment
	A	79.280	5	0 (control)
	B	61.540	5	1 g/kg
C	B	47.920	5	2 g/kg
C		32.760	5	4 g/kg

Figure 15.7 SAS output for Example 15.6.

Example 15.7 Roommate Satisfaction

How satisfied are college students with dormitory roommates? The article "Roommate Satisfaction and Ethnic Identity in Mixed-Race and White University Roommate Dyads" (*Journal of College Student Development* [1998]: 194–199) investigated differences among randomly assigned African American/white, Asian/white, Hispanic/white, and white/white roommate pairs. The researchers used a one-way ANOVA to analyze scores on the Roommate Relationship Inventory to see whether a difference in mean score existed for the four types of roommate pairs. They reported "significant differences among the means ($P < .01$). Follow-up Tukey [intervals] . . . indicated differences between White dyads ($M = 77.49$) and African American/White dyads ($M = 71.27$). . . . No other significant differences were found."

Although the mean satisfaction score for the Asian/white and Hispanic/white groups were not given, they must have been between 77.49 (the mean for the white/white pairs) and 71.27 (the mean for the African American/white pairs). (If they had been larger than 77.49, they would have been significantly different from the African American/white pairs mean, and if they had been smaller than 71.27, they would have been significantly different from the white/white pairs mean.) An underscoring consistent with the reported information is

White/White   Hispanic/White and Asian/White   African-American/White

▪

Exercises 15.23–15.34 Turn to the Expanded Answers Section for answers not shown next to exercises.

15.23 Leaf surface area is an important variable in plant gas-exchange rates. The article "Fluidized Bed Coating of Conifer Needles with Glass Beads for Determination of Leaf Surface Area" (*Forest Science* [1980]: 29–32) included an analysis of dry matter per unit surface area (mg/cm^3) for trees raised under three different growing conditions. Let μ_1, μ_2, and μ_3 represent the true mean dry matter per unit surface area for the growing conditions 1, 2, and 3, respectively. The given 95% simultaneous confidence intervals are based on summary quantities that appear in the article:

Difference	$\mu_1 - \mu_2$	$\mu_1 - \mu_3$	$\mu_2 - \mu_3$
Interval	$(-3.11, -1.11)$	$(-4.06, -2.06)$	$(-1.95, .05)$

Which of the following four statements do you think describes the relationship between μ_1, μ_2, and μ_3? Explain your choice.

a. $\mu_1 = \mu_2$, and μ_3 differs from μ_1 and μ_2.
b. $\mu_1 = \mu_3$, and μ_2 differs from μ_1 and μ_3.
c. $\mu_2 = \mu_3$, and μ_1 differs from μ_2 and μ_3.
d. All three μ's are different from one another.

15.24 The degree of success at mastering a skill often depends on the method used to learn the skill. The article "Effects of Occluded Vision and Imagery on Putting Golf Balls" (*Perceptual and Motor Skills* [1995]: 179–186) reported on a study involving the following four learning methods: (1) visual contact and imagery, (2) nonvisual contact and imagery, (3) visual contact, and (4) control. There were 20 subjects randomly assigned to each method. The following summary information on putting performance score was reported:

Method	1	2	3	4	
$\bar{x}$	16.30	15.25	12.05	9.30	$\bar{\bar{x}} = 13.23$
s	2.03	3.23	2.91	2.85	

a. Is there sufficient evidence to conclude the mean putting performance score is not the same for the four methods? $F = 25.98$, P-value < 0.001, reject H_0
b. Calculate the 95% T–K intervals, and then use the underscoring procedure described in this section to identify significant differences among the learning methods.

15.25 The article referenced in Exercise 15.24 reported on a study involving the following four learning methods: (1) visual contact and imagery, (2) nonvisual contact and imagery, (3) visual contact, and (4) control. There were 20 subjects assigned to each method. Calculate the 99% T–K intervals, indicate which methods differ significantly from one another, and summarize the results by underscoring. How do the 99% T–K intervals compare to the 95% T–K intervals of Exercise 15.24?

15.26 ● The accompanying data resulted from a flammability study in which specimens of five different fabrics were tested to determine burn times.

1	17.8	16.2	15.9	15.5	
2	13.2	10.4	11.3		
Fabric **3**	11.8	11.0	9.2	10.0	
4	16.5	15.3	14.1	15.0	13.9
5	13.9	10.8	12.8	11.7	

MSTr = 23.67
MSE = 1.39
$F = 17.08$
P-value = .000

The accompanying output gives the T–K intervals as calculated by MINITAB. Identify significant differences and give the underscoring pattern.

Individual error rate = 0.00750
Critical value = 4.37
Intervals for (column level mean) – (row level mean)

	1	2	3	4
	1.938			
2	7.495			
	3.278	−1.645		
3	8.422	3.912		
	−1.050	−5.983	−6.900	
4	3.830	−0.670	−2.020	
	1.478	−3.445	−4.372	0.220
5	6.622	2.112	0.772	5.100

15.27 ▼ Sample mean chlorophyll concentrations for the four Jerusalem artichoke varieties introduced in Exer-

cise 15.9 were .30, .24, .41, and .33, with corresponding sample sizes of 5, 5, 4, and 6, respectively. In addition, MSE = .0130. Calculate the 95% T–K intervals, and then use the underscoring procedure described in this section to identify significant differences among the varieties.

15.28 Do lizards play a role in spreading plant seeds? Some research carried out in South Africa would suggest so ("Dispersal of Namaqua Fig (*Ficus cordata cordata*) Seeds by the Augrabies Flat Lizard (*Platysaurus broadleyi*)," *Journal of Herpetology* [1999]: 328–330). The researchers collected 400 seeds of this particular type of fig, 100 of which were from each treatment: lizard dung, bird dung, rock hyrax dung, and uneaten figs. They planted these seeds in batches of 5, and for each group of 5 they recorded how many of the seeds germinated. This resulted in 20 observations for each treatment. The treatment means and standard deviations are given in the accompanying table.

Treatment	n	$\bar{x}$	s
Uneaten figs	20	2.40	.30
Lizard dung	20	2.35	.33
Bird dung	20	1.70	.34
Hyrax dung	20	1.45	.28

a. Construct the appropriate ANOVA table, and test the hypothesis that there is no difference between mean number of seeds germinating for the four treatments.
b. Is there evidence that seeds eaten and then excreted by lizards germinate at a higher rate than those eaten and then excreted by birds? Give statistical evidence to support your answer. H_0: $\mu_2 = \mu_3$, H_a: $\mu_2 > \mu_3$, $t = 6.56$, P-value ≈ 0, reject H_0

15.29 The article "Growth Response in Radish to Sequential and Simultaneous Exposures of NO_2 and SO_2" (*Environmental Pollution* [1984]: 303–325) compared a control group (no exposure), a sequential exposure group (plants exposed to one pollutant followed by exposure to the second four weeks later), and a simultaneous-exposure group (plants exposed to both pollutants at the same time). The article states, "Sequential exposure to the two pollutants had no effect on growth compared to the control. Simultaneous exposure to the gases significantly reduced plant growth." Let $\bar{x}_1$, $\bar{x}_2$, and $\bar{x}_3$ represent the sample means for the control, sequential, and simultaneous groups, respectively. Suppose that $\bar{x}_1 > \bar{x}_2 > \bar{x}_3$. Use the given information to construct a table where the sample means are listed in increasing order, with those that are judged not to be significantly different underscored.

15.30 The nutritional quality of shrubs commonly used for feed by rabbits was the focus of a study summarized in the article "Estimation of Browse by Size Classes for Snowshoe Hare" (*Journal of Wildlife Management* [1980]: 34–40). The energy content (cal/g) of three sizes (4 mm or less, 5–7 mm, and 8–10 mm) of serviceberries was studied. Let μ_1, μ_2, and μ_3 denote the true energy content for the three size classes. Suppose that 95% simultaneous confidence intervals for $\mu_1 - \mu_2$, $\mu_1 - \mu_3$, and $\mu_2 - \mu_3$ are $(-10, 290)$, $(150, 450)$, and $(10, 310)$, respectively. How would you interpret these intervals?

15.31 The accompanying underscoring pattern appeared in the article "Effect of SO_2 on Transpiration, Chlorophyll Content, Growth, and Injury in Young Seedlings of Woody Angiosperms" (*Canadian Journal of Forest Research* [1980]: 78–81). Water loss of plants (*Acer saccharinum*) exposed to 0, 2, 4, 8, and 16 hours of fumigation was recorded, and a multiple comparison procedure was used to detect differences among the mean water losses for the different fumigation durations. How would you interpret this pattern?

Duration of fumigation	16	0	8	2	4
Sample mean water loss	27.57	28.23	30.21	31.16	36.21

15.32 ● Samples of six different brands of diet or imitation margarine were analyzed to determine the level of physiologically active polyunsaturated fatty acids (PAPUFA, in percent), resulting in the data shown in the accompanying table. (The data are fictitious, but the sample means agree with data reported in *Consumer Reports*.)

Imperial	14.1	13.6	14.4	14.3	
Parkay	12.8	12.5	13.4	13.0	12.3
Blue Bonnet	13.5	13.4	14.1	14.3	
Chiffon	13.2	12.7	12.6	13.9	
Mazola	16.8	17.2	16.4	17.3	18.0
Fleischmann's	18.1	17.2	18.7	18.4	

a. Test for differences among the true average PAPUFA percentages for the different brands. Use $\alpha = .05$.
b. Use the T–K procedure to compute 95% simultaneous confidence intervals for all differences between means and interpret the resulting intervals.

15.33 Exercise 15.8 presented the accompanying summary information on satisfaction levels for employees on three different work schedules:

$$n_1 = 24 \qquad n_2 = 24 \qquad n_3 = 20$$
$$\bar{x}_1 = 6.60 \qquad \bar{x}_2 = 5.37 \qquad \bar{x}_3 = 5.20$$

MSE was computed to be 2.028. Calculate the 95% T–K intervals, and identify any significant differences.

15.34 ● Consider the accompanying data on plant growth after the application of different types of growth hormone.

	1	13	17	7	14
	2	21	13	20	17
Hormone	3	18	14	17	21
	4	7	11	18	10
	5	6	11	15	8

a. Carry out the *F* test at level $\alpha = .05$.
b. What happens when the T–K procedure is applied? (*Note*: This "contradiction" can occur when H_0 is "barely" rejected. It happens because the test and the multiple comparison method are based on different distributions. Consult your friendly neighborhood statistician for more information.)

Bold exercises answered in back ● Data set available online but not required ▼ Video solution available

Activity 15.1 Exploring Single-Factor ANOVA

Working with a partner, consider the following:
1. Each of the four accompanying graphs shows a dotplot of data from three separate random samples. For each of the four dotplots, indicate whether you think that the basic assumptions for single-factor ANOVA are plausible. Write a sentence or two justifying your answer.

Case 1

Case 2

Case 3

Case 4

2. Each of the three accompanying graphs shows a dot-plot of data from three separate random samples. For each of the three dotplots, indicate whether you think that the three population means are probably not all the same, you think that the three population means might be the same, or you are unsure whether the population means could be the same. Write a sentence or two explaining your reasoning.

Case A

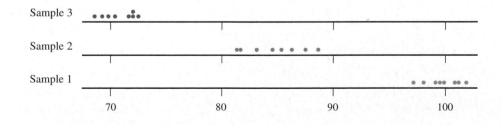

Case B

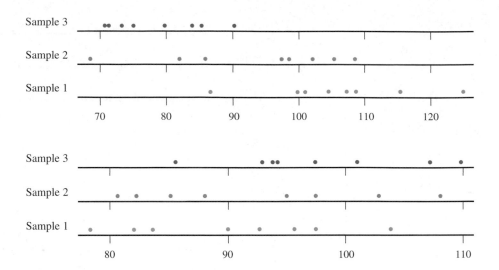

Case C

3. Sample data for each of the three cases in Step 2 are shown in the accompanying table. For each of the three cases, carry out a single-factor ANOVA. Is the result of the F test in each case consistent with your answers in Step 2? Explain.

| | Case A | | | Case B | | | Case C | |
Sample 1	Sample 2	Sample 3	Sample 1	Sample 2	Sample 3	Sample 1	Sample 2	Sample 3
99.7	91.3	69.3	104.2	81.9	71.7	82.3	82.4	94.2
98.0	82.0	72.1	107.0	105.4	79.7	97.4	87.5	109.8
101.4	83.6	71.7	88.6	98.4	70.9	83.7	97.3	94.9
99.2	84.8	69.9	99.6	108.4	76.6	103.6	102.6	85.8
101.0	86.5	68.8	124.3	102.1	85.3	78.6	94.8	97.4
101.8	91.5	70.7	100.7	68.9	90.3	90.1	81.3	101.0
99.5	81.8	72.7	108.3	85.8	84.2	92.8	85.2	93.0
97.0	85.5	72.2	116.5	97.5	74.6	95.5	107.8	107.1

Summary of Key Concepts and Formulas

Term or Formula	Comment
Single-factor analysis of variance (ANOVA)	A test procedure for determining whether there are significant differences among k population or treatment means. The hypotheses tested are $H_0: \mu_1 = \mu_2 = \cdots = \mu_k$ versus H_a: at least two μ's differ.
Treatment sum of squares: $$\text{SSTr} = n_1(\bar{x}_1 - \bar{\bar{x}})^2 + \cdots + n_k(\bar{x}_k - \bar{\bar{x}})^2$$	A measure of how different the k sample means $\bar{x}_1$, $\bar{x}_2, \ldots, \bar{x}_k$ are from one another; associated df $= k - 1$.
Error sum of squares: $$\text{SSE} = (n_1 - 1)s_1^2 + \cdots + (n_k - 1)s_k^2$$	A measure of the amount of variability within the individual samples; associated df $= N - k$, where $N = n_1 + \cdots + n_k$.

Term or Formula	**Comment**
Mean square	A sum of squares divided by its df. For single-factor ANOVA, MSTr = SSTR/$(k - 1)$ and MSE = SSE/$(N - k)$.
$F = \dfrac{\text{MSTr}}{\text{MSE}}$	The test statistic for testing H_0: $\mu_1 = \mu_2 = \cdots = \mu_k$ in a single-factor ANOVA. When H_0 is true, F has an F distribution with numerator df $= k - 1$ and denominator df $= N - k$.
SSTo = SSTr + SSE	The fundamental identity in single-factor ANOVA, where SSTo = total sum of squares = $\sum(x - \bar{\bar{x}})^2$.
Tukey–Kramer multiple comparison procedure	A procedure for identifying significant differences among the μ's once the hypothesis H_0: $\mu_1 = \mu_2 = \cdots = \mu_k$ has been rejected by the ANOVA F test.

Graphing Calculator Exploration

Exploration 15.1 Single Factor ANOVA

Single-factor ANOVA is at the limit of the capability of a graphing calculator unless you have a programmable calculator and *reliable* programs. Nevertheless, for small ANOVA problems, the calculator can be used effectively when access to a computer is limited.

By now you should be familiar with the capabilities of your calculator, so we don't need to go into a great deal of detail here. Begin by entering the data, one treatment per List. To illustrate, we will use the data from Exercise 15.5 on box compression strength. Put the data from box type 1 into List1, data from box type 2 into List2, etc. Boxplots and normal probability plots (discussed in the Graphing Calculator Explorations of Chapter 4) can be used to assess the plausibility of the ANOVA assumptions.

After entering the data, navigate the menus to the ANOVA screen. Screens will differ from one calculator to another, but two common versions are shown in Figure 15.8.

Figure 15.8   Two common ANOVA screens.

```
ANOVA
How Many: 4
List1 : List1
List2 : List2
List3 : List3
List4 : List4
Execute
```

```
ANOVA
(List1, List2, List3,
List4)
```

Press either Execute or the Enter key to initiate the ANOVA computations on the calculator. The results should look like one of the two "larger-than-life" displays shown in Figure 15.9.

Figure 15.9 Results of ANOVA computations.

```
ANOVA
F = 25.0943
p = 5.5254499E-7
xpσn − 1 = 41.1333
    Fdf = 3
    SS = 127374.755
    MS = 1691.94875
    Edf = 20
    SS = 33838.9750
    MS = 1691.94875
```

```
One-way ANOVA
 F = 25.0942894
 p = 5.5254499E-7
Factor
    df = 3.0000000
    SS = 127374.755
    MS = 42458.2515
Error
    df = 20.0000000
    SS = 33838.9750
    MS = 1691.94875
    Sxp = 41.1333046
```

From the information provided by the calculator, you can reproduce the single-factor ANOVA table.

Source of Variation	df	Sum of Squares	Mean Square	F
Treatments	3	127374.755	42458.2515	25.094
Error	20	33838.9750	1691.94875	
Total	23	???		

Check to make sure you understand the correspondence between the calculator output and the ANOVA table. Notice specifically that the total sum of squares does not appear on the calculator screen. It must be calculated by summing the treatment and error sums of squares. The quantity labeled Sxp is $\sqrt{MSE}$, an estimate of the assumed common population standard deviation σ.

The Tukey–Kramer multiple comparison procedure could be performed by combining the information from statistics calculated from the individual lists and the calculator output, but these procedures are not usually built-in on the calculator. If you anticipate performing many single-factor ANOVAs on your calculator, you might consider downloading a calculator program from the web. Freeware varies in quality, so be sure to check such a program against a known correct solution!

Appendix A

Appendix Statistical Tables

Table 1 Random Numbers

Row																				
1	4	5	1	8	5	0	3	3	7	1	2	8	4	5	1	1	0	9	5	7
2	4	2	5	5	8	0	4	5	7	0	7	0	3	6	6	1	3	1	3	1
3	8	9	9	3	4	3	5	0	6	3	9	1	1	8	2	6	9	2	0	9
4	8	9	0	7	2	9	9	0	4	7	6	7	4	7	1	3	4	3	5	3
5	5	7	3	1	0	3	7	4	7	8	5	2	0	1	3	7	7	6	3	6
6	0	9	3	8	7	6	7	9	9	5	6	2	5	6	5	8	4	2	6	4
7	4	1	0	1	0	2	2	0	4	7	5	1	1	9	4	7	9	7	5	1
8	6	4	7	3	6	3	4	5	1	2	3	1	1	8	0	0	4	8	2	0
9	8	0	2	8	7	9	3	8	4	0	4	2	0	8	9	1	2	3	3	2
10	9	4	6	0	6	9	7	8	8	2	5	2	9	6	0	1	4	6	0	5
11	6	6	9	5	7	4	4	6	3	2	0	6	0	8	9	1	3	6	1	8
12	0	7	1	7	7	7	2	9	7	8	7	5	8	8	6	9	8	4	1	0
13	6	1	3	0	9	7	3	3	6	6	0	4	1	8	3	2	6	7	6	8
14	2	2	3	6	2	1	3	0	2	2	6	6	9	7	0	2	1	2	5	8
15	0	7	1	7	4	2	0	0	0	1	3	1	2	0	4	7	8	4	1	0
16	6	6	5	1	6	1	8	1	5	5	2	6	2	0	1	1	5	2	3	6
17	9	9	6	2	5	3	5	9	8	3	7	5	0	1	3	9	3	8	0	8
18	9	9	9	6	1	2	9	3	4	6	5	6	4	6	5	8	2	7	4	0
19	2	5	6	3	1	9	8	1	1	0	3	5	6	7	9	1	4	5	2	0
20	5	1	1	9	8	1	2	1	1	6	9	8	1	8	1	9	9	1	2	0
21	1	9	8	0	7	4	6	8	4	0	3	0	8	1	1	0	6	2	3	2
22	9	7	0	9	6	3	8	9	9	7	0	6	5	4	3	6	5	0	3	2
23	1	7	6	4	8	2	0	3	9	6	3	6	2	1	0	7	7	3	1	7
24	6	2	5	8	2	0	7	8	6	4	6	6	8	9	2	0	6	9	0	4
25	1	5	7	1	1	1	9	5	1	4	5	2	8	3	4	3	0	7	3	5
26	1	4	6	6	5	6	0	1	9	4	0	5	2	7	6	4	3	6	8	8
27	1	8	5	0	2	1	6	8	0	7	7	2	6	2	6	7	5	4	8	7
28	7	8	7	4	6	5	4	3	7	9	3	9	2	7	9	5	4	2	3	1
29	1	6	3	2	8	3	7	3	0	7	2	4	8	0	9	9	9	4	7	0
30	2	8	9	0	8	1	6	8	1	7	3	1	3	0	9	7	2	5	7	9
31	0	7	8	8	6	5	7	5	5	4	0	0	3	4	1	2	7	3	7	9
32	8	4	0	1	4	5	1	9	1	1	2	1	5	3	2	8	5	5	7	5
33	7	3	5	9	7	0	4	9	1	2	1	3	2	5	1	9	3	3	8	3
34	4	7	2	6	7	6	9	9	2	7	8	7	5	5	5	2	4	4	3	4
35	9	3	3	7	0	7	0	5	7	5	6	9	5	4	3	1	4	6	6	8
36	0	2	4	9	7	8	1	6	3	8	7	8	0	5	6	7	2	7	5	0
37	7	1	0	1	8	4	7	1	2	9	3	8	0	0	8	7	9	2	8	6
38	9	7	9	4	4	5	3	1	9	3	4	5	0	6	3	5	9	6	9	8
39	0	4	2	5	0	0	9	9	6	4	0	6	9	0	3	8	3	5	7	2
40	0	7	1	2	3	6	1	7	9	3	9	5	4	6	8	4	8	8	0	6
41	3	5	6	6	2	4	4	5	6	3	7	8	7	6	5	2	0	4	3	2
42	6	6	8	5	5	2	9	7	9	3	3	1	6	9	5	9	7	1	1	2
43	9	5	0	4	3	1	1	7	3	9	2	7	7	4	7	0	3	1	2	8
44	5	1	7	8	9	4	7	2	9	2	8	9	9	8	0	6	3	7	2	1
45	1	6	3	9	4	1	3	2	1	1	8	5	6	3	4	1	9	3	1	7
46	4	4	8	6	4	0	3	8	3	8	3	5	9	5	9	4	8	3	9	4
47	7	7	6	6	4	5	4	4	8	4	4	0	3	9	8	5	2	0	2	3
48	2	5	6	6	3	7	0	6	5	6	9	0	1	9	5	2	6	9	1	2
49	9	4	0	4	7	5	3	2	8	7	2	7	4	9	3	9	6	5	5	6
50	7	3	1	5	6	6	5	0	3	5	3	7	2	8	6	2	4	1	8	7

Table 1 ■ Random Numbers   815

Table 1 Random Numbers (*Continued*)

Row																				
51	7	5	8	2	8	8	8	7	6	4	1	1	0	2	3	1	9	3	6	0
52	3	3	6	0	9	1	1	0	3	2	7	8	2	0	5	3	4	8	9	8
53	0	2	9	6	9	8	9	3	8	1	5	3	9	9	7	0	7	7	1	6
54	8	5	9	6	2	9	6	8	2	1	2	4	7	0	6	8	3	4	6	1
55	5	4	7	6	1	0	0	1	0	4	6	1	4	1	5	0	9	6	5	5
56	5	0	3	6	4	1	9	8	4	4	1	2	0	2	5	1	8	1	2	1
57	0	2	6	3	7	5	1	1	6	6	0	5	8	1	2	3	3	6	1	3
58	3	8	1	6	3	8	1	4	5	2	9	4	2	5	7	3	2	3	1	8
59	9	1	5	6	0	6	5	6	6	3	6	2	3	0	0	0	1	8	5	9
60	5	3	5	6	3	9	5	4	7	3	6	6	7	5	0	1	5	6	7	3
61	9	6	6	4	5	7	7	6	1	5	4	4	8	0	6	5	7	6	3	0
62	6	3	0	6	7	9	5	5	4	6	2	2	8	4	4	0	0	9	9	8
63	8	5	8	3	5	2	0	6	6	0	0	6	0	6	3	0	1	7	0	5
64	3	8	2	4	9	0	9	2	6	2	9	5	1	9	1	9	0	8	3	3
65	1	4	4	1	1	7	4	6	3	6	5	6	5	5	7	7	0	3	5	8
66	5	9	9	5	3	7	2	5	1	7	1	1	0	7	1	0	9	2	8	8
67	8	7	1	7	5	2	5	6	8	7	9	9	1	3	9	6	4	9	3	0
68	6	7	2	3	1	4	9	2	1	7	0	8	6	7	8	9	9	4	7	4
69	2	3	2	8	7	0	9	7	1	1	1	2	8	2	9	1	0	6	7	7
70	2	9	5	7	8	4	7	9	0	3	6	9	2	0	6	0	6	2	6	8
71	4	8	9	8	3	2	7	6	9	1	9	8	6	9	5	2	4	9	9	9
72	1	5	6	5	7	7	5	4	3	4	3	8	1	8	9	9	4	4	1	1
73	1	8	1	1	7	2	8	5	5	8	9	9	9	6	2	0	1	6	6	7
74	5	7	7	0	9	5	5	6	8	6	8	2	2	6	0	5	5	1	8	7
75	1	8	6	0	5	4	8	3	4	5	3	5	8	7	7	7	8	5	7	0
76	2	6	6	7	9	4	2	2	8	7	4	3	4	9	6	1	9	4	3	9
77	3	6	6	4	5	7	8	3	0	2	8	4	6	7	2	1	4	5	2	3
78	0	7	8	0	1	2	1	1	3	4	2	1	6	9	3	3	5	4	0	4
79	8	3	6	0	5	7	7	9	1	5	8	8	4	9	5	7	2	2	7	6
80	5	3	6	9	0	6	3	8	7	5	9	5	9	7	4	2	5	6	2	9
81	0	9	3	7	7	2	8	6	4	3	2	9	4	8	2	9	9	6	9	9
82	9	4	7	4	0	0	0	3	5	4	6	6	2	6	2	3	6	1	1	4
83	5	5	4	1	7	8	6	4	2	3	2	9	8	4	6	3	8	3	0	3
84	5	3	0	0	5	4	8	0	7	4	7	6	2	1	1	2	1	2	6	9
85	3	3	0	9	3	2	9	4	0	5	5	4	8	7	5	7	5	3	8	8
86	3	0	5	7	1	9	5	8	0	0	4	5	3	0	3	0	2	7	6	7
87	5	0	8	6	0	8	1	6	2	0	8	6	5	4	0	7	2	9	1	0
88	3	6	4	7	8	2	3	5	7	9	8	5	2	7	6	9	0	2	4	9
89	9	0	4	4	9	1	6	8	5	2	8	9	0	7	5	7	2	5	1	8
90	9	5	2	6	9	3	9	6	5	1	8	8	7	8	2	0	4	4	7	9
91	9	4	5	7	0	3	4	6	4	2	5	4	8	6	1	1	9	1	8	8
92	8	1	1	8	0	5	4	2	8	5	3	3	3	0	1	1	4	4	8	3
93	6	9	4	7	8	3	3	9	1	2	5	0	1	2	3	0	1	1	2	5
94	0	0	6	8	8	7	2	4	4	7	6	6	0	3	4	7	5	6	8	2
95	5	3	3	9	3	8	4	9	1	9	1	7	8	4	5	2	2	5	4	4
96	2	5	6	2	7	6	0	3	8	1	4	4	2	6	8	3	6	3	2	8
97	7	4	3	7	9	6	8	6	2	8	3	8	4	2	2	0	7	0	5	3
98	1	9	0	8	8	0	1	2	2	2	7	5	6	5	5	7	8	7	2	6
99	2	4	8	0	2	5	2	7	0	5	9	6	6	1	5	8	7	9	7	5
100	4	1	7	8	6	7	1	1	5	8	9	4	8	9	8	3	0	9	0	7

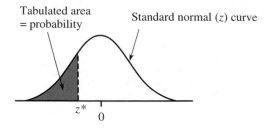

Tabulated area = probability

Standard normal (z) curve

z^* 0

Table 2 Standard Normal Probabilities (Cumulative z Curve Areas)

z^*	.00	.01	.02	.03	.04	.05	.06	.07	.08	.09
−3.8	.0001	.0001	.0001	.0001	.0001	.0001	.0001	.0001	.0001	.0000
−3.7	.0001	.0001	.0001	.0001	.0001	.0001	.0001	.0001	.0001	.0001
−3.6	.0002	.0002	.0001	.0001	.0001	.0001	.0001	.0001	.0001	.0001
−3.5	.0002	.0002	.0002	.0002	.0002	.0002	.0002	.0002	.0002	.0002
−3.4	.0003	.0003	.0003	.0003	.0003	.0003	.0003	.0003	.0003	.0002
−3.3	.0005	.0005	.0005	.0004	.0004	.0004	.0004	.0004	.0004	.0003
−3.2	.0007	.0007	.0006	.0006	.0006	.0006	.0006	.0005	.0005	.0005
−3.1	.0010	.0009	.0009	.0009	.0008	.0008	.0008	.0008	.0007	.0007
−3.0	.0013	.0013	.0013	.0012	.0012	.0011	.0011	.0011	.0010	.0010
−2.9	.0019	.0018	.0018	.0017	.0016	.0016	.0015	.0015	.0014	.0014
−2.8	.0026	.0025	.0024	.0023	.0023	.0022	.0021	.0021	.0020	.0019
−2.7	.0035	.0034	.0033	.0032	.0031	.0030	.0029	.0028	.0027	.0026
−2.6	.0047	.0045	.0044	.0043	.0041	.0040	.0039	.0038	.0037	.0036
−2.5	.0062	.0060	.0059	.0057	.0055	.0054	.0052	.0051	.0049	.0048
−2.4	.0082	.0080	.0078	.0075	.0073	.0071	.0069	.0068	.0066	.0064
−2.3	.0107	.0104	.0102	.0099	.0096	.0094	.0091	.0089	.0087	.0084
−2.2	.0139	.0136	.0132	.0129	.0125	.0122	.0119	.0116	.0113	.0110
−2.1	.0179	.0174	.0170	.0166	.0162	.0158	.0154	.0150	.0146	.0143
−2.0	.0228	.0222	.0217	.0212	.0207	.0202	.0197	.0192	.0188	.0183
−1.9	.0287	.0281	.0274	.0268	.0262	.0256	.0250	.0244	.0239	.0233
−1.8	.0359	.0351	.0344	.0336	.0329	.0322	.0314	.0307	.0301	.0294
−1.7	.0446	.0436	.0427	.0418	.0409	.0401	.0392	.0384	.0375	.0367
−1.6	.0548	.0537	.0526	.0516	.0505	.0495	.0485	.0475	.0465	.0455
−1.5	.0668	.0655	.0643	.0630	.0618	.0606	.0594	.0582	.0571	.0559
−1.4	.0808	.0793	.0778	.0764	.0749	.0735	.0721	.0708	.0694	.0681
−1.3	.0968	.0951	.0934	.0918	.0901	.0885	.0869	.0853	.0838	.0823
−1.2	.1151	.1131	.1112	.1093	.1075	.1056	.1038	.1020	.1003	.0985
−1.1	.1357	.1335	.1314	.1292	.1271	.1251	.1230	.1210	.1190	.1170
−1.0	.1587	.1562	.1539	.1515	.1492	.1469	.1446	.1423	.1401	.1379
−0.9	.1841	.1814	.1788	.1762	.1736	.1711	.1685	.1660	.1635	.1611
−0.8	.2119	.2090	.2061	.2033	.2005	.1977	.1949	.1922	.1894	.1867
−0.7	.2420	.2389	.2358	.2327	.2296	.2266	.2236	.2206	.2177	.2148
−0.6	.2743	.2709	.2676	.2643	.2611	.2578	.2546	.2514	.2483	.2451
−0.5	.3085	.3050	.3015	.2981	.2946	.2912	.2877	.2843	.2810	.2776
−0.4	.3446	.3409	.3372	.3336	.3300	.3264	.3228	.3192	.3156	.3121
−0.3	.3821	.3783	.3745	.3707	.3669	.3632	.3594	.3557	.3520	.3483
−0.2	.4207	.4168	.4129	.4090	.4052	.4013	.3974	.3936	.3897	.3859
−0.1	.4602	.4562	.4522	.4483	.4443	.4404	.4364	.4325	.4286	.4247
−0.0	.5000	.4960	.4920	.4880	.4840	.4801	.4761	.4721	.4681	.4641

Table 2 ■ Standard Normal Probabilites (Cumulative *z* Curve Areas)   **817**

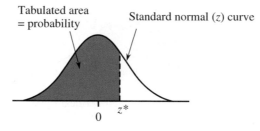

Tabulated area = probability   Standard normal (*z*) curve

0   *z**

Table 2 Standard Normal Probabilities (*Continued*)

*z**	.00	.01	.02	.03	.04	.05	.06	.07	.08	.09
0.0	.5000	.5040	.5080	.5120	.5160	.5199	.5239	.5279	.5319	.5359
0.1	.5398	.5438	.5478	.5517	.5557	.5596	.5636	.5675	.5714	.5753
0.2	.5793	.5832	.5871	.5910	.5948	.5987	.6026	.6064	.6103	.6141
0.3	.6179	.6217	.6255	.6293	.6331	.6368	.6406	.6443	.6480	.6517
0.4	.6554	.6591	.6628	.6664	.6700	.6736	.6772	.6808	.6844	.6879
0.5	.6915	.6950	.6985	.7019	.7054	.7088	.7123	.7157	.7190	.7224
0.6	.7257	.7291	.7324	.7357	.7389	.7422	.7454	.7486	.7517	.7549
0.7	.7580	.7611	.7642	.7673	.7704	.7734	.7764	.7794	.7823	.7852
0.8	.7881	.7910	.7939	.7967	.7995	.8023	.8051	.8078	.8106	.8133
0.9	.8159	.8186	.8212	.8238	.8264	.8289	.8315	.8340	.8365	.8389
1.0	.8413	.8438	.8461	.8485	.8508	.8531	.8554	.8577	.8599	.8621
1.1	.8643	.8665	.8686	.8708	.8729	.8749	.8770	.8790	.8810	.8830
1.2	.8849	.8869	.8888	.8907	.8925	.8944	.8962	.8980	.8997	.9015
1.3	.9032	.9049	.9066	.9082	.9099	.9115	.9131	.9147	.9162	.9177
1.4	.9192	.9207	.9222	.9236	.9251	.9265	.9279	.9292	.9306	.9319
1.5	.9332	.9345	.9357	.9370	.9382	.9394	.9406	.9418	.9429	.9441
1.6	.9452	.9463	.9474	.9484	.9495	.9505	.9515	.9525	.9535	.9545
1.7	.9554	.9564	.9573	.9582	.9591	.9599	.9608	.9616	.9625	.9633
1.8	.9641	.9649	.9656	.9664	.9671	.9678	.9686	.9693	.9699	.9706
1.9	.9713	.9719	.9726	.9732	.9738	.9744	.9750	.9756	.9761	.9767
2.0	.9772	.9778	.9783	.9788	.9793	.9798	.9803	.9808	.9812	.9817
2.1	.9821	.9826	.9830	.9834	.9838	.9842	.9846	.9850	.9854	.9857
2.2	.9861	.9864	.9868	.9871	.9875	.9878	.9881	.9884	.9887	.9890
2.3	.9893	.9896	.9898	.9901	.9904	.9906	.9909	.9911	.9913	.9916
2.4	.9918	.9920	.9922	.9925	.9927	.9929	.9931	.9932	.9934	.9936
2.5	.9938	.9940	.9941	.9943	.9945	.9946	.9948	.9949	.9951	.9952
2.6	.9953	.9955	.9956	.9957	.9959	.9960	.9961	.9962	.9963	.9964
2.7	.9965	.9966	.9967	.9968	.9969	.9970	.9971	.9972	.9973	.9974
2.8	.9974	.9975	.9976	.9977	.9977	.9978	.9979	.9979	.9980	.9981
2.9	.9981	.9982	.9982	.9983	.9984	.9984	.9985	.9985	.9986	.9986
3.0	.9987	.9987	.9987	.9988	.9988	.9989	.9989	.9989	.9990	.9990
3.1	.9990	.9991	.9991	.9991	.9992	.9992	.9992	.9992	.9993	.9993
3.2	.9993	.9993	.9994	.9994	.9994	.9994	.9994	.9995	.9995	.9995
3.3	.9995	.9995	.9995	.9996	.9996	.9996	.9996	.9996	.9996	.9997
3.4	.9997	.9997	.9997	.9997	.9997	.9997	.9997	.9997	.9997	.9998
3.5	.9998	.9998	.9998	.9998	.9998	.9998	.9998	.9998	.9998	.9998
3.6	.9998	.9998	.9999	.9999	.9999	.9999	.9999	.9999	.9999	.9999
3.7	.9999	.9999	.9999	.9999	.9999	.9999	.9999	.9999	.9999	.9999
3.8	.9999	.9999	.9999	.9999	.9999	.9999	.9999	.9999	.9999	1.0000

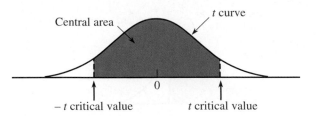

Central area   *t* curve

0

$-t$ critical value t critical value

Table 3 *t* Critical Values

		.80	.90	.95	.98	.99	.998	.999
Central area captured:		80%	90%	95%	98%	99%	99.8%	99.9%
Confidence level:								
	1	3.08	6.31	12.71	31.82	63.66	318.31	636.62
	2	1.89	2.92	4.30	6.97	9.93	23.33	31.60
	3	1.64	2.35	3.18	4.54	5.84	10.21	12.92
	4	1.53	2.13	2.78	3.75	4.60	7.17	8.61
	5	1.48	2.02	2.57	3.37	4.03	5.89	6.86
	6	1.44	1.94	2.45	3.14	3.71	5.21	5.96
	7	1.42	1.90	2.37	3.00	3.50	4.79	5.41
	8	1.40	1.86	2.31	2.90	3.36	4.50	5.04
	9	1.38	1.83	2.26	2.82	3.25	4.30	4.78
	10	1.37	1.81	2.23	2.76	3.17	4.14	4.59
	11	1.36	1.80	2.20	2.72	3.11	4.03	4.44
	12	1.36	1.78	2.18	2.68	3.06	3.93	4.32
	13	1.35	1.77	2.16	2.65	3.01	3.85	4.22
	14	1.35	1.76	2.15	2.62	2.98	3.79	4.14
	15	1.34	1.75	2.13	2.60	2.95	3.73	4.07
	16	1.34	1.75	2.12	2.58	2.92	3.69	4.02
Degrees of	17	1.33	1.74	2.11	2.57	2.90	3.65	3.97
freedom	18	1.33	1.73	2.10	2.55	2.88	3.61	3.92
	19	1.33	1.73	2.09	2.54	2.86	3.58	3.88
	20	1.33	1.73	2.09	2.53	2.85	3.55	3.85
	21	1.32	1.72	2.08	2.52	2.83	3.53	3.82
	22	1.32	1.72	2.07	2.51	2.82	3.51	3.79
	23	1.32	1.71	2.07	2.50	2.81	3.49	3.77
	24	1.32	1.71	2.06	2.49	2.80	3.47	3.75
	25	1.32	1.71	2.06	2.49	2.79	3.45	3.73
	26	1.32	1.71	2.06	2.48	2.78	3.44	3.71
	27	1.31	1.70	2.05	2.47	2.77	3.42	3.69
	28	1.31	1.70	2.05	2.47	2.76	3.41	3.67
	29	1.31	1.70	2.05	2.46	2.76	3.40	3.66
	30	1.31	1.70	2.04	2.46	2.75	3.39	3.65
	40	1.30	1.68	2.02	2.42	2.70	3.31	3.55
	60	1.30	1.67	2.00	2.39	2.66	3.23	3.46
	120	1.29	1.66	1.98	2.36	2.62	3.16	3.37
z critical values	∞	1.28	1.645	1.96	2.33	2.58	3.09	3.29

Table 4 ▪ Tail Areas for *t* Curves   **819**

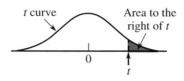

Table 4 Tail Areas for *t* Curves

df *t*	1	2	3	4	5	6	7	8	9	10	11	12
0.0	.500	.500	.500	.500	.500	.500	.500	.500	.500	.500	.500	.500
0.1	.468	.465	.463	.463	.462	.462	.462	.461	.461	.461	.461	.461
0.2	.437	.430	.427	.426	.425	.424	.424	.423	.423	.423	.423	.422
0.3	.407	.396	.392	.390	.388	.387	.386	.386	.386	.385	.385	.385
0.4	.379	.364	.358	.355	.353	.352	.351	.350	.349	.349	.348	.348
0.5	.352	.333	.326	.322	.319	.317	.316	.315	.315	.314	.313	.313
0.6	.328	.305	.295	.290	.287	.285	.284	.283	.282	.281	.280	.280
0.7	.306	.278	.267	.261	.258	.255	.253	.252	.251	.250	.249	.249
0.8	.285	.254	.241	.234	.230	.227	.225	.223	.222	.221	.220	.220
0.9	.267	.232	.217	.210	.205	.201	.199	.197	.196	.195	.194	.193
1.0	.250	.211	.196	.187	.182	.178	.175	.173	.172	.170	.169	.169
1.1	.235	.193	.176	.167	.162	.157	.154	.152	.150	.149	.147	.146
1.2	.221	.177	.158	.148	.142	.138	.135	.132	.130	.129	.128	.127
1.3	.209	.162	.142	.132	.125	.121	.117	.115	.113	.111	.110	.109
1.4	.197	.148	.128	.117	.110	.106	.102	.100	.098	.096	.095	.093
1.5	.187	.136	.115	.104	.097	.092	.089	.086	.084	.082	.081	.080
1.6	.178	.125	.104	.092	.085	.080	.077	.074	.072	.070	.069	.068
1.7	.169	.116	.094	.082	.075	.070	.066	.064	.062	.060	.059	.057
1.8	.161	.107	.085	.073	.066	.061	.057	.055	.053	.051	.050	.049
1.9	.154	.099	.077	.065	.058	.053	.050	.047	.045	.043	.042	.041
2.0	.148	.092	.070	.058	.051	.046	.043	.040	.038	.037	.035	.034
2.1	.141	.085	.063	.052	.045	.040	.037	.034	.033	.031	.030	.029
2.2	.136	.079	.058	.046	.040	.035	.032	.029	.028	.026	.025	.024
2.3	.131	.074	.052	.041	.035	.031	.027	.025	.023	.022	.021	.020
2.4	.126	.069	.048	.037	.031	.027	.024	.022	.020	.019	.018	.017
2.5	.121	.065	.044	.033	.027	.023	.020	.018	.017	.016	.015	.014
2.6	.117	.061	.040	.030	.024	.020	.018	.016	.014	.013	.012	.012
2.7	.113	.057	.037	.027	.021	.018	.015	.014	.012	.011	.010	.010
2.8	.109	.054	.034	.024	.019	.016	.013	.012	.010	.009	.009	.008
2.9	.106	.051	.031	.022	.017	.014	.011	.010	.009	.008	.007	.007
3.0	.102	.048	.029	.020	.015	.012	.010	.009	.007	.007	.006	.006
3.1	.099	.045	.027	.018	.013	.011	.009	.007	.006	.006	.005	.005
3.2	.096	.043	.025	.016	.012	.009	.008	.006	.005	.005	.004	.004
3.3	.094	.040	.023	.015	.011	.008	.007	.005	.005	.004	.004	.003
3.4	.091	.038	.021	.014	.010	.007	.006	.005	.004	.003	.003	.003
3.5	.089	.036	.020	.012	.009	.006	.005	.004	.003	.003	.002	.002
3.6	.086	.035	.018	.011	.008	.006	.004	.004	.003	.002	.002	.002
3.7	.084	.033	.017	.010	.007	.005	.004	.003	.002	.002	.002	.002
3.8	.082	.031	.016	.010	.006	.004	.003	.003	.002	.002	.001	.001
3.9	.080	.030	.015	.009	.006	.004	.003	.002	.002	.001	.001	.001
4.0	.078	.029	.014	.008	.005	.004	.003	.002	.002	.001	.001	.001

(Continued)

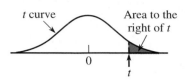

t curve

Area to the right of *t*

0

t

Table 4 Tail Areas for *t* Curves (*Continued*)

df *t*	13	14	15	16	17	18	19	20	21	22	23	24
0.0	.500	.500	.500	.500	.500	.500	.500	.500	.500	.500	.500	.500
0.1	.461	.461	.461	.461	.461	.461	.461	.461	.461	.461	.461	.461
0.2	.422	.422	.422	.422	.422	.422	.422	.422	.422	.422	.422	.422
0.3	.384	.384	.384	.384	.384	.384	.384	.384	.384	.383	.383	.383
0.4	.348	.347	.347	.347	.347	.347	.347	.347	.347	.347	.346	.346
0.5	.313	.312	.312	.312	.312	.312	.311	.311	.311	.311	.311	.311
0.6	.279	.279	.279	.278	.278	.278	.278	.278	.278	.277	.277	.277
0.7	.248	.247	.247	.247	.247	.246	.246	.246	.246	.246	.245	.245
0.8	.219	.218	.218	.218	.217	.217	.217	.217	.216	.216	.216	.216
0.9	.192	.191	.191	.191	.190	.190	.190	.189	.189	.189	.189	.189
1.0	.168	.167	.167	.166	.166	.165	.165	.165	.164	.164	.164	.164
1.1	.146	.144	.144	.144	.143	.143	.143	.142	.142	.142	.141	.141
1.2	.126	.124	.124	.124	.123	.123	.122	.122	.122	.121	.121	.121
1.3	.108	.107	.107	.106	.105	.105	.105	.104	.104	.104	.103	.103
1.4	.092	.091	.091	.090	.090	.089	.089	.089	.088	.088	.087	.087
1.5	.079	.077	.077	.077	.076	.075	.075	.075	.074	.074	.074	.073
1.6	.067	.065	.065	.065	.064	.064	.063	.063	.062	.062	.062	.061
1.7	.056	.055	.055	.054	.054	.053	.053	.052	.052	.052	.051	.051
1.8	.048	.046	.046	.045	.045	.044	.044	.043	.043	.043	.042	.042
1.9	.040	.038	.038	.038	.037	.037	.036	.036	.036	.035	.035	.035
2.0	.033	.032	.032	.031	.031	.030	.030	.030	.029	.029	.029	.028
2.1	.028	.027	.027	.026	.025	.025	.025	.024	.024	.024	.023	.023
2.2	.023	.022	.022	.021	.021	.021	.020	.020	.020	.019	.019	.019
2.3	.019	.018	.018	.018	.017	.017	.016	.016	.016	.016	.015	.015
2.4	.016	.015	.015	.014	.014	.014	.013	.013	.013	.013	.012	.012
2.5	.013	.012	.012	.012	.011	.011	.011	.011	.010	.010	.010	.010
2.6	.011	.010	.010	.010	.009	.009	.009	.009	.008	.008	.008	.008
2.7	.009	.008	.008	.008	.008	.007	.007	.007	.007	.007	.006	.006
2.8	.008	.007	.007	.006	.006	.006	.006	.006	.005	.005	.005	.005
2.9	.006	.005	.005	.005	.005	.005	.005	.004	.004	.004	.004	.004
3.0	.005	.004	.004	.004	.004	.004	.004	.004	.003	.003	.003	.003
3.1	.004	.004	.004	.003	.003	.003	.003	.003	.003	.003	.003	.002
3.2	.003	.003	.003	.003	.003	.002	.002	.002	.002	.002	.002	.002
3.3	.003	.002	.002	.002	.002	.002	.002	.002	.002	.002	.002	.001
3.4	.002	.002	.002	.002	.002	.002	.002	.001	.001	.001	.001	.001
3.5	.002	.002	.002	.001	.001	.001	.001	.001	.001	.001	.001	.001
3.6	.002	.001	.001	.001	.001	.001	.001	.001	.001	.001	.001	.001
3.7	.001	.001	.001	.001	.001	.001	.001	.001	.001	.001	.001	.001
3.8	.001	.001	.001	.001	.001	.001	.001	.001	.001	.000	.000	.000
3.9	.001	.001	.001	.001	.001	.001	.000	.000	.000	.000	.000	.000
4.0	.001	.001	.001	.001	.000	.000	.000	.000	.000	.000	.000	.000

Table 4 ■ Tail Areas for *t* Curves   **821**

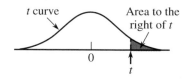

t curve   Area to the right of *t*

0   *t*

Table 4 Tail Areas for *t* Curves (*Continued*)

df *t*	25	26	27	28	29	30	35	40	60	120	∞(= z)
0.0	.500	.500	.500	.500	.500	.500	.500	.500	.500	.500	.500
0.1	.461	.461	.461	.461	.461	.461	.460	.460	.460	.460	.460
0.2	.422	.422	.421	.421	.421	.421	.421	.421	.421	.421	.421
0.3	.383	.383	.383	.383	.383	.383	.383	.383	.383	.382	.382
0.4	.346	.346	.346	.346	.346	.346	.346	.346	.345	.345	.345
0.5	.311	.311	.311	.310	.310	.310	.310	.310	.309	.309	.309
0.6	.277	.277	.277	.277	.277	.277	.276	.276	.275	.275	.274
0.7	.245	.245	.245	.245	.245	.245	.244	.244	.243	.243	.242
0.8	.216	.215	.215	.215	.215	.215	.215	.214	.213	.213	.212
0.9	.188	.188	.188	.188	.188	.188	.187	.187	.186	.185	.184
1.0	.163	.163	.163	.163	.163	.163	.162	.162	.161	.160	.159
1.1	.141	.141	.141	.140	.140	.140	.139	.139	.138	.137	.136
1.2	.121	.120	.120	.120	.120	.120	.119	.119	.117	.116	.115
1.3	.103	.103	.102	.102	.102	.102	.101	.101	.099	.098	.097
1.4	.087	.087	.086	.086	.086	.086	.085	.085	.083	.082	.081
1.5	.073	.073	.073	.072	.072	.072	.071	.071	.069	.068	.067
1.6	.061	.061	.061	.060	.060	.060	.059	.059	.057	.056	.055
1.7	.051	.051	.050	.050	.050	.050	.049	.048	.047	.046	.045
1.8	.042	.042	.042	.041	.041	.041	.040	.040	.038	.037	.036
1.9	.035	.034	.034	.034	.034	.034	.033	.032	.031	.030	.029
2.0	.028	.028	.028	.028	.027	.027	.027	.026	.025	.024	.023
2.1	.023	.023	.023	.022	.022	.022	.022	.021	.020	.019	.018
2.2	.019	.018	.018	.018	.018	.018	.017	.017	.016	.015	.014
2.3	.015	.015	.015	.015	.014	.014	.014	.013	.012	.012	.011
2.4	.012	.012	.012	.012	.012	.011	.011	.011	.010	.009	.008
2.5	.010	.010	.009	.009	.009	.009	.009	.008	.008	.007	.006
2.6	.008	.008	.007	.007	.007	.007	.007	.007	.006	.005	.005
2.7	.006	.006	.006	.006	.006	.006	.005	.005	.004	.004	.003
2.8	.005	.005	.005	.005	.005	.004	.004	.004	.003	.003	.003
2.9	.004	.004	.004	.004	.004	.003	.003	.003	.003	.002	.002
3.0	.003	.003	.003	.003	.003	.003	.002	.002	.002	.002	.001
3.1	.002	.002	.002	.002	.002	.002	.002	.002	.001	.001	.001
3.2	.002	.002	.002	.002	.002	.002	.001	.001	.001	.001	.001
3.3	.001	.001	.001	.001	.001	.001	.001	.001	.001	.001	.000
3.4	.001	.001	.001	.001	.001	.001	.001	.001	.001	.000	.000
3.5	.001	.001	.001	.001	.001	.001	.001	.001	.000	.000	.000
3.6	.001	.001	.001	.001	.001	.001	.000	.000	.000	.000	.000
3.7	.001	.001	.000	.000	.000	.000	.000	.000	.000	.000	.000
3.8	.000	.000	.000	.000	.000	.000	.000	.000	.000	.000	.000
3.9	.000	.000	.000	.000	.000	.000	.000	.000	.000	.000	.000
4.0	.000	.000	.000	.000	.000	.000	.000	.000	.000	.000	.000

Table 5 Curves of $\beta = P$(Type II Error) for t Tests

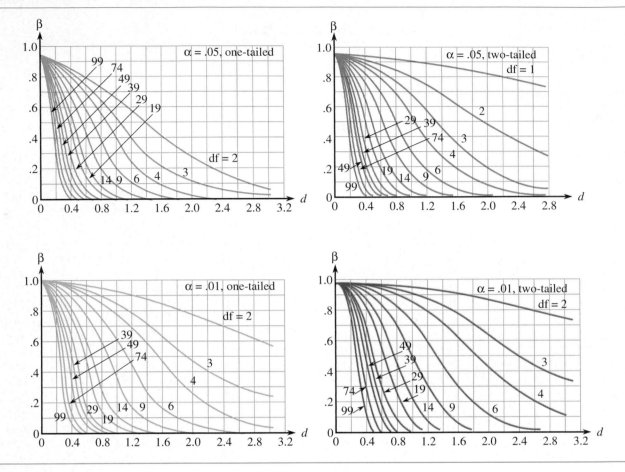

Table 6 ■ Values That Capture Specified Upper-Tail *F* Curve Areas   **823**

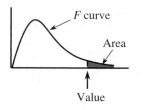

F curve

Area

Value

Table 6 Values That Capture Specified Upper-Tail *F* Curve Areas

| df$_2$ | Area | \multicolumn{10}{c}{df$_1$} |
		1	2	3	4	5	6	7	8	9	10
1	.10	39.86	49.50	53.59	55.83	57.24	58.20	58.91	59.44	59.86	60.19
	.05	161.40	199.50	215.70	224.60	230.20	234.00	236.80	238.90	240.50	241.90
	.01	4052.00	5000.00	5403.00	5625.00	5764.00	5859.00	5928.00	5981.00	6022.00	6056.00
2	.10	8.53	9.00	9.16	9.24	9.29	9.33	9.35	9.37	9.38	9.39
	.05	18.51	19.00	19.16	19.25	19.30	19.33	19.35	19.37	19.38	19.40
	.01	98.50	99.00	99.17	99.25	99.30	99.33	99.36	99.37	99.39	99.40
	.001	998.50	999.00	999.20	999.20	999.30	999.30	999.40	999.40	999.40	999.40
3	.10	5.54	5.46	5.39	5.34	5.31	5.28	5.27	5.25	5.24	5.23
	.05	10.13	9.55	9.28	9.12	9.01	8.94	8.89	8.85	8.81	8.79
	.01	34.12	30.82	29.46	28.71	28.24	27.91	27.67	27.49	27.35	27.23
	.001	167.00	148.50	141.10	137.10	134.60	132.80	131.60	130.60	129.90	129.20
4	.10	4.54	4.32	4.19	4.11	4.05	4.01	3.98	3.95	3.94	3.92
	.05	7.71	6.94	6.59	6.39	6.26	6.16	6.09	6.04	6.00	5.96
	.01	21.20	18.00	16.69	15.98	15.52	15.21	14.98	14.80	14.66	14.55
	.001	74.14	61.25	56.18	53.44	51.71	50.53	49.66	49.00	48.47	48.05
5	.10	4.06	3.78	3.62	3.52	3.45	3.40	3.37	3.34	3.32	3.30
	.05	6.61	5.79	5.41	5.19	5.05	4.95	4.88	4.82	4.77	4.74
	.01	16.26	13.27	12.06	11.39	10.97	10.67	10.46	10.29	10.16	10.05
	.001	47.18	37.12	33.20	31.09	29.75	28.83	28.16	27.65	27.24	26.92
6	.10	3.78	3.46	3.29	3.18	3.11	3.05	3.01	2.98	2.96	2.94
	.05	5.99	5.14	4.76	4.53	4.39	4.28	4.21	4.15	4.10	4.06
	.01	13.75	10.92	9.78	9.15	8.75	8.47	8.26	8.10	7.98	7.87
	.001	35.51	27.00	23.70	21.92	20.80	20.03	19.46	19.03	18.69	18.41
7	.10	3.59	3.26	3.07	2.96	2.88	2.83	2.78	2.75	2.72	2.70
	.05	5.59	4.74	4.35	4.12	3.97	3.87	3.79	3.73	3.68	3.64
	.01	12.25	9.55	8.45	7.85	7.46	7.19	6.99	6.84	6.72	6.62
	.001	29.25	21.69	18.77	17.20	16.21	15.52	15.02	14.63	14.33	14.08
8	.10	3.46	3.11	2.92	2.81	2.73	2.67	2.62	2.59	2.56	2.54
	.05	5.32	4.46	4.07	3.84	3.69	3.58	3.50	3.44	3.39	3.35
	.01	11.26	8.65	7.59	7.01	6.63	6.37	6.18	6.03	5.91	5.81
	.001	25.41	18.49	15.83	14.39	13.48	12.86	12.40	12.05	11.77	11.54
9	.10	3.36	3.01	2.81	2.69	2.61	2.55	2.51	2.47	2.44	2.42
	.05	5.12	4.26	3.86	3.63	3.48	3.37	3.29	3.23	3.18	3.14
	.01	10.56	8.02	6.99	6.42	6.06	5.80	5.61	5.47	5.35	5.26
	.001	22.86	16.39	13.90	12.56	11.71	11.13	10.70	10.37	10.11	9.89

(Continued)

Table 6 Values That Capture Specified Upper-Tail *F* Curve Areas (*Continued*)

df₂	Area	1	2	3	4	5	6	7	8	9	10
											df₁
10	.10	3.29	2.92	2.73	2.61	2.52	2.46	2.41	2.38	2.35	2.32
	.05	4.96	4.10	3.71	3.48	3.33	3.22	3.14	3.07	3.02	2.98
	.01	10.04	7.56	6.55	5.99	5.64	5.39	5.20	5.06	4.94	4.85
	.001	21.04	14.91	12.55	11.28	10.48	9.93	9.52	9.20	8.96	8.75
11	.10	3.23	2.86	2.66	2.54	2.45	2.39	2.34	2.30	2.27	2.25
	.05	4.84	3.98	3.59	3.36	3.20	3.09	3.01	2.95	2.90	2.85
	.01	9.65	7.21	6.22	5.67	5.32	5.07	4.89	4.74	4.63	4.54
	.001	19.69	13.81	11.56	10.35	9.58	9.05	8.66	8.35	8.12	7.92
12	.10	3.18	2.81	2.61	2.48	2.39	2.33	2.28	2.24	2.21	2.19
	.05	4.75	3.89	3.49	3.26	3.11	3.00	2.91	2.85	2.80	2.75
	.01	9.33	6.93	5.95	5.41	5.06	4.82	4.64	4.50	4.39	4.30
	.001	18.64	12.97	10.80	9.63	8.89	8.38	8.00	7.71	7.48	7.29
13	.10	3.14	2.76	2.56	2.43	2.35	2.28	2.23	2.20	2.16	2.14
	.05	4.67	3.81	3.41	3.18	3.03	2.92	2.83	2.77	2.71	2.67
	.01	9.07	6.70	5.74	5.21	4.86	4.62	4.44	4.30	4.19	4.10
	.001	17.82	12.31	10.21	9.07	8.35	7.86	7.49	7.21	6.98	6.80
14	.10	3.10	2.73	2.52	2.39	2.31	2.24	2.19	2.15	2.12	2.10
	.05	4.60	3.74	3.34	3.11	2.96	2.85	2.76	2.70	2.65	2.60
	.01	8.86	6.51	5.56	5.04	4.69	4.46	4.28	4.14	4.03	3.94
	.001	17.14	11.78	9.73	8.62	7.92	7.44	7.08	6.80	6.58	6.40
15	.10	3.07	2.70	2.49	2.36	2.27	2.21	2.16	2.12	2.09	2.06
	.05	4.54	3.68	3.29	3.06	2.90	2.79	2.71	2.64	2.59	2.54
	.01	8.68	6.36	5.42	4.89	4.56	4.32	4.14	4.00	3.89	3.80
	.001	16.59	11.34	9.34	8.25	7.57	7.09	6.74	6.47	6.26	6.08
16	.10	3.05	2.67	2.46	2.33	2.24	2.18	2.13	2.09	2.06	2.03
	.05	4.49	3.63	3.24	3.01	2.85	2.74	2.66	2.59	2.54	2.49
	.01	8.53	6.23	5.29	4.77	4.44	4.20	4.03	3.89	3.78	3.69
	.001	16.12	10.97	9.01	7.94	7.27	6.80	6.46	6.19	5.98	5.81
17	.10	3.03	2.64	2.44	2.31	2.22	2.15	2.10	2.06	2.03	2.00
	.05	4.45	3.59	3.20	2.96	2.81	2.70	2.61	2.55	2.49	2.45
	.01	8.40	6.11	5.18	4.67	4.34	4.10	3.93	3.79	3.68	3.59
	.001	15.72	10.66	8.73	7.68	7.02	6.56	6.22	5.96	5.75	5.58
18	.10	3.01	2.62	2.42	2.29	2.20	2.13	2.08	2.04	2.00	1.98
	.05	4.41	3.55	3.16	2.93	2.77	2.66	2.58	2.51	2.46	2.41
	.01	8.29	6.01	5.09	4.58	4.25	4.01	3.84	3.71	3.60	3.51
	.001	15.38	10.39	8.49	7.46	6.81	6.35	6.02	5.76	5.56	5.39
19	.10	2.99	2.61	2.40	2.27	2.18	2.11	2.06	2.02	1.98	1.96
	.05	4.38	3.52	3.13	2.90	2.74	2.63	2.54	2.48	2.42	2.38
	.01	8.18	5.93	5.01	4.50	4.17	3.94	3.77	3.63	3.52	3.43
	.001	15.08	10.16	8.28	7.27	6.62	6.18	5.85	5.59	5.39	5.22

Table 6 ■ Values That Capture Specified Upper-Tail *F* Curve Areas   **825**

Table 6 Values That Capture Specified Upper-Tail *F* Curve Areas (*Continued*)

df_2	Area	df_1									
		1	2	3	4	5	6	7	8	9	10
20	.10	2.97	2.59	2.38	2.25	2.16	2.09	2.04	2.00	1.96	1.94
	.05	4.35	3.49	3.10	2.87	2.71	2.60	2.51	2.45	2.39	2.35
	.01	8.10	5.85	4.94	4.43	4.10	3.87	3.70	3.56	3.46	3.37
	.001	14.82	9.95	8.10	7.10	6.46	6.02	5.69	5.44	5.24	5.08
21	.10	2.96	2.57	2.36	2.23	2.14	2.08	2.02	1.98	1.95	1.92
	.05	4.32	3.47	3.07	2.84	2.68	2.57	2.49	2.42	2.37	2.32
	.01	8.02	5.78	4.87	4.37	4.04	3.81	3.64	3.51	3.40	3.31
	.001	14.59	9.77	7.94	6.95	6.32	5.88	5.56	5.31	5.11	4.95
22	.10	2.95	2.56	2.35	2.22	2.13	2.06	2.01	1.97	1.93	1.90
	.05	4.30	3.44	3.05	2.82	2.66	2.55	2.46	2.40	2.34	2.30
	.01	7.95	5.72	4.82	4.31	3.99	3.76	3.59	3.45	3.35	3.26
	.001	14.38	9.61	7.80	6.81	6.19	5.76	5.44	5.19	4.99	4.83
23	.10	2.94	2.55	2.34	2.21	2.11	2.05	1.99	1.95	1.92	1.89
	.05	4.28	3.42	3.03	2.80	2.64	2.53	2.44	2.37	2.32	2.27
	.01	7.88	5.66	4.76	4.26	3.94	3.71	3.54	3.41	3.30	3.21
	.001	14.20	9.47	7.67	6.70	6.08	5.65	5.33	5.09	4.89	4.73
24	.10	2.93	2.54	2.33	2.19	2.10	2.04	1.98	1.94	1.91	1.88
	.05	4.26	3.40	3.01	2.78	2.62	2.51	2.42	2.36	2.30	2.25
	.01	7.82	5.61	4.72	4.22	3.90	3.67	3.50	3.36	3.26	3.17
	.001	14.03	9.34	7.55	6.59	5.98	5.55	5.23	4.99	4.80	4.64
25	.10	2.92	2.53	2.32	2.18	2.09	2.02	1.97	1.93	1.89	1.87
	.05	4.24	3.39	2.99	2.76	2.60	2.49	2.40	2.34	2.28	2.24
	.01	7.77	5.57	4.68	4.18	3.85	3.63	3.46	3.32	3.22	3.13
	.001	13.88	9.22	7.45	6.49	5.89	5.46	5.15	4.91	4.71	4.56
26	.10	2.91	2.52	2.31	2.17	2.08	2.01	1.96	1.92	1.88	1.86
	.05	4.23	3.37	2.98	2.74	2.59	2.47	2.39	2.32	2.27	2.22
	.01	7.72	5.53	4.64	4.14	3.82	3.59	3.42	3.29	3.18	3.09
	.001	13.74	9.12	7.36	6.41	5.80	5.38	5.07	4.83	4.64	4.48
27	.10	2.90	2.51	2.30	2.17	2.07	2.00	1.95	1.91	1.87	1.85
	.05	4.21	3.35	2.96	2.73	2.57	2.46	2.37	2.31	2.25	2.20
	.01	7.68	5.49	4.60	4.11	3.78	3.56	3.39	3.26	3.15	3.06
	.001	13.61	9.02	7.27	6.33	5.73	5.31	5.00	4.76	4.57	4.41
28	.10	2.89	2.50	2.29	2.16	2.06	2.00	1.94	1.90	1.87	1.84
	.05	4.20	3.34	2.95	2.71	2.56	2.45	2.36	2.29	2.24	2.19
	.01	7.64	5.45	4.57	4.07	3.75	3.53	3.36	3.23	3.12	3.03
	.001	13.50	8.93	7.19	6.25	5.66	5.24	4.93	4.69	4.50	4.35
29	.10	2.89	2.50	2.28	2.15	2.06	1.99	1.93	1.89	1.86	1.83
	.05	4.18	3.33	2.93	2.70	2.55	2.43	2.35	2.28	2.22	2.18
	.01	7.60	5.42	4.54	4.04	3.73	3.50	3.33	3.20	3.09	3.00
	.001	13.39	8.85	7.12	6.19	5.59	5.18	4.87	4.64	4.45	4.29

(Continued)

Table 6 Values That Capture Specified Upper-Tail *F* Curve Areas (*Continued*)

df$_2$	Area	df$_1$ 1	2	3	4	5	6	7	8	9	10
30	.10	2.88	2.49	2.28	2.14	2.05	1.98	1.93	1.88	1.85	1.82
	.05	4.17	3.32	2.92	2.69	2.53	2.42	2.33	2.27	2.21	2.16
	.01	7.56	5.39	4.51	4.02	3.70	3.47	3.30	3.17	3.07	2.98
	.001	13.29	8.77	7.05	6.12	5.53	5.12	4.82	4.58	4.39	4.24
40	.10	2.84	2.44	2.23	2.09	2.00	1.93	1.87	1.83	1.79	1.76
	.05	4.08	3.23	2.84	2.61	2.45	2.34	2.25	2.18	2.12	2.08
	.01	7.31	5.18	4.31	3.83	3.51	3.29	3.12	2.99	2.89	2.80
	.001	12.61	8.25	6.59	5.70	5.13	4.73	4.44	4.21	4.02	3.87
60	.10	2.79	2.39	2.18	2.04	1.95	1.87	1.82	1.77	1.74	1.71
	.05	4.00	3.15	2.76	2.53	2.37	2.25	2.17	2.10	2.04	1.99
	.01	7.08	4.98	4.13	3.65	3.34	3.12	2.95	2.82	2.72	2.63
	.001	11.97	7.77	6.17	5.31	4.76	4.37	4.09	3.86	3.69	3.54
90	.10	2.76	2.36	2.15	2.01	1.91	1.84	1.78	1.74	1.70	1.67
	.05	3.95	3.10	2.71	2.47	2.32	2.20	2.11	2.04	1.99	1.94
	.01	6.93	4.85	4.01	3.53	3.23	3.01	2.84	2.72	2.61	2.52
	.001	11.57	7.47	5.91	5.06	4.53	4.15	3.87	3.65	3.48	3.34
120	.10	2.75	2.35	2.13	1.99	1.90	1.82	1.77	1.72	1.68	1.65
	.05	3.92	3.07	2.68	2.45	2.29	2.18	2.09	2.02	1.96	1.91
	.01	6.85	4.79	3.95	3.48	3.17	2.96	2.79	2.66	2.56	2.47
	.001	11.38	7.32	5.78	4.95	4.42	4.04	3.77	3.55	3.38	3.24
240	.10	2.73	2.32	2.10	1.97	1.87	1.80	1.74	1.70	1.65	1.63
	.05	3.88	3.03	2.64	2.41	2.25	2.14	2.04	1.98	1.92	1.87
	.01	6.74	4.69	3.86	3.40	3.09	2.88	2.71	2.59	2.48	2.40
	.001	11.10	7.11	5.60	4.78	4.25	3.89	3.62	3.41	3.24	3.09
∞	.10	2.71	2.30	2.08	1.94	1.85	1.77	1.72	1.67	1.63	1.60
	.05	3.84	3.00	2.60	2.37	2.21	2.10	2.01	1.94	1.88	1.83
	.01	6.63	4.61	3.78	3.32	3.02	2.80	2.64	2.51	2.41	2.32
	.001	10.83	6.91	5.42	4.62	4.10	3.74	3.47	3.27	3.10	2.96

Table 7 Critical Values of q for the Studentized Range Distribution

Error df	Confidence level	\multicolumn{8}{c}{Number of populations, treatments, or levels being compared}
		3	4	5	6	7	8	9	10
5	95%	4.60	5.22	5.67	6.03	6.33	6.58	6.80	6.99
	99%	6.98	7.80	8.42	8.91	9.32	9.67	9.97	10.24
6	95%	4.34	4.90	5.30	5.63	5.90	6.12	6.32	6.49
	99%	6.33	7.03	7.56	7.97	8.32	8.61	8.87	9.10
7	95%	4.16	4.68	5.06	5.36	5.61	5.82	6.00	6.16
	99%	5.92	6.54	7.01	7.37	7.68	7.94	8.17	8.37
8	95%	4.04	4.53	4.89	5.17	5.40	5.60	5.77	5.92
	99%	5.64	6.20	6.62	6.96	7.24	7.47	7.68	7.86
9	95%	3.95	4.41	4.76	5.02	5.24	5.43	5.59	5.74
	99%	5.43	5.96	6.35	6.66	6.91	7.13	7.33	7.49
10	95%	3.88	4.33	4.65	4.91	5.12	5.30	5.46	5.60
	99%	5.27	5.77	6.14	6.43	6.67	6.87	7.05	7.21
11	95%	3.82	4.26	4.57	4.82	5.03	5.20	5.35	5.49
	99%	5.15	5.62	5.97	6.25	6.48	6.67	6.84	6.99
12	95%	3.77	4.20	4.51	4.75	4.95	5.12	5.27	5.39
	99%	5.05	5.50	5.84	6.10	6.32	6.51	6.67	6.81
13	95%	3.73	4.15	4.45	4.69	4.88	5.05	5.19	5.32
	99%	4.96	5.40	5.73	5.98	6.19	6.37	6.53	6.67
14	95%	3.70	4.11	4.41	4.64	4.83	4.99	5.13	5.25
	99%	4.89	5.32	5.63	5.88	6.08	6.26	6.41	6.54
15	95%	3.67	4.08	4.37	4.59	4.78	4.94	5.08	5.20
	99%	4.84	5.25	5.56	5.80	5.99	6.16	6.31	6.44
16	95%	3.65	4.05	4.33	4.56	4.74	4.90	5.03	5.15
	99%	4.79	5.19	5.49	5.72	5.92	6.08	6.22	6.35
17	95%	3.63	4.02	4.30	4.52	4.70	4.86	4.99	5.11
	99%	4.74	5.14	5.43	5.66	5.85	6.01	6.15	6.27
18	95%	3.61	4.00	4.28	4.49	4.67	4.82	4.96	5.07
	99%	4.70	5.09	5.38	5.60	5.79	5.94	6.08	6.20
19	95%	3.59	3.98	4.25	4.47	4.65	4.79	4.92	5.04
	99%	4.67	5.05	5.33	5.55	5.73	5.89	6.02	6.14
20	95%	3.58	3.96	4.23	4.45	4.62	4.77	4.90	5.01
	99%	4.64	5.02	5.29	5.51	5.69	5.84	5.97	6.09
24	95%	3.53	3.90	4.17	4.37	4.54	4.68	4.81	4.92
	99%	4.55	4.91	5.17	5.37	5.54	5.69	5.81	5.92
30	95%	3.49	3.85	4.10	4.30	4.46	4.60	4.72	4.82
	99%	4.45	4.80	5.05	5.24	5.40	5.54	5.65	5.76
40	95%	3.44	3.79	4.04	4.23	4.39	4.52	4.63	4.73
	99%	4.37	4.70	4.93	5.11	5.26	5.39	5.50	5.60
60	95%	3.40	3.74	3.98	4.16	4.31	4.44	4.55	4.65
	99%	4.28	4.59	4.82	4.99	5.13	5.25	5.36	5.45
120	95%	3.36	3.68	3.92	4.10	4.24	4.36	4.47	4.56
	99%	4.20	4.50	4.71	4.87	5.01	5.12	5.21	5.30
∞	95%	3.31	3.63	3.86	4.03	4.17	4.29	4.39	4.47
	99%	4.12	4.40	4.60	4.76	4.88	4.99	5.08	5.16

(Continued)

Table 8 Upper-Tail Areas for Chi-Square Distributions

Right-tail area	df = 1	df = 2	df = 3	df = 4	df = 5
>0.100	< 2.70	< 4.60	< 6.25	< 7.77	< 9.23
0.100	2.70	4.60	6.25	7.77	9.23
0.095	2.78	4.70	6.36	7.90	9.37
0.090	2.87	4.81	6.49	8.04	9.52
0.085	2.96	4.93	6.62	8.18	9.67
0.080	3.06	5.05	6.75	8.33	9.83
0.075	3.17	5.18	6.90	8.49	10.00
0.070	3.28	5.31	7.06	8.66	10.19
0.065	3.40	5.46	7.22	8.84	10.38
0.060	3.53	5.62	7.40	9.04	10.59
0.055	3.68	5.80	7.60	9.25	10.82
0.050	3.84	5.99	7.81	9.48	11.07
0.045	4.01	6.20	8.04	9.74	11.34
0.040	4.21	6.43	8.31	10.02	11.64
0.035	4.44	6.70	8.60	10.34	11.98
0.030	4.70	7.01	8.94	10.71	12.37
0.025	5.02	7.37	9.34	11.14	12.83
0.020	5.41	7.82	9.83	11.66	13.38
0.015	5.91	8.39	10.46	12.33	14.09
0.010	6.63	9.21	11.34	13.27	15.08
0.005	7.87	10.59	12.83	14.86	16.74
0.001	10.82	13.81	16.26	18.46	20.51
<0.001	>10.82	>13.81	>16.26	>18.46	>20.51

Right-tail area	df = 6	df = 7	df = 8	df = 9	df = 10
>0.100	<10.64	<12.01	<13.36	<14.68	<15.98
0.100	10.64	12.01	13.36	14.68	15.98
0.095	10.79	12.17	13.52	14.85	16.16
0.090	10.94	12.33	13.69	15.03	16.35
0.085	11.11	12.50	13.87	15.22	16.54
0.080	11.28	12.69	14.06	15.42	16.75
0.075	11.46	12.88	14.26	15.63	16.97
0.070	11.65	13.08	14.48	15.85	17.20
0.065	11.86	13.30	14.71	16.09	17.44
0.060	12.08	13.53	14.95	16.34	17.71
0.055	12.33	13.79	15.22	16.62	17.99
0.050	12.59	14.06	15.50	16.91	18.30
0.045	12.87	14.36	15.82	17.24	18.64
0.040	13.19	14.70	16.17	17.60	19.02
0.035	13.55	15.07	16.56	18.01	19.44
0.030	13.96	15.50	17.01	18.47	19.92
0.025	14.44	16.01	17.53	19.02	20.48
0.020	15.03	16.62	18.16	19.67	21.16
0.015	15.77	17.39	18.97	20.51	22.02
0.010	16.81	18.47	20.09	21.66	23.20
0.005	18.54	20.27	21.95	23.58	25.18
0.001	22.45	24.32	26.12	27.87	29.58
<0.001	>22.45	>24.32	>26.12	>27.87	>29.58

Table 8 ■ Upper-Tail Areas for Chi-Sqare Distributions **829**

Table 8 Upper-Tail Areas for Chi-Square Distributions (*Continued*)

Right-tail area	df = 11	df = 12	df = 13	df = 14	df = 15
>0.100	<17.27	<18.54	<19.81	<21.06	<22.30
0.100	17.27	18.54	19.81	21.06	22.30
0.095	17.45	18.74	20.00	21.26	22.51
0.090	17.65	18.93	20.21	21.47	22.73
0.085	17.85	19.14	20.42	21.69	22.95
0.080	18.06	19.36	20.65	21.93	23.19
0.075	18.29	19.60	20.89	22.17	23.45
0.070	18.53	19.84	21.15	22.44	23.72
0.065	18.78	20.11	21.42	22.71	24.00
0.060	19.06	20.39	21.71	23.01	24.31
0.055	19.35	20.69	22.02	23.33	24.63
0.050	19.67	21.02	22.36	23.68	24.99
0.045	20.02	21.38	22.73	24.06	25.38
0.040	20.41	21.78	23.14	24.48	25.81
0.035	20.84	22.23	23.60	24.95	26.29
0.030	21.34	22.74	24.12	25.49	26.84
0.025	21.92	23.33	24.73	26.11	27.48
0.020	22.61	24.05	25.47	26.87	28.25
0.015	23.50	24.96	26.40	27.82	29.23
0.010	24.72	26.21	27.68	29.14	30.57
0.005	26.75	28.29	29.81	31.31	32.80
0.001	31.26	32.90	34.52	36.12	37.69
<0.001	>31.26	>32.90	>34.52	>36.12	>37.69

Right-tail area	df = 16	df = 17	df = 18	df = 19	df = 20
>0.100	<23.54	<24.77	<25.98	<27.20	<28.41
0.100	23.54	24.76	25.98	27.20	28.41
0.095	23.75	24.98	26.21	27.43	28.64
0.090	23.97	25.21	26.44	27.66	28.88
0.085	24.21	25.45	26.68	27.91	29.14
0.080	24.45	25.70	26.94	28.18	29.40
0.075	24.71	25.97	27.21	28.45	29.69
0.070	24.99	26.25	27.50	28.75	29.99
0.065	25.28	26.55	27.81	29.06	30.30
0.060	25.59	26.87	28.13	29.39	30.64
0.055	25.93	27.21	28.48	29.75	31.01
0.050	26.29	27.58	28.86	30.14	31.41
0.045	26.69	27.99	29.28	30.56	31.84
0.040	27.13	28.44	29.74	31.03	32.32
0.035	27.62	28.94	30.25	31.56	32.85
0.030	28.19	29.52	30.84	32.15	33.46
0.025	28.84	30.19	31.52	32.85	34.16
0.020	29.63	30.99	32.34	33.68	35.01
0.015	30.62	32.01	33.38	34.74	36.09
0.010	32.00	33.40	34.80	36.19	37.56
0.005	34.26	35.71	37.15	38.58	39.99
0.001	39.25	40.78	42.31	43.81	45.31
<0.001	>39.25	>40.78	>42.31	>43.81	>45.31

Table 9 Binomial Probabilities

$n = 5$

							π						
x	0.05	0.1	0.2	0.25	0.3	0.4	0.5	0.6	0.7	0.75	0.8	0.9	0.95
0	.774	.590	.328	.237	.168	.078	.031	.010	.002	.001	.000	.000	.000
1	.203	.329	.409	.396	.360	.259	.157	.077	.029	.015	.007	.000	.000
2	.022	.072	.205	.263	.309	.346	.312	.230	.132	.088	.051	.009	.001
3	.001	.009	.051	.088	.132	.230	.312	.346	.309	.263	.205	.072	.022
4	.000	.000	.007	.015	.029	.077	.157	.259	.360	.396	.409	.329	.203
5	.000	.000	.000	.001	.002	.010	.031	.078	.168	.237	.328	.590	.774

$n = 10$

							π						
x	0.05	0.1	0.2	0.25	0.3	0.4	0.5	0.6	0.7	0.75	0.8	0.9	0.95
0	.599	.349	.107	.056	.028	.006	.001	.000	.000	.000	.000	.000	.000
1	.315	.387	.268	.188	.121	.040	.010	.002	.000	.000	.000	.000	.000
2	.075	.194	.302	.282	.233	.121	.044	.011	.001	.000	.000	.000	.000
3	.010	.057	.201	.250	.267	.215	.117	.042	.009	.003	.001	.000	.000
4	.001	.011	.088	.146	.200	.251	.205	.111	.037	.016	.006	.000	.000
5	.000	.001	.026	.058	.103	.201	.246	.201	.103	.058	.026	.001	.000
6	.000	.000	.006	.016	.037	.111	.205	.251	.200	.146	.088	.011	.001
7	.000	.000	.001	.003	.009	.042	.117	.215	.267	.250	.201	.057	.010
8	.000	.000	.000	.000	.001	.011	.044	.121	.233	.282	.302	.194	.075
9	.000	.000	.000	.000	.000	.002	.010	.040	.121	.188	.268	.387	.315
10	.000	.000	.000	.000	.000	.000	.001	.006	.028	.056	.107	.349	.599

Table 9 ■ Binomial Probabilities **831**

Table 9 Binomial Probabilities (*Continued*)

n = 15

							π						
x	0.05	0.1	0.2	0.25	0.3	0.4	0.5	0.6	0.7	0.75	0.8	0.9	0.95
0	.463	.206	.035	.013	.005	.000	.000	.000	.000	.000	.000	.000	.000
1	.366	.343	.132	.067	.030	.005	.000	.000	.000	.000	.000	.000	.000
2	.135	.267	.231	.156	.092	.022	.004	.000	.000	.000	.000	.000	.000
3	.031	.128	.250	.225	.170	.064	.014	.002	.000	.000	.000	.000	.000
4	.004	.043	.188	.225	.218	.126	.041	.007	.001	.000	.000	.000	.000
5	.001	.011	.103	.166	.207	.196	.092	.025	.003	.001	.000	.000	.000
6	.000	.002	.043	.091	.147	.207	.153	.061	.011	.003	.001	.000	.000
7	.000	.000	.014	.040	.081	.177	.196	.118	.035	.013	.003	.000	.000
8	.000	.000	.003	.013	.035	.118	.196	.177	.081	.040	.014	.000	.000
9	.000	.000	.001	.003	.011	.061	.153	.207	.147	.091	.043	.002	.000
10	.000	.000	.000	.001	.003	.025	.092	.196	.207	.166	.103	.011	.001
11	.000	.000	.000	.000	.001	.007	.041	.126	.218	.225	.188	.043	.004
12	.000	.000	.000	.000	.000	.002	.014	.064	.170	.225	.250	.128	.031
13	.000	.000	.000	.000	.000	.000	.004	.022	.092	.156	.231	.267	.135
14	.000	.000	.000	.000	.000	.000	.000	.005	.030	.067	.132	.343	.366
15	.000	.000	.000	.000	.000	.000	.000	.000	.005	.013	.035	.206	.463

n = 20

							π						
x	0.05	0.1	0.2	0.25	0.3	0.4	0.5	0.6	0.7	0.75	0.8	0.9	0.95
0	.358	.122	.012	.003	.001	.000	.000	.000	.000	.000	.000	.000	.000
1	.377	.270	.058	.021	.007	.000	.000	.000	.000	.000	.000	.000	.000
2	.189	.285	.137	.067	.028	.003	.000	.000	.000	.000	.000	.000	.000
3	.060	.190	.205	.134	.072	.012	.001	.000	.000	.000	.000	.000	.000
4	.013	.090	.218	.190	.130	.035	.005	.000	.000	.000	.000	.000	.000
5	.002	.032	.175	.202	.179	.075	.015	.001	.000	.000	.000	.000	.000
6	.000	.009	.109	.169	.192	.124	.037	.005	.000	.000	.000	.000	.000
7	.000	.002	.055	.112	.164	.166	.074	.015	.001	.000	.000	.000	.000
8	.000	.000	.022	.061	.114	.180	.120	.035	.004	.001	.000	.000	.000
9	.000	.000	.007	.027	.065	.160	.160	.071	.012	.003	.000	.000	.000
10	.000	.000	.002	.010	.031	.117	.176	.117	.031	.010	.002	.000	.000
11	.000	.000	.000	.003	.012	.071	.160	.160	.065	.027	.007	.000	.000
12	.000	.000	.000	.001	.004	.035	.120	.180	.114	.061	.022	.000	.000
13	.000	.000	.000	.000	.001	.015	.074	.166	.164	.112	.055	.002	.000
14	.000	.000	.000	.000	.000	.005	.037	.124	.192	.169	.109	.009	.000
15	.000	.000	.000	.000	.000	.001	.015	.075	.179	.202	.175	.032	.002
16	.000	.000	.000	.000	.000	.000	.005	.035	.130	.190	.218	.090	.013
17	.000	.000	.000	.000	.000	.000	.001	.012	.072	.134	.205	.190	.060
18	.000	.000	.000	.000	.000	.000	.000	.003	.028	.067	.137	.285	.189
19	.000	.000	.000	.000	.000	.000	.000	.000	.007	.021	.058	.270	.377
20	.000	.000	.000	.000	.000	.000	.000	.000	.001	.003	.012	.122	.358

Table 9 Binomial Probabilities (*Continued*)

n = 25

							π						
x	0.05	0.1	0.2	0.25	0.3	0.4	0.5	0.6	0.7	0.75	0.8	0.9	0.95
0	.277	.072	.004	.001	.000	.000	.000	.000	.000	.000	.000	.000	.000
1	.365	.199	.023	.006	.002	.000	.000	.000	.000	.000	.000	.000	.000
2	.231	.266	.071	.025	.007	.000	.000	.000	.000	.000	.000	.000	.000
3	.093	.227	.136	.064	.024	.002	.000	.000	.000	.000	.000	.000	.000
4	.027	.138	.187	.118	.057	.007	.000	.000	.000	.000	.000	.000	.000
5	.006	.065	.196	.164	.103	.020	.002	.000	.000	.000	.000	.000	.000
6	.001	.024	.163	.183	.148	.045	.005	.000	.000	.000	.000	.000	.000
7	.000	.007	.111	.166	.171	.080	.015	.001	.000	.000	.000	.000	.000
8	.000	.002	.062	.124	.165	.120	.032	.003	.000	.000	.000	.000	.000
9	.000	.000	.030	.078	.134	.151	.061	.009	.000	.000	.000	.000	.000
10	.000	.000	.011	.042	.091	.161	.097	.021	.002	.000	.000	.000	.000
11	.000	.000	.004	.019	.054	.146	.133	.044	.004	.001	.000	.000	.000
12	.000	.000	.002	.007	.027	.114	.155	.076	.011	.002	.000	.000	.000
13	.000	.000	.000	.002	.011	.076	.155	.114	.027	.007	.002	.000	.000
14	.000	.000	.000	.001	.004	.044	.133	.146	.054	.019	.004	.000	.000
15	.000	.000	.000	.000	.002	.021	.097	.161	.091	.042	.011	.000	.000
16	.000	.000	.000	.000	.000	.009	.061	.151	.134	.078	.030	.000	.000
17	.000	.000	.000	.000	.000	.003	.032	.120	.165	.124	.062	.002	.000
18	.000	.000	.000	.000	.000	.001	.015	.080	.171	.166	.111	.007	.000
19	.000	.000	.000	.000	.000	.000	.005	.045	.148	.183	.163	.024	.001
20	.000	.000	.000	.000	.000	.000	.002	.020	.103	.164	.196	.065	.006
21	.000	.000	.000	.000	.000	.000	.000	.007	.057	.118	.187	.138	.027
22	.000	.000	.000	.000	.000	.000	.000	.002	.024	.064	.136	.227	.093
23	.000	.000	.000	.000	.000	.000	.000	.000	.007	.025	.071	.266	.231
24	.000	.000	.000	.000	.000	.000	.000	.000	.002	.006	.023	.199	.365
25	.000	.000	.000	.000	.000	.000	.000	.000	.000	.001	.004	.072	.277

Appendix B
References

Chapter 1

Hock, Roger R. *Forty Studies That Changed Psychology: Exploration into the History of Psychological Research*. New York: Prentice-Hall, 1995.

Moore, David. *Statistics: Concepts and Controversies*, 6th ed. New York: W. H. Freeman, 2006. (A nice, informal survey of statistical concepts and reasoning.)

Peck, Roxy, ed. *Statistics: A Guide to the Unknown*, 4th ed. Belmont, CA: Duxbury Press, 2006. (Short, nontechnical articles by a number of well-known statisticians and users of statistics on the application of statistics in various disciplines and subject areas.)

Utts, Jessica. *Seeing Through Statistics*, 3rd ed. Belmont, CA: Duxbury Press, 2005. (A nice introduction to the fundamental ideas of statistical reasoning.)

Chapter 2

Cobb, George. *Introduction to the Design and Analysis of Experiments*. New York: Springer-Verlag, 1998. (An interesting and thorough introduction to the design of experiments.)

Freedman, David, Robert Pisani, and Roger Purves. *Statistics*, 3rd ed. New York: W. W. Norton, 1997. (The first two chapters contain some interesting examples of both well-designed and poorly designed experimental studies.)

Lohr, Sharon. *Sampling Design and Analysis*. Belmont, CA: Duxbury Press, 1996. (A nice discussion of sampling and sources of bias at an accessible level.)

Moore, David. *Statistics: Concepts and Controversies*, 6th ed. New York: W. H. Freeman, 2001. (Contains an excellent chapter on the advantages and pitfalls of experimentation and another chapter in a similar vein on sample surveys and polls.)

Scheaffer, Richard L., William Mendenhall, and Lyman Ott. *Elementary Survey Sampling*, 5th ed. Belmont, CA: Duxbury Press, 1996. (An accessible yet thorough treatment of the subject.)

Sudman, Seymour, and Norman Bradburn. *Asking Questions: A Practical Guide to Questionnaire Design*. San Francisco: Jossey-Bass, 1982. (A good discussion of the art of questionnaire design.)

Chapter 3

Chambers, John, William Cleveland, Beat Kleiner, and Paul Tukey. *Graphical Methods for Data Analysis*. Belmont, CA: Wadsworth, 1983. (This is an excellent survey of methods, illustrated with numerous interesting examples.)

Cleveland, William. *The Elements of Graphing Data*, 2nd ed. Summit, NJ: Hobart Press, 1994. (An informal and informative introduction to various aspects of graphical analysis.)

Freedman, David, Robert Pisani, and Roger Purves. *Statistics*, 3rd ed. New York: W. W. Norton, 1997. (An excellent, informal introduction to concepts, with some insightful cautionary examples concerning misuses of statistical methods.)

Moore, David. *Statistics: Concepts and Controversies*, 6th ed. New York: W. H. Freeman, 2006. (A nonmathematical yet highly entertaining introduction to our discipline. Two thumbs up!)

Chapter 4

Chambers, John, William Cleveland, Beat Kleiner, and Paul Tukey. *Graphical Methods for Data Analysis*. Belmont, CA: Wadsworth, 1983. (This is an excellent survey of methods, illustrated with numerous interesting examples.)

Cleveland, William. *The Elements of Graphing Data*, 2nd ed. Summit, NJ: Hobart Press, 1994. (An informal and informative introduction to various aspects of graphical analysis.)

Freedman, David, Robert Pisani, and Roger Purves. *Statistics*, 3rd ed. New York: W. W. Norton, 1997. (An excellent, informal introduction to concepts, with some insightful cautionary examples concerning misuses of statistical methods.)

Moore, David. *Statistics: Concepts and Controversies*, 6th ed. New York: W. H. Freeman, 2006. (A nonmathematical yet highly entertaining introduction to our discipline. Two thumbs up!)

Chapter 5

Neter, John, William Wasserman, and Michael Kutner. *Applied Linear Statistical Models*, 4th ed. New York: McGraw-Hill, 1996. (The first half of this book gives a comprehensive treatment of regression analysis without overindulging in mathematical development; a highly recommended reference.)

Chapter 6

Devore, Jay L. *Probability and Statistics for Engineering and the Sciences*, 7th ed. Pacific Grove, CA: Brooks/Cole, 2007. (The treatment of probability in this source is more comprehensive and at a somewhat higher mathematical level than ours is in this textbook.)

Mosteller, Frederick, Robert Rourke, and George Thomas. *Probability with Statistical Applications*. Reading, MA: Addison-Wesley, 1970. (A good introduction to probability at a modest mathematical level.)

Chapter 8

Freedman, David, Robert Pisani, Roger Purves. *Statistics*, 3rd ed. New York: W. W. Norton, 1997. (This book gives an excellent informal discussion of sampling distributions.)

Chapter 9

Devore, Jay L. *Probability and Statistics for Engineering and the Sciences*, 7th ed. Pacific Grove, CA: Brooks/Cole, 2007. (This book gives a somewhat general introduction to confidence intervals.)

Freedman, David, Robert Pisani, and Roger Purves, *Statistics*, 3rd ed. New York: W. W. Norton, 1997. (This book contains an informal discussion of confidence intervals.)

Chapter 10

The books by Freedman et al. and Moore listed in previous chapter references are excellent sources. Their orientation is primarily conceptual, with a minimum of mathematical development, and both sources offer many valuable insights.

Chapter 11

Devore, Jay. *Probability and Statistics for Engineering and the Sciences*, 7th ed. Belmont, Calif.: Duxbury Press, 2007. (Contains a somewhat more comprehensive treatment of the inferential material presented in this and the two previous chapters, although the notation is a bit more mathematical than that of the present textbook.)

Chapter 12

Agresti, Alan, and B. Finlay. *Statistical Methods for the Social Sciences*, 3rd ed. Englewood Cliffs, NJ: Prentice-Hall, 1997. (This book includes a good discussion of measures of association for two-way frequency tables.)

Everitt, B. S. *The Analysis of Contingency Tables*. New York: Halsted Press, 1977. (A compact but informative survey of methods for analyzing categorical data.)

Mosteller, Frederick, and Robert Rourke. *Sturdy Statistics*. Reading, Mass.: Addison-Wesley, 1973. (Contains several readable chapters on the varied uses of the chi-square statistic.)

Chapter 13

Neter, John, William Wasserman, and Michael Kutner. *Applied Linear Statistical Models*, 4th ed. New York: McGraw-Hill, 1996. (The first half of this book gives a comprehensive treatment of regression analysis without overindulging in mathematical development; a highly recommended reference.)

Chapter 14

Additional reference is listed at the end of Chapter 5.

Kerlinger, Fred N., and Elazar J. Pedhazur, *Multiple Regression in Behavioral Research*. Austin, Texas: Holt, Rinehart & Winston, 1973. (A readable introduction to multiple regression.)

Chapter 15

Miller, Rupert. *Beyond ANOVA: The Basics of Applied Statistics*. New York: Wiley, 1986. (This book contains a wealth of information concerning violations of basic assumptions and alternative methods of analysis.)

Winer, G. J., D. R. Brown, and K. M. Michels, *Statistical Principles in Experimental Design*, 3rd edition. Boston: McGraw-Hill, 1991. (This book contains extended discussion of ANOVA with many examples worked out in great detail.)

Chapter 16

Conover, W. J. *Practical Nonparametric Statistics*, 3rd ed. New York: Wiley, 1999. (An accessible presentation of distribution-free methods.)

Daniel, Wayne. *Applied Nonparametric Statistics*, 2nd ed. Boston: PWS-Kent, 1990. (An elementary presentation of distribution-free methods, including the rank-sum test discussed in Section 16.1.)

Mosteller, Frederick, and Richard Rourke. *Sturdy Statistics*. Reading, Mass.: Addison-Wesley, 1973. (A readable, intuitive development of distribution-free methods, including those based on ranks.)

Expanded Answers to Exercises

Chapter 1

1.1 Descriptive statistics is the branch of statistics consisting of those methods for summarizing the values in a data set. Inferential statistics refers to procedures that allow conclusions to be drawn about a population based on information from a sample.

1.2 A population is the entire collection of objects or individuals about which information is desired. A sample is a subset of the population selected for study in some prescribed manner.

1.4 The sample is the 2121 children that were in the study; the population of interest is all children between the ages of 1 and 4.

1.5 The population is the entire student body—15,000 students. The sample consists of the 200 students interviewed.

1.6 The population consists of all 7000 property owners. The sample consists of the 500 property owners surveyed.

1.7 The population consists of all single-family homes in Northridge. The sample consists of the 100 homes selected for inspection.

1.8 The population consists of all 2006 Mazda 6s. The sample consists of the six Mazdas of this type selected for testing.

1.9 The population consists of all 5000 bricks in the lot. The sample consists of the 100 bricks selected for inspection.

1.10 b. categorical **d.** numerical (continuous)

1.11 a. categorical **b.** categorical

1.14 For example: **a.** General Motors, Toyota, Aston Martin, Ford, Jaguar, . . . **b.** 3.23, 2.92, 4.0, 2.8, . . . **c.** 2, 0, 1, 4, 3, . . . **d.** 49.2, 48.84, 50.3, 50.23, . . . **e.** 10, 15.5, 17, 3, 6.5, . . .

1.15 a. Gender, Brand, and Telephone Area Code **b.** Number of previous motorcycles owned

1.16

Box office (in millions of dollars)

Most summer movies have box office sales of between $50 million and $150 million. There is a small cluster of three films that have sales of about $200 million. The two top box office totals for the summer of 2002 were noticeably higher.

1.17

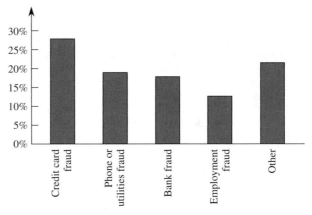

The most common single type of identity theft is credit card fraud, with 28% of the total complaints, followed by phone and utilities and bank fraud, both just under 20%. Employment fraud is less at 13%.

1.18 a. Categorical. **b.** No, because a dotplot is used for numerical data.

c.

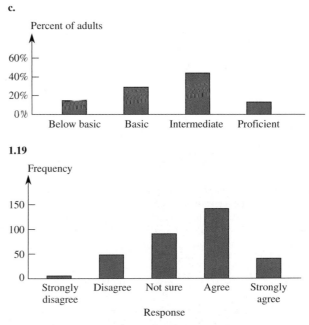

1.19

1.20 a.

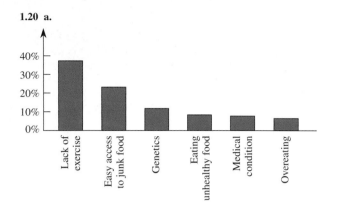

b. The categories "Easy access to junk food'" "Eating unhealthy food," and "Overeating" could all be combined into a single category. They can all be described as "Poor eating habits."

1.21

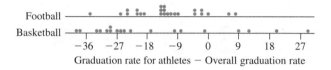

For both sports, there are few universities where the overall graduation rate is lower than that of the scholarship athletes. At the schools where athletes do better than the overall rate, there are more schools where the basketball players do well; however, there are some schools where the basketball players graduate at a much poorer rate than overall.

1.22

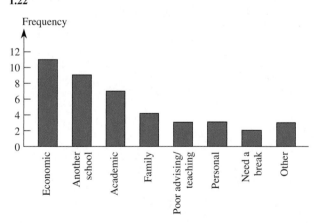

The most common reasons that college seniors have for leaving during their senior year before graduation seem to be for academic reasons or financial reasons rather than personal reasons.

1.23 a.

Grade	Frequency	Relative Frequency
A+	11	0.306
A	10	0.278
B	3	0.083
C	4	0.111
D	4	0.111
F	4	0.111
Total	**36**	**1.000**

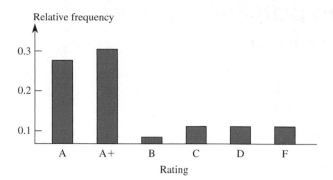

Two-thirds of the beaches are rated B or above.

1.24

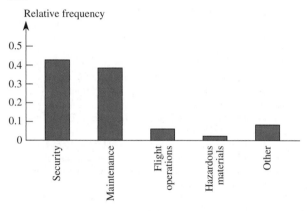

The two most common violations are security and maintenance. Each is responsible for more than twice the number of fines imposed for the rest of the violations put together (18%).

1.25

Sleepy on the Job?	Relative Frequency
Not at all	0.31
Few days each month	0.40
Few days each week	0.22
Daily occurrence	0.07

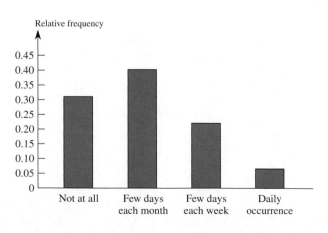

1.26

Type of family	Percentage
Nonfamily	29
Married with children	27
Married without children	29
Single-parent family	15
Total	100

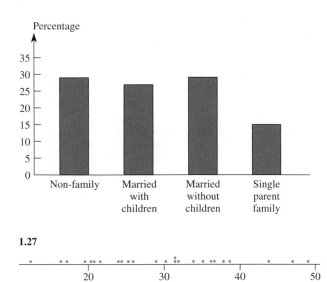

1.27

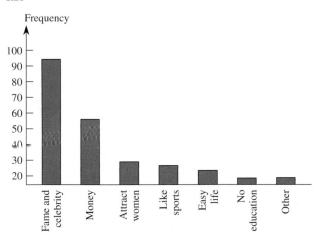

Most business schools have acceptance rates between 15% and 40%. There are three schools that have acceptance rates of more than 40%.

1.28

1.29

Sport	Frequency	Rel. Freq.
Touch football (TF)	38	0.226
Soccer (SO)	24	0.143
Basketball (BK)	19	0.113
Baseball/Softball (BA)	11	0.065
Jogging/Running (JR)	11	0.065
Bicycling (BI)	11	0.065
Volleyball (VO)	7	0.042
Others (OT)	47	0.280
	$n = 168$	0.999

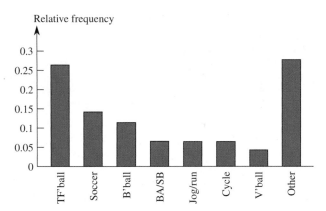

1.30

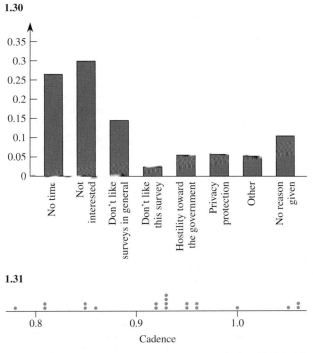

1.31

The display suggests that a representative value is around .93. The 20 observations are quite spread out around this value. There is a gap that separates the three smallest and the three largest cadence values from the rest of the data.

Chapter 2

2.3 Better educated, wealthier, and more active people tend to have better health care, which may provide an alternative explanation for reduced risk.

2.4 a. Observational study; group assignment not determined by researchers.

2.5 No cause-and-effect conclusion can be made on the basis of an observational study.

2.6 a. No, observational study using volunteers.   **b.** No, no reason to think that sample is representative of *Oprah* watchers.

2.7 a. The definition of affluent Americans was household income over $75,000 and simple random sample.

2.8 Observational study—potential confounding variables include diet and lifestyle of those who live in the South.

2.9 No, it is an observational study and cause-and-effect conclusions cannot be made on the basis of an observational study.

2.10 Answers will vary, and could include drawing names from a hat, using random numbers to select from numbered list, etc.

2.11 Number names on list, use random number generator to select 30 numbers between 1 and 500. Signatures corresponding to the 30 selected numbers constitute the random sample.

2.12 Number cases, use random numbers to select 50.

2.13 Stratified sampling would be worthwhile if, within the resulting strata, the elements are more homogeneous with respect to textbook cost than the population as a whole.
a. Worthwhile   **b.** Worthwhile   **c.** Not worthwhile

2.14 Method B would be preferable as the rows may differ in growing conditions.
b. No, the probability of selection is higher for part-time students.

2.15 b. Stratified random sampling   **d.** Simple random sample

2.16 a. Answers will vary.   **b.** No, the probability of selection is higher for part-time students.

2.17 Using a convenience sample may introduce selection bias.

2.18 Volunteers tend to differ from the population of interest.

2.19 a. Use three-digit random numbers to select pages; count words on these pages.   **b.** It would be reasonable to stratify by chapter (some topics more "wordy" than others) or to stratify by chapter text, exercises, graphing calculator explorations, index, tables, etc.   **c.** Choose one of the first *k* pages at random and the select every *k*th page thereafter.
d. Randomly choose *k* chapters and count the words on every page in the selected chapters.   **e.** Answers will vary.   **f.** Answers will vary.

2.20 Selection bias has been eliminated by random sampling; the sample of 1000 registered voters is a good representation of the registered voters in California even though it consists of a small fraction of the population.

2.21 Randomization is used in selecting the households and available subjects in each family. However, the sampling plan eliminates households with no telephone and may overrepresent households with many telephones.

2.22 No, nonresponse rate was high (1128/1260 [89.5%]) and the sample excludes all who do not have Internet.

2.23 Small sample of people who chose to attend the workshop; probably not representative of the population of interest.

2.24 There is no reason to think that the strata for Schemes 1 or 3 would be more homogeneous than the population as a whole. In Scheme 2 however, the strata may tend to be more similar with respect to income, which may be related to support for a sales tax increase.

2.25 No, volunteer sample

2.26 Only 2000 completed surveys were returned. The wording of the survey is important; leaving the state may have nothing to do with a doctor's view on managed care.

2.27 Different subpopulations may have used the web site and the telephone.

2.28 Response bias and selection bias.

2.29 One possibility would be to select a random sample of students at the school; there are other reasonable answers.

2.30 Selection bias; the studies looked only at women who had sisters, mothers, and grandmothers who had breast cancer.

2.31 Response variable is time to exit the parking space. Extraneous factors include location of the parking spaces, time of day, etc.

2.32 a. Strength of binding   **b.** Type of glue   **c.** Number of pages in the book, type of cover (paperback vs. hard cover); other possible extraneous factors include paper type, etc.

2.33 a. Select a random sample of students from the school. Choose two comparable IQ tests, Test 1 and Test 2. Randomly assign subjects to one of two groups and administer Test 1 to the first group and Test 2 to the second group. Record the IQ scores. All listen to Mozart piano sonata. Then administer Test 2 to the first group and Test 1 to the second group. Record the IQ scores and compute differences.   **b.** Direct control for extraneous variables such as time of day, Mozart selection, etc. Two IQ tests are used to eliminate the possibility that scores may be higher when retaking the same test.   **c.** This design has "student" as a blocking factor to "block out" the effects of differences in IQ scores across students.   **d.** Random assignment to groups should create "equivalent" experimental groups.

2.34 The researchers must randomize the order in which the cyclists try the three different drinks.

2.35 a. Blocking

2.36 It is important to create comparable groups; if allowed to choose groups, higher IQ subjects might tend to choose to take the test without distractions, lower IQ subjects might choose distractions.

2.37 Other factors could have contributed to the difference in the pregnancy rate.

2.38 To ensure that our experiment doesn't favor one experimental condition over another; for example to avoid having communities from the healthier neighborhoods all in the same group.

2.39 Gender is a useful blocking variable if men and women of the same status tend to differ with respect to "rate of talk."

2.40 A placebo treatment is included to assess whether the treatment really has an effect or whether subjects are responding merely because they think the treatment will help. A control group allows us to separate a treatment effect from an effect that might be due to other intervening factors.

2.41 People may have their own personal beliefs about the effectiveness of various treatments, which can influence their response.

2.42 Answers will vary. See solutions manual for examples.

2.43 a. To have comparable groups   **b.** A control group is needed for comparison. For example, it may be the social aspect of meeting other people that creates the physical changes.

2.44 a. Subjects were volunteers, not randomly selected.   **b.** Observational study; can't draw cause-and-effect conclusions.   **c.** Design must include random assignment to treatments.

2.45 If either the dog handlers or the experimenters knew which patients had cancer, they might give physical clues (consciously or unconsciously) that the dogs might pick up.

2.46 a. To assess whether the treatment really has an effect over and above a possible psychological response.   **b.** To create comparable experimental groups.

2.47 a. No, the judges probably believed that a fancy seafood restaurant would have better chowder than Denny's and this may have influenced the rating.

2.48 Describes a placebo effect; describes the importance of including a placebo when evaluating new medications.

2.49 a. Randomly divide into two groups of 50, one group gets PH50 and the other group gets a placebo nasal spray. Assess before and after the treatment, measure improvement.   **b.** A placebo treatment is needed to see if any improvement is due to PH50 or to any other influence that occurs during the time of testing.   **c.** Blinding of subjects is strongly recommended.

2.50 a. Wouldn't expect a sheep to have a psychological response that would affect kidney function!   **b.** Would allow comparison of amalgam filling and resin filling treatments.   **c.** Reduce the possibility of harm to human subject (kidney damage).

2.51 a. "Forests are being destroyed"   **b.** "Vanishing tropical forests"   **c.** "Manmade extinction of animal and plant species"   **d.** "Destruction of tropical forests"

2.57 b. Parents of children with ADD may be more likely to let child watch TV if it has a calming effect.

2.58 b. No, can't draw cause-and-effect conclusions from an observational study.

2.59 For example, it may be the young, who are more likely to be single, that are at higher risk.

2.60 a. Yes, because of the random assignment of children to treatment.   **b.** No, children were not randomly selected.

2.61 Possibilities include nonresponse bias and selection bias (eliminating anyone without an address).

2.62 Sample consisted only of women, sample consisted of volunteers, all participants from a single university.

2.63 This experiment uses replication (many doctors) and direct control (same gown, identical gestures, etc.). Not mentioned in article, but would need random assignment of doctors into the experimental groups to ensure good design.

2.64 There may be potential confounding factors such as environmental conditions (availability of cigarettes, smoke in the air, etc.). If the researchers were careful to control such factors, the conclusion would be valid.

2.65 a. Gender, age, weight, lean body mass, and capacity to lift weights were all dealt with by direct control.   **b.** Yes. Knowing whether they were receiving creatine might have influenced the effort a subject put into his workout.   **c.** Yes, especially if the trainer is the one taking the measurements. If the trainer knows which men are receiving creatine it might influence the way in which he works with the subjects.

2.66 a. Alcohol consumption may vary by province.

2.67 a. Treatments are standing and squatting; response variable is amount of tip   **b.** Extraneous factors include table location and how busy the restaurant is.   **c.** Blocking could be used to control for factors such as table location, time of day, etc.   **d.** For example, economic status   **e.** Randomization (flipping the coin to determine which treatment is received) evens out the potential effects of other extraneous factors.

2.68 Answers will vary. See solutions manual for an example answer.

2.69 Answers will vary depending on proposed design. Typically temperature is dealt with through blocking or randomization.

2.70 Answers will vary. See solutions manual for an example answer.

Chapter 3

3.1

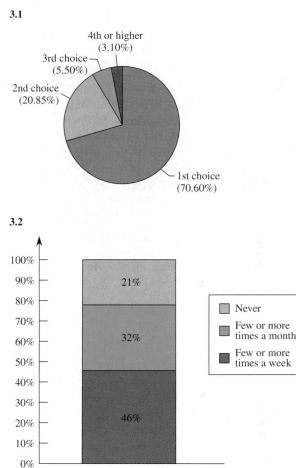

3.2

3.3

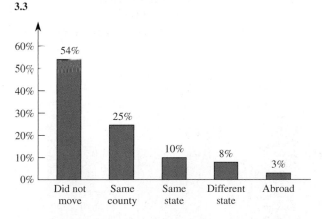

It is clear in the graph that most Americans did not move between the years 1995 and 2000, and if they did, they stayed in the same county. Comparisons of categories with similar relative frequencies can be difficult with a pie chart.

3.4

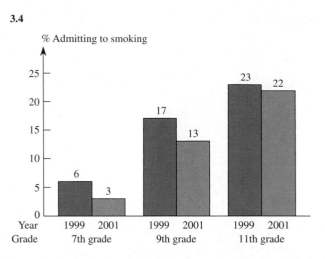

It appears that in all three grades the percentage of students that admitted to smoking tobacco was less in the year 2001 than it was in 1999. However, the reduction was greater in 7th graders (6% down to 3%) and in 9th graders (17% down to 13%) than in 11th graders, which was at the highest of the three levels in 1999 at 23% and only decreased by 1% to 22% in 2001.

3.5

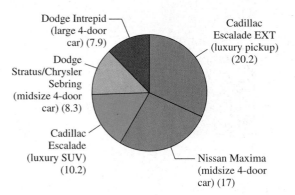

3.6

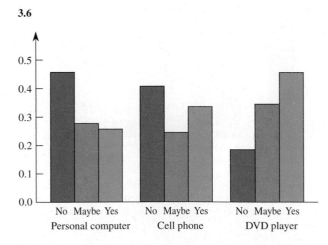

3.7 a.

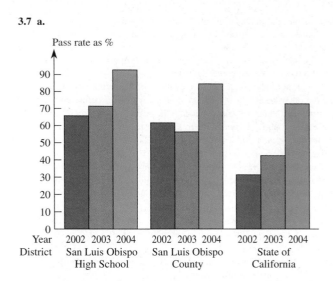

b. The pass rate increased each year for San Luis Obispo High School and the State of California from 2002 to 2004 with a sharp rise in 2004. However, in San Luis Obispo County, there was a drop in the pass rate in 2003, followed by a sharp increase in 2004 when a pass in the exam was needed for graduation.

3.8 a.

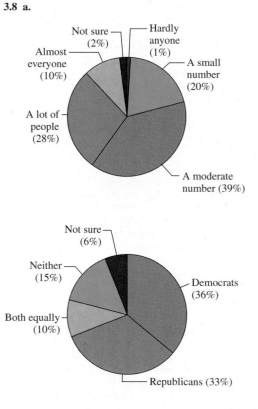

b.

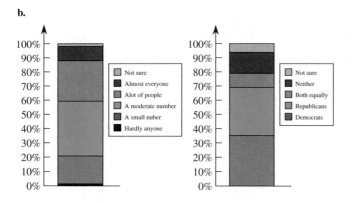

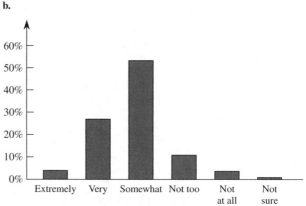

c. It's very clear from both graphs that "a moderate number" and "a lot of people" are the two biggest categories in the corruption question, with the "a small number" category being the only other large category. It is also clear that the "Democrat," "Republican," and "Neither" categories are the largest in the ethical question. However, because the answers in the corruption question are ordered, it can be clearly seen in the segmented bar chart that the most popular answers are in the middle range of the answers, a fact that is not so obvious in the pie chart.

3.9 a.

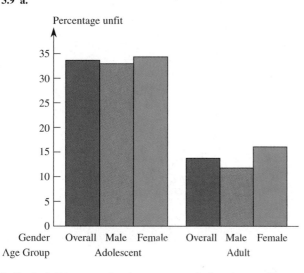

b. For both age groups, females are more unfit than the overall age group, and men are less unfit. However, this difference is much less marked in adolescents who on the whole are much more unfit than their older counterparts.

3.10 a.

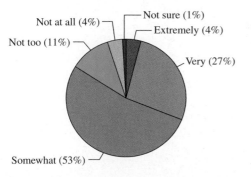

c. Both graphs clearly show that "Very" and "Somewhat" are the top two categories. However, as the answer choices have an order, it is easier to see in the bar chart that the popular answers are in the more favorable end of the categories.

3.11 a.

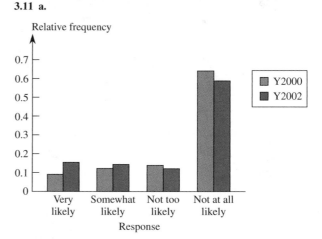

The proportion of Christmas shoppers who are "very likely" or "somewhat likely" to use the Internet has increased from 2000 to 2002, and the proportion who are "not too likely" or "not at all likely" has decreased. However it should be noted that the vast majority of Christmas shoppers (71%) are hesitant to do their Christmas shopping on-line.

b.

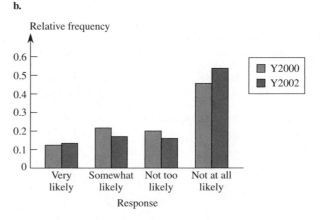

The proportion of people who felt strongly about mail order catalogues ("very likely" or "not at all likely") increased over the two-year period while the proportion of those who weren't too sure decreased.

3.12 a. A bar chart would be a better choice. There are eight categories, which may be too many for an effective pie chart.

b.

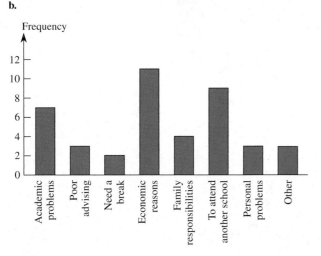

c. Most students left for economic reasons, to attend another school, or for academic problems

3.13 a. There are too many categories for this pie chart to be effective.

b.

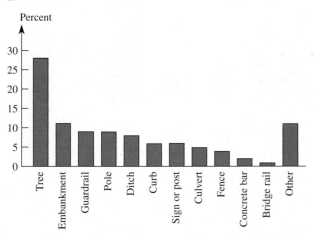

The bar chart is much more effective than the pie chart. It is very easy to compare the hazards, even for the categories that have small percentages.

3.14 a. The number of people killed in highway work zones has varied between 650 and 850 a year in a cyclical way. There were two peaks, in 1994 and 1999 when over 800 people were killed and two troughs, in 1992 and 1997 when less than 700 were killed.

3.15 a.

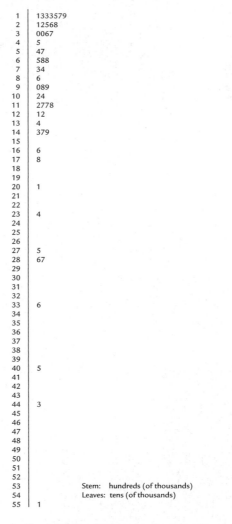

b. The distribution is skewed to the right with most of the states having values at the lower end of the scale. Forty out of the 50 states have fewer than 1,500,000 people who have smoked in the past month. There are some outliers at the high end of the distribution. **c.** No, it does not indicate that tobacco is necessarily a problem in these states. NY, CA, and TX are the three most heavily populated states in the United States and even if they have the same proportion of smokers as others states, they will have a higher *number* of smokers because of the greater population. **d.** No, it would be better to use the proportion of the population of each state that smoked during the past month. That way, the population of the state would not affect the result.

3.16

Calorie Content (cal/100ml) of 26 Brands of Light Beer

1	9
2	23788999
3	001112233459
4	0123

Stem: Tens
Leaf: Ones

Most brands of light beer have calorie contents of between 27 and 35 cal/100ml. Four of the brands have a lower calorie content (between 19 and 23 cal/100ml) and another four have a higher calorie content (between 39 and 43 cal/100ml).

3.17

```
         Calorie Content (cal/100ml) of 26 Brands of Light Beer

1L
1H  |  9
2L  |  123
2H  |  78899
3L  |  00111223345
3H  |  9
4L  |  0123                                    Stem: Tens
4H  |                                          Leaf: Ones
```

3.18 a.

```
         % of Fully Credentialed Teachers in CA Counties

                                             Stem: Tens
7  |  5                                       Leaf: Ones
8  |  0334555556788899
9  |  0011112234444555555566777777777788888899
10 |  0
```

Most counties in California have over 90% of their teachers fully credentialed. Los Angeles County is the lowest with only 75% and Alpine is the only County where 100% of their teachers are fully credentialed.

b.

```
         % of Fully Credentialed Teachers in CA Counties

7L  |                                         Stem: Tens
7H  |  5                                       Leaf: Ones
8L  |  0334
8H  |  555556788899
9L  |  0011112234444
9H  |  5555555566777777777788888899
10L |  0
```

We can now see that there are only three counties with less then 85% of their teachers fully credentialed.

3.19 a.

```
Very Large Urban        Large Urban Areas
Areas
                  1  |  023478
                  2  |  369
             8    3  |  0033589
            99    4  |  0366
          1178    5  |  012355
          0379    6  |
             2    7  |
                  8  |              Stem: Tens
             3    9  |              Leaf: Ones
```

b. Not necessarily. Philadelphia is a larger urban area than Riverside, CA, but has less extra travel time. However, *overall*, taking into account all the urban areas mentioned, or if we were to calculate the average or typical value for each type of area, then we would find that *on the whole*, the larger the urban area, the greater the extra travel time.

3.20 a.

```
         % Increase in Population 1990 to 2000

0  |  013444555567888899999
1  |  00000011234444578
2  |  00113368
3  |  01
4  |  0
5  |                                          Stem: Tens
6  |  6                                        Leaf: Ones
```

b. Forty-eight of the states have an increase in population of 31% or less, and most of these are under 12%. There are two states that have a much larger percentage of increase: Nevada (66%), and Arizona (40%).

c.

```
% Increase in Population 1990 to 2000
        WEST      |      EAST
   998880    0    |  134445555678999
     4430    1    |  0000011244578
    83100    2    |  136
       10    3    |
        0    4    |
            5    |              Stem: Tens
        6    6    |              Leaf: Ones
```

The states that show a large percentage increase in population are in the West. There are five states in the West (out of 19) that have a percentage increase greater than the maximum increase in the East.

3.21

```
         High School Dropout Rates 1997–1999 by State

0f  |  555
0s  |  666667777777
0*  |  88888889999999999
1.  |  00011111
1t  |  22223333
1f  |                                         Stem: Tens
1s  |  77                                      Leaf: Ones
```

3.22 a.

Class Interval	Frequency	Relative Frequency
$0–<$3	7	.1373
$3–<$6	3	.0588
$6–<$9	15	.2941
$9–<$12	11	.2157
$12–<$15	8	.1569
$15–<$18	4	.0784
$18–<$21	3	.0588
Total	51	

b.

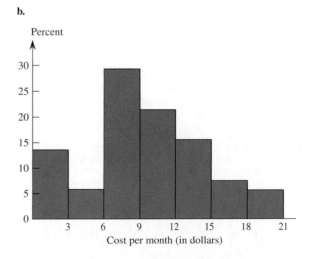

Most of the states have plans in the midrange of cost with a few states with very expensive plans. More states have plans at the cheaper end than there are states at the expensive end of the distribution.

3.23 a.

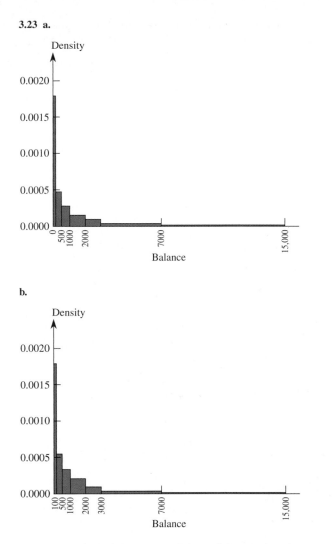

b.

c. The results for both the survey and the credit bureau show that most college student debt is less than $2000. However, the results don't agree exactly; the credit card bureau had more students with very large debt than the survey data suggested. Maybe some of the students with very large debts were reluctant to admit to the scale of their problems!
d. Only 132 out of 1260 students replied, an 89.5% nonresponse rate. It may be that students with a large debt would be reluctant to respond, so this sample was not representative of all students.

3.24 a.

Number of Impairments	Frequency	Relative Frequency
0	100	.4167
1	43	.1792
2	36	.1500
3	17	.0708
4	24	.1000
5	9	.0375
6	11	.0458
	$n = 240$	1.0000

3.24 c. $1 - .7459 = .2541$

3.25 a.

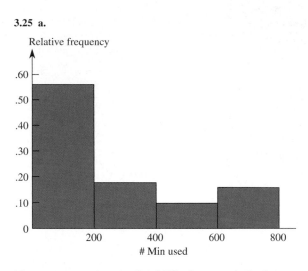

Most men average between 0 and 200 mins a month. Far fewer average between 400 and 800 mins a month.

b.

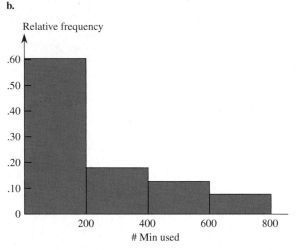

The distribution for men and women is similar in that most women average between 0 and 200 mins as do men. Fewer women average 600 to 800 minutes than men.
3.26 a. If a commute is longer than 45 minutes, then the time is often rounded to the nearest 15 or 30 minutes.
b.

Commute Time	Frequency	Relative Frequency	Density
0 to <5	5200	.0518	.0104
5 to <10	18200	.1813	.0363
10 to <15	19600	.1952	.0390
15 to <20	15400	.1534	.0390
20 to <25	13800	.1375	.0307
25 to <30	5700	.0568	.0275
30 to <35	10200	.1016	.0114
35 to <40	2000	.0199	.0203
40 to <45	2000	.0199	.0040
45 to <60	4000	.0398	.0027
60 to <90	2100	.0209	.0007
90 to <120	2200	.0219	.0007

c.

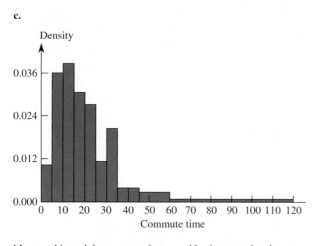

Most working adults commute between 10 minutes and an hour to work. However, there are a still a surprising number (look at the *area*) that commute between one and two hours to work.

d.

Commute Time	Relative Frequency	Cumulative Relative Frequency
0 to <5	.0518	.0518
5 to <10	.1813	.2331
10 to <15	.1952	.4283
15 to <20	.1534	.5817
20 to <25	.1375	.7192
25 to <30	.0568	.7760
30 to <35	.1016	.8776
35 to <40	.0199	.8975
40 to <45	.0199	.9174
45 to <60	.0398	.9572
60 to <90	.0209	.9781
90 to <120	.0219	1.0000

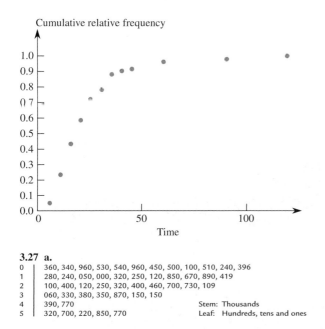

3.27 a.

```
0 │ 360, 340, 960, 530, 540, 960, 450, 500, 100, 510, 240, 396
1 │ 280, 240, 050, 000, 320, 250, 120, 850, 670, 890, 419
2 │ 100, 400, 120, 250, 320, 400, 460, 700, 730, 109
3 │ 060, 330, 380, 350, 870, 150, 150
4 │ 390, 770                          Stem: Thousands
5 │ 320, 700, 220, 850, 770           Leaf: Hundreds, tens and ones
```

The stem-and-leaf display suggests that a typical or representative value is in the stem 2 row, perhaps around 2230. There are no gaps in the display. The shape of the display is not symmetric.

b.

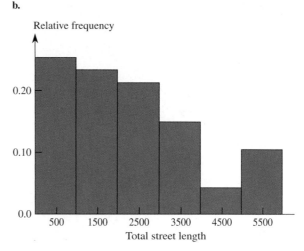

There is a longer upper tail; the histogram is positively skewed.

3.28 a. We don't know the width of the last interval "100,000 or more" and the widths of the intervals are unequal.

b.

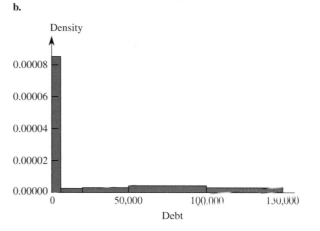

c. Only 42.7% of medical students who have completed their residencies have educational debts of under $5,000. 41.8% of them have educational debts of between $50,000 and $150,000. It seems that medical students finishing their residency have either a very large debt or a relatively small debt with only 15.5% of them having a debt of between $5,000 and $50,000.

3.30 a. and b.

Classes	Frequency	Relative Frequency	Cumulative Relative Frequency
0 to <6	2	.0225	.0225
6 to <12	10	.1124	.1349
12 to <18	21	.2360	.3709
18 to <24	28	.3146	.6855
24 to <30	22	.2472	.9327
30 to <36	6	.0674	1.0001 ≈ 1.0
	$n = 50$	1.0001	

c. (Rel. Freq. for $12 - <18$) = (Cum. Rel. Freq. for <18) − (Cum. Rel. Freq. for <12) = .3709 − .1349 = .2360.

e.

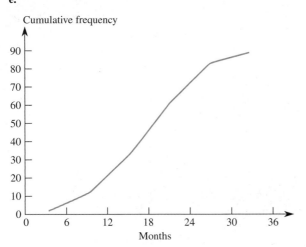

Cumulative frequency

i. Approximately 20 months   **ii.** Approximately 29 months

3.31 a.

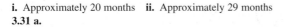

Class Intervals	Frequency	Rel. Freq.	Density
.15–<.25	8	.02192	0.2192
.25–<.35	14	.03836	0.3836
.35–<.45	28	.07671	0.7671
.45–<.50	24	.06575	1.3150
.50–<.55	39	.10685	2.1370
.55–<.60	51	.13973	2.7946
.60–<.65	106	.29041	5.8082
.65–<.70	84	.23014	4.6028
.70–<.75	11	.03014	0.6028
	$n = 365$	1.00001	

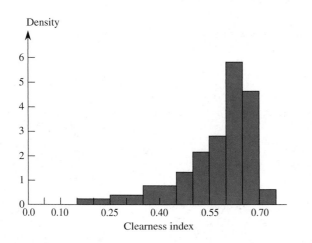

3.32 Almost all the differences are positive, indicating that the runners slow down. The graph is positively skewed. A typical difference value is about 150. About .02 of the runners ran the late distance more quickly that the early distance.

3.33 i. Symmetric

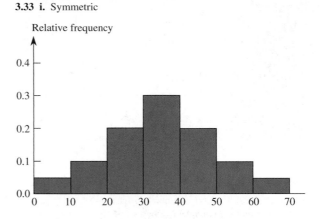

ii. Positively skewed

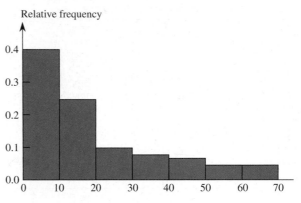

iii. Bimodal and approximately symmetric

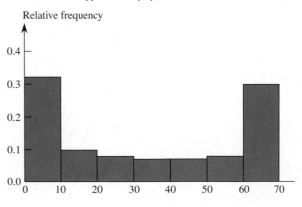

iv. Bimodal

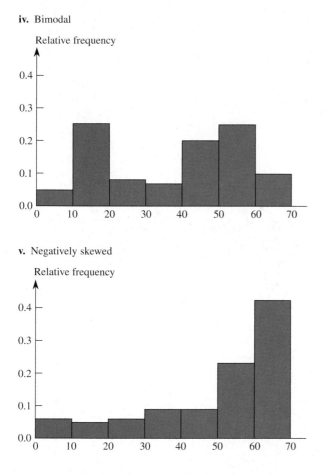

v. Negatively skewed

3.34 For example (there are other possible answers):

Class Intervals	Part a.	Part b.	Part c.	Part d.
100–<120	5	5	35	20
120–<140	10	7	15	10
140–<160	40	10	10	4
160–<180	10	15	5	25
180–<200	5	33	5	11

3.35 a. The graft weight ratio tends to decrease as recipient body weight increases, but the relationship does not look linear.   **b.** People with low body weight tend to be small people and it is possible their livers may be smaller than the liver of an average person. Conversely, people with high weight tend to be large people and their livers may be larger than the liver of an average person. Therefore, we would expect the graft weight ratio to be large for low weight people and small for high weight people.

3.36 In general, there appears to be a positive relationship between dropout rate and poverty rate. On average, as the poverty rate increases, the dropout rate increases.

3.37 a.

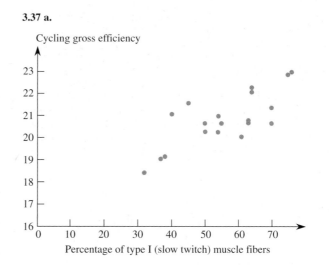

There is a tendency for y to increase as x does; larger values of gross efficiency tend to be associated with larger values of percentage type I fibers. **b.** There are several observations that have identical x-values yet different y-values; the value of y is *not* determined solely by x.

3.38 a.

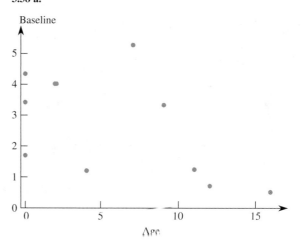

b.

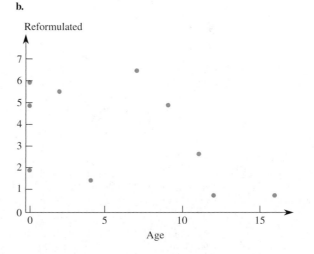

c. There appears to be a negative relationship between age and emissions. As age increases, emissions decrease.   **d.** The two scatterplots have very similar patterns. As engine age increases beyond age = 8 years, there is a definite decrease in emissions in both plots. For age = 0, there are several different *y*-values, which indicates that emissions are not solely determined by age.

3.39 There are several observations that have identical or nearly identical *x*-values yet different *y*-values. Therefore, the value of *y* is not determined solely by *x*, but also by various other factors. There appears to be a general tendency for *y* to decrease in value as *x* increases in value. There are two data points that are far removed from the remaining data points.

3.40 a. There is not a deterministic relationship between *x* and *y;* there are two data points, (100, 222) and (100, 241), which have the same *x*-value but different *y*-values.

b.

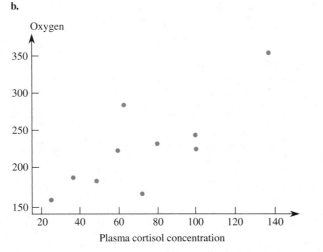
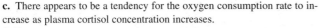

c. There appears to be a tendency for the oxygen consumption rate to increase as plasma cortisol concentration increases.

3.41 a.

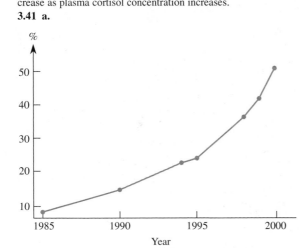

b. There has been an increase in the ownership of computers over time since 1985. At first the increase was slow and then from 1995 ownership has been increasing at a more rapid rate.

3.42 a.

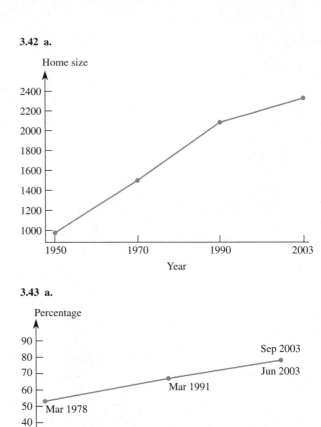

3.43 a.

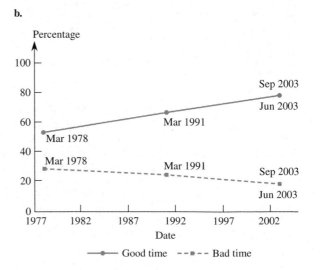

b.

c. The time series plot gives a better trend over time as it shows a time scale, which the bar chart does not.

3.44 a.

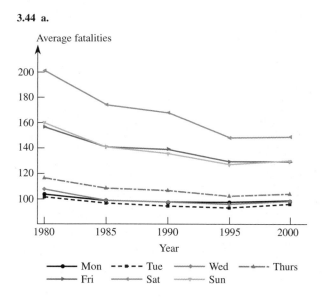

b. There are many more fatalities on a Friday, Saturday, and Sunday. Most people work Monday to Friday and may drive more on at the weekend. **c.** There has been a steady decrease in the average number of fatalities between 1980 and 2000, more so on Fridays to Sundays than Mondays to Thursdays. This could be because most states introduced the compulsory use of front, and then rear seat belts in cars during this time. **3.45** In both 2001 and 2002 the box office sales dropped in Weeks 2 and 6 and in the last two weeks of the summer. The seasonal peaks occurred during Weeks 4, 9, and 13.

3.46

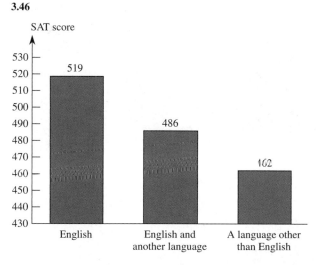

3.47

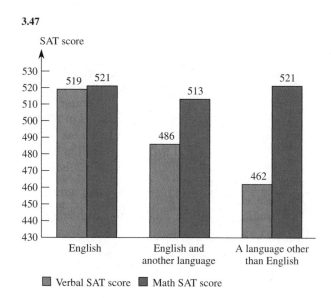

For those students who have English as a first language there is very little difference between the average Verbal and Math scores. For students who speak English and another language, they do less well in both subjects, but it is more noticeable in the Verbal scores. Those students who speak a language other than English as their first language score on average about 55 points less than those who speak only English. However, their average Math scores are as high as the English-only speaking students.

3.48

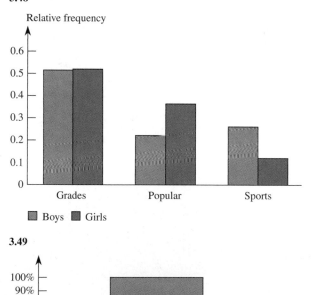

3.49

3.50

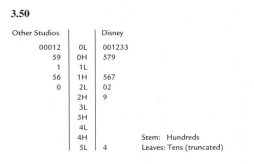

Other Studios		Disney
00012	0L	001233
59	0H	579
1	1L	
56	1H	567
0	2L	02
	2H	9
	3L	
	3H	
	4L	
	4H	Stem: Hundreds
	5L	4 Leaves: Tens (truncated)

The first impression is that there are many films where the tobacco exposure time is small, both for Disney films and for those made by other studios. Disney has made more films with longer exposure to tobacco and there is an obvious outlier in the Disney data.

3.51 a.

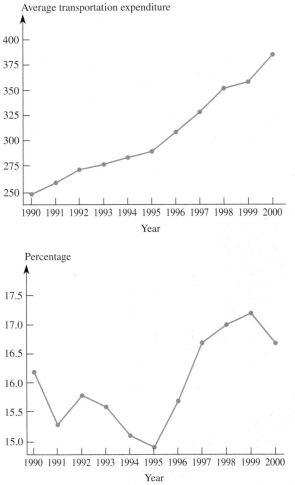

b. The time-series plots from Part (a) do agree with the statement. It is clear from the first graph that the actual expenditure has been increasing. Although the percentage of household expenditure looks volatile, in the 10 years of this study, it has varied from 14.9% to 17.2%, and is small compared with the increase in actual expenditure.

3.52

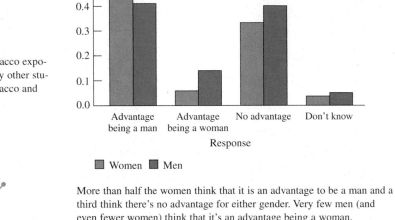

More than half the women think that it is an advantage to be a man and a third think there's no advantage for either gender. Very few men (and even fewer women) think that it's an advantage being a woman.

3.53 a.

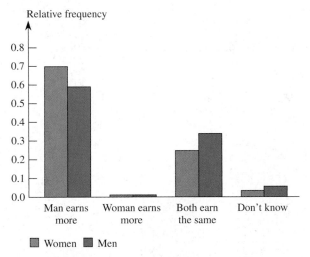

b.

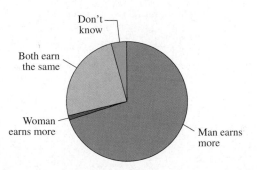

Men

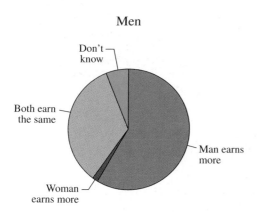

c. It's much easier to compare the differences using the comparative bar chart. The categories are next to each other and any difference is easy to see. Comparing the size of the slice for two pie charts is more difficult. **d.** The majority of both men and women think that men earn more money for the same work. Very few men or women think that a woman earns more. More women than men think that a man earns more or earns the same as a woman.

3.54 a.

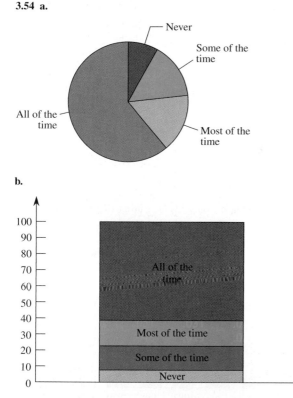

b.

c. It is easier to see the difference in the proportions with a pie chart; however, it is easier to estimate the percentage of each response with a segmented bar chart.

3.55 a.

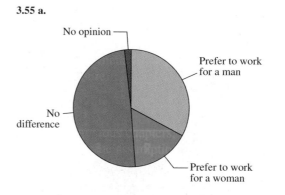

Given the choice, most people don't care whether they work for a man or a woman. Of those that do have an opinion, most would prefer to have a male boss.

b.

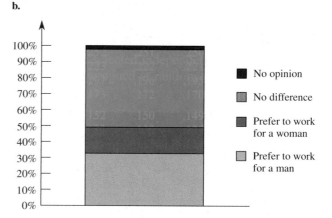

3.56 a. 2.2 liters/min **b.** In the row with stem of 8. The leaf of 9 would be placed to the right of the other leaves. **c.** A large number of flow rates are between 6.0 and 8.0. Perhaps 6.9 or 7.0 could be selected as a typical flow rate. **d.** There appears to be quite a bit of variability in the flow rates. While there are a large number of flow rates in the 6.0 to 9.0 range, the flow rates appear to vary quite a bit in relation to one another. **e.** The distribution is not symmetric. Taking 7.0 as a typical value, the smaller flow rates are spread from 2.2 to 7.0, while the larger flow rates are spread from 7.0 to 18.9 (a larger spread). **f.** The value 18.9 appears to be somewhat removed from the rest of the data and hence is an outlier.

3.57

Class Interval	Frequency	Rel. Freq.
30−<50	2	.04
50−<70	4	.08
70−<90	4	.08
90−<110	11	.22
110−<130	12	.24
130−<150	3	.06
150−<170	5	.10
170−<190	4	.08
190−<210	3	.06
210−<230	1	.02
230−<250	1	.02
n = 50		1.00

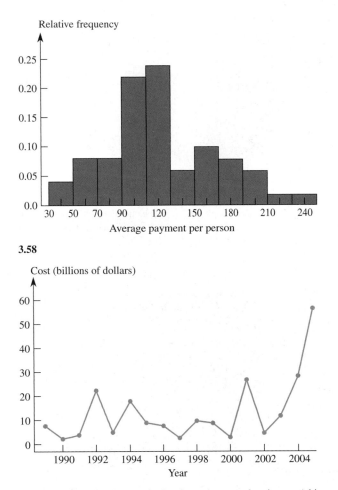

3.58

The peaks coincide with a major hurricane (or several major ones) hitting the mainland causing loss of buildings and life. The top five costliest hurricanes were: Katrina (2005), Andrew (1992), Charley (2004), Ivan (2004), and Wilma (2005).

3.59

```
0 | 675890050
1 | 8615216706312284318087877
2 | 56907869061971
3 | 7550001505
4 | 5678230
5 |
6 | 71                    Stem: Tens
7 | 0                     Leaf: Ones
```

The data values are concentrated between 0 and 40, with a few larger values. Overall, the plot appears to be skewed to the right.

3.60 a.

Day of Week	Frequency
Sunday	109
Monday	73
Tuesday	97
Wednesday	95
Thursday	83
Friday	107
Saturday	100
	$n = 664$

b. .4759   **c.** If a murder were no more likely to be committed on some days than on other days, the proportion of murders on a specific day would be $1/7 = .1429$. So, for three days the proportion would be $3(.1429) = .4287$. Since the proportion for the weekend is .4759, there is some evidence to suggest that a murder is more likely to be committed on a weekend day than on a nonweekend day.

3.61

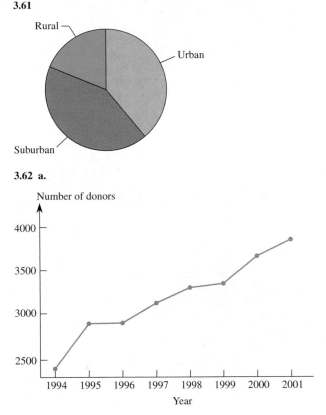

3.62 a.

The number of transplants from a living relative has been increasing steadily from 1994 to 2001.

b.

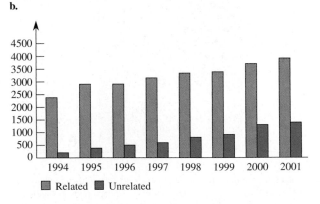

There are many more related donors than unrelated donors used for kidney transplants. However the use of nonrelated donors is rising at a faster rate than that of related donors.

3.63 a. For the first histogram, it appears that more frozen meals have sodium content around 800 mg. However, the second histogram suggests that sodium content is fairly uniform from 300 mg to 900 mg and then drops off above 900 mg.   **b.** Using the first histogram, the proportion of observations that are less than 800 is approximately .667. Using the second histogram, the proportion of observations that are less than 800 is also approximately .667.

3.64 a.

Class	Frequency	Rel. Freq.
.175–<.225	4	.0727
.225–<.275	2	.0364
.275–<.325	16	.2909
.325–<.375	15	.2727
.375–<.425	9	.1636
.425–<.475	6	.1091
.475–<.525	2	.0364
.525–<.575	0	.0000
.575–<.625	1	.0182
	$n = 55$	1.0000

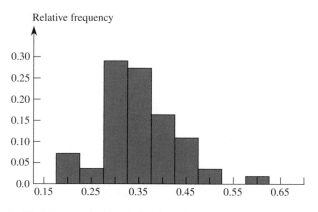

Relative frequency

b. The histogram is skewed slightly in the positive direction. The majority of the observations are in the center.

Chapter 4

4.1 Mean = 970.7, median = 738. They are different because outliers at the high end of the data set influence the mean more than the median.
4.2 a. The mean is larger for this data set because there is an outlier at the high end that will affect the mean, but not the median.   **c.** The median. The distribution is skewed to the right and with the outlier at the high end, the mean would be much higher than the typical value.
d. The top 20 newspapers in the country is not a random sample of daily U.S. newspapers and therefore not representative of the population.
4.3 a.

% of Copper Content for a Sample of Bidri Artifacts

copper content %

b. Mean = 3.654%, median = 3.35% **c.** An 8% trimmed mean will drop the outlier of 10.1%; the 8% trimmed mean will be smaller than the mean.
4.4 b. It is possible that the lowest score, 14%, is from DC. If this is eliminated, the mean is 23.6%, which rounds to 24%. However, the mean of the data set to the nearest percent is 23%, not 24%.

4.5 c. These 20 days had the highest number of fatalities; it is not a random sample of the 365 days of the year and cannot be used to generalize to the rest of the year.
4.6 a. It suggests that the distribution for angioplasty wait times is skewed to the left and the distribution for bypass wait times is skewed to the left.   **b.** No. The median is equivalent to "50% completed within" and so must be a shorter time than the "90% completed within."
4.7 a. No, we have no knowledge of the costs of any intermediate plans.
b. The number of plans is not given in the data set.   **c.** Iowa, Minnesota, Montana, Nebraska, North and South Dakota, and Wyoming.
4.8 a.

Lowest Premium Medicare Drug Plan Cost by State

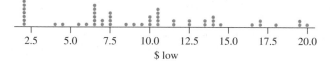

$ low

b. ii. The mean would be about the same as the median.
d.

Highest Premium Medicare Drug Plan Cost by State

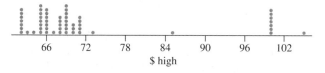

$ high

4.9 Expensive houses do not pull the median price up. Sales of very expensive houses will increase the mean price, just as the sales of inexpensive houses will decrease the mean price. If there are more sales of either very high- or very low-priced homes, this will affect the median price. More higher-priced sales than lower-priced sales, the median increases; more lower-priced sales, the median price decreases.
4.10 This statement could be correct if there were a small group of residents with very large wages. This group would make the average wage very large and thus a large percentage (perhaps even as large as 65%) could have wages less than the average wage.
4.12 c. There need to be 25(.8) = 20 successes in all. Since there are 7 in the first 10, there need to be 13 successes in the 15 added observations.
4.14 The total for the 19 papers is 19(70) = 1330. To have a class average of 71, the total of all 20 papers would have to be 20(71) = 1420. Thus, the last paper would have to be a 1420 − 1330 = 90. To have a class average of 72, the total of all 20 papers would have to be 20(72) = 1440. Thus the last paper would have to be a 1440 − 1330 = 110. (An impossibility, since the maximum score possible is 100.)
4.16 Answers will vary. For example: Set 1: 2, 3, 7, 11, 12; $\bar{x} = 7$ and $s = 4.528$ Set 2: 5, 6, 7, 8, 9; $\bar{x} = 7$ and $s = 1.581$; Set 1: 2, 3, 4, 5, 6; $\bar{x} = 4$ and $s = 1.581$ Set 2: 4, 5, 6, 7, 8; $\bar{x} = 6$ and $s = 1.581$.
4.17 It would have a large standard deviation. Some parents would spend much more than the average and some parents would either not be able to afford to spend that much (or not be willing to) or would rather spread it over the school year.
4.18 Standard deviation is a measure of spread and so is volatility in the stock market. A small standard deviation means the stock stays close to its mean value and a large standard deviation means that it varies—sometimes it is worth a lot more than its mean, sometimes a lot less—in other words, it is more volatile.

4.19 b. Smaller. There is less variability in the data for Memorial Day.
c. There is less variability for holidays that always occur on the same days of the week. Memorial Day, Labor Day, and Thanksgiving Day have standard deviations of 18.2, 17.7, and 15.3 respectively whereas New Year's Day, July 4th, and Christmas Day have standard deviations of 50.1, 47.1, and 52.4.
4.20 a. i. The lower quartile must be lower than the median of 14.
b. iii. The upper quartile is between 13 and 42.
4.21 a. The sample sizes for Los Osos and Morro Bay are different so the average for the combined areas is $\dfrac{(606456)(114) + (511866)(123)}{(114 + 123)}$.
b. Because Paso Robles has the bigger range of values, it is likely to have the larger standard deviation. **c.** Because the difference in the high prices is so much greater than the difference in the low prices, it would suggest that the distribution of house prices in Paso Robles is right skewed and the mean price is higher than the median, suggesting the median would be lower in Paso Robles than in Grover Beach.
4.22 a. $s^2 = \dfrac{35.07321}{9} = 3.897$, $s = \sqrt{3.897} = 1.974$ **b.** Because these were the 10 hospitals with the lowest cost-to-charge ratios, the variability of the ratios of *all* the hospitals would be greater than for these 10. Therefore, the standard deviation for all the hospitals in California would be larger than 1.974. **c.** The data from these 10 hospitals are, by definition, the lowest cost-to-charge ratio of all the hospitals in California. They are not representative of all the hospitals in California and so should not be used to draw conclusions about this population.
4.24 b. Standard deviation = \$186.236, iqr = \$155 **c.** There is more variability in the repair cost for moderately priced midsize cars. Both the standard deviation and interquartile range values are higher than the corresponding values for the inexpensive midsize cars. **d.** Mean repair cost for moderately priced midsize cars = \$348.9, mean repair cost for inexpensive midsize cars = \$298.36 **e.** On average, the repair cost is higher for a moderately priced midsize car compared to an inexpensive midsize car. The variability (or spread) of the individual costs is greater for the moderately priced cars.
4.26 a. Mean = 2965.2, standard deviation = 542.6, iqr = 602 **b.** The iqr for the sodium content for chocolate pudding is 602 and the iqr for sodium content in catsup is 1300; there is less variability in sodium content in chocolate pudding.
4.27 a. 1st quartile = 51, 3rd quartile = 69, interquartile range = 18
b. The iqr for inpatient cost-to-charge ratio in Example 4.9 is 14. There is more variability in cost-to-charge ratios for outpatient services.
4.28 a. For sample 1, $\bar{x} = 7.81$ and $s = .39847$, for sample 2, $\bar{x} = 49.68$ and $s = 1.73897$. **b.** For sample 1, $CV = (100)(.39847)/7.81 = 5.10$, for sample 2, $CV = (100)(1.73897)/49.68 = 3.50$. Even though the first sample has a smaller standard deviation than the second, the variation relative to the mean is greater.
4.29 a. The mean (22.4) is greater than the median (18); the distribution is more likely to be positively skewed.

b.

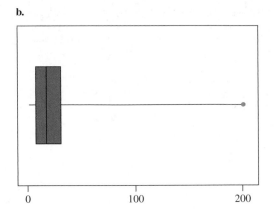

c. The maximum (205) is more than 3 times the interquartile range (3 × 24 = 72) above the upper quartile. Therefore there is at least one extreme outlier.
4.30

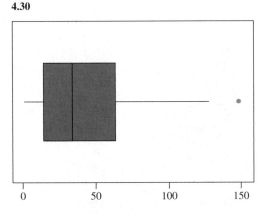

Half the states in the continental U.S. have nitrous oxide emissions of less than 34,000 tons. There is one state that has a much larger quantity of emissions than any of the other states The distribution looks positively skewed, meaning that most states have emission values at the lower end of the range.
4.31 a. Mean = 59.85 hours, median = 55.0 hours. As the mean is greater than the median, the distribution is likely to be skewed to the right.
b.

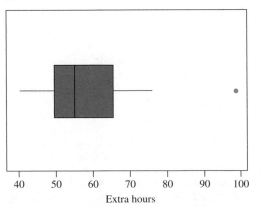

Extra hours

The number of extra commuter hours for urban areas is centered around 55 hours and the distribution is skewed to the right. There is a major outlier, Los Angeles at 98 hours. Excluding the outlier, the range is from 40 hours to 75 hours.

4.32 a. The iqr because there are two outliers (one low and one high) that would inflate the standard deviation.   **b.** The iqr = $94.0 - 81.5 = 12.5$. The mild outliers lie between $Q3 + 1.5(iqr)$ and $Q3 + 3(iqr)$ and $Q1 - 1.5(iqr)$ and $Q1 - 3(iqr)$ or $94 + 18.75 = 112.75$ and $95 + 37.5 = 132.5$ and $81.5 - 18.75 = 62.75$ and $81.5 - 37.5 = 44$. There is one mild outlier at the lower end: Firefighter (67). The extreme outliers are greater than $Q3 + 3(iqr)$ or less than $Q1 - 3(iqr)$, i.e., >132.5 or <44. There are two extreme outliers, one on the high end: Students (152) and one on the low end: Farmer (43).

c.

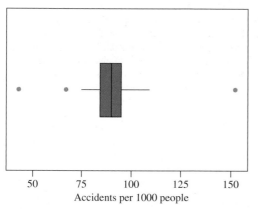

d. Apart from authors of statistical text books, it appears that farmers and firemen should be offered a professional discount as they seem to have far fewer accidents than other occupations.

4.33 a. The mean blood lead level (4.93) is higher than the median (3.6). This is due to the unusually large values in the data set.

b.

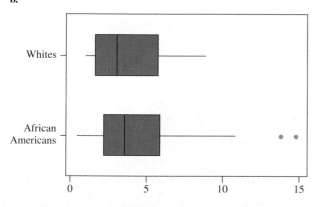

Overall, African Americans had a higher blood lead level than whites and also had greater variability. There were two values from the African Americans that had much higher blood lead levels than the others of the same ethnicity.

4.34.

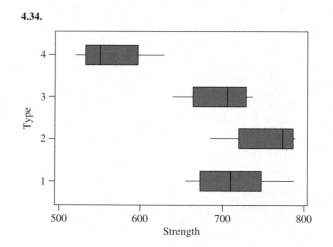

The median marker for Type 1 is centered in the box whereas the medians for Type 2 and Type 3 are closer to the higher end of their boxes and the median for Type 4 is closer to the lower end of its box. The widths of the boxes are approximately equal indicating similar variability in the middle half of the data for each box type. There are no outliers for any of the boxplots.

4.35 a. Excited delirium: median = 0.4, Q1 = 0.1, Q3 = 2.8, iqr = 2.7
No excited delirium: median = 1.6, Q1 = 0.3, Q3 = 7.9, iqr = 7.6
b. There are two mild and two extreme outliers for the Excited Delirium sample. There are no outliers in the No Excited Delirium sample.

c.

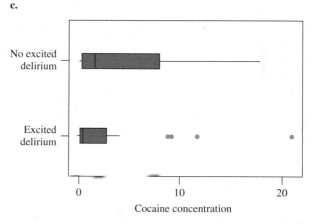

The median marker for each boxplot is located at the lower end of the box. The boxplot for the No Excited Delirium is much wider than for Excited Delirium indicating much more variability in the middle half of the data. Both boxplots have short whiskers on the lower end and long whiskers on the higher end. The Excited Delirium sample has two mild outliers and two extreme outliers. There are no outliers for the No Excited Delirium sample.

4.36 a. 40 is one standard deviation above the mean; 30 is one standard deviation below the mean; 25 is two standard deviations below the mean; 45 is two standard deviations above the mean.   **c.** No more than 11%
d. Approximately .15%

4.38 a.

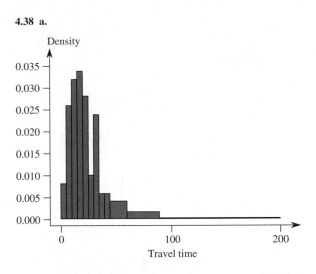

Travel time

b. Most people take around 20 minutes to get to work, although the time varies from less than 5 minutes to more than 90 minutes. The distribution is positively skewed, meaning very few people take a long time to get to work. **c.** No, it wouldn't be appropriate to use the Empirical Rule for this distribution. One of the conditions for the Empirical Rule is a normal distribution and this distribution is skewed. **d.** According to the Empirical Rule approximately 95% of the data lie within two standard deviations of the mean. That implies that approximately $(100 - 95)/2 = 2.5\%$ of the data will lie below the $27 - 2(24) = -21$ mins. This is clearly impossible. **e. i.** At least 75% **ii.** At least 21%

f. Between Times Chebyshev's Actual
  0 and 75 mins at least 75% Between 93% and 98%
  0 and 47 mins at least 21% Between 87% and 93%

4.39 a. At least 75% **b.** (2.90, 70.94) **c.** If the distribution of NO_2 values was normal, approximately 2.5% of the values would be less than -9.64 (two standard deviations below the mean). Since NO_2 concentration can't be negative, the distribution shape must not resemble a normal curve.

4.40 For these data, the mean minus one standard deviation is $3.15 - 6.09 = -2.94$, a negative number. This suggests that the distribution of annual wine consumption is positively skewed and the Empirical Rule should not be used.

4.42 Out of all the people who took the verbal section, 17% scored higher than she did and 83% scored at or below her score. For the math section, 94% of those who took the exam scored at or below her score and only 6% scored higher than she did.

4.43 b. Approximately 5% **c.** Approximately 13.5% **d.** Because the histogram is well approximated by a normal curve.

4.44 a.

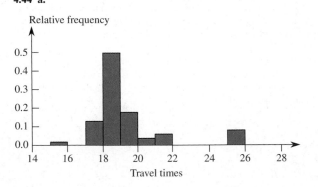

Travel times

b. i. 21 **ii.** 15 **iii.** Approximately 21.67, because the cumulative relative frequency up to 21 is 86%, up to 22 is 92%, and 21.67 is two-thirds of the way across the class of 21–<22. **iv.** The 95th percentile will be in the class 25–<26. The cumulative relative frequency at 25 is .92. The 95th percentile will be 3/8 of the way across the 25–<26 class; it is approximately 25.375. **v.** The 10th percentile will be 8/13 of the way across the 17–<18 class; it is approximately 17.615.

4.45 The proportion is at most 0.16.

4.46 a. A z-score of $+2.2$ indicates that this person scored 2.2 standard deviations above the mean. Roughly 95% of the scores are within 2 standard deviations of the mean, so this person scored in the upper 2.5% of the class. **b.** A z-score of $-.4$ indicates that this person scored .4 standard deviations below the mean. **c.** A z-score of -1.8 indicates that this person scored 1.8 standard deviations below the mean. **d.** A z-score of 1.0 indicates that this person scored 1 standard deviation above the mean. Roughly 68% of the scores are within 1 standard deviation of the mean. This means that about $(1 - .68)/2 = 16\%$ scored higher than the person. Another way of stating this is that this person scored at the 84th percentile. **e.** A z-score of 0 indicates that this person scored at the mean. Because a normal curve is symmetric, the mean is also the median. Thus half of the scores were higher than this person's score and half were lower.

4.47 One standard deviation below the mean is a negative number and the number of answers changed from right to wrong cannot be negative. Therefore the distribution cannot be well described by a normal curve. From Chebyshev's Rule, the proportion who changed at least six answers from right to wrong is at most 0.106

4.49 a.

Class	Frequency	Rel Freq.	Cum. Rel. Freq.
5 −< 10	13	.26	.26
10 −< 15	19	.38	.64
15 −< 20	12	.24	.88
20 −< 25	5	.10	.98
25 −< 30	1	.02	1.00
	$n = 50$	1.00	

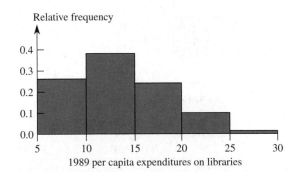

1989 per capita expenditures on libraries

4.50 The z-score associated with the first stimulus value is $(4.2 - 6.0)/1.2 = -1.5$ and for the second stimulus value it is $(1.8 - 3.6)/.8 = -2.25$. Since the z-score associated with the second stimulus reading is more extreme (-2.25 is less than -1.5), you are reacting more quickly (relatively) to the second stimulus than to the first.

4.51 The supervisors will be paid the median: $4286. The mean (more favorable to the taxpayers): $3968.67.

4.52 a.

32	55
33	49
34	
35	6699
36	34469
37	03345
38	9
39	2347
40	23
41	
42	4

Stem: Tens
Leaf: Ones

The stem-and-leaf display suggests that the mean and median will
be fairly close to each other.   **b.** Mean = 370.69, median = 369.5
c. The largest value, 424, could be increased by any arbitrary amount
without affecting the sample median.
4.53 a. The mean is $1,110,766, which is larger than the median, which
is $275,000. There are probably a few baseball players who earn a lot
of money (in the millions of dollars) while the bulk of baseball players
earn in the hundreds of thousands of dollars. The outliers have a greater
influence on the mean value making the value larger than the median.
b. If the population was all 1995 salaries and all the salaries were
used to compute the mean then the reported mean was the population
mean μ.
4.54 a. The fact that only half the homes in this county cost less than
$278,380 is the definition of a median. Statement is OK.   **b.** The me-
dian is not the midpoint of the range. It is the price at which 50% of the
homes cost less and 50% of the houses cost more.   **c.** If there are no
houses costing below $300,000 and unless at least half the homes in this
county cost exactly $300,000, Walker is correct; the median will be
above $300,000.
4.56 In general, adding the same number to each observation has no effect on the variance or standard deviation.
4.57 $s = 180.846$. The standard deviation for the new data set is 10 times
larger than the standard deviation for the original data set.
4.58 b.

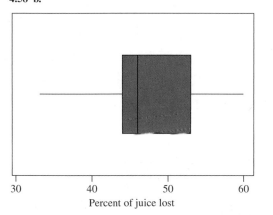

Percent of juice lost

The median line is somewhat closer to the lower edge of the box suggesting a bit of positive skewness in the middle half of the data. The
lower whisker is slightly longer than the upper whisker giving the impression of a slight negative skewness in the extremes of the data.
4.59 a. Median = 20.88, Quartile 1 = 18.09, Quartile 3 = 22.20
b. Interquartile range = 4.11; there are two outliers, 35.78 and 36.73.

c.

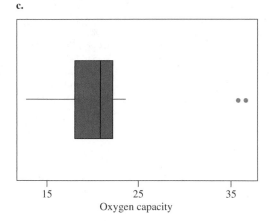

Oxygen capacity

Most of the subjects have an oxygen capacity of between 20 and 25
mL/kg/min. There are two subjects with much higher oxygen capacities
of over 35 mL/kg/min.
4.60 Mean = 11.61, median = 10.05. The sample mean is somewhat
larger than the sample median because of the outliers 15.2, 16.2, and
20.4. The sample median is more representative of a typical value since
it is not influenced by the mentioned outliers.
4.61 a. $\bar{x} = 192.57$, median = 189   **b.** $\bar{x} = 189.71$, median = 189 (unchanged)   **d.** If the largest observation is 204, new trimmed mean =
185. If the largest observation is 284, the trimmed mean is the same as
in (c).
4.63

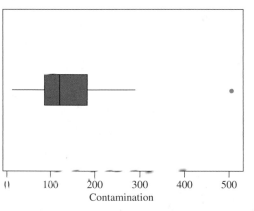

Contamination

The median line is not at the center of the box so there is a slight asymmetry in the middle half of the data. The upper whisker is slightly longer
than the lower whisker. There is one extreme outlier, 511.
4.65

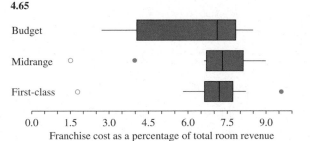

Franchise cost as a percentage of total room revenue

The median franchise cost is about the same for each type of hotel. The variability of franchise cost differs substantially for the three types. With the exception of a couple of outliers, first-class hotels vary very little in the cost of franchises, while budget hotels vary a great deal in regard to franchise costs. Ignoring the outliers, the distribution of franchise costs for first-class hotels is quite symmetric; midrange hotels have a distribution of franchise costs that is slightly skewed to the right; while budget hotels have a distribution of franchise costs that is skewed to the left.

4.67 Since the mean is larger than the median, this suggests that the distribution of values is positively skewed or has some outliers with very large values.

4.68 a. Mean $= 22.15$, standard deviation $= 11.366$ **b.** 10% trimmed mean $= 19.4375$. It is a better measure of location for this data set since it eliminates a very large value (69) from the calculation. **c.** Upper quartile $= 20.5$, lower quartile $= 18$, iqr $= 2.5$ **d.** 25 and 28 are mild outliers and 69 is an extreme outlier

e.

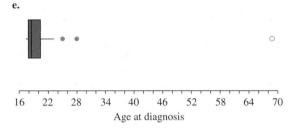

Age at diagnosis

4.69 c. -0.5 **d.** 97.5 **e.** The value 40 is three standard deviations below the mean, so only about $.5(.3\%)$ or $.15\%$ of the scores are below 40. There would not be many scores below 40.

Chapter 5

5.1 a. Positive **b.** Negative **g.** Negative **h.** No correlation

5.2 The statement is incorrect. The correlation coefficient measures the extent to which x and y are linearly related. They may have a strong non-linear relationship and yet have a correlation of zero.

5.3 For example:

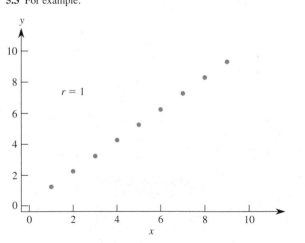

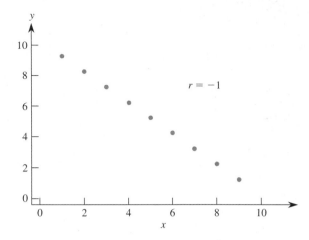

5.4 Association does not imply a causal relationship. For example, it could be age, or the amount they entertain, or even the age of their children that has a more important effect on their drinking habits rather than the amount they earn.

5.5 a. $r = 0.94$. The correlation is strong and positive. **b.** Increasing sugar consumption doesn't cause or lead to higher rates of depression; it may be another reason that causes an increase in both. For instance, a high sugar consumption may indicate a need for comfort food for a reason that also causes depression. **c.** These countries may not be representative of any other countries. It may be that only these countries have a strong positive correlation between sugar consumption and depression rate and other countries may have a different type of relationship between these factors. It is therefore not a good idea to generalize these results to other countries.

5.6 a.

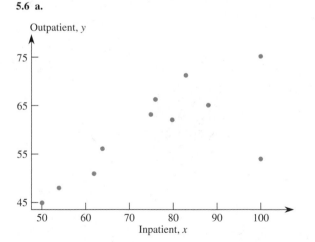

$r = 0.73$. There appears to be a reasonably strong positive linear relationship between the cost-to-charge ratio for inpatient and outpatient services at these Oregon hospitals. **b.** There is one hospital, Harney District, that has a lower outpatient cost-to-charge ratio or higher inpatient cost-to-charge ratio than the other 10 hospitals. **c.** If this observation was removed, the remaining points would all be much closer to a line and so the correlation coefficient would be greater. The relationship would be stronger.

5.8 b.

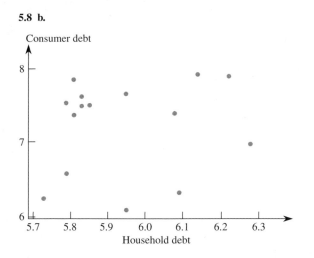

The scatterplot supports the correlation coefficient that indicates a very weak, or no linear relationship between consumer and household debts.

5.9 a. $r = 0.335$. There is a weak positive relationship between timber sales and the amount of acres burned in forest fires.   **b.** No. Correlation does not imply a causal relationship.

5.10 a.

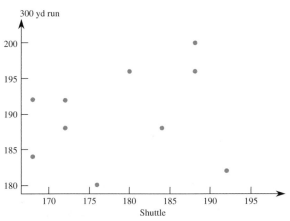

There does not appear to be any linear relationship between peak heart rate during a shuttle run and peak heart rate during a 300 yard run.
b. $r = 0.2096$; The value of .2096 suggests at best a very weak linear relationship between the two variables. The conclusion is consistent with the one of Part (a).   **c.** The value of r does not depend on which variable is labeled x. So switching the labels will not change the value of r.
5.11 a. There appears to be a strong, positive correlation.   **b.** $r = 0.9366$ indicates a strong positive linear relationship.   **c.** If x and y had a correlation coefficient of 1, all the points would lie on a straight line, but the line would not necessarily have slope 1 and intercept 0.

5.12 a.

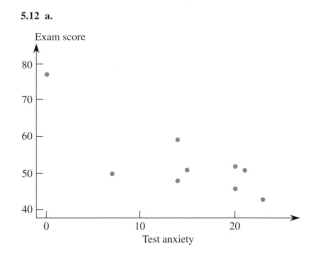

There are several observations that have identical x-values yet different y-values, so the value of y is not determined solely by x. There is one data point that is far removed from the remaining data points. There may be a tendency for exam scores to decrease as test anxiety increases.
b. There appears to be a moderate negative linear relationship.   **c.** $r = -0.7878$; consistent with answer to Part (b).   **d.** No, correlation measures the extent of association, but association does not imply causation.
5.13 $r = 0.39$; there is a weak positive linear relationship.
5.15 No. An r value of -0.085 indicates, at best, a very weak relationship between support for environmental spending and degree of belief.
5.16 The sample correlation coefficient would be closest to $-.9$. This is because there is an almost perfect negative linear relationship between speed and time required to travel a fixed distance. That is, as speed increases, time required to traverse the fixed distance decreases.
5.17 a.

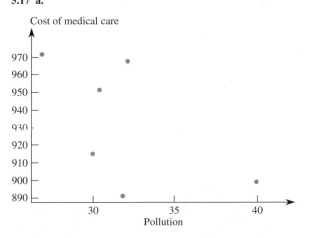

There appears to be a weak negative relationship.

5.18 a.

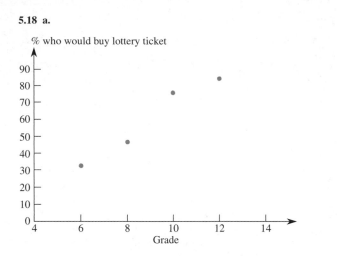

% who would buy lottery ticket

Grade

There appears to be positive linear relationship between the grade level and the percentage who said they were more likely to purchase a lottery ticket.

5.19 a. The dependent variable is the number of fruit and vegetable servings per day. The independent variable is the number of hours of TV viewed per day. **b.** Negative, because as the number of hours of TV watched *increases*, the number of servings of fruit and vegetables *decreases*.

5.20 a. For lower values of patient to nurse ratios, nurse job satisfaction might be low because up to a point, the more patients a nurse has to look after, the more interesting the job would be. After a certain number, however, the job would get difficult to do well and might get frustrating. The relationship might be nonlinear. **b.** Patient satisfaction is probably related to the amount of attention received. The higher the patient to nurse ratio, the less personal attention would be received, so the relationship would be negative. **c.** Quality of care probably declines as the number of patients a nurse must care for increases. The relationship would be negative.

5.21 a.

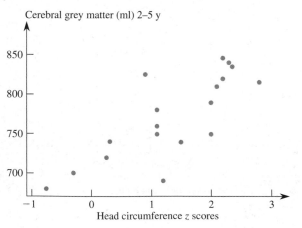

Cerebral grey matter (ml) 2–5 y

Head circumference z scores

b. $r = .7863$ **c.** $\hat{y} = 714.145 + 42.521x$ **e.** The least-squares line was calculated using values of z-scores of between -0.75 and 2.8 and therefore is only valid for values in this range.

5.22 a. Intercept $= -147$, slope $= 6.175$. On average, a 1 cm increase in snout-vent length is associated with a 6.175 increase in clutch size.

5.23 a.

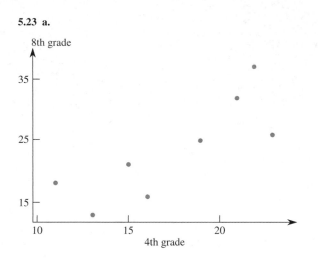

8th grade

4th grade

There appears to be a positive relationship between the percentages of public school students who were at or above proficient level in mathematics in the 4th grade in 1996 and in the 8th grade in 2000. There are two states with slightly unusual results: CA and UT.

5.24 a. $\hat{y} = 0.2508 + 0.7052x$ **b.** On average, a 1.0 increase in high school GPA is associated with an increase of 0.7052 in first-year college GPA.

5.25 a.

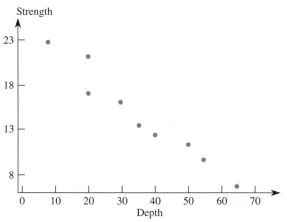

Strength

Depth

Yes, as the carbonation depth increases, the strength of the concrete decreases. **b.** $\hat{y} = 24.39 - 0.275x$ **d.** The least-squares line was calculated using values of "depth" of between 8 mm and 65 mm and therefore is only valid for values in this range.

5.26 It certainly seems that the sooner the paramedics get there, the higher your chances of survival. The slope of the least-squares line is -9.30, which means that for every extra minute, on average, the survival rate decreases by 9.30%.

5.28 a. $r = 0.70$ There is a moderately strong positive linear relationship between sale price and property size. **b.** $r = -0.333$ There is a very weak negative linear relationship (if any!) between sale price and land/building ratio. **c.** Size because it has a correlation coefficient much closer to $|1|$.

5.29 a. $\hat{y} = 59.910 + 27.462x$ **b.** 554.234 **c.** No, because the data used to obtain the least-squares equation were from steeply sloped plots, so it would not make sense to use it to predict runoff sediment from gradually sloped plots. You would need to use data from gradually sloped plots to create a least-squares regression equation to predict runoff sediment from gradually sloped plots.

5.30 a. Slope $= 244.9$, intercept $= -275.1$ **d.** No. When shell height (x) equals 1, the equation would result in a predicted breaking strength of -30.2. It is impossible for breaking strength to be a negative value, so the equation results in a predicted value which is not meaningful.

5.31 a.

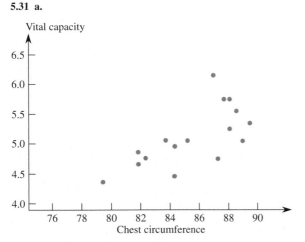

The graph shows a moderate positive linear relationship between chest capacity and vital capacity.

b.

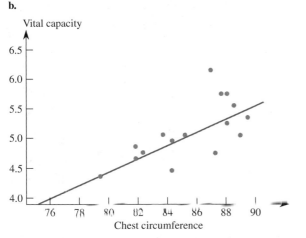

e. No—there are two data points whose x-values are 81.8 but they have different y-values.

5.32 It is dangerous to use the least-squares line to obtain predictions for x-values outside the range of those contained in the sample, because there is no information in the sample about the relationship that exists between x and y beyond the range of the data.

5.34 The denominators of b and of r are always positive numbers. The numerator of b and r is $\sum(x - \bar{x})(y - \bar{y})$. Since both b and r have the same numerator and positive denominators, they will always have the same sign.

5.35 b. Both the slope and the y intercept are changed by the multiplicative factor c.

5.36 a.

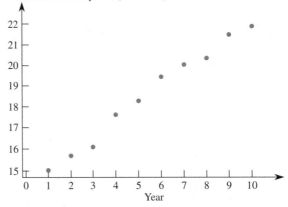

$\hat{y} = 14.213 + 0.7903x$. The number of transplants has increased steadily over time.

b.

x	y	$\hat{y}$	$y - \hat{y}$
1	15	15.0036	$-.0036$
2	15.7	15.7939	$-.0939$
3	16.1	16.5842	$-.4842$
4	17.6	17.3745	.2254
5	18.3	18.1648	.1351
6	19.4	18.9552	.4448
7	20	19.7455	.2545
8	20.3	20.5358	$-.2358$
9	21.4	21.3261	$-.0739$
10	21.8	22.1164	$-.3164$

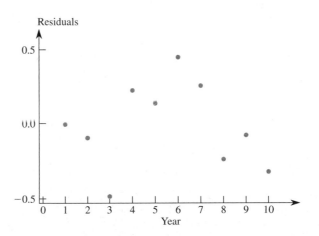

There does appear to be curvature in the residual plot, which indicates that the relationship between year and number of transplants may be better described by a curve rather than a line.

5.37 a.

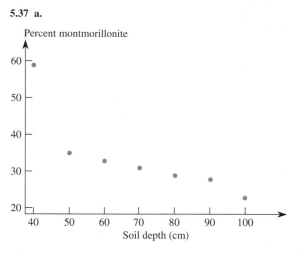

b.

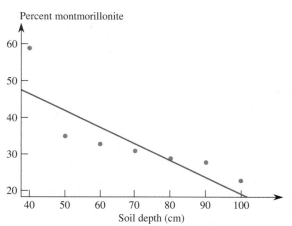

There are three points with large residuals, those points with x-values of 40, 50, and 60.

c.

x	y	$\hat{y}$	$y - \hat{y}$
40	58	46.5	11.5
50	34	42.0	−8.0
60	32	37.5	−5.5
70	30	33.0	−3.0
80	28	28.5	−0.5
90	27	24.0	3.0
100	22	19.5	2.5

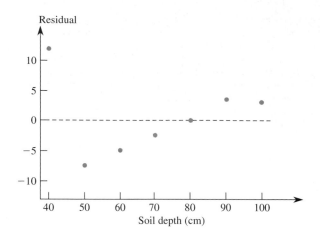

The residuals for small x-values and large x-values are positive, while the residuals for the middle x-values are negative.

5.38 a.

x	y	$\hat{y}$	Residuals
30.0	915	941.47	−26.47
31.8	891	933.03	−42.03
32.1	968	931.62	36.38
26.8	972	956.48	15.52
30.4	952	939.59	12.41
40.0	899	894.56	4.44

b. $r = -1.581$; there is a moderately strong negative linear relationship.

c.

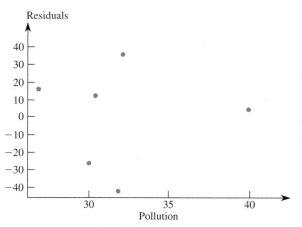

It appears that the areas with the high and low pollution have smaller residuals than the areas with pollution rates in the middle range. This might warrant further investigation.

d. The observation is influential. With this observation deleted, the equation of the regression line is $\hat{y} = 974 - 1.35x$, which is quite different than the line based on the complete data set.

5.39 a. $\hat{y} = 94.33 - 15.388x$

b.

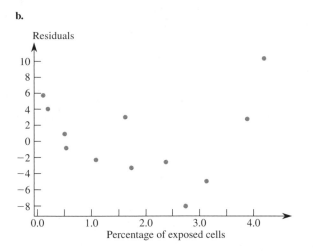

Residuals

There is a curved pattern in the plot, suggesting the linear model may not be appropriate.

5.40 a. The least-squares line with observation 11 is $\hat{y} = -1.1 + 1.29x$. Without observation 11 the least-squares line is $\hat{y} = -17.59 + 1.59x$. **b.** Yes, residual = 31.44, which is quite large when $s_e = 12.185$.   **c.** No, the least-squares line with observation 5 is $\hat{y} = -1.1 + 1.29x$; without observation 5: $\hat{y} = 5.26 + 1.16x$ (not a lot of difference in the slope).

5.41 a. 15.4%   **b.** No. $r^2 = 16\%$, so only 16% of the observed variation in first year college grades can be attributed to the approximate linear relationship between first year college grades and SAT II score. The least-squares line does not effectively summarize the relationship between SAT score and first year college grades.

5.42 a. 76.64% of the observed variability in clutch size can be explained by an approximate linear relationship between clutch size and snout vent length.   **b.** $s_e = 29.25$; this is a typical deviation.

5.43 a.

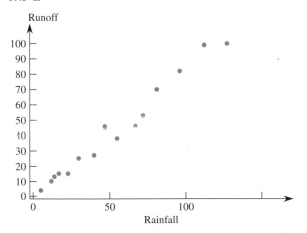

Runoff

There does appear to be a positive linear relationship between rainfall volume and runoff volume.

d.

Rainfall	Runoff	Residual
5	4	0.99344
12	10	1.20463
14	13	2.55068
17	15	2.06976
23	15	−2.89208
30	25	1.31911
40	27	−4.95062
47	46	8.26057
55	38	−6.35522
67	46	−8.2789
72	53	−5.41376
81	70	4.14348
96	82	3.73888
112	99	7.50731
127	100	−3.89728

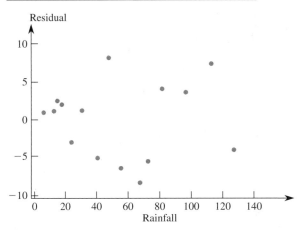

Residual

Yes, the variability of the residuals appears to be increasing with x, indicating that a linear regression may not be appropriate.

5.44 a. $\hat{y} = 1.339 - 0.008x$ **c.** With r^2 close to 0, the linear relationship between perceived stress and telomere length accounts for a very small proportion of variability in telomere length.

5.45 a. $\hat{y} = 914.5$, residual = −21.5   **b.** The typical amount that average SAT score deviates from the least-squares line is 53.7.   **c.** Only about 16% of the observed variation in average SAT scores can be attributed to the approximate linear relationship between average SAT scores and expenditure per pupil.

5.46 a. $\hat{y} = 184.31$, residual = −19.31

5.47 a.

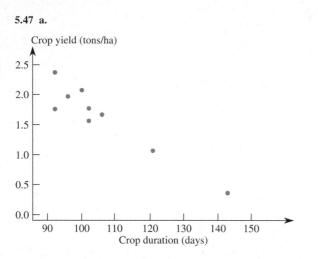

Crop yield (tons/ha)

Yes, there is a reasonably strong linear relationship.   **b.** The least-squares equation with the point deleted is $\hat{y} = 5.1151 - .0333x$. The deletion of this point does not greatly affect the equation of the line. For the full data set, the least-squares line is $\hat{y} = 5.20683 - .03421x$. The deletion of this point does not have a large effect.   **c.** For the full data set, $r^2 = .8829$. For the data set with the point (143, .3) deleted, $r^2 = .6774$. The value of r^2 becomes smaller when the point (143, 0.3) is deleted. The reason for this is that in the equation $r^2 = 1 - \text{SSResid/SSTo}$, the value of SSTo is lowered by dropping the point (143, 0.3) but the value of SSResid remains about the same.

5.48 $r^2 = .9512$; 95.12% of the variability in hardness can be explained by the linear relationship between hardness and elapsed time.

5.49 a. -20.8 **b.** $r = -0.755$ **c.** 11.64

5.50 a. r^2 could be large, indicating that the least-squares line is a big improvement over the line $\hat{y} = \bar{y}$, but even with a big improvement, the prediction errors for the least-squares line could still be large.
b. r^2 could be small, indicating that the least-squares line is a not much of an improvement over the line $\hat{y} = \bar{y}$, even when s_e is small. This would happen when points are clustered fairly tightly around the horizontal line $\hat{y} = \bar{y}$.
c. When r^2 is large and s_e is small, then not only has a large proportion of the total variability in y been explained by the linear association between x and y, but the typical error of prediction is small.

5.51 a. When $r = 0$, then $s_e = s_y$. The least-squares line in this case is a horizontal line with intercept of $\bar{y}$.   **b.** When r is close to 1 in absolute value, then s_e will be much smaller than s_y.   **c.** 1.5

5.52 a.

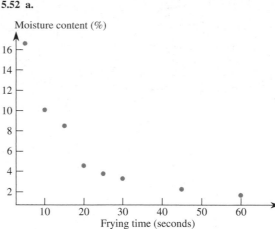

Moisture content (%)

Because of the substantial curvature in the plot, a straight line would not provide an effective summary of the relationship.
b.

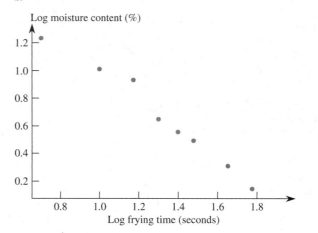

Log moisture content (%)

The plot of the transformed variables suggests that the relationship could be modeled by a straight line.   **c.** The coefficient of determination between y' and x' is .973. This suggests that a least-squares line might effectively summarize the relationship.

5.53 a.

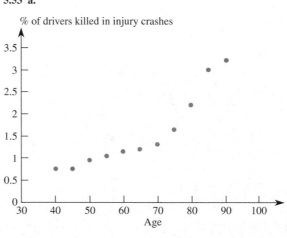

% of drivers killed in injury crashes

5.53 b. Figure 5.31 suggests moving x up or y down; one reasonable choice is $y' = \dfrac{1}{y}$.   **c.** For the transformation suggested in Part (b) (there are other possible answers):

Age x	Fatality Rate y	$y' = 1/y$
40	0.75	1.33
45	0.75	1.33
50	0.95	1.05
55	1.05	0.95
60	1.15	0.87
65	1.2	0.83
70	1.3	0.77
75	1.65	0.61
80	2.2	0.45
85	3	0.33
90	3.2	0.31

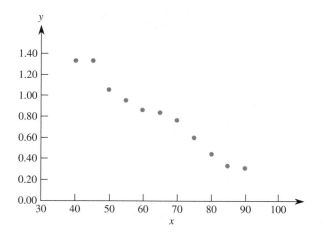

d. Yes, the plot is much more linear.
5.54 a.

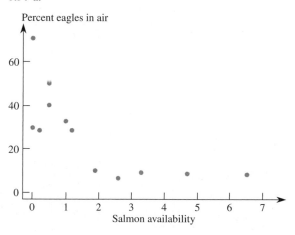

The relationship does not appear to be linear.   **b.** The plot looks like section 3 of figure 5.31, which suggests going down the ladder on y and/or x. Both $\sqrt{x}$ and $\sqrt{y}$ are down the ladder.

c.

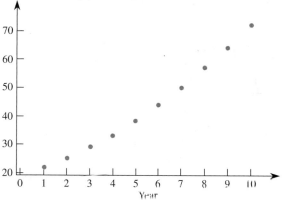

Yes, the plot is straighter than the plot in Part (a).   **d.** Since there are x observations whose values are 0, both $\log(x)$ and $1/x$ cannot be employed. Another transformation that might be helpful in straightening the plot is cube root of x and cube root of y.
5.55 a. $r = -0.717$ **b.** $r = -0.835$ for the transformed data; the transformation appears to have been successful.
5.56 a.

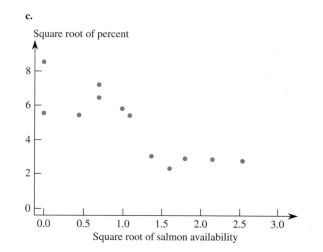

From 1990 to 1999 the number of people waiting for a transplant has increased. Each year, the number of people added to the waiting list increases.   **b.** For example, with $y' = \sqrt{y}$:

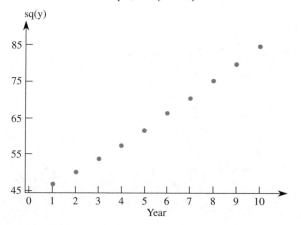

c. $\sqrt{\hat{y}} = 4.12 + 0.427x$, the predicted number of patients waiting for an organ transplant in 2000 (year 11) is $\sqrt{\hat{y}} = 4.12 + 0.427(11) = 8.817 \Rightarrow y = 2.269$. As y is measured in thousands, we predict 2269 patients awaiting transplant surgery in 2000. **d.** We are assuming the relationship between year and the number awaiting transplant stays the same outside the range of x-values in the given data range. The further from the data range a prediction is going to be made, the less accurate it may be. 2010 is further away from the data used to create the least-squares line and we don't know if the relationship between the two variables is still the same. We would be less confident to make a prediction if the year was 2010.

5.57 a.

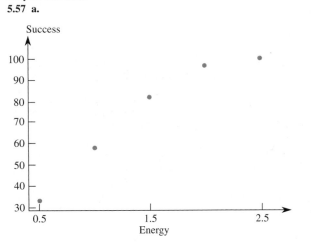

The relationship appears to be nonlinear. **b.** $\hat{y} = 22.48 + 34.36x$.

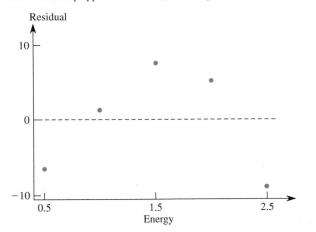

The curved pattern in residual plot confirms that the relationship is nonlinear. **c.** The value of r^2 is higher and the residuals are smaller for the log transformation. **d.** $\hat{y} = 62.417 + 101.07\log(x)$

5.58 a.

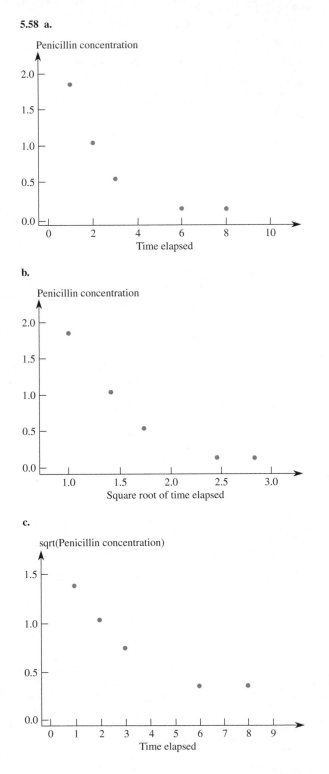

d.

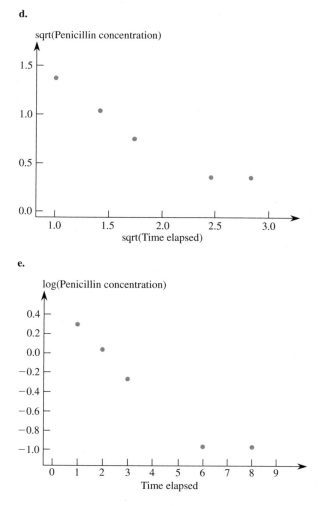

e.

The transformations of (d) or (e) appear to do the best job of straighten-ing the plot.
5.59 a.

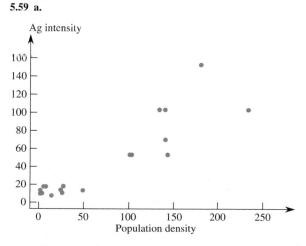

The plot does show a positive relationship.

b.

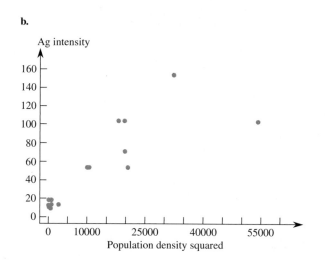

The plot appears straighter, but the variability of y increases as x increases.
c.

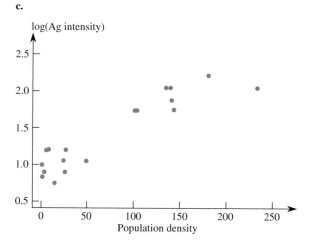

The plot also looks linear, but the variability in y does not appear to be increasing with x.
d.

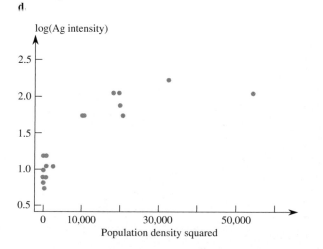

The plot has a curvature in the opposite direction as the plot in Part (a).

5.60

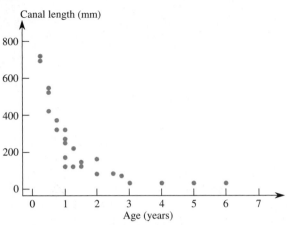

Canal length (mm)

The relationship between age and canal length is not linear, but curvilinear. Transforming to $1/x$ produces a scatterplot that is much straighter than the plot above.

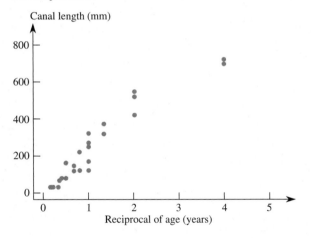

Canal length (mm)

5.62 a.

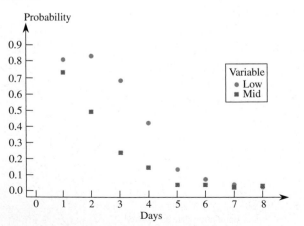

Probability

Variable
● Low
■ Mid

The plots have the S-shape characteristic of a logistic plot.

b.

Days	Proportion	$\dfrac{p}{(1-p)}$	$y' = \ln\!\left(\dfrac{p}{(1-p)}\right)$
1	0.75	3	1.098612
2	0.67	2.030303	0.708185
3	0.36	0.5625	−0.57536
4	0.31	0.449275	−0.80012
5	0.14	0.162791	−1.81529
6	0.09	0.098901	−2.31363
7	0.06	0.06383	−2.75154
8	0.07	0.075269	−2.58669

The resulting best fit line is: $y' = a + bx = 1.513 - 0.587x$, where y is the proportion of eggs hatched and $x =$ the days of exposure. The negative slope mean that the value of $b < 0$, indicating that the curve starts near 1 for small x values and then decreases as x increases.

5.63 b. $\ln\!\left(\dfrac{p}{1-p}\right) = 7.537 - 0.0079x$ **c.** 0.632

5.64 a.

Concentration	0.10	0.15	0.20	0.30	0.50	0.70	0.95
Proportion killed	.2083	.25	.4464	.6078	.8298	.9623	.9608

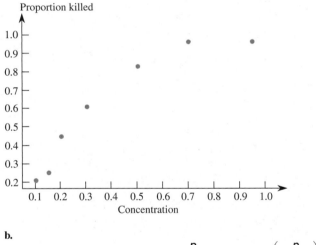

Proportion killed

b.

Concentration	Proportion	$\dfrac{p}{(1-p)}$	$y' = \ln\!\left(\dfrac{p}{(1-p)}\right)$
0.1	0.2083	0.263105	−1.3352
0.15	0.25	0.333333	−1.09861
0.2	0.4464	0.806358	−0.21523
0.3	0.6078	1.54972	0.438074
0.5	0.8298	4.875441	1.58421
0.7	0.9623	25.5252	3.239666
0.95	0.9608	24.5102	3.19909

The resulting best fit line is $y' = a + bx = -1.559 + 5.767x$, where y is the proportion of mosquitoes killed and $x =$ the concentration of pesticide. The positive slope, $b > 0$, shows that as the concentration of the pesticide increases, the proportion of the mosquitoes killed also increases.

5.65 a.

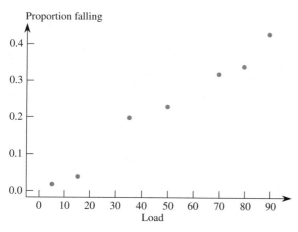

Proportion falling

b. $\ln\left(\dfrac{p}{1-p}\right) = -3.579 + 0.0397x$; a positive slope indicates that as the load increases, the proportion failing increases.   **c.** 0.232
5.66 a. $r = -0.981$; there appears to be a strong negative linear relationship.
b.

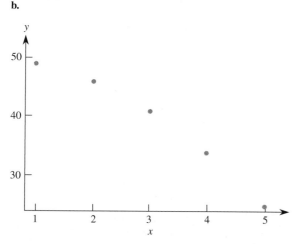

There is some curvature to the plot. This illustrates the importance of looking at both numerical and graphical descriptions of the relationship.
5.68 a. $y = 4027.083 - 577.895x$ **b.** -577.895 is the estimate of the average change in myoglobin level associated with a one unit increase in finishing time.   **c.** $\hat{y} = -596.077$; this is clearly unreasonable since myoglobin level cannot be negative.
5.69 $r^2 = .6561$; 65.61% of the variability in age is explained by the linear relationship. $s_e = 13.27$; a typical prediction error is approximately 13.27.

5.70 a. The least-squares line is $\hat{y} = 32.08 + 0.5549x$.

x	y	Predicted	Residual
15	23	40.4048	−17.4048
19	52	42.6245	9.3755
31	65	49.2837	15.7163
39	55	53.7231	1.2769
41	32	54.8330	−22.8330
44	60	56.4978	3.5022
47	78	58.1626	19.8374
48	59	58.7175	0.2825
55	61	62.6020	−1.6020
65	60	68.1513	−8.1513

b. SSResid $=1635.68$, $r^2 = 0.2859$ **c.** Only 28.59% of the observed variation in age is explained by the linear relationship between percentage of root transparent dentine and age. Also, $s_e = 14.3$, so a typical prediction error is quite large. The least-squares line does not give very accurate predictions.
5.71 a. $\hat{y} = 94.33 - 15.38x$ **b.** SSTo $= 5534.62$, SSResid $= 285.82$
d. $s_e = 5.346$; a typical prediction error would be about 5.35%.
5.72 a.

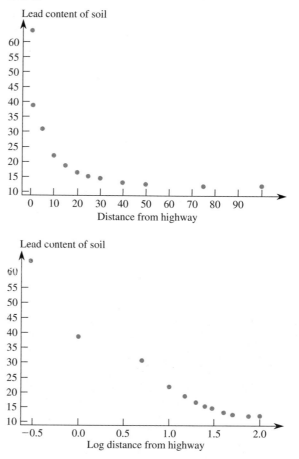

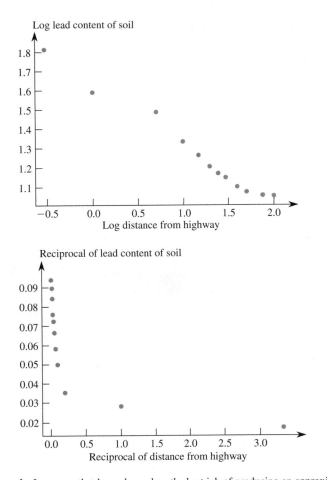

Log lead content of soil

Reciprocal of lead content of soil

b. It appears that log x, log y does the best job of producing an approximate linear relationship; the least-squares equation for predicting $y' = \log y$ from $x' = \log x$ is $\hat{y}' = 1.61867 - .31646x'$. When $x = 25$, $x' = 1.39764$, $\hat{y}' = 1.61867 - .31646(1.39764) = 1.17628$ and $\hat{y} = 10^{1.17628} = 15.0064$.

5.74 a. The plot does not appear linear, but it is difficult to say because there is one unusual point, $(51.3, 49.3)$. **c.** The value of r^2 is not very large and the value of s_e is 4.70, which is large relative to the size of the y-values in the sample. A straight line is not very effective in summarizing the relationship. **d.** $\hat{y} = 36.4175 - .1978x$, $r^2 = .027$. When the point $(51.3, 49.3)$ is deleted from the data set, there is little evidence of a linear relationship. This observation is very influential.

5.75 a. $\hat{y} = 13.9617 + 3.1703x$

b.

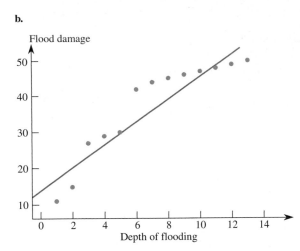

The relationship may be better described by a curve. **d.** No, relationship appears to be nonlinear and 18 is outside the range of the x values in the data set.

5.76 b. For example, if $y = 1$ when $x = 6$, then $r = .509$. (any value greater than .973 for y will work) **c.** For example, if $y = -1$ when $x = 6$, then $r = -.509$ (any value less than $-.973$ for y will work).

5.77 a.

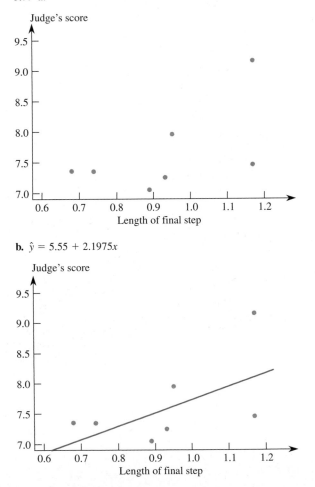

b. $\hat{y} = 5.55 + 2.1975x$

5.78 a.

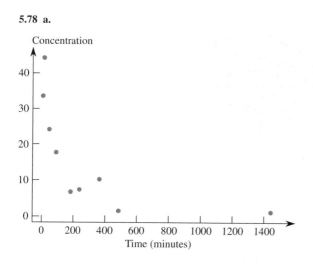

b. Based on the plot in Part (a) and Figure 5.34, a transformation going down the ladder on *x* or *y* is suggested. The transformation log(time) will produce a reasonably straight plot.

Chapter 6

6.1 A chance experiment is any activity or situation in which there is uncertainty about which of two or more possible outcomes will result.
6.2 The collection of all possible outcomes of a chance experiment; {(H, H), (H, T), (T, H), (T, T)}.
6.3 b.

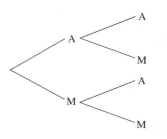

c. *B* = {AA, AM, MA}, *C* = {AM, MA}, *D* = {MM}. *D* is a simple event.
6.4 a. *A* = {Head oversize, Prince oversize, Slazenger oversize, Wimbledon oversize, Wilson oversize}   **b.** *B* = {Wimbledon midsize, Wimbledon oversize, Wilson midsize, Wilson oversize}   **c.** *Not B* = {Head midsize, Head oversize, Prince midsize, Prince oversize, Slazenger midsize, Slazenger oversize}   **d.** *B or C* = {Head midsize, Head oversize, Prince midsize, Prince oversize, Wimbledon midsize, Wimbledon oversize, Wilson midsize, Wilson oversize}   **e.** *B and C* = {Wilson midsize, Wilson oversize}

f.

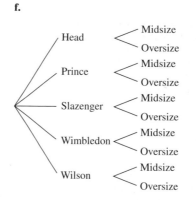

6.5 a.

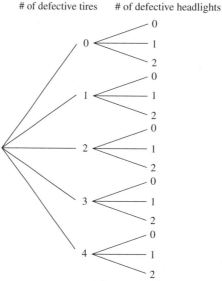

b. A^C = {(2, 0), (2, 1) (2, 2) (2, 3), (2, 4)}; $A \cup B$ = {(0, 0), (0, 1), (0, 2), (0, 3), (0, 4), (1, 0), (1, 1), (1, 2), (1, 3), (1, 4), (2, 0) (2, 1)}; $A \cap B$ = {(0, 0), (0, 1), (1, 0), (1, 1)}
6.6 a.

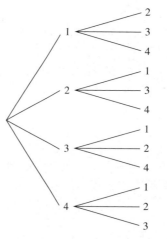

b. A = {(1,2), (1,4), (2,1), (2,3), (2,4), (3,2), (3,4), (4,1), (4,2), (4,3), (4,4)} where (i, j) means that copy i was selected for the 2-hour service and copy j was selected for the overnight loan.   **c.** B = {(1,3), (1,4), (2,3), (2,4), (3,1), (3,2), (4,1), (4,2)} where (i, j) means that copy i was selected for the 2-hour service and copy j was selected for the overnight loan.

6.7 a.

First book   Second book   Third book
selected   selected   selected

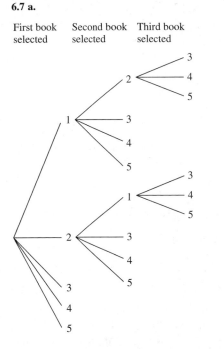

6.8 a. For example, N, DN, DDN, DDDN, DDDDN.   **b.** There are an infinite number of outcomes in the sample space.   **c.** E = {DN, DDDN, DDDDDN, ...}. This event consists of all outcomes in which the first n batteries selected are all defective with n being an odd number, followed by a nondefective battery.

6.10 a. The 27 possible outcomes are (1,1,1), (1,1,2), (1,1,3), (1,2,1), (1,2,2), (1,2,3), (1,3,1), (1,3,2), (1,3,3), (2,1,1), (2,1,2), (2,1,3), (2,2,1), (2,2,2), (2,2,3), (2,3,1), (2,3,2), (2,3,3), (3,1,1), (3,1,2), (3,1,3), (3,2,1), (3,2,2), (3,2,3), (3,3,1), (3,3,2), (3,3,3).
c. B = {(1,2,3), (1,3,2), (2,1,3), (2,3,1), (3,1,2), (3,2,1)}
d. C = {(1,1,1), (1,1,3), (1,3,1), (1,3,3), (3,1,1), (3,1,3), (3,3,1), (3,3,3)}
e. B^c = {(1,1,1), (1,1,2), (1,1,3), (1,2,1), (1,2,2), (1,3,1), (1,3,3), (2,1,1), (2,1,2), (2,2,1), (2,2,2), (2,2,3), (2,3,2), (2,3,3), (3,1,1), (3,1,3), (3,2,2), (3,2,3), (3,3,1), (3,3,2), (3,3,3)}
C^c = {(1,1,2), (1,2,1), (1,2,2), (1,2,3), (1,3,2), (2,1,1), (2,1,2), (2,1,3), (2,2,1), (2,2,2), (2,2,3), (2,3,1), (2,3,2), (2,3,3), (3,1,2), (3,2,1), (3,2,2), (3,2,3), (3,3,2)}
$A \cup B$ = {(1,1,1), (2,2,2), (3,3,3), (1,2,3), (1,3,2), (2,1,3), (2,3,1), (3,1,2), (3,2,1)}
$A \cap B$ = the empty set.
$A \cap C$ = {(1,1,1), (3,3,3)}

6.11 a.

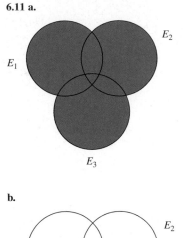

b.

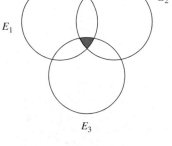

c.

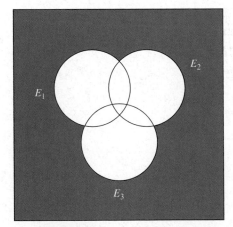

d.

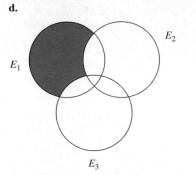

e.

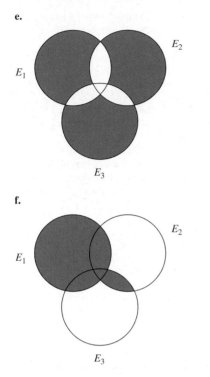

f.

6.12 a. $(A \cup B)^C$

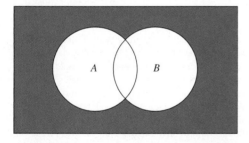

$A^C \cap B^C$

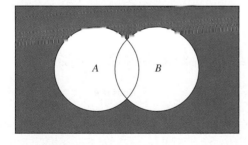

We see that $(A \cup B)^C = A^C \cap B^C$.

b. $(A \cap B)^C$

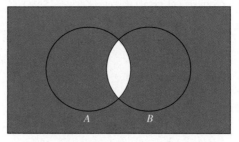

$A^C \cup B^C$

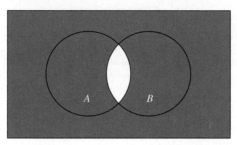

We see that $(A \cap B)^C = A^C \cup B^C$.

6.13 a. $A = \{(C,N), (N,C), (N,N)\}$; $P(A) = 0.19$

6.15 a. In the long run, 1% of all people who suffer cardiac arrest in New York City survive.

6.16 The article sounds as though you will fly safely for 26,000 years and crash on the very last day. Obviously no one lives that long. In fact, there is a very small risk with each flight.

6.18 a. In the long run, 35% of all customers who purchase a tennis racket at this store will buy one with a grip size of $4\frac{1}{2}$ inches.

6.21 No, it is not true. The addition rule is for disjoint events. The events "midsize" and "$4\frac{3}{8}$ inch grip" are not disjoint since P(midsize AND $4\frac{3}{8}$ inch grip) = 0.10 is greater than zero.

6.23 a. The 24 outcomes (including the outcome (2, 4, 3, 1)) are
(1, 2, 3, 4), (1, 2, 4, 3), (1, 3, 2, 4), (1, 3, 4, 2), (1, 4, 2, 3), (1, 4, 3, 2)
(2, 1, 3, 4), (2, 1, 4, 3), (2, 3, 1, 4), (2, 3, 4, 1), (2, 4, 1, 3), (2, 4, 3, 1)
(3, 1, 2, 4), (3, 1, 4, 2), (3, 2, 1, 4), (3, 2, 4, 1), (3, 4, 1, 2), (3, 4, 2, 1)
(4, 1, 2, 3), (4, 1, 3, 2), (4, 2, 1, 3), (4, 2, 3, 1), (4, 3, 1, 2), (4, 3, 2, 1)
b. (1, 2, 4, 3), (1, 4, 3, 2), (1, 3, 2, 4), (4, 2, 3, 1), (3, 2, 1, 4), (2, 1, 3, 4), 0.25

6.24 a. (C,D,P), (C,P,D), (D,C,P), (D,P,C), (P,C,D), (P,D,C). Each outcome would be assigned the probability of 1/6.

6.25 a. Suppose the students selected are from Mathematics and Physics. We abbreviate this outcome by {M, P}. With this notation, the 10 possible outcomes are {B, C}, {B, M}, {B, P}, {B, S}, {C, M}, {C, P}, {C, S}, {M, P}, {M, S}, {P, S}. **c.** 4/10 = 0.4

6.27 a. $P(O_1) = P(O_3) = P(O_5) = 1/9$, $P(O_2) = P(O_4) = P(O_6) = 2/9$.
c. $P(O_1) = 1/21$; $P(O_2) = 2/21$; $P(O_3) = 3/21$; $P(O_4) = 4/21$; $P(O_5) = 5/21$; $P(O_6) = 6/21$; P(odd) = 9/21 = 0.4286, P(at most 3) = 6/21 = 0.2857

6.31 The statement in the article implies the following two conditions: (1) $P(D|Y^C) > P(D|Y)$ and (2) $P(Y) > P(Y^C)$. This claim is consistent with the information given in Part (a) but not with any of the others. In Part (b) and Part (c), condition (1) is violated; in Part (d), condition (2) is violated; in Part (e) and Part (f), both conditions are violated.

6.32 It appears to be true that most basketball players are over 6 feet tall, but many people over 6 feet tall are not basketball players. So it is reasonable to expect $P(A|B) > P(B|A)$.

6.35 b.

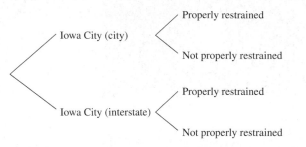

Iowa City (city) — Properly restrained / Not properly restrained

Iowa City (interstate) — Properly restrained / Not properly restrained

6.36 f. The probabilities in Parts (c), (d), and (e) differ in that the probability that a randomly selected individual from that group perceives smoking as very harmful decreases if he or she is a former smoker and decreases even more if he or she has never smoked. This is probably not a surprise.

6.37 b. 0.2 **e.** The probabilities in Part (c) and Part (d) are not equal because the events "wearing a seatbelt" and "being female" are not independent. It is clearly more likely adults will regularly use a seatbelt if they are female than if they are male.

6.38 a. i. $\frac{100}{600} = 0.167$ **ii.** $\frac{41}{600} = 0.068$ **iii.** $\frac{41}{172} = 0.241$

iv. $\frac{500}{600} = 0.8333$ **v.** $\frac{41}{100} = 0.41$ **vi.** $\frac{18}{100} = 0.18$

b. The probability that a randomly selected senior uses a seatbelt is much lower than for a randomly selected person in the 18 to 24 year age group.

6.39 a. i. 0.76 **ii.** 0.717 **iii.** 0.803 **iv.** Treatment B has the higher survival rate. **b. i.** 0.583 **ii.** 0.6 **iii.** 0.5 **iv.** Treatment A has the higher survival rate. **c. i.** 0.878 **ii.** 0.95 **iii.** 0.85 **iv.** Treatment A has the higher survival rate. **d.** Women seem to respond well to both treatments. Men seem to respond very well to treatment A but not so well to treatment B. When the data is combined, the small quantity of data collected for men on treatment B—only 20% of those men who had treatment A, becomes "lost" in the rest of the data.

6.40 Suppose the probability that a driver who does not live close to a restaurant will have an accident is x. So $P(A|R^C) = x$. If a driver who does live close to a restaurant is 30% more likely to have an accident, then $P(A|R) = 1.3x$. The difference between these is $0.3x$, which divided by $x\, P(A|R^C)$ will give 0.3. The 4th choice is the correct one.

6.41 a. Approximately 85% of calls are for medical assistance.
d. 0.1275 **f.** Probably not. There are likely to be several calls related to the same event—several reports of the same accident or fire that would be received close together in time.

6.42 a. In the long run, 30 out of every 100 accidents reported to the police that result in a death were a single-vehicle rollover. **b.** In the long run, 54 out of every 100 accidents reported to the police that result in a death were a frontal collision. **d.** No. Because $P(F|D) \neq P(F)$, F and D are not independent. **e.** No, F is the event that the selected accident was a frontal collision. R^C is the event that the selected accident was not a single-vehicle rollover. These are not the same. There are accidents that are one of these, or both, or neither.

6.43 No, a randomly selected adult is more likely to experience pain daily if she is a woman. The events are dependent.

6.44 They are dependent because the probability of TB for a recent immigrant (0.0075) is not the same as the unconditional probability of TB (0.0006).

6.45 $P(\textit{favors gun control}|\textit{female})$ and $P(\textit{favors gun control})$ are not equal, the two events are dependent.

6.46 They are dependent. $P(\textit{begins smoking again}|\textit{uses nicotine aid})$ and $P(\textit{begins smoking again})$ are not equal.

6.50 c. 0.7143 **d.** No, the probability that a student has both a Master-Card and a Visa is not the product of the two individual probabilities.

6.52 b. no, $P(F) \neq P(F|C)$. **c.** no, $P(F) \neq P(F|O)$.

6.53 a. The expert assumed that there is a 1 in 12 chance for each valve to be in any one of the clock positions and that the positions of the two valves are independent. **b.** The assumption of independence is questionable because the wheels turn in a synchronized manner. The correct probability would be higher than 1/144.

6.54 b. 0.56 **c.** 0.06

6.55 a. The events E_1 and E_2 are dependent, because the probability of E_2 differs depending on whether E_1 occurs or not. **c.** $P(E_2|E_1) = 39/4999$ and $P(E_2|\text{not } E_1) = 40/4999$. They are nearly equal.

6.56 $P(B_1 \cap S) = 0.12$, $P(B_1 \cap M) = 0.2$, $P(B_1 \cap L) = 0.08$, $P(B_2 \cap S) = 0.18$, $P(B_2 \cap M) = 0.3$, $P(B_2 \cap L) = 0.12$

6.57 a. 0.00391 **b.** 0.00383. This is very close to the value of the probability in Part (a).

6.62 $E \cap F$ is the event that a randomly selected voter has signed the recall petition and also votes in the recall election, 0.08.

6.63 The likelihood of using a cell phone is different for each type of vehicle; for instance if you drive a van or SUV, you are more likely to use a cell phone than if you drive a pickup truck.

6.64 a. 0.523, 0.477, 0.018, 0.03 **b.** 0.6032, this is the probability that a randomly selected gun purchase background check resulted in a blocked sale.

6.67 a. In four consecutive years, there are three years of 365 days (1095 nonleap days) and one year of 366 days (365 nonleap days and one leap day)—a total of 1461 days, of which one, Feb 29th, is a leap day. Hence a leap day occurs once in 1461 days. **b.** If babies are induced or born by C-section, they are less likely to be born during weekends or on holidays so they are not equally likely to be born on the 1461 days. **c.** 1 in 2.1 million is the probability of picking a mother and baby randomly and finding *both* to be a leap day baby $(1/1461)^2$. The probability of a leap year baby becoming a leap year mom means that the mom's birthday is already fixed, so the hospital spokesperson's probability is too small.

6.69 b. 0.08325 **e.** The benefit of this retest scheme is that it reduces the rate of false positives; the disadvantage is that it increases the risk of false negatives.

6.70 a. It is possible. For example (there are other possibilities), consider

	Works Full-time	Does Not Work Full-time	Total
Uses drugs	70	30	100
Does not use drugs	805	7095	7900
Total	875	7125	8000

c. No, need to know at least one of the cell entries in the table given in Part (a).

6.74 a.

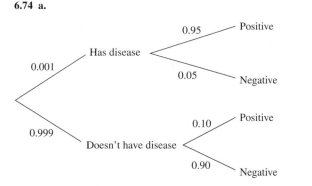

d. 0.00942. The reason this probability is so small is that the incidence rate of the disease is so small.

6.75 a. 0.85 **b.** 0.19

6.76 a.

Probation	High School GPA			
	2.5–<3.0	3.0–<3.5	3.5 and Above	Total
Yes	0.10	0.11	0.06	0.27
No	0.09	0.27	0.37	0.73
Total	0.19	0.38	0.43	1.00

d. They are not independent; the probability a student has a GPA 3.5 or above and is on academic probation is not equal to the product of the two individual probabilities.

6.78 i. 0.4876

6.79 The simulation results will vary from one simulation to another. The following probabilities were obtained from a simulation using a computer with 100,000 trials:

a. 0.71293 **b.** 0.02201

6.81 The simulation results will vary from one simulation to another. The approximate probability should be around 0.8504.

6.82 The simulation results will vary from one simulation to another. The approximate probability should be around 0.8468.

6.83 a. The simulation results will vary from one simulation to another. The approximate probability should be around 0.6504 **b.** The decrease in the probability of on-time completion for Jacob made the biggest change in the probability that the project is completed on time.

6.84 a.

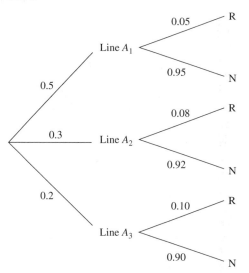

6.87 c. Select a random digit. If it is a 0, 1, or 2, then award A one point. If it is a 3 or 4, then award B one point. If it is 5, 6, 7, 8, or 9, then award A and B one-half point each. Repeat the selection process until a winner is determined or the championship ends in a draw (5 points each). Replicate this entire procedure a large number of times. The estimate of the probability that A wins the championship would be the ratio of the number of times A wins to the total number of replications. **d.** It would take longer if no points for a draw are awarded. It is possible that a number of games would end in a draw and so more games would have to be played in order for a player to earn 5 points.

6.88 a. For example, choose a random digit. If it is a 1, 2, 3, 4, 5, 6, 7 or 8, then seed 1 defeats seed 4. If it is a 9 or 0, then seed 4 defeats seed 1. **b.** For example, choose a random digit. If it is a 1, 2, 3, 4, 5 or 6, then seed 2 defeats seed 3. If it is a 7, 8, 9, or 0, then seed 3 defeats seed 2. **c.** For example, if seeds 1 and 2 have won in the first round, then they are competing in game 3. Select a random digit. If it is a 1, 2, 3, 4, 5 or 6, then seed 1 defeats seed 2 in game 3. If it is a 7, 8, 9, or 0, then seed 2 defeats seed 1 in game 3. If seeds 1 and 3 have won in the first round, then they are competing in game 3. Select a random digit. If it is a 1, 2, 3, 4, 5, 6, or 7, then seed 1 defeats seed 3 in game 3. If it is an 8, 9 or 0, then seed 3 defeats seed 1 in game 3. If seeds 2 and 4 have won in the first round, then they are competing in game 3. Select a random digit. If it is a 1, 2, 3, 4, 5, 6, or 7, then seed 2 defeats seed 4 in game 3. If it is an 8, 9, or 0, then seed 4 defeats seed 2 in game 3. If seeds 3 and 4 have won in the first round, then they are competing in game 3. Select a random digit. If it is a 1, 2, 3, 4, 5, or 6, then seed 3 defeats seed 4 in game 3. If it is a 7, 8, 9, or 0, then seed 4 defeats seed 3 in game 3. **d.** Answers will vary. See solutions manual for one example. **e.** Answers will vary. See solutions manual for one example. **f.** Answers will vary. **g.** The answers for Parts (e) and (f) differ because they are based on different random selections and different number of trials. Generally, the larger the number of trials, the better the estimate is. So one would believe that the estimate from Part (f) is better than the estimate of Part (e).

6.94 R_1 and R_2 are not independent events because $P(R_2|R_1) = 1139/2516$ and this is not equal to $P(R_2|R_1^C) = 1140/2516$. However, these two probabilities are very nearly equal, so from a practical point of view these two events may be regarded as independent (approximately).

6.95 a. 0.6222

Chapter 7

7.2 d. Discrete

7.4 All values in the interval 0 to $\sqrt{2}$, continuous

7.7 b. $-3, -2, -1, 1, 2, 3$ **c.** $0, 1, 2$

7.8 e. 0.90, 0.65; they are different because the first probability includes the probability associated with 3 and 6 courses.

7.9 b. About 20% of the cartons have exactly one broken egg. **d.** 0.85; $P(y < 2)$ is less than $P(y \leq 2)$ because y is discrete and $(y \leq 2)$ includes $y = 2$.

7.10 c.

Value of x	Probability
0	1/6
1	4/6
2	1/6

7.12 a. (1, 2), (1, 3), (1, 4), (1, 5), (2, 3), (2, 4), (2, 5), (3, 4), (3, 5), (4, 5)

b.

Value of x	Probability
0	3/10
1	6/10
2	1/10

7.13 Results will vary. See solutions manual for the results from one possible simulation.

7.14

Value of x	Probability
0	0.027
1	0.189
2	0.441
3	0.343

7.15 a. $P(x = 0) = 0.4096$, $P(x = 1) = 0.4096$, $P(x = 2) = 0.1536$, $P(x = 3) = 0.0256$, $P(x = 4) = 0.0016$.

7.16

Value of w (in dollars)	Probability
1	3/10 = 0.3
10	3/10 = 0.3
25	4/10 = 0.4

7.17 a. Smallest y value: 1, resulting from outcome S. Second smallest value: 2, resulting from outcome FS. **b.** Positive integers: 1, 2, 3, . . . **c.** $P(y = 1) = 0.7$, $P(y = 2) = (0.3)(0.7)$, $P(y = 3) = (0.3)^2(0.7)$, $P(y = 4) = (0.3)^3(0.7)$, $P(y = 5) = (0.3)^4(0.7)$. In general, $P(y) = (0.3)^{y-1}(0.7)$.

7.19 $P(y = 0) = 0.16$, $P(y = 1) = 0.33$, $P(y = 2) = 0.32$, $P(y = 3) = 0.19$

7.20 a.

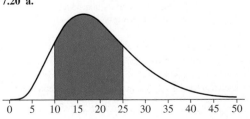

b.

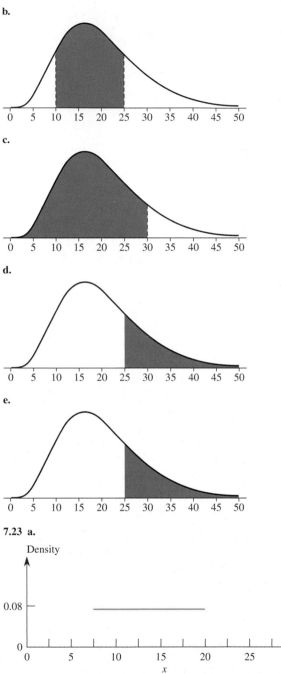

c.

d.

e.

7.23 a.

7.25 b. 0.25

7.26 a. Area under the density curve = $(0.5)(40)(0.05) = 1$.

7.29 a. $\mu_y = 0.56$; the long-run average number of broken eggs per carton is 0.56. **b.** 65% **c.** Because the probability distribution indicates that there are more cartons with 0 or 1 broken eggs than 2, 3, or 4 broken eggs.

7.32 b. 1 and 2 are the only values that are more than 2 standard deviations from the mean; 0.05.

7.35 Under the royalty plan, the mean amount received would be $10,550, which exceeds the flat payment. This makes the royalty plan sound like a good idea. However, with the royalty plan, the probability of receiving over $10,000 is only 0.25 which makes the flat payment look more favorable. It all depends on how confident the author was!

7.36 Probability distribution 1:

x	1	2	3	4	5
$p(x)$	0.5	0	0	0	0.5

Probability distribution 2:

x	1	2	3	4	5
$p(x)$	0.2	0.2	0.2	0.2	0.2

The mean for distribution 1 is $(1)(0.5) + (5)(0.5) = 3$. The standard deviation is 3.9452. The mean for distribution 2 is $(1)(0.2) + (2)(0.2) + (3)(0.2) + (4)(0.2) + (5)(0.2) = 3$. The standard deviation is 1.41421. Thus the two distributions have the same mean but quite different standard deviations.

7.37 a. Discrete, there are only 6 possible values.

7.40 a. 16.38, 3.9936

7.41 a. Because if $y < 0$, the peg wouldn't fit in the hole!   **b.** 0.003   **d.** Yes, it's reasonable; the hole and the peg are made at different times using different tools.   **e.** With a standard deviation larger than a mean, it seems fairly likely to obtain a negative value of y so it would seem a relatively common occurrence to find a peg that was too big to fit in the pre-drilled hole.

7.42 a. 0; this means that the average score for students who are just guessing is zero, which seems fair.

7.43 a. $2.8, $1.29   **b.** $0.7, $0.78

7.44 a. 3.5, 2.9167, 1.7078   **b.** 3.5, 2.9167, 1.7078   **e.** Depends on how much risk you are willing to take. The variability in winnings would be much greater for game 2, which means the potential for greater winnings is with game 2, but there is also the potential for greater losses.

7.45 a. Six outcomes: SFFFFF, FSFFFF, FFSFFF, FFFSFF, FFFFSF, FFFFFS.   **b.** 10 successes: 184,756 outcomes; 15 successes: 15,504 outcomes; 5 successes: 15,504 outcomes.

7.46 a. 0.0486

7.47 a. $p(4) = 0.24576$.

7.50

x = Number of Female Puppies in a Litter of Size 5	$p(x)$
0	.03125
1	.15625
2	.31250
3	.31250
4	.15625
5	.03125

7.51 d. 25 failing is more than 3 standard deviations below the mean, so would be a very surprising result.

7.53 If the graphologist was guessing, $P(x \geq 6) = 0.377$

7.55 b. 0.484   **c.** 17.5, 2.2913

7.56 a. Binomial with $n = 100$ and $\pi = 0.2$.   **b.** 20   **c.** 16, 4   **d.** It is highly unlikely; a score of 50 is 7.5 standard deviations away from the mean.

7.57 Since the sampling is done without replacement and the sample size is more than 5% of the population size, the number of invalid signatures in that situation cannot be considered to have a binomial distribution.

7.58 c. 0.846, 0.846; they are larger because these values of π are closer to 0.5.   **d.** It is more likely that a fair coin will not be judged unfair, but there is also a greater risk of judging an unfair coin to be fair.

7.59 c. The error probability in Part (a) decreases whereas the error probability in Part (b) increases.

7.60 d. If 90% favor the ban, $P(x < 20) = 0.033$. It would be very unusual to see fewer than 20 people favoring the ban if the "at least 90%" assertion is true.

7.61 a. Geometric

7.64 e. 0.6887

7.69 a. 1.88   **b.** 2.33

7.73 d. Weights less than 1836 grams or weights greater than 5018 grams.   **e.** The birth weight distribution would be normal with mean 7.5663 lbs and standard deviation 1.06263 lbs. The probability that the birth weight is greater than 7 lbs. is still 0.7019.

7.79 c. Yes, the probability is about 0.001.

7.81 Since this plot appears to look like a straight line, it is reasonable to conclude that a normal distribution provides an adequate description of the steam rate distribution.

7.82 This plot is not very straight, suggesting that the cadmium concentration distribution is not normal.

7.83

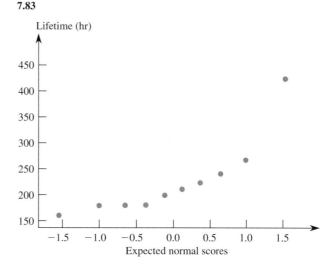

Since the graph exhibits a curve, we cannot conclude that the variable component lifetime be adequately modeled by a normal distribution.

7.84

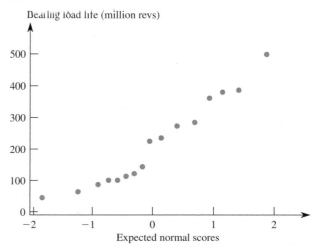

Since the plot is fairly straight, normality is plausible.

7.85

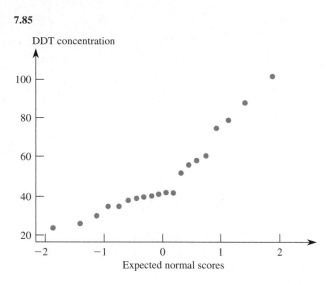

DDT concentration

Although the graph seems to follow a straight line for values of DDT concentration below 45 and above 45, they don't appear to be the *same* straight line. A normal distribution may not be an appropriate model for this population.

7.86

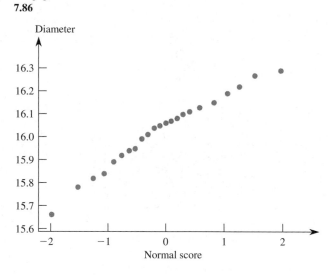

Diameter

The plot is quite straight; normality is plausible.

7.87

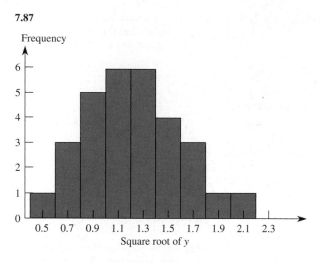

Frequency

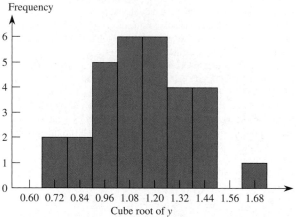

Frequency

Cube root transformation is more symmetric

7.88 a.

Class	Frequency	Relative Frequency
0–<300	8	0.186047
300–<600	12	0.27907
600–<900	5	0.116279
900–<1200	5	0.116279
1200–<1500	4	0.093023
1500–<1800	2	0.046512
1800–<2100	4	0.093023
2100–<2400	1	0.023256
2400–<2700	2	0.046512

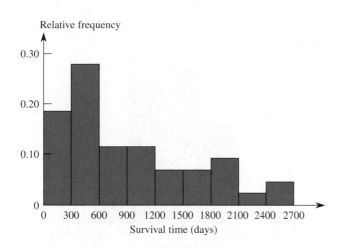

Relative frequency

Survival time (days)

c. Many of the commonly used statistical inference procedures are valid only when the sample data are from a population distribution that is at least approximately normal. If such inference methods are to be used to draw conclusions from the survival data of this problem then a transformation should be considered so that, on the transformed scale, the data are approximately normally distributed.

7.89 a.

Class	Frequency	Relative Frequency
0–<100	22	0.22
100–<200	32	0.32
200–<300	26	0.26
300–<400	11	0.11
400–<500	4	0.04
500–<600	3	0.03
600–<700	1	0.01
700–<800	0	0.00
800–<900	1	0.01

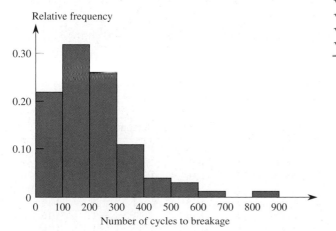

Relative frequency

Number of cycles to breakage

7.90 a.

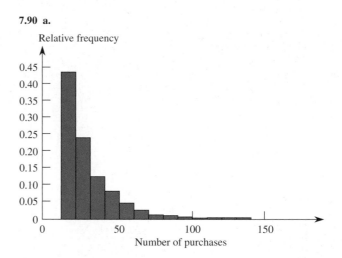

Relative frequency

Number of purchases

The histogram is positively skewed.

b.

Class Interval	Frequency	Rel. Frequency Frequency $= \dfrac{\text{Frequency}}{2071}$	Class Interval Width	Density $= \dfrac{\text{Rel. Frequency}}{\text{Class Width}}$
$\sqrt{10}$–<$\sqrt{20}$	904	0.436504	1.30986	0.333245
$\sqrt{20}$–<$\sqrt{30}$	500	0.241429	1.00509	0.240207
$\sqrt{30}$–<$\sqrt{40}$	258	0.124577	0.84733	0.147024
$\sqrt{40}$–<$\sqrt{50}$	167	0.080637	0.74651	0.108019
$\sqrt{50}$–<$\sqrt{60}$	94	0.045389	0.6749	0.067253
$\sqrt{60}$–<$\sqrt{70}$	56	0.02704	0.62063	0.043569
$\sqrt{70}$–<$\sqrt{80}$	26	0.012554	0.57767	0.021733
$\sqrt{80}$–<$\sqrt{90}$	20	0.009657	0.54256	0.017799
$\sqrt{90}$–<$\sqrt{100}$	13	0.006277	0.51317	0.012232
$\sqrt{100}$–<$\sqrt{110}$	9	0.004346	0.48809	0.008904
$\sqrt{110}$–<$\sqrt{120}$	7	0.00338	0.46636	0.007248
$\sqrt{120}$–<$\sqrt{130}$	6	0.002897	0.4473	0.006477
$\sqrt{130}$–<$\sqrt{140}$	6	0.002897	0.43041	0.006731
$\sqrt{140}$–<$\sqrt{150}$	3	0.001449	0.41529	0.003488
$\sqrt{150}$–<$\sqrt{160}$	0	0.000000	0.40166	0.000000
$\sqrt{160}$–<$\sqrt{170}$	2	0.000966	0.38929	0.002481

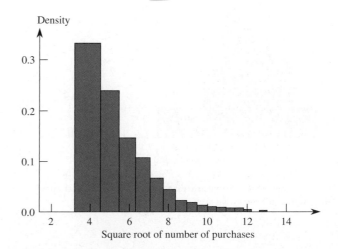

Density

Square root of number of purchases

The histogram is less skewed, but it is still far from symmetric.

7.91 a.

0	56667889999
1	000011122456679999
2	0011111368
3	11246
4	18
5	
6	8
7	
8	
9	39 HI: 448

b.

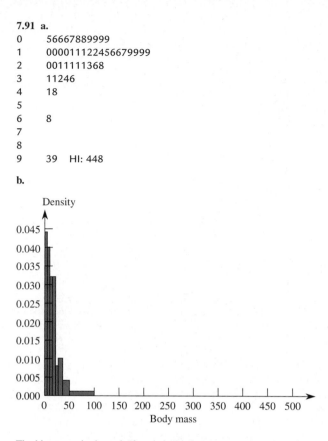

Body mass

The histogram is skewed. If statistical inference procedures based on normality assumptions are to be used to draw conclusions from the body mass data of this problem, then a transformation should be considered so that, on the transformed scale, the data are approximately normally distributed.

c.

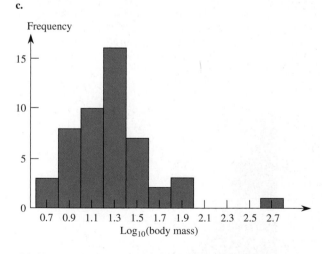

Log_{10}(body mass)

This histogram is more symmetric than the one for the untransformed data, but it is still somewhat skewed.

d.

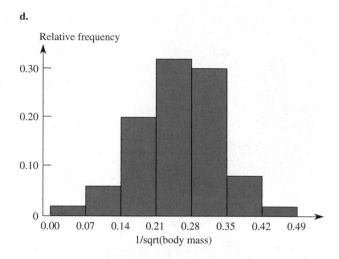

1/sqrt(body mass)

This histogram does resemble a normal curve.

7.92 The transformation appears to have been more successful for Mn than for Cu and Zn. The histogram of log(Zn) shows a hint of bimodality. In all three cases the histograms appear to be more symmetric in the transformed scale than in the original scale.

7.93 c. 0.7557

7.97 b. 0.8739

7.100 a. A normal approximation would not be appropriate because $n\pi < 10$. **b.** A normal approximation would be appropriate in this case; $P(x \geq 20) \approx 0.87046$.

7.102 a. No, because sampling is without replacement and more than 5% of the population is sampled. **c.** No, the variance doubles. The standard deviation increases by a factor of $\sqrt{2}$.

7.104 e. 0.65

7.105 a. 2.64, 1.53961

7.106 c. In order for y to take the value 5, there must be only 1 satisfactory battery among the first four batteries selected and the fifth battery selected must be a satisfactory one. The four outcomes that satisfy this requirement are SUUUS, USUUS, UUSUS, UUUSS. Hence $p(5) = 0.02048$.

d. $p(y) = \binom{y-1}{1}(.8)^2(1-.8)^{y-1} = (y-1)(.8)^2(1-.8)^{y-1}$

for $y = 2, 3, 4, \ldots$

7.108 b. Yes, $P(y < 25) = 0.00001545$.

7.113 b. About 69% of adult women would be excluded due to the height requirement.

7.114 a.

Outcome	y	p	Outcome	y	p
SSSS	4	1/16	FSSF	2	1/16
SSSF	3	1/16	FSFS	1	1/16
SSFS	2	1/16	FFSS	2	1/16
SFSS	2	1/16	SFFF	1	1/16
FSSS	3	1/16	FSFF	1	1/16
SSFF	2	1/16	FFSF	1	1/16
SFSF	1	1/16	FFFS	1	1/16
SFFS	1	1/16	FFFF	0	1/16

The probability distribution of y is given below. Its mean is 1.6875.

y	$p(y)$
0	1/16
1	7/16
2	5/16
3	2/16
4	1/16

b.

Outcome	y	p	Outcome	y	p
SSSS	4	0.1296	FSSF	2	0.0576
SSSF	3	0.0864	FSFS	1	0.0576
SSFS	2	0.0864	FFSS	2	0.0576
SFSS	2	0.0864	SFFF	1	0.0384
FSSS	3	0.0864	FSFF	1	0.0384
SSFF	2	0.0576	FFSF	1	0.0384
SFSF	1	0.0576	FFFS	1	0.0384
SFFS	1	0.0576	FFFF	0	0.0256

The probability distribution of y is given below. Its mean is 2.0544.

y	$p(y)$
0	0.0256
1	0.3264
2	0.3456
3	0.1728
4	0.1296

c.

Outcome	z	p	Outcome	z	p
SSSS	4	1/16	FSSF	2	1/16
SSSF	3	1/16	FSFS	1	1/16
SSFS	2	1/16	FFSS	2	1/16
SFSS	2	1/16	SFFF	3	1/16
FSSS	3	1/16	FSFF	2	1/16
SSFF	2	1/16	FFSF	2	1/16
SFSF	1	1/16	FFFS	3	1/16
SFFS	2	1/16	FFFF	4	1/16

The probability distribution of y is given below. Its mean is 2.375.

z	$p(z)$
1	2/16
2	8/16
3	4/16
4	2/16

7.115 a.

w	$p(w)$
0	6/36
1	10/36
2	8/36
3	6/36
4	4/36
5	2/36

7.116

x	$p(x)$
1	12/24
2	8/24
3	4/24

7.117 c.

x	$p(x)$
4	$(0.6)^4 \quad\quad + (0.4)^4 = 0.1552$
5	$\binom{4}{3}(0.6)^4(0.4) + \binom{4}{3}(0.6)(0.4)^4 = 0.26880$
6	$\binom{5}{3}(0.6)^4(0.4)^2 + \binom{5}{3}(0.6)^2(0.4)^4 = 0.29952$
7	$\binom{6}{3}(0.6)^4(0.4)^3 + \binom{6}{3}(0.6)^3(0.4)^4 = 0.27648$

7.118

y	$p(y)$
0	0.1552
1	0.26880
2	0.29952
3	0.27648

7.119 $P(x = 0) = 0.216$, $P(x = 1) = 0.432$, $P(x = 2) = 0.288$, $P(x = 3) = 0.064$.

7.123 b. 0.8760

7.124 d. Yes, the probability of a pregnancy having a duration of at least 310 days is only 0.0030.

Chapter 8

8.1 A statistic is computed from the observations in a sample, whereas a population characteristic is a quantity that describes the whole population.

8.2 μ is the mean of the population whereas $\bar{x}$ is the sample mean. Similarly, σ is the standard deviation of the population whereas s is the sample standard deviation.

8.3 a. Population characteristic **c.** Population characteristic

8.4 b. Answers will vary. See solutions manual for an example.
c. Answers will vary. See solutions manual for an example.

8.6 Histogram for samples of size 10 would be centered in the same place but would be less spread out than for samples of size 5.

8.7 a.

Sample	Sample Mean	Sample	Sample Mean
1, 2	1.5	3, 1	2.0
1, 3	2.0	3, 2	2.5
1, 4	2.5	3, 4	3.5
2, 1	1.5	4, 1	2.5
2, 3	2.5	4, 2	3.0
2, 4	3.0	4, 3	3.5

The sampling distribution of $\bar{x}$ based on $n = 2$, when sampling is done *without replacement*, is

Value of $\bar{x}$	1.5	2.0	2.5	3.0	3.5
Probability	2/12	2/12	4/12	2/12	2/12

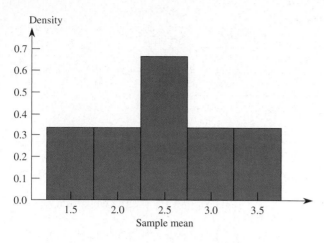

Density

b.

Sample	Sample Mean	Sample	Sample Mean
1, 1	1.0	3, 1	2.0
1, 2	1.5	3, 2	2.5
1, 3	2.0	3, 3	3.0
1, 4	2.5	3, 4	3.5
2, 1	1.5	4, 1	2.5
2, 2	2.0	4, 2	3.0
2, 3	2.5	4, 3	3.5
2, 4	3.0	4, 4	4.0

The sampling distribution of $\bar{x}$ based on $n = 2$, when sampling is done *with replacement*, is

Value of $\bar{x}$	1.0	1.5	2.0	2.5	3.0	3.5	4.0
Probability	1/16	2/16	3/16	4/16	3/16	2/16	1/16

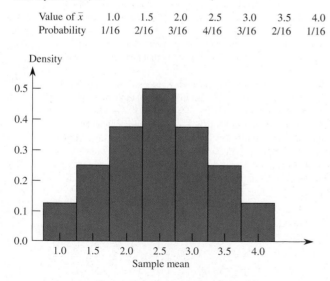

c. They are similar in that they are both symmetric about the population mean of 2.5. Also, both have largest probability at 2.5. They are different in that the values in Part (a) cover a smaller range, 1.5 to 3.5, than in Part (b), 1.0 to 4.0.

8.9

Statistic #1:

Value	2.67	3.00	3.33	3.67
Probability	0.1	0.4	0.3	0.2

Statistic #2:

Value	3	4
Probability	0.7	0.3

Statistic #3:

Value	2.5	3.0	3.5
Probability	0.1	0.5	0.4

Statistic #1 is the only one that is unbiased, but, because it has a large standard deviation, there is a good chance that an estimate would be far away from the value of the population mean. For this population, Statistic #3 may be a better choice because it is more likely to produce an estimate that is close to the population mean.

8.12 σ represents the standard deviation of the population being sampled, while $\sigma_{\bar{x}}$ represents the standard deviation of the sampling distribution of $\bar{x}$. Similarly, μ represents the mean value of the population being sampled, while $\mu_{\bar{x}}$ represents the mean value of the sampling distribution of $\bar{x}$.

8.15 b. $n = 20$: $\mu_{\bar{x}} = 2$, $\sigma_{\bar{x}} = 0.179$; $n = 100$: $\mu_{\bar{x}} = 2$, $\sigma_{\bar{x}} = 0.08$; a sample of size 100 would be most likely to result in a value close to μ.

8.16 c. Total weight limit of 2500 lbs will be exceeded if $\bar{x} > 156.25$.

8.17 a. 0.8185, 0.0013

8.18 b. Approximately 95% of the time, $\bar{x}$ will be within $2\,\sigma_{\bar{x}} = 2(1) = 2$ of μ. Approximately 0.3% of the time, $\bar{x}$ will be further than $3\sigma_{\bar{x}} = 3(1) = 3$ from μ.

8.20 a.

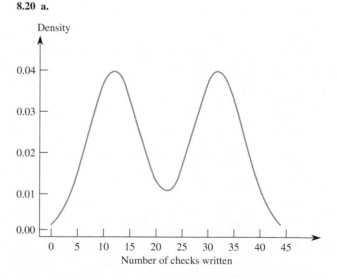

Density

b. Approximately normal with mean 22.0 and standard deviation 1.65

8.22 a. Approximately normal with mean 3 and standard deviation 2.9625 **b.** 2.9625 to 3.0375

8.24 For $\pi = 0.65$, $n = 30$, 50, 100, and 200; for $\pi = 0.2$, $n = 50$, 100, and 200

8.25 c. If $n = 200$, μ_p does not change, but σ_p becomes 0.01804.
d. Yes, $n\pi$ and $n(1 - \pi)$ are now both > 10.

8.26 a. $\mu_p = 0.15$, $\sigma_p = 0.0357$ **b.** Yes, $n\pi$ and $n(1 - \pi)$ are both > 10 **c.** If $n = 200$, μ_p does not change, but σ_p becomes 0.02525.
d. Yes, $n\pi$ and $n(1 - \pi)$ are both > 10. **e.** 0.9761

8.27 a. $\mu_p = 0.005$, $\sigma_p = 0.007$ **b.** Since $n\pi < 10$, the sampling distribution of p cannot be assumed to be approximated by a normal curve.
8.28 b. $\mu_p = 0.3$, $\sigma_p = 0.0229$
8.29 a. $\pi = 0.5$: $\mu_p = 0.5$, $\sigma_p = 0.0333$, $\pi = 0.6$: $\mu_p = 0.6$, $\sigma_p = 0.0327$; in both cases, the sampling distribution of p is approximately normal. **b.** $\pi = 0.5$: $P(p \geq 0.6) = 0.0013$, $\pi = 0.6$: $P(p \geq 0.6) = 0.5$
c. The probability for $\pi = 0.5$ would decrease and the probability for $\pi = 0.6$ would stay the same.
8.33 a. Approximately normal with a mean of 50 lbs and a standard deviation of 0.1 lb
8.36 b. If $\pi = 0.4$ for your state, it would be very unlikely to observe such a sample.

Chapter 9

9.1 Statistic II because it is unbiased and it has a smaller variability than the other two.
9.2 An unbiased statistic is generally preferred over a biased statistic, because there is no long run tendency for the unbiased statistic to overestimate or underestimate the true population value. One might choose a biased statistic over an unbiased one if the bias of the first is small and if its sampling distribution has a small variance.
9.8 a. $\bar{x} = 20.6526$
9.9 c. One possibility is a trimmed mean. An 8.3% trimmed mean = 11.73.
9.10 a. 120.6
9.11 a. 95% confidence level
9.14 a. As the confidence level increases, the width of the confidence interval for π increases. **b.** As the sample size increases, the width of the confidence interval for π decreases. **c.** As the value of p gets farther from 0.5, either larger or smaller, the width of the confidence interval for π decreases.
9.15 a. (0.38, 0.46); we can be 99% confident that the true proportion of adult Americans who made plans in May 2005 based on an incorrect weather report is between .38 and .46.
9.16 c. The estimated standard deviation of the sampling distribution of p is larger when p is 0.48 than when $p = 0.37$ so the width of the confidence interval will be wider.
9.22 a. (0.494, 0.546) **b.** (0.6252, 0.6748) **c.** It is estimated that about half of all adult Americans think lying is never justified, yet more than 60% of them think that it is OK to lie if it avoids hurting someone's feelings. Obviously there are some adults who consider "lying to avoid hurting someone's feelings" as not really a lie!
9.23 c. Increased confidence level in Part (a), increased sample proportion in Part (a).
9.24 a. (0.891, 0.909) **b.** The sample was large ($np > 10$ and $n(1 - p) > 10$, which in this case, they are), and that the respondents were randomly selected.
9.25 (0.2169, 0.4755); must assume that it is reasonable to regard this sample as representative of young adults with pierced tongues.
9.26 With $p = .36$ and $n = 1004$, the bound on error is
$$1.96\sqrt{\frac{(.36)(.64)}{1004}} \approx .03.$$
9.27 With $p = .25$ and $n = 1002$, the bound on error is
$$1.96\sqrt{\frac{(.25)(.75)}{1002}} \approx .03.$$
9.28 (0.597, 0.643); must assume that the sample is representative of Australian children.

9.32 b. (114.4, 115.6) is the 90% interval and (114.1, 115.9) is the 99% interval because the width of the interval increases as the confidence level increases.
9.34 a. Narrower, because the confidence level is lower. **b.** Incorrect; 95% refers to the percentage of all possible samples that result in an interval that includes μ.
9.35 A 95% confidence interval is (262.045, 350.135). Other confidence levels are possible.
9.36 a. The two groups "12 to 23 months" and "24 to 35 months" would both have the same variability as each other but greater variability than the "less than 12 months group" because the interval width for both groups is 0.4, which is wider than the "less than 12 months" group with an interval width of 0.2. **b.** The group "less than 12 months" would have the largest sample size because the interval width for that group is narrower than for the other groups. **c.** It would be a 99% confidence level because if everything else remains constant, an increase in the confidence level results in an increase in the interval width. The new confidence interval is wider.
9.37 (26.72. 30.28); with a confidence level of 90%, we estimate the mean daily commute time for working residents of Calgary to be between 26.72 min and 30.28 min.
9.38 a. The standard deviation could be larger than the mean if the distribution is extremely skewed (in this case, positive skew). **b.** No, because expense must be nonnegative and here the mean is smaller than the standard deviation. **c.** Yes, because the sample size is large.
9.39 c. The new FAA recommendations are above the upper bounds for both summer and winter.
9.41 b. If we took many samples using this sampling method and obtained confidence intervals from each of the samples, 90% of them would capture the true mean airborne time for Flight 448.
9.46 b. No, the confidence interval is an estimate of the mean, not individual values.
9.47 No. The sample size is small and there are two outliers in the data set, indicating that the assumption of a normal population distribution is not reasonable.
9.50 For 90% confidence level: $n = \left[\frac{(1.645)\sigma}{B}\right]^2$; for 98% confidence level: $n = \left[\frac{(2.33)\sigma}{B}\right]^2$.
9.52 303, 385, unless it were very costly to obtain observations, the larger sample size would be preferred.
9.54 b. No, even if the mean is nonzero, there could be some individual 0s in the population meaning there are some students who don't lie to their mothers.
9.55 c. The sample size is much larger.
9.61 a. (14.21, 16.99) **b.** CBS: (10.51, 13.29), FOX: (10.31, 13.09), NBC: (9.61, 12.39)

Chapter 10

10.1 $\bar{x}$ is a statistic and hypotheses are always expressed in terms of a population characteristic.
10.2 a. Does not comply, the null and alternative hypothesis cannot be identical. **b.** Does not comply, the hypothesized value must be the same in both hypotheses. **d.** Does not comply, the hypothesized values are not the same and the alternative hypothesis cannot include the equal case.
10.3 The burden of proof is placed on the contractor to show that the mean weld strength is greater than 100 lb/in^2.
10.4 a. The second pair of hypotheses; strong evidence (or lack of strong evidence) can only be claimed for the alternative hypothesis.

10.5 Failing to reject the null hypothesis is not the same as proving it to be true or even being able to claim strong support in favor of the null hypothesis.

10.6 The second pair of hypotheses, because the change will be made only if there is evidence that more than 60% favor the change. The alternative hypothesis should be $H_a: \pi > 0.6$.

10.12 a. Type I error **b.** A Type I error occurs when screening indicates cancer but patient does not have cancer. Consequences might be unnecessary follow-up tests or possible treatment. **c.** A Type II error occurs when a patient has cancer but this is not detected by the screening procedure. Consequences might be that the patient does not receive timely treatment.

10.13 a. A Type I error is concluding the symptoms are due to disease when they are really due to child abuse. A Type II error is concluding the symptoms are due to child abuse when they are really due to disease.

10.14 a. A Type I error would be deciding that a particular man is not the father when in fact he is the father. A Type II error would be deciding that a particular man is the father when in fact he is not the father.

10.16 a. A Type I error consists of saying that the manufacturer's claim is not true ($\pi < 0.9$), when in fact the manufacturer is correct in its claim. A Type II error occurs if the manufacturer's claim is incorrect but the consumer agency fails to detect it. A Type I error would result in false accusation of false advertising. A Type II error would enable the manufacturer to continue false advertising without penalty. **b.** A small value for α, such as 0.01, because a Type I error has serious consequences for the agency.

10.17 a. A Type I error is returning to the supplier a shipment that is not of inferior quality. A Type II error is accepting a shipment of inferior quality. **b.** The calculator manufacturer would consider a Type II error more serious, since it would end up producing defective calculators. **c.** The supplier would consider a Type I error more serious because it would lose the profits from perfectly good printed circuits.

10.18 A Type I error is concluding that the discharged water is too hot when it really is not too hot. A Type II error is concluding that the mean temperature of discharged water is not too hot when it really is too hot. A Type II error may be the more serious because it could lead to damage to the river ecosystem.

10.19 a. We would probably not complain if the defective rate was below 10%; we would want to determine whether the defective rate is greater than is claimed by the manufacturer. **b.** A Type I error is concluding that the defect rate exceeds 10%, when in fact, it isn't. This would lead to unjustified charges of false advertising. A Type II error is concluding that there isn't enough evidence to conclude that the defective rate exceeds 10% when in fact, it is. This would allow the manufacturer to continue making an incorrect claim.

10.20 a. The second pair of hypotheses, because we would want to be convinced that the mercury concentration is acceptable.

10.21 a. Failed to reject H_0 **c.** Since the null hypothesis is initially assumed to be true, and is rejected only if there is strong evidence against it, if we fail to reject the null hypothesis, we have not "proven" it to be true. We can say only that there is not strong evidence against it.

10.22 b. A Type I error is changing to robots when in fact they are not superior to humans. A Type II error is not changing to robots when in fact they are superior to humans. **c.** Since a Type I error would result in a substantial monetary loss to the company and also the loss of jobs for employees, a small α should be used.

10.23 a. A P-value of 0.0003 indicates that these test results are very unlikely to occur if H_0 is true. The more plausible explanation for the occurrence of these results is that H_0 is false. **b.** A P-value of 0.35

indicates that these results are quite likely to occur (consistent) if H_0 is true. The data do not cast reasonable doubt on the validity of H_0.

10.35 a. $z = -1.90$, P-value $= 0.0287$, reject H_0 **b.** $z = -0.6$, P-value $= 0.2743$, fail to reject H_0 **c.** The denominator of the test statistic is different because the sample sizes are different. With a small sample the difference between the sample proportion of 0.47 and the hypothesized proportion of 0.5 could plausibly be attributed to chance, whereas when the sample size is larger this difference is no longer likely to be attributable to chance.

10.44 The value of 38% is a population value. No test is necessary to conclude that the population proportion is less than 0.40.

10.48 a. P-value $= 2$(area under the 9 df t curve to the right of 0.73); $2(0.251) > P$-value $> 2(0.222) \Rightarrow 0.502 > P$-value > 0.444 **b.** P-value $=$ area under the 10 df t curve to the right of $-0.5 = 1 -$ area under the 10 df t curve to the right of $0.5 = 1 - 0.314 = 0.686$ **c.** P-value $=$ area under the 19 df t curve to the left of $-2.1 =$ area under the 19 df t curve to the right of $2.1 = 0.025$ **d.** P-value $=$ area under the 19 df t curve to the left of $-5.1 =$ area under the 19 df t curve to the right of $5.1 = 0$ **e.** P-value $= 2$(area under the 39 df t curve to the right of 1.7) $\approx 2(0.048) = 0.096$

10.53 $t = 14.8$, P-value ≈ 0, reject H_0; to generalize, we would need to assume that the sample could be considered as representative of the population of interest.

10.55 With large samples, it is possible to be convinced that the observed difference is not attributable to chance and still have a relatively small difference from a practical point of view.

10.56 a. $t = 1.26$, P-value ≈ 0.1038, fail to reject H_0 **c.** The sample standard deviation in Part (b) was smaller, resulting in a larger value of the test statistic.

10.58 a. $t = -10.58$, P-value ≈ 0, reject H_0 **b.** Both questions looked at the mean number of hours per week that the teens spent online. However for 10.57 Part (b), the data were reported by the teens themselves, and for 10.58 Part (a), the data were reported by their parents. Obviously the parents grossly underestimate the amount of time the teens spend online!

10.59 The authors are saying that the positive effects of music on pain intensity found in the study are statistically significant, but not practically significant from a clinical point of view.

10.63 a. Since the boxplot is nearly symmetric with no outliers and the normal probability plot is reasonably straight, the t test is appropriate. **b.** The sample median is slightly less than 245 and because of the near symmetry and variability, a mean of 240 is plausible.

10.65 a. When the significance level is held fixed, increasing the sample size will increase the power of the test. **b.** When the sample size is held fixed, increasing the significance level will increase the power of the test.

10.66 b. A Type I error involves concluding that the water being discharged from the power plant has a mean temperature in excess of 150°F when, in fact, the mean temperature is not greater than 150°F. A Type II error is concluding that the mean temperature of water being discharged is 150°F or less when, in fact, the mean temperature is in excess of 150°F. **c.** 0.0359 (area under the normal curve to the right of 1.8)

d.

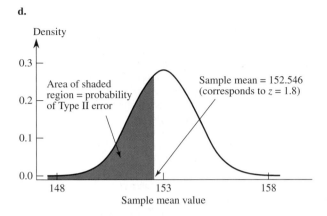

Density

Area of shaded region = probability of Type II error

Sample mean = 152.546 (corresponds to $z = 1.8$)

Sample mean value

10.67 c. It would be smaller.

10.70 a. i. Approximately 0.85 **ii.** Approximately 0.55 **iii.** Approximately 0.10 **iv.** Approximately 0

10.76 a. $z = -1.56$, P-value $= 0.0594$, fail to reject H_0 (at $\alpha = 0.05$), reject H_0 (at $\alpha = 0.10$)

10.82 c. Yes, because if the P-value is greater than 0.10, it is also greater than 0.05 or 0.01.

10.85 $z = 2.50$, P-value $= 0.0124$, at $\alpha = 0.05$: reject H_0, but fail to reject H_0 at $\alpha = 0.01$

10.89 a. If the distribution were normal, the Empirical Rule implies that about 16% of the values would be less than the value that is one standard deviation below the mean. Since the standard deviation is larger than the mean, this value is negative. Caffeine consumption can't be negative, so it is not plausible that the distribution is normal. However, since the sample size is large ($n = 47$), the t test can still be used.

10.94 a. Answers will vary based on simulation results.

Chapter 11

11.2 a. H_0: $\mu_1 - \mu_2 = 10$, H_a: $\mu_1 - \mu_2 > 10$

11.3 a. Assumptions are that the population distributions are approximately normal and that the samples are independently selected and can be regarded as random samples. **b.** $t = 2.24$, df $= 33$, P-value ≈ 0.032, reject H_0, the conclusion is consistent with the paper.

11.4 a. Because the standard deviation is nearly as large as the mean, the population distribution will not fit the 68-95-99.8 rule for normal distributions without the number of hours spent on the Internet becoming negative; clearly impossible. The population distribution is skewed to the right. **b.** Yes, because the sample sizes are large.

11.6 a. Must assume that the samples are independently selected and can be regarded as random samples from the populations of interest. Since the sample sizes are large, no other assumptions are necessary.

11.7 a. H_0: $\mu_1 = \mu_2$ vs. H_0: $\mu_1 < \mu_2$, where μ_1 is the mean payment for all claims not involving errors and μ_2 is the mean payment for all claims involving errors. **b.** With sample sizes so large (515 and 889) the df will be large and the bottom row of the t table can be used. With a P-value of 0.004 and a lower tail test, the value of the test statistic must have been between 2.58 and 3.09. $t = 2.65$ is the best answer.

11.11 a. $t = 10.4$, df $= 82$, P-value ≈ 0, reject H_0 **b.** $t = 16.3$, df $= 62$, P-value ≈ 0, reject H_0 **c.** $t = -4.69$, df $= 85$, P-value ≈ 0, reject H_0 **d.** Although it appears that the male monkeys spent time playing with the "boy" toys and the female monkeys spent more time playing with the "girl" toys, there was a difference in the "neutral" toy as well. Any difference in the time spent playing with these toys may have nothing to do with the gender choices—male monkeys may like a

certain color better, or shiny surfaces or some other underlying factor. This is an observational study and as such, causation cannot be shown. **e.** One of the conditions of a two-sample t test is that the two samples are independent. In this example, if the mean percentage of the time spent playing with the police car increased, the mean percentage of the time spent playing with the doll would have to decrease. Therefore one affects the other, they are not independent, and this violates one of the conditions of the test.

11.12 b. i. $t = 27.39$, df $= 78$, P-value ≈ 0, reject H_0 **ii.** $t = 9.50$, df $= 78$, P-value ≈ 0, reject H_0 **iii.** $t = -11.18$, df $= 78$, P-value ≈ 0, reject H_0 **c.** It appears that it is the gender of the passenger that has the greater effect than the gender of the driver on the teen drivers speed. Although the amount by which the speed limit was exceeded was higher for a male teen driver than a female teen driver, if they have a passenger, a male passenger is associated with a greater average number of miles per hour over the speed limit than a female passenger regardless of the gender of the driver.

11.14 a. The sample sizes are too small and we don't know if the response distributions are approximately normal. **b.** The sample sizes are now large enough (>30) to use a two-sample t test.

11.15 a. $t - 6.56$, df $- 201$, P-value ≈ 0, reject H_0

11.16 a. $t = 1.77$, df $= 37$, P-value ≈ 0.08, fail to reject H_0 **b.** $t = -1.28$, df $= 29$, P-value $\approx .204$, fail to reject H_0

11.17 $(0.643, 1.357)$ Because the interval does not include zero, there is evidence that the mean score for males is higher than the mean score for females.

11.18 Assuming it is reasonable to regard the samples as representative, yes. Boxplots are approximately symmetric and there are no outliers, so the assumption of normality is reasonable.

11.19 a. $t = 2.93$, df $= 54$, P-value ≈ 0.006, reject H_0 **b.** $t = 0.15$, df $= 68$, P-value ≈ 0.842, fail to reject H_0

11.28 b. Two sample $t = 9.78$, df $= 22$, P-value ≈ 0, reject H_0, in this case the conclusion is the same.

11.29 a. If possible, treat each patient with both drugs, with one drug used on one eye and the other drug used on the other eye. Determine which eye will receive the new drug at random. If this treatment method is not possible, then request that the ophthalmologist pair patients according to their eye pressure so that the two people in a pair have approximately equal eye pressure. Then select one patient in each pair at random to receive the new drug, and treat the other patient in each pair with the standard treatment. **b.** Both procedures in Part (a) would result in paired data. **c.** Randomly assign subjects to one of the two treatment groups. This experiment is probably not as informative as a paired experiment with the same number of subjects.

11.30 Take n pieces of pipe and cut each into two pieces, resulting in n pairs of pipe. Coat one piece in each pair with coating 1 and the other piece with coating 2. Then put both pipes from a pair into service where they are buried at the same depth, orientation, in the same soil type, etc. After the specified length of time, measure the depth of corrosion penetration for each piece of pipe.

11.34 a. $t = -17.38$, df $= 15$, P-value ≈ 0, reject H_0 (for $\alpha = .05$) **b.** $t = 2.44$, df $= 13$, P-value ≈ 0.032, reject H_0 (for $\alpha = .05$)

11.35 c. Yes, the confidence interval from Part (a) (for pitchers) lies entirely above the confidence interval from Part (b) (for position players).

11.37 a. $t = 4.32$, df $= 23$, P-value ≈ 0, reject H_0 **b.** $t = 1.66$, df $= 23$, P-value ≈ 0.055, fail to reject H_0 **c.** A mean difference of 8.82 is more unusual when there is no difference and the standard deviation of the difference distribution is small.

11.38 a. $t = -1.15$, df $= 6$, P-value ≈ 0.853, fail to reject H_0 **b.** No, the seven schools in the sample are not representative of all schools in California.

11.41 It is not necessary to use an inference procedure since complete information on all 50 states is available. Inference is necessary only when a sample is selected from some larger population.

11.42 a. The data are paired because the same group of girls is used for each sample.

11.43 a. (6.702, 20.548)   **c.** Because the two intervals overlap, it is possible that the mean change is the same for men and women.

11.47 (.0184, .0816), 0 is not included in the interval indicating that the proportion of Americans age 12 and older who own an MP3 player was significantly higher in 2006 than in 2005.

11.51 a. $(-0.401, -0.225)$

11.52 a. Because otherwise it is not possible to "blind" the subjects as to which treatment they are receiving.

11.54 a. Yes, the sample proportion for Americans could have been as large as 0.09, because the test statistic using this proportion is $z = 1.17$, which does not lead to the rejection of the hypothesis of no difference. **b.** No, the sample proportion for Americans could not have been as large as 0.10, because the value of the test statistic is then $z = 2.83$ and the difference would have been judged as significant.

11.58 a. $(-0.0561, 0.0695)$ **b.** With 90% confidence, it is estimated that the true proportion of patients developing diabetes may be as much as 0.0695 more in the insulin group than in the control group; but it also may be as much as 0.0561 less in the insulin group as in the control group.   **c.** Because 0 is in the interval, it is possible that there is no difference in the proportion of the patients developing diabetes in the two groups. The proposed treatment doesn't appear very effective.

11.60 The decision was based on an analysis of all the soldiers that served in the Gulf War (the population). Because a census was taken, no inference procedure was necessary.

11.61 a. $t = 11.95$, df = 18, P-value ≈ 0, reject H_0 **b.** $t = -68.8$, df = 10, P-value ≈ 0, reject H_0

11.62 For small prey: $t = 1.27$, df = 7, P-value ≈ 0.23, fail to reject H_0; for medium prey: $t = 1.83$, df = 10, P-value ≈ 0.102, fail to reject H_0; for large prey: $t = -18.08$, df = 12, P-value ≈ 0, reject H_0.

11.64 c. No, would need to know the sample means and standard deviations.

11.65 a. $(-0.529, 0.329)$ No, because the interval includes 0.

11.69 a. This would be the case if alcohol consumption varies quite a bit from week to week for the individuals in the samples.

11.80 $(-0.2743, -0.0823)$ The interval does not contain 0. The proportion of children with decayed teeth is lower for children who drink fluorinated water by somewhere between 0.0823 and 0.2743.

11.82 b. $t = 0.1398$, df = 18, P-value ≈ 0.890, fail to reject H_0

11.86 a. $t = 0.56$, df = 95, P-value ≈ 0.5795, fail to reject H_0 **b.** $t = 0.1306$, df = 92, P-value ≈ 0.8964, fail to reject H_0

11.88 The large sample z test for testing a difference of two proportions requires that the samples be independent. In this survey, the same people answered the initial question and the revised question, and so the two samples are not independent. There is no procedure from this chapter that can be used to answer the question posed.

Chapter 12

12.1 a. $0.020 < P$-value < 0.025, fail to reject H_0 **b.** $0.040 < P$-value < 0.045, fail to reject H_0 **c.** $0.035 < P$-value < 0.040, fail to reject H_0 **d.** P-value < 0.001, reject H_0 **e.** P-value > 0.10, fail to reject H_0

12.3 a. $X^2 = 19.0$, P-value < 0.001, reject H_0 **b.** If $n = 40$, the chi-squared test should not be used because the expected cell count for one of the categories (nut type 4) would be less than 5.

12.5 a. $X^2 = 21.81$, P-value < 0.001, reject H_0 **b.** If 50% of all the smokers in this population start smoking between the ages of 16 to 20,

and if all 5 years in this age range are equally likely, there should be about 10% starting to smoke for each of the ages 16, 17, 18, 19, 20. This is equivalent to 20% of all the smokers starting between the ages of 16 and 17 and 30% of all the smokers starting between the ages of 18 and 20.

12.7 b. The proportion for each month would reflect the number of days in the month out of the 366 days in 2004. April, June, September, and November have 30 days (30/366 = .082), February has 29 (29/366 = .079), and the rest have 31 (31/366 = .085).

12.10 $X^2 = 25.48$, P-value < 0.001, reject H_0

12.11 $X^2 = 0.931$, P-value > 0.100, fail to reject H_0

12.12 a. $X^2 = 4.63$, P-value > 0.100, fail to reject H_0

12.15 c. 15

12.17 $X^2 = 5.204$, P-value > 0.100, fail to reject H_0

12.19 a. $X^2 = 34.544$, P-value < 0.001, reject H_0 **b.** Menopausal women who are in the large HMO

12.29 b. Yes, because there is no evidence of an association between gender and workaholism type.

12.31 One cell has an expected count of less than five. The chi-squared approximation is probably not reasonable.

12.35 $X^2 = 9.84$, $0.010 < P$-value < 0.005, reject H_0

12.36 The data are from a census, not a sample.

12.38 $X^2 = 4.035$, P-value > 0.100, fail to reject H_0

12.39 a. The proportions falling into each category appear to be dissimilar. For instance, only 17.62% of the subjects in the "never smoked" category consumed one drink a day, whereas 34.97% of those in the "currently smokes" category consume one drink per day. Similar discrepancies are seen for other categories as well.   **b.** $X^2 = 980.068$, P-value < 0.001, reject H_0 **c.** The results are consistent with our observations in Part (a).

12.40 a. $X^2 = 27.61$, P-value < 0.001, reject H_0

12.42 $X^2 = 10.090$, $0.005 < P$-value < 0.010, reject H_0

12.45 a. Homogeneity

12.46 a. $X^2 = 8216.5$, P-value < 0.001, reject H_0 **b.** Answers will vary.   **c.** $X^2 = 10.748$, P-value > 0.100 (or P-value = 0.465), fail to reject H_0

12.47 $X^2 = 5.1$, P-value > 0.100, fail to reject H_0

Chapter 13

13.1 a. $y = -5.0 + 0.017x$
b.

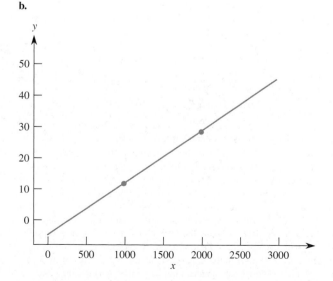

13.4 a. The estimate of the change in mean fuel efficiency associated with an increase of 1 horsepower is a decrease of 0.15 mpg.
b. It is the estimate of the mean fuel efficiency in mpg of all small cars with 100 horsepower. It is also the predicted value of the fuel efficiency of a particular small car which has 100 horsepower.   **c.** $\hat{y} = 44.0 - 015(300) = 44 - 45 = -1$. The predicted fuel efficiency is negative 1, which is an impossible value. This situation has occurred because the fitted equation has been used to make predictions outside its range of validity. Small cars do not have 300 horsepower engines. The fitted equation is valid only for the range of x values for which data have been collected.   **d.** 68% percent of the variability of fuel efficiency of small cars can be explained by (or attributed to) the linear relationship between horsepower and fuel efficiency.   **e.** It is the magnitude of a typical deviation from the least squares line, i.e., it is the typical difference between actual fuel efficiency and the estimated fuel efficiency.
13.6 a. $y = \alpha + \beta x$ is the equation of the population regression line and $\hat{y} = a + bx$ is the equation of the least squares line (the estimated regression line).   **b.** The quantity b is a statistic. It is the slope of the estimated regression line. The quantity β is a population characteristic. It is the slope of the population regression line. The quantity b is an estimate of β.   **c.** $\alpha + \beta x^*$ is the true mean y value for $x = x^*$ so $\alpha + \beta x^*$ is a population characteristic. $a + bx^*$ is a point estimate of the mean y value when $x = x^*$ or $a + bx^*$ is a point estimate of an individual y value to be observed when $x = x^*$. The quantity $a + bx^*$ is a statistic.
d. σ represents the standard deviation of the random deviation e. It is the typical deviation about the true regression line; s_e is an estimate of σ.
13.7 b. $s_e = 1.5923$
13.8 a.

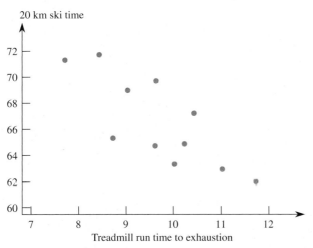

There is a general relationship that as treadmill run time to exhaustion increases, the 20 km ski time decreases. The plot does suggest that the simple linear regression model may be useful.

b. $\hat{y} = 88.7956 - 2.3335x$

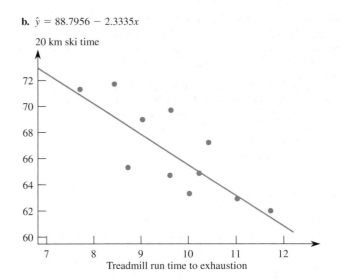

c. -2.3335 **e.** No, 15 minutes is outside the range of the data.
f. 0.6342, about 63.42% of the variability in the ski time values can be explained by the linear relationship between treadmill run time and ski time.
g. 2.1874, the estimated magnitude of a typical difference between actual ski time and predicted ski time is 2.1874.
13.11 a.

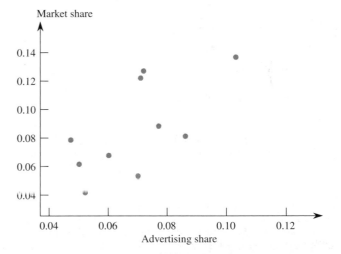

There seems to be a general tendency for y to increase at a constant rate as x increases. However, there is also quite a bit of variability in the y values.   **b.** $\hat{y} = -0.002274 + 1.246712x$. When $x = 0.09$, the predicted value is 0.1099.   **c.** $r^2 = 0.436$. This means that 43.6% of the total variability in market share can be explained by the linear relationship with advertising share.
13.12 σ is the standard deviation of the random error term. It describes how much the points are spread out around the true regression line. σ_b is the standard deviation of the statistic b. It describes how much the value of b varies from sample to sample. s_b is an estimate of σ_b based on the sample data.
13.14 Yes. The x values given in the problem are more spread out than those in Example 13.3, which would result in a larger value for S_{xx}. This results in a smaller value for s_b.

13.17 b. $\hat{y} = 786.62$. Residual is $= -29.62$.

13.18 a. Yes. Based on the P-value of 0.000, the null hypothesis of no useful linear relationship ($\beta = 0$) would be rejected. **b.** A 95% confidence interval is $(-0.161, 0.092)$. Other confidence levels are also possible.

13.19 a. $t = 2.8302$, P-value $= 0.008$, reject H_0, and conclude that the model is useful. **b.** (4.294, 25.706). Based on this interval, the estimated change in mean average SAT score associated with an increase of $1000 in expenditures per child is between 4.294 and 25.706.

13.20 From the MINITAB output, $t = 1.79$, P-value $= 0.111$, fail to reject H_0, and conclude that there is not convincing evidence that the model is useful.

13.21 a. $\hat{y} = -96.6711 + 1.595x$ **b.** $t = 27.17$, P-value ≈ 0, reject H_0, and conclude that the model is useful. **c.** The mean response time for individuals with closed-head injuries is 1.48 times as great as for those with no head injury.

13.22 A 99% confidence interval for β is $(-0.069, -0.043)$. Other confidence levels are possible.

13.23 a. $t = 6.53$, P-value ≈ 0, reject H_0, and conclude that the model is useful.

13.24 a. $t = 2.28$, P-value $= 0.062$, fail to reject H_0, there is not convincing evidence that the model is useful.

13.26 $H_0: \beta = 20$, $t = 0.45$, P-value $= 0.670$, fail to reject H_0, the data does not contradict prior belief.

13.27

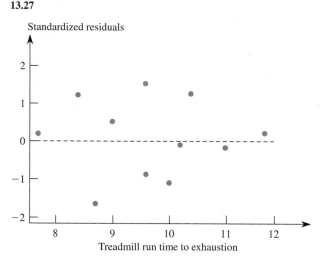

The standardized residual plot does not exhibit any unusual features.

13.28 a. $\hat{y} = 0.939 + 0.873x$

b. It is clear that the data value with min-width $= 1.20$ and max-width $= 4.40$ is an outlier, and including this point inflates SSResid. Eliminating this point results in $\hat{y} = 0.703 + 0.918x$.

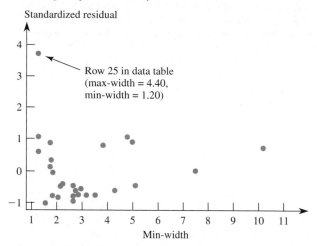

c. 0.918 estimates the increase in average max-width of the container associated with a 1 cm increase in min-width. The intercept is not meaningful because a food container cannot have a width of 0.

d.

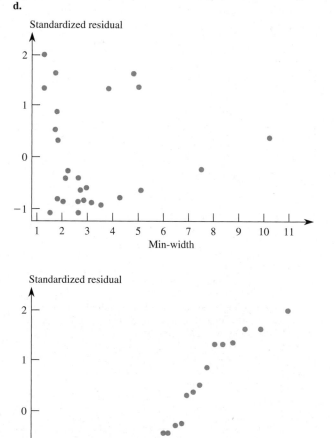

The residual plot raises concerns about the assumption of constant variance and the normal probability plot is curved, suggesting that the assumption of normality of the error distribution may not be reasonable.
13.29 a. The assumptions required in order that the simple linear regression model be appropriate are **i.** The distribution of the random deviation e at any particular x value has mean value 0. **ii.** The standard deviation of e is the same for any particular value of x. **iii.** The distribution of e at any particular x value is normal. **iv.** The mean value of vigor is a linear function of stem density. **v.** The random deviations associated with different observations are independent of one another.
b.

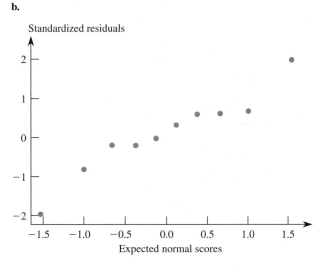

The plot is reasonably straight. The assumption of normality is plausible.
c.

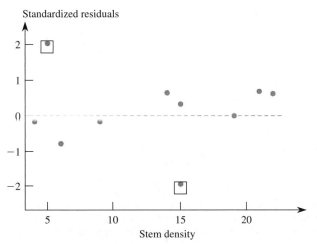

There are two residuals that are unusually large. **d.** There is a pattern in the residual plot; negative residuals are for the small x values and the positive residuals are for the large x values.

13.30 a.

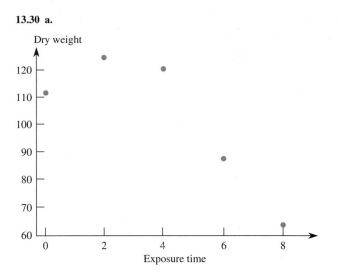

A linear model is not appropriate; the relationship would be better modeled by a curve.
b.

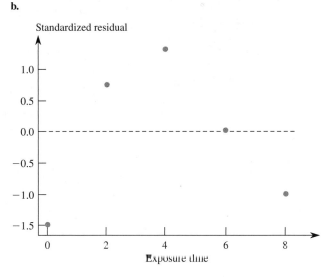

The residual plot shows a curved pattern, confirming the observation in Part (a).

13.31 a.

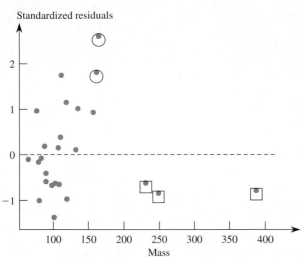

There are several large residuals and potentially influential observations. **b.** The residuals associated with the potentially influential observations are all negative. Without these three, there appears to be a positive trend to the standardized residual plot. The plot suggests that the simple linear regression model might not be appropriate. **c.** There does not appear to be any pattern in the plot that would suggest that it is unreasonable to assume that the variance of y is the same at each x value.

13.32 a.

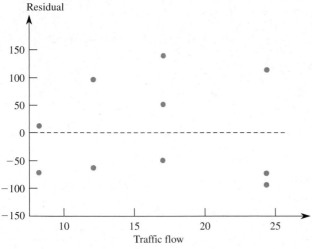

A linear regression model appears to be appropriate.

b.

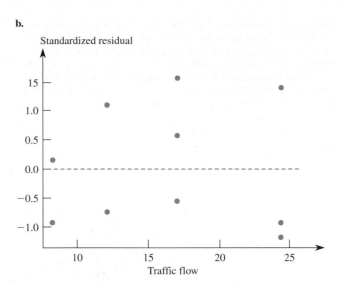

The plot exhibits the same pattern as the residual plot of Part (a). Only the scale on the vertical axis is different.

13.33

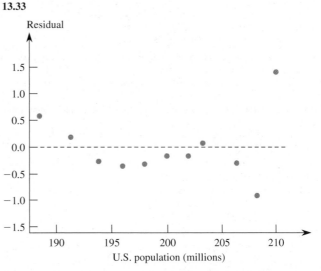

The residuals are positive in 1963 and 1964, then they are negative from 1965 through 1972. This pattern in the residual plot casts doubt on the appropriateness of the simple linear regression model.

13.34 A confidence interval is an interval estimate of a population mean, whereas a prediction interval is an interval of plausible values for a single future observation. A 95% prediction interval is one for which 95% of all possible samples would yield interval limits capturing the future observation.

13.35 If the request is for a confidence interval for β the wording would likely be "estimate the *change* in the average y value associated with a one unit increase in the x variable." If the request is for a confidence interval for $\alpha + \beta x^*$ the wording would likely be "estimate the *average y* value when the value of the x variable is x^*."

13.36 b. 4.038, since 3 is the same distance from $\bar{x} = 2.5$ as 2 is, the estimated standard deviation will be the same.

13.38 a. (6.551, 6.570) **b.** (6.560, 6.616)

13.39 a. (6.4722, 6.6300)

13.40 a. $r^2 = 0.63$ **b.** P-value for the model utility test is 0.000, reject H_0 and conclude that the model is useful.   **d.** (4.529, 10.806)

13.41 a. $\hat{y} = -133.02 + 5.919x$ **b.** $s_b = 1.127$ **c.** $t = 5.25$, P-value ≈ 0, reject H_0 **e.** Shouldn't use regression equation to predict for $x = 105$ because it is outside the range of the data.

13.42 a. $t = 13.532$, P-value ≈ 0, reject H_0 and conclude that the model is useful.   **c.** (5.709, 6.647)

13.43 a. $\hat{y} = 2.78551 + 0.04462x$ **b.** $t = 5.25$, P-value ≈ 0, reject H_0, and conclude that the model is useful   **c.** (3.671, 4.577)   **d.** Since 4.1 is in the interval, it is plausible that a box that has been on a shelf 30 days will not be acceptable.

13.44 Since 17 is farther away from $\bar{x} = 19.21$ than is 20, the confidence interval with $x^* = 17$ would be wider. The same would be true for a prediction interval.

13.45 a. $\hat{y} = 30.019 + 0.1109x$ **d.** The interval of Part (b) is for the mean blood level of all people who work where the air lead level is 100. The interval of Part (c) is for a single randomly selected individual who works where the air lead level is 100.

13.46 b. For $x^* = 45$, (1.809, 2.106) is a 95% confidence interval. This interval and the 95% interval in Part (a) for $x^* = 35$ are a set of simultaneous intervals with a confidence level of at least 90%.   **c.** The simultaneous confidence level would be at least 98%.   **d.** The simultaneous confidence level would be at least 58%.

13.47 a. Prediction interval for $x = 35$ is (1.9519, 3.0931). Prediction interval for $x = 45$ is (1.3879, 2.5271).

13.49 r is the sample correlation coefficient and measures the strength of the linear relationship in the sample. r is an estimate of ρ, the population correlation coefficient.

13.50 a. No, the sample points might fall exactly on a straight line even though not all population pairs do.   **b.** Yes. If $\rho = 1$, then r will be 1 for every sample, since a perfect linear relation in the population implies that all (x, y) pairs in the sample will lie on the same line.

13.51 $t = 2.07$, P-value ≈ 0.036, reject H_0.

13.52 $t = 2.84$, P-value ≈ 0.0032, reject H_0.

13.53 a. $t = -6.175$, P-value ≈ 0, reject H_0.   **b.** No, $r^2 = 0.0676$.

13.54 a. $t = 7.75$, P-value ≈ 0, reject H_0.   **b.** $r^2 = 0.822$

13.55 $r = 0.574$, $t = 1.85$, P-value ≈ 0.1, fail to reject H_0

13.56 a. The null hypothesis would be rejected. There is evidence that the correlation coefficient is not 0.   **b.** The small P-value just indicates that $\rho \neq 0$, but does not necessarily imply that ρ is far from 0.

13.57 $t = 2.2$, P-value $= 0.028$, reject H_0. The result of the test is significant, indicating that we are convinced that $\rho \neq 0$ (largely because of the very large sample size). However r is quite close to 0, indicating a very weak linear relationship.

13.58 a. -0.640 **b.** 80.7   **c.** $t = -7.05$, P-value ≈ 0, reject H_0 and conclude that there is a useful linear relationship.

13.59 a. Yes, $t = -3.40$ and for either a one- or two-tailed test P-value is less than 0.05.   **b.** Answers will vary.   **c.** Probably not. $r^2 = (-0.18)^2 = 0.0324$, so only about 3.24% of the observed variability in sense of humor can be explained by the linear regression model.

13.60 a. $t = 8.24$, P-value ≈ 0, reject H_0 and conclude that there is a positive linear relationship.   **b.** 25.058

13.61 a. $t = -3.95$, P-value $= 0.003$, reject H_0 and conclude that the linear model is useful.   **b.** A 95% confidence interval is $(-3.669, -0.998)$. Other confidence levels are possible.   **c.** A 95% prediction interval is (60.276, 70.646). Other confidence levels are possible.   **d.** Because 11 is farther from $\bar{x}$.

13.62 a. $t = 15.26$, P-value $= 0.000$, reject H_0 and conclude that the linear model is useful.   **b.** $H_0: \beta \geq 0.8$, $H_a: \beta < 0.8$, $t = -1.48$, P-value ≈ 0.086, fail to reject H_0 **c.** A 95% confidence interval is (139.08, 156.64).

Other confidence levels are possible.   **d.** The confidence interval would be wider because 250 is farther from $\bar{x}$.

13.63 a. $t = -6.09$, P-value ≈ 0, reject H_0, and conclude that there is a linear relationship.   **b.** A 95% confidence interval is (2.346, 3.844). Other confidence levels are possible.   **c.** The predicted trail length when soil hardness is 10 is -2.58. Since trail length cannot be negative, the predicted value makes no sense.

13.64 a.

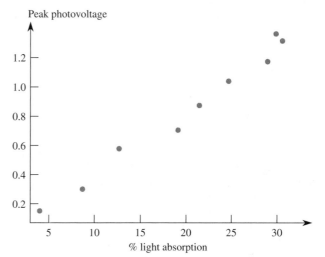

The relationship looks linear.   **b.** $\hat{y} = -0.08259 + 0.044649x$ **c.** $r^2 = .983$ **d.** $\hat{y} = 0.7702$, residual $= -0.0902$ **e.** Yes, $t = 20.00$, P-value ≈ 0, reject H_0, and conclude that there is a useful linear relationship.   **f.** A 95% confidence interval is (0.039, 0.050). Other confidence levels are possible.   **g.** A 95% confidence interval is (0.762, 0.859). Other confidence levels are possible.

13.65 a. $H_0: \beta = -0.5$, $H_a: \beta \neq -0.5$, $t = 0.49$, P-value ≈ 0.624, fail to reject H_0, the data do not contradict the researchers' prior belief.   **b.** A 95% confidence interval is (47.083, 54.099). Other confidence levels are possible.

13.66 a. $\hat{y} = -1.7521 + 0.68485x$ **b.** $s_e = 0.337$, $H_0: \alpha = 0$, $H_a: \alpha \neq 0$, $t = -5.20$, P-value ≈ 0, reject H_0 **c.** $(-2.531, -0.974)$. No, because 0 is not included in the interval.

13.67 $H_0: \beta = \beta'$, $H_a: \beta \neq \beta'$, $t = -1.05$, P-value ≈ 0.322, fail to reject H_0

13.68 A linear model is not appropriate for data sets 2, 3, and 4.

13.69 When the point is included in the computations, the slope will be negative and much more extreme (farther from 0) than if the point is excluded from the computations.

13.70 a. $\hat{y} = 57.964 + 0.0357x$ **b.** $t = 0.023$, P-value ≈ 1, fail to reject H_0

c.

x	y	Predicted	Residual
1	74	58.00	16.00
2	54	58.04	−4.04
3	52	58.07	−6.07
4	51	58.11	−7.11
5	52	58.14	−6.14
6	53	58.18	−5.18
7	58	58.21	−0.21
8	71	58.25	12.75

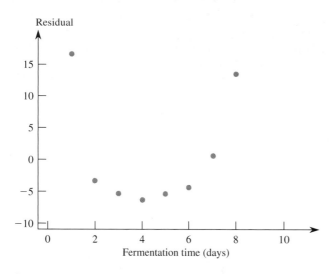

d. The residual plot has a distinct curved pattern which indicates that a simple linear regression model is not appropriate.

13.71 Since the P-value of 0.0076 is smaller than most reasonable levels of significance, the percentage raise does appear to be linearly related to productivity.

13.72 a. $t = -0.06$, P-value $= 0.922$, fail to reject H_0 **b.** $t = 4.35$, P-value ≈ 0, reject H_0 **c.** $a = 3.0781$, $b = 0.01161$ **d.** (2.9778, 4.3392) **e.** (3.6197, 4.161)

13.73 Even though the P-value is small and $r^2 = 0.65$, the variability about the least-squares line appears to increase as soil depth increases. If this is so, a residual plot would be "funnel shape," opening up toward the right. One of the conditions for valid application of the procedures described in this chapter is that the variability of the y's be constant. It appears that this requirement may not be true in this instance.

13.74 a. The e_i's are the deviations of the observations from the population regression line, whereas the residuals are the deviations of the observations from the estimated regression line. **b.** The simple linear regression model states that $y = \alpha + \beta x + e$. Without the random deviation e, the equation implies a deterministic model, whereas the simple linear regression model is probabilistic. **c.** The quantity b is a statistic. Its value is known once the sample has been collected, and different samples result in different b values. Therefore, it does not make sense to test hypotheses about b. Only hypotheses about a population characteristic can be tested. **d.** If $r = +1$ or -1, then each point falls exactly on the regression line and SSResid would equal zero. A true statement is that SSResid is always greater than or equal to zero. **e.** The sum of the residuals must equal zero. Thus, if they are not all exactly zero, at least one must be positive and at least one must be negative. They cannot all be positive. Since there are some positive and no negative values among the reported residuals, the student must have made an error. **f.** SSTo $= \Sigma(y - \bar{y})^2$ must be greater than or equal to SSResid $= \Sigma(y - \hat{y})^2$. Thus, the values given must be incorrect.

13.75 a. When $r = 0$, then $s_e \approx s_y$. The least squares line in this case is a horizontal line with intercept of $\bar{y}$. **b.** When r is close to 1 in absolute value, then s_e will be much smaller than s_y.

Chapter 14

14.1 A deterministic model does not have the random deviation component e, while a probabilistic model does contain such a component.

14.2 a. (mean y value for fixed values of x_1, x_2, x_3) $= 30 + 0.90x_1 + 0.08x_2 - 4.5x_3$ **c.** The average change in acceptable load associated

with a 1-cm increase in left lateral bending, when grip endurance and trunk extension ratio are held fixed, is 0.90 kg. **d.** The average change in acceptable load associated with a 1 N/kg increase in trunk extension ratio, when grip endurance and left lateral bending are held fixed, is −4.5 kg. **f.** $P(13.5 < y < 33.5) =$

$$P\left(\frac{13.5 - 23.5}{5} < z < \frac{33.5 - 23.5}{5}\right) = 0.9544$$

14.3. $y = \beta_0 + \beta_1 x_1 + \beta_2 x_2 + e$. An interaction term is not included in the model because it is given that x_1 and x_2 make independent contributions.

14.4 b. The value of x_4 is now 0 and the predicted score decreases to 1.03. **d.** If all other variables remain the same, the average increase in mean ecology score associated with an increase in income of $1000 is 0.01. **e.** Ideology and social class are categorical (qualitative) variables. An appropriate way of incorporating these variables in a regression model is to define indicator or dummy variables. Each of these variables takes on five different values, so four dummy variables need to be defined for each of ideology and social class.

14.5 a. 103.11 **b.** 96.87 **c.** $\beta_1 = -6.6$; 6.6 is the expected decrease in yield associated with a one-unit increase in mean temperature when the mean percentage of sunshine remains fixed. $\beta_2 = -4.5$; 4.5 is the expected decrease in yield associated with a one-unit increase in mean percentage of sunshine when mean temperature remains fixed.

14.6 c. $\beta_2 = -1.40$. The expected decrease in error percentage associated with a one-unit increase in character subtense when level of backlight, viewing angle, and level of ambient light remain fixed, is equal to 1.40. $\beta_3 = 0.02$. The expected increase in error percentage associated with a one-unit increase in viewing angle when character subtense, and level of ambient light remain fixed is 0.02.

14.7 a.

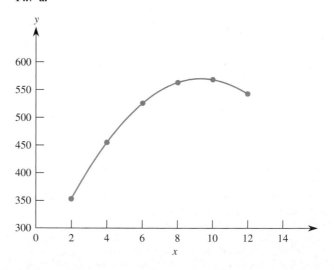

b. The mean chlorine content is higher for $x = 10$ than for $x = 8$.
c. When the degree of delignification increases from 8 to 9 the mean chlorine content increases by 7. Mean chlorine content decreases by 1 when degree of delignification increases from 9 to 10.

14.8 a. mean $y = 21.09 + 0.653x_1 + 0.0022x_2 - 0.0206x_1^2 + 0.00004x_2^2$
b. May 6 is 16 days after April 20. The values of x_1 and x_2 are 16 and 41180 respectively and the estimated mean yield is 67948.56.

14.9 a. For $x_1 = 30$, $y = 4.8 + 0.8x_2$; for $x_1 = 20$, $y = 3.8 + 0.8x_2$; for $x_1 = 10$, $y = 2.8 + 0.8x_2$.

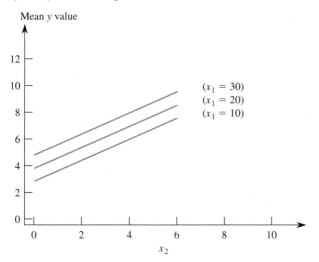

b. For $x_2 = 60$, $y = 49.8 + 0.1x_2$; for $x_2 = 55$, $y = 45.8 + 0.1x_2$; for $x_2 = 50$, $y = 41.8 + 0.1x_2$.

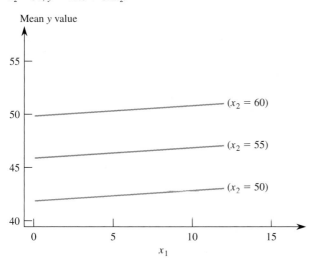

c. The parallel lines in each graph are attributable to the lack of interaction between the two independent variables.

d. For $x_1 = 30$, $y = 4.8 + 1.7x_2$; for $x_1 = 20$, $y = 3.8 + 4.1x_2$; for $x_1 = 10$, $y = 2.8 + 1.1x_2$.

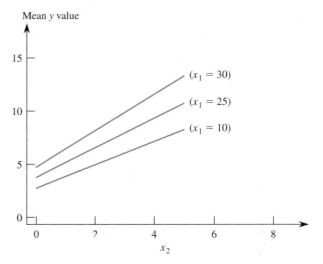

For $x_2 = 60$, $y = 49.8 + 1.9x_1$; for $x_2 = 55$, $y = 45.8 + 1.75x_1$; for $x_2 = 50$, $y = 41.8 + 1.6x_1$.

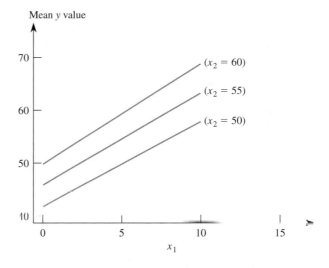

Because there is an interaction term, the lines are not parallel.

14.10 a. mean $y = \alpha + \beta_1 x_1 + \beta_2 x_2 + \beta_3 x_3$ where x_2 is 1 if intake setting is medium and 0 if it is not medium and x_3 is 1 if intake setting is high and 0 if it is not high. β_1 is the expected increase in particulate matter concentration associated with a one-unit increase in flue temperature when intake setting remains fixed. β_2 is the expected increase in particulate matter concentration associated with a change in intake setting from low to medium when flue temperature remains fixed. β_3 is the expected increase in particular matter concentration associated with a change in intake setting from low to high when flue temperature remains fixed.

b. $x_1 \cdot x_2$ and $x_1 \cdot x_3$

14.11 a. $y = \alpha + \beta_1 x_1 + \beta_2 x_2 + \beta_3 x_3 + e$

b. $y = \alpha + \beta_1 x_1 + \beta_2 x_2 + \beta_3 x_3 + \beta_4 x_1^2 + \beta_5 x_2^2 + \beta_6 x_3^2 + e$

c. $y = \alpha + \beta_1 x_1 + \beta_2 x_2 + \beta_3 x_3 + \beta_4 x_1 x_2 + e$, $y = \alpha + \beta_1 x_1 + \beta_2 x_2 + \beta_3 x_3 + \beta_4 x_1 x_3 + e$, $y = \alpha + \beta_1 x_1 + \beta_2 x_2 + \beta_3 x_3 + \beta_4 x_2 x_3 + e$

d. $y = \alpha + \beta_1 x_1 + \beta_2 x_2 + \beta_3 x_3 + \beta_4 x_1^2 + \beta_5 x_2^2 + \beta_6 x_3^2 + \beta_7 x_1 x_2 + \beta_8 x_1 x_3 + \beta_9 x_2 x_3 + e$

14.12 a. -64.6241 **b.** No, because when an interaction term is present, changing the value of one independent variable will also change the value of the interaction term. Thus, you cannot change one model predictor while holding all the rest fixed.

14.13 a. Three dummy variables would be needed to incorporate a nonnumerical variable with four categories. For example, you could define $x_3 = 1$ if the car is a subcompact and 0 otherwise, $x_4 = 1$ if the car is a compact and 0 otherwise, and $x_5 = 1$ if the car is a midsize and 0 otherwise. The model equation is then $y = \alpha + \beta_1 x_1 + \beta_2 x_2 + \beta_3 x_3 + \beta_4 x_4 + \beta_5 x_5 + e$.
b. For the variables defined in Part (a), $x_6 = x_1 x_3$, $x_7 = x_1 x_4$, and $x_8 = x_1 x_5$ are the additional predictors needed to incorporate interaction between age and size class.

14.14 c. The relationship between the volume of a cylindrical can and its width and height is a deterministic one. So, in principle, there is no need for regression. However, if the cans are not exactly cylindrical (they never are), then a suitably chosen regression model can be useful.
d. Yes

14.15 a. $b_1 = -2.18$ is the estimated change in mean fish intake associated with a one-unit increase in water temperature when the values of the other predictor variables are fixed. $b_4 = 2.32$ is the estimated change (increase) in mean fish intake associated with a one-unit increase in speed when the value of the other predictor variables are fixed.

14.16 a. $0.01 < P\text{-value} < 0.05$ **b.** $P\text{-value} > 0.10$
c. $P\text{-value} = 0.01$ **d.** $0.001 < P\text{-value} < 0.01$

14.17 b. $0.001 < P\text{-value} < 0.01$ **c.** $0.01 < P\text{-value} < 0.05$

14.18 $F = 24.41$, $P\text{-value} < 0.001$, reject H_0, and conclude that the model is useful.

14.19 $F = 34.286$, $P\text{-value} < 0.001$, reject H_0, and conclude that the model is useful.

14.20 $F = 3.5$, $0.01 < P\text{-value} < 0.05$, reject H_0, and conclude that the model is useful.

14.21 a. $\hat{y} = 86.85 - 0.12297x_1 + 5.090x_2 - 0.07092x_3 + 0.001538x_4$
b. $F = 64.4$, $P\text{-value} < 0.001$, reject H_0, and conclude that the model is useful.   **c.** $R^2 = 0.908$. This means that 90.8% of the variation in the observed tar content values has been explained by the fitted model. $s_e = 4.784$. This means that the typical distance of an observation from the corresponding mean value is 4.784.

14.23 b. $F = 10.21$, $0.001 < P\text{-value} < 0.01$, reject H_0, and conclude that the model is useful.

14.24 a. $\hat{y} = 1.44 - 0.0523 \text{ length} + 0.00397 \text{ speed}$ **c.** $F = 24.02$, $P\text{-value} \approx 0$, reject H_0, and conclude that the model is useful.
d. $\hat{y} = 1.59 - 1.40\left(\dfrac{\text{length}}{\text{speed}}\right)$ **e.** The model in Part (a); the model in Part (a) has $R^2 = 0.75$ and $R^2 \text{ (adj)} = 0.719$, whereas the model in Part (d) has $R^2 = 0.543$ and $R^2 \text{ (adj)} = 0.516$.

14.25 a. $\hat{y} = -859 + 23.7 \text{ minwidth} + 226 \text{ maxwidth} + 225 \text{ elongation}$ **b.** Adjusted R^2 takes into account the number of predictors used in the model whereas R^2 does not do so.   **c.** $F = 16.03$, $P\text{-value} < 0.001$, reject H_0, and conclude that the model is useful.

14.26 a. $F = 23.05$, $P\text{-value} < 0.001$, reject H_0, and conclude that the model is useful.   **b.** It is a bit surprising, but with a large sample size it is possible to be convinced that the model is useful (better than the model $\hat{y} = \bar{y}$) even when only a small proportion of variability in y is explained by the model.   **c.** $R^2 = 0.16$ and adjusted $R^2 = 0.16$. To two decimal places, they are equal; this is attributable to the large sample size.

14.27 a. SSResid $= 390.4347$, SSTo $= 1618.2093$, SSRegr $= 1227.7746$ **b.** $R^2 = 0.759$. This means that 75.9% of the variation in observed shear strength values has been explained by the fitted model.

c. $F = 5.039$, $0.01 < P\text{-value} < 0.05$, reject H_0, and conclude that the model is useful.

14.28 a. $F = 6.00$, $0.001 < p\text{-value} < 0.01$, reject H_0, and conclude that the model is useful.   **b.** $R^2 = .4898$, $s_e = 0.894$. About 49% of the variability in error percentage has been explained by the fitted model. A typical deviation from the mean value corresponding to any specified set of values for the predictors is estimated to be 0.894.   **c.** The relatively large value of s_e and the moderate value of R^2 suggest that the estimated regression equation may not provide sufficiently good predictions of error rate.

14.29 $F = 96.64$, $P\text{-value} < 0.001$, reject H_0 and conclude that the model is useful.

14.30 As long as the number of predictors variables is less than 10, the corresponding model would be judged useful at level of significance 0.05. A high R^2 value can often be obtained simply by including a great many predictors in the model, even though the actual population relationship between y and the predictors is weak.

14.31 See solutions manual for sample computer output.

14.32 a. $\hat{y} = 76.4 - 7.3x_1 + 9.6x_2 - 0.91x_3 + 0.0963x_4 - 13.5x_1^2 + 2.80x_2^2 + 0.0280x_3^2 - 0.000320x_4^2 + 3.75x_1x_2 - 0.750x_1x_3 + 0.142x_1x_4 + 2.00x_2x_3 - 0.125x_2x_4 + 0.00333x_3x_4$ **b.** $F = 8.76$, $p\text{-value} \approx 0$, reject H_0, and conclude that the model is useful.   **c.** $R^2 = 0.885$, $s_e = 0.3529$. About 88.5% of the variability in observed brightness has been explained by the fitted model. A typical deviation from the mean value corresponding to any specified set of values for the predictors is estimated to be 0.3529.

14.33 $\hat{y} = 35.8 - 0.68x_1 + 1.28x_2$, $F = 18.95$, $P\text{-value} < 0.001$, reject H_0, and conclude that the model is useful.

Chapter 15

15.2 a. $0.001 < P\text{-value} < 0.01$ **b.** $0.05 < P\text{-value} < 0.10$
c. $0.01 < P\text{-value} < 0.05$ **d.** $P\text{-value} < 0.001$

15.3 a. $H_0: \mu_1 = \mu_2 = \mu_3 = \mu_4$, H_a: At least two of the four μ_i's are different.   **b.** $0.01 < P\text{-value} < 0.05$, fail to reject H_0

15.7 a.

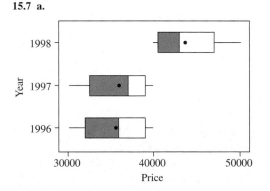

1996 and 1997 boxplots are similar, but the 1998 boxplot values are higher. The boxplot for 1998 also has a longer tail on the upper end than the other two boxplots.

15.8 a. numerator df $= 2$, denominator df $= 65$

15.12 b. If each technique was used on 45 subjects, then the total number of subjects in the study is $3(45) = 135$. The denominator degrees of freedom $df_2 = 135 - 3 = 132$. The critical value decreases as the denominator df increases. Since the computed F of 4.12 exceeds the 0.05 level critical value when the denominator df $= 30$, it will exceed the 0.05 level critical value when the denominator df $= 132$. Hence, the null hypothesis would be rejected and the conclusion is the same as in Part (a).

15.14 a. See solutions manual for detailed computations.
15.15

Source of variation	Degrees of Freedom	Sum of Squares	Mean Square	F
Treatments	3	75081.72	25027.24	1.70
Error	16	235419.04	14713.69	
Total	19	310500.76		

$F = 1.70$, P-value > 0.10, fail to reject H_0
15.16

Source of Variation	Degrees of Freedom	Sum of Squares	Mean Square	F
Treatments	2	152.18	76.09	5.56
Error	71	970.96	13.675	
Total	73	1123.14		

$F = 5.56$, $0.001 < P$-value < 0.01, reject H_0
15.17

Source of Variation	Degrees of Freedom	Sum of Squares	Mean Square	F
Treatments	3	29.304	9.768	2.56
Error	418	1595.088	3.816	
Total	421	1624.392		

$F = 2.56$, $0.05 < P$-value < 0.10, fail to reject H_0
15.20

Source of Variation	Degrees of Freedom	Sum of Squares	Mean Square	F
Treatments	2	1.348	0.674	3.51
Error	18	3.457	0.192	
Total	20	4.805		

$F = 3.51$, $0.05 < P$-value < 0.10, fail to reject H_0
15.21 a. See solutions manual for detailed computations.
15.23 Since the intervals for $\mu_1 - \mu_2$ and $\mu_1 - \mu_3$ do not contain zero, μ_1 and μ_2 are judged to be different and μ_1 and μ_3 are judged to be different. Since the interval for $\mu_2 - \mu_3$ contains zero, μ_2 and μ_3 are not judged to be different. Hence, statement (iii) is the correct choice.
15.24 b. $\mu_1 - \mu_2$: $(-1.284, 3.384)$, $\mu_1 - \mu_3$: $(1.916, 6.584)$, $\mu_1 - \mu_4$: $(4.666, 9.334)$, $\mu_2 - \mu_3$: $(0.866, 5.534)$, $\mu_2 - \mu_4$: $(3.616, 8.284)$, $\mu_3 - \mu_4$: $(0.416, 5.084)$; We would conclude that μ_1 is not significantly

different from μ_2, but that all other means differ from one another. Ordering the sample means from smallest to largest:

Method	4	3	2	1
$\bar{x}$	9.30	12.05	15.25	16.30

15.25 $\mu_1 - \mu_2$: $(-1.81, 3.91)$, $\mu_1 - \mu_3$: $(1.39, 7.11)$, $\mu_1 - \mu_4$: $(4.14, 9.86)$, $\mu_2 - \mu_3$: $(0.34, 6.06)$, $\mu_2 - \mu_4$: $(3.09, 8.81)$, $\mu_3 - \mu_4$: $(-0.11, 5.61)$; μ_1 differs from μ_3, μ_1 differs from μ_4, μ_2 and differs from μ_3, and μ_2 differs from μ_4.

Method	4	3	2	1
$\bar{x}$	9.30	12.05	15.25	16.30

15.26 μ_1 differs from μ_2; μ_1 differs from μ_3; μ_1 differs from μ_5; μ_2 differs from μ_4; μ_3 differs from μ_4; μ_4 differs from μ_5.

Fabric	3	2	5	4	1
$\bar{x}$	10.5	11.63	12.3	14.96	16.35

15.27 $\mu_1 - \mu_2$: $(-0.1465, 0.2665)$, $\mu_1 - \mu_3$: $(-0.3190, 0.1090)$, $\mu_1 - \mu_4$: $(-0.2277, 0.1677)$, $\mu_2 - \mu_3$: $(-0.3890, 0.0490)$, $\mu_2 - \mu_4$: $(-0.2877, 0.1077)$, $\mu_3 - \mu_4$: $(-1.308, 2.908)$. No differences are detected.
15.28 a. $F = 45.6432$, P-value < 0.001, reject H_0
15.29

Group	Simultaneous	Sequential	Control
Mean	$\bar{x}_3$	$\bar{x}_2$	$\bar{x}_1$

15.30 The interval for $\mu_1 - \mu_2$ contains zero, and hence μ_1 and μ_2 are judged not different. The intervals for $\mu_1 - \mu_3$ and $\mu_2 - \mu_3$ do not contain zero, so μ_1 and μ_3 are judged to be different and μ_2 and μ_3 are judged to be different. Hence, μ_3 is different from the other two means.
15.31 The mean water loss when exposed to 4 hours fumigation is different from all other means. The mean water loss when exposed to 2 hours fumigation is different from that for levels 16 and 0, but not 8. The mean water losses for duration 16, 0, and 8 hours are not different from one another.
15.32 a. $F = 79.264$, P-value < 0.001, reject H_0 **b.** See solutions manual for the required 15 T-K intervals. The resulting underscoring pattern is

Brand	2	4	3	1	5	6
Mean	12.8	13.1	13.825	14.1	17.14	18.1

15.33 $\mu_1 - \mu_2$: $(0.24, 2.22)$, $\mu_1 - \mu_3$: $(0.36, 2.44)$, $\mu_2 - \mu_3$: $(-0.87, 1.21)$; μ_1 differs from μ_2 and μ_3, but μ_2 and μ_3 do not differ.
15.34 a. $F = 3.30$, $0.01 < P$-value < 0.05, reject H_0 **b.** See solutions manual for the required 10 T-K intervals. No significant differences are determined using the T-K method.

Index

A

Addition rule, general, 323
Additive probabilistic model, 690
Adjusted coefficient of multiple determination, 768
All subsets method, *14-14*
Alternative hypotheses, 526
Analysis of variance (ANOVA)
    assumptions, 787
    cautions and limitations, 15–23
    computations, *15-25–26*
    distribution-free, *16-23–16-28*
    multiple comparisons procedure, 800–6
    notation, 785
    randomized block experiment, *15-25–26*
    single-factor, 784–90, *15-25*
    single-factor F test, 790–3
    summarizing, 793–4
    Tukey-Kramer (T-K) procedure, 800–1, 804
    two-factor, *15-9–19*
ANOVA table, 793
Assumption of independence, 319
Automatic selection of variables, *14-14*

D

Backward elimination method, *14-9–21*
β and power for the t test, 567–70
Bar charts, 16–17
    comparative, 76–7
    other uses, 81–3
    segmented, 80–1
Bayes' rule (Bayes theorem), 330–2
Bias
    measurement, 33–4
    nonresponse, 33–4
    response, 33–4
    selection, 33–4
    social desirability, 59
Biased statistics, 478
Bimodal distribution, 109
Binomial distribution, 386–92
Binomial probability distributions
    normal approximation to, 426–8

properties, 386
sampling without replacement, 317–20
tables, 390, 830–2
Binomial random variables, 386–92
Bivariate data, 13
    categorical, 660
    cautions and limitations, 266–7
    communicating results, 264
    interpreting analysis of, 264–5
    scatterplots, 117–22
    time-series plots, 122–4
    See also correlation
Bivariate normal distributions, 734
Blocking, 44, 46, 47
Blocks, *15-1*
Bound on error of estimation, 491
Boxplots
    on calculators, 196–7
    modified, 171
    skeletal, 170

C

Calculators. *See* Graphing Calculator Explorations
Categorical data, 12
    bivariate data set, 13
    frequency distributions, 15
    graphical displays, 77–81
    homogeneity tests, 660–7
    independence tests, 660–1, 667–71
    in multiple regression, 757–8
    relative frequency, 15
    sample proportion of successes, 155, 462
    univariate, 12, 647–56
    See also chi-square tests
Causation
    correlation and, 207
    drawing conclusions about, 44–5
    See also experiments
Censuses, 32–3
Center, measures of. *See* mean; median; trimmed mean
Central Limit Theorem, 457
Chance experiments, 279–82

Charts. *See* graphs
Chebyshev's Rule, 177–9
Chi-square distributions, 652, *16-25*, *16-35*, 828–9
Chi-square tests, 647–56
    assumptions, 677–9
    cautions and limitations, 679–80
    communicating results, 677
    goodness-of-fit, 651, 652–6, 685–6
    for homogeneity, 661–7, 677, 686–7
    for independence, 667–71, 677, 686–7
    interpreting, 677
CI. *See* confidence interval
Classes, 101
Class intervals, 101–5
Clusters, 38–9
Cluster sampling, 38–9
Coefficient of determination, 228–31
Coefficient of multiple determination, 767
Combined estimate of common population proportion, 621
Comparative bar charts, 76–7
Comparative experiments, 42–7
Comparative stem-and-leaf displays, 91–2
Completely randomized design, *16-23–26*
Complete second-order model, 755
Conditional probability, 302–10
Confidence interval (CI), 521–2, *14-2*
    assumptions, 509–11
    cautions and limitations, 511–13
    definition, 483
    for difference between two population or treatment means, 595–7, 613–14
    for difference between two proportions, 624
    for estimated mean value in simple linear regression, 726–7
    for population mean, 522–3
    for population proportion, 521–2
    for regression coefficients, 760
    for the slope of the population regression line, 704
    general form of a, 490–1
    Graphing Calculator Explorations, 642–3
    interpreting, 624–6
    for mean y value, 727, *14-8*
    modified, 489

Italicized hyphenated entries (14-, 15-, 16-) pertain to sections of the book found at www.cengage.com/statistics/peck

Italicized hyphenated entries (14-, 15-, 16-) pertain to sections of the book found at www.cengage.com/statistics/peck

Italicized hyphenated entries (14-, 15-, 16-) pertain to sections of the book found at www.cengage.com/statistics/peck

Italicized hyphenated entries (14-, 15-, 16-) pertain to sections of the book found at www.cengage.com/statistics/peck

Italicized hyphenated entries (14-, 15-, 16-) pertain to sections of the book found at www.cengage.com/statistics/peck

Tabulated area = probability

Standard normal (z) curve

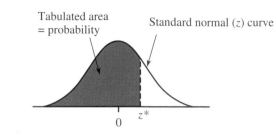

0   z*

z*	.00	.01	.02	.03	.04	.05	.06	.07	.08	.09
0.0	.5000	.5040	.5080	.5120	.5160	.5199	.5239	.5279	.5319	0.5359
0.1	.5398	.5438	.5478	.5517	.5557	.5596	.5636	.5675	.5714	0.5753
0.2	.5793	.5832	.5871	.5910	.5948	.5987	.6026	.6064	.6103	0.6141
0.3	.6179	.6217	.6255	.6293	.6331	.6368	.6406	.6443	.6480	0.6517
0.4	.6554	.6591	.6628	.6664	.6700	.6736	.6772	.6808	.6844	0.6879
0.5	.6915	.6950	.6985	.7019	.7054	.7088	.7123	.7157	.7190	0.7224
0.6	.7257	.7291	.7324	.7357	.7389	.7422	.7454	.7486	.7517	0.7549
0.7	.7580	.7611	.7642	.7673	.7704	.7734	.7764	.7794	.7823	0.7852
0.8	.7881	.7910	.7939	.7967	.7995	.8023	.8051	.8078	.8106	0.8133
0.9	.8159	.8186	.8212	.8238	.8264	.8289	.8315	.8340	.8365	0.8389
1.0	.8413	.8438	.8461	.8485	.8508	.8531	.8554	.8577	.8599	0.8621
1.1	.8643	.8665	.8686	.8708	.8729	.8749	.8770	.8790	.8810	0.8830
1.2	.8849	.8869	.8888	.8907	.8925	.8944	.8962	.8980	.8997	0.9015
1.3	.9032	.9049	.9066	.9082	.9099	.9115	.9131	.9147	.9162	0.9177
1.4	.9192	.9207	.9222	.9236	.9251	.9265	.9279	.9292	.9306	0.9319
1.5	.9332	.9345	.9357	.9370	.9382	.9394	.9406	.9418	.9429	0.9441
1.6	.9452	.9463	.9474	.9484	.9495	.9505	.9515	.9525	.9535	0.9545
1.7	.9554	.9564	.9573	.9582	.9591	.9599	.9608	.9616	.9625	0.9633
1.8	.9641	.9649	.9656	.9664	.9671	.9678	.9686	.9693	.9699	0.9706
1.9	.9713	.9719	.9726	.9732	.9738	.9744	.9750	.9756	.9761	0.9767
2.0	.9772	.9778	.9783	.9788	.9793	.9798	.9803	.9808	.9812	0.9817
2.1	.9821	.9826	.9830	.9834	.9838	.9842	.9846	.9850	.9854	0.9857
2.2	.9861	.9864	.9868	.9871	.9875	.9878	.9881	.9884	.9887	0.9890
2.3	.9893	.9896	.9898	.9901	.9904	.9906	.9909	.9911	.9913	0.9916
2.4	.9918	.9920	.9922	.9925	.9927	.9929	.9931	.9932	.9934	0.9936
2.5	.9938	.9940	.9941	.9943	.9945	.9946	.9948	.9949	.9951	0.9952
2.6	.9953	.9955	.9956	.9957	.9959	.9960	.9961	.9962	.9963	0.9964
2.7	.9965	.9966	.9967	.9968	.9969	.9970	.9971	.9972	.9973	0.9974
2.8	.9974	.9975	.9976	.9977	.9977	.9978	.9979	.9979	.9980	0.9981
2.9	.9981	.9982	.9982	.9983	.9984	.9984	.9985	.9985	.9986	0.9986
3.0	.9987	.9987	.9987	.9988	.9988	.9989	.9989	.9989	.9990	0.9990
3.1	.9990	.9991	.9991	.9991	.9992	.9992	.9992	.9992	.9993	0.9993
3.2	.9993	.9993	.9994	.9994	.9994	.9994	.9994	.9995	.9995	0.9995
3.3	.9995	.9995	.9995	.9996	.9996	.9996	.9996	.9996	.9996	0.9997
3.4	.9997	.9997	.9997	.9997	.9997	.9997	.9997	.9997	.9997	0.9998
3.5	.9998	.9998	.9998	.9998	.9998	.9998	.9998	.9998	.9998	0.9998
3.6	.9998	.9998	.9999	.9999	.9999	.9999	.9999	.9999	.9999	0.9999
3.7	.9999	.9999	.9999	.9999	.9999	.9999	.9999	.9999	.9999	0.9999
3.8	.9999	.9999	.9999	.9999	.9999	.9999	.9999	.9999	.9999	1.0000

Standard normal
probabilities
(cumulative z curve
areas)

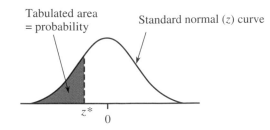

Tabulated area
= probability

Standard normal (z) curve

z^* 0

z^*	.00	.01	.02	.03	.04	.05	.06	.07	.08	.09
−3.8	.0001	.0001	.0001	.0001	.0001	.0001	.0001	.0001	.0001	.0000
−3.7	.0001	.0001	.0001	.0001	.0001	.0001	.0001	.0001	.0001	.0001
−3.6	.0002	.0002	.0001	.0001	.0001	.0001	.0001	.0001	.0001	.0001
−3.5	.0002	.0002	.0002	.0002	.0002	.0002	.0002	.0002	.0002	.0002
−3.4	.0003	.0003	.0003	.0003	.0003	.0003	.0003	.0003	.0003	.0002
−3.3	.0005	.0005	.0005	.0004	.0004	.0004	.0004	.0004	.0004	.0003
−3.2	.0007	.0007	.0006	.0006	.0006	.0006	.0006	.0005	.0005	.0005
−3.1	.0010	.0009	.0009	.0009	.0008	.0008	.0008	.0008	.0007	.0007
−3.0	.0013	.0013	.0013	.0012	.0012	.0011	.0011	.0011	.0010	.0010
−2.9	.0019	.0018	.0018	.0017	.0016	.0016	.0015	.0015	.0014	.0014
−2.8	.0026	.0025	.0024	.0023	.0023	.0022	.0021	.0021	.0020	.0019
−2.7	.0035	.0034	.0033	.0032	.0031	.0030	.0029	.0028	.0027	.0026
−2.6	.0047	.0045	.0044	.0043	.0041	.0040	.0039	.0038	.0037	.0036
−2.5	.0062	.0060	.0059	.0057	.0055	.0054	.0052	.0051	.0049	.0048
−2.4	.0082.	.0080	.0078	.0075	.0073	.0071	.0069	.0068	.0066	.0064
−2.3	.0107	.0104	.0102	.0099	.0096	.0094	.0091	.0089	.0087	.0084
−2.2	.0139	.0136	.0132	.0129	.0125	.0122	.0119	.0116	.0113	.0110
−2.1	.0179	.0174	.0170	.0166	.0162	.0158	.0154	.0150	.0146	.0143
−2.0	.0228	.0222	.0217	.0212	.0207	.0202	.0197	.0192	.0188	.0183
−1.9	.0287	.0281	.0274	.0268	.0262	.0256	.0250	.0244	.0239	.0233
−1.8	.0359	.0351	.0344	.0336	.0329	.0322	.0314	.0307	.0301	.0294
−1.7	.0446	.0436	.0427	.0418	.0409	.0401	.0392	.0384	.0375	.0367
−1.6	.0548	.0537	.0526	.0516	.0505	.0495	.0485	.0475	.0465	.0455
−1.5	.0668	.0655	.0643	.0630	.0618	.0606	.0594	.0582	.0571	.0559
−1.4	.0808	.0793	.0778	.0764	.0749	.0735	.0721	.0708	.0694	.0681
−1.3	.0968	.0951	.0934	.0918	.0901	.0885	.0869	.0853	.0838	.0823
−1.2	.1151	.1131	.1112	.1093	.1075	.1056	.1038	.1020	.1003	.0985
−1.1	.1357	.1335	.1314	.1292	.1271	.1251	.1230	.1210	.1190	.1170
−1.0	.1587	.1562	.1539	.1515	.1492	.1469	.1446	.1423	.1401	.1379
−0.9	.1841	.1814	.1788	.1762	.1736	.1711	.1685	.1660	.1635	.1611
−0.8	.2119	.2090	.2061	.2033	.2005	.1977	.1949	.1922	.1894	.1867
−0.7	.2420	.2389	.2358	.2327	.2296	.2266	.2236	.2206	.2177	.2148
−0.6	.2743	.2709	.2676	.2643	.2611	.2578	.2546	.2514	.2483	.2451
−0.5	.3085	.3050	.3015	.2981	.2946	.2912	.2877	.2843	.2810	.2776
−0.4	.3446	.3409	.3372	.3336	.3300	.3264	.3228	.3192	.3156	.3121
−0.3	.3821	.3783	.3745	.3707	.3669	.3632	.3594	.3557	.3520	.3483
−0.2	.4207	.4168	.4129	.4090	.4052	.4013	.3974	.3936	.3897	.3859
−0.1	.4602	.4562	.4522	.4483	.4443	.4404	.4364	.4325	.4286	.4247
−0.0	.5000	.4960	.4920	.4880	.4840	.4801	.4761	.4721	.4681	.4641